Springer Collected Works in Mathematics

More information about this series at http://www.springer.com/series/11104

Moscow State University, about 1947

Israel M. Gelfand

Collected Papers I

Editors
Semen G. Gindikin
Victor W. Guillemin
Alexandr A. Kirillov
Bertram Kostant
Shlomo Sternberg

Reprint of the 1987 Edition

Springer

Author
Israel M. Gelfand (1913 – 2009)
Department of Mathematics
Rutgers State University of New Jersey
New Brunswick, NJ
USA

Editors
Semen G. Gindikin
Department of Mathematics
Rutgers State University of New Jersey
New Brunswick, NJ
USA

Victor W. Guillemin
Massachusetts Institute of Technology
Cambridge, MA
USA

Alexandr A. Kirillov
University of Pennsylvania
Philadelphia, PA
USA

Bertram Kostant
Massachusetts Institute of Technology
Cambridge, MA
USA

Shlomo Sternberg
Harvard University
Cambridge, MA
USA

ISSN 2194-9875
Springer Collected Works in Mathematics
ISBN 978-3-662-43735-3 (Softcover)
978-3-540-13619-4 (Hardcover)

Library of Congress Control Number: 2012954381

Mathematics Subject Classification (2010): 22E46, 22E65, 28C20, 33A35, 46F25, 53C65, 55N, 57Q, 57R, 60G20, 90D40, 92A09, 94A17

Springer Heidelberg New York Dordrecht London

Printed on acid-free paper

Springer-Verlag GmbH Berlin Heidelberg is part of Springer Science+Business Media
(www.springer.com)

IZRAIL M. GELFAND

COLLECTED PAPERS

VOLUME I

Edited by

S.G. Gindikin V.W. Guillemin A.A. Kirillov
B. Kostant S. Sternberg

SPRINGER-VERLAG
BERLIN HEIDELBERG NEW YORK
LONDON PARIS TOKYO

Professor Izrail M. Gelfand
Member of the Academy of Sciences of the USSR
A.N. Belozersky Laboratory of Molecular Biology and
Bioorganic Chemistry, Building "A", Moscow State University
Moscow, GSP-234, 119899, USSR

Editors:

Professor Semen G. Gindikin
A.N. Belozersky Laboratory of Molecular Biology and
Bioorganic Chemistry, Building "A", Moscow State University
Moscow, GSP-234, 119899, USSR

Professor Victor W. Guillemin
Department of Mathematics, Massachusetts Institute of Technology
Cambridge, MA 02139, USA

Professor Aleksandr A. Kirillov
Moskovskii Universitet, Mehmat, Moscow 117234, USSR

Professor Bertram Kostant
Department of Mathematics, Massachusetts Institute of Technology
Cambridge, MA 02139, USA

Professor Shlomo Sternberg
Department of Mathematics, Harvard University, Science Center
One Oxford Street, Cambridge, MA 02138, USA

Mathematics Subject Classification (1980):
05B 30, 05B 35, 06B 15, 06D 05; 10D; 12A, 12B; 17B 65; 22E 30; 22E 35; 26E 46; 26E 65; 32N 15; 33A 35; 33A 45; 35Q 20; 43A 83; 43A 90; 46B 20, 46F, 46J, 46L, 47B, 47e, 47F 05, 49F; 53C 65, 58E, 58F 17, 58H; 60G 20, 65P 05, 65Q 05; 68D 27; 68G 05; 70H, 81C, 81G 20, 81G 30, 94A 17

ISBN 3-540-13619-3 Springer-Verlag Berlin Heidelberg New York
ISBN 0-387-13619-3 Springer-Verlag New York Berlin Heidelberg

Library of Congress Cataloging-in-Publication Data.
Gel'fand, I.M. (Izrail' Moiseevich).
[Works, 1987]. Collected papers / Izrail M. Gelfand: edited by S.G. Gindikin ... [et al.].
p. cm. In English, French, and German. Bibliography: v. 1, p.
ISBN 0-387-13619-3 (U.S.)
1. Mathematics – Collected works. 2. Gel'fand, I.M. (Izrail' Moiseevich)
3. Mathematicians – Soviet Union – Biography. I. Gindikin, S.G. (Semen Grigor'evich)
II. Title. QA3.G38 1987 510-dc19 87-32254 CIP

Printed in Germany
Printing: Druckhaus Beltz, Hemsbach/Bergstr.; binding: J. Schäffer GmbH & Co. KG, Grünstadt
2141/3140-543210

Preface

I am very grateful to Springer-Verlag for publishing my three volumes of Collected Papers. I was asked to write a survey of the papers included: I do not think that the author has this right. It seems to me that whatever a man achieves in science comes from above. For this reason, he cannot be the judge of his own work. But I hope that at the end of the third volume I will be able to write what I have tried to do in these papers and did not succeed or did not fully succeed to do. It may be that some of these problems remain interesting today. From the bibliography you can see that a lot of papers are written together with some of my colleagues and friends. I want to express my deep gratitude to them. I very much enjoyed working with them and I learned a lot from my contact with them.

I am very grateful to all my friends who did a lot of work, especially in translating, correcting and generally editing these three volumes. Without them it would not have been possible to get this edition ready.

Thanks also very much to Professors S.G. Gindikin, V.W. Guillemin, A.A. Kirillov, B. Kostant and S. Sternberg.

Moscow, September 1987 I. M. Gelfand

Table of contents

Part I. Survey lectures and articles of general content

1. Some aspects of functional analysis and algebra 3
2. On some problems of functional analysis 27
3. Some questions of analysis and differential equations 37
4. Integral geometry and its relation to the theory of group representations 56
5. On elliptic equations 65
6. Automorphic functions and the theory of representations 76
7. The cohomology of infinite dimensional Lie algebras; some questions of integral geometry 88

Part II. Banach spaces and normed rings

1. Sur un lemme de la théorie des espaces linéaires 107
2. Abstrakte Funktionen und lineare Operatoren 113
3. On one-parametrical groups of operators in a normed space 163
4. On normed rings 169
5. To the theory of normed rings. II. On absolutely convergent trigonometrical series and integrals 172
6. To the theory of normed rings. III. On the ring of almost periodic functions 175
7. On the theory of characters of commutative topological groups (with D.A. Rajkov) 177
8. Normierte Ringe 181
9. Über verschiedene Methoden der Einführung der Topologie in die Menge der maximalen Ideale eines normierten Ringes (mit G.E. Shilov) 202
10. Ideale und primäre Ideale in normierten Ringen 216
11. Zur Theorie der Charaktere der Abelschen topologischen Gruppen . . 223
12. Über absolut konvergente trigonometrische Reihen und Integrale . . 225
13. On the imbedding of normed rings into the ring of operators in Hilbert space (with M.A. Najmark) 241
14. Commutative normed rings (with D.A. Rajkov and G.E. Shilov) . . . 258
15. Normed rings with an involution and their representations (with M.A. Najmark) 364

Part III. Differential equations and mathematical physics

1. Eigenfunction expansions for equations with periodic coefficients . . 401
2. On the determination of a differential equation from its spectral function (with B.M. Levitan) 405
3. On a simple identity for eigenvalues of second order differential operator (with B.M. Levitan) 457
4. Solution of quantum field equations (with R.A. Minlos) 462
5. On the structure of the regions of stability of linear canonical systems of differential equations with periodic coefficients (with V.B. Lidskij) 466
6. Eigenfunction expansions for differential and other operators (with A.G. Kostyuchenko) 505
7. On identities for eigenvalues of a second order differential operator . . 510
8. Some problems in the theory of quasilinear equations 518
9. On a theorem of Poincaré (with I.I. Piatetski-Shapiro) 605
10. Fractional powers of operators and Hamiltonian systems (with L.A. Dikij) 610
11. A family of Hamiltonian structures related to nonlinear integrable differential equations (with L.A. Dikij) 625
12. Asymptotic behaviour of the resolvent of Sturm-Liouville equations and the algebra of the Korteweg-de Vries equations (with L.A. Dikij) 647
13. The resolvent and Hamiltonian systems (with L.A. Dikij) 684
14. Integrable nonlinear equations and the Liouville theorem (with L.A. Dikij) 697
15. Hamiltonian operators and algebraic structures related to them (with I.Ya. Dorfman) 707
16. The Schouten bracket and Hamiltonian operators (with I.Ya. Dorfman) 722
17. Hamiltonian operators and the classical Yang-Baxter equation (with I.Ya. Dorfman) 726

Essays in honour of Izrail M. Gelfand

On his fiftieth birthday (by M.I. Vishik, A.N. Kolmogorov, S.V. Fomin, and G.E. Shilov) 735

On his sixtieth birthday (by S.G. Gindikin, A.A. Kirillov, and D.B. Fuks; O.V. Lokytsievskij and N.N. Chentsov; M.B. Berkinblit, Yu.M. Vasil'ev, and M.L. Shik) 753

On his seventieth birthday (by N.N. Bogolyubov, S.G. Gindikin, A.A. Kirillov, A.N. Kolmogorov, S.P. Novikov, and L.D. Faddeev) 812

Some remarks on I.M. Gelfand's works (by V.W. Guillemin and S. Sternberg) 831

Tentative table of contents for volumes II and III 837

Bibliography 843

Acknowledgements 883

Part I

Survey lectures and articles of general content

1.

Some aspects of functional analysis and algebra

Proc. Int. Congr. Math. 1954, Amsterdam **1** (1957) 253–276. Zbl. **79**:326

1. Introduction
2. Representations of Groups
3. Generalized Functions
4. Generalized Functions and Representations of Groups
5. Differential Operators
6. Analysis in Functional Spaces

1. *Introduction.*

Functional analysis, which has become an independent branch of mathematics at the beginning of this century, occupies one of the central positions in contemporary mathematics. This is explained on the one hand by the fact that functional analysis made use of the main classical methods of analysis and algebra, and on the other hand by the rôle which functional analysis plays in contemporary physical science, especially in quantum physics.

The study of mathematical problems connected with quantum mechanics was a turning point in the development of functional anlysis itself, and at the present time to a great extent it determines the main paths for its development.

It can be said without exaggeration that contemporary functional analysis represents a new and serious step in the development of mathematics.

In the last few years a number of new branches have arisen in functional analysis. Although relatively recently (about 20 to 30 years ago) functional analysis was thought of mainly as the theory of linear normed spaces, at the present time that theory, which is important and, roughly speaking, is completed, cannot even be considered as one of the basic branches of functional analysis.

In general, functional analysis is still far from being completed, but the basic tendencies in its development are considerably clearer now than they were some 15 to 20 years ago.

It is of course impossible to discuss in this paper all the basic questions of functional analysis. Therefore, we will limit ourselves to the consideration of a few selected problems. Although at a first glance the problems considered differ in character, common to all of them is, for instance, the fact that the development of each of these branches is closely connected with and is stimulated by the development of quantum mechanics and the quantum field theory. Perhaps not all the problems which will be discussed lie on the main path of development of functional analysis. However we hope that they will at least aid in determing that path.

2. *Representations of groups.*

One of the main branches of functional analysis in which the cooperation of analytic and algebraic methods typical for that field of mathematics, is clearly seen, is the theory of representations of groups.

The theory of representations developed as a branch of algebra in the study of representations of finite groups. The connection between the theory of representations and analysis was discovered after the study of representations of compact topological groups had begun. From that time on the theory of representations, both in the character of the problems with which it was concerned and in its methods, developed essentially as a branch of functional analysis.

It is sufficient to point out, for instance, the connection of the theory of representations with the almost periodic functions, the spherical functions, or the generalized spherical functions, which arise in the representations of the group of rotations of threedimensional space [1], [2].

The development of the theory of representations of groups was stimulated first by the quantum theory and later by the quantum field theory. In particular, from these theories it became clear that the theory of representations is one of the basic mathematical methods for the study of symmetries (invariancies) occurring in physics.

The analytic character of that theory is, naturally, most clearly seen in the representations of Lie groups. We shall limit ourselves to the best known case of semi-simple Lie groups [5], [7], which includes such important groups as, for example, the group of unimodular matrices, the orthogonal group, the Lorentz group, etc.

An important example of representation is the so-called regular representation. This representation acts in the space of all square integrable functions on a group and consists in making correspond to each element h the linear transformation T_h given by the formula: $T_h\varphi(g) = \varphi(gh)$. Decomposing this representation into irreducible ones, in the case of a compact group one obtains all the irreducible representations of the group under consideration. In the case of semi-simple Lie groups the decomposition of the regular representation gives rise to an important class of irreducible representations, the so-called basic series of irreducible representations. However (and in this respect locally compact groups differ essentially from compact ones), not all the irreducible representations of the semi-simple Lie groups are contained in this class.

The reason for the appearance of the so-called supplementary series may be illustrated by a simple example. Let us consider two similar groups: the group of the rotations of a sphere, i.e. the group of the motions of a surface of constant positive curvature and the group of the motions of a surface of constant negative curvature (Lobachevsky planes).

All irreducible representations of rotation groups are given by spherical functions and can be obtained as follows. Let us consider a Laplace operator on the surface of a sphere (the second differential Beltrami parameter). The eigenfunctions of this operator corresponding to a given eigenvalue (i.e. spherical functions) form the space in which the required irreducible unitary representations act. Let us now consider the Laplace operator in the space of the square integrable functions on a Lobachevsky plane. Its spectrum consists of all numbers from $-\infty$ to 0, and spectral expansion in this case also gives irreducible representations, more explicitely the main series of representations of the group of the motions of the Lobachevsky plane. All the eigenfunctions decrease in this case as $e^{-\sqrt{k}r}$, where $-k$ is the curvature of the Lobachevsky plane. However, if we should only require that the eigenfunctions considered be bounded, then more eigenfunctions might arise, corresponding to the supplementary part of the spectrum — the segment from 0 to k. Such a difference between the case of square integrability and the case of boundedness is connected with the fact that the circumference of a circle on a Lobachevsky plane increases with the radius r as $e^{-\sqrt{k}r}$, and, therefore, the class of square integrable functions is much narrower than the class of bounded functions. These supplementary eigenfunctions do not occur in the expansion of square integrable functions, and give rise to a supplementary series of irreducible representations. More exactly, eigenfunctions depending only on the radius are positive definite functions, and with their aid supplementary series of representations may be realized. This example shows that for locally compact homogeneous spaces, in contrast to compact ones, the collection of those invariant elementary positive definite functions, which are required for the Plancherel expansion theorem (the expansion of square integrable functions into an integral analogous to the Fourier integral) is essentially different from the collection of all invariant elementary positive definite functions.

Apparently, supplementary series of representations appear in those cases where there is such a neighbourhood U of the unity in the group, that the measure of U^n increases as a geometric progression.

Before exposing the results related to the representations of the basic series it is desirable to make a few remarks on the so-called dimensions of functional spaces. As is well known, spaces of square integrable functions of any number of variables are isomorphic. However, as far as we know, in concrete questions of analysis it never happens that the space of ,,all'' the functions of a certain number of variables is effectively transformed one-to-one into the space of ,,all'' the functions of another number of variables. It may be said without exaggeration that for analysis the isomorphism of all Hilbert spaces has no more importance than, let us say, for algebraic geometry the fact that a curve and a, surface are sets of equal power.

We shall attempt to determine in a given space of functions the number of variables occurring in the functions of the space where the representations of the semi-simple Lie groups are defined. It is possible here to make a comparison with fundamental Burnside theorems on finite groups, according to which the number of nonequivalent representations of a finite group G is equal to the number of classes of conjugated elements in a given group and the sum of the squares of the dimensions of all (nonequivalent) representations is equal to the order of the group. The proof of these theorems is based on the decomposition of regular representations into irreducible ones. Similar facts hold for representations of the Lie groups if the dimensions of functional spaces are conceived in the corresponding way. We shall illustrate this taking as an example the group of complex unimodular matrices with determinant equal to 1.

An element of this group is determined by the values of $n^2 - 1$ complex parameters. A class of conjugated elements is in general determined by the values of $n - 1$ parameters (it consists of all the matrices with a given set of eigenvalues $\lambda_1, \lambda_2, \ldots, \lambda_n$, $\lambda_1 \lambda_2 \ldots \lambda_n = 1$). The analogue of the first Burnside theorem learns that the representations of the groups mentioned above are determined by $n - 1$ complex (i.e. $2n - 2$ real) parameters. And the representations of the basic series are, in fact, determined by the values of $2n-2$ parameters ($n - 1$ integral ones and $n - 1$ real ones).

Let us determine now in a functional space the number of variables occurring in the functions of the space on which the irreducible representations are defined. The regular representation acts in the space of the functions on the group, i.e. in the space of functions of $n^2 - 1$ complex parameters; it is decomposed into irreducible representations. Let l represent the dimension of the functional space on which an irreducible representation of a given group acts. The analogue of Burnside's second theorem in this case will be expressed by the relationship $n^2 - 1 = n - 1 + 2l$, from which it follows that $l = \frac{n(n-1)}{2}$, i.e. that each of the irreducible representations is realized in a space of functions depending on $\frac{n(n-1)}{2}$ parameters. This is what really occurs. It is highly improbable that there exists a realization of these representations in a space of ,,all'' the functions depending on a different number of variables, which in some way is natural. It would be interesting to develop a theory in which these considerations would become exact.

Let us now return to the description of the irreducible representations of semi-simple Lie groups. We shall limit ourselves to the case of the group of complex unimodular matrices, as the picture in that case is typical for the general case. The representation of the group of second order matrices operates in the space of functions of a complex variable z subject to fractional linear transformations $\frac{\alpha z+\beta}{\gamma z+\delta}$ given by the second order matrices $\begin{pmatrix}\alpha & \beta \\ \gamma & \delta\end{pmatrix}$. In constructing the representations of the group of matrices of the nth order it is important, in the first place, to define correctly the conception of fractional linear transformation for this case. The conventional definition of fractional linear transformations is unsatisfactory, since it would lead to representations in the space of functions of $n-1$ variables and not in the space of functions of $\frac{n(n-1)}{2}$ variables, as should occur in accordance with what has been said above. The generalization of the fractional linear transformations necessary in this case is as follows: let us take a point in the n-dimensional projective space, a straight line passing through this point, a plane passing through this straight line, etc. We shall call such a combination a *generalized linear element*. It is not difficult to check that in the $(n-1)$-dimensional projective space the generalized linear element is determined by the values of $\frac{n(n-1)}{2}$ parameters. The representations of the group under consideration are obtained as follows: a measure is introduced in some way in the space of the linear elements, and in the space of the functions of "z" which are square integrable with this measure the operator T_g corresponding to the element "g" is determined by means of the formula: $T_g f(z) = f(zg)\,\alpha(zg)$, where "$zg$" is the image of the linear element "z" in the projective transformation with the matrix "g", and $\alpha(zg)$ is a fixed function, specific for each representation. The function $\alpha(zg)$ is determined from the requirement that $Tg_1\, Tg_2$ should be equal to $Tg_1\, g_2$ and the requirement that the representation should be unitary. The basic series differ from the supplementary one in the manner of choosing the scalar product.

It should be noted that the function α is determined apart from equivalences by the character of the group of diagonal matrices, and for an arbitrary semi-simple group, by the character of a similar (Cartan) subgroup. This character itself is a generalization of Cartan's highest weight for the case of infinite-dimensional representation.

Considering the space of functions dependent on linear elements "with omissions" (for example now a point, then a plane, etc.) we obtain a degenerated series of representations. Let us assume $z_{n_1 n_2 \dots n_k}$ to be the respective "incomplete" linear element, $n_1, n_2, \dots, n_k$ being the dimensions of the manifolds comprising the given linear element. Each set of dimensions has as degenerated series of its own.

The spaces of linear elements "z" and $z_{n_1 n_2 \dots n_k}$, as well as the respective representation series may be constructed in a similar way for any semi-simple Lie group. It is possible to establish which of these representations are equivalent.

Continuing to develop the procedures expounded in [5], [6] and in [70], M. A. Naimark proved that the representations described here are all representations of classical complex groups.

The problem of the equivalence of representations can be solved using the classical theory of characters which, and this was a surprise at the time of discovery, is fully applicable to the case of infinite-dimensional representations. It is interesting to note that the formulae of the theory of characters are, in this case, not at all more complicated and, in some cases, are simpler than the formulae of finite-dimensional representations. The existence of characters for complex semi-simple groups was proved by direct calculation on the basis of the apparent type of representation for these groups which was already known at that time [5]. Harish Chandra [7] and Godement [13] proved remarkable theorems about the existence of characters, for any semi-simple group, complex or not.

Interesting new questions arise in studying the representations of real semi-simple Lie groups. When considering, for example, any of the real forms of the group of complex unimodular nth order matrices, the same calculation of the number of parameters given above shows that irreducible unitary representations should be realized in a space of functions depending on $\frac{n(n-1)}{2}$ real parameters $\left(\text{i.e. in a functional space of real dimension } \frac{n(n-1)}{2}\right)$. In order to obtain actually these representations, let us again consider the space "Z" of linear elements which correspond to a complex group. The space "Z" decomposes into parts, which are transitive with respect to the real form. One of these parts is a manifold of real dimension $\frac{n(n-1)}{2}$. On this manifold a functional space is constructed, in which representations of the real group are assigned in a fashion similar to the one described above for a complex group.

For the transitive components whose real dimension is $\frac{n(n-1)}{2} + r$, where $r > 0$, the representation is realized in a space of functions which are arbitrary functions of $\frac{n(n-1)}{2} - r$ real parameters and depend analytically on r complex parameters. The transitive manifold of the highest possible dimension $\left(\frac{n(n-1)}{2}\right.$ complex variables, i.e. $n(n-1)$ real variables$\left.\right)$ is of special interest. The corresponding irreducible representations of the real group are constructed in a space of functions which are analytic in all $\frac{n(n-1)}{2}$ parameters. A similar construction may be used for degenerated series as well. In doing so one should consider the decomposition of the respective spaces of "incomplete linear elements" into transitive components. In this manner we can describe, for example, the representations of the real unimodular group [11] previously described by Bargman [1] for $n = 2$.

The analyticity with respect to some of the variables becomes manifest in the theory of representations in the following manner. Every transitive homogeneous space is a space of the classes of conjugate elements with respect to some subgroup. Functions on a homogeneous space may be considered as functions on a group G, which are constant on classes of conjugate elements with respect to some subgroup K. In passing from the group G to its real form the intersection of this real form with the subgroup K or with a conjugate of it may be partially imaginary. The requirement of constancy on the classes of conjugate elements is replaced by the requirement of analyticity, just as constancy on straight lines parallel to the straight line $x = y$ (i.e. the fulfilment of the equation $\frac{\partial f}{\partial x} = \frac{\partial f}{\partial y}$) becomes analyticity of f (i.e. the fulfilment of the Cauchy-Riemann conditions $\frac{\partial f}{\partial x} = -i\frac{\partial f}{\partial y}$), when that line is replaced by the straight line $x = iy$. The regular way of determining for which parameters analyticity is required is to write the condition that the functions be constant on the classes of conjugate elements, in the complex case, as $X_j f = 0$, where f is a function on the group, and X_j are Lie operators corresponding to infinitesimal transformations in the subgroup. By introducing the parameters in the required way we find that f is independent of some of them and satisfies the Cauchy-Riemann condition for pairs of others. Of special interest in connection with the theory of automorphic functions are the transitive manifolds of the highest dimensions already mentioned, on which the functions are analytic in all parameters. General automorphic functions should be defined as functions on these manifolds, invariant with respect to some discrete subgroup of the corresponding semi-simple group. The study of such automorphic functions would probably be of interest.

For the special case of a unimodular matrix group of the second order a somewhat different connection with the theory of automorphic functions was discovered by S. V. Fomin and the author of this report in a paper [12] on the theory of dynamic systems. Automorphic functions are obtained in this paper by decomposing into irreducible components the representation acting in the space of the functions on G, constant on the classes of conjugate elements with respect to a certain discrete subgroup. In each of the irreducible representations of the discrete series contained in this representation there is a certain automorphic form determining uniquely the given irreducible representation. The full decomposition of the representation of the group of second order matrices mentioned above into irreducible ones, as well as the analogous decomposition for the case when the representation acts in a space of cosets of the discrete subgroup of the semi-simple group G, would without doubt be of interest.

3. *Generalized functions.*

In one way or another generalized functions have been considered in mathematics and its applications for rather a long time.

A substantial contribution to the formation of the concept of generalized functions was made in the works of Hadamard [15], and later of M. Riesz [16], on finite parts of divergent integrals. Generalized functions (Dirac's δ-function, etc.) were systematically used in quantum mechanics beginning with the nineteen twenties. The general concept of generalized functions as functionals was developed by S. L. Sobolev in connection with his investigations on equations of the hyperbolic type. The treatment of generalized functions as linear functionals on some space of sufficiently smooth functions is the most convenient and natural one.

Much was done in developing the theory of generalized functions by L. Schwartz [18] who combined and systematized the material at hand and presented it from a unified standpoint. The appearance of his book fostered the penetration of these concepts into various branches of mathematics. Schwartz has also introduced, with the aid of generalized functions, the notion of the Fourier transform of functions which do not grow faster than a certain power.

It is, however, necessary to generalize the definition of the Fourier transform for a broader class of functions. This can be seen from the following example which is traditional for the Fourier method. In considering the heat equation $\frac{\partial u}{\partial t} = \frac{\partial^2 u}{\partial x^2}$, as was shown by A. N. Tichonov, the class of functions, which do not increase faster than $e^{|x|^2}$, is the natural class of functions in which the existence and uniqueness of the solution of the Cauchy problem is ensured. However, the various generalizations of the definition of the Fourier integral, given by Bochner [20], Carleman [19] and Schwartz [18], allow the Cauchy problem to be solved only for the class of functions which do not increase faster than a certain power of x. A further generalization of the Fourier transform requires a further development of the idea of generalized function [60], which, incidentally, is important from other points of view as well. Starting with Schwartz's definition we shall generalize the definition of Fourier transform so that it shall fit some classes of fast growing functions [60]. It is important that in order to define the Fourier transform we must, as a rule, use analytic functions.

By the term generalized functions we shall understand functionals over some topological space S of infinitely differentiable functions which approach zero faster than any power of x. We shall call this space the basic space. The Fourier transforms of all the functions of the basic space also form a basic space $\tilde{S}$, to which we shall refer as the dual of S.

If we are given the generalized function T, i.e. the functional $T(\varphi)$ over some basic space S, then in generalizing Schwartz's definition we shall introduce the Fourier transform $\tilde{T}$ as a functional over $\tilde{S}$ determined by the equation $\tilde{T}(\tilde{\varphi}) = T(\varphi)$ $(\varphi \in S, \tilde{\varphi} \in \tilde{S})$. The space Z_p^p of all entire analytic functions $\varphi(z)$ of order of growth not higher than p and of order of decrease along the real axis not lower than p: $|\varphi(z)| \leqq C_1 e^{c_2|z|^p}$, $|\varphi(x)| < C_3 e^{-c_4|x|^p}$ may serve as an example of a basic space. It is possible to mention a number of other spaces, for example the space K of all finite infinitely differentiable functions introduced by Schwartz, and its dual space Z^1 of entire analytic functions $\varphi(z)$ of order of growth not higher than 1, decreasing along each straight line parallel to the axis of x, faster than any power of x:

$$\sup_{-\infty < x < \infty} |x^n \varphi(x + iy)| \leqq C_n e^{c_n|y|} \quad (n = 1, 2, \ldots).$$

For applying the Fourier transform method to a problem (for example to a given type of differential equations) it is important to select the basic space (and, consequently, the set of generalized functions). For instance, if we have a system of differential equations of order p,

$$\frac{\partial u_i}{\partial t} = \sum P_{ik} \left(\frac{\partial}{\partial x_1}, \frac{\partial}{\partial x_2}, \ldots, \frac{\partial}{\partial x_n} \right) u_k,$$

where the P_{ik} represent polynomials of degree not higher than p, then by introducing the basic space of analytic functions Z_q^q $(p < q < p + \varepsilon)$ we can, with the help of the Fourier transform, prove the uniqueness of the solution of the Cauchy problem for the class of functions which do not grow faster than $e^{|z|^{p'-\varepsilon}}$ $\left(\frac{1}{p} + \frac{1}{p'} = 1\right)$. One of the principal points in the proof of this theorem, the fact that $\tilde{Z_p^p} = Z_{p'}^{p'}$, is based on the Phragmén-Lindelöf theorem. Thus, the use of a complex variable plays an important part in this case and this is evidently connected with the very essence of the problem. And indeed from the general theorem formulated here follows the uniqueness of the solution of the Schroedinger equation $\frac{\partial u}{\partial t} = i \triangle u$ in the class of functions which do not grow faster than $e^{|z|^{2-\varepsilon}}$. While for the equation of heat transfer the proof of uniqueness in the same class of functions may be obtained in numerous ways, it is hard to conceive how one could prove this theorem for the Schroedinger equation without using the theory of functions of a complex variable.

Lack of time does not permit us to discuss other applications of generalized functions. We shall only emphasize once more the following two points:

1) In order to be able to use the method of generalized functions it is important that for each problem, or class of problems, one can construct the corresponding basic space. A "universal" basic space valid for all problems does not exist, and it is senseless to attempt to construct it.

2) For a broad class of problems in which generalized functions are used it is necessary to pass to the complex area. This implies the consideration of a basic space consisting of analytic functions (or a space consisting of the solutions of some differential equations).

Another significant aspect of the theory of generalized functions should be mentioned, viz. the calculation of divergent integrals and series. The calculation of divergent integrals is carried out by Hadamard, essentially by dropping the divergent part (or, as a physicist would put it, by the "regularization" of the given divergent integral). M. Riesz obtains the convergence of some types of integrals by applying the method of analytic continuation. It must be underlined that we are speaking here not of summing oscillating integrals, but of giving finite values to integrals which really tend to $+\infty$.

In connection with this group of questions the following two problems are of interest:

I. Let $P(x_1, x_2, \ldots, x_n)$ be a polynomial. Consider the area in which $P > 0$. Let $\varphi(x_1, \ldots, x_n)$ be an infinitely many differentiable function equal to zero outside a certain finite area. We shall examine the functional

$$(P^\lambda \cdot \varphi) = \int_{P>0} P^\lambda(x_1, \ldots, x_n)\varphi(x_1, \ldots, x_n)dx_1 \ldots dx_n.$$

It is necessary to prove that this is a meromorphic function of λ (it would be natural to call it a ζ-function of the given polynomial), whose poles are located in points forming several arithmetical progressions, as well as to calculate the residues of this function.

II. The second problem may be stated as follows: What should one understand by a rational function (as a generalized function) and how might one find its Fourier transform? We shall consider a rational function as a functional on Z^1, i.e. we shall understand the value of this functional to be an integral along some surface parallel at infinity to the real subspace of Z^1 (different surfaces produce different functionals). The Fourier transform of such a functional is the functional on K (a space of finite functions), dual with respect to Z^1. The general case of these Fourier transforms has been studied little. For the special case where the denominator of the rational function is a polynomial of the so called hyperbolic type, they were investigated, in connection with hyperbolic differential equations, by Herglotz [22], Petrovski [69], Garding [23] and Leray [24]. In connection with elliptic equations Fourier transforms of rational functions with "elliptic" denominators were studied by Shapiro [25], Lopatinski [26] and Bochner [27].

4. *Generalized Functions and Representations of Groups.*

The development of the theory of representations of Lie groups in the last few years shows that here, too, generalized functions are a convenient and useful apparatus. The use of generalized functions has, in particular, been proved useful in the investigation of the equivalence and irreducibility of representations (Brouchat [21], Mackey [30] and Mautner [37]). Generalized functions may also be employed conveniently in assigning scalar products which fit the supplementary series (see [6], § 18).

One of the specific questions of the theory of representations, where generalized functions can be used conveniently, is the so-called Plancherel theorem which gives the expansion of the function $f(g)$ on the Lie group into the analogue of the Fourier integral [9, 10]. This is equivalent to the calculation of $f(e)$ (e is the unit of the group), if for each character $\chi(g)$ of this group $\int f(g)\chi(g)dg$ is known.

For classical Lie groups, from the values of this integral, we can easily find the integral of the function $f(g)$, taken in any class of conjugate elements in general position. Thus in the case of compact groups $f(e)$ is determined. For instance, for the group of unitary matrices the conjugate element class is the set of all matrices with given eigenvalues. When all these eigenvalues tend to 1, the respective classes of conjugate elements tend to the unit, just like concentric spheres $x^2 + y^2 + z^2 = c$ shrink into their centre when $c \to 0$.

In the case of a noncompact group, for instance a group of all nth order matrices with determinant 1, the class of conjugate elements in general position is again the set of all matrices with given unequal eigenvalues. In this case, however, when the eigenvalues approach 1, the corresponding class does not reduce to a single matrix at all, but approaches the set of all matrices with the only eigenvalue 1, just as hyperboloids $x^2 + y^2 - z^2 = c$, when $c \to 0$, do not reduce to the origin, but become a cone. The unit matrix itself constitutes a rather complicated singular point in the manifold of all matrices, which have all eigenvalues equal to 1, and it is, therefore, far from obvious how one can find $f(e)$ if one knows the integral of the function on the class of conjugate elements. This problem may be solved with the aid of generalized functions by applying the method due to M. Riesz [9, 10].

In order to explain the essential idea of the method, we shall illustrate it by a simple model problem. Let f be a finite, sufficiently smooth function. Let I_c denote the integral of the function f on the hyperboloid $x^2 + y^2 - z^2 = c$. The problem is to calculate $f(0,0,0)$, if, for each c, I_c is known. Note that for this purpose the integral $\int f(x,y,z)(x^2 + y^2 - z^2)^\lambda \, dxdydz$ can be calculated if one knows only I_c, as on each hyperboloid the second factor is constant. Now, if λ approaches $-\frac{3}{2}$, the integral will approach $f(0,0,0)$. Since on the other hand the same integral can be written in terms of I_c, we obtain an expression for $f(0,0,0)$ in terms of I_c. The general problem stated above for groups is solved in a similar way.

The following problem is closely connected with the one examined. Let $P(x,y,z, \ldots)$ be a polynomial and $f(x,y,z, \ldots)$ a finite function. It is necessary to find the values of f in singular points, the integrals of f along singular lines of the surfaces $P = C$, etc., if the integrals of f over all surfaces of constant level of the polynomial P are known.

Let us consider a few problems arising in connection with the application of the theory of generalized functions to representations.

It is well-known from the theory of representations of compact groups that finding the representations of a compact group is equivalent to finding the representations of the centre of the group ring consisting of those functions which are constant on the classes of conjugate elements, i.e. which satisfy the condition $f(g) = f(g_0 g g_0^{-1})$ for all $g, g_0 \in G$. The product (convolution) of such functions is determined by the formula: $(f_1 \cdot f_2)(g) = \int f_1(g g_1^{-1}) f_2(g_1) \, d\mu(g_1)$.

The representation of the centre of the group ring is given by the formula $f(g) \to \int f(g)\chi(g)dg$ where $\chi(g)$ is a character of the group.

As mentioned above, characters for arbitrary classical groups also exist and also determine uniquely the representation. It would be interesting, in this case, too, to construct the centre of the group ring and thus to obtain characters.

The direct determination of the centre is hindered by the fact that any function belonging to the centre is constant on the classes of conjugate elements, and, consequently, as a rule, is not integrable; therefore the direct calculation of the convolution leads to divergent integrals. A more general problem of the same type is the problem of the construction of a general theory of spherical functions. It is known that general spherical functions are connected with the group of the transformations of a manifold in the same manner as the conventional spherical functions are connected with the group of the rotations of a sphere. Instead of a group of transformations of a manifold one may speak of a group and a given stationary subgroup. Then the representations of the group prove to be connected in a natural way with the ring of functions which are constant on two-sided cosets of the group on this subgroup [3].

So far only the case has been studied, where the respective stationary group is compact [31, 13, 70]. In the general case the functions of the ring, as a rule, are not integrable and the calculation of their convolution leads to divergent integrals. It is quite probable that the application of the theory of generalized functions will allow the construction of a theory of spherical functions for noncompact Lie subgroups as well.

For compact Lie groups there are two ways of proving that the representations constructed are all the possible representations of this group. One of them consists in using the theory of characters, the other in using the Cartan theory of highest weights. We have already mentioned the possibility of applying the first of them to local compact Lie groups. It is interesting that the wider use of the second method is also connected with the theory of spherical functions on noncompact stationary subgroups. For instance, in the case when G is a group of matrices with determinant 1 the subgroup Z of triangular matrices with all diagonal elements equal to one is such a subgroup.

It seems to me that, if the theory of generalized functions is used correctly, the main problems of the theory of representations of locally compact Lie groups will be not more complicated than for compact groups.

5. *Differential Operators.*

The theory of differential equations is one of the sources of modern functional analysis. In 1910 H. Weyl published one of the first works on the spectral properties of differential operators. Several related concepts such as the eigenfunction and the eigenvalue came into use at the beginning of the last century (Fourier method). A comprehensive exposition of the main questions in differential operator theory should require at least a separate report. Therefore, we shall occupy ourselves only with a few special questions.

At the present time the main progress in differential operator theory has been achieved in the following topics grouped somewhat arbitrarily:

1) Differential operators and boundary value problems
2) Spectral theory of differential operators
3) Inverse problems.

To the most important topics of the theory of differential equations belong boundary value problems for various types of equations. The connection between these problems and spectral theory is clearly indicated in the works of Friedrichs. In the last few years new important results have been obtained by Vishik [32, 33, 34, 35]. Unfortunately, due to lack of time, we cannot discuss here these works in detail.

We shall now treat more in detail the spectral theory of differential operators. Let us consider first of all selfadjoint differential operators. The existence of a general theorem for the spectral expansion of an arbitrary selfadjoint operator in differential operator theory cannot be considered as a final solution to the problem, as it is desirable to obtain this expansion as an expansion by ordinary or generalized eigenfunctions and not as an abstract resolution of the unity E_λ.

The expansion by eigenfunctions over a finite interval, in the case of ordinary differential equations, has been known for a long time. In 1910 Weyl considered the case of an expansion over an infinite interval for a differential operator of the second order. The spectral expansion has further been extended to the case of ordinary selfadjoint differential operators of higher orders by M. G. Krein [28] and Kodaira [29], and to the case of partial elliptic differential equations by A. J. Powsner [65] who used Carleman's results for this purpose. Mautner [37] presented an interesting study filling the gap between general spectral expansions and expansions by eigenfunctions. In this study restrictions are imposed on an operator in functional space, allowing the spectral expansion to be accomplished by functions. From Mautner's results it is possible to obtain for elliptic operators the expansions by eigenfunctions (see also Garding [30]).

In the field of partial differential operators progress has been made only for elliptic operators. A solution to the problem of a spectral expansion of an arbitrary (e.g. hyperbolic) operator has not been found as yet. Generalized functions do not give any new results for elliptic operators. However, eigenfunctions should be sought for as generalized functions in the case of hyperbolic differential operators.

Let us consider the problem of a spectral expansion of a dynamic system. Let a system of differential equations be given in a certain analytic manifold:

$$\frac{dx_i}{dt} = X_i(x_1, x_2, \ldots, x_n),$$

where the X_i are analytic functions. In the manifold there exists an invariant integral (invariant volume).

This system may be compared with the partial differential operator

$$Z = X_1 \frac{\partial}{\partial x_1} + X_2 \frac{\partial}{\partial x_2} + \ldots + X_n \frac{\partial}{\partial x_n}$$

(which can be called the Lie operator for this system), characterizing an infinitely small displacement along the trajectory. If a Hilbert space consisting of square integrable functions is considered, the operator iL will be a selfadjoint operator in this space having, in accordance with the general theory, a certain spectral expansion. The rather trivial case where the spectrum of this operator is a pure point spectrum has been investigated completely. As far as we know, the continuous spectrum in a spectral expansion has been investigated only for the so-called geodesic streams in manifolds of constant negative curvature [12]. However, even in this case, the investigation of the corresponding eigenfunctions has not yet been completed. Here it is necessary to prove the existence of a complete system of generalized eigenfunctions for any analytic dynamic system. In all known examples of transitive dynamic systems with a continuous spectrum the multiplicity of this spectrum is infinite. At the same time, the eigenfunctions (considered in the usual way and not as generalized functions) are only constant here. It is interesting to determine whether other generalized eigenfunctions (corresponding let us say to the eigenvalue $\lambda = 0$) exist in such systems.

A second problem in the field of spectral expansion of differential operators is as follows. It is well-known that the theory of eigenvalues is equivalent to the study of a pencil of quadratic forms $A + \lambda E$. The more general problem concerning the study of the pencil $A + \lambda B$, where $B(u,u)$ is a positive definite quadratic form, in the general theory of linear operators (of finite dimensions or not), is equivalent to this one, as we have only to choose a new scalar product. However, in the case of differential operators, where the theorem for the spectral expansion of the unit operator is not the final result of the whole theory but rather a leading thread, the problem for the quadratic differential form $A(u,u) + \lambda B(u,u)$ is a new one. An interesting example of such a study is found in the publications of Sobolev [38] and Alexandryan [39]. There a problem is considered which is essentially equivalent to the study of quadratic forms

$$\int\int \frac{\partial^2 u}{\partial x^2} u dx dy + \lambda \int\int \left\{ \left(\frac{\partial u}{\partial x} \right)^2 + \left(\frac{\partial u}{\partial y} \right)^2 \right\} dx dy$$

for functions equal to zero on the boundary of a region G.

An interesting and important problem in the spectral theory of differentia operators is the study of the asymptotic behaviour of the eigenfunctions and eigenvalues. Numerous results have been obtained recently in the study of partial differential equations by B. M. Levitan [41, 42]. The investigation of the asymptotic behaviour of the eigenfunctions and eigenvalues of differential operators is closely related to the so-called quasi-classical approximation in quantum mechanics.

Selfadjoint operators were considered above. Much less progress has been made in the spectral theory of non-selfadjoint operators, although it sphysical application is of great value. In this field we have the early works of Tamarkin and Hille where the case of a finite interval of a straight line is considered and an important work of Carleman where the existence of eigenfunctions for the elliptic differential operator of the second order is proved. The results obtained by M. V. Keldysh [40] in 1951 are an important contribution to this field. He proved that the system of eigenfunctions for an elliptic equation together with the so-called allied functions (the analogue of the basis with respect to which the matrix takes the Jordanform) is complete. In the works of M. V. Keldysh a wide use has been made of the theory of functions of a complex variable and also of his theorems on the general theory of operators (see also Browder [43, 44]).

These results may be considered as the beginning of a new stage in spectral theory, since here, for the first time, a theorem on the completeness of eigenfunctions has been given that goes beyond the field of selfadjoint operators.

A study by M. S. Lifshitz [45] should be mentioned where the possibility of transforming a selfadjoint and completely continuous operator into the triangular form has been proved. M. S. Lifshitz makes use of the methods of the theory of functions of a complex variable and in particular of Potapov's theorem [46], when dealing with the expansion of bounded analytic matric functions in a product, similar to the Blaschke product.

Hardly anything has been done in the spectral theory of non-selfadjoint operators in the case of an infinite region. In this case it is not even clear under what conditions for the operator a spectral expansion is possible. Let us consider the simple example of a differential equation of the first order:

$$Ly = i\frac{dy}{dx} + a(x)y,$$

where $a(x)$ represents a certain complex function. In this example we shall use the following most conservative definition of a spectrum. Let us consider the operator over a finite interval $(-N,N)$ having, for instance, periodic boundary conditions. Then a discrete spectrum $\lambda(N)$ is obtained. Now let $N \to \infty$. The limit of the set $\lambda(N)$ for $N \to \infty$ will be called the spectrum of the operator L along the whole axis. The eigenvalues of the operator L, with periodic boundary conditions, in the interval $(-N, N)$ are equal to

$$\lambda_n^N = \frac{1}{2N}\int_{-N}^{N} a(s)ds + \pi\frac{n}{N}.$$

We can see, for instance, that, if the mean value of the imaginary part of the function $a(x)$ in the interval $(-N, N)$ approaches ∞ as $N \to \infty$, then the whole spectrum goes to infinity as $N \to \infty$. Consequently, in this case (e.g. for the operator $i\frac{dy}{dx} + ix^2y$) there is no spectrum in the finite part of the plane and the very problem of the spectral expansion for such an operator has no meaning.

It is hardly possible to have a spectral expansion over an infinite interval, for example in a Hilbert space for the operator $i\frac{d^2y}{dx^2} + ixy$, which is the "Fourier transform" of the above mentioned operator. As far as we know, the only spectral expansion of non-selfadjoint differential operators in an infinite region is the expansion of the differential operators of the second order shown in the work of Naimark [47].

Finally we shall say something about inverse problems. So-called inverse problems (see also [48, 49]) have become of great importance in connection with certain problems of mathematical physics and especially of quantum mechanics. Let us examine for example Schroedinger's equation

$$-\psi''(r) + V(r)\psi = k^2\psi$$

with the conditions that $\psi(0) = 0$, $\psi'(0) = 1$. Assuming that the potential $V(r) = 0$ for $r \geqq a$ we have, if r is sufficiently large, $\psi(r) = \varrho(k) \sin(kr + \eta(k))$ where $\eta(k)$ is the phase shift. The inverse problem in this case is to find the potential from $\eta(k)$ or $\varrho(k)$. Where there is no discrete spectrum, the potential is determined uniquely by the phase shift, as was proved by Levinson [50]. At about the same time, similar solutions to inverse problems in geophysics were obtained by A. N. Tichonov [51].

The uniqueness of the solution to the inverse problem for an arbitrary

$V(r)$ has been proved by Marchenko [52], who considered, instead of the phase, the so-called spectral function $\varrho(\lambda)$, which in the preceding particular case coincides with the multiplier $\varrho(k)$.

Further topics in the field of inverse problems for equations of the second order are the following questions:

a. What can serve as spectral function and in particular what point set can serve as a spectrum for a differential operator;

b. In which way can the potential be reproduced effectively by $\varrho(k)$?

All these problems were solved by M. G. Krein [53, 54, 55], B. M. Levitan and the author [56]. In these studies the authors derived necessary and sufficient conditions for $\varrho(k)$ being the spectral function of a differential operator. The determination of the coefficients of the equation by means of $\varrho(k)$ was reduced to the solution of a certain Fredholm integral equation [56]. These results make it also possible to penetrate into the nature of the spectrum of differential operators. It follows that:

1) For any closed set in a finite interval there exists a differential operator the spectrum of which in this interval coincides with the given set.

2) Any sequence fulfilling certain asymptotic conditions can be the sequence of eigenvalues for a Sturm-Liouville operator in a finite interval (with corresponding boundary conditions).

The above mentioned integral equation for inverse problems was applied to problems on dispersion theory in quantum mechanics in the works of Jost and Kohn [57]. It was used by Corinalesi [58] in the case of a relativistic particle.

Using the same integral equation M. G. Krein derived important formulas in order to determine $V(r)$ by means of $\varrho(\lambda)$ for a large class of functions $\varrho(\lambda)$, and in particular for all rational functions.

So far we have considered the one-dimensional case of an inverse problem. Little has been done for the case of several independent variables. It would be most natural to consider in this case a problem similar to that of determining $V(r)$ by means of the phase.

Let us consider the equation:

$$-\Delta u + V(x,y,z)u + \varkappa^2 u = 0.$$

Assume that $V=0$ outside a certain finite region and that only equations without any discrete spectrum are to be considered. Let us examine the solution of this equation and also a normal derivative of the solution on a sufficiently large closed surface. The function u and its normal derivative can be divided into two components, u_1 and u_2 say, representing convergent and divergent waves. We may consider these waves (both outgoing and ingoing) without taking into account the derivatives of u_1 and u_2, because the values of the normal derivatives are determined by the values of the function itself. Let $S(k)$ be the operator which to each outgoing wave associates the corresponding ingoing wave. The problem is to determine whether the potential $V(r)$ can be defined by means of the operator $S(k)$. Instead of the operator $S(k)$ the operator $R(\lambda)$ introduced by Wigner can be considered. It establishes a relationship between the function on the surface and its normal derivative [67, 68].

In quantum mechanics this problem can be treated in terms of dispersion theory in the same way as in the one-dimensional case. We mention also an interpretation of this problem in the field of optics. A light wave coming from infinity is dispersed in the points of inhomogeneity defined by a certain function V. We observe a dispersed wave. Is it possible to reproduce the function V using the data, obtained by varying in any manner the oncoming waves and observing the corresponding dispersed waves? If it is possible, how can we do it?

A possible variant of the inverse problem is as follows. Let us examine the external boundary condition problem for the equation

$$-\Delta u + V(x_1, x_2, \ldots, x_n)u = \lambda u,$$

with the boundary condition $\frac{\partial u}{\partial n} = 0$. Let us consider the resolution of the unity $E(\lambda)$ corresponding to this operator. $E(\lambda)$ is an integral operator with a kernel $\varrho(P,Q,\lambda)$. If we consider this kernel for points P and Q lying on the boundary of the region, we get the function $\varrho(S_1,S_2,\lambda)$, which is an analogue of the function $\varrho(\lambda)$ in the one-dimensional problem. Now one has to find the potential $V(r)$ knowing $\varrho(S_1,S_2,\lambda)$. Some results for this second problem were obtained by Berezanski [59].

6. *Analysis in Functional Spaces.*

Although the branches of functional analysis mentioned above are comparatively new and are continuing to develop rapidly, they nevertheless have already taken a definite form and acquired, so to say, a "personal" character. This is not true for the group of questions covered by the last section of this report. We shall deal here with problems and procedures which are just beginning to appear; it is, however, possible that in the future they will occupy a central place in functional analysis as a whole. There are a number of physical problems in which, apart from difficulties of physical character, other difficulties arise, due to the absence of a sufficiently general, adequate mathematical apparatus. Some questions of quantum electrodynamics, the theory of turbulence, etc. are of this type. Lately such a mechanism has begun to take shape. It might be called analysis in functional spaces. We shall illustrate the questions and methods arising here on a problem of quantum electrodynamics. We shall not use any data of quantum electrodynamics, but shall only consider the simplest model equation. On the basis of ideas which are to be found in the well-known paper by V.A. Fok [17], Schwinger [62] developed quantum electrodynamics with the aid of functionals in a space (referred to as the space of external sources). In doing so the equations of quantum electrodynamics become a comparatively complicated system of integral equations and lead to relationships between functionals and their variational derivatives. This system of equations could be simplified much and reduced to a linear differential equation in the functional space, or to a system of such equations [66]. We shall demonstrate this procedure by using a very simple equation, the so-called Thiring quantum equation

$$(-\Box + \varkappa^2)\psi = \lambda\psi^3 + I,$$

where

$$\Box = \frac{\partial^2}{\partial x_1^2} + \frac{\partial^2}{\partial x_2^2} + \frac{\partial^2}{\partial x_3^2} - \frac{\partial^2}{\partial t^2}$$

and $\psi(x_1,x_2,x_3,x_4)$, as usual in quantum theory, designate certain operators. Following the method of Schwinger we add certain scalar functions $I(x)$ (called external sources) to this equation and reduce it to a system of ordinary differential equations (not containing operators) in the following way. Let us assume that at the moment $t = -\infty$ the external sources $I(x)$ are absent and let e_0 designate the state of vacuum without external sources. If then external sources are introduced, the state e_0 will alter in some way, and at $t = +\infty$ will change into some state e_0^1. Let us assume that Z is equal to the scalar product of the wave functions e_0 and e_0^1 ($|Z|^2$ indicates the probability that the vacuum has remained a vacuum). Z is a linear functional of the external sources I and satisfies the following differential equation:

$$\lambda \frac{\delta^2 Z}{\delta I(x)^2} + (\Box - \varkappa^2)\frac{\delta Z}{\delta I(x)} = I(x)Z. \tag{1}$$

Here $\frac{\delta Z}{\delta I(x)}$ is the so-called variational-functional derivative, i.e. the limit of the ratio of the increase of Z and $\int \delta I(x)dx$, where the variation $\delta I(x)$ is concentrated in an infinitesimal neighbourhood of the fixed point x_0.

The equation (1) has a similar form as the Airy equation, which is obtained from it, if Z is considered, not as a functional, but as a function and the variational-functional derivatives are replaced by usual derivatives.

It seems natural to call the solutions of the functional equation (1) Airy functionals. They can be obtained in the following way.

Let us consider in the four-dimensional space of variables x_1,x_2,x_3,x_4 a cube of side L, divide it into cells of side τ and assume that Z depends only on the values of I at the joints of these cells. The equation (1) reduces to a system of partial differential equations

$$\frac{1}{\omega^2}\frac{\partial^2 Z}{\partial I_i^2} + \frac{1}{\omega}\sum_k R_{ik}\frac{\partial Z}{\partial I_k} = I_i Z,$$

where ω is the volume of a cell and R^{ik} is the matrix obtained by replacing the operator $\square - \varkappa^2$ by the corresponding difference operator.

We shall solve this equation with the aid of a Laplace transform, i.e. we shall search a solution of the form

$$Z(I_1, I_2, \ldots, I_n) = \int \xi(s_1, \ldots, s_n) e^{i\Sigma I_k s_k} ds_1 \ldots ds_n, \tag{2}$$

where the integration is extended over the surface obtained as the direct product of n contours C_k chosen in the planes of the corresponding complex variables s_k. Thus we obtain the following expression for $\xi(s_1, s_2, \ldots, s_n)$:

$$\xi(s^1, \ldots, s_n) = e^{i\Sigma\frac{s_k^3}{3} + \Sigma R_{ik} s_i s_k}.$$

In order that the integral (2) be different from zero it is necessary that each of the contours C_k extend into infinity. At the same time we must ensure the convergence of the integral (2). This problem can be solved for each of the contours C_k along which one integrates. It is easy to verify that for the convergence of the integral (2) it is necessary and sufficient that each of the contours C_k extend into infinity within one of three angles, I, II and III say. In order to obtain a nonzero integral, one must take the contour C_k in the union of the angles I and II, I and III, or II and III. In the latter case the value of the integral is equal to the sum of its values in the two former cases, so there will be only two independent integrals. In integrating along all the n contours we obtain in this way 2^n independent solutions. Linear combination of the solutions furnishes the general solution of the equation (1). However, if we should now pass to the limit by letting the dimensions of the cells tend to zero ($\tau \to 0$), i.e. if we go over from indices $1, 2, \ldots, n$ to a continuous variable x, then in the limit we shall obtain only two solutions (if in different planes we take the contours in different pairs of angles, then there will be no solution in the limiting case).

If the functional Z has been found, the calculation of various quantum-mechanical effects reduces to the determination of some functional derivatives of Z.

This treatment of quantum-mechanical problems is closely connected with the theory constructed by Feynman [63] on the basis of other considerations. Similar methods are also used by Edwards and Peierls [64].

For a rigorous justification of the method of solution given above it is necessary to be able to answer a number of other questions. For instance, one must justify the transition to the limit for $n \to \infty$. It would be still better to obtain the solution directly by integrating in the functional space. It is desirable to find the asymptotic behaviour of the Airy functionals (in a similar way as was done for the Airy functions). All these specific mathematical questions are of unquestionable interest not only for quantum electrodynamics, where they arise, but for many other fields as well. Questions which are very close to those just discussed, arise, for example, in the theory of turbulence in the interpretation given in a recent work by E. Hopf [61]. Let $u_\alpha(x)$ be the velocity field of a certain liquid. Let us introduce the functions $y^\alpha(x)$ and assume that $Z = \langle e^{\int y^\alpha u_\alpha dx} \rangle$ (here the simbol $\langle\,\rangle$ indicates taking the mean). Z represents a functional of $y(x)$. On the basis of the equations of hydrodynamics E. Hopf has given a linear equation for Z in functional derivatives. The similarity with quantum electrodynamics found here is of interest. For instance, the correlation between velocities in different points is calculated, as quantum-mechanical effects, by means of functional derivatives of Z. Here questions of the kind of integration in functional spaces may also play an essential rôle.

It should be noted that some of the concepts of analysis in functional spaces do exist already rather a long time (f.i. N. Wiener's concept of measure in a functional space). However, in the near future, they may occupy a considerably more significant place than they have so far.

1. Bargman V., Annals of Math., *48*, 568-640 (1947).
2. Гельфанд И. М., Шапиро З. Я., Успехи мат. Наук, т. XII, в. 1 (47), 3-117 (1952).
3. Гельфанд И. М., Райков Д. А., Мат. сборник, *13* (55), 301-316 (1943).
4. Mautner F. Y., Proc. Nat. Acad. Sci. USA, *34*, 52-54 (1948).
5. Гельфанд И. М., Наймарк М. А., Известия Ак. наук СССР, серия математическая, *12*, No. 5, 445-480 (1948).
6. Гельфанд И. М., Наймарк М. А., Труды Мат. ин-та АН СССР им. Стеклова, т. 36, 1-288 (1950).
7. Harish-Chandra, Trans. Amer. Math. Soc., *75*, No. 2, 185-243 (1953).
8. Гельфанд И. М., Граев М. И., Доклады Ак. наук СССР, *86*, No. 3, 461-463 (1952).
9. Гельфанд И. М., Граев М. И., Доклады Ак. наук СССР, *92*, No. 2, 221-224 (1953).
10. Гельфанд И. М., Граев М. И., Доклады Ак. наук СССР, *92*, No. 3, 461-464 (1953).
11. Гельфанд И. М., Граев М. И., Известия Ак. наук СССР, серия матем., *17*, вып. 3, 189-248 (1953).
12. Гельфанд И. М., Фомин С. В., Доклады Ак. наук, *76*, No. 6, 771-774 (1951), Успехи матем. наук, т. XII, вып. 1 (47), 118-136 (1951).

13. Godement R., Trans. Amer. Math. Soc., *73*, 496-556 (1952).
14. Godement R., Annals of Math., *59*, No. 1, 63-85 (1954).
15. Hadamard J., Le problème de Cauchy et les équations aux dérivées partielles linéaires hyperboliques, Paris, 1932.
16. Riesz M., Acta Math., *81*, 1-123, No. 1-2 (1949).
17. Фок В. А., Sow. Phys., *6* (1934).
18. Schwartz L., Théorie des distributions, v. I, II, Paris, 1951.
19. Carleman T., L'intégrale de Fourier et questions qui s'y rattachent. Leçons Uppsala, 1944.
20. Bochner S., Vorlesungen über Fouriersche Integrale, Leipzig, 1932.
21. Brouchat C. R., Acad. Sci. Paris, *237*, 1478-1480 (1953).
C. R. Acad. Sci. Paris, *238*, 38-40, 437-439, 550-553 (1954).
22. Herglotz G., Acad., Wiss. Leipzig, Math. Phys. Klasse, *78*, 93-196, 287-318 (1923).
23. Garding L., Acta Math., *85*, 1-62 (1951).
24. Leray J., C. R. Acad. Sci. Paris, *234*, 1112-1115 (1952).
25. Шапиро З. Я., Известия Ак. наук СССР, серия матем., *17*, No. 6 (1953).
26. Лопатинский Я. Б., Доклады Ак. наук СССР, *71*, No. 3, 433-436 (1950).
27. Bochner S., Annals of Math., *57*, No. 1, 32-56 (1953).
28. Крейн М. Г., Доклады Ак. наук СССР, *74*, 9-12 (1950).
29. Kodaira K., Amer., J. Math., *72*, 502-544 (1950).
30. Mackay G. W., a) Ann. of Math., *55*, No. 1, 101-139 (1952).
b) Ann. of Math., *58*, No. 2, 193-221 (1953).
31. Гельфанд И. М., Доклады Ак. Наук СССР, *70*, No. 1, 5-8 (1950).
32. Вишик М. И., Доклады Ак. наук СССР, *64*, 433-437 (1949).
33. Вишик М. И., а) Доклады Ак. наук СССР, *93*, No. 2, 225-228 (1953).
б) Доклады Ак. наук СССР, *93*, No. 1, 9-12 (1953).
34. Вишик М. И., Мат. сборник, *29*, (71) No. 3, 615-676 (1951).
35. Вишик М. И., Труды Моск. Мат. общ-ва, т. I, 187-246 (1952).
36. Garding L., Comptes rendus du Douzième Congrès Math. Scandinaves, 44-55 (1953).
37. Mautner F. Y., Proc. Nat. Acad. Sci. U.S.A., *39*, No. 1 (1953).
38. Александрян Р. А., Доклады Ак. наук СССР, *73*, No. 5, 869-872 (1950).
39. Соболев С. Л., а) Доклады Ак. Наук СССР, *81*, No. 6, 1007-1009 (1951).
б) Доклады Ак. наук СССР, *82*, 205-208 (1952).
в) Мат. сборник, *1* (43) (1936).
40. Келдыш М. В., Доклады Ак. наук СССР, *77*, No. 1, 11-14 (1951).
41. Левитан Б. М., а) Доклады Ак. наук СССР, *90*, No. 1, 133-135 (1953).
б) Доклады Ак. наук СССР, *94*, No. 2, 179-182 (1954).
42. Левитан Б. М., Известия Ак. наук СССР, серия матем., *17*, No. 4, 331-364, No. 5, 473-484 (1953).
43. Browder F. E., Proc. Nat. Acad. Sci. U.S.A., *39*, 179-184 (1953).
44. Browder F. E., Proc. Nat. Acad. Sci. U.S.A., *39*, 433-439 (1953).
45. Лифшиц М. С., Мат. сборник, *34*, вып, 1, 145-199 (1954).
46. Потапов, Доклады Ак. наук СССР, *72*, No. 5, 849-852 (1950).

47. Наймарк М. А., Труды Моск. Матем. общ-ва, т. III, 181-270 (1954).
48. Амбарцумян В. А., Zeits. f. Physik, *53*, 690 (1929).
49. Borg G., C. R. Dixième Congrès Math. Scandinaves 1946, Copenhagen, 172-180 (1947).
50. Levinson N., Bull. Am. Math. Soc., *55*, No. 5 (1949).
51. Тихонов А. Н., Доклады Ак. наук СССР, *69*, 797-800 (1949).
52. Марченко В. А., Доклады Ак. наук СССР, *72*, No. 3, 457-463 (1950).
53. Крейн М. Г., Доклады Ак. наук СССР, *76*, No. 1, 21-24 (1951).
54. Крейн М. Г., Доклады Ак. наук СССР, *76*, No. 3, 345-348 (1951).
55. Крейн М. Г., а) Доклады Ак. наук СССР, *93*, 617-620, 767-770 (1953).
б) Доклады Ак. наук СССР, *94*, No. 6, 987-990 (1954).
56. Гельфанд И. М., Левитан Б. М., Известия Ак. наук СССР, серия матем., *15*, 309-360 (1951).
57. Yost R., Kohn W., Kgl. Danske Vidensk. Selskab, *27*, No. 9 (1953).
58. Corinaldesi E., Il. Nuovo Cimento, XI, No. 5 (1954).
59. Березанский Ю. М., Доклады Ак. наук СССР, *93*, No. 4, 591-594 (1953).
60. Гельфанд И. М., Шилов Г. Е., Успехи мат. наук, XIII, No. 6 (58), 3-54 (1953).
61. Hopf E., Journal Rat. Mech. and Analys, *1*, 87-123 (1952).
62. Schwinger J., Proc. Nat. Ac. Sci. U.S.A., *37*, No. 7 (1951).
63. Feynman R. P., Phys. Rev., *80*, No. 3, 440-457 (1950).
64. Edwards, Peierls R., Proc. Roy. Soc. A., *224*, No. 156, 24-33 (1954).
65. Повзнер А. Я., Мфт. сборник, *32* (74), 109-156 (1953).
66. Гельфанд И. М., Минлос Р., Доклады Ак. наук СССР, *95*, No. 2, 209-212 (1954).
67. Wigner E., Ann. of Math., *53*, No. 1, 36-67 (1951).
68. Wigner E., Neuman J., Ann. of Math., *59*, No. 3, 418-433 (1954).
69. Петровский И. Г., Мат. сборник, *7* (59), No. 3, 289-384 (1945).
70. Гельфанд И. М., Наймарк М. А., Труды Моск. Мат. общ-ва, т. I, 423-475 (1952).

2.

On some problems of functional analysis[1]

Usp. Mat. Nauk **11** (6) (1956) 3–12 [Transl., II. Ser., Am. Math. Soc. **16** (1960) 315–324]
Zbl. **100**: 321

How widely the ideas and methods of functional analysis have spread nowadays is attested by this conference. Originating in the needs of quantum mechanics and certain branches of mathematics itself (integral equations, the calculus of variations), functional analysis in the nineteen-thirties was basically the theory of linear normed spaces and of linear operators in these spaces. At the present time the domain of applications of functional analysis has been much extended. Its methods are widely used in the theory of differential equations, in numerical methods, in the theory of probability and so forth, on the one hand, and in problems of theoretical physics on the other hand. It is natural that the very concept "functional analysis" has at the same time undergone an essential modification.

It would be difficult to determine at the present time the future lines of development of functional analysis. Only one thing is clear, that just as originally, so also presently functional analysis will develop in close connection with its applications. Two regions, in our opinion, will exert the strongest influence of all on the future course of development of functional analysis. The first of these is hydrodynamics (the problem of the flow of a viscous fluid, the theory of a compressible gas, the theory of turbulence). In the second place there is theoretical physics, more precisely the quantum theory of fields and the theory of elementary particles. This region of physics, it is true, itself now stands at a crossroad, and it is not clear what the nature of its future development will be, but however it develops, one thing is clear: it and functional analysis are to be fellow-travellers.

There is yet another region of science which may substantially influence the future development of mathematics and, in particular, functional analysis. I may provisionally associate this region with high-speed calculating machines, the theory of information and suchlike questions. However at present it is hard to foresee how the development of these regions will bear on functional analysis.

In the present report I want to indicate certain topics which, in my opinion, offer interest for the further development of the methods of functional analysis. In this I make no claim either to treat such topics exhaustively, or that just these topics are the most important ones. The choice of questions to be considered here has been dictated in the first instance by my personal tastes. In the second place,

1) The review articles which are published in the present issue of the Uspehi Mat. Nauk constitute an account of the corresponding review lectures which were read at the All-Union conference on functional analysis and its applications, which was held in January 1956 in Moscow (for information about the conference see Uspehi Mat. Nauk 11, no. 3, 249, (1956)). The remaining lectures will be published in Uspehi Mat. Nauk 12, no. 1, (1957).

it did not seem to me expedient to speak about branches of functional analysis of an established nature, which are now being worked on sufficiently intensively. For this reason I do not dwell, for example, on the theory of self-adjoint and non self-adjoint differential operators, on the topological methods of functional analysis or on that group of questions which owes its origin to the classical problem of moments, on which M. G. Kreĭn is working so successfully. And finally, I have attempted to select those questions whose solution by methods of functional analysis seems already to me to be a more or less realistic task at the present time.

§1. **Linear topological spaces.** At the present time the development of analysis and the theory of generalised functions shows that the framework of linear normed spaces has become restrictive for functional analysis, and that it is necessary to study linear topological spaces, i.e., topological spaces with linear operations which are continuous in the given topology. The introduction of such spaces is dictated not by love of generality, but simply by the fact that the spaces which are the most interesting and necessary for concrete problems of analysis are not normed. Thus, the entire theory of generalised functions may be developed as an application to analysis of linear topological spaces of functions and their adjoints.

As it has turned out, the most natural class of linear topological spaces for analysis is given by the so-called denumerably normed spaces, introduced already before the war in the works of Mazur and Orlicz and then, independently, in the works of the French school. These spaces are suitable firstly for the reason that one can carry over to them practically unaltered the existing function-space technique for Banach spaces, and secondly since the class of these spaces is sufficiently wide for applications.

A linear topological space is termed a denumerably normed linear topological space if the topology in it is defined by a denumerable set of norms $||x||_1 \leq \leq ||x||_2 \leq \cdots$. A simple example of a denumerably normed space is given by infinitely differentiable functions $\phi(t)$ on an interval, if one puts

$$||\phi||_n = \max \sum_{i=0}^{n} |\phi^{(i)}(t)|.$$

This space enjoys an important property which is not available for an ordinary normed (infinite-dimensional) space, namely that every bounded set in it is compact. Such denumerably normed linear topological spaces are called perfect. Grothendieck introduced a narrower and apparently fairly essential class of linear topological spaces, namely the so-called nuclear spaces. According to Grothendieck a denumerably normed space Φ is termed nuclear, if for every sequence of functionals $f_n \in \Phi'$, such that for any $\phi \in \Phi$ the series $\Sigma\, |(f_n, \phi)|$ converges, there also converges for some m the series $\Sigma\, ||f_n||_m$. Of course the above definition of Grothendieck is still ill formulated, and the axiomatics of nuclear spaces or of classes of spaces close to them require further "elucidation". Grothendieck showed that a topology in nuclear spaces could be given by means of a denumerable number of Hilbert norms.

Nuclear spaces are essential, for example, in constructing a complete system of eigen-functionals (generalised eigen-functions) of a self-adjoint operator.

Another problem, where such a definition is essential, is the problem of constructing a measure in linear topological spaces. Let in fact E be a linear topological space and E' its adjoint. Let further $\phi_1, \cdots, \phi_n$ be elements of E. We fix in E' a measure of the strips

$$a_i < F(\phi_i) < b_i \ (i = 1, 2, \cdots, n).$$

We demand of the measures of the strips that they be made to agree among themselves, and that the measure should depend continuously on $\phi_1, \cdots, \phi_n$. The problem is to continue this measure to a measure of all closed sets in E'. One can show without difficulty that for an arbitrary linear topological space (for example, for a Hilbert space) such a continuation is impossible. However it is probably possible for all sufficiently good linear topological spaces. It is likely that a sufficient condition for the possibility of such a continuation is the nuclearity of the space E.

An important question for the theory of arbitrary infinite-dimensional spaces is that of their approximation by finite-dimensional ones. A good approximation would play an essential role, for example, in the theory of measure in functional spaces and in numerical methods of analysis.

It is inadmissible in normed spaces to speak of their approximation by finite-dimensional ones, since the unit sphere in such spaces deviates from any finite-dimensional space by unity. In denumerably-normed spaces one can consider the sphere in one norm, and define the deviation from finite-dimensional subspaces by means of another norm. We may accordingly speak of this or that rapidity of approximation to a denumerably-normed space by finite-dimensional ones. Here "good" spaces are distinguished precisely by this speed of approximation. It would be very interesting to classify all linear topological spaces by just this speed of approximation.

G. M. Adel'son-Vel'skiĭ has given a construction of a spectrum (directed system) of finite-dimensional spaces, which has as its limit an arbitrary perfect linear topological space. This definition is the following. Let there be a partially ordered set of finite-dimensional spaces R_α. In each of these spaces is introduced a sequence of norms $||x||_n^\alpha$. If $\beta > \alpha$, then there exists a linear mapping of R_β on to R_α, enjoying the property of transitivity, and such that under the mapping the corresponding norms can only decrease. One designates as a point of the limit space a sequence x_α (with respect to α) of elements of R_α, such that $\|x\|_n < C_n$, and such that if $\alpha < \beta$, then x_α is the image of x_β.[1)]

Apart from the above problem of the continuation of measures in sufficiently good linear topological spaces, we would like to indicate the following two questions, which are closely connected with each other.

1. To give a definition of the "speed of approximation" to an infinite-dimensional space by finite-dimensional spaces and hence to deduce properties of well-approximable spaces. It is quite possible that it is just these spaces that are the most useful for applications.

2. To give an inner determination of the dimensionality of linear topological spaces of functions, understood as the number of arguments on which these functions depend (for the case of sufficiently good spaces of functions, for example for spaces of infinitely differentiable functions). This problem is closely connected with the problem of the speed of approximation to linear topological spaces by finite-dimensional spaces. For example, differentiable functions of one variable can be approximated to by their values at n equidistant points with a precision of order $\frac{1}{n}$, while functions of two variables can be approximated to by their values at the same number of points with a precision of order $\frac{1}{\sqrt{n}}$.[2)] This shows that a space of functions of one variable can be "more rapidly" approximated to by finite-dimensional spaces (space of their values at a finite number of points), than a space of functions of two variables. One should remark that from the point of view of the theory of Hilbert spaces there is no difference between spaces of various numbers of variables. But this merely indicates that the Hilbert space apparatus is not sufficiently delicate for analysis. A graphic illustration of this is given by the solution of various problems of the theory of the representations of the classical Lie groups. It is usually possible from a priori considerations, connected with certain analogues of Burnside's theorems, to determine the dimensionality of the spaces of functions in which the representations are realised. It would be very interesting if one could give a rigorous content to these considerations, which today have only a suggestive character.

§II. Linear partial differential equations. Linear equations have been studied

1) We remark that the approximation of functional spaces by finite-dimensional spaces which is encountered in numerical analysis is an approximation of just such a type. For nowadays in using methods of functional analysis in problems of numerical analysis one usually tries to replace infinite-dimensional spaces by their finite-dimensional subspaces (for example, the space of continuous functions by the space of broken-linear functions with angular points having given abscissas). However it seems more natural to replace spaces by their finite-dimensional factor-spaces (which is precisely the case in the construction of G. M. Adel' son-Vel' skii).

2) After this lecture and before the manuscript was put into final form there appeared the very interesting and important work of A. N. Kolmogorov, in which is investigated the approximation to spaces of functions of various numbers of variables by finite-dimensional spaces.

for a long time, and many problems of this theory may be rightly termed classical. A real possibility has now appeared of bringing to a close the theory of linear differential equations and of giving a complete answer to all questions in this region. Firstly, the study of the equations themselves has been substantially improved. Secondly, there is the apparatus of linear topological spaces which is extremely useful for these equations.

It is not necessary to give some sort of appreciation of the existing situation in this area, and so I will simply confine myself to listing certain questions which appear to me not without interest.

1. The characterisation of boundary problems and their solubility conditions for elliptic equations. This problem has confronted the theory of differential equations for a long time now, and M. I. Višik has obtained important results here. I would like to indicate a possible approach to its solution, arising from the theory of generalised functions. As is known, in theorems on the solubility of boundary problems the main difficulty usually is formed by the proof of the fact that the available solution does indeed satisfy the boundary condition in some sense. The behaviour of the function on approach to the boundary of the region plays in these questions the same role as behaviour at infinity in the theory of generalised functions.

In the works of G. E. Šilov and myself the investigation of the uniqueness and the existence of a solution, with reference to the behaviour of the function at infinity, were obtained with the aid of the theory of generalised functions by the construction of various linear spaces. In a completely analogous way it is necessary, probably, to construct spaces of functions with various types of behaviour on approach to the boundary in order to solve boundary problems.

Apparently, in the correct formulation of a boundary problem for the equation $Lu = f$ there must figure a pair of linear topological spaces (of functions u and functions f), in whose definition the boundary properties demanded on the solutions are already taken into account. Here it is required that the operator $Lu = f$ should be continuous on this pair of spaces and should have a continuous inverse (to within a finite-dimensional operator). From each of the spaces one must require that they should contain all sufficiently good finite functions (i.e., functions which vanish in a neighbourhood of a boundary strip). [1] In addition, one must require that

1) Such a requirement in the case of the equation $\Delta u = f$ admits, in the role of the first space, a space of functions which vanish on the boundary, and excludes, for example, a space of functions which vanish on the boundary together with their derivatives, since in this latter case the space of functions f does not contain all finite functions. (The equation $\Delta u = f$ has a solution which vanishes on the boundary for every finite f, but may fail to have a solution if the requirement is imposed that u should vanish on the boundary together with its derivatives.)

the spaces be linear ones, for which the above requirements hold.

The enumeration of correct problems is in fact the enumeration of such pairs of spaces. It is to be wished, of course, that such a solution should make it possible for us to obtain hitherto known boundary problems as a special case, and not only such traditional problems as those of Dirichlet or Neumann, but also such as the generalised problem of Tricomi, for example.

2. While on the subject of general boundary problems, I would like now to formulate some concrete problems. We consider an elliptic equation $Lu = 0$ and boundary conditions $L_k u|_\Gamma = f_k$, where Γ is the boundary of the region and the L_k are differential operators. The problem is termed a Fredholm one if it is soluble for all right-hand sides which satisfy a finite number of conditions. Let their number be m and let also the homogeneous problem have a finite number n of linearly independent conditions.* The number $m-n$ is called the index of the problem. The question is, how to calculate the index even in the case of the equation $\Delta u = 0$ in space of three dimensions with some differential boundary conditions. Here an important role is played by the topological properties of the system of coefficients in the boundary conditions. Can this index in three-dimensional space be finite and other than zero? This problem, but only for the case of the equation $\Delta u = 0$ in the plane, is solved by methods of the theory of analytic functions in works of I. N. Vekua.

3. Considering now the question of the solution of Cauchy's problem, one can indicate here two problems. The first is the solution of Cauchy's problem for equations with coefficients which have surfaces of discontinuity. For example, if we have a linear hyperbolic equation, for which the coefficients of the higher derivatives are discontinuous, then not for all coefficients does there hold the uniqueness and solubility of the Cauchy problem. It is not very hard to indicate conditions on the coefficients on the surfaces of discontinuity, under which the Cauchy problem always has a solution, and moreover a unique one. However the existence and uniqueness theorem for such equations is itself not proved.

The other question is the enumeration of all correct boundary conditions for evolutionary equations. Here we term an evolutionary equation one for which the Cauchy problem is correct, such as, for example, hyperbolic or parabolic equations.[1)]

* *Translator's note:* "Solutions" should perhaps replace the word "conditions".

1) In a recent work of John "On incorrect problems for partial differential equations" it is proved that the Cauchy problem for the backward heat-conduction equation $\frac{\partial u}{\partial t} = -\frac{\partial^2 u}{\partial t^2}$ (the classical example of an incorrect problem) is correct in the class of positive solutions (and so generally in the class of solutions bounded from below. M. M. Lavrent'ev and E. M. Landis showed that the Cauchy problem for the Laplace equation, and general elliptic equations of the second order, is likewise correct, if we stipulate in advance that the solution should not exceed a constant M. This fact probably holds good for arbitrary equations (Footnote continued on next page.)

4. The following question is that of the proof of the uniqueness of the solution of Cauchy's problem for elliptic equations with continuous coefficients. The classical theory of Carleman solves this question for an equation of the second order with two independent variables. By means of the principle of the maximum E. M. Landis obtained a solution for elliptic equations of the second order with any number of variables. Precisely on account of this approach, however, there is little hope of extending his proof to the general case. The point is that the use of the principle of the maximum and the related technique of barriers are essentially restricted to the study of one second-order equation, and at the present time one must either give up equations of higher order and systems of elliptic equations or else, what would of course be extremely interesting, give some genuine extension to them of the principle of the maximum and the technique of barriers. Very interesting results in this direction, based on other principles, are due to M. M. Lavrent'ev.

§III. **Quasi-linear equations.** In contrast to linear equations, the theory of general non-linear equations, and even quasi-linear equations, is still at present completely undeveloped, although in fact very many applied problems of hydromechanics and other regions of physics and mechanics lead to them. For certain classes of quasi-linear equations, theorems have been proved on the solubility of Cauchy's problem in the small, but these are still not theorems which essentially describe the solution and which can be useful for applications. In these theorems it is a question of sufficiently smooth solutions while, as is generally known, the equations of a compressible gas, for example, under certain conditions undergo a discontinuity. * It is extremely important to construct a theory to include these discontinuous solutions. The only problem of which I shall speak here is the following: to prove the theorem of existence, uniqueness and stability of Cauchy's problem for quasi-linear equations in the class of piece-wise continuous functions. Here one must be able to "follow-up" the solution for just as long as it does not cease to exist, in fact does not go off to infinity.

More precisely, it is a question of a system of quasi-linear equations

$$\frac{\partial u_i}{\partial t} = \sum_i \frac{\partial F_i(x_1, \cdots, x_k; u_1, \cdots, u_n, t)}{\partial x_j} + G_i$$

$$(i = 1, \cdots, n).$$

We assume in addition that the system of equations is hyperbolic. On the surfaces of discontinuity of the solutions we shall assume that there hold relations which are obtained by integration of the above system. The question is posed of the

$\frac{\partial u}{\partial t} = p(\frac{\partial}{\partial x})\, u$, which by the way, it would be interesting to prove. All this, perhaps, indicates a certain provisional character in the current concept of correctness.

* *Translator's note:* The meaning is probably that the solutions of the equations undergo discontinuities.

existence and uniqueness of the solution of this system, if the initial data are, for example, continuous. Here it is permissible to consider discontinuous solutions also. But not all discontinuous solutions are suitable here. If we admitted all discontinuous solutions, then, as is shown by the example of so-called shock waves of rarefaction type in hydrodynamics, the uniqueness theorem would not hold. A requirement which picks out the necessary class of solutions, is the requirement of the stability of this solution with respect to small perturbations. The equations of hydrodynamics enjoy an important peculiarity, on which I should like to dwell. To explain this, let us consider, for example, the case of equations with two independent variables

$$\frac{\partial u_i}{\partial t} + \frac{\partial F_i(u_1, \cdots, u_n)}{\partial x} = 0$$

$$(i = 1, 2, \cdots, n).$$

If there is a solution with a line of discontinuity, then, at a point of discontinuity, integrating our system, we get $n - 1$ conditions. However we need altogether $2n$ conditions to determine the solution, determining the n values u_i^- to the left of the discontinuity and the n values u_i^+ to the right of the discontinuity. The requirement of stability does indeed have just the effect, that at a point of discontinuity the common number of conditions introduced along the characteristics is equal to $n + 1$, of which p conditions are from one side and $n - p + 1$ from the other. The equations of hydromechanics are characterised by the property that for them as the solutions evolve in time this number p does not change. In the general case these conditions may be formulated in terms of the eigen-values of the matrices

$$\left\|\frac{\partial F_i(u^+)}{\partial u_k}\right\| \text{ and } \left\|\frac{\partial F_i(u^-)}{\partial u_k}\right\|$$

and the values of the functions $F_i(u_1, \cdots, u_n)$ themselves.

It would be very interesting to formulate these conditions properly and to study the class of equations which specifically arises. For the case of a single equation

$$\frac{\partial u}{\partial t} + \frac{\partial F(u)}{\partial x} = 0$$

this condition is the condition that the function $F(u)$ be convex.

If the study of Cauchy's problem for quasi-linear equations has as its origin the theory of a compressible gas, another important section of hydrodynamics, the theory of a viscous fluid, is also linked with important mathematical considerations. Here one may indicate 1° the theory of the boundary layer (or, mathematically, of a small parameter before the higher derivatives) and 2° the investigation of the causes of the non-existence of a solution of the problem of laminar flow around a body for large Reynolds numbers. These two problems are, as a matter of fact, closely connected.

§IV. **Measures in functional spaces.** The importance of developing a theory of measure is evident from the following example. In constructing the quantum mechanics of a system with n degrees of freedom one investigates the Schrödinger equation in space of n dimensions, that is to say one considers self-adjoint differential operators in n-dimensional space. If again we had to construct a logically consistent quantum theory of fields, then, since a field has infinitely many degrees of freedom, we should in this case need to have a logically consistent theory of self-adjoint differential operators in infinite-dimensional space. But the concept of self-adjointness in n-dimensional space is founded on the concept of Euclidean measure, which is lacking in infinite-dimensional space.

The most natural measure in functional space is perhaps the so-called Wiener measure. Wiener obtained his measure more than 30 years ago from considerations of probability theory. One may arrive at it from geometrical considerations. It is clear from considerations of similarity, that in passing to an infinite-dimensional space similar sets have measures of which the ratio is infinite. Hence in infinite-dimensional space there are no grounds for expecting a good concept of measure. If however we confine ourselves to sets located on the surface of a sphere, then considerations of similarity fall away and there are no counter-examples to the construction of a measure. If we consider the mean-value of functions, defined on a sphere, and increase the number of dimensions, at the same time increasing its radius proportionately to the square root of the number of dimensions, then in the limit from the mean we get the integral with respect to the Wiener measure. More precisely, suppose we are given a functional $F(\phi)$, which is to be integrated with respect to the Wiener measure. Expanding the function ϕ in a Fourier series with respect to the orthonormal functions ϕ_n, $\phi = \sum_{n=1}^{\infty} c_n \phi_n$, we may form a sequence of linear functionals F_n, where F_n is a functional in n-dimensional space, acting according to the formula

$$F_n(c_1, \cdots, c_n) = F(c_1, \cdots, c_n, 0, \cdots).$$

The integral with respect to the Wiener measure is obtained by taking the mean-value of the functional F_n over a sphere of radius $a\sqrt{n}$ and passing to the limit as $n \to \infty$.

Apart from the quantum theory of fields, integration and analysis in functional spaces are interesting for the theory of stochastic processes, the theory of information, statistical physics. They, of course, demand by no means merely one Wiener measure which, being the simplest, does not in the least exhaust the whole wide field of applications of the theory of measure in functional spaces. The development of this branch of analysis may accordingly turn out to be very appropriate for future progress.

§V. **Rings of type II_1.** These rings were introduced as rings of operators in Hilbert space. One considers weakly closed rings, which together with each operator contain also its adjoint. The centre of the ring consists of operators which are multiples of the identity. In n-dimensional space every such ring is isomorphic to the full matrix ring. In the infinite-dimensional case there is a series of such rings. Certain of them are again isomorphic to the full ring of operators. However the most beautiful, "perfect" structures are the so-called rings of type II_1. Thus, to each of the operators of a II_1 ring one can, in a reasonable way, define a trace and a determinant; the trace of a projection operator is a number lying between 0 and 1, which may be called the dimension of the subspace determined by this differential operator, and so on. In spite of the great elegance of this theory, in applications these things are encountered only in the theory of the representations of discrete groups.

However the beauty and harmony of the theory of II_1 rings makes us think that this theory must undoubtedly find a wide field of applications. There is reason for thinking that these rings are most naturally realised as rings of operators on spaces with Wiener measure. Furthermore, much as the analogous finite-dimensional rings (I_n) are the rings of all operators in a finite-dimensional space, the question arises of finding such a "space", in which a II_1 should be the ring of "all" "operators".

§VI. **Hypergeometric functions.** It is known that almost all the special functions of one variable to be met with in mathematical physics may be obtained from the general hypergeometric function of Gauss by a suitable choice of parameters. These same functions appear as elements of representations of the simplest classical groups, namely the groups of rotations of the sphere and of the Lobacevskii plane. This connection lies in the nature of the matter, since the special functions make their appearance by way of considerations connected with this or that invariance of a problem under transformations of a space. Hence it is natural to construct the theory of hypergeometric functions of several variables, relying on results and methods of the theory of the representations of compact or locally compact Lie groups. It is thus necessary so to construct the theory of hypergeometric functions that it should contain the theory of general spherical functions, connected with the representations of semi-simple groups.

Translated by:
F. V. Atkinson

3.

Some questions of analysis and differential equations*

Usp. Mat. Nauk **14** (3) (1959) 3–19 [Transl., II. Ser., Am. Math. Soc. **26** (1963) 201–219]
Zbl. **91**:88

Lately I have had to prepare two survey lectures on analysis. One for the Moscow Mathematical Society (1957), the other for the International Congress of Mathematicians in Edinburgh (1958). In the process of preparation of these lectures, in particular the latter, it was natural for me to think about various problems which it would be interesting to solve, and though the content of these two lectures is completely represented by the problems discussed below, it seemed to me to be appropriate to present them separately. I have with regret not included those problems prepared by me for the Edinburgh lecture which are connected with the theory of representations of groups. This has to do with the fact that the problem of posing them is closely bound up with a number of concrete results in that direction, so that the presentation could not have been made without those results. But this would have changed the paper partly into a paper on the theory of representations, which did not seem appropriate to me.

The author expresses his deep thanks to the direction of the Moscow Mathematical Society and to the organization committee of the Edinburgh Congress, for the honor of an invitation which served as the stimulus for thinking about the considerations presented below.

Certain first questions, which will be discussed below, are connected with the theory of partial differential equations. Hence we shall begin with a number of general remarks connected with differential equations.

In the theory of partial differential equations it is possible to take two different points of view. The first of these consists in taking an interest not in arbitrary equations, but rather in equations describing definite physical processes. From this point of view partial differential equations may describe physical processes of two different sorts: stationary (as, for example, stationary distributions of temperature, periodic oscillations and so forth), and evolutionary (for example, the diffusion of waves or heat). These two types of physical process are described by equations of different types.

This point of view has the advantage that the problems it leads to are of "high quality." The deficiency of such an approach is however evident. Indeed, partial differential equations do not have to arise immediately from physical problems. They may also appear in further transformations and modifications of

* In the preparation of this article a great deal of assistance was given me by S. V. Fomin. My conversations with him were of essential importance to me in the preparation of the final text.

such problems, as may be illustrated in a series of examples.

The second possible point of view here consists in constructing a general theory of partial differential equations independent of the type, not immediately connected with definite physical representations (much as in analysis, where one considers one or another class of functions, where we do not try to see right away where these functions enter in physics, since these classes may be obtained in further transformations and modifications of the problems). This second point of view is best represented in the significant and deep paper of L. Gårding at the Edinburgh Congress. Undoubtedly, for mathematicians the second point of view has a great attraction. It is regrettable that in the theory of equations of arbitrary type one has only partial, far from final, results.

At the present time there is no foundation for a strong preference for either of the two points of view discussed above. Each of them has its advantages and deficiencies.

1. Various definitions of ellipticity. In connection with the study of definite types of equations, in particular those which describe stationary process, several variants of the definition of ellipticity of systems of differential equations are of interest.

Consider, for example, a system with constant coefficients of the form

$$\sum_{i,k=1}^{n} A_{ik}^{pq} \frac{\partial^2 u^{(p)}}{\partial x_i \partial x_k} = f_q \quad (p, q = 1, 2, \cdots, n). \tag{A}$$

For such a system the following definition of ellipticity is known* (due to I. G. Petrovskiĭ). The system (A) is said to be weakly elliptic if the matrix

---

* For the introduction of one or another classification system for equations it is natural to require the following. Suppose that the system is for example of the form

$$\sum_k a_{ijk} \frac{\partial u_i}{\partial x_k} = 0 \quad (i, j = 1, 2, \cdots, m;\ k = 1, 2, \cdots, n). \tag{B}$$

We shall consider for this system the following transformations:

1) the replacement of equations (B) by independent linear combinations of them (in the same number);

2) linear substitution for the unknown functions u_i;

3) linear substitution of the independent variables x_k.

Then it is natural to require that systems obtained by one or another of the transformations of the indicated type should be of the same type.

The Petrovskiĭ definition of ellipticity satisfies this requirement, and from this point of view, is an invariant definition of ellipticity.

In connection with what has been said, it is natural to pose the following problem on the classification of trajectories in the space of the systems of equations. Consider in the space of systems of type (B) a group of mappings, generated by mappings of types 1), 2), and 3). This group is evidently representable in the form of the product of the three matrix groups

$$G = A \times B \times C,$$

corresponding to the three types of mappings. Here the space of the systems of equations

$\|\Sigma A_{ik}^{pq} \xi_i \xi_k\|$ is not degenerate for any set $\xi_1, \cdots, \xi_n$, among which not all are null.

A number of definitions of semiboundedness are quite close to the concept of ellipticity. For example, M. I. Višik has introduced the following definition.

The system (A) is said to be *strongly elliptic,** if the symmetric part of the matrix $\Sigma A_{ik}^{pq} \xi_i \xi_k$, for any non-null $\xi_1, \cdots, \xi_n$, is positive definite (for any choice of the scalar product).

Another possible definition of semiboundedness: the system (A) is said to be *semibounded,* if all eigenvalues of the matrix $\Sigma A_{ik}^{pq} \xi_i \xi_k$ lie in the same halfplane.

For strongly elliptic systems in the sense of Višik, one has solvability and uniqueness for the Dirichlet problem.**

It is likely that for weakly elliptic systems whose characteristic polynomials do not have multiple roots, the Dirichlet problem is solvable and unique, up to finite-dimensional subspaces, i.e., the homogeneous Dirichlet problem may have only a finite number of independent solutions, and for solvability of the nonhomogeneous problem it suffices to impose a finite number of conditions on the right side of the equations.*** One may show that the set of weakly elliptic systems

---

decomposes into trajectories. We shall call a trajectory *regular,* if one of its neighborhoods is a fibering. The trajectory will be called *degenerate,* if it is a manifold of a smaller number of variables than a regular trajectory. The remaining trajectories are called *singular.* The problem consists in characterizing (in algebraic terms) the differences among the various types of trajectories and, in particular, to find a condition for the regularity of a trajectory. It is easy to verify that in the case of two independent variables, i.e., of systems of the type

$$\frac{\partial u_j}{\partial t} = \Sigma\, a_{jk} \frac{\partial u_k}{\partial x},$$

the classification of trajectories comes down to the theory of elementary divisors. Indeed, a system belongs to a regular trajectory, if the matrix

$$A = \|a_{jk}\|$$

may be brought to diagonal form. It belongs to a singular trajectory if it can be brought to Jordan form (not diagonal), while the various Jordan squares correspond to the various eigenvalues. Finally, the system belongs to a degenerate trajectory if in the Jordan form of the matrix there are several squares which belong to the same eigenvalue.

* This term has entered the literature; however we shall more correctly speak not of ellipticity, but of semiboundedness.

** It is necessary to observe, however, that there are no particular grounds for attaching exclusive significance to the Dirichlet problem. Indeed, in the equations of the theory of elasticity for the displacement vector, the most natural boundary conditions—the given loadings on the boundary—lead to a quite different boundary problem.

*** If the characteristic polynomial has multiple roots, then this assertion is not true. For example, for the system

$$\left.\begin{aligned} \frac{\partial^2 u}{\partial x^2} - \frac{\partial^2 u}{\partial y^2} - 2\,\frac{\partial^2 v}{\partial x \partial y} = 0, \\ \frac{\partial^2 v}{\partial x^2} - \frac{\partial^2 v}{\partial y^2} + 2\,\frac{\partial^2 u}{\partial x \partial y} = 0, \end{aligned}\right\}$$

which is weakly elliptic (but not strongly elliptic), the Dirichlet problem either has no solution or else has an infinite number of solutions. Its characteristic polynomial has multiple roots (Bicadze).

is not connected (i.e., in it one may find two systems which cannot be brought into one another by a continuous variation of the coefficients in a way which preserves ellipticity). For systems of two equations with two independent variables, this set decomposes into two components. One of them contains the system $\Delta u = 0$, $\Delta v = 0$, and the other the system mentioned in the footnote,* and also other systems, for which the Dirichlet problem is not unique (see Višik [1]).

Problem 1. Into how many connected components does the set of weakly elliptic systems of type (A) decompose?

Problem 2. How does one (algebraically) characterize the component which contains the system $\Delta u_i = 0$?

Problem 3. If the system belongs to the same component as $\Delta u_i = 0$, then is the Dirichlet problem for it always solvable and unique?

2. Boundary problems for elliptic systems. Suppose given some elliptic system

$$P\left(\frac{\partial}{\partial x_1}, \dots, \frac{\partial}{\partial x_n}\right) u = 0,$$

where P is a matrix whose elements are polynomials, and u is a set of functions. Suppose that for this system there are given on the boundary of the region D several boundary conditions "of differentiable type", i.e., which connect the values of the function and its derivatives at the same point. Some of these boundary conditions may be given on the whole boundary of the region in which the boundary problem is posed, and some may be given on portions of a smaller number of dimensions than the whole boundary.

Problem. To enumerate all "nice" boundary problems (i.e., those for which there is unicity up to a finite-dimensional subspace, and correctness). For the equation $\Delta u = 0$ with two independent variables this problem was considered by I. N. Vekua [2]. Already for the equation $\Delta u = 0$ in three-dimensional space the problem indicated has significant interest. The analogous problem may be posed also for equations of other types.

3. Equations with discontinuous coefficients. Conjugate solutions of two different equations. For hyperbolic systems with sufficiently smooth coefficients the theorem on existence and uniqueness of the solution of the Cauchy problem is known (Petrovskiĭ, Leray). In these theorems the assumed smoothness of the coefficients appears as essential. Indeed, consider for example the equation

$$\frac{\partial u}{\partial t} + a(x)\frac{\partial u}{\partial x} = 0,$$

where $a(x)$ is a function at some points having no derivative. The Cauchy problem

* See footnote *** of preceding page.

for this equation may have no solution at all (for instance for $a(x) = -\sqrt{|x|}$), or else have an infinite number of solutions (for example, for $a(x) = +\sqrt{|x|}$). However, from the point of view of applications, it is important to clarify the conditions which ensure the existence and uniqueness of solutions in the case of coefficients which are not only not smooth, but even discontinuous. Clearly, one must impose in this case definite conditions on the discontinuities of the coefficients, as well as on the behavior of the solution at the points of discontinuity of the coefficients (for example, continuity of the solution at these points).

Below, in speaking of discontinuous coefficients, we shall understand this in the classical sense, i.e., suppose that these discontinuities lie on a finite number of smooth surfaces.

For definiteness we consider the system of equations

$$\frac{\partial u}{\partial t} = A \frac{\partial u}{\partial x} \tag{1}$$

($u = (u_1, \cdots, u_n)$, $x = (x_1, \cdots, x_n)$) with a discontinuous matrix of coefficients A (for example, $A = A^+$ for $x_1 > 0$ and $A = A^-$ for $x_1 < 0$).

Suppose that the matrices A^+ and A^- are such that each of the systems

$$\frac{\partial u}{\partial t} = A^+ \frac{\partial u}{\partial x} \quad \text{and} \quad \frac{\partial u}{\partial t} = A^- \frac{\partial u}{\partial x} \tag{2}$$

is hyperbolic (i.e., the eigenvalues of the matrices A^+ and A^- are real). We shall seek solutions of problem (1), continuous for $x_1 = 0$. For the existence of such a solution it is necessary that certain additional conditions be satisfied. Namely, let (r^+, r^-) be the index * of problem (1) at $x_1 = 0$, and n the number of unknown functions. Then for the existence of a continuous solution the following equality must be satisfied:

$$r^+ + r^- = n. \tag{3}$$

This equality shows the possibility of "gluing together" the solutions on one curve of discontinuity. Further questions arise in the presence of several such curves. Suppose, for example, that the coefficients of the system

$$\frac{\partial u}{\partial t} = A_1 \frac{\partial u}{\partial x_1} + A_2 \frac{\partial u}{\partial x_2}, \tag{4}$$

where A_1 and A_2 are matrices, are discontinuous along three rays issuing from one point, and suppose that on each of these rays there is satisfied an equation of type (3). Does it follow from this that there exists a solution continuous on the whole space?

---

* The concept of index was presented in [3]. In the present case r^+ is just the number of negative eigenvalues of the matrix A^+, and r^- the number of positive eigenvalues of the matrix A^-.

As we have already said above, for hyperbolic systems with sufficiently smooth coefficients the theorems of existence and uniqueness for the Cauchy problem are valid without any kind of additional conditions. In the case of continuous, but not differentiable coefficients, the existence or the uniqueness of the solution may not hold. It is interesting to clear up the question of what conditions assure the existence and uniqueness of the solution in the case of continuous but not differentiable coefficients. In the case of two independent variables, such a condition, it is possible, is the regularity of the system of characteristics.*

The problem of discovering a solution, continuous in the whole space, of a system of equations with discontinuous coefficients, is a special case of the following problem on the *conjugacy of two systems.*

Suppose that in some region G and outside of it there are given two different equations, and that on the boundary of the region G there are given certain "conjugacy conditions", for example the continuity of the solution, or a discontinuity of a definite type. One seeks solutions satisfying the corresponding equations inside G and outside G, and, on the boundary, the conjugacy conditions.

Example. Consider the diffusion of gas in a canal, surrounded by a porous medium. In the canal the motion of the gas is described by the wave equation

$$\frac{\partial^2 p}{\partial t^2} = a \Delta p,$$

and outside of it by the diffusion equation

$$\frac{\partial p}{\partial t} = b \Delta p.$$

On the boundary it is necessary to require the continuity of the solution.

The study of two equations, given in two parts of the space and connected by one or another "conjugacy conditions" on the boundary, is dictated by real physical problems. On the other hand, it is possible that such an approach to the study of the "usual" boundary problems will also show itself to be useful, since the setting of one or another boundary conditions may be considered as a limiting case of problems with conjugacy of two equations.

4. Correct boundary problems for "evolutionary" systems. We shall call the system

$$\frac{\partial u}{\partial t} = L\left[\frac{\partial}{\partial x_i}, x\right] u \tag{5}$$

"evolutionary" (i.e., describing some nonstationary process) if the Cauchy problem for it has a unique solution in the entire space.

---

* It is possible that in the case of a large number of independent variables these additional conditions may turn out to be the regularity of the family of bicharacteristics, whose important role in the Cauchy problem was exposed by Leray in his significant paper [4].

Problem. Describe the boundary condition which must be considered for the system (5) (i.e., conditions for which the corresponding Cauchy problem is correct, its solution exists and is unique in some region of the spaces of variables).

For the solution of this problem, in every case when it is limited by boundary conditions of local type, i.e., establishing a connection between the values of the function and its derivatives in a given point, one may sketch the following approach.

Consider first the statement of the problem for the case when the coefficients of the equations are constant, and the solution is sought in a half-space of the x-space. Then for variable (but sufficiently smooth) coefficients and an arbitrary (smooth) region we proceed as follows: taking an arbitrary point of the boundary of the region in the x-space and passing through it a tangent hyperplane, we "freeze" the coefficients of the equations and the boundary operator. We thus obtain the special case mentioned above (the half-space problem). If the problem thus obtained is correct, then we shall say that the original problem is correct in the point in question.

Question. Is it true that a boundary problem is correct if and only if it is correct in every point, or will there arise "obstructions" inside the region?

Here by correctness at the point we mean gross (i.e., with respect to variations of the coefficients of the equation, the plane, and the boundary conditions).

5. Energy integral. In hyperbolic systems an important role is played by the so-called "energy integral", i.e., positive definite quadratic functionals of the solution, nonincreasing as time increases.

For systems with constant coefficients the energy integrals always exist (Petrovskiĭ, Leray). It is interesting to determine whether there exists an energy integral for any correct boundary problem (i.e., when, aside from the initial data, we also impose boundary conditions), and including also the case of discontinuous conditions.

The first step here would be the resolution of the question for constant coefficients and a half-space.

All that is proved at present is that the energy integral exists for a hyperbolic system of type (5) with discontinuous coefficients for any correct problem in the case of a system of equations with two variables x and t [5].*

An undoubted interest lies in the question as to whether one can obtain any existence theorem for hyperbolic equations by using the energy integral, if there are given arbitrary, correct boundary conditions.

6. General boundary problems. General definition of correctness. The ques-

* In this paper we have analyzed the algebraic side of the question of energy integrals for systems with constant coefficients.

tion as to whether one may pose boundary problems for one or another types of systems of differential equations, has been investigated little enough at the present. A more or less exhaustive analysis exists only for the equation

$$\Delta u = 0.$$

At the same time a series of questions, for example certain problems of the theory of stochastic processes, require the consideration, for the differential equations, of rather peculiar boundary conditions, differing from the boundary problems usually considered.

An important source from which general boundary problems appear is the theory of the extension of selfadjoint operators.

The question involving general boundary problems has two aspects:

1) the enumeration of possible "general boundary conditions" for regions with sufficiently smooth boundaries;

2) the investigation of correct boundary problems for arbitrary regions.

This last problem was considered in ample detail by Wiener and Keldyš for the Laplace equation (applied to the Dirichlet problem).

In connection with the investigation of various boundary problems it is important to formulate in general terms the definition of correctness of the boundary problem.

Consider the equation

$$Lu = f,$$

where L is some operator. We shall say that a correct boundary problem is given for this equation if there are given two spaces U and F, consisting respectively of functions u and f, defined in some region G, while:

1°. The spaces U and F contain all "finite" (i.e., equal to zero in some neighborhood of the boundary of the region G) sufficiently smooth functions.

2°. U and F are sufficiently nice, for example nuclear topological spaces, and the operator L is continuous on U, carries it into F and has a continuous inverse on F, up to a finite-dimensional one.

3°. U is the minimal space having properties 1° and 2°.

7. **Quasilinear equations.** In recent times, in connection with problems of hydrodynamics and a series of other questions, considerable significance has been acquired by quasilinear equations, i.e., equations linear in the derivatives, but not linear in the unknown functions themselves. Certain questions connected with quasilinear equations were formulated and published in a preceding issue of Uspehi Matematičeskih Nauk. These lectures [3], properly speaking, are in fact nothing other than a detailed statement of two or three problems on quasilinear equations.

8. Asymptotic solutions of hyperbolic equations in dynamical systems. The investigation of dynamical systems, defined by a system of differential equations

$$\frac{dx_i}{dt} = F(x_1, \cdots, x_n) \qquad (i = 1, 2, \cdots, n), \tag{6}$$

is closely connected with the study of the corresponding partial differential equation

$$\frac{\partial u}{\partial t} + F_1(x_1, \cdots, x_n)\frac{\partial u}{\partial x_1} + \cdots + F_n(x_1, \cdots, x_n)\frac{\partial u}{\partial x_n} = 0, \tag{7}$$

whose characteristics are the integral curves of the system (6). The passage from the system (6) to the equation (7) means that instead of the dynamical system itself, given on an n-dimensional manifold Ω, one considers the time-dependent function $f(t, x_1, \cdots, x_n)$ on that manifold.

The consideration of a certain space of functions on a dynamical system (namely, the space of functions with integrable square) underlies the so-called spectral method of the theory of dynamical systems * due to Koopman. In terms of functions given on a dynamical system, one may formulate, for example, such properties of this system as ergodicity, the presence of mixing, the invariance of the measure given on the dynamical system, and so forth. Thus the theory of dynamical systems may to a significant degree be interpreted as a theory relative to a linear partial differential equation of first order. More precisely speaking, inasmuch as the theory of dynamical systems consists in the first place of the investigation of the asymptotic behavior of solutions (for large t), the corresponding problem for equation (7) also consists in the study of the asymptotics of its solution as $t \to \infty$. But in such a form the problem does not need to be set up for one equation only. It makes sense also for hyperbolic systems ** of linear partial differential equations, i.e., the replacement of the desired dynamical system by several families of characteristics. So we shall consider a system of form

$$\frac{\partial u_j}{\partial t} + \Sigma\, a^j_{ik}\frac{\partial u_i}{\partial x_k} = 0 \qquad (j = 1, 2, \cdots, n), \tag{8}$$

each solution of which has the form of a vector function

$$u(t, x_1, \cdots, x_n). \tag{9}$$

We shall sketch a possible approach to the extension of a number of concepts of

* The study of some object by considering one or another system of functions on that object is a rather widespread method quite convenient in many cases.

** It is natural here to consider just hyperbolic systems, i.e., those for which there exists a system of real characteristics—as in the desired dynamical system.

the theory of dynamical systems to the solution of the system (8). An invariant measure on a dynamical system, defined by equations (6), may be defined as a measure having the property that, in the space of functions on Ω integrable with their squares with respect to that measure, the differential operator

$$i \sum F_k \frac{\partial}{\partial x_k}$$

is self-adjoint. As a generalization of the concept of the invariant measure of a system of type (8), one may therefore consider a scalar product (which may be dependent also on the derivatives of the function u) in the space of vector-functions (9), such that with respect to it the differential operator

$$i \sum a_{pk}^{(j)} \frac{\partial}{\partial x_k}$$

(the infinitesimal-displacement operator in t) becomes self-adjoint.

In a compact dynamical system, as is known, there always exists an invariant measure. Does this assertion remain valid on passing to systems of type (8)?

One says that a dynamical system (with invariant measure μ) has (strong) mixing, if for any two functions f and g, summable with respect to it with their squares, there exists a limit for the expression (f_t, g) as t approaches infinity. It is clear that this definition carries over at once to vector functions, i.e., to systems of type (8).

It would be interesting to carry over to systems of partial differential equations all the fundamental concepts and results of the theory of dynamical systems, in particular the spectral theory.

9. "Analytic" topology and differential equations. At the very time of its inception topology developed in continuous connection with analysis. Indeed, problems of analysis (the integration of entire differentials in an n-dimensional space and the qualitative investigation of differential equations) led Poincaré to the creation of the fundamental concepts of topology.

The analytic origins and the analytic orientation of the fundamental concepts of topology are evident, for example, in the fact that in topology three-dimensional manifolds do not play any exclusive role in comparison with general n-dimensional manifolds.

"Analytic" topology (i.e., topology in its connection with analysis) may be defined as the science of the discrete, in particular of the integer, invariants of analysis.* As we have already said, topological concepts arose and to a certain degree developed in connection with various questions of analysis. A particularly

* If one may describe in one phrase the content of such an extensive part of mathematics.

essential role here was played by the theory of algebraic functions and Abelian integrals. Nevertheless, throughout the well-known period, topology developed in an essentially "autonomous" way, as a theory of invariants of homeomorphisms, and achieved significant results. Thus, invariants arising in topology are connected in analysis basically with the problem of the solution of algebraic equations (algebraic functions) and with the problem of integration of these functions (Abelian integrals). Very few of the essential invariants were obtained in connection with ordinary differential equations (in the complex domain and with rational coefficients, in order that these invariants should be more classical if possible), and perhaps, nothing at all is known in connection with partial differential equations.

New and important connections between analysis and topology were established in the papers of I. G. Petrovskiĭ and E. M. Landis (Mat. Sb. (N.S.) 43 (85) (1957), 147–148 and 37 (79) (1955), 209–250). In these papers, for the differential equation

$$\frac{dy}{dx} = \frac{P(x, y)}{Q(x, y)} \tag{10}$$

in the complex domain, where P and Q are polynomials, an integer-valued characteristic was introduced, which for the case of the real equation (10) turns out to be connected with the number of limit cycles of the equation (10). This invariant is connected with the study of the behavior in the large of solutions in the complex domain; the clarification of the general (topological) character of the concepts which arise, and, in particular, the carryover of this invariant of Petrovskiĭ-Landis to systems of differential equations with rational coefficients appears to be very important and interesting.

10. Canonical form of non-selfadjoint operators, in particular nuclear (kernel) operators. At a time when the theory of self-adjoint linear operators has been successfully carried over to infinite self-adjoint operators, the general theory of linear operators in Hilbert space (or linear topological space), i.e., the infinite analogue of the Jordan form, is still lacking. Among the operators studied, and those not completely, are only operators which differ very little from self-adjoint operators. One of the first important results here was the paper of M. V. Keldyš on the completeness of the system of eigenfunctions and adjoint functions for the operator $L_1 + iL_2$, where L_1^{-p} has for some $p > 0$ a finite trace, $L_1^* = L_1$, and the operator $L_1^{-1}L_2$ is completely continuous. Keldyš's result was strengthened by V. B. Lidskiĭ, who showed the completeness of a system of eigenfunctions and adjoint functions of the operator $L_1 + iL_2$, where $L_1^* = L_1$, L_1^{-1} being a Hilbert-Schmidt operator, $L_2 = L_2^*$ a semibounded operator. The general theory of operators was taken up by Dunford, who obtained the analogue of the spectral decomposition for non-selfadjoint operators (a continuous direct sum of Jordan matrices) under definite assumptions on the rate of growth of the resolvent of the operator on entering the spectrum.

In the study of the "normal" form of operators in Hilbert space, interesting results were obtained by M. S. Livšic and his followers. Livšic proved in [6] that the operator

$$C = A + iB, \tag{11}$$

where $A = A^*$ and B is a completely continuous operator with a finite trace, may be brought by an orthogonal transformation to triangular form; in the infinite case, naturally, the triangular form also a "continuous" triangular form, which is a Volterra operator (for more precise definitions see the work of Livšic). Further Sahnovič [7] established conditions for the affine equivalence of operators, brought into triangular form. Thus, in these papers Livšic and Sahnovič solved, in essence, the question of the normal form of operators of type (11).

It remains, however, to be emphasized that these results, which are essentially affine, were established with the use of metric methods not particular to them.

Perhaps, as the first, one ought to solve the problem of the Jordan form for nuclear operators (in the sense of Grothendieck), doing it by affine methods.* This problem requires, probably, a further interesting development of normed rings, and in the case that the problem is solved in linear topological spaces, also the development of the theory of topological rings.

11. Topology in the space of representations and the factors of II_1. Let $x \longrightarrow T(x)$ be a representation of some ring R, which we shall assume a normed and separable ring with involution. For example, by R we may understand the group ring of a countable discrete group. In the set of continuous representations of the ring R we introduce a topology, giving neighborhoods in the following way.

We choose in the ring R in any way a finite number of elements $x_1, x_2, \cdots, x_k$, and in the space in which there operates a given irreducible representation $T^{(0)}(x)$, a finite system of vectors $\phi_1^{(0)}, \cdots, \phi_n^{(0)}$, and we construct a scalar product

$$(T^{(0)}(x_\alpha)\,\phi_i^{(0)}, \phi_j^{(0)}).$$

We shall say that the representation $T(x)$ belongs to the neighborhood $V(T^{(0)}; \phi_1^{(0)}, \cdots, \phi_n^{(0)}; x_1, \cdots, x_k; \epsilon)$ of the representation $T^{(0)}(x)$, if, in the space in which the representation $T(x)$ operates, one may find a system of vectors $\phi_1, \cdots, \phi_n$ such that

$$|(T(x_\alpha)\,\phi_i, \phi_j) - (T^{(0)}(x_\alpha)\,\phi_i^{(0)}, \phi_j^{(0)})| < \epsilon$$

for all $i, j = 1, \cdots, l, \quad \alpha = 1, \cdots, k$.

* It is probable that for the problem at hand this is a more reasonable class of operators than the class of all completely continuous operators.

The topology thus defined in the space of representations, generally speaking, is not Hausdorff. Indeed, consider the ring of matrices of second order

$$A(t) = \begin{bmatrix} a(t), & b(t) \\ c(t), & d(t) \end{bmatrix}, \quad 0 \leq t \leq 1,$$

in which the functions a, b, c, d are continuous functions of t, while $b(0) = c(0) = 0$. Then the relation $A(t) \to A(t_0)$ for each $t_0 > 0$ gives a continuous representation, and the relations $A(t) \to a(0)$ and $A(t) \to d(0)$ are two one-dimensional representations. It is easy to verify that these two one-dimensional representations do not have nonintersecting neighborhoods.

We shall call representations contiguous, if any pair of their neighborhoods intersect. We shall say further that representations are connected to each other, if one may join them by a chain of representations, each two successive links of which are contiguous. For algebraic Lie groups representations which are contiguous to each other form a discrete group (for example, for the Lorentz group the finite-dimensional (spinor) representations are contiguous to each other, and so are their "tails", i.e., representations given by integral-valued choices of the parameters of the representation). On the opposite side, in rings of type II_1 all representations, from all appearances, are either contiguous or connected to each other. This of course makes much more difficult the problem of classification of rings of type II_1 (one ought to mention that probably one would do better to consider not the rings II_1 as weakly closed rings, but those rings from which they arise by weak closure, as, for example, the group rings of certain discrete groups).* One must here, however, emphasize that the principal problem is not the problem of classification of all factors, but rather the problem of making them easily accessible and to find their real applications in analysis or physics. On the possibility of a connection of factors with Wiener measure I have already given some indications in my report in Uspehi Matematičeskih Nauk XI, issue 6 (1956), 3–12.

One of those concrete problems, which we may indicate here, consists in the following. Zinger considered the following concrete ring. Suppose given a dynamical system Ω with an invariant measure. Consider the product $\Omega \times I$ of the space Ω with the numerical axis I and the space of functions $f(x, t)$ with integrable square on that product. In this space we define the following operator:

$$U(f) = \int_{-\infty}^{+\infty} \phi(s) f(x_s, t + s)\, ds,$$

* An interesting "gluing" plays a role in another problem of classification of rings of continuous complex functions, in which convergence is uniform convergence. As was shown in the important and instructive examples of Goffman and Zinger, published in this issue of Uspehi Matematičeskih Nauk, the structure of such rings is probably determined by the choice of rings of analytic functions in which points are glued together in a definite way.

where the function $\phi(s)$ is some finite continuous function on the numerical axis. (If in the dynamical system the time is discrete, then in place of the integral we consider the sum.) Further, we introduce operators A, writing

$$Af = a(x) f(x, t),$$

where $a(x)$ is a function on Ω. We consider the ring generated by operators of type U and A. What are the types of these rings for the following dynamical systems?

1°. The dynamical system Ω is a torus with the usual motions on it.

2°. The dynamical system Ω is a dynamical system on a surface of negative curvature. (A spectral study of this dynamical system will be found in [8].)

12. Asymptotics in equations with analytic data. In a series of problems of asymptotic character an essential role is frequently played by the analyticity of the initial data and the coefficients. Consider for example the problem of the so-called barrier dispersion. Suppose that a stream of particles of a given impulse moves in the direction of the x axis and suppose that in its path there is a potential barrier. According to the rules of quantum mechanics, a certain quantity of particles will reflect, even though its height is less than the energy of the moving particles. It is required to find the portion of reflected particles, more precisely the asymptotic behavior of it as $E \longrightarrow \infty$. Here, if one has barrier dispersion, then this asymptotic behavior depends on the analytic properties of the potential. Indeed, in the case that the potential is an analytic function, the coefficient of reflection increases exponentially, and in the case that it has a discontinuity of any order, the dependence will be a power.

As another example we consider the problem of an adiabatic invariant pendulum, whose oscillations are described by the equation

$$y'' + a(\epsilon t) y = 0,$$

where $a(-\infty) = a$, $a(+\infty) = b$, and ϵ is a parameter tending to zero. Thus this is a pendulum whose length slowly (adiabatically) varies. The problem consists in determining the precision to which the adiabatic invariant is maintained. It turns out that this precision again depends on the analyticity of the function $a(t)$ and will be different in it depending on whether or not the function $a(t)$ is analytic.

The examples already introduced show that in problems on asymptotics, the habit of requiring correctness of the problem may lose its meaning. If the process goes on for a long time, then the very small perturbations may have an essential significance. Analogously, it is perfectly natural that the asymptotics may depend on the large "smoothnesses" of the functions, and the problem of its whereabouts may therefore be incorrect.

As to general problems relative to this point we shall consider two: the Cauchy problem and the Dirichlet problem.

1°. We consider some or other *system* of equations:

$$\frac{\partial u}{\partial t} = Lu, \tag{12}$$

where L is a matrix of arbitrary differential operators in the variable x, and consider for this equation the Cauchy problem, not assuming its correctness. We add to the right side some or other terms, such that for the new equation the Cauchy problem will be correct. That is, we consider the equation

$$\frac{\partial u}{\partial t} = \epsilon L_1 u + Lu \tag{13}$$

(sometimes the same result may be attained by adding higher derivatives in t). It is required to find the asymptotic solution of the Cauchy problem for equation (13) as ϵ approaches zero. Two cases are possible: 1) the usual problem, when the limiting Cauchy problem is correct, and 2) when the limiting Cauchy problem is not correct. In the latter case, the asymptotics are delicate, while, as in the problems indicated above, an essential role is played by the analytic properties of the initial data and the coefficients of the operator L. Indeed, the analyticity of the coefficients and the initial data guarantee, in view of the Kowalevski theorem, the existence of a solution for some interval of variation of t. It is natural to expect that this solution will be a limit for the solution of the perturbed equation. One may expect further that if the initial data are not analytic (having analytic coefficients), then the order of convergence as $\epsilon \to 0$ will depend essentially on the analytic character of the initial data.

Consider the following example. Suppose we have the equation

$$\frac{\partial u}{\partial t} = -\frac{\partial^2 u}{\partial x^2}.$$

We first add the term $-\epsilon \dfrac{\partial^4 u}{\partial x^4}$, i.e., we consider the equation

$$\frac{\partial u}{\partial t} = -\frac{\partial^2 u}{\partial x^2} - \epsilon \frac{\partial^4 u}{\partial x^4}; \tag{14}$$

analogously we could have considered the equation

$$\frac{\partial u}{\partial t} - \epsilon \frac{\partial^2 u}{\partial t^2} = -\frac{\partial^2 u}{\partial x^2}.$$

It is required to find the asymptotic solution of the problem of equation (14) as $\epsilon \to 0$.

2°. Consider some region D and the equation

$$Lu = f.$$

We add to L, for example, the operator $\epsilon\Delta^p$, where the power p is sufficiently high that it exceeds the order of the equation L, and set up for the equation

$$(L + \epsilon\Delta^p)\, u = f$$

a Dirichlet problem, i.e., we give on the boundary of the region a function and a sufficient number of normal derivatives. It is required to find the asymptotic behavior as $\epsilon \longrightarrow 0$.

We indicate the following example, connected with the theory of membranes. Consider the equation

$$\epsilon\Delta^4 u + \left(\frac{\partial^2}{\partial x^2} - \frac{\partial^2}{\partial y^2}\right)^2 u = f.$$

On the boundary of the region is given a function and normal derivatives of the necessary order. It is required to find the asymptotic solution as $\epsilon \longrightarrow 0$. This asymptotic behavior depends, of course, on the solvability of the limit problem of Cauchy.

13. Vacuum measure. In order to look into mathematical questions arising in the quantum theory of fields, it is useful to sharpen a number of fundamental concepts. In quantum mechanics one considers wave functions. The square of the modulus of a wave function gives us the probability density of detection of a particle in a given position, and the integral of the square of the wave function over a region the probability of finding the particle in that region. That integral is taken over Euclidean measure in a space whose dimension is determined by the number of degrees of freedom of the system. Thus, as the number of degrees of freedom increases we must integrate over spaces of ever higher numbers of dimensions, and for systems with infinitely many degrees of freedom (e.g., fields), we would have to integrate over a space with an infinite number of variables. The equations of the theory of fields may be written as the Schrödinger equation in an infinite-dimensional space. (See, for example, [9].) There, in order to attach a meaning to a wave function in the quantum theory of fields we must indicate over what measure we are going to integrate in an infinite-dimensional space. Inasmuch as there is no ordinary Euclidean measure in an infinite-dimensional space, we have to introduce a measure into the corresponding infinite-dimensional space in some other way, namely, we replace the fields in question by a system with a finite number n of degrees of freedom, and write down for this the Schrödinger equation. Then we consider the eigenfunction of this equation corresponding to the smallest eigenvalue. This eigenfunction never vanishes, and therefore the square of its modulus, integrated over the volume of the n-dimensional space, may be taken to be the measure in this n-dimensional space. The limit of this measure as the number of degrees of freedom increases will be called the vacuum measure in the infinite-dimensional space. In this case it is a Gaussian measure (i.e., the measure given by the Gaussian distribution). Apparently, all the fundamental difficulties of the contemporary methods of the quantum theory of fields are connected with the difficulty of estimations of this measure. In particular, the methods of the theory of perturbations are essentially connected with the use, in the role of this measure, of the vacuum measure of free fields. Inasmuch as it is improbable that the pure vacuum measure is absolutely continuous with respect to the vacuum measure of free fields, it is not astonishing that its use leads to a series of difficulties. Of course, all kinds of regularizations help, but only very relatively.

The problems which we may now formulate are the following.

First, do we need to prove the existence of a vacuum measure even in the simplest case of the Thirring equation:

$$\Box \psi + k^2 \psi = \lambda \psi^3.$$

Such a measure must be given in the space of generalized functions $\phi(x)$, as is shown by the example of free fields, where the measure gives a generalized Gaussian measure.

The second and very important question is how can effective methods be developed in quantum theory, beginning from the hypothesis of the existence of a vacuum measure for mutually interacting fields?

14. Quasi-invariant measures. The measure $\mu(P)$, given in a linear space, is called quasi-invariant if for each vector x the measure $\mu(P + x)$ is absolutely continuous over the measure $\mu(P)$. In an n-dimensional space each quasi-invariant measure may be represented in the form $\int f(x)\, dx$, where the function $f(x)$ is positive and almost everywhere different from zero. In an infinite-dimensional space, there are not only no invariant measures, but as Mitjagin and Gorin proved, not even quasi-invariant measures. However one may introduce the notion of quasi-invariant measure, consisting in a certain way of modifying it. To do this we proceed as follows. We consider a certain nuclear linear topological space Φ. Suppose that there is given in Φ a positive definite quadratic form, giving in Φ the scalar product. Then one may suppose that we have

$$\Phi \subset H \subset \Phi',$$

where H is a Hilbert space gotten by adjoining the scalar product, and Φ' is adjoint to Φ. A quasi-invariant measure may now be defined as follows: we shall require that the measure should be defined in Φ', and quasi-invariance will be taken only relative to motions on elements from Φ. The Wiener measure is an example of quasi-invariant measure. One may construct further many other examples of quasi-invariant measure and find, for example, conditions which say when

Gaussian measure becomes quasi-invariant. The problem consists in describing all quasi-invariant measures in the triple $\Phi \subset H \subset \Phi'$ in the sense described above. It still remains to be noted that the three questions indicated here on quasi-invariant measure, on vacuum measure, and the question on rings of type II are closely related to one another.

In order to give some indications concerning this connection, we observe, for example, that the vacuum measure, from all appearances, is quasi-invariant. I would like in these terms to give some short suggestions concerning the important paper of Segal [10].

Suppose that we have a space Φ, which is a nuclear space. We consider, as before, $\Phi \subset H \subset \Phi'$, and construct an infinite-dimensional Lie group as follows. The elements of this group will be triples (ϕ, ψ, α), where $\phi, \psi \in \Phi$, and α is a real number. We give the product by the formula

$$(\phi, \psi, \alpha)(\phi_1, \psi_1, \alpha_1) = (\phi + \phi_1, \psi + \psi_1, \beta),$$

where $\beta = \alpha\alpha_1 e^{2i}(\psi, \phi_1)$. Then the question of irreducible (unitary) representations of this group is connected with the finding of a quasi-invariant measure in $\Phi \subset H \subset \Phi'$. Since, on the other hand, this group is connected with commutator relations in the quantum theory of fields, which arise as commutator relations in the Schrödinger equation relative to spaces with a vacuum measure, then the connection between all of these is established without doubt.

BIBLIOGRAPHY

[1] M. N. Višik, *On strongly elliptic systems of differential equations*, Mat. Sb. (N.S.) **29** (71) (1951), 615–676. (Russian)

[2] I. N. Vekua, *New methods for solving elliptic equations*, OGIZ, Moscow, 1948. (Russian)

[3] I. M. Gel'fand, *Some problems in the theory of quasilinear equations*, Uspehi Mat. Nauk **14** (1959), no. 2 (86), 87–158. (Russian)

[4] J. Leray, *Problème de Cauchy*. I, Bull. Soc. Math. France **85** (1957), 38–429, and also successive parts II (ibid. **86** (1958), 75–96), III (ibid. **87** (1959), 81–180), etc.

[5] K. I. Babenko and I. M. Gel'fand, *Remarks on hyperbolic systems*, Naučn. Dokl. Vysš. Skoly Fiz.-Mat. **1958**, no. 1, 12–18. (Russian)

[6] M. S. Livšic, *On spectral decomposition of linear non-selfadjoint operators*, Mat. Sb. (N.S.) **34** (76) (1954), 145–199. (Russian)

[7] L. A. Sahnovič, *On reduction of Volterra operators to the simplest form and on inverse problems*, Izv. Akad. Nauk SSSR Ser. Mat. **21** (1957), 235–262. (Russian)

[8] I. M. Gel'fand and S. V. Fomin, *Geodesic flows on manifolds of constant negative curvature*, Uspehi Mat. Nauk **8** (1952), no. 1 (47), 118–137. (Russian)

[9] I. M. Gel'fand and A. M. Jaglom, *Integration in functional spaces and its application in quantum physics*, ibid. **11** (1956), no. 1(67), 77–114. (Russian)

[10] I. E. Segal, *Distributions in Hilbert space and canonical systems of operators*, Trans. Amer. Math. Soc. **88** (1958), 12–42.

Translated by:
J. M. Danskin, Jr.

4.

Integral geometry and its relation to the theory of group representations

Usp. Mat. Nauk **15** (2) (1960) 155–164 [Russ. Math. Surv. **15** (2) (1960) 143–151]
Zbl. **119**:177

Part of a report, prepared by me for the Edinburgh Mathematical Congress, has already been published [1]. This account suffered, however, from several substantial defects. In the first place, a number of questions bound up with group representations were omitted, while, secondly, the discussion of a number of problems was not continued sufficiently far. The present paper is a first supplement to [1]. The next number of "Uspekhi" will contain a detailed exposition of the questions touched on in [1] relating to elliptic equations, while other questions in the theory of group representations and in the theory of automorphic functions will be published later.

A number of problems of integral geometry (in particular, those discussed in this paper as examples 2, 4, 5, 6 and 8) have been formulated and solved by the author in collaboration with M.I. Graev [5], [6].

The author expresses his thanks to A.A. Kirillov for his assistance in preparing this paper for publication.

Let X be some space and in it let there be given certain manifolds which we shall suppose to be analytic and dependent analytically on parameters $\lambda_1, \ldots, \lambda_k$:

$$\mathfrak{M} = \mathfrak{M}(\lambda) = \mathfrak{M}(\lambda_1, \ldots, \lambda_k).$$

With a function $f(x)$ in X we associate its integrals over these manifolds

$$I(\lambda) = \int_{\mathfrak{M}(\lambda)} f(x)\, dx. \tag{1}$$

We then ask whether it is possible to determine $f(x)$, knowing its integrals over the $\mathfrak{M}(\lambda)$. More precisely, we are interested in the following questions:

1. When is the correspondence defined by formula (1), one-one, $f(x)$ being an "arbitrary" function on X? If the correspondence is one-one, then it would be desirable to have a formula expressing $f(x)$ in terms of the $I(\lambda)$.

2. Which functions of λ can be represented as an integral of type (1), $f(x)$ being an "arbitrary" function on X?

When we speak of "arbitrary" functions in this connection we have in mind some class of functions singled out by conditions on their smoothness or their rate of decrease at infinity, but containing locally all infinitely differentiable functions[1]. Such a class might be, for instance, the class of bounded infinitely differentiable functions or the class of infinitely differentiable functions rapidly decreasing at infinity.

---

[1] That is, every point has a neighbourhood U with the property that if φ is any infinitely differentiable function whatsoever on $\bar{U}$, then the chosen class contains a function ψ, agreeing with φ on U.

It is natural to expect that if a function of n variables is to be uniquely determined, then we must know a function of the same number of variables. More precisely, if an "arbitrary" function $f(x)$ is to correspond to an "arbitrary" function $I(\lambda)$, then the dimension of the space should necessarily coincide with the number of parameters λ_i. Unfortunately, a theorem of this kind on the invariance of dimension has not yet been proved.

From our point of view the solution of the problems which we have posed is precisely the basic content of integral geometry. Until now the field of study covered by the name "integral geometry" has been a very poor one—principally the calculation of invariant measures for some homogeneous space or other, where the problems which arise are more or less trivial. (Cf. [7], [8].)

The problem of integral geometry arises in a natural way in the theory of representations of Lie groups [5] in connection with the following problem.

Let X be a homogeneous space operated on by a group of transformations G. Then G has a representation in the space E of functions $f(x)$ on X: to every element $g \in G$ there corresponds the operator T_g where $T_g f(x) = f(xg)$. (xg is the point of the space to which x is mapped by the transformation g.) One should like to resolve this representation into its irreducible components.

The general form of the solution of this problem has been found by I.M. Gel'fand and M.I. Graev [5] for the case when G is a complex semisimple Lie group. In the space X certain submanifolds, called orispheres, are singled out. A transformation of the group G sends every orisphere into another orisphere. Therefore the group G can be considered as acting on the space of orispheres X'. It appears that for the representation of G defined by its action on the space E' of functions on X' the resolution into irreducible components is, generally speaking, very easy. We have still to show the connection between the spaces E and E'. Consider the transformation α: $E \to E'$, where to an arbitrary function $f(x)$ on X there corresponds a function φ on X':

$$\varphi(\mathfrak{M}) = \int_{\mathfrak{M}} f(x)\, dx, \qquad (1')$$

where $\mathfrak{M}$ is an orisphere. In this way there arises a problem of integral geometry—to determine whether α is invertible and which are the functions $\varphi(\mathfrak{M})$ on X' which can be defined by a formula of type (1'). If α is invertible, then the problem reduces to finding the formula of inversion.

We shall now consider several examples and formulate several problems in integral geometry.

1. As a classical example we have Radon's problem of the representation of functions in n-dimensional space by their integrals over hyperplanes.

Let $f(x) = f(x_1, \ldots, x_n)$ be a bounded infinitely differentiable function. We set

$$I(a, p) = \int_{(a,x)=p} f(x)\, d\sigma_a,$$

where $(a, x) = \sum_{j=1}^{n} a_j x_j$, and $d\sigma_a$ is the element of area on the hyperplane.

The integral $I(a, p)$ is a function of the hyperplane $(a, x) = p$. It depends on n parameters, namely the numbers a_j and p, determined uniquely up to a constant multiplier. Then we have the following inversion formulae [2]: for even n

$$f(x) = \frac{(-1)^{\frac{n}{2}}(n-1)!}{2(2\pi)^{n-1}} \int_{\Omega} \left(\int_{-\infty}^{\infty} \frac{I(a, p)}{|p-(a, x)|^n} dp \right) d\omega;$$

while, for odd n

$$f(x) = \frac{(-1)^{\frac{n-1}{2}}}{2(2\pi)^{n-1}} \int_{\Omega} \left(\frac{\partial^{n-1} I(a, p)}{\partial p^{n-1}} \bigg|_{p=(a, x),\ \Sigma a_j^2=1} \right) d\omega\ ^{1}).$$

Here Ω is the unit sphere $\Sigma a_j^2 = 1$, and $d\omega$ the element of area. In the first integral the interior integral is understood to be its regularised value.

The one-oneness of the correspondence between the functions $f(x)$ and the functions $I(a, p)$ follows from Plancherel's formula, which for even n takes the form

$$\int |f(x)|^2 dx = \frac{1}{2(2\pi)^n} \int_{-\infty}^{\infty} \int_{\Omega} \left| \frac{\partial^{\frac{n-1}{2}} I(a, p)}{\partial p^{\frac{n-1}{2}}} \right|^2 d\omega\, dp.$$

Even this classical existence problem is not fully solved. For example we can pose the following question, the answer to which would be analogous to the theorem of Paley-Wiener: which functions $I(a, p)$ correspond to bounded infinitely differentiable functions? Or: which functions $I(a, p)$ correspond to rapidly decreasing infinitely differentiable functions?

In place of the functions $f(x)$ we can consider positive measures satisfying the demand that the measure of the whole space be finite. The problem then becomes: which functions of halfspace correspond to measures?

2. A problem, analogous to Radon's problem, can be formulated for n-dimensional Minkowski space, where the square of the distance between the points $(x_1, x_2, \ldots, x_n)$ and $(x_1 + dx_1, x_2 + dx_2, \ldots, x_n + dx_n)$ is given by the formula $ds^2 = dx_1^2 - dx_2^2 - \ldots - dx_n^2$. For simplicity we consider the case of three-dimensional Minkowski space. Let the integrals of the function $f(x)$ in Minkowski space be given over all spacelike planes, i.e. over planes $a_1x_1 + a_2x_2 + a_3x_3 = p$, where $a_1^2 - a_2^2 - a_3^2 > 0$. Naturally, the knowledge of these integrals is insufficient for the redetermination of $f(x)$ (in Radon's problem the integrals of the function were given over every plane). It seems that it is sufficient to add only the integrals of $f(x)$ over isotropic lines, i.e. lines along which $ds^2 = 0$[2]. We observe that the isotropic lines are

[1] Alternatively, there is a single formula (cf. [2]):

$$f(x) = \frac{(n-1)!}{(2\pi i)^n} \int_{\Omega} \left(\int_{-\infty}^{+\infty} \frac{I(a, p)\, dp}{[p-(a, x)-i0]^n} \right) d\omega.$$

[2] Since the metric on an isotropic line is absent, it is necessary to be precise about what we understand by an integral over an isotropic line. We define measure on an isotropic line by the condition that it be invariant under displacements. Then the measure on the line is uniquely determined up to a constant multiplier.

parallel to the generators of the "light" (or isotropic) cone $x_1^2 - x_2^2 - x_3^2 = 0$. Therefore, in place of the integrals over the isotropic lines we could give the integrals on the hyperboloids of one sheet $(x_1 - a_1)^2 - (x_2 - a_2)^2 - (x_3 - a_3)^2 = -c^2$, generated by the isotropic lines. Giving these integrals is sufficient because the norms of the integrals over different isotropic lines are related to one another[1].

Suppose, therefore, that we are given the integrals of $f(x)$ over the spacelike planes

$$I_1(a, p) = \int_{(a, x)=p} f(x)\, d\sigma,$$

where $(a, x) = a_1x_1 + a_2x_2 + a_3x_3$, $a_1^2 - a_2^2 - a_3^2 > 0$, and the integrals over the hyperboloids[2]

$$I_2(a, c) = \int_{-(x_1-a_1)^2+(x_2-a_2)^2+(x_3-a_3)^2=c^2} f(x)\, d\sigma.$$

Then the formula of inversion takes the following form:

$$f(x) = C_1 \int_{a_1^2-a_2^2-a_3^2=1} \left.\frac{\partial^2 I_1(a, p)}{\partial p^2}\right|_{p=(a, x)} d\omega + C_2 \int_0^\infty I_2(x, c)\, c^{-3}\, dc.$$

The second integral here must be understood to be its regularised value or, precisely, the value of the analytic function $F(\lambda) = \int_0^\infty I_2(x, c)\, c^\lambda\, dc$ for $\lambda = -3$ (it can be shown that $F(\lambda)$ is regular at $\lambda = -3$).

3. The following problem is closely related to integral geometry. Let $P(x, y)$ be a polynomial such that the curve $P(x, y) = 0$ has a singular point (x_0, y_0) but the curves $P(x, y) = c$ do not have singularities for sufficiently small $c > 0$. Suppose given the integrals of certain functions $\varphi(x, y)$ along the level curves of P

$$I(c) = \int_{P=c} \varphi(x, y)\, d\omega,$$

where the differential form $d\omega$ on the curve $P = c$ is defined by the equality $dx\, dy = dP\, d\omega$. Knowing the integrals $I(c)$ we have to determine the values of $\varphi(x, y)$ at (x_0, y_0). This problem has a simple solution when the second differential of P

$$d^2P = \alpha\, dx^2 + 2\beta\, dx\, dy + \gamma\, dy^2$$

---

[1] We observe that functions on the set of isotropic lines, defined as integrals of functions over these lines, are bound by certain relations. In fact their integrals over each of the family of generators of the hyperboloid $(x_1 - a_1)^2 - (x_2 - a_2)^2 - (x_3 - a_3)^2 = -c$ coincide.

[2] The element of area $d\sigma$ of the surface $x_1 = \varphi(x_2, x_3)$ in Minkowski space is given by the formula

$$d\sigma = \sqrt{1 - \left(\frac{\partial x_1}{\partial x_2}\right)^2 - \left(\frac{\partial x_1}{\partial x_3}\right)^2}\, dx_2\, dx_3.$$

is nondegenerate at the singularity, i.e. $\alpha\gamma - \beta^2 \neq 0$. In fact, if $\alpha\gamma - \beta^2 = \Delta > 0$ at the singularity, then

$$\varphi(x_0, y_0) = \frac{\sqrt{\Delta}}{\pi} \lim_{c \to 0} (\lambda + 1) \int_0^{\varepsilon} I(c) c^{\lambda} dc = \frac{\sqrt{\Delta}}{\pi} \lim_{c \to 0} I(c);$$

while, if $\alpha\gamma - \beta^2 = \Delta < 0$, then

$$\varphi(x_0, y_0) = -\frac{\sqrt{-\Delta}}{2} \lim_{c \to 0} (\lambda + 1)^2 \int_0^{\varepsilon} I(c) c^{\lambda} dc.$$

Here we suppose $\varepsilon > 0$ to be so small, that there are no singularities on the curves $P = c$ for $0 < c \leqslant \varepsilon$.

There is an analogous formula for the case of an arbitrary number of independent variables. In addition, it is possible to solve the analogous problem for the case of the so-called reduced singularities [2].

4. In the two-dimensional complex space C^2 of points (z_1, z_2) consider the one-dimensional complex straight lines which do not pass through the point (0, 0). The lines themselves form a complex manifold. Each of them can be given by equations

$$\begin{vmatrix} z_1 & z_2 \\ w_1 & w_2 \end{vmatrix} = 1.$$

We take as measure on $\begin{vmatrix} z_1 & z_2 \\ w_1 & w_2 \end{vmatrix} = 1$ the "area" of the triangle with vertices the points (0, 0), (z_1, z_2), $(z_1 + dz_1, z_2 + dz_2)$, i.e. we set

$$ds^2 = |z_1 dz_2 - z_2 dz_1|^2 = \left|\frac{dz_2}{w_2}\right|^2 = \left|\frac{dz_1}{w_1}\right|^2.$$

With every bounded function f on C^2 we associate the function $\varphi(w_1, w_2)$ on the space of lines on C^2:

$$\varphi(w_1, w_2) = \int\limits_{\left|\begin{smallmatrix} z_1 & z_2 \\ w_1 & w_2 \end{smallmatrix}\right| = 1} f(z_1, z_2)\, ds^2 = \int f\left(z_1, \frac{z_1 w_2 - 1}{w_1}\right) \frac{dz_1\, d\bar{z}_1}{w_1 \bar{w}_1}.$$

It appears that f can be redetermined from the $\varphi(w_1, w_2)$ in the following manner:

$$f(z_1, z_2) = c \int\limits_{\left|\begin{smallmatrix} z_1 & z_2 \\ w_1 & w_2 \end{smallmatrix}\right| = 1} \left(\frac{\partial^2 \varphi(\lambda w_1, \lambda w_2)}{\partial\lambda\, \partial\bar{\lambda}} \bigg|_{\lambda = 1} \right) |w_1 dw_2 - w_2 dw_1|^2.$$

This problem can also be considered as the problem of determining a function in four-dimensional real space when its integrals over certain two-dimensional submanifolds are known.

5. Consider the Beltrami-Klein interpretation of Lobachevskii space. Lobachevskii space is realised as the interior of the unit sphere in three-dimensional Euclidean space; displacements appear as projective transformations leaving the unit sphere fixed. Straight lines in the Lobachevskii

sense are ordinary straight lines; parallel straight lines are lines intersecting on the unit sphere. The surfaces, orthogonal in the Lobachevskii sense to a fixed family of parallel straight lines, are called orispheres. In the representation of Lobachevskii space which we are considering these are spheres touching the unit sphere. Orispheres in Lobachevskii space carry a Euclidean geometry; for this reason they are the natural analogues of the planes of three-dimensional Euclidean space. It appears that it is possible to determine a bounded continuous differentiable function $f(x)$ in Lobachevskii space, if its integrals are known on all the orispheres. The formula of inversion is analogous to the formula of inversion given in example 1 for the case of three-dimensional Euclidean space. We fix the point x_0 and consider all the orispheres which are at a distance λ from x_0. Every such orisphere is given by the family of straight lines orthogonal to it, or, what comes to the same thing, by a point ω on the unit sphere Ω. We denote $I_{x_0}(\omega, \lambda)$ the integral of f over this orisphere. Then the inversion formula has the following form (cf. [5]):

$$f(x_0) = c \int_{\Omega} \left(\frac{\partial^2 I_{x_0}(\omega, \lambda)}{\partial \lambda^2} \bigg|_{\lambda=0} \right) d\omega.$$

6. A more complicated example. Consider the exterior X of the unit sphere Ω in three-dimensional Euclidean space. X is a homogeneous space with respect to the group of projective transformations leaving Ω fixed. In X we can introduce an invariant metric, which is indefinite. We consider two-dimensional submanifolds in X analogous to the orispheres of Lobachevskii space. They can be determined, for example, as those surfaces on which Euclidean geometry is realized. By analogy with Lobachevskii space we call these surfaces orispheres. In the representation given these will be certain surfaces of the second order touching Ω. It appears that for the redetermination of a function given on X it is necessary to have knowledge of its integrals not only over all the orispheres, but also over all the isotropic lines (i.e. over the lines along which $ds = 0$). In the representation given these are straight lines touching Ω.

In this case the formula of inversion is found as follows (cf. [6]). We construct the cone with vertex at the point x_0, touching the sphere along a circle (see diagram). From x_0 we draw the geodesics inside one of the sheets of the cone, on them we lay off segments of length λ and then we construct orispheres orthogonal to the corresponding geodesics. Every such orisphere is given by the geodesics orthogonal to it, or, what comes to the same thing, by the point ω of "intersection" of the geodesics with the sphere Ω. We denote by $I_{x_0}(\omega, \lambda)$ the integral of $f(x)$ over this orisphere. It appears that

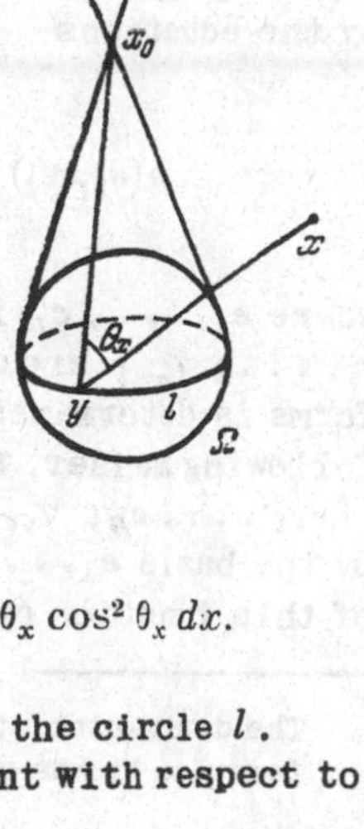

$$f(x_0) = C_1 \int_{\Omega_{x_0}} \left(\frac{\partial I_{x_0}(w, \lambda)}{\partial \lambda^2} \bigg|_{\lambda=0} \right) d'\omega + C_2 \int_{X \setminus K} f(x) \sin^{-3} \theta_x \cos^2 \theta_x \, dx.$$

The first integral is taken over the domain on Ω bounded by the circle l. The integration is with respect to the measure $d'\omega$ invariant with respect to transformations leaving x_0 fixed. The second integral is taken over the exterior of the cone; θ_x is the angle between the isotropic lines, passing through x and touching Ω at y on 1, and the corresponding generator of the cone (see diagram). In this way the exterior of the cone is divided into isotropic lines along which θ_x is constant so that for the calculation of this integral it is sufficient to know the integrals of $f(x)$ over the isotropic lines.

7. In complex three-dimensional space suppose given the integrals of a bounded continuous function $f(z_1, z_2, z_3)$ over all the "straight lines"

$$z_1 = \frac{1}{\lambda} + w_2 z_2,$$
$$z_3 = \lambda - w_1 z_2,$$

passing through the "hyperbola" $z_2 = 0$, $z_1 z_3 = 1$:

$$\varphi(w_1, w_2; \lambda) = \int f\left(\frac{1}{\lambda} + w_2 z, z, \lambda - w_1 z\right) dz\, d\bar{z};$$

then the formula of inversion takes the form[1] :

$$f(z_1, z_2, z_3) = \left.\frac{\partial^2 I(\lambda)}{\partial\lambda\, \partial\bar{\lambda}}\right|_{\lambda=1},$$

where

$$I(\lambda) = -\frac{1}{8\pi^2}\int \varphi\left(w, \frac{(z_2 w + z_3) z_1 - 1}{(z_2 w + z_3) z_2}; (z_2 w + z_3)\lambda\right) |z_2 w + z_3|^{-2}\, dw\, d\bar{w}.$$

Notice that here $\varphi(w_1, w_2; \lambda)$ is a non-arbitrary function of three complex variables. It satisfies the condition $I(\lambda) = I(\lambda^{-1})$ (cf. [4]).

It would be of interest to solve the following more general problem: to determine the function $f(z_1, z_2, z_3)$ knowing its integrals over all the complex lines which intersect a given algebraic curve.

8. Consider the space of positive definite Hermitian forms p of order n with discriminant 1. In this space the group G of complex unimodular matrices acts transitively: the form with matrix x transforms into the form with matrix g^*xg. In this space we give the name of orispheres to surfaces defined by the equations

$$p(e_1, e_1) = e^{\tau_1}, \ldots, \begin{vmatrix} p(e_1, e_1) & \ldots & p(e_1, e_{n-1}) \\ \cdot & \cdot\ \cdot\ \cdot & \cdot \\ p(e_{n-1}, e_1) & \ldots & p(e_{n-1}, e_{n-1}) \end{vmatrix} = e^{\tau_{n-1}},$$

where $e_1, \ldots, e_n$ is an arbitrary basis in the linear space C^n and $\tau_1, \ldots, \tau_{n-1}$ are arbitrary numbers. A bounded function f on the space of forms is determined if its integrals are known on all the orispheres in the following manner. Let p_0 be a fixed form. Then we denote by $I(e_1, \ldots, e_n; \tau_1, \ldots, \tau_{n-1})$ the integral of f over the orisphere defined by the basis $e_1, \ldots, e_n$ and the numbers $\tau_1, \ldots, \tau_{n-1}$. We now take the mean of this function over all bases in which p_0 has the form $p_0 = \bar{x}_1\bar{x}_1 + \ldots + \bar{x}_n\bar{x}_n$

[1] The derivation of the Plancherel formula for the Lorentz group is based on this formula. It was obtained in [4] in an implicit form.

and denote the function so obtained by $I_{p_0}(\tau_1, \ldots, \tau_{n-1})$. It turns out (cf. [5]) that

$$f(p_0) = LI_{p_0}(\tau_1, \ldots, \tau_{n-1})|_{\tau_1=\ldots=\tau_{n-1}=0},$$

where

$$L = c \prod_{2\leqslant p<q\leqslant n} \left(\frac{\partial}{\partial\tau_p} - \frac{\partial}{\partial\tau_q}\right)^2 \prod_{p=2}^{n} \frac{\partial^2}{\partial\tau_p^2}.$$

9. Let G be a group and Z a subgroup. With a function $f(g)$ on G we associate its integrals $\varphi(g_1^{-1}, g_2)$ over all possible sets $g_1^{-1} Z g_2$.

A subgroup Z will be called *resolvent* if we are able to determine an "arbitrary" function $f(g)$ knowing its integrals $\varphi(g_1^{-1}, g_2)$ over sets of the form $g_1^{-1} Z g_2$.

It has been shown [3], that the maximal nilpotent subgroups of a complex semisimple Lie group[1] are resolvent. For all groups in [3] and [9] formulae are obtained representing $f(g)$ by integrals φ.

We give the result for the case where G is the group of complex matrices of order n with determinant equal to 1 (cf. [3] and [9]). The resolvent Z of G consists in this case of all the triangular matrices $\zeta = \|\zeta\|_{ij}$, where $\zeta_{ij} = 0$ for $i > j$, and $\zeta_{ii} = 1$. We introduce the subgroup Z of triangular matrices $z = \|z_{ij}\|$, where $z_{ij} = 0$ for $i < j$ and $z_{ii} = 1$. Then an arbitrary set $g_1^{-1} Z g_2$ coincides with some set $z_1^{-1} \delta Z z_2$ where $z_1, z_2 \in Z$ and δ is a diagonal matrix: $\delta = [e^{\tau_1 + i\varphi_1}, \ldots, e^{\tau_n + i\varphi_n}]$.

We define the integral of $f(g)$ over the set $z_1^{-1} \delta Z z_2$ by the formula

$$\varphi(z_1, z_2, \delta) = \beta^{-1}(\delta) \int f(z_1^{-1}\delta\zeta z_2)\, d\mu(\zeta).$$

Here $d\mu(\zeta)$ means the product of the differentials of the real and imaginary parts of the matrix ζ, and

$$\beta(\delta) = |\delta_2|^4 |\delta_3|^8 \ldots |\delta_n|^{4n-4} = e^{4\tau_2 + 8\tau_3 + \ldots + (4n-4)\tau_n}.$$

The value of $f(g)$ at the point $g = e$ (the unit matrix) is determined by the following formula:

$$f(e) = cL \int \varphi(z, z, \delta)\, d\mu(z)|_{\delta=e},$$

where c is a numerical multiplier and

$$L = \prod_{2\leqslant p<q\leqslant n} \left[\left(\frac{\partial}{\partial\tau_p} - \frac{\partial}{\partial\tau_q}\right)^2 + \left(\frac{\partial}{\partial\varphi_p} - \frac{\partial}{\partial\varphi_q}\right)^2\right] \prod_{p=2}^{n} \left(\frac{\partial^2}{\partial\tau_p^2} + \frac{\partial^2}{\partial\varphi_p^2}\right).$$

(By applying a transformation of the group of displacements to this it is easy to express the value of $f(g)$ for arbitrary g in terms of $\varphi(z_1, z_2\, \delta)$.)

It would be of interest to give a description of the resolvent for other classes of groups and to solve the corresponding problem of integral geometry for them. In particular, it is of interest to study the simple groups of

---

[1] These subgroups are conjugate to each other.

finite order from this point of view.

We observe that the resolvent of a group G of finite order can be characterized in terms of the rank of a subgroup. We say that the subgroup Z of a group G of finite order has rank n, if amongst the characteristic functions of the subsets $g_1^{-1}Zg_2$ of G there are exactly n which are linearly independent. (Obviously the rank of a subgroup does not exceed the order of the whole group G since an arbitrary characteristic function of a set is the sum of functions equal to 1 at one of the points of the set and equal to 0 for the remaining elements of G.) Then we have the following simple assertion: the formula

$$\varphi(g_1^{-1}Zg_2) = \sum_{\zeta \in Z} f(g_1^{-1}\zeta g_2),$$

where Z is a subgroup of G, permits of inversion for those, and only those, subgroups Z whose rank coincides with the order of G (i.e. has the greatest possible value).

The examples which we have given exhaust almost all the problems of integral geometry (in the sense indicated at the beginning of this paper) which have been solved up to the present time.

Amongst the unsolved problems we mention the generalisation of problem 8 to the space of indefinite matrices with given signature, and the analogue of problem 7 for real space, which can be formulated in the following way: to determine a function in three-dimensional real space, if its integrals are known over all the lines which intersect an arbitrary curve of the second order (or, more generally, any algebraic curve). It would be interesting to obtain any results at all concerning the invariance of dimension in a sufficiently general case.

Received 3rd December, 1959.

REFERENCES

[1] Gel'fand, I.M. Some problems of analysis and differential equations. Uspekhi Mat. Nauk, 14 (1959), 3-20.

[2] Gel'fand, I.M. and Shilov, G.E. *Obobshchennye funktsii i deistviya nad nimi.* (Generalised functions and operations on them.) Gosudarstv. Izdat. Fiz.-Mat. Lit., Moscow, 1958.

[3] Gel'fand, I.M. and Naimark, M.A. Unitary representations of the classical groups. Trudy Mat. Inst. Steklov, 36 (1950).

[4] Gel'fand, I.M. and Naimark, M.A. Unitary representations of the Lorentz group. Izv. Akad. Nauk SSSR. Ser. Mat., 11 (1947) 411-504.

[5] Gel'fand, I.M. and Graev, M.I. The geometry of homogeneous spaces, group representations in homogeneous spaces and questions in integral geometry related to them. I. Trudy Moskov. Mat. Obshch., 8 (1959) 321-390.

[6] Gel'fand, I.M. and Graev, M.I. The resolution into irreducible components of representations of the Lorentz group in the spaces of functions defined on symmetric spaces. Dokl. Akad. Nauk SSSR., 127 (1959) 250-253.

[7] Blaschke, V. Lektsii po integral'noi geometrii. (Vorlesungen über Integralgeometrie) Uspekhi Mat. Nauk, 5 (1938) 97-149.

[8] Santaló, L.A. *Vvedenie v integral'nuyu geometriyu.* Izdat, Inostr. Lit., Moscow, 1956. (cf. Introduction to Integral Geometry Paris, Act. Sci et Ind., (No. 1198.), 1953.)

[9] Gel'fand, I.M. and Graev, M.I. The analogue of Plancherel's formula for classical groups. Trudy Moskov. Mat. Obshch. 4 (1955) 375-404.

Translated by I. R. Porteous.

5.

On elliptic equations

Usp. Mat. Nauk **15** (3) (1960) 121–132 [Russ. Math. Surv. **15** (3) (1960) 113–123]
Zbl. **95**:78

This paper, like the note on integral geometry in the last number of the "Uspekhi", is an addendum to my paper [1].

The main idea of the paper is contained in § 2, where we pose the problem of describing linear elliptic equations and their boundary problems in topological terms. The most important of the properties in the large of the solutions of these equations and problems are preserved under small deformations of the problem and must therefore be, in some sense, homotopy invariants. The discovery and study of these invariants is the right way to sort out the whole multiplicity of boundary problems for elliptic equations and to classify these problems.

§ 1 is introductory, and the reader may skim through it lightly if he wishes. In this section we introduce the basic definitions and notations and, in preparation for the discussion of general boundary problems for elliptic equations in § 2, we advance some general considerations about the nature of these problems.

In the preparation of this paper M. S. Agranovich and Z. Ya. Shapiro have given me considerable help, and I take this opportunity of expressing my thanks to them.

§ 1. Introduction

We write the system of partial differential equations to be considered in the form:

$$A\left(x, \frac{\partial}{\partial x}\right) u(x) = f(x). \tag{1}$$

Here $x = (x_1, \ldots, x_n)$ is a point of n-dimensional real space, and A is a square matrix of order m, whose elements

$$A_{ik}\left(x, \frac{\partial}{\partial x}\right) = A_{ik}\left(x_1, \ldots, x_n; \frac{\partial}{\partial x_1}, \ldots, \frac{\partial}{\partial x_n}\right)$$

are polynomials in the $\dfrac{\partial}{\partial x_j}$ with real or complex coefficients depending on x; $u(x)$ and $f(x)$ are m-dimensional vectors

$$u(x) = (u_1(x), \ldots, u_m(x)), \quad f(x) = (f_1(x), \ldots, f_m(x)).$$

We shall suppose that (1) holds in the closure $\bar{\Omega}$ of a domain Ω of the x-space and that it is *elliptic* in $\bar{\Omega}$ in the sense of I. G. Petrovskii. This means that if A^0 is the principal part of A, i.e.

$$A\left(x, \frac{\partial}{\partial x}\right) = A^0\left(x, \frac{\partial}{\partial x}\right) + A'\left(x, \frac{\partial}{\partial x}\right),$$

where A° contains only derivatives of the highest order p, and A' only derivatives of order $< p$, then

$$\det A^0 (x, \sigma) \neq 0 \tag{2}$$

for all $x \in \Omega$ and all real vectors $\sigma = (\sigma_1, \ldots, \sigma_n)$ with $\Sigma\sigma_j^2 \neq 0$.

We shall also suppose that on the boundary Γ of Ω we have the boundary condition

$$B\left(x, \frac{\partial}{\partial x}\right) u(x)|_\Gamma = g(x). \tag{3}$$

Here B is a rectangular matrix with m columns, whose elements, like those of A, are polynomials in the $\frac{\partial}{\partial x_i}$ with coefficients depending on x. It is natural to assume that mp is even and that B has $\frac{1}{2}mp$ rows; we shall say more about this later..

If nothing is said to the contrary, we shall assume that the coefficients in A and B are continuous and that the boundary Γ of Ω is smooth, i.e. it has a tangent hyperplane everywhere.

* * *

We shall say that the boundary problem defined by (1) and (3) is *normally soluble* or *soluble apart from a finite-dimensional subspace*, if the following two conditions are satisfied: (i) the non-homogeneous problem is soluble when a finite number of conditions are imposed on the right-hand side, viz. the right-hand side is orthogonal to certain given functions; (ii) the homogeneous problem has a finite number of linearly independent solutions.

It is well known that many problems for elliptic equations with variable coefficients can be solved by reducing the equations to ones with constant coefficients. It seems likely that a boundary problem for an equation with variable coefficients in an arbitrary domain will always be normally soluble provided that all the boundary problems arising by reductions of this kind are normally soluble. We shall formulate this hypothesis more precisely in a moment.

Let us fix some point $x^\circ \in \overline{\Omega}$. Imagine that we are examining the neighbourhood of this point under a microscope, with greater and greater magnification. In other words, supposing for simplicity that x° coincides with the origin, we make the similarity transformation

$$y = \lambda x, \tag{4}$$

and then let λ tend to infinity. Under the transformation (4) the problem defined by (1) and (3) becomes

$$A\left(\frac{y}{\lambda}, \lambda\frac{\partial}{\partial y}\right) u_\lambda(y) = f\left(\frac{y}{\lambda}\right), \tag{5}$$

$$B\left(\frac{y}{\lambda}, \lambda\frac{\partial}{\partial y}\right) u_\lambda(y)|_\Gamma = g\left(\frac{y}{\lambda}\right), \tag{6}$$

where

$$u_\lambda(y) = (u/\lambda)\,^{1}).$$

We now examine what happens to our problem when $\lambda \to \infty$. If x° is an interior point of Ω, the boundary moves off to infinity and the boundary condition disappears with it. Dividing all the equations of the system (5) by λ^p, we see that in the limit we obtain the system

$$A^0\left(x^0, \frac{\partial}{\partial x}\right)u(x) = 0 \tag{7}$$

(we have replaced y by x and denoted the unknown function again by $u(x)$). This system has constant coefficients and is homogeneous with respect to differentiation; ellipticity is obviously preserved.

If x° lies on the boundary Γ, then in the limit Ω becomes a half-space (bounded by the hyperplane Σ touching Ω at x°). In the interior of this half-space (1) goes over into (7). On the bounding hyperplane Σ we have a new boundary condition

$$C\left(x^0, \frac{\partial}{\partial x}\right)u(x)\,|_\Sigma = h. \tag{8}$$

This is obtained from (6) by proceeding to the limit after first dividing each equation by the highest power of λ in it. The matrix C therefore has constant coefficients, and in each row there remain only derivatives of a fixed order (the highest); h will be a vector of constant components, vanishing for all the equations in which this fixed order is greater than zero.

We now examine the *limiting problem*, arising in the way we have just described when $\lambda \to \infty$. If this problem corresponds to an interior point x° of Ω, it is simply a matter of satisfying in the whole space the system

$$A^0\left(x^0, \frac{\partial}{\partial x}\right)u(x) = f(x), \tag{9}$$

where the right-hand side has, say, finite and reasonably smooth components. If $x^\circ \in \Gamma$, then (9) is required to hold in a half-space bounded by the tangent hyperplane Σ of Ω at x°, and on Σ we have the boundary condition

$$C\left(x^0, \frac{\partial}{\partial x}\right)u(x)\,|_\Sigma = 0. \tag{10}$$

It is natural to seek a solution of the limiting problem in the class of functions bounded together with their derivatives up to a certain order.

We shall say that the original problem

---

[1] Here we are regarding the unknown functions as "dimensionless". We have done this for simplicity; more generally we might suppose that $u_{i\lambda}(y) = \lambda^{k_i} u(y/\lambda)$. It is not difficult to give an example of a case where we do have to make such an assumption. We reduce the equation $A(x, \frac{\partial}{\partial x})\,u(x) = f(x)$ $(m=1)$ to a system by putting $u_1(x) = u(x)$, $u_2(x) = \frac{\partial u(x)}{\partial x_1}$; if we then make our similarity transformation, putting $u_1(y) = u_1(y/\lambda)$, we have to put $u_{2\lambda}(y) = \lambda u_2(y/\lambda)$.

$$A\left(x, \frac{\partial}{\partial x}\right) u(x) = f(x), \tag{1}$$

$$B\left(x, \frac{\partial}{\partial x}\right) u(x)\Big|_{\Gamma} = g(x) \tag{3}$$

is *regular at* x° if the corresponding limiting problem is normally soluble. Furthermore, we shall say that the original problem is *regular* if it is regular at every point of Ω and of its boundary Γ.

It is natural to expect that the following general proposition will hold; *if the problem defined by* (1) *and* (3) *is regular, then it is normally soluble.* In other words, *if all the limiting problems corresponding to the interior and boundary points of the domain are uniquely soluble apart from a finite-dimensional subspace, then the same holds for the original problem.* (The requirement of regularity may possibly have to be replaced by the more severe restriction of *stable regularity*, viz. that all sufficiently near problems are also regular.)

A number of authors have proved this assertion, completely or partially, under various supplementary conditions; see the papers by Z.Ya. Shapiro [2], Ya.B. Lopatinskii [3], F. Browder [4], M. Schechter [5] and the references given there. But no such assertion has been established in complete generality.

We remark that the regularity of the problem at an interior point of Ω is a consequence of the condition of ellipticity and therefore does not constitute an additional condition.

* *
*

We now take up the question: what boundary conditions (3) are permissible in a normally soluble boundary problem?

We begin with the limiting problem, taking it in the form

$$A\left(\frac{\partial}{\partial x}\right) u(x) = 0 \qquad (x_n > 0), \tag{11}$$

$$B\left(\frac{\partial}{\partial x}\right) u(x)\Big|_{x_n=0} = g(x_1, \ldots, x_{n-1}), \tag{12}$$

where we assume the function g to be finite. The matrix A is homogeneous with respect to differentiation, and so is each row of B. We take the Fourier transform with respect to $(x_1, \ldots, x_{n-1})$, obtaining the system of ordinary differential equations

$$A\left(-i\sigma', \frac{d}{dx_n}\right) v(\sigma', x_n) = 0 \qquad (x_n > 0) \tag{13}$$

with the boundary condition

$$B\left(-i\sigma', \frac{d}{dx_n}\right) v(\sigma', x_n)\Big|_{x_n=0} = h(\sigma'), \tag{14}$$

where $\sigma' = (\sigma_1, \ldots, \sigma_{n-1})$. We seek a solution of (11)-(12) in the class of functions bounded together with their derivatives up to a certain order. The solution of (13)-(14) for each σ' must be a linear combination of those vectors of a fundamental system of solutions of (13) that remain bounded as

x_n increases. These vectors are of the form

$$(c_1 x_n^k e^{-i\sigma_n x_n}, \ldots, c_m x_n^k e^{-i\sigma_n x_n}), \tag{15}$$

where σ_n is a root of the equation

$$\det A(\sigma', \sigma_n) = 0, \tag{16}$$

lying in the *lower* half-plane, k is a non-negative integer smaller than the multiplicity of this root, and $c = c(\sigma') = (c_1, \ldots, c_m)$ is a constant vector satisfying the system of equations

$$A(\sigma', \sigma_n)\, c(\sigma') = 0$$

and thus defined only up to a multiplicative constant. To satisfy the boundary condition (14), we must form a linear combination of the vectors (15) with coefficients depending on σ', and then find these coefficients. The number of these coefficients is equal to the number of roots σ_n of (16) that lie in the lower half-plane, each root being counted a number of times equal to its multiplicity. This number must clearly be equal to the number of rows in the matrix B of condition (12).

Because of the ellipticity condition, equation (16) has degree mp in σ_n and its roots are non-real for real $\sigma' \neq 0$. If the coefficients in the matrix A are real, these roots are evenly divided between the upper and lower half-planes. It is not difficult to show that this also holds for complex coefficients if $n \geqslant 3$ (Ya.B. Lopatinskii [3]). We exclude from consideration the case of plane systems with complex (non-real) coefficients. We can then infer that mp is *even* and that *the number of rows in the matrix B must be $\frac{1}{2}mp$.* Obviously, this condition must also be satisfied by the original problem defined by (1) and (3).

Continuing the argument, we can obtain conditions *in algebraic form* for the limiting problem to be normally soluble. For clarity we shall do this in the case when $m = 1$ (one equation) and the roots $\sigma_n^{(\mu)}$ of (16), for real $\sigma' = 0$, are all distinct.

We write the boundary condition (12) in more detail as follows:

$$B_\nu\left(\frac{\partial}{\partial x}\right) u(x)\Big|_{x_n=0} = g_\nu(x_1, \ldots, x_{n-1}) \quad \left(\nu = 1, \ldots, \frac{mp}{2}\right) \tag{17}$$

Substituting a linear combination of the vectors (15) $(k = 0)$ in the equations

$$B_\nu\left(-i\sigma', \frac{d}{dx_n}\right) v\Big|_{x_n=0} = h(\sigma'),$$

we see that the condition

$$\det\{B_\nu(\sigma', \sigma_n^{(\mu)})\} \neq 0 \qquad \left(\mu, \nu = 1, 2, \ldots, \frac{mp}{2}\, \sigma' \neq 0\right) \tag{18}$$

must be satisfied (we recall that the polynomials B_ν are homogeneous). In the general case (when $m \geqslant 1$ and (16) may have multiple roots) an analogous condition can be obtained. We have thus found a necessary condition for the limiting problem (11)-(12) to be normally soluble. For the original problem defined by (1) and (3) to be regular this condition must be satisfied by the limiting problem at every boundary point of the domain. In the case when the polynomials B_ν all have the same degree, this condition is called the *ellipticity condition of the boundary problem.* In the papers cited above, Z.Ya. Shapiro, Ya.B. Lopatinskii, M. Schechter and F. Browder use various modifications or reinforcements of this ellipticity condition.

Z.Ya. Shapiro [2] considers the case of a homogeneous system in three-dimensional space, with constant coefficients. The general boundary problem for a half-plane is solved in explicit form (in the class of functions increasing no faster than a power of the distance); for a bounded domain he reduces it to an integral equation for which the Fredholm theory holds, whence it follows that the problem is soluble when a finite set of conditions is imposed on the right-hand side. Ya.B. Lopatinskii [3] obtained (under stronger algebraic conditions) an analogous result for systems with variable coefficients in a space with an arbitrary number of dimensions, under the hypothesis that the highest derivative in the boundary conditions(3) is of lower order than that in the system (1). This last hypothesis is retained by F. Browder [4] and M. Schechter [5], who consider the case of a single equation ($m = 1$) and strengthen Lopatinskii's results. In particular, Browder establishes normal solubility for general boundary problems in a bounded domain Ω in the class $L^p(\Omega)$, the boundary condition and the system (1) itself being satisfied in a generalised sense.

We also mention a paper by L. Hörmander [6]: he shows that the ellipticity condition on the problem is necessary and sufficient for every solution of the homogeneous problem (11)-(12) ($g \equiv 0$) in a half-space to have an analytic continuation across the bounding hyperplane.

We remark that in the case of equations with *discontinuous* coefficients or conditions we can still ask whether normal solubility of the general boundary problem follows from the normal solubility of the limiting problems. However, the form of the limiting problems at points of discontinuity becomes more complicated. Suppose that there is a discontinuity in the coefficients of the matrix A of the system

$$A\left(x, \frac{\partial}{\partial x}\right)u(x) = f(x), \tag{1}$$

along some smooth curve in the interior of the (plane) domain Ω. At points of this curve the limiting problem reduces to the construction of a solution satisfying one elliptic equation with constant coefficients on one side of a straight line (the tangent to the discontinuity curve) and another such system on the other side. In each of these half-planes, as we have already indicated, the number of conditions that have to be imposed on the boundary in order to determine the solution uniquely in the half-plane is $\frac{1}{2}mp$, where m is the number of unknown functions in the system (1) and p is its order. We must therefore have mp conditions in all. These conditions may be derived by requiring, for instance, that the solution and its derivatives up to some order are to be continuous across the discontinuity curve; in this connection see [7]. The condition that the solution be smooth may also be replaced by some other requirement.

If three discontinuity curves meet at an interior point of the domain, then the limiting problem may consist, for example, in determining a solution, continuous together with its derivatives up to some order, and satisfying three distinct equations in three angles with their common vertex at this point. Here we must have mp conditions on each of the three bounding rays; we then have to find out whether this is sufficient for normal solubilty or whether some supplementary conditions are needed.

It would also be interesting to find out what normal solubility of the limiting problem implies at points of discontinuity of the coefficients in the matrix B of the boundary condition

$$B\left(x, \frac{\partial}{\partial x}\right)u(x)\Big|_{\Gamma} = g(x). \tag{3}$$

Analogous questions arise in the study of *singular* problems. The so-called oblique derivative problem may be taken as typical; for the Laplace equation in the plane, say, we have to find a solution such that at each point of the boundary the derivative in a prescribed direction (varying with the particular point) has a given value. Obviously regularity is violated at points where this direction is tangent to the boundary. It is natural to begin by considering the case when such violations of regularity are confined to a manifold of lower dimensions than that of the boundary. We have to find out what supplementary conditions have to be imposed on the limiting problems at points where regularity is violated for the problem as a whole to be uniquely soluble apart from a finite-dimensional subspace.

To investigate this, it is useful to "straighten out" the boundary by means of appropriate coordinate transformations at points where regularity is violated. As a result, if the coefficients in the boundary condition were constant, they will in general become variable. In the limiting problem these variable coefficients will be replaced, not by constants, but, it appears, by linear functions. The same applies to the coefficients in the equations of the system.

Problems for equations with linear coefficients can be studied with the help of the Laplace transform. An original and very interesting operational calculus has been applied to the study of these problems by J. Leray [8].

§ 2. Homotopy invariants of elliptic equations

We consider, in a given domain Ω, a class of systems of the form

$$A\left(x, \frac{\partial}{\partial x}\right) u(x) = f(x) \tag{1}$$

where the number of variables, the number of unknown functions and the order of the highest derivative in the equations are all fixed[1], and the coefficients may be arbitrary continuous functions. We shall call this class a *class of systems of given structure.* We require the ellipticity condition to be satisfied; this rejects a certain set of systems, and the set of systems remaining breaks up into connected components. In the matrix B of the boundary condition

$$B\left(x, \frac{\partial}{\partial x}\right) u(x)\Big|_{\Gamma} = g(x) \tag{2}$$

we fix the order of the highest derivative, and we again allow the coefficients to be arbitrary continuous functions. The elliptic systems (1) of given structure combined with an arbitrary boundary condition (2) form a class of problems, which we shall call a *class of problems of given structure*[2].

We now require in addition that the *regularity condition of the problem* be satisfied. (We may have to require stable regularity, i.e. the regularity of all problems in a neighbourhood of each problem considered). In this way we reject some of the problems of given structure, and the set of problems remaining again breaks up into connected components. The set of boundary problems rejected by the regularity condition seems always to be of lower dimension than the set of all boundary problems.

---

[1] Other conditions may be adjoined to these; e.g. it may be assumed that all the equations are homogeneous with respect to differentiation.

[2] We remark that in the case of system on a closed manifold (e.g. the Laplace equation on the sphere) the question of boundary conditions does not arise.

Two problems (or two systems, if no boundary conditions are present) are said to be *homotopic* if they lie in the same connected component, i.e. one can be obtained from the other by continuous variation of the coefficients in the matrices A and B, the conditions of *ellipticity* and *regularity* being maintained throughout.

Everything we know about elliptic equations supports the view that the most essential global properties of elliptic systems (1) and their solutions are unaltered by small deformations of the system or the problem within a given structure. It was just for this reason that we introduced the concept of homotopy.

If all the essential non-local properties of the solutions of elliptic systems are homotopy invariants, it becomes important to discover these invariants. This is the language in which the essential properties of the solutions can be expressed. We have no doubt that the converse also holds: all the homotopy invariants must have a meaning for the equations themselves from the classical point of view.

It is natural to begin by trying to discover how many connected components there are in the set of elliptic systems with constant coefficients[1] and how each components can be described algebraically.

Thus there are two important questions here: firstly *to find all homotopy invariants of elliptic problems* (i.e. equations with boundary conditions) and, secondly, *to discover what these invariants mean in terms of the solutions of the equations*.

We emphasise that, if the condition of regularity of the problem is expressed in an algebraic form (see § 1, formula (18)), then the first of these questions becomes one of pure algebraic topology.

It is to be expected, in particular, that one homotopy invariant will be the *index* of the problem, its "degree of non-Fredholmness", i.e. the difference between the number of linearly independent solutions of the given homogeneous problem and the corresponding number for the adjoint homogeneous problem (or, equivalently, the minimum number of conditions on the right-hand side of the non-homogeneous problem under which it has a solution).

In papers by I.N. Vekua, F.D. Gakhov, A.I. Vol'pert, B.V. Boyarskii and others (see [10], [11], [12], [13] and the references given there) the index has been evaluated for various classes of problems in the plane. It would be desirable to discover whether there are non-trivial examples of regular boundary problems with non-zero index in the case of three or more dimensions[2].

We now formulate the problem of homotopy invariants in a somewhat different way. Suppose for simplicity that all the left-hand sides of the equations of the system

$$A\left(x, \frac{\partial}{\partial x}\right) u(x) = f(x) \tag{1}$$

and of the boundary condition

$$B\left(x, \frac{\partial}{\partial x}\right) u(x) \Big|_{\Gamma} = g(x) \tag{2}$$

---

[1] For systems of two equations in the plane this question has been considered by B.V. Boyarskii [9].

[2] Note added in proof. In answer to a question posed in [14], A.I. Vol'pert has constructed such an example in three dimensions.

are homogeneous with respect to differentiation. In order to describe the problem, it is then sufficient to describe its set of limiting problems (see § 1). Consider the auxiliary set S of the limiting problems associated with the boundary points of the domain. A point of the space S is a triple (a, b, c), where a is a homogeneous equations of given structure with constant coefficients, b is the half-space in which the problem is to be solved, and c is a boundary condition of given structure, compatible with a and b. We denote the sets of all a, b and c by $\mathfrak{A}$, $\mathfrak{B}$ and $\mathfrak{C}$. respectively. Our space S can then be considered as a fibre space with base $\mathfrak{A} \times \mathfrak{B}$ and fibre $\mathfrak{C}$. Obviously $\mathfrak{A}$ and $\mathfrak{E}$ are subsets of finite-dimensional spaces (their elements have as coordinates the components of the corresponding matrices) and $\mathfrak{B}$ is a sphere, provided that the domain Ω is convex and finite - we confine ourselves to this case for the present. Suppose that the domain Ω and the system (1) are given; then to every point x of the boundary Γ of Ω corresponds an equation a and a half-space b. This correspondence defines a mapping of Γ into the base of the fibre space S:

$$x \longrightarrow (a, b). \tag{3}$$

Here the mapping $x \rightarrow a$ has a continuous extension into the interior of the domain. Specifying a boundary condition can be interpreted as lifting the mapping (3) into the fibre space itself, i.e. defining a mapping

$$x \longrightarrow (a, b, c), \tag{4}$$

whose projection on the base coincides with (3).

We have to investigate the homotopy types of the mappings $\Gamma \rightarrow S$. Now Γ is essentially a sphere; we are therefore interested in the homotopy group $\pi_{n-1}(S)$ of S.

It is obvious that the projection on $\mathfrak{A}$ of each mapping of Γ into S will belong to a single connected component of $\mathfrak{A}$. If we take two mappings of Γ into S whose projections on $\mathfrak{A}$ lie in distinct components, then we cannot deform one mapping continuously into the other. This means that we must begin by discovering how many connected components $\mathfrak{A}$ has and how they are to be described. After this question has been settled, we single out one particular component of $\mathfrak{A}$. When this has been done, $\mathfrak{A}$ by itself cannot provide any homotopy invariants, and we can replace $\mathfrak{A}$ by a point. We shall now be dealing with a single fixed equation, and the base of the fibre space reduces to $\mathfrak{B}$. We remark that, in the case considered above, of a convex domain, the mapping of Γ on $\mathfrak{B}$ is the trivial mapping of an n-dimensional sphere on an n-dimensional sphere. All the mappings of Γ into S in which we are interested project into this. Thus we do not need the whole group $\pi_{n-1}(S)$, but only the part of it that is defined by mappings of the type indicated.

* *
*

As an example[1] we consider the first order system, with two independent variables

$$A(x)\frac{\partial u}{\partial x} + E\frac{\partial u}{\partial y} = f(x, y), \tag{5}$$

[1] This example has been worked out by Z.Ya. Shapiro.

where E is the unit matrix. We assume a boundary condition of the form

$$B(x)\, u(x)\,|_{\Gamma} = 0. \tag{6}$$

We suppose that the matrices A and B are real, that m is even and that B has ½m rows. The ellipticity condition turns out to be

$$\det\{A - \lambda E\} \neq 0$$

for real λ, i.e. A has no real eigenvalues.

We are interested in how to describe the global properties of problems of this kind. In other words, we are interested in the homotopy invariants of these problems. In this case we shall show that there is a single homotopy invariant, and we shall do this without appealing to the theory of differential equations.

In the first place, it is not difficult to show that in this case the set $\mathfrak{A}$ is connected. By a small change in the elements of A (if necessary), we can ensure that its eigenvalues are distinct for all x. Next, for each value of x we can find a non-singular real matrix $T(x)$ such that

$$T(x)\, A(x)\, T^{-1}(x)$$

is in a canonical form; this will be a matrix consisting of blocks of order two, with non-zero blocks only on the principal diagonal; these will be of the form

$$\begin{pmatrix} \alpha & -\beta \\ \beta & \alpha \end{pmatrix},$$

where $\beta \neq 0$. Any such block can obviously be continuously deformed into one of the following blocks:

$$\begin{pmatrix} 0 & -1 \\ 1 & 0 \end{pmatrix} \text{ and } \begin{pmatrix} 0 & 1 \\ -1 & 0 \end{pmatrix}. \tag{7}$$

Thus we have obtained the result that every system (5) can be continuously deformed into a fixed system, which splits up into ½m Cauchy-Riemann equations (with right-hand sides). We can therefore restrict ourselves to the consideration of this systems

We shall therefore suppose that A is a matrix consisting of 2×2 blocks, with non-zero blocks only on the principal diagonal, these being of the form (7).

For this system it can be shown that the necessary algebraic condition for the problem to be regular (cf. relation (18) in § 1) can be formulated as follows. Let $b_1, \ldots, b_m$ be the columns of the matrix B in (6). We form the complex square matrix of order ½m

$$(b_1 + ib_2, \ldots, b_{m-1} + ib_m) = B^*(x).$$

The necessary condition for regularity is then that

$$\det B^*(x) \neq 0 \tag{8}$$

at every point of the boundary Γ of Ω. Now let Ω be a bounded convex domain and suppose that (8) holds. Since Γ is compact, we can normalise the matrix $B^*(x)$ by the condition

$$|\det B^*(x)| = 1.$$

The boundary condition now reduces to prescribing a continuous mapping of Γ, which may be thought of as a circle, into the space of non-singular complex matrices of order $\frac{1}{2}m$ with determinant of absolute value 1. We denote this space by $\mathfrak{C}$. We are therefore interested in the fundamental group $\pi_1(\mathfrak{C})$.

We remark that, as x describes Γ, $\det B^*(x)$ returns to its initial value, but its argument may change by an integral multiple of 2π. It is not difficult to verify that, if there is no change in the argument, then the mapping of Γ into $\mathfrak{C}$ is null-homotopic. Hence $\pi_1(\mathfrak{C})$ consists of classes K_j of mappings of Γ into $\mathfrak{C}$ where K_j contains those mappings under which the argument of $\det B^*(x)$ changes by $2\pi j$ ($j = 0, \pm 1, \pm 2, \ldots$) as x describes Γ.

Thus the number j is the single homotopy invariant of the problem, and all the global properties of the problems considered must be expressible in terms of this number. In particular, it must be related to the index, provided that the index is actually a homotopy invariant.

This conclusion is confirmed by the results of A.I. Vol'pert and others, who have shown that a condition equivalent to (8) is sufficient for the problem to be normally soluble, and have given an explicit formula for the index.

Received 14th January, 1960.

REFERENCES

[1] I.M. Gel'fand: Some problems in analysis and in differential equations. *Uspekhi Mat. Nauk*, N.S. 14, No. 3 (1959), 3-20. (Russian).

[2] Z.Ya. Shapiro: On general boundary problems for equations of elliptic type. *Izv. Akad. Nauk SSSR, Ser. Mat.* 17 (1953), 539-562. (Russian).

[3] Ya.B. Lopatinskki: On a method of reducing boundary problems for a system of differential equations of elliptic type to a regular integral equation. *Ukrain. Mat. Zh.* 5 (1953), 123-151. (Russian).

[4] F.E. Browder: Estimates and existence theorems for elliptic boundary value problems. *Proc. Nat. Acad. Sci. U.S.A.* 45 (1959), 365-372.

[5] M. Schechter: General boundary value problems for elliptic partial differential equations. *Comm. Pure Appl. Math.* 12 (1959), 457-486.

[6] L. Hörmander: On the regularity of solutions of boundary problems. *Acta Math.* 99 (1958), 225-264.

[7] I.M. Gel'fand: Some problems in the theory of quasilinear equations. *Uspekhi Mat. Nauk*, N.S. 14, No. 2 (1959), 87-158. (Russian).

[8] J. Leray: Le problème de Cauchy pour une équation linéaire à coefficients polynomiaux. *C. R. Acad. Sci. Paris*, 242 (1956), 953-959.

[8] B.V. Boyarskii: On the first boundary problem for systems of equations of elliptic type of the second order in the plane. *Bull. Acad. Polon. Sci., Sér. Sci. Math. Astr. Phys.* 7 (1959), 565-570. (Russian).

[10] F.D. Gakhov: *Kraevye zadachi* (Boundary problems). Gosudarstv. Izdat. Fiz.-Mat. Lit., Moscow, 1958.

[11] I.N. Vekua: *Obobshchennye analiticheskie funktsii* (Generalised analytic functions). Gosudarstv. Izdat. Fiz.-Mat. Lit., Moscow, 1959.

[12] A.I. Vol'pert: Boundary problems for elliptic systems of higher order in the plane. *Dokl. Akad. Nauk SSSR*, N.S. 127 (1959), 739-741. (Russian).

[13] B.V. Boyarskii: Some boundary problems for systems of $2n$ equations of elliptic type in the plane. *Dokl. Akad. Nauk SSSR*, N.S. 124 (1959), 15-18. (Russian).

[14] I.M. Gel'fand: On some problems of functional analysis. *Uspekhi Mat. Nauk*, N.S. 11, No. 6 (1956), 3-12.

Translated by F. Smithies.

6.

Automorphic functions and the theory of representations[1]

Proc. Int. Congr. Math. Stockholm (1962) 74–85. Zbl. **138**:71

1. Introduction

According to F. Klein the investigation of geometrical objects may be reduced to the study of properties invariant under a group of transformations, i.e. to the study of homogeneous spaces.

By a homogeneous space X we mean any manifold acted upon by a group of transformations which is a Lie group. We can therefore study in a unified manner the whole class of the most symmetrical objects, aesthetically perfect, such as sphere, Lobachevsky space, Grassmann's manifold, the space of positively-definite matrices, etc.

It is rather remarkable that we can reach the same aesthetic perfection in the study of the set of functions $f(x)$ defined on the homogeneous space X. A transformation $x \to xg$ in X gives rise to a linear operator $T_g f(x) = f(xg)$ in the function space. So we are led to a representation of the group G, since the product of the transformations T_{g_1} and T_{g_2} corresponds to the product of the elements g_1 and g_2 of G.

Roughly speaking the problem is to decompose the function space into minimal invariant subspaces, or, which is the same, to decompose the representation into irreducible representations and to study invariant families involved.

The compact case of this problem (i.e. compact X and G) was investigated by H. Weyl and E. Cartan who took as model the rotation group. The invariant families for the rotation group consist of spherical functions. The least invariant systems of functions, arising in the general case, we shall also call the spherical functions on X. But only after one rejects the compactness condition and passes over to infinite-dimensional representations can one fully appreciate the importance and interest of this problem.

In this report I should like to tell about some results obtained by me in collaboration with my friends I. I. Pyatezki-Shapiro and M. I. Graev. Some of the results were inspired by important works of Selberg and Godement. It should be said that only through systematic employment of the theory of infinite-dimensional representations can one obtain complete understanding of these results.

Each homogeneous space X is associated with a group G and its subgroup Γ (stationary subgroup) consisting of those elements of G which leave immobile some fixed point $x_0 \in X$. The first part of the report is devoted mainly to the consideration of the case in which this subgroup Γ is a discrete group and the space X has finite volume. The functions relating to this case we shall call automorphic functions. Thus automorphic functions are spheri-

(1) The author dedicates this report to his friends, L. Schwartz, M. Morse, J. Leray, O. A. Oleinik, J. G. Petrovskiy, L. Bers, R. Courant, P. Lax, whose kind assistance has greatly helped him.

cal functions, associated with a discrete stationary subgroup. Automorphic functions in this sense were introduced in 1950 in a paper by S. V. Fomin and myself [3]. The ordinary automorphic functions are included among these if we take the group of real 2×2-matrices.

Permit me to cite another interesting and important example. Let us be given an arbitrary Riemann surface with a Riemann metric on it. Consider the manifold X the points of which are taken to be linear elements (i.e. points with directions) of the initial surface. Suppose that the universal covering space of our Riemann surface is the upper halfplane. Then the group G acting on X will be the group of all linear fractional transformations with real coefficients and the stationary subgroup Γ will be one of its discrete subgroups. If the original Riemann manifold is compact, i.e. if it is associated with an algebraic function, then the resulting representation of G in $L_2(X)$ can be decomposed into a countable direct sum of representations. Every irreducible representation of the group G under consideration can be characterised by a set of indices which we shall call the number of the representation. The numbers of the representations involved in the decomposition are invariants of the Riemann surface. They appear to form a complete system of invariants, and so it may happen that this question is connected with the classical problem of moduli in which such remarkable success was attained recently by Bers and Ahlfors. We can prove a weaker result: every continuous deformation of Riemann surface, leaving unchanged these invariants, is the identity.

So in the first half of the report we consider some general questions about the decomposition of a representation in a function space $L_2(X)$ on a homogeneous space X with discrete stationary subgroup. The characterization of the resulting representations and the method of decomposition lead us to very interesting functions which we call Zeta-functions of the given homogeneous space. This system of functions is closely connected with the classical Zeta-function. In the case of the group of real 2×2-matrices we can reduce these functions, using the duality theorems, to the functions, which were introduced by Selberg from particular considerations.

Now we wish to say a word about the methods. M. I. Graev and I have proposed [1] the so-called method of horyspheres for the investigation of representations in $L_2(X)$ in the case in which Γ is continuous, but not discrete as in the case discussed above. This method can be applied with suitable modifications also to the case considered here. This method leads us to a series of theorems concerning the structure of the representations and to the Zeta-function of a homogeneous space. But the general idea of the horysphere method is not sufficient. It is necessary to study the operators $\int F(g)T_g dg$ in detail, where $F(g)$ is a finite function. This part of our work reminds one strongly of the theory of the S-matrix in quantum mechanics. The analogy is intrinsic. For example the Zeta-functions of a homogeneous space are quite analogous to the Heisenberg S-matrix.

In the second part of the report we consider analogous questions concerning the structure of representations in $X = G/\Gamma$ in the case of a semisimple Lie group over a finite field (in Chevalley-Dickson sense). So it is necessary to describe the representations of Lie groups over finite fields. These questions could serve as the theme of a separate report and we are forced to give only a short account of them. We obtain here a interesting system of Zeta-functions for a homogeneous space above a finite field.

The author has chosen as the theme of his report developments that are in the very earliest stages.

We hope that the many interesting problems and relationships which arise will compensate for the unavoidable lack of a description clear and comprehensible in every detail.

We hope that a reasonable bourbakisation of all facts given in this report and of other facts and problems of the theory of representations will lead to the creation of a domain of algebraic functional analysis in which these facts will be the main examples.

2. The case of a compact homogeneous space

If G is a semisimple Lie group, Γ a discrete subgroup and X the space G/Γ of left cosets, then evidently to each element $g \in G$ there corresponds a movement in X transforming x to xg. Denote by $L_2(X)$ the set of all functions on X with integrable square. To each element $g \in G$ we make correspond the unitary shift operator T_g in $L_2(X): T_g f(x) = f(xg)$. The operators T_g form a representation of the group G.

The main problem is to decompose this representation into irreducible ones.

The operators $T_\varphi = \int_G \varphi(g) T_g dg$ where $\varphi(g)$ is a finite function on G play an important role in the theory of infinite-dimensional representations. In our situation the use of operators T_φ is based upon the fact that for every continuous finite function $\varphi(g)$ the operator T_φ is an integral operator in X with the kernel

$$K(x_1, x_2) = \sum_{\gamma \in \Gamma} \varphi(g_1^{-1} \gamma g_2), \tag{1}$$

where g_1 and g_2 are representatives of cosets x_1 and $x_2 \in X$. It is easy to verify that the series for the function $K(x_1, x_2)$ converges uniformly in every compact domain. Thus if X is compact the operator T_φ is completely continuous for every finite continuous function $\varphi(g)$.

It is easy to prove the following general proposition.

If the unitary representation $g \to T_g$ of Lie group G in a space H is such that operator T_φ is completely continuous for every finite continuous function $\varphi(g)$, then H can be decomposed into a countable sum of irreducible representations (unitary) of the group G, where the multiplicity of each irreducible representation is finite.

It follows at once from this proposition that when X is compact the representation in $L_2(X)$ is decomposable into a countable sum of irreducible unitary representations of the group G. This fact does not exhaust all information about the irreducible representations contained in $L_2(X)$ which can be obtained by the use of the operators T_φ. There is a formula which gives in a reasonable sense a complete description of all irreducible unitary representations contained in $L_2(X)$.

Let $H_1, H_2, \ldots$ be irreducible non-equivalent representations contained in $L_2(X)$; $N_1, N_2, \ldots$, their multiplicities. Let $\pi_k(g)$ be the character of the irreducible representation H_k. The existence of such characters was proved by Godement and Harish Chandra. Then the following important formula

$$\sum_{m=1}^{\infty} \int_{F_m} \varphi(g^{-1}\gamma_m g)\, dg = \sum_{k=1}^{\infty} N_k \int \varphi(g)\, \pi_k(g)\, dg \tag{2}$$

is valid, where $\gamma_1, \gamma_2, \ldots$ is a sequence of mutually non conjugate elements of the group Γ and F_m is a fundamental domain for the centralizer Γ_m of the element γ_m. This formula makes it often feasible to find out what representations enter in $L_2(X)$, and with what multiplicities. A particular case of this formula (namely the case in which the functions $\varphi(g)$ are invariant under U—the maximal compact subgroup of the semisimple group) can be essentially reduced by the use of the so-called duality theorem [2] to the previously known formula of Selberg [4].

Formula (2) may be used to investigate the asymptotic distribution of "numbers" of the representations, occurring in $L_2(X)$. The leading term of the asymptotic expansion will be obtained if we choose the function $\varphi(g)$ to be concentrated in a decreasing sequence of neighbourhoods of the unit of the group. Now we give following example.

Consider the asymptotic behaviour not of all representations but only of the so-called representations of class I. By definition these representations are the representations entering in the decomposition of $L_2(G/U)$ where U is a maximal compact subgroup of G.

These representations are given in the following manner. Let $\mathfrak{A}$ be Cartan subalgebra of the symmetric space G/U and let $\mathfrak{A}^+$ be the cone of the positive vectors in $\mathfrak{A}$ (i.e. vectors for which $(\alpha, \alpha) \geqslant 0$ for all positive roots α). Then every representation of class I is uniquely determined by some vector in $\mathfrak{A}^+$. If now X is a compact homogeneous space G/Γ then the representations of class I entering into $L_2(X)$ form a countable set of points in $\mathfrak{A}^+$. Their asymptotic distribution is given by the following formula. Denote by α the positive roots of the symmetric space G/U and by v_α the multiplicities of the roots of G/U. Then the formula

$$N(B_n) \sim C_G C_\Gamma \int_{B_n} \prod_{\alpha > 0} (\varrho, \alpha)^{v_\alpha}\, d\varrho \tag{3}$$

is valid, where B_n runs through an increasing sequence of subregions in $\mathfrak{A}^+$; C_G is a constant depending only on G; C_Γ is the volume of space $X = G/\Gamma$ and $N(B_n)$ is the number of irreducible representations entering in $L_2(X)$ indices of which belong to B_n. The proof is based on some results of F. I. Karpelevich and S. G. Gindikin.

Apparently a similar formula is valid for the other types of irreducible representations.

Since for the group of 2×2-matrices as was said, the "numbers" of the representations are invariants of the Riemann surface, the terms of the asymptotic expansion are also such. It would be very interesting if a few leading terms of this expansion were to play the role of moduli. The next terms of asymptotics demand apparently more averaging than the mere evaluation of $N(B_n)$.

Classical automorphic forms enter here in natural way. Among the representations of the group G there are isolated ones, i.e. those occurring in isolated form already in $L_2(G)$. Among the isolated ones there are representations which are realizable by analytic functions [11, 12, 13]. These isolated representations, if they do occur in the decomposition of $L_2(X)$, are connected with automorphic forms in natural way. It is interesting to observe that there are still other isolated representations (Graev [6]). For every isolated representation one can deduce from the spur formula an explicit formula for the multiplicity N_k. In the case of classical automorphic forms such a formula was obtained by Hirzebruch [14] who used the Riemann-Roch theorem and by Selberg [4] who used a method close to ours.

3. Horyspheres and regular subgroups

The method of horyspheres, which has been elaborated elsewhere, mainly in the paper [1], is effective when X is not compact. In general the method is as follows. Let X be a homogeneous space acted upon by a group G. With the space X one can associate a space Ω whose elements are taken to be surfaces in X which we shall call horyspheres (the definition of horysphere will be given later). To each function defined on X we make correspond its integrals over horyspheres, then the representation in the space of functions on X maps homomorphically onto a representation in the space of functions on Ω. As a consequence the problem of decomposing the representation in the space of functions on X is reduced to the following two problems: (1) to find the kernel of the homomorphism; (2) to decompose the obtained set of functions on Ω into irreducible representations. In some important cases, discussed in [1], the kernel is zero and induced representation on Ω can be given a simple description. But in our case the kernel of homomorphism is not zero and the description is not trivial. Nevertheless the method of horyspheres is very productive in this situation.

In particular it makes it possible to separate quite effectively the series of representations occurring continuously in $L_2(X)$ from the series occurring discretely, or in other words to separate the discrete spectrum from the continuous one.

Let X be an arbitrary homogeneous space; then the set of horyspheres in X is not a homogeneous space itself and has not in general a Hausdorff structure. The structure of the space the points of which are transitive components in the space of horyspheres (i.e. the sets of horyspheres which can be carried one into another by the movements of the group) plays a fundamental role in the description of the spectrum of irreducible representations in $L_2(X)$.

Now we proceed to give the exact definitions. Let G be a real semisimple Lie group and $g(t)$ a one-parameter subgroup. The set Z consisting of all $z \in G$ such that

$$\lim_{t\to\infty} g(-t)zg(t) = 1$$

is called the horyspherical subgroup associated with the subgroup $g(t)$. For example if G is the group of real 2×2-matrices then every horyspheric subgroup is conjugate to the subgroup of all matrices of the form $\begin{pmatrix}1 & \zeta\\ 0 & 1\end{pmatrix}$.

If G is the group of real matrices of order $\mathbf{n}$, then there exist as many non-conjugate horyspherical subgroups as there are different partitions of the the number $\mathbf{n}:\mathbf{n}=k_1+\ldots+k_s$ with positive integers k. (Partitions that are distinguished by the order are to be considered as the different ones.)

Now we proceed to define the horyspheres. Let X be a homogeneous space of a group G.

The orbits of horyspherical groups will be called horyspheres in X. The horysphere will be called compact if the set of points of which it consists, is compact.

Now we proceed to describe the class of discrete subgroups Γ which we shall be concerned with. First of all here belong these for which the volume of G/Γ is finite. It is quite possible that all such subgroups satisfy the assumptions formulated below. For the groups of second order this follows from the results of Siegel. It would be very interesting to prove the assertion in the general case.

Now we introduce the following definition. A set $Y \subset X$ will be called a cylindrical set, if it can be split into mutually non-intersecting compact horyspheres which can be carried one into another by movements. Thus in a cylindrical set all the horyspheres are of the form $x_0 Zg$ where Z is some fixed horyspherical subgroup.

Now we formulate the following fundamental definition.

A discrete subgroup Γ of the semisimple group G will be called regular if the factorspace G/Γ has a finite covering by regular bounded cylindrical sets and the intersection of each pair of them is compact (the definition of regular bounded cylindrical sets is given below).

It is easy to verify that the factor-space G/Γ for each regular discrete subgroup Γ of semisimple group G has finite volume.

We shall observe finally that except for the group of real 2×2-matrices all presently known examples of irreducible discrete subgroups of semisimple Lie groups such that the factor space has finite volume are arithmetical groups that are constructed by the well-known construction of Borel.

We shall call linear algebraic group G every group consisting of all complex $n \times n$-matrices the elements of which satisfy given polynomial relations. In the following we shall assume that the coefficients of these polynomials are rational numbers. We denote by G_z the set of all matrices in G whose elements are integers and whose determinant is equal to 1. In a similar way G_R is the set of matrices in G whose elements are real numbers. It is easy to see that G_z is a discrete subgroup of the group G_R.

A. Borel and Harish Chandra [8] have proved that the volume of the factor-space G_R/G_z is finite, if G_R is semisimple group.

Apparently it can be proved by their methods that G_z is a discrete regular subgroup of G_R.

Now we proceed to define the regular bounded cylindrical sets. It is not difficult to see that for every cylindrical set Y there exists a horyspherical subgroup Z and a set $S \subset G$, whose image in G/Γ is Y, with the following properties:

(1) For every $g \in S$ and $z \in Z$ there exists $\delta \in \Delta = \Gamma \cap Z$ such that $\delta zg \in S$.

(2) If $g_1, g_2 \in S$ and $g_1 g_2^{-1} \in \Gamma$ then $g_1 g_2^{-1} \in \Delta$.

The first condition means that Y consists of horyspheres. The second that these horyspheres do not intersect.

If the set Y is compact then one can choose S so that it will have also the following properties:

(3) There exists a neighbourhood of unity U_2 such that if $g_1^{-1} \gamma g_2 \in U$, where g_1, $g_2 \in S$, then $\gamma \in \Delta$.

(This condition is more strong than condition 2.)

(4) For every $z \in Z$ there exists a compact neighbourhood of unity U_z such that $g^{-1} zg \in U_z$ for any $g \in S$.

Condition (4) plays in this theory the fundamental role.

We shall agree to call bounded any cylindrical set Y for which there exist sets S and Z with the properties (1)–(4). Bounded cylindrical sets are in general non-compact and this seems to be the geometrical reason for the existence in semisimple groups of discrete subgroups such that the factor-space G/Γ has finite volume and at the same time is non-compact.

Let Y be a bounded cylindrical set. It can be shown that all elements $g \in S$ are representable in the form $g = zat$, where a belongs to a certain accompanying subgroup A of Z,(1) and t belongs to a certain compact set T in G. We shall agree to call a normal subgroup $\tilde{Z}$ of Z allowable, or simply allowable, if the intersection of the Lie algebra of the group $\tilde{Z}$ with every root subspace is either vacuous or contains the root subspace.

A bounded cylindrical set will be called regular if

(5) For each allowable subgroup $\tilde{Z}$ of Z (including the group Z itself) the factor-space $\tilde{Z}/(\tilde{Z} \cap \Gamma)$ is compact.

Thus finally the regular bounded cylindrical sets are characterized by the existence of the sets S,Z,A for which the conditions (1)–(5) hold.

4. The separation of the continuous spectrum of representation from the discrete one

We shall suppose as formerly that we have a homogeneous space $X = G/\Gamma$, where the discrete group Γ is regular and G is a real semisimple group.

By the use of the horysphere method we can decompose the space $L_2(X)$ into a direct orthogonal sum of two spaces. One of them, which we denote $L_2^0(X)$, is decomposable into a discrete direct sum of irreducible representations. The other is decomposable into representations of continuous spectra. However there may be representations which enter into the second space discretely, the so-called representations of the complementary series. They get into the second space because they are, so to speak, the analytical continuation of the continuous spectrum involved. One more justification of this fact is that they hit exactly at the singular points of Zeta-functions corresponding to the given continuous series. It is very much like the complementary discrete spectrum of quantum mechanics that hits exactly at the zeros of the S-matrix.

The decomposition into the direct sum is carried out in the following way. To every function $f(x) \in L_2(X)$ we make correspond its integrals over *compact* horyspheres and we denote by $L_2^0(X)$ the space of functions which have their integral over any compact horysphere equal to zero; $L_2'(X)$ is the orthogonal complement of $L_2^0(X)$.

Then *the space $L_2^0(X)$ can be decomposed into a countable number of irreducible representations*. The proof is based on a study of asymptotic properties of the kernel of the integral operator $\int \varphi(g) T_g dg$ at the regular cylindrical sets.

---

(1) A commutative subgroup A is called accompanying if each element of A is semisimple and if A is generated by oneparametrical subgroups $g_1(t), \ldots, g_n(t)$ to each of it Z is associated.

5. Functions in the space of horyspheres

We divide the set of compact horyspheres into transitive families Ω_i (i.e. sets of horyspheres which can be carried one into another by movements). When Γ is regular there will be only a finite number of such families. The families Ω_i are partially ordered in a natural way, namely $\Omega_i < \Omega_j$ if each horysphere of Ω_i can be split into horyspheres of Ω_j.

Now we take one of the transitive horysphere families Ω_i and make correspond to each function $f(x) \in L_2(X)$ its integrals over horyspheres of Ω_i. We obtain a function $\check{f}(\omega)$ defined on Ω_i. The inner product defined in $L_2(X)$ is carried in a natural way into the set of these functions. Thus in $L_2(\Omega_i)$ there is defined a quadratic form. To this form there corresponds an operator M defined by

$$\int_{\Omega_i} f\check{\varphi}\, d\omega = [\check{f}, M\varphi],$$

where $[\check{f}_1, \check{f}_2]$ is the inner product inherited from $L_2(X)$. This operator plays a fundamental role. It is permutable with the movements in Ω_i. As a consequence this operator, by the decomposition of representation in $L_2(\Omega_i)$ into irreducible ones, will be, in every system of representations equivalent to the given one, a matrix whose order coincides with the number of equivalent representations. In the general case this number is less than or equal to the order of Weyl group. Thus after the decomposition there will arise a matrix of order equal to the order of Weyl group. (For a group of $n \times n$-matrices this order is equal to $n!$). This matrix depending on the "number" of the representation is by definition the Zeta-function of the space X. These functions, associated with the "number" of representation, are closely connected with such functions as the Riemann Zeta-function and its generalizations. It is not excluded that a deeper development of this theory will throw light on the blank spaces in the theory of the classical Zeta-function. The results just formulated give us a chance to describe the representations involved. It could be done in detail but lack of time prevents us from doing so. We merely point out that the spectrum of the representation of the component, associated with any Ω_i, has Lebesgue type and multiplicity 1. The indices of the representations fill up several linear subspaces of Cartan algebra. The description of the allowable subspaces is carried out by induction. The dimension of these spaces is easy to calculate. It is equal to the dimension of a group which is associated with Ω_i and which consists of all the homeomorphisms permutable with the movements of G in Ω_i.

6. Example: the matrix group G and the subgroup Γ of matrices with integer elements

As an example we take the subgroup Γ of integer elements of the real $n \times n$-matrix group. Each transitive family of compact horyspheres is defined by a partition $n = n_1 + \ldots + n_k$. The maximal family Ω_1 is defined by the partition $n = 1 + \ldots + 1$ and consists of orbits of the triangular group. Every irreducible representation of G can be defined by an index which is a vector in a space of dimension $\mathbf{n}-1$. The set of indices of all the irreducible representations involved in the decomposition of $L_2(X)$, where $X=G/\Gamma$, is called the spectrum of $L_2(X)$. To each Ω_i there corresponds a spectrum consisting of not more than a countable number of linear subspaces. To Ω_1 itself there corresponds a spectrum of multiplicity 1 filling up the whole space. The spectrum of $L_2(X)$ is the union of the subspaces mentioned above and of a countable number of points.

Now we shall exhibit the Zeta-function corresponding to the homogeneous space G/Γ. We shall restrict ourselves to the Zeta-function of highest dimension. The Weyl group in this case is the symmetric group of degree $\mathbf{n}$. Thus the Zeta-function is a matrix of order $\mathbf{n}!$. We shall write down the first row because the rank of the matrix is equal to 1 and all other rows are easily expressed in terms of the first. (Each representation associated with Ω_1 enters in $L_2(X)$ only one time!) So we have: if σ is a permutation of the symmetric group and $x=(x_1,\ldots,x_{\mathbf{n}})$ is the "number" of a representation $(x_1+\ldots+x_{\mathbf{n}}=0)$, then

$$\zeta_\sigma(x)=\prod_{\substack{i>j\\ \sigma(i)<\sigma(j)}} B\left(\frac{1}{2},\frac{x_i-x_j}{2}\right)\frac{\zeta(x_i-x_j+1)}{\zeta(x_i-x_j)}, \tag{5}$$

where $\sigma=(\sigma_1,\ldots,\sigma_{\mathbf{n}})$.

It is not difficult to write this formula in standard root notation. Written in this form it can be generalized.

The functional equation for the Zeta-function which is a consequence of the general theory of representations, is of the form

$$\zeta_{\sigma_1\sigma_2}(x)=\zeta_{\sigma_2}(x)\,\zeta_{\sigma_1}(\sigma_2 x). \tag{6}$$

7. The representations of finite Chevalley groups

The success of the theory of representations depends after all on a happy construction of irreducible representations. In the case of the complex semisimple Lie groups a simple construction of irreducible representations was given (in 1947–1950 by M. A. Naimark and the author [9] (later we shall speak about the construction). However, if we go to other classes of semisimple groups—the real semisimple groups, the matrix groups over finite fields etc.—the construction of Naimark and author does not yield all the representations but only a small part of them.

Now we shall discuss the classification of the irreducible representations of semisimple matrix groups over a finite field. We shall deal with the groups which were considered in the well-known papers by Dickson [15] and Chevalley [16]. Included here for example are the unimodular matrix groups over finite fields, the matrix groups leaving invariant some quadratic form etc.

Already in the last century in one of the first papers on the theory of representations Frobenius found the characters of the unimodular group of 2×2-matrices over a finite field. Later (in 1928) Hecke gave the construction for half of the irreducible representations of this group. The construction for the other groups was absent.

Now we proceed to construct the irreducible representations. In the case of a complex semisimple Lie group G the construction of the principal series of irreducible representations is as follows. We consider the homogeneous space G/Z of the right cosets of G with respect to a maximal nilpotent subgroup Z. We call this space the principal affine space associated with G. The set of homeomorphisms of G/Z permutable with the movements form a commutative group which we shall call the homothety group. A function defined on G/Γ will be called homogeneous if it is multiplied by a constant under a homothety.

The irreducible representations of the principal series are realized in spaces of homogeneous functions on G/Z; the operator of representation is defined as the shift operator. We have already noted that this construction when transferred to the simple Dickson–Chevalley groups over a finite field, will give us only a small part of all irreducible representations. Now we shall explain how to construct *all* irreducible representations. Again we consider a representation in the space of functions on G/Z where Z is a maximal nilpotent subgroup. An operator of the representation will be defined not as mere shift but as a shift operator multiplied by a fixed function depending on a point of the space and on an element of the group

$$T_\chi(g)f(z) = f(zg)\,\alpha(z,g).$$

It is easy to show that the function $\alpha(z,g)$ is defined essentially by a one-dimensional representation $\chi(z)$ of the group Z. For brevity we shall call the $\chi(z)$ simply characters. It can be shown that by the decomposition of these representations we shall obtain all the irreducible representations. This follows from the fact that in every representation of G there exists a vector which is an eigenvector with respect to the element of Z. Now we introduce a partial ordering into the set of characters: let $\chi_1 < \chi_2$ if from $\chi_2(s) = 1$ it follows $\chi_1(s) = 1$ for each $s \in Z$ such that $s = e^{E_\alpha t}$ where α is a root of G; the maximal characters $\chi(s)$ will be called the characters of general position. It turns out that the representations $T^\chi(g)$ corresponding to the characters χ of general position do contain each irreducible representation of G not more than one time. The irreducible representations entering into these $T_\chi(g)$ we shall call the principal representations of G, all others will be called degenerate. The number of all irreducible representations of G is a polynomial of k, where k is the order of field under consideration. It turns out that the number of degenerate representation is a polynomial of lower degree. In this sense we can say that the principal representations are almost all irreducible representations of G. It can be shown that the dimensions of the principal representations of G can be expressed as polynomials in k of degree N, where N is the dimension of group Z, and that the dimensions of the representations of the degenerate series are expressed as polynomial of lower degree. As an illustration we give a formula for the dimensions of the different principal representations of the unimodular matrix group of the $\mathbf{n}$th order. The principal representations are split into several series. Each series is defined by a partition $\mathbf{n} = \mathbf{n}_1 + \dots + \mathbf{n}_l$ of $\mathbf{n}$ into sum of positive integers. The dimensions of the representations of this series are equal to(1)

(1) (Added in proof.): There is a very interesting paper of I. A. Green. *Trans. Amer. Math. Soc.* (1955).

$$\frac{(k-1)(k^2-1)\dots(k^l-1)}{(k^{n_1}-1)\dots(k^{n_s}-1)}, \tag{7}$$

where k is order of the field under consideration. There are exceptions to this formula, the so-called singular representations. The number of these divided by the total number of representations goes to zero with increasing k.

Now we shall explain how to split into series the principal representations of the arbitrary group G. We combine into one series the principal representations contained in the same $T_\chi(g)$ and with the same multiplicities. It seems that the representations belonging to same series are "contrived in the same manner"; their dimensions coincide; formulae for the characters are written in the same way; when realized they give rise to the same special functions.

Since the principal representations enter in $T_\chi(g)$ of general position with the multiplicity 1, the realization of the representation is comparatively easy. It is sufficient to consider the ring of operators, permutable with the representation $T_\chi(g)$, where χ is a character of general position. This ring is commutative (!), and so it decomposes into a direct sum of complex fields. Each summand defines one of the principal representations of G. It should be noted that the procedure leads to interesting classes of special functions on G. Thus in the case of the unimodular group of second order the one-dimensional components of the ring of operators permutable with $T_\chi(g)$ can be expressed as certain sums known as the Kloosterman sums. It is remarkable that the summation is carried sometimes over a certain "contour" in a quadratic extension of the field. Depending on the set of summation we obtain representations of one or another principal series.

8. Zeta-functions associated with homogeneous space over a finite field

The representations of the group G over a finite field being constructed we have the possibility to define the Zeta-function. Here we restrict ourselves to the Zeta-function associated with principal representations with $\chi\equiv 1$; we shall not carry out the "analytic continuation" to the other series. Let G be a Dickson-Chevalley group over a finite field, let Γ be any of its subgroups and let Z be a maximal nilpotent subgroup. In order to construct the Zeta-function we consider the functions $\varphi(g)$ constant on the cosets of G with respect to Z, i.e. the functions satisfying the equation $\varphi(\zeta g)=\varphi(g)\chi(\zeta)$. The operator M acting in the space of these functions is given by

$$M\varphi=\psi(g)=\sum_{\substack{\gamma\in\Gamma\\ \zeta\in Z}}\varphi(\gamma\zeta g)\chi'(\zeta). \tag{8}$$

Then $\psi(\zeta g)=\psi(g)\chi(\zeta)$. The operator M is permutable with the movement T_{g_0} of the group: $T_{g_0}\varphi(g)=\varphi(gg_0)$. So after the decomposition of the representation into irreducible ones it will be given by a matrix which depends on the "number" of the representation and has order equal to the order of Weyl group. By the definition the matrix is the Zeta-function of group G respect to the subgroup Γ. It depends on the index π of representation which is a multiplicative character of the Cartan subgroup. We shall give the explicit form of this function. For this purpose we construct the function $k(g)$ equal for each g to the number of ways in which g can be written in the form $g=\zeta_1\gamma\zeta_2$, where $\gamma\in\Gamma$ and ζ_1 and ζ_2 are in Z. Each g is of the form

$g = \zeta' \delta s \zeta''$, where δ is in the Cartan subgroup and s is in the Weyl group; therefore $k(g)$ depends only on δ and s: $k(g) = k(\delta, s)$. The Fourier transform of $k(\delta, s)$ with respect to δ is now the Zeta-function:

$$\zeta_s(\pi) = \sum_\delta k(\delta, s)\, \pi(\delta).$$

We shall not write down the form of this function for the representations of other principal series.

9. Representations associated with groups over other fields

We shall not consider the case of groups associated with the field of p-adic numbers which were studied by Mautner and Bruhat. We hope that the indicated constructions are applicable to this case and make it possible to describe all representations.

Very interesting is the problem of studying the representations of the Chevalley groups over the field of algebraic functions over the field of complex numbers. Here it is not evident how to define a representation. Apparently the following definition is the most natural. We require T_g to be an operator defined every where except a manifold of lower dimension. We require T_g to depend continuously on g and to satisfy the equation $T_{g_1} T_{g_2} = T_{g_1 g_2}$ everywhere except a manifold of lower dimension. These groups are examples of infinite-dimensional groups for which the problem of constructing their irreducible representations is quite real and interesting. Also in this case the constructions indicated above give many important types of representations. It is possible, however, to exhibit the examples showing that the construction does not exhaust all degenerate representations.

REFERENCES

[1]. Гельфанд, И. М. & Граев, И. М. И., *Труды Моск. Матем. общ.*, 8 (1959), 323–390.

[2]. Гельфанд, И. М. & Пятецкий-Шациро, И. М. И., *Успехи матем. наук*, XIV 2 (1959), 172–194.

[3]. Гельфанд, И. М. & Фомин, И. С. В., *Успехи матем. наук*, 7:1 (47) (1952), 118–137.

[4]. Selberg, A., *Tata Inst. of Fund. Res., Bombay*, 1960, 147–164.

[5]. Selberg, A., *J. Ind. Math. Soc.*, 20 (1956), 47–87.

[6]. Граев, М. И. *Докл. А. Н. С.С.С.Р.*, 113 (1957), 966–969.

[7]. Selberg, A., *Analytic Functions*, vol. 2, 1960.

[8]. Borel, A. & Harish-Chandra, *Ann. Math.*, 75 3 (1962).

[9]. Гельфанд, И. М. & Наймарк, И. М. А., *Труды матем. ин-та А.Н. С.С.С.Р.*, 36 (1950), 1–288).

[10]. Harish-Chandra, *Proc. Nat. Acad. Sci. USA.*, 45 (1959), 570–573.

[11]. Bargman, *Ann. Math.*, 48 (1947), 568–640.

[12]. Гельфанд, И. М. & Граев, И. М. И., *Изв. А.Н. С.С.С.Р, сер. матем.*, 17 (1953), 189–238.

[13]. Граев, М. И., *Труды Моск. матем общ.*, VII (1959).

[14]. Hirzebruch, *Analytic Functions*, vol. 2, 1960.

[15]. Dickson, L., *Linear Groups*.

[16]. Chevalley, C., *Tôhoku Math. J.*, 7 (1955), 14–66.

7.

The cohomology of infinite dimensional Lie algebras; some questions of integral geometry

Int. Congr. Math. 1970, Nice **1** (1970) 95–111. Zbl. **239**: 58004

This report is concerned with certain results and problems arising in the theory of the representation of groups. In the last twenty years much has been achieved in this field and—most important—its almost boundless possibilities have become apparent.

Indeed, its problems, touching on the interests of algebraic geometry, on many questions of the algebraic number theory, analysis, quantum field theory and geometry, as well as its inner symmetry and beauty have resulted in the growing popularity of the theory of representations.

It is impossible to list even briefly its main achievements, and this is not the aim of this communication. Nevertheless, one cannot omit mentioning the outstanding papers by Harish-Chandra, Selberg, Langlands, Kostant, A. Weil, which considerably advanced the development of the theory of representations and opned up new relationships; and, since we do not go into these questions, we will not be able to touch upon many of the deep notions and results of the theory of representations.

We feel that the methods which have arisen in the theory of representation of groups may be used in a considerably more general non-homogeneous situation. We will give some examples:

1. The proof of the fact that the spectrum of a flow on symmetric spaces of constant negative curvature is a Lebesgue spectrum [1] was based on methods of the theory of representations, namely the decomposition of representations into irreducible ones. One of the most useful methods of decomposing representations into irreducible representations is the orisphere method [5]. In the works of Sinai, Anosov, Margulis [2], [3], [4], only the orispheres are considered and groups symmetries are left out. This rendered possible the study of the spectrum of dynamic systems in a considerably more general situation.

2. The theorem of Plancherel and the method of orispheres gives rise to the consideration of more general problems of integral geometry, taking place in a non-homogeneous situation [5], [6], [7], [8], [9], [10], [11], [12], [13], [14], [15], [16], [17].

3. If we have a manifold and its mapping, the study of distributions " constant on the inverse image of each point " of this mapping is an extremely interesting problem, special examples of which were studied in the homogeneous situation (functions in

four-dimensional space, invariant relative to the Lorenz group, functions constant on classes of conjugate elements of a semi-simple Lie group [18], etc.). There are various aspects of this problem which are considerably more interesting and important than may seem at first glance. Of course, the main interest of the problem is the study of these distributions at singularies of the mapping. To be more precise, suppose X is a manifold (C^∞-analytical, algebraic) and $\mathscr{G}$ is some (perhaps infinitely dimensional) Lie algebra of smooth vector fields. One wishes to describe the space of unvariant distributions.

A more natural statement of the problem is obtained by replacing the distributions by generalised sections of a vector bundle which vary according to a given finite dimensional representation. Unfortunately consideration of length prevent me from giving a series of existing examples. Those examples are particularly interesting when X has only a finite number of orbits relative to $\mathscr{G}$. For interesting example in the non-homogeneous situation see [34].

4. The theory of representation of groups makes the consideration of interesting examples possible and shows the importance of studying the ring of all the regular differential operators on those algebraic manifolds which are homogeneous spaces. It is quite natural to wish to describe the structure of the ring R of regular differential operators on any algebraic manifold. Perhaps, as in [19], [20], it would be helpful to consider the quotient skew-field of the ring R. Another interesting problem is the description of the involutions of this ring R.

In this report I would like to tell about certain problems which were studied by my friends and myself while thinking about questions connected with representation theory.

I. Representations of semisimple Lie algebras.

0. Suppose $\mathscr{G}$-is a semisimple Lie algebra. The study of representations is essentially the study of a category of $\mathscr{G}$-modules. The choice of the particular category of $\mathscr{G}$-modules considered in the algebraic problems of the theory of representations is essential. Suppose $\mathscr{f}$ is a fixed subalgebra of $\mathscr{G}$. $\mathscr{G}$-the module will be called $(\mathscr{g}, \mathscr{f})$ finite iff 1° it is a finitely generated $\mathscr{U}(\mathscr{G})$-module and 2° as an $\mathscr{U}(\mathscr{G})$-module it is the algebraic direct sum of finite dimensional irreducible representations of $\mathscr{f}$ and in this decomposition each of the irreducible representation appears only a finite number of times.

The following two cases are very interesting:

1° $\mathscr{G}$ is a real semisimple algebra, $\mathscr{f}$ is the subalgebra corresponding to the maximum compact subgroup. The corresponding $(\mathscr{g}, \mathscr{f})$-modules were considered by V. A. Ponamaryov and the author and were called by them " Harish-Chandra modules ".

2° $\mathscr{G}$ is a real Lie algebra, $\mathscr{f}$ is a Cartan subalgebra or, more generally, the semisimple part of the parabolic subalgebra.

1. Let us consider in more detail the category of Harish-Chandra modules in the case when $\mathscr{G}$ is the algebra of a complex semisimple Lie group.

Further, each module is the direct sum of modules on each of which the Laplace operators have only one eigen-value.

Consider an example. Suppose G is a simply connected Lie group over the algebra $\mathcal{G}$, B – its Borel subgroup, $\mathcal{N}$ is a unipotent radical of B, H – a Cartan subgroup. Consider the indecomposable finite dimensional representation ρ of the group H. Note that since $H = C^* \times C^* \times \ldots \times C^*$ the question of the finite dimensional representations of H is reduced to the determination of a finite number of pairwise commutative matrices. Let us extend this representation ρ of the group H to a representation of the group B and consider, further, the representation of the group G induced by this representation B. The representation thus obtained will be called a Jordan representation. In the case when ρ is of dimension one, we obtain the well-known representation of the principal series. Thus we have constructed, using the representation of the group H, a representation of the group G. Note that the description of the canonical form of the representation of H is in some sense an unsolvable problem if the rang of H is greater than 1 [21].

If we consider the representation of the algebra $\mathcal{G}$ thus constructed only on the space of vectors which vary over the finite dimensional representation of the maximum compact subgroup, we will obtain Harish-Chandra modules. Apparently the following hypothesis holds: at the points of general position all the indecomposable Harish-Chandra modules are all Jordan representations (*).

For $SL(2, \mathbb{C})$ this statement follows from work of Zhelobenko. The most interesting is the study of Harish-Chandra modules at singular points. Of course, the problem of listing all the Jordan modules is already a badly stated (unsolvable) problem, since it is based on the classification of systems of pairwise commutative matrices. However, it is not clear whether it is possible to solve this problem at a singular point, considering the Jordan modules as given. If such a solution were possible, it would have exceptional interest.

The problem of describing Harish-Chandra modules was completely solved by V. A. Ponomaryov and the author for the Lie algebra of the group $SL(2, \mathbb{C})$ [22], [23], [24]. Then these representations were constructed as a group representation (and not only as an algebra representation) by M. I. Graev and the authors cited above [25].

The classification of indecomposable Harish-Chandra modules is carried out in two stages.

1. The problem is reduced to a problem in linear algebra.
2. The linear algebra problem obtained for $SL(2, \mathbb{C})$ generalises the problem of describing the canonical form of pairs of matrices A, B such that $AB = BA = 0$. To solve this problem we apply the Maclane relation theory, which allows us to use the relations $A^\#$ and $B^\#$, inverse to the degenerate operators A and B, as well as the monomials $A^{\# k_1} B^{\# k_2} A^{\# k_3} \ldots$.

The Harish-Chandra modules at a singular point may be divided into two classes.

(*) To be more precise, each HARISH-CHANDRA module is decomposed into direct sum of submodules on which the Laplace operators have precisely one eigen-value. The set of eigen-values thus obtained is called singular if the representation of the fundamental series with the same eigen-values of the Laplace operator are reducible. The points of general position will be exactly the non-singular points.

The modules of first class are uniquely defined by any set of natural numbers, the modules of the second class are determined by any set of natural numbers together with one complex number λ. It is thus interesting to note that at singular points the module space is not discrete. The most convenient canonical form of Harish-Chandra modules are given in [25].

In the case of $SL(2, \mathbb{R})$ the problem of classifying Harish-Chandra modules is easily reduced to a problem in linear algebra; explicitely the category of Harish-Chandra modules at a given singular point is isomorphic to the following category of diagrams in the category of finite dimensional linear spaces:

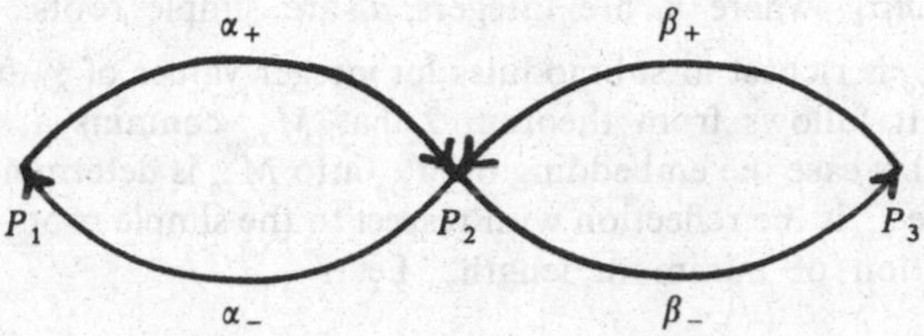

with the condition $\alpha_+\alpha_- = \beta_+\beta_- = \gamma$, where γ is nilpotent. The question of the classification of the objects of this category is aparantly solvable but leads to considerable difficulties.

CONJECTURE. — The category of Harish-Chandra modules for any semisimple group with given eigen-values of Laplace operators is equivalent to a certain category of diagrams in the category of finite dimensional linear spaces.

2. This and the following section of the report summarise some results of I. N. Bernstein, S. I. Gel'fand and the author.

Suppose $\mathscr{G}$ is a semisimple Lie algebra over $\mathbb{C}$, b is its Borel subalgebra, $\mathfrak{u}$ is a radical and $\mathfrak{f}$ is a Cartan subalgebra. Consider the following category $\mathcal{O}$. Its objects are $(\mathscr{G}, \mathfrak{f})$ – finite modules M, satisfying the following condition: for every vector $\xi \in M$ the space $\mathscr{U}(u)\xi$ is finite dimensional. This category is most useful for the application of the theory of highest weights. In this category, let us chose a class of objects which will be called elementary. All the others will be constructed from them and their factor modules by step by step extensions.

Suppose χ is a linear functional over $\mathfrak{f}$. Denote by $M_\chi \mathscr{U}(\mathscr{G})$-module, generated by f_χ, with the relations $nf_\chi = 0$ and $hf_\chi = (\chi - \rho, h).f_\chi$ for all $h \in \mathfrak{f}$ and $n \in \mathfrak{u}$. Here ρ denotes the half-sum of the positive roots. By studying the modules M_χ we get extensive information on the representation of the algebras $\mathscr{G}$, including finite dimensional ones. We now state a few theorems on M_χ modules and their morphisms.

THEOREM 1 (Verma). — Let the modules M_{χ_1} and M_{χ_2} be given. Two cases are possible:

1° $$\mathrm{Hom}\,(M_{\chi_1}, M_{\chi_2}) = 0;$$

and

2° $$\mathrm{Hom}\,(M_{\chi_1}, M_{\chi_2}) \approx \mathbb{C},$$

then any non-trivial homomorphism M_{χ_1} into M_{χ_2} is an embedding.

To state the next theorem we must introduce a partial ordering in the Weyl group W. Suppose $s_1, s_2 \in W$. We shall say that $s_1 > s_2$ iff there exist reflexions $\sigma_1, \ldots, \sigma_r$ in W such that $s_1 = \sigma_1 \ldots \sigma_r s_2$ and $l(\sigma_{i+1} \ldots \sigma_r s_2) = l(\sigma_{i-1} \ldots \sigma_r s_2) + 1, i = 1, \ldots, r$, where $l(s)$ is the length of the element $s \in W$.

THEOREM 2. — Let M_{χ_1} and M_{χ_2} be given. M_{χ_1} imbeds into M_{χ_2} if and only if,

1. There exists such an χ that Re χ lies in the positive Weyl chamber and such a pair of elements $s_1, s_2 \in W$, $s_1 > s_2$ that $\chi_1 = s_1\chi$, $\chi_2 = s_2\chi$.

2. $\chi_1 - \chi_2 = \Sigma n_i\alpha_i$, where n_i are integers, α_i are simple roots.

The module M_{χ_0} is richest in submodules for integer values of χ_0 from the positive Weyl chamber. It follows from theorem 2 that M_{χ_0} contains a submodule $M_{s\chi_0}$ for all $s \in W$. In this case the embedding of $M_{s\chi_0}$ into M_{χ_0} is determined in the following way. Suppose s_{α_i} is the reflection with respect to the simple roots α_i, $s = s_{\alpha_i} \ldots s_{\alpha_k}$ is the decomposition of minimum length. Let

$$\chi_i = s_{\alpha_i} s_{\alpha_{i+1}} \ldots s_{\alpha_k} \chi_0 .$$

Then

$$f_{s\chi_0} = a f_{\chi_0},$$

where

$$a = E_{-\alpha_1}^{\frac{(\chi_2 - \chi_1, \alpha_1)}{(\alpha_1, \alpha_1)}} . E_{-\alpha_2}^{\frac{(\chi_3 - \chi_2, \alpha_2)}{(\alpha_2, \alpha_2)}} \ldots E_{-\alpha_k}^{\frac{(\chi_0 - \chi_k, \alpha_k)}{(\alpha_k, \alpha_k)}}$$

Since the minimum representation s in the form of the product of s_{α_i} is not unique, whereas the injection $M_{s\chi_0}$ into M_{χ_0} is uniquely determined, the theorem gives relations between " chains " of the type described. In the general case the embedding is more complicated.

The relations between $M_{s\chi_0}$ may easily be shown by the following commutative diagram. The vertices of the diagram are numbered by the elements s of the Weyl group and correspond to the modules $M_{s\chi_0}$. If $s_1 < s_2$, then an arrow going from s_2 to s_1 is drawn. The mapping is defined by the embedding of $M_{s_2\chi_0}$ into $M_{s_1\chi_0}$. We obtain a commutative diagram. It is not difficult, using this diagram, to get in particular, a resolution of the finite dimensional representation by free $\mathcal{U}(u)$-modules.

The finite dimensional representation with highest weight $\chi_0 - \rho$ is of the form

$$M = M_{\chi_0} / \sum_{s \neq l} M_{s\chi_0} .$$

The theorems stated above and this diagram contain, in this case, the formulas of Kostant, Weyl's formulas for characters, the Borel-Weil theorem and the Harish-Chandra theorem concerning the left ideals of enveloping algebras.

3. *The ring of differential operators on the principal affine space and the generalisation of the Segal-Bargman representation to any compact group.*

Suppose G is a complex semisimple Lie group, $\mathcal{N}$ is the maximum unipotent subgroup, H — a Cartan subgroup. The manifold $A = \mathcal{N} \backslash G$ is called the principal affine space of the group G. It is an algebraic quasi affine manifold. It is interesting to consider the ring $\mathscr{D}$ of regular differential operators on A. Suppose $f(g)$ ranges over all the regular algebraic functions (polynomials) on the group G. We will give a method allowing to construct for any such function a differential operator on A. Since H normalises $\mathscr{N}$, the transformation $g \to hg$ may be carried over to A (left translations [5]). Using these left translations we can assign to every element of the Lie algebra $\not h$ of the group H a differential operator on A. The commutative ring of differential operators on A generated by these operators will be denoted, following [20], by W_u. Suppose π is the natural map of G into A. Denote by π^* extension of the functions over A to functions over G induced by π. The operation π_*, mapping the functions on G into functions on A is less obvious and supplements, in our case, the operation of averaging the function over the subgroup. The construction of π_* is carried out in the following way.

Suppose $f(g)$ is a regular algebraic function on G. Consider it as the linear combination of matrix elements of finite dimensional irreducable representations in the basis of weight vectors H. Threw out all the elements of this sum except the summands corresponding to those matrix elements whose first index is the highest weight of the corresponding representations. Denote by $\pi_* f$ the function thus obtrained.

Suppose f is a fixed function on G. Define the operator f in the functions by the formula

$$\bar{f}(\varphi) = \pi_*(f\pi_*(\varphi)).$$

THEOREM 1. — There exists an element $w \in W_u$ such that $w_0 f$ is a regular differential operator on A. Conversely, every regular differential operator on A may be represented in the form $\Sigma w_i \,.\, \bar{f}_i$, $w_i \in W_u$ where f_i are functions on G.

Suppose $\mathscr{K}$ is the quotient field of the W_u ring, $\mathscr{F}(G)$ is the ring of regular algebraic functions on G. The map constructed in theorem 1 may be expanded to the map

$$i : \mathscr{D} \bigotimes_{W_u} \mathscr{K} \to \mathscr{F}(G) \bigotimes_{\mathbf{C}} \mathscr{K}$$

THEOREM 2. — i is a linear space isomorphism over $\mathscr{K}$, compatible with the right translations by elements of G.

Note that the fact of the existence of an isomorphism of the spaces above was obtained earlier in a joint paper of A. A. Kirillov and the author [20].

For the group $SU(2)$ there exists an extremely useful realisation of the whole series of representations of this group due to Segal and Bargmann. This realisation is in the Hilbert space of analytic functions of two complex variables, square, integrable with weight $e^{-|z_1|^2-|z_2|^2}$. We will point out a generalisation of this construction for any compact Lie group.

Suppose K is a simply connected compact Lie group of rang r, G — its complexification, A — the principal affine space of the group G. Introduce the weight function $e^{-H(a)}$, $a \in A$. Suppose ρ_i is the i'-th fundamental representation of G, let ξ_i denote the vector of highest weight in ρ_i. Put

$$H_i(g) = (\rho_i(g)\xi_i\,, \rho_i(g)\xi_i),$$

where (,) is the scalar product in the space of the representation ρ_i invariant relative to K. It is clear that $H_i(g)$ is a function on A and we can then put

$$H(a) = \sum_{i=1}^{r} H_i(a).$$

Now consider the analytic functions on A which are square integrable with weight $e^{-H(a)}$. Call the Hilbert space of all these functions a " generalised Segal-Bargmann space ". The group K thus obtained acts on it in a natural way and the unitary representation thus obtained contains every irreducible one exactly once. Let us call any operator with polynomial regular algebraic coefficients a " differential operator on A ".

CONJECTURE. — The operator conjugate (in the generalised Segal-Bargmann space) with a regular differential operator is again a regular differential operator.

The involutions which arise in the ring of regular differential operators are far from trivial. Thus, for the case of $SU(n)$ the operator, say, conjugate with multiplication by a simple first order function, is a differential operator of the $(n-1)$-st order. The techniques developed in the previous section apparently will turn out to be very useful in the study of the ring of differential operators on A, in particular, for the proof of the conjecture stated aboce. The fact of the matter is that the construction of the involution itself is most conveniently carried out in the terms developed there. Using this method the conjecture was checked for $SU(3)$.

We state another problem. Let the real form of the group G be given. Its unitary representation naturally gives rise to an involution in the enveloping algebra $\mathcal{U}(\mathcal{G})$. We must find all the possible extensions of this involution from $\mathcal{U}(\mathcal{G})$ to the ring of all the regular differential operators on A. In the simpliest examples these extended involutions correspond to series of unitary representations (of real groups) contained in the regular one. It would be interesting to list the involutions in the ring of regular differential operators on any quasiaffine algebraic manifold.

It would also be interesting to consider the factor space of the group G, not only over the maximal unipotent group, but also over any orispherical subgroup.

II. Integral geometry.

In this paragraph I will only consider one elementary example [17]. The derivation of the Plancherel formula for $G = GL(n, \mathbb{C})$ is based on the following problem in integral geometry. Denote by $\mathcal{N} \in G$ the set of all the upper triangular matrices with units on the diagonal. Suppose the function $f(x)$, $x \in G$ is given. Let

$$\varphi(x_1, x_2) = \int_{\mathcal{N}} f(x_1^{-1} z x_2) dz,$$

where x_1 and x_2 are any matrices. The problem is: given $\varphi(x_1, x_2)$ find $f(x)$. It suffices to solve the problem when $x = e$ is the unit matrix. We can assume that the fonction f is given on $\mathbb{C}^{n^2}$ and the equation $y = x_1^{-1} z x_2$ for fixed x_1 and x_2 defines in $\mathbb{C}^{n^2}$ a plane of dimension $\dfrac{n(n-1)}{2}$.

Now replace our problem with the following, at first glance meaningless, problem. Consider the space $H_{n^2,k}\left(k = \dfrac{n(n-1)}{2}\right)$ of all the k-dimensional planes in $\mathbb{C}^{n^2}$. For all $h \in H_{n^2,k}$ consider the function

$$\varphi(h) = \int_h f(x)dx.$$

We must now recover $f(x)$. In the paper [10] this problem is solved in the following manner. Using the function φ and its derivatives construct a differential (k, h) form $\mathscr{H}\varphi$ on the Grassman manifold $G_{n^2,k}$ of k-dimensional planes containing the point x. This form $\mathscr{H}\varphi$ is closed and the value of $f(x)$ is equal to $\int_{\gamma_0} \mathscr{H}\varphi$, where γ_0 is any cycle homologic to the set of all k-dimensional planes containing the point x and lying in a fixed $k+1$-dimensional plane passing through the point (Euler's cycle (*)). As to the integral over the other k-dimensional cycles in the basis of Schubert cells in $G_{n^2,k}$, it is equal to zero.

In our case the function $\varphi(x_1, x_2)$ is known not on the whole manifold $H_{n^2,k}$ but only on a certain submanifold. The submanifold of $H_{n^2,k}$ will from now on be called the " complex of k-dimensional planes ". The complex is called permissible if the form $\mathscr{H}\varphi$ on this complex is determined by the values of the function φ on this complex only. In the case when φ is given on a permissible complex we can recover $f(x)$ by using the formula

$$f(x) = C_\gamma \int_\gamma \mathscr{H}\varphi,$$

where γ is a cycle lying in the complex; thus to find C_γ it suffices to decompose the cycle γ over the Schubert cell basis. In our case the complex will consist of planes of the form $h_{x_1,x_2} = \{ y/y = x_1^{-1}zx_2 \}$ and has dimension n^2. It turns out to be permissible. The set of these planes of this complex which contain the point e has the necessary dimension k and forms a cycle. The coefficient of the Euler cycle is equal to $n!$ Considering the form $\mathscr{H}\varphi$ only on the complex, we will obtain the classical inversion formula

$$f(e) = [(2i)^k\pi^{2k}n!]^{-1} \int \prod_{q<p} \left(\frac{\partial}{\partial\delta_p} - \frac{\partial}{\partial\delta_q}\right)\left(\frac{\partial}{\partial\bar{\delta}_p} - \frac{\partial}{\partial\bar{\delta}_q}\right) \times \varphi(\mathscr{C}^{-1}\delta\mathscr{C})\,|_{\delta=e} \bigwedge_{q<p} d\mathscr{C}_{qp} \bigwedge_{q<p} d\overline{\mathscr{C}}qp.$$

Apparently one can obtain the Paley-Wiener theorem for $GL(n, \mathbb{C})$, in a similar manner; in other words, obtain conditions on φ, which imply the decrease of f at infinity. To do this we embed $GL(n, \mathbb{C})$ not into $\mathbb{C}^{n^2}$ but into $\mathbb{CP}^{n^2}$ and consider the problem as a projective problem of integral geometry (see [15]). Since in this case we can recover $f(x)$ in the points at infinity as well, the Paley-Wiener conditions will consist in the

(*) Note that other problems of integral geometry give rise to integration over other cycles in $G_{n^2,k}$; see, for example [16].

following: the function f' and its derivatives (recovered by using φ) must be equal to zero at all the points of infinity.

III. Cohomology of infinite algebras.

0. *This part of the report contains results obtained jointly by D. B. Fuks and the author.*

We know how difficult it is to describe any reasonable category of representations. On the other hand, the problem of determining cohomology groups is a sumpler one. Here we list results about the cohomology of Lie algebras of vector spaces, which show that these cohomologies are reasonable, are not equal to zero and are not infinite dimensional.

Recall that the cohomology $H^*(\mathcal{G}; M) = \sum_q H^q(\mathcal{G}; M)$ of the topological algebra $\mathcal{G}$ with coefficients in the $\mathcal{G}$-module is defined as the cohomology of the complex $C(\mathcal{G}; M) = \{ c^q(\mathcal{G}; M), d^q(\mathcal{G}; M) \}$ where $c^q(\mathcal{G}; M)$ is the space of continuous skew-symetric q-linear functionals on $\mathcal{G}$ ranging over M, and the differential $d^q = d^r(\mathcal{G}; M)$ is defined by the formula

$$(d^q L)(\xi_1, \dots, \xi_{q+1}) = \sum_{1 \leq s < t \leq q+1} (-1)^{s+t-1} L([\xi_s, \xi_t], \xi_1, \dots, \hat{\xi}_s, \dots, \hat{\xi}_t, \dots, \xi_{q+1})$$
$$- \sum_{1 \leq s \leq q} (-1)^s \xi_s L(\xi_1, \dots, \hat{\xi}_s, \dots, \xi_{q+1}).$$

If M is a ring, and the operators on $\mathcal{G}$ are its differentials, then the complex $C(\mathcal{G}; M)$ has a natural multiplicative structure.

1. *Problems and examples.*

The main example of an infinitely dimensional Lie algebra will be the algebra of smooth vector fields on a smooth manifold.

Suppose M is a closed orientable connected smooth (*) manifold. Denote by $\mathfrak{A}(M)$ the Lie algebra of smooth tangent vector fields on M with Poisson brackets for commuting. The first of the problems considered is a follows. Define the cohomology ring $\mathfrak{H}^*(M) = H^*(\mathfrak{A}(M); \mathbb{R})$ of the algebra $\mathfrak{A}(M)$ with coefficients in the unit representation, i. e., in the field $\mathbb{R}$ of real numbers with a trivial $\mathfrak{A}(M)$-module structure. This ring obviously is a differential invariant of the manifold M. Looking ahead we shall say that the space $H^q(\mathfrak{A}(M); \mathbb{R})$ will turn out to be finite dimensional for any q (see [28]). The problem of computing the ring $\mathfrak{H}^*(M)$ is not as of yet completely solved.

We would like to point out the difference between the method of constructing invarients of manifolds by using objects of differential geometry (the Lie algebra of vector fields) and the usual method of constructing differential invariants. Whereas usually the differential form representing a Pontryagin of Chern class on the manifold X is built up from the individual object (by using the metric) on the manifold,

(*) By smooth we always mean of class C^∞.

in our case the invariants are constructed using the infinite dimensional set of all smooth vector fields on the manifold.

As an example consider the case when M is the circle S^1. We can show that the ring $H^*(S^1)$ is generated by a two-dimensional generator a and a three-dimensional generator, the two being related only by the skewsymetry condition.

Further the generators $a \in \mathfrak{H}^2(S^1)$, $\mathscr{C} \in \mathfrak{H}^3(S^1)$ are represented by cocycles $A \in C^2(\mathfrak{A}(S^1); \mathbb{R})$, $B \in C^3(\mathfrak{A}(S^1); \mathbb{R})$ given by the formulas

$$A(f, g) = \int_{S^1} \begin{vmatrix} f'(x) & f''(x) \\ g'(x) & g''(x) \end{vmatrix} dx$$

$$B(f, g, h) = \int_{S^1} \begin{vmatrix} f(x) & f'(x) & f''(x) \\ g(x) & g'(x) & g''(x) \\ h(x) & h'(x) & h''(x) \end{vmatrix} dx$$

When the dimension of the manifold M increases the ring $\mathfrak{H}^*(M)$ becomes considerably richer; thus the ring $\mathfrak{H}^*(S^2)$ has 10 generators, and the ring $\mathfrak{H}^*(S^1 \times S^2)$, 20 generators (see [29]).

The cohomology of the Lie algebra of smooth vector fields is intimately connected with the cohomology of Lie algebras of formal vector fields. By a formal vector field at the point O of the space R^n we mean a linear combination of the form $\Sigma p_i(x_1, \ldots, x_n)e_i$ where $e_1, \ldots, e_n$ are the standard basis vectors of the space R^n and $p_i(x_1, \ldots, x_n)$, the formal power series with real coefficients in the coordinates $x_1, \ldots, x_n$ of the space. The set of formal vector fields is, in an obvious sense, a linear topological space, and a natural commutation operation transforms it into a topological Lie algebra. This algebra is denoted by W_n.

2. *The algebra of formal vector fields. The cohomology of the algebra W_n with coefficients in, R.*

In order to state the final result it is necessary to describe a certain auxilliary topological space X_n $(n = 1, 2, \ldots)$. Suppose $\mathcal{N} \geqslant 2n$ and let $p_i E(N, n) \to G(N, n)$ be the canonical $U(n)$ bundle over the (complex) Grasman manifold $G(\mathcal{N}, n)$. The usual (W-complex of the manifold $G(\mathcal{N}, n)$ has the following property: the $2n$-th skeleton $[G(\mathcal{N}, n)]_{2n}$ does not depend on $\mathcal{N}$ when $\mathcal{N} \geqslant 2n$. The inverse image of the set $[G(\mathcal{N}, n)]$ under the map p will be denoted by X_n.

The space X_1 is a three-dimensional sphere, the other spaces do not have such a simply visualised description. We have the following.

THEOREM 2.1. — For all q, n there is an isomorphism

$$H^q(W_n; R) = H^q(X_n; R).$$

Multiplication in the ring $H^*(W_n; R)$ (as well as in the ring $H^*(X_n; R)$) is trivial, i. e., the product of any two elements of positive dimension is equal to zero.

The cohomology of the space X_n may be computed by using standard topological methods. For example, it is trivial for $0 < q \leq 2n$ and for $q > n(n + 2)$.

Theorem 2.1 is the central result of the article [30]. Its proof uses a somewhat modified version of the Serre-Hoschild spectral sequence [31] corresponding to the subalgebra of the algebra W_n, generated by the elements $x_i e_j$ (*); this subalgebra is isomorphic to $\mathscr{G}l(n, R)$. Beginning with the second member, this spectral sequence turns out to be isomorphic to the Leray-Serre spectral sequence of the bundle $X_n \to [G(\mathcal{N}, n)]_{2n}$ with fibre $U(n)$.

It turns out also that each element $\alpha \in H^q(W_n; \mathbb{R})$ is represented by such a cocycle $A \in C^q(W_n; \mathbb{R})$, that $A(\xi_1, \ldots, \xi_q)$ depends only on the 2-jets of formal vector fields $\xi_1, \ldots, \xi_q$ (see [30]).

To study the cohomology of W_n with coefficients in other modules (and to describe those modules) it is important to know the structure of the subalgebras

$$\ldots \subset L_k \subset \ldots \subset L_0 \subset W_n$$

where L_k consists of vector spaces whose components are series without terms of power less than or equal to K.

The relation between the cohomology of the algebras W_n and L_0. The following general fact is easily generalised to the case of the cohomology of infinite dimensional Lie algebras.

Suppose B is an subalgebra of Lie algebra A; M – some B-module; $\hat{M}$ – an induced A-module (i. e. $\hat{M} = \mathrm{Hom}_{[B]}(M, [A])$ where $[A]$, $[B]$ are enveloping algebras for A, B). Then

$$H^*(A; \hat{M}) = H^*(B; M).$$

We will apply this statement in the case when M is a tensor representation of the algebra L_0 (i. e. a finite dimensional representation obtained from the representation of the algebra $\mathscr{G}l(n; R)$ by means of the projection $L_0 \to L_0/L_1 = \mathscr{G}l(n; R)$). At the same time the induced representation $\hat{M}$ of the algebras W_n is none other than the space of the corresponding formal tensor fields. For example, if $M = R$ is the unit representation of the algebra L_0, then $\hat{M}$ is the space $F(R^n)$ of formal power series in n variables with the natural action of the algebra W_n; if M is the space $\Lambda^r(R^n)'$ of skewsymetric r-linear forms in R^n, then $\hat{M}$ is the space Ω^r of formal exterior differential forms of r order in R^n.

The cohomology of the algebra W_n with coefficients in the spaces of formal exterior differential forms. The space

$$H^*(W_n, \Omega^*) = \sum_{q,\mathfrak{A}} H^q(W_n; \Omega^r)$$

is obviously a bigraduated algebra (over R), isomorphic, as we just found out, to $H^*(L_0; \Lambda^*(R^n)')$.

THEOREM 2.2. — The bigraduated ring $H^*(W_n; \Omega^*) = H^*(L_0; \Lambda^*(R^n)')$ is multiplicatively generated by $2n$ generators

$$\rho_i \in H^{2i-1}(L_0; \Lambda^0(R^n)') \qquad (i = 1, \ldots, n)$$
$$\tau_i \in H^i \quad (L_0; \Lambda^i(R_n)') \qquad (i = 1, \ldots, n)$$

These generators are connected only by the following relations $\rho_i \rho_k = -\rho_k \rho_i$; $\rho_k \tau_i = \tau_i \rho_k$; $\tau_i \tau_k = \tau_k \tau_i$; $\tau_1^{i_1} \tau_2^{i_2} \ldots \tau_n^{i_n} = 0$ if $i_1 + 2i_2 + \ldots + ni_n > n$.

(*) $i, j = 1, \ldots, n$.

In particular, the ring $H^*(L_0; R) = H^*(L_0; \Lambda^0(R^n)') = H^*(W_n; F(R^n))$ is an exterior algebra in generators of dimension $1, 3, 5, \ldots, 2n-1$. i. e.

$$H^*(W_n; F(R^n)) = H^*(gl(n, R); R).$$

Moreover,

$$H^q(L_0; \Lambda^r(R^n)') = \begin{cases} 0 & \text{where} \quad q < r \\ H^r(L_0; \Lambda^r(R^n)') \otimes H^{q-r}(gl(n, R); R) & \text{where} \quad q \geqslant r \end{cases}$$

while the dimension of the space $H^r(L_0, \Lambda^r(R^n)')$ is equal to the number of ways in which the number r may be represented as the sum of natural numbers.

The computation of the cohomology of L_0 with coefficients in the tensor representation reduces to the computation of the cohomology of the algebra L_1 with coefficients in $\mathbb{R}$. In a similar way for jets, to the cohomology of L_k with coefficients in $\mathbb{R}$.

Apparently the following statement holds.

CONJECTURE. — For any n the spaces $H^q(L_k; R)$ are finite dimensional.

For $n=1$ the dimension of the space $H^q(L_k; R)$ equals $C_q^{k-1} + C_{q+1}^{k-1}$, $q, k=0, 1, \ldots$).

Using previously mentioned results to compute the cohomology of the algebra L_0 with tensor coefficients we can deduce that the classes of cohomology of the algebra W_n (even W_1) with coefficients in tensor fields is not always representable by cocycles depending only on 2-jets of their arguments (in contrast with the cases of constant and skewsymetric coefficients).

We have been unsuccessful, so far, in computing the cohomology $H^*(A, \mathbb{R})$ for other Cartan algebras. Note that all these cohomologies are connected with very important standard complexes. For this complex consists of the polynomials $P(\alpha_1, \ldots, \alpha_q); (\beta_1, \ldots, \beta_q)$, $\alpha_i \in \mathbb{R}^n$, $\beta_i \in (\mathbb{R}^n)'$; the polynomial P is skewsymetric under the simultaneous interchange of α_i, β_i with α_j, β_j. The differential is given by the formula

$$\begin{aligned} dP(\alpha_1, \ldots, \alpha_{q+1}; \beta_1, \ldots, \beta_{q+1}) \\ = \Sigma(-1)^{s+t}(\alpha_s, \beta_t) - (\alpha_t, \beta_s))P(\alpha_f + \alpha_t, \alpha_f, \ldots, \hat{\alpha}_f, \ldots, \hat{\alpha}_t, \ldots; \\ \beta_f + \beta_t, \beta_1, \ldots, \hat{\beta}_f, \ldots, \hat{\beta}_t, \ldots, \beta_{q+1}). \end{aligned}$$

Usually, the infinite dimensional Lie algebras which arise in the formal theory are factor subcomplexes of this complex.

3. *The algebra of smooth vector fields. Cohomology with coefficients in R.*

Suppose M is a compact connected orientable smooth n-dimensional manifold without boundary, $\mathfrak{A}(M)$ – the Lie algebra of smooth tengent yest fields on M. In the standard complex $C(M) = \{ C^q(M) = C(\mathfrak{A}(M); R)d^q \}$ we introduce a filtration $0 = C_0(M) \subset C_1(M) \subset \ldots \subset C(M)$ where $C_k(M) = \{ C_k^q(M) \}$ is a subcomplex of the complex $C(M)$, defined in the following way. A cochain $L \in C^q(M)$ belongs to $C_k^q(M)$ if it equals zero on any C^q the vector fields $\xi_1, \ldots, \xi_q$ such that for any k points of the manifold M one of the fields $\xi_1, \xi_2, \ldots, \xi_q$ equals zero in the neighbourhood of each of these points. For example, $C_0^q(M) = 0$; $C_1^q(M)$ consists of such cochains L that $L(\xi_1, \ldots, \xi_q) = 0$ when the supports of the fields $\xi_1, \ldots, \xi_q$ are pairwise non-intersecting; $C^a_{q-1}(M)$ consists of such cochains L that $L(\xi_1,\ldots,\xi_q)=0$ when the supports of the fields $\xi_1,\ldots,\xi_q$ have no common intersection to all of them; $C^a_k(M)=C^a(M)$ when $k\geqslant q$. It is clear that $C_k(M)$ for all k is a subcomplex of the complex $C(M)$ and that $C^q_k(M)C^r_l(M)\subset C^{q+r}_{k+l}(M)$.

To compute the cohomology of the factor complex $C_k(M)/C_{k-1}(M)$ we have defined a spectral sequence, the first term of which may be expressed by using the cohomology of the manifold M and the algebra W_n. A special role is played by the complex $C_1(M)$. This complex we shall call a diagonal complex.

Conjecture. — The image of the cohomology of the diagonal complex $C_1(M)$ in $\mathfrak{H}^*(M)$ under the embedding $C_1(M)\to C(M)$ multiplicatively generates all of the ring $\mathfrak{H}^*(M)$. In particular the ring $\mathfrak{H}^*(M)$ is always finitely generated.

Remark. — This is true for the second term of the spectral sequence.

Let us describe a spectral sequence which converges to the cohomology of the diagonal complex. It arises in connection with two different filtrations of the diagonal complex of the manifold. In order to describe the first filtration, note that the q-cochains of the diagonal complex $C_1(M)$ are determined by distributions (more precisely, by the generalized sections of a certain fibre bundle) on M^q which are supported by the diagonal. The m-th term $C^q_{1,m}$ of the first filtration consists of those distributions which have an order (relative to Δ) less than or equal to m.

To define the second filtration fix a triangulation of the manifold

$$M=M_n\supset M_{n-1}\supset\ldots\supset M_0$$

where M_i is the i-dimensional skeleton, and the m-th term $C_{1,m}$ of the filtration consider those q-cochains which are realised by distributions whose support is $M_m\subset\Delta$.

Knowing the cohomology of W_n can construct a spectral sequence which allows us to compute the cohomology of the diagonal complex.

Theorem 3.1. — There exists a spectral sequence $\mathscr{E}=\{E_r^{p,q},d_r^{p,q}\}$ which converges to the cohomology of the diagonal complex $\mathfrak{H}^*(M)$ such that

$$E_r^{p,q}=H^{p+n}(M)\otimes H^q(W_n;R);$$

$E_r^{p,q}$, in particular, can be different from zero only when $-n\leqslant p\leqslant 0$.

Let us clarify the operation of " globalizing " the formal cohomology: construct a mapping of the space $E_r^{-r,q+r}=H^{n-r}(M)\otimes H^{q+r}(W_n,R)$ into $C^q_1(M)$. This mapping is not uniqual determined: it depends on the choice of the system of local coordinates on M. Suppose $\Gamma=\{U_1,\ldots,U_{\mathcal{N}}\}$ is a coordinate covering of M with coordinates $y_{k_1},\ldots,y_{k_n}$ on U_i and $\{\rho_i\}$ is a decomposition of unity consistent with this covering. In order to construct the element $\mathscr{I}(a\otimes\Psi)(a\in H^{n+r}(W_n,R),\ \Psi\in H^{n-v}(M))$ find a cochain $\alpha\in C^{n+v}(W_n;R)$ representing the closed form ω from the class Ψ. Set

$$\mathscr{I}(a_n\otimes\Psi)(\xi_1,\ldots,\xi_q)=\int_M\omega\Lambda[\sum_{k=1}^{N}\rho_k\varphi(\alpha,U_k;\xi_1,\ldots,\xi_q)]$$

where $\varphi(\alpha, U_i; \xi_1, \ldots, \xi_q)$ is a form on U_k, which equals

$$\sum_{1 \leqslant i < \ldots < i_r \leqslant n} \alpha(\xi_1(u, U_k), \ldots, \xi_q(u, U_i), e_{k,i_1 \ldots} e_{k,i_r}) \times dy_i \wedge \ldots \wedge dy_{iv}$$

at the point $u \in U_i$, where the ξ_i are considered as a formal field in the neighbourhood of the point u under the coordinates y_{k_i}. The theorem is proved in [29] (statement 1.4).

The cohomology with coef cients in the spaces of smooth sections of smooth vector bundles. Suppose A is a finite dimensional $GL(n, R)$ module and suppose M is a smooth connected manifold (we do not assume M either orientable, or compact, or without boundary). Denote by α the vector bundle over M with fiber isomorphic to A, induced by the tangent bundle and by means of the representation of the group $CL(n, R)$ in A. By $\mathscr{A}$ denote the space of smooth sections of the fiber bundle α. The space $\mathscr{A}$ has an obvious $\mathfrak{A}(M)$ module structure. Our goal is the study of the cohomology of the algebra $\mathfrak{A}(M)$ with coefficients in the $\mathfrak{A}(M)$ module $\mathscr{A}$.

In the complex $C(M; A) = \{ C^a(\mathfrak{A}(M); \mathscr{A}); d^a \}$ we will introduce a filtration similar to the one considered above for $C(M)$. We shall say that the cocycle $L \in C^q(\mathfrak{A}(M); \mathscr{A})$ has filtration no greater than k if the section $L(\xi_1, \ldots, \xi_q)$ of the bundle α is equal to zero for any point $x \in M$ with the following property: for any points $x_1, \ldots, x_k \in M$ one of the vector fields $\xi_1, \ldots, \xi_q$ equals zero in the neighbourhood of each of the points $x_1, \ldots, x_k, x$.

The space of q-dimensional cocycles which have filtration no greater than k is denoted by $C_k^q(\mathfrak{A}(M); \mathscr{A})$. It is clear that $C_k(M; \mathscr{A}) = \{ C_k^q(\mathfrak{A}(M); \mathscr{A}) \}$ is a subcomplex of the complex $C(M; \mathscr{A})$.

The subcomplex $C_0(M; \mathscr{A})$ is called " diagonal ". We denote it by $C_\Delta(M; \mathscr{A})$.

THEOREM 3.5. — We have the following spectral sequence $\{ E_r^{p,q}, d_r^{p,q} \}$ which converges to $\mathfrak{H}_\Delta^*(M; \mathscr{A})$ and is such that $E_r^{p,q} = H^p(M; R) \otimes H^q(L_0; A)$. In the multiplicative case the spectral sequence is a multiplicative one and the isomorphism considered above is an isomorphism of rings.

CONJECTURE. — $H_\Delta^*(M, \mathscr{A}) = H(T_M, R) \otimes \mathrm{Hom}_{CL(n)}(A, H^*(L, R))$ where T is the principal $U(n)$ bundle over M induced by the complexification of the tangent bundle.

This conjecture has been proved in the case when $A = \Lambda^q$ is the exterior power of the standard representation. The case $q = 0$ was independently studied by Locik [33].

In the end of this part of the report I would like to introduce a general concept of formal differential geometry. It arises when one formalises and generalises the methods of construction of Pontryagin and Chern classes (by means of metrics and connections); also in the expression of the index of a differential operator in terms of the symbol and the metric of the manifold.

Suppose we have an algebra W_n of formal vector fields. Consider the jet space and, in it, a invariant algebraic submanifold X. Examples of such manifolds are the space of all symmetric tensors of rang 2, the set of all affine connections.

Let us define the complex $\Omega(X)$. Any rational map of X into the complex of formal differential forms will be called a chain of $\Omega(X)$, the differential will be obtained by differentiation in the image. Set $\Omega(X) = \mathrm{Hom}(X, \Omega)$, where Ω is the complex of formal differential forms, and call the maps of the rational cohomology of $\Omega(X)$-generalised Chern classes. It can be shown, in the case when X is the manifold of symmetric tensors of rang 2, that they coincide with Pontryagin classes ($q < n$).

REFERENCES

[1] I. M. Gel'fand, S. V. Fomin. — Geodesic flows on surfaces of constant negative curvature (Геодезические потоки на поверхностях постоянной отрицательной кривизны). Успехи математически ких наук, т. ХУI, 1 (1952), pp. 118-137.

[2] Y. G. Sinai. — Classical dynamic systems with countable Lebesgue spectrum (2) (Классические динамические системы со счетнократным лебеговым спектром). Изв. АН СССР, сер. матем. 30, 1 (1966), pp. 15-68.

[3] D. V. Anosov. — Geodesic flows on closed Riemann manifolds of negative curvature (Геодезические потоки на замкнутых римановых многообразиях отрицательной кривизны). Труды Матем. ин-та АН СССР, 90 (1967), pp. 1-210.

[4] G. A. Margulis. — On some applications of ergodic theory to the study of manifolds of negative curvature (О некоторых применениях эргодической теории к изучению многообразий отрицательной кривизны). Функц. анализ, 3, вып. 4 (1969), pp. 89-90.

[5] I. M. Gel'fand, M. I. Graev. — The geometry of homogeneous spaces, representation theory in homogeneous spaces and related questions of integral geometry (Геометрия однородных пространств, представления групп в однородных пространствах и связанные с ними вопросы интегралвной геометрии). Труды Моск. матем. об-ва, 8 (1959), pp. 321-390.

[6] I. M. Gel'fand. — Automorphic functions and the theory of representations (Автоморфные функции и теория представлений). *Proceedings of the International Congress of Mathematicians*, Stockholm (1962), pp. 74-85.

[7] I. M. Gel'fand, M. I. Graev, I. I. Pyatetskii-Shapiro. — Representation theory and automorphic functions (Теория представлений и автоморфные функции). Изд. « Наука » (1966), pp. 1-512.

[8] Harish-Chandra. — *Eisenstein series over finite fields*. Stone Jubilee volume, Springer.

[9] Helgason. — *Seminare Bourbaki* (1965).

[10] I. M. Gel'fand, M. I. Graev, Z. Y. Shapiro. — Integral geometry on κ-dimensional planes (Интегральная геометрия на κ-мерных плоскостях). Функц. анал., т. 1, вып. 1 (1967), pp. 15-31.

[11] I. M. Gel'fand, M. I. Graev. — Permissible complexes of planes in C^n (Допустимые комилексы прямых в пространстве C^n). Функц. анализ, т. 2, вып. 3 (1968).

[12] —, —, Z. Y. Shapiro. — Differential forms and integral geometry (Дифференциальные формы и интегральн ная геометрия). Функц. анализ, т. 3, вып. 2 (1969).

[13] Langlands. — *On the functional equations satisfied by Eisenstein series* (1967).

[14] I. M. Gel'fand, M. I. Graev, N. Y. Vilenkin. — Integral geometry and related questions of representation theory (Интегральная геометрия и связанные с ней вопросы теории представлений). Гос. изд. физ.-мат. лит. (1962), pp. 1-656.

[15] —, —, Z. Y. — Shapiro. Integral geometry in projective space (Интегральная геометрия в проективном пространстве). Функц. анализ, т. 4, вып. 1 (1970), pp. 14-32,

[16] —, —, —. — Problems of integral geometry related with a pair of Grassman manifolds (Задачи интегральной геометрии, связанные с парой грассмановых многообразий). ДАН СССР, 193, N° 2 (1970).

[17] —, —. — Complexes of κ-dimensional planes and the Plancherel formula for GL(n, ℂ) (Комплексы κ-мерных плоскостей и формула Планшереля для GL(n, ℂ)). ДАН СССР, 179 N° 3 (1968).

[18] Harish-Chandra. — Invariant distributions on Lie algebras. *Am. J. Math.*, LXXXVI, 2 (1964), pp. 272-309.

[19] I. M. Gel'fand, A. A. Kirillov. — Sur les corps liés aux algèbres enveloppantes des algèbres de Lie (О телах, связанных с обертывающими алгебрами алгебр Ли). Institut des hautes études scientifiques, Publications Mathématiques, (1966).

[20] —, —. — Structure of the Lie skew-field, related to a semisimple split Lie algebra (Структура тела Ли, связанного с полупростой расщепимой алге брой Ли). Функц. анализ, 3, вып. 1 (1969).

[21] I. M. Gel'fand, V. A. Ponomarev. A remark on the classification of pairs of commuting linear operators in finite dimensional space (Замечания о классификации пары коммутирующих линейных преобразований в конечномерном пространстве). Функц. анализ, 3, вып. 4 (1969), pp. 81-82.

[22] —, —. — Undecomposable representations of the Lorenz group (Неразложимые представления группы Лоренца). Успехи матем. наук, XXIII, вып. 2 (140), 1968 г.

[23] —, —. — The category of Harish-Chandra modules over the Lie algebra of the Lorenz group (Категория модулей Хариш-Чандры над алгеброй Ли группы Лоренца). ДАН СССР, N° 2, 176 (1967), pp. 243-246.

[24] —, —. — The classification of undecomposable infinitesimal representations of the Lorenz group. (Классификация неразложимых инфинитезимальных представлений группы Лоренца). ДАН СССР, N° 3, 176 (1967).

[25] —, —, M. I. Graev. — The classification of linear representations of the group SL (2, C) (Классификация линейных представлений группы SL (2, C)). ДАН СССР, N° 5, 194 (1970).

[26] I. N. Bernstein, I. M. Gel'fand, S. I. Gel'fand. — Differential operators on the principal affine space (Дифференциальные операторы на основном пространстве). ДАН СССР, N° 6, 195 (1970).

[27] V. Bargmann. — On a Hilbert space of analytic functions and an associated integral transform. I, II. *Com. Pure and Appl. Math.*, XIV, 3 (1961), pp. 187-214 and XX, 1 (1967), pp. 1-101.

[28] I. M. Gel'fand, D. B. Fuks. — The cohomology of the Lie algebra of tangent vector fields of a smooth manifolds, I (Когомологии алгебры Ли касательных векторных полей гладкого многообразия). Функц. анализ, 3, вып. 3, pp. 32-52.

[29] —, —. — The cohomology of the Lie algebra of tangent vector fields of a smooth manifolds, II (Когомологии алгебры Ли касательных векторных полей гладкого многообразия). Функц. анализ, вып. 2, 4 (1970), pp. 23-32.

[30] —, —. — The cohomology of the Lie algebra of formal vector fields (Когомологии алгебры Ли формальных векторных полей). Изв. АН СССР, сер. мат., 34 (1970), pp. 322-337.

[31] I. P. Serre, C. Hochshild. — Cohomology of group extensions. *Trans. Am. Math. Soc.*, 74 (1953), pp. 110-134.

[32] I. M. Gel'fand, D. B. Fuks. — The cohomology of the Lie algebra of vector fields with non-trivial coefficients (Когомологии алгебры Ли векторных полей с нетривиальными коэффициентами). Функц. анализ, вып. 3, 4 (1970).

[33] M. V. Locik. — Concerning the cohomology of infinite dimensional Lie algebras of vector fields (О когомологиях бесконечномерных алгебр Ли векторных полей). Функц. анал., вып. 2, 4 (1970), pp. 43-53.

[34] I. N. Bernstein, S. I Gel'fand. — The function ρ^λ meromorphic (Мероморфность функции ρ^λ). Функц. анал., вып. 1, 3 (1969).

University of Moscow,
Department of Mathematics,
Moscow V 234 (U. R. S. S.)

Part II

Banach spaces and normed rings

1.

Sur un lemme de la théorie des espaces linéaires

Izv. Nauchno-Issled. Inst. Mat. Khar'kov Univ., Ser. 4, **13** (1936) 35–40
Zbl. **14**:162

Le but de cette note est de démontrer un lemme présentant un petit extrait de ma dissertation.

Ce lemme permet de simplifier beaucoup les démonstrations des propositions diverses en les fondant sur le même principe. Pour expliquer, comment ça se fait, je donne plusieurs exemples Pour démontrer le lemme je fais emploi d'une méthode que m'a communiqué A. I. Plessner dans une conversation à propos d'un autre sujet.

§ 1.

Soit donnée un espace linéaire E du type (B).[1])

Si à tout point x de E on fait correspondre un nombre réel $f(x)$, on dit que $f(x)$ est une fonctionnelle dans E.

Lemme: *Soit $p(x)$ une fonctionnelle dans E jouissant des propriétés* [2]):

1° $p(x) \geqslant 0$
2° $p(x+y) \leqslant p(x)+p(y)$
3° $p(\alpha x) = |\alpha|\, p(x)$.

Si en outre $p(x)$ est une fonctionnelle semi-continue inférieurement, c'est-à-dire à tout x_0 et $\varepsilon > 0$ il correspond un tel $\delta > 0$, que $p(x_0) - p(x) < \varepsilon$ pour $\| x - x_0 \| < \delta$, il existe un tel $M > 0$, que pour tout $x \subset E$

$$p(x) \leqslant M \| x \| . \tag{1}$$

Démonstration: D'après la propriété 3° de $p(x)$, il suffit de démontrer, que $p(x)$ est bornée dans la sphère $K_1 = K(1, 0)$ ($\| x \| \leqslant 1$).

Admettons le contraire, c'est-à-dire que $p(x)$ n'est pas bornée dans K_1. Cela posé, $p(x)$ est non bornée dans toute autre sphère $K(\delta, x_0)$. En effet,

---

[1]) Sous un *espace linéaire du type* (B) on entend un ensemble E d'éléments x, y, z..., pour lequel les conditions suivantes sont remplies:

1°. Pour ses éléments sont définies les opérations d'addition et de multiplication par un nombre réel, conformement aux règles d'algèbre

2°. A tout élément x est attribué un nombre réel $\| x \|$, — la *norme* de cet élément, — de manière, que: a) $\| x \| > 0$ pour $x \neq 0$; $\| 0 \| = 0$; b) $\| \alpha x \| = |\alpha| \cdot \| x \|$ pour tout α réel et c) $\| x + y \| \leq \| x \| + \| y \|$

3°. L'ensemble est *complet*, c'est-à-dire pour toute suite d'éléments $x_1, x_2, \ldots x_n, \ldots$ telle que $\lim\limits_{m, n \to \infty} \| x_n - x_m \| = 0$ il existe toujours un élément x tel que $\lim\limits_{n \to \infty} \| x - x_n \| = 0$.

Les éléments de E seront dits „points" de E. Sous une *sphère* $K(\delta, x_0)$ dans E au centre x_0 et de rayon $\delta > 0$ on entend l'ensemble de tous éléments x tels que $\| x - x_0 \| \leq \delta$. Il est aisé de voir que la condition 3° est equivalente à la suivante:

3°$_1$ Pour toute suite de sphères $K_1 \supset K_2 \supset K_3 \supset \ldots$ dans E dont les rayons tendent vers zéro il existe un et seulement un point x appartenant à toutes sphères K_i $(i = 1, 2, 3, \ldots)$.

Naturellement, une suite $x_1, x_2, \ldots, x_n, \ldots$ d'éléments de E sera dit *convergente* vers x ce qu'on écrit par $\lim\limits_{n \to \infty} x_n = x$, si $\lim\limits_{n \to \infty} \| x - x_n \| = 0$. Cela posé, la condition 3° peut être regardée comme l'exigence de ce, que le théorème de Bolzano-Cauchy reste vrai dans E. (Cf. Banach, Théorie des opérations linéaires, Warsowie 1932).

[2]) Une telle fonctionnelle sera dite *convexe*.

admettant, que $p(x) < C$ pour $\| x - x_0 \| \leqslant \delta$, on aura $p(x-x_0) \leqslant p(x) + p(-x_0) = p(x) + p(x_0) < 2C$ pour $\| x - x_0 \| \leqslant \delta$; donc, en posant $\frac{1}{\delta}(x - x_0) = y$, on aura $p(y) < \frac{2C}{\delta}$ pour tout $y \subset K_1$, ce qu'est en contradiction avec l'hypothèse.

Cela établi, prenons dans K_1 un élément x tel que $p(x_1) > 1$. La fonctionelle $p(x)$ étant semi-continue, il existe une sphère $K(\delta_1, x) = K_2 \subset K_1$ telle que $p(x) > 1$ pour tout $x \subset K_2$. D'autre part, $p(x)$ étant aussi non bornée dans K_2, choisissons un point $x_2 \subset K_2$ tel que $p(x_2) > 2$; alors il existe une sphère $K(\delta_2, x_2) = K_3 \subset K_2$ tel que $p(x) > 2$ pour tout $x \subset K_3$ etc. En continuant infiniment ce procédé, on arrive à une suite de sphères $K_1 \supset K_2 \supset K_3 \supset \ldots$ telle que $p(x) > n$ pour tout $x \subset K_{n+1}$ $(n = 1, 2, \ldots)$. En outre, on peut toujours supposer, que $\lim\limits_{n\to\infty} \delta_n = 0$. Or, il existe un point x_0 appartenant à toutes sphères K_i. Par conséquent, on a en ce point l'inégalité absurde

$$p(x_0) > n \qquad (n = 1, 2, 3, \ldots).$$

Le lemme est donc démontré.

R e m a r q u e: Avant de passer aux applications du lemme, remarquons, que la propriété 2° avec l'inégalité (1) entraîne $|p(x) - p(x_0)| \leqslant p(x - x_0) \leqslant M \| x - x_0 \|$; donc, $\lim x_n = x$ entraîne $\lim p(x_n) = p(x)$. Ainsi il résulte du lemme que:

Si la fonctionnelle convexe est semi-continue inférieurement, elle est aussi continue.

§ 2.

T h é o r è m e 1. *Soit $p_1(x), p_2(x), \ldots, p_n(x), \ldots$ une suite de fonctionnelles convexes et continues dans E. Si en outre cette suite est bornée pour tout $x \subset E$, la fonctionnelle $p(x) = \sup p_n(x)$ est aussi continue et convexe dans E.*

D é m o n s t r a t i o n: La convexité de $p(x)$ est évidente. D'après le lemme précédent, il ne reste qu'à montrer, que $p(x)$ est semi-continue inférieurement. Soit x_0 un point quelconque de E. Choisissons un tel N que

$$p(x_0) - p_N(x_0) < \frac{\varepsilon}{2} \tag{2}$$

Le nombre N étant choisi, prenons un tel $\delta > 0$, que

$$|p_N(x_0) - p_N(x)| < \frac{\varepsilon}{2} \qquad (\text{pour } \| x - x_0 \| < \delta) \tag{3}$$

Mais alors en vertu de (2) et (3)

$$p(x_0) - p(x) < p_N(x_0) + \frac{\varepsilon}{2} - \sup p_n(x) \leqslant p_N(x_0) + \frac{\varepsilon}{2} - p_N(x) < \varepsilon$$

$$(\text{pour } \| x - x_0 \| < \delta),$$

c. q. f. d.

C o r o l l a i r e 1. *Soient $f_1(x), f_2(x), \ldots$ une suite de fonctionnelles linéaires*[3])

---

[3]) Une fonctionnelle $f(x)$ définie dans E est dite *linéaire*, s'il jouit de deux propriétés suivantes:

$$1^0, f(\alpha x + \beta y) = \alpha f(x) + \beta f(y),$$
$$2^0, \lim_{n\to\infty} f(x_n) = f(x), \text{ si } \lim_{n\to\infty} x_n = x.$$

On prouve aisément, que la deuxième propriété (la continuité de f) est équivalente à la propriété de $f(x)$ d'être bornée, c'est-à-dire, de satisfaire pour un $M > 0$ à l'inégalité

$$|f(x)| \leq M \| x \| \text{ pour tout } x \text{ de } E$$

Le plus petit possible des nombres M pour lequel cette inégalité a encore lieu est dit la *norme* de $f(x)$ et sera désigné par $\| f \|$ (Cf. Banach, loc. cit. p. 54).

36

dans E; si pour tout $x \subset E$ cette suite est bornée, la suite de normes $\| f_1 \|$ $\| f_2 \|, \ldots$ *est bornée*[4]).

En effet, en posant $p_n(x) = |f_n(x)|$ $(n = 1, 2, \ldots)$ et $p(x) = \sup p_n(x)$, on aura $\| f_n \| \leqslant \underset{\|x\| \leq 1}{\mathrm{Max}}\, p(x)$ $(n = 1, 2, 3, \ldots)$.

Corollaire 2. *Soient $f_1(x), f_2(x), \ldots, f_n(x), \ldots$ une suite de fonctionnelles linéaires dans E; si pour tout $x \subset E$ la série $\sum_{i=1}^{\infty} |f_i(x)|^p$ $(p \geqslant 1)$ converge il existe un tel $M > 0$ que*

$$\sum_{i=1}^{\infty} |f_i(x)|^p \leqslant M^p \| x \|^p.$$

En effet, posons $p_n(x) = \sqrt[p]{\sum_{i=1}^{n} |f_i(x)|^p}$ $(n = 1, 2, \ldots)$ et $p(x) = \lim_{n \to \infty} p_n(x) = \sup p_n(x)$. Alors, d'après le lemme,

$$p(x) = \sqrt[p]{\sum_{i=1}^{\infty} |f_i(x)|^p} \leqslant M \| x \|,$$

c. q. f. d.

Corollaire 3. *Soient $f_1(x), f_2(x), \ldots, f_n(x), \ldots$ une suite de fonctionnelles linéaires dans E; si pour tout $x \subset E$ $\sum_{i=1}^{\infty} |f_i(x)|$ converge, la fonctionnelle $f(x) = \sum_{i=1}^{\infty} f_i(x)$ est linéaire dans E.*

En effet, d'après le corollaire précedent, $|f(x)| \leqslant p(x) = \sum_{i=1}^{\infty} |f_i(x)| \leqslant M \| x \|$.

§ 3.

1. Le corollaire 3 permet de démontrer très simplement le théorème de Landau[5]):

Si la série $\sum_{i=1}^{\infty} a_i \xi_i$ converge pour toute suite $\xi_1, \xi_2, \ldots$ telle que $\sum_{i=1}^{\infty} |\xi_i|^p < \infty$ $(p > 1)$, la série $\sum_{i=1}^{\infty} |a_i|^q$, où $\frac{1}{p} + \frac{1}{q} = 1$, converge.

En effet, désignons par $l^{(p)}$ l'espace dont les points sont des suites $x = (\xi_1, \xi_2, \xi_3, \ldots)$ telles, que $\sum_{i=1}^{\infty} |\xi_i|^p < \infty$ et où $\| x \| = \sqrt[p]{\sum_{i=1}^{\infty} |\xi_i|^p}$. En posant $f_i(x) = a_i \xi_i$ pour $x = (\xi_1, \xi_2, \ldots)$ on obtient ainsi une suite de fonctionnelles linéaires $f_1(x)$ $f_2(x), \ldots$ vérifiant évidemment les conditions du corollaire 3. D'où, la fonctionnelle

$$f(x) = \sum_{i=1}^{\infty} f_i(x) = \sum_{i=1}^{\infty} a_i \xi_i$$

---

[4]) Voir Banach, loc. cit. p. 80 et S. Mazur — W. Orlicz, Studia Mathematica, t. IV p. 153.
[5]) *Landau* Über einen Konvergenzsatz (Nachrichten Ges. Wiss. Göttingen, 1 07, p. 25—27); voir aussi F. Riesz, Les systèmes d'équations linéaires à une infinité d'inconnues, p. 48.

37

est linéaire dans $l^{(p)}$ et par suite, d'après la proposition bien connue,

$$\sum_{i=1}^{\infty} |a_i|^q < \infty,$$

c. q. f. d.

Le corollaire 2 contient comme un cas particulier la proposition de Hellinger - Toeplitz[6]) et de plus sa généralisation suivante:

Si la matrice infinie $\| a_{ik} \|_1^{\infty}$ *fait correspondre au chaque point* $x = (\xi_1, \xi_2, \ldots)$ *de* $l^{(p)}$ *un point* $y = (\eta_1, \eta_2, \ldots)$ *de* $l^{(p_1)}$ *par les formules* $\eta_i = \sum_{k=1}^{\infty} a_{ik} \xi_k \; (i = 1, 2, \ldots)$, *alors l'opérateur correspondant* $y = F(x)$ *est borné, c'est-à-dire, il existe un tel* $M > 0$ *que* $\| y \| \leqslant M \| x \|$.

Pour la démonstration il suffit de remarquer, que

$$\| y \| = \sqrt[p_1]{\sum_{i=1}^{\infty} |\eta_i|^{p_1}} = \sqrt[p_1]{\sum_{i=1}^{\infty} |f_i(x)|^{p_1}},$$

où $f_i(x) = \eta_i = \sum_{k=1}^{\infty} a_{ik} \xi_k \; (i = 1, 2, 3, \ldots)$ sont, d'après le théorème de Landau des fonctionnelles linéaires dans $l^{(p)}$.

2. Au moyen du théorème 1 on établit presque immédiatement le théorème de F. Riesz[7]):

Si pour toute fonction $x(t)$ $(0 \leqslant t \leqslant 1)$ *sommable de p-ième puissance* $(p > 1)$, *l'intégrale* $\int_0^1 x(t) \psi(t)\, dt$, *existe,* $\psi(t)$ *étant* $(0 \leqslant t \leqslant 1)$ *une fonction donnée, la fonction* $\psi(t)$ *est sommable de q-ième puissance, où* $\frac{1}{p} + \frac{1}{q} = 1$.

En effet, envisageons l'espace $L^{(p)}$ de toutes fonctions $x = x(t)$ $(0 \leqslant t \leqslant 1)$ sommable de p-ième puissance où

$$\| x \| = \sqrt[p]{\int_0^1 |x(t)|^p \, dt}.$$

Posons

$$p_n(x) = \int_{\mathfrak{M}_n} |x(t) \psi(t)| \, dt,$$

où $\mathfrak{M}_n$ est l'ensemble de tous points t pour lesquels: $|\psi(t)| < n$. D'après le théorème 1, la fonctionnelle

$$p(x) = \sup p_n(x) = \lim p_n(x) = \int_0^1 |x(t) \psi(t)| \, dt$$

est continue et, par suite, la fonctionnelle

$$f(x) = \int_0^1 x(t) \psi(t) \, dt$$

est linéaire, car $|f(x)| \leqslant p(x) < M \| x \|$. Mais cela étant,[8])

$$\int_0^1 |\psi(t)|^q \, dt < \infty,$$

c. q. f. d.

---

[6]) Mathematische Annalen (69), 1910, p. 289 — 330.
[7]) *F. Riesz,* Mathem Annalen (69), 1910, v. p. 475.
[8]) Voir Banach, loc. cit. p. 64.

38

De même on démontre:

Si l'intégrale $\int_0^1 x(t)\psi(t)\,dt$, *où* $\psi(t)$ *est une fonction donnée, existe pour tout* $x(t)$ *sommable, alors vrai* $\max|\psi(t)|<\infty$.

§ 4.

1. Le théorème 1 permet d'établir aussi la proposition suivante:

Théorème 2. *Soit E un espace séparable et soit* $f_t(x)$ *une fonctionnelle dans E dépendant d'un paramètre réel* $t\,(a\leqslant t\leqslant b)$. *Si pour tout* $x\subset E$ *l'intégrale* $\int_0^1 |f_t(x)|^p\,dt\ (p\geqslant 1)$ *existe, il existe un tel M, que*

$$\int_a^b |f_t(x)|^p\,dt \leqslant M^p \|x\|^p.$$

Démonstration. Rappelons d'abord que l'espace E est dit *séparable*, s'il existe dans E une suite $x_1, x_2, x_3, \ldots$ partout dense dans E.

Envisageons l'ensemble $\mathfrak{M}_n$ de tous points t pour lesquels $\|f_t\|^p<n$. Cet ensemble est mesurable, car il est le produit des ensembles

$$\mathfrak{M}_n^{(k)}=\mathfrak{M}(|f_t(x_k)|^p<n) \qquad (k=1,2,\ldots).$$

Le fonctionnelles

$$p_n(x)=\sqrt[p]{\int_{\mathfrak{M}_n}|f_t(x)|^p\,dt} \qquad (n=1,2,\ldots)$$

sont évidemment convexes et continues; donc, d'après le théorème 1, la fonctionelle

$$p(x)=\sup p_n(x)=\lim p_n(x)=\sqrt[p]{\int_a^b |f_t(x)|^p\,dt}$$

est convexe et continue et, par suite, il existe un tel $M>0$ que $p(x)\leqslant M\|x\|$, c. q. f. d.

Le théorème démontré permet d'introduire la notion de l'intégrale de Lebesque d'une fonctionnelle linéaire dépendant d'un paramètre réel $(a\leqslant t\leqslant b)$. Nous disons que f_t est sommable *faiblement*, si pour tout $x\subset E$ l'expression $f_t(x)$ est une fonction réelle sommable du paramètre t. Si f_t est sommable *faiblement*, la fonctionelle

$$f(x)=\int_a^b f_t(x)\,dt$$

est linéaire dans E. En effet, d'après le théorème démontré, il existe un tel $M>0$, que

$$|f(x)|\leqslant\int_a^b |f_t(x)|\,dt\leqslant M\ \|x\|.$$

Or, la fonctionnelle $f(x)$ nous appelons l'intégrale de la fonction $f_t(x)$, dont toutes les valeurs sont les fonctionnelles dans l'espace E.

2. Appelons la fonctionnelle $f_t(x)$, $(a\leqslant t\leqslant b)$ *fonction à variation faiblement bornée*, si pour tout $x\subset E$ $f_t(x)$ est une fonction de t à variation bornée dans (a, b). Démontrons le

39

Théorème 3: *Si la fonctionnelle linéaire $f_t(x)$ $(a \leqslant t \leqslant b)$ est une fonction de t à variation faiblement bornée, la fonctionnelle $p(x) = \operatorname{Var} f_t(x)$ est convexe et continue*[9]).

Démonstration: La convexité de $p(x)$ est évidente. Soit S une subdivision quelconque de l'intervalle (a, b) au moyen des points $t_0 = a < t_1 < < t_2 < \ldots < t_n = b$. Posons

$$p_S(x) = \sum_{i=1}^{n} |f_{t_i}(x) - f_{t_{i-1}}(x)|.$$

La fonctionnelle $p_S(x)$ est évidemment convexe et continue. D'autre part, on a

$$p(x) = \sup p_S(x) = \operatorname{Var} f_t(x).$$

Le théorème 1 évidemment restant vrai, si l'on y remplace la suite $\{p_i(x)\}$ par un ensemble $\{p_S(x)\}$ quelconque, on voit que le théorème en question est une conséquence de celui-ci.

3. Soit x_t une fonction de variable réel $t\,(a \leqslant t \leqslant b)$ dont les valeurs sont les points de l'espace E.

Une telle fonction x_t sera dite à *variation faiblement bornée*, si pour toute fonctionnelle linéaire $f(x)$ dans E, $f(x_t)$ est une fonction à variation bornée. Démontrons le.

Théorème 4: *Si $x_t\,(a \leqslant t \leqslant b)$ est une fonction à variation faiblement bornée, il existe un tel $M > 0$, que*

$$\operatorname{Var} f(x_t) \leqslant M \, \| f \| \,.$$

Démonstration: Ce théorème se réduit au théorème précédent. En effet, envisageons l'espace E' de toutes fonctionnelles f définies dans E. L'espace E' est aussi un espace linéaire. On peut toujours règarder $f(x_t)$ comme $F_t(f)$ c'est-à-dire, comme une fonctionnelle dans E' dépendant du paramètre $t\,(a \leqslant t \leqslant b)$. Vu que $\operatorname{Var} f(x_t) = \operatorname{Var} F_t(f)$, il ne reste qu'appliquer le théorème précédent à la fonctionnelle $p(f) = \operatorname{Var} F_t(f)$.

[9]) Ce théorème permet d'introduire la notion de l'intégrale de Stieltjes

$$\int_a^b \alpha(t)\, df_t(x).$$

2.

Abstrakte Funktionen und lineare Operatoren

Mat. Sb., Nov. Ser. **4** (46) (1938) 235–284. Zbl. **20**: 367

Die vorliegende Arbeit stellt den Inhalt einer im Jahre 1935 öffentlich verteidigten Dissertation dar. Die seitdem veröffentlichten Resultate von Birkhoff, Dunford, Clarkson u. a. fallen teilweise mit den hier erhaltenen Resultaten zusammen.

Die Arbeit besteht aus zwei Teilen. Im ersten Teil werden abstrakte Funktionen untersucht [d. h. Funktionen x_t, die jeder Zahl t ein Element eines linearen (B) Raumes zuordnen].

In § 6 wird eine Definition der Funktion mit beschränkter Variation gegeben, die sich von der Bochnerschen Definition unterscheidet. Dieser Begriff knüpft ebenso naturgemäss an unsere Definition des Lebesgueschen Integrals einer abstrakten Funktion an, wie die Bochnersche Definition der Funktion mit beschränkter Variation an seine Definition des Integrals anknüpft.

Streng genommen führen wir zwei Definitionen der Funktion mit beschränkter Variation ein, nämlich die der Funktion mit schwacher und die der Funktion mit starker Variation. Die zweite Definition gestattet uns die allgemeine Form des linearen vollstetigen Operators zu finden, der den Raum stetiger Funktionen auf einen beliebigen linearen Raum abbildet. Die erste Definition macht es möglich, die allgemeine Form eines beliebigen linearen Operators zu finden, der den Raum stetiger Funktionen auf einen beliebigen linearen Raum abbildet (Teil II, § 8). Der Begriff einer abstrakten Funktion mit stark beschränkter Variation ist mit der Orliczschen Definition der unbedingten Konvergenz von Reihen verbunden. Hier ist dieser Begriff in einer etwas umgewandelten und der weiteren Anwendung angepassten Form dargelegt (§ 4).

In § 7, Teil I, geben wir die Definition des Lebesgueschen Integrals einer abstrakten Funktion. Diese Definition umfasst sowohl die Bochnersche als auch die Birkhoffsche Definitionen.

Schliesslich werden in § 9 des ersten Teils die Bedingungen für den Fall gegeben, in dem die Ableitung der abstrakten Funktion fast überall existiert, wenn nur die Werte der Funktion einem regulären Raum angehören (d. h. einem solchen Raum, der mit seinem zweiten konjugierten zusammenfällt). Das ermöglicht es, im Besonderen, die notwendige und hinreichende Bedingung dafür zu finden, dass eine Funktion ein unbestimmtes Integral im Bochnerschen Sinne ist. Andere Bedingungen, die erfüllt sein müssen, damit eine abstrakte

Funktion fast überall eine Ableitung besitze, haben Birkhoff, Clarkson, Dunford und Morse gegeben.

Im zweiten Teil dieser Arbeit werden auf Grund der Ergebnisse des ersten Teils die allgemeinen Formen verschiedener linearer Operatoren gefunden. In der Untersuchung gehen wir hierbei folgendermassen vor: es wird gesucht die allgemeine Form des linearen Operators, der einen gewissen konkreten Raum auf einen beliebigen Raum abbildet (oder umgekehrt).

Aus den Resultaten des zweiten Teils möchte ich besonders den Satz des § 2 hervorheben (die Bedingung für die Kompaktheit von Mengen in einem beliebigen linearen Raum).

Eine kurze Zusammenfassung dieser Arbeit erschien in den Comptes rendus de l'Académie des Sciences de l'U. R. S. S., **XVII**, 5, (1937), 243–248.

Inhaltsverzeichnis

Seite

I. Teil. Abstrakte Funktionen.
§ 1. (B) messbare Funktionen 236
§ 2. Nach Lebesgue messbare Funktionen 238
§ 3. Ein Hilfssatz 240
§ 4. Unbedingte Konvergenz 241
§ 5. Volladditive Mengenfunktionen 246
§ 6. Funktionen mit beschränkter Variation 246
§ 7. Das Lebesguesche Integral 253
§ 8. Das Stieltjessche und das Lebesgue-Stieltjessche Integral 259
§ 9. Differentiation abstrakter Funktionen 260
II. Teil. Allgemeine Form linearer Operatoren.
§ 1. Die allgemeine Form der linearen Operatoren, die einen beliebigen linearen Raum auf den Raum (*C*) der stetigen Funktionen abbilden . . . 266
§ 2. Die Bedingungen der Kompaktheit von Mengen in linearen Räumen . . 268
§ 3. Die allgemeine Form des stetigen und des vollstetigen Operators, der $l^{(1)}$ auf *E* abbildet 269
§ 4. Die allgemeine Form des linearen stetigen und vollstetigen Operators, der einen beliebigen linearen Raum auf $l^{(1)}$ abbildet 271
§ 5. Die allgemeine Form des linearen Operators, der (*c*), den Raum aller konvergenten Folgen, auf einen beliebigen linearen Raum *E* abbildet . . . 272
§ 6. Die allgemeine Form eines Operators, der $L^{(1)}$ auf einen beliebigen linearen Raum abbildet 275
§ 7. Die allgemeine Form des Operators, der einen beliebigen linearen Raum *E* auf *M* abbildet 279
§ 8. Die allgemeine Form des linearen Operators, der (*C*) auf einen beliebigen Raum *E* abbildet 280
§ 9. Die allgemeine Form der Operatoren, die *E* auf (*V*) abbilden 283

I. TEIL

ABSTRAKTE FUNKTIONEN

§ 1. (B) messbare Funktionen

Die abstrakte Funktion x_t wird schwach stetig genannt, wenn für ein beliebiges lineares Funktional f die reelle Funktion fx_t nach t stetig ist. fx ist der Wert des linearen Funktionals f im Punkt x des linearen Raumes E.

Die Funktion x_t wird stark stetig genannt, wenn bei $t_n \longrightarrow t$ $\lim x_{t_n} = x_t$ im Sinne der starken Konvergenz ist.

Man beachte, dass, wenn die Funktion x_t schwach stetig und die Menge ihrer Werte kompakt ist, x_t auch stark stetig ist.

Die Funktion x_t wird nach Baire eine Funktion erster Klasse in bezug auf die starke Konvergenz genannt, wenn sie ein starker Limes (für jedes t) einer Folge von stark stetigen Funktionen ist.

Ebenso wie für reelle Funktionen ist die punkthafte Unstetigkeit (in bezug auf die starke Konvergenz) eine notwendige und hinreichende Bedingung dafür, dass x_t eine Funktion erster Klasse nach Baire ist.

Jede schwach stetige Funktion besitzt Punkte starker Stetigkeit, wenn nur die Werte der Funktion einem separablen Raum angehören. Genauer gesagt gilt der

Satz 1. *Ist E separabel, so ist jede schwach stetige Funktion, deren Werte E angehören, eine Funktion nicht höherer als erster Klasse nach Baire in bezug auf die starke Konvergenz.*

Beweis. Beweisen wir zuerst den Satz unter der Annahme, dass E eine Basis $x_1, x_2, \ldots, x_n, \ldots$ [1] besitzt.

Die schwach stetige Funktion kann folgendermassen dargestellt werden:

$$x_t = \sum_{i=1}^{\infty} f_i x_t \cdot x_i;$$

dabei sind $f_i x_t$ nach Definition stetige Funktionen von t. D. h., dass die Funktion $\sum_{i=1}^{n} f_i x_t \cdot x_i$ stark stetig ist, und folglich ist x_t als starker Limes stark stetiger Funktionen eine Funktion erster Klasse. Befreien wir uns nun von der Forderung, dass E eine Basis hat. Hierzu betten wir E in den Raum (C) [2] der stetigen Funktionen ein. Da in (C) eine Basis vorhanden ist [3], ist die Funktion x_t in bezug auf (C) eine Funktion erster Klasse. Aber die Eigenschaft einer Funktion der ersten Klasse anzugehören hängt nicht von dem umfassenden Raum ab, sondern nur von dem minimalen abgeschlossenen Teilraum, der die Werte der Funktion x_t enthält. (Tatsächlich, damit x_t eine Funktion erster Klasse sei, ist notwendig und hinreichend, dass sie auf jeder perfekten Menge punkthaft unstetig ist. Das letztere aber ist eine innere Eigenschaft der Funktion und hängt von dem umfassenden Raum nicht ab.) Folglich gehört x_t zur ersten Klasse in bezug auf E, w. z. b. w.

Die stark (B) messbaren abstrakten Funktionen können ähnlich wie die reellen (B) messbaren Funktionen definiert werden: *als Klasse der stark* (B) *messbaren Funktionen bezeichnen wir die minimale Gesamtheit der Funktionen,*

---

[1] Wir sagen, dass der Raum E eine Basis besitzt, wenn es eine solche abzählbare Elementenfolge $x_1, \ldots, x_n, \ldots$ gibt, dass ein jedes Element des Raumes E eindeutig in einer der beiden Formen $x = \sum_{i=1}^{\infty} a_i x_i$ oder $x = \sum_{i=1}^{\infty} f_i x \cdot x_i$ dargestellt werden kann (f_i ist eine Folge von Funktionalen, siehe Banach, Théorie des opérations linéaires, S. 111).

[2] Banach hat bewiesen, dass ein beliebiger separabler Raum einem Teilraum des Raumes der stetigen Funktionen äquivalent ist. Siehe Banach, l. c., S. 187.

[3] Schauder, Mathematische Zeitschrift, **26**, (1927), 48—49.

die alle stark stetigen Funktionen und gleichzeitig mit jeder stark konvergenten Funktionenfolge auch deren Limes enthält. Die Funktion wird schwach (B) *messbar genannt, wenn für ein beliebiges f die reelle Funktion* fx_t (B) *messbar ist.*

Man kann den folgenden Satz beweisen:

Satz 2. *In separablen Räumen fallen die Begriffe der starken und schwachen* (B) *Messbarkeit zusammen.*

§ 2. Nach Lebesgue messbare Funktionen

Definitionen. 1. *Die Funktion* x_t *wird stark messbar nach Lebesgue genannt, wenn es eine Folge stark stetiger Funktionen gibt, die für fast all t gegen* x_t *konvergiert* [4].

2. *Die Funktion* x_t *wird schwach messbar genannt, wenn für ein beliebiges f die reelle Funktion* fx_t *messbar nach Lebesgue (in bezug auf t) ist.*

Hilfssatz 1. *Ist* x_t *stark messbar, so ist* $|x_t|$ *eine messbare reelle Funktion* [4].

Beweis. Ist $x_t^{(n)}$ stark stetig, so ist $|x_t^{(n)}|$ stetig. Aus

$$\lim_{n\to\infty} x_t^{(n)} = x_t \quad \text{für fast alle } t$$

folgt

$$\lim_{n\to\infty} |x_t^{(n)}| = |x_t| \quad \text{für fast alle } t.$$

Folglich ist $|x_t|$ messbar.

Hilfssatz 2. *Gehören die Werte der Funktion* x_t *dem abgeschlossenen linearen Teilraum des Raumes E an, und ist* x_t *in E stark messbar, so ist* x_t *auch in bezug auf den Teilraum stark messbar. Auch die Umkehrung gilt.*

Mit anderen Worten: die Eigenschaft der Funktion x_t messbar zu sein hängt nur vom minimalen Teilraum ab, der die Werte von x_t enthält.

Beweis. Zu beweisen ist nur der erste Teil des Satzes. D. h. wir haben eine Folge stark stetiger Funktionen zu finden, deren Werte dem gegebenen Teilraum angehören, und die stark gegen x_t konvergieren. Bemerken wir dazu, dass die (C)-Eigenschaft [5] auch für stark messbare abstrakte Funktionen gültig bleibt. Infolgedessen gibt es für ein beliebiges n auf dem Segment (a, b) eine perfekte Punktmenge P_n mit einem Mass, das sich von dem Mass des ganzen Segments um weniger als $\frac{1}{n^2}$ unterscheidet, auf der x_t stark stetig ist. Auf P_n setzen wir nun $x_t^{(n)} = x_t$ und interpolieren $x_t^{(n)}$ in den angrenzenden Intervallen linear. Es ist dann leicht zu sehen, dass $x_t^{(n)}$ die gesuchte Folge ist.

Satz 1. *Ist E separabel, so fallen die Begriffe der starken und der schwachen Messbarkeit der Funktion* x_t *(deren Werte E angehören) zusammen.*

Beweis. Es ist offensichtlich, dass jede stark messbare Funktion auch schwach messbar ist. Beweisen wir nun das Umgekehrte zuerst unter der

---

[4] Siehe Bochner, Fund. Math., **XX**, (1933).

[5] Die (C)-Eigenschaft bedeutet Folgendes: ist die Funktion x_t auf der Menge P messbar, so existiert eine perfekte Teilmenge der Menge P, auf welcher x_t stetig ist, und deren Mass sich durch eine beliebig kleine Zahl vom Masse der Menge P unterscheidet.

Annahme, dass E eine Basis hat [1]. Die Funktion

$$x_t^{(n)} = \sum_{i=1}^{n} x_i f_i x_t$$

ist stark messbar, und da $x_t = \lim_{n\to\infty} x_t^{(n)}$ ist, so ist x_t auch stark messbar.

Von dieser Annahme können wir uns folgendermassen befreien:

Betten wir E in den Raum der stetigen Funktionen (C) [2] ein. Dann ist x_t stark messbar in bezug auf (C) [da in (C) eine Basis vorhanden ist [3]], und also, nach Hilfssatz 2, auch in bezug auf E, w. z. b. w.

In konjugierten Räumen können die folgenden beiden Definitionen der schwachen Messbarkeit gegeben werden:

1. *Die abstrakte Funktion f_t, deren Werte $\overline{E}$ angehören, wird schwach messbar genannt, wenn für ein beliebiges x aus E die reelle Funktion $f_t x$ messbar ist.*

2. *Die abstrakte Funktion f_t wird schwach messbar genannt, wenn für ein beliebiges ξ aus $\overline{\overline{E}}$ die reelle Funktion ξf_t messbar ist.*

Beweisen wir nun den

Satz 2. *Ist der Raum $\overline{E}$ separabel, so stimmen in ihm die Definitionen 1 und 2 überein.*

Beweis. Es ist nur zu beweisen, dass 2 aus 1 folgt [6]. Das ergibt sich aber unmittelbar aus dem

Hilfssatz 3 (Banach). *Ist E separabel, so ist ein beliebiges Element ξ aus $\overline{\overline{E}}$ ein schwacher Limes der Folge $x_1, x_2, \ldots, x_n, \ldots$ der Elemente aus E, d. h. für ein beliebiges f gilt die Gleichung $\lim_{n\to\infty} f x_n = \xi f$.*

Beweis des Hilfssatzes 3. Es sei $f_1, \ldots, f_k, \ldots$ eine überall dichte Menge in $\overline{E}$. Konstruieren wir das Element x_n so, dass $f_k x_n = \xi f_k$ für $k = 1, \ldots, n$, und $|x_n| \leq |\xi| + 1$. Solch ein Element existiert auf Grund des folgenden Satzes von Helly:

Es ist eine endliche Zahl linearer Funktionale $f_1, \ldots, f_n$ gegeben. Damit ein Element x existiert, dessen Norm kleiner als $M + \varepsilon$ ist (ε ist eine beliebige positive Zahl), und für welches $f_i x = c_i$ gilt (c_i sind gegebene Zahlen), ist notwendig und hinreichend, dass folgende Bedingung erfüllt ist: für beliebige reelle h_i gilt die Ungleichung

$$\left|\sum_{i=1}^{n} h_i c_i\right| \leq M \left|\sum_{i=1}^{n} h_i f_i\right|.$$

Diese Bedingung ist in unserem Falle erfüllt, da an Stelle der c_i hier ξf_i tritt, und folglich

$$\left|\sum_{i=1}^{n} h_i c_i\right| = \left|\sum_{i=1}^{n} \xi f_i h_i\right| \leq |\xi| \cdot \left|\sum_{i=1}^{n} h_i f_i\right|$$

ist.

[6] Da einem jeden Element x aus E ein solches Element ξ aus $\overline{\overline{E}}$ entspricht, dass für alle f die Gleichung $\xi f = f x$ stattfindet.

Die Folge der Elemente $x_1, \ldots, x_n, \ldots$ konvergiert schwach gegen ξ. In der Tat haben wir: $|x_n| \leqslant |\xi| + 1$ und $\lim_{n \to \infty} f_k x_n = \xi f_k$ für ein beliebiges Element unserer überall dichten Menge.

Jetzt ist Satz 2 leicht zu beweisen. Es ist zu zeigen, dass für ein beliebiges ξ die reelle Funktion ξf_t summierbar ist. Dazu konstruieren wir eine Elementenfolge $x_1, \ldots, x_n, \ldots$, die schwach gegen ξ konvergiert. Dann ist $\lim_{n \to \infty} f x_n = \xi f$, d. h. $\lim_{n \to \infty} f_t x_n = \xi f_t$. Da die $f_t x_n$ gemäss Definition 1 messbare Funktionen sind, so besitzt auch ξf_t dieselbe Eigenschaft.

§ 3. Ein Hilfssatz

Der Darlegung weiterer Resultate schicken wir den folgenden Hilfssatz voraus, den wir oft anwenden werden.

Hilfssatz [7]. *Im linearen Raum E sei ein Funktional $p(x)$ mit folgenden Eigenschaften gegeben:*

1°) $p(x) \geqslant 0$,

2°) $p(x+y) \leqslant p(x) + p(y)$,

3°) $p(ax) = |a| p(x)$.

Wenn $p(x)$ ein von unten halbstetiges Funktional ist [d. h. wenn es für ein beliebiges ε ein solches δ gibt, dass für alle x mit $|x - x_0| < \delta$ $p(x_0) - p(x) < \varepsilon$ ist], *dann ist es stetig und in der Einheitssphäre beschränkt.*

Also ist nach 3°) die Ungleichung $p(x) \leqslant M|x|$ erfüllt, wo M eine passende Konstante ist.

Beweis. Beweisen wir zuerst, dass $p(x)$ in der Einheitssphäre beschränkt ist. Nehmen wir das Gegenteil an: $p(x)$ sei in der Einheitssphäre nicht beschränkt. Dann ist $p(x)$ in einer beliebigen Sphäre $K(\delta, x_0)$ mit dem Zentrum x_0 und dem Radius δ nicht beschränkt. In der Tat, wäre für $|x - x_0| < \delta$ $p(x) \leqslant k$, so hätten wir

$$p(x - x_0) \leqslant p(x) + p(-x_0) \leqslant k + p(-x_0),$$

und folglich

$$p(y) \leqslant \frac{1}{\delta}(k + p(-x_0)) \quad \left[y = \frac{1}{\delta}(x - x_0)\right],$$

wenn

$$|y| \leqslant 1.$$

Dies in Betracht ziehend, wählen wir in der Einheitssphäre ein Element x_1, für das $p(x_1) > 1$ ist. Infolge der Halbstetigkeit von $p(x)$ gibt es eine solche Sphäre $K(\delta_1, x_1)$, dass $p(x) > 1$ für alle x, für die $|x - x_1| < \delta_1$ ist. Aus dem Obigen folgt, dass es in dieser Sphäre einen Punkt x_2 gibt, für den $p(x_2) > 2$ ausfällt. Deshalb gibt es auch eine Sphäre $K(\delta_2, x_2)$, die innerhalb der ersten liegt, und in der $p(x) > 2$ gilt. Setzen wir diesen Prozess weiter fort, und wählen wir δ_n so, dass $\lim \delta_n = 0$ ist, so erhalten wir eine Folge von ineinandergeschachtelten Kugeln, die sich auf einen Punkt zusammenziehen. Aber dann ist $p(x)$ in diesem Punkt sinnlos, was der Voraussetzung widerspricht.

---

[7] Siehe Orlicz, Studia Math., I, (1929), 9.

Die Stetigkeit von $p(x)$ kann jetzt sehr leicht bewiesen werden: da für alle x mit $|x| \leqslant 1$ $p(x) \leqslant M$ gilt, so ist $|p(x) - p(x_0)| \leqslant M|x - x_0|$, und folglich ist $p(x)$ stetig.

Anmerkung. Geometrisch lautet dieser Hilfssatz folgendermassen: besitzt ein abgeschlossener, konvexer, Null enthaltender Körper auf jedem Strahl aus dem Anfangspunkt des Koordinatensystems mehr als einen Punkt, so ist der Nullpunkt ein innerer Punkt.

§ 4. Unbedingte Konvergenz

1. *Wir nennen die Reihe* $x_1 + x_2 + \ldots + x_n + \ldots$ *unbedingt konvergent, wenn für ein beliebiges Funktional* f *die Reihe* $fx_1 + \ldots + fx_n + \ldots$ *absolut konvergiert* [8].

Einer unbedingt konvergenten Reihe kann man nicht immer eine Summe aus demselben Raum zuschreiben. Das ist im allgemeinen nur in schwach vollständigen Räumen möglich.

Satz 1. *Konvergiert die Reihe* $fx_1 + \ldots + fx_n + \ldots$ *unbedingt, so gibt es ein solches* M, *dass*

$$|fx_1| + \ldots + |fx_n| + \ldots \leqslant M|f|$$

ist.

Beweis. Dieser Satz folgt aus dem Hilfssatz in § 3. Dazu untersuchen wir $\bar{E}$ und nehmen $p(f) = \sum_{i=1}^{\infty} |fx_i|$ an. Zeigen wir, dass $p(f)$ allen Forderungen des Hilfssatzes genügt. Die Eigenschaften 1°), 2°), 3°) sind offensichtlich. Man hat nur die Halbstetigkeit von $p(f)$ zu beweisen, d. h. zu beweisen, dass es für jedes f_0 und ε ein solches δ gibt, dass $p(f_0) - p(f) < \varepsilon$ ausfällt für jedes f, für das $|f - f_0| < \delta$ ist. Schliesslich wählen wir n derart, dass

$$\sum_{i=n+1}^{\infty} |f_0 x_i| < \frac{\varepsilon}{2}$$

ist. Da $\sum_{i=1}^{n} |fx_i|$ ein stetiges Funktional von f ist, so gibt es eine Umgebung des Punktes f_0, in der die Schwankung des Funktionals kleiner als $\frac{\varepsilon}{2}$ ist. In dieser Umgebung ist dann $p(f_0) - p(f) < \varepsilon$. In der Tat:

$$p(f_0) - \sum_{i=1}^{n} |f_0 x_i| < \frac{\varepsilon}{2},$$

$$\sum_{i=1}^{n} |f_0 x_i| - \sum_{i=1}^{n} |fx_i| < \frac{\varepsilon}{2},$$

$$\sum_{i=1}^{n} |fx_i| - p(f) < 0,$$

und die gesuchte Ungleichung folgt durch Addition der drei letzten Ungleichungen. Also genügt $p(f)$ den Bedingungen des Hilfssatzes des § 3, und es gibt ein solches M, dass $p(f) \leqslant M|f|$ ist, w. z. b. w.

[8] Siehe Orlicz, Studia Math., I, S. 241—248. Unsere Definition unterscheidet sich etwas von der Orliczschen. Sie fallen nur in shwach vollständigen Räumen zusammen.

Der Satz 1 zeigt uns, was unter der Summe einer unbedingt konvergenten Reihe zu verstehen ist. Wir haben

$$\left|\sum_{i=1}^{\infty} fx_i\right| \leqslant \sum_{k=1}^{\infty} |fx_k| \leqslant M|f|;$$

folglich ist $\sum\limits_{k=1}^{\infty} fx_k$ ein lineares Funktional in bezug auf $\overline{E}$, d. h. $\sum\limits_{k=1}^{\infty} x_k$ ist ein gewisses Element ξ aus $\overline{\overline{E}}$. Dieses Element betrachten wir als die Summe der Reihe $\sum\limits_{i=1}^{\infty} x_i$. Wenn E schwach vollständig ist, so ist ξ ein Element aus E selbst, da die Folge der Teilsummen der Reihe schwach konvergiert [6]. Weisen wir noch auf das folgende Merkmal der unbedingten Konvergenz einer Reihe hin:

Satz 2 [9]. *Damit die Reihe $\sum\limits_{i=1}^{\infty} x_i$ unbedingt konvergiere, ist die Existenz eines solchen M notwendig und hinreichend, dass $\left|\sum\limits_{i=1}^{n} \varepsilon_i x_i\right| \leqslant M$ für beliebige n und $\varepsilon_i = \pm 1$ gilt.*

Beweis. Dieser Satz folgt unmittelbar aus dem vorhergehenden Satz 1. Beweisen wir zuerst die Notwendigkeit. Wenn die Reihe $\sum\limits_{i=1}^{\infty} x_i$ unbedingt konvergiert, gibt es ein solches M, dass $\sum\limits_{i=1}^{\infty} |fx_i| \leqslant M|f|$ ist für ein beliebiges f. Dann ist aber

$$\left|f \sum_{i=1}^{n} \varepsilon_i x_i\right| \leqslant \sum_{i=1}^{n} |fx_i| \leqslant M|f|;$$

und da dies für ein beliebiges f gilt, so ist

$$\left|\sum_{i=1}^{n} \varepsilon_i x_i\right| \leqslant M.$$

Beweisen wir nun das Umgekehrte. Wir haben

$$\sum_{i=1}^{n} |fx_i| = \sum_{i=1}^{n} \varepsilon_i fx_i = f \sum \varepsilon_i x_i \leqslant M|f|$$

$$(\varepsilon_i = \operatorname{sign} fx_i),$$

und folglich konvergiert die Reihe $\sum\limits_{i=1}^{\infty} |fx_i|$ für jedes f, w. z. b. w.

Vollständig analog dem Satz 2 wird der folgende Satz bewiesen:

Satz 2a [9]. *Für die unbedingte Konvergenz der Reihe $x_1 + \ldots + x_n + \ldots$ ist die Existenz eines solchen M notwendig und hinreichend, dass für eine beliebige Folge der Indizes $n_1 < n_2 < \ldots < n_k < \ldots$ die Ungleichung*

$$|x_{n_1} + \ldots + x_{n_k}| < M$$

gilt.

---

[9] Diese Sätze wurden von Orlicz gegeben.

2. Im konjugierten Raum können zwei Definitionen der unbedingten Konvergenz gegeben werden:

1°. *Die Reihe* $\sum_{i=1}^{\infty} f_i$ *heisst unbedingt konvergent, wenn für ein beliebiges* x *aus* E *die Reihe* $\sum_{i=1}^{\infty} |f_i x|$ *konvergiert.*

2°. *Die Reihe* $\sum_{i=1}^{\infty} f_i$ *heisst unbedingt konvergent, wenn für ein beliebiges* ξ *aus* $\overline{\overline{E}}$ *die Reihe* $\sum_{i=1}^{\infty} |\xi f_i|$ *konvergiert.*

Bei der ersten Definition wird $\overline{E}$ als ein Raum von Funktionalen betrachtet, in der zweiten als ein Raum von Elementen.

Satz 3. *Die Definitionen 1° und 2° stimmen überein.*

Beweis. Für jede dieser Definitionen gilt der Satz 2. In der Tat kann man den Beweis der Sätze 1 und 2 wörtlich wiederholen, wenn man im ersten Falle $p(x) = \sum_{i=1}^{\infty} |f_i x|$ und im zweiten $p(\xi) = \sum_{i=1}^{\infty} |\xi f_i|$ annimmt. Folglich ist bei jeder der beiden Definitionen der unbedingten Konvergenz die Existenz eines solchen M notwendig und hinreichend, dass $\left|\sum_{i=1}^{n} \varepsilon_i f_i\right| \leq M \quad (\varepsilon_i = \pm 1)$ ist. Dadurch ist bewiesen, dass die beiden Definitionen übereinstimmen.

Anmerkung. Aus Satz 3 folgt, dass wenn E zu einem gewissen anderen Raum konjugiert ist, die Summe der unbedingt konvergenten Reihe ein Element desselben Raumes ist.

3. Unbedingte Konvergenz im starken Sinne. *Wir sagen, dass die Reihe* $\sum_{i=1}^{\infty} x_i$ *unbedingt im starken Sinne konvergiert, wenn die Reihen* $\sum_{i=1}^{\infty} |f x_i|$ *gleichmässig in bezug auf alle* f *mit* $|f| \leq 1$ *konvergieren;* oder, mit anderen Worten, wenn es für jedes ε ein solches n gibt, dass $\sum_{k=n+1}^{\infty} |f x_k| < \varepsilon$ für jedes f mit $|f| \leq 1$.

Aus der Definition der unbedingten Konvergenz im starken Sinne folgt, dass jede unbedingt im starken Sinne konvergente Reihe stark konvergiert, und folglich ihre Summe ein Element desselben Raumes ist. In der Tat: eine Reihe konvergiere unbedingt im starken Sinne. Dann existiert für ein gegebenes ε ein solches n, dass für $m > n$ $\sum_{k=n+1}^{m} |f x_k| < \varepsilon$ für jedes f mit $|f| \leq 1$ ist. Folglich ist, für dieselben f, $\left|f \sum_{k=n+1}^{m} x_k\right| < \varepsilon$ und also auch $\left|\sum_{k=n+1}^{m} x_k\right| < \varepsilon$.

Satz 4. *Damit die Reihe* $\sum_{k=1}^{\infty} x_k$ *unbedingt im starken Sinne konvergiere, ist die gleichmässige Konvergenz der Reihengesamtheit* $\sum_{i=1}^{\infty} \varepsilon_i x_i$ $(\varepsilon_i = \pm 1)$ *in bezug auf alle* ε_i $(\varepsilon_i = \pm 1)$ *notwendig und hinreichend.*

Beweis. Die Reihe $\sum_{i=1}^{\infty} x_i$ konvergiere unbedingt im starken Sinne. Dann gibt es für jedes ε ein solches n, dass $\sum_{i=n}^{\infty} |fx_i| < \varepsilon$ ist für jedes f mit $|f| \leqslant 1$. Nun ist aber

$$|f \sum_{i=n}^{m} \varepsilon_i x_i| \leqslant \sum_{i=n}^{m} |fx_i| < \varepsilon,$$

und da dies für ein beliebiges f mit $|f| \leqslant 1$ gilt, so ist auch $\left| \sum_{i=n}^{m} \varepsilon_i x_i \right| < \varepsilon$. Das Umgekehrte ist ebenfalls leicht zu beweisen. Für ein beliebiges ε gibt es ein solches n, dass für $m \geqslant n$ $\left|\sum_{i=n}^{m} \varepsilon_i x_i\right| < \varepsilon$ $(\varepsilon_i = \pm 1)$ ist. Also haben wir

$$\sum_{i=n}^{m} |fx_i| = \sum_{i=n}^{m} \varepsilon_i fx_i = f. \sum_{i=n}^{m} \varepsilon_i x_i \leqslant |f| \, \varepsilon \quad (\varepsilon_i = \operatorname{sign} fx_i),$$

und folglich konvergiert die Reihe unbedingt im starken Sinne, w. z. b. w.

Satz 4a. *Damit die Reihe* $\sum_{i=1}^{\infty} x_i$ *unbedingt im starken Sinne konvergiere, ist die gleichmässige Konvergenz der Reihen* $x_{n_1} + \ldots + x_{n_k} + \ldots$ *in bezug auf die Indizes* $n_1, n_2, \ldots, n_k, \ldots$ *notwendig und hinreichend.*

Das ist folgendermassen zu verstehen: unabhängig von der Wahl der Folge $n_1, \ldots, n_k, \ldots$ gibt es für ein beliebiges ε ein solches N, dass $\left| \sum_{n_k > N} x_{n_k} \right| < \varepsilon$ ist. Dieser Satz wird analog dem vorhergehenden bewiesen.

Satz 5. *Damit die Reihe* $x_1 + \ldots + x_n + \ldots$ *unbedingt im starken Sinne konvergiere, ist notwendig und hinreichend, dass die Punktmenge* $\sum_{i=1}^{n} \varepsilon_i x_i$ *kompakt ist, wobei* $\varepsilon_i = \pm 1$ *ist und* n *eine natürliche Zahl.* ε_i *und* n *dürfen alle möglichen zulässigen Werte annehmen.*

Beweis. Wir zeigen die Notwendigkeit dieser Bedingung auf Grund des folgenden Satzes, der auf S. 268 bewiesen wird:

Damit die Punktmenge R des linearen Raumes E kompakt ist, ist die gleichmässige Konvergenz auf R einer jeden gegen Null schwach konvergenten Folge linearer Funktionale notwendig und hinreichend, d. h. für jedes ε muss ein solches N existieren, dass $|f_n x| < \varepsilon$ ausfällt für alle x aus R und alle $n > N$.

Die Reihe $\sum_{i=1}^{\infty} x_i$ konvergiere also unbedingt im starken Sinne. Es ist zu beweisen, dass die Menge $\sum_{i=1}^{N} \varepsilon_i x_i$ $(\varepsilon_i = \pm 1)$ kompakt ist. f_n konvergiere schwach gegen Null. Dann ist für ein beliebiges x $\lim_{n \to \infty} f_n x = 0$ und

$$\left| f_n \sum_{i=1}^{N} \varepsilon_i x_i \right| \leqslant \sum_{i=1}^{\infty} |f_n x_i| \,. \tag{1}$$

Da f_n schwach konvergiert, gibt es ein solches M, dass $|f_n| \leqslant M$. Nach der Definition der unbedingt starken Konvergenz gibt es ein solches k_0, dass

$\sum_{i=k_0}^{\infty} |f_n x_i| < \frac{\varepsilon}{2}$ für jedes n ist. Wegen $\lim_{n\to\infty} f_n x_i = 0$ existiert ein solcher Index n_0, dass für jedes $n > n_0$ und jedes $i = 1, 2, \ldots, k_0 - 1$

$$|f_n x_i| < \frac{\varepsilon}{2k_0}$$

ausfällt. Dann ist $\sum_{i=1}^{\infty} |f_n x_i| < \varepsilon$ für $n > n_0$, und folglich sind [nach (1)] die Voraussetzungen des zitierten Satzes erfüllt. Also ist die Menge $\sum_{i=1}^{n} \varepsilon_i x_i$ kompakt.

Satz 5a. *Damit die Reihe $\sum_{i=1}^{\infty} x_i$ unbedingt im starken Sinne konvergiere, ist es notwendig und hinreichend, dass die Punktmenge $\sum_{i=1}^{k} x_{n_i}$ kompakt ist, wobei $n_1, n_2, \ldots, n_k, \ldots$ eine beliebige anwachsende Folge ganzer Zahlen ist.*

Beweis. Dieser Satz folgt unmittelbar aus dem vorhergehenden Satz 5. Hierzu bemerken wir zunächst, dass, wenn eine gegebene Menge kompakt ist, die Menge jener Punkte, die die Summen oder Differenzen von Punkten erster Menge darstellen, ebenfalls kompakt ist. Die Notwendigkeit der Voraussetzung ist jetzt daraus ersichtlich, dass jeder Punkt $\sum_{i=1}^{k} x_{n_i}$ die halbe Summe zweier Punkte der Form $\sum_{n=1}^{n_k} \varepsilon_n x_n$ mit entsprechend gewählten n_k darstellt. Diese Voraussetzung ist aber auch hinreichend, da man jeden Punkt der Form $\sum_{n=1}^{N} \varepsilon_n x_n$ als Differenz zweier Punkte der Form $\sum x_{n_i}$ mit entsprechend gewählten n_i darstellen kann.

Folglich könnte die unbedingte und die unbedingt starke Konvergenz einer Reihe noch folgendermassen definiert werden:

Die Reihe $x_1 + \ldots + x_n + \ldots$ wird unbedingt konvergent genannt, wenn die Menge der Summen der Art $\sum_{i=1}^{n_k} x_i$ beschränkt ist ($n_1, \ldots, n_k, \ldots$ bedeutet dabei eine beliebige anwachsende Folge natürlicher Zahlen).

Die Reihe $x_1 + \ldots + x_n + \ldots$ wird unbedingt konvergent im starken Sinne genannt, wenn die Menge $\sum_{i=1}^{k} x_{n_i}$ kompakt ist.

Beispiel. Die unbedingte und die unbedingt starke Konvergenz im Raume stetiger Funktionen $[x_n = \varphi_n(t)]$.

Es kann gezeigt werden, dass in (C) die unbedingte Konvergenz der Reihe $\sum_{n=1}^{\infty} x_n$ der Konvergenz der Funktionenreihe $\sum |\varphi_n(t)|$ und der Beschränktheit ihrer Teilsummen äquivalent ist.

Die unbedingte Konvergenz im starken Sinne der Reihe $\sum x_n$ ist der gleichmässigen Konvergenz der Reihe $\sum |\varphi_n(t)|$ äquivalent.

§ 5. Volladditive Mengenfunktionen

Definition 1. *Wenn jeder Menge P eines gewissen Borelschen Mengenkörpers ein Punkt $x(P)$ des linearen Raumes entspricht, und wenn aus der Gleichung $P = P_1 + \ldots + P_n + \ldots$* ($P_n$ sind paarweise elementenfremd) *die Gleichung $x(P) = x(P_1) + \ldots + x(P_n) + \ldots$ im Sinne der schwachen Konvergenz folgt, so wird $x(P)$ eine abstrakte volladditive Mengenfunktion genannt*[10]. (Wir werden der Kürze halber auch einfach volladditive Mengenfunktion sagen).

Mit anderen Worten: *die Funktion $x(P)$ wird eine volladditive Mengenfunktion genannt, wenn $fx(P)$ für ein beliebiges lineares Funktional f eine reelle volladditive Mengenfunktion ist.*

Definition 2. *Ist die Wertmenge einer volladditiven Funktion $x(P)$ kompakt, so wird sie volladditiv im starken Sinne genannt.*

Satz 1. *Jede volladditive Mengenfunktion ist beschränkt*[9].

Beweis. Für jedes Funktional f ist die Funktion $fx(P)$ eine reelle volladditive Mengenfunktion, und folglich ist sie beschränkt. Setzen wir $p(f) = \sup |fx(P)|$. Dann ist $p(f)$ ein Funktional, welches allen Voraussetzungen des Hilfssatzes aus § 3 genügt. [Die Eigenschaften 1°), 2°), 3°) sind leicht nachzuprüfen. Die Halbstetigkeit folgt daraus, dass sich jedes $fx(P)$, und folglich auch $\sup |fx(P)|$, bei kleiner Änderung von f wenig ändert.]

Folglich ist $p(f)$ in der Einheitssphäre beschränkt. Es gibt also ein solches M, dass $|fx(P)| \leqslant M$ für jedes f mit $|f| \leqslant 1$. Das bedeutet, dass für ein beliebiges P $|x(P)| \leqslant M$ ist, w. z. b. w.

Satz 2. *Ist $x(P)$ eine volladditive Mengenfunktion im starken Sinne, so besteht für ein System paarweise elementenfremder Mengen $P_1, P_2, \ldots, P_n, \ldots$ die Gleichung $x(P) = x(P_1) + \ldots + x(P_n) + \ldots$ im Sinne der unbedingt starken Konvergenz.*

Beweis. Nach dem Satz 5a, § 4 genügt es zu beweisen, dass die Punktmenge $\sum_{i=1}^{k} x(P_{n_i})$ kompakt ist. Das folgt aber unmittelbar aus der Definition der starken Additivität.

§ 6. Funktionen mit beschränkter Variation

Definition 1. *Die Funktion x_t wird eine Funktion mit beschränkter Variation genannt, wenn für ein beliebiges Funktional f die reelle Funktion fx_t von beschränkter Variation ist.*

Satz 1. *Ist x_t eine Funktion mit beschränkter Variation, so gibt es ein solches M, dass $\operatorname{var}_t[fx_t] \leqslant M|f|$ bei beliebigem f.*

Beweis. $\operatorname{var}_t[fx_t]$ ist ein Funktional bezüglich f. Bezeichnen wir es mit $p(f)$. Um den Satz zu beweisen, ist nach dem Hilfssatz aus § 3 zu zeigen, dass $p(f)$ den Voraussetzungen dieses Hilfssatzes genügt. Es ist leicht nachzuprüfen, dass die Bedingungen 1°), 2°), 3°) erfüllt sind. Beweisen wir nun die Halbstetigkeit von $p(f)$, d. h., dass es für beliebige f_0 und ε ein solches

[10] Die Birkhoffsche Definition, [19], der vollen Additivität unterscheidet sich von der unseren nur dadurch, dass sie sich auf den Orliczschen Begriff der unbedingten Konvergenz stützt.

δ gibt, dass für jedes f mit $|f-f_0|<\delta$ die Ungleichung $\text{var}_t[f_0x_t]-$ $-\text{var}_t[fx_t]<\varepsilon$ besteht. Das beweisen wir folgendermassen.

Es gibt eine solche Zerlegung des Intervalls $a\leqslant t\leqslant b$ in Teilintervalle, dass

$$\left|\text{var}[f_0x_t]-\sum_{i=1}^{n}|f_0x_{t_{i+1}}-f_0x_{t_i}|\right|<\frac{\varepsilon}{2}$$

ist. Man kann nun ein solches δ wählen, dass

$$|f_0x_{t_{i+1}}-fx_{t_{i+1}}|<\frac{\varepsilon}{2n}\quad(i=0,1,\ldots,n)$$

ist, wenn $|f_0-f|<\delta$.

Dann ist

$$\left|\sum_{i=1}^{n}|fx_{t_{i+1}}-fx_{t_i}|-\sum_{i=1}^{n}|f_0x_{t_{i+1}}-f_0x_{t_i}|\right|<\frac{\varepsilon}{2},$$

und folglich

$$\text{var}_t[f_0x_t]-\sum_{i=1}^{n}|fx_{t_{i+1}}-fx_{t_i}|<\varepsilon.$$

Da aber

$$\text{var}_t[fx_t]=\sup\sum_{i=1}^{n}|fx_{t_{i+1}}-fx_{t_i}|$$

ist, gilt um so mehr $\text{var}_t[f_0x_t]-\text{var}_t[fx_t]<\varepsilon$, w. z. b. w.

S a t z 2. *Damit x_t eine Funktion mit beschränkter Variation sei, ist die Existenz eines solchen M notwendig und hinreichend, dass*

$$\left|\sum_{i=1}^{n}\varepsilon_i[x_{t_{i+1}}-x_{t_i}]\right|\leqslant M$$

für beliebige $\varepsilon_i=\pm1$ und jede Zerlegung des Intervalls (a,b) in Teilintervalle ist.

B e w e i s. Beweisen wir zuerst, dass diese Bedingung hinreichend ist. Aus

$$\left|\sum_{i=1}^{n}\varepsilon_i(x_{t_{i+1}}-x_{t_i})\right|\leqslant M$$

folgt

$$\left|\sum_{i=1}^{n}\varepsilon_i f(x_{t_{i+1}}-x_{t_i})\right|=\left|f\sum_{i=1}^{n}\varepsilon_i(x_{t_{i+1}}-x_{t_i})\right|\leqslant M|f|\quad(\varepsilon_i=\pm1),$$

d. h. fx_t ist eine reelle Funktion mit beschränkter Variation.

Ist umgekehrt x_t eine Funktion mit beschränkter Variation, so folgt aus dem Satz 1 dieses Paragraphen, dass $\text{var}_t[fx_t]\leqslant M|f|$ ist, d. h.

$$\left|f\sum_{i=1}^{n}\varepsilon_i(x_{t_{i+1}}-x_{t_i})\right|\leqslant\sum_{i=1}^{n}|f(x_{t_{i+1}}-x_{t_i})|\leqslant\text{var}_t[fx_t]\leqslant M|f|.$$

Dies gilt aber für ein beliebiges f, und also ist

$$\left|\sum\varepsilon_i(x_{t_{i+1}}-x_{t_i})\right|\leqslant M.$$

Satz 2a. *Damit* x_t *eine Funktion mit beschränkter Variation sei, ist die Existenz eines solchen* M *notwendig und hinreichend, dass*

$$\left|\sum_{i=1}^{n}(x_{t_i}-x_{s_i})\right|\leqslant M$$

für beliebige n *und* $a\leqslant s_1<t_1\leqslant s_2<\ldots\leqslant b$ *ist.*

Der Beweis dieses Satzes unterscheidet sich nicht wesentlich von dem des Satzes 2.

In konjugierten Räumen können zwei Definitionen der Funktion mit beschränkter Variation gegeben werden:

1°. f_t *wird eine Funktion mit beschränkter Variation genannt, wenn für ein beliebiges* ξ *aus* $\overline{\overline{E}}$ *die Funktion* ξf_t *von beschränkter Variation (nach* t*) ist.*

2°. f_t *wird eine Funktion mit beschränkter Variation genannt, wenn für ein beliebiges* x *aus* E *die Funktion* $f_t x$ *von beschränkter Variation ist.*

Beweisen wir nun den

Satz 3. *Die Definitionen 1° und 2° stimmen überein.*

Beweis. Es kann wie früher gezeigt werden, dass mit jeder der beiden Definitionen die Behauptung des Satzes 2 gültig bleibt. Folglich stimmen die beiden Definitionen überein. Man beachte, dass eine Funktion mit beschränkter Variation beschränkt ist.

In der Tat haben wir $\operatorname{var}[fx_t]\leqslant K|f|$, d. h.

$$|fx_t-fx_0|\leqslant K|f|,$$

und also

$$|x_t|\leqslant K+|x_0|.$$

Beispiel. Funktionen mit beschränkter Variation im Raum (C) der stetigen Funktionen.

Die Werte von x_t gehören (C) an. Das bedeutet, dass jedem fixierten t eine stetige Funktion entspricht. Mit anderen Worten ist x_t eine Funktion zweier Veränderlicher $K(s, t)$, die für jedes festgewählte t nach s stetig ist.

Damit x_t eine abstrakte Funktion mit beschränkter Variation sei, ist notwendig und hinreichend

1) dass $K(s, t)$ bei jedem s eine Funktion mit beschränkter Variation für t ist, und

2) dass es ein solches M gibt, dass $\operatorname{var}_t[K(s, t)]\leqslant M$ ist.

Wir gehen nun zur weiteren Betrachtung der Funktionen mit beschränkter Variation über.

Satz 4. *Ist* x_t *eine Funktion mit beschränkter Variation, und strebt* $t_n\to t$*, wobei* $t_n<t$ *bleibt, so konvergiert* x_{t_n} *schwach.*

Beweis. fx_t ist eine reelle Funktion mit beschränkter Variation. Folglich strebt $fx_{t_n}\to fx_t$, und da f beliebig ist, ist der Satz hiermit bewiesen.

Anmerkungen. 1°. Wenn E ausserdem ein schwach vollständiger Raum ist, so existieren in jedem Punkt x_{t+0} und x_{t-0} (im Sinne der schwachen Konvergenz).

2°. Wir haben bewiesen, dass die schwache Konvergenz in jedem Punkt stattfindet.

Führen wir jetzt ein Beispiel an, wo die starke Konvergenz nicht besteht.

Betrachten wir im Raume (c)[11] die folgende abstrakte Funktion

$$x_t = \{\varphi_n(t)\} = \begin{cases} 1, \text{ wenn } 0 \leqslant t \leqslant \frac{1}{n}, \\ 0, \text{ wenn } \frac{1}{n} < t \leqslant 1. \end{cases}$$

Für $t_n \longrightarrow 0$ konvergiert x_{t_n} schwach, die starke Konvergenz findet aber nicht statt.

3°. Wenn die Werte einer Funktion mit beschränkter Variation eine kompakte Menge bilden, so existiert in jedem Punkt ein starker Limes von rechts und von links.

Satz 5. *Ist E separabel, so hat f_t — eine Funktion mit beschränkter Variation, deren Werte $\overline{E}$ angehören, — nicht mehr als eine abzählbare Menge von Punkten schwacher Unstetigkeit.*

Beweis. $x_1, x_2, \ldots, x_n, \ldots$ sei eine überall dichte Menge in E. Bezeichnen wir mit P_n die Punktmenge der Unstetigkeiten der reellen Funktion $f_t x_n$. Ausserhalb der Menge $P = P_1 \dot{+} P_2 \dot{+} \ldots \dot{+} P_n \dot{+} \ldots$ ist die Funktion f_t schwach stetig. In der Tat, wenn t_0 der Menge P nicht angehört, konvergiert für eine beliebige Folge $t_k \longrightarrow t_0$ die Folge $f_{t_k} x_n$ gegen $f_{t_0} x_n$.

Die $|f_{t_n}|$ sind aber gleichmässig beschränkt (nach Satz 2 dieses Paragraphen), und dies ist für die schwache Konvergenz hinreichend.

Eine Funktion mit beschränkter Variation kann sogar in keinem Punkte eine schwache Ableitung besitzen.

Das kann durch ein einfaches Beispiel gezeigt werden: betrachten wir im Hilbertschen Raum die abstrakte Funktion

$$x_t = \varphi_t = \varphi_t(s) = \begin{cases} 1, \text{ wenn } s \leqslant t, \\ 0 \quad \text{"} \quad s > t; \end{cases}$$

x_t ist eine Funktion mit beschränkter Variation, die in keinem Punkt eine Ableitung besitzt.

Definition 2. *Wir sagen, dass x_t eine Funktion mit stark beschränkte Variation ist, wenn:*

1°) *x_t eine Funktion mit beschränkter Variation ist, und*

2°) *die Menge aller reellen Funktionen $f x_t$ mit $|f| \leqslant 1$ kompakt ist.*

Die Konvergenz in der Menge dieser reellen Funktionen bedeutet dabei, dass die volle Variation der Differenz zweier Funktionen zu Null strebt, d. h., dass *man aus einer jeden unendlichen Funktionenfolge*

$$f_1 x_t, \ldots, f_n x_t, \ldots, |f_n| \leqslant 1,$$

eine solche Teilfolge $f_{n_1} x_t, f_{n_2} x_t, \ldots, f_{n_k} x_t, \ldots$ auswählen kann, dass $\lim\limits_{k, l \to \infty} \operatorname{var}[f_{n_k} x_t - f_{n_l} x_t] = 0$ ist.

Satz 6. *Damit x_t eine Funktion mit stark beschränkter Variation sei, ist notwendig und hinreichend, dass die Summenmenge*

$$\sum_{i=0}^{n} \varepsilon_i [x_{t_{i+1}} - x_{t_i}] \qquad (1)$$

---

[11] (c) ist der Raum der konvergenten Folgen.

kompakt ist, wobei $\varepsilon_i = \pm 1$, $a = t_0 \leqslant t_1 \leqslant \ldots \leqslant t_{n+1} \leqslant b$, *und* ε_i *und* t_i *beliebige zulässige Werte annehmen dürfen.*

Beweis. Beweisen wir zunächst, dass für jede Funktion mit stark beschränkter Variation die Summenmenge (1) auf Grund des folgenden Merkmals der Kompaktheit, das im zweiten Teil dieser Arbeit bewiesen wird (§ 2, Satz 1), kompakt ist:

Damit die Punktmenge R des linearen Raumes E kompakt sei, ist die gleichmässige Konvergenz einer jeden gegen Null schwach konvergenten Folge linearer Funktionale $f_1, \ldots, f_n, \ldots$ auf dieser Menge notwendig und hinreichend, d. h. es muss für jedes ε ein solches N existieren, dass für beliebige $x \subset R$ und $n > N$ $|f_n x| < \varepsilon$ ist.

Wir haben also aus der schwachen Konvergenz von f_n gegen Null ihre gleichmässige Konvergenz auf einer Elementenmenge der Art (1) zu beweisen. Um die Behauptung zu beweisen, genügt es, wegen

$$\left| f_n \sum_{i=0}^{n} \varepsilon_i (x_{t_{i+1}} - x_{t_i}) \right| \leqslant \mathrm{var}_t [f_n x_t], \tag{2}$$

zu beweisen, dass $\lim \mathrm{var}_t [f_n x_t] = 0$ ist. Da f_n schwach konvergiert, gibt es ein solches M, dass $|f_n| \leqslant M$ gilt. Also ist, nach der Definition einer Funktion mit schwach beschränkter Variation, die Menge der reellen Funktionen $f_n x_t$ im Sinne der totalen Variation kompakt, und da infolge der schwachen Konvergenz für jedes t $\lim\limits_{n\to\infty} f_n x_t = 0$ gilt, ist $\lim\limits_{n\to\infty} \mathrm{var}_t [f_n x_t] = 0$. (Andernfalls müsste wegen der Kompaktheit eine von Null verschiedene Funktion existieren, gegen die eine Teilfolge aus $f_n x_t$ konvergieren würde, und folglich könnte $\lim\limits_{n\to\infty} f_n x_t = 0$ nicht bestehen.)

Beweisen wir nun, dass diese Voraussetzung auch hinreichend ist. Die Menge $\sum\limits_{i=0}^{n} \varepsilon_i (x_{t_{i+1}} - x_{t_i})$ sei also kompakt. Dann ist die Wertmenge der x_t ebenfalls kompakt. In der Tat, nehmen wir $n = 2$, $t_0 = a$, $t_1 = t$, $t_2 = b$, $\varepsilon_1 = +1$, $\varepsilon_2 = -1$ an, so wird die Elementenmenge $2x_t - x_a - x_b$ kompakt, und folglich ist auch die Wertmenge der x_t kompakt. Also ist der minimale lineare Teilraum, der Werte von x_t enthält, separabel. Nach dem Satz über die Fortsetzung linearer Funktionale hängt aber die Eigenschaft einer Funktion, eine Funktion mit stark beschränkter Variation zu sein, nur von dem minimalen linearen Teilraum ab, der die Werte der Funktion enthält. Es genügt also den Satz für einen separablen Raum zu beweisen.

E sei also ein separabler Raum. Es ist zu beweisen, dass eine unendliche Funktionenmenge $f_n x_t$ mit $|f_n| \leqslant 1$ eine im Sinne der totalen Variation konvergente Teilfolge enthält. Da $\bar{E}$ schwach kompakt ist[12], kann aus der gegebenen Menge der Funktionen die schwach konvergente Teilfolge $f_1, \ldots, f_n, \ldots, \lim f_n = f$ (im Sinne der schwachen Konvergenz), gewählt werden.

---

[12] Banach, l. c., S. 123.

Dann konvergiert, da die Menge $\sum_{i=0}^{n} \varepsilon_i (x_{t_{i+1}} - x_{t_i})$ kompakt ist, gemäss dem in diesem Beweis bereits zitierten Satz, die Folge $(f_k - f) \sum_{i=0}^{n} \varepsilon_i (x_{t_{i+1}} - x_{t_i})$ mit $n \to \infty$ gleichmässig gegen Null, unabhängig von den ε_i und der Zerlegung in Teilintervalle. Folglich existiert für ein gegebenes ε ein solches N, dass für alle $k > N$

$$\left| (f_k - f) \sum_{i=0}^{n} \varepsilon_i (x_{t_{i+1}} - x_{t_i}) \right| < \varepsilon$$

ausfällt unabhängig von den ε_i und der Zerlegung in Teilintervalle. Setzen wir $\varepsilon_i = \operatorname{sign} (f_k - f)(x_{t_{i+1}} - x_{t_i})$ und bestimmen die obere Schranke für alle Zerlegungen in Teilintervalle, so erhalten wir

$$\operatorname{var}_t [f_k x_t - f x_t] < \varepsilon$$

für alle $k > N$. Also kann aus jeder unendlichen Funktionenmenge $f x_t$ mit $|f| \leqslant 1$ eine Teilfolge der im Sinne der totalen Variation konvergierenden Funktionen ausgewählt werden, w. z. b. w.

Satz 6a. *Damit x_t eine Funktion mit stark beschränkter Variation sei, ist notwendig und hinreichend, dass die Summenmenge $\sum_{i=1}^{n} (x_{t_i} - x_{s_i})$ kompakt ist, wobei $(s_1, t_1), (s_2, t_2), \ldots, (s_n, t_n)$ ein beliebiges System paarweise disjunkter Intervalle bedeutet.*

Wie der Satz 5a, § 4, wird auch dieser Satz aus Satz 5, § 4, abgeleitet.

Folgerungen aus Satz 6. 1°. Die Werte der Funktion x_t mit stark beschränkter Variation bilden eine kompakte Menge.

2°. Ist x_t eine Funktion mit stark beschränkter Variation, so existiert in jedem Punkt ein starker linker und rechter Limes.

Beweis. Dies folgt aus der vorangehenden Anmerkung, dem Satz 4 dieses Paragraphen und der Anmerkung 3° zu Satz 4.

3°. Jede Funktion mit stark beschränkter Variation hat eine nicht mehr als abzählbare Punktmenge starker Unstetigkeiten.

Beweis. Diese Anmerkung folgt aus Satz 5 dieses Paragraphen, de Anmerkung 1° zu Satz 6 und der Anmerkung 3° zu Satz 4. Da in Satz 5 die Separabilität von E gefordert wird, betrachten wir den minimalen, die Werte von x_t enthaltenden, Teilraum.

Satz 7. *$x(P)$ sei eine volladditive Mengenfunktion [die Mengen mögen auf dem Intervall (a, b) liegen und (B) messbar sein]. P_t sei das Intervall (a, t). Dann ist die abstrakte Funktion $x_t = x(P_t)$ eine Funktion mit beschränkter Variation.*

Beweis. $fx(P)$ ist eine reelle volladditive Funktion. Folglich ist $fx(P_t) = fx_t$ eine reelle Funktion mit beschränkter Variation. Da dies für ein beliebiges f gilt, ist x_t definitionsgemäss eine Funktion mit beschränkter Variation.

Satz 8. *$x(P)$ sei eine volladditive Mengenfunktion im starken Sinne. P_t sei das Intervall (a, t). Dann ist $x_t = x(P_t)$ eine Funktion mit stark beschränkter Variation.*

Beweis. Die Werte von $x(P)$ bilden eine kompakte Menge. Folglich bilden die Summen $\sum_{i=1}^{n} (x_{t_i} - x_{s_i})$, wo $s_1 \leqslant t_1 \leqslant \ldots \leqslant s_n \leqslant t_n$, ebenfalls eine kompakte Menge, und nach Satz 6a ist x_t eine Funktion mit stark beschränkter Variation.

Wir wollen nun die folgenden Sätze 7a und 8a beweisen, die im gewissen Sinne die Umkehrungen der Sätze 7 und 8 darstellen:

Satz 7a. *Ist E ein schwach vollständiger Raum, so entspricht jeder Funktion x_t mit stark beschränkter Variation eine volladditive Mengenfunktion mit Werten aus E. Ist E kein schwach vollständiger Raum, so gehören die Werte dieser volladditiven Mengenfunktion im allgemeinen dem $\overline{\overline{E}}$ an.*

Beweis. Wir beweisen zuerst die letzte Behauptung des Satzes 7a.

x_t sei also eine Funktion mit beschränkter Variation. Dann ist fx_t eine reelle Funktion mit beschränkter Variation. Folglich entspricht ihr eine gewisse volladditive Mengenfunktion $\varphi(P, f)$, die, natürlich, von der Wahl des Funktionals f abhängt.

Es ist klar, dass

$$\varphi(P, f_1 + f_2) = \varphi(P, f_1) + \varphi(P, f_2),$$
$$\varphi(P, af) = a\varphi(P, f)$$

gilt. $\varphi(P, f)$ ist ein stetiges Funktional von f, denn wir haben

$$|\varphi(P, f)| \leqslant \text{var}_t[\,x_t] \leqslant M|f|.$$

Also kann $\varphi(P, f)$ folgendermassen dargestellt werden:

$$\varphi(P, f) = \xi(P) f,$$

wo $\xi(P)$ eine abstrakte volladditive Funktion ist, deren werte $\overline{\overline{E}}$ angehören [nach der Definition der Funktion $\varphi(P, f)$ gilt die Gleichung

$$(P_1 \dotplus P_2 \dotplus \ldots \dotplus P_n \dotplus \ldots) f = \xi(P_1) f + \ldots + \xi(P_n) f + \ldots$$

für ein beliebiges f und paarweise elementenfremde Mengen $P_1, P_2, \ldots, P_n, \ldots$; das bedeutet aber, dass $\xi(P)$ im schwachen Sinne volladditiv ist].

Satz 8a. *Für jede Funktion mit stark beschränkter Variation kann eine ihr entsprechende stark volladditive Mengenfunktion konstruiert werden.*

Nach dem vorhergehenden Satz entspricht einer Funktion mit stark beschränkter Variation eine volladditive Funktion $\xi(P)$ mit Werten aus $\overline{\overline{E}}$. Es ist vorerst zu beweisen, dass diesen Punkten gewisse Punkte aus E entsprechen. Hierzu reicht es zu zeigen, dass aus $f_n \to 0$ (schwach) $\xi(P) f_n \to 0$ folgt. Aber

$$|\xi(P) f_n| \leqslant \text{var}_t [f_n x_t], \tag{1}$$

und die rechte Seite strebt zu Null, da für jedes t $f_n x_t \to 0$, und, ausserdem, nach der Definition der Funktionen mit stark beschränkter Variation, die Funktionenfolge $f_n x_t$ kompakt ist.

Beweisen wir nun die Kompaktheit der Wertmenge $x(P)$ [statt $\xi(P)$ schreiben wir jetzt $x(P)$]. Hierzu ist es, nach Satz 1, § 2 des zweiten Teils dieser

Arbeit, hinreichend, die gleichmässige Konvergenz von $f_n x(P)$ in bezug auf die Wertmenge der $x(P)$ für jede zu Null schwach konvergierende Folge f_n zu zeigen. Das aber folgt wieder aus (1).

§ 7. Das Lebesguesche Integral

Wir führen jetzt den Begriff des Lebesgueschen Integrals einer abstrakten Funktion x_t ein.

Definition 1. *Die Funktion x_t, deren Werte E angehören, wird summierbar genannt, wenn für ein beliebiges lineares Funktional f die reelle Funktion fx_t summierbar ist.*

Definition 2. *Die Funktion f_t wird (*) summierbar genannt, wenn für ein beliebiges x die reelle Funktion $f_t x$ summierbar ist.*

In $\bar{E}$ ist auch die Definition 1 anwendbar. Sie sieht dann natürlich so aus: die Funktion f_t wird summierbar genannt, wenn für ein beliebiges ξ aus $\bar{\bar{E}}$ die Funktion ξf_t summierbar ist.

Folgender Hilfssatz ist klar:

Hilfssatz 1. *Jede summierbare Funktion f_t ist (*) summierbar.*

Beschäftigen wir uns jetzt damit, was man als Integral einer summierbaren Funktion bezeichnen kann. Diese Frage zu beantworten hilft der folgende

Satz 1. *Wenn x_t summierbar ist, so gibt es ein solches M, dass für ein beliebiges f die Ungleichung*

$$\int_a^b |fx_t|\,dt \leqslant M|f|$$

besteht.

Den Satz 1 werden wir etwas später beweisen.

Satz 1a. *Wenn x_t summierbar und E separabel ist, so gibt es ein solches M, dass für ein beliebiges f die Ungleichung*

$$\int_a^b |fx_t|\,dt \leqslant M|f|$$

gilt[13].

Beweis. Nach dem Hilfssatz aus § 3 reicht es zu zeigen, dass

$$P(f) = \int_a^b |fx_t|\,dt$$

ein halbstetiges Funktional von f ist. Es ist also zu beweisen, dass es für beliebige f_0 und ε ein solches δ gibt, dass für jedes f mit $|f - f_0| < \delta$ die Ungleichung

$$\int_a^b |f_0 x_t|\,dt - \int_a^b |fx_t|\,dt < \varepsilon$$

[13] Der Beweis des Satzes 1 unterscheidet sich von den Beweisen der Sätze 1a und 1b nur deshalb, weil die Menge P_N in diesem Falle nicht notwendigerweise messbar zu sein braucht.

besteht. Bezeichnen wir mit P_N die Menge jener t, für die $|x_t| \leq N$ ist; N kann man so wählen, dass

$$\int_a^b |f_0 x_t|\, dt - \int_{P_N} |f_0 x_t|\, dt < \frac{\varepsilon}{2}$$

ausfällt. Da $|f - f_0| < \delta$ ist, gilt

$$|f x_t - f_0 x_t| < N|f - f_0| < N\delta$$

für alle t, die P_N angehören, d. h.

$$\int_a^b |f_0 x_t|\, dt - \int_a^b |f x_t|\, dt < \frac{\varepsilon}{2} + \int_{P_N} |f_0 x_t|\, dt - \int_{P_N} |f x_t|\, dt < \frac{\varepsilon}{2} + N\delta(b-a).$$

Nimmt man $\delta = \frac{\varepsilon}{2N(b-a)}$, so erhält man die gesuchte Ungleichung. Die Messbarkeit der Menge P_N folgt aus der Separabilität von E. Folglich ist nach Satz 1, § 2, x_t stark messbar, und also $|x_t|$ nach dem Hilfssatz 1, § 2, messbar.

Satz 1b. *Wenn E separabel und f_t (*) summierbar ist, so gibt es ein solches M, dass*

$$\int_a^b |f_t x|\, dt \leq M|x|$$

gilt.

Beweis. Der Beweis unterscheidet sich von dem vorhergehenden nur dort, wo die Messbarkeit der Menge P_N jener t, für die $|f_t| \leq N$ ist, bewiesen wird. Das kann man aber folgendermassen zeigen: $x_1, \ldots, x_n, \ldots$ sei eine in der Einheitssphäre des Raumes E überall dichte Menge; dann ist

$$|f_t| = \sup_{n=1,2,\ldots} |f_t x_n|,$$

und da $f_t x_n$ eine messbare Funktion ist, ist die Menge $P_N^{(n)}$ jener t, für welche $|f_t x_n| \leq N$, messbar. P_N, als der Durchschnitt aller $P_N^{(n)}$, ist also auch messbar.

Hilfssatz 2. *Wenn $\varphi_1(t), \ldots, \varphi_n(t), \ldots$ eine Folge summierbarer Funktionen ist, für welche*

1) *fast überall* $\sup\limits_{n=1,2,\ldots} |\varphi_n(t)| < +\infty$ *und*

2) $\sup \int_a^b |\varphi_n \quad dt = +\infty$,

so existiert eine solche Folge $c_1, \ldots, c_n, \ldots$ mit $\sum |c_n| < +\infty$, dass die Funktion

$$\psi(t) = \sum_{n=1}^{\infty} c_n \varphi_n(t)$$

nicht summierbar ist.

Beweis. Betrachten wir den Raum (m), dessen Elemente f beschränkte Folgen sind: $f = (a_1, \ldots, a_n, \ldots)$, und $|f| = \sup\limits_{n=1,2,\ldots} |a_n|$. Er ist zu dem Raum $l^{(1)}$ konjugiert. [Elemente des Raumes $l^{(1)}$ sind absolut konvergente Folgen $x = (c_1, \ldots, c_n, \ldots)$. Die Norm ist dabei

$$|x| = \sum_{n=1}^{\infty} |c_n|.]$$

Es sei eine abstrakte Funktion $f_t = (\varphi_1(t), \ldots, \varphi_n(t), \ldots)$ gegeben, wo die $\varphi_n(t)$ den Bedingungen des Hilfssatzes genügen. Ihre Werte gehören (m) an, und sie ist nicht (*) summierbar. In der Tat, wenn wir $x_k = (0, \ldots, 1, 0, \ldots)$ setzen, so sehen wir, dass die Integrale

$$\int_a^b |f_t x_k| \, dt = \int_a^b |\varphi_k(t)| \, dt$$

in k nicht beschränkt sind, was dem Satz Ib widerspricht. Folglich gibt es ein solches Element x, dass $f_t x$ nicht summierbar ist, d. h. dass es eine solche Folge $c_1, \ldots, c_n, \ldots$ mit $\sum |c_n| < +\infty$ gibt, dass die Funktion $\psi(t) = \sum c_n \varphi_n(t)$ nicht summierbar ist.

Satz 1. *Ist die Funktion x_t summierbar, so gibt es ein solches M, dass*

$$\int_a^b |f x_t| \, dt \leqslant M |f|$$

für beliebige f ist.

Beweis. Zum Beweis benötigen wir lineare Räume, die wir jetzt konstruieren werden. P sei eine Menge gewisser Elemente $\alpha, \beta, \ldots$. Bilden wir den Raum $l_p^{(1)}$ der reellen Funktionen $\varphi(\alpha)$, die auf der Menge P definiert sind und den folgenden Bedingungen genügen:

1) $\varphi(\alpha)$ ist nicht mehr als in einer abzählbaren Punktmenge von Null verschieden.

2) $\sum |\varphi(\alpha)| < +\infty$, wo die Summe sich über alle α, für die $\varphi(\alpha) \neq 0$, erstreckt. Als Betrag $|\varphi|$ des Elementes $\varphi(\alpha)$ nehmen wir $|\varphi| = \sum |\varphi(\alpha)|$.

Ein so definierter Raum ist ein (B) Raum, wie man leicht einsieht. Zu ihm ist der Raum M_P aller reellen Funktionen $\varphi(\alpha)$, die auf P definiert und beschränkt sind, konjugiert. Die Norm des Elementes $\varphi(\alpha)$ ist $|\varphi| = \sup_{\alpha \subset P} |\varphi(\alpha)|$. Konjugiert zu M_P ist der Raum, dessen Elemente $\psi(R)$ additive (im allgemeinen nicht volladditive) Mengenfunktionen von R sind[14] (R ist eine Teilmenge von P). Dabei muss die Norm folgendermassen definiert werden: $|\varphi| = \operatorname{var} \varphi(R)$. Wir gehen nun zum Beweis des Satzes über.

Jeder lineare Raum E ist ein Teilraum eines gewissen Raumes M_P, wo P eine passend gewählte Menge ist. In der Tat: betrachten wir eine Menge linearer Funktionale f, die in der Einheitssphäre des Raumes E überall dicht ist. Diese Menge nehmen wir als die Menge P. Dann entspricht jedem Element x eine auf P beschränkte Funktion, nähmlich die Gesamtheit der Werte fx, wo x die Menge P durchläuft. Die Norm des Elementes ist in diesem Falle

$$|x| = \sup_{f \subset P} |fx|.$$

Da die Definition der Summierbarkeit einer abstrakten Funktion x_t eine innere Eigenschaft ist, die (nach dem Satz von der Fortsetzung eines linearen Funktionals) nur von dem minimalen linearen Teilraum abhängt, der die Werte von x_t enthält, genügt es den Satz 1 für summierbare Funktionen im Raume M_P zu beweisen. Es sei also $x_t = \varphi(t, \alpha)$ eine abstrakte Funktion, deren Werte M_P

[14] Fichtenholz et Kantorovitch, Studia mathematica, 5, (1935), 69—98.

angehören. Die Summierbarkeit von x_t bedeutet folgendes: für eine beliebige additive Mengenfunktion $\psi(R)$ $(R \subset P)$ ist $\int \varphi(t, \alpha)\, d_\alpha \psi(R)$[15] eine summierbare Funktion von t. Zeigen wir, dass $\varphi(t, \alpha)$ bei jedem fixierten α summierbar ist, und dass es ein solches M gibt, dass

$$\int_a^b |\varphi(t, \alpha)|\, dt \leqslant M$$

für jedes α gilt[16].

Es ist klar, dass für jedes α die Funktion $\varphi(t, \alpha)$ nach t summierbar ist. In der Tat, ordnen wir jeder Funktion $\varphi(\alpha)$ ihre Werte in einem gewissen fixierten Punkte α zu. Das ist ein lineares Funktional im Raume M_P. Da das lineare Funktional einer abstrakten Funktion nach der Definition eine summierbare Funktion ist, ist auch $\varphi(t, \alpha)$ nach t summierbar. Jetzt muss die Existenz eines solchen M, dass

$$\int_a^b |\varphi(t, \alpha)|\, dt \leqslant M,$$

bewiesen werden. Nehmen wir das Gegenteil an. Dann gibt es eine solche Folge $\alpha_1, \ldots, \alpha_n, \ldots$, dass

$$\int_a^b |\varphi(t, \alpha_n)|\, dt > n.$$

In diesem Falle existiert nach dem Hilfssatz 2 dieses Paragraphen eine solche Folge $c_1, \ldots, c_n, \ldots$, dass $\sum |c_n| < +\infty$ und die Funktion

$$\chi(t) = \sum_{n=1}^{\infty} c_n \varphi(t, \alpha_n)$$

nicht summierbar ist. Betrachten wir jetzt das Element $\varphi(\alpha)$ des Raumes $l_P^{(1)}$:

$$\varphi(\alpha_n) = c_n, \quad \varphi(\alpha) = 0, \text{ wenn } \alpha \neq \alpha_1, \text{ oder } \alpha \neq \alpha_2, \ldots$$

Dann ist die Funktion $\sum_\alpha \varphi(\alpha)\varphi(t, \alpha)$ nicht summierbar (die Summe erstreckt sich über alle α, für die $\varphi(\alpha) \neq 0$ ist).

Folglich ist $\varphi(t, \alpha)$ nicht (*) summierbar und nach dem Hilfssatz 1 dieses Paragraphen auch schlechthin nicht summierbar. Wir kamen also zu einem Widerspruch und haben damit die Existenz eines solchen M bewiesen, dass

$$\int_a^b |\varphi(t, \alpha)|\, dt \leqslant M$$

ist. Beweisen wir jetzt die Ungleichung (1). Es ist zu zeigen, dass

$$\int_a^b \left| \int_P \varphi(t, \alpha)\, d_\alpha \psi(R) \right| dt \leqslant M \operatorname{var}_P \psi(R),$$

[15] Da es möglich ist, ein jedes lineare Funktional im Raume M_P in der Form $\int \varphi(\alpha)\, d_\alpha \psi(R)$ darzustellen [$\varphi(\alpha)$ ist ein Punkt des Raumes M_P, $\psi(R)$ — eine additive Mengenfunktion]. Das Integral wird hier im Sinne von Fréchet verstanden.

[16] Es kann gezeigt werden, dass für die Summierbarkeit der abstrakten Funktion x_t, deren Werte M_P angehören, diese Voraussetzungen auch hinreichend sind.

wo die Variation der Funktion $\psi(R)$ bezüglich der Menge P genommen wird. Aber

$$\int_a^b \left| \int_P \varphi(t, \alpha)\, d_\alpha \psi(R) \right| dt \leqslant \int_a^b \int_P |\varphi(t, \alpha)|\, d_\alpha \operatorname{var}_R \psi(R_1)\, dt =$$

$$= \int_P \int_a^b |\varphi(t, \alpha)|\, dt\, d_\alpha \operatorname{var}_R \psi(R_1) \leqslant \int_P M\, d_\alpha \operatorname{var}_R \psi(R_1) = M \operatorname{var}_P \psi(R),$$

und wenn man dies in die Sprache der Theorie der abstrakten Funktionen übersetzt, so erhält man die Ungleichung (1).

Definition 3. *Das Lebesguesche Integral einer abstrakten Funktion* x_t *wird ein solches Element* ξ *des Raumes* $\overline{\overline{E}}$ *genannt, dass für ein beliebiges Funktional* f *die Gleichung*

$$\xi f = \int_a^b f x_t\, dt$$

besteht.

Satz 2. *Jede summierbare Funktion besitzt ein Integral.*

Beweis. $\int_a^b f x_t\, dt$ ist ein lineares Funktional bezüglich f. In der Tat ist nach Satz 1

$$\left| \int_a^b f x_t\, dt \right| \leqslant \int_a^b |f x_t|\, dt \leqslant M \; f \,.$$

Bezeichnen wir $\int_a^b f x_t\, dt$, als ein Funktional bezüglich f, mit ξf. Dann ist ξ das Integral von x_t.

Anmerkung. Es ist möglich, dass $\int_a^b x_t\, dt$ schon ein Element aus E und nicht nur aus $\overline{\overline{E}}$ ist.

Führen wir ein Beispiel an, wo das Integral nur ein Element aus $\overline{\overline{E}}$, aber nicht aus E ist. Betrachten wir den Raum (c_0), dessen Elemente x gegen Null konvergente Folgen sind:

$$x = (a_1, \ldots, a_n, \ldots), \text{ wo } |x| = \sup_{n=1,2,\ldots} |a_n| \text{ und } \lim a_n = 0.$$

Konstruieren wir in (c_0) die folgende abstrakte Funktion:

$$x_t = \varphi_n(t), \text{ wo } \varphi_{2n}(t) \equiv 0 \text{ und } \varphi_{2n+1}(t) = \begin{cases} 0 & \text{für } t = 0, \\ 2n & \;\;\text{"}\;\; 0 < t \leqslant \frac{1}{2n}, \\ 0 & \;\;\text{"}\;\; t > \frac{1}{2n}. \end{cases}$$

Es ist leicht nachzuprüfen, dass

$$\int_a^b x_t\, dt = \xi_0$$

ist, wo $\xi_0 = (1, 0, 1, \ldots)$ nicht dem Raum (c_0) angehört, sondern nur dem Raum (m) der beschränkten Folgen, der der zweite zu (c_0) konjugierte Raum ist.

Satz 3. *Ist der gegebene Raum E zu einem anderen konjugiert, so ist das Integral einer summierbaren Funktion ein Element desselben Raumes.*

Beweis. In der Tat, ist f_t summierbar, so ist das Integral

$$\int_a^b |f_t x|\, dt \leqslant M|x|,$$

wo M eine passend gewählte Konstante ist. Folglich ist $\int_a^b f_t x\, dt$ ein lineares Funktional bezüglich x, d. h. ein Element aus $\bar{E}$.

Satz 4[17]. *Wenn* $\int_a^b |x_t|\, dt$ *existiert, und* x_t *schwach messbar ist, so ist* x_t *summierbar, und* $\int_a^b x_t\, dt$ *ist ein Element aus* E *(und nicht nur aus* $\bar{\bar{E}}$*).*

Beweis. Zunächst *ist* x_t summierbar, denn für ein beliebiges f ist $|fx_t| \leqslant$ $\leqslant |f|\,|x_t|$, und da fx_t messbar und durch eine summierbare Funktion beschränkt ist, so ist fx_t summierbar.

Beweisen wir jetzt, dass $\int_a^b x_t\, dt$ ein Element aus E ist. Es sei

$$\int_a^b x_t\, dt = \xi.$$

Es ist die Existenz eines solchen Elementes x zu beweisen, dass für ein beliebiges f die Gleichung $\xi f = fx$ besteht. Dazu reicht es die schwache Stetigkeit von ξf bezüglich f zu zeigen, d. h. aus der schwachen Konvergenz von f_n gegen 0 die Konvergenz von ξf_n gegen 0 abzuleiten[18]. Die Folge f_n konvergiere also schwach gegen 0. Dann gibt es ein solches M, dass $|f_n| \leqslant M$ ist. Ferner ist

$$\xi f_n = \int_a^b f_n x_t\, dt, \quad |f_n x_t| \leqslant M|x_t|,$$

und dies bedeutet, dass $f_n x_t$ eine konvergente Folge summierbarer Funktionen ist, die durch die summierbare Funktion $M|x_t|$ beschränkt sind. Folglich ist es gestattet, zum Limes unter dem Integralzeichen überzugehen, und so erhalten wir $\lim_{n\to\infty} \xi f_n = \xi f$, w. z. b. w.

Satz 5. *Ist E ein separabler schwach vollständiger Raum, so ist* $\int_a^b x_t\, dt$ *ein Element aus* E (x_t ist dabei eine beliebige summierbare Funktion).

---

[17] Dieser Satz zeigt, dass eine jede im Bochnerschen Sinne integrierbare Funktion auch in unserem Sinne integrierbar ist.

[18] Birkhoff und Bochner definieren die Funktion mit beschränkter Variation als eine Funktion, die der Bedingung (A) (siehe S. 261) genügt. Unsere Definition der Funktion mit beschränkter Variation unterscheidet sich von der Birkhoffschen.

Beweis. Betrachten wir die Funktion x_t^N, die folgendermassen definiert ist:

$$x_t^N = \begin{cases} 0, & \text{wenn } |x_t| > N, \\ x_t & \text{„} \quad |x_t| \leq N; \end{cases}$$

x_t^N ist schwach messbar (siehe den Beweis des Satzes 1a), und also, nach Satz 1, § 2, ist x_t stark messbar. Nach Hilfssatz 1, § 2, folgt daraus, dass die Funktion $|x_t^N|$ messbar ist. Da sie auch beschränkt ist, existiert das Integral $\int_a^b |x_t^N|\,dt$, das, nach Satz 4 dieses Paragraphen, ein Element aus E ist. Für jedes f gilt die Gleichung

$$\lim_{N\to\infty} \int_a^b f x_t^N\,dt = \int_a^b f x_t\,dt,$$

und das bedeutet, dass $\int_a^b x_t\,dt$, als schwacher Limes der Elementenfolge $\int_a^b x_t^N\,dt$ aus E, auch ein Element aus E ist.

Wir gehen nun zu weiteren Eigenschaften des Lebesgueschen Integrals über.

Satz 6. *Ist E separabel, und fallen die unbestimmten Integrale zweier Funktionen $f_t^{(1)}$ und $f_t^{(2)}$ zusammen, so fallen auch die Funktionen selbst fast überall zusammen.*

Beweis. $x_1, x_2, \ldots, x_n, \ldots$ sei eine in E überall dichte Menge. Bezeichnen wir mit P_n jene Menge, auf der $f_t^{(1)} x_n \neq f_t^{(2)} x_n$ ist. Dann fallen $f_t^{(1)} x_n$ und $f_t^{(2)} x_n$ bei beliebigem n zusammen, wenn nur t der Menge $P = P_1 \dot{+} \ldots \dot{+} P_n \dot{+} \ldots$ nicht angehört. Da die Folge $x_1, \ldots, x_n, \ldots$ überall dicht in E ist, so ist $f_t^{(1)} = f_t^{(2)}$ ausserhalb P, w. z. b. w.

Man kann ein Beispiel anführen, wo die unbestimmten Integrale zweier abstrakter Funktionen zusammenfallen, dagegen die Funktionen selbst in keinem Punkte zusammenfallen.

Satz 7. *Das unbestimmte Integral ist eine Funktion mit beschränkter Variation.*

Beweis. Es sei $\xi_t = \int_a^t x_s\,ds$. Dann ist $\xi_t f = \int_a^t f x_s\,ds$. Da die Funktion $\xi_t f$ das unbestimmte Integral einer reellen Funktion ist, so besitzt sie eine beschränkte Variation, und da f beliebig ist, hat auch ξ_t eine beschränkte Variation.

Analog lässt sich beweisen, dass $\int_P x_t\,dt$ eine volladditive Mengenfunktion ist.

§ 8. Das Stieltjessche und das Lebesgue-Stieltjessche Integral

Das Stieltjessche Integral wird folgendermassen definiert:

Definition. *x_t sei eine abstrakte und $\varphi(t)$ eine reelle Funktion. Das Integral $\int_a^b \varphi(t)\,dx_t$ wird ein solches Element ξ aus $\overline{\overline{E}}$ genannt, dass für ein beliebiges f die Gleichung $\xi f = \int_a^b \varphi(t)\,df x_t$ gilt.*

Satz 1. *Ist $\varphi(t)$ stetig, und besitzt x_t eine beschränkte Variation, so existiert das Integral $\int_a^b \varphi(t)\, dx_t$.*

Beweis. $\int_a^b \varphi(t)\, dfx_t$ existiert, da fx_t eine reelle Funktion mit beschränkter Variation ist. Ferner,

$$\left| \int_a^b \varphi(t)\, dfx_t \right| \leqslant \max |\varphi(t)| \cdot \mathrm{var}_t [fx_t],$$

oder, nach Satz 1, § 6,

$$\left| \int_a^b \varphi(t)\, dfx_t \right| \leqslant M \cdot \max |\varphi(t)| \cdot |f|,$$

wo M eine passend gewählte Konstante ist. Folglich ist $\int_a^b \varphi(t)\, df_t$ ein lineares Funktional bezüglich f. Man könnte diesen Satz auch anders beweisen, und zwar könnte man die Existenz des schwachen Limes des Ausdrucks

$$\sum_{i=1}^{n} \varphi(t_i)(x_{t_{i+1}} - x_{t_i})$$

zeigen, wenn die Intervallenlängen zu Null streben.

Daraus folgt

Satz 2. *Ist E schwach vollständig, so ist*

$$\int_a^b \varphi(t)\, dx_t$$

ein Element aus E (und nicht nur aus $\overline{\overline{E}}$).

Wir sagen, dass die Funktion $\varphi(t)$ messbar bezüglich x_t (einer Funktion mit beschränkter Variation) ist, wenn $\varphi(t)$ messbar bezüglich einer beliebigen reellen Funktion fx_t ist. Ist speziell $\varphi(t)$ (B) messbar, so ist sie messbar bezüglich einer beliebigen Funktion mit beschränkter Variation.

Analog dem Satz 1 kann man den folgenden Satz beweisen:

Satz 3. *Besitzt die Funktion x_t eine beschränkte Variation, und ist $\varphi(t)$ beschränkt und bezüglich x_t messbar, so existiert das Integral*

$$\int_a^b \varphi(t)\, dx_t.$$

Satz 4. *Besitzt die Funktion x_t eine stark beschränkte Variation, und ist $\varphi(t)$ beschränkt, so ist $\int_a^b \varphi(t)\, dx_t$ ein Element aus E (und nicht nur aus $\overline{\overline{E}}$).*

§ 9. Differentiation abstrakter Funktionen

Der starke Limes des Ausdruckes $\frac{x_{t+h} - x_t}{h}$ bei $h \to 0$ wird die starke Ableitung der abs rakten Funktion x_t im Punkte t genannt.

Der schwache Limes desselben Ausdrucks wird die schwache Ableitung von x_t genannt.

Wir sagen, dass *die abstrakte Funktion x_t der Bedingung* (A) *genügt, wenn es ein solches M gibt, dass für eine beliebige Zerlegung* $(t_0, \ldots, t_n)$ *des Intervalls* (a, b) *in Teilintervalle die Ungleichung*

$$\sum_{i=0}^{n} |x_{t_{i+1}} - x_{t_i}| \leqq M \tag{A}$$

besteht.

Das Hauptresultat dieses Paragraphen bilden die folgenden Sätze:

Satz 1[19]. *Ist E ein regulärer Raum (d. h. wenn es für ein beliebiges ξ aus $\bar{\bar{E}}$ ein solches x aus E gibt, dass für jedes f die Gleichung $\xi f = fx$ besteht), so besitzt eine jede abstrakte Funktion, die der Bedingung* (A) *genügt, fast überall eine starke Ableitung.*

Satz 2. *Ist E separabel, so besitzt eine abstrakte Funktion f_t, deren Werte $\bar{E}$ angehören, und die der Bedingung* (A) *genügt, fast überall eine schwache Ableitung.*

(Siehe auch Hilfssatz 5 und Satz 1a dieses Paragraphen.)

Wir skizzieren zuerst den Plan der Beweise der Sätze 1 und 2: Punkt 1 enthält einige einleitende Bemerkungen. In Punkt 2 wird bewiesen, dass eine Funktion x_t, die der Bedingung (A) genügt, fast überall eine schwache Ableitung hat. Hiernach beweisen wir in Punkten 3 und 4, dass bei der nachträglichen Forderung der „absoluten Stetigkeit" der Funktion x_t diese das Integral ihrer Ableitung im Bochnerschen Sinne[4] ist. Daraus ergibt sich, dass x_t fast überall eine starke Ableitung besitzt. In Punkt 5 befreien wir uns dann von dieser nachträglichen Forderung. Punkt 6 ist einigen Gegenbeispielen gewidmet. In Punkt 7 werden charakteristische Eigenschaften des unbestimmten Integrals im Bochnerschen Sinne angeführt.

1. Eine Funktion x_t, die der Bedingung (A) genügt, bestimmt im Raume E eine gewisse Kurve. Die Bedingung (A) bedeutet, dass diese Kurve rektifizierbar ist. Die Länge s der Kurve x_t ist dabei die obere Schranke des Ausdrucks

$$\sum_{i=0}^{n} | x_{t_{i+1}} - x_{t_i} |$$

bei allen möglichen Zerlegungen des Intervalls (a, b) in Teilintervalle. Wir sagen, dass die Funktion x_t der Lipschitzschen Bedingung genügt, wenn es ein solches M gibt, dass $| x_{t_1} - x_{t_2} | \leqq M | t_2 - t_1 |$ bei beliebigen t_1 und t_2 ist. Von einer jeden Funktion mit beschränkter Variation kann man durch Einführung einer neuen unabhängigen Veränderlichen zu einer Funktion, die der Lipschitzschen Bedingung genügt, übergehen. Hierzu genügt es als die neue unabhängige Veränderliche die

---

[19] Die Differenzierbarkeit der Funktion x_t, die der Bedingung (A) genügt, und deren Werte dem Hilbertschen Raum angehören, hat Birkhoff bewiesen [Birkhoff, Trans. of Amer. Math. Soc., **38**, (1935), 357—378]. Clarkson hat die Differenzierbarkeit von x_t unter der Voraussetzung, dass E ein gleichmässig konvexer Raum ist, bewiesen. Clarkson, ibid., **40**, 396—414; Dunford a. Morse, ibid., 415—420. In der Arbeit von Dunford und Morse wurde die Differenzierbarkeit von x_t unter anderen Voraussetzungen untersucht.

Bogenlänge $s=s(t)$ der Kurve x_t von Punkt $t=a$ bis zu einem Punkt mit gegebenem t zu nehmen. In der Tat: da die Sehnenlänge kleiner als die Bogenlänge ist, so ist $|x_{s'}-x_s|\leqslant|s'-s|$, wo $s=s(t)$ eine monotone Funktion von t ist.

2. Hilfssatz 1. *Jede Funktion, die der Bedingung* (A) *genügt, genügt in fast allen Punkten der Lipschitzschen Bedingung (mit einer Konstante K_t, die von t abhängt).*

Beweis. Nehmen wir das Gegenteil an. Das bedeutet, dass es eine solche Menge P mit $\operatorname{mes} P=a>0$ existiert, dass für ein beliebiges N und jedes t, das P angehört, eine Zahlenfolge $t_N^{(m)}$ ($\lim\limits_{m\to\infty} t_N^{(m)}=t$) gefunden werden kann, für die $|x_{t_N^{(m)}}-x_t|>N|t_N^{(m)}-t|$ ist.

Nach dem Satz von Vitali kann man aus dem Segmentensystem $(t_N^{(m)}, t)$, das die Menge P bedeckt, eine endliche Anzahl paarweise einander nicht überdeckender Segmente auswählen, die eine Teilmenge Q der Menge P mit $\operatorname{mes} Q>\frac{a}{2}$ bedecken. Bezeichnen wir diese Segmente mit (t_i, s_i). Dann ist

$$\sum|x_{t_i}-x_{s_i}|>\sum_{i=1}^{n} N|t_i-s_i|>N\frac{a}{2}.$$

Da N beliebig ist, so widerspricht dies der Bedingung (A). Wir sind nun imstande den obenformulierten Satz 2 zu beweisen.

Es sei f_t eine abstrakte Funktion, die der Bedingung (A) genügt. Bezeichnen wir mit P_0 die Menge jener t, in welchen f_t der Lipschitzschen Bedingung nicht genügt. Nach dem Hilfssatz 1 ist $\operatorname{mes} P_0=0$. Betrachten wir jetzt in E eine abzählbare überall dichte Menge $x_1, x_2, \ldots, x_n, \ldots$ Die Funktion $f_t x$ ist für ein beliebiges x eine reelle Funktion mit beschränkter Variation. Tatsächlich ist

$$\sum_{i=1}^{n}|f_{t_{i+1}}x-f_{t_i}x|\leqslant|x|\sum|f_{t_{i+1}}-f_{t_i}|\leqslant M|x|.$$

Folglich besitzt die Funktion $f_t x_n$ für fast alle t eine Ableitung. Bezeichnen wir mit P_n die Menge jener t, für welche $f_t x_n$ keine Ableitung besitzt.

Die Funktion x_t besitzt eine schwache Ableitung ausserhalb der Punktmenge $P=P_0\dotplus P_1\dotplus\ldots\dotplus P_n\dotplus\ldots$, denn gehört t der Punktmenge P nicht an, so existiert, erstens, der Limes des Ausdrucks $\frac{f_{t+h}x_n-f_t x_n}{h}$ bei $h\to 0$ und, zweitens, ein solches $K(t)$, dass für alle genügend kleinen h

$$\left|\frac{f_{t+h}-f_t}{h}\right|\leqslant K(t)$$

ausfällt. Für $h\to 0$ konvergiert also, nach dem Satz von der schwachen Konvergenz [20], der Ausdruck $\frac{f_{t+h}-f_t}{h}$ schwach.

3. Hilfssatz 2. *Ist E separabel, so ist die Funktion f_t (mit Werten aus $\bar{E}$), die der Bedingung* (A) *genügt, und für die $f_t x$ eine total stetige Funktion von t bei beliebigem x ist, das unbestimmte Integral ihrer Ableitung.*

[20] Banach, l. c., S. 123.

B e w e i s. Nach Satz 3 existiert $\frac{d}{dt} f_t$. Es sei $\frac{d}{dt} f_t = g_t$; g_t ist eine (*) summierbare Funktion, da für jedes x die Gleichung

$$\frac{d}{dt} f_t x = g_t x \tag{1}$$

fast überall besteht, und $f_t x$ absolut stetig ist. Folglich existiert $\int_a^t g_s\, ds$. Aber infolge (1) ist die Differenz

$$\int_a^t g_s x\, ds - f_t x \tag{2}$$

für jedes x konstant. Die Differenz zweier linearer Funktionale (2) ist auch ein lineares Funktional, und folglich ist

$$f_t = \int_a^t g_s\, ds - f_0,$$

wo f_0 ein fixiertes Funktional ist.

H i l f s s a t z 3. *Ist E separabel, und genügt f_t (mit Werten aus $\bar{E}$) der Bedingung* (A), *so existiert*

$$\int_a^b \left| \frac{d}{dt} f_t \right| dt:$$

B e w e i s. Die Funktion f_t bestimmt in $\bar{E}$ eine rektifizierbare Kurve. Bezeichnen wir mit $s(t)$ die Kurvenlänge von Punkt f_a bis zum Punkt f_t; dann ist

$$|f_{t_2} - f_{t_1}| \leqslant |s(t_2) - s(t_1)|,$$

und folglich

$$\left| \frac{f_{t_2} - f_{t_1}}{t_2 - t_1} \right| \leqslant \left| \frac{s(t_2) - s(t_1)}{t_2 - t_1} \right|.$$

Aber

$$\left| \lim_{t_2 \to t_1} \frac{f_{t_2} - f_{t_1}}{t_2 - t_1} \right| \leqslant \overline{\lim_{t_2 \to t_1}} \left| \frac{f_{t_2} - f_{t_1}}{t_2 - t_1} \right|$$

(Limes im Sinne der schwachen Konvergenz), und also gilt fast überall

$$\left| \frac{d}{dt} f_t \right| \leqslant \left| \frac{d}{dt} s(t) \right|.$$

Da $s(t)$ eine monotone Funktion ist, so ist $\frac{d}{dt} s(t)$ summierbar.

Beweisen wir, dass $\left| \frac{d}{dt} f_t \right|$ messbar ist. Das ergibt sich aus der Gleichung

$$\left| \frac{d}{dt} f_t \right| = \sup_{n=1,2,\ldots} \left| \frac{d}{dt} f_t x_n \right|,$$

wo $x_1, x_2, \ldots, x_n, \ldots$ eine in der Einheitssphäre des Raumes E überall dichte Menge ist.

Da $\left| \frac{d}{dt} f_t \right|$ messbar und durch eine summierbare Funktion von oben beschränkt ist, ist sie auch summierbar.

4. Hilfssatz 4. *Ist E separabel, so ist es für die Existenz des Integrals $\int_a^b x_t dt$ im Bochnerschen Sinne*[19] *notwendig und hinreichend, dass*

1°) *das Integral* $\int_a^b |x_t| dt$ *existiert,*

2°) *für jedes f die reelle Funktion fx_t messbar ist* [*es ist hinreichend, dass die Bedingung* 2°) *für eine überall dichte Menge von Funktionalen erfüllt ist*].

Beweis. Die Notwendigkeit ist offenbar. Um zu beweisen, dass diese Bedingung hinreichend ist, muss gezeigt werden, dass x_t stark messbar ist. Das aber wurde in Satz 1, § 2, behauptet.

Hilfssatz 5. *Ist $\bar{E}$ separabel, genügt f_t der Bedingung* (A), *und ist dabei die Funktion $f_t x$ für ein beliebiges x totalstetig, so besitzt die Funktion f_t fast überall eine starke Ableitung.*

Beweis. Ist $\bar{E}$ separabel, so ist auch E separabel[21]. Folglich hat f_t gemäss Satz 3 dieses Paragraphen fast überall eine schwache Ableitung $\frac{d}{dt} f_t = g_t$. Diese Ableitung ist im Bochnerschen Sinne integrierbar, da $g_t x$ für jedes x messbar ist, und also, nach Satz 2, § 2, auch die Funktion ξf_t für jedes ξ messbar ist. Folglich ist g_t nach Hilfssatz 4 dieses Paragraphen im Bochnerschen Sinne integrierbar, und zwar ist

$$f_t = \int_a^t g_s\, ds.$$

Aber das Bochnersche Integral hat fast überall eine starke Ableitung[22]. Folglich besitzt f_t fast überall eine starke Ableitung.

Hilfssatz 6. *Ist E ein regulärer Raum, und ist die Funktion x_t, die der Bedingung* (A) *genügt, derart, dass für jedes f die Funktion fx_t totalstetig ist, so besitzt x_t fast überall eine starke Ableitung.*

Beweis. Die Behauptung ist offenbar, wenn E separabel ist. In der Tat, ein regulärer Raum ist zu seinem konjugierten konjugiert, d. h. die Bedingungen des Hilfssatzes 5 sind erfüllt.

Im allgemeinen Fall gehen wir folgendermassen vor. Die Werte von x_t gehören einem separablen Teilraum des Raumes E an. Aber A. I. Plessner hat bewiesen, dass jeder Teilraum eines regulären Raumes ebenfalls regulär ist. Somit ist alles auf den oben erledigten Fall reduziert.

5. Hilfssatz 7. *Führt eine monotone Funktion $s = s(t)$ die Menge P der Werte t auf die Menge Q mit* $\operatorname{mes} Q = 0$ *der Werte s über, so hat $s(t)$ eine Ableitung, die fast überall auf P Null ist.*

Diesen Hilfssatz aus der Theorie der Funktionen der reellen Veränderlichen werden wir hier nicht beweisen.

Wir gehen jetzt zum Beweis des am Anfang dieses Paragraphen formulierten Satzes 1 über:

---

[21] Banach, l. c., S. 189.

[22] Bochner, Fund. Math., **XX**, (1933).

Satz 1. *Ist E regulär, so besitzt jede Funktion x_t (mit Werten aus E), die der Bedingung* (A) *genügt, fast überall eine starke Ableitung.*

Beweis. Gemäss den Anmerkungen zu Punkt 1 dieses Paragraphen, kann man durch Einführung einer neuen unabhängigen Veränderlichen zu einer Funktion, die der Lipschitzschen Bedingung genügt, übergehen. Da die Funktion x_s der Lipschitzschen Bedingung genügt, ist die Funktion fx_s bei beliebigem f absolut stetig, und also hat nach Hilfssatz 6 dieses Paragraphen die Funktion x_s fast überall eine starke Ableitung. Da $\frac{d}{dt}x_t = \frac{d}{ds}x_{t(s)}\frac{ds}{dt}$ und s eine monotone Funktion von t ist ($\frac{ds}{dt}$ existiert also für fast alle t), brauchen wir nur die Wertmenge der s (vom Mass Null) zu untersuchen, auf welcher $\frac{dx_{t(s)}}{ds}$ nicht existiert. Beweisen wir, dass $\frac{dx_t}{dt}$ auf dieser Menge für fast alle t existiert und gleich Null ist.

In der Tat, nach Hilfssatz 7 besteht auf dieser Menge für fast alle t die Gleichung $\frac{ds(t)}{dt} = 0$. Aber

$$\left|\frac{x(t_2) - x(t_1)}{t_2 - t_1}\right| \leqslant \frac{s(t_2) - s(t_1)}{t_2 - t_1},$$

und folglich existiert $\frac{dx_t}{dt}$ auf dieser Menge fast überall und ist gleich Null. Hiermit ist der Satz bewiesen.

Der Beweis des Satzes 2 kann analog dem des Satzes 1 geführt werden. Den Satz 1 kann man auch folgendermassen formulieren:

Satz 1a. *Jede im regulären Raum rektifizierbare Kurve besitzt fast überall eine Tangente.*

6. Führen wir jetzt ein Beispiel einer Funktion an, *die der Bedingung* (A) *genügt und keine Ableitung besitzt.*

Definieren wir im Raum $L^{(1)}$ summierbarer Funktionen eine abstrakte Funktion durch

$$x_t = K(s, t) = \begin{cases} 1, & \text{wenn } s < t, \\ 0 & \text{,, } s \geqslant t, \end{cases}$$
$$0 \leqslant s \leqslant 1, \quad 0 \leqslant t \leqslant 1.$$

Es ist leicht nachzuprüfen, dass diese Funktion weder eine schwache noch eine starke Ableitung hat. Hieraus folgt eine Anmerkung, die nicht ohne Interesse sein dürfte:

der Raum $L^{(1)}$ und jeder separable Raum, der einen $L^{(1)}$ isomorphen Teil enthält, ist keinem konjugierten Raum isomorph.

Konstruieren wir jetzt das Beispiel einer Funktion, die zwar eine schwache, aber keine starke Ableitung besitzt. Hierzu betten wir $L^{(1)}$ in den Raum V der Funktionen mit beschränkter Variation ein. Da V zu einem separablen Raum, nämlich dem Raum der stetigen Funktionen, konjugiert ist, so hat nach Satz 3 dieses Paragraphen die im ersten Teil dieses Punktes konstruierte abstrakte Funktion fast überall eine schwache Ableitung. Diese Ableitung kann nicht stark sein, da sie sonst auch im vorhergehenden Beispiel existieren müsste.

7. Schliesslich führen wir für den Fall der regulären Räume die notwendige und hinreichende Bedingung dafür an, dass x_t das unbestimmte Integral einer Funktion im Bochnerschen Sinne ist.

Satz 3. *Damit die Funktion x_t das unbestimmte Integral einer Funktion sei (E ist dabei regulär), ist es notwendig und hinreichend, dass die folgenden Bedingungen erfüllt sind:*

1°) *Bei beliebiger Zerlegung des Intervalls (a, b) in Teilintervalle $(a \leq t_1 \leq \ldots \leq t_n \leq b)$ gilt die Ungleichung*

$$\sum_{i=0}^{n-1} |x_{t_{i+1}} - x_{t_i}| \leq M,$$

wo M eine passend gewählte Konstante ist.

2°) *Ist* $\lim \sum |t_k - s_k| = 0$, *so ist*

$$\lim \left(\sum |x_{t_k} - x_{s_k}|\right) = 0.$$

Die Bedingung 2°) kann durch eine schwächere ersetzt werden, nämlich durch

2°a) *Für jedes f ist die abstrakte Funktion fx_t absolut stetig.*

Der Beweis dieses Satzes ist im Wesentlichen in dem Beweis des Hilfssatzes 5 dieses Paragraphen enthalten.

II. TEIL

ALLGEMEINE FORM LINEARER OPERATOREN

§ 1. Die allgemeine Form der linearen Operatoren, die einen beliebigen linearen Raum auf den Raum (C) der stetigen Funktionen abbilden [1]

Betrachten wir den Raum (C), dessen Elemente stetige, auf dem Intervall (a, b) definierte, Funktionen sind. Die Norm wird hierbei als $\max|\varphi(t)|$ definiert. $\varphi(t) = U(x)$ sei ein linearer, den Raum E auf (C) abbildender Operator. Betrachtet man den fixierten Wert t, so entspricht jedem Element x des Raumes E eine Zahl $\varphi(t)$. Folglich ist bei fixiertem t $\varphi(t) = U(x)$ ein Funktional und zwar ein lineares Funktional in bezug auf x. In der Tat: seine Additivität ist offensichtlich; die Stetigkeit ergibt sich daraus, dass einer konvergenten Elementenfolge eine konvergente Folge stetiger Funktionale entspricht. Folglich kann der Operator folgendermassen dargestellt werden:

$$\varphi(t) = U(x) = f_t x. \tag{1}$$

Betrachten wir nun die Bedingung, die f_t auferlegt werden muss, damit $f_t x$ eine stetige Funktion sei. Diese Bedingung lautet naturgemäss: wenn $t_n \to t$, dann $f_{t_n} x \to f_t x$. Mit anderen Worten ist f_t eine schwach stetige abstrakte Funktion (siehe Teil I, § 1).

Diese Bedingung ist auch hinreichend, denn erstens ordnet der Operator $\varphi(t) = f_t(x)$ unter dieser Voraussetzung jedem Element x eine stetige Funktion $\varphi(t)$ zu, und zweitens existiert $\sup|f_t|$ (da sonst eine konvergente Folge $t_n \to t$

[1] Die Resultate dieses Paragraphen hat im Wesentlichen schon Radon erhalten. Radon untersuchte die allgemeine Form des (C) auf (C) abbildenden Operators. Siehe auch Clarkson (Anm. [19] zum ersten Teil).

existieren müsste, für die $|f_{t_n}| > n$ wäre, und das würde der schwachen Konvergenz der Folge f_{t_n} für $t_n \to t$ widersprechen). Folglich ist

$$\max|\varphi(t)| \leqslant \max|f_t| \cdot |x|, \tag{2}$$

und das bedeutet, dass der Operator (1) stetig ist. Zeigen wir noch, dass die Norm des Operators gleich $\sup|f_t|$ ist:

$$|U| = \sup_{|x| \leqslant 1} |U(x)| = \sup_{|x| \leqslant 1} \max_{a \leqslant t \leqslant b} |f_t x| = \sup_{a \leqslant t \leqslant b} \sup_{|x| \leqslant 1} |f_t x| = \sup_{a \leqslant t \leqslant b} |f_t|.$$

Auf diese Weise erhalten wir den folgenden

Satz 1. *$\varphi(t) = f_t x$ ist die allgemeine Form des linearen Operators, der den Raum E auf (C) abbildet. Dabei ist f_t eine schwach stetige abstrakte Funktion, deren Werte $\bar{E}$ angehören. Die Norm des Operators ist gleich* $\sup|f_t|$.

Finden wir jetzt die allgemeine Form des vollstetigen Operators, der E auf (C) abbildet.

Satz 2. *$\varphi(t) = f_t x$ ist die allgemeine Form des vollstetigen Operators, der den Raum E auf (C) abbildet. Dabei ist f_t eine abstrakte stark stetige Funktion, deren Werte $\bar{E}$ angehören. Die Norm des Operators ist gleich* $\sup|f_t|$.

Beweis. $\varphi(t) = U(x)$ sei ein vollstetiger Operator. Die Elementenmenge $|x| \leqslant 1$ muss in eine in (C) kompakte Menge stetiger Funktionen $\varphi(t)$ übergehen. Folglich sind nach dem Arzelàschen Satz diese stetigen Funktionen gleichartig stetig, d. h.

$$\lim_{h \to 0} \sup_{|x| \leqslant 1} |\varphi(t+h) - \varphi(t)| = \lim_{h \to 0} \sup_{|x| \leqslant 1} |f_{t+h}x - f_t x| = \lim_{h \to 0} |f_{t+h} - f_t| = 0.$$

Das bedeutet, dass die abstrakte Funktion stark stetig ist. Da die abstrakte Funktion f_t und folglich auch $|f_t|$ stetig sind, so kann man in diesem Falle statt $\sup|f_t|$ auch $\max|f_t|$ schreiben.

Es sei nun umgekehrt der Operator $\varphi(t) = f_t x$ gegeben, wo f_t eine stark stetige Funktion ist. Dann geht die Einheitssphäre $|x| \leqslant 1$ in eine Menge gleichartig stetiger Funktionen über [da $|\varphi(t+h) - \varphi(t)| = |f_{t+h}x - f_t x| \leqslant |f_{t+h} - f_t| \cdot |x|$], und folglich ist der Operator vollstetig. Der Satz ist bewiesen.

Beispiele. Führen wir jetzt zur Erläuterung die allgemeine Form der linearen Operatoren an, die konkrete lineare Räume auf (C) abbilden.

1. Die allgemeine Form des linearen Operators, der $L^{(p)}$ $(p > 1)$ auf (C) abbildet. Um die allgemeine Form dieses Operators zu finden, nützen wir die allgemeine Form des linearen Funktionals in $L^{(p)}$ und die Bedingung der schwachen Konvergenz aus. Die allgemeine Form des linearen Operators, der $L^{(p)}$ auf (C) abbildet, ist

$$\varphi(t) = \int_a^b K(s, t)\, \psi(s)\, ds,$$

wo $\psi(s)$ ein Element aus $L^{(p)}$ und $\varphi(t)$ ein Element aus (C) ist.

Der Kern $K(s, t)$ genügt den folgenden Bedingungen:

1°) Es gibt ein solches M, dass $\int_a^b |K(s, t)|^q\, ds \leqslant M$ ist $\left(\frac{1}{p} + \frac{1}{q} = 1\right.$.

2°) Für ein beliebiges u ist $\int_0^u K(s, t)\,ds$ eine stetige Funktion von t.

Die Norm des Operators ist $\sup \sqrt[q]{\int_a^b |K(s, t)|^q\,ds}$.

2. Die allgemeine Form des vollstetigen Operators, der $L^{(p)}$ auf (C) abbildet. Die allgemeine Form vollstetiger Operatoren, die $L^{(p)}$ auf (C) abbilden, lautet:

$$\varphi(t) = \int_a^b K(s, t)\,\psi(s)\,ds,$$

wo $K(s, t)$ der Bedingung

$$\lim_{t_1 \to t_2} \int_a^b |K(s, t_1) - K(s, t_2)|^q\,ds = 0 \quad \left(\frac{1}{p} + \frac{1}{q} = 1\right)$$

genügt.

Analog kann die allgemeine Form der Operatoren, die einen beliebigen Raum E auf (C) abbilden, gefunden werden.

Hierzu genügt es, die Bedingung für die schwache Konvergenz in diesem Raum zu kennen.

§ 2. Die Bedingungen der Kompaktheit von Mengen in linearen Räumen

Beweisen wir nun den folgenden für uns wichtigen Satz als eine Anwendung der Resultate des vorhergehenden Paragraphen:

Satz 1. *Damit die Punktmenge M des linearen Raumes E kompakt sei, ist die gleichmässige Konvergenz gegen* 0 *einer jeden zu Null schwach konvergenten Folge linearer Funktionale notwendig und hinreichend.*

Beweis. Wir beweisen zuerst, dass diese Bedingung hinreichend ist. Nehmen wir zunächst an, dass E separabel ist. Die Menge M ist beschränkt. Wäre dies nämlich nicht der Fall, so müsste eine zu M gehörende Folge $x_1, \ldots, x_n, \ldots$ mit $|x_n| > n$ existieren. Diese Annahme ihrerseits würde die Existenz einer Folge $f_1, \ldots, f_n, \ldots$ solcher Funktionale nach sich ziehen, für die $|f_n| = \frac{1}{n}$ und $f_n x_n = 1$ ist. Das aber widerspricht der Voraussetzung des Satzes.

Da der Raum E separabel ist, so ist er einem Teil von (C) isometrisch (siehe Banach, loc. cit., S. 185). Nach dem Satz des vorigen Paragraphen kann diese Übereinstimmung folgendermassen dargestellt werden: $\varphi(t) = f_t x$. Hierbei entspricht der Menge M eine gewisse Menge stetiger Funktionen. Aus der gleichmässigen Konvergenz der Folge $f_{t_n} x$ auf M gegen f_t $(t_n \to t)$ folgt, dass die M entsprechende Menge stetiger Funktionen gleichartig stetig ist. Da sie nach dem obengesagten auch beschränkt ist, so ist sie nach dem Arzeläschen Satz kompakt Folglich ist unsere Menge stetiger Funktionen, und also auch M, kompakt.

Da jede abzählbare Menge $x_1, \ldots, x_n, \ldots$ in einem separablen Teilraum des Raumes E eingebettet werden kann, ist der erste Teil des Satzes bewiesen.

Die Notwendigkeit kann folgendermassen bewiesen werden. Auf jeder kompakten Menge M konvergiert eine konvergente Folge linearer Operatoren gleichmässig. In der Tat: $U_n(x)$ konvergiere in jedem Punkt gegen Null. Dann sind die Normen der Operatoren gleichmässig beschränkt. Es ist die gleichmässige Konvergenz auf M zu beweisen. Wäre dies nicht der Fall, so gäbe es ein solches a, dass

$$|U_n(x_n)| > a \tag{1}$$

ist (x_n gehören M an). Da M kompakt ist, kann eine gegen ein gewisses Element x_0 konvergente Teilfolge x_{n_k} ausgeschieden werden. Für alle $k > k_0$ gelte $|x_{n_k} - x_0| < \varepsilon$, und für alle $n > n_0$ sei $|U_n(x_0)| < \varepsilon$. Dann ist für hinreichend grosse k

$$|U_{n_k}(x_{n_k})| \leqslant |U_{n_k}(x_{n_k} - x_0)| + |U_{n_k}(x_0)| \leqslant |U_{n_k}|\,|x_{n_k} - x_0| + \varepsilon \leqslant k\varepsilon + \varepsilon.$$

Wenn wir nun ε genügend klein wählen, so geraten wir in einen Widerspruch mit (1).

Aus dem angeführten Merkmal der Kompaktheit kann folgender Satz abgeleitet werden:

Satz 2. *Gegeben sei ein linearer Operator $y = U(x)$, der den Raum E in den Raum E_1 überführt. Der konjugierte Operator, $f = U^*(g)$ (der $\bar{E}_1$ auf $\bar{E}$ abbildet), überführe jede schwach konvergente Folge in eine stark konvergente. Dann ist der Operator $y = U(x)$ [und folglich auch der Operator $f = U^*(g)$] vollstetig.*

Beweis. Es ist zu beweisen, dass die Elementenmenge $y = U(x)$, $|x| \leqslant 1$, kompakt ist. g_n sei eine gegen Null schwach konvergente Folge von Funktionalen. Da

$$g_n y = g_n U(x) = U^*(g_n)\, x$$

ist, und da für alle Elemente unserer Menge $|x| \leqslant 1$ gilt, so folgt aus der vorausgesetzten starken Konvergenz von $U^*(g_n)$ die gleichmässige Konvergenz von $g_n y$ für alle in Betracht kommenden y, d. h., nach Satz 1, auch die Kompaktheit dieser Menge.

Anmerkung. In der obigen Formulierung des Satzes 2 wird unter der schwachen Konvergenz der Folge g_n die schwache Konvergenz von g_n als Funktionale und nicht als Elemente verstanden. Andernfalls ist der Satz einfach falsch.

Beispiel. Die Abbildung des Raumes (C) auf (m).

Folgerung. Der Satz, der aus Satz 1 durch Vertauschung der Worte „Funktional" und „Element" entsteht (mit anderen Worten der zu Satz 1 duale Satz), ist falsch.

§ 3. Die allgemeine Form des stetigen und des vollstetigen Operators, der $l^{(1)}$ auf E abbildet [2]

Die allgemeine Form des Operators zu finden, der $l^{(1)}$ auf den linearen Raum E abbildet, ist eine einfache Aufgabe.

[2] $l^{(1)}$ ist der Raum, dessen Elemente Folgen $x = (c_1, \ldots, c_n, \ldots)$ sind, für die

$$\sum_{n=1}^{\infty} |c_n| < +\infty$$

gilt. Die Norm von x wird dabei als $\sum_{n=1}^{\infty} |c_n|$ definiert.

$y = U(x)$ sei ein linearer, den Raum $l^{(1)}$ auf E abbildender Operator; $x = (c_1, \ldots, c_n, \ldots)$ sei ein Element des Raumes $l^{(1)}$. Das Element $x_n = (0, \ldots, 0, 1, 0, \ldots)$ gehe in ein Element y_n des Raumes E über.

Als Bilder der Elemente der Einheitssphäre bilden die Elemente y_n eine beschränkte Menge: $|y_n| \leqslant M$. Das Element $x = (c_1, \ldots, c_n, \ldots) = \sum_{n=1}^{\infty} c_n x_n$ geht in das Element $\sum c_n y_n$ über. Auf diese Weise erhalten wir den

Satz I. *Ein beliebiger linearer Operator $y = U(x)$, der $l^{(1)}$ auf E abbildet, kann folgendermassen dargestellt werden:*

$$y = \sum_{n=1}^{\infty} c_n y_n, \tag{1}$$

wo $y_1, \ldots, y_n, \ldots$ eine beliebige Folge beschränkter Elemente aus E ist. Umgekehrt ist jeder Operator der Art (1) linear und bildet den Raum $l^{(1)}$ auf E ab.

Satz II. *Jeder vollstetige Operator $y = U(x)$, der $l^{(1)}$ auf den Raum E abbildet, kann folgendermassen dargestellt werden:*

$$y = \sum_{n=1}^{\infty} c_n y_n, \tag{1}$$

wo $x = (c_1, \ldots, c_n, \ldots)$ ein Element aus $l^{(1)}$ und y ein Element aus E ist; $y_1, y_2, \ldots, y_n, \ldots$ bilden eine kompakte Menge. Umgekehrt ist jeder Operator der Form (1) vollstetig, wenn die Folge $y_1, \ldots, y_n, \ldots$ kompakt ist.

Beweis. Beweisen wir zunächst den direkten Satz. Die Elemente $x_n = (0, \ldots, 1, 0, \ldots)$ liegen auf der Einheitssphäre. Folglich bilden ihre Bilder y_n eine kompakte Menge, w. z. b. w.

Es sei jetzt umgekehrt y_n eine kompakte Menge. Beweisen wir, dass der Operator vollstetig ist. Es ist zu beweisen, dass das Bild der Einheitssphäre, d. h. die Gesamtheit der Elemente $\sum c_n y_n$ (mit $\sum |c_n| \leqslant 1$), eine kompakte Menge bildet. Nach Satz 1 des § 2 reicht es hierzu zu zeigen, dass jede gegen Null schwach konvergente Folge von Funktionalen auf dieser Menge gleichmässig konvergiert. $f_1, \ldots, f_k, \ldots$ sei eine Folge von Funktionalen. Da y_n nach der Voraussetzung eine kompakte Menge bilden, so gibt es gemäss Satz 1, § 2, für jedes ε ein solches N, dass für alle $k > N$ $|f_k y_n| < \varepsilon$ ist. Aber dann ist

$$\left| f_k \sum_{n=1}^{\infty} c_n y_n \right| = \left| \sum_{n=1}^{\infty} c_n f_k y_n \right| \leqslant \varepsilon \sum |c_n| \leqslant \varepsilon,$$

und folglich ist die Konvergenz auf der Elementengesamtheit $\sum c_n y_n$ ($\sum |c_n| \leqslant 1$) gleichmässig, und diese Menge kompakt, w. z. b. w.

Beispiel[3]. Durch unmittelbare Anwendung des vorhergehenden Satzes erhalten wir:

Die allgemeine Form des Operators, der $l^{(1)}$ auf $l^{(1)}$ abbildet, ist

$$b_k = \sum_{i=1}^{\infty} a_{ik} c_i \quad (\sum |b_k| < +\infty, \ \sum |a_k| < +\infty),$$

[3] Dieses Resultat hat Lorenz angegeben [Comptes rendus de l'Académie des Sciences de l'U. R. S. S., I, N. 2—3, (1935)].

wo die Matrix $\| a_{ik} \|$ der folgenden Bedingung genügt: es gibt ein solches M, dass $\sum_{k=1}^{\infty} | a_{ik} | \leqslant M$ für alle i ist.

Die allgemeine Form des linearen vollstetigen Operators, der $l^{(1)}$ auf $l^{(1)}$ abbildet, ist $b_k = \sum_{i=1}^{\infty} a_{ik} c_i$. Die Matrix $\| a_{ik} \|$ genügt dabei der Bedingung: für jedes ε gibt es ein solches $N(\varepsilon)$, dass $\sum_{i=N(\varepsilon)}^{\infty} | a_{ik} | < \varepsilon$ bei beliebigen k ist. Der letzte Satz kann aus der Bedingung für die Kompaktheit der Mengen in $l^{(1)}$ erhalten werden.

§ 4. Die allgemeine Form des linearen stetigen und vollstetigen Operators, der einen beliebigen linearen Raum auf $l^{(1)}$ abbildet

$y = U(x)$ sei ein linearer Operator, der einen Punkt x aus E dem Punkt y aus $l^{(1)}$ zuordnet $[y = (a_1, \ldots, a_n, \ldots)]$. Ist n fixiert, so ist a_n ein lineares Funktional von x: $a_n = f_n x$. Da $\sum | a_n | < +\infty$ ist, so gilt auch für ein beliebiges x: $\sum | f_n x | < +\infty$. Folglich konvergiert die Reihe $\sum f_n$ unbedingt (Teil I, § 4).

Umgekehrt entspricht einer beliebigen unbedingt konvergenten Reihe $\sum f_n$ ein linearer den Raum E auf $l^{(1)}$ abbildender Operator. In der Tat: nehmen wir $a_n = f_n x$ an, so entspricht jedem Element x ein Element des Raumes $l^{(1)}$. Die Additivität dieser Abbildung ist offensichtlich, ihre Stetigkeit folgt aus den Sätzen 1 und 3, § 4, Teil I. In der Tat ist $\sum | a_n | = | \sum f_n x | \leqslant M | x |$. Es ist unschwer zu zeigen, dass die Norm des Operators gleich der oberen Schranke der Zahlenmenge $| \sum_{n=1}^{k} \varepsilon_n f_n |$ $(\varepsilon_n = \pm 1)$ ist. Somit erhalten wir den folgenden

Satz 1. *Jeder lineare Operator $y = U(x)$, der einen beliebigen linearen Raum E auf $l^{(1)}$ abbildet [x ist ein Element aus E, $y = (a_1, \ldots, a_n, \ldots)$ ist ein Element aus $l^{(1)}$], kann folgendermassen dargestellt werden:*

$$a_n = f_n x, \tag{1}$$

wo 1) f_n ein lineares Funktional in E ist, und 2) die Reihe $\sum_{n=1}^{\infty} f_n$ unbedingt konvergiert. Die Operatornorm ist die obere Schranke der Zahlen

$$\left| \sum_{n=1}^{k} \varepsilon_n f_n \right| \quad (\varepsilon_n = \pm 1,\ k - \textit{beliebig.})$$

Umgekehrt ist jeder Ausdruck der Form (1) *ein linearer Operator in E*

Finden wir nunmehr die allgemeine Form des vollstetigen Operators, der E auf $l^{(1)}$ abbildet.

Satz 2. *Jeder lineare vollstetige Operator, der einen beliebigen linearen Raum E auf $l^{(1)}$ abbildet, kann folgendermassen dargestellt werden:*

$$c_n = f_n x, \tag{2}$$

wo f_n ein lineares Funktional in E, x ein Element aus E und $y = (c_1, \ldots, c_n, \ldots)$ ein Element aus $l^{(1)}$ ist. Für die Vollstetigkeit des Operators ist *hierbei die unbedingte starke Konvergenz der Reihe* $\sum_{n=1}^{\infty} f_n$ *notwendig und hinreichend* (Teil I, § 4).

Beweis. Da der vollstetige Operator gleichzeitig auch stetig ist, so kann er nach Satz 1 dieses Paragraphen folgendermassen dargestellt werden: $c_n = f_n x$, wo die Reihe $\sum_{n=1}^{\infty} f_n$ unbedingt konvergiert. Der Operator ist vollstetig. Folglich ist die Punktmenge $(f_1(x), \ldots, f_n(x), \ldots)$ aus $l^{(1}$ für alle x mit $|x| \leqslant 1$ kompakt.

Wir nützen nun das folgende leicht zu beweisende Merkmal der Kompaktheit in $l^{(1)}$ aus. Damit eine Punktmenge M aus $l^{(1)}$ kompakt sei, ist die gleichmässige Konvergenz der Reihen $\sum |c_n|$ bezüglich der Menge M notwendig und hinreichend; $(c_1, \ldots, c_n, \ldots)$ sind dabei Punkte aus M. Hieraus erhalten wir: damit der Operator $y = U(x)$ vollstetig sei, ist die gleichmässige Konvergenz der Reihen $\sum_{n=1}^{\infty} |f_n x|$ für jedes x mit $|x| \leqslant 1$ notwendig und hinreichend. Folglich ist Satz 2 bewiesen, da (Teil I, § 4, S. 243) das obengesagte ja die Definition der unbedingten Konvergenz im starken Sinne ist.

Somit führt die Untersuchung der allgemeinen Form des Operators, der einen gegebenen linearen Raum E auf $l^{(1)}$ abbildet, zur Untersuchung der unbedingten Konvergenz in $\bar{E}$.

Beweisen wir nun den folgenden

Satz 3. *Jeder lineare Operator* $y = U(x)$, *der einen regulären Raum* E *auf* $l^{(1)}$ *abbildet, ist vollstetig.*

Beweis. Ist E regulär, so ist auch $\bar{E}$ regulär. Nach Satz 4 des folgenden Paragraphen heisst das, dass die starke und die schwache unbedingte Konvergenz in $\bar{E}$ übereinstimmen. Folglich ist der Operator $y = U(x)$ (gemäss den Sätzen 1 und 2 dieses Paragraphen) vollstetig.

§ 5. Die allgemeine Form des linearen Operators, der (c), den Raum aller konvergenten Folgen, auf einen beliebigen linearen Raum E abbildet

Der lineare Operator $y = U(x)$ ordne jedem Element $x = (c_1, \ldots, c_n, \ldots)$ ein Element y des schwach vollständigen Raumes E zu. Bezeichnen wir mit y_n das Bild des Elementes $x_n = (0, \ldots, 0, 1, 0, \ldots)$. Dem Element $x = (\varepsilon_1, \ldots$ $\ldots, \varepsilon_k, 0, \ldots)$ $(\varepsilon_i = \pm 1)$ entspricht das Element $y = \sum_{i=1}^{k} \varepsilon_i y_i$ $(\varepsilon_i = \pm 1)$. Da $|x| = 1$, so ist $\left|\sum_{i=1}^{k} \varepsilon_i y_i\right| \leqslant |U|$, d. h. für beliebige $\varepsilon_i = \pm 1$ und k gilt die Ungleichung

$$\left|\sum_{i=1}^{k} \varepsilon_i y_i\right| \leqslant |U|. \tag{1}$$

Folglich konvergiert die Reihe $\sum_{i=1}^{\infty} y_i$ nach Satz 2, § 4, Teil I, unbedingt. Demnach konvergiert, was leicht aus der Definition der unbedingten Konvergenz folgt (S. 241), auch die Reihe $\sum_{i=1}^{\infty} c_i y_i$ unbedingt. Da nach Voraussetzung E ein schwach vollständiger Raum ist, so ist $\sum_{i=1}^{\infty} c_i y_i$ ein Element aus E (im allgemeinen ist es ein Element aus $\overline{\overline{E}}$). Ausserdem gilt die Ungleichung

$$\left|\sum_{i=1}^{\infty} c_i y_i\right| \leqslant \sup_{i=1,2,\ldots} |c_i| \cdot \sup\left|\sum_{i=1}^{n} \varepsilon_i y_i\right|$$

(sup wird nach allen n und allen möglichen $\varepsilon_i = \pm 1$ genommen).

In der Tat gilt für ein beliebiges lineares Funktional fy die Ungleichung

$$\left|f\sum_{i=1}^{n} c_i y_i\right| \leqslant \sum_{i=1}^{n} |c_i f y_i| \leqslant \sup_{i=1,2,\ldots} |c_i| \cdot \sum_{i=1}^{n} |fy_i| =$$

$$= \sup|c_i| \cdot \sum_{i=1}^{n} \varepsilon_i f y_i = \sup|c_i| \cdot f\sum_{i=1}^{n} \varepsilon_i y_i \leqslant \sup|c_i| \cdot |f| \cdot \left|\sum_{i=1}^{n} \varepsilon_i y_i\right| \leqslant$$

$$\leqslant |f| \cdot \sup|c_i| \cdot \sup\left|\sum_{i=1}^{n} \varepsilon_i y_i\right| \quad (\varepsilon_i = \operatorname{sign} fy_i),$$

und also gilt

$$\left|\sum_{i=1}^{n} c_i y_i\right| \leqslant \sup|c_i| \cdot \sup\left|\sum \varepsilon_i y_i\right| \quad (\varepsilon_i = \pm 1).$$

Da das Element $\sum_{i=1}^{\infty} c_i y_i$ ein schwacher Limes der Folge $\sum_{i=1}^{n} c_i y_i$ ist, so erhalten wir

$$\left|\sum_{i=1}^{\infty} c_i y_i\right| \leqslant \sup|c_i| \cdot \sup\left|\sum_{i=1}^{n} \varepsilon_i y_i\right| \quad (\varepsilon_i = \pm 1,\ n \text{ — beliebig}), \tag{2}$$

denn die Norm des schwachen Limes einer Folge ist kleiner oder gleich dem oberen Limes der Norm der Elementenfolge.

Umgekehrt zeigen dieselben Überlegungen, dass jeder unbedingt konvergenten Reihe, deren Werte einem schwach vollständigen Raum angehören, ein (c) auf E abbildender Operator entspricht. Somit erhielten wir den folgenden

Satz 1. *Die allgemeine Form des linearen Operators, der* (c) *auf den schwach vollständigen Raum* E *abbildet, ist* $y = \sum c_i y_i$, *wo* $x = (c_1, \ldots, c_n, \ldots)$ *ein Element aus* (c) *und* $y_1, \ldots, y_n, \ldots$ *eine gewisse fixierte Folge aus* E *ist, und zwar eine solche, dass die Reihe* $\sum_{n=1}^{\infty} y_n$ *unbedingt konvergiert. Die Operatornorm ist gleich der oberen Schranke der Zahlen* $\left|\sum_{i=1}^{k} \varepsilon_i y_i\right|$ ($\varepsilon_i = \pm 1$, k *— eine beliebige ganze Zahl). Das letztere ergibt sich aus dem Vergleich der Ungleichungen* (1) *und* (2).

Da die Reihe $\sum c_n y_n$ auch für eine beliebige beschränkte Folge c_n konvergiert, und da die Ungleichung (2) für solche Folgen gültig bleibt, so erhalten wir die folgende

Anmerkung. Jeder lineare Operator, der (c) auf einen schwach vollständigen Raum E abbildet, kann in den Raum (m) fortgesetzt werden, d. h. das

Definitionsgebiet eines Operators kann bis (m) ausgedehnt werden. [(c) ist ja ein Teilraum von (m).]

Es ist auch leicht die allgemeine Form des vollstetigen Operators, der (c) auf einen beliebigen linearen Raum E abbildet, zu finden. Mit den obigen Bezeichnungen haben wir: da die Elemente $(\varepsilon_1, \ldots, \varepsilon_n, 0, \ldots)$ in der Einheitssphäre liegen, so bilden ihre Bilder, die Elemente $\sum_{i=1}^{n} \varepsilon_i y_i$, eine kompakte Menge. Folglich konvergiert nach Satz 5, § 4, Teil I, die Reihe $\sum y_n$ unbedingt im starken Sinne. (Die schwache Vollständigkeit des Raumes E ist hier nicht notwendig, da die Summe einer unbedingt im starken Sinne konvergenten Reihe ein Element des Raumes E ist.)

Zeigen wir jetzt, dass umgekehrt jeder im starken Sinne unbedingt konvergenten Reihe ein vollstetiger Operator entspricht. Es ist also zu zeigen, dass eine Menge der Elemente der Form $\sum_{i=1}^{\infty} c_i y_i$, $\sup |c_i| \leqslant 1$, kompakt ist, wenn die Reihe $\sum_{i=1}^{\infty} y_i$ unbedingt im starken Sinne konvergiert.

$f_1, \ldots, f_n, \ldots$ sei eine gegen Null schwach konvergente Folge von Funktionalen. Wir beweisen, dass diese Folge auf der Menge M von Punkten der Form $\sum_{i=1}^{\infty} c_i y_i$, $\sup |c_i| \leqslant 1$, gleichmässig konvergiert. Zunächst haben wir

$$\left| f_n \left(\sum_{i=1}^{\infty} c_i y_i \right) \right| \leqslant \sum_{i=1}^{\infty} |f_n c_i y_i| \leqslant \sum_{i=1}^{\infty} |f_n y_i|;$$

nach der Definition der unbedingten Konvergenz (§ 4, Nr. 3, Teil I), konvergieren aber die Reihen $\sum_{i=1}^{\infty} |f_n y_i|$ gleichmässig bezüglich n, und also ist

$$\lim_{n \to \infty} \sum_{i=1}^{\infty} |f_n y_i| = \sum_{i=1}^{\infty} \lim_{n \to \infty} |f_n y_i| = 0.$$

Folglich konvergiert jede gegen Null schwach konvergente Folge linearer Funktionale auf M gleichmässig, und das bedeutet gemäss Satz 1, § 2, Teil II, dass M kompakt ist. Wir erhalten somit den folgenden

Satz 2. *Die allgemeine Form eines linearen vollstetigen Operators,* $y = U(x)$, *der* (c) *auf einen beliebigen Raum* E *abbildet, ist* $y = \sum_{i=1}^{\infty} c_i y_i$ [$x = (c_1, \ldots, c_n, \ldots)$ *ist ein Element des Raumes* (c), y *ein Element aus* E, $y_1, \ldots, y_n, \ldots$ *einesolche Elementenfolge aus* E, *dass die Reihe* $\sum_{n=1}^{\infty} y_n$ *unbedingt im starken Sinne konvergiert*].

Hieraus folgt, dass ein jeder vollstetige Operator, der (c) auf einen beliebigen linearen Raum abbildet, in (m) fortgesetzt werden kann.

Satz 3. *Ist* E *regulär, so ist ein jeder lineare Operator* $y = U(x)$, *der* (c) *auf* E *abbildet, vollstetig.*

Beweis. Der zu $y = U(x)$ konjugierte Operator bildet E auf $l^{(1)}$ ab. Ein linearer Operator führt eine schwach konvergente Elementenfolge in eine ebenfalls schwach konvergente Folge über (Hilfssatz 1, § 8, Teil II). Da die starke und die schwache Konvergenz der Elemente in $l^{(1)}$ zusammenfallen, so führt unser Operator eine schwach konvergente Elementenfolge in eine stark konvergente über. $\overline{E}$ ist jedoch regulär, d. h. die Bedingungen des Satzes 2, § 2, Teil II, sind erfüllt, denn in $\overline{E}$ stimmt die schwache Konvergenz von Funktionalen mit der von Elementen überein. Folglich ist der Operator $y = U(x)$ vollstetig.

Satz 4. *Ist E regulär, so fallen in ihm die starke und die schwache unbedingte Konvergenz zusammen* (vgl. Orlicz, Studia Mathematica, Bd. 1, S. 241—248).

Beweis. Dieser Satz folgt unmittelbar aus den Sätzen 1, 2 und 3 dieses Paragraphen.

§ 6. Die allgemeine Form eines Operators, der $L^{(1)}$ auf einen beliebigen linearen Raum abbildet [4]

Betrachten wir zuerst den Fall, in dem E ein regulärer Raum ist.

Satz 1. *Die allgemeine Form eines linearen Operators* $y = U(x)$, *der* $L^{(1)}$ *auf einen linearen Raum E abbildet, ist*

$$y = \int_a^b \varphi(t)\, y_t\, dt, \tag{1}$$

wobei $\varphi(t)$ *ein Element aus* $L^{(1)}$, y *ein Element aus E und* y_t *eine abstrakte, messbare, fast überall beschränkte Funktion von t ist. Die Norm dieses Operators ist gleich*

$$|U| = \operatorname{vrai\,max}_t |y_t|.$$

Beweis. Beweisen wir zunächst, dass ein jeder Operator der Form (1) ein linearer stetiger Operator ist.

Ist $\varphi(t)$ summierbar, so ist $\varphi(t) y_t$ im Bochnerschen Sinne integrierbar. In der Tat: $\varphi(t) y_t$ ist messbar, und zwar stark messbar, da E regulär ist (§ 2, Teil I). Das Integral $\int_a^b |\varphi(t) y_t|\, dt$ existiert, weil fast überall $|\varphi(t) y_t| \leqslant K |\varphi(t)|$ ($K = \operatorname{vrai\,max} |y_t|$) ist, und folglich $|\varphi(t) y_t|$ summierbar ist. D. h., dass (1) für eine beliebige summierbare Funktion $\varphi(t)$ einen Sinn hat. Um die Stetigkeit des Operators (1) zu beweisen, reicht es seine Beschränktheit zu zeigen. Aber

$$\left| \int_a^b \varphi(t)\, y_t\, dt \right| \leqslant \int_a^b |\varphi(t)|\, |y_t|\, dt \leqslant \operatorname{vrai\,max} |y_t| \cdot |x| \tag{2}$$

[das Element $\varphi(t)$ wird bald durch x, bald durch $\varphi(t)$ bezeichnet].

Beweisen wir nun, dass, umgekehrt, jeder lineare Operator, der $L^{(1)}$ auf E abbildet, durch (1) dargestellt wird. $y = U(x)$ sei ein linearer Operator, der $L^{(1)}$ auf E abbildet, und $\varphi(t)$ sei eine summierbare Funktion — ein Element des Raumes $L^{(1)}$. Bezeichnen wir mit $\chi_t(s)$ eine Funktion, die folgendermassen in

[4] N. Dunford [Trans. of Amer. Math. Soc., 40, (1937), 474—493] hat Sätze erhalten, die den Sätzen 1 und 2 analog sind.

$L^{(1)}$ definiert ist:

$$\varkappa_t(s) = \begin{cases} 1, & \text{wenn } s > t, \\ 0 & \text{"} \quad s \leqslant t \end{cases} \quad (0 \leqslant s \leqslant 1,\ 0 \leqslant t \leqslant 1).$$

Bei jedem fixierten t entspricht dem Element $\varkappa_t(s)$ des Raumes $L^{(1)}$ ein Element y_t des Raumes E. Folglich entspricht der Funktion $\varkappa_t(s)$ eine abstrakte Funktion y_t aus E.

Dem Element $\frac{1}{t_2 - t_1}[\varkappa_{t_2}(s) - \varkappa_{t_1}(s)]$ (t_2 und t_1 sind fixiert) entspricht der Punkt $\frac{1}{t_2 - t_1}(y_{t_2} - y_{t_1})$ des Raumes E. Da das Element $\frac{1}{t_2 - t_1}[\varkappa_{t_2}(s) - \varkappa_{t_1}(s)]$ auf der Einheitssphäre in $L^{(1)}$ liegt, so ist

$$\left| \frac{1}{t_2 - t_1}(y_{t_2} - y_{t_1}) \right| \leqslant |U|,$$

oder

$$|y_{t_2} - y_{t_1}| \leqslant |U| \, |t_2 - t_1|.$$

Also hat die abstrakte Funktion y_t fast überall eine starke Ableitung y'_t (Satz 1, § 8, Teil I), deren Integral im Bochnerschen Sinne sie darstellt. Hierbei ist

$$\text{vrai max} |y'_t| \leqslant |U|. \tag{3}$$

Zeigen wir jetzt, dass der Operator $y = U(x)$ in der Form $y = \int_a^b \varphi(t) y'_t \, dt$, wo y eine summierbare, fast überall beschränkte Funktion ist, dargestellt werden kann.

Die Norm dieses Operators ist gleich vrai max $|y'_t|$ [$x = \varphi(t)$ ist ein Element des Raumes $L^{(1)}$]. In der Tat besteht für jede Funktion der Art $\varphi(s) = \varkappa_{t_2}(s) - \varkappa_{t_1}(s)$ die Gleichung

$$U(x) = \int_a^b \varphi(t) y'_t \, dt$$

[da nämlich einerseits $U(\varkappa_{t_2}(s) - \varkappa_{t_1}(s)) = y_{t_2} - y_{t_1}$, und andererseits

$$\int_a^b [\varkappa_{t_2}(s) - \varkappa_{t_1}(s)] y'_s \, ds = \int_{t_1}^{t_2} y'_s \, dt = y_{t_2} - y_{t_1}$$

ist].

Also gilt Gleichung (1) auch für eine beliebige Treppenfunktion. Da der linke und der rechte Teil dieser Gleichung stetige Operatoren sind, und da diese Gleichung für eine in $L^{(1)}$ überall dichte Elementenmenge gilt, so gilt sie auch für ein beliebiges $\varphi(t)$, w. z. b. w.

Der Vergleich der Ungleichungen (2) und (3) ergibt

$$|U| = \text{vrai} \max_{a \leqslant t \leqslant b} |y'_t|.$$

Satz 2. *Die allgemeine Form des Operators, der $L^{(1)}$ auf den zu einem separablen Raum konjugierten Raum E abbildet, ist $y = \int_a^b \varphi(t) y_t \, dt$ [y ist ein Element aus E, $\varphi(t)$ — eine summierbare Funktion — ist ein Element aus $L^{(1)}$,*

y_t ist eine summierbare, fast überall beschränkte, abstrakte Funktion. Das Integral ist im Sinne des § 7, I. Teil, zu verstehen].

Der Beweis unterscheidet sich vom Beweis des vorhergehenden Satzes nur dadurch, dass hier die Existenz einer schwachen Ableitung (auf Grund des Hilfssatzes 2, § 9, I. Teil) statt der einer starken benutzt werden muss. Also existiert das Integral (auf Grund des Satzes 4, § 7).

Satz 3:

$$y=\int_a^b \varphi(t)\, y_t\, dt \tag{4}$$

ist die allgemeine Form des vollstetigen Operators, der $L^{(1)}$ auf einen beliebigen linearen Raum E abbildet. Die Werte von y_t bilden eine kompakte Menge, y_t ist schwach messbar [$\varphi(t)$ ist ein Element des Raumes $L^{(1)}$, y_t ist eine abstrakte Funktion mit Werten aus E].

Beweis. Zeigen wir vorerst, dass ein jeder Operator der Form (4) vollstetig ist. Dazu ist es, gemäss § 2, Teil II, ausreichend zu zeigen, dass jede schwach konvergente Folge von Funktionalen f_n gleichmässig bezüglich aller y mit $|x|\leqslant 1$ konvergiert (d. h. bezüglich aller y, deren entsprechende $\varphi(t)$ in der Einheitssphäre des Raumes $L^{(1)}$ liegen). Dadurch wird dann die Kompaktheit des Bildes der Einheitssphäre gezeigt.

Es sei $f_n \longrightarrow 0$ (schwach). Dann ist

$$|f_n \int_a^b \varphi(t)\, y_t\, dt| \leqslant \int_a^b |\varphi(t)|\, |f_n y_t|\, dt \leqslant \operatorname{vrai\,max} |f_n y_t|. \tag{5}$$

Aber die Menge von Elementen y_t ist kompakt. Folglich strebt $f_n y_t$ gemäss Satz I, § 2, Teil II, und deshalb auch $f_n y$, gleichmässig zu Null, w. z. b. w.

Zeigen wir jetzt, dass ein jeder vollstetige Operator, der $L^{(1)}$ auf E abbildet, in der Form (4) dargestellt werden kann.

Die Elementenmenge der Art $\frac{1}{t_2-t_1}(y_{t_2}-y_{t_1})$ ist kompakt, da sie dem Bilde der Einheitssphäre angehört (hier benutzen wir die Bezeichnungen des Beweises des Satzes 1 dieses Paragraphen). Die Funktion y_t besitzt eine Ableitung. Um dies zu zeigen, betten wir den separablen, die Werte von y_t enthaltenden, Teil des Raumes E in (m) ein. Da[5] (m) zu einem separablen Raum konjugiert ist, so besitzt y_t gemäss Satz 3, § 9, Teil I, fast überall eine schwache Ableitung [im Sinne der Konvergenz in (m)]. Aber die Elementenmenge der Form $\frac{1}{t_2-t_1}[y_{t_2}-y_{t_1}]$ ist kompakt. Demnach fällt diese Ableitung mit der starken Ableitung zusammen. Aber die Kompaktheit der Menge bleibt bei Hinzufügung ihrer Häufungspunkte erhalten. Folglich ist die Wertmenge von y'_t kompakt.

Wird $L^{(1)}$ auf $L^{(1)}$ abgebildet, so bedeutet dieses Resultat folgendes (den folgenden Satz hat Dunford bewiesen[4]):

---

[5] Siehe z. B. den Anfang des Beweises des Satzes 1, § 7, Teil I, dieser Arbeit.

Satz 4. *Die allgemeine Form des linearen vollstetigen Operators, der $L^{(1)}$ auf $L^{(1)}$ abbildet, ist*

$$\psi(s) = \int_a^b K(s, t)\varphi(t)\,dt,$$

wo

1°) $\int_a^b |K(s, t)|\,ds$ *für fast alle t einen Sinn hat,*

2°) $\lim\limits_{h\to 0} \operatorname{vrai\,max} \int_a^b |K(s, t+h) - K(s, t)|\,ds = 0.$

Den Beweis dieses Satzes erhält man unmittelbar aus dem vorhergehenden Satz, wenn nur die Bedingung der Kompaktheit der Mengen in $L^{(1)}$ ausgenützt wird.

Die allgemeine Form des vollstetigen, $L^{(1)}$ auf $L^{(p)}$ (bei $p > 1$) abbildenden Operators ist

$$\psi(s) = \int_a^b K(s, t)\varphi(t)\,dt,$$

wo $K(s, t)$ den nachstehenden Bedingungen genügt:

1°) $\int_a^b |K(s, t)|^p\,ds$ ist für fast alle t definiert,

2°) $\lim\limits_{h\to 0} \operatorname{vrai\,max} \int_a^b |K(s, t+h) - K(s, t)|^p\,dt = 0$[4].

Die allgemeine Form des stetigen, $L^{(1)}$ auf $L^{(1)}$ abbildenden Operators kann folgendermassen erhalten werden.

Betten wir $L^{(1)}$ in den Raum V der Funktionen mit beschränkter Variation ein. Da (V) zum separablen Raum (C) konjugiert ist, so ist die allgemeine Form des Operators, der $L^{(1)}$ auf V abbildet,

$$\varkappa(s) = \int_a^b K(s, t)\varphi(t)\,dt.$$

Dabei ist $K(s, t)$ messbar und

$$\operatorname{vrai\,max}_t \operatorname{var}_s [K(s, t)] < +\infty. \tag{6}$$

Die Norm des Operators ist gleich der linken Seite der Relation (6). Es kann gezeigt werden, dass für die absolute Stetigkeit von $\varkappa(s)$ die absolute Stetigkeit der Funktion

$$\int_0^u K(s, t)\,dt \tag{7}$$

bezüglich der Veränderlichen s notwendig und hinreichend ist. Jetzt erhalten wir die allgemeine Form des gesuchten Operators:

$$\psi(s) = \frac{d}{ds}\int_a^b K(s, t)\varphi(t)\,dt,$$

wo $K(s, t)$ den Relationen (6) und (7) genügt.

Die allgemeine Form des stetigen Operators $y = U(x)$, der $L^{(1)}$ auf M abbildet, ist

$$\psi(s) = \int_a^b K(s, t)\, \varphi(t)\, dt,$$

wo

1°) $K(s, t)$ als Funktion zweier Veränderlicher messbar ist, und

2°) $K(s, t)$ durch eine Konstante M überall beschränkt ist, ausser, vielleicht, einer Menge vom ebenen Masse Null.

Die Norm des Operators ist

$$|U| = \underset{\substack{a \leqslant s \leqslant b \\ a \leqslant t \leqslant b}}{\operatorname{vrai\,max}} |K(s, t)|.$$

vrai max wird in bezug auf die beiden Veränderlichen verstanden.

Der Beweis folgt unmittelbar aus dem Grundsatz dieses Paragraphen (da M zu $L^{(1)}$ konjugiert ist) und aus dem Satz von Fubini. Der Satz von Fubini lautet: der Durchschnitt einer ebenen Menge vom Masse Null mit fast allen parallelen Geraden ist eine Menge, deren lineares Mass gleich Null ist.

§ 7. Die allgemeine Form des Operators, der einen beliebigen linearen Raum E auf M abbildet

[M ist der Raum der fast überall beschränkten Funktionen $\varphi(s)$. Die Norm von $\varphi(s)$ ist $|\varphi| = \underset{a \leqslant s \leqslant b}{\operatorname{vrai\,max}} |\varphi(s)|$.]

E sei ein regulärer oder separabler Raum. Bezeichnen wir mit $\varphi = U(x)$ einen Operator, der E auf M abbildet [$x \subset E$, $\varphi = \varphi(s) \subset M$].

Bei festgewähltem t ist $\int_a^t \varphi(s)\, ds$ ein lineares Funktional bezüglich x. Bezeichnen wir dieses Funktional mit $g_t x$. Es genügt der Ungleichung

$$|g_{t_2} x - g_{t_1} x| = \left| \int_{t_1}^{t_2} \varphi(s)\, ds \right| \leqslant |t_1 - t_2|\, \|\varphi\| \leqslant |t_2 - t_1|\, |U| \cdot |x|.$$

Da aber diese Ungleichung bei einem beliebigen x gültig ist, so ist

$$|g_{t_2} - g_{t_1}| \leqslant |U|\, |t_2 - t_1|.$$

Nach dem Satz 3, § 9, Teil I, hat g_t fast überall eine schwache Ableitung f_t (ist E regulär, so besitzt g_t eine starke Ableitung).

Da andererseits die Gleichung

$$\varphi(t) = \lim_{t_1 \to t} \frac{1}{t_1 - t} \int_t^{t_1} \varphi(s)\, ds$$

für fast alle t besteht, so ergibt sich der

Satz 1. *Die allgemeine Form des linearen Operators $\varphi = U(x)$, der einen beliebigen regulären oder separablen Raum E auf M abbildet, ist $\varphi(t) \doteq f_t x$, wo f_t den folgenden Bedingungen genügt:*

1) *f_t ist schwach messbar,*

2) $\operatorname{vrai\,max} |f_t| < +\infty$.

Wir haben bereits die Notwendigkeit dieser Bedingungen gezeigt. Dass sie hinreichend sind, ist offensichtlich. Im Falle des regulären Raumes könnte man diese Resultate aus denen des vorhergehenden Paragraphen auf Grund des folgenden unschwer zu beweisenden Satzes erhalten:

Satz 2. *E_2 sei ein regulärer Raum. Dann ist der Operator, der $\overline{E}_2$ auf $\overline{E}_1$ abbildet, zu dem Operator, der E_1 auf E_2 abbildet, konjugiert.*

§ 8. Die allgemeine Form des linearen Operators, der (C) auf einen beliebigen Raum E abbildet

Bevor wir uns mit der Frage der allgemeinen Form linearer Operatoren, die (C) auf einen beliebigen Raum E abbilden, befassen, beweisen wir zwei Hilfssätze.

Hilfssatz 1. *Jeder stetige Operator $y = U(x)$ führt eine schwach konvergente Folge in eine wiederum schwach konvergente über.*

Beweis. $y = U(x)$ sei ein gewisser Operator. fx sei ein lineares Funktional bezüglich x und gy ein lineares Funktional bezüglich y. Bezeichnen wir mit U^* den zu U konjugierten Operator. Nach der Definition ist $gU(x) = U^*(g)x$ für beliebige g und x. x_n sei eine gegen x schwach konvergente Folge. Dann ist $gU(x_n) = U^*(g)x_n$, und folglich konvergiert $gU(x_n)$ bei beliebigem g. Das aber bedeutet die schwache Konvergenz von $U(x_n)$.

Hilfssatz 2. *Führt der Operator $y = U(x)$ den Raum (C) in einen schwach vollständigen Raum E über, so kann sein Definitionsbereich fortgesetzt werden, und zwar kann man diesen Operator im Raume aller stückweise stetigen Funktionen definieren [mit derselben Definition der Norm, wie im Raum (C) der stetigen Funktionen].*

Beweis. Jede beschränkte stückweise stetige Funktion kann als Limes einer Folge stetiger Funktionen dargestellt werden. Wählen wir dabei die Folge stetiger Funktionen $x_1, x_2, \ldots, x_n, \ldots$ so, dass ihre Normen gegen die Norm von x konvergieren. Folglich konvergiert $x_1, x_2, \ldots, x_n, \ldots$ schwach. Gemäss dem oben bewiesenen Hilfssatz konvergiert die Elementenfolge $U(x_n) = y_n$ schwach. y sei ihr schwacher Limes (er existiert auf Grund der schwachen Vollständigkeit von E).

Nehmen wir $y = U(x)$ an. Diese Definition ist eindeutig. Der Operator, der auf diese Weise definiert ist, ist additiv. Ferner ist

$$|y| = |\lim U(x_n)| \leqslant \overline{\lim} |U(x_n)| \leqslant |U \ \overline{\lim} | x_n | = |U| \cdot |x|$$

[lim $U(x_n)$ — im Sinne der schwachen Konvergenz]. Folglich ist der Operator stetig und hat dieselbe Norm, wie im Raume (C).

Satz 1. *Die allgemeine Form des linearen Operators $y = U(x)$, der den Raum (C) auf einen beliebigen schwach vollständigen linearen Raum E abbildet, ist*

$$y = \int_a^b \varphi(t)\, dy_t, \tag{1}$$

wo y_t eine abstrakte Funktion mit beschränkter Variation ist [$\varphi(t) = x$ — ein Element aus (C) — ist eine stetige Funktion; y ist ein Element aus E].

Beweis. Beweisen wir vorerst, dass jede Transformation der Form (1) ein linearer Operator ist. Gemäss dem Satz 2, § 8, Teil I, ist $\int_a^b \varphi(t)\,dy_t$ ein Element aus E (im allgemeinen ein Element aus $\bar{\bar{E}}$). Folglich ist $y = \int_a^b \varphi(t)\,dy_t$ ein additiver Operator. Es bleibt nur seine Stetigkeit (Beschränktheit) zu zeigen. Wir haben

$$|fy| = \left|\int_a^b \varphi(t)\,dfy_t\right| \leqslant \max|\varphi(t)|\,\mathrm{var}_t[fy_t];$$

da y_t eine Funktion mit beschränkter Variation ist, so gibt es ein solches M, dass $\mathrm{var}_t[fy_t] \leqslant M|f|$ ist. Endgültig erhalten wir

$$|fy| \leqslant \max|\varphi(t)|\,M|f| = M|x|\,|f|.$$

Da diese Ungleichung bei beliebigem f besteht, so ist $|y| \leqslant M|x|$, und das bedeutet, dass der Operator beschränkt ist.

Diese Behauptung bleibt für den Raum aller stückweise stetigen Funktionen gültig.

Zeigen wir jetzt, dass ein jeder Operator in der Form (1) dargestellt werden kann. Nach Hilfssatz 2 kann man den Operator im Raum der stückweise stetigen Funktionen definieren.

Betrachten wir die Funktion $\varkappa_t(s) = \begin{cases} 0 & \text{bei } s \leqslant t, \\ 1 & \text{„} \ s > t. \end{cases}$ Bei fixiertem t ist sie ein Element des Raumes der stückweise stetigen Funktionen. Diesem Element ordnet der Operator $y = U(x)$ ein bestimmtes Element y_t des Raumes E zu.

Bei veränderlichem t ist y_t eine gewisse abstrakte Funktion. Zeigen wir, dass diese Funktion eine beschränkte Variation besitzt. In der Tat haben wir

$$U\Big(\sum_{i=1}^{n} \varepsilon_i\,[\varkappa_{t_{i+1}}(s) - \varkappa_{t_i}(s)]\Big) = \sum_{i=1}^{n} \varepsilon_i\,(y_{t_{i+1}} - y_{t_i})$$

$$(\varepsilon_i = \pm 1,\ t_1 \leqslant t_2 \leqslant \ldots \leqslant t_{n+1}).$$

Folglich ist

$$\Big|\sum_{i=1}^{n} \varepsilon_i\,[y_{t_{i+1}} - y_{t_i}]\Big| \leqslant |U|$$

(da die Norm des Elementes $\sum_{i=1}^{n} \varepsilon_i\,[\varkappa_{t_{i+1}}(s) - \varkappa_{t_i}(s)]$ gleich 1 ist). Nach dem Satz 2, § 6, Teil I bedeutet dies die Beschränktheit der Variation von y_t. Zeigen wir nun, dass man den Operator $y = U(x)$ in Form $y = \int_a^b \varphi(t)\,dy_t$ darstellen kann [x und $\varphi(t)$ sind ein und dasselbe Element]. Da nach dem oben Bewiesenen y_t eine beschränkte Variation besitzt, so ist gemäss dem ersten Teil des Beweises dieses Satzes (Anmerkung zu Satz 1) das Integral $\int_a^b \varphi(t)\,dy_t$ ein linearer Operator im Raume der stückweise stetigen Funktionen. Es sei

$\varphi(s) = \varkappa_t(s)$. Dann ist $\int_a^b \varkappa_t(s)\,dy_s = y_t$, wenn nur der Wert $s = t$ ein Stetigkeitspunkt der Funktion y_t ist.

Stetigkeitspunkte von y_t sind alle Punkte mit der möglichen Ausnahme einer abzählbaren Menge (Anmerkung zu Satz 5, § 6).

Andererseits ist $y_t = U(\varkappa_t(s))$, nach der Definition von y_t. Demnach gilt die Relation

$$U(\varphi) = \int_a^b \varphi(s)\,dy_s$$

für alle Treppenfunktionen, deren Unstetigkeitspunkte einer gewissen abzählbaren Menge nicht angehören. Da man mit solchen Funktionen eine beliebige stetige Funktion approximieren kann, so ist hiermit der Satz bewiesen. Ist E kein schwach vollständiger Raum, so kann man doch behaupten, dass jeder lineare Operator, der (C) auf E abbildet, folgendermassen dargestellt werden kann:

$$y = \int_a^b \varphi(t)\,d\xi_t$$

(ξ_t ist dabei eine abstrakte Funktion mit beschränkter Variation, deren Werte $\overline{\overline{E}}$ angehören). Ist ξ_t eine beliebige Funktion mit beschränkter Variation, so braucht das Element $\int_a^b \varphi(t)\,d\xi'_t$ (ξ_t ist stetig) nicht unbedingt dem Raum E anzugehören.

Beschäftigen wir uns jetzt mit der allgemeinen Form der vollstetigen Operatoren, die (C) auf einen beliebigen linearen Raum abbilden. Beweisen wir vorerst den folgenden

Hilfssatz 3. *Führt ein vollstetiger Operator $y = U(x)$ den Raum (C) in den Raum E über, so kann man ihn derart fortsetzen, dass er auch im Raume aller stückweise stetigen Funktionen vollstetig ist* (siehe Hilfssatz 2 dieses Paragraphen).

Beweis. Stellen wir eine beliebige stückweise stetige Funktion x, wie im Hilfssatz 2 dieses Paragraphen, als Limes einer Folge stetiger Funktionen $x_1, \ldots, x_n, \ldots$ dar. Diese Folge wählen wir dabei so, dass die Normen von x_n gegen die Norm von x konvergieren. Gemäss Hilfssatz 1 dieses Paragraphen konvergiert $U(x_n)$ schwach. Da der Operator vollstetig, und folglich die Menge $U(x_1), \ldots, U(x_n), \ldots$ kompakt ist, konvergiert $U(x_n)$ stark gegen ein gewisses y. Setzen wir $y = U(x)$. Dieser Operator ist stetig (Hilfssatz 2). Da man die Punkte des Bildes der neuen Einheitskugel durch Zufügung der Häufungspunkte zum Bilde der ersten Einheitskugel bekommt, so ist das Bild der neuen Einheitskugel kompakt, und dies bedeutet, dass der Operator vollstetig bleibt.

Satz 2. *Die allgemeine Form des linearen vollstetigen Operators $y = U(x)$, der (C) auf einen beliebigen linearen Raum abbildet, ist*

$$y = \int_a^b \varphi(t)\,dy_t, \tag{1}$$

wobei $\varphi(t)$ ein Element aus (C), y ein Element aus E und y_t eine Funktion mit stark beschränkter Variation ist (§ 6, Teil I).

Beweis. Beweisen wir zuerst, dass ein jeder Operator der Form (1) vollstetig ist. Es muss gezeigt werden, dass das Bild der Einheitssphäre kompakt ist, d. h., dass die Elementenmenge $\int_a^b \varphi(t)\, dy_t$, wo $|\varphi(t)| \leqslant 1$, kompakt ist. [Die Stetigkeit des Operators (1) wurde schon im Satz 1 dieses Paragraphen bewiesen.]

Nach Satz 1, § 2, Teil II, reicht es zu zeigen, dass für eine jede gegen Null schwach konvergente Folge von Funktionalen $f_1, \ldots, f_n, \ldots$ das Integral $\int_a^b \varphi(t)\, df_n y_t$ in bezug auf alle $\varphi(t)$ mit $|\varphi(t)| \leqslant 1$ gleichmässig zu Null strebt.

Da $|f_n|$ beschränkt sind, ist nach der Definition der Funktion mit stark beschränkter Variation die Folge reeller Funktionen $f_n y_t$ im Sinne der Konvergenz im Raume der Funktionen mit beschränkter Variation kompakt. Da $f_n y_t$ infolge der schwachen Konvergenz von f_n gegen Null auch gegen Null konvergiert, ist $\underset{n\to\infty}{\mathrm{var}}\,[f_n y_t] \longrightarrow 0$. Weiterhin ist

$$\left|\int_a^b \varphi(t)\, df_n y_t\right| \leqslant \max|\varphi(t)|\, \mathrm{var}_t\,[f_n y_t] = \mathrm{var}_t\,[f_n y_t].$$

Hieraus ist ersichtlich, dass die Konvergenz gleichmässig ist. Beweisen wir jetzt, dass ein jeder vollstetige Operator in der Form (1) dargestellt werden kann. Gemäss Hilfssatz 3 kann man den Operator in den Raum aller stückweise stetigen Funktionen fortsetzen. Wenn wir hier die Bezeichnungen des Satzes 1 dieses Paragraphen beibehalten, so sehen wir, dass die Elementenmenge $\sum_{i=1}^{n} \varepsilon_i (y_{t_{i+1}} - y_{t_i})$ kompakt ist ($\varepsilon_i = \pm 1$, $t_1 \leqslant \ldots \leqslant t_n \leqslant t_{n+1}$ ist eine beliebige Zerlegung des Intervalls).

y_t ist, nach Satz 6 des § 6, Teil I, eine Funktion mit stark beschränkter Variation.

Weitere Überlegungen stellen eine fast wörtliche Wiederholung der Betrachtungen dar, die zu dem Satz 1 dieses Paragraphen führten.

Der Unterschied besteht nur im folgenden: die Existenz nicht mehr als einer abzählbaren Menge von Unstetigkeitspunkten wird aus der Folgerung 3° des Satzes 6, § 6, Teil I, erhalten.

§ 9. Die allgemeine Form der Operatoren, die *E* auf (*V*) abbilden

(V) ist der Raum, dessen Elemente Funktionen $y = \varphi(t)$ mit beschränkter Variation sind, für die $\varphi(a) = 0$ ist. Die Norm wird als $\mathrm{var}_t[\varphi(t)]$ definiert.

Satz 1. *Die allgemeine Form des linearen Operators $y = U(x)$, der E auf V abbildet, ist*

$$\varphi(t) = f_t x \tag{1}$$

[$\varphi(t)$ *und y sind ein und dasselbe Element*], *wo f_t eine abstrakte Funktion mit beschränkter Variation* (siehe § 6, Teil I) *und Werten aus E ist, für die $f_a = 0$ ist.*

Beweis. Zunächst ist es klar, dass eine jede Transformation der Form (1) ein linearer stetiger Operator ist, der E auf V abbildet.

In der Tat ergibt sich aus der Definition der abstrakten Funktion mit beschränkter Variation, dass jedem Element x des Raumes E eine Funktion mit beschränkter Variation entspricht. Die Additivität dieser Zuordnung ist offenbar. Die Stetigkeit aber ist aus folgendem ersichtlich: gemäss Satz 1 und Satz 3, § 6, Teil I, gibt es ein solches M, dass $|y| = \mathrm{var}_t[f_t x] \leqslant M|x|$ für ein beliebiges x besteht. Zeigen wir jetzt, dass ein jeder Operator $y = U(x)$ in der Form (1) dargestellt werden kann.

Es sei der Operator $\varphi(t) = U(x)$ gegeben. Jedem x entspricht eine Funktion mit beschränkter Variation. Bei festgewähltem t entspricht einem jeden x eine Zahl. Diese Zuordnung, wie leicht einzusehen ist, ist stetig und additiv. Folglich ist $\varphi(t)$ bei jedem fixierten t ein lineares Funktional von x, d. h. $\varphi(t) = f_t x$. Da $f_t x$ für jedes x eine Funktion mit beschränkter Variation ist, so ist auch f_t eine abstrakte Funktion mit beschränkter Variation.

Satz 2. *Die allgemeine Form des vollstetigen Operators, der E auf V abbildet, ist*

$$\varphi(t) = f_t x, \tag{1}$$

wo f_t eine abstrakte Funktion mit stark beschränkter Variation ist, für die $f_a = 0$ gilt.

Der Beweis ist gemäss Satz 1 dieses Paragraphen und der Definition der abstrakten Funktion mit stark beschränkter Variation (Teil I, § 6) offensichtlich.

(Поступило в редакцию 28/IV 1938 г.)

3.

On one-parametrical groups of operators in a normed space

Dokl. Akad. Nauk SSSR **25** (1939) 713–718. Zbl. **22**:358

(Communicated by A. N. Kolmogorov, Member of the Academy, 3. XI. 1939)

1. Suppose we are given a family of operators U_t in a linear space E (t — a real number) satisfying the following conditions:

1°. U_t is a weakly continuous function of t, i. e. $(U_t x, f)$ is for every x and f a continuous function of t; (y, f) or (f, y) means the value of the linear functional f at the point y of the space E^*.

2°. $U_{t+s} = U_t \cdot U_s$ for arbitrary numbers t and s and $U_0 = 1$.

3°. $|U_t| \leqslant M$ (in the sequel we shall for simplicity assume that $M = 1$. This can be always achieved by replacing the norm in the space by $|x|_1 = \sup_t |U_t x|$ which is equivalent to it).

We shall prove that such a family U_t of operators may be represented in the form $U_t = e^{tA}$ and shall study the properties of the operator A (the operator A will be, in general, an unbounded operator). In the case, when E is a Hilbert space and U_t is a family of unitarian operators, this theorem was proved by Stone ([1]). The problem which is solved in the present paper was proposed by A. N. Kolmogorov.

2. Construction of the Operator A. It is natural to construct the operator A in the following way. Suppose that for some x there exists

$$\lim_{h \to 0} \frac{U_h x - x}{h}.$$

Then put

$$Ax = \lim_{h \to 0} \frac{U_h x - x}{h}. \tag{1}$$

Let us prove that the limit (1) exists for an everywhere dense set of elements x of the space E.

Consider the following operator:

$$C = \int_{-\infty}^{+\infty} U_s \, d\varphi(s), \tag{2}$$

* See § 5, where condition 1° is weakened.

where $\varphi(s)$ is a function of bounded variation on the real axis. The equality (2) has the following sense: for every x and f (f is a functional, x an element of the space E)

$$(Cy, f) = \int_{-\infty}^{+\infty} (U_s y, f)\, d\varphi(s).$$

The existence of the operator C in the case when E is a separable or a regular space is proved in a previous paper of the author (²). The proof of the existence of the operator C in the general case will be given elsewhere.

It is easily seen that $|C| \leqslant \mathrm{var}_s [\varphi(s)]$ and

$$\frac{U_h - 1}{h} C = \int_{-\infty}^{+\infty} U_s d \frac{\varphi(s-h) - \varphi(s)}{h}.$$

We choose $\varphi(s)$ differentiable and such that

$$\lim_{h \to 0} \mathrm{var}_s \left[\frac{\varphi(s-h) - \varphi(s)}{h} - \varphi'(s) \right] = 0. \tag{3}$$

Then there exists $\lim\limits_{h \to 0} \frac{U_h - 1}{h} C$ (the limit being understood here in the sence of the convergence of operators by the norm) and is equal to

$$C_1 = \int_{-\infty}^{+\infty} U_s d\varphi'(s).$$

In fact,

$$\left| \frac{U_h - 1}{h} C - C_1 \right| = \left| \int_{-\infty}^{+\infty} U_s d \left[\frac{\varphi(s-h) - \varphi(s)}{h} - \varphi'(s) \right] \right| \leqslant$$

$$\leqslant \mathrm{var}_s \left[\frac{\varphi(s-h) - \varphi(s)}{h} - \varphi'(s) \right].$$

Consequently, for an arbitrary element

$$x = Cy, \tag{4}$$

the limit of the expression $\frac{U_h x - x}{h}$ exists (by the limit is meant here the strong limit).

We collect all elements x which may be obtained from the formula (4) with at least one admittable C [i. e. a C defined by the formula (2) with a function $\varphi(s)$ satisfying the equality (3)]. We shall prove that the set of these x is everywhere dense in E.

Assume the contrary: then by Hahn's theorem there exists a linear functional $f \neq 0$ such that

$$\int_{-\infty}^{+\infty} (U_t y, f)\, d\varphi(t) = 0$$

for any y and any function $\varphi(t)$ satisfying the relation (3). Consequently, $(U_t y, f) = 0$ for any y and t; in particular, putting $t = 0$, we have $(y, f) = 0$ for any y, i. e. $f = 0$, and we obtain a contradiction.

Thus, we have proved that *on an everywhere dense set of elements* x *there exists a limit in the sense of strong convergence of the expression* $\frac{U_h x - x}{h}$ *for* $h \to 0$. *Put*

$$Ax = \lim_{h \to 0} \frac{U_h x - x}{h}$$

for all those x, *for which the limit of the right-hand side exists, be it even in the sense of weak convergence.*

3. *Investigation of the Properties of the Operator* A. We prove in the first place that *the operator* A *is closed*. Before we proceed to the proof, we make a simple remark which will be used in the sequel.

If Ax *is defined, then the expression* $AU_t x$ *is also defined and we have the equality*

$$AU_t x = U_t A x = \frac{d}{dt} U_t x. \tag{5}$$

This follows directly from the definition of A, if we observe that

$$\frac{U_{t+h} x - U_t x}{h} = U_t \frac{U_h x - x}{h} = \frac{U_h [U_t x] - U_t x}{h}.$$

We pass now to the proof of the closedness of the operator A.

An operator is said to be closed, if $x_n \to x$ *and* $Ax_n \to y$ *imply that* Ax *is defined and that* $y = Ax$.

Let $x_n \to x$ and $Ax_n \to y$. According to (5)

$$U_t x_n - U_0 x_n = \int_0^t \frac{d}{ds} U_s x_n \, ds = \int_0^t U_s A x_n \, ds.$$

Letting $n \to \infty$, we obtain

$$U_t x - x = \int_0^t U_s y \, ds$$

(we have the right to pass to the limit under the sign of the integral, since $|U_s A x_n - U_s y| \leqslant |Ax_n - y|$), or

$$\frac{U_t x - x}{t} = \frac{1}{t} \int_0^t U_s y \, ds.$$

For $t \to 0$ the weak limit of the right-hand side exists and is equal to y. Consequently, Ax exists and $Ax = y$, q. e. d.

Let us now show that the spectrum of the operator A is situated on the imaginary axis. To this end we prove first the following

Lemma 1.

$$|(A - \lambda 1) x| \geqslant |\sigma| \cdot |x|, \tag{6}$$

where $\lambda = \sigma + i\tau$. *In the general case this estimation cannot be improved.*

Proof. Suppose first that $\tau = 0$. It is evidently sufficient to prove our assertion for $|x| = 1$. Suppose that for some x with $|x| = 1$ (6) is not satisfied, i. e. that $|(A - \lambda 1) x| \leqslant a < |\sigma|$. According to (5)

$$\frac{d}{dt} U_t x = U_t A x = U_t (A - \sigma 1) x + \sigma U_t x. \tag{7}$$

There exists an f, $|f| = 1$, such that $(f, x) = 1$. Applying f to both sides of (7) and putting $(U_t x, f) = \varphi(t)$, we obtain

$$\frac{d}{dt} \varphi(t) = \sigma \varphi(t) + \psi(t)$$

where $\psi(t)=(U_t(A-\sigma 1)x, f)$, and consequently, $|\psi(t)| \leqslant a < |\sigma|$. Thus, we have $|\varphi'(t)-\sigma\varphi(t)| \leqslant a < |\sigma|$ and $\varphi(0)=(U_0 x, f)=1$. Let $\varphi(t)=\varphi_1(t)+i\varphi_2(t)$. Then obviously

$$|\varphi_1'(t)-\sigma\varphi_1(t)| \leqslant a < |\sigma| \quad \text{and} \quad \varphi_1(0)=\varphi(0)=1.$$

Let $\sigma > 0$, $\varphi_1'(t) \geqslant \sigma\varphi_1(t)-a$; put $\sigma\varphi_1 - a = \rho(t)$. Then

$$\rho'(t) \geqslant \sigma\rho(t) \quad \text{and} \quad \rho(0)=\sigma - a > 0. \tag{8}$$

The function $\rho(t)$ remains positive on the interval $(0, \infty)$. In fact, let t_0 be the leftmost of the points of the interval $(0, \infty)$ at which $\rho(t_0)=0$. Then $\rho(t_0)-\rho(0)=t_0\rho'(t_1)$, where $0 < t_1 < t_0$, and consequently $\rho'(t_1) < 0$; on the other hand, $\rho'(t_1) \geqslant \sigma\rho(t_1) > 0$. Thus, there exists no point on $(0, \infty)$ at which $\rho(t)$ vanishes, and consequently $\rho(t)$ does not change its sign. From (8) follows $\frac{\rho'(t)}{\rho(t)} \geqslant \sigma$, i. e. $\rho(t) \geqslant \rho(0)e^{\sigma t}$; consequently, $\rho(t)$, and therefore also $\varphi(t)=(U_t x, f)$, is unbounded. But this contradicts to the boundedness of $|U_t|$. Thus, for $\tau=0$ the lemma is proved.

If $\lambda=\sigma+i\tau$, we put $U_t^{(1)}=e^{-i\tau t}U_t$. Then

$$A^{(1)}x=\lim_{\tau\to 0}\frac{e^{-i\tau t}U_t x - e^{-i\tau t}x}{t}=Ax-i\tau x.$$

Applying the already obtained result to $A^{(1)}x$, we obtain

$$|Ax-\lambda x| \geqslant |\sigma|\cdot|x|.$$

Denote by Ω_A the domain of definition of the operator A.

Lemma 2. $A-\lambda 1$ $(\lambda=\sigma+i\tau,\ \sigma \neq 0)$ *maps* Ω_A *on the whole space.*

Proof. We prove first that the aggregate of elements $Ax-\lambda x$ $(x\in\Omega_A)$ is everywhere dense in E. Assume the contrary. Then there exists $f\neq 0$ such that $(Ax-\lambda x, f)=0$ for any $x\in\Omega_A$, i. e. $(Ax, f)=\lambda(x, f)$ for $x\in\Omega_A$. Since together with x also $U_t x\in\Omega_A$, $(AU_t x, f)=\lambda(U_t x, f)$, i. e. $\frac{d}{dt}(U_t x, f)=\lambda(U_s x, f)$. Consequently, $(U_t x, f)=(x, f)e^{\lambda t}$; since $e^{\lambda t}$ is unbounded on the real axis and $(U_t x, f)$ is bounded, $(x, f)=0$ for $x\in\Omega_A$, i. e. $f=0$, and we obtain a contradiction.

We prove now that the set $(A-\lambda 1)x$, $x\in\Omega_A$, is closed. Let $y_n=Ax_n-\lambda x_n$ and $y_n\to y$. According to Lemma 1, the x_n also have a limit: $\lim x_n=x$. Since $x_n\to x$ and $Ax_n-\lambda x_n\to y$, $Ax_n\to\lambda x+y$, and since A is a closed operator, Ax is defined and $Ax=\lambda x+y$. Thus, $y=(A-\lambda 1)x$ belongs also to our set, i. e. we have proved its closedness.

From Lemmas 1 and 2 follows that the spectrum of the operator A is situated on the imaginary axis and that it is impossible to extend the domain of its definition without the set of its spectral points becoming the whole plane.

4. In the present paragraph we prove that for any h we have $e^{Ah}=U_h$; e^{hA} we define as follows:

$$e^{hA}x=\sum_{n=0}^{\infty}\frac{h^n}{n!}A^n x. \tag{9}$$

Let us show that the right-hand side of (9) has sense for an everywhere dense set of elements x.

Consider the aggregate of infinitely differentiable functions $\varphi(t)$ such that

$$\lim_{N\to\infty}\operatorname{var}\,[\varphi(t+h)-\varphi_N(t)]=0, \tag{10}$$

where

$$\varphi_N(t) = \sum_{n=0}^{N-1} \varphi^{(n)}(t) \frac{h^n}{n!}$$

(such will be, for instance, all functions of the form $\frac{P(t)}{a^2+t^2}$ where $P(t)$ is a trigonometrical polynomial, $a > |h|$). Then, for $x = \int_{-\infty}^{+\infty} U_t y \, d\varphi(t)$, $A^n x$ are defined for every n (the proof is obtained in the following way: we apply the arguments of § 2 to $Ax = \int_{-\infty}^{+\infty} U_t y \, d\varphi'(t)$, which proves the existence of $A^2 x$, etc.). In the same way as in § 2 we prove that the set of these x is everywhere dense in E. The series $\sum_{n=0}^{\infty} \frac{A^n x}{n!} h^n$ converges. In fact, $\sum_{n=0}^{N-1} \frac{A^n x}{n!} h^n = \int_{-\infty}^{+\infty} U_t y \, d\varphi_N(t)$ and consequently, according to (10), for $N \to \infty$ it converges to $\int_{-\infty}^{+\infty} U_t y \, d\varphi(t+h)$.

Let us now prove that if for any x the series $\sum_{n=0}^{\infty} \frac{A^n x}{n!} h^n$ converges, then its sum is equal to $U_h x$ [it is sufficient to demand that

$$\frac{|A^n x|}{n!} h_n \to 0; \tag{11}$$

as we shall see later, hereform already follows that the series (9) converges].

According to (5), § 3, $U_t A^n x = \frac{d^n}{dt^n} U_t x$. Consequently,

$$\left| U_{t+h} x - U_t \sum_{n=0}^{N-1} A^n x \frac{h^n}{n!} \right| = \left| U_{t+h} x - \sum_{n=0}^{N-1} \frac{d^n}{dt^n} U_t x \frac{h^n}{n!} \right| =$$

$$= \left| \frac{1}{N!} \int_t^{t+h} (t-s+h)^{N-1} \frac{d^N}{dt^N} U_s x \, ds \right|$$

(the last equality is a corollary of Taylor's formula with the integral form of the residual term), i. e., according to (11),

$$\left| U_{t+h} x - U_t \sum_{n=0}^{N-1} A^n x \frac{h^n}{n!} \right| = \left| \frac{1}{N!} \int_t^{t+h} (t-s+h)^{N-1} U_s A^N x \, ds \right| \leqslant$$

$$\leqslant \frac{|h|^N}{N!} |U_s A^N x| \leqslant \frac{|h|^N}{N!} |A^N x| \to 0. \tag{12}$$

Putting in (12) $t=0$, we obtain:

$$U_h x = \sum_{n=0}^{\infty} A^n x \frac{h^n}{n!},$$

q. e. d.

5. In the present paragraph will be proved that from the weak measurability of the function U_t follows ist strong continuity.

Theorem. *Let U_t be weakly measurable, i. e. let $(U_t x, f)$ be a measurable function of t for any x and f. Then $U_t x$ is a strongly continuous function of t for every x, i. e. $t_n \to t$ implies $|U_{t_n} x - U_t x| \to 0$.*

Proof. In § 2 has been proved that $\lim\limits_{h\to 0} \frac{U_h x - x}{h}$ exists for an everywhere dense set of elements x (this proof depended only on the weak measurability of U_t). Consequently, for elements of this everywhere dense set from $h \to 0$ follows $|U_h x - x| \to 0$, and consequently also from $t + h \to t$ follows $|U_{t+h} x - U_t x| \to 0$.

Let y be an arbitrary element. Then

$$|U_{t+h} y - U_t y| \leqslant |U_{t+h} y - U_{t+h} x| + |U_{t+h} x - U_t x| + |U_t x - U_t y| \leqslant$$
$$\leqslant 2|y - x| + |U_h x - x|.$$

Consequently, choosing x such that $|y - x| < \frac{\varepsilon}{4}$ and then δ such that for $|h| < \delta$, $|U_h x - x| < \frac{\varepsilon}{2}$, we obtain $|U_{t+h} y - U_t y| < \varepsilon$, q. e. d.

Corollary. If for $h \to 0$ the expression $\frac{U_h x - x}{h}$ weakly converges, then it converges also strongly.

Proof. Let

$$\lim \frac{U_h x - x}{h} = Ax,$$

$$\frac{U_h x - x}{h} = \frac{1}{h} \int_0^h U_t Ax \, dt. \tag{13}$$

Since the expression under the sign of the integral is strongly continuous, for $h \to 0$ there exists a strong limit of the right-hand side, and consequently also of the left-hand side of formula (13), q. e. d.

V. A. Steklov Mathematical Institute.
Academy of Sciences of the USSR.

Received
5. XI. 1939.

REFERENCES

[1] M. Stone, Linear transformations in Hilbert space. [2] I. Gelfand, Recueil Mathém., 4 (46), 2 (1938).

4.

On normed rings

Dokl. Akad. SSSR **23** (1939) 430–432. Zbl. **21**:294

(Communicated by A. N. Kolmogoroff, Member of the Academy, 27. III. 1939)

Definition 1. The set R of elements $x, y, \ldots$ is called a normed ring, if: a) it is a complete normed space with multiplication by complex numbers; b) together with the elements x and y exists their product xy possessing the usual algebraical properties and continuous at least with respect to one of its factors; c) it contains a unit element e.

It may be shown that in R exists a norm equivalent to a given one and possessing the following properties:

$$\|xy\| \leqslant \|x\| \cdot \|y\|, \quad \|e\| = 1.$$

In what follows we consider only commutative rings.

Definition 2. A set of elements $x, y, \ldots \subset R$ is called an ideal I, if it possesses the following properties: a) if $x \in I$, $y \in I$, then $\alpha x + \beta y \in I$, where α and β are arbitrary complex numbers; b) if $x \in I$ and z is an arbitrary element of the ring R, then $xz \in I$.

The whole ring and the nul ideal, i. e. the ideal consisting of only one element 0, are trivial ideals. Every ideal different from these two we shall call non-trivial.

Definition 3. A non-trivial ideal not contained in any other non-trivial ideal we call a maximal ideal.

Theorem 1. *Every maximal ideal is closed* (*i. e. the set of elements contained in this ideal is closed in* R).

Theorem 2. *Every non-trivial ideal is contained in a maximal ideal.*

The ring of residues R/I, where I is a closed ideal, may be again made a normed ring. As is known, the elements of R/I are classes X, Y, $\ldots$, consisting of such elements x', x'', $\ldots$ that $x' - x'' \in I$. The norm of the class X is defined in the following way: $\|X\| = \inf |x|$, where $x \in X$. It may be shown that in the ring of residues all axioms of a normed ring are satisfied.

Theorem 3. *The ring of residues to a maximal ideal is the corpus of complex numbers.*

Thus, to every element $x \in R$ and every maximal ideal M there corresponds a complex number $x(M)$, namely the number, which is correlated to x by the homomorphism $R \sim R/M$. Hence follows

Theorem 4. *To every element* $x \in R$ *there corresponds a function*

$x(M)$ defined on the set $\mathfrak{M}$ of all maximal ideals. To the sum of elements corresponds the sum of functions, to the product of elements—the product of functions.

The function $x(M)$ is bounded. In fact, $|x(M)| \leqslant \|x\|$. It may be shown that $\sup |x(M)| = \lim_{n\to\infty} \sqrt[n]{\|x^n\|}$. Hence we obtain

Theorem 5. *If in the ring R there are no elements different from zero, for which $\sqrt[n]{\|x^n\|} \to 0$, then our ring is isomorphic to the ring of functions defined on the set of maximal ideals.*

We shall show that these functions $x(M)$ are continuous. To this end we must topologize the set $\mathfrak{M}$ of maximal ideals. We shall define the topology by a system of neighbourhoods.

Definition 4. A set K of elements of the ring R is called a set of generators of the ring R, if the smallest closed ring containing K is the whole ring R.

Definition 5. By the neighbourhood of the maximal ideal M_0 we shall mean the set of maximal ideals satisfying the inequalities

$$|x_i(M) - x_i(M_0)| < \varepsilon, \quad (i = 1, 2, \ldots, n),$$

where ε and n are arbitrary and $x, \ldots, x_n$ are arbitrary elements of the set K of generators of the ring R.

The so obtained topology in $\mathfrak{M}$ does not depend on the choice of K. The set $\mathfrak{M}$ of maximal ideals in this topology proves to be a bicompact Hausdorff space and the functions $x(M)$ are continuous on it. As the result we obtain the following fundamental

Theorem 6. *Every abstract normed ring R may be homomorphically mapped in the ring of continuous functions defined on the Hausdorff bicompact space of maximal ideals of the ring R. For the isomorphism it is necessary and sufficient that the conditions of Theorem 5 should be satisfied.*

In some cases the topological properties of the set $\mathfrak{M}$ may be stated more precisely, namely:

1. If the ring R is separable, then the set $\mathfrak{M}$ is metrisable.
2. If the ring R is generated by a system of n generators, then the set $\mathfrak{M}$ is homeomorphic to a compact subspace of the n-dimensional complex space.

Theorem 7. *If in the ring R for each element x there is a complex-conjugated element, i. e. such an element y that $y(M) = \overline{x(M)}$, whatever be M from $\mathfrak{M}$, then each function continuous on the set $\mathfrak{M}$ is the limit of a uniformly convergent sequence of functions corresponding to the elements of the ring R*.*

Hence it follows that if in the ring R a norm is introduced in such a way that from the uniform convergence of the functions $x_n(M)$ follows the convergence of the norms of the elements x_n (for which it is, for instance, sufficient to demand that $\|x^2\| = \|x\|^2$), then the ring R consists of all functions continuous on $\mathfrak{M}$ (we assume that the conditions of theorem 7 are satisfied in the ring R).

Thus, in many cases the aggregate of functions $x(M)$ uniquely determines the ring. This follows from

---

* This theorem was proved by the author together with G. Šilov. The proof will be given in their common paper.

Theorem 8. *Let there be given two normed rings R_1 and R_2. Let the intersection of all maximal ideals in each of them be the nul ideal. Then from the algebraic isomorphism of the rings R_1 and R_2 follows their continuous isomorphism.*

The applications of the above results will be given elsewhere. A detailed exposition will be published in the «Recueil Mathématique de Moscou».

Institute of Mathematics.
Moscow State University.

Received
27. III. 1939.

5.

To the theory of normed rings. II.
On absolutely convergent trigonometrical series and integrals

Dokl. Akad. Nauk SSSR **25** (1939) 570–572. Zbl. **22**:357

(Communicated by A. N. Kolmogorov, Member of the Academy, 10. X. 1939)

N. Wiener(1), Cameron(2), Pitt(3), and N. Wiener and Pitt (4) proved some interesting theorems on absolutely convergent trigonometrical series and Fourier integrals. In the present note we indicate a general method of obtaining theorems of this type. As a direct application of this method we obtain and generalize the theorems of the above authors.

Let us illustrate this method in all details on the simplest of Wiener's theorems:

Theorem 1. *If a function $f(t)$ can be developed into an absolutely convergent Fourier series and does not vanish, then $\frac{1}{f(t)}$ may be also developed into an absolutely convergent Fourier series.*

The proof is based on the following simple

Lemma. In order that an element x of a normed ring R should have an inverse element, it is necessary and sufficient that it should not belong to any maximal ideal of the ring R.

If x has an inverse element, then it does not belong to any non-trivial ideal and, a fortiori, to any maximal ideal.

Suppose that x has no inverse element. Then the aggregate of elements xy, where y runs through the whole R, forms an ideal not coinciding with the whole ring; by Theorem 2 of my preceding note(5) it is contained in some maximal ideal of R.

Consider the ring R formed by all functions $f(t)=\sum_{n=-\infty}^{\infty} a_n e^{int}$ such that $\sum_{n=-\infty}^{\infty} |a_n| < \infty$; under the sum and the product of elements of the ring R we understand the sum and the product (in the usual sense) of functions, and put the norm $\|f(t)\| = \sum_{n=-\infty}^{\infty} |a_n|$. It is easily verified that all axioms of a normed ring are satisfied.

Let M be an arbitrary maximal ideal of R. By Theorem 4 (5) to every element $x \in R$ we correlate a number $x(M)$. Let a be the number corresponding to the function e^{it}; then to the function e^{-it} corresponds the number $\frac{1}{a}$. Since $|a| \leqslant \|e^{it}\| = 1$ and $\frac{1}{|a|} \leqslant \|e^{-it}\| = 1$, we have $|a|=1$,

i. e. $a=e^{it_0}$ and $\frac{1}{a}=e^{-it_0}$: Hence follows that to every trigonometrical polynomial $\sum_{n=-N}^{n=N} a_n e^{int}$ corresponds the number $\sum_{n=-N}^{n=N} a_n e^{int_0}$ which is the value of this polynomial at the point $t=t_0$.

Since the homeomorphism $R \to \frac{R}{M}$ is continuous, to every function $f(t) \in R$ corresponds the number $f(t_0)$. The maximal ideal M consists of all functions $f(t)$ to which corresponds the number 0; in other words, the maximal ideal consists of all functions vanishing at a given point. Consequently, the sentence «$f(t)$ does not vanish anywhere» means: «$f(t)$ does not belong to any maximal ideal». By the Lemma, the element $f(t)$ has an inverse element.

By the same method may be proved the following proposition:

Theorem 2. *Let there be given a formal power series* $\sum_{-\infty}^{\infty} a_n x^n$ *such that* $\sum_{-\infty}^{\infty} |a_n| \alpha_n < \infty$, *where* $\alpha_{m+n} \leqslant C\alpha_m \alpha_n$, $\alpha_k > 0$. *In order that there should exist a series* $\sum_{-\infty}^{\infty} b_n x^n$ *with coefficients satisfying the condition* $\sum_{-\infty}^{\infty} |b_n| \alpha_n < \infty$ *such that*

$$\sum_{-\infty}^{\infty} a_n x^n \sum_{-\infty}^{\infty} b_n x^n = 1,$$

it is necessary and sufficient that the function $\varphi(r, t) = \sum_{-\infty}^{\infty} a_n r^n e^{int}$ *should not vanish in the domain* $r_1 \leqslant r \leqslant r_2$, $0 \leqslant t \leqslant 2\pi$, *where*

$$r_1 = \lim_{n \to +\infty} \alpha_n^{\frac{1}{n}}, \quad r_2 = \lim_{n \to -\infty} \alpha_n^{\frac{1}{n}}. \tag{1}$$

Remark 1. For $r_1 \leqslant r \leqslant r_2$ the series $\sum_{-\infty}^{\infty} a_n r^n e^{int}$ converges.

Remark 2. The limits (1) exist.

Before formulating a similar theorem for Fourier-Stieltjes integrals, we introduce the following notations.

Let $f(t)$ be a function having on every finite segment a bounded variation; then $f(t) = g(t) + h(t) + s(t)$, where $g(t)$ is absolutely continuous on every finite segment, $h(t)$ is a step-function and $s(t)$ is the singular component, i. e. a continuous function of bounded variation possessing almost everywhere a derivative equal to zero.

Theorem 3. *If* $f(t)$ *is such that*

1) $\int_{-\infty}^{\infty} \alpha(t) |df(t)| < \infty$,

where $\alpha(t+s) \leqslant C\alpha(t)\alpha(s)$, $\alpha(t) > 0$, $\alpha(t)$ *is continuous,*

2) $\left| \int_{-\infty}^{\infty} e^{\rho t + i\lambda t} df(t) \right| \geqslant C > 0$

for all t and all ρ *satisfying the inequality*

$$\lim_{\lambda\to-\infty} \frac{\ln\alpha(\lambda)}{\lambda} \leqslant \rho \leqslant \overline{\lim_{\lambda\to-\infty}} \frac{\ln\alpha(\lambda)}{\lambda}$$

and

$$3)\quad \inf\left|\int_{-\infty}^{\infty} e^{\rho\lambda+i\lambda t}\,dh(t)\right| > \int_{-\infty}^{\infty} \alpha(t)\,|\,ds(t)\,|,$$

then there exists a function $\varphi(t)$ *of bounded variation on every finite segment such that*

$$1)\quad \int_{-\infty}^{\infty} \alpha(t)\,|\,d\varphi(t)\,| < \infty$$

and

$$2)\quad \int_{-\infty}^{\infty} f(s)\,d\varphi(t-s) = \begin{cases} 1 & \text{for } t>0, \\ 0 & \text{for } t\leqslant 0. \end{cases}$$

In the case $\alpha(t)\equiv 1$ this theorem is proved by Wiener and Pitt ([4]).

Finally, by the same methods we may prove the following general proposition.

Theorem 4. *Let G be an abelian group with the elements* $t, s, \ldots$, $\alpha(t)$ — *a positive function defined on G and satisfying the condition* $\alpha(t+s)\leqslant C\alpha(t)\alpha(s)$, $f(s)$ — *a function defined on G, differing from zero on a set of points of G at most enumerable and such that*

$$1)\quad \sum_{t\in G} \alpha(t)\,|\,f(t)\,| < \infty,$$

$$2)\quad \sum_{t\in G} f(t)\,e^{\rho(t)}\,\chi(t) \neq 0$$

for any character $\chi(t)$ *of the group G and any function* $\rho(t)$ *satisfying the conditions*: $\rho(t+s)=\rho(t)+\rho(s)$, $\rho(t)$ *is real and* $\rho(t)\leqslant \ln\alpha(t)$.

In this case there exists a function $g(t)$ *different from zero on a set of points of G at most enumerable such that*

$$1)\quad \sum_{t\in G} \alpha(t)\,|\,g(t)\,| < \infty$$

and

$$2)\quad \sum_{s\in G} g(t-s)\,f(s) = \begin{cases} 0 & \text{for } t\neq 0, \\ 1 & \text{for } t=0. \end{cases}$$

Theorem 1 follows from Theorem 4 if for the group G we take the additive group of integers and put $\alpha(t)=1$.

Remark. From the course of the proof of Theorem 4 follows the existence of an additive function $\rho(t)\leqslant \ln\alpha(t)$. In other words, if we denote $\ln\alpha(t)$ by $\beta(t)$, we obtain the following assertion:

Let on an abelian group $\mathfrak{G}$ be defined a function $\beta(t)$ such that $\beta(t+s)\leqslant\rho(t)+\beta(s)$ and $\beta(0)=0$. Then there exists a function $\rho(t)\leqslant\beta(t)$ such that $\rho(t+s)=\rho(t)+\rho(s)$.

Steklov Mathematical Institute.
Academy of Sciences of the USSR.

Received
17.X.1939.

REFERENCES

[1] N. Wiener, Annals of Math., 33, 1—100 (1932). [2] R. H. Cameron, Duke Math. Journal, 3, 662 (1937). [3] H. R. Pitt, Journal of Math. and Phys., M. I. T., 16, 191 (1938). [4] N. Wiener and H. R. Pitt, Duke Math. Journal, 4, 420 (1938). [5] I. Gelfand, C. R. Acad. Sci. URSS. XXIII, No. 5 (1939).

6.

To the theory of normed rings. III. On the ring of almost periodic functions

Dokl. Akad. Nauk SSSR **25** (1939) 573–574. Zbl. **22**: 357

(Communicated by A. N. Kolmogorov, Member of the Academy, 10. X. 1939)

The aggregate R of all almost periodic functions $x(t)$ in the sense of Bohr forms, according to the definitions of my earlier note ([1]), is a normed ring, if the operations of addition and multiplication are understood in the usual sense and the norm is defined by the condition

$$\|x(t)\| = \sup_{-\infty<t<\infty} |x(t)|.$$

Observe that according to Theorem 8 of my note ([1]) just referred to, this is the unique (up to an equivalence) way of norming the ring R.

By the fundamental theorem of Bohr the system of functions $\{e^{2\pi i\lambda t}\}$ is a system of generators of the ring R.

Let M be an arbitrary maximal ideal of the ring R. Under the homeomorphism $R \to \frac{R}{M}$ to the function $e^{2\pi i\lambda t}$ corresponds a complex number $\varphi(\lambda)$, $|\varphi(\lambda)| \leqslant \|e^{2\pi i\lambda t}\| = 1$. From the equality $e^{2\pi i\lambda t} \cdot e^{2\pi i(-\lambda)t} = 1$ follows $|\varphi(\lambda)| \cdot |\varphi(-\lambda)| = 1$, whence $|\varphi(\lambda)| = 1$; from the equality $e^{2\pi i(\lambda+\mu)t} = e^{2\pi i\lambda t} \cdot e^{2\pi i\mu t}$ follows that $\varphi(\lambda+\mu) = \varphi(\lambda)\varphi(\mu)$. The function $\varphi(\lambda)$ defines thus a homeomorphical mapping of the additive group K of real numbers into the group H of rotations of a circumference; as is known, such a mapping is called a character of the discrete group K. Let us show that also the converse is true: to every character $\varphi(\lambda)$ of the discrete group K corresponds a maximal ideal M of the ring R such that under the homeomorphism $R \to \frac{R}{M}$ the function $e^{2\pi i\lambda t}$ is transformed into $\varphi(\lambda)$.

Lemma. Let there be given a polynomial in n variables $P(x_1, x_2, \ldots, x_n)$. If the real numbers $\lambda_1, \lambda_2, \ldots, \lambda_n$ are linearly independent, then

$$\max_{|x_k|=1,\ k=1,2,\ldots,n} |P(x_1, x_2, \ldots, x_n)| = \sup_{-\infty<t<\infty} |P(e^{2\pi i\lambda_1 t}, e^{2\pi i\lambda_2 t}, \ldots, e^{2\pi i\lambda_n t})|. \quad (1)$$

For the proof observe that the polynomial on the right-hand side of (1) is a uniformly continuous function of its arguments; for any given $\varepsilon > 0$ we may find such a $\delta > 0$ that a variation of every argument by a quantity not exceeding δ changes the value of the polynomial by not more than ε. The function e^{iu} is uniformly continuous for $-\infty < u < \infty$ and hence there exists such an $\eta > 0$ that $|u_1 - u_2| < \eta$ implies $|e^{iu_1} - e^{iu_2}| < \delta$. Let $x_1^0 = e^{2\pi i a_1}$, $x_2^0 = e^{2\pi i a_2}, \ldots, x_n^0 = e^{2\pi i a_n}$ be arbitrary values ot the variables $x_1, x_2, \ldots, x_n$. By Kronecker's theorem we may find a real number t_0 and integers $m_1, m_2, \ldots, m_n$ such that the following inequalities will be satisfied:

$$|a_k - \lambda_k t_0 - m_k| < \frac{\eta}{2\pi}\ (k = 1, 2, \ldots, n), \text{ or } |(2\pi a_k - 2\pi m_k) - 2\pi\lambda_k t_0| < \eta,$$

whence

$$|e^{2\pi i a_k - 2\pi i m_k} - e^{2\pi i \lambda_k t_0}| = |e^{2\pi i a_k} - e^{2\pi i \lambda_k t_0}| < \delta,$$

$$|P(x_1^0, x_2^0, \ldots, x_n^0) - P(e^{2\pi i \lambda_1 t_0}, e^{2\pi i \lambda_2 t_0}, \ldots, e^{2\pi i \lambda_n t_0})| < \varepsilon. \qquad (2)$$

The equality (1) is an obvious corollary of the inequality (2).

Let now be given an arbitrary character $\varphi(\lambda)$ of the group K. Correlate to the trigonometrical polynomial $P(e^{2\pi i \lambda_1 t}, e^{2\pi i \lambda_2 t}, \ldots, e^{2\pi i \lambda_n t})$, where $\lambda_1, \lambda_2, \ldots, \lambda_n$ are arbitrary real numbers, the quantity

$$P(\varphi(\lambda_1), \varphi(\lambda_2), \ldots, \varphi(\lambda_n)). \qquad (3)$$

From the condition $\varphi(\lambda+\mu) = \varphi(\lambda)\varphi(\mu)$ immediately follows that to the sum and the product of trigonometrical polynomials corresponds the sum and the product of the quantities (3). From the lemma proved above follows that in the case when $\lambda_1, \lambda_2, \ldots, \lambda_n$ are linearly independent,

$$|P(\varphi(\lambda_1), \varphi(\lambda_2), \ldots, \varphi(\lambda_n)) \leqslant \| P(e^{2\pi i \lambda_1 t}, e^{2\pi i \lambda_2 t}, \ldots, e^{2\pi i \lambda_n t}) \|.$$

Thus, to a uniformly convergent sequence of such polynomials corresponds a convergent sequence of quantities (3). Since a linear combination $\sum c_k e^{2\pi i \mu_k t}$ with arbitrary μ_k may be represented as a polynomial $P(e^{2\pi i \lambda_1 t}, e^{2\pi i \lambda_2 t}, \ldots, e^{2\pi i \lambda_n t})$ with linearly independent λ_k, we may, applying the fundamental theorem of Bohr, extend the correspondence between the polynomials

$$P(e^{2\pi i \lambda_1 t}, e^{2\pi i \lambda_2 t}, \ldots, e^{2\pi i \lambda_n t})$$

and the complex numbers (3) to all functions $x(t) \in R$. To the sum and the product of elements $x(t)$ will correspond the sum and the product of numbers; hence easily follows that those functions $x(t)$, to which corresponds the number 0, form a maximal ideal $M \subset R$.

According to the general theory, R is homeomorphically mapped into the ring of continuous functions defined on the set of all its maximal ideals; since here the conditions of Theorem 7 ([1]) are satisfied, we conclude the following

Fundamental Theorem. *The ring* R *of all almost periodic functions in the sense of Bohr is isomorphic to the ring of all continuous functions defined on the group of characters of the additive group of real numbers.*

The topology on the group of characters is introduced according to ([1]) thus: the neighbourhood of the character $\varphi_0(\lambda)$ is called the aggregate of all characters $\varphi(\lambda)$ satisfying the following inequalities:

$$|\varphi(\lambda_k) - \varphi_0(\lambda_k)| < \varepsilon \quad (k = 1, 2, \ldots, n),$$

where $\varepsilon > 0$; $\lambda_1, \lambda_2, \ldots, \lambda_n$ (n — an integer) are arbitrary fixed numbers. The group of characters topologized in this manner becomes a bicompact Hausdorff space *.

The characters $\varphi(\lambda)$ having the form $e^{2\pi i \lambda t_0}$ (continuous characters) compose in the group of all characters an everywhere dense subgroup isomorphic to the additive group of real numbers with the following topology: the neighbourhood of the number t_0 is formed by all numbers t satisfying the inequalities

$$|t - t_0 - n\lambda_i| < \delta_i \quad (i = 1, 2, \ldots, m;\ n = 0, \pm 1, \pm 2, \ldots),$$

where λ_i, δ_i (m—an integer) are arbitrary fixed numbers.

Steklov Mathematical Institute.
Academy of Sciences of the USSR.

Received
17.X. 1939.

REFERENCES

[1] I. Gelfand, C. R. Acad. Sci. URSS, XXIII, No. 5 (1939).

* This topologization of the group of characters (as the set of maximal ideals) coincides with the usual topologization of the group of characters introduced by van Kampen.

7.

(with D. A. Rajkov)

On the theory of characters of commutative topological groups

Dokl. Akad. Nauk SSSR **28** (1940) 195–198. Zbl. **24**:120

(Communicated by A. N. Kolmogoroff, Member of the Academy, 13. V. 1940)

Let G be a commutative topological group (with the second axiom of enumerability or locally bicompact), for all sets X of whose elements an exterior measure $\mu(X)$ is defined which satisfies the following conditions:

I_1) $0 \leqslant \mu(X) \leqslant \infty$. I_2) $\mu(X) \leqslant \Sigma\mu(X_n)$, if $X \subset \Sigma X_n$ (the sum being finite or enumerable). I_3) $\mu(\Lambda) = 0$, Λ being the empty set.

II_1) All open sets are measurable and their measures are greater than zero. II_2) $\mu(X) = \inf \mu(U)$, where U are open sets covering X. II_3) There exists a neighbourhood of the zero of the group having a finite measure.

III_1) The measure is invariant: $\mu(X+g) = \mu(X)$ for all X and all $g \in G$. III_2) The measure is symmetric: $\mu(-X) = \mu(X)$.

IV) The measure is continuous with respect to translations: if $\mu(X) < \infty$, then for every $\varepsilon > 0$ there exists a neighbourhood U_ε of the zero of the group such that $\mu(X) - \mu(X \cap X+g) < \varepsilon$ for all $g \in U_\varepsilon$.

The aggregate of all summable (with respect to the measure μ) complex functions $x(g)$ forms a Banach space L with the usual addition and multiplication by complex numbers and with the norm

$$\|x\| = \int |x(g)|\, dg.$$

We may also define in L the multiplication («Faltung») of elements: in virtue of Fubini's theorem, for arbitrary $x, y \in L$ the function

$$x * y(g) = \int x(g-h)\, y(h)\, dh = \int y(g-h)\, x(h)\, dh = y * x(g)$$

exists for almost all g and belongs to L, and

$$\|x * y\| \leqslant \|x\|\, \|y\|.$$

From IV it follows that if $y \in L$ is bounded, then $z = x * y$ is continuous and bounded for arbitrary $x \in L$.

Two cases are possible:

1°. Every point $g \in G$ has a positive measure. In this case the group is discrete. The function $e(g)$ equal to 1 for $g = 0$ and to 0 for $g \neq 0$ is the unit with respect to multiplication introduced in L. L is a normed ring (1).

2°. Every point $g \in G$ has the measure 0. L does not contain the unit; in fact, the bounded discontinuous functions contained in L yield

in «Faltung» with an arbitrary function from L, continuous functions. In order to extend L to a normed ring R we adjoin formally the unit e and count as elements of the ring R the sums $\mathfrak{z} = \lambda e + x(g)$, where λ are complex numbers and $x \in L$. The algebraical operations are defined naturally; the norm $\|\mathfrak{z}\|$ we define as $|\lambda| + \|x\|$. The number, into which the element $\mathfrak{z}$ passes under the homeomorphism $R \to R/M$ of the ring R by its maximal ideal M, we shall denote by $(\mathfrak{z}, M)$. The aggregate of all functions from L forms in the ring R a maximal ideal; denote it by M_0. We have $(\mathfrak{z}, M_0) = \lambda$, in particular $(x, M_0) = 0$ for all $x \in L$.

The considerations which follow are carried out for the ring R, however, they remain valid with corresponding simplifications also in the case 1°, when L is itself a normed ring with a unit.

Theorem 1. *Every maximal ideal M of the ring R different from M_0 generates a continuous character of the group G.*

Proof. Since $M \neq M_0$, there exists $x \in L$ such that $(x, M) \neq 0$. Put

$$\chi(h) = \frac{(x(g+h), M)}{(x(g), M)}. \tag{1}$$

We have $\chi(0) = 1$, $|\chi(h)| \leqslant \frac{\|x\|}{|(x, M)|} = C$; further,

$$|\chi(a+h) - \chi(a)| \leqslant \frac{\|x(g+a+h) - x(g+a)\|}{|(x, M)|}$$

and since, in virtue of IV and III, the numerator is continuous with respect to h and does not depend on a, $\chi(h)$ is a uniformly continuous function. In virtue of the fact that the measure is invariant,

$$x(g+a+b) * x(g) = x(g+a) * x(g+b)$$

whence

$$(x(g+a+b), M)(x(g), M) = (x(g+a), M)(x(g+b), M).$$

Dividing by $[x(g), M]^2$, we obtain

$$\chi(a+b) = \chi(a)\chi(b). \tag{2}$$

Hence follows that $|\chi(h)| \leqslant 1$. In fact, if $|\chi(h_0)| = 1 + \varepsilon > 1$, then $|\chi(nh_0)| = (1+\varepsilon)^n$, which contradicts to the boundedness of $\chi(h)$. Putting in (2) $b = -a$, we obtain $1 = \chi(a)\chi(-a)$, whence $|\chi(h)| \equiv 1$. Thus, $\chi(h)$ is a continuous character of the group G. From the equality $x(g+h) * y(g) = x(g) * y(g+h)$ it follows that

$$\chi(h)(y(g), M) = (y(g+h), M) \quad \text{for all } y \in L \tag{3}$$

i. e. that χ does not depend on the choice of x.

Theorem 2. *Every measurable character $\chi(h)$ of the group G generates a maximal ideal of the ring R different from M_0.*

Proof. To every $\mathfrak{z} = \lambda e + x(g)$ we correlate the number

$$(\mathfrak{z}, \chi) = \lambda + \int x(g)\overline{\chi}(g)\,dg.$$

To the addition and multiplication of the elements of the ring will correspond the usual multiplication and addition of these numbers (as regards addition this is obvious, whereas for multiplication it may be verified by means of Fubini's theorem). Hence follows that the elements $\mathfrak{z} \in R$, for which $(\mathfrak{z}, \chi) = 0$, form a maximal ideal M_χ of the ring R. Taking $\mathfrak{z} = x(g)\chi(g)$, where $\int x(g)\,dg \neq 0$, we obtain $(\mathfrak{z}, \chi) \neq 0$ and, consequently, $M_\chi \neq M_0$.

Theorem 3. *A measurable character coincides with the continuous character generated by the maximal ideal generated by it. Thus, every measurable character is continuous.*

Proof. Let $(x, \chi) \neq 0$. We have

$$(x(g+h), \chi) = \int x(g+h)\overline{\chi}(g)\,dg = \chi(h)\int x(g+h)\overline{\chi}(g+h)\,dg =$$
$$= \chi(h)\int x(g)\overline{\chi}(g)\,dg = \chi(h)\,(x(g), \chi),$$

whence

$$\chi(h) = \frac{(x(g+h), \gamma)}{(x(g), \chi)}.$$

Theorem 4. *Every continuous character is generated by only one maximal ideal.*

Proof. Let the maximal ideal M generate the character $\chi(g)$. If M does not coincide with the maximal ideal M_χ generated by this character, then there exists a function $x \in L$ such that

$$(x, M) \neq \int x(h)\overline{\chi}(h)\,dh.$$

Then for an arbitrary function $y \in L$ with $(y, M) \neq 0$ we obtain

$$(x, M)(y, M) = \left(\int y(g-h)\,x(h)\,dh, M\right) \neq$$
$$\neq \int x(h)\,(y(g), M)\,\overline{\chi}(h)\,dh = \int (y(g-h)\,x(h), M)\,dh.$$

Thus, it suffices to prove that for arbitrary $x, y \in L$ holds the equality

$$\left(\int y(g-h)\,x(h)\,dh, M\right) = \int (y(g-h)\,x(h), M)\,dh. \tag{4}$$

But the linear functional (x, M) defines in G a complex «measure» which, in virtue of the inequality $|(x_X(g), M)| \leqslant \|x_X(g)\| = \mu(X)$ [$x_X(g)$ is the characteristic function of the set X], satisfies the conditions which are sufficient for (4) to be considered as a consequence of Fubini's theorem.

Corollary [Generalization of Wiener's Theorem(²)]. *In order that the element $\zeta = \lambda e + x(g)$ should have in the ring R an inverse, it is necessary and sufficient that for all continuous characters of the group G should hold the inequality*

$$\lambda\left(\lambda + \int x(g)\,\chi(g)\,dg\right) \neq 0.$$

Theorem 5. *The ring R does not contain generalized nilpotent elements.*

The proof of theorem 5 is based on the following

Lemma. *A bounded Hermite-symmetrical function $x \in L$ with a positive norm is not a generalized nilpotent element of the ring R.*

Proof of the Lemma. By condition, $x(-g) = \overline{x}(g)$. Put $x^{(0)}(g) = e$, $x^{(n)}(g) = x^{(n-1)}(g) * x(g)$ $(n = 1, 2, \ldots)$. The $x^{(n)}(g)$ are Hermite-symmetrical and for $n \geqslant 2$ continuous. Were $x(g)$ a generalized nilpotent element, we would have $\|\lambda^n x^{(n)}\| \to 0$ for arbitrary λ. Let us show that $\|\lambda^{2n-1} x^{(2n-1)}\| \to \infty$ for all sufficiently large λ. At first

$$|\lambda^n x^{(n)}(0)| = |\lambda| \left|\int x(-g)\,\lambda^{n-1} x^{(n-1)}(g)\,dg\right| \leqslant$$
$$\leqslant |\lambda| \sup|x(g)|\,\|\lambda^{n-1} x^{(n-1)}\|. \tag{5}$$

Put $x^{(2n)}(0) = c_n$. We have $c_n = \int |x^n(g)|^2\,dg$. Hence $c_1 > 0$ and, in virtue of the continuity of $x^{(2)}(g)$, also $c_2>0$, since otherwise $c_1=x^{(2)}(0)=0$. Further, by Schwarz' inequality,

$$c_n^2=\left|\int x^{(n-1)}(-g)\,x^{(n+1)}(g)\,dg\right|^2\leqslant$$
$$\leqslant\int|x^{(n-1)}(g)|^2\,dg\int|x^{(n+1)}(g)|^2\,dg=c_{n-1}c_{n+1}.$$

Thus, we have

$$0<\frac{c_2}{c_1}\leqslant\frac{c_3}{c_2}\leqslant\ldots\leqslant\frac{c_{n+1}}{c_n}\leqslant\ldots,$$

and consequently $\lambda^{2n}c_n\to\infty$ for all $\lambda>\lambda_0\geqslant 0$. Then, in virtue of (5), also $\|\lambda^{2n-1}x^{(2n-1)}\|\to\infty$.

Proof of Theorem 5. If $\mathfrak{z}=\lambda e+x(g)$ is a generalized nilpotent element, then $\lambda=(\mathfrak{z}, M_0)=0$ and $\mathfrak{z}=x(g)$. Let $\|x\|>0$. It is easily shown that there exists a bounded function $u\in L$ such that $y(g)=x*u(g)\not\equiv 0$, and moreover, $\Re y(0)\neq 0$. $y(g)$ is continuous and bounded. Consider the function $z(g)=y(g)+\bar{y}(-g)$. We have $z(-g)=\bar{z}(g)$ and $z(0)=2\Re y(0)\neq 0$. Since $z(g)$ is continuous, $\|z\|>0$. Thus, in virtue of the lemma, $z(g)$ cannot be a generalized nilpotent element. But then the same is true also with respect to $y(g)$, since, together with $y(g)$, $\bar{y}(-g)$ would also be a generalized nilpotent element. But in this case $\mathfrak{z}$ as well could not be a generalized nilpotent element.

Theorem 6. *The group G possesses a sufficient set of continuous characters: for every $g_0\in G$ there exists a continuous character $\chi(g)$ of the group G such that $\chi(g_0)\neq 1$.*

Proof. Choose a function $x(g)\in L$ such that $\|x(g+g_0)-x(g)\|>0$. Since, by theorem 5, $x(g+g_0)-x(g)$ is not a generalized nilpotent element, we can find such a maximal ideal M that $(x(g+g_0), M)\neq(x(g), M)$. From (1) then follows that

$$\chi(g_0)=\frac{(x(g+g_0), M)}{(x(g), M)}\neq 1.$$

Theorems 1—4 show that the set X of all continuous characters of the group G may be considered as the set of all maximal ideals of the ring R different from M_0. Let $\mathfrak{M}$ be the bicompact topological space of all maximal ideals of the ring R [([1]), definition 5]. Then $\mathfrak{M}-M_0=X$ is a locally bicompact space (with the second axiom of enumerability)*. The neighbourhoods of the character χ_0 are defined by arbitrary functions $x_k(g)\in L$ $(k=1,\ldots,n;\ n=1, 2,\ldots)$ and arbitrary $\varepsilon>0$ as the aggregates of all characters χ satisfying the inequalities

$$\left|\int x_k(g)\,\bar{\chi}(g)\,dg-\int x_k(g)\,\bar{\chi}_0(g)\,dg\right|<\varepsilon\quad(k=1,\ldots,n).$$

It may be shown that if G is a locally bicompact group, then this topology coincides with the usual one [([3]), definition 34] and X forms a topological group.

Steklov Institute of Mathematics.
Academy of Sciences of the USSR.

Received
14. V. 1940.

REFERENCES

[1] I. M. Gelfand, C. R. Acad. Sci. URSS, XXIII, No. 5 (1939). [2] I. M. Gelfand, ibid., XXV, No. 7 (1939). [3] Л. С. Понтрягин, Непрерывные группы (1938).

* In the case, when G is discrete, M_0 is absent and X coincides with $\mathfrak{M}$.

8.

Normierte Ringe[1]

Mat. Sb., Nov. Ser. 9 (51) (1941) 3–23. Zbl. 24:320

§ 1. Die Axiomatik

Die Gesamtheit R der Elemente $x, y, \ldots$ wird ein n o r m i e r t e r R i n g genannt, wenn:

(α) R ein linearer, normierter, vollständiger Raum im Sinne von Banach ist.

(β) In R die Operation der Multiplikation der Elemente definiert ist, die die gewöhnlichen algebraischen Eigenschaften

$$x(\lambda y + \mu z) = \lambda xy + \mu xz,$$
$$x(yz) = (xy)z$$

besitzt.

(γ) In R eine Einheit vorhanden ist, d. h. ein solches Element e, dass $ex = x$ für ein beliebiges Element $x \in R$, dabei $\|e\| \neq 0$.

(δ) Die Operation der Multiplikation bezüglich des jeden der Faktoren stetig ist: wenn $x_n \to x$, so $x_n y \to xy$, und wenn $y_n \to y$, so $xy_n \to xy$.

Im Folgenden werden wir die Operation der Multiplikation als kommutativ betrachten; die Ergebnisse des §1 hängen aber von der Kommutativität nicht ab.

Das Axiom (δ) werden wir im Folgenden in der folgenden Form benutzen:

(δ') $\quad\|x\cdot y\| \leqslant \|x\|\cdot\|y\|$ und $\|e\| = 1$.

A priori würde es scheinen, dass (δ') eine wesentliche Verstärkung von (δ) ist; wir werden aber zeigen, dass in der Tat (δ') und (δ) äquivalent sind. Es gilt nämlich der folgende

S a t z 1. *Für jeden normierten Ring R kann man einen isomorphen und homöomorphen Ring R' finden, in dem das Axiom (δ') Platz hat.*

Es sei Q der Ring aller linearen Operatoren über R (der als linearer Raum betrachtet wird). Jedem Element x können wir einen linearen Operator $A_x \in Q$ zuordnen, der folgendermassen definiert ist:

$$A_x y = x \cdot y.$$

Offenbar gehen verschiedene Elemente in verschiedene Operatoren über, die Summe und das Produkt von Elementen gehen in die Summe und das Produkt von Operatoren über, das Element e geht in den Einheitsoperator E über.

Somit existiert in Q ein Unterring R', der R isomorph ist. Finden wir, durch welche Eigenschaft werden die linearen Operatoren $A \in R'$ charakterisiert.

[1] Ein Abriss des Inhaltes dieser Arbeit (ausser den letzten zwei Paragraphen) ist in den C. R. de l' Ac. d. Sc. de l' U. R. S. S., **XXIII**, N. 5, (1939) veröffentlicht.

Wenn $A \in R'$, so ist $A = A_x$, und daher

$$A(yz) = x \cdot yz = (xy)z = Ay \cdot z$$

für beliebige y und z aus R. Zeigen wir, dass wenn ein linearer Operator $A \in Q$ die Eigenschaft $A(yz) = Ay \cdot z$ (für beliebige y und z aus R) besitzt, so $A \in R'$. In der Tat, sei $Ae = x$; dann ist

$$Ay = A(ey) = Ae \cdot y = xy$$

für beliebiges $y \in R$.

Aus diesem Kriterium folgt leicht, dass R' ein abgeschlossener Unterring von Q ist. Es mögen die $A_n \in R'$ zu $A \in Q$ konvergieren. Dann ist

$$A(x \cdot y) = \lim A_n(xy) = \lim (A_n x) y = Ax \cdot y,$$

woraus, nach dem Bewiesenen, $A \in R'$ folgt. Also ist R' ein vollständiger Raum. Es sei $\|e\| = a$. Dann ist für $A_x \in R'$

$$\|A_x\| = \sup_{\|y\| \leq 1} \|A_x y\| = \sup_{\|y\| \leq 1} \|x \cdot y\| \geq \frac{1}{a} \|x \cdot e\| = \frac{1}{a} \|x\|.$$

Dies bedeutet, dass die Abbildung von R' auf R stetig ist. Da die Räume R und R' vollständig sind, so folgt aus der Stetigkeit der Abbildung von R' auf R, dass, nach einem bekannten Satz von Banach, die inverse Abbildung stetig ist; so bekommen wir, dass R R' homöomorph ist. Aber für die Elemente $A \in R'$ als lineare Operatoren ist das Axiom (δ') offenbar erfüllt.

F o l g e r u n g 1. Das Produkt xy ist nach beiden Faktoren gleichzeitig stetig: wenn $x_n \to x$ und $y_n \to y$, so $x_n y_n \to xy$.

F o l g e r u n g 2. In absolut konvergenten Reihen kann man, ohne die Summe zu ändern, die Summanden beliebig umstellen; solche Reihen kann man multiplizieren (addieren), wobei die erhaltenen Reihen zu dem Produkt (der Summe) der Reihen konvergieren.

§ 2. Vorläufige Tatsachen über Ideale

H i l f s s a t z 1. *Wenn $x \in R$ und $\|e - x\| < 1$, so besitzt x ein inverses Element.*

Zum Beweis betrachten wir die Reihe

$$e + (e - x) + (e - x)^2 + \ldots$$

Da $\|(e - x)^n\| \leq \|e - x\|^n$ ist, konvergiert diese Reihe; es sei y ihre Summe. Bilden wir das Produkt von y und $x = e - (e - x)$ und benutzen die Folgerung 2 aus Satz 1, so bekommen wir:

$$yx = e + (e - x) + (e - x)^2 + \ldots - (e - x) - (e - x)^2 - \ldots = e.$$

Folglich ist $y = x^{-1}$.

Somit besitzt die Einheit eine Umgebung $U(e)$, die aus Elementen besteht, welche inverse Elemente besitzen. Es ist leicht einzusehen, dass eine solche Umgebung für jedes Element x existiert, das ein inverses Element besitzt. Es sei $xy = e$, und sei $U(x)$ eine solche Umgebung von x, dass $U(x)y \subset U(e)$; ist z ein beliebiges Element von $U(x)$, so besitzt $z \cdot y$ ein inverses Element w,

$zyw = e$, woraus folgt, dass das Element z auch ein inverses Element besitzt. Folglich ist die Menge V der Elemente, die inverse Elemente besitzen, in R offen.

Hilfssatz 2. *x^{-1} ist eine stetige Funktion von x auf V. Mit anderen Worten, wenn die $x_n \in V$ zu $x \in V$ konvergieren, so*

$$x_n^{-1} \to x^{-1}.$$

Sei $x = e$; dann ist die Folge x_n^{-1} beschränkt, da für $\|x_n - e\| < \frac{1}{2}$

$$\|x_n^{-1}\| = \|e + (e - x_n) + \ldots\| \leq 1 + \frac{1}{2} + \frac{1}{4} + \ldots = 2$$

ist.

Sei $K = \sup \|x_n^{-1}\|$. Dann ist

$$\|e - x_n^{-1}\| = \|x_n^{-1}(x_n - e)\| \leq K\|x_n - e\| \to 0,$$

und folglich $x_n^{-1} \to e$.

Im allgemeinen Falle konvergiert die Folge $x_n x^{-1}$ zu $xx^{-1} = e$, und daher ist

$$(x_n x^{-1})^{-1} = x x_n^{-1} \to e, \quad x_n^{-1} \to x^{-1}.$$

Die Gesamtheit I der Elemente $x \in R$ nennen wir ein Ideal (oder ein nichttriviales Ideal), wenn

1°. Aus $x \in I$, $y \in I$ folgt $px + qy \in I$, wo p und q beliebige Elemente des Ringes sind.

2°. $I \neq R$.

Offenbar kann I kein Element x aus V enthalten, denn andernfalls würde es ein beliebiges Element $y \in R$ enthalten:

$$y = yx^{-1} \cdot x,$$

was der Bedingung 2° widersprechen würde. In jedem Ring bildet das Element 0 offenbar ein Ideal; wir werden es das Nullideal nennen und (0) bezeichnen.

Hilfssatz 3. *Wenn der Ring R keine Ideale besitzt, die von dem Nullideal verschieden sind, so existiert für jedes $x \neq 0$ das inverse Element x^{-1}; mit anderen Worten, R ist in diesem Falle ein Körper.*

Betrachten wir die Gesamtheit I der Elemente yx, wo $x \neq 0$ ist, und y den ganzen Ring R durchläuft. Da $I \neq 0$ (wegen $x \in I$) ist, so ist $I = R$. Dies bedeutet, insbesondere, dass ein $y \in R$ derart existiert, dass $xy = e$ ist. Somit besitzt x ein inverses Element.

Wir haben stets $I \subset R - V$; da $R - V$ abgeschlossen ist, so haben wir auch $\bar{I} \subset R - V$. Offenbar ist $\bar{I}$ auch ein Ideal. Wir bekommen somit den

Satz 2. *Die Abschliessung $\bar{I}$ eines nichttrivialen Ideals I ist wieder ein nichttriviales Ideal.*

Es sei ein Ideal I gegeben. Ein Element $x \in R$ heisst kongruent $y \in R$ nach dem Ideal I, $x \sim y$, wenn $x - y \in I$; da die Kongruenzbeziehung symmetrisch, reflektiv und transitiv ist, so zerfällt R in Klassen untereinander nach dem Ideal I kongruenter Elemente; führen wir in natürlicher Weise die Operationen der Addition, Subtraktion und Multiplikation dieser Klassen ein, so erhalten wir den Restklassenring R/I. Dieser Ring besitzt die Einheit, deren Rolle die $e \in R$

enthaltende Klasse spielt. Die Null dieses Ringes ist die Klasse, die von allen $x \in I$ gebildet wird.

Hilfssatz 4. *Ist I ein abgeschlossenes Ideal, so ist R/I ein normierter Ring.*

Bezeichnen wir die Elemente von R/I durch $X, Y, \ldots$ und setzen $\|X\| = \inf_{x \in X} \|x\|$. Zeigen wir, dass alle Axiome der Norm erfüllt sind:

1°. $\|\lambda X\| = |\lambda| \|X\|$ — ist offenbar.

2°. $\|X+Y\| = \inf_{z \in X+Y} \|z\| \leqslant \inf_{x \in X, y \in Y} \|x+y\| \leqslant \inf_{x \in X, y \in Y} \{\|x\| + \|y\|\} \leqslant$
$\leqslant \inf_{x \in X} \|x\| + \inf_{y \in Y} \|y\| = \|X\| + \|Y\|$.

3°. In ähnlicher Weise $\|X \cdot Y\| = \|X\| \cdot \|Y\|$.

4°. Es sei $\|X\| = 0$, d. h. es existiere eine solche Folge $x_n \in X$, dass $x_n \to 0$. Wenn x ein beliebiges Element aus X ist, so ist $x - x_n = y_n \in I$; da $x = \lim y_n$ ist, so $x \in \bar{I} = I$. Hieraus folgt, dass die ganze Klasse X in I enthalten ist und folglich mit ihm zusammenfällt, d. h. die Null des Ringes R/I ist.

5°. Es sei E die Einheit von R/I. Da $e \in E$, so ist $\|E\| \leqslant 1$; wäre $\|E\| < 1$, so würde ein Element $x \in E$ existieren derart, dass $\|x\| < 1$ ist. Dann würde, nach Lemma 1, das Element $e - x$ ein inverses Element besitzen, was wegen $e - x \in I$ unmöglich ist.

6°. Die Vollständigkeit. Es möge $\|X_n - X_m\| \to 0$; wählen wir eine Unterfolge X'_n derart, dass die Reihe $\sum \|X'_n - X'_{n+1}\|$ konvergiert. Für ein beliebiges $x_1 \in X'_1$ wird sich ein $x_2 \in X'_2$ derart finden, dass

$$\|x_2 - x_1\| < 2\|X'_2 - X'_1\|,$$

ein $x_3 \in X'_3$ — derart, dass

$$\|x_3 - x_2\| < 2\|X'_3 - X'_2\|,$$

usw. Die Punkte x_n bilden eine Fundamentalfolge; also

$$x_n \to x, \quad X'_n \to X \ni x.$$

Folglich auch $X_n \to X$.

§ 3. Normierte Körper

Es sei R ein vollständiger normierter Raum, und es möge jeder komplexen Zahl λ aus einem Bereich $\mathfrak{G}$ der komplexen Ebene ein Element $x \in R$ entsprechen. Die so definierte abstrakte Funktion $x(\lambda)$ heisst analytisch, wenn für jedes $\lambda \in \mathfrak{G}$ der Grenzwert

$$\lim_{h \to 0} \frac{x(\lambda+h) - x(\lambda)}{h}$$

im Sinne der Konvergenz nach der Norm existiert.

Es sei f ein beliebiges lineares Funktional aus R; dann ist $f(x(\lambda))$ eine gewöhnliche analytische Funktion von λ. Dieser Umstand erlaubt uns eine Reihe von Eigenschaften gewöhnlicher analytischer Funktionen auf abstrakte analytische Funktionen zu übertragen.

1. Der Liouvillesche Satz. Wenn $x(\lambda)$ für alle λ definiert und der Norm nach beschränkt ist, so ist $x(\lambda) \equiv x$, wo x ein konstantes Element von R

ist. In der Tat, haben wir für beliebiges f nach dem gewöhnlichen Liouvilleschen Satz $f[x(\lambda)] = \text{const}$, und folglich ist auch $x(\lambda)$ konstant, denn sollte $x(\lambda)$ auch nur zwei Werte $x(\lambda_1) = x_1$ und $x(\lambda_2) = x_2$ annehmen, so würde nach einem Satz von Hahn ein lineares Funktional f existieren derart, dass $f(x_1) \neq f(x_2)$ ist, was unmöglich ist.

2. Integrale. Es sei in dem Bereich $\mathfrak{G}$ eine rektifizierbare Kurve Γ gegeben. Dann existiert

$$y = \int_\Gamma x(\lambda)\, d\lambda.$$

Es seien $\lambda_0, \lambda_1, \ldots, \lambda_n$ beliebige Punkte auf Γ. Betrachten wir die Summe

$$S = \sum_{k=0}^{n} x(\lambda_k)(\lambda_k - \lambda_{k-1}).$$

Es sei nun $\lambda_{00}, \ldots, \lambda_{10}, \ldots, \ldots, \lambda_{n-1, p_{n-1}}, \lambda_n$ eine andere Gesamtheit von Punkten auf Γ, die die erste enthält, derart, dass $\lambda_{00} = \lambda_0$, $\lambda_{10} = \lambda_1, \ldots$; schätzen wir die Differenz zwischen den entsprechenden Summen ab. Sie ist gleich

$$\left\| \sum (x(\lambda_k) - x(\lambda_{k,l}))(\lambda_{k,l+1} - \lambda_{k,l}) \right\| \leq \omega(x)\Gamma,$$

wo $\omega(x)$ die grösste der Differenzen $\|x(\lambda_k) - x(\lambda_{k,l})\|$ bedeutet, und Γ die Länge der Kurve ist. Es sei

$$\delta = \sup_{\lambda \in \Gamma} \min \rho(\lambda, \lambda_k).$$

Wir behaupten, dass die Summen S einen Grenzwert besitzen, wenn $\delta \to 0$. Sei, in der Tat, ein beliebiges $\varepsilon > 0$ vorgegeben. Finden wir ein $\delta > 0$ derart, dass aus $|\lambda' - \lambda''| < \frac{\delta}{2}$ die Ungleichung $\|x(\lambda') - x(\lambda'')\| < \varepsilon$ folgt, und schätzen wir die Abweichung voneinander aller Summen mit einem gegebenen $\frac{\delta}{2}$ ab. Es mögen den Punktfolgen $\{\mu_k\}$ und $\{\lambda_k\}$ die Summen S_2 und S_1 entsprechen. Wenn die Folge $\{\mu_k\}$ die Folge $\{\lambda_k\}$ enthält, so ist $\|S_2 - S_1\| < \varepsilon\Gamma$. Im entgegengesetzten Falle bilden wir die Folge $\nu_1, \nu_2, \ldots, \nu_k, \ldots$ — die Vereinigung der Folgen $\{\lambda_k\}$ und $\{\mu_k\}$. Für die entsprechende Summe haben wir

$$\|S_3 - S_1\| < \varepsilon\Gamma, \quad \|S_3 - S_2\| < \varepsilon\Gamma,$$

woraus

$$\|S_2 - S_1\| < 2\varepsilon\Gamma$$

folgt. Also besitzen die Summen S für $\delta \to 0$ einen Grenzwert, den wir das Integral nennen.

3. Der Satz von Cauchy. Wenn die Funktion $x(\lambda)$ im Inneren und auf der Grenze einer geschlossenen Kurve Γ analytisch ist, so gilt

$$y = \int_\Gamma x(\lambda)\, d\lambda = 0.$$

In der Tat, für ein beliebiges lineare Funktional f gilt, nach dem gewöhnlichen Satz von Cauchy, $f(y) = 0$; hieraus folgt $y = 0$.

Die Integralformel von Cauchy

$$x(\lambda) = \frac{1}{2\pi i} \int_{\Gamma} \frac{x(\xi)\, d\xi}{\xi - \lambda}$$

wird analog bewiesen.

Als Folge der Integralformel von Cauchy erhalten wir die Existenz aller Ableitungen der Funktion $x(\lambda)$ und die Entwicklung in die Taylorsche Reihe:

$$x(\lambda) = x(\lambda_0) + x'(\lambda_0)(\lambda - \lambda_0) + \ldots,$$

die in jedem Kreise konvergiert, in welchem die Funktion $x(\lambda)$ analytisch ist.

Wir beweisen nun den folgenden

Satz 3. *Ein normierter Körper ist dem Körper T aller komplexen Zahlen isomorph* [2].

Mit anderen Worten, kann jedes Element x in der Form $x = \lambda e$ erhalten werden, wo λ eine komplexe Zahl ist.

Nehmen wir den Gegenteil an: es möge für ein Element x die Differenz $x - \lambda e$ für kein λ verschwinden. Dann existiert für jedes λ $(x - \lambda e)^{-1}$. Zeigen wir, dass dies eine analytische Funktion von λ ist.

Der Ausdruck

$$\frac{1}{h}\left[(x - (\lambda + h)e)^{-1} - (x - \lambda e)^{-1}\right] = -\left[(x - \lambda e - he)^{-1}\right](x - \lambda e)^{-1}$$

hat, infolge von Hilfssatz 2, einen Grenzwert, nämlich $-(x - \lambda e)^{-2}$. Für grosse λ ist

$$|(x - \lambda e)^{-1}| \cdot |\lambda^{-1}| \left| \left(\frac{x}{\lambda} - e\right)^{-1} \right| \to 0,$$

wiederum nach Hilfssatz 2; da $(x - \lambda e)^{-1}$ in jedem Kreise infolge der Stetigkeit beschränkt ist, so ist $(x - \lambda e)^{-1}$ auch für alle λ beschränkt. Wenden wir den Liouvilleschen Satz an, so finden wir: $(x - \lambda e)^{-1}$ ist konstant und folglich gleich Null, da es Null zum Grenzwert für $\lambda \to \infty$ hat; aber dann ist auch $e = (x - \lambda e)(x - \lambda e)^{-1} = 0$, was unmöglich ist.

§ 4. Maximale Ideale

Ein Ideal $M \subset R$ heisst maximal, wenn es nicht ein Teil eines anderen Ideals ist.

Satz 4. *Jedes maximale Ideal M ist abgeschlossen.*

Im entgegengesetzten Falle wäre M, nach Satz 2, ein Teil des Ideals $\overline{M}$, und folglich könnte es nicht maximal sein.

Satz 5. *Jedes Ideal I ist in einem maximalen Ideal enthalten.*

Dieser Satz wird leicht mit Hilfe der transfiniten Induktion bewiesen. Es sei $x_1, x_2, \ldots, x_\omega, \ldots, x_\alpha, \ldots = \{x_\alpha\}$ eine totalgeordnete Folge aller Elemente von R. Jedem nichtmaximalen Ideal A kann ein Ideal $A^+ \supset A$ folgendermassen zugeordnet werden: betrachten wir die Menge aller Elemente $x \in \{x_\alpha\}$ derart, dass die Gesamtheit der Elemente $a + px$ $(a \in A, p \in R)$ ein Ideal $\supset A$ bildet; nach der Annahme ist diese Menge nicht leer und besitzt daher ein

[2] Dieser Satz wurde zuerst von Mazur bewiesen. Sein Beweis ist von dem unseren verschieden.

erstes Element x_A. Setzen wir $A^+ = \{a + px_A\}$ und konstruieren eine transfinite Folge von Idealen I_α in folgender Weise: wir setzen $I_0 = I$; es seien die I_α für alle $\alpha < \beta$ bereits konstruiert; gehört β der ersten Klasse an, d. h. existiert $\alpha = \beta - 1$, so setzen wir $I_\beta = I_\alpha^+$; wenn β der zweiten Klasse angehört, so setzen wir

$$I_\beta = \sum_{\alpha < \beta} I_\alpha.$$

Diese Folge hat eine Mächtigkeit, die die Mächtigkeit von R nicht übertrifft; daher muss diese Folge ein letztes Glied besitzen, welches eben das gesuchte maximale Ideal darstellt.

Für verschiedene Anwendungen ist der folgende einfache Satz von Wichtigkeit:

Satz 6. *Dafür, dass ein Element $x \in R$ ein inverses Element besitze, ist es notwendig und hinreichend, dass x keinem maximalen Ideal angehöre.*

Wenn x ein inverses Element besitzt, so gehört x keinem Ideal an; umsomehr also gehört x keinem maximalen Ideal an.

Wenn x kein inverses Element besitzt, so bildet die Gesamtheit der Elemente $\{x \cdot y\}$, wo y alle Elemente von R durchläuft, ein Ideal; nach Satz 5 gehört es einem maximalen Ideal an.

§ 5. Funktionen auf maximalen Idealen

Es sei M ein maximales Ideal des Ringes R, und R/M — der Restklassenring.

Satz 7. *R/M ist dem Körper der komplexen Zahlen isomorph.*

Nach den Sätzen 3 und 4 und dem Hilfssatz 3 genügt es zu zeigen, dass R/M keine Ideale, die von dem Nullideal verschieden sind, enthält. Dies ist aber klar: wäre in R/M ein Ideal I, das von dem Nullideal verschieden ist, vorhanden, so würden wir, wenn wir alle Elemente aller I bildenden Klassen zusammenfassen, ein Ideal in R erhalten, das M enthält, was der Maximalität des letzteren widerspricht.

Es gilt auch die inverse Behauptung: wenn der Restklassenring nach dem Ideal R/M dem Körper der komplexen Zahlen isomorph ist, so ist M ein maximales Ideal. Denn anderenfalls würden wir ein Ideal $M' \supset M$ betrachten, und alle Klassen in R/M, die Elemente von M' enthalten, zusammenfassen; dabei würden wir ein Ideal im Körper der komplexen Zahlen erhalten, was unmöglich ist.

Der Satz 7 erlaubt uns jedem Element $x \in R$ eine komplexe Zahl $x(M)$ zuzuordnen, nämlich diejenige, der bei dem Isomorphismus $R/M \approx T$ die Restklasse entspricht, die das Element x enthält.

Bei fixiertem x und variierendem M erhalten wir eine Funktion $x(M)$, die auf der Menge $\mathfrak{M}$ aller maximalen Ideale des Ringes R definiert ist.

Diese Funktionen besitzen die folgenden Eigenschaften:

(α) Wenn $x = x_1 + x_2$ ist, so ist $x(M) = x_1(M) + x_2(M)$.

(β) Wenn $x = x_1 \cdot x_2$ ist, so ist $x(M) = x_1(M) \cdot x_2(M)$.

(γ) $e(M) = 1$.

(δ) $|x(M)| \leq \|x\|$.

(ε) Wenn $M_1 \neq M_2$ ist, so existiert ein $x \in R$ derart, dass $x(M_1) \neq x(M_2)$ ist.

Diese Eigenschaften sind nichts anderes als einfach Formulierungen der Tatsache, dass die Abbildung $R \to R/M$ ein Homomorphismus ist.

(ζ) Wenn $x(M)$ nicht verschwindet, so existiert ein $y \in R$ derart, dass

$$y(M) = \frac{1}{x(M)}$$

ist. Dies ist eine Folgerung aus dem Satz 6.

§ 6. Das Radikal eines normierten Ringes

Wir stellen uns die Aufgabe zu finden, für welche Elemente $x \in R$ die Funktion $x(M) = 0$ ist; mit anderen Worten, welche Elemente allen maximalen Idealen angehören.

Definition. Ein Element $x \in R$ heisst ein verallgemeinertes nilpotentes Element, wenn $\sqrt[n]{\|x^n\|} \to 0$. Es ist klar, dass gewöhnliche nilpotente Elemente (d. h. solche, für die für ein bestimmtes n $x^n = 0$ ist) auch verallgemeinerte nilpotente Elemente sind; die inverse Behauptung gilt im allgemeinen nicht. Die Gesamtheit der verallgemeinerten nilpotenten Elemente heisst das Radikal des Ringes.

Satz 8. *Der Durchschnitt aller maximalen Ideale fällt mit der Menge aller verallgemeinerten nilpotenten Elemente zusammen.*

Es möge x einem bestimmten maximalen Ideal M_0 nicht angehören; dann ist $x(M_0) \neq 0$. Da

$$\|x^n\| \geq |x^n(M_0)| = |x^n(M_0)|$$

ist, haben wir

$$\sqrt[n]{\|x^n\|} > |x(M_0)|;$$

somit ist x kein verallgemeinertes nilpotentes Element. Es möge nun x allen maximalen Idealen angehören. Dann gehört $e - \lambda x$ für beliebiges λ keinem maximalen Ideal an und besitzt daher ein inverses Element; die analytische Funktion

$$(e - \lambda x)^{-1}$$

ist also ganz und, nach § 3, kann in eine beständig konvergente Taylorsche Reihe entwickelt werden:

$$(e - \lambda x)^{-1} = e + \lambda x + \lambda^2 x^2 + \ldots$$

Da $\lambda^n x^n$ ein Glied einer konvergenten Reihe ist, so haben wir für $n \to \infty$

$$\|\lambda^n x^n\| \to 0.$$

Hieraus folgt

$$\|x^n\| = \frac{\|\lambda^n x^n\|}{|\lambda|^n} < \frac{1}{|\lambda|^n}$$

für genügend grosse n, und also

$$\overline{\lim_{n \to \infty}} \sqrt[n]{\|x^n\|} \leq \frac{1}{|\lambda|};$$

da aber λ beliebig ist, so erhalten wir

$$\lim \sqrt[n]{\|x^n\|} = 0.$$

Es gilt auch ein allgemeinerer Satz, der das max $|x(M)|$ mit den Normen der Potenzen von x verbindet:

Satz 8′. *Für jedes* $x \in R$ *existiert*

$$\lim \sqrt[n]{\| x^n \|}$$

und ist dem max $|x(M)|$ *gleich.*

Setzen wir max $|x(M)| = a$. Dann besitzt das Element $x - \mu e$ ein inverses für alle μ mit $|\mu| > a$, denn es wird durch die Funktion $x(M) - \mu$ dargestellt, die nicht Null wird. Setzen wir $\lambda = \frac{1}{\mu}$; dann ist $(e\mu - x)^{-1} = \lambda(e - \lambda x)^{-1}$ eine analytische Funktion im Kreise $|\lambda| < \frac{1}{a}$. Entwickeln wir sie in die Taylorsche Reihe:

$$\lambda(e - \lambda x)^{-1} = \lambda(e + \lambda x + \ldots).$$

Da das allgemeine Glied einer konvergenten Reihe zu Null strebt, so erhalten wir:

$$\| \lambda^n x^n \| \to 0, \quad \| x^n \| = \frac{1}{|\lambda|^n} \| \lambda^n x^n \| \leqslant \frac{1}{|\lambda|^n}$$

für genügend grosse n; hieraus folgt

$$\overline{\lim} \sqrt[n]{\| x^n \|} \leqslant \frac{1}{|\lambda|}.$$

Für $|\lambda| \to \frac{1}{a}$ erhalten wir

$$\overline{\lim} \sqrt[n]{\| x^n \|} \leqslant a. \tag{1}$$

Da $\| x \| \geqslant a$, $\| x^n \| \geqslant a^n$, gilt $\sqrt[n]{\| x^n \|} \geqslant a$ für alle n, woraus

$$\underline{\lim} \sqrt[n]{\| x^n \|} \geqslant a \tag{2}$$

folgt. Vergleichen wir (1) und (2), so finden wir:

$$\lim \sqrt[n]{\| x^n \|} = a = \max |x(M)|.$$

§ 7. Die Topologisierung der Menge $\mathfrak{M}$

Da wir beabsichtigen die Funktionen $x(M)$ in stetige Funktionen umzuwandeln, führen wir in der Menge $\mathfrak{M}$ eine Topologie ein.

Eine Umgebung $U(M_0)$ des Punktes $M_0 \in \mathfrak{M}$ definieren wir als die Gesamtheit aller $M \in \mathfrak{M}$, für die die Ungleichungen

$$|x_i(M) - x_i(M_0)| < \varepsilon \quad (i = 1, 2, \ldots, n)$$

erfüllt sind. Somit wird die Umgebung durch die Angabe von ε und von n beliebigen Elementen aus R definiert.

Zeigen wir, dass die Axiome des topologischen Raumes erfüllt sind.

1°. Jeder Punkt besitzt eine Umgebung und ist in ihr enthalten.

2°. Der Durchschnitt zweier Umgebungen enthält eine dritte Umgebung.

3°. Wenn $M_1 \subset U(M_0)$, so existiert ein $U(M_1) \subset U(M_0)$.

Es ist klar, dass die ersten zwei Axiome erfüllt sind. Um zu zeigen, dass auch das dritte Axiom erfüllt ist, finden wir ein solches ε', dass der Kreis vom

Radius ε' mit dem Mittelpunkt in dem Punkt $x_i(M_1)$ sich im Inneren des Kreises vom Radius ε mit dem Mittelpunkt in dem Punkt $x_i(M_0)$ befindet; es ist dann offenbar, dass

$$U(M_1) = \{|x_i(M) - x_i(M_1)| < \varepsilon'\}$$

der gestellten Bedingung genügt.

Satz 9. *Der topologische Raum $\mathfrak{M}$ ist bikompakt und genügt den Axiomen von Hausdorff.*

Der Beweis, den wir hier anführen, stützt sich auf den Satz von Tychonoff über die Bikompaktheit des topologischen Produktes bikompakter Räume [3]. Einen Beweis, der sich auf den Tychonoffschen Satz nicht stützt, kann man in [4] finden.

Ordnen wir jedem Element $x \in R$ einen Kreis Q_x der komplexen Ebene vom Radius $\|x\|$ zu und betrachten das topologische Produkt Ω aller dieser Kreise. Nach dem Satz von Tychonoff ist Ω bikompakt. Jedem $M_0 \in \mathfrak{M}$ ordnen wir den Punkt aus Ω zu, der durch die Zahlen $x(M_0)$ bestimmt ist. Aus (ε), § 5, folgt, dass $\mathfrak{M}$ eineindeutig auf einen Teil von Ω abgebildet wird; die Topologie in $\mathfrak{M}$ fällt mit der auf dem Bild von $\mathfrak{M}$ in Ω induzierten zusammen; somit ist $\mathfrak{M}$ ein homöomorpher Teil von Ω. Um die Bikompaktheit von $\mathfrak{M}$ zu zeigen, genügt es nun seine Abgeschlossenheit in dem bikompakten Raum Ω zu zeigen.

Es möge $\Lambda = \{\lambda_x\}$, wo $\lambda_x \in Q_x$, ein Grenzpunkt in Ω für die Menge $\mathfrak{M}$ sein; konstruieren wir ein maximales Ideal $M_0 \in \mathfrak{M}$ derart, dass $x(M_0) = \lambda_x$ für jedes x ist; dadurch wird gerade gezeigt, dass $\Lambda \in \mathfrak{M}$. Zeigen wir, dass $\lambda_{x+y} = \lambda_x + \lambda_y$ ist. Zu diesem Ende betrachten wir die Umgebung des Punktes Λ, die durch die Punkte $x \cdot y$ und $x + y$ und eine beliebig kleine Zahl ε bestimmt ist. Da Λ ein Grenzpunkt für $\mathfrak{M}$ ist, so wird sich in ihrer Umgebung ein Punkt $M \in \mathfrak{M}$ finden, d. h. für ein bestimmtes M wird

$$|\lambda_x - x(M)| < \varepsilon,$$
$$|\lambda_y - y(M)| < \varepsilon,$$
$$|\lambda_{x+y} - x(M) - y(M)| < \varepsilon$$

sein. Daher ist $|\lambda_{x+y} - \lambda_x - \lambda_y| < \varepsilon$, und da ε beliebig war, $\lambda_{x+y} = \lambda_x + \lambda_y$. Wir erhalten, dass die Korrespondenz $x \to \lambda_x$ ein Homomorphismus des Ringes R in den Körper der komplexen Zahlen ist. In derselben Weise zeigen wir, dass $\lambda_{ie} = i$ ist. Folglich existiert ein maximales Ideal $M_0 \in \mathfrak{M}$ derart, dass

$$x(M_0) = \lambda_x$$

für beliebiges $x \in R$ ist, und der Satz ist bewiesen.

Beachten wir ferner, dass alle Funktionen $x(M)$ automatisch stetig werden: um eine Umgebung $V(M_0)$ zu finden, in der die Funktion $x(M)$ um nicht mehr als ε variiert, genügt es die Menge

$$V(M_0) = \{|x(M) - x(M_0)| < \varepsilon\}$$

zu betrachten, die nach Definition eine Umgebung ist.

---

[3] A. Tychonoff, Math. Annalen, **102**, (1929).

[4] Siehe I. Gelfand und G. Šilov, Über verschiedene Methoden der Einführung der Topologie in die Menge der maximalen Ideale eines normierten Ringes (in diesem Heft des Recueil mathématique, S. 31).

Wir bekommen so den

Satz 10. *Jeder normierte Ring R kann homomorph in den Ring* $C(\mathfrak{M})$ *aller auf dem bikompakten Hausdorffschen Raum* $\mathfrak{M}$ *stetigen Funktionen abgebildet werden. Dafür, dass diese Abbildung von R in* $C(\mathfrak{M})$ *isomorph sei, ist es notwendig und hinreichend, dass die folgende Bedingung erfüllt sei: aus* $\sqrt[n]{\|x^n\|} \to 0$ *für* $x \in R$ *folgt* $x = 0$.

Der folgende Satz zeigt die Notwendigkeit der eingeführten Topologie.

Satz 10'. *Sei* $\mathfrak{M}$ *in irgendeiner Weise so topologisiert, dass*

(α) $\mathfrak{M}$ *bikompakt ist,*

(β) *die Funktionen* $x(M)$ *stetig sind.*

Bezeichnen wir den topologischen Raum in der neuen Topologie durch $\mathfrak{M}_1$ *in der alten — durch* $\mathfrak{M}_0$. *Dann ist* $\mathfrak{M}_0$ *homöomorph* $\mathfrak{M}_1$.

Die Abbildung von $\mathfrak{M}_0$ auf $\mathfrak{M}_1$, die jedem $M \in \mathfrak{M}_0$ dasselbe M in $\mathfrak{M}_1$ zuordnet, ist eineindeutig. Infolge von (β) ist das Bild in $\mathfrak{M}_1$ einer jeden in $\mathfrak{M}_0$ offenen Menge auch in $\mathfrak{M}_1$ eine offene Menge; folglich ist die Abbildung $\mathfrak{M}_1 \to \mathfrak{M}_0$ stetig. Nach einem bekannten topologischen Satz ist, infolge von (α), die Abbildung stetig in beiden Richtungen, d. h. $\mathfrak{M}_0$ ist homöomorph $\mathfrak{M}_1$.

Da wir die Anwendungen dieser Theorie im Auge haben, bemerken wir, dass es nicht notwendig ist bei der Angabe der Topologie alle Elemente des Ringes zu benutzen: man kann sich auf die erzeugenden Elemente beschränken. Die Gesamtheit $K \subset R$ heisst die Gesamtheit der Erzeugenden von R, wenn der kleinste abgeschlossene, K enthaltende Unterring mit R zusammenfällt.

Satz 11. *Die Gesamtheit der Umgebungen*

$$U(M_0) = \{|x_i(M) - x_i(M_0)| < \varepsilon\}, \qquad i = 1, 2, \ldots, n,$$

wo ε *und* n *beliebig sind, und* $x_1, x_2, \ldots, x_n$ — *Elemente von* K *sind, ist ein definierendes System von Umgebungen in* $\mathfrak{M}$.

Wir müssen zeigen, dass in jeder Umgebung

$$U_1(M_0) = \{|x_i(M) - x_i(M_0)| < \varepsilon\}, \qquad i = 1, 2, \ldots, n,$$

$$x_1, x_2, \ldots, x_n \in R,$$

eine Umgebung aus dem definierenden System enthalten ist.

Finden wir n Polynome

$$P_1(x_{11}, x_{12}, \ldots, x_{1n_1}), \ldots, P_n(x_{n1}, x_{n2}, \ldots, x_{nn_n})$$

in den Elementen $x_{ik} \in K$, die der Norm nach um weniger als $\frac{\varepsilon}{3}$ von den Elementen $x_1, x_2, \ldots, x_n$ verschieden sind. Für jedes Polynom finden wir ein solches δ, dass bei der Abweichung der Argumente $x_{k1}(M), \ldots, x_{kn_k}(M)$ von den Werten $x_{k1}(M_0), \ldots, x_{kn_k}(M_0)$ um weniger als δ der Wert des Polynoms sich um nicht mehr als $\frac{\varepsilon}{3}$ ändert.

Betrachten wir die Umgebung

$$U(M_0) = \{ | x_{ki}(M) - x_{ki}(M_0) | < \delta \}, \qquad i = 1, 2, \ldots, n_k, \quad k = 1, 2, \ldots, n.$$

Wir behaupten, dass $U(M_0) \subset U_1(M_0)$ ist. In der Tat, für diejenigen M, für welche unsere Ungleichungen erfüllt sind, haben wir

$$| P_k(M) - P_k(M_0) | < \frac{\varepsilon}{2};$$

da

$$| x_k(M) - P_k[x(M)] | < \frac{\varepsilon}{3}$$

und

$$| x_k(M_0) - P_k[x(M_0)] | < \frac{\varepsilon}{3}$$

ist, so finden wir $| x_k(M) - x_k(M_0) | < \varepsilon$, w. z. b. w.

Satz 12. *Wenn R separabel ist, so ist $\mathfrak{M}$ metrisierbar (und folglich, nach Satz 9, ist ein Kompakt).*

Es ist leicht einzusehen, dass das abzählbare System von Umgebungen

$$\{ | x_{i_k}(M) - \lambda_{i_k} | < r_s \}, \quad k = 1, 2, \ldots, n,$$

wo die r_s und λ_{i_k} rationale Zahlen sind, und die $x_{i_1}, x_{i_2}, \ldots, x_{i_n}$ aus der abzählbaren dichten Menge genommen sind, die die Basis des Raumes $\mathfrak{M}$ bildet, ein definierendes System ist. Hieraus folgt nach einem Satz von Urysohn, dass $\mathfrak{M}$ metrisierbar ist. Überdies kann man unmittelbar die Entfernung mittels der Formel

$$\rho(M_1, M_2) = \sum_{n=1}^{\infty} \frac{1}{2^n} \frac{| x_n(M_1) - x_n(M_2) |}{1 + | x_n(M_1) - x_n(M_2) |}$$

definieren.

Diese Bedingung ist nicht notwendig. Um eine notwendige und hinreichende Bedingung zu erhalten, betrachten wir den Ring aller auf dem Kompakt stetigen Funktionen: wenn man in ihm die Konvergenz nicht der Norm nach, sondern die gleichmässige Konvergenz betrachtet, so existiert immer eine abzählbare überall dichte Menge. Da in der Definition der Umgebungen nicht die Norm der Funktionen wesentlich ist, sondern ihre Werte, so folgt die Metrisierbarkeit von $\mathfrak{M}$ aus der Separabilität von R bezüglich der gleichmässigen Konvergenz auf $\mathfrak{M}$ der Funktionen $x(M)$.

Somit haben wir den

Satz 13. *Führen wir in dem Ring R eine neue Norm*

$$| x | = \lim \sqrt[n]{\| x^n \|}$$

ein. Die Separabilität von $\mathfrak{M}$ in der neuen Norm stellt eine notwendige und hinreichende Bedingung der Metrisierbarkeit von R dar.

(Beachten wir, dass mit der neuen Normierung R, im allgemeinen, nicht mehr ein vollständiger Raum sein wird.)

Satz 14. *Wenn R eine endliche Anzahl von Erzeugenden $x_1, x_2, \ldots, x_n$ hat, so ist $\mathfrak{M}$ eine abgeschlossene und beschränkte Untermenge des n-dimensionalen komplexen Raumes.*

Die Funktionen $x_1(M), \; x_2(M), \ldots, x_n(M), \ldots$ bilden $\mathfrak{M}$ eindeutig und stetig auf eine bestimmte abgeschlossene und beschränkte Untermenge $\mathfrak{M}'$ des

n-dimensionalen komplexen Raumes ab. Zeigen wir, dass diese Abbildung eineindeutig ist; dann wird sie auch in beiden Richtungen stetig sein, und folglich $\mathfrak{M}$ wird $\mathfrak{M}'$ homöomorph sein. Es mögen zwei Punkte M_1 und M_2 in einen Punkt von $\mathfrak{M}'$ abgebildet werden. Dies bedeutet, dass $x_k(M_1) = x_k(M_2)$ $(k = 1, 2, \ldots, n)$ ist. Dann wird aber für alle Polynome in $x_1, x_2, \ldots, x_n$ und folglich auch für alle ihre Grenzwerte, d. h. für alle $x \in R$, $x(M_1) = x(M_2)$ sein. Also folgt aus $x \in M_1$, d. h. $x(M_1) = 0$, $x \in M_2$, d. h. M_1 und M_2 fallen zusammen.

§ 8. Reelle Ringe

Wir haben bewiesen, dass jeder normierte Ring in den Ring aller stetigen komplexen auf der Menge $\mathfrak{M}$ definierten Funktionen $x(M)$ abgebildet werden kann. Wir stellen nun die Frage, wie der Ring R beschaffen sein soll, damit diese Funktionen reell werden.

Wenn die Funktion $x(M)$ reell ist, so ist $x^2(M)$ nicht negativ, $x^2(M) + 1$ ist positiv und folglich besitzt eine inverse Funktion in dem Ring. Auf Grund dieser Tatsache können wir die folgende Definition einführen:

Ein Ring heisst r e e l l, wenn

(α) Die Multiplikation von Elementen mit Zahlen ist nur für den Fall definiert, wo mit reellen Zahlen multipliziert wird (und nicht, wie früher, mit komplexen).

(β) Für jedes $x \in R$ $x^2 + e$ besitzt ein inverses Element.

S a t z 15. *Wenn R ein reeller Ring ist, so sind alle Funktionen $x(M)$ reell.*

Es genügt zu zeigen, dass der Restklassenring R/M, wo M ein maximales Ideal ist, dem Körper der reellen Zahlen isomorph ist. Da R/M ein Körper ist, so hat die Bedingung (β) für ihn die folgende Form:

(β') Für beliebiges $x \in R/M$ ist $x^2 + e \neq 0$.

Hieraus folgt auch, dass $x^2 + y^2 \neq 0$ ist, was auch die x und y aus R/M seien, wenn nur $x \neq 0$ oder $y \neq 0$ ist: in der Tat, wenn, z. B., $x \neq 0$ ist, es ist $x^2 + y^2 = x^2 \left(e + \left(\frac{y}{x}\right)^2\right) \neq 0$, da im Körper R/M keine Nullteiler vorhanden sind.

Bilden wir aus den Elementen von R/M die Gesamtheit T der Elemente

$$x + iy$$

und definieren in T die üblichen Operationen der Addition und Multiplikation unter der Annahme, dass $i^2 = -e$ ist. Insbesondere ist die Division immer möglich, da

$$\frac{1}{x + iy} = \frac{x - iy}{x^2 + y^2}$$

ist, wo der Nenner auf der rechten Seite nach dem Bewiesenen von Null verschieden ist, sowie die Multiplikation mit komplexen Zahlen $\lambda + i\mu$. Definieren wir die Norm in der folgenden Weise:

$$\|x + iy\| = \sup_\alpha \|x \cos \alpha + y \sin \alpha\| \text{ [5]}.$$

[5] Auf die Möglichkeit einer solchen Wahl der Norm wurde ich von M. Krein aufmerksam gemacht.

Es ist leicht einzusehen, dass alle Axiome der Norm erfüllt sind. Somit ist T ein normierter Körper; nach Satz 3 ist T dem Körper der komplexen Zahlen isomorph. Betrachten wir, wie werden bei diesem Isomorphismus die Elemente von R/M transformiert. Sei $x \to \lambda + i\mu$; dann $(x - \lambda e)^2 + (\mu e)^2 \to 0$, d. h. $(x - \lambda e)^2 + (\mu e)^2$ ist die Null von R/M. Nach dem Bewiesenen ist $\lambda e = x$, $\mu = 0$. Somit ist R/M dem Körper der reellen Zahlen isomorph.

§ 9. Die Beziehung zwischen R und $C(\mathfrak{M})$

Wir werden zeigen, dass jeder normierte Ring R in den Ring $C(\mathfrak{M})$ aller auf dem bikompakten Raum $\mathfrak{M}$ stetigen Funktionen homomorph abgebildet werden kann. Stellen wir die Frage, in welcher Beziehung der Ring der Funktionen $x(M)$ zu dem Ring aller stetigen Funktionen steht; insbesondere, welchen Bedingungen muss man R unterwerfen, damit R mit $C(\mathfrak{M})$ zusammenfalle.

Satz 16. *Wenn für jedes Element $x \in R$ ein Element $y \in R$ gefunden werden kann derart, dass $x(M) = \overline{y(M)}$ für beliebiges $M \in \mathfrak{M}$ [$\overline{y(M)}$ ist die zu $y(M)$ komplex-konjugierte Funktion], so bilden die Funktionen des Ringes R eine überall dichte Untermenge (im Sinne der gleichmässigen Konvergenz) in dem Ring $C(\mathfrak{M})$. Mit anderen Worten, jede auf $\mathfrak{M}$ stetige Funktion ist die Grenzfunktion einer gleichmässig konvergenten Folge von Funktionen $x(M)$, $x \in R$.*

Für den Beweis siehe [4].

Folgerung 1. Die Behauptung des Satzes gilt immer für einen reellen Ring R (§ 9), wenn man unter $C(\mathfrak{M})$ den Ring aller reellen stetigen Funktionen auf $\mathfrak{M}$ versteht.

Folgerung 2. Dafür, dass $R \equiv C(\mathfrak{M})$ sei, ist es notwendig und hinreichend, dass aus der gleichmässigen Konvergenz der Funktionen $x(M)$ die Konvergenz der Elemente x nach der Norm in R folge.

Die Notwendigkeit dieser Bedingung folgt aus Satz 17, und das Hinreichen — aus Satz 16.

Die bequemste für die Anwendungen ist die folgende hinreichende Bedingung

Folgerung 3. Wenn für beliebiges $x \in R$

$$\|x^2\| = \|x\|^2$$

ist, so ist $R = C(\mathfrak{M})$.

In der Tat, nach Satz 8′ ist

$$\max|x(M)| = \lim \sqrt[2^n]{\|x^{2^n}\|} = \lim \sqrt[2^n]{\|x\|^{2^n}} = \lim \|x\| = \|x\|,$$

woraus folgt, dass die gleichmässige Konvergenz nach sich die Konvergenz der Norm nach zieht.

In Folgerungen 2 und 3 wird angenommen, dass die Bedingungen des Satzes 16 erfüllt sind.

§ 10. Die Beziehung zwischen dem algebraischen und dem stetigen Isomorphismus

Es sei R ein Ring, in dem der Durchschnitt der maximalen Ideale das Nullideal ist (d. h. aus $\sqrt[n]{\|x^n\|} \to 0$ $x = 0$ folgt). Nach Satz 8 ist ein solcher Ring einem bestimmten Ring von stetigen Funktionen auf dem bikompakten Hausdorffschen Raum $\mathfrak{M}$ isomorph.

Satz 17. *Es sei R_1 ein normierter Ring, der dem Ring R algebraisch isomorph ist. Dann ist er auch topologisch isomorph, d. h. die Konvergenz der Elemente $x \in R$ ist äquivalent der Konvergenz der entsprechenden Elemente $y \in R_1$.*

Der Sinn und die Bedeutung dieser Behauptung liegen im folgengen. Um einen normierten Ring anzugeben, muss man, erstens, einen Vorrat an Elementen mit ihren algebraischen Eigenschaften und, zweitens, die Norm dieser Elemente vorgeben. Der Satz 17 zeigt, dass für den Funktionenring wird die Norm, abgesehen von einer Äquivalenz, schon durch den Vorrat an Elementen mit ihren algebraischen Eigenschaften in vollkommen eindeutiger Weise bestimmt.

Beweis. Offenbar ist der Ring R_1 auch ein Funktionenring auf derselben Menge $\mathfrak{M}$ (aber, möglicherweise, mit einer anderen Topologie), denn aus dem algebraischen Isomorphismus folgt die Korrespondenz zwischen den maximalen Idealen von R und R_1; wenn, insbesondere, $y \in R_1$ allen maximalen Idealen von R_1 angehört, so gehört das entsprechende $x \in R$ allen maximalen Idealen von R. Laut unserer Bedingung ist $x=0$, woraus auch $y=0$ folgt. Ferner werden die entsprechenden Elemente durch gleiche Funktionen dargestellt: wenn $x-\lambda e \in M$, so $y-\lambda e \in M_1$, woraus $x(M)=y(M)$ $(=\lambda)$ folgt.

Somit können wir annehmen, dass wir mit einem und demselben Ring R der Funktionen $x(M)$ zu tun haben, in dem zwei Normen, $\|x\|_1$ und $\|x\|_2$, eingeführt sind. Wir haben zu zeigen, dass diese Normen äquivalent sind.

Führen wir in R eine neue Norm $\|x\|_3=\|x\|_1+\|x\|_2$ ein. Zeigen wir dass R bezüglich dieser Norm vollständig ist. Es sei $\|x_n-x_m\|_3 \to 0$; dann $\|x_n-x_m\|_1 \to 0$ und $\|x_n-x_m\|_2 \to 0$. Da R auch bezüglich der ersten und der zweiten Norm vollständig ist, so existieren die Grenzwerte $x=\lim_1 x_n$ und $x'=\lim_2 x_n$. Aber die Konvergenz der Norm nach zieht die gleichmässige Konvergenz nach sich; deshalb werden x und x' durch eine und dieselbe Funktion dargestellt, und folglich fallen sie zusammen: $x'=x$. Dann haben wir

$$\|x-x_n\|_3=\|x-x_n\|_1+\|x-x_n\|_2 \to 0,$$

d. h. $x=\lim_3 x_n$. Nach dem bereits zitierten Satz von Banach (siehe § 1) ist die dritte Norm äquivalent der ersten und der zweiten, woraus auch die Äquivalenz dieser letzten folgt.

§ 11. Die Zerlegung des Ringes in eine direkte Summe von Idealen

Satz 18. *Dafür, dass ein Ring in eine direkte Summe von Idealen, die Ringe mit Einheiten sind, zerlegbar sei, ist es notwendig und hinreichend, dass die Menge der maximalen Ideale nicht zusammenhängend sei. Dabei wird vorausgesetzt, dass mit x in den Ring R auch ein solches Element y eingeht, dass $x(M)=\overline{y(M)}$ ist.*

Beweis. Es sei $\mathfrak{M}=\mathfrak{M}_1+\mathfrak{M}_2$, wo $\mathfrak{M}_1$ und $\mathfrak{M}_2$ zwei abgeschlossene sich nicht überschneidende Mengen von maximalen Idealen sind. Definieren wir die Funktion $\varphi(M)$ in folgender Weise: $\varphi(M)=1$, wenn $M \in \mathfrak{M}_1$, und $\varphi(M)=0$, wenn $M \in \mathfrak{M}_2$. Die Funktion $\varphi(M)$ ist stetig. Folglich existiert, nach Satz 16, ein Element $x \in R$ derart, dass $|x(M)-\varphi(M)|<\frac{1}{4}$ für jedes maximale Ideal M

ist. Die Werte der Funktion $x(M)$ liegen in zwei Kreisen vom Radius $\frac{1}{4}$ mit den Mittelpunkten in den Punkten 0 und 1. Folglich existiert für $|\lambda| = \frac{1}{2}$ $(x - \lambda e)^{-1}$ [nach der Eigenschaft (ζ), § 5]. Setzen wir

$$e_1 = \frac{1}{2\pi i} \int_C (x - \lambda e)^{-1} d\lambda,$$

wo C der Kreis vom Radius $\frac{1}{2}$ um den Punkt 0 ist. Zeigen wir, dass $e_1^2 = e_1$ ist. Wenn wir den Radius des Kreises etwas vergrössern, so wird sich das Integral nicht ändern, da die Funktion $(x - \lambda e)^{-1}$ in einer Umgebung des Kreises $|\lambda| = \frac{1}{2}$ analytisch ist. Folglich können wir schreiben:

$$e_1^2 = \frac{1}{2\pi i} \int_C (x - \lambda e)^{-1} d\lambda \cdot \frac{1}{2\pi i} \int_{C_1} (x - \mu e)^{-1} d\mu,$$

wo C_1 ein etwas grösserer Kreis ist. Dann haben wir

$$e_1^2 = -\frac{1}{4\pi^2} \int_C (x - \lambda e)^{-1} d\lambda \cdot \int_{C_1} (x - \mu e)^{-1} d\mu = -\frac{1}{4\pi^2} \int_C \int_{C_1} (x - \lambda e)^{-1} (x - \mu e)^{-1} d\lambda \, d\mu =$$

$$= -\frac{1}{4\pi^2} \int_C \int_{C_1} \frac{(x - \lambda e)^{-1}}{\lambda - \mu} d\lambda \, d\mu - \frac{1}{4\pi^2} \int_C \int_{C_1} \frac{(x - \mu e)^{-1}}{\mu - \lambda} d\lambda \, d\mu =$$

$$= \frac{1}{2\mu i} \int_C \left[\frac{1}{2\pi i} \int_{C_1} \frac{d\mu}{\lambda - \mu} \right] (x - \lambda e)^{-1} d\lambda - \frac{1}{2\pi i} \int_{C_1} \left[\frac{1}{2\pi i} \int_C \frac{d\lambda}{\mu - \lambda} \right] (x - \mu e)^{-1} d\mu =$$

$$= \frac{1}{2\pi i} \int_C (x - \lambda e)^{-1} d\lambda,$$

da

$$\frac{1}{2\pi i} \int_{C_1} \frac{d\mu}{\lambda - \mu} = -1, \quad \frac{1}{2\pi i} \int_C \frac{d\lambda}{\mu - \lambda} = 0$$

ist. Zerlegen wir nun den Ring R folgendermassen in eine direkte Summe: $R = Re_1 + R(e - e_1)$. Es ist leicht einzusehen, dass die Ideale Re_1 und $R(e - e_1)$ sich nicht überschneiden. Finden wir die Werte der Funktion $e_1(M)$:

$$e_1(M) = \frac{1}{2\pi i} \int_C (x(M) - \lambda)^{-1} d\lambda,$$

und folglich $e_1(M) = 1$ für $M \in \mathfrak{M}_1$ und $e_1(M) = 0$ für $M \in \mathfrak{M}_2$, d. h. $e_1 \in M$, wenn $M \in \mathfrak{M}_2$, und $e_1 \overline{\in} M$, wenn $M \in \mathfrak{M}_1$. Somit ist das Ideal Re_1 in allen maximalen Idealen des Systems $\mathfrak{M}_2$ und in keinem maximalen Ideal des Systems $\mathfrak{M}_1$ enthalten.

Re_1 und $R(e - e_1)$ sind normierte Ringe. Als Einheit des Ringes Re_1 dient e_1, und als Einheit des Ringes $R(e - e_1)$ dient $e - e_1$. Finden wir eine Beziehung zwischen den maximalen Idealen der Ringe Re_1 und R. Es sei ein maximales Ideal M' des Ringes Re_1 gegeben. Wie wir bereits gesehen haben, enthalten alle maximalen Ideale des Systems $\mathfrak{M}_2$ das Ideal M'. Wir behaupten, dass überdies genau ein maximales Ideal des Systems $\mathfrak{M}_1$ existiert, das M' enthält. Dieses Ideal M wird durch die Formel

$$M = M' + R(e - e_1)$$

bestimmt. Beweisen wir dies. Da $M \supset M'$ nach Voraussetzung dem System $\mathfrak{M}_1$ angehört, so enthält es $R(e-e_1)$ und folglich auch $M' + R(e-e_1)$. Es genügt somit zu zeigen, dass $M' + R(e-e_1)$ ein maximales Ideal ist. Nun fällt, erstens, dieses Ideal mit dem ganzen Ring nicht zusammen, denn nehmen wir den Gegenteil an, so haben wir $e \in M' + R(e-e_1)$, und folglich $e_1 \in M'$, was unmöglich ist. Zweitens, ist dieses Ideal ein maximales, denn jedes Element $x \in R$ kann in folgender Weise dargestellt werden:

$$x = \lambda e + y,$$

wo $y \in M' + R(e-e_1)$. In der Tat: $x = xe_1 + x(e-e_1)$. Aber $xe_1 \in Re_1$, und folglich haben wir $xe_1 = \lambda e_1 + y_1$, wo $y_1 \in M'$. Also ist $x = \lambda e_1 + y_1 + x(e-e_1)$, wo $y = y_1 + x(e-e_1) \in M' + R(e-e_1)$. Somit entspricht jedem maximalen Ideal $M' \subset Re_1$ ein einziges maximales Ideal $M \in \mathfrak{M}_1$. Umgekehrt, jedem maximalen Ideal $M \in \mathfrak{M}_1$ entspricht ein maximales Ideal $M' \subset Re_1$. M' wird durch die Formel $M' = M \cap Re_1$ bestimmt. Dieser Durchschnitt ist, wie man leicht einsieht, ein Ideal. Es ist ein maximales Ideal, denn jedes Element $xe_1 \in Re_1$ kann in folgender Weise dargestellt werden:

$$xe_1 = x(M)e_1 + (x - x(M)e)e_1,$$

wo das Element $(x - x(M)e)e_1 \in M \cap Re_1$. Somit ist eine eineindeutige Korrespondenz zwischen der Menge $\mathfrak{M}'$ der maximalen Ideale des Ringes Re_1 und der Menge $\mathfrak{M}_1$ festgestellt. Man kann auch beweisen, dass diese Korrespondenz ein Homöomorphismus ist.

Wenn der Ring R in eine direkte Summe zerlegbar ist, so ist $\mathfrak{M}$ nicht zusammenhängend. Es sei $R = R_1 + R_2$. Es seien ferner e_1 und e_2 die Einheiten der Ideale R_1 und R_2. Es ist bekannt, dass

$$e_1^2 = e_1, \quad e_2^2 = e_2, \quad e_1 + e_2 = e$$

ist. Folglich ist $e_1^2(M) = e_1(M)$, d. h. die Funktion $e_1(M)$ kann nur die Werte 0 und 1 annehmen. Da $\mathfrak{M}$ zusammenhängend ist, so ist entweder $e_1(M) = 1$, oder $e_1(M) = 0$. Es sei $e_1(M) = 0$ [andernfalls ist $e_2(M) = 0$]. Nach Satz 8', § 6, ist

$$\max |e_1(M)| = \lim \sqrt[n]{\|e_1^n\|}.$$

Da $e_1(M) = 0$ ist, und $e_1^n = e_1$, so folgt hieraus, dass $e_1 = 0$ ist. Also ist eine nichttriviale Zerlegung des Ringes R in eine direkte Summe von Idealen unmöglich.

Man beachte, dass wir eigentlich folgendes bewiesen haben. Wenn $\mathfrak{M} = \mathfrak{M}_1 + \mathfrak{M}_2$ ist, wo $\mathfrak{M}_1$ und $\mathfrak{M}_2$ zwei abgeschlossene, sich nicht überschneidende Mengen sind, so kann der Ring R in eine direkte Summe von Ringen zerlegt werden, beim ersten von welchen als die Menge von maximalen Idealen $\mathfrak{M}_1$ dient, und beim zweiten $\mathfrak{M}_2$ (in dem Sinne, in welchem es beim Beweis des Satzes dieses Paragraphen erklärt wurde).

Folgerung. Wenn in dem Ringe R nur eine endliche Anzahl von maximalen Idealen existiert, so kann R in eine direkte Summe von Ringen $R_1, \ldots, R_n$ zerlegt werden, in jedem von welchen nur ein einziges maximales Ideal vorhanden ist (primäre Ringe).

§ 12. Analytische Funktionen von Elementen des Ringes

In einem normierten Ring sind mit jedem Element x alle Polynome in x enthalten und, allgemeiner, auch alle „ganzen Funktionen" von x, d. h. alle Elemente der Form $\sum_0^\infty c_n x^n$, wo die Reihe $\sum_0^\infty c_n \zeta^n$ eine ganze analytische Funktion der komplexen Veränderlichen ζ darstellt. In der Tat, in diesem Falle wird die Reihe $\sum_0^\infty c_n x^n$ durch die konvergente Reihe $\sum_0^\infty |c_n| \| x \|^n$ majorisiert. Somit entspricht jeder ganzen analytischen Funktion $f(\zeta)$ eine abstrakte analytische Funktion $f(x)$, die für alle Elemente des Ringes definiert ist. Ferner haben wir gesehen, dass mit jedem Element x, für das $x(M)$ auf keinem maximalen Ideal zu Null wird, in dem Ring auch das Element x^{-1} enthalten ist. Somit entspricht der analytischen Funktion $\frac{1}{\zeta}$, die im Punkte $\zeta = 0$ einen Pol besitzt, eine abstrakte analytische Funktion x^{-1}, die für alle Elemente x definiert ist, für die die Menge der Werte von $x(M)$ den Punkt 0 nicht enthält. In genau derselben Weise entspricht, allgemein, jeder rationalen Funktion $Q(\zeta)$ der komplexen Veränderlichen ζ eine abstrakte analytische Funktion $Q(x)$, die für alle Elemente x definiert ist, für die die Menge der Werte von $x(M)$ keinen Pol der Funktion $Q(\zeta)$ enthält.

Es entsteht nun naturgemäss die Frage über die Erweiterung dieser Korrespondenz, die für ganze und rationale Funktionen festgestellt wurde, auf alle analytischen Funktionen. Diese Frage wird im positiven Sinne durch den folgenden Satz beantwortet:

Satz 19. *Es sei D ein beschränkter abgeschlossener Bereich in der Ebene der komplexen Veränderlichen ζ, A_D der algebraische Ring aller analytischen Funktionen, die in D regulär sind, mit den gewöhnlichen Operationen der Addition und Multiplikation; es sei $\mathfrak{D}$ die Gesamtheit aller Elemente x des Ringes derart, dass die Menge der Werte von $x(M)$ in A_D vollständig enthalten ist, und $\Gamma_{\mathfrak{D}}$ der algebraische Ring aller Funktionen mit Werten aus dem Ring, die auf $\mathfrak{D}$ definiert sind. Es existiert eine isomorphe Abbildung $f(\zeta) \to f(x)$ des Ringes A_D auf einen Teil $A_{\mathfrak{D}}$ des Ringes $\Gamma_{\mathfrak{D}}$, bei welcher $f(\zeta) = \zeta$ in $f(x) = x$ übergeht, und eine Folge von Funktionen $f_n(\zeta)$, die in irgendeinem D enthaltenden offenen Bereich gleichmässig konvergent ist, in eine Folge von Funktionen $f_n(x)$ übergeht, die für jedes $x \in \mathfrak{D}$ der Norm nach konvergent ist.*

Es gibt nur eine solche Abbildung, und diese wird durch die Formel von Cauchy[6]

$$f(x) = \frac{1}{2\pi i} \int_\Gamma (\zeta e - x)^{-1} f(\zeta)\, d\zeta \qquad (1)$$

[6] Die Anwendung des Integrals von Cauchy in ähnlichen Fragen (Untersuchung von Operatoren) findet sich schon bei F. Riesz in seinem Buch über die Theorie der unendlichen Systeme von Gleichungen mit einer unendlichen Anzahl von Unbekannten.

gegeben, wo als Integrationsweg eine beliebige rektifizierbare geschlossene Kurve genommen wird, die ganz in dem Regularitätsbereich der Funktion $f(\zeta)$ gelegen ist und von D eine positive Entfernung hat.

Beweis. Wir werden zuerst zeigen, dass wenn die in dem Satz behauptete Abbildung überhaupt existiert, so muss $f(\zeta)$ in die Funktion $f(x)$ übergehen, die durch die Formel (1) definiert ist, und werden sodann zeigen, dass die durch die Formel (1) gegebene Abbildung in der Tat allen gestellten Bedingungen genügt.

Es möge also die uns interessierende Abbildung existieren. Da ζ in x übergeht, so ist der Homomorphismus $f(\zeta) \to f(x)$ nicht trivial. Daher geht die Funktion $f(\zeta) \equiv 1$ in $f(x) \equiv e$ über. Daraus folgt, dass die Funktion $\frac{1}{f(\zeta)}$, wenn sie zusammen mit $f(\zeta)$ in A_D enthalten ist, in $f^{-1}(x) = (f(x))^{-1}$ übergeht. Es sei nun ζ_0 ein beliebiger, zu D nicht angehörender, Punkt. Dann ist die Funktion $(\zeta - \zeta_0)^{-1}$ in A_D enthalten und muss, nach dem vorherigen, in die Funktion $(x - \zeta_0 e)^{-1}$ übergehen. Da nach der Bedingung $x \in \mathfrak{D}$ und $\zeta_0 \overline{\in} D$, so existiert diese letztere Funktion *.

Betrachten wir nun eine beliebige Funktion $f(\zeta) \in A_D$. Da $f(\zeta)$ in dem beschränkten abgeschlossenen Bereich D regulär ist, so existiert eine rektifizierbare geschlossene Kurve Γ, die ganz in dem Regularitätsbereich der Funktion $f(\zeta)$ liegt und von D eine positive Entfernung hat. Nach der Formel von Cauchy haben wir für alle Punkte $\zeta \in \mathfrak{D}$

$$f(\zeta) = \frac{1}{2\pi i} \int_\Gamma \frac{f(\lambda)\, d\lambda}{\lambda - \zeta}.$$

Der Ausdruck auf der rechten Seite dieser Gleichung ist der Grenzwert der Folge von Summen der Form

$$f_n(\zeta) = \frac{1}{2\pi i} \sum_{k=0}^{n} \frac{f(\lambda_k)(\lambda_{k+1} - \lambda_k)}{\lambda_k - \zeta} \qquad (\lambda_k \in \Gamma),$$

die in dem ganzen Bereich Γ gleichmässig konvergent ist. Einer jeden solchen Summe entspricht die abstrakte Funktion

$$f_n(x) = \frac{1}{2\pi i} \sum_{k=0}^{n} f(\lambda_k)(\lambda_k e - x)^{-1}(\lambda_{k+1} - \lambda_k).$$

Nach der Bedingung müssen diese Funktionen für jedes $x \in \mathfrak{D}$ der Norm nach zu einem Grenzwert konvergieren. Sie sind aber Integralsummen für das Integral $\int_\Gamma x(\lambda)\, d\lambda$ der Funktion $x(\lambda) = \frac{1}{2\pi i} f(\lambda)(\lambda e - x)^{-1}$ der komplexen Veränderlichen λ,

---

* Damit die Funktion $(x - \zeta_0 e)^{-1}$ für alle $\zeta_0 \overline{\in} D$ existiere, müssen wir die betrachteten Elemente x auf den Bereich $\mathfrak{D}$ beschränken.

Man beachte, dass $\mathfrak{D}$ nicht leer ist. In der Tat, sei λ_0 ein innerer Punkt des Bereiches $\mathfrak{D}$, so dass ein $\rho > 0$ derart existiert, dass ein Kreis vom Radius ρ um den Punkt λ_0 ganz in D enthalten ist. Dann gilt

$$x' = \frac{(\lambda - \lambda_0)x}{\|x\|} + \lambda_0 e \in \mathfrak{D}$$

für alle Elemente x des betrachteten Ringes und alle λ, die in dem angegebenen Kreis liegen.

welche infolge der Wahl von Γ existiert und der Norm nach stetig ist. Wie früher hervorgehoben wurde, existiert ein solches Integral im Sinne der Konvergenz nach der Norm. Dabei hängt es nicht von der Wahl von Γ ab, solange Γ den gestellten Bedingungen genügt, da $x(\lambda)$ eine abstrakte analytische Funktion der Veränderlichen λ ist, die ausserhalb D regulär ist, und für solche Funktionen gilt, wie wir gesehen haben, der Satz von Cauchy.

Somit sind wir zu dem Schluss gekommen, dass jeder Funktion $f(\zeta) \in A_D$ eine abstrakte Funktion

$$f(x) = \frac{1}{2\pi i} \int_\Gamma \frac{f(\lambda)\, d\lambda}{\lambda e - x}$$

entsprechen muss, wobei diese Funktion tatsächlich existiert und für alle $x \in \mathfrak{D}$ eindeutig definiert ist.

Zeigen wir nun, dass die Abbildung $f(\zeta) \rightarrow f(x)$, die durch die Formel (1) gegeben ist, tatsächlich die in dem Satz behaupteten Eigenschaften besitzt.

Beweisen wir, dass diese Abbildung ein Homomorphismus ist. Es braucht nur bewiesen zu werden, dass das Produkt in das Produkt übergeht. Mit anderen Worten, wir haben zu zeigen, dass

$$\frac{1}{2\pi i} \int_\Gamma \frac{f(\lambda)\, g(\lambda)\, d\lambda}{\lambda e - x} = \frac{1}{2\pi i} \int_\Gamma \frac{f(\lambda)\, d\lambda}{\lambda e - x} \cdot \frac{1}{2\pi i} \int_\Gamma \frac{g(\lambda)\, d\lambda}{\lambda e - x} \tag{2}$$

gilt, wo Γ eine rektifizierbare geschlossene Kurve bedeutet, die ganz in dem gemeinsamen Regularitätsbereich der Funktionen $f(\zeta)$ und $g(\zeta)$ enthalten ist und von D eine positive Entfernung hat. Ersetzen wir in dem zweiten Integral auf der rechten Seite von (2) Γ durch ein Γ enthaltendes und von ihm eine positive Entfernung besitzendes Γ_1, das noch ganz in dem Regularitätsbereich der Funktionen $f(\zeta)$ und $g(\zeta)$ liegt, so erhalten wir

$$\frac{1}{2\pi i} \int_\Gamma \frac{f(\lambda)\, d\lambda}{\lambda e - x} \cdot \frac{1}{2\pi i} \int_\Gamma \frac{g(\lambda)\, d\lambda}{\lambda e - x} = -\frac{1}{4\pi^2} \int_\Gamma \int_{\Gamma_1} \frac{f(\lambda)\, g(\mu)}{(\lambda e - x)(\mu e - x)}\, d\lambda\, d\mu =$$

$$= \frac{1}{4\pi^2} \int_\Gamma \int_{\Gamma_1} \frac{f(\lambda) g(\mu)}{\mu - \lambda} \left(\frac{1}{\lambda e - x} - \frac{1}{\mu e - x} \right) d\lambda\, d\mu = \frac{1}{2\pi i} \int_\Gamma \frac{f(\lambda)}{\lambda e - x} \left(\int_{\Gamma_1} \frac{1}{2\pi i} \frac{g(\mu)\, d\mu}{\mu - \lambda} \right) d\lambda +$$

$$+ \frac{1}{4\pi^2} \int_{\Gamma_1} \frac{g(\mu)}{\mu e - x} \left(\int_\Gamma \frac{f(\lambda)\, d\lambda}{\mu - \lambda} \right) d\mu.$$

Da die Funktion $\frac{f(\lambda)}{\mu - \lambda}$ in dem abgeschlossenen Bereich der Ebene der komplexen Veränderlichen λ, der von Γ begrenzt wird, regulär ist, so finden wir nach dem Satz von Cauchy, dass der zweite Summand auf der rechten Seite gleich Null ist. Ferner, da λ ein innerer Punkt des Bereiches ist, der von Γ_1 begrenzt wird, so ist das innere Integral in dem ersten Summanden nach der Formel von Cauchy gleich $g(\lambda)$. Somit kommen wir zu der Gleichung (2).

Zeigen wir nun, dass ζ in x übergeht, d. h. dass

$$\frac{1}{2\pi i} \int \frac{\zeta\, d\zeta}{\zeta e - x} = x$$

gilt. Da die Funktion $f(\zeta)=\zeta$ in der ganzen Ebene regulär ist, so können wir für Γ einen Kreis von beliebig grossem Radius um den Nullpunkt nehmen. Wählen wir diesen Kreis so, dass die Reihe

$$\zeta(\zeta e-x)^{-1}=e+\zeta^{-1}x+\zeta^{-2}x^2+\ldots$$

absolut konvergiert. Dann erhalten wir nach gliedweiser Integration

$$\frac{1}{2\pi i}\int\frac{\zeta\,d\zeta}{\zeta e-x}=\frac{1}{2\pi i}\int_{\Gamma}(e+\zeta^{-1}x+\ldots)\,d\zeta=x.$$

Es möge nun $f_n(\zeta)\to f(\zeta)$ gleichmässig in dem offenen Bereich D' gelten, der D enthält. Nehmen wir Γ im Inneren von D' so liegend, dass es von D eine positive Entfernung hat, so erhalten wir

$$\|f_n(x)-f(x)\|=\left\|\frac{1}{2\pi i}\int_{\Gamma}\frac{f(\zeta)-f_n(\zeta)}{\zeta e-x}\,d\zeta\right\|\leqslant$$

$$\leqslant\max_{\zeta\in\Gamma}|f(\zeta)-f_n(\zeta)|\cdot\frac{1}{2\pi}\int_{\Gamma}\|(\zeta e-x)^{-1}\|\,d\zeta\to 0.$$

Es bleibt uns zu zeigen, dass der Homomorphismus $f(\zeta)\to f(x)$ ein Isomorphismus ist, d. h. dass $f(x)$ identisch Null nur in dem Fall ist, wenn $f(\zeta)\equiv 0$ ist.

Aber wenn $f(\zeta)\not\equiv 0$ ist, so existiert ein abgeschlossener Bereich $D_1\subset D$, in dessen allen Punkten $f(\zeta)\neq 0$ ist. Dann ist $\frac{1}{f(\zeta)}\in A_{D_1}$, und die Funktion $f^{-1}(x)$ ist für alle x definiert, für die die Menge der Werte von $x(M)$ in D_1 enthalten ist. Nach dem Bewiesenen haben wir $f(x)\cdot f^{-1}(x)=e$, und folglich $f(x)\neq 0$ für alle angegebenen Elemente x.

Somit ist der Beweis des Satzes 19 geleistet. Aus der Formel (1) erhalten wir noch den folgenden

Satz 20. *Wenn $f(\zeta)$ in dem Bereich der Werte der Funktion $x(M)$ regulär ist, so existiert im Ringe ein solches Element y, dass $y(M)=f(x(M))$ für alle maximalen Ideale M ist.*

Beweis. Da $x(M)$ eine stetige und beschränkte Funktion ist, die auf einem bikompakten Raum definiert ist, so ist die Gesamtheit ihrer Werte eine beschränkte abgeschlossene Menge F in der komplexen Ebene. Wie wir gesehen haben, existiert das Integral (1) und nimmt einen und denselben Wert an für jedes Γ, das ganz in dem Regularitätsbereich der Funktion $f(\zeta)$ enthalten ist und von F eine positive Entfernung hat. Dieses Integral ist gerade das Element des Ringes, das die im Satz angegebene Eigenschaft besitzt. In der Tat, bei fixiertem M und veränderlichem x ist $x(M)$ ein lineares Funktional von x, und also haben wir

$$f(x)(M)=\frac{1}{2\pi i}\int_{\Gamma}(\lambda e-x)^{-1}(M)f(\lambda)\,d\lambda=\frac{1}{2\pi i}\int_{\Gamma}\frac{f(\lambda)\,d\lambda}{\lambda-x(M)}=f(x(M)).$$

Mathematisches Institut der Akademie
der Wissenschaften der U. d. S. S. R.

(Поступило в редакцию 27/VI 1940 г.)

9.

(mit G. E. Shilov)

Über verschiedene Methoden der Einführung der Topologie in die Menge der maximalen Ideale eines normierten Ringes

Mat. Sb., Nov. Ser. **9** (51) (1941) 25–38. Zbl. **24**:321

§ 1. Die erste Methode der Einführung der Topologie

1. Es sei K ein normierter Ring [1] und $\mathfrak{M}$ die Menge aller seinen maximalen Ideale. Ein maximales Ideal M heisst ein B e r ü h r u n g s p u n k t der Untermenge $\mathfrak{A} \in \mathfrak{M}$, wenn es den Durchschnitt aller maximalen Ideale, die in $\mathfrak{A}$ eingehen, enthält; die Gesamtheit aller Berührungspunkte von $\mathfrak{A}$ bildet die A b s c h l i e s s u n g $\overline{\mathfrak{A}}$ *.

Es ist leicht die Erfüllung der Axiome des topologischen Raumes nachzuweisen: wenn $M_0 \in \overline{\mathfrak{A}+\mathfrak{B}}$, so ist, z.B., $M_0 \in \overline{\mathfrak{A}}$, d. h. $M_0 \supset \prod_{M \in \mathfrak{A}} M \supset \prod_{M \in \mathfrak{A}+\mathfrak{B}} M$, d. h. $M_0 \in \overline{\mathfrak{A}+\mathfrak{B}}$; wenn aber $M_0 \overline{\in} \overline{\mathfrak{A}}+\overline{\mathfrak{B}}$, so bedeutet dies, dass Elemente $x \in \prod_{M \in \mathfrak{A}} M$ und $y \in \prod_{M \in \mathfrak{B}} M$ existieren, die zu M_0 nicht gehören; ihr Produkt $xy \in \prod_{M \in \mathfrak{A}+\mathfrak{B}} M$ gehört auch nicht zu M_0 (weil der Restklassenring R/M_0 ein Körper ist und keine Nullteiler enthält), und hieraus folgt, dass M_0 kein Berührungspunkt von $\mathfrak{A}+\mathfrak{B}$ ist. Somit ist $\overline{\mathfrak{A}}+\overline{\mathfrak{B}}=\overline{\mathfrak{A}+\mathfrak{B}}$. Ferner ist klar, dass $\mathfrak{A} \subset \overline{\mathfrak{A}}$ und $\overline{\overline{\mathfrak{A}}}=\overline{\mathfrak{A}}$ ist; endlich setzen wir definitionsgemäss $\overline{0}=0$.

2. Jeder einzelne Punkt $M \in \mathfrak{M}$ ist offenbar eine abgeschlossene Menge; folglich ist $\mathfrak{M}$ ein T_1-Raum. Wie das Beispiel 1 (am Ende dieses Paragraphen) zeigt, ist $\mathfrak{M}$ im allgemeinen Falle kein T_2-Raum.

Die Menge $\mathfrak{A}$ der Form $\{x(M)=0\}$ ist abgeschlossen; ihre Berührungspunkte müssen das Element x enthalten und müssen daher $\mathfrak{A}$ angehören. Die offenen Mengen — Ergänzungen zu den solchen abgeschlossenen — bilden ein vollständiges System von Umgebungen: wenn eine beliebige offene Menge $\mathfrak{O}$ und ihre abgeschlossene Ergänzung $\mathfrak{F}$ gegeben sind, so existiert für einen beliebigen Punkt $M_0 \in \mathfrak{O}$ ein $x \in K$ derart, dass $x(M_0) \neq 0$, $x(M)=0$ für alle $M \in \mathfrak{F}$ ist, denn andernfalls wäre das maximale Ideal M_0 ein Berührungspunkt von $\mathfrak{F}$; wenn wir die Umgebung $\{x(M) \neq 0\}$ betrachten, so sehen wir, dass sie ganz in $\mathfrak{O}$ eingeschlossen ist.

---

* Diese Topologie wurde schon in einem anderen Zusammenhang von Stone [2] (in Booleschen Ringen) und in der Arbeit von I. Gelfand und A. Kolmogoroff [3] betrachtet.

3. Zeigen wir nun, dass der topologische Raum $\mathfrak{M}$ bikompakt ist. Es sei ein beliebiges System $\{\mathfrak{F}_\alpha\}$ von abgeschlossenen Mengen gegeben. Jedem $\mathfrak{F}_\alpha$ ordnen wir das Ideal $I_\alpha = \prod_{M \in \mathfrak{F}_\alpha} M$ zu; betrachten wir das Ideal I, das von allen endlichen Summen von Elementen, die aus den Idealen I_α genommen sind, gebildet wird. Nehmen wir an, dass I in einem maximalen Ideal M_0 enthalten ist. Dieses Ideal M_0 enthält für gegebenes α alle Elemente des Ideals I_α und gehört deshalb zu der abgeschlossenen Menge $\mathfrak{F}_\alpha$; hieraus folgt, dass $M_0 \in \prod_\alpha \mathfrak{F}_\alpha$. Es sei nun gegeben, dass der Durchschnitt $\prod_\alpha \mathfrak{F}_\alpha$ leer ist; dann kann das Ideal I in keinem maximalen Ideal enthalten sein und fällt daher mit dem ganzen Ring K zusammen. Folglich gibt eine bestimmte Summe $\sum_{i=1}^{n} x_i$, wo $x_i \in I_{\alpha_i}$, die Einheit. Bilden wir den endlichen Durchschnitt $\prod_{i=1}^{n} \mathfrak{F}_{\alpha_i}$; wir behaupten, dass er leer ist. In der Tat, wenn ein maximales Ideal in allen Mengen $\mathfrak{F}_{\alpha_i}$ enthalten wäre, so würde es alle x_i enthalten, und mit ihnen also auch die Einheit, was unmöglich ist. Somit erhalten wir:

Aus jedem System von abgeschlossenen Mengen $\{\mathfrak{F}_\alpha\}$ mit leerem Durchschnitt kann man ein endliches Untersystem $\{\mathfrak{F}_{\alpha_i}\}$ auswählen, dessen Durchschnitt auch leer ist.

Dies ist aber eine der Definitionen der Bikompaktheit.

Beispiel. Betrachten wir den Ring A, der von allen Funktionen $\varphi(z)$ gebildet wird, die im Kreise $|z| \leqslant 1$ stetig und in seinem Inneren analytisch sind. Unter der Norm des Elements $\varphi(z)$ werden wir $\max_{|z| \leqslant 1} |\varphi(z)|$ verstehen.

Es ist bekannt, dass eine jede solche Funktion der Grenzwert einer Folge von Polynomen ist, die im Kreise $|z| \leqslant 1$ gleichmässig konvergiert. Alle Funktionen, die in einem gegebenen Punkt z_0, $|z_0| \leqslant 1$, verschwinden, bilden ein maximales Ideal (§ 2, n°. 4); zeigen wir, dass A keine anderen maximalen Ideale besitzt. In der Tat, sei M ein beliebiges maximales Ideal; da $|z(M)| \leqslant \|z\| = 1$ ist, so existiert ein Punkt z_0, $|z_0| \leqslant 1$, derart, dass für das ihr entsprechende maximale Ideal M_0 $z(M_0) = z(M)$ gilt. Die Erzeugende z hat, folglich, gleiche Werte auf zwei maximalen Idealen; dann werden aber auch alle Elemente des Ringes als Grenzwerte von Polynomen in der Erzeugenden auf diesen maximalen Idealen gleiche Werte haben: insbesondere zieht $\varphi(M) = 0$ $\varphi(M_0) = 0$ nach, d. h. $M \subset M_0$ und, infolge der Maximalität, $M = M_0$. Betrachten wir nun, was hier unsere Methode der Topologisierung ergibt.

Jede unendliche Menge $\mathfrak{F}$, die einen Grenzpunkt (im gewöhnlichen Sinne) innerhalb des Kreises $|z| \leqslant 1$ enthält, liegt in $\mathfrak{M}$ überall dicht. In der Tat, ein Berührungspunkt einer solchen Menge ist ein beliebiges maximales Ideal, das alle Funktionen $\varphi(z) \in A$ enthält, die auf $\mathfrak{F}$ gleich Null sind; aber da nach dem klassischen Einzigkeitssatz der analytischen Funktionen nur eine Funktion — nämlich die Null des Ringes A — in allen Punkten von $\mathfrak{F}$ gleich Null sein kann, so genügt ein beliebiges maximales Ideal dieser Bedingung, und folglich ist $\overline{\mathfrak{F}} = \mathfrak{M}$. Jeder Bereich enthält somit jeden inneren Kreis vollständig mit der Ausnahme von, vielleicht, einer endlichen Anzahl seiner Punkte; folglich können keine zwei Punkte in einander nicht überschneidende Umgebungen eingeschlossen werden.

§ 2. Bikompakte Erweiterungen eines topologischen Raumes

4. Es sei der normierte Ring K als der Ring der Funktionen auf einer Menge S mit gewöhnlichen Operationen der Addition und Multiplikation gegeben.

Betrachten wir die Menge aller Funktionen $x(t) \in K$, die in einem fixierten Punkt $a \in S$ den Wert Null annehmen. Offenbar ist diese Menge M_a ein Ideal in K. Wenn wir zu ihm eine beliebige in ihm nicht enthaltene Funktion $y(t) \in K$ hinzufügen, so erhalten wir ein Ideal (M_a, y), das eine von Null verschiedene Konstante

$$y(a) = y(t) - [y(t) - y(a)]$$

enthält und also mit dem ganzen Ring K zusammenfällt. Folglich ist M_a ein maximales Ideal und also auch ein abgeschlossenes (siehe [1], Satz 4). Einem beliebigen Element $x \in K$ entspricht eine Zahl $x(M_a)$ (siehe [1], § 5); zeigen wir, dass $x(M_a) = x(a)$ ist.

In zwei Sonderfällen, nämlich für $x = e$, wenn $x(a) = x(M_a) = 1$, und $x(a) = 0$, wenn $x \in M_a$, $x(M_a) = 0$ ist, ist dies klar; im allgemeinen Falle betrachten wir die Funktion $x - x(a)e$: wir haben

$$[x - x(a)e](M_a) = 0, \quad x(M_a) - x(a)e(M_a) = 0, \quad x(M_a) = x(a).$$

5. Wollen wir erreichen, dass zwei verschiedenen Punkten zwei verschiedene maximale Ideale entsprechen, so muss der Ring K der folgenden evidenten Bedingung (die notwendig und hinreichend ist) unterworfen werden: für beliebige zwei verschiedene Punkte a und b existiert ein $x(t) \in K$ derart, dass $x(a) \neq x(b)$ ist. Wenn diese Bedingung erfüllt ist, stellt sich heraus, dass S ein Teil von $\mathfrak{M}$ ist, der dabei überall dicht gelegen ist: der Durchschnitt aller maximalen Ideale, die den Punkten von S entsprechen, besteht aus einer Funktion $x(t) \equiv 0$, die als die Null des Ringes K in jedem maximalen Ideal enthalten ist; folglich ist jedes maximale Ideal ein Berührungspunkt der Menge S.

6. Es sei nun S ein topologischer Raum. Betrachten wir, welcher Bedingung muss K genügen, damit bei der angegebenen Einbettung S einem Teil von $\mathfrak{M}$ homöomorph ist. Nehmen wir an, dass dies der Fall ist; betrachten wir eine beliebige abgeschlossene Menge $A \subset S$ und die entsprechende Menge $\mathfrak{A}$ der maximalen Ideale von K. Da nach der Bedingung $\mathfrak{A}$ in S abgeschlossen ist, so wird sich für jedes maximale Ideal $M_a \subset S$, das in $\mathfrak{A}$ nicht enthalten ist, ein Element x finden, derart, dass $x(M_a) \neq 0$, $x(\mathfrak{A}) = 0$ ist, denn andernfalls wäre M_a ein Berührungspunkt von $\mathfrak{A}$. Zu den Funktionen übergehend, erhalten wir: für eine beliebige abgeschlossene Menge $A \subset S$ und einen beliebigen zu ihr nicht gehörenden Punkt a muss in K eine Funktion $x(t)$ existieren derart, dass $x(a) \neq 0$, $x(A) = 0$ ist. Es ist leicht einzusehen, dass diese Bedingung auch hinreichend ist, wenn die Funktionen des Ringes K auf S stetig sind. In der Tat, wenn a ein Berührungspunkt von A ist, so ist jede Funktion $x(t) \in K$, die auf A Null ist, auch in a Null; dies aber bedeutet gerade, dass das maximale Ideal M_a ein Berührungspunkt von $\mathfrak{A}$ ist. Umgekehrt, wenn a kein Berührungspunkt von A ist, so existiert laut unserer Bedingung eine Funktion $x(t) \in K$, die auf A gleich Null ist und in a von Null verschieden ist; also ist das Ideal M_a kein Berührungspunkt von $\mathfrak{A}$.

Bekanntlich wird ein topologischer Raum **vollständig regulär** genannt, wenn für eine beliebige abgeschlossene Menge A und einen beliebigen zu ihr nicht gehörenden Punkt a eine auf S stetige Funktion $x(t)$ existiert derart, dass $x(a) \neq 0$, $x(A) = 0$ ist.

Es ist klar, dass die oben angeführten Betrachtungen zu einer neuen Definition des vollständig regulären Raumes führen: *ein topologischer Raum S heisst vollständig regulär, wenn er auf eine homöomorphe Weise in die Menge $\mathfrak{M}$ aller maximalen Ideale des Ringes $K = C(S)$ aller reellen auf S stetigen Funktionen $x(t)$ eingebettet werden kann.*

7. Dafür, dass bei dieser Einbettung der vollständig reguläre Raum S vollständig mit $\mathfrak{M}$ zusammenfalle, ist es notwendig, dass S bikompakt sei (n°. 3). Diese Bedingung ist auch hinreichend: zeigen wir, der Arbeit [3] folgend, dass wenn S bikompakt ist, für jedes maximale Ideal $M \subset C(S)$ ein Punkt a existiert, in dem alle Funktionen des Ideals M den Wert Null annehmen, und, folglich, dass M aus allen Funktionen $x(t) \in C(S)$ besteht, die in a gleich Null sind. Lassen wir das Gegenteil zu, so können wir für jeden Punkt $a \in S$ eine Funktion $x_a(t) \in M$ derart angeben, dass $x_a(a) \neq 0$ ist. Betrachten wir nun eine Umgebung $U(a)$ des Punktes a, in der $x_a(t)$ nicht verschwindet. Infolge der Bikompaktheit von S kann man aus der erhaltenen Überdeckung eine endliche Überdeckung $U(a_1)$, $U(a_2), \ldots, U(a_n)$ auswählen. Bilden wir nun die Funktion $y(t) = \sum_{k=1}^{n} x_{a_k}^2(t)$; sie wird auf S in keinem Punkt zu Null, und daher existiert $\frac{1}{y(t)}$ und gehört zu $C(S)$; da das Ideal M $y(t)$ enthält, muss es auch $1 = y(t) \cdot \frac{1}{y(t)}$ enthalten, was unmöglich ist. Wenn also S bikompakt ist, so ist

$$S = \mathfrak{M}.$$

8. Man beachte, dass bei der Topologisierung der Menge $\mathfrak{M}$ (§ 1) wir uns ausschliesslich auf die algebraischen Eigenschaften des Ringes K stützten; wenn also zwei normierte Ringe algebraisch isomorph sind, so werden die entsprechenden topologischen Räume $\mathfrak{M}_1$ und $\mathfrak{M}_2$ topologisch identisch sein. Wenn man, insbesondere, die Ringe $C(S_1)$ und $C(S_2)$ aller auf zwei vollständig regulären bikompakten Räumen S_1 und S_2 stetigen Funktionen nimmt, so sind diese Räume S_1 und S_2 zueinander homöomorph, denn nach dem Bewiesenen ist $S_1 = \mathfrak{M}_1 = \mathfrak{M}_2 = S_2$.

9. Wenn S ein vollständig regulärer, aber im allgemeinen kein bikompakter Raum ist, so ist, wie wir gesehen haben, $\mathfrak{M}$ ein bikompakter Raum, der S als eine überall dichte Untermenge enthält. Jede Funktion $x(t) \in C(S)$ kann natürlich von S auf das ganze $\mathfrak{M}$ wie $x(M)$ fortgesetzt werden [wir haben bereits gesehen, dass für das dem Punkte $a \in S$ entsprechende maximale Ideal M_a $x(M_a) = x(a)$ ist]; wir werden nun zeigen, dass bei dieser Fortsetzung auf $\mathfrak{M}$ alle Funktionen **stetig** bleiben. Dem Beweis dieser Tatsache schicken wir zwei einfache Lemmas voraus.

Lemma 1. *Es sei A eine Untermenge von S und $\bar{A}$ ihre Abschliessung in $\mathfrak{M}$. Dann sind die Werte einer beliebigen Funktion $x(M) \in C(S)$ auf $\bar{A}$ unter ihren Werten auf A und, vielleicht, einigen Grenzwerten derselben enthalten.*

Wenn eine Funktion $x(M)$ in allen Punkten von A gleich Null ist, so wird sie auch, nach der Definition der Abschliessung, in allen Punkten von $\bar{A}$ gleich Null sein.

Wenn $x(M)$ in allen Punkten von A gleich 1 ist, so ist $y(M) = x(M) - 1$ in allen Punkten von A gleich Null und folglich auch in allen Punkten von $\bar{A}$ gleich Null, woraus folgt, dass $x(M)$ in allen Punkten von $\bar{A}$ gleich 1 ist.

Es sei nun die Funktion $x(M)$ derart, dass Null weder unter ihren Werten auf A, noch unter den Grenzwerten dieser Werte enthalten ist; zeigen wir, dass dasselbe auch auf $\bar{A}$ der Fall sein wird. In dem Ringe $C(S)$ ist eine Funktion $y(M)$ enthalten derart, dass auf A $y(M) = \frac{1}{x(M)}$ ist; diese Funktion kann, z. B., folgendermassen definiert werden: in den Punkten $M \in S$, in denen $|x(M)| \geqslant \alpha$ ist (wo $\alpha = \inf_{M \in A} |x(M)| > 0$), setzen wir $y(M) = \frac{1}{x(M)}$ und in den übrigen $y(M) = \frac{x(M)}{\alpha^2}$. Das Produkt $xy(M)$ ist gleich 1 auf A und folglich, nach dem Vorhergehenden, gleich 1 auch auf $\bar{A}$; wegen der Beschränktheit von $y(M)$ ($|y(M)| \leqslant \|y\|$) kann die Funktion $x(M)$ auf $\bar{A}$ weder verschwinden, noch sich unbegrenzt Null nähern.

Nun ist die Behauptung des Lemmas evident: sollte eine Funktion $y(M)$ auf $\bar{A}$ einen Wert λ annehmen, der weder ihr Wert auf A, noch ein Grenzwert dieser Werte ist, so würde $y(M) - \lambda$ auf $\bar{A}$ den Wert Null annehmen, der weder ihr Wert auf A, noch ein Grenzwert dieser Werte ist, — was, wie wir gesehen haben, unmöglich ist.

Lemma 2. *Es sei U eine Umgebung der Art $\{x(M) \neq 0\}$ (n°. 2). Es gilt die Einschliessungsbeziehung $U \subset \overline{US}$.*

In der Tat, bezeichnen wir mit CU die Ergänzung von U zu $\mathfrak{M}$, d. h. die Menge $\{x(M) = 0\}$. Dann ist $S = US + CU \cdot S$, $\bar{S} = \overline{US} + \overline{CUS}$; da aber $\bar{S} = \mathfrak{M} \supset U$ ist, so gilt $U \subset \overline{US} + \overline{CU \cdot S}$. Aber, nach Lemma 1, haben wir in allen Punkten von $\overline{CU \cdot S}$ $x(M) = 0$, so dass $U \subset \overline{US}$ ist.

Satz. *Alle Funktionen $x(M)$ sind auf $\mathfrak{M}$ stetig.*

Es seien ein Punkt $M_0 \in \mathfrak{M}$, eine Funktion $y(M) \in C(S)$ und eine positive Zahl ε gegeben; wir haben eine solche Umgebung U des Punktes M_0 zu finden, in deren allen Punkten $|y(M) - y(M_0)| < \varepsilon$ gilt. Es sei A die Menge derjenigen Punkte von S, in denen $|y(M) - y(M_0)| \geqslant \varepsilon$ ist. Nach Lemma 1 enthält $\bar{A}$ nicht den Punkt M_0. Da das System von Umgebungen der Form $\{x(M) \neq 0\}$ ein vollständiges System ist (n°. 2), so existiert eine Umgebung $U = \{x(M) \neq 0\}$ des Punktes M_0, die in $\mathfrak{M} - \bar{A}$ enthalten ist. Diese Umgebung ist die gesuchte: aus $U\bar{A} = 0$ folgt, dass $US \cdot \bar{A}S = US \cdot A = 0$ ist, d. h. dass in allen Punkten von US $|y(M) - y(M_0)| < \varepsilon$ ist; nach Lemma 1 ist $|y(M) - y(M_0)| < \varepsilon$ in allen Punkten von $\overline{US}$, insbesondere, nach Lemma 2, in allen Punkten der Umgebung U.

Folgerung. *Der Raum $\mathfrak{M}$ ist ein Hausdorffscher Raum.*

In der Tat, für zwei gegebene Punkte M_0 und M_1 kann man eine Funktion $x(M)$ konstruieren derart, dass $x(M_0) = 0$, $x(M_1) = 1$ ist. Da $x(M)$ stetig ist, so sind die Mengen $\{|x(M)|<\frac{1}{2}\}$ und $\{|x(M)|>\frac{1}{2}\}$ offen. Die erste Menge enthält den Punkt M_0, die zweite — den Punkt M_1, und diese Mengen überschneiden sich nicht.

10. Es sei eine beliebige bikompakte Erweiterung des Raumes S gegeben, d. h. ein bikompakter Hausdorffscher Raum Q, der S als eine überall dichte Menge enthält. Wir werden zeigen, dass Q ein stetiges Bild des von uns konstruierten Raumes $\mathfrak{M}$ ist. Mit anderen Worten, $\mathfrak{M}$ ist in einem bestimmten Sinne die grösste bikompakte Erweiterung von S.

Betrachten wir den Ring $C(Q)$ aller auf Q stetigen Funktionen. Da jede Funktion, die auf Q stetig ist, auch auf $S \subset Q$ stetig ist, so ist der Ring $C(Q)$ ein abgeschlossener Unterring des Ringes $C(S)$. Es sei M ein Punkt der Menge $\mathfrak{M}$; wählen wir aus dem maximalen Ideal M alle Funktionen aus, die dem Ringe $C(Q)$ angehören. Diese Funktionen bilden, offenbar, in $C(Q)$ ein Ideal M'. Fügen wir zu ihm eine beliebige in ihm nicht enthaltene Funktion $y(t) \in C(Q)$ hinzu, so erhalten wir ein Ideal (M', y), das eine von Null verschiedene Konstante $y(M) = y(t) - [y(t) - y(M)]$ enthält und also mit $C(Q)$ zusammenfällt; folglich ist das Ideal M' ein maximales Ideal in $C(Q)$ und entspricht einem bestimmten Punkt $t' \in Q$ (n°. 7). Wir erhalten so eine Abbildung von $\mathfrak{M}$ in Q; es ist leicht einzusehen, dass diese Abbildung stetig ist; wenn der Punkt M ein Berührungspunkt der Menge $\mathfrak{A} \in \mathfrak{M}$ ist, so werden alle auf $\mathfrak{M}$ stetigen Funktionen, die auf $\mathfrak{A}$ gleich Null sind, auch in M gleich Null sein; da es insbesondere auch für Funktionen aus $C(Q)$ gilt, so ist das Bild des Punktes M ein Berührungspunkt des Bildes der Menge $\mathfrak{A}$. Daher ist das Bild der Menge $\mathfrak{M}$ in Q ein bikompakter Raum, der S enthält (bei unserer Abbildung gehen die Punkte von S offenbar in sich über) und, folglich, mit Q zusammenfällt; somit ist die Behauptung bewiesen.

Die von Čech [4] konstruierte Erweiterung $\beta(S)$ ist auch in diesem Sinne die grösste Erweiterung von S; der bewiesene Satz erlaubt nun zu schliessen, dass $\beta(S) = \mathfrak{M}$ ist.

§ 3. Die zweite Methode der Einführung der Topologie

11. Es sei K ein normierter Ring, der der folgenden Bedingung genügt:

(*) *Für jedes Element* $x \in K$ *kann ein Element* $y \in K$ *derart gefunden werden, dass* $x(M) = \overline{y(M)}$ *für beliebiges* $M \in \mathfrak{M}$ *ist* (der Strich oben bedeutet den Übergang zu der konjugiert-komplexen Grösse).

In diesem Paragraphen führen wir in der Menge $\mathfrak{M}$ der maximalen Ideale eines solchen Ringes eine neue Topologie ein, die, im allgemeinen, von der in § 1 eingeführten verschieden ist.

Es sei nämlich eine Menge $\mathfrak{A} \in \mathfrak{M}$ gegeben; ein maximales Ideal M_0 werden wir einen Berührungspunkt von $\mathfrak{A}$ nennen, wenn es in der Abschliessung der Menge aller Elemente von K enthalten ist, die in allen maximalen Idealen von $\mathfrak{A}$ enthalten sind. Es ist leicht einzusehen, dass diese Definition der folgenden äquivalent ist M_0 heisst ein Berührungspunkt von $\mathfrak{A}$, wenn für jedes $x \in M_0$ $\inf_{M \in \mathfrak{A}} |x(M)| = 0$ ist. In der Tat, wenn M_0 ein Berührungspunkt von $\mathfrak{A}$ nach der zweiten Definition ist, so existiert für jedes $x \in M_0$ eine Folge von maximalen Idealen $M_n \in \mathfrak{A}$ derart, dass $x\,(M_n) = \lambda_n \to 0$. Dann ist

$$x = \lim_{n \to \infty} (x - \lambda_n e), \quad x - \lambda_n e \in M_n.$$

Wenn, umgekehrt, M_0 kein Berührungspunkt von $\mathfrak{A}$ nach der zweiten Definition ist, so existiert ein Element $y \in M_0$ derart, dass $\inf\limits_{M \in \mathfrak{A}} |y(M)| = a > 0$ ist. Dieses Element kann nicht ein Grenzwert von Elementen $x \in M \in \mathfrak{A}$ sein, denn sie alle werden zu Null in mindestens einem Punkte von $\mathfrak{A}$, während die Konvergenz der Elemente $x \in K$ muss, infolge der Ungleichung $|x(M)| \leqslant \|x\|$ ([1], § 5), die gleichmässige Konvergenz der entsprechenden Funktionen $x(M)$ nach sich ziehen.

Zeigen wir, dass die Axiome der Abschliessung erfüllt sind. Es sei $M_0 \in \overline{\mathfrak{A}} + \overline{\mathfrak{B}}$; dann ist, z. B., $M_0 \in \overline{\mathfrak{A}}$; für jedes $x_0 \in M_0$ ist $\inf\limits_{M \in \mathfrak{A}} |x(M)| = 0$, und umsomehr $\inf\limits_{M \in \mathfrak{A}+\mathfrak{B}} |x(M)| = 0$, so dass $M_0 \in \overline{\mathfrak{A}+\mathfrak{B}}$. Wenn $M_0 \overline{\in} \overline{\mathfrak{A}} + \overline{\mathfrak{B}}$, so haben wir $M_0 \overline{\in} \overline{\mathfrak{A}}$ und $M_0 \overline{\in} \overline{\mathfrak{B}}$, d. h. es existieren in M_0 Elemente y und z derart, dass

$$\inf_{M \in \mathfrak{A}} |y\ (M)| > 0, \quad \inf_{M \in \mathfrak{B}} |z\ (M)| > 0;$$

das Element $\bar{y}y + \bar{z}z$ besitzt ein $\inf\limits_{M \in \mathfrak{A}+\mathfrak{B}} |(\bar{y}y + \bar{z}z)\ (M)| > 0$ [hier benutzen wir wesentlich die Bedingung (*), die von uns auf den Ring K auferlegt war] und ist in M_0 enthalten, so dass M_0 kein Berührungspunkt von $\mathfrak{A} + \mathfrak{B}$ ist; somit ist $\overline{\mathfrak{A}} + \overline{\mathfrak{B}} = \overline{\mathfrak{A} + \mathfrak{B}}$. Ferner ist es klar, dass $\mathfrak{A} \subset \overline{\mathfrak{A}}$ und $\overline{\overline{\mathfrak{A}}} = \overline{\mathfrak{A}}$ ist; endlich setzen wir definitionsgemäss $\overline{0} = 0$.

12. Da jedes maximale Ideal abgeschlossen ist ([1], Satz 4), so ist jeder Punkt $M \in \mathfrak{M}$ eine abgeschlossene Menge; folglich ist $\mathfrak{M}$ ein T_1-Raum.

Eine Menge $\mathfrak{A} \in \mathfrak{M}$ der Form $\{|x(M)| \geqslant \imath\}$ ist abgeschlossen; ein Punkt $M_0 \in \mathfrak{M}$, in dem $x(M_0) = b$, wo $|b| < |a|$ ist, kann nicht ein Berührungspunkt von $\mathfrak{A}$ sein, denn $M_0 \ni x - be$, $\inf\limits_{M \in \mathfrak{A}} |x - be| > 0$. Folglich sind die Mengen $\{\,|x(M)| < a\,\}$ offen; es ist leicht zu zeigen, dass sie ein vollständiges System von Umgebungen bilden: wenn eine beliebige offene Menge $\mathfrak{O}$ und ihre abgeschlossene Ergänzung $\mathfrak{F}$ gegeben sind, so existiert für jeden Punkt $M_0 \in \mathfrak{O}$ ein solches $x \in K$, dass $x(M_0) = 0$, $\inf\limits_{M \in \mathfrak{F}} |x(M)| = a \neq 0$ ist, denn andernfalls wäre das maximale Ideal M_0 ein Berührungspunkt von $\mathfrak{F}$; betrachten wir die Umgebung $\{\,|x(M)| < a\}$, so sehen wir, dass sie vollständig in $\mathfrak{O}$ eingeschlossen ist. Es ist klar, dass alle Funktionen $x(M)$ stetige Funktionen sind; hieraus folgt in derselben Weise, wie in n°. 9, § 2, dass $\mathfrak{M}$ ein Hausdorffscher Raum ist.

13. Zeigen wir schliesslich, dass $\mathfrak{M}$ bikompakt ist. Zum Beweis benutzen wir die Ergebnisse des § 2; die Eigenschaft (*) nutzen wir nicht wesentlich aus.

Es sei eine abgeschlossene Menge $\mathfrak{F} \in \mathfrak{M}$ und ein zu ihr nicht gehörender Punkt M_0 gegeben; wie wir bereits wissen, existiert eine Umgebung $U = \{\,|x(M)| < a\}$ des Punktes M_0, die in $\mathfrak{M} - \mathfrak{F}$ eingeschlossen ist. Setzen wir $\varphi(M) = 0$ ausserhalb U und $\varphi(M) = a - |x(M)|$ in U, so erhalten wir eine auf $\mathfrak{M}$ stetige

Funktion, die auf $\mathfrak{F}$ Null ist und in U von Null verschieden ist; somit ist $\mathfrak{M}$ vollständig regulär, und die ganze Theorie der bikompakten Erweiterungen, die in § 2 konstruiert werden, ist auf $\mathfrak{M}$ anwendbar. Die Menge $\mathfrak{M}'$ der maximalen Ideale des Ringes $C(\mathfrak{M})$, topologisiert nach der Methode des § 1, ist nämlich eine bikompakte Erweiterung von $\mathfrak{M}$. Wir werden zeigen, dass $\mathfrak{M}$ ein stetiges Bild von $\mathfrak{M}'$ ist. In der Tat, wählen wir aus dem maximalen Ideal $M' \subset C(\mathfrak{M})$ alle zu K gehörenden Funktionen aus, so erhalten wir in derselben Weise, wie in n°. 10, ein maximales Ideal M des Ringes K, d. h. einen Punkt der Menge $\mathfrak{M}$; est ist klar, dass bei dieser Abbildung $\mathfrak{M}$ in sich übergeht. Ausserdem, wenn $M' \in \mathfrak{M}'$ in $M \in \mathfrak{M}$ übergeht, so gilt für eine beliebige Funktion $x \in K$: $x(M') = x(M)$. Somit ist ein vollständiges Urbild der offenen Menge $\{|x(M)| < a\} \subset \mathfrak{M}$ die Menge $\{|x(M')| < a\} \subset \mathfrak{M}'$, die, infolge der Stetigkeit von $x(M')$ auf $\mathfrak{M}'$ (n°. 9), selbst offen ist; dies aber bedeutet gerade die Stetigkeit der Abbildung $\mathfrak{M}' \rightarrow \mathfrak{M}$. Hieraus folgt, dass $\mathfrak{M}$ als ein stetiges Bild des bikompakten $\mathfrak{M}'$ bikompakt ist.

Bemerkung. Es ist leicht zu zeigen, dass die hier eingeführte Topologie als das Resultat der Anwendung der allgemeinen Topologie ([1], S. 11) auf den Ring, der die Eigenschaft (*) besitzt, entsteht. Dank gerade dieser Eigenschaft enthält der Durchschnitt einer endlichen Anzahl von Mengen $\{|x_i(M)| < a_i\}$, $i = 1, 2, \ldots, n$, der eine elementare Umgebung in der allgemeinen Topologie ist, eine Menge der Form $\{|x(M)| < a\}$.

14. Es sei der Ring K als ein Ring der stetigen Funktionen (im allgemeinen, nicht aller) auf einem topologischen Raum S gegeben. Wenn die Bedingung des n°. 5 erfüllt ist, so kann S in die Menge $\mathfrak{M}$ aller maximalen Ideale des Ringes K eingebettet werden; betrachten wir, welche Bedingung auf K auferlegt werden muss, damit S in $\mathfrak{M}$ auf eine homöomorphe Weise eingebettet werden kann. Nehmen wir an, dass es dem so ist; betrachten wir eine beliebige abgeschlossene Menge $A \subset S$ und die entsprechende Menge $\mathfrak{A}$ der maximalen Ideale von K. Da nach der Bedingung $\mathfrak{A}$ in S abgeschlossen ist, so wird sich für ein beliebiges maximales Ideal $M_0 \subset S$, das in $\mathfrak{A}$ nicht enthalten ist, ein Element $x \in K$ derart finden, dass $x(M_0) = 0$, $\inf\limits_{M \in \mathfrak{A}} |x(M)| > 0$ ist; andernfalls wäre M_0 ein Berührungspunkt von $\mathfrak{A}$. Gehen wir zu Funktionen über, so erhalten wir: für jede beliebige abgeschlossene Menge A und jeden nicht zu ihr gehörenden Punkt a muss in K eine Funktion $x(t)$ existieren derart, dass $x(a) = 0$, $\inf\limits_{t \in A} |x(t)| > a$ ist. Es ist leicht einzusehen, dass diese Bedingung auch hinreichend ist. Wenn K der Ring aller auf S stetigen Funktionen ist, so fällt diese Bedingung mit der ähnlichen Bedingung zusammen, die wir in n°. 6 erhalten haben; im allgemeinen Falle ist dies aber nicht der Fall.

§ 4. Der Zusammenhang zwischen dem Ring der Funktionen $x(M)$ und dem Ring aller stetigen Funktionen

15. In § 3 haben wir festgestellt, dass ein normierter Ring K, der die Eigenschaft (*) besitzt, in den Ring der stetigen Funktionen auf dem Hausdorffschen bikompakten Raum $\mathfrak{M}$ seiner maximalen Ideale abgebildet werden kann.

Im allgemeinen braucht, natürlich, nicht jede auf $\mathfrak{M}$ stetige Funktion das Bild eines Elementes $x \in K$ zu sein. Wir werden aber in diesem Paragraphen zeigen, dass *jede auf* $\mathfrak{M}$ *stetige Funktion der Grenzwert* (im Sinne der gleichmässigen Konvergenz) *von Funktionen ist, die Bilder von Elementen von* K *sind.*

Beim Beweis werden wir uns auf das folgende Grundprinzip stützen:

Es sei auf einer Menge S *eine Familie* K *von reellen beschränkten Funktionen* $x(t)$ *gegeben, die alle Konstanten enthält und einen algebraischen Ring mit den gewöhnlichen Operationen der Addition und Multiplikation darstellt. Die Untermengen von* S *der Form* $U_x^a = \{\, |x(t)| < a\}$ *werden wir ausgewählt nennen. Dafür, dass eine Funktion* $y(t)$, *die auf* S *definiert ist, der Grenzwert einer gleichmässig konvergenten Folge von Funktionen aus* K *sei, ist es notwendig und hinreichend, dass für jedes* $\varepsilon > 0$ *eine endliche Überdeckung von* S *durch ausgewählte Untermengen* $U_{x_1}^{a_1}, U_{x_2}^{a_2}, \ldots, U_{x_n}^{a_n}$ *existiere, die die folgenden Eigenschaften besitzt:*

(1) *für jede* $U_{x_i}^{a_i}$ *existiert eine Zahl* l_i *derart, dass aus* $|x_i(t)| < a_i$ $|y(t) - l_i| \leqslant \frac{\varepsilon}{2}$ *folgt,*

(2) *ausserdem, aus* $|y(t) - l_i| > 2\varepsilon$ *folgt* $|x_i(t)| > a_i + \delta_i$, *wo* $\delta_i > 0$ *ist.*

16. Lemma. *Auf einer abgeschlossenen Menge* P, *die in einem Würfel* Q *im* n*-dimensionalen Raume gelegen ist, sei eine Funktion* $\varphi(x)$ *gegeben, die in jedem Punkte* $x \in P$ *eine Schwankung besitzt, die eine vorgegebene Grösse* $\varepsilon > 0$ *nicht übertrifft. Dann existiert eine im Würfel* Q *definierte stetige Funktion* $f(x)$, *die sich auf* P *von* $\varphi(x)$ *um nicht mehr als* 2ε *unterscheidet.*

Beweis. Nach einem bekannten Satz von Baire existiert eine Zahl $\eta > 0$ derart, dass aus $\rho(x', x'') < \eta$ $(x', x'' \in P)$

$$|\varphi(x') - \varphi(x'')| \leqslant 2\varepsilon$$

folgt. Führen wir eine simpliziale Zerlegung des Würfels Q in eine endliche Anzahl von n-dimensionalen Simplexen durch, deren Durchmesser $\frac{\eta}{2}$ nicht übertrifft. Es sei x_0 eine Ecke irgendeines Simplexes und $\tilde{x}_0$ — der zu ihr nächste Punkt auf P (wenn es solcher mehrere gibt, so nehmen wir einen beliebigen von ihnen); setzen wir $f(x_0) = \varphi(\tilde{x}_0)$. Nachdem wir so $f(x)$ in den Ecken der Simplexe definiert haben, betrachten wir einen beliebigen Punkt $x \in Q$; es sei $(x_0, x_1, \ldots, x_n)$ der Simplex, der x enthält, und $(\tau_1, \tau_2, \ldots, \tau_n)$ — die baryzentrischen Koordinaten des letzteren in diesem Simplex, $\tau_i \geqslant 0$, $\sum\limits_i \tau_i = 1$. Setzen wir $f(x) = \sum\limits_i \tau_i f(x_i)$; es ist evident, dass bei dieser Definition $f(x)$ in Q stetig wird. Es sei nun $x \in P$, $(x_0, x_1, \ldots, x_n)$ — der Simplex, der x enthält, $x_i \in P$ — derjenige Punkt, den wir bei der Definition des Wertes der Funktion $f(x)$ im Punkte x_i benutzten. Da $\rho(x, x_i) < \frac{\eta}{2}$ ist, so ist

$$\rho(x, \tilde{x}_i) \leqslant \rho(x, x_i) + \rho(x_i, \tilde{x}_i) \leqslant 2\rho(x, x_i) < \eta,$$
$$|\varphi(x) - f(x_i)| = |\varphi(x) - \varphi(\tilde{x}_i)| < 2\varepsilon,$$
$$|\varphi(x) - f(x)| = |\varphi(x) - \sum_i \tau_i f(x_i)| = |\sum_i \tau_i (\varphi(x) - f(x_i))| < 2\varepsilon \sum_i \tau_i = 2\varepsilon,$$

w. z. b. w.

17. Der Beweis des Grundprinzips. Die Notwendigkeit der Bedingung ist offenbar: wenn $x(t) \in K$ derart ist, dass $\sup |x(t) - y(t)| < \varepsilon$ ist, so kann S durch eine endliche Anzahl von ausgewählten Untermengen $\{ |x(t) - n\varepsilon| < \varepsilon \}$ überdeckt werden, in jeder von welchen $y(t)$ um nicht mehr als 4ε variiert. Es möge $y(t)$ der gestellten Bedingung genügen; zeigen wir, dass für beliebiges $\varepsilon > 0$ ein $x(t) \in K$ existiert derart, dass $\sup |x(t) - y(t)| < 4\varepsilon$ ausfällt. Es sei

$$U_1 = \{ |x_1(t)| < a_1 \}, \; U_2 = \{ |x_2(t)| < a_2 \}, \; \ldots, \; U_n = \{ |x_n(t)| < a_n \}$$

eine endliche Überdeckung von S durch ausgewählte Untermengen, auf jeder von welchen $y(t)$ um nicht mehr als $\frac{\varepsilon}{4}$ variiert; es sei t_k ein beliebiger fixierter Punkt aus U_k. Die Funktionen $x_1(t), x_2(t), \ldots, x_n(t)$ bilden S auf eine beschränkte Menge P_0 des n-dimensionalen Raumes; bezeichnen wir diese Abbildung durch $x = \omega(t)$. In der umgekehrten Richtung wird sie, selbstverständlich, im allgemeinen nicht eindeutig sein; aber wir können behaupten, dass falls $\omega(t_0) = x_0$ ist und $t_0 \in U_k$, so folgt aus $\omega(t) = x_0$ auch $t \in U_k$: in der Tat, die Bilder der zu U_k nicht angehörenden Punkte haben ihre k-te Koordinate $\geqslant a_k$ und können daher nicht mit dem Bilde eines Punktes aus U_k zusammenfallen.

Setzen wir $\varphi(x) = y(t_k)$, wo k so gewählt ist, dass die ausgewählte Untermenge U_k mindestens ein (und, folglich, auch jedes) Urbild des Punktes x enthält. Es ist leicht einzusehen, dass die so definierte Funktion in jedem Punkte eine Schwankung hat, die ε nicht übertrifft: sei $x_p \to x$, wo $x_p \in P_0$; da die k-te Koordinate von x kleiner als a_k ist, so wird, von einer bestimmten Nummer an, die k-te Koordinate von x_p auch kleiner als a_k sein; folglich sind die Urbilder von x_p in U_k enthalten; wenn daher $\varphi(x_p) = y(t_p)$ ist, so wird, von einer bestimmten Nummer p an, $U_p \cap U_k \neq 0$ sein, woraus

$$|\varphi(x) - \varphi(x_p)| = |y(t_k) - y(t_p)| < 2 \cdot \frac{\varepsilon}{4} = \frac{\varepsilon}{2}$$

folgt. Also haben wir für die Schwankung von $\varphi(x)$:

$$\overline{\lim}\, |\varphi(x_p) - \varphi(x_q)| \leqslant 2 \cdot \frac{\varepsilon}{2} = \varepsilon.$$

Aus der Bedingung (2) folgt, dass die Funktion $\varphi(x)$ kann man ohne Vergrösserung der Schwankung auf die Abschliessung $\bar{P}_0 = P$ erweitern. Betrachten wir einen beliebigen Würfel Q, der P enthält; nach dem Lemma existiert eine auf Q stetige Funktion $f(x)$, die auf P sich von $\varphi(x)$ um nicht mehr als 2ε unterscheidet. Nach dem Satz von Weierstrass kann ein Polynom $R(x_1, x_2, \ldots, x_n)$ konstruiert werden, das sich von $f(x)$ auf Q um weniger als ε unterscheidet. Betrachten wir genau dasselbe Polynom $R(x_1(t), x_2(t), \ldots, x_n(t))$; dies ist ein Element von K, und wir behaupten, dass es sich in allen Punkten von S von $y(t)$ um nicht mehr als 4ε unterscheidet. In der Tat:

$$\begin{aligned} |y(t) - R[x_1(t), x_2(t), \ldots, x_n(t)]| &= |y(t) - R[\omega(t)]| \leqslant \\ \leqslant |y(t) - \varphi(\omega(t))| + |\varphi(\omega(t)) - f(\omega(t))| &+ |f(\omega(t)) - R[\omega(t)]| \leqslant \\ \leqslant \varepsilon + 2\varepsilon + \varepsilon &= 4\varepsilon, \end{aligned}$$

w. z. b. w.

18. Satz. *Es sei K ein normierter Ring, der die Eigenschaft* (*) *besitzt* (n°. 11), *und* $\mathfrak{M}$ *— die Menge seiner maximalen Ideale, so topologisiert, wie es in § 3*

angegeben war; jede auf $\mathfrak{M}$ stetige Funktion $y(M)$ ist der Grenzwert einer gleichmässig konvergenten Folge von Funktionen $x(M)$ — Bildern von Elementen des Ringes K.

Es genügt zu zeigen, dass jede reelle stetige Funktion ein Grenzwert der Funktionen $x(M)$ ist. Es sei K' die Familie aller reellen Funktionen $x(M)$. Betrachten wir für jeden Punkt $M \in \mathfrak{M}$ eine Umgebung $U(M)$ so klein, dass die Funktion $y(M)$ in ihr um nicht mehr als ε variiert; diese Umgebungen können aus der Basis $\{U = \{|x(M)| < a\}\}$ (n°. 12) genommen werden, wobei man, überdies, die Funktionen $x(M)$ reell annehmen darf, denn man kann nötigenfalls $x(M)$ durch $x(M)\overline{x(M)}$ ersetzen. Die Bikompaktheit von $\mathfrak{M}$ ausnutzend, kann man eine endliche Anzahl solcher Umgebungen aussondern, die das ganze $\mathfrak{M}$ überdecken. Somit sind die Bedingungen der Anwendbarkeit des Grundprinzips erfüllt; wenden wir es an, so erhalten wir das Gewünschte.

§ 5. Sätze über Ideale und Unterringe des Ringes der stetigen Funktionen

In diesem Paragraph zeigen wir, dass die von M. Stone [2] aufgestellten Sätze über Ringe stetiger Funktionen und Ideale in ihnen auch Folgerungen der §§ 3 und 4 sind. Im Folgenden werden wir mit $C(S)$ den Ring aller — reellen und komplexen — auf S stetigen Funktionen bezeichnen. Es ist leicht einzusehen, dass alle Ergebnisse des § 2 auch für diesen erweiterten Ring gültig bleiben.

19. Wie wir gesehen haben (n°. 10), entspricht jeder bikompakten Erweiterung Q eines vollständig regulären Raumes S ein bestimmter abgeschlossener Unterring $K \subset C(S)$, nämlich die Gesamtheit aller auf Q stetigen Funktionen, oder, mit anderen Worten, die Gesamtheit aller Funktionen, die von S auf Q stetig fortgesetzt werden können [wenn $x(t)$ auf Q stetig fortgesetzt werden kann, so ist dies, wegen der Dichtigkeit von S in Q, nur auf eine einzige Weise möglich].

Der Unterring K besitzt die folgende Eigenschaft: für eine beliebige abgeschlossene Menge $\mathfrak{F} \subset S$ und einen beliebigen nicht zu ihr angehörenden Punkt $a \in S$ existiert eine Funktion $x(t) \in K$ derart, dass $x(a) = 0$, $|x(t)| \geqslant \alpha > 0$ für $t \in \mathfrak{F}$ ist. In der Tat, die Bedingung des n°. 14 ist für den Ring K als den Ring der Funktionen auf dem Raum $Q = \mathfrak{M}_k$ (n°. 7) erfüllt (n°. 13), und folglich ist Q homöomorph $\mathfrak{M}_k$, topologisiert nach der Regel des § 3; da S in $\mathfrak{M}_k = Q$ homöomorph eingebettet ist, so erhalten wir, die Bedingung des n°. 14 formulierend, das Gewünschte.

Es ist natürlich den Unterring $K \subset C(S)$, der diese Eigenschaft besitzt, vollständig regulär zu nennen. Zeigen wir, dass jedem vollständig regulären Unterring $K \subset C(S)$, der zusammen mit jeder Funktion ihre komplex konjugierte enthält, eine bestimmte bikompakte Erweiterung Q entspricht, auf die solche und nur solche Funktionen aus $C(S)$ stetig fortgesetzt werden können, die zu K gehören.

20. Lemma 1. *Wenn $x(t) \in K$ eine reelle Funktion ist, wobei $\inf_{t \in S} |x(t)| = l > 0$ ist, so ist $\frac{1}{x(t)} \in K$.*

Es sei L die obere Grenze der Werte der Funktion $|x(t)|$. Die Menge der Werte $y = x(t)$ verteilt sich auf zwei Segmente $[-L, -l]$ und $[l, L]$; betrachten wir y als die unabhängige Veränderliche, so finden wir nach dem Weierstrassschen Satz ein Polynom

$P(y)$, das sich von $\frac{1}{y}$ auf $[-L, -l]+[l, L]$ um weniger als ε unterscheidet; dann unterscheidet sich auch die Funktion $P[x(t)] \in K$ von $\frac{1}{x(t)}$ auf S um weniger als ε; da ε beliebig und K abgeschlossen ist, so finden wir:

$$\frac{1}{x(t)} \in K,$$

w. z. b. w.

Lemma 2. *Der Unterring K besitzt die Eigenschaft* (*) *des § 3.*

Es sei $x(t)$ eine reelle Funktion; zeigen wir, dass sie auf jedem maximalen Ideal $M \in \mathfrak{M}_k$ einen reellen Wert $x(M)$ annimmt. Wäre für ein $M_0 \in \mathfrak{M}_k$ $x(M_0) = \lambda + i\mu$, $\mu \neq 0$, so würde die reelle Funktion $[x(t) - \lambda]^2 + \mu^2$ auf M_0 den Wert Null haben; aber, $\inf\{[x(t) - \lambda]^2 + \mu^2\} \geqslant \mu^2 > 0$, woraus nach Lemma 1 folgt, dass $[x(t) - \lambda]^2 + \mu^2$ eine inverse Funktion hat und also zu keinem maximalen Ideal angehören kann. Es sei nun $x(t)$ eine beliebige Funktion und $y(t) = \overline{x(t)}$ ihre komplex konjugierte; zeigen wir, dass auch auf einem beliebigen $M \in \mathfrak{M}$ $x(M) = \overline{y(M)}$ ist. Dies folgt daraus, dass $x(t) + y(t)$ und $\frac{1}{i}(x(t) - y(t))$ reelle Funktionen sind und also, nach dem Bewiesenen, auf jedem maximalen Ideal reelle Werte haben; wenn aber $z_1 + z_2$ und $\frac{1}{i}(z_1 - z_2)$, wo z_1 und z_2 komplexe Zahlen sind, reell sind, so sind diese komplexen Zahlen konjugiert.

21. Es möge nun der Unterring K den gestellten Bedingungen genügen. Nach Lemma 2 kann die Menge $\mathfrak{M}_k$ der maximalen Ideale von K nach der Regel des § 3 topologisiert werden. Dies wird ein bikompakter Raum, der nach der Bedingung (siehe n°. 14) S als einen homöomorphen Unterraum enthält. Zeigen wir, dass S in $\mathfrak{M}_k$ überall dicht gelegen ist; für den Beweis genügt es zu zeigen, dass wenn $x(t)$ derart ist, dass $\inf_{t \in S} |x(t)| = l > 0$ ist, so haben wir $\frac{1}{x(t)} \in K$, und folglich $x(t)$ zu keinem maximalen Ideal angehören kann.

Für eine reelle Funktion $x(t)$ folgt dies aus Lemma 1; wenn nun $x(t) = \lambda(t) + i\mu(t)$ ist, so haben wir:

$$\inf|\lambda^2(t) + \mu^2(t)| = \inf|x(t) \cdot \overline{x(t)}| = l^2 > 0,$$

$$\frac{1}{\lambda^2(t) + \mu^2(t)} \in K, \quad \frac{1}{x(t)} = \frac{\lambda(t) - i\mu(t)}{\lambda^2(t) + \mu^2(t)} \in K.$$

Somit ist $\mathfrak{M}_k$ eine bikompakte Erweiterung von S. Jede Funktion $x(t) \in K$ ist auf dem ganzen $\mathfrak{M}_k$ (n°. 12) stetig, und folglich kann jede Funktion $x(t)$ stetig auf $\mathfrak{M}_k$ fortgesetzt werden. Eine beliebige auf $\mathfrak{M}_k$ stetige Funktion $y(t)$ ist ein Grenzwert von Funktionen aus K (n°. 18) und infolge der Abgeschlossenheit von K gehört zu diesem letzten; die Behauptung ist somit vollständig bewiesen.

22. Wenn ein vollständig regulärer Raum S selbst bikompakt ist, so fällt er mit allen seinen bikompakten Erweiterungen zusammen; somit fällt jeder vollständig reguläre Unterring K, der zusammen mit jeder Funktion ihre komplex konjugierte enthält, mit dem ganzen Ring $C(S)$ zusammen.

Man kann sogar noch mehr behaupten: wenn ein solcher Unterring K vollständig Hausdorffsch ist, d. h. wenn für zwei beliebige Punkte $t_1, t_2 \in S$ eine Funktion $x(t) \in K$ derart existiert, dass $x(t_1) \neq x(t_2)$ ist, so ist $K = C(S)$. Es genügt zu zeigen, dass K ein vollständig regulärer Unterring ist. Es sei $\mathfrak{F} \subset S$ eine abgeschlossene Menge und $a \overline{\in} \mathfrak{F}$. Laut Bedingung kann man für jeden Punkt $b \in \mathfrak{F}$ eine solche Funktion $x_b(t) \in K$ finden, dass $x_b(a) = 0$, $x_b(b) \neq 0$ ist. Umgeben wir jeden Punkt $b \in \mathfrak{F}$ mit einer so kleinen Umgebung $U(b)$, dass die Funktion $x_b(t)$ in $U(b)$ nicht zu Null wird; da $\mathfrak{F}$ als eine abgeschlossene Untermenge des bikompakten Raumes S selbst bikompakt ist, so kann man eine endliche Anzahl solcher Umgebungen $U(b_1), U(b_2), \ldots, U(b_n)$ aussondern, die das ganze $\mathfrak{F}$ überdecken; die Funktion $x(t) = \sum_{i=1}^{n} x_{b_i}(t) \overline{x_{b_i}(t)}$ wird auf $\mathfrak{F}$ von Null verschieden und infolge der Bikompaktheit von $\mathfrak{F}$ besitzt auf $\mathfrak{F}$ ein Minimum, das von Null verschieden ist.

23. Die Gesamtheit aller Funktionen $x(t) \in C(S)$, die auf einer gegebenen abgeschlossenen Menge $\mathfrak{F} \subset S$ zu Null werden, bildet offenbar ein abgeschlossenes Ideal in $C(S)$; wir behaupten, dass wenn S bikompakt ist, der Ring $C(S)$ keine anderen abgeschlossenen Ideale enthält.

Nehmen wir zuerst an, dass das betrachtete Ideal I zusammen mit jeder Funktion ihre komplex konjugierte enthält. Es sei $\mathfrak{F}$ die maximale abgeschlossene Menge, auf der alle Funktionen $x(t) \in I$ zu Null werden; zeigen wir, dass I aus allen Funktionen besteht, die auf $\mathfrak{F}$ zu Null werden.

Fügen wir zu I alle Konstanten hinzu, so erhalten wir einen abgeschlossenen Unterring $K \subset C(S)$. Es sei S' der topologische Raum, der aus S durch Identifizierung aller Punkte von $\mathfrak{F}$ entsteht; offenbar ist S' auch bikompakt. Jeder Funktion $x(t) \in K$ entspricht eine stetige Funktion $x'(t) \in C(S')$. Es ist leicht einzusehen, dass der erhaltene Unterring $K' \subset C(S')$ vollständig Hausdorffsch ist; folglich enthält er alle auf S' stetigen Funktionen. Dies aber bedeutet gerade, dass I alle Funktionen $x(t) \in C(S)$ enthält, die auf $\mathfrak{F}$ gleich Null sind, denn einer jeden solchen Funktion entspricht eine auf S' stetige Funktion, die in dem Bild der Menge $\mathfrak{F}$ gleich Null ist.

Es bleibt nun zu zeigen, dass in jedem abgeschlossenen Ideal zusammen mit jeder Funktion auch ihre komplex konjugierte enthalten ist.

Es sei $\mathfrak{F}$ die Menge derjenigen Punkte $t \in S$, in denen $x(t) = 0$ ist. Die zwei abgeschlossenen Mengen $\mathfrak{F}_1 = \{|x(t)| \leq \varepsilon\} \supset \mathfrak{F}$, $\mathfrak{F}_2 = \{|x(t)| \geq 2\varepsilon\}$ überschneiden sich nicht; wegen der vollständigen Regularität des bikompakten Raumes S existiert eine stetige Funktion $\alpha(t)$, die auf $\mathfrak{F}_1$ gleich 0, auf $\mathfrak{F}_2$ gleich 1 ist, und in den übrigen Punkten die Zwischenwerte annimmt; das Produkt $\alpha(t)\overline{x(t)}$ stellt eine Funktion dar, die I angehört [da $\alpha(t)\overline{x(t)} = \beta(t)x(t)$ ist, wo $\beta(t)$ stetig ist] und sich von $\overline{x(t)}$ um weniger als 4ε unterscheidet; da ε beliebig ist, so erhalten wir $x(t) \in I$, w. z. b. w.

Literatur

1. I. Gelfand, Normierte Ringe (in diesem Heft des Recueil mathématique, S. 3—23).

2. M. Stone, Applications of the theory of Boolean tings to general topology, Trans. of Amer. Math. Soc., **41**, (1937).

3. I. Gelfand and A. Kolmogoroff, On rings of continuous functions on topological spaces, C. R. de l'Ac. des Sc. de l'U. R. S. S., n. s., **XXII**, N. 1, (1939).

4. E. Čech, On bicompact spaces, Ann. of Math., **38**, (1937), 823—845; P. Alexandroff, Réc. math., **5 (47) : 2**, (1939), 403—423; Wallman, Ann. of Math., **39**, (1938), 112—127.

Mathematisches Institut der
Akademie der Wissenschaften der
U. d. S. S. R.

(Поступило в редакцию 27/VI 1940 г.)

10.

Ideale und primäre Ideale in normierten Ringen

Mat. Sb., Nov. Ser. **9** (51) (1941) 41–48. Zbl. **24**:322

Ein abgeschlossenes Ideal I des Ringes R heisst primär, wenn es nur in einem maximalen Ideal des Ringes R enthalten ist. Diese Bedingung ist der folgenden äquivalent: in dem Restklassenring R/I ist jedes Element x, das kein inverses besitzt, ein verallgemeinertes nilpotentes Element, d. h. $\sqrt[n]{\|x^n\|} \to 0$.

Zeigen wir, dass diese Bedingungen äquivalent sind.

Aus der ersten folgt die zweite. In der Tat: I ist nur in einem maximalen Ideal enthalten. Folglich enthält R/I nur ein maximales Ideal. Wenn das Element $x \in R/I$ kein inverses besitzt, so geht es in ein nichttriviales Ideal ein, also auch in ein maximales ([1], Satz 5). Die Gesamtheit der verallgemeinerten nilpotenten Elemente ist der Durchschnitt von maximalen Idealen ([1], Satz 8). Folglich, da in R/I nur ein maximales Ideal vorhanden ist, ist jedes Element, das kein inverses besitzt, ein verallgemeinertes nilpotentes Element.

Aus der zweiten folgt die erste. Die Gesamtheit der verallgemeinerten nilpotenten Elemente des Ringes R/I bildet ein Ideal, das in allen maximalen Idealen enthalten ist ([1], Satz 8). Dieses Ideal in R/I ist das einzige maximale Ideal, denn jedes Element, das in dieses Ideal nicht eingeht, besitzt ein inverses und folglich in kein nichttriviales Ideal eingehen kann. Somit ist in R/I nur ein einziges maximales Ideal vorhanden, und folglich existiert in R nur ein maximales Ideal, das I enthält.

Wir werden eine allgemeine Methode angeben, die es in vielen Fällen gestattet die primären Ideale eines Ringes R zu finden.

Satz 1. *Es sei x ein verallgemeinertes nilpotentes Element des Ringes R, und sei*

$$\|(e-\lambda x)^{-1}\| \leqslant \frac{cr^n}{\cos^n\varphi} \qquad (\lambda = re^{i\varphi})$$

für alle genügend grosse r und alle $\varphi \neq \frac{\pi}{2}$. Dann ist $x^{n+1}=0$.

Beweis. Betrachten wir ein Polynom $P(\lambda)$ der Ordnung n, dessen Nullstellen in der unteren Halbebene liegen. Dann ist die Funktion $\Phi(\lambda) = \frac{(e-\lambda x)^{-1}}{P(\lambda)}$ auf der reellen Achse beschränkt. Ausserdem ist es leicht nachzuprüfen, dass

$$\lim_{r\to\infty} \frac{1}{r} \int_0^\pi \log^+ |\Phi(\lambda)| \sin\varphi \, d\varphi = 0$$

ist. Folglich ist, nach dem Satz von Phragmén-Lindelöf [2] (in der Form, die ihm von Nevanlinna gegeben wurde), $\Phi(\lambda)$ in der oberen Halbebene beschränkt, und also haben wir für $0 \leqslant \varphi \leqslant \pi$: $\|(e - \lambda x)^{-1}\| \leqslant C_1 r^n$. Ähnlich haben wir für $-\pi \leqslant \varphi \leqslant \pi$: $\|(e - \lambda x)^{-1}\| \leqslant C_2 r^n$. Somit ist $\|(e - \lambda x)^{-1}\| \leqslant C r^n$ in der ganzen Ebene. Aber dann ist die analytische Funktion $(e - \lambda x)^{-1}$ ein Polynom von nicht höher als n-ter Ordnung. Folglich ist $x^{n+1} = 0$, denn x^{n+1} ist der Koeffizient von λ^{n+1} in der Reihe $\sum\limits_{n=0}^{\infty} \lambda^n x^n = (e - \lambda x)^{-1}$.

Bemerkung. Wir haben auf die Funktion $(e - \lambda x)^{-1}$ die Ergebnisse der Theorie der gewöhnlichen analytischen Funktionen angewendet. Die Legalität dieses Verfahrens kann am einfachsten folgendermassen festgestellt werden: für jedes fixierte λ ist $(e - \lambda x)^{-1}$ ein Element eines linearen Raumes (des Ringes R). Wenn wir ein lineares Funktional f nehmen, so ist $f[(e - \lambda x)^{-1}]$ eine gewöhnliche analytische Funktion. Unsere Überlegungen kann man als auf diese analytische Funktion bezogen betrachten. Variieren wir f, so erhalten wir das gewünschte Ergebnis.

Wir stellen uns nun die folgende Aufgabe: die primären Ideale zu finden, die in einem maximalen Ideal M_0 enthalten sind. Wir werden dabei die Untersuchung der primären Ideale mit der Geschwindigkeit des Wachstums der Funktion $(x - \lambda e)^{-1}$ in der Umgebung des Punktes $\lambda_0 = x(M_0)$ verbinden. In der Mehrheit der konkreten Fälle ist es bedeutend bequemer die Grösse von $(x - \lambda e)^{-1}$ nicht mit $|\lambda - \lambda_0|$, sondern mit $d(\lambda)$ zu vergleichen, wo $d(\lambda)$ der Abstand des Punktes λ von der Menge der Werte der Funktion $x(M)$ ist.

Satz 2. *Es mögen $x_1, x_2, \ldots$* * *ein System von Erzeugenden des Ringes R bilden. Es seien die Werte der Funktionen $x_i(M)$ reell. Sei ferner M_0 ein fixiertes maximales Ideal, und es möge für jede der Erzeugenden die Beziehung*

$$|(x_k - \lambda e)^{-1}| = o\left(\frac{1}{|\mathfrak{J}(\lambda - \lambda_k^{(0)})|^{n_k}}\right)$$

in einer Umgebung des Punktes $\lambda_k^{(0)} = x_k(M_0)$ erfüllt sein. Dann existiert ein kleinstes der primären Ideale I des Ringes R, die in M_0 enthalten sind. Dieses Ideal wird von den Elementen $(x_k - \lambda_k^{(0)} e)^{n_k - 1}$ erzeugt. Sind, insbesondere, alle $n_k \leqslant 2$, so ist das einzige primäre Ideal, das in M_0 enthalten ist, M_0 selbst.

Beweis. Betrachten wir den Restklassenring $\tilde{R} = R/I$, wo I ein beliebiges primäres in M_0 enthaltenes Ideal ist. In R ist nur ein maximales Ideal, nämlich das Bild des Ideals M_0, vorhanden. Folglich existiert im Ringe $\tilde{R}$ $(\tilde{x}_k - \lambda\tilde{e})^{-1}$ für jedes $\lambda \neq \lambda_k^{(0)} - x_k(M_0)$ (wir bezeichnen mit $\tilde{y}$ das Bild des Elementes $y \in R$ beim Homomorphismus $R \to R/I$). Da beim Übergang zum Restklassenring die Norm sich nur verkleinern kann, so ist in einer bestimmten Umgebung von $\lambda_k^{(0)}$

$$\|(\tilde{x}_k - \lambda\tilde{e})^{-1}\| = o\left(\frac{1}{|\mathfrak{J}(\lambda - \lambda_k^{(0)})|^{n_k}}\right). \tag{1}$$

* Die Indexe bei den x sind nur bequemlichkeitshalber eingeführt: wir setzen nicht voraus, dass das System der Erzeugenden abzählbar ist.

Setzen wir $\mu=\frac{1}{\lambda-\lambda_k^{(0)}}$ und $\tilde{x}_k-\lambda_k^{(0)}\tilde{e}=\tilde{y}_k$. Die Beziehung (1) kann folgendermassen umgeschrieben werden:

$$\left\|\left(\tilde{y}_k-\frac{1}{\mu}e\right)^{-1}\right\|=o\left(\frac{1}{\left|\mathcal{G}\left(\frac{1}{\mu}+\lambda_0\right)\right|^{n_k}}\right)$$

für genügend grosse $|\mu|$, oder

$$\|(\tilde{e}-\mu\tilde{y}_k)^{-1}\|=o\left(\frac{1}{|\mu|\left|\mathcal{G}\left(\frac{1}{\mu}+\lambda_0\right)\right|^{n_k}}\right),$$

d. h.

$$\|(\tilde{e}-\mu\tilde{y})^{-1}\|=o\left(\frac{r^{n_k-1}}{|\cos^{n_k}\varphi|}\right), \tag{2}$$

wo $\mu=re^{i\left(\varphi+\frac{\pi}{2}\right)}$ ist.

Folglich ist nach Satz 1 $y^{n_k+1}=0$, d. h.

$$(\tilde{e}-\mu\tilde{y}_k)^{-1}=\sum_{p=0}^{n_k}\mu^p\tilde{y}_k^p.$$

Wäre $\tilde{y}^{n_k-1}\neq 0$, so würde der Ausdruck $(\tilde{e}-\mu\tilde{y}_k)^{-1}$ eine höhere Ordnung haben, als die Beziehung (2) erlaubt. Somit ist $\tilde{y}_k^{n_k-1}=0$, und folglich geht $(x_k-\lambda_k^{(0)}e)^{n_k-1}$ beim Homomorphismus $R\to R/I$ in Null über, d. h. $(x_k-\lambda_k^{(0)}e)^{n_k-1}\in I$. Somit haben wir bewiesen, dass das Ideal I, das von den Elementen $(x_k-\lambda_k^{(0)}e)^{n_k-1}$ erzeugt wird, in jedem primären Ideal $I\subset M_0$ enthalten ist. Zeigen wir, dass das Ideal I auch primär ist. Nehmen wir an, dass es in irgendeinem maximalen Ideal M_1 enthalten ist. Beim Homomorphismus $R\to R/I$ geht $(x_k-\lambda_k^{(0)}e)^{n_k-1}$ in Null über. Folglich ist $x_k(M_1)=\lambda_k^{(0)}=x_k(M_0)$. Aber aus $x_k(M_1)=x_k(M_0)$ folgt, dass für jedes Element $x\in R$ $x(M_1)=x(M_0)$ ist. Also folgt aus $x(M_0)=0$, d.h. $x\in M_0$, $x\in M_1$, d. h. $M_1=M_0$. Folglich ist M_0 das einzige maximale Ideal, in dem I enthalten ist, d. h. I ist ein primäres Ideal.

Zeigen wir nun, dass wenn alle $n_k\leqslant 2$ sind, jedes primäre Ideal $I\subset M_0$ mit M_0 zusammenfällt. Beim Homomorphismus $R\to R/I$ geht $x_k-\lambda_k^{(0)}e$ in Null über. Folglich ist $\tilde{x}_k=\lambda_k^{(0)}\tilde{e}$. Da die $\tilde{x}_k$ ein System von Erzeugenden des Ringes R/I bilden, so ist R/I der Körper der komplexen Zahlen. Folglich ist das Ideal I ein maximales.

B e m e r k u n g. Wir haben angenommen, dass die Funktionen $x_k(M)$ reelle Werte annehmen. Es ist klar, dass diese Bedingung abgeschwächt werden kann. Es würde genügen zu verlangen, dass in einer Umgebung des Punktes $\lambda_k^{(0)}=x_k(M_0)$ ein Segment existiert, das durch den Punkt $\lambda_k^{(0)}$ durchgeht, auf welchem

$$\|(x_k-\lambda e)^{-1}\|=o\left(\frac{1}{|\lambda-\lambda_k^{(0)}|^{n_k}}\right)$$

ist, und dass bei $t\to 0$

$$\frac{1}{t}\int_0^{2\pi}\log^+\|(x-\lambda e)^{-1}\|\cdot|\sin\varphi|\,d\varphi\to 0 \qquad (\lambda=\lambda_k^{(0)}+te^{i\varphi}).$$

Der Satz 2 gestattet uns das kleinste primäre Ideal $J \subset M_0$ zu finden. Wie kann man die übrigen primären Ideale finden? Um diese Frage zu klären, nehmen wir an, dass in dem Ring R (der den Bedingungen des Satzes 2 genügt) ein endliches System von Erzeugenden $x_1, x_2, \ldots, x_n$ existiert.

Nehmen wir das kleinste primäre Ideal J, das M_0 angehört, und bilden R/J. In R/J gibt es eine endliche Anzahl von Erzeugenden $\tilde{x}_1, \tilde{x}_2, \ldots, \tilde{x}_n$. Anstatt $\tilde{x}_k$ nehmen wir besser $\tilde{y}_k = \tilde{x}_k - \lambda_k^{(0)} \tilde{e}$. Daraus, dass $\tilde{y}_k^{n_k - 1} = 0$ ist, folgt, dass R/J ein hyperkomplexes System mit einer endlichen Anzahl von Einheiten ist. Jedem primären Ideal $I \supset J$ entspricht ein Ideal in R/J und umgekehrt.

Somit ist die Aufgabe der Auffindung der primären Ideale $I \in M_0$ in diesem Falle auf die Aufgabe der Auffindung aller Ideale in einem hyperkomplexen System zurückgeführt. Man kann zeigen, dass R/J ein primärer Ring ist. Umgekehrt, kann jedes beliebige hyperkomplexe System, das ein primärer Ring ist, auf diese Weise erhalten werden.

Wenn R ein Ring mit einer Erzeugenden ist, so ist der Ring R/J offenbar ein hyperkomplexes System mit den Erzeugenden $e, \tilde{y}, \ldots, \tilde{y}^{n-1}$. Die Ideale in R/J sind Hauptideale über den Elementen $\tilde{y}, \tilde{y}^2, \ldots, \tilde{y}^{n-1}$.

Gehen wir zu dem Ring R über, so erhalten wir den folgenden

S a t z 3. *Es sei R ein Ring mit einer Erzeugenden x. Es möge $x(M)$ nur reelle Werte annehmen und in einer Umgebung von $\lambda_0 = x(M_0)$*

$$\|(x - \lambda e)^{-1}\| = o\left(\frac{1}{|\mathfrak{I}(\lambda - \lambda_k^{(0)})|^n}\right)$$

sein. Dann existieren nicht mehr als $n-1$ primäre Ideale, die in M_0 enthalten sind. Jedes von ihnen ist das Hauptideal über dem Element $(x - \lambda_0 e)^k$ $(k = 1, 2, \ldots, n-1)$.

B e m e r k u n g. Das Hauptideal über dem Element z ist das kleinste z enthaltende abgeschlossene Ideal.

D e f i n i t i o n. *Ein normierter Ring R ist ein N-Ring, wenn jedes abgeschlossene Ideal des Ringes R der Durchschnitt maximaler Ideale ist.*

Ein solcher Ring ist, z. B., der Ring der stetigen Funktionen auf einem bikompakten Raum ([3], n°. 23). Wir werden hinreichende Bedingungen dafür aufstellen, dass ein Ring R ein N-Ring ist.

Zunächst beweisen wir das folgende

L e m m a. *Dafür, dass ein Ring R ein N-Ring sei, ist es notwendig und hinreichend, dass in keinem Restklassenring von R von Null verschiedene verallgemeinerte nilpotente Elemente vorhanden seien.*

B e w e i s. Es sei ein Ideal L gegeben. Wir haben zu zeigen, dass es der Durchschnitt maximaler Ideale ist. In R/L gibt es keine verallgemeinerten nilpotenten Elemente. Folglich ist der Durchschnitt der maximalen Ideale des Ringes R/L das Nullideal. Die Urbilder in R der maximalen Ideale des Ringes R/L sind maximale Ideale des Ringes R. Ihr Durchschnitt ist ein vollständiges Urbild der Null, d. h. das Ideal L.

Umgekehrt, sei das Ideal L der Durchschnitt der maximalen Ideale. Dann ist in R/L der Durchschnitt aller maximalen Ideale das Nullideal, d. h. in R/L gibt es keine verallgemeinerten nilpotenten Elemente, die von Null verschieden sind.

Satz 4. *Es gebe in dem Ring R für jedes Element x sein komplex-konjugiertes. Es sei für jedes Element x, für das die Funktion $x(M)$ reell ist, und x^{-1} nicht existiert, die Beziehung*

$$\|(x-\lambda e)^{-1}\| = o\left(\frac{1}{|\mathfrak{J}(\lambda)|^2}\right)$$

bei $\lambda \to 0$, $\mathfrak{J}(\lambda) \neq 0$, erfüllt. Dann ist R ein N-Ring.

Beweis. Infolge des Lemmas genügt es zu zeigen, dass in einem beliebigen Restklassenring keine von Null verschiedenen verallgemeinerten nilpotenten Elemente vorhanden sind.

Es sei $\tilde{x}$ ein verallgemeinertes nilpotentes Element des Ringes R/L (wir bezeichnen mit $\tilde{x}$ das Bild des Elementes x des Ringes R). Dann ist $\tilde{\bar{x}}$ auch ein verallgemeinertes nilpotentes Element des Ringes R/L. In der Tat, dass $\tilde{x}$ ein verallgemeinertes nilpotentes Element ist, bedeutet, dass für $M \supset L$ $x(M) = 0$ ist. Aber dann ist auch $\bar{x}(M) = 0$ für $M \supset L$, d. h. $\tilde{\bar{x}}$ ist ein verallgemeinertes nilpotentes Element. Die Elemente

$$\tilde{x}_1 = \frac{\tilde{x} + \tilde{\bar{x}}}{2} \quad \text{und} \quad \tilde{x}_2 = \frac{\tilde{x} - \tilde{\bar{x}}}{2}$$

sind auch verallgemeinerte nilpotente Elemente. Zeigen wir, dass $\tilde{x}_1 = \tilde{x}_2 = 0$ ist. Hieraus wird folgen, dass auch $\tilde{x} = 0$ ist. Zeigen wir dies für $\tilde{x}_1$.

Die Funktion $x_1(M)$ ist reell, und folglich haben wir

$$\|(x_1 - \lambda e)^{-1}\| = o\left(\frac{1}{|\mathfrak{J}(\lambda)|^2}\right).$$

Da in dem Restklassenring die Norm nur kleiner sein kann, so ist auch

$$\|(\tilde{x}_1 - \lambda\tilde{e})^{-1}\| = o\left(\frac{1}{|\mathfrak{J}(\lambda)|^2}\right).$$

Setzen wir $\lambda = \frac{1}{\mu}$, so erhalten wir

$$\|(\tilde{e} - \mu\tilde{x}_1)^{-1}\| = o\left(\frac{r}{\cos^2\varphi}\right) \quad \left(\mu = re^{i\left(\varphi + \frac{\pi}{2}\right)}\right).$$

Folglich ist nach Satz 1 $\tilde{x}_1 = 0$; ähnlich beweisen wir, dass $\tilde{x}_2 = 0$ ist, d. h. $\tilde{x} = 0$.

Untersuchen wir nun die primären Ideale in einigen Ringen. Wir werden voraussetzen, dass in dem Ring R zwei Erzeugende x und x^{-1} vorhanden sind. Jedes maximale Ideal wird dabei durch die Zahl $x(M)$ bestimmt.

Nehmen wir an, dass die Menge der Zahlen $x(M)$ auf dem Einheitskreise liegt, d. h. dass $|x(M)| = 1$ ist. Dies ist der Beziehung

$$\lim_{n \to \pm\infty} \sqrt[n]{\|x^n\|} = 1$$

äquivalent. In der Annahme, dass die $a_n = \|x^n\|$ nicht zu schnell wachsen, werden wir imstande sein die primären Ideale des Ringes R zu finden.

Satz 5. *Wenn die a_n derart sind, dass*

$$\left.\begin{aligned} &\lim_{r\to 1}(1-r)^k \sum_{n=0}^{\infty} a_n r^n = 0 \\ \textit{und} \quad & \\ &\lim_{r\to 1}(1-r)^k \sum_{n=0}^{\infty} a_{-n} r^n = 0 \end{aligned}\right\} \quad (3)$$

ist, so sind in jedem maximalen Ideal M_0 nicht mehr als $k-1$ primäre Ideale $I_1, \ldots, I_{k-1}$ enthalten. Das Ideal I_l wird durch das Element $(x-\lambda_0 e)^l$ erzeugt, wo $\lambda_0 = x(M_0)$ ist.

Beweis. Betrachten wir $(e-\lambda x)^{-1}$. Es sei $|\lambda| < 1$. Dann ist

$$\|(e-\lambda x)^{-1}\| \leqslant \sum_{n=0}^{\infty} |\lambda|^n |x^n| \leqslant \sum_{n=0}^{\infty} a_n r^n,$$

wo $r = |\lambda|$ ist. Es sei $|\lambda| > 1$. Dann ist

$$\|(e-\lambda x)^{-1}\| = o\left(\frac{1}{(1-r)^k}\right).$$

Somit haben wir

$$\|(e-\lambda x)^{-1}\| = o\left(\frac{1}{(1-r)^k}\right).$$

Aus Satz 3 (siehe auch die Bemerkung zu Satz 2) folgt, dass es nicht mehr als $k-1$ primäre Ideale existieren, die in M_0 enthalten sind, und dass jedes von ihnen durch das Element

$$(x-\lambda_0 e)^l \quad (1 \leqslant l \leqslant k)$$

erzeugt wird.

Wir haben verlangt, dass $\sum a_n r^n$ die Ordnung $o\left(\frac{1}{(1-r)^k}\right)$ hat. Der Satz gilt nicht mehr, wenn man o durch O ersetzt. Als Beispiel betrachten wir den Ring aller Funktionen $f(t)$, die auf einem Umkreis definiert sind und eine stetige Ableitung der Ordnung k besitzen. Die Norm in diesem Ringe definieren wir folgendermassen:

$$\|f(t)\| = \max|f(t)| + \ldots + \frac{1}{k!}\max|f^{(k)}(t)|.$$

Die Erzeugenden dieses Ringes sind $x = e^{it}$ und $x^{-1} = e^{-it}$. Wir haben

$$a_n = \|x^n\| \sim c|n|^k \quad (n = \ldots, -1, 0, +1, \ldots),$$

$$\sum a_n r^n \sim c \sum n^k r^n \sim \frac{c_1}{(1-r)^{k+1}},$$

wo $c_1 \neq 0$ ist. Nach Satz 5 sind in diesem Ring nicht mehr als $k+1$ primäre Ideale vorhanden, die in M_0 enthalten sind. Es wäre ihrer nicht mehr als k, wenn man in der Bedingung des Satzes 5 o durch O ersetzen könnte. Es ist aber leicht zu zeigen, dass in diesem Falle $k+1$ primäre Ideale existieren, die M_0

angehören; $I_0 = M_0$ besteht nämlich aus allen Funktionen, die im Punkte $\lambda_0 = x(M_0)$ zu Null werden, I_1 besteht aus allen Funktionen, die zusammen mit ihrer Ableitung in diesem Punkte zu Null werden, usw.

Literatur

1. I. Gelfand, Normierte Ringe (in diesem Heft des Recueil mathématique, S. 3—23).

2. E. Lindelöf, Acta Soc. Fennicae, **50**, N. 5, (1922). Einen einfachen Beweis siehe bei Alfors [Trans. of Amer. Math. Soc., **41**, (1937), 1].

3. I. Gelfand und G. Šilov, Über verschiedene Methoden der Einführung der Topologie in die Menge der maximalen Ideale eines normierten Ringes (in diesem Heft des Recueil mathématique, S. 37).

Mathematisches Institut der
Akademie der Wissenschaften
der U. d. S. S. R.

(Поступило в редакцию 27/VI 1940 г.)

11.

Zur Theorie der Charaktere der Abelschen topologischen Gruppen

Mat. Sb., Nov. Ser. **9** (51) (1941) 49–50. Zbl. **24**: 323

Es sei ein kommutativer normierter Ring R und eine beschränkte Menge $\mathfrak{G} \subset R$ derart gegeben, dass $\mathfrak{G}$ eine Gruppe bildet (als Operation in $\mathfrak{G}$ dient die Operation der Multiplikation, die in R definiert ist). Wir werden zeigen, dass in $\mathfrak{G}$ genügend viele stetige Charaktere vorhanden sind (der Sinn der Worte „genügend viele" wird weiter unten definiert).

Satz 1. *Es sei $x = e - y$, wo y ein verallgemeinertes nilpotentes Element ist. Es sei $\|x^n\| \leq M$ für $n = 0, \pm 1, \ldots$ Dann ist $y = 0$.*

Beweis. Betrachten wir $(e - \lambda x)^{-1}$. Für $|\lambda| < 1$ haben wir

$$(e - \lambda x)^{-1} = \sum_{n=0}^{\infty} \lambda^n x^n.$$

Für $|\lambda| > 1$

$$(e - \lambda x)^{-1} = -\sum_{n=1}^{\infty} \lambda^{-n} x^{-n}.$$

Die respektiven Reihen konvergieren, da $\|x^n\| \leq M$ ist. Es gilt die Abschätzung:

a) für $|\lambda| < 1$

$$\|(e - \lambda x)^{-1}\| \leq \sum_{n=0}^{\infty} |\lambda^n| \, \|x^n\| \leq \frac{M}{1 - |\lambda|},$$

b) für $|\lambda| > 1$

$$\|(e - \lambda x)^{-1}\| \leq \sum_{n=1}^{\infty} |\lambda^{-n}| \, \|x^{-n}\| \leq \frac{M}{|\lambda| - 1}.$$

Setzen wir $\lambda = \frac{\mu}{\mu - 1}$. Dann ist

$$(e - \lambda x)^{-1} = -(\mu - 1)(e - \mu y)^{-1}$$

und folglich

$$\|(e - \mu y)^{-1}\| \leq \frac{M}{\left| |\mu| - |\mu - 1| \right|},$$

wenn $|\mu| \neq |\mu - 1|$ ist, oder

$$\|(e - \mu y)^{-1}\| \leq \frac{M \left[|\mu| + |\mu - 1| \right]}{|1 - 2r\cos\varphi|} = O\left(\frac{1}{\cos\varphi}\right),$$

wo $\mu = re^{i\varphi}$ ist. Also ist, nach Satz 1 der vorhergehenden Arbeit, $y^2 = 0$. Aber in diesem Falle $x^n = (e - y)^n = e - ny$, denn $\|x^n\|$ beschränkt ist, $y = 0$.

Bemerkung. In derselben Weise kann man zeigen, dass, wenn

$$\|e+y\|=\|e-y\|=1$$

ist, $y=0$ ist. Hier wird schon y a priori nicht als ein verallgemeinertes nilpotentes Element vorausgesetzt.

Nehmen wir irgendein maximales Ideal M des Ringes R. Dann entspricht jedem Element $x \in \mathfrak{G} \subset R$ eine komplexe Zahl $x(M)$, $|x(M)|=1$. In der Tat: die Folge x^n, $n=0, \pm 1, \ldots$, ist beschränkt (denn $x^n \in \mathfrak{G}$ und $\mathfrak{G}$ ist nach Voraussetzung beschränkt). Folglich, da $|x^n(M)| \leqslant \|x\|^n$ ist, ist auch $x^n(M)$ eine beschränkte Zahlenfolge, also $|x(M)|=1$. Es sei x ein fixiertes Element aus $\mathfrak{G}$, das von e verschieden ist. Zeigen wir, dass ein M derart existiert, dass $x(M) \neq 1$ ist. Nehmen wir den Gegenteil an, d. h. nehmen wir an, dass für jedes M $x(M)=1$ ist; dann gehört das Element $e-x=y$ zu allen maximalen Idealen, d. h. ist ein verallgemeinertes nilpotentes Element. Da ausserdem die Folge x^n beschränkt ist, so ist nach Satz 1 $y=0$, d. h. $x=e$, was einen Widerspruch ergibt. Jede für $x \in \mathfrak{G}$ definierte Funktion $x(M)$ ist ein Charakter der Gruppe $\mathfrak{G}$. Wir haben bewiesen, dass es Charaktere gibt, die von dem Charakter $\equiv 1$ verschieden sind.

Zeigen wir, dass es genügend viele Charaktere gibt, d. h. dass für zwei verschiedene x_1 und x_2 ein M existiert derart, dass $x_1(M) \neq x_2(M)$ ist. Setzen wir das Gegenteil voraus. Dann nimmt $x_1 x_2^{-1}$ auf allen maximalen Idealen den Wert 1 an, d. h. $x_1 x_2^{-1} - e = y$ gehört allen maximalen Idealen an. Dieselben Überlegungen, wie oben, wiederholend, erhalten wir, dass $x_1 x_2^{-1} = e$, d. h. $x_1 = x_2$ ist, was einen Widerspruch ergibt. Somit haben wir den

Satz 2. *Es sei R ein normierter Ring. Es möge die Menge $\mathfrak{G} \in R$ beschränkt sein und eine Gruppe bilden. Dann existieren stetige nichttriviale Charaktere der Gruppe $\mathfrak{G}$. Dieser Charaktere gibt es genügend viele, d. h. für zwei beliebige Elemente x_1 und x_2 derart, dass $x_1 \neq x_2$ ist, existiert ein stetiger Charakter, der auf diesen Elementen verschiedene Werte annimmt.*

Mathematisches Institut der
Akademie der Wissenschaften
der U. d. S. S. R.

(Поступило в редакцию 27/VI 1940 г.)

12.

Über absolut konvergente trigonometrische Reihen und Integrale

Mat. Sb., Nov. Ser. 9 (51) (1941) 51–66. Zbl. 24:323

§ 1. N. Wiener [1], R. H. Cameron [2], N. Wiener und H. R. Pitt [3] haben einige sehr interessante Sätze über absolut konvergente trigonometrische Reihen und Integrale bewiesen, die wesentliche Anwendungen in den Sätzen des Tauberschen Typs besitzen.

Das Ziel der vorliegenden Arbeit besteht in der Angabe einer allgemeinen Methode zur Deduktion solcher Sätze, sowie im Beweis einer Reihe von allgemeineren Sätzen. Um die Grundideen am deutlichsten auseinanderzusetzen, beweisen wir zuerst mit unserer Methode den einfachsten der Wienerschen Sätze.

Betrachten wir den folgenden Ring R_1. Die Elemente des Ringes sind die Funktionen

$$f(t) = \sum_{n=-\infty}^{+\infty} a_n e^{int}$$

derart, dass $\sum_{n=-\infty}^{+\infty} |a_n| < \infty$ ist. Als die Norm der Funktion nehmen wir

$$\|f\| = \sum_{n=-\infty}^{+\infty} |a_n|.$$

Die Addition und die Multiplikation in dem Ringe definieren wir als die gewöhnlichen Addition und Multiplikation von Funktionen. Es ist leicht nachzuprüfen, dass alle Axiome des normierten Ringes (siehe „Normierte Ringe", § 1 *), das Axiom (δ') einschliessend, erfüllt sind.

Der Wienersche Satz besteht im folgenden:

S a t z 1. *Wenn $f(t)$ zu R_1 angehört und nirgends zu Null wird, so gehört $\frac{1}{f(t)}$ auch zu R_1.*

B e w e i s. Finden wir die maximalen Ideale des Ringes R_1. Es sei M ein maximales Ideal. Nach Satz 7, § 5, N. R., entspricht jedem Element f beim Homomorphismus $R \to R/M$ eine bestimmte Zahl. Bezeichnen wir mit a die Zahl, die der Funktion e^{it} entspricht, $|a| \leqslant \|e^{it}\| = 1$ nach (β), § 5, N. R. Der Funktion e^{-it} entspricht die Zahl $\frac{1}{a}$. Folglich ist auch $\left|\frac{1}{a}\right| \leqslant 1$, und also haben wir $|a| = 1$, d. h.

* In diesem Heft des Recueil mathématique, S. 3—23. Im folgenden werden wir diese Arbeit als N. R. zitieren.

a kann folgendermassen geschrieben werden: $a=e^{it_0}$. Der Funktion e^{int} entspricht die Zahl e^{int_0} (da das Produkt in das Produkt übergeht). Folglich geht $\sum_{n=-N}^{N} a_n e^{int}$ in $\sum_{n=-N}^{N} a_n e^{int_0}$ über, d. h. jedes Polynom geht beim Homomorphismus in den Wert dieses Polynoms im Punkte t_0 über. Da unser Homomorphismus stetig ist, so geht jede Funktion $f(t)$ des Ringes in den Wert dieser Funktion im Punkte t_0 über. Somit besteht jedes maximale Ideal M aus denjenigen Funktionen, die beim Homomorphismus in die Null übergehen, d. h. aus solchen Funktionen, für die $f(t_0)=0$ gilt. Somit bedeutet der Satz „$f(t)$ wird nirgends zu Null", dass „$f(t)$ in keinem maximalen Ideal enthalten ist". Nach Satz 6, **N. R.**, besitzt also das Element f ein inverses.

§ 2. Beim Beweis des Satzes 1 haben wir viele Eigenschaften des Ringes R_1 nicht benutzt. Es ist deshalb natürlich die durchgeführten Überlegungen zu verallgemeinern. Betrachten wir die Menge R der formalen Potenzreihen $\sum a_n x^n$, wo a_n eine beliebige Folge komplexer Zahlen ist, die der Bedingung

$$\sum_{n=-\infty}^{+\infty} |a_n| \alpha_n < +\infty \tag{1}$$

genügt (die α_n sind fixierte positive Zahlen). Finden wir die Bedingungen, denen die α_n genügen müssen, damit in der Menge R zusammen mit zwei beliebigen Potenzreihen auch ihr Produkt enthalten wäre.

Satz 2. *Dafür, dass zusammen mit zwei beliebigen Potenzreihen auch ihr Produkt der Bedingung* (1) *genüge, ist es notwendig und hinreichend, dass eine Zahl c existiere derart, dass*

$$\alpha_{m+n} \leqslant c\alpha_m \alpha_n \tag{2}$$

ist.

Beweis. Zeigen wir die Notwendigkeit der Bedingung (2). Wenn wir

$$\left\| \sum_{-\infty}^{+\infty} a_n x^n \right\| = \sum_{-\infty}^{+\infty} |a_n| \alpha_n$$

setzen, so können wir R als einen normierten linearen Raum betrachten. Es sei

$$y=\sum_{-\infty}^{+\infty} a_n x^n, \quad z=\sum_{-\infty}^{+\infty} b_n x^n,$$

wo y und $z \in R$. Nach der Bedingung konvergiert die Reihe $\sum_{n=-\infty}^{+\infty} a_k b_{n-k}$ für eine beliebige Folge a_k, die der Bedingung (1) genügt. Für fixierte b_k ist das Funktional $\sum_{k=-N}^{+N} |a_k b_{n-k}|$ bezüglich y stetig. Folglich ist

$$\sum_{k=-\infty}^{+\infty} |a_k b_{n-k}| = \sup_N \sum_{k=-N}^{+N} |a_k b_{n-k}|,$$

als die obere Grenze von stetigen Funktionalen, ein halbstetiges Funktional. Also ist es stetig und beschränkt — dies nach dem folgenden Lemma, das wir noch in diesem Beweis benutzen werden:

Wenn $p(y)$ ein halbstetiges konvexes Funktional ist, das in einem vollständigen normierten Raum definiert ist, so ist $p(y)$ stetig und genügt der Bedingung

$$p(y) \leqslant c\|y\| \quad [4].$$

Zeigen wir, dass

$$\|yz\| \leqslant c\|y\|\cdot\|z\|$$

ist. Bei fixiertem z ist $\|yz\|$ ein konvexes Funktional bezüglich y. Ausserdem haben wir

$$\|yz\| = \sum_{n=-\infty}^{+\infty} a_n \Big| \sum_{k=-\infty}^{+\infty} a_k b_{n-k} \Big| = \sup_N \sum_{n=-N}^{+N} \Big| \sum_{k=-\infty}^{+\infty} a_k b_{n-k} \Big|.$$

Folglich ist $\|yz\|$, als die obere Grenze von stetigen Funktionalen, ein halbstetiges Funktional, und also, nach dem zitierten Lemma, ein stetiges Funktional, das nach y in der Einheitssphäre $\|y\| \leqslant 1$ beschränkt ist. Somit existiert

$$p(z) = \sup_{\|y\| \leqslant 1} \|yz\|.$$

Das Funktional $p(z)$ ist, als die obere Grenze von stetigen Funktionalen, ein halbstetiges Funktional. Folglich, nach dem zitierten Lemma, ist $p(z)$ ein beschränktes Funktional, d. h. $p(z) \leqslant c\|z\|$. Setzen wir $y = x^n$, $z = x^m$, so erhalten wir

$$a_{m+n} \leqslant c a_m a_n.$$

Dass die Bedingung (2) hinreichend ist, wird durch direktes Nachrechnen gezeigt. Die Rechnung zeigt auch, dass

$$\|yz\| \leqslant c\|y\|\cdot\|z\|$$

ist. In dem Ring R kann man eine Norm einführen, die der ursprünglichen äquivalent ist, und für die $\|yz\|_1 \leqslant \|y\|_1 \cdot \|z\|_1$ gilt. Zu diesem Ende genügt es zu setzen

$$\|y\|_1 = \sup_m \frac{\sum_{n=-\infty}^{+\infty} a_{m+n} \,|\, a_n \,|}{a_m};$$

man beachte, dass

$$\|x^n\|_1 = \sup_m \frac{a_{m+n}}{a_n}$$

ist.

Den in Satz 1 betrachteten Ring R_1 kann man aus der Klasse der normierten Ringe durch folgende Bedingungen herausheben (diese Bedingungen zeigen, übrigens, dass dieser Ring besonders interessant ist).

Satz 3. *Es sei R ein normierter Ring mit zwei Erzeugenden x und x^{-1}, der den folgenden Bedingungen genügt:*

1°) $\|\sum_{-N}^{N} a_n x^n\|$ *hängt nur von den* $|a_n|$ *ab, und*

2°) $\|x^n\| = 1$.

Dann fällt der Ring R mit dem R_1 des Satzes 1 zusammen.

Beweis. Da in dem Ring zwei Erzeugende x und x^{-1} vorhanden sind, so wird jedes maximale Ideal M vollständig durch die Zahl bestimmt, in die x beim

Homomorphismus $R \to R/M$ übergeht. Genau so, wie im Satz 1, wird gezeigt, dass, wenn x in a übergeht, $a = e^{i\varphi}$ ist. Bei diesem Homomorphismus geht $\sum_{-N}^{N} a_n x^n$ in $\sum_{-N}^{N} a_n e^{in\varphi}$ über. Da in R maximale Ideale existieren, so existiert ein solches φ, dass das Element

$$\sum_{-N}^{N} a_n x^n - \sum_{-N}^{N} a_n e^{in\varphi} \cdot e$$

kein inverses besitzt. Wir haben zu zeigen, dass

$$\|\sum_{-N}^{N} a_n x^n\| = \sum_{-N}^{N} |a_n|$$

ist. Erstens ist es klar, dass

$$\|\sum_{-N}^{N} a_n x^n\| \leqslant \sum_{-N}^{N} |a_n| \cdot \|x^n\| \leqslant \sum_{-N}^{N} |a_n|$$

ist. Nehmen wir an, dass

$$\|\sum_{-N}^{N} a_n x^n\| < \sum_{-N}^{N} |a_n|$$

ist. Infolge der Homogenität der Ungleichung darf man annehmen, dass $\sum_{-N}^{N} |a_n| = 1$ ist. Es sei also

$$\sum_{-N}^{N} |a_n| = 1, \quad \|\sum_{-N}^{N} a_n x^n\| < 1.$$

Nach der Bedingung 1°) ist auch

$$\|\sum_{-N}^{N} a_n e^{-in\varphi} x^n\| < 1.$$

Folglich besitzt, nach Lemma 1, § 2, N. R., das Element $e - \sum_{-N}^{N} |a_n| e^{-in\varphi} x^n$ ein inverses. Dies ist aber unmöglich, denn beim Homomorphismus nach M das Element $e - \sum_{-N}^{N} |a_n| e^{-in\varphi} x^n$ in $1 - \sum_{-N}^{N} |a_n| = 0$ übergeht.

Somit haben wir bewiesen, dass

$$\|\sum_{-N}^{N} a_n x^n\| = \sum_{-N}^{N} |a_n|$$

ist. Da es in dem Ring nur zwei Erzeugende x und x^{-1} gibt, so folgt hieraus, dass jedes Element des Ringes in der Form $\sum_{n=-\infty}^{+\infty} a_n x^n$ dargestellt werden kann, wobei

$$\|\sum_{-\infty}^{+\infty} a_n x^n\| = \sum_{-\infty}^{+\infty} |a^n|$$

ist.

§ 3. Wir stellen nun die folgende Aufgabe. Es sei eine Potenzreihe $\sum_{-\infty}^{+\infty} a_n x^n$ gegeben, deren Koeffizienten der Bedingung (1) und die in der Bedingung (1) vorkommenden α_n der Bedingung (2) genügen. Was sind die notwendigen und hinreichenden Bedingungen dafür, dass eine Reihe $\sum_{-\infty}^{+\infty} b_n x^n$ derart existiere, dass das Produkt dieser Reihe und der Reihe $\sum_{-\infty}^{+\infty} a_n x^n$ gleich $x^0 = e$ wäre?

Satz 4. *Es sei eine formale Potenzreihe* $\sum_{-\infty}^{+\infty} a_n x^n$ *derart gegeben, dass*

$$\sum_{-\infty}^{+\infty} |a_n| \alpha_n < +\infty \tag{1}$$

ist, wo

$$\alpha_{m+n} \leqslant c\alpha_m \cdot \alpha_n, \quad \alpha_n > 0. \tag{2}$$

Dafür, dass eine Reihe $\sum_{-\infty}^{+\infty} b_n x^n$, *deren Koeffizienten der Bedingung*

$$\sum \alpha_n |b_n| < +\infty$$

genügen, derart existiere, dass $\sum a_n x^n \cdot \sum b_n x^n = e$ *ist, ist es notwendig und hinreichend, dass die Funktion*

$$\varphi(r, t) = \sum_{-\infty}^{+\infty} a_n r^n e^{int}$$

in dem Bereich $r_1 \leqslant r \leqslant r_2$, $0 \leqslant t \leqslant 2\pi$, *nicht verschwinde, wo*

$$r_1 = \lim_{n \to -\infty} \alpha_n^{\frac{1}{n}}, \quad r_2 = \lim_{n \to +\infty} \alpha_n^{\frac{1}{n}}$$

ist.

Bemerkungen: 1°. Für $r_1 \leqslant r \leqslant r_2$ ist die Reihe, die die Funktion $\varphi(r, t)$ darstellt, konvergent.

2°. Der Grenzwert $\lim \alpha_n^{\frac{1}{n}}$ existiert.

Beweis. Finden wir die maximalen Ideale des Ringes R aller Reihen, die der Bedingung (1) genügen. Es sei M ein maximales Ideal. Beim Homomorphismus $R \to R/M$ geht x in re^{it} über. Dann geht x^n in $r^n e^{int}$ über. Dabei ist $|r^n e^{int}| \leqslant \|x^n\|_1$, und folglich

$$r^n \leqslant \sup_m \frac{\alpha_{m+n}}{\alpha_m},$$

d. h.

$$r \leqslant \lim_{n \to +\infty} \left(\sup_m \frac{\alpha_{m+n}}{\alpha_m} \right)^{\frac{1}{n}} = r_2$$

und

$$r \geqslant \lim_{n \to -\infty} \left(\sup_m \frac{\alpha_{m+n}}{\alpha_m} \right)^{\frac{1}{n}} = r_1$$

(für die Existenz dieser Grenzwerte siehe **N. R.**, Satz 8', oder Pólya und Szegö, Aufgaben und Lehrsätze aus der Analysis. I, Nr. 98).

Da

$$\frac{a_n}{a_0} \leqslant \sup_m \frac{a_{m+n}}{a_m} \leqslant c a_n$$

ist, so haben wir

$$\lim_{n\to+\infty} \left(\sup_m \frac{a_{m+n}}{a_m}\right)^{\frac{1}{n}} = \lim_{n\to+\infty} a_n^{\frac{1}{n}}.$$

Folglich geht x in re^{it} über, wo $r_1 \leqslant r \leqslant r_2$ ist. Dabei geht x^n in $r^n e^{int}$ und $\sum_{-N}^{+N} a_n x^n$ in $\sum_{-N}^{+N} a_n r^n e^{int}$ über. Da der Homomorphismus stetig ist, so geht $\sum_{-\infty}^{+\infty} a_n x^n$ in $\sum_{-\infty}^{+\infty} a_n r^n e^{int}$ über (und hieraus folgt schon die Konvergenz dieser letzteren Reihe). Somit besteht das maximale Ideal M aus allen Elementen $\sum_{-\infty}^{+\infty} a_n x^n$ derart, dass $\sum_{-\infty}^{+\infty} a_n r^n e^{int} = 0$ ist, wo r und t fixierte Zahlen sind. Wenn also $\varphi(r, t)$ in dem betrachteten Bereich nicht zu Null wird, so gehört $\sum_{-\infty}^{+\infty} a_n x^n$ zu keinem maximalen Ideal und folglich besitzt ein inverses Element.

Die Notwendigkeit der Bedingung folgt daraus, dass für jede r und t, die dem betrachteten Bereich angehören, ein Homomorphismus existiert, der x in re^{it} überführt.

§ 4. Betrachten wir nun die Gesamtheit aller Funktionen $f(t)$ derart, dass

$$\int_{-\infty}^{+\infty} |f(t)|\, \alpha(t)\, dt < +\infty \tag{3}$$

ist. Dabei ist $\alpha(t)$ eine stetige Funktion, die den Bedingungen

1°) $\alpha(t) > 0$ und

2°) $\alpha(s+t) \leqslant \alpha(s)\,\alpha(t)$

genügt.

Die Faltung zweier Funktionen f_1 und f_2 definieren wir folgendermassen:

$$f_3 = f_1 \times f_2,$$

wo

$$f_3(s) = \int_{-\infty}^{+\infty} f_2(s-t)\, f_1(t)\, dt$$

ist. Die Gesamtheit der Funktionen, die der Bedingung (3) genügen, besitzt die Eigenschaft, dass zusammen mit zwei beliebigen Funktionen in diese Gesamtheit auch ihre Faltung und ihre Summe eingehen. Somit bildet die Gesamtheit dieser Funktionen einen Ring.

Setzen wir

$$\|f\| = \int_{-\infty}^{+\infty} \alpha(t)\, |f(t)|\, dt,$$

so wird unser Ring ein normierter Ring, wobei $\|f_1 \times f_2\| \leqslant \|f_1\| \cdot \|f_2\|$ sein wird. In diesem normierten Ring gibt es keine Einheit. Wir erweitern ihn deshalb durch Hinzufügung des Symbols e und werden als Elemente des erweiterten Ringes R die Symbole $\lambda e + f$ betrachten und zwar mit der folgenden Multiplikationsregel:

$$(\lambda e + f_1) \times (\mu e + f_2) = \lambda \mu e + \lambda f_2 + \mu f_1 + f_1 \times f_2.$$

Die Norm in dem erweiterten Ring führen wir als

$$\|\lambda e + f\| = |\lambda| + \|f\|$$

ein.

Finden wir die maximalen Ideale in diesem Ring. Es sei M ein maximales Ideal. Betrachten wir die Funktion

$$f_a(t) = \begin{cases} 1 & \text{für } 0 \leqslant t \leqslant a, \\ 0 & \text{„ } t \,\overline{\in}\, [0, a]. \end{cases}$$

Es möge beim Homomorphismus $R \to R/M$ der Funktion $f_a(t)$ die Zahl $\varphi(a)$ entsprechen. Zeigen wir, dass $\varphi(a)$ eine Ableitung besitzt, die der Beziehung

$$\varphi'(a+b) = \varphi'(a)\,\varphi'(b)$$

genügt. Setzen wir

$$f(t) = \frac{f_{a+\Delta a}(t) - f_a(t)}{\Delta a} \times f_b(t).$$

Das Diagramm der Funktion $f(t)$ ist daneben dargestellt. Wir haben

$$\|f - [f_{a+b} - f_a]\| = \int_a^{a+\Delta a} \alpha(t)\,dt + \int_{a+b}^{a+b+\Delta a} \alpha(t)\,dt,$$

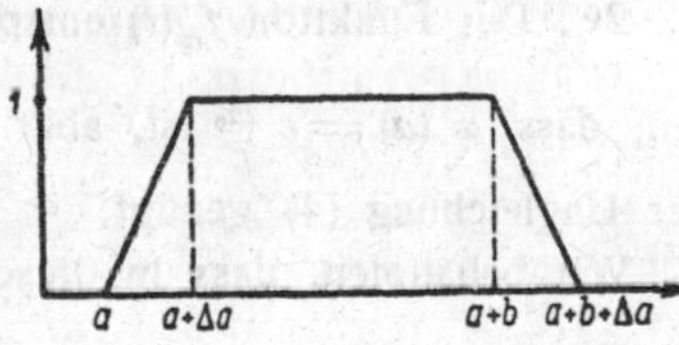

d. h. für $\Delta a \to 0$ strebt das Element f zu $f_{a+b} - f_a$. Beim Homomorphismus $R \to R/M$ geht f in

$$\frac{\varphi(a+\Delta a) - \varphi(a)}{\Delta a} \cdot \varphi(b)$$

und $f_{a+b} - f_a$ in $\varphi(a+b) - \varphi(a)$ über. Da der Homomorphismus stetig ist, so existiert der Grenzwert des Ausdruckes

$$\frac{\varphi(a+\Delta a) - \varphi(a)}{\Delta a} \cdot \varphi(b)$$

und ist gleich $\varphi(a+b) - \varphi(a)$.

Es sind zwei Fälle möglich:

1°. $\varphi(b) = 0$. In diesem Fall ist alles klar.

2°. Für ein bestimmtes b_0 ist $\varphi(b_0) \neq 0$. Dann besitzt $\varphi(a)$ eine Ableitung, denn

$$\frac{\varphi(a+\Delta a) - \varphi(a)}{\Delta a} \cdot \varphi(b_0) \to \varphi(a+b_0) - \varphi(a).$$

Wir erhalten somit $\varphi'(a)\,\varphi(b) = \varphi(a+b) - \varphi(a)$. Differenzieren wir diese Beziehung nach b, so erhalten wir, dass

$$\varphi'(\alpha)\,\varphi'(b) = \varphi'(a+b)$$

ist. $\varphi'(a)$ ist eine stetige Funktion, da $\varphi'(a) = \frac{\varphi(a+b)-\varphi(a)}{\varphi(b_0)}$ ist. Hieraus folgt: $\varphi'(a) = e^{iua}$, wo u eine Konstante ist.

Betrachten wir die Zahl u. Da bei einem Homomorphismus die Norm sich nur verkleinern kann, so ist

$$\left|\frac{\varphi(a+\Delta a)-\varphi(a)}{\Delta a}\right| \leqslant \left\|\frac{f_{a+\Delta a}-f_a}{\Delta a}\right\| = \frac{1}{\Delta a}\int_a^{a+\Delta a} \alpha(t)\,dt.$$

Folglich haben wir $|\varphi'(a)| < \alpha(t)$, oder $e^{-a\mathfrak{J}(u)} \leqslant \alpha(a)$, d. h.

$$\mathfrak{J}(u) \leqslant \frac{\log\alpha(-a)}{a},$$

wenn $a < 0$, und

$$\mathfrak{J}(u) \geqslant -\frac{\log\alpha(a)}{a},$$

wenn $a > 0$. Folglich ist

$$\lim_{a\to+\infty}\frac{\log\alpha(a)}{-a} \leqslant \mathfrak{J}(u) \leqslant \lim_{a\to+\infty}\frac{\log\alpha(-a)}{a}. \tag{4}$$

Somit sind beim Homomorphismus $R \to R/M$ zwei Fälle möglich:

1°. $\varphi(b) = 0$. Dann geht jede Funktion [als der Grenzwert linearer Kombinationen der Funktionen $f_a(t)$] in Null über, und das maximale Ideal besteht aus allen $\lambda e + f$, für die $\lambda = 0$ ist.

2°. Der Funktion $f_a(t)$ entspricht die Zahl $\varphi(a) = \frac{e^{iau_0}-1}{a}$ [wir haben bewiesen, dass $\varphi'(a) = e^{iau_0}$ ist, aber da $\varphi(0) = 0$ ist, so ist $\varphi(a) = \frac{e^{iau_0}-1}{a}$], wo u_0 der Ungleichung (4) genügt.

Wir behaupten, dass in diesem Falle der Funktion f die Zahl

$$\int_{-\infty}^{+\infty} f(t)\,e^{itu_0}\,dt \tag{5}$$

entspricht. In der Tat, dies gilt für die Funktionen $f_a(t)$, folglich auch für ihre lineare Kombinationen, und da der Homomorphismus stetig ist, auch für die Grenzwerte der linearen Kombinationen, d. h. für beliebige Funktionen des Ringes.

Somit besteht das maximale Ideal aus allen Elementen $\lambda e + f$ derart, dass $\lambda + \int_{-\infty}^{+\infty} f(t)\,e^{itu_0}\,dt = 0$ ist, und wird folglich durch die Zahl u_0 bestimmt. Zeigen wir, dass, umgekehrt, die Gesamtheit der Funktionen, die dieser Bedingung genügen, wo u_0 der Ungleichung (4) genügt, ein maximales Ideal bestimmt.

Erstens hat das Integral (5) einen Sinn. In der Tat, die Bedingung (4) ist äquivalent der folgenden:

$$|e^{iu_0t}| \leqslant \alpha(t)$$

[denn

$$\lim_{a\to\infty}\frac{\log\alpha(-a)}{a} = \inf_{a\geqslant 0}\frac{\log\alpha(-a)}{a};$$

siehe Pólya und Szegö, Aufgaben und Lehrsätze aus der Analysis. I, Nr. 98] Folglich konvergiert das Integral (5) und dabei absolut, denn es existiert $\int_{-\infty}^{+\infty} a(t)\,|f(t)|\,dt$.

Ordnen wir dem Element $\lambda e + f$ die Zahl $\lambda + \int_{-\infty}^{+\infty} f(t)\, e^{i a_0 t}\, dt$ zu. Diese Zuordnung ist ein Homomorphismus, denn der Faltung zweier Funktionen entspricht das Produkt ihrer Fouriertransformierten. Aber jeder Homomorphismus auf dem Körper der komplexen Zahlen bestimmt ein maximales Ideal.

Den Satz 6, § 4, **N. R.**, S. 9 anwendend, erhalten wir somit den

Satz 5. *Es sei eine positive stetige Funktion $a(t)$ gegeben, die der Bedingung $a(t+s) \leqslant a(t)\,a(s)$ genügt. Es sei $f(t)$ derart gegeben, dass $\int_{-\infty}^{+\infty} |f(t)|\, a(t)\, dt < +\infty$ ist. Betrachten wir die Funktion*

$$F(u) = a + \int_{-\infty}^{+\infty} f(t)\, e^{itu}\, dt,$$

die in dem Streifen

$$\lim_{t \to +\infty} \frac{\log a(t)}{-t} \leqslant \mathfrak{I}(u) \leqslant \lim_{t \to +\infty} \frac{\log a(-t)}{t}$$

definiert ist; dafür, dass die Funktion $\frac{1}{F(u)}$ (in demselben Streifen) in der Form

$$\frac{1}{F(u)} = b + \int_{-\infty}^{+\infty} f_1(t)\, e^{iut}\, dt$$

dargestellt werden könne, wo

$$\int_{-\infty}^{+\infty} |f_1(t)|\, a(t)\, dt < +\infty$$

ist, ist es notwendig und hinreichend, dass $F(u)$ in diesem Streifen nicht zu Null werde, und dass $a \neq 0$ sei.

Bemerkungen: 1°. Für den Fall $a(t) \equiv 1$ war dieser Satz von N. Wiener bewiesen.

2°. Die Bedingung $a(t+s) = a(t)\,a(s)$ kann durch eine schwächere, nämlich $a(t+s) \leqslant c\, a(t)\, a(s)$, ersetzt werden.

3°. Interessante Sonderfälle des Satzes erhält man, wenn man $a(t) = |t|^k + 1$, $k \geqslant 0$, oder

$$a(t) = \{|t|^k + 1\}\{|\log|t+1||\}^s, \quad s \geqslant 0, \quad k \geqslant 0,$$

setzt.

§ 5. Gehen wir nun zu der Betrachtung des folgenden Ringes über. Nehmen wir eine Funktion $a(t)$, die den folgenden Bedingungen genügt:

1°) $a(t) > 0$,

2°) $a(t+s) \leqslant a(t)\, a(s)$,

3°) $\alpha(0)=1$,

4°) $\log\alpha(t)$ ist auf dem Intervall $-1\leqslant t\leqslant+1$ beschränkt.

Als Elemente des Ringes R werden wir die $x=\sum_{-\infty}^{+\infty} a_n e^{i\lambda_n t}$ nehmen, wo die a_n derart sind, dass

$$\sum_{-\infty}^{+\infty} |a_n|\,\alpha(\lambda_k)<+\infty \tag{6}$$

(die Reihe $\sum a_n e^{i\lambda_n t}$ selbst setzen wir nicht als konvergent voraus). Als das Element xy definieren wir die Reihe, die durch Multiplikation von Reihen erhalten wird, die den Elementen x und y entsprechen. Die Norm des Elementes x setzen wir gleich dem Ausdruck auf der linken Seite von (6).

Es ist leicht zu zeigen, dass das Produkt von Reihen, die der Bedingung (6) genügen, auch der Bedingung (6) genügt, und dass $\|xy\|\leqslant\|x\|\cdot\|y\|$ ist.

Beschreiben wir die maximalen Ideale in diesem Ring R. Es möge beim Homomorphismus $R\to R/M$ das Element des Ringes $x=e^{i\lambda t}$ in die Zahl $\varphi(\lambda)$ übergehen. Da $\|x\|=\alpha(\lambda)$ ist, so ist $|\varphi(\lambda)|\leqslant\alpha(\lambda)$. Ausserdem haben wir $\varphi(\lambda+\mu)=\varphi(\lambda)\cdot\varphi(\mu)$. Folglich ist $\varphi^n(\lambda)=\varphi(n\lambda)$, d. h.

$$\log|\varphi(\lambda)|\leqslant\frac{\log\alpha(n\lambda)}{n} \quad \text{für } n>0$$

und

$$\log|\varphi(\lambda)|\geqslant\frac{\log\alpha(n\lambda)}{n} \quad \text{für } n<0.$$

Somit ist

$$\lim_{n\to-\infty}\frac{\log\alpha(n\lambda)}{n}\leqslant\log|\varphi(\lambda)|\leqslant\lim_{n\to+\infty}\frac{\log\alpha(n\lambda)}{n}.$$

Setzen wir

$$\lim_{n\to-\infty}\frac{\log\alpha(n\lambda)}{n}=\beta_1(\lambda),\qquad \lim_{n\to+\infty}\frac{\log\alpha(n\lambda)}{n}=\beta_2(\lambda),$$

so haben wir

$$\beta_1(\lambda)\leqslant\log|\varphi(\lambda)|\leqslant\beta_2(\lambda).$$

Also ist $\log|\varphi(\lambda)|$ eine additive Funktion von λ, die auf dem Intervall (0, 1) beschränkt ist [die Beschränktheit von $\log|\varphi(\lambda)|$ folgt aus der Beschränktheit von $\beta_2(\lambda)$; $\beta_2(\lambda)$ ist beschränkt, weil

$$\beta_2(\lambda)=\lim_{n\to+\infty}\frac{\log\alpha(n\lambda)}{n}\leqslant\lim_{n\to\infty}\frac{n\log\alpha(\lambda)}{n}=\log\alpha(\lambda)$$

ist, und $\log \alpha(\lambda)$ ist auf $(0, 1)$ beschränkt nach der Bedingung 4°)]. Folglich haben wir

$$\log |\varphi(\lambda)| = \rho\lambda,$$

wo

$$\lim_{n\to-\infty} \frac{\log \alpha(n)}{n} \leqslant \rho \leqslant \lim_{n\to+\infty} \frac{\log \alpha(n)}{n}, \tag{7}$$

d. h. $\varphi(\lambda) = \psi(\lambda) e^{\lambda\rho}$, wo

$$|\psi(\lambda)| = 1, \quad \psi(\lambda+\mu) = \psi(\lambda)\psi(\mu) \tag{8}$$

ist.

Umgekehrt, entspricht jeder Funktion $e^{\rho\lambda}\psi(\lambda)$, wo ρ eine fixierte Zahl ist, die der Bedingung (7) genügt, und $\psi(\lambda)$ eine Funktion ist, die der Bedingung (8) genügt, ein bestimmtes maximales Ideal.

Um dies zu beweisen, konstruieren wir den Homomorphismus, der der Funktion $e^{\lambda\rho}\psi(\lambda)$ entspricht. Ordnen wir dem Element $\sum_{k=1}^{\infty} a_k e^{i\lambda_k t}$ die Zahl $\sum_{k=1}^{\infty} a_k e^{\rho\lambda_k}\psi(\lambda_k)$ zu. Zeigen wir, dass die Reihe $\sum_{k=1}^{\infty} a_k e^{\rho\lambda_k}\psi(\lambda_k)$ absolut konvergiert.

Da $\lim_{\lambda\to\infty} \frac{\log \alpha(\lambda)}{\lambda}$ existiert und gleich $\inf_{\lambda\geqslant 0} \frac{\log \alpha(\lambda)}{\lambda}$ ist (siehe Pólya und Szegö, Aufgaben und Lehrsätze aus der Analysis. I, Nr. 98), so ist $\rho \leqslant \frac{\log \alpha(\lambda)}{\lambda}$ für $\lambda > 0$ und $\rho \geqslant \frac{\log \alpha(\lambda)}{\lambda}$ für $\lambda < 0$. Somit ist $e^{\rho\lambda} \leqslant \alpha(\lambda)$, und also folgt aus der Konvergenz der Reihe $\sum |a_k| \alpha(\lambda_k)$ die Konvergenz der Reihe $\sum |a_k| e^{\rho\lambda_k}$, usw.

Unter den maximalen Idealen des Ringes R die einfachsten sind diejenigen, für die $\varphi(\lambda)$ stetig ist. Sie haben die folgende Form:

$$\varphi(\lambda) = e^{\rho\lambda} e^{i\lambda t_0}, \tag{9}$$

wo ρ der Bedingung (7) genügt. Wenn man die Menge aller maximalen Ideale topologisiert (siehe N. R., § 7), so erweist sich, dass die Menge der maximalen Ideale der Form, welche von den Funktionen der Form (9) beschrieben wird, in der Gesamtheit aller maximalen Ideale eine überall dichte Menge bildet.

Beweisen wir dies. Betrachten wir ein maximales Ideal, das von der Funktion $e^{\lambda\rho}\psi(\lambda)$ beschrieben wird. Bilden wir mit Hilfe der Erzeugenden $e^{i\lambda t}$ ein System seiner Umgebungen und beweisen, dass in einer beliebigen solchen Umgebung ein maximales Ideal der Form (9) vorhanden ist. Eine Umgebung des maximalen Ideals wird durch eine endliche Anzahl von Erzeugenden $e^{i\lambda_1 t}, \ldots, e^{i\lambda_n t}$ und eine Zahl ε bestimmt. In diese Umgebung gehen maximale Ideale ein, die von Funktionen beschrieben werden, die der Bedingung

$$|e^{\lambda_k\rho}\psi_0(\lambda_k) - \varphi(\lambda_k)| < \varepsilon \quad (k = 1, 2, \ldots, n) \tag{10}$$

genügen. Es ist leicht zu zeigen, dass es genügt sich auf solche Umgebungen zu beschränken, für welche zwischen den $\lambda_1, \lambda_2, \ldots, \lambda_n$ keine Beziehung der Art $\sum a_k \lambda_k = 0$ mit rationalen a_k existiert. Es sei $\varphi_0(\lambda_k) = e^{2\pi i \alpha_k}$; nach dem Kroneckerschen Satz existiert ein t_0 derart, dass

$$|t_0 \lambda_k - p_k| < \delta \quad (k = 1, 2, \ldots, n) \tag{10'}$$

ist, wo die p_k ganze Zahlen sind und δ eine vorgegebene Zahl ist. δ kann man so klein wählen, dass aus (10') (10) für $\varphi(\lambda) = e^{i\lambda t_0} e^{\lambda \rho}$ folgt. Unsere Behauptung ist somit bewiesen.

Satz 6. *Es sei*

$$\sum_{k=1}^{\infty} |a_k| \, \alpha(\lambda_k) < +\infty.$$

Es sei ferner

$$\left| \sum_{k=1}^{\infty} a_k e^{(\rho + ti)\lambda_k} \right| \geqslant c > 0 \tag{11}$$

für alle t *und für ein* ρ, *das der Bedingung* (7) *genügt.* [*Die Konvergenz dieser Reihe ist eine Folgerung der Bedingungen* (6) *und* (7).] *Dann existiert eine Funktion* $\sum b_k e^{(\rho + it)\mu_k}$ *derart, dass* $\sum b_k \alpha(\mu_k) < +\infty$ *und*

$$\sum a_n e^{(\rho + it)\lambda_k} \cdot \sum b_n e^{(\rho + it)\mu_k} = 1$$

ist für alle t *und* ρ, *das der Bedingung* (7) *genügt.*

Es gilt auch die inverse Behauptung.

Beweis. Beim Homomorphismus nach dem maximalen Ideal, das von der Funktion $e^{\lambda\rho + i\lambda t_0}$ beschrieben wird, geht das Element y unseres Ringes in die Zahl $\sum a_k e^{\lambda_k \rho + i\lambda_k t_0}$ über. Folglich entspricht dem Element y, über das in der Bedingung des Satzes die Rede ist, eine Funktion $y(M)$ derart, dass

$$|y(M)| \geqslant c > 0$$

ist. Da die Menge der maximalen Ideale der Form (9) in der Menge aller maximalen Ideale überall dicht ist, so gilt $|y(M)| \geqslant c$ auch für ein beliebiges maximales Ideal M [denn die Funktion $y(M)$ ist stetig]. Folglich gehört y keinem maximalen Ideal an, d. h. besitzt ein inverses Element.

§ 6. Betrachten wir nun einen Ring R, der die vorigen als Unterringe enthält. Als Elemente des Ringes R werden wir Funktionen $f(s)$ annehmen, die den folgenden Bedingungen genügen:

a) $f(s)$ ist für alle s definiert und ist auf jedem endlichen Intervall von beschränkter Schwankung,

b) $\int\limits_{-\infty}^{+\infty} \alpha(s) \, |d f(s)| < +\infty$, wo $\alpha(s)$ eine Funktion ist, die den folgenden Bedingungen genügt:

1°) $\alpha(s + t) \leqslant \alpha(s) \, \alpha(t)$,

2°) $\alpha(s) > 0$,

3°) $\alpha(0)=1$,

4°) $\alpha(s)$ ist stetig.

Die Multiplikation im Ring definieren wir folgendermassen: wir werden sagen, dass $f=f_1\times f_2$ ist, wenn

$$f(t)=\int_{-\infty}^{+\infty} f_1(t-s)\,d f_2(s)$$

ist. Setzen wir

$$\| f \|=\int_{-\infty}^{+\infty} |d f(t)|.$$

Es ist leicht ersichtlich, dass $\| f_1\times f_2\| \leqslant \| f_1\|\cdot\| f_2\|$ ist. Die Rolle der Einheit spielt in diesem Ringe die Funktion

$$e(t)=\begin{cases}1 & \text{für } t>0,\\ 0 & \ ,, \quad t<1.\end{cases}$$

Setzen wir

$$f(t)=h(t)+g(t)+s(t),$$

wo

$g(t)$ absolut stetig ist,

$h(t)$ eine Treppenfunktion ist,

$s(t)$ eine stetige Funktion mit beschränkter Schwankung ist, die fast überall eine verschwindende Ableitung besitzt.

Satz 7. *Ist $f(t)$ eine Funktion unseres Ringes, und ist*

$$\left|\int_{-\infty}^{+\infty} e^{i\lambda t+\rho t}\,d f(t)\right| \geqslant c>0$$

für alle λ und alle ρ, die den Ungleichungen

$$\lim_{\lambda\to-\infty}\frac{\log\alpha(\lambda)}{\lambda}\leqslant\rho\leqslant\lim_{\lambda\to+\infty}\frac{\log\alpha(\lambda)}{\lambda} \qquad (12)$$

genügen, und gilt schliesslich

$$\inf_{\rho,\lambda}\left|\int_{-\infty}^{+\infty} e^{\rho t+i t\lambda}\,dh(t)\right| > \int_{-\infty}^{+\infty}\alpha(t)\,|\,ds(t)|,$$

so existiert im Ring eine Funktion $f_1(t)$ derart, dass

$$f(t)\times f_1(t)=e(t)\ *$$

ist.

Beweis. Wir haben zu zeigen, dass das Element $f(t)$ im Ringe ein inverses besitzt. Betrachten wir den Homomorphismus unseres Ringes nach dem maximalen Ideal M: $R\to R/M$. Dabei können zwei Fälle vorkommen:

* Im Falle $\alpha(t)\equiv 1$ gehört dieser Satz N. Wiener und H. R. Pitt [3].

1) nicht alle absolut stetige Funktionen gehen in M ein,

2) alle absolut stetige Funktionen kommen in M vor.

Betrachten wir den ersten Fall. Es sei $g_1(t)$ eine absolut stetige Funktion, die in M nicht eingeht. Die Gesamtheit aller absolut stetigen Funktionen unseres Ringes ist auch ein Ring. Die Homomorphismen dieses Ringes auf den Körper der komplexen Zahlen wurden beim Beweis des Satzes 5 betrachtet. $g_1(t)$ geht bei diesem Homomorphismus $R \rightarrow R/M$ in $\int\limits_{-\infty}^{+\infty} g_1'(t)\, e^{\rho t + i\lambda t}\, dt \neq 0$ über, wo ρ und λ fixiert sind, wobei ρ den Ungleichungen (12) genügt. $g_2(t) = g_1(t) \times f(t)$ ist auch absolut stetig. Also geht diese Funktion in

$$\int\limits_{-\infty}^{+\infty} e^{\rho t + i\lambda t}\, dg_2(t)$$

über. Da das Fourier-Stieltjessche Integral der Funktion $g_2(t)$ gleich dem Produkt der Fourierschen Integrale der Funktionen $g_1(t)$ und $f(t)$ ist, so haben wir

$$\int\limits_{-\infty}^{+\infty} e^{\rho t + i\lambda t}\, dg_2(t) = \int\limits_{-\infty}^{+\infty} e^{\rho t + i\lambda t}\, dg_1(t) \cdot \int\limits_{-\infty}^{+\infty} e^{\rho t + i\lambda t}\, df(t).$$

Andererseits geht $g_2(t)$ in das Produkt derjenigen Zahlen über, in die $g_1(t)$ und $f(t)$ übergehen. Folglich geht $f(t)$ in

$$\int\limits_{-\infty}^{+\infty} e^{\rho t + i\lambda t}\, df(t)$$

über. Da nach der Bedingung des Satzes diese Zahl von Null verschieden ist, so gehört $f(t)$ nicht zu den maximalen Idealen des ersten Typs.

Betrachten wir nun die maximalen Ideale M des zweiten Typs. In jeden von ihnen gehen alle absolut stetige Funktionen ein, d. h. beim Homomorphismus nach einem solchen M geht $g(t)$ in Null über. Um das Bild der Funktion $f(t) = g(t) + h(t) + s(t)$ zu finden, genügt es, folglich, das Bild der Funktion $h(t) + s(t)$ zu finden.

Die Homomorphismen des Ringes der Funktionen $h(t)$ wurden beim Beweis des Satzes 6 betrachtet. Aus dem Beweis des Satzes 6 ist ersichtlich, dass wenn $\left| \int\limits_{-\infty}^{+\infty} e^{\rho t + i\lambda t}\, dh(t) \right| \geqslant a$ für alle den Ungleichungen (12) genügenden ρ und λ ist, so geht $h(t)$ beim Homomorphismus nach einem beliebigen maximalen Ideal in eine Zahl über, die dem absoluten Betrage nach grösser oder gleich a ist; da nach Voraussetzung

$$\left| \inf \int\limits_{-\infty}^{+\infty} e^{\rho t + i\lambda t}\, dh(t) \right| \geqslant \| s \|$$

ist, so geht also beim Homomorphismus $h(t)$ in eine Zahl $\geqslant \|s\|$ und $s(t)$ — in eine Zahl kleiner als $\|s\|$ über. Folglich geht $h(t)+s(t)$ nicht in Null über. Folglich geht auch $f(t)=g(t)+h(t)+s(t)$ nicht in Null über. Folglich gehört f zu keinem maximalen Ideal und besitzt also ein inverses Element.

Zusatz. Der Satz 4 kann auf beliebige Abelsche Gruppen verallgemeinert werden.

Satz 8. *Es sei* $\mathfrak{G}$ *eine Abelsche Gruppe; t, s,... — die Elemente der Gruppe* $\mathfrak{G}$*;* $\alpha(t)$ *— eine Funktion, die auf der Gruppe definiert ist und den Bedingungen*

1°) $\alpha(t+s) \leqslant \alpha(t)\,\alpha(s)$,

2°) $\alpha(t) > 0$,

3°) $\alpha(0) = 1$

genügt; $f(t)$ *— eine Funktion, die auf* $\mathfrak{G}$ *definiert ist, auf einer nicht mehr als abzählbaren Menge von Punkten von Null verschieden ist und derart, dass*

$$\sum_{t \in \mathfrak{G}} \alpha(t)\,|f(t)| < +\infty \tag{13}$$

und

$$\sum_{t \in \mathfrak{G}} f(t)\, e^{\rho(t)} \chi(t) \neq 0$$

für einen beliebigen Charakter $\chi(t)$ *der Gruppe* $\mathfrak{G}$ *und eine beliebige Funktion* $\rho(t)$*, die den Bedingungen*

$$\rho(t+s) = \rho(t) + \rho(s), \quad \rho(t) \leqslant \log \alpha(t)$$

genügt.

Dann existiert eine Funktion $g(t)$*, die von Null auf einer nicht mehr als abzählbaren Menge von Punkten verschieden ist, derart, dass*

a) $$\sum_{t \in \mathfrak{G}} \alpha(t)\,|g(t)| < +\infty$$

und

b) $$\sum_{s \in \mathfrak{G}} f(s)\, g(t-s) = \begin{cases} 0 & \text{für } t \neq 0, \\ 1 & \text{„} \quad t = 0 \end{cases}$$

ist.

Wenn wir für $\mathfrak{G}$ die Gruppe der ganzen Zahlen nehmen, so erhalten wir den Satz 4.

Der Satz 8 wird fast genau in derselben Weise, wie Satz 4, bewiesen.

Beachten wir, dass beim Beweis des Satzes 7 gleichzeitig die Existenz einer additiven Funktion $\rho(t) \leqslant \log \alpha(t)$ bewiesen wird.

Aus dem Satz 20, § 12, **N. R.** ergeben sich Sätze über analytische Funktionen von Elementen des Ringes, die die Sätze von Lévy [5] und Wiener-Pitt [3] verallgemeinern.

Literatur

1. N. Wiener, Tauberian Theorem, Annals of Math., **33**, (1932).

2. R. H. Cameron, Duke Math. Journal, **3**, (1937), 682—688.

3. N. Wiener and H. R. Pitt, ibid., **4**, N. 2, (1938), 420—436.

4. I. Gelfand, Abstrakte Funktionen und lineare Operatoren, I. Teil, § 3, Recueil math., **4** (**46**): **2**, (1938), 235—284.

5. P. Lévy, Sur la convergence absolue des séries de Fourier, Compositio Math., I, (1934).

Mathematisches Institut der
Akademie der Wissenschaften
der U. d. S. S. R.

(Поступило в редакцию 27/VI 1940 г.)

13.
(with M. A. Najmark)

On the embedding of normed rings into the ring of operators in Hilbert space

Mat. Sb., Nov. Ser. **12** (54) (1942) 197–213. Zbl. **60**:270

§ 1. Fundamental notions

This paper is devoted to the investigation of a class of normed rings. A set R is called n o r m e d r i n g (cf. [3]) if

(α) R is a linear normed complete space in the sense of Banach [1];

(β) an (in general non-commutative) operation of multiplication is defined in R with the ordinary properties (λ, μ are complex numbers)

$$x(\lambda y + \mu z) = \lambda xy + \mu xz, \quad x(yz) = (xy)z;$$

(γ) for any two elements $x, y \in R$

$$\|xy\| \leq \|x\| \cdot \|y\|;$$

(δ) R possesses a unit *i. e.* an element e such that $ex = xe = x$ for all $x \in R$; moreover $\|e\| = 1$.

A set I_l of elements $x \in R$ will be called l e f t i d e a l, if

1° $x \in I_l$, $y \in I_l$ imply $px + qy \in I_l$ for all $p, q \in R$;

2° $I_l \neq R$.

In a similar manner we define the r i g h t i d e a l s I_r. A set I will be called t w o - s i d e d i d e a l, when it is a left and a right ideal. A normed ring is called s i m p l e if it does not contain any two-sided ideals.

Repeating the argument of [3], p. 5, we can state the following facts:

I. *An element* $x \in R$ *possesses a left (right) inverse element, if and only if it does not belong to any left (right) ideal.*

II. *The closure of a left (right) ideal is a left (right) ideal.*

A left (right) ideal will be called m a x i m a l, when it is not a rpoper subset of a left (right) ideal. The proposition II implies:

III. *Every maximal left (right) ideal is closed.*

Further we have:

IV. *Every left (right) ideal is contained in a maximal left (right) ideal.*

Hence by I and IV:

V. *An element* $x \in R$ *possesses a left (right) inverse element, if and only if it does not belo ng any maximal left (right) ideal.*

A normed ring R will be called an $*$-ring if to every $x \in R$ there corresponds an element $x^* \in R$ satisfying the following conditions *:

1′. $(\lambda x + \mu y)^* = \bar{\lambda} x^* + \bar{\mu} y^*$;

2′. $x^{**} = x$;

3′. $(xy)^* = y^* x^*$;

4′. $\|x^* x\| = \|x^*\| \cdot \|x\|$;

5′. $\|x^*\| = \|x\|$ **;

6′. $x^* x + e$ possesses a two-sided inverse element for all $x \in R$.

From $xe = ex = e$ we obtain by 3′

$$e^* x^* = x^* e^* = e^*. \tag{1}$$

Since by 2′ x^* runs through the whole R when x does so (1) means that e^* is a unit in R. As R possesses only one unit, we have

$$e^* = e. \tag{2}$$

Every closed subring R of an $*$-ring is clearly also an $*$-ring, if $x \in R$ implies $x^* \in R$. As an example of an $*$-ring we point out the set B of all bounded operators in Hilbert spaces *** $\mathfrak{H}$ where x^* is the operator adjoint to x and $\|x\|$ denotes the norm of x. In fact, 1′—3′, 5′—6′ express the well known properties of adjoint operators in Hilbert space, so that it remains to prove 4′. By 5′ ($|\xi| = |\eta| = 1$; $\xi, \eta \in \mathfrak{H}$)

$$\|x^* x\| = \sup_{\xi, \eta} (x^* x \xi, \eta) | \geqslant \sup_{\xi} |(x^* x \xi, \xi)| = \sup_{\xi} \|x \xi\|^2 = \|x\|^2 = \|x\| \cdot \|x^*\|;$$

on the other hand, by (γ)

$$\|x^* x\| \leqslant \|x^*\| \cdot \|x\|.$$

B is thus an $*$-ring; hence every closed subring R of B is also an $*$-ring, if $x \in R$ implies $x^* \in R$.

The main purpose of this paper is the proof of the following.

Theorem 1. *Every normed $*$-ring can be isomorphically mapped onto a closed subring R_1 of the set B of all bounded operators in a Hilbert space $\mathfrak{H}$ in such a manner that, if $x \in R$ and $X \in R_1$ correspond to each other, then $\|x\| = \|X\|$ and x^*, X^* also correspond to each other by this mapping.*

§ 2. Some lemmas

In order to prove this theorem we establish some lemmas that are of independent interest.

* Here and below λ, μ are complex numbers and $\bar{\lambda}$, $\bar{\mu}$ the conjugate numbers.

** The authors suppose the last two axioms to be corollaries of 1′—4′, but they have not succeeded in proof of this fact. We also note that the axioms 4′, 5′ may be replaced by the axiom: $\|x^* x\| = \|x\|^2$. For (γ) § 1 implies $\|x\|^2 = \|x^* x\| \leqslant \|x^*\| \cdot \|x\|$, hence $\|x\| \leqslant \|x^*\|$. Replacing x by x^* gives $\|x^*\| \leqslant \|x\|$, hence $\|x^*\| = \|x\|$.

*** Hilbert spaces considered in this paper are not supposed to be separable.

L e m m a 1. *Let R be a normed commutative ring and let an operation be de defined in R putting in correspondence to every element $x \in R$ an element x^* under $\in R$ the following conditions:*

$1'$. $(\lambda x+\mu y)^* = \bar{\lambda} x^* + \bar{\mu} y^*$;

$2'$. $x^{**} = x$;

$3'$. $(xy)^* = x^* y^*$;

$4'$. $\|x^* x\| = \|x^*\| \cdot \|x\|$.

Then R can be isomorphically mapped onto the ring of all complex-valued continuous functions x (M) over a bicompact topological space $\mathfrak{M}$ in such a manner that, if $x \in R$ and x (M) correspond to each other, then

$$\|x\| = \max_{M \in \mathfrak{M}} |x(M)|$$

and the function corresponding to x^* is the conjugate function $\overline{x(M)}$

P r o o f. We first prove that in R

$$\|x^2\| = \|x\|^2.$$

By $2'-4'$

$$\|x^{*2}\| \cdot \|x^2\| = \|x^{*2} x^2\| = \|(x^* x)^* (x^* x)\| = \|(x^* x)^*\| \cdot \|x^* x\| = \|x^* x\|^2 = = \|x^*\|^2 \cdot \|x\|^2; \tag{1}$$

on the other hand, by (γ) § 1

$$\|x^{*2}\| \leqslant \|x^*\|^2, \quad \|x^2\| \leqslant \|x\|^2$$

hence (1) can take place if and only if

$$\|x^{*2}\| = \|x^*\|^2, \quad \|x^2\| = \|x\|^2.$$

In virtue of Theorem 8′ and Theorem 10 in [3] R is therefore isomorphic to a subring C_1 of the set $C(\mathfrak{M})$ of all complex-valued continuous functions $x(M)$ over the set $\mathfrak{M}$ of all maximal ideals M in R. Moreover

$$\|x\| = \max_{M \in \mathfrak{M}} \|x(M)\|$$

if x and $x(M)$ correspond to each other.

As G. Šilov [4] has proved, $\mathfrak{M}$ contains a unique minimal closed set $\mathfrak{M}_0$ on which every function $x(M) \in C_1$ attains the maximum of its modulus. Let M_0 be an element of $\mathfrak{M}_0$; we prove that $M_0^* = M_0$, where M^* denotes the set of all x^*, when x ranges through M. We first notice that the mapping $TM = M^*$ is a homeomorphism of $\mathfrak{M}$ into itself. In fact, every neighbourhood $U(M_1)$ is given by inequalities of the form

$$|x_k(M) - x_k(M_1)| < \varepsilon, \quad k = 1, 2, \dots, n, \quad \varepsilon > 0, \tag{2}$$

so that $U(M)$ is defined by $n, x_1, \dots, x_n$ and $\varepsilon > 0$. On the other hand, by the definition of $x(M)$

$$x = x(M)e + m, \quad m \in M,$$

hence

$$x^* = \overline{x(M)}e + m^*,$$

i. e.,

$$x^*(M^*) = \overline{x(M)}. \tag{3}$$

We can therefore write (2) in the form

$$|x_k^*(M^*) - x_k^*(M_1^*)| = |\overline{x_k(M)} - \overline{x_k(M_1)}| = |x_k(M) - x_k(M_1)| < \varepsilon$$

We see that T maps $U(M_1)$ onto a neighbourhood $V(M_1^*)$ defined by n, $x_1^*, \dots, x_n^*$, ε. Hence T is a homeomorphism.

Now suppose $M_0 \neq M_0^*$; there exist then two neighbourhoods $U(M_0)$, $U(M_0^*)$ without common element. By the continuity of T, there exists in $x(M_0)$ such a neighbourhood $V(M_0)$ that its T-map $(V(M_0))^*$ lies in $U(M_0^*)$. Thus $V(M_0)$ and $(V(M_0)^*$ have no element in common.

By the definition of $\mathfrak{M}_0$ there exists such a function $x_0(M) \in C_1$ that $|x_0(M)|$ does not attain its maximum on $\mathfrak{M}_0 - V(M_0)$. Hence it attains this maximum on $V(M_0)$. By (3) we have

$$|x^*(M)| = |\overline{x^*(M)}| = |x^{**}(M^*)| = |x(M^*)|. \tag{4}$$

Hence $|x_0^*(M)|$ does not attain its maximum on $T\mathfrak{M}_0 - (V(M_0))^*$, but it attains this maximum on $(V(M_0))^*$. On the other hand, it follows from (4) that $T\mathfrak{M}_0$ is also a minimal closed set, on which every function $x(M) \in C_1$ attains the maximum of its modulus. By the uniqueness of such a set we have $T\mathfrak{M}_0 = \mathfrak{M}_0$. Hence, by the construction of $x_0(M)$,

$$\max_{M \in \mathfrak{M}_0} |x_0^*(M) x_0(M)| = \max_{M \in \mathfrak{M}_0} (|x_0^*(M)| \cdot |x_0(M)|) < \max_{M \in \mathfrak{M}_0} |x_0^*(M)| \cdot \max_{M \in \mathfrak{M}} |x_0(M),$$

i. e.,

$$\|x_0^* x_0\| < \|x^*\| \cdot \|x_0\|.$$

This inequality contradicts 4′, and thus $M^* = M$ for all $M \in \mathfrak{M}_0$.

Now consider all $x(M) \in C_1$ for $M \in \mathfrak{M}_0$ only. By the definition of $\mathfrak{M}_0$ this contraction of the domain of $x(M)$ does not change $\max|x(M)|$, so that after this contraction R and C_1 remain isomorphic and $\|x\| = \max_{M \in \mathfrak{M}_0} |x(M)|$. Moreover by (3)

$$x^*(M) = \overline{x(M^*)} = \overline{x(M)} \quad \text{for } M \in \mathfrak{M}_0,$$

so that $x(M) \in C_1$ implies $\overline{x(M)} \in C_1$. By Theorem 16 in [3] C_1 is the set of all continuous functions over C_1; on the other hand, C_1 and R are continuously isomorphic, hence $\mathfrak{M}_0 = \mathfrak{M}$, $C_1 = C(\mathfrak{M})$, q. e. d.

The lemma just proved implies 5′, 6′ § 1. In fact,

$$\|x^*\| = \max_{M \in \mathfrak{M}} |\overline{x(M)}| = \max_{M \in \mathfrak{M}} |x(M)| = \|x\|.$$

Further $\frac{1}{|x(M)|^2 + 1}$ is a continuous function over $\mathfrak{M}$, consequently $(x^*x + e)^{-1}$ exists.

2. We shall now deduce some corollaries from Lemma 1.

Let R be an arbitrary normed $*$-ring; an element $h \in R$ will be called Hermitian, if $h^* = h$. Every element $x \in R$ can be uniquely represented in the form $x = h_1 + ih_2$, where $h_1, h_2 \in R$ are Hermitian. In fact, if $x = h_1 + ih_2$ is any such representation, then $x^* = h_1 - ih_2$, whence

$$h_1 = \frac{x + x^*}{2}, \quad h_2 = \frac{x - x^*}{2i}. \tag{5}$$

Such representation thus is unique. Conversely h_1, h_2 in (5) are Hermitian and $x = h_1 + ih_2$.

Now let $h \in R$ be an Hermitian element and R_1 the minimal closed subring of R containing h. This ring R_1 is evidently commutative, hence it satisfies the conditions of Lemma 1. Thus R_1 is isomorphic to the set $C(\mathfrak{M})$. Denote by $h(M)$ the function corresponding to h; then $h = h^*$ means that $h(M)$ is real. Hence for any non-real λ

$$\frac{1}{h(M) - \lambda}$$

is a function belonging to $C(M)$, *i. e.* the two-sided inverse $(h - \lambda e)^{-1}$ exists. If we call the spectrum of x the set of all λ's, for which the two-sided $(x - \lambda e)^{-1}$ does not exist, we have the following result:

Corollary 1. *The spectrum of every Hermitian element is real.*

Further we have

$$\left| \frac{1}{1 + ih(M)} \right| \leqslant 1,$$

whence

$$\max_{M \in \mathfrak{M}} \left| \frac{1}{1 + ih(M)} \right| \leqslant 1,$$

i. e.

$$|(e + ih)^{-1}| \leqslant 1. \tag{5}$$

An Hermitian element h will be called positive if its spectrum consists of non-negative numbers*. Evidentely in this case $h(M) \geqslant 0$, hence $\sqrt{h(M)}$ belongs to $C(\mathfrak{M})$ and is also $\geqslant 0$. Denote by g the element of R_1 corresponding to $\sqrt{h(M)}$; then g is positive and $g^2 = h$. Thus, we have

Corollary 2. *Every positive element h can be represented in the form $h = g^2$, where g is also positive.*

We denote g by $\sqrt{h}$ or by $h^{1/2}$.

Corollary 3. *If h_1 is positive and*** h_1^{-1} *exists, then* $(h_1 + ih_2)^{-1}$ *exists also for any Hermitian h_2.*

In fact,

$$h_1 + ih_2 = h_1^{1/2}(e + ih_1^{-1/2} h_2 h_1^{-1/2}) h_1^{1/2};$$

but, $h_1^{-1/2} h_2 h_1^{-1/2}$ being Hermitian, $(e + ih_1^{-1/2} h_2 h_1^{-1/2})^{-1}$ exists, hence $(h_1 + ih_2)^{-1}$ exists also***.

* The spectrum of h will then be said to be positive.

** Here and below x^{-1} always means the two-sided inverse element.

*** So far we have only used the axioms 1'—5' (p. 198).

If in 6′ § 1 we replace x by $\frac{1}{\sqrt{\varepsilon}}x$ $(\varepsilon > 0)$ we obtain that $\left(\frac{1}{\varepsilon}x^*x+x\right)^{-1}$ exists; hence $(x^*x+\varepsilon e)^{-1}=\frac{1}{\varepsilon}\left(\frac{1}{\varepsilon}x^*x+e\right)^{-1}$ exists also, *i. e.* x^*x *is positive for all* $x \in R$.

Corollary 4. *If* h_1, h_2 *are positive and* h_1^{-1} *exists, then the spectrum of* $h_1 h_2$ *is positive.*

Put $x = h_2^{1/2} h_1^{1/2}$; then

$$h_1^{1/2} h_2 h_1^{1/2} = x^* x$$

is positive, *i. e.* the spectrum of $h_1^{1/2} h_2 h_1^{1/2}$ is positive. On the other hand, the correspondence $y = h_1^{1/2} x h_1^{1/2}$ being a ring-automorphism, it does not change the spectrum. Hence, the spectrum of

$$h_1 h_2 = h_1^{1/2} (h_1^{1/2} h_2 h_1^{1/2}) h_1^{-1/2}$$

is also positive.

Corollary 5. *The sum of two positive elements is positive.*

Let h_1, h_2 be positive; we have to prove, that $h_1 + h_2$ is also positive, *i. e.* that $(h_1+h_2+\varepsilon h)^{-1}$ exists for all $\varepsilon > 0$. As $(h_1+\varepsilon e)^{-1}$ exists, we have

$$h_1+h_2+\varepsilon e=(h_1+\varepsilon e)\,[e+(h_1+\varepsilon e)^{-1}h_2].$$

By Corollary 4 the spectrum of $(h_1+\varepsilon e)^{-1}h_2$ is positive, hence $[e+(h_1+\varepsilon e)^{-1}h_2]^{-1}$ exists. Thus $(h_1+h_2+\varepsilon e)^{-1}$ exists also.

3. A set R will be called algebraic normed ring, if all the axioms (α)—(δ) § 1 are satisfied for R with the eventual exception of completeness of R. Hence an algebraic normed ring is a normed ring if and only if it is complete.

Lemma 2. *Let* R *be a commutative ring satisfying the conditions of Lemma* 1, *and let* R_1 *be an algebraic normed ring containing no generalized nilpotent elements**. *If* R *and* R_1 *are algebraically isomorphic, then* R_1 *is complete, and every algebraic isomorphism between* R *and* R_1 *is continuous.*

Proof. Let $\tilde{R}_1$ be the minimal complete ring containing R_1 and N a maximal ideal in $\tilde{R}_1$. This ideal determines a homomorphism of $\tilde{R}$ into the field K of all complex numbers, hence a homomorphism of R_1 into K. Let now R_1 and R be isomorphic. Then we have also a homomorphism of R into K, which determines a maximal ideal M in R. Thus to every maximal ideal N in $\tilde{R}_1$ there corresponds a maximal ideal M in R. Moreover if x and y are elements of R and R_1 which correspond to each other by the given isomorphism φ between R and R_1 [*i. e.* $x=\varphi(y)$], then by the definition of M

$$x(M)=y(N). \tag{6}$$

* An element $x \neq 0$ is called generalized nilpotent element if $\lim_{n\to\infty} \sqrt[n]{\|x^n\|}=0$ (cf. [3] § 6).

This correspondence between maximal ideals in $\tilde{R}$ and R, that we denote by $M=\psi(N)$, is one-to-one. In fact, if

$$\psi(N_1)=\psi(N_2)=M,$$

then N_1, N_2 determine the same homomorphism of R_1 into K, hence by continuity also the same homomorphism of $\tilde{R}_1$ into K. But this can be only if $N_1=N_2$.

Now prove that $M=\psi(N)$ is continuous, so that, $\mathfrak{M}$ being bicompact, this correspondence is a homeomorphism.

Any neighbourhood $U(M_0)$ of a maximal ideal $M_0 \in \mathfrak{M}$ can be represented as the set of all $M \in \mathfrak{M}$ satisfying the inequalities

$$|x_k(M)-x_k(M_0)|<\varepsilon, \quad k=1, 2, \ldots, n; \quad \varepsilon>0, \quad x_k \in R, \tag{7}$$

where $\varepsilon>0$, n, x_k, $k=1, \ldots, n$, are fixed for this $U(M_0)$. Let $M_0=\psi(N_0)$, $x_k=\varphi(y_k)$, $k=1, 2, \ldots, n$; denote by $V(N_0)$ the neighbourhood of N_0 determined by

$$|y_k(N)-y_k(N_0)|<\varepsilon, \quad k=1, 2, \ldots, n. \tag{8}$$

But in virtue of (6) for $M=\psi(N_0)$ we have

$$y_k(N)=x_k(M), \quad y_k(N)=x_k(M_0),$$

so that (8) implies (7), *i. e.* $M=\psi(N) \in U(M_0)$, as soon as $N \in V(N_0)$. This means that ψ is continuous.

Denote by $\mathfrak{M}_1$ the ψ-map of $\mathfrak{N}$; as a topological map of the bicompact $\mathfrak{N}$ it is also bicompact and therefore closed in $\mathfrak{M}$. Suppose $\mathfrak{M}_1 \neq \mathfrak{M}$; there exists then a function $x(M) \in C(\mathfrak{M})$, satisfying the conditions

$$x(M) \not\equiv 0, \quad x(M)=0 \quad \text{for all } M \in \mathfrak{M}_1. \tag{9}$$

Let x be the corresponding element of R; then (9) can be written in the form

$$x \neq 0, \quad x \in M \quad \text{for all } M \in \mathfrak{M}_1. \tag{10}$$

Denote by y the corresponding element in R_1, so that $x=\varphi(y)$. Then (10) imply

$$y \neq 0, \quad y \in N \quad \text{for all } y \in \mathfrak{N}.$$

By Theorem 8 in [3] this means that y is a generalized nilpotent element in R_1.

The contradiction so obtained shows that $\mathfrak{M}_1=\mathfrak{M}$, so that ψ is a homeomorphism of $\mathfrak{N}$ into $\mathfrak{M}$. Hence using (6) we can consider R as the ring $C(\mathfrak{N})$ of all continuous functions over $\mathfrak{N}$; by (6) we then have $R=R_1$. On the other hand, $R \supseteq \tilde{R}_1 \supseteq R_1$, hence $R_1=\tilde{R}_1$. As R_1 is complete, by Theorem 17 in [3] every isomorphism between R and R_1 is continuous.

Corollary 6. *Let R, R_1 be two normed $*$-rings* and let $y=\varphi(x)$ be an isomorphic mapping of R into R_1 satisfying the condition $y^*=\varphi(x^*)$. Then $\|y\|=\|x\|$, so that the φ-map of R is closed in R_1.*

* In this corollary condition 6' will not be used.

Proof. Let h be an Hermitian element in R and $g=\varphi(h)$ the corresponding element in R_1. Let further R' be the minimal closed subring of R containing h, and $R'_1=\varphi(R')$—the φ-map of R'. Evidentely R', R'_1 are commutative and R' satisfies all the conditions of Lemma 1. Using the same argument as on p. 199 we get: $\|y^2\|=\|y\|^2$ for all $y\in R'_1$. Thus R'_1 contains no generalized nilpotent elements. By Lemma 2 R'_1 is complete and φ is a continuous isomorphism between R' and R'_1. If $\mathfrak{M}$, $\mathfrak{N}$ are the sets of all maximal ideals in R', R'_1 respectively, then for $y=\varphi(x)$, $x\in R'$, $y\in R'_1$

$$\|y\|=\max_{N\in\mathfrak{N}}|y(N)|=\max_{M\in\mathfrak{M}}|x(M)|=\|x\|,$$

in particular,

$$\|h\|=\|g\|=\|\varphi(h)\|. \tag{11}$$

We see that (11) holds for any Hermitian h. Now let x be any element of R and $y=\varphi(x)$ the corresponding element of R_1. Then x^*x is Hermitian and $y^*y=\varphi(x^*x)$. Hence by (11)

$$\|x^*x\|=\|y^*y\|,$$

i. e., by 4′, 5′ § 1

$$\|x\|^2=\|y\|^2, \quad \|x\|=\|y\|.$$

3. Proof of Theorem 1

Let us now proceed to the proof of Theorem 1. First we construct the Hilbert space. Let M be a maximal left ideal in R; we divide R into equivalence-classes modulo M, the totality of which will be denoted by $\mathfrak{H}'$. These equivalence-classes will be later considered as elements of a Hilbert space. If a is an element of R, ξ—an equivalence-class from $\mathfrak{H}'$, then $x_1, x_2\in\xi$ imply $x_1-x_2\in M$, hence $a(x_1-x_2)\in M$, *i. e.* ax_1, ax_2 belong to the same equivalence-class modulo M. We denote this equivalence-class by $A\xi$ and write $A=\varphi(a)$. It is clear that A is a linear transformation in $\mathfrak{H}'$ and that the correspondence φ between a and A so obtained is a homomorphism.

We shall now define a scalar product in $\mathfrak{H}'$. To this purpose we first construct in R a functional $f(x)$ satisfying the following conditions:

1° $f(\lambda x)=\lambda f(x)$;
2° $f(x+y)=f(x)+f(y)$;
3° $f(x^*)=\overline{f(x)}$;
4° $f(x)\geqslant 0$ if x is Hermitian and positive;
5° $f(x)=1$;
6° $f(x)=0$ for $x\in M$.

We denote by H the set of all Hermitian elements in R and by P—the set of all elements h of H representable in the form $h=m_1+m^*_2$ m_1, $m_2\in M$, *i. e.* of those belonging to the linear sum $M+M^*$. All elements of P can also be written in the form $h=m+m^*$, $m\in M$.

In fact, $h=m_1+m_2^*$; m_1, $m_2\in M$ implies $h=(m_1+m_2)^*=m_2+m_1^*$, hence

$$h=\frac{m_1+m_2}{2}+\left(\frac{m_1+m_2}{2}\right)^*=m+m^*,$$

where $m=\frac{m_1+m_2}{2}$.

For no positive element $h\in P$ can h^{-1} exist. In fact, if

$$h=m+m^*$$

is a positive element of P and h^{-1} exists, put $m=h_1+ih_2$; h_1, $h_2\in H$. Then $h_1=\frac{1}{2}h$ hence h_1 is positive and h_1^{-1} exists. By corollary 3 this implies that m^{-1} exists also, which is impossible.

The same is true for the closure $\overline{P}$ of P. If $h\in\overline{P}$ is a positive element and h^{-1} exists, we choose an element g in P in such a way that

$$\|g-h\|<\|h^{-1}\|^{-1}$$

and write

$$g=h+(g-h)=h^{1/2}[e+h^{-1/2}(g-h)h^{-1/2}]h^{1/2}.$$

As

$$\|h^{-1/2}(g-h)h^{1/2}\|\leqslant\|h^{-1/2}\|^2\cdot\|g-h\|=\|h^{-1}\|\cdot\|g-h\|<1,$$

the element

$$e+h^{-1/2}(g-h)h^{-1/2}$$

is positive and $[e+h^{-1/2}(g-h)h^{-1/2}]^{-1}$ exists. Put

$$e+h^{-1/2}(g-h)h^{-1/2}=h_1^2,$$

where h_1 is Hermitian and positive. We obtain that

$$g\in P,\quad g=h^{1/2}h_1^2h^{1/2}=(h_1h^{1/2})^*(h_1h^{1/2})$$

is positive and

$$g^{-1}=h^{-1/2}h_1^{-2}h_1^{-1/2}$$

exists. The obtained contradiction proves our statement also for $\overline{P}$.

From this fact we easily deduce that $\overline{P}$ lies at the distance 1 from e. Otherwise we would have

$$h=e+x,\quad h\in\overline{P},\quad x\in H,\quad \|x\|<1,$$

whence follows that h is positive and h^{-1} exists, which is impossible. It follows by a well known theorem of Banach ([1], p. 57, Lemma) that there exists a linear functional defined on the set

$$E=\{\lambda e+x,\ x\in\overline{P},\ \lambda \text{ real}\}$$

such that

$$f(e)=x,\quad f(x)=0\qquad\text{for } x\in\overline{P}.$$

We prove that it satisfies also 4°. Evidently,

$$f(\lambda e+x)=\lambda\qquad\text{for } x\in\overline{P}.$$

Let us prove that $\lambda \geqslant 0$ if $\lambda e + x$ is positive. If λ were negative, $\lambda = -\varepsilon$, $\varepsilon > 0$ for a positive $h = \lambda e + x$, $x \in P$, then

$$x = h - \lambda e = h + \varepsilon e$$

should be a positive element and x^{-1} should exist. This contradiction proves our statement. Now consider the set $\mathfrak{K}$ of all positive elements of H. It possesses the following properties:

1^{**} $x \in \mathfrak{K}$, $\lambda \geqslant 0$ imply $\lambda x \in \mathfrak{K}$;

2^{**} $x \in \mathfrak{K}$, $x \neq 0$ imply $-\overline{x} \in \mathfrak{K}$;

3^{**} $x, y \in \mathfrak{K}$ implies $x + y \in \mathfrak{K}$ (Corollary 5).

According to the terminologie of M. Krein [5] $\mathfrak{K}$ is a cone. It contains the sphere $\| x - e \| < 1$, hence contains inner points. By a theorem due to M. Krein [5] $f(x)$ can be extended over the whole H in such a manner that it remains linear and non-negative on $\mathfrak{K}$.

For an arbitrary element $x = h_1 + ih_2$ of R we put

$$f(x) = f(h_1) + if(h_2).$$

Evidently $f(x)$ satisfies $1^* - 6^*$.

Now, for any $\xi, \eta \in \mathfrak{H}'$ we put

$$(\xi, \eta) = f(y^* x),$$

where $x \in \xi$, $y \in \eta$. The value of (ξ, η) does not depend on the choice of the elements $x \in \xi$, $y \in \eta$. In fact, if, *e. g.*, x_1 is any other element of ξ, then $x_1 - x \in M$, hence, $y^* x_1 - y^* x \in M$ and by 6^*

$$f(y^* x_1 - y^* x) = 0, \quad f(y^* x_1) = f(y^* x).$$

We now prove that (ξ, η) possesses all the properties of the scalar product. By $1^* - 3^*$

$$(\eta, \xi) = f(x^* y) = f((y^* x)^*) = \overline{f(y^* x)} = \overline{(\xi, \eta)}; \quad x \in \xi, \quad y \in \eta;$$

$$(\xi_1 + \xi_2, \eta) = f(y^*(x_1 + x_2)) = f(y^* x_1) + f(y^* x_2) = (\xi_1, \eta) + (\xi_2, \eta);$$

$$x_1 \in \xi_1, \quad x_2 \in \xi_2, \quad y \in \eta;$$

$$(\lambda \xi, \eta) = f(y^* \lambda x) = \lambda f(y^* x) = \lambda (\xi, \eta); \quad x \in \xi, \quad y \in \eta.$$

Further, $x^* x$ being positive, we have by 4^*

$$(\xi, \xi) = f(x^* x) \geqslant 0; \quad x \in \xi.$$

Thus it remains to prove that $(\xi, \xi) = 0$ implies $\xi = M$ (M plays here the rôle of the null-element). From the properties of (ξ, η) already established follows Schwarz inequality

$$|(\xi, \eta)|^2 \leqslant (\xi, \xi)(\eta, \eta).$$

Consequently $(\xi_0, \xi_0) = 0$ implies $(\xi, \eta) = 0$ for any $\eta \in \mathfrak{H}'$, *i. e.*, $f(y^* x_0) = 0$ for any $y \in R$. y being arbitrary, we may replace y by y^* that gives

$$f(yx^*) = 0 \quad \text{for all } y \in R, \ x_0 \in \xi_0. \tag{1}$$

Denote by I the set of all elements $x \in R$ satisfying the equality

$$g(yx) = 0 \quad \text{for all } y \in R.$$

By (1)
$$\xi_0 \subseteqq I. \tag{2}$$
We shall prove that I is a left ideal containing M. In fact,
$$x_1, x_2 \in I; \quad p, q \in R,$$
imply
$$f(y(px_1+qx_2)) = f((yp)x_1)+f((yq)x_2)=0$$
for all $y \in R$, *i. e.* $px_1+qx_2 \in I$. Further, $I \neq R$, because for $y=e$, $x=e$
$$f(yx)=f(x)=1.$$
Hence $x \in I$ and we see that I is a left ideal. If $x \in M$, then $yx \in M$ for all $y \in R$, and by 6*
$$f(yx)=0,$$
i. e. $x \in I$. Thus $M \subseteqq I$. M being maximal, this implies $M=I$, whence by (2) $\xi_0 = M$. Thus (ξ, η) possesses all the properties of the scalar product.

If now $\mathfrak{H}'$ is not complete with respect to (ξ, η) we extend it to the minimal complete space $\mathfrak{H}$ containing $\mathfrak{H}'$. Clearly $\mathfrak{H}$ is a Hilbert space. We have seen that to every $a \in R$ there corresponds a linear operator $A=\varphi(a)$ in $\mathfrak{H}'$. This operator is bounded with respect to the norm $|\xi| = (\xi, \xi)^{1/2}$. In order to prove this we first note that
$$\left\| \frac{a^*a}{\|a\|^2+\varepsilon} \right\| < 1,$$
whence
$$[(\|a\|^2+\varepsilon)e-a^*a]^{-1} = \frac{1}{\|a\|^2+\varepsilon}\left[e-\frac{a^*a}{\|a\|^2+\varepsilon}\right]^{-}$$
exists for any $\varepsilon > 0$. This means that $\|a\|^2 e - a^*a$ is positive. We put $\|a\|^2 e - a^*a = g^2$, where g is positive Hermitian. Then for any $x \in R$
$$x^*(\|a\|^2 e - a^*a)x = x^*g^2x = (gx)^*(gx)$$
is positive and by 4*
$$f(x^*(\|a\|^2 e - a^*a)x) \geqslant 0,$$
i. e.
$$f(x^*a^*ax) \leqslant \|a\|^2 f(x^*x).$$
If $x \in \xi$, the last inequality can be written in the form
$$(A\xi, A\xi) \leqslant \|a\|^2(\xi, \xi),$$
i. e. A is bounded in $\mathfrak{H}'$. It can therefore be uniquely extended to a bounded operator in $\mathfrak{H}$. This operator we also denote by A and write also $A=\varphi(a)$. Evidently $\|A\| \leqslant \|a\|$, where $\|A\|$ is the norm of the operator, and φ remains to be a homomorphism. If $B=\varphi(a^*)$, then for $\xi, \eta \in \mathfrak{H}'$, $x \in \xi$, $y \in \eta$
$$(A\xi, \eta) = f(y^*ax) = f((a^*y)^*x) = (\xi, B\eta).$$
Thus we have
$$(A\xi, \eta) = (\xi, B\eta). \tag{3}$$
In virtue of the continuity of the scalar product and by the definition of A, B in $\mathfrak{H}$ (3) is also valid for any $\xi, \eta \in \mathfrak{H}$, *i. e.*
$$B=A^*, \quad \varphi(a^*) = [\varphi(a)]^*.$$

If now φ is an isomorphism, then all the conditions of Corollary 6 are satisfied; we have only to take for R_1 the set of all bounded operators in $\mathfrak{H}$. By this corollary $\|a\|=\|A\|$ and our theorem is proved. If R is simple, then φ is an isomorphism. In fact, let I be the set of all $x\in R$ such that $\varphi(x)=0$. This set I is a two-sided ideal in R, because $x, y\in I$; $a, b\in R$ imply

$$\varphi(ax+by)=\varphi(a)\varphi(x)+\varphi(b)\varphi(y)=0,$$
$$\varphi(xa+yb)=\varphi(x)\varphi(a)+\varphi(y)\varphi(b)=0,$$

and $\varphi(e)=1$, where 1 is the operator defined by $1\cdot\xi=\xi$. Thus, R being simple, $I=(0)$, *i. e.* φ is an isomorphism and our theorem is proved in the case of a simple ring R.

Let now R be any $*$-ring. For every maximal ideal we construct as above the corresponding Hilbert space and denote it now by $\mathfrak{H}_M$. Then we take the direct sum

$$\mathfrak{H}=\sum_M \mathfrak{H}_M$$

of all these spaces, *i. e.*, the set of all complexes $\xi=\{\xi_M\}$, $\xi_M\in\mathfrak{H}_M$, with $\sum_M|\xi_M|^2<+\infty$* where the operations are defined by

$$\lambda\xi=\{\lambda\xi_M\},\qquad \xi+\eta=\{\xi_M+\eta_M\},\qquad (\xi,\eta)=\sum_M(\xi_M,\eta_M)$$

for

$$\xi=\{\xi_M\},\qquad \eta=\{\eta_M\}.$$

To every element $a\in R$ there corresponds a bounded operator A_M; in $\mathfrak{H}_M$. Write $A_M=\varphi_M(a)$. As we have shown

$$\|A_M\|\leqslant\|a\|. \tag{4}$$

Put

$$A\{\xi_M\}=\{A_M\xi_M\},$$

where $A_M=\varphi_M(a)$. A is obviously a linear operator. In virtue of (4)

$$|A\{\xi_M\}|^2=\sum_M|A_M\xi_M|^2\leqslant\|a\|^2\sum_M|\xi_M|^2=\|a\|^2|\{\xi_M\}|^2,$$

i. e. A is bounded. We write $A=\varphi(a)$; evidently $\varphi(a^*)=[\varphi(a)]^*$.

Suppose that φ is an isomorphism. Then by Corollary 6 $\|a\|=\|\varphi(a)\|=\|A\|$ and our theorem is proved. It remains to show that φ is really an isomorphism.

Consider the set I of all $a\in R$ such that $\varphi(a)=0$. It is sufficient to show that $I=(0)$. $a\in I$ is equivalent to $\varphi(a)=0$ or $A_M=\varphi_M(a)=0$

* Whence evidently follows that in every such complex only enumerably many ξ's do not vanish.

for all M. This means further: $A_M\xi_M = M$ for all $\xi_M \in \mathfrak{H}_M$ and all M, *i. e.* $ax \in M$ for all $x \in R$ and all M. In particular, for $x=e$ we get: $a \in M$ for all M, *i. e.* $a \in \Pi M$, where ΠM is the intersection of all M. But $\Pi M = 0$. In fact as an intersection of left ideals it is also a left ideal, hence $a \in \Pi M$ implies $a^*a \in \Pi M$. If $\lambda \neq 0$ belongs to the spectrum of a^*a, then $a^*a - \lambda e$ belongs to some of the M, say M_0. On the other hand, a^*a, as an element of ΠM belongs to M_0 too; hence $\lambda e \in M_0$, which is impossible. Thus we see that the spectrum of a^*a consists only of $\lambda = 0$, and we obtain $a^*a = 0$, $a = 0$ that completes the proof of the theorem.

It follows from the theorem just proved that all rings discussed by S. W. P. Steen [10] can be considered as rings of operators in Hilbert space. Consequently, the results of Steen follow immediately from the corresponding results of F. J. Murray and J. v. Neumann ([6], [7], [9]).

The residue ring of an $*$-ring is an $*$-ring itself. Hence follows that if R is a closed $*$-subring of the ring of operators in Hilbert space, then any its residue ring can be imbedded into the ring of operators in Hilbert space.

§ 4. Weakly closed rings

We have proved that every $*$-ring can be considered as a subring of the ring B of all bounded operators in a Hilbert space $\mathfrak{H}$. Let us now consider the case, when this subring is weakly closed (cf. [7]). If $\mathfrak{S}$ is a subset of B, denote by $\mathfrak{S}^p$ the set of all projections from $\mathfrak{S}$. We then have:

Lemma 3. *If I is a left ideal in R, then $I^p = (0)$ implies $I = (0)$.*

Proof. Let $A \in I$, $A \neq 0$. Since R is weakly closed, A can be represented in the form

$$A = UH,$$

where H is positive Hermitian, U—partially isometric and $U, H \in R$ (cf. [8], Theorem 7 and [6] I, Lemma 4.4.2). Moreover $U^*UH = H$, whence

$$H = U^*A \in I.$$

Since $H = 0$ implies $A = 0$, we must have $H \neq 0$. Let $E(\lambda)$ $-\infty < \lambda < +\infty$, be the resolution of the identity corresponding to H; H being positive, $E(\lambda) = 0$ for $\lambda < 0$. Moreover the equality $E(\varepsilon) = 1$ does not hold for all $\varepsilon > 0$, because otherwise we would have $H = 0$. Thus for some $\varepsilon > 0$

$$E(\varepsilon) \neq 1. \tag{1}$$

As $1 - E(\varepsilon) \in R$, we have also $H_\varepsilon = [1 - E(\varepsilon)]H \in I$. The operator $B = [1 - E(\varepsilon)]H + E(\varepsilon)$ is an element of R and possesses a bounded inverse. Hence

$$B^{-1}H_\varepsilon \in I. \tag{2}$$

On the other hand, it is obvious that $B^{-1}=H_\varepsilon^{-1}$ on $[1-E(\varepsilon)]\mathfrak{H}$ and $B^{-1}=1$ on $E(\varepsilon)\mathfrak{H}$. Hence $B^{-1}H_\varepsilon=1$ on $[1-E(\varepsilon)]\mathfrak{H}$ and $B^{-1}H_\varepsilon=0$ on $E(\varepsilon)\mathfrak{H}$, *i. e.* $B^{-1}H_\varepsilon=1-E(\varepsilon)$, so that by (1) and (2) $1-E(\varepsilon)\in I$, $1-E(\varepsilon)\neq 0$ contrary to the hypothesis: $I^p=(0)$.

Lemma 4. *Let R be a factor*, I a two-sided ideal in R and P an element of I^P. Then every projection Q equivalent to P (mod R) is also an element of I^P.*

Proof. By the definition of equivalence modulo R a partially isometric operator $U\in R$ exists, satisfying

$$U^*U=P, \quad UU^*=Q.$$

Hence, I being a two-sided ideal,

$$Q=Q^2=UU^*UU^*=UPU^*\in I.$$

Corollary 7. *If R is a factor in the separable Hilbert space and I is a two-sided ideal in P, then I does not contain infinite projections.*

Proof. Let P be an infinite projection from I; then P is equivalent to 1 (mod R) (cf. [6] I, Lemma 7. 2. 1), and by Lemma 4, $1\in I$ which is impossible.

Corollary 8. *Every factor of class* III *in the separable Hilbert space is a simple ring.*

Proof. Let I be a two-sided ideal in R. Since all projections in R different from zero are all infinite, Corollary 7 gives $I^P=(0)$, whence by Lemma 3, $I=(0)$.

Corollary 9. *Every factor R of finite class is simple.*

Proof. Let $D(P)$ be the relative dimension in R (cf. [6] I, Theorem VII) and $I\neq(0)$ a two-sided ideal in R. Then there exists an element $P_0\in I^p$, $P_0\neq 0$, and $D(P_0)>0$. Choose an integer n such that $\frac{1}{n}\leqslant D(P_0)$. In virtue of the well known properties of $D(P)$ a projection** $Q\leqslant P_0$ exists satisfying the equality $D(Q)=\frac{1}{n}$. Moreover

$$Q=QP_0\in I.$$

On the other hand, there exist n mutually orthogonal projections $Q_1,\dots,Q_n$ equivalent to Q such that $Q_1+Q_2+\dots+Q_n=1$. By Lemma 4 all Q_j as well as their sum 1 belong to I that contradicts to the definition of the ideal.

Lemma 5. *Every simple weakly closed ring is a factor.*

Proof. Denote by R' the set of all elements of B that commute with every element of R. If Z is the centrum of R, we have:

$$Z=R\cap R',$$

whence Z is also a weakly closed ring.

* A weakly closed ring is called a factor (cf. [6] I, p. 138). if its centrum consists only of elements $\lambda 1$.

** $Q\leqslant P$ means $Q\mathfrak{H}\subseteq P\mathfrak{H}$, *i. e.* $QP=PQ=Q$.

Let P be an element of Z^P. Then the set

$$I=\{AP,\ A\in R\}$$

is a two-sided ideal, provided that it does not coincide with R. In fact, if $A_1, A_2, B_1, B_2 \in R$ then

$$B_1(A_1P)+B_2(A_2P)=(B_1A_1+B_2A_2)P\in I,$$
$$(A_1P)B_1+(A_2P)B_2=(A_1B_1+A_2B_2)P\in I.$$

But, R being simple, we have either $I=(0)$, or $I=R$. In the first case $AP=0$ for all $A\in P$, hence $P=1\cdot P=0$. In the second case all elements of R have the form AP, in particular, $1=AP$ for some A. Then $1\cdot(1-P)=AP(1-P)=0$, $P=1$. We see that, Z^P consists only of 0 and 1, consequently $Z=\{\lambda 1\}$ (cf. [7] Theorem 2).

Theorem 2. *Every factor R of class I_∞ or II_∞ in the separable Hilbert space is not simple. It possesses only one* non-trivial *two-sided ideal closed in the uniform topology, which coincides with the smallest two-sided ideal I_0 containing all finite projections and closed in the uniform topology.*

Proof. Let I be the set of all such $A\in R$ that $\overline{A\mathfrak{H}}$ is finite*; I is a two-sided ideal. In fact**

$$\overline{(A+B)\mathfrak{H}}\subseteq\overline{A\mathfrak{H}}+\overline{B\mathfrak{H}},$$

hence $A\in I$, $B\in I$ imply $A+B\in I$. We have further

$$\overline{AB\mathfrak{H}}\subseteq\overline{A\mathfrak{H}}.$$

hence $A\in I$, $B\in R$ imply $AB\in I$. The equality $D(\overline{A\mathfrak{H}})-D(\overline{A^*\mathfrak{H}})$ (cf. [6] I, Lemma 6.2.1) shows that $A\in I$ implies $A^*\in I$. Consequently, if $B\in\mathfrak{B}$, $A\in I$, then

$$A^*\in I,\quad A^*B^*\in I,\quad BA=(A^*B^*)^*\in I.$$

Since, on the other hand, $1\bar{\in} I$ the set I is a two-sided ideal $\neq(0)$. We see that R is not simple.

Let now I_1 be an arbitrary non-trivial two-sided ideal closed in the uniform topology. According to Lemma 3 I_1^P contains elements different from zero; by Corollary 7 all these elements must be finite. Suppose that $P\in I_1^P$, $P\neq 0$ and Q is any projection from R satisfying the condition $D(Q)\leqslant D(P)$. Then***

$$Q\sim P'\leqslant P\quad(\text{mod } R).$$

Since $P'=PP'\in I_1$, Lemma 4 gives: $Q\in I_1$. Now let Q_0 be any finite projection from R; then Q_0 can be represented in the form

$$Q_0=P_1+P_2+\ldots+P_n+Q,$$

where P_i, Q, $i=1, 2, \ldots, n$, are mutually orthogonal, $P_i\sim P$ (mod R) and $D(Q)<D(P)$.

* $\overline{A\mathfrak{H}}$ denotes the closure of $A\mathfrak{H}$.

** $\mathfrak{M}+\mathfrak{N}$ denotes the set of all $\xi+\eta$, $\xi\in\mathfrak{M}$, $\eta\in\mathfrak{N}$.

*** $Q\sim P'$ (mod R) denotes that Q and P' are equivalent with respect to R.

By Lemma 4 $P_i \in I$; moreover, as we have just proved, $Q \in I$, hence $Q_0 \in I$. Thus, I_1 contains all finite projections and therefore

$$I_0 \subseteq I_1. \tag{3}$$

Further we notice that $A \in I_1$ implies $A^* \in I_1$. In fact, if $A \in I_1$, $A = UH$, where H is positive Hermitian and U—partially isometric, then $H = U^*A \in I_1$, whence $A^* = HU^* \in I_1$. Therefore if we write

$$A = H_1 + iH_2, \quad H_1 = \frac{A + A^*}{2}, \quad H_2 = \frac{A - A^*}{2i},$$

we see that $A \in I_1$ implies H_1, $H_2 \in I_1$. Let now $E_1(\lambda)$ be the resolution of the identity corresponding to H_1 Repeating the argument used on p. 209—210 we obtain that $E_1(\Delta) \in I_1$ for any closed interval Δ, which does not contain zero. By Corollary 7 every such $E_1(\Delta)$ is finite, hence $E_1(\Delta) \in I_0$. Since H_1 can be represented as the limit

$$H_1 = \lim \sum_{k=1}^{n} \lambda_k E_1(\Delta_k), \quad 0 \overline{\in} \Delta_k, \quad k = 1, \ldots, n$$

in the sense of the uniform topology, we have $H_1 \in I_0$. Analogously $H_2 \in I_1$, so that $A = H_1 + iH_2 \in I_0$. We have proved that $A \in I_1$ implies $A \in I_0$, *i. e.*, $I_1 \subseteq I_0$. Combining this result with (3) we finally get: $I_0 = I_1$, that completes the proof of the theorem.

R e m a r k. Consider the quotient-ring $\frac{R}{I_0}$ of equivalence-classes modulo I_0. It is evidently simple and is also an, *-ring. By Theorem 1 it can therefore be considered as a ring of operators in a Hilbert space. If, in particular, $R = B$, then the finiteness of a projection P means that $P\mathfrak{H}$ is finite-dimensional in the usual sense. Consequently I_0 coincides with the set of all completely-continuous operators, so that Theorems 1 and 2 contain, in particular, the results of J. W. Calkin [2].

Combining Corollaries 8, 9, Lemma 5 and Theorem 2 we obtain:

T h e o r e m 3. *The only simple weakly closed rings in the separable Hilbert space are the factors of the classes* I_n, II_1 *and* III.

Literature

1. S. B a n a c h, Théorie des opérations linéaires, Warszawa, 1932.

2. J. W. C a l k i n, A quotient ring over the ring of bounded operators in Hilbert space. I, Bull. Amer. Math. Soc., **46**, No 5, (May 1940), 400.

3. I. G e l f a n d, Normierte Ringe, Recueil Math. (Mat. Sbornik), **9** (**51**): **1**, (1941).

4. I. G e l f a n d, D. R a i k o v and G. Š i l o v, Normed ring, Uspekhi Mat. Nauk, vol. 11 (Russ.).

5. M. K r e i n, Sur les fonctionnelles positives additives dans les espaces linéaires normés, Communications de l'Institut d. Sc Math. et Méc. d. Kharkoff, série 4, t. XIV, (1937).

6. F. J. M u r r a y and J. v. N e u m a n n, On rings of operators. I, Annals of Math., **37**, (1936); II, Trans. Amer. Math. Soc., **41**, (1937).

7. J. v. N e u m a n n, Zur Algebra der Funktionaloperatoren, Math. Annalen, **102**, (1929).

8. J. v. Neumann, Über adjungierte Funktionaloperatoren, Annals of Math., 33, (1932).

9. J. v. Neumann, On rings of operators. III, Annals of Math., 41, (1940).

10. S. W. P. Steen, Introduction to the theory of operators. I, Proc. London Math. Soc. (2), 41, (1936), 361—392; II, ibid., 43, (1937), 529—543; III, ibid, 44, (1938), 398—411; IV, Proc. Cambridge Philos. Soc., 35, (1939), 562—578; V, ibid., 36, (1940), 139—149.

Поступило в редакцию 22/VIII 1941 г.

14.

(with D. A. Rajkov and G. E. Shilov)

Commutative normed rings

Usp. Mat. Nauk 1 (2) (1946) 48–146 [Transl., II. Ser., Am. Math. Soc. 5 (1957) 115–220]
Zbl. 201:457

Nomed rings, that is, linear normed spaces in which a continuous multiplication satisfying the usual algebraic restrictions is defined, are actually encountered in many mathematical problems. The present work is devoted to commutative normed rings. A theory of non-commutative nomed rings, the construction of which would doubtless have great value, has not been accomplished up to now.

This paper consists of a principal division, devoted to the general theory of commutative nomed rings, and four appendices.

Fundamental in the theory of commutative nomed rings is the concept of maximal ideals, that is, ideals contained in no other ideals but the whole ring. The basic role of this concept is related to the fact that it gives an invariant form to the definition of the spectrum of an operator. In this paper it is shown that the quotient ring by a maximal ideal is the field of complex numbers. From this it follows that each nomed ring can be mapped homomorphically into a ring of complex funcplex (defined on the set of maximal ideals). The kernel of this homomorphism consists of those elements x for which $\lim \sqrt[n]{\| x^n \|} = 0$ ("generalized nilpotent elements"). A topology can be introduced into the set of maximal ideals under which it becomes a compact Hausdorff space, and the functions which are the images of elements of the ring are continuous. The set of values of the function corresponding to an element x coincides with the spectrum of x, i.e., the set of numbers λ for which $(x-\lambda^e)^{-1}$ does not exist.

The "canonical" representation as a function ring established here for a ring without radical is often useful and interesting for rings presented as function rings with ordinary algebraic operations. The point is, roughly speaking, that the representation of the ring which we are considering here defines the functions on the largest set where they can be defined and topologizes this set in the way most naturally dictated by the functions of the ring.

Appendix I is devoted to rings, closed under unifom convergence, of real-valued continuous functions. In particular, compact extensions of topological spaces are discussed here.

Appendix II is devoted to the study of functions of bounded variation on the line, with multiplication defined by "convolution", and to some related rings. In it is explained the close relation between these rings and the theory of Fourier integrals. Among the partial results established are generalizations of known theorems of Wiener and his students and disciples on Fourier-Stieltjes transforms of assorted classes of functions of bounded variation.

Appendix III, developing some ideas noted in the preceding Appendix, gives an application of the theory of commutative rings to the theory of characters of locally compact groups and the construction of harmonic analysis on these groups.

Appendix IV studies the relation between maximal ideals in a given ring and those in rings which contain it. The fundamental result is the determination of a class of maximal ideals ("the boundary") which can be extended to every larger ring.

Appendices I, II, and IV are independent of each other. Appendix III does not depend on Appendices I and IV.

The text in small print can, if it be desired, be left unread, for the basic text is not formally based on it. In order not to interrupt the exposition, the papers cited have been collected in the bibliography at the end of this paper (pages 217 - 220) Numbers in square brackets refer to papers listed in this bibliography *.

Added in proof. This paper was written in 1940. The circumstances of war time greatly delayed its publication. Not wishing to postpone longer the appearance of the paper, the authors were obliged to refrain from discussing the greater part of the results published after this was written. At the end of this paper can be found a supplementary list of papers on normed rings which appeared in the years 1940 to 1945.

CONTENTS

General theory of commutative normed rings

§ 1. Definitions and examples of normed rings 117
§ 2. Maximal ideals 120
§ 3. Abstract analytic functions 126
§ 4. Functions on maximal ideals. Radical of a ring 128
§ 5. Topologization of the set of maximal ideals 133
§ 6. The ring of functions $x(M)$ 139
§ 7. Rings with involution 143
§ 8. Real rings 146
§ 9. Analytic functions of elements of a ring 151
§10. Decomposition of a ring into a direct sum of ideals 155
§11. Relations between algebraic and topological homomorphisms .. 157

Appendix I. Some topological problems in the theory of rings.

§12. A new method of topologizing the set of maximal ideals 160

* Editorial note: In the original text substantial parts were set in small type in order to indicate that they could be omitted without interfering with understanding the remainder. In order to avoid complications in vari-typing, this has not been followed.

§13. Compact extensions of completely regular spaces 162
§14. Another proof of the compactness of the set of maximal ideals in the weak topology 165
§15. Classification of all closed subrings of the ring $C(S)$ 166
Appendix II. Rings with convolution.
§16. The ring of absolutely integrable functions 168
§17. The ring of functions of bounded variation. 176
Appendix III. Group rings of commutative locally compact groups and the theory of characters.
§18. Group rings 193
§19. Relation between maximal ideals of a group ring and characters of the group. 198
§20. Uniqueness theorem for the Fourier transform and sufficiency of the set of characters 202
§21. Positive definite functions 205
Appendix IV. Extensions of maximal ideals.
§22. Generalized zero divisors 208
§23. Extension of maximal ideals of a symmetric ring 210
§24. Extension of maximal ideals in the general case 211

GENERAL THEORY OF COMMUTATIVE NORMED RINGS

§1. Definitions and examples of normed rings

Definition 1. A *normed ring* is a set R of elements $x, y, \cdots$, satisfying the following conditions:

a) *R is a linear normed complete space, with scalars from the complex number field.*

b) *In R is defined an associative multiplication, commuting with multiplication by complex numbers, distributive with respect to addition, and continuous in each variable.* Furthermore we shall assume hereafter that *multiplication is commutative* *.

c) *In R there exists an identity e with respect to multiplication.*

Remarks. Each ring which satisfies conditions a) and b) but is without an identity can be enlarged to a normed ring by formally supplying an identity e, that is, by constructing the ring of formal sums $\lambda e + x$, where λ runs over all complex numbers and x over all elements of the given ring; addition and multiplication are uniquely determined: $(\lambda e+x) + (\mu e+y) = (\lambda+\mu)e + (x+y)$, $\mu(\lambda e+x) = \mu\lambda e + \mu x$, $(\lambda e+x)(\mu e+y) = \lambda\mu e + (\mu x+\lambda y+xy)$; the norm may be defined, for example, by the

* In deriving facts independent of this assumption we shall not use it.

formula $\| \lambda e + x \| = |\lambda| + \| x \|$.

Let us display some examples of normed rings.

1. The ring C of all complex functions defined and continuous on the segment $[0, 1]$, with the norm $\| x(t) \| = \max_{0 \le t \le 1} |x(t)|$, and with ordinary multiplication.

2. The ring D_n of all complex functions defined and possessing a continuous nth derivative on the segment $[0, 1]$, with norm

$$\| x(t) \| = \sum_{k=0}^{n} \max_{0 \le t \le 1} |x^{(k)}(t)| \tag{1}$$

and with ordinary multiplication.

3. The ring W of all absolutely convergent trigonometric series $\sum_{n=-\infty}^{\infty} c_n e^{int}$, with norm $\| \sum_{n=-\infty}^{\infty} c_n e^{int} \| = \sum_{n=-\infty}^{\infty} |c_n|$, and with ordinary multiplication.

4. The ring $l^{(n)}$ of all polynomials $\sum_{k=0}^{n} a_k x^k$ of nth degree with complex coefficients, with norm $\| \sum_{k=0}^{n} a_k x^k \| = \sum_{k=0}^{n} |a_k|$ and multiplication

$$\sum_{k=0}^{n} a_k x^k \sum_{l=0}^{n} b_l x^l = \sum_{m=0}^{n} \left(\sum_{p=0}^{m} a_{m-p} b_p \right) x^m,$$

that is, ordinary multiplication except that powers of x beginning with the $(n+1)$ st are thrown away.

5. The ring $l^{(a)}$ of all absolutely continuous functions on the segment $[0,1]$, with norm $\| x(t) \| = \int_0^1 |x(t)| \, dt$ and multiplication

$$(x * y)(t) = \int_0^1 x(t - \tau) y(\tau) \, d\tau,$$

and formal adjunction of an identity.

6. The ring A of all functions regular in the circle $|\xi| < 1$ and continuous in $|\xi| \le 1$, with $\| x(\xi) \| = \max_{|\xi| \le 1} |x(\xi)|$ and with ordinary multiplication.

It is easily verified that in examples 1 and 3-6 the norm satisfies yet another condition:

$$\| xy \| \le \| x \| \, \| y \|. \tag{2}$$

Thus in example 5 we have

$$\begin{aligned} \| x * y \| &= \int_0^1 \left| \int_0^t x(t-\tau) y(\tau) \, d\tau \right| dt \le \int_0^1 \left(\int_0^t |x(t-\tau)| \, |y(\tau)| \, d\tau \right) dt = \\ &= \int_0^1 \left(\int_\tau^1 |x(t-\tau)| \, dt \right) |y(\tau)| \, d\tau = \int_0^1 \left(\int_0^{1-\tau} |x(t)| \, dt \right) |y(\tau)| \, d\tau \le \\ &\le \int_0^1 |x(t)| \, dt \int_0^1 |y(\tau)| \, d\tau = \| x \| \, \| y \|. \end{aligned}$$

However in the second example, with $n \ge 2$, relation (2), generally speaking, is not valid; thus for $x(t) \equiv t$ we have: $\| x(t) \| = 2$, $\| x^2(t) \| = 5 > \| x(t) \|^2$. But if in place of (1) we use in D_n the norm

$$\|x(t)\| = \sum_{k=0}^{n} \frac{\max\limits_{0\le t\le 1} |x^{(k)}(t)|}{k!}, \qquad (3)$$

then the inequality (2), as is not difficult to show, is also verified. But these norms (1) and (3) are *topologically equivalent*. The possibility of such renorming is a general property of normed rings.

Theorem 1. *For each normed ring R there can be found a topological and algebraic isomorphism of it with a ring R' satisfying the conditions*

b') $\|xy\| \le \|x\| \, \|y\|$ *and* $\|e\| = 1$.

Proof. To each element x of the ring R corresponds an operator A_x of multiplication by x: $A_x y = xy$. In view of property b) this operator is linear. The operators A_x generate, in the ring Q of all linear operators mapping the normed space R into itself, a subring R' with identity (the identity operator E corresponds to the identity e of the ring R).

Let us show that the ring R' is a normed ring under the norm $\|A_x\| = \sup\limits_{\|y\|\le 1} \|xy\|$. All that is needed for the proof is the completeness of R', that is, the closure of R' in Q.

In view of the associativity of multiplication we have $A_x(yz) = x(yz) = (xy)z = A_x y \cdot z$. It is not difficult to see that this property is characteristic for elements of R'. Indeed, if an operator A is such that for every y and z the equation $A(yz) = Ay \cdot z$ holds, then setting $Ae = x$ we have $Ay = A(ey) = Ae \cdot y = xy$; that is, A is the operation of multiplication by x.

Suppose now that the operators $A_n \in R'$ converge strongly to some operator A, that is, that $A_n x$ converges to Ax in the norm of the space R for each x in R. Then by continuity of multiplication we have $A(xy) = \lim A_n(xy) = \lim A_n x \cdot y = Ax \cdot y$ and, by what was proved above, we have also that $A \in R'$. Hence R' is closed in Q not only in the sense of uniform but also in the sense of strong convergence of operators.

It is evident that the rings R and R' are algebraically isomorphic. Let us show that the isomorphism is also topological. We have

$$\|A_x\| = \sup_{\|y\|\le 1} \|xy\| \ge \left\| x \frac{e}{\|e\|} \right\| = \left\| \frac{x}{e} \right\| \, *$$

or

$$\|x\| \le \|e\| \, \|A_x\|. \qquad (4)$$

Hence the mapping $A_x \to x$ of the space R' onto the space R is continuous. But, since both of these spaces are complete, by a well known theorem of Banach

* It is easily seen that $e \neq 0$ and that therefore $\|e\| > 0$.

[1, pages 40-41]* the inverse operation $x \to A_x$ is also continuous. With this, then, is proved the topological isomorphism of the rings R and R', and with it also the condition of the theorem, for R' satisfies the condition b′). At the same time we have showed that *each normed ring is topologically and algebraically isomorphic to a ring of operators in a linear normed space.*

Remark. If the condition b′) is satisfied by the ring R, then R and R' are *isometric*. Relation (4) in this case gives $\| x \| \leq \| A_x \|$. On the other hand, by (2) we have $\| A_x \| = \sup_{\|y\| \leq 1} \| xy \| \leq \| x \| \sup_{\|y\| \leq 1} \| y \| = \| x \|$. Combining both inequalities we have

$$\| A_x \| = \| x \|.$$

Corollary 1. *The product xy is continuous in both variables simultaneously.*

The proof follows from the inequality (2).

Definition 2. A series $x_1 + x_2 + \cdots + x_n + \cdots$ is called *absolutely convergent* if the series of norms

$$\| x_1 \| + \| x_2 \| + \cdots + \| x_n \| + \cdots$$

is convergent.

Corollary 2. *With absolutely convergent series of elements of a normed ring one can operate just as with absolutely convergent series of numbers.*

In what follows this point we shall assume that the norm satisfies the condition b′).

§ 2. Maximal ideals

Let us denote by O the set of all elements x for which there exists an inverse element x^{-1}. Let us show that *O is an open set.*

First of all, *each element x such that $\| e - x \| < 1$ has an inverse element x^{-1}*. Indeed consider the series

$$e + (e-x) + (e-x)^2 + \cdots .$$

Since $\| (e-x)^n \| \leq \| e-x \|^n$, this series is absolutely convergent, and therefore represents and element of R. Multiplication by $x = e - (e-x)$ and application of Corollary 2, § 1, yields

$$e + (e-x) + (e-x)^2 + \cdots - (e-x) - (e-x)^2 - \cdots = e.$$

Consequently, $y = e + (e-x) + (e-x)^2 + \cdots$ is indeed an inverse element, x^{-1}, whose existence we have proved.

Now suppose that x is an arbitrary element of O. Let us denote by $U_0(e)$ the neighborhood $\| e - z \| < 1$ of the identity element e, which we have proved is con-

* See also F. Hausdorff, *Mengenlehre*, de Gruyter, Berlin, 1935, Ergänzungen, pp. 283-284, or Lyusternik, Uspehi Mat. Nauk 1(1936), pp. 131-133.

tained in O. Since $xx^{-1} = e$, by continuity of multiplication there is a neighborhood $U(x)$ of the element x such that $U(x)x^{-1} \subset U_0(e)$. This means that zx^{-1} has an inverse element for each z in $U(x)$: $zx^{-1}(zx^{-1})^{-1} = e$. But then $x^{-1}(zx^{-1})^{-1}$ is an inverse element for z; that is, along with x the set O contains a neighborhood of it, $U(x)$.

Lemma. *x^{-1} is a continuous function of x on the set O.*

Proof. It is necessary for us to prove that if $x \in O$ and $x_n \to x$, then also $x_n^{-1} \to x^{-1}$. It is easily verified that

$$x^{-1} - x_n^{-1} = \{e - [e - x^{-1}(x - x_n)]^{-1}\} x^{-1}.$$

Hence

$$\| x^{-1} - x_n^{-1} \| \leq \| e - [e - x^{-1}(x - x_n)]^{-1} \| \, \| x^{-1} \| =$$

$$= \| x^{-1} \| \, \| \sum_{m=1}^{\infty} x^{-m}(x - x_n)^m \| \leq \sum_{m=1}^{\infty} \| x^{-1} \|^{m+1} \, \| x - x_n \|^m,$$

and this also proves that $x_n^{-1} \to x^{-1}$ when $x_n \to x$.

Definition 3. A set I of elements of a normed ring R will be called an *ideal* if it satisfies the following conditions:

a) If $x \in I$ and $y \in I$, then $x + y \in I$.

b) If $x \in I$, then $zx \in I$ for all z in R.

c) $I \neq R$.

An example of an ideal I in the ring C of example 1 of §1 is the set of all functions of C which vanish on the segment $[0, \frac{1}{2}]$.

No element having an inverse can belong to any ideal. Indeed, if $x \in I$, then under the existence of x^{-1} we would find for each element z in R that $z = (zx^{-1})x \in I$, that is, that I coincides with R, in contradiction with c) of Definition 3.

On the other hand, each element without an inverse belongs to some ideal, the set of all elements zx, where z runs over all of R.

Hence, *in order that an element of the ring R have an inverse it is necessary and sufficient that it belong to no ideal.* In particular, *if the ring R has no ideals except the null ideal (that is, the ideal containing only 0), then R is a field.*

It is not difficult to show that the closure $\bar{I}$ of an ideal I satisfies the conditions a) and b) of Definition 3. Since, in addition, each ideal I is contained in the closed set $R - O$, its closure $\bar{I}$ is contained in $R - O$, so is not R. Hence, *the closure of an ideal is again an ideal.*

Definition 4. A *maximal ideal* is an ideal contained in no other ideal.

Let us find all maximal ideals in the ring C of example 1, §1.

A maximal ideal in the ring C is the set of all functions in C which vanish *at any fixed point of the segment* $[0, 1]$.

The set M_r of all functions $x(t) \in C$ for which $x(r) = 0$ is an ideal in the ring C. Let $y(t)$ be some function of C not vanishing at r. We need to show that there is no ideal in C containing both $y(t)$ and M_r. But this follows by representing each function $z(t)$ of C in the form

$$z(t) = \frac{z(r)}{y(r)}\, y(t) + \left(z(t) - \frac{y(t)}{y(r)}\, z(r)\right)$$

where $z(t) - \frac{y(t)}{y(r)} z(r) \in M_r$, and the first term is a multiple of $y(t)$ *.

Suppose now that M is any maximal ideal of the ring C. Let us show that all functions belonging to this maximal ideal vanish at some fixed point of the segment $[0, 1]$. If this were not the case, then for each point r of $[0, 1]$ could be found a function $x_r(t) \in M$ such that $x_r(r) \neq 0$, and hence that $|x_r(t)| > \delta_r > 0$ in some interval about r. By the lemma of Borel-Lebesgue there exists a finite set of these intervals covering the whole segment $[0, 1]$. Let $r_1, \cdots, r_n$ be the corresponding points. The function

$$x(t) = x_{r_1}(t)\overline{x_{r_1}(t)} + \cdots + x_{r_n}(t)\overline{x_{r_n}(t)} = |x_{r_1}(t)|^2 + \cdots + |x_{r_n}(t)|^2$$

belongs to M. But, on the other hand, $x(t) > \min\limits_{1 \leq k \leq n} \delta_{r_k}^2 > 0$, which means that in C there is an inverse function $\frac{1}{x(t)}$, and in this case, as we have seen, $x(t)$ can not belong to any ideal, and in particular can not belong to the maximal ideal M. This contradiction shows that there exists a point r such that $x(r) = 0$ for all $x(t)$ in M. But then M, which is maximal in M_r, the set of all functions in C which vanish at r.

Just the same method shows that the set of all absolutely convergent series $\sum_{n=-\infty}^{\infty} c_n e^{int}$ the sum of which vanishes at a point r is a maximal ideal in the ring W of example 3, §1. However, if we wish to follow through the proof given above for the converse assertion, we will arrive at the point where from the fact that the sum of an absolutely convergent series $\sum_{n=-\infty}^{\infty} c_n e^{int}$ is different from 0 for every t we can derive the conclusion that $(\sum_{n=-\infty}^{\infty} c_n e^{int})^{-1}$ also belongs to the ring W, that is, is represented by an absolutely convergent trigonometric series. This assertion is valid and is the content of a theorem of Wiener [29, p. 14; 30, p. 91] **; however we shall prove it below, basing the proof on the fact that the maximal ideals in the ring W have the given form.

Remark. Let R be an arbitrary normed ring consisting of (not necessary all) continuous functions $x(t)$ defined on a compact set F with ordinary addition and multiplication. C is of this type; W is also, if the functions appearing in it are

* This yields a proof that in each ring of functions (with ordinary multiplication) the set of all functions which vanish at a fixed point is a maximal ideal.

** See also Zygmund, *Trigonometrical series*, p. 140.

defined not on the line but on the circumference of radius 1. Examining the reasoning used above, we see that the set of all functions in R vanishing at a point of F always forms a maximal ideal in the ring R^*; in order that the converse assertion be true it is sufficient that R have the following properties:

a) If a function $x(t)$ is in R, so is its complex conjugate function $\overline{x(t)}$.

b) If a function $x(t)$ which never vanishes is in R, so is its reciprocal function $\frac{1}{x(t)}$.

In order that this establish a one-to-one correspondence between maximal ideals of R and points of F it is still necessary that the ring R satisfy the following "separation condition".

c) For each pair of distinct points $t_1 \neq t_2$ of the set F there is in R a function $x(t)$ such that $x(t_1) \neq x(t_2)$.

Condition b) is not only sufficient (along with condition a)) but also necessary. Condition a) is not necessary, as we shall show in § 4 by determining the maximal ideals of the ring A of example 6, § 1.

It is evident that each maximal ideal is closed; if it were not, then its closure would contain it as a proper subset, which would mean it was not maximal.

Theorem 2. *Each ideal I is contained in a maximal ideal.*

The proof of this will proceed by transfinite induction. Let $x_1, x_2, \cdots, x_\omega, \cdots, x_\alpha, \cdots$ be a well ordering of the set R. To each non-maximal ideal I can be matched an ideal $I^+ \supset I$ in the following way: since I is non-maximal there exists x in R, $x \overline{\in} I$, satisfying the condition that the set of all elements $j + rx$, where $j \in I$, $r \in R$, is an ideal; let x_I be the first such element in the sequence $\{x_\alpha\}$; then set $I^+ = \{j + rx_I\}$. Now construct a transfinite sequence of ideal I_α in the following way: set $I_0 = I$ and suppose that I_α are constructed for every $\alpha < \beta$; if β is of the first class, that is, if $\beta - 1$ exists, then set $I_\beta = I^+_{\beta-1}$; if β is of the second class, then set $I_\beta = \Sigma_{\alpha<\beta} I_\alpha$. This sequence has cardinal number not exceeding that of the ring R, and therefore has a final form which, therefore, is a maximal ideal containing the ideal I.

In combination with the earlier remarks on the existence of inverse elements it is easy to derive from this

Theorem 3. *In order that an element of a commutative normed ring R have an inverse element it is necessary and sufficient that it belong to no maximal ideal. In particular, if the ring has no non-null maximal ideals, then it is a field.*

Elements x and $y \in R$ are called *equivalent with respect to the ideal I* if $x - y \in I$. Since this relation is reflexive, symmetric, and transitive, the ring R

* This is independent of compactness of F or continuity of $x(t)$ (see footnote * of the preceding page).

is decomposed into classes of equivalent elements. Define the sum (product) of two classes X, Y to be the class containing the sums (products) of elements x, y from X, Y, and denote by λX (λ a complex number) the class containing the elements λx, where $x \in X$; this gives the ring R/I of equivalence classes of R with respect to I. The zero element of this ring is the class containing all elements of I, and the identity E is the class containing the identity e of the ring R.

Introduce in R/I the norm

$$\|X\| = \inf_{x \in X} \|x\|. \tag{1}$$

Theorem 4. *If the ideal I is closed, then R/I is a normed ring.*

Proof.

1) $\|\lambda X\| = |\lambda| \, \|X\|$.

Obvious.

2) $\|X+Y\| \le \|X\| + \|Y\|$.

We have:

$$\|X+Y\| = \inf_{z \in X+Y} \|z\| = \inf_{x \in X, y \in Y} \|x+y\| \le \inf_{x \in X, y \in Y} \{\|x\| + \|y\|\} =$$

$$= \inf_{x \in X} \|x\| + \inf_{y \in Y} \|y\| = \|X\| + \|Y\|.$$

3) $\|XY\| \le \|X\| \, \|Y\|$.

We have:

$$\|XY\| = \inf_{z \in XY} \|z\| \le \inf_{x \in X, y \in Y} \|xy\| \le \inf_{x \in X, y \in Y} \|x\| \, \|y\| =$$

$$= \inf_{x \in X} \|x\| \inf_{y \in Y} \|y\| = \|X\| \, \|Y\|.$$

4) $\|E\| = 1$.

First of all, since $e \in E$, then $\|E\| \le 1$. Let y be an arbitrary element of E. We have $y = e + x$, where $x \in I$. If $\|y\|$ were less than 1, then, by the results of §2, x would have an inverse, and therefore could not be in any ideal I. Hence $\|E\| \ge 1$, and, therefore, $\|E\| = 1$.

5) If $\|X\| = 0$, then X is the null class.

In view of (1) there is a sequence $x_n \in X$ such that $x_n \to 0$ as $n \to \infty$. Let x be an arbitrary element of the class X. We have $x - x_n \in I$, and, since $x_n \to 0$, $x = \lim_{n\to\infty}(x - x_n) \in \bar{I}$. But by assumption I is closed, $I = \bar{I}$. Hence, X coincides with I, that is, the null class.

6) R/I is complete under the norm (1).

Suppose that $\{X_n\}$ is a fundamental sequence of classes, that is, that $\|X_n - X_m\| \to 0$ as $m, n \to \infty$. Then from it may be chosen a subsequence $\{X_{n_k}\}$ such that the series $\sum_k \|X_{n_{k+1}} - X_{n_k}\|$ converges. By (1) for an arbitrary element x_1 of X_{n_1} can be found an element x_2 of X_{n_2} such that $\|x_2-x_1\|<2\|X_{n_2}-X_{n_1}\|$; then for this x_2 can be found an $x_3 \in X_{n_3}$ such that $\| x_3-x_2 \| < 2 \| X_{n_3}-X_{n_2}\|$, and so on. It is clear that the sequence $\{x_n\}$ is fundamental and, consequently, converges to some element $x \in R$. But then the sequence $\{X_{n_k}\}$, and consequently the sequence $\{X_n\}$, converges to the class X containing x.

Remark 1. The homomorphic mapping of the ring R onto the quotient ring R/I by the closed ideal I, determined if an element $x \in R$ is carried to its class X, is an open continuous mapping*. In fact, let $U \subset R$ be an open sphere with center zero: $U = \{ \|x\| < \delta \}$, and let U' be the image of U in R/I. By definition of the norm in the quotient ring this image consists of those and only those cosets $X \in R/I$ for which $\|X\| < \delta$; consequently U' is an open set in R/I. In just this way it can be proved that the image of every open sphere in R is an open set in R/I. Since the open spheres in R form a defining system of neighborhoods, it follows that each open set in R has an open image in R/I. On the other hand, suppose that $F' \subset R/I$ is closed, that F is the complete inverse image of F', that $x_1 (\in X_1)$, $x_2(\in X_2), \cdots, x_n(\in X_n), \cdots$ is a fundamental sequence in F, that $x \in R$ is its limit, and that $x \in X$. Since $\| X-X_n \| \leq \| x-x_n \|$, then $X = \lim_{n\to\infty} x_n$, and, consequently, belongs to F'; but then $x \in F$, so F is closed.

Remark 2. Between the closed ideals $J \supset I$ in the ring R and the closed ideals J' in R/I there is a one-to-one correspondence in which each ideal J corresponds to its image J' in R/I.

Indeed, in view of the continuity of the mapping $R \to R/I$, the complete inverse image J of each closed ideal J' of the ring R/I is a closed ideal of the ring R which evidently contains I, and also $J_1' \neq J_2'$ implies $J_1 \neq J_2$. Conversely, the image J' of each ideal J containing I is an ideal of the ring R/I; under this mapping J is the complete inverse image of the ideal J' since J contains along with any x each element equivalent to x with respect to I. Since $R \to R/I$ is an open mapping, and since under open mappings from closure of a complete inverse image follows closure of the image, the image J' of each closed ideal J containing I is a closed ideal in R/I.

Theorem 5. *The quotient ring R/M of a commutative normed ring R by a maximal ideal M is a field.*

Proof. By Theorems 2 and 3, it suffices to show that the ring R/M contains no non-null ideals. But if there were in R/M such an ideal J, then its inverse image in R would be an ideal containing M as a proper subset, in contradiction with maximality of the ideal M.

* That is, the image of each open set in R is an open set in R/I (mapping is open) and the complete inverse image of each closed set in R/I is a closed set in R (mapping is continuous).

Let us note that, in view of Theorem 4, R/M is a normed ring, for, as we saw above, a maximal ideal is always closed.

It is not difficult to see that the converse of Theorem 5 is also true.

Theorem 6. *If the ring R/I of cosets by the ideal I is a field then I is a maximal ideal. Here it need not be assumed a priori that the ideal I is closed.*

Proof. If there were in the ring R an ideal J containing I as a proper subset, then its image in the ring R/I would be a non-null ideal, which is impossible since by assumption R/I is a field.

Let us consider the quotient ring of the ring C by a maximal ideal M. Since M consists of all functions in C which vanish at some point r (see pp. 121, 122) each coset X consists of all those functions $x(t) \in C$ having at this point some one value λ_X. Then $\lambda_{X+Y} = \lambda_X + \lambda_Y$, $\lambda_{XY} = \lambda_X \lambda_Y$, $\lambda_{\mu X} = \mu \lambda_X$. Also $\| X \| = | \lambda_X |$, since if $x(t) \in X$, then $\| x(t) \| \geq | x(r) | = | \lambda_X |$, but, on the other hand, the function $x(t) \equiv \lambda_X$ belongs to the class X. Hence C/M is isomorphic to the field of complex numbers.

Later we shall see that this property is shared by all commutative normed rings. To prove this we shall use methods from the theory of analytic functions.

§3. Abstract analytic functions

Definition 5. A function $x(\lambda)$ defined in some domain D of the plane of the complex variable λ and taking values in a normed ring R will be called *analytic* in D if for each $\lambda \in D$ it is strongly differentiable, that is, if the ratio

$$\frac{x(\lambda+h)-x(\lambda)}{h} \tag{1}$$

converges in the norm to some limit $x'(\lambda)$ as $|h| \to 0$.

Then, if the inverse element $x(\lambda) = (z - \lambda e)^{-1}$ exists for $\lambda = \lambda_0$ (and then by the lemma of §2 it exists for all λ sufficiently close to λ_0), it is an analytic function of λ in some neighborhood of λ_0. Indeed,

$$\frac{(z-(\lambda+h)e)^{-1} - (z-\lambda e)^{-1}}{h} = (z-(\lambda+h)e)^{-1}(z-\lambda e)^{-1}$$

and in view of the lemma of §2, the product on the right side converges to $(z-\lambda e)^{-2}$ when $|h| \to 0$.

If the function $x(\lambda)$ is analytic in D and f is an arbitrary linear functional defined on the space R, then $f\{x(\lambda)\}$ is an ordinary analytic function in D. Indeed, from the strong convergence of the ratio (1) follows also the convergence of the ratio

$$\frac{f\{x(\lambda+h)\} - f\{x(\lambda)\}}{h},$$

that is, the differentiability of the function $f\{x(\lambda)\}$.

The basic results of the theory of ordinary analytic functions can be extended to our abstract analytic functions, and the first of all are the theorem and formula of Cauchy. For these we need to define contour integration of abstract functions.

Let Γ be an oriented arc of a rectifiable curve in the plane of the complex variable λ and let $x(\lambda)$ be an abstract function defined and continuous in the norm on Γ. We define the integral $\int_\Gamma x(\lambda)\,d\lambda$ in the usual way:

$$\int_\Gamma x(\lambda)\,d\lambda = \lim_{\max|\lambda_{k+1}-\lambda_k|\to 0} \sum_{k=0}^{n-1} x(\lambda_k')(\lambda_{k+1}-\lambda_k),$$

where $\lambda_0, \lambda_1, \cdots, \lambda_{n-1}, \lambda_n$ are points in order along the curve, λ_k' is an arbitrary point between λ_k and λ_{k+1}, and the limit is taken in the sense of strong convergence. The existence and uniqueness of this limit follow from the rectifiability of the curve Γ and the uniform continuity of $x(\lambda)$ on Γ, and is proved in the ordinary way. From the definition of the integral it is clear that

$$f\{\int_\Gamma x(\lambda)\,d\lambda\} = \int_\Gamma f\{x(\lambda)\}\,d\lambda \tag{2}$$

for each linear functional f.

Theorem of Cauchy. *If $x(\lambda)$ is analytic in the closed region bounded by the simple rectifiable curve Γ, then*

$$\int_\Gamma x(\lambda)\,d\lambda = 0.$$

Proof. Set $\int_\Gamma x(\lambda)\,d\lambda = y$. By (2) for each linear functional f we have $f\{y\} = \int_\Gamma f\{x(\lambda)\}\,d\lambda$. Consequently, by the theorem of Cauchy for ordinary analytic functions, $f\{y\} = 0$ for each f. But then $y = 0$, because by the theorem of Hahn [1, p. 55] for each $y \neq 0$ there exists a linear functional f such that $f\{y\} \neq 0$.

In an analogous way is proved the

Integral Formula of Cauchy. *If the function $x(\lambda)$ is analytic in the closed region bounded by the simple closed curve Γ, then for every interior point of this region it can be represented in the form*

$$x(\lambda) = \frac{1}{2\pi i}\int_\Gamma \frac{x(\xi)\,d\xi}{\xi-\lambda}. \tag{3}$$

From the formula (3) follows in the usual way that *an abstract analytic function is indefinitely differentiable, and in a neighborhood of each point of regularity $\lambda = \lambda_0$ can be expanded in a strongly convergent Taylor's series*

$$x(\lambda) = x(\lambda_0) + x'(\lambda_0)(\lambda-\lambda_0) + \frac{x''(\lambda_0)}{2!}(\lambda-\lambda_0)^2 + \cdots,$$

for which the radius of convergence is the distance from λ_0 to the nearest singularity of $x(\lambda)$.

As an example (which we shall need later) let us determine the radius of the largest circle with center at $\lambda = 0$ inside which $(e-\lambda x)^{-1}$ exists. This function is differentiable, that is analytic, in all of the plane where it exists. Therefore its Taylor's series

$$\sum_{n=0}^{\infty} \lambda^n x^n \tag{4}$$

must converge absolutely in the interior of some circle. Conversely, the function $(e-\lambda x)^{-1}$ evidently exists inside the circle of absolute convergence of the series (4) and coincides with its sum there. But the circle of absolute convergence of the series (4) is

$$|\lambda| < \frac{1}{\lim\limits_{n\to\infty} \sqrt[n]{\|x^n\|}} \; *.$$

Hence the desired radius is $\dfrac{1}{\lim\limits_{n\to\infty} \sqrt[n]{\|x^n\|}}$.

In an analogous way it may be shown that the largest radius ρx of a circular neighborhood of a point $x \in O$ lying completely in O is $\dfrac{1}{\lim\limits_{n\to\infty} \sqrt[n]{\|x^{-n}\|}}$.

§4. Functions on maximal ideals. Radical of a ring.

By use of the results of the preceding section we can now describe finally the quotient ring of a commutative normed ring by a maximal ideal, first discussed in §2.

Theorem 7. *A normed field is always isomorphic to the field of complex numbers.*

Proof. We need to show that each element x of the normed field R can be represented in the form $x = \lambda e$, where λ is some complex number and e is the identity element. Suppose, on the contrary, that there is an element x such that $x - \lambda e$ is zero for no λ. Then the same will be true for x^{-1} and, since R is a field, both $(x-\lambda e)^{-1}$ and $(x^{-1}-\lambda e)^{-1}$ must exist for all λ. In this case, by §3, their Taylor's series $\sum_{n=0}^{\infty} x^{-n-1}\lambda^n$ and $\sum_{n=0}^{\infty} x^{n+1}\lambda^n$ converge absolutely in the whole plane, and, in particular, at $\lambda = 1$. But then it would follow simultaneously that $x^n \to 0$ and $x^{-n} \to 0$, which is impossible since $x^n x^{-n} = e$.

Remark. Theorem 7 can also be proved by means of the theorem of Liouville,

* From the inequality (2) of §1 it follows easily that $\lim\limits_{n\to\infty} \sqrt[n]{\|z^n\|}$ exists for each $z \in R$. Indeed, in view of that inequality, setting $\|z^n\| = a_n$ we have $a_n = a_{mk+l} \le a_k^m a_l$ $(0 \le l < k)$, whence $a_n^{\frac{1}{n}} \le a_k^{\frac{1}{k} - \frac{1}{kn}} a_l^{\frac{1}{n}}$. Fixing k and letting $n \to \infty$ we get $\overline{\lim}\, a_n^{\frac{1}{n}} \le a_k^{\frac{1}{k}}$. Now let $k \to \infty$ to get $\overline{\lim}\, a_n^{\frac{1}{n}} \le \underline{\lim}\, a_n^{\frac{1}{n}} = \lim a_n^{\frac{1}{n}}$. At the time we see that $\sqrt[k]{\|z^k\|} \ge \lim\limits_{n\to\infty} \sqrt[n]{\|z^n\|}$ for all k. Note that of the properties of the sequence of norms $\|z^n\| = a_n$ (≥ 0) we used only that $a_{m+n} \le a_n a_m$.

which for abstract analytic functions reads:

If an abstract analytic function is regular in the whole λ *plane and uniformly bounded in norm, then* $x(\lambda) \equiv x_0$ *where* x_0 *is some fixed element of the ring* R.

The proof of this theorem uses the same idea as that given in § 3 to prove the theorem and formula of Cauchy. By the theorem of Liouville in the ordinary case we have for each linear functional f that $f\{x(\lambda)\} \equiv$ const. But then $x(\lambda) \equiv$ const, for if $x(\lambda)$ takes two distinct values $x(\lambda_1)$ and $x(\lambda_2)$, then by the theorem of Hahn there exists a linear functional f such that $f\{x(\lambda_1)\} \neq f\{x(\lambda_2)\}$.

Now suppose that $(x-\lambda e)^{-1}$ exists for all λ. We have

$$\|(x-\lambda e)^{-1}\| = |\lambda^{-1}|\, \|(e-\lambda^{-1}x)^{-1}\|.$$

By the lemma of § 2 the second factor of the right side tends to 1 as $|\lambda| \to \infty$, so $\|(x-\lambda e)^{-1}\| \to 0$. From this it follows that the function $(x-\lambda e)^{-1}$ is bounded in the whole plane; then by the theorem of Liouville $(x-\lambda e)^{-1} \equiv$ const. Since $(x-\lambda e)^{-1} \to 0$ as $|\lambda| \to \infty$, we have that $(x-\lambda e)^{-1} \equiv 0$. But this is impossible.

From Theorems 5 and 7 it is easy to derive the following important theorem:

Theorem 8. *The quotient ring* R/M *of a commutative normed ring* R *by a maximal ideal* M *is isomorphic to the complex field.*

In this way the maximal ideal M determines a homomorphism of the ring R onto the field of complex numbers, under which all elements of any one of the classes making up R/M are carried to that complex number which corresponds to this class under the isomorphism between R/M and the complex field.

It can also be shown from Theorem 6 that, conversely, *every non-trivial algebraic homomorphism of the ring* R *into the field of complex numbers is determined be a maximal ideal.* Indeed, the kernel of the homomorphism is an ideal in R whose quotient ring is isomorphic to the field of complex numbers; by Theorem 6 this ideal (consisting of all elements of the ring which are carried to zero) is maximal.

Let us denote by $x(M)$ the number attached to the element $x \in R$ under the homomorphic mapping of R into the field of complex numbers determined by the maximal ideal M. For each fixed x we define for variable M a function $x(M)$ on the set $\mathfrak{M}$ of all maximal ideals of the ring R. This function evidently has the following properties:

a) If $x = x_1 + x_2$, then $x(M) = x_1(M) + x_2(M)$.

b) If $x = x_1 x_2$, then $x(M) = x_1(M)\, x_2(M)$.

c) If $x_2 = \lambda x_1$, then $x_2(M) = \lambda x_1(M)$.

d) $e(M) \equiv 1$.

e) If $x \in M_0$, then $x(M_0) = 0$, and, conversely, if $x(M_0) = 0$, then $x \in M_0$.

f) If $M_1 \neq M_2$, then there is an $x \in R$ such that $x(M_1) \neq x(M_2)$.

In addition to these

g) $|x(M)| \leq \|x\|$.

$x(M)$ is that number λ_X which under the isomorphism of R/M onto the field of complex numbers corresponds to the class X containing x. Since $X = \lambda_X E$, $\|X\| = |\lambda_X| \, \|E\| = |\lambda_X|$. Using the definition of norm in the quotient ring we have $|x(M)| = |\lambda_X| = \inf_{z \in X} \|z\| \leq \|x\|$.

By e) we can now formulate Theorem 3 in the following way: *In order that an element x of the ring R have an inverse it is necessary and sufficient that the function $x(M)$ vanish nowhere in $\mathfrak{M}$.*

As an illustration of the results just derived we determine the maximal ideals of the rings W (example 3 of § 1) and A (example 6 of § 1).

1. Let the element e^{it} of the ring W be carried by the homomorphism determined by the ideal M to a number a, so e^{-it} is carried to a^{-1}. By g) we have $|a| \leq \|e^{it}\| = 1$ and $|a^{-1}| \leq \|e^{-it}\| = 1$, so $a = e^{it_0}$ $(0 \leq t_0 \leq 2\pi)$. Hence e^{it} is carried into e^{it_0}, but then $\sum_{n=-\infty}^{\infty} c_n e^{int} \in W$ is carried to $\sum_{n=-\infty}^{\infty} c_n e^{int_0}$. Then M consists of all series carried to zero, that is, of all functions $\sum_{n=-\infty}^{\infty} c_n e^{int}$ vanishing at the point t_0.

On the basis of this result we can now prove the theorem of Wiener discussed in § 2 (page 122). Indeed suppose that the sum of the absolutely convergent series $\sum_{n=-\infty}^{\infty} c_n e^{int}$ never vanishes. By what has already been proved this means that as an element of W it belongs to no maximal ideal, but then, by Theorem 3, there is in W an inverse element, that is, $\dfrac{1}{\sum_{n=-\infty}^{\infty} c_n e^{int}}$ is also expandable in an absolutely convergent trigonometric series.

2. Just as in the discussion of the ring C (pp. 121, 122) it can be shown that the set of all functions of A which vanish at a point ζ_0 of the circle $|\zeta| \leq 1$ is a maximal ideal in A. Let us show that the converse is valid. Let M_0 be a maximal ideal of A and let ζ_0 be the number into which the function $x(\zeta) \equiv \zeta$ is carried by the homomorphism determined by this maximal ideal. The function ζ is a *generator* of the ring A; all functions in A are limits of uniformly convergent sequences of polynomials. From this it follows that for each element $x(\zeta) \in A$ we have $x(M_0) = x(\zeta_0)$ and that M_0 consists of those $x(\zeta) \in A$ which vanish at ζ_0. Notice that the ring A does not satisfy the condition a) of page 123.

Remark. It follows from our investigation of the maximal ideals in the rings C, W, and A that between the points t of the domain of definition of the functions $x(t)$ belonging to the ring and the maximal ideals of the ring there is a one-to-one correspondence under which $x(t) \equiv x(M)$. We will therefore identify, in these rings, the maximal ideals with the corresponding points.

Between the function $x(M)$ and the *spectrum* of the element x, that is, the set of all those λ for which $(x-\lambda e)^{-1}$ does not exist, we have a simple relation.

Theorem 9. *The set of values taken on by the function $x(M)$ coincides with the spectrum of the element x.*

Proof. If $x(M_0)=\lambda_0$, then $(x-\lambda_0 e)(M_0)=0$, so $x-\lambda_0 e \in M_0$, and therefore $(x-\lambda_0 e)^{-1}$ does not exist. Conversely, if $(x-\lambda_0 e)^{-1}$ does not exist, it follows that $(x-\lambda_0 e)(M)$ vanishes for some maximal ideal M_0, that is, $x(M_0)=\lambda_0$.

The ring R is mapped homomorphically on the ring of image functions $x(M)$. The kernel of this homomorphic mapping is the set of all elements $x \in R$ for which $x(M)=0$, that is, which belong to all the maximal ideals of the ring. Suppose x is such an element. By Theorem 9 $(x-\lambda e)^{-1}$ exists for all λ except $\lambda=0$, and, consequently, the function $(e-\lambda x)^{-1}$ is integral. But this means that the radius of convergence of the Taylor's series $\Sigma_{n=0}^{\infty} \lambda^n x^n$, which equals $\dfrac{1}{\lim\limits_{n\to\infty}\sqrt[n]{\|x^n\|}}$ (see page 128) is infinite, that is,

$$\lim_{n\to\infty}\sqrt[n]{\|x^n\|}=0. \tag{1}$$

Conversely, if the relation (1) is satisfied by an element x of the ring R, then the Taylor's series for $(e-\lambda x)^{-1}$ is absolutely convergent for all λ, which means that $(x-\lambda e)^{-1}$ exists for all λ except $\lambda=0$, so, by Theorem 9, $x(M)\equiv 0$.

Therefore, the kernel of the homomorphism of the ring R into the ring of functions $x(M)$ consists of those and only those elements $x \in R$ which satisfy the condition (1).

Definition 6. An element $x \in R$ which satisfies the condition (1) is called *generalized nilpotent* *, and the set of all generalized nilpotent elements is called the *radical* of the ring.

The result just derived can now be formulated in the following way:

Theorem 10. *The intersection of all maximal ideals coincides with the radical of the ring.*

A simple example of a ring with radical, i.e., a ring in which there are generalized nilpotent elements, different from zero, is the ring $I^{(n)}$ of example 4, §1. In it $x^{n+1}=0$ so every element without a constant term is nilpotent, and therefore generalized nilpotent. The unique maximal ideal of this ring is the set of all polynomials without constant term.

A non-trivial example of a ring with radical is the ring $I^{(a)}$ of example 5, §1.

* It is evident that ordinary nilpotent elements, i.e., elements x satisfying the condition $x^n=0$ for some n, are also generalized nilpotents. The converse, generally speaking, is false.

The ring $I^{(a)}$ has as a generator the function $x_0(t) \equiv t$ *. Indeed, $x_0^n = \frac{t^{n-1}}{(n-1)!}$, so x_0 generates all polynomials, and these are dense in the norm in the space of absolutely integrable functions on the interval $[0, 1]$. But $\|x_0^n\| = \frac{1}{(n-1)!}\int_0^1 t^{n-1}dt = \frac{1}{n!}$ and therefore $\sqrt[n]{\|x_0^n\|} \to 0$ as $n \to \infty$. Hence x_0 is a generalized nilpotent element. Since it is a generator, all elements of $I^{(a)}$ not involving the identity are generalized nilpotents. The unique maximal ideal of the ring $I^{(a)}$ is the set of all elements not involving the adjoined identity.

The rings considered earlier, C, W, and A, are examples of rings without generalized nilpotent elements different from zero. Indeed, for these rings we established a one-to-one correspondence between maximal ideals and points of the set on which the functions making up the ring are defined; then the value of an element x at a maximal ideal M coincides with the value of the function $x(t)$ at the corresponding element t. Hence in these rings the only element vanishing on every maximal ideal is the identically zero function.

Let us now consider the ring of hermitian-symmetric kernels $k(s, t)$, $k(s, t) = \overline{k(t, s)}$, defined and continuous on the square $0 \leq s,\ t \leq 1$, with norm $\|h(s, t)\| = \max_{0 \leq s,\ t \leq 1} |h(s, t)|$ and multiplication

$$g(s, t) * h(s, t) = \int_0^1 g(s, r)\, h(r, t)\, dr,$$

with an identity adjoined formally. The absence of a generalized nilpotent element is equivalent here to the theorem on existence for each hermitian-symmetric kernel $h(s, t)$ of a characteristic number. Indeed, if $g(s, t)$ is a generalized nilpotent element, so is the hermitian-symmetric kernel,

$$g(s, t) * \overline{g(t, s)} = \int_0^1 g(s, r)\, \overline{g(t, r)}\, dr,$$

which is equal to zero only if $g(s, t) \equiv 0$. But, as is well known, a characteristic number of a hermitian-symmetric kernel is a singular point of the Neumann series of the kernel, i.e., a singular point of the function $(e - \lambda h)^{-1}$. If h were a generalized nilpotent element, then this function would have no singularities, that is, h would have no characteristic numbers, in contradiction with well-known results in the theory of integral equations.

The relation (1) which holds for generalized nilpotent elements is a special case of the following more general relation:

Theorem 11. *For each* $x \in R$

$$\sup_{M \in \mathfrak{M}} |x(M)| = \lim_{n \to \infty} \sqrt[n]{\|x^n\|}, \tag{2}$$

where on the left side $\mathfrak{M}$ *denotes the set of all maximal ideals of the ring* R.

Proof. Let us set $\sup_{M \in \mathfrak{M}} |x(M)| = a$. Then by Theorem 9 the element $x - \mu e$

* The identity of the ring is not counted as a generator.

has an inverse for every $|\mu| > a$ and therefore the function $(e-\lambda x)^{-1}$, where $\lambda = \frac{1}{\mu}$, is analytic in the circle $|\lambda| < \frac{1}{a}$. Hence it follows that $\frac{1}{a}$ is not less than the radius of convergence $\frac{1}{\overline{\lim\limits_{n\to\infty}} \sqrt[n]{\|x^n\|}}$ of the Taylor's series of this function, or

$$\overline{\lim_{n\to\infty}} \sqrt[n]{\|x^n\|} \leq a. \tag{3}$$

On the other hand, since for all n, $\|x^n\| \geq \sup\limits_{M\in\mathfrak{M}} |x^n(M)| = a^n$, then $\sqrt[n]{\|x^n\|} \geq a$, and therefore

$$\underline{\lim_{n\to\infty}} \sqrt[n]{\|x^n\|} \geq a. \tag{4}$$

Combination of the inequalities (3) and (4) yields the relation (2). The existence of $\lim\limits_{n\to\infty} \sqrt[n]{\|x^n\|}$ is derived here as a side result.

As an illustration of Theorem 11 let us consider the ring W of example 3, §1. Earlier (p. 130) we found that there exists a one-to-one correspondence between the maximal ideals of this ring and the points of the interval $0 \leq t < 2\pi$, under which $x(M) = x(t)$ for all $x \in W$. Consequently $\sup\limits_{M\in\mathfrak{M}} |x(M)| = \max |x(t)|$. Therefore from application of Theorem 11 to this case follows that *the maximum modulus of the sum of an absolutely convergent trigonometrical series is equal to the limit of the nth root of the sum of the absolute values of the coefficients of the series associated with the nth powers of the given series.* Applied to the ring considered on page 132, Theorem 11 shows that *the modulus of the first characteristic number of a hermitian-symmetric kernel is equal to the limit of the inverse of the nth root of the maximum modulus of the nth iterated kernel.*

§5. Topologization of the set of maximal ideals

The desire to make all the functions $x(M)$ continuous leads to the following topologization of the set $\mathfrak{M}$ of maximal ideals of a ring R.

Definition 7. Suppose that $x_1, \cdots, x_n$ are arbitrary elements of the ring R and that ϵ is an arbitrary positive number. By the neighborhood $[M_0; x_1, \cdots, x_n; \epsilon]$ *of a minimal ideal* M_0 we shall mean the set of all maximal ideals M which satisfy the inequalities

$$|x_1(M_0) - x_1(M)| < \epsilon, \cdots, |x_n(M_0) - x_n(M)| < \epsilon.$$

With this definition of neighborhoods $\mathfrak{M}$ becomes a Hausforff topological space. Indeed, first of all, each maximal ideal evidently has neighborhoods and belongs to each of its neighborhoods. Furthermore, the intersection of the neighborhoods

$$[M; x_1, \cdots, x_n; \epsilon_1] \quad \text{and} \quad [M; x_{n+1}, \cdots, x_{n+m}; \epsilon_2]$$

of the maximal ideal M contains the neighborhood

$$[M; x_1, \cdots, x_n, x_{n+1}, \cdots, x_{n+m}; \min(\epsilon_1, \epsilon_2)].$$

Finally, if $M' \neq M$, then, by the "separation property" f) (p. 129) there is an element $x \in R$ such that $x(M') \neq x(M)$. Evidently the neighborhoods $[M; x; \frac{\epsilon}{2}]$ and $[M'; x; \frac{\epsilon}{2}]$ do not intersect if $\epsilon < |x(M') - x(M)|$.

Continuity of the functions $x(M)$ follows immediately from the definition of neighborhoods; in order to find a neighborhood of the maximal ideal M in which the values of a function x differ by less than ϵ from its value at M it suffices to stay in the set $[M; x; \epsilon]$ but this is a neighborhood.

It is useful to observe that for the rings C, W, and A the topology introduced here into the set $\mathfrak{M}$ coincides with the topology of the domain of the given functions belonging to the ring (that is, with the topology of the segment for C, circumference for W, and circle for A). For this it is necessary to show that in each new neighborhood is contained an old one and in each old neighborhood is contained a new. The first of these follows from the fact that the functions $x(t)$ belonging to the ring are continuous, so the sets $[t_0; x_1, \cdots, x_n; \epsilon]$ are open. For the proof of the converse assertion note that in each of the rings under consideration there is a function vanishing only at a given point t_0. Let U be an arbitrary neighborhood of the point t_0 and let $x(t)$ be a function with this property. If there were for each ϵ a point in $[t_0; x; \epsilon]$ lying outside U, then this would imply that the minimum of the function $x(t)$ on the complement of U equals zero. But since $x(t)$ is continuous and the domain of the function in any of the three rings is compact, $x(t)$ attains its minimum, that is, it would vanish at some point distinct from t_0, which is contrary to assumption. Hence for some ϵ the neighborhood $[t_0; x; \epsilon]$ is entirely contained in the old neighborhood U, as was required.

Theorem 12. *The space $\mathfrak{M}$ of maximal ideals of a commutative normed ring R when topologized in accordance with Definition 7 is compact.*

Proof. To each element x of the ring R corresponds the circle Q_x in the complex plane which has center 0 and radius $\|x\|$. Let Q be the topological product of all these circles, that is, the points of the space are functions $\{\lambda_x\}$ (x runs over all of R) such that $\lambda_x \in Q_x$, and the neighborhoods of a point $\{\lambda_x^{(0)}\}$ are determined by arbitrary finite sets of elements $x_1, \cdots, x_n \in R$ and arbitrary positive ϵ as the set of all those points $\{\lambda_x\}$ whose coordinates at the points $x_1, \cdots, x_n$ satisfy the inequalities

$$|\lambda_{x_1} - \lambda_{x_1}^{(0)}| < \epsilon, \cdots, |\lambda_{x_n} - \lambda_{x_n}^{(0)}| < \epsilon$$

(the remaining coordinates are completely arbitrary). Since every Q_x is compact, by the well-known theorem of Tychonov [27] Q is also compact.

In view of the inequality $|x(M)| \leq \|x\|$, to each maximal ideal M corresponds a point $\{\mu_x\} \in Q$, where $\mu_x = x(M)$. Moreover, by the "separation property" f) (p. 129) to two distinct maximal ideals M and M' correspond points $\{\mu_x\}$ and

$\{\mu'_x\}$ differing from each other in at least one coordinate, and therefore distinct. In this way $\mathfrak{M}$ is mapped one-to-one on a subset of the space Q. From the definitions of the topologies in $\mathfrak{M}$ and in Q it follows that this mapping is a homeomorphism. In order to show the compactness of $\mathfrak{M}$ it suffices to show that $\mathfrak{M}$ is closed in Q.

Let $\Lambda = \{\lambda_x\}$ be a limit point of the set $\mathfrak{M}$. We show that $\Lambda \in \mathfrak{M}$, that is, that there is a maximal ideal M_0 such that $\lambda_x = x(M_0)$ for all $x \in R$. For this we show that $\lambda_{x+y} = \lambda_x + \lambda_y$, $\mu\lambda_x = \lambda_{\mu x}$, $\lambda_{xy} = \lambda_x\lambda_y$. We restrict ourselves to a proof of the last of these; the proofs of the others follow the same pattern. Consider the neighborhood of the point Λ determined by the elements e, x, y, and xy, and an arbitrary positive number ϵ. Since Λ is a limit point of $\mathfrak{M}$, there is in this neighborhood a point $M \in \mathfrak{M}$, that is, for some M we have

$$|\lambda_e - e(M)| = |\lambda_e - 1| < \epsilon, \ |\lambda_x - x(M)| < \epsilon, \ |\lambda_y - y(M)| < \epsilon,$$

$$|\lambda_{xy} - (xy)(M)| = |\lambda_{xy} - x(M)\,y(M)| < \epsilon.$$

But then

$$\begin{aligned} |\lambda_{xy} - \lambda_x\lambda_y| &\le |\lambda_{xy} - x(M)y(M)| + |x(M)[y(M) - \lambda_y]| + |\lambda_y[x(M) - \lambda_x]| \le \\ &\le |\lambda_{xy} - x(M)y(M)| + \|x\|\,|y(M) - \lambda_y| + |\lambda_y|\,|x(M) - \lambda_x| < \\ &< \epsilon(1 + \|x\| + |\lambda_y|), \end{aligned}$$

and, since ϵ is arbitrary, also

$$\lambda_{xy} = \lambda_x\lambda_y, \ \lambda_e = 1.$$

Therefore the correspondence $x \to \lambda_x$ is a homomorphism of the ring R onto the field of complex numbers; the kernel of this mapping is a maximal ideal M_0 such that $x(M_0) = \lambda_x$ for all $x \in R$. With this the theorem is verified *.

Remark. We arrived at a definition of a topology in the space $\mathfrak{M}$ of maximal ideals starting from the desire to make all the functions $x(M)$ continuous on $\mathfrak{M}$. The following proposition shows that this condition uniquely determines the topology in $\mathfrak{M}$ in which $\mathfrak{M}$ is compact.

Theorem 12′. *Let $\mathfrak{M}'$ be the space of maximal ideals of a commutative normed ring topologized in any way satisfying* 1) *$\mathfrak{M}'$ is compact, and* 2) *the functions $x(M)$ are continuous on $\mathfrak{M}'$. Then $\mathfrak{M}'$ is homeomorphic to $\mathfrak{M}$, where $\mathfrak{M}$ is the space of maximal ideals topologized according to Definition* 7.

Proof. The mapping of $\mathfrak{M}'$ on $\mathfrak{M}$ which makes correspond to each maximal ideal $M \in \mathfrak{M}'$ the same ideal in $\mathfrak{M}$ is one-to-one. As a consequence of condition 2) the inverse image in $\mathfrak{M}'$ of each neighborhood, and hence of each open set, in

* In Appendix I (page 165) a proof of Theorem 12 will be given which does not require the theorem of Tyhonov on the compactness of the topological product of compact spaces.

$\mathfrak{M}$ is open in $\mathfrak{M}'$. Therefore, this mapping of $\mathfrak{M}'$ on $\mathfrak{M}$ is continuous. But, since $\mathfrak{M}'$ and $\mathfrak{M}$ are both compact (condition 1)), by a well-known theorem of topology the inverse mapping is also continuous, that is, $\mathfrak{M}'$ is homeomorphic to $\mathfrak{M}$.

These theorems contain as special cases the earlier results (page 134) that for the rings C, W, and A the topologies of the set of maximal ideals coincide with the original topologies in the domains of definition of the corresponding functions.

In connection with the results derived in §4 Theorem 12 gives:

Theorem 13. *Each commutative normed ring can be mapped homomorphically into some ring of continuous functions on a compact space, so that the kernel of the homomorphism is the radical of the ring. Hence if the radical of the ring contains only 0, this mapping is an isomorphism.*

Is there some new formulation of this result applied to rings of functions with ordinary algebraic operations? We answer this question in the following way: Let us find how the set of maximal ideals is determined naturally by the domain of the functions making up the ring.

First, it is clear that each point t_0 of the initial domain of definition of the functions determines a maximal ideal: The mapping $x(t) \to x(t_0)$ is a homomorphism of the ring onto the field of complex numbers; by its very definition $x(M_0)$, where M_0 is the maximal ideal determined by t_0, is equal to $x(t_0)$. But in order that two different points t_1 and t_2 determine two distinct maximal ideals it is necessary that these points be "separated" by the ring: that is, that in the ring there exists a function $x(t)$ such that $x(t_1) \neq x(t_2)$. If two points t_1 and t_2 are not separated by the ring, then under the correspondence all points t which correspond to the same maximal ideal are identified, that is carried to one maximal ideal. Thus we identify point differing by integral multiples of 2π under the mapping of points of the line to the corresponding maximal ideals of the ring W; in this way the line is carried to the circumference of radius 1, the natural domain of the functions of the ring W since all have period 2π.

But this is not the only change which can affect the initial domain of definition of the functions determining the ring under the passage to the maximal ideals; it may also be enlarged. Consider, for example, the ring C, and consider it as made up of functions defined only on the rational points of the segment $[0, 1]$. Since the rational numbers are everywhere dense in $[0, 1]$, the functions in the ring C, because they are continuous, are completely determined by their values on these points. Hence the set of maximal ideals of the ring just described can not be the set of rationals; the maximal ideals correspond to the points of the whole segment $[0, 1]$. Passing from the given function ring to that on the set of maximal ideals consists here of extending the domain of definition to its natural limit, in topological terms passing from the space of rational numbers (with the topology induced by the segment $[0, 1]$) to its *compact extension**.

Let us consider in this connection the ring B of continuous almost periodic functions of Bohr defined on the line with norm $\| x(t) \| = \sup_{-\infty<t<+\infty} | x(t) |$ and ordinary algebraic operations. The points of the line are separated by the elements of this ring and therefore none of them are identified under the correspondence with maximal ideals. However the set of maximal ideals is not exhausted by the points of the line; if this were not the case it would follow from the non-vanishing of an almost periodic function that its reciprocal would also be an almost periodic funcfion, but this is denied by the example of the function $2 - \sin x - \sin \lambda x$ with an irrational λ; it never vanished, but its lower bound is zero, so its reciprocal function is unbounded and, therefore, does not belong to the ring B.

In these last two examples the points of the initial domain of definition of the functions were *everywhere dense* in the compact space of maximal ideals and the construction of the maximal ideals had a character of a topological closure of the initial domain**. However, extensions of another kind are possible. Consider, for example, the ring A. Since the functions belonging to this ring are completely determined by their values on the boundary of the circle $|\zeta| \leq 1$, it is possible to consider the ring A as a ring of functions given a priori not on the whole circle but only on the circumference $|\zeta| = 1$. Moreover, if we introduce as a norm the maximum modulus on the circumference, we regain the old norm, because the maximum modulus of a function regular inside the circle $|\zeta| \leq 1$ is attained on the boundary of the circle. But the set of maximal ideals of the ring A is not exhausted by the set of points of the circumference $|\zeta| = 1$, and, still worse, these points do not make up an everywhere dense subset of the compact of maximal ideals, which coincides, as we know, with the circle. Here the passage from the a priori domain of definition of the functions of the ring to the space of maximal ideals has a character not of a topological but of an "analytical" extension of this domain.

To give the topology in the set of maximal ideals does not require the use of all elements of the ring.

Definition 8. A set K of elements of a ring R is called a *set of generators* of the ring if the smallest closed subring with identity which contains the set K is the whole ring R.

Remark. The identity is not counted as a generator..

* See Appendix I.

** In the second example the topology induced on the line $-\infty < t < \infty$ by the bicompact space $\mathfrak{M}$ is different from the ordinary topology of the line. See page 179.

Theorem 14. *The set $\{U\}$ of neighborhoods $[M; x_1, \cdots, x_n; \epsilon]$, where the x_i are chosen from a system K of generators of the ring R, is a determining system of neighborhoods in $\mathfrak{M}$.*

Proof. We need to show that in each neighborhood $[M; y_1, \cdots, y_n; \epsilon]$ there is a neighborhood from $\{U\}$. Let

$$P_1(x_{11}, x_{12}, \cdots, x_{1k_1}), \cdots, P_n(x_{n1}, x_{n2}, \cdots, x_{nk_n})$$

be polynomials in the elements $x_{ik} \in K$ $(1 \le i \le n,\ 1 \le k \le k_i)$ differing in norm from the corresponding elements $y_1, \cdots, y_n$ by less than $\frac{\epsilon}{3}$ and let δ be such that $[M; x_{11}, \cdots, x_{nk_n}; \delta]$ is contained in $[M; P_1, \cdots, P_n; \frac{\epsilon}{3}]$. We assert that the neighborhood $[M; x_{11}, \cdots, x_{nk_n}; \delta]$ (which belongs to the system $\{U\}$) is contained in $[M; y_1, \cdots, y_n; \epsilon]$. Indeed, for all M' in this neighborhood and all $i = 1, \cdots, n$ we have

$$|y_i(M') - y_i(M)| \le |y_i(M') - P_i\{x_{ik}(M')\}| + |P_i\{x_{ik}(M')\} - P_i\{x_{ik}(M)\}| +$$
$$+ |P_i\{x_{ik}(M)\} - y_i(M)| \le$$
$$\le 2\,\|P_i - y_i\| + |P_i\{x_{ik}(M')\} - P_i\{x_{ik}(M)\}| \le$$
$$\le 2\frac{\epsilon}{3} + \frac{\epsilon}{3} = \epsilon.$$

Remark. *If the ring R has a countable set of generators $x_1, x_2, \cdots, x_n, \cdots$, or, equivalently, if R is separable, then the second axiom of denumerability is satisfied in $\mathfrak{M}$* (and by a theorem of Urysohn [28] *$\mathfrak{M}$ is metrizable*). To see this, suppose that $r_1, r_2, \cdots, r_n, \cdots$ is the sequence of all complex rational numbers (that is, numbers of the form $r + is$, where r and s are real rational numbers). We assert that a determining system of neighborhoods in $\mathfrak{M}$ is the countable set of open sets of the form

$$|x_{n_k}(M) - r_{m_k}| < \frac{1}{p} \quad (k = 1, 2, \cdots, l;\ p \text{ an integer}). \qquad (*)$$

By Theorem 14 it suffices to prove that each set $[M_0; x_{n_1}, \cdots, x_{n_l}; \epsilon]$ contains such a set. But for this we need only take, in (*), $p > \frac{2}{\epsilon}$ and $|r_{m_k} - x_{n_k}(M_0)| < \frac{1}{p}$ $(k = 1, 2, \cdots, l)$.

Of special interest is the case in which the ring has a finite set of generators.

Theorem 15. *If the ring R has a finite set of generators $x_1, \cdots, x_n$, then $\mathfrak{M}$ is homeomorphic to a closed bounded subset of the n-dimensional complex space.*

Proof. The functions $x_1(M), \cdots, x_n(M)$ map $\mathfrak{M}$ continuously on some closed closed subset $\mathfrak{M}'$ of n-dimensional complex space. Let us show that this mapping is one-to-one; since $\mathfrak{M}$ and $\mathfrak{M}'$ are compact it will have a continuous inverse mapping; that is, that $\mathfrak{M}$ is homeomorphic to $\mathfrak{M}'$.

Suppose that two points M_1 and M_2 are mapped to the same point of $\mathfrak{M}'$. This means that $x_1(M_1) = x_1(M_2), \cdots, x_n(M_1) = x_n(M_2)$. Then for each polynomial P in the $x_1, \cdots, x_n$ we will have $P(M_1) = P(M_2)$, and, because $x_1, \cdots, x_n$ generate the ring, the equation $x(M_1) = x(M_2)$ will hold for all $x \in R$. But in this case by the separation property (property f), page 129) $M_1 = M_2$.

With this the theorem is proved. From it follows, in particular, that *if R is a ring with one generator, then $\mathfrak{M}$ is homeomorphic to some closed bounded subset of the complex plane.*

§6. The ring of functions $x(M)$

It is natural to raise the following question: How big is the set of functions $x(M)$? And, first of all, are there sufficiently many to approximate all continuous functions defined on $\mathfrak{M}$?

For the ring A this last question has a negative answer: As we have seen, the space of maximal ideals of this ring coincides with the circle $|\zeta| \leq 1$; and since A is closed under uniform convergence on this circle, a non-analytic function continuous on the circle $|\zeta| \leq 1$ can not be uniformly approximated by functions from A.

We show that the last question has a positive answer for an important class of normed rings.

Definition 9. We call an element $x^* \in R$ *conjugate* to the element x if $x^*(M) = \overline{x(M)}$ for all maximal ideals M^*. We call the element $x \in R$ *real* if $x(M)$ is real for every maximal ideal M. The ring R is called *symmetric* if it contains a conjugate element to each of its elements.

The ring A is not symmetric: $x(\zeta)$ and $\overline{x(\zeta)}$ are simultaneously analytic in the circle $|\zeta| < 1$ only if $x(\zeta) = \text{const}$.

We denote by R' the set of all functions $x(M)$, where $x \in R$.

Theorem 16. *If R is a symmetric ring, then each function $f(M)$ continuous on $\mathfrak{M}$ is a limit of a uniformly convergent sequence of functions $x(M) \in R'$.*

We note first that in view of the symmetry of the ring R each function $x(M)$ in R' can be decomposed in R' into real and imaginary parts:

$$x(M) = \frac{x(M) + x^*(M)}{2} + i\,\frac{x(M) - x^*(M)}{2i} = \Re x(M) + i\Im x(M).$$

Then if $x(M_1) \neq x(M_2)$ at least one of the analogous inequalities must hold for $\Re x(M)$ and $\Im x(M)$, so that distinct maximal ideals are already separated by the real functions R'. Also from nearness of $x(M)$ to $f(M)$ follows nearness of $\Re x(M)$ to $\Re f(M)$ and $\Im x(M)$ to $\Im f(M)$, and conversely. Hence Theorem 16 reduces to the following theorem:

Theorem 16′. *Suppose that on a compact space $\mathfrak{M}$ is given a set K of real,*

* It is evident that the difference of two elements both conjugate to the same element x is a generalized nilpotent element of the ring, and conversely, adding to x^* an arbitrary generalized nilpotent element yields another element conjugate to x.

continuous functions $x(M)$ containing all the constant functions, closed under the ordinary addition and multiplication operations, and so large that for each two points $M_1 \neq M_2$ there is a function $x(M) \in K$ such that $x(M_1) \neq x(M_2)$. Then each continuous real function defined on $\mathfrak{M}$ is the limit of a uniformly convergent sequence of functions from K.

Proof of Theorem 16′. Denote by $C(\mathfrak{M})$ the space of all continuous real functions on $\mathfrak{M}$ with the norm $\| f(M) \| = \max | f(M)|$ and by K^* the closure of the set K in $C(\mathfrak{M})$. We need to show that $K^* \equiv C(\mathfrak{M})$. We split the proof into a sequence of steps.

1. If $f(M) \in K^*$, then $| f(M) | \in K^*$.

Because $f(M)$ is a continuous function defined on a compact space, it is bounded; $| f(M) | \leq a$. We have

$$| f(M) | = \sqrt{a^2-[a^2-f^2(M)]} = a \sqrt{1-\left(1-\frac{f^2(M)}{a^2}\right)} =$$

$$= a \left\{ 1- \sum_{n=1}^{\infty} \frac{1\cdot 1\cdot 3 \cdots (2n-3)}{2\cdot 4\cdot 6 \cdots 2n} \left(1-\frac{f^2(M)}{a^2}\right)^n \right\},$$

where the series is uniformly convergent, because $0 \leq 1 - \frac{f^2(M)}{a^2} \leq 1$. Hence $|f(M)|$ is a limit of a uniformly convergent sequence of polynomials from K^*, that is, it is itself in K^*.

2. If $f(M), g(M), \cdots, h(M)$ belong to K^* so do
$$\max [f(M), g(M), \cdots, h(M)] \text{ and } \min [f(M), g(M), \cdots, h(M)]$$
belong to K^*.

Evidently it suffices to consider the case of two functions. But

$$\max [f(M), g(M)] = \frac{f(M)+g(M)+|f(M)-g(M)|}{2}$$

and

$$\min [f(M), g(M)] = \frac{f(M)+g(M)-|f(M)-g(M)|}{2}$$

so the validity of our assertion follows from the result of the preceding step.

3. For each pair of points $M_1 \neq M_2$ there exists in K^* a non-negative function not exceeding 1 which equal 1 at M_1 and equals 0 in some neighborhood of M_2.

By assumption there exists a function $x(M) \in K$ such that $x(M_1) \neq x(M_2)$. Set $y(M) = \frac{x(M)-x(M_2)}{x(M_1)-x(M_2)}$. Evidently $y(M_1) = 1$ and $y(M_2) = 0$. By continuity $y(M) < \epsilon < 1$ in some neighborhood of M_2. The function $\phi(M) = \frac{\max[y(M)-\epsilon, 0]}{1-\epsilon}$ equals 1 at M_1 and 0 in the same neighborhood of the point M_2 and is non-negative; hence $\psi(M) = \min[\phi(M), 1]$ satisfies all the requirements.

4. Suppose that F is a closed set in $\mathfrak{M}$ and that M_1 is a point not belonging

to it. Then in K^* there is a non-negative function not exceeding 1 which is equal to 1 at M_1 and 0 on F.

Indeed, to each point $M_2 \in F$ there corresponds a function, found in the preceding step, and a neighborhood in which the function vanishes. Since $\mathfrak{M}$ is compact, there exists a finite number n of these neighborhoods covering all of F. Let $\psi_1(M), \cdots, \psi_n(M)$ be the corresponding functions. Then the function $\psi(M) = \min[\psi_1(M), \cdots, \psi_n(M)]$ clearly satisfies our condition.

5. Suppose that F_1 and F_2 are two non-intersecting closed sets. Then there is in K^* a non-negative function not exceeding 1 which equals 1 on F_1 and 0 on F_2.

Indeed by the result of the preceding step to each point M_1 of F_1 there is some non-negative function $\phi(M)$ not exceeding 1 equal to 1 at this point and to 0 on F_2. By continuity $\phi(M) > 1 - \epsilon > 0$ in some neighborhood of the point M_1. Then the function $\psi(M) = \frac{\min[\phi(M), 1-\epsilon]}{1-\epsilon}$ will be 0 on F_2 and 1 in some neighborhood of M_1. Letting M_1 run over F_1 gives a set of neighborhoods covering the compact set F_1. Choose from these a finite covering $U_1, \cdots, U_n$ and let $\psi_1(M), \cdots, \psi_n(M)$ be the corresponding functions. Then the function $\psi(M) = \max[\psi_1(M), \cdots, \psi_n(M)]$ will satisfy all our requirements.

6. $K^* = C(\mathfrak{M})$.

Let $f(M)$ be an arbitrary real continuous function on $\mathfrak{M}$. Without loss of generality we can assume that $\min f(M) = 0$ and $\max f(M) = 1$. Let n be any natural number. Denote by P_k the set of those points M at which $f(M) \le \frac{k}{n}$ and by Q_k the set of points of M at which $f(M) \ge \frac{k+1}{n}$ $(k = 0, 1, \cdots, n-1)$. By continuity of the function $f(M)$ all of these sets are closed. Also $P_0 \subset P_1 \subset \cdots \subset P_{n-1}$, $Q_0 \supset Q_1 \supset \cdots \supset Q_{n-1}$, and P_k does not intersect Q_k. Let $\psi_k(M)$ be the function from K^* constructed for the sets Q_k and P_k as in the preceding step; that is, equal to 1 on Q_k and 0 on P_k and included between 0 and 1. We set $\psi(M) = \frac{1}{n}\sum_{k=0}^{n-1} \psi_k(M)$. We assert that $|f(M) - \psi(M)| \le \frac{1}{n}$. Indeed, suppose that at a point M the function $f(M)$ lies between $\frac{k}{n}$ and $\frac{k+1}{n}$; $\frac{k}{n} \le f(M) \le \frac{k+n}{n}$, so that M is contained in Q_{k-1} and P_{k+1}. Then $\psi_0(M) = \psi_1(M) = \cdots = \psi_{n-1}(M) = 1$, $\psi_{k+1}(M) = \psi_{k+2}(M) = \cdots = \psi_{n-1}(M) = 0$, and since $0 \le \psi_k(M) \le 1$, also $\frac{k}{n} \le \psi(M) \le \frac{k+1}{n}$; consequently $f(M)$ and $\psi(M)$ belong to the same interval of length $\frac{1}{n}$, which means that $|f(M) - \psi(M)| \le \frac{1}{n}$. Since n is arbitrary and K^* is closed, $f(M) \in K^*$. Therefore $K^* = C(\mathfrak{M})$, which proves Theorem 16′.

With this Theorem 16 is also proved.

As an illustration of Theorem 16′ consider on the topological product $S \times T$ of compact, completely regular spaces S and T the algebraic ring generated by the functions of the form $\phi(s)\psi(t)$, where $\phi(s)$ and $\psi(t)$ are real, continuous functions defined respectively on the spaces S and T. The elements of this ring are, evidently, defined and continuous on the compact space $S \times T$. Moreover, since S is completely regular, for each two points $s_1 \neq s_2$ there exists a continuous function $\phi(s)$ such that $\phi(s_1) \neq \phi(s_2)$, and the same is true for T. Therefore for each two points $(s_1, t_1) \neq (s_2, t_2)$ of the space $S \times T$ a function in K can be found which takes different values at these two points. Hence the set K satisfies all the conditions of Theorem 16′, and we conclude: Each real continuous kernel $k(s, t)$ given on $S \times T$ can be approximated uniformly by "degenerate" kernels of the form $\Sigma_{k=1}^{n} \phi_k(s)\psi_k(t)$, where $\phi_k(s)$ and $\psi_k(t)$ are real continuous functions defined respectively on S and T.

Remark. The condition that the set K of Theorem 16′ contain the products of its elements was used only in step 1 of the proof of the theorem. Hence we have actually proved the following proposition:

Suppose that the set K of continuous functions $x(M)$ defined on the compact set $\mathfrak{M}$ 1) contains the function $x(M) \equiv 1$, 2) along with two functions contains also their sum, 3) along with $x(M)$ contains also $\lambda x(M)$, where λ is an arbitrary complex number, 4) along with $x(M)$ contains $|x(M)|$, and 5) for each two points $M_1 \neq M_2$ contains a function $x(M)$ such that $x(M_1) \neq x(M_2)$. Then each real continuous function defined on $\mathfrak{M}$ is the limit of a uniformly convergent sequence of functions from K.

From Theorem 16 follows immediately

Theorem 17. *If from uniform convergence of functions $x(M)$ determined by elements of a symmetric ring R follows the convergence in the norm of R of the corresponding elements of R, then R is isomorphic to the ring of all continuous complex functions on $\mathfrak{M}$ with norm $\|f(M)\| = \max|f(M)|$.*

Corollary. *If for each element x of a symmetric ring R one has $\|x^2\| = \|x\|^2$, then R is isomorphic to the ring of all continuous complex functions on $\mathfrak{M}$.*

Indeed, from this equation and Theorem 11 it follows that

$$\max|x(M)| = \lim_{n\to\infty} \sqrt[2^n]{\|x^{2^n}\|} = \lim_{n\to\infty} \sqrt[2^n]{\|x\|^{2^n}} = \lim_{n\to\infty} \|x\| = \|x\|,$$

and from this it follows that the uniform convergence of functions $x(M)$ is equivalent to convergence of the corresponding elements x in norm.

Now suppose that S is an arbitrary topological space and that R is some normed ring of continuous bounded functions on it, with ordinary algebraic operations and with norm $\|x\| = \sup_{t \in S}|x(t)|$, and assume also that R contains with each function $x(t)$ its complex conjugate function $\overline{x(t)}$. Obviously here $\|x^2\| = \|x\|^2$. Therefore R is isomorphic to the ring of all continuous functions of the compact space $\mathfrak{M}$ of all maximal ideals of R. The relation between the spaces $\mathfrak{M}$ and S will be discussed in Appendix I.

§7. Rings with involution

Definition 10. By an *involution* we shall mean an operation attaching to each element x of a commutative normed ring R a unique element x' such that a) $(x')' = x$, b) $(\lambda x + \mu y)' = \bar{\lambda} x' + \bar{\mu} y'$, c) $(xy)' = x'y'$, and d) $(e + xx')^{-1}$ exists for each $x \in R$. A ring R in which such an involution is defined will be called a *ring with involution.* A linear functional defined on a ring with involution R will be called *positive* if $f(xx') \geq 0$ for all $x \in R$. We shall say that R possesses *sufficiently many positive linear functionals* if for each element $x_0 \in R$, different from 0, there exists a positive linear functional f_0 such that $f_0(x_0) \neq 0$.

Each ring with involution is symmetric, because the element x' *is conjugate (in the sense of Definition* 9) *to* x: $x' = x^*$.

Let us show first that *if* $x = x'$, *then the function* $x(M)$ *is real.* By Theorem 9 for this it suffices to show that $(x - (\lambda + i\mu)e)^{-1}$ (where λ and μ are real) exists for all $\mu \neq 0$. But $(x - \lambda e)' = x' - \bar{\lambda} e' = x - \lambda e$, because use in c) of Definition 10 of $x = e$, $y = e'$ along with condition a) shows that $e = e'$. Therefore

$$(x - (\lambda + i\mu)e)(x - (\lambda - i\mu)e) = (x - \lambda e)^2 + \mu^2 e = (x - \lambda e)(x - \lambda e)' + \mu^2 e,$$

and since by condition d) the right side has an inverse, when $\mu \neq 0$

$$(x - (\lambda + i\mu)e)^{-1} = (x - (\lambda - i\mu)e)\,((x - \lambda e)^2 + \mu^2 e)^{-1}$$

also exists.

Suppose now that x is an arbitrary element of the ring. Set $y = x + iAe$ where $A > \max |x(M)|$. We have

$$\Re y(M) = \Re x(M), \quad \Im y(M) = \Im x(M) + A \neq 0,$$
$$\Re y'(M) = \Re x'(M), \quad \Im y'(M) = \Im x'(M) - A.$$

Because $(y + y')' = y' + y$, by what was just proved, the sum $y(M) + y'(M)$ is real, that is, $\Im y'(M) = -\Im y(M)$, which means also that $\Im x'(M) = -\Im x(M)$. Furthermore, because $(yy')' = yy'$, the product $y(M)y'(M)$ is real, that is $\Im y(M)[\Re y'(M) - \Re y(M)] = 0$; since $\Im y(M) \neq 0$, it follows from this that $\Re y'(M) = \Re y(M)$, that is, also $\Re x'(M) = \Re x(M)$. This shows that $x'(M) = x(M)$.

On the basis of this result we shall hereafter denote the involution by the symbol * of the conjugate.

Let Q' be any commutating set of bounded hermitian linear operators defined in Hilbert space H, and containing the identity operator E. The smallest set containing Q' which is closed under the operations of multiplication by complex numbers, addition, multiplication, passing to the conjugate operator, and taking limits in norm is a commutative normed ring Q. We show that taking the conjugate operator is an involution in Q. Suppose that $A \in Q$ and that $\|A\| = a$. It is only needful to show that $(AA^* + E)^{-1}$ exists and belongs to Q. For this, evidently, it suffices to show that $\left\|\frac{a^2E-AA^*}{a^2+1}\right\| < 1$ because then

$$(AA^*+E)^{-1} = \frac{1}{a^2+1}\sum_{n=0}^{\infty}\left(\frac{a^2E-AA^*}{a^2+1}\right)^n.$$

But since $(AA^*f, AA^*f) = (A^*AA^*fA^*f) \leq \|A^*A\|(A^*f, A^*f) = a^2(A^*f, A^*f)$, then

$$\|a^2E-AA^*\|^2 = \sup\frac{((a^2E-AA^*)f, (a^2E-AA^*)f)}{(f,f)} =$$

$$= \sup\frac{a^4(f,f)-2a^2(A^*f,A^*f)+(AA^*f,AA^*f)}{(f,f)} \leq \sup\frac{a^4(f,f)-a^2(A^*f,A^*f)}{(f,f)} \leq a^4,$$

from which it follows that

$$\left\|\frac{a^2E-AA^*}{a^2+1}\right\| \leq \frac{a^2}{a^2+1} < 1.$$

Each maximal ideal M in the commutative normed ring R determines in R a linear functional defined by the equation

$$M(x) = x(M)$$

for fixed M and varying x. Indeed, by properties a), c), and g) of the functions $x(M)$ (pages 129, 130) we have

$$M(x_1+x_2) = M(x_1) + M(x_2);\ M(\lambda x) = \lambda M(x);\ |M(x)| \leq \|x\|.$$

Moreover, in view of the "separation property" f) distinct maximal ideals determine distinct linear functionals. The maximal ideal M itself is the "hyperplane" $M(x) = 0$.

Evidently each maximal ideal M in a ring R with involution when regarded as a linear functional is positive. To prove this, $M(x^*) = \overline{M(x)}$ so $M(xx^*) = |M(x)|^2 \geq 0$ for all $x \in R$.

Theorem 18. *Each positive linear functional f defined on a ring with involution R can be regarded as a positive linear functional on the space of functions $x(M)$ determined uniquely by a positive linear functional on the space of all continuous complex functions on the space $\mathfrak{M}$ of maximal ideals of the ring R. In this statement positive functional in a space of functions has its ordinary meaning.*

Proof. From positivity of the expression

$$f((x+\lambda y)(x+\lambda y)^*) = f(xx^*) + \lambda f(yx^*) + \overline{\lambda} f(xy^*) + |\lambda|^2 f(yy^*),$$

considered as a quadratic form in the variable λ, it follows that

$$f(yx^*) = \overline{f(xy^*)}$$

and

$$|f(xy^*)|^2 \leq f(xx^*)f(yy^*);$$

in particular, when $y = e$,

$$f(x^*) = \overline{f(x)} \tag{1}$$

and

$$|f(x)|^2 \leq f(e)\, f(xx^*). \tag{2}$$

Setting $xx^* = z$, the second term of the inequality (2) combined with the fact that $(z^m)^* = z^m$ yields

$$|f(x)| \leq f(e)^{\frac{1}{2}} f(z)^{\frac{1}{2}} \leq f(e)^{\frac{1}{2}+\frac{1}{4}} f(z^2)^{\frac{1}{4}} \leq \cdots \leq f(e)^{\frac{1}{2}+\frac{1}{4}+\cdots+\frac{1}{2^n}} f(z^{2^{n-1}})^{\frac{1}{2^n}}.$$

Since $|f(y)| \leq |f|\, \|y\|$, it follows from this that

$$|f(x)| \leq f(e)^{1-\frac{1}{2^n}}\, |f|^{\frac{1}{2^n}}\, \|z^{2^n-1}\|^{\frac{1}{2^n}}.$$

Taking the limit here as $n \to \infty$ and applying the result of Theorem 11, we get

$$|f(x)| \leq f(e) \max_{M \in \mathfrak{M}} |z(M)|^{\frac{1}{2}},$$

that is,

$$|f(x)| \leq f(e) \max_{M \in \mathfrak{M}} |x(M)|. \tag{3}$$

The inequality (3) shows that f is a linear functional in the space of functions $x(M)$, where the norm is defined as the maximum modulus. Let us show that this functional is positive; that is, that $f(x) \geq 0$ if $x(M) \geq 0$ for all $M \in \mathfrak{M}$. We note first of all that if $x(M)$ is real for all $M \in \mathfrak{M}$, then $f(x)$ is real. In fact, since $x^*(M) \equiv \overline{x(M)} \equiv x(M)$, x^* can differ from x only by a generalized nilpotent element. But by the inequality (3) the functional f vanishes on generalized nilpotent elements. Therefore $f(x^*) = f(x)$; that is, in view of equation (1) $\overline{f(x)} = f(x)$. Suppose now that $x(M) \geq 0$ for all maximal ideals M. Denote $\max_{M \in \mathfrak{M}} x(M) = \alpha$. Since

$$0 \leq \alpha - x(M) = (\alpha e - x)(M) \leq \alpha$$

and since, by what has just been proved, $f(\alpha e - x)$ is real, then applying (3) we get

$$\alpha f(e) - f(x) = f(\alpha e - x) \leq f(e) \max_{M \in \mathfrak{M}} (\alpha e - x)(M) \leq \alpha f(e).$$

This shows that $f(x) \geq 0$.

By Theorem 16 the completion of the space of all $x(M)$ with norm defined by the maximum modulus is the set of all continuous complex functions defined on $\mathfrak{M}$. From this it follows that the functional f is determined by a linear functional in that complete space. From the same Theorem 16 it is easily shown that each non-negative real continuous function defined on $\mathfrak{M}$ is uniformly approximable by non-negative real functions $x(M)$ $(x \in R)$. Combined with the preceding sentences this shows that the functional f is positive on the space of all continuous complex functions defined on $\mathfrak{M}$. With this Theorem 18 is completely proved.

From Theorem 18 and theorems about representation of positive linear functionals [16] immediately follows

Theorem 19. *Each positive linear functional f defined in a ring with involution R can be represented, in only one way, in the form*

$$f(x) = \int_{\mathfrak{M}} x(M)\, d\Phi(M),$$

*where Φ is a non-negative, completely additive, regular * set function on the field of Borel sets of the space $\mathfrak{M}$ of maximal ideals of the ring R.*

Theorem 20. *In order that a commutative normed ring R be symmetric and contain no generalized nilpotent elements it is necessary and sufficient that R be a ring with involution and that R possess sufficiently many positive linear functionals.*

Proof. The necessity of the conditions of the theorem is evident. Indeed, the absence from a symmetric ring R of generalized nilpotent elements means that the operation $x \to x^*$ is single valued in R (and is, evidently, an involution) and also means that for each element $x_0 \in R$ different from zero there is a maximal ideal M_0, which determines a positive linear functional, such that $M_0(x_0) = x_0(M_0) \neq 0$. Conversely, suppose that x_0 is an arbitrary element different from zero, of a ring with involution R, and suppose that f_0 is a positive linear functional such that $f_0(x_0) \neq 0$. Then by the inequality (3) we have

$$\max_{M \in \mathfrak{M}} |x_0(M)| \geq \left\| \frac{f_0(x_0)}{f_0(e)} \right\| > 0,$$

which shows that x_0 is not a generalized nilpotent element.

The ring Q considered on pp. 143, 144 satisfies the conditions of Theorem 20. Indeed, each element of the Hilbert space H, on which the operators A of Q are defined, determines on Q a linear functional $f(A) = (Af, f)$; this functional is positive: $f(AA^*) = (AA^*f, f) = (A^*f, A^*f) \geq 0$; moreover for each $A_0 \neq 0$ there exists an element $f_0 \in H$ such that $(A_0 f_0, f_0) \neq 0$. Hence, by Theorem 20, there are in Q no generalized nilpotent elements different from zero.

Another application of Theorem 20 will be made in Appendix III.

§8. Real rings

Definition 11. A *commutative normed ring* is a set $\mathfrak{R}$ of elements $x, y, \cdots$, satisfying the following conditions:

a) $\mathfrak{R}$ is a linear normed complete space with multiplication only by real numbers.

b) In $\mathfrak{R}$ there is defined an associative and commutative multiplication, commuting with multiplication by real numbers, distributive with respect to addition,

* A function defined on the field of Borel sets is called *regular* if the value of its total variation on each Borel set equals the lower bound of the values of its total variation on the open sets containing this Borel set.

and continuous in each variable.

c) In $\mathfrak{R}$ there is an identity e with respect to multiplication.

d) $(x^2+e)^{-1}$ exists for every $x \in \mathfrak{R}$.

Just as in §1 it can be shown that it is possible to renorm $\mathfrak{R}$ so that it also satisfies the condition

b′) $\|xy\| \le \|x\| \, \|y\|$ and $\|e\| = 1$.

As an example of a real normed ring one can consider any ring K of continuous, bounded, real functions on any topological space, with ordinary algebraic operations, with norm $\|x\| = \sup_{t \in S} |x(t)|$, containing all constant functions, and complete under its norm.

Clearly, we need only check the validity of d) of Definition 11; that is, we need to show that along with a function $x(t)$ K also contains the function $\frac{1}{x^2(t)+1}$.

$x(t)$ is bounded, $|x(t)| \le C$. Hence

$$\frac{1}{x^2(t)+1} = \frac{1}{(C^2+1)-[(C^2+1)-(x^2(t)+1)]} = \frac{1}{C^2+1} \cdot \frac{1}{1-\frac{C^2-x^2}{C^2+1}} =$$

$$= \frac{1}{C^2+1} \sum_{n=0}^{\infty} \left(\frac{C^2-x^2}{C^2+1}\right)^n,$$

where the series is uniformly convergent because $\left|\frac{C^2-x^2}{C^2+1}\right| \le \frac{C^2}{C^2+1} < 1$. Since every partial sum of this series is in K, in view of the fact that K is closed under uniform limits, the sum of the series, i.e., the function $\frac{1}{x^2(t)+1}$ belongs to K.

Another important example of a real normed ring is presented by the set $\mathfrak{Q}$ of all Hermitian operators in the ring Q considered on pages 143, 144.

In general the set of all real elements of an arbitrary commutative normed ring R (that is, the elements x which correspond to real functions $x(M)$) is a real ring in the sense of Definition 11.

Let us remark that not every normed ring of real functions $x(t)$ is a real ring in the sense of Definition 11. Thus, consider the ring of real functions on the segment $[-1, 1]$ analytically extendable to the circle $|t+ui| \le 1$, with ordinary algebraic operations and with norm $\|x\| = \max_{|t+ui| \le 1} |x(t+ui)|$. It satisfies conditions a), b), and c) of Definition 11 but does not satisfy d). To see this observe that the function $x(t) = t$ belongs to this ring but $\frac{1}{t^2+1}$ has poles at the points $\pm i$, and therefore can not belong to the ring.

Theorem 21. *Each commutative real normed ring $\mathfrak{R}$ can be imbedded with preservation of norm in a normed ring with involution R so that to each element*

$x \in \mathfrak{R}$ corresponds a real function $x(M)$ on the set of maximal ideals of the ring R and each element of R which determines a real function on the maximal ideals differs from an element of $\mathfrak{R}$ only by a generalized nilpotent element.

Proof. We construct the ring R whose existence is asserted by the theorem out of the formal sums $x + iy$, where $x, y \in \mathfrak{R}$, and i is the imaginary unit. Then we say that $x + iy = 0$ only in case $x = y = 0$. The algebraic operations are defined naturally:

$$(x+iy) + (u+iv) = (x+u) + i(y+v),$$

$$(x+iy)(u+iv) = (xu-yv) + i(xv+yu).$$

Evidently the identity element of $\mathfrak{R}$ will serve as an identity for R. Define the norm in the following way

$$\| x+iy \| = \max \| x \cos \alpha + y \sin \alpha \|. \tag{1}$$

It is clear that the norm of an element $x \in \mathfrak{R}$ is preserved by this.

Let us show that all the conditions which a norm should satisfy are fulfilled.

1) $\| (\lambda+i\mu)(x+iy) \| = \sqrt{\lambda^2+\mu^2}\, \| x+iy \| = | \lambda+i\mu |\, \| x+iy \|$ (λ and μ are real).

Indeed, setting $\dfrac{\lambda}{\sqrt{\lambda^2+\mu^2}} = \cos \phi$, $\dfrac{\mu}{\sqrt{\lambda^2+\mu^2}} = \sin \phi$, we get

$$\| (\lambda+i\mu)(x+iy) \| = \max_\alpha \| \sqrt{\lambda^2+\mu^2}\, [x \cos (\alpha-\phi) + y \sin (\alpha-\phi)] \| =$$

$$= \sqrt{\lambda^2+\mu^2}\, \| x+iy \|.$$

2) $\| (x+iy)+(u+iv) \| \le \| x+iy \| + \| u+iv \|$.

Obvious.

3) $\| (x+iy)(u+iv) \| \le 2\, \| x+iy \|\, \| u+iv \|$; hence multiplication is continuous.

Indeed, we have

$$\| (x+iy)(u+iv) \| = \max_\alpha \| (xu-yv) \cos 2\alpha + (xv+yu) \sin 2\alpha \| =$$

$$= \max_\alpha \| (x \cos \alpha + y \sin \alpha)(u \cos \alpha + v \sin \alpha) -$$

$$- (x \sin \alpha - y \cos \alpha)(u \sin \alpha - v \cos \alpha) \| \le$$

$$\le \max_\alpha \| x \cos \alpha + y \sin \alpha \| \max_\alpha \| u \cos \alpha + v \sin \alpha \| +$$

$$+ \max_\alpha \| x \sin \alpha - y \cos \alpha \| \max_\alpha \| u \sin \alpha - v \cos \alpha \| =$$

$$= 2\, \| x+iy \|\, \| u+iv \|,$$

for, setting $\alpha = \beta + \frac{\pi}{2}$, we get $\max_\alpha \| x \sin \alpha - y \cos \alpha \| = \max_\beta \| x \cos \beta + y \sin \beta \| = \| x+iy \|$ and analogously $\max_\alpha \| u \sin \alpha - v \cos \alpha \| = \| u+iv \|$.

4) If $\| x+iy \| = 0$, then $x + iy = 0$, i.e., $x = y = 0$.

Indeed, from (1) follows $\| x+iy \| \ge \| x \cos 0 + y \sin 0 \| = \| x \|$ and

$\| x+iy \| \geq \left\| x \cos \frac{\pi}{2} + y \sin \frac{\pi}{2} \right\| = \| y \|$. Hence, if $\| x+iy \| = 0$ then $\| x \| = \| y \| = 0$, that is, $x = y = 0$.

5) R is complete under the norm (1).

Indeed, if $\| (x_n + iy_n) - (x_m + iy_m) \| \to 0$, then by what has just been proved $\| x_n - x_m \| \to$ and $\| y_n - y_m \| \to 0$; by completeness of $\mathfrak{R}$ this implies the existence of $x, y \in \mathfrak{R}$ such that $x_n \to x$ and $y_n \to y$. Then

$$\| (x_n + iy_n) - (x+iy) \| = \max_a \| (x_n - x) \cos a + (y_n - y) \sin a \| \leq$$
$$\leq \| x_n - x \| + \| y_n - y \| \to 0.$$

Therefore R is a normed ring*. Let us show that the operation $x+iy \to x-iy$ is an involution. If $z = z'$, that is, $z \in \mathfrak{R}$, then just as on page 143 it can be shown that the function $z(M)$ is real. Therefore, $x^2 + y^2$ has in R a non-negative spectrum for every $x, y \in \mathfrak{R}$ and hence

$$(e + (x+iy)(x-iy))^{-1} = (e + x^2 + y^2)^{-1}$$

exists for every $x + iy \in R$. The validity of the other properties of an involution is evident. Hence R is a ring with involution.

Finally, if the function $(x+iy)(M) = x(M) + iy(M)$ is real for all M, then, by the reality of the functions $x(M)$ and $y(M)$, we have $y(M) \equiv 0$, which is to say that $x + iy$ differs from the element x of $\mathfrak{R}$ only by a generalized nilpotent element iy. With this Theorem 21 is completely proved.

Since in §§ 1 and 2 we used nowhere the possibility of multiplying the complex numbers, all the results of those sections preserve their validity in real rings. In particular, in a commutative real normed ring $\mathfrak{R}$ there exists maximal ideals, and the ring of cosets by a maximal ideal is a field.

Theorem 22. *The quotient ring of a commutative real normed ring $\mathfrak{R}$ by a maximal ideal is the field of real numbers.*

Proof. We shall establish a one-to-one correspondence between maximal ideals M' of the ring $\mathfrak{R}$ and maximal ideals M of the larger ring R of Theorem 21 in such a way that for each element $x \in \mathfrak{R}$ the equation $x(M') = x(M)$ is satisfied. Since by the preceding theorem the functions $x(M)$ are real, this proves Theorem 22.

Suppose that M is some maximal ideal of the ring R. Attaching to each element $x \in \mathfrak{R}$ the corresponding number $x(M)$ gives a homomorphism of the ring $\mathfrak{R}$ into the field of (real) numbers. This mapping is non-trivial, since the identity element e ($\in \mathfrak{R}$) is carried to the number 1. By Theorem 6 this mapping determines

* We remark that so far we have not used condition d) of Definition 11. Hence we have shown that each ring $\mathfrak{R}$ satisfying the first three of the conditions of Definition 11 can be embedded with norm preserved in some normed ring in the sense of Definition 1.

in $\mathfrak{R}$ some maximal ideal M', for which for all $x \in \mathfrak{R}$ we have $x(M') = x(M)$. In particular an element $x \in \mathfrak{R}$ belongs to M' if and only if $x(M) = 0$, i.e., if and only if x belongs to M. Hence M' is the intersection of M with $\mathfrak{R}$. Let us show that $\mathfrak{R}$ contains no other kind of maximal ideal. Suppose that M' is a maximal ideal of the ring $\mathfrak{R}$. The set $M' + iM'$ of elements $x + iy$, where x and y belong to M' is an ideal in R. This ideal is maximal, because making correspond to each element $u + iv \in R$ the number $u(M') + iv(M')$ we define a non-trivial homomorphism of the ring R into the field of complex numbers, and the ideal $M' + iM'$ is the kernel of this homomorphism. Evidently M' is the intersection of the maximal ideal $M' + iM'$ with $\mathfrak{R}$. Since $M' + iM'$ is the smallest ideal of R containing M', M' is contained in no other maximal ideal M of the ring R. In this way the maximal ideals of the rings R and $\mathfrak{R}$ are put in one-to-one correspondence, and Theorem 22 is proved.

Remark. From the last part of Theorem 21 it is clear that the real ring $\mathfrak{R}$ coincides with the set of all real elements of the ring R constructed in that theorem if and only if $\mathfrak{R}$ has no generalized nilpotent elements different from zero. Therefore, if $\mathfrak{R}$ is the set of all real elements of some symmetric ring R_1 which contains generalized nilpotent elements different from zero, then R *does not coincide with* R_1. The cause of this is clearer if we remark that if $x (\neq 0)$ is a generalized nilpotent of R_1, so is ix, which therefore belongs to $\mathfrak{R}$, but then $x + i \cdot ix$ (which equals zero in R_1) *is not zero in* R because in R $\|x + i(ix)\| = \sup \|x(\cos\alpha + i\sin\alpha)\| = \|x\|$.

If there are no generalized nilpotent elements in $\mathfrak{R}$ different from zero, then R_1 and R are algebraically isomorphic. By Theorem 29 of § 11 it follows that they are also topologically isomorphic. Hence a real normed ring without generalized nilpotents different from zero can not be determined by two distinct symmetric rings.

All the results of § 5 can easily be carried over to commutative real normed rings. In particular, the set $\mathfrak{M}'$ of all maximal ideals of a real normed ring $\mathfrak{R}$ topologized as in Definition 7 is compact. Since the elements of $\mathfrak{R}$ generate the symmetric ring R of Theorem 21 and take on corresponding ideals of these two rings the same values, the topology in $\mathfrak{M}'$ coincides with the topology of the set $\mathfrak{M}$ of all maximal ideals of the ring R. From Theorems 16 - 16' if follows that the *functions* $x(M')$ *determined by elements of a real normed ring* $\mathfrak{R}$ *are everywhere dense in the space* $C(\mathfrak{M}')$ *of all continuous real functions on the compact space* $\mathfrak{M}'$ *of its maximal ideals.*

Remark. From the coincidence for an element x of a real normed ring $\mathfrak{R}$ of the function $x(M')$ with the function $x(M)$ $(M \in \mathfrak{M})$ follows also for real normed rings with the validity of the formula

$$\max_{M\in\mathfrak{M}'} |x(M')| = \lim_{n\to\infty} \sqrt[n]{\|x^n\|}. \tag{2}$$

As an illustration we show that *the real normed ring* $\mathfrak{Q}$ *of commuting bounded Hermitian operators, described on page 147, is isomorphic to the ring of all continuous real functions on the compact space* $\mathfrak{M}'$ *of its maximal ideals.*

Since the functions $A(M')$ $(A \in \mathfrak{Q}, M' \in \mathfrak{M}')$ are everywhere dense in $C(\mathfrak{M}')$, for the proof of our assertion it suffices to show that from uniform convergence of functions $A(M')$ follows convergence in the norm of $\mathfrak{Q}$ of the corresponding operators. But by formula (2) this follows (see the corollary of Theorem 17) from the equation

$$\|A^2\| = \|A\|^2, \tag{3}$$

which holds for all bounded Hermitian operators *.

§9. Analytic functions of elements of a ring

A normed ring always contains along with any one of its elements x all polynomials, and, more generally, all "integral" functions of x, that is, all elements of the form $\sum_{n=0}^{\infty} c_n x^n$, where the series $\sum_{n=0}^{\infty} c_n \zeta^n$ represents an integral analytic function of the complex variable ζ. Indeed, in this case the series $\sum_{n=0}^{\infty} c_n x^n$ is majorized by the convergent series $\sum_{n=0}^{\infty} |c_n| \, \|x\|^n$. Hence to each integral analytic function $f(\zeta)$ corresponds an "abstract" analytic function $f(x)$ defined for all elements of the ring. Furthermore, we see that for each x such that the set of values $x(M)$ does not include the number zero for any maximal ideal the ring contains also an inverse element x^{-1}. Hence corresponding to the analytic function $\frac{1}{\zeta}$ which has a pole at the point $\zeta = 0$, is the function x^{-1} which is defined for the set of all elements x whose spectrum does not contain the point 0. In the same way to each rational function $Q(\zeta)$ of the complex variable ζ corresponds an abstract "analytic" function $Q(x)$ defined for all elements x whose spectrum does not contain a pole of $Q(\zeta)$.

It is natural to raise the question of extension of this correspondence established for integral and rational functions to all analytic functions. This problem is given a positive solution in the following theorem.

Theorem 23. *Suppose that* D *is a bounded closed domain in the plane of the complex variable* ζ, *that* A_D *is the ring of all analytic functions regular in* D *(with ordinary operations of addition and multiplication), that* $\mathfrak{D}$ *is the set of all*

* We display a proof of (3). On the one hand $\|A^2\| \le \|A\|^2$, since generally, $\|AB\| \le \|A\| \, \|B\|$. On the other hand, by Schwartz's inequality $|(A^2 f, f)|^2 \le (A^2 f, A^2 f)\,(f, f)$, so $\|A\|^2 = \sup \frac{(Af, Af)}{(f, f)} = \sup \left| \frac{(A^2 f, f)}{(f, f)} \right| \le \sqrt{\sup \frac{(A^2 f, A^2 f)}{(f, f)}} = \|A^2\|$. Comparison of these inequalities gives (3).

elements of the ring whose spectrum is contained in D, and that $F_{\mathfrak{D}}$ is the algebraic ring of all abstract functions with values in the ring defined on $\mathfrak{D}$. Then there exists an isomorphism $f(\zeta) \to f(x)$ of the ring A_D on a subset $A_{\mathfrak{D}}$ of the ring $F_{\mathfrak{D}}$, which carries the function $f(\zeta) \equiv \zeta$ into $f(x) \equiv x$ and carries a sequence of functions $f_n(\zeta)$ uniformly convergent on some region containing D to a sequence of functions $f_n(x)$ converging in the norm for each $x \in \mathfrak{D}$.

This mapping is unique and is given by the Cauchy formula

$$f(x) = \frac{1}{2\pi i} \int_\Gamma (\zeta e - x)^{-1} f(\zeta)\, d\zeta, \tag{1}$$

where the integration is carried out along a rectifiable contour Γ which is entirely contained in the domain of regularity of the function $f(\zeta)$, which surrounds the region D, and which passes around this region at a positive distance.

Proof. We prove first that if a mapping exists which satisfies the conditions of the theorem, then the function $f(\zeta)$ must be carried to the function $f(x)$ defined by formula (1), and then show that the mapping by formula (1) has indeed all the required properties.

Now suppose that a mapping such as interests us exists. Since ζ is carried to x, the homomorphic mapping $f(\zeta) \to f(x)$ is non-trivial (i.e., does not take everything to zero). Then the function $f(\zeta) \equiv 1$ is carried to $f(x) \equiv e$. Hence it follows that if $\frac{1}{f(\zeta)}$ and $f(\zeta)$ both belong to A_D, the former is carried into $f^{-1}(x) - [f(x)]^{-1}$. Now let ζ_0 be an arbitrary point not belonging to D. Then the the function $(\zeta - \zeta_0)^{-1}$ is in A_D and is carried into the function $(x - \zeta_0 e)^{-1}$. Since by assumption $x \in \mathfrak{D}$ and $\zeta_0 \,\overline{\in}\, D$, this last function actually exists*.

Consider now an arbitrary function $f(\zeta) \in A_D$. Since $f(\zeta)$ is regular in the bounded closed region D, there is a rectifiable contour lying wholly in the domain of regularity of the function $f(\zeta)$ and passing around D at a positive distance. By Cauchy's formula, for all points $\zeta \in D$ we have

$$f(\zeta) = \frac{1}{2\pi i} \int_\Gamma \frac{f(\lambda)\, d\lambda}{\lambda - \zeta}.$$

The right side of this equation is a limit of a sequence of sums of the form

$$f_n(\zeta) = \frac{1}{2\pi i} \sum_{k=0}^{n} \frac{f(\lambda_k)(\lambda_{k+1} - \lambda_k)}{\lambda_k - \zeta} \qquad (\lambda_k \in \Gamma),$$

uniformly convergent in the whole domain D. To each such sum corresponds an

* The requirement that the function $(x - \zeta_0 e)^{-1}$ exists for all $\zeta_0 \,\overline{\in}\, D$ depends on restricting the element x considered to $\mathfrak{D}$.

We remark that $\mathfrak{D}$ is not empty; let λ_0 be an interior point of D, so there exists $\rho > 0$ such that the circle of radius ρ and center λ_0 is entirely in D. Then $x' = \frac{(\lambda - \lambda_0)x}{\|x\|} + \lambda_0 e \in \mathfrak{D}$ for all elements x of the ring being considered and all λ in the given circle.

abstract function

$$f_n(x) = \frac{1}{2\pi i} \sum_{k=0}^{n} f(\lambda_k)(\lambda_k e - x)^{-1} (\lambda_{k+1} - \lambda_k).$$

By assumption, these functions must converge in norm for each $x \in \mathfrak{D}$ to some limit. But these are approximating sums for the integral $\int_\Gamma x(\lambda)\, d\lambda$ of the function $x(\lambda) = \frac{1}{2\pi i} f(\lambda)(\lambda e - x)^{-1}$ of the complex variable λ, which, by the choice of the contour Γ, exists and is continuous in norm. As was shown in §3, such an integral actually exists in the sense of norm convergence. Moreover, it does not depend on the choice of the contour Γ satisfying the requirements, because $x(\lambda)$ is an abstract analytic function of the variable λ regular outside D, and for such functions, as we saw in §3, Cauchy's theorem is valid.

Therefore, we have reached the conclusion that to each function $f(\zeta) \in A_D$ must correspond the abstract function

$$f(x) = \frac{1}{2\pi i} \int_\Gamma (\lambda e - x)^{-1} f(\lambda)\, d\lambda, \tag{1}$$

for this function exists and is uniquely determined for all $x \in \mathfrak{D}$.

Let us show that the mapping $f(\zeta) \to f(x)$ by formula (1) actually satisfies the conditions of the theorem.

Let us show that this mapping is a homomorphism. The proof requires only that products be carried to products. In other words, we must prove the equation

$$\frac{1}{2\pi i} \int_\Gamma (\lambda e - x)^{-1} f(\lambda)\, g(\lambda)\, d\lambda = \frac{1}{2\pi i} \int_\Gamma (\lambda e - x)^{-1} f(\lambda)\, dx \cdot \frac{1}{2\pi i} \int_\Gamma (\lambda e - x)^{-1} g(\lambda) d\lambda, \tag{2}$$

where Γ is a rectifiable contour completely contained in the domains of regularity of the functions $f(\zeta)$ and $g(\zeta)$ and surrounding D at a positive distance. In the second integral on the right hand side replace the contour Γ by a contour Γ_1 outside Γ at a positive distance and entirely in the domain of regularity of both functions $f(\zeta)$ and $g(\zeta)$ to get

$$\frac{1}{2\pi i} \int_\Gamma (\lambda e - x)^{-1} f(\lambda)\, d\lambda \cdot \frac{1}{2\pi i} \int_{\Gamma_1} (\lambda e - x)^{-1} g(\lambda)\, d\lambda =$$

$$= -\frac{1}{4\pi^2} \int_\Gamma \int_{\Gamma_1} (\lambda e - x)^{-1} (\mu e - x)^{-1} f(\lambda)\, g(\mu)\, d\lambda\, d\mu =$$

$$= -\frac{1}{4\pi^2} \int_\Gamma \int_{\Gamma_1} \frac{f(\lambda) g(\mu)}{\mu - \lambda} [(\lambda e - x)^{-1} - (\mu e - x)^{-1}]\, d\lambda\, d\mu =$$

$$= \frac{1}{2\pi i} \int_\Gamma (\lambda e - x)^{-1} f(\lambda) \left(\frac{1}{2\pi i} \int_{\Gamma_1} \frac{g(\mu)\, d\mu}{\mu - \lambda}\right) d\lambda +$$

$$+ \frac{1}{4\pi^2} \int_{\Gamma_1} (\mu e - x)^{-1} g(\mu) \left(\int_\Gamma \frac{f(\lambda) d\lambda}{\mu - \lambda}\right) d\mu.$$

Since the function $\frac{f(\lambda)}{\mu - \lambda}$ is regular in the closed region of the λ plane bounded by the contour Γ, by Cauchy's theorem the second integral on the right is zero. Furthermore, since λ in an interior point of the region in the plane of the variable μ bounded by the contour Γ_1, the formula of Cauchy shows that the inner integral in the first term equals $g(\lambda)$. This proves the equality (2).

Let us show now that ζ is carried to x; that is, that

$$\frac{1}{2\pi i}\int_\Gamma (\zeta e - x)^{-1}\zeta\, d\zeta = x.$$

Since the function $f(\zeta) \equiv \zeta$ is regular in the whole plane, a circle with center 0 and arbitrarily large radius can be chosen as the contour Γ. Choose it so that the series

$$\zeta(\zeta e - x)^{-1} = e + \zeta^{-1}x + \zeta^{-2}x^2 + \cdots$$

is absolutely convergent on it. Then termwise integration gives

$$\frac{1}{2\pi i}\int_\Gamma (\zeta e - x)^{-1}\zeta\, d\zeta = \frac{1}{2\pi i}\int_\Gamma (e + \zeta^{-1}x + \zeta^{-2}x^2 + \cdots)\, d\zeta = x\frac{1}{2\pi i}\int_\Gamma \zeta^{-1} d\zeta = x.$$

Now suppose that $f_n(\zeta) \to f(\zeta)$ uniformly on an open region D' containing D. Choose for Γ a contour surrounding D at a positive distance and lying entirely in D', to get

$$\| f(x) - f_n(x) \| = \left\| \frac{1}{2\pi i}\int_\Gamma (\zeta e - x)^{-1}[f(\zeta) - f_n(\zeta)]\, d\zeta \right\| \le$$

$$\le \max_{\zeta \in \Gamma} |f(\zeta) - f_n(\zeta)| \frac{1}{2\pi i}\int_\Gamma \|(\zeta e - x)^{-1}\|\, |d\zeta| \to 0 \quad (n \to \infty).$$

It remains to be proved that the homomorphism $f(\zeta) \to f(x)$ is an isomorphism; that is, that $f(x)$ is identically zero only in case $f(\zeta) \equiv 0$.

But if $f(\zeta) \not\equiv 0$, then there is a closed region $D_1 \subset D$ in all of which $f(\zeta) \neq 0$. Then $\frac{1}{f(\zeta)} \in A_{D_1}$ and the function $\frac{1}{f}(x)$ is defined for all x with spectrum contained in D_1. By the results above $f(x)\frac{1}{f}(x) = e$, so $f(x) \neq 0$ for all such elements x.

With this the proof of Theorem 23 is complete. From formula (1) we derive the following proposition:

Theorem 24. *If $f(\zeta)$ is regular on the set of values of the function $x(M)$ then there is in the ring an element y such that $y(M) = f(x(M))$ for all maximal ideals M.*

Proof. Since $x(M)$ is a continuous and bounded function defined on a bicompact space, the set of its values is a bounded closed set F in the complex plane. As we saw, the integral (1) exists (and has the same value) for every contour Γ lying entirely in the domain of regularity of the function $f(\zeta)$ and surrounding F at a positive distance. This integral is an element of the ring satisfying the properties required for the theorem. To see this, $x(M)$ for fixed M and varying x is a linear functional of x. Therefore, by formula (2) of § 3, we have

$$f(x)(M) = \frac{1}{2\pi i} \int_\Gamma (\lambda e - x)^{-1}(M) f(\lambda)\, d\lambda = \frac{1}{2\pi i} \int_\Gamma \frac{f(\lambda)\, d\lambda}{\lambda - x(M)} = f(x(M)).$$

This theorem contains as a special case the following generalization, due to P. Lévy [15], of the theorem of Wiener discussed on pages 122 and 130*. *Suppose that the Fourier series of the function $x(t)$ is absolutely convergent and that the values of $x(t)$ are in the circle $|\zeta - \zeta_0| < \rho$. If $f(\zeta)$ is a complex function of a complex variable ζ regular in all of this circle, then the Fourier series of the function $f(x(t))$ is also absolutely convergent.*

§ 10. Decomposition of a ring into a direct sum of ideals

Definition 12. A normed ring R is called a *direct sum* of two of its ideals I_1 and I_2 if:

a) The intersection $I_1 \cap I_2$ contains only the element 0.

b) Each element $x \in R$ can be represented as a sum, $x = x_1 + x_2$, where $x_1 \in I_1$ and $x_2 \in I_2$.

From condition a) it follows easily that the representation in b) is unique. In fact, if $x = x_1 + x_2 = y_1 + y_2$, then $0 = (x_1 - y_1) + (x_2 - y_2)$. Since 0 and $x_1 - y_1$ belong to I_1, also $x_2 - y_2 \in I_1$; but, on the other hand, $x_2 - y_2 \in I_2$, so $x_2 - y_2 = 0$, $x_2 = y_2$, and, hence, also $x_1 = y_1$.

Suppose that R is the direct sum of its ideals I_1 and I_2. Consider the decomposition of the identity $e = e_1 + e_2$, $e_1 \in I_1$, $e_2 \in I_2$. Since $e_1 e_2 \in I_1 \cap I_2$, then $e_1 e_2 = 0$, and $e = e^2 = e_1^2 + e_2^2$, from which and from the uniqueness of the decomposition follows that $e_1^2 = e_1$, $e_2^2 = e_2$. The element e_1 in the ideal I_1, just as does e_2 in I_2, plays the role of an identity: if $x_1 \in I_1$, then $x_1 e_2 = 0$ and then $x_1 e_1 = x_1(e - e_2) = x_1$. The ideals I_1 and I_2 are closed; if $x_n \subset I_1$ and $x_n \to x$, then $x_n = x_n e_1 \to x e_1$, and consequently $x = x e_1 \in I_1$. Hence, *I_1 and I_2 are themselves normed rings with corresponding identities e_1 and e_2.*

Suppose, conversely, that there is an element $e_1 \in R$ different from zero and the identity with the property that $e_1^2 = e_1$. Consider the element $e_2 = e - e_1$; evidently

$$e_1 e_2 = e_1(e - e_1) = e_1 - e_1^2 = 0 \text{ and } e_2^2 = e_2(e - e_1) = e_2.$$

Let I_1 and I_2 be the ideals generated by the elements e_1 and e_2. Let $x \in I_1 \cap I_2$; this means that $x = x' e_1 = x'' e_2$, and from this $x = x(e_2 + e_1) = x' e_1 e_2 + x'' e_2 e_1 = 0$. Therefore the intersection $I_1 \cap I_2$ contains only 0. Moreover each element $x \in R$ can be represented as a sum $x = x e_1 + x e_2$, where $x e_1 \in I_1$, $x e_2 \in I_2$. We see that the conditions a) and b) of Definition 12 are satisfied, so R is the direct sum of the ideals I_1 and I_2.

* See also Zygmund, *Trigonometrical series*, p. 140.

Theorem 25. *If the commutative normed ring R is decomposed into a direct sum of two of its ideals I_1 and I_2, then the space $\mathfrak{M}(R)$ of maximal ideals of the ring R is decomposed into the sum of two non-intersecting closed sets F_1 and F_2 corresponding to the spaces of maximal ideals of the rings I_1 and I_2.*

Proof. Suppose that e_1 and e_2 are the identities of the rings I_1 and I_2. It follows from the equations $e_1^2 = e_1$, $e_2^2 = e_2$, that the functions $e_1(M)$ and $e_2(M)$ take on only the values 0 and 1. Let F_1 be the set of all those maximal ideals for which $e_1(M) = 1$, and F_2 the set of all those maximal ideals where $e_2(M) = 1$. These set are closed. Since $e_1 + e_2 = e$, also $e_1(M) + e_2(M) = 1$, and therefore F_1 and F_2 do not intersect and their sum is $\mathfrak{M}(R)$. Therefore, to the decomposition of the ring R into a direct sum of the ideals I_1 and I_2 corresponds a decomposition of the space $\mathfrak{M}(R)$ of maximal ideals of the ring R into a sum of two closed non-intersecting sets: $\mathfrak{M}(R) = F_1 + F_2$, where

$$F_1 = \{ e_1(M) = 1 \}, \quad F_2 = \{ e_2(M) = 1 \}.$$

Let us show that $F_1 = \mathfrak{M}(I_1)$, $F_2 = \mathfrak{M}(I_2)$. To this end, suppose that $M_1 \in F_1$. Attaching to each element $x_1 \in I_1$ the number $x_1(M_1)$ we get a homomorphism of the ring I_1 into the field of complex numbers, non-trivial because e_1 is carried to 1. Hence, each maximal ideal $M_1 \in F_1$ of the ring R determines some maximal ideal of the ring I_1. Two distinct ideals M_1, $M_1' \in F_1$ determine distinct maximal ideals of the ring I_1, because for each element $x \in R$ we have $M_1(x) = M_1(xe_1)$ and $M_1'(x) = M_1'(xe_1)$, and hence M_1 and M_1' must differ for some element $xe_1 \in I_1$. Conversely, suppose M' is a maximal ideal of the ring I_1. Attach to each element $x \in R$ the number $x_1(M')$, where $x_1 = xe_1$ to get a nontrivial homomorphism of the ring R into the field of complex numbers. The maximal ideal defined by this homomorphism must belong to F_1, because for each maximal ideal $M_2 \in F_2$ we have $x_1(M_2) = 0$. Hence, each maximal ideal M' of the ring I_1 determines a maximal ideal M_1 of the ring R, so that, evidently, distinct maximal ideals $M' \subset I_1$ determine distinct maximal ideals $M_1 \subset R$, and if M' determines M_1, then M_1, by the earlier method, determines precisely the ideal M'. This establishes a one-to-one correspondence between F_1 and $\mathfrak{M}(I_1)$. From the equations $x(M') = x_1(M_1)$ it follows that the topological spaces F_1 and $\mathfrak{M}(I_1)$ are homeomorphic. The equivalence of F_2 with $\mathfrak{M}(I_2)$ is established in the same manner.

Theorem 26. *Suppose that R is a symmetric ring and that $\mathfrak{M}(R)$ can be represented as a sum of two disjoint closed sets F_1 and F_2. Then R can be decomposed into a direct sum of two ideals I_1 and I_2, which are rings with identities and which have the sets F_1 and F_2 as their sets of maximal ideals. Moreover the ideals I_1 and I_2 are uniquely determined by the sets F_1 and F_2.*

Proof. For the proof of the first assertion it suffices, by Theorem 25, to establish the existence of an element e_1 such that

$$\text{a) } e_1^2 = e_1, \quad \text{b) } \{e_1(M) = 1\} = F_1.$$

The function $f(M)$ which equals 1 on F_1 and 0 on F_2 is continuous; by Theorem 16 there exists an element $x \in R$ such that $|x(M) - f(M)| < \frac{1}{3}$. Hence all the values of $x(M)$ on F_1 are included in the circle $|1-\lambda| < \frac{1}{3}$ and all the values of $x(M)$ on F_2 lie in the circle $|\lambda| < \frac{1}{3}$. Denote by D the set of these circles. The element x belongs to the region $\mathfrak{D} \subset R$ corresponding to D (see §9). The function $e_1(\lambda)$ which equals 1 in the circle $|1-\lambda| < \frac{1}{2}$ and equals 0 in the circle $|\lambda| < \frac{1}{2}$ is analytic in D so it belongs to the ring A_D. We set

$$e_1 = e_1(x) = \frac{1}{2\pi i} \int_{|1-\lambda| = \frac{1}{2}} (\lambda e - x)^{-1} d\lambda .$$

Since $(e_1(\lambda))^2 = e_1(\lambda)$, by equation (2) of §9 $e_1^2 = e_1$. Moreover, $e_1(M)$ is 1 on F_1 and 0 on F_2, so that e_1 is not 0 or e. This proves the first part of the theorem.

For the proof of the second assertion it suffices to prove the following: if e_1 and e_1' are such that $e_1^2 = e_1$, $e_1'^2 = e_1'$ and $\{e_1(M) = 1\} = \{e_1'(M) = 1\}$, then $e_1' = e_1$.

Since the function $e_1(M) - e_1'(M)$ is zero everywhere in $\mathfrak{M}(R)$, the powers of $e_1 - e_1'$ must tend to zero. But, as is simple to verify, $(e_1 - e_1')^{2n+1} = e_1 - e_1'$ for each integer $n \geq 0$. From this it follows that $e_1 = e_1'$.

Remark. In the proof of Theorem 26 we used the symmetry of the ring R only to establish the existence in R of an element x such that $|x(M)|$ was greater than some constant in F_1 and less than that constant in F_2. Therefore the result of Theorem 26 still holds if in place of symmetry of the ring one assumes directly the existence of such an element.

The concept of direct sum of ideals obviously generalizes to the case of an arbitrary finite number of terms $I_1, I_2, \cdots, I_n$; each element $x \in R$ must be represented uniquely as a sum $x_1 + x_2 + \cdots + x_n$, where $x_i \in I_i$ $(i = 1, 2, \cdots, n)$.

Theorem 27. *A ring R with a finite number of maximal ideals can be decomposed into a direct sum of a finite number of ideals, each a ring with one maximal ideal.*

This theorem follows from Theorem 26 since a ring satisfying this condition is symmetric. Indeed, suppose that x is an arbitrary element of R and $\alpha_i = x(M_i)$ $(i = 1, 2, \cdots, n)$ are its values on the maximal ideals. Consider a polynomial $P(\alpha)$ which at the points $\alpha_1, \alpha_2, \cdots, \alpha_n$ takes the corresponding values $\bar{\alpha}_1, \bar{\alpha}_2, \cdots, \bar{\alpha}_n$; the element $y = P(x)$ is evidently complex conjugate to x.

§11. Relation between algebraic and topological homomorphisms

If R is a *ring without radical* (for brevity we speak thus of a ring whose radical contains only the element 0), then, by Theorem 13, R is isomorphically embeddable in the ring of functions $x(M)$ on its set of maximal ideals. Therefore we also call rings without radical *function rings*.

In the present section we consider relations between topological and algebraic properties of function rings.

Theorem 28. *Suppose that R_1 and R_2 are two normed rings of complex functions defined on the same set, that this set serves both rings as the space of maximal ideals, and that R_1 is contained in R_2. Then each sequence of functions $x_1, x_2, \cdots, x_n, \cdots$ from R_1 which converges in the norm of R_1 converges also in the norm of R_2.*

Proof. Denote the norm of a function $x(M)$ ($x \in R_1$) in the ring R_1 by $\|x\|_1$, and that in the ring R_2 by $\|x\|_2$. Introduce into the ring R_1 the new norm

$$\|x\| = \max\{\|x\|_1, \|x\|_2\}. \tag{1}$$

It is easily verified that all the conditions to which a norm is subject are satisfied. Let us show that R_1 is a complete space under this new norm. Suppose that x_1, $x_2, \cdots, x_n, \cdots$ is a sequence of functions of R_1 fundamental in the sense of the norm (1). Clearly it is fundamental in both the earlier senses. Suppose that x and x' are its limits in these two norms. Since the sequence of functions $x_n(M)$ is uniformly convergent to $x(M)$ and $x'(M)$, we have $x(M) \equiv x'(M)$. But R_2 has no generalized nilpotents, so $x = x'$. We have

$$\|x - x_n\| = \max\{\|x - x_n\|_1, \|x - x_n\|_2\} \to 0 \ (n \to \infty).$$

Consequently, x is the limit of the sequence $x_1, x_2, \cdots, x_n, \cdots$ in the norm (1). Thus R_1 is complete in this norm.

Evidently convergence in R_1 under the norm (1) carries with it convergence in the norm of R_1. Since R_1 is complete in both norms, by a theorem of Banach on the bicontinuity of a continuous one-to-one mapping between complete spaces [1, pp. 40-41*] convergence to a limit in the norm of R_1 is equivalent to convergence in the norm (1), which implies convergence in the norm of R_2. This proves the theorem.

As an illustration of Theorem 28 let us prove the following proposition:

Suppose that a normed ring R consists of infinitely differentiable functions $x(t)$ defined on the segment $0 \le t \le 1$. Then there exists a sequence of positive numbers $m_0, m_1, \cdots, m_n, \cdots$ such that for each $x(t) \in R$

$$\max_{0 \le t \le 1} |x^{(n)}(t)| < Cm_n \tag{2}$$

where the constant C depends only on the function and not on the number n.

Proof. It suffices to show that for each n there is a positive number m_n

* See also footnote on page 120.

such that from $x(t) \in R$, $\|x\| \leq 1$ follows $\max_{0 \leq t \leq 1} |x^{(n)}(t)| \leq m_n$. Suppose that for some n no such m_n could be found; then there would exist a sequence of functions $x_1(t), x_2(t), \cdots, x_k(t), \cdots$ in R such that $\|x_k\| \leq 1$ for all k and $\max_{0 \leq t \leq 1} |x_k^{(n)}(t)| > k$. The functions $y_k(t) = \frac{1}{k} x_k(t)$ converge in norm to zero. Since R is evidently contained in the ring D_n, by Theorem 28 the $y_k(t)$ must tend to zero in the norm of D_n, which is impossible since $\max_{0 \leq t \leq 1} |y_k^{(n)}(t)| > 1$.

Corollary. *The ring D_∞ of all infinitely differentiable functions is not normable.*

Proof. Suppose that D_∞ is normed and that $m_0, m_1, \cdots, m_n, \cdots$ is the sequence of numbers from the proof of the preceding proposition. It is possible to construct an infinitely differentiable function $y(t)$ such that $\max_{0 \leq t \leq 1} |y^{(n)}(t)| > nm_n$. But for this function the inequalities (2) can not all hold, in contradiction with the preceding proof.

From Theorem 28 follows simply

Theorem 29. *An algebraic isomorphism of commutative normed rings without radical is always a topological isomorphism; that is, convergence of elements of one ring is equivalent to convergence of the sequence of corresponding elements in the other.*

The meaning and value of this proposition is found in the following. To define a normed ring it is necessary to give, first, the supply of elements and their algebraic properties, and, second, the norm. Theorem 28 shows that for function rings the norm is determined up to equivalence by giving the elements and the algebraic properties.

From Theorem 29 follows, in particular, that the condition formulated in Theorem 17, sufficient for isomorphism of a symmetric ring with the ring of all continuous functions on its space of maximal ideals, is also necessary. Hence, the algebraic ring of all continuous functions on a compact Hausdorff space S can be normed, up to equivalence, in just one way; all possible norms in it are equivalent in the norm $\|x(s)\| = \max_{s \in S} |x(s)|$.

Theorem 30. *Each automorphism of a commutative normed ring R without radical is continuous.*

Proof. Suppose an automorphism of the ring R is given, that is, a one-to-one mapping of the ring on itself such that if x is carried to x', then λx is carried to $\lambda x'$ for each complex number λ, and if x, y are carried to x', y', then $x + y$ is carried to $x' + y'$ and xy to $x'y'$. Now we consider this automorphism as an algebraic isomorphism between the ring R of functions x and the ring R' of functions x'. By Theorem 29 the ring R and R' are topologically isomorphic;

that is, convergence of a sequence $x_1, x_2, \cdots, x_n, \cdots$ is equivalent to convergence of the corresponding sequence $x_1', x_2', \cdots, x_n', \cdots$. But this is just continuity of the automorphism.

Suppose that R is a symmetric ring without radical. In this case the element x^* conjugate to x is uniquely defined. The mapping $x \to x^*$ of the ring on itself differs from an automorphism only in that λx goes not to λx^* but to $\bar{\lambda} x^*$. Theorem 28 shows that this mapping is continuous; that is, from the convergence of x_n to x follows convergence of x_n^* to x^* *. In order to prove this it suffices to introduce in R, along with the old norm $\|x\|$, a new norm $|x| = \|x^*\|$.

APPENDIX I

Some topological problems in the theory of rings

§12. A new method of topologizing the set of maximal ideals

Let R be a normed ring and $\mathfrak{M} = \mathfrak{M}(R)$ its set of maximal ideals. We shall introduce here a topology in $\mathfrak{M}$ which, generally speaking, is different from that introduced in §5 which we shall call the *weak* topology hereafter.

Definition. A maximal ideal M will be called a *cluster point* of a subset $\mathfrak{A} \subset \mathfrak{M}$ if it contains the intersection of all the maximal ideals in $\mathfrak{A}$; the set of all cluster points of a set $\mathfrak{A}$ will be called its closure $\overline{\mathfrak{A}}$.

We verify the axioms of a topological space:

1. $\overline{\mathfrak{A} + \mathfrak{B}} = \overline{\mathfrak{A}} + \overline{\mathfrak{B}}$.

a) $\overline{\mathfrak{A}} + \overline{\mathfrak{B}} \subset \overline{\mathfrak{A} + \mathfrak{B}}$. In fact, if $M_0 \in \overline{\mathfrak{A}} + \overline{\mathfrak{B}}$, then M_0 belongs to at least one of the sets $\overline{\mathfrak{A}}, \overline{\mathfrak{B}}$, for example $\overline{\mathfrak{A}}$; consequently $M_0 \supset \prod_{M \in \mathfrak{A}} M \supset \prod_{M \in \mathfrak{A}+\mathfrak{B}} M$ and hence $M_0 \in \overline{\mathfrak{A} + \mathfrak{B}}$.

b) $\overline{\mathfrak{A}} + \overline{\mathfrak{B}} \supset \overline{\mathfrak{A} + \mathfrak{B}}$. Indeed, if $M_0 \,\overline{\in}\, \overline{\mathfrak{A}} + \overline{\mathfrak{B}}$, then there exist elements $x \in \prod_{M \in \mathfrak{A}} M$ and $y \in \prod_{M \in \mathfrak{B}} M$ not belonging to M_0, so that $x(M_0) \neq 0$ and $y(M_0) \neq 0$; then also $xy(M_0) \neq 0$, that is, the product xy does not belong to M_0. But $xy \in \prod_{M \in \mathfrak{A}+\mathfrak{B}} M$; therefore M_0 is not a cluster point of $\mathfrak{A} + \mathfrak{B}$.

2. $\mathfrak{A} \subset \overline{\mathfrak{A}}$. Obvious.

3. $\overline{\overline{\mathfrak{A}}} = \overline{\mathfrak{A}}$. Obvious.

4. We will define the closure of the empty set to be empty.

Each single point $M \in \mathfrak{M}$ is evidently a closed set; consequently $\mathfrak{M}$ is a T_1-space. In general it need not be a T_2-space.

Indeed, let us consider what happens when we introduce this topology into

* The converse assertion (from convergence of x_n^* to x^* follows convergence of x_n to x) is already contained in this, since for each $z \in R$ we have $z^{**} = z$.

the set $\mathfrak{M}$ of the ring A. As we know (see page 130) $\mathfrak{M}(A)$ is the set of all points of the circle $|z| \leq 1$. Let F be an infinite subset having a limit point (in the usual sense) in the interior of the circle; we find the points of accumulation of F. By our definition these are points which carry to zero each function $\phi(z) \in A$ which vanishes on F. But by the uniqueness theorem for analytic functions only one function –the identically zero function– equals zero for all points of F. Hence every point of the circle is a cluster point for F and $\bar{F} = \mathfrak{M}$. From this it follows that each domain $O \subset \mathfrak{M}$, if it is not empty, contains the entire interior of the circle with the possible exception of a finite set of points. Therefore not every pair of points can be included in disjoint neighborhoods so $\mathfrak{M}(A)$ in its new topology is not a T_2- space. From this example it is evident that, generally speaking, this topology is different from weak. Hence, from just this it follows that, generally speaking, the functions $x(M)$ are not all continuous in the new topology, for were they continuous $\mathfrak{M}(A)$ would be a T_2- space.

The sets $\mathfrak{A}$ of the form $\{x(M) = 0\}$ are closed since each cluster point of such a set must contain the element x and must belong to $\mathfrak{A}$. The open sets complementary to such closed sets form a complete system of neighborhoods. In fact, if O is an arbitrary open set and F is its complement, then for each $M_0 \in O$ there exists $x \in R$ such that $x(M_0) \neq 0$ and $x(M) = 0$ for all $M \in F$, because the maximal ideal M_0 is not a cluster point of F. The neighborhood $\{x(M) \neq 0\}$ is contained in O.

Theorem 1. *The space* $\mathfrak{M}$ *topologized in this way is compact.*

Proof. Suppose that $\{F_\alpha\}$ is an arbitrary system of closed sets. To each F_α corresponds an ideal $I_\alpha = \prod_{M \in F_\alpha} M$. Consider the set I of all finite sums of elements taken from the ideals I_α. Either I coincides with R or else it is an ideal and is, therefore, contained in some maximal ideal M_0. For each α this ideal M_0 contains all the elements of I_α and therefore belongs to the closed set F_α; hence $M_0 \in \prod_\alpha F_\alpha$.

Now suppose that it is given that the intersection $\prod_\alpha F_\alpha$ is empty; then I can not be contained in any maximal ideal, and therefore is the whole ring R. In this case some sum $\sum_{i=1}^n x_i$ equals e ($x_i \in I_{\alpha_i}$, $i = 1, 2, \cdots, n$). We assert that the finite intersection $\prod_{i=1}^n F_{\alpha_i}$ is empty. Indeed, if some maximal ideal belonged to all the sets F_{α_i}, $i = 1, 2, \cdots, n$, then it would contain all the points $x_i \in I_{\alpha_i}$, and with them also the identity, which is impossible.

Thus we see that from each system of closed sets F of the space $\mathfrak{M}$ whose intersection is empty a finite subset can be selected, also with empty intersection. But this is just compactness of the space $\mathfrak{M}$.

§13. Compact extensions of completely regular spaces

Let us apply our method of topologization of the set $\mathfrak{M}$ to the ring $C(S)$ of all real bounded continuous functions defined in some topological space S, with norm $\|x\| = \sup_{t \in S} |x(t)|$ and ordinary algebraic operations. $C(S)$ is a real normed ring (see page 147). To each point $a \in S$ corresponds a maximal ideal $M_a \in \mathfrak{M}$, consisting of all functions vanishing at $t = a$, so that $x(M_a) = x(a)$ for all $x \in C(S)$. If, in addition, for each pair of distinct points a and b there is a function $x(t) \in C(S)$ such that $x(a) \neq x(b)$, then distinct points yield distinct maximal ideals. Here S is dense in $\mathfrak{M}$, because the intersection of all the maximal ideals corresponding to points of the space S is the single function $x(t) \equiv 0$, which, as the zero element of the ring $C(S)$, belongs to every maximal ideal; consequently each maximal ideal is a cluster point of S.

Let us find those properties of the domain S necessary that its topology agree with that induced on S by $\mathfrak{M}$; that is, that S be homeomorphic to a subset of $\mathfrak{M}$. Let us suppose that this holds; let us consider a closed set $A \subset S$ and the corresponding set $\mathfrak{A}$ of maximal ideals. Since by assumption $\mathfrak{A}$ is closed in S, then for each maximal ideal $M_a \in S$ but not in $\mathfrak{A}$ there is an element x such that $x(M_a) \neq 0$ and $x(M) = 0$ for all $M \in \mathfrak{A}$; otherwise M_a would be a cluster point of the set $\mathfrak{A}$. Returning to the functions defined on the set S, we get for each closed set $A \subset S$ and each point a in S but not in A a function $x(t)$ in $C(S)$ such that $x(a) \neq 0$ and $x(t) = 0$ for all $t \in A$. As is well known, a space S with property is called completely regular.

Conversely, if S is completely regular, then it is homeomorphic to a subset of $\mathfrak{M}$. In fact, suppose that A is a subset of S and that $\mathfrak{A}$ is the corresponding set of maximal ideals; suppose further that $a \,\overline{\in}\, A$ and that M_a is the maximal ideal corresponding to a. If a is a cluster point of the set A, then each function $x(t) \in C(S)$ which vanishes on all of A also vanishes at a, so that M_a is a cluster point of $\mathfrak{A}$. If a is not a cluster point of A, so that $a \,\overline{\in}\, \overline{A}$, then be the complete regularity of the space S there is a function $x(t) \in C(S)$ vanishing at all points of $\overline{A}$ but different from zero at a; therefore M_a is not a cluster point of $\mathfrak{A}$.

We have proved the following proposition:

Theorem 2. *A topological space S is homeomorphic to a subset of the space $\mathfrak{M}(C(S))$ if and only if it is completely regular.*

Suppose that S is a regular space. Each function $x(t) \in C(S)$ has a natural extension from S to all of $\mathfrak{M}$, the function $x(M)$; we prove next that under this assumption all these functions are continuous.

Lemma 1. *Suppose that A is some subset of the space S and that $\overline{A}$ is its closure in $\mathfrak{M}$. Then the values of any function $x(M) \in C(S)$ on $\overline{A}$ are its values on*

A and, perhaps, some limit points of these.

Proof. If the function $x(M)$ vanishes at all points of the set A, then, by definition of closure, it vanishes at all points of the set $\overline{A}$. If the function $x(M) = 1$ at all points of A, then $y(M) = x(M) - 1$ vanishes at all points of A, and hence of $\overline{A}$; hence $x(M)$ is one at all points of $\overline{A}$.

Now suppose that $\alpha = \inf_{M \in A} |x(M)| > 0$; let us show that the same if true for $\overline{A}$. A function $y(M)$ can be found in $C(S)$ which coincides on A with $\frac{1}{x(M)}$. It can be defined, for example, thus: At each point $M \in S$ at which $|x(M)| \geq \alpha$ set $y(M) = \frac{1}{x(M)}$, and at all other points set $y(M) = \frac{x(M)}{\alpha^2}$. The continuity of this function is easily verified. The product $x(M)y(M)$ equals 1 on A and consequently, by the preceding, on $\overline{A}$; and since $|y(M)| \leq \frac{1}{\alpha}$, then $\inf_{M \in \overline{A}} |x(M)| \geq \alpha$.

After this the assertion of the lemma is almost evident. If there were a function $y(M)$ which took on $\overline{A}$ a value λ not one of its values or a limit of its values on A, then $y(M) - \lambda$ would take on $\overline{A}$ the value 0 which would not be a value or limit of its values on A, which we have just seen is impossible.

Lemma 2. *Suppose that U is a neighborhood in $\mathfrak{M}$ of the form $\{x(M) \neq 0\}$. Then $U \subset \overline{U \cap S}$.*

Proof. Set $CU = \mathfrak{M} - U = \{x(M) = 0\}$. Then $S = U \cap S + CU \cap S$, $\overline{S} = \overline{U \cap S} + \overline{CU \cap S}$. But since $\overline{S} = \mathfrak{M} \supset U$, then $U \subset \overline{U \cap S} + \overline{CU \cap S}$. But by Lemma 1 $x(M) = 0$ in all of $\overline{CU \cap S}$, so that $U \subset \overline{U \cap S}$.

Theorem 3. *All functions $y(M)$ of the ring $C(S)$ are continuous on $\mathfrak{M}$.*

Proof. Suppose given a point $M_0 \in \mathfrak{M}$, a function $y(M) \in C(S)$, and a number $\epsilon > 0$; we need to find such a neighborhood U of M_0 that at each of its points $|y(M) - y(M_0)| < \epsilon$. Suppose that A is the set of all points M of S at which $|y(M) - y(M_0)| \geq \frac{\epsilon}{2}$. By Lemma 1 $M_0 \overline{\in} \overline{A}$. Since the system of all neighborhoods $\{x(M) \neq 0\}$ is complete, there exists a neighborhood $U = \{x(M) \neq 0\}$ of the point M_0 which is contained in $\mathfrak{M} - \overline{A}$. This is the desired neighborhood. From $U \cap \overline{A} = 0$ follows that $(U \cap S) \cap (\overline{A} \cap S) = (U \cap S) \cap A = 0$; that is, that in all the points of $U \cap S$ $|y(M) - y(M_0)| < \frac{\epsilon}{2}$; then by Lemma 1 $|y(M) - y(M_0)| \leq \frac{\epsilon}{2} < \epsilon$ for all points of $\overline{U \cap S}$ and, in particular, by Lemma 2, at all points of U.

Corollary 1. *$C(S)$ is isomorphic to $C(\mathfrak{M})$.*

Indeed, each function continuous on $\mathfrak{M}$ is continuous on $S \subset \mathfrak{M}$; we just showed that the converse is true. Therefore $C(S)$ and $C(\mathfrak{M})$ are algebraically isomorphic. Since S is dense in $\mathfrak{M}$, $C(S)$ and $C(\mathfrak{M})$ are also isometric.

Corollary 2. *The space $\mathfrak{M}(C(S))$ is Hausdorff.*

Indeed, for each two points M_0 and M_1 there exists a function $x(M)$ such that $x(M_0) = 0$, $x(M_1) = 1$. Since $x(M)$ is continuous, the sets $\{|x(M)| < \frac{1}{2}\}$ and

and $\{|x(M)| > \frac{1}{2}\}$ are open. The first contains M_0, the second M_1, and the intersection is empty.

Theorem 4. *Suppose that an arbitrary compact extension of S is given, that is, a compact Hausdorff space Q containing a dense subset homeomorphic to S. Then there exists a continuous mapping of the space $\mathfrak{M}(C(S))$ onto Q under which the points of S are carried to themselves; in other words, in some sense $\mathfrak{M}$ is the largest bicompact extension of the space S.*

Proof. Consider the ring $C(Q)$. Since each function continuous on Q is continuous on $S \subset Q$, the ring $C(Q)$ is a closed subring of $C(S)$. By the well known theorem of Urysohn* the ring $C(Q)$ has the "separation property"; for each two distinct points of the space Q there is a continuous function taking distinct values at them. Also, if the function $x(t) \in C(Q)$ does not vanish on Q, then its modulus has a positive lower bound on Q, and therefore $\frac{1}{x(t)} \in C(Q)$. Using this property just as on pp. 121, 122 we show that the set of maximal ideals of the ring $C(Q)$ coincides with the set Q. Again by the theorem of Urysohn Q is completely regular. Therefore by Theorem 2 the set of maximal ideals of the ring $C(Q)$ is homeomorphic to Q. Suppose that M is a point of $\mathfrak{M}(C(S))$; choosing out of the maximal ideal M all functions belonging to $C(Q)$ we get, evidently, a maximal ideal $M' \in \mathfrak{M}(C(Q))$, that is, a point of the set Q. Indeed, attaching to each function $x(t) \in C(Q)$ its values $x(M)$ on the maximal ideals M, we get a homomorphism of the ring $C(Q)$ into the field of real numbers, which is non-trivial because $x(t) = 1$ is carried to the number 1. The kernel of this homomorphism is a maximal ideal in $C(Q)$ and coincides with the set of all functions of $C(Q)$ which belong to M. We have constructed a mapping $M \to M'$ of the space $\mathfrak{M}(C(S))$ into the space Q which, evidently, leaves the points of S fixed; it remains only to show that this mapping is continuous, and then, in view of the bicompactness of a continuous image of a compact space and the everywhere density of S in Q, it will follow that the image of $\mathfrak{M}(C(S))$ is all of Q.

But the continuity of the mapping is evident; if M is a cluster point of $\mathfrak{A} \subset \mathfrak{M}(C(S))$, then all continuous functions on $\mathfrak{M}$ which vanish on $\mathfrak{A}$ also vanish at M; since this is true, in particular, for the functions in $C(Q)$, the image of M is a cluster point of the image of $\mathfrak{A}$. With this our assertions are completely proved.

The construction of Čech [4] of a compact extension $\beta(S)$ also gives a largest extension in this sense of the space S; the proposition just proved allows us to conclude that $\beta(S)$ coincides with $\mathfrak{M}(C(S))$, since it is not difficult to show that a maximal extension is unique (this follows, for example, from the results of §15).

* See, for example, F. Hausdorff, *Mengenlehre*, de Gruyter, Berlin, 1935, pp. 231-232.

§ 14. Another proof of the compactness of the set of maximal ideals in the weak topology

By adding to the ring $C(S)$ all complex valued, continuous, bounded functions on S, we derive a new ring; denote it by $C_1(S)$. All functions $x(t) + iy(t) \in C_1(S)$ can be extended continuously to $\mathfrak{M}(C(S))$ and, by Corollary 1 of Theorem 3 of the preceding section, generates thus the whole ring of complex continuous functions. Since $\mathfrak{M}(C(S))$ is compact, it must coincide with the whole set of maximal ideals of the ring $C_1(S)$.

Now suppose that R is an arbitrary commutative normed ring, and that $\mathfrak{M}(R)$ is the set of all its maximal ideals. Introduce in $\mathfrak{M}(R)$ the "weak" topology (see Definition 7 on page 133): A neighborhood $U(M_0)$ is defined by an arbitrary number $\epsilon > 0$ and elements $x_1, x_2, \cdots, x_n$ of R to be the set of all maximal ideals $M \subset R$ which satisfy the inequalities

$$|x_i(M) - x_i(M_0)| < \epsilon \quad (i = 1, 2, \cdots, n).$$

We show here, without using the theorem of Tyhonov on the compactness of the topological product of compact spaces, that *with this topology the space* $\mathfrak{M}(R)$ *becomes a compact Hausdorff space* (Theorem 12 of §5).

That the space is Hausdorff, that is, that each two of its points are separated by disjoint neighborhoods, follows easily from the definition of neighborhoods. But more than that, the space $\mathfrak{M}(R)$ is completely regular. In fact, let F be an arbitrary closed set and M_0 a point not belonging to it. Then there is a neighborhood $U(M_0)$

$$\{|x_i(M) - x_i(M_0)| < \epsilon\} \quad (i = 1, 2, \cdots, n),$$

containing no points of F. Consider the function

$$f(M) = \sum_{i=1}^{n} |x_i(M) - x_i(M_0)|.$$

This function is continuous and at all points of the set F takes values not less than ϵ; at the point M_0 it vanishes. Now set

$$f_1(M) = \epsilon - \min[f(M), \epsilon].$$

It is clear that $f_1(M)$ is continuous, equals 0 everywhere on F, and is not zero at M_0. Hence $\mathfrak{M}(R)$ is completely regular.

To prove compactness of $\mathfrak{M}(R)$ set $\mathfrak{M}(R) = S$ and consider the ring $C_1(S)$. The space $\mathfrak{M}(C_1(S))$ of its maximal ideals, topologized as on page 160, contains a subset homeomorphic to S and is its compact extension. We show that S is a continuous image of $\mathfrak{M}(C_1(S))$. Choose from the maximal ideal $M_1 \subset C_1(S)$ the functions belonging to R; these functions form some maximal ideal M in R. Define a mapping of $\mathfrak{M}(C_1(S)) \to S$ by attaching to each maximal ideal $M_1 \subset C_1(S)$ the related maximal ideal $M \subset R$. Under this mapping the points of S are carried into themselves, so the image covers all of S. Moreover, if M_1 is carried to M, then for each $x \in R$ $x(M_1) = x(M)$. Therefore, the complete inverse image of the set $\{ |x_i(M) - x_i(M_0)| < \epsilon; i = 1, 2, \cdots, n\}$ in $\mathfrak{M}(R)$ is the set $\{ |x_i(M_1) - x_i(M_0)| < \epsilon; i = 1, 2, \cdots, n\}$ in $\mathfrak{M}(C_1(S))$, which, by continuity of the functions $x(M_1)$, is open in $\mathfrak{M}(C_1(S))$. But this is continuity of the mapping $M_1 \to M$. Hence $\mathfrak{M}(R) = S$ is compact, because it is the continuous image of a compact space $\mathfrak{M}(C_1(S))$.

The following problem arises naturally: Under what conditions do the two topologies we have introduced into the set of maximal ideals of a commutative normed ring coincide? It is easily proved that this occurs then and only then when for each closed (in the weak topology) set $\mathfrak{A} \subset \mathfrak{M}$ and each point M_0 not in $\mathfrak{A}$ can be found a function $x(M)$ $(x \in R)$ such that $x(M_0) \neq 0$ and $x(M) = 0$ for all $M \in \mathfrak{A}$. Indeed, suppose that this condition, which we shall call the *regularity condition*, is satisfied. Then inside every "weak" neighborhood $U = \{ |x_i(M) - x_i(M_0)| < \epsilon; i = 1, 2, \cdots, n\}$ can be found a neighborhood in the new sense $\{ y(M) \neq 0\}$, for we can find such a $y(M)$ by taking $y(M) = 0$ in $\mathfrak{M} - U$ and different from 0 at M_0. On the other hand, the sets $\{ y(M) \neq 0\}$, which, as we saw, from a basis for neighborhoods in the new topology, are all open in the weak topology. Hence the topologies coincide. Conservely, if it is known that the topologies coincide, then in the "weak" neighborhood $\{ |x_i(M) - x_i(M_0)| < \epsilon; i = 1, 2, \cdots, n\}$, separating a point M_0 from a given closed set $\mathfrak{A}$ can be found a neighborhood in the new sense $\{ y(M) \neq 0\}$ containing M_0; evidently $y(M_0) \neq 0$ and $y(M) = 0$ for all $M \in \mathfrak{A}$, so the condition of regularity is fulfilled.

A ring which satisfies the condition of regularity is called *regular*.

§15. Classification of all closed subrings of $C(S)$

The weak topology of the space of maximal ideals of a real ring can be given by a determining system of neighborhoods of the following form:

$$\{ |x(M)| < \epsilon\} \quad (x \in R), \tag{1}$$

since in each neighborhood $[M_0; x_1, \cdots, x_n; \delta]$ is contained a neighborhood (1), where $x = \sum [x_k - x_k(M_0)]^2$ and $\epsilon = \delta^2$.

The same topology is found if cluster points of a subset $\mathfrak{A} \subset \mathfrak{M}(R)$ are defined in the following way: A maximal ideal M_0 is a cluster point of the set $\mathfrak{A}$ if for each $x \in M_0$ $\inf_{M \in \mathfrak{A}} |x(M)| = 0$*. Indeed, if M_0 satisfies this last condition, then in every neighborhood of M_0, which we may think of as $\{ |x(M)| < \epsilon\}$, where $x \in M_0$, contains a point of $\mathfrak{A}$, and consequently M_0 is indeed a cluster point of $\mathfrak{A}$. Conversely, if M_0 does not have a given property, then there exists $x \in M_0$ such that $|x(M)| \geq c > 0$ for all $M \in \mathfrak{A}$, so the neighborhood $\{ |x(M)| < c\}$ of the point M_0

* Or alternatively: A maximal ideal M_0 is a cluster point of the set $\mathfrak{A}$ if it is contained in the closure (in the ring norm) of the union of the maximal ideals $M \in \mathfrak{A}$.

contains no point of $\mathfrak{A}$, and, consequently, M_0 is no cluster point of $\mathfrak{A}$.

Suppose now that K is a real normed ring made up of continuous functions (generally speaking, not all of them) defined on a topological space S; let us explain those conditions under which S, which, as we know, is a subset of $\mathfrak{M}(K)$, be homeomorphic to this part, that is, that the original topology in S be that induced by $\mathfrak{M}(K)$.

Let us assume that this holds. Consider an arbitrary closed set $A \subset S$ and the corresponding set $\mathfrak{A}$ of maximal ideals of the ring K. Since by assumption A is closed in S, for each maximal ideal $M_0 \in S$, but not in $\mathfrak{A}$, can be found an $x \in K$ such that $x(M_0) = 0$, $\inf\limits_{M \in \mathfrak{A}} |x(M)| > 0$; no such M_0 can be a cluster point for $\mathfrak{A}$ (see above). Returning to the functions, we get: *For each closed set $A \subset S$ and each point a in S but not in A, there is in K a function $x(t)$ such that $x(a) = 0$, $\inf\limits_{t \in A} |x(t)| > 0$.*

It is easily verified that this condition, which we shall call condition (*), is also sufficient.

For this, suppose that (*) is satisfied. Let A be a subset of the space S, and let $\mathfrak{A}$ be the corresponding set of maximal ideals; suppose also that $a \,\overline{\in}\, A$ and that $M_a \,\overline{\in}\, \mathfrak{A}$ is the corresponding maximal ideal. If M_a is not a cluster point of $\mathfrak{A}$, there is a function $x(M)$ such that $x(M_a) = 0$, $|x(M)| \geq c > 0$ for all $M \in \mathfrak{A}$. Then the neighborhood $\{ |x(t)| < c \}$ of the point a contains no point of the set A, so the point a is not a cluster point of the set A. Conversely, if a is not a cluster point of A, so that $a \,\overline{\in}\, A$, then there is $x(t) \in K$ such that $x(a) = 0$, $\inf\limits_{t \in A} |x(t)| > 0$; but then M_a is not a cluster point of $\mathfrak{A}$.

Theorem 5. *For each closed subring K of the ring $C(S)$ which satisfies the condition (*) there exists a compact extension Q of the space S such that K is isomorphic to the ring $C(Q)$. Conversely, if the subring K of the ring $C(S)$ is isomorphic to $C(Q)$, where Q is some compact extension of the space S, then K satisfies condition (*). In this case a function in $C(Q)$ coincides on the set S with the function in K which corresponds to it, and a function which is not in K can not be extended to a function in $C(Q)$.*

Proof. Suppose that the closed subring $K \subset C(S)$ satisfies the condition (*); then the compact space $Q = \mathfrak{M}(K)$ topologically contains S. We show that S is dense in Q, that is, that each point M_0 of $Q - S$ is a cluster point of S.

Suppose the contrary; there is a maximal ideal M_0 not a cluster point of S. Then we can find such an element $x \in M_0$ that $\inf\limits_{t \in S} |x(t)| = l > 0$. Let $L = \sup\limits_{t \in S} |x(t)|$. The values of the function $x(t)$ are contained in the two intervals $[-L, -l]$ and $[l, L]$. By the theorem of Weierstrass a polynomial $P(x)$ can be found which, everywhere in these two intervals, differs from the continuous function $\frac{1}{x}$ by less than ϵ, where ϵ is an arbitrary positive number. Then the polynomial $P(x(t))$ differs from $\frac{1}{x(t)}$ by less than ϵ everywhere in S. Since $P(x(t)) \in K$ and ϵ is arbitrary and K is closed, $\frac{1}{x(t)} \in K$. But this contradicts the supposition that $x \in M_0$, and consequently has no inverse in K.

In consequence of this the functions $x(M)$ make up on $Q = \mathfrak{M}(K)$ a ring which is also closed under uniform convergence. Applying Theorem 16′ we get $K = C(Q)$.

Now suppose that $K = C(Q)$; we show that S satisfies condition (*). Suppose that $A \subset S$ is closed and that $a \overline{\in} A$. Since S is topologically imbedded in Q, the closure $\overline{A}$ of A in Q does not contain a. By Urysohn's theorem there is a function $x(t)$ continuous on Q, equal to 1 on $\overline{A}$ and 0 at a. Considering it only in S we have the function we needed in K.

It is now simple to classify all closed subrings of the ring $C(S)$. Let K be an arbitrary closed subring; introduce in S a new topology in which by definition a point a is a cluster point of a set $A \subset S$ if from $x(a) = 0$ and $x \in K$ follows $\inf_{t \in A} |x(t)| = 0$. It is easily verified that all the topological axioms are satisfied and that the new topological space S' is a continuous image of the space S. Moreover all functions $x(t) \in K$ are continuous on S' and make up a ring on it satisfying condition (*). By Theorem 5 of this section, K is isomorphic to $C(Q')$, where Q' is some compact extension of the space S'.

Also conversely if Q' is a compact extension of some continuous image S' of the space S, then the functions of the ring $C(Q')$ determine a closed subring of the ring $C(S)$.

Let us formulate this result in final form.

Theorem 6. *To each closed subring K of the ring $C(S)$, where S is an arbitrary topological space, corresponds some compact extension Q' of some continuous image S' of S so that K is isomorphic to $C(Q')$, and conversely.*

APPENDIX II.

Rings with convolution

§16. The ring of absolutely integrable functions

In this Appendix an important class of commutative normed rings connected with the theory of Fourier integrals will be considered.

Denote by L_1 the space of all measurable, absolutely integrable complex functions of a real variable $t(-\infty < t < \infty)$ with norm

$$\|x\| = \int_{-\infty}^{\infty} |x(t)|\, dt.$$

A multiplication of elements, called convolution, can be defined in L_1 by the formula

$$(x^*y)(t) = \int_{-\infty}^{\infty} x(t-\tau)y(\tau)\,d\tau. \tag{1}$$

With the aid of the theorem of Fubini it is easy to show that *the integral on the right-hand side of formula* (1) *exists for almost all* t, *is absolutely integrable as a function of* t, *and satisfied* $\|x^*y\| \le \|x\|\,\|y\|$. Indeed,

$$\int_{-\infty}^{\infty} \left| \int_{-\infty}^{\infty} x(t-\tau)y(\tau)\,d\tau \right| dt \le \int_{-\infty}^{\infty} \left(\int_{-\infty}^{\infty} |x(t-\tau)|\,|y(\tau)|\,d\tau \right) dt,$$

and by the theorem of Fubini

$$\int_{-\infty}^{\infty} \left(\int_{-\infty}^{\infty} |x(t-\tau)|\,|y(\tau)|\,d\tau \right) dt = \int_{-\infty}^{\infty} \left(\int_{-\infty}^{\infty} |x(t-\tau)|\,dt \right) |y(\tau)|\,d\tau,$$

where convergence of one of these two iterated integrals carries with it convergence of the other. But the integral on the right hand side converges and equals

$$\int_{-\infty}^{\infty} |x(t)|\,dt \int_{-\infty}^{\infty} |y(\tau)|\,d\tau = \|x\|\,\|y\|.$$

From this our assertion follows. It is easily verified that *convolution is associative and commutative.*

L_1 *contains no identity under convolution.* Indeed, it is easily shown that *the result of convolution of a bounded function with an arbitrary absolutely integrable function is a continuous function.* In fact, suppose that $|x(t)| \le C$, $y(t)$ is an arbitrary function from L_1, and $z = x^*y$. Then

$$|z(t+h) - z(t)| = \left| \int_{-\infty}^{\infty} y(t+h-\tau)x(\tau)\,dt - \int_{-\infty}^{\infty} y(t-\tau)x(\tau)\,d\tau \right| \le$$

$$\le C \int_{-\infty}^{\infty} |y(t+h-\tau) - y(t-\tau)|\,d\tau = C \int_{-\infty}^{\infty} |y(\tau+h) - y(\tau)|\,d\tau.$$

But the last integral tends to zero as $h \to 0$; consequently, $z(t)$ is a uniformly continuous function of t. Now if L_1 contained a unit under convolution, then each bounded function in L_1 would have to coincide almost everywhere with some continuous function, specifically, with that continuous function resulting from convolution with the identity. But L_1 contains bounded discontinuous functions which differ from each continuous function on sets of positive measure; a simple example is a function equal to 1 in some interval and equal to 0 outside it. We have reached a contradiction and shown that there is no unit in L_1.

Denote by $V^{(a)}$ the normed ring obtained by formal adjunction of a unit to L_1, Then $V^{(a)}$ consists of elements $\mathfrak{z} = \lambda e + x(t)$, where e is the adjoined identity, λ is an arbitrary complex number, and $x(t)$ is an arbitrary element of L_1, and where $\|\mathfrak{z}\| = |\lambda| + \|x\|$. It is clear that L_1 is a maximal ideal in $V^{(a)}$. We denote this maximal ideal by M_∞; below we shall show that in the space of maximal ideals of the ring $V^{(a)}$ it plays the role of an infinitely distant point.

Let us find the other maximal ideals of $V^{(a)}$. Let s be an arbitrary real number. For each element $\mathfrak{z} = \lambda e + x(t) \in V^{(a)}$ we set

$$\mathfrak{Z}(s) = \lambda + \int_{-\infty}^{\infty} x(t)\, e^{ist} dt.$$

It is easily verified that the mapping $\mathfrak{z} \to \mathfrak{Z}(s)$ is a homomorphism of the ring $V^{(a)}$ into the field of complex numbers. To prove this we need only show that to the product of elements of the ring corresponds the product of the corresponding numbers. For this it suffices to show that to the convolution of functions $x(t), y(t) \in L_1$ corresponds the product of their Fourier transforms

$$X(s) = \int_{-\infty}^{\infty} x(t)\, e^{ist} dt, \quad Y(s) = \int_{-\infty}^{\infty} y(t)\, e^{ist} dt.$$

But after all

$$\int_{-\infty}^{\infty} (x*y)(t)\, e^{ist} dt = \int_{-\infty}^{\infty} \Big(\int_{-\infty}^{\infty} x(t-\tau)\, y(\tau)\, d\tau \Big) e^{ist} dt =$$
$$= \int_{-\infty}^{\infty} \Big(\int_{-\infty}^{\infty} x(t-\tau) e^{is(t-\tau)} y(\tau) e^{is\tau} d\tau \Big) dt = \int_{-\infty}^{\infty} \Big(\int_{-\infty}^{\infty} x(t-\tau) e^{is(t-\tau)} dt \Big) y(\tau) e^{is\tau} d\tau =$$
$$= \int_{-\infty}^{\infty} x(t) e^{ist} dt \int_{-\infty}^{\infty} y(\tau) e^{is\tau} d\tau.$$

The maximal ideals are generated by the mappings $\mathfrak{z} \to \mathfrak{Z}(s_0)$, i.e., the set of all elements $\mathfrak{z} = \lambda e + x(t) \in V^{(a)}$ such that

$$\mathfrak{Z}(s_0) = \lambda + X(s_0) = \lambda + \int_{-\infty}^{\infty} x(t) e^{is_0 t} dt = 0,$$

we denote by M_{s_0}. Then, $\mathfrak{z}(M_{s_0}) = \mathfrak{Z}(s_0)$. Two maximal ideals M_{s_1} and M_{s_2} are distinct if $s_1 \neq s_2$, because it is easy to construct a function $x(t) \in L_1$ such that $X(s_1) \neq X(s_2)$. Moreover, all these maximal ideals are distinct from M_∞, because $x(M_\infty) = 0$ for all x in L_1 but for each s_0 there can be found a function $x(t) \in L_1$ such that $x(M_{s_0}) \neq X(s_0) \neq 0$.

We show that *the maximal ideals* M_s *and* M_∞ *exhaust the set of maximal ideals of the ring* $V^{(a)}$.

We set

$$h_a(t) = \begin{cases} 1 \text{ if } 0 \le t < a, \\ 0 \text{ for all other values of } t. \end{cases}$$

Evidently the characteristic function of a half-open interval $[a_1, a_2)$ equals the difference $h_{a_1}(t) - h_{a_2}(t)$. Hence it follows that *the functions* $h_a(t)$ *generate* L_1, *and therefore also* $V^{(a)}$.

Lemma 1. *A closed ideal which contains an absolutely integrable function*

$x(t)$ contains all the translated functions $x(t-\lambda)$; moreover,

$$x(t-\lambda) = \lim_{\tau\to 0} x(t) * \frac{h_{\lambda+\tau}(t) - h_\lambda(t)}{\tau}, \tag{2}$$

where the limit is taken in the sense of norm convergence.

Proof. The functions $\frac{h_{\lambda+\tau}(t)-h_\lambda(t)}{\tau}$ are bounded in norm (all have norm 1). From this it follows that to verify condition (2) it suffices to prove it for functions $x(t) = h_a(t)$, which, as was noted above, generate $V^{(a)}$. But for $x(t) = h_a(t)$ the product appearing under the limit sign in (2) is easily calculated; a graph of it appears in the sketch. The difference between this product and $h_a(t-\lambda)$ has norm τ, as it equals the sum of the areas of the triangles ABC and $A_1B_1C_1$. Therefore, the limit of the product as $\tau \to 0$ is the function $h_a(t-\lambda)$.

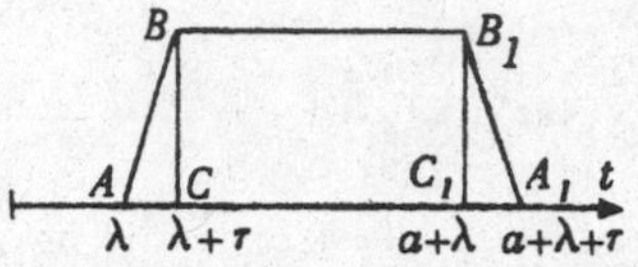

Theorem 1. *If the maximal ideal M of the ring $V^{(a)}$ is distinct from the maximal ideal M_∞, then there exists a real number s such that for each $\mathfrak{z} = \lambda e + x(t) \in V^{(a)}$*

$$\mathfrak{z}(M) = \lambda + \int_{-\infty}^{\infty} x(t) e^{ist} dt; \tag{3}$$

in other words, M coincides with the maximal ideal M_s.

Proof. Let M be a maximal ideal of the ring $V^{(a)}$ distinct from M_∞. Under the homomorphic mapping of $V^{(a)}$ onto $V^{(a)}/M$ a function $h_a(t)$ is carried to some complex number $\phi(a)$. Consequently the function $h_a(t-\lambda) = h_{\lambda+a}(t) - h_\lambda(t)$ is carried to the number $\phi(\lambda+a) - \phi(\lambda)$ and the limit formula (2) in which one sets $x(t) = h_a(t)$ is carried to the formula

$$\phi(\lambda+a) - \phi(\lambda) = \lim_{\tau\to 0} \phi(a) \frac{\phi(\lambda+\tau)-\phi(\lambda)}{\tau}. \tag{4}$$

If $\phi(a)$ vanished identically, then we should have $h_a(t) \in M$ for all a; since linear combinations of the functions $h_a(t)$ are dense in $L_1 = M_\infty$, it would follow from this that $M_\infty \subset M$, i.e., $M_\infty = M$, which contradicts the hypotheses. Hence $\phi(a)$ does not vanish identically. In this case it follows from relation (4) that $\phi'(\lambda)$ exists for every λ and has the following properties:

(α) $\phi(\lambda+a) - \phi(\lambda) = \phi(a)\phi'(\lambda)$.

(β) $\phi'(\lambda)$ is continuous; this follows from (α) by use of a point where $\phi(a)$ is non-zero and the continuity of $\phi(\lambda)$.

(γ) $|\phi'(\lambda)| \le 1$; indeed we saw earlier that the function $\frac{h_{\lambda+\tau}(t)-h_\lambda(t)}{\tau}$ has norm 1, so the fraction $\frac{\phi(\lambda+\tau)-\phi(\lambda)}{\tau}$ does not exceed 1 in absolute value; clearly this property is preserved during passage to the limit $\phi'(\lambda)$.

(δ) $\phi'(0) = 1$; first it is evident that $\phi(0) = 0$; therefore, if in the relation (α) we set $\lambda = 0$ we get $\phi(a) = \phi(a)\phi'(0)$, from which our assertion follows because $\phi(a) \not\equiv 0$.

Differentiating (α) with respect to a we get

$$\phi'(\lambda+a)=\phi'(\lambda)\,\phi'(a). \tag{5}$$

In particular, when $a=-\lambda$ we get from this and (δ) that

$$1=\phi'(0)=\phi'(\lambda)\,\phi'(-\lambda).$$

Hence, in view of (γ), it follows that

$$|\phi'(\lambda)|\equiv 1 \text{ for all } \lambda. \tag{6}$$

The relations (5) and (6) together show that $\phi'(\lambda)$ is a (continuous) homomorphism of the additive group of real numbers into the multiplicative group of complex numbers on the unit circumference. Such a mapping always has the form $e^{is\lambda}$, where s is some real number. We have, consequently,

$$\phi'(\lambda)=e^{is\lambda},$$

and since $\phi(0)=0$ also

$$h_\lambda(M)=\phi(\lambda)=\int_0^\lambda e^{ist}dt=\int_{-\infty}^{\infty} h_\lambda(t)\,e^{ist}dt.$$

The equation (3) is valid, therefore, for functions $\mathfrak{z}=h_\lambda(t)$. Since these functions generate $V^{(a)}$, the equation is valid for all $\mathfrak{z}\in V^{(a)}$ and Theorem 1 is proved. From it, by Theorems 3 and 24 of §§2 and 9, follow:

Corollary 1. *In order that an element* $\mathfrak{z}=\lambda e+x(t)\in V^{(a)}$ *have an inverse in the ring* $V^{(a)}$ *it is necessary and sufficient that the expression*

$$\lambda\Big(\lambda+\int_{-\infty}^{\infty} x(t)\,e^{ist}dt\Big)$$

be different from zero for every s.

Corollary 2. *If* $X(s)$ *is the Fourier transform of a function* $x(t)\in L_1$ *and* $F(z)$ *is an analytic function regular in the closure of the set of values of* $X(s)$ *and vanishing at the point* $z=0$, *then* $Y(s)=F[X(s)]$ *is also the Fourier transform of some function* $y(t)\in L_1$.

Let us determine the structure of the topological space $\mathfrak{M}(V^{(a)})$. Indeed, let us show that $\mathfrak{M}(V^{(a)})$ *is the projective line*, that is, is homeomorphic to the circumference. The projective line is defined by adjoining to the real line $-\infty<s<\infty$ an infinitely distant point, in such a way that the neighborhoods of ordinary points remain the same and the neighborhoods of the point at infinity are defined as the sets of all those s (including the point at infinity) satisfying the inequality $|s|>A$, for all $A>0$. We show that *if the maximal ideal* M_s *is identified with the corresponding point* s *of the real line* $-\infty<s<\infty$, *and the maximal ideal* M_∞ *is identified with the point at infinity, then the topology defined in* $\mathfrak{M}(V^{(a)})$ *as a space of maximal ideals coincides with the topology of the projective line.*

By the compactness of the projective line, for this it suffices in view of Theorem 12′ §5, to show that $\mathfrak{z}(M)=\lambda+x(M)$ considered as a function on the projective line is continuous for all $\mathfrak{z} = \lambda e + x(t) \in V^{(a)}$. Since $x(M_s) = X(s)$ and $x(M_\infty) = 0$, it is necessary only to show that the Fourier transform $X(s)$ of an absolutely integrable function $x(t)$ is continuous and tends to zero as $|s| \to \infty$. But these properties are satisfied by the Fourier transform $H_a(s)$ of the function $h_a(t)$, as can be seen from the equation

$$H_a(s) = \int_{-\infty}^{\infty} h_a(t) e^{ist} dt = \int_0^a e^{ist} dt = \frac{e^{isa} - 1}{is}.$$

But since the functions $h_a(t)$ generate the ring $V^{(a)}$ an application of the elementary inequality $|X(s)| \leq \|x\|$ easily shows that the desired property holds for all $X(s)$.

Of importance in the theory of Fourier integrals is the theorem of uniqueness, according to which *an absolutely integrable function is uniquely determined by its Fourier transform.* By Theorem 1 this is nothing more than the assertion that in the ring $V^{(a)}$ there are no generalized nilpotent elements $x(t) \in L_1$ distinct from zero. Since an element $\mathfrak{z} = \lambda e + x(t)$ for which $\lambda \neq 0$ is evidently not a generalized nilpotent (because $\mathfrak{z}(M_\infty) = \lambda \neq 0$), we see that *the theorem of uniqueness for Fourier transforms of absolutely integrable functions is equivalent to the theorem about absence of a radical in* $V^{(a)}$. In Appendix III this theorem will be proved for a large class of rings including the ring $V^{(a)}$.

The results just described can be generalized almost without changing the argument followed.

Let $\alpha(t)$ be a positive continuous function defined on $-\infty < t < \infty$ and such that

$$\alpha(t_1 + t_2) \leq \alpha(t_1) + \alpha(t_2) \tag{7}$$

for every real t_1 and t_2, and

$$\alpha(0) = 1. \tag{8}$$

It is easily verified that the set $L_{<\alpha(t)>}$ of all functions $x(t)$, $-\infty < t < \infty$, satisfying the condition

$$\|x\| = \int_{-\infty}^{\infty} |x(t)|\, \alpha(t)\, dt < \infty,$$

form a normed ring with the ordinary linear operations and with convolution with multiplication. There is no identity in these rings. The normed ring derived by adjoining an identity to $L_{<\alpha(t)>}$ shall be denoted by $V^{(a)}_{<\alpha(t)>}$. We determine the form of the maximal ideals in the ring $V^{(a)}_{<\alpha(t)>}$. Like $V^{(a)}$, the ring $V^{(a)}_{<\alpha(t)>}$ is also generated by the functions

$$h_a(t) = \begin{cases} 1 \text{ if } 0 \leq t < a, \\ 0 \text{ for all values of } t. \end{cases}$$

Let M be a maximal ideal of the ring $V^{(a)}_{<\alpha(t)>}$ not coinciding with the set $M_\infty = L_{<\alpha(t)>}$, which is, evidently, a maximal ideal in the ring $V^{(a)}_{<\alpha(t)>}$. Denote by $\phi(a)$ the number to which the homomorphism $V^{(a)}_{<\alpha(t)>} \to V^{(a)}_{<\alpha(t)>}/M$ carries the element $h_a(t)$. Repeating the argument used for the proof of Theorem 1 we reach the relation

$$\phi'(\lambda+a) = \phi'(\lambda)\,\phi'(a), \tag{9}$$

corresponding to (5), and the inequality

$$|\phi'(\lambda)| \leq \alpha(\lambda), \tag{10}$$

replacing the inequality (γ). From (9) and continuity of the function $\phi'(\lambda)$ it follows that

$$\phi'(a) = e^{iza}, \tag{11}$$

where $z = \sigma + i\tau$ is some complex number; inequality (10) implies that

$$e^{-\tau a} \leq \alpha(a) \tag{12}$$

for every real a, or, what amounts to the same thing, $\tau \leq \dfrac{\ln \alpha(a)}{-a}$ when $a < 0$ and $\tau \geq \dfrac{\ln \alpha(a)}{-a}$ when $a > 0$. From this it follows that

$$\tau_1 = \lim_{a\to+\infty} \frac{\ln \alpha(a)}{-a} = \sup_{a>0} \frac{\ln \alpha(a)}{-a} \leq \tau \leq \inf_{a<0} \frac{\ln \alpha(a)}{-a} = \lim_{a\to-\infty} \frac{\ln \alpha(a)}{-a} = \tau_2.^* \tag{13}$$

Thus z lies in the strip defined by the inequalities (13). Since the generators $h_a(t)$ under the homomorphism $V^{(a)}_{<\alpha(t)>} \to V^{(a)}_{<\alpha(t)>}/M$ are carried to

$$\phi(a) = \int_0^a \phi'(t)\,dt = \int_{-\infty}^{\infty} h_a(t)\,e^{izt}dt,$$

every element $\lambda e + x(t) \in V^{(a)}_{<\alpha(t)>}$ is carried to

$$\lambda + \int_{-\infty}^{\infty} x(t)\,e^{izt}dt, \tag{14}$$

where, by (12), the integral in (14) is absolutely convergent.

Conversely, substituting each element $\lambda e + x(t) \in V^{(a)}_{<\alpha(t)>}$ into formula (14), where $z = \sigma + i\tau$ is an arbitrary complex number in the strip $\tau_1 \leq \tau \leq \tau_2$, we get a homomorphic mapping of the ring $V^{(a)}_{<\alpha(t)>}$ into the field of complex numbers which defines a maximal ideal $M \subset V^{(a)}_{<\alpha(t)>}$.

Therefore the maximal ideals $M \neq M_\infty$ fill the strip $\tau_1 \leq \Im z \leq \tau_2$; the maximal ideal M_∞ completes it to a compact space. Evidently there are propositions analogous to the Corollaries 1 and 2 of Theorem 1.

Now let us consider the set $L'_{<\alpha(t)>}$ of all functions $x(t)$ $(0 \leq t < \infty)$ satisfying the condition

$$\|x\| = \int_0^\infty |x(t)|\,\alpha(t)\,dt < \infty,$$

where the function $\alpha(t)$ is defined only for $t \geq 0$ and satisfies the conditions (7)

* See the footnote on page 128.

and (8) for these values of t. This set becomes a normed linear space with ordinary linear operations and with multiplication defined by the formula

$$(x*y)(t) = \int_0^t x(t-s)\,y(s)\,ds.$$

The normed ring formed by adjoining an identity to $L'_{<\alpha(t)>}$ we denote by $l_{<\alpha(t)>}$.

The same argument with which we determined the maximal ideals of the ring $V^{(a)}_{<\alpha(t)>}$ leads here to the following result: The set of maximal ideals of the ring $l^{(a)}_{<\alpha(t)>}$ is representable as a half plane completed in the natural way to a compact space $\Im z \geq \tau_1 = \lim\limits_{a\to+\infty} \dfrac{\ln \alpha(a)}{-a}$, by the maximal ideal $M_\infty = L'_{<\alpha(t)>}$. (We remark that for sufficiently rapid increase of $\alpha(t)$, that is, in case $\lim\limits_{a\to+\infty} \dfrac{\ln \alpha(a)}{-a} = \infty$, the half plane degenerates and M_∞ is the only maximal ideal in the space $l_{<\alpha(t)>}$; in this case those and only those elements $\lambda e + x(t) \in l_{<\alpha(t)>}$ for which $\lambda \neq 0$ have inverses.)

We apply this result in the deduction of the following theorem of Tauberian type:

Suppose that $\alpha(t)$ is a function defined for $t > 0$ and satisfying the conditions (7) *and* (8), *and that $F(t)$ is a function measurable and bounded on every finite interval of the positive axis. Let $x_0(t) \in L'_{<\alpha(t)>}$ be such that*

$$\int_0^\infty x_0(t)\,e^{izt}dt \neq -1 \tag{15}$$

for all $z = \sigma + i\tau$ with $\tau \geq \lim\limits_{a\to+\infty} \dfrac{\ln \alpha(a)}{-a}$, and that

$$\Big(F(t) + \int_0^t x_0(t-s)\,F(s)\,ds\Big)\,\alpha(t) \to 0 \text{ as } t\to\infty. \tag{16}$$

Then also

$$F(t)\,\alpha(t) \to 0 \text{ as } t \to \infty.$$

Proof. To each element $\mathfrak{z} = \lambda e + x(t) \in l_{<\alpha(t)>}$ corresponds an operation $\mathfrak{z}*$ carrying each function $\Phi(t)$ $(0 \leq t < \infty)$ continuous and bounded on each finite interval to the function

$$\mathfrak{z} * \Phi(t) = \lambda\Phi(t) + \int_0^t x(t-s)\,\Phi(s)\,ds,$$

which is also measurable and bounded in every finite interval. Evidently $\mathfrak{z}_1*(\mathfrak{z}_2*\Phi(t)) = (\mathfrak{z}_1*\mathfrak{z}_2)*\Phi(t)$. By definition of the ring $l_{<\alpha(t)>}$ the operation $\mathfrak{z}*$ is the operation of multiplication by $\mathfrak{z}$. It is easily shown that from $\Phi(t)\,\alpha(t) \to 0$ (as $t\to\infty$) it follows that $(\mathfrak{z}*\Phi(t))\,\alpha(t) \to 0$ (as $t \to \infty$) for each $\mathfrak{z} \in l_{<\alpha(t)>}$. By condition (15) the element $x_0 + e$ belongs to no maximal ideal of the ring $l_{<\alpha(t)>}$ and therefore it has in $l_{<\alpha(t)>}$ an inverse. Furthermore, condition (16) can be written in the form $[(e + x_0) * F(t)]\,\alpha(t) \to 0$. But then by associativity

$$(e+x_0)^{-1} * [(e+x_0) * F(t)]\,\alpha(t) = [(e+x_0)^{-1} * (e+x_0)] * F(t)\,\alpha(t) = F(t)\,\alpha(t) \to 0.$$

This theorem in the particular case $\alpha(t) \equiv 1$ was first proved by Paley and Wiener [18, pages 59-60].

§ 17. The ring of functions of bounded variation

Suppose that $f(t)$ is a function of bounded variation on the line $-\infty < t < \infty$. As is well known, $f(t)$ can be represented as the difference of two nondecreasing functions of bounded variation. Therefore at each point t the right and left hand limits, $f(t+0)$ and $f(t-0)$, exist; in addition, $f(+\infty)$ and $f(-\infty)$ exist. The function $f(t+0)$ is evidently continuous on the right and also has bounded variation. Since a function of bounded variation has at most a countable set of points of discontinuity, $f(t+0)$ coincides, except perhaps on a countable set, with $f(t)$. The function $f(t+0)$ has no removable points of discontinuity; all of its points of discontinuity are of the second kind.. From here on when speaking of a function of bounded variation we will assume that it is continuous on the right, i.e., that it satisfies the condition $f(t) = f(t+0)$.

We denote by $V^{(b)}$ the linear space of all complex functions $f(t)$ with bounded variation on $-\infty < t < \infty$, satisfying the condition $f(-\infty) = 0$ and continuous on the right, with norm

$$\|f\| = \operatorname{Var} f.$$

The space $V^{(b)}$ is complete. In it can be defined an operation of multiplication, called convolution, by the formula

$$(f_1 * f_2)(t) = \int_{-\infty}^{\infty} f_1(t-\tau)\, df_2(\tau),$$

where the integral is taken in the sense of Riemann-Stieltjes. Indeed, it is not difficult to verify that *if* $f_1, f_2 \in V^{(b)}$, *then also* $f_1 * f_2 \in V^{(b)}$ *and* $\|f_1 * f_2\| \le \|f_1\|\,\|f_2\|$. With the help of Fubini's theorem it can be shown that *convolution is associative and commutative.* Evidently the function

$$\epsilon(t) = \begin{cases} 0 & \text{if } t < 0, \\ 1 & \text{if } t \ge 0, \end{cases}$$

is an identity for convolution, for which $\|\epsilon\| = 1$. Hence $V^{(b)}$ *is a commutative normed ring* in the sense of Definition 1 of § 1.

Let $f(t) \in V^{(b)}$ and let $\{\lambda_k\}$ be the sequence of all points of discontinuity of the function $f(t)$ (infinite, finite, or empty). The function

$$h(t) = \sum_{\lambda_k \le t} h_k,$$

where for brevity we have set $f(\lambda_k+0) - f(\lambda_k-0) = h_k$ and where the sum is taken over all those points of discontinuity λ_k not lying to the right of t, is called the jump function of $f(t)$. We have

$$\operatorname{Var} h = \Sigma \mid f(\lambda_k+0)-f(\lambda_k-0) \mid \leq \operatorname{Var} f < \infty.$$

The difference $f(t) - h(t) = c(t)$ is called the continuous part of $f(t)$. In it can be found a singular part, that is, a summand $s(t)$ all the variation of which is concentrated on some set of measure zero in such a way that the difference $c(t) - s(t) = g(t)$ is absolutely continuous, that is, for each $\epsilon > 0$ there exists $\delta > 0$ such that $\Sigma \mid g(t_{k+1}) - g(t_k) \mid < \epsilon$ for each finite system of nonoverlapping intervals (t_k, t_{k+1}) the sum of whose lengths is less than δ; $s(t)$ is a continuous function of bounded variation with a derivative almost everywhere equal to zero. Therefore each function $f(t) \in V^{(b)}$ can be represented in a unique way as a sum

$$f(t) = g(t) + h(t) + s(t),$$

where $g(t)$ is absolutely continuous, $h(t)$ is a jump function, and $s(t)$ is a singular function. Under this condition

$$\|f\| = \|g\| + \|h\| + \|s\|.$$

Finally, any of these components may vanish, that is, actually be missing.

An absolutely continuous function $g(t) \in V^{(b)}$ is characterized by the following property: It has almost everywhere a derivative $g'(t)$ which is absolutely integrable and $g(t)$ is its integral, i.e., $g(t) = \int_{-\infty}^{t} g'(\tau)\,d\tau$. Hence it follows that $\operatorname{Var} g = \int_{-\infty}^{\infty} |g'(t)|\,dt$. Moreover, the convolution of absolutely continuous functions corresponds to the convolution of their derivatives as functions of $V^{(a)}$, that is, if $g(t) = \int_{-\infty}^{\infty} g_1(t-\tau)\,dg_2(\tau)$, then $g'(t) = \int_{-\infty}^{\infty} g_1'(t-\tau)\,g_2'(\tau)\,d\tau$. All of this shows that by matching to each element $\lambda e + x(t) \in V^{(a)}$ the function $\lambda \epsilon(t) + \int_{-\infty}^{t} x(\tau)\,d\tau$ we define an isometric and isomorphic mapping of the ring $V^{(a)}$ into the ring $V^{(b)}$; in other words *the ring $V^{(a)}$ is a subring of the ring $V^{(b)}$*.

The jump functions $h(t)$ also form a ring; we denote it by H.

The problem of finding the maximal ideals of the ring H is solved in the following theorem.

Theorem 2. *Between the maximal ideals of the ring H and the characters $\chi(t)$ of the discrete additive group Γ of all real numbers there is a one-to-one correspondence under which a function $h(t) = \sum_{\lambda_k \leq t} h_k \in H$ takes at the maximal ideal M corresponding to a character χ the value*

$$\int_{-\infty}^{\infty} \chi(t)\,dh(t) = \Sigma\, h_k\, \chi(\lambda_k).$$

The set of maximal ideals M_s corresponding to the continuous characters $\chi(t) = e^{ist}$, that is, the set of points of the line $-\infty < s < \infty$, is everywhere dense in the space $\mathfrak{M}(H)$.

We recall that the characters of a discrete commutative group G are the functions $\chi(g)$ $(g \in G)$ taking complex values of modulus 1, and such that $\chi(g_1+g_2) = \chi(g_1)\chi(g_2)$ for all $g_1, g_2 \in G$. It is easy to verify that each character of the additive group of real numbers which is continuous in the usual topology has the form $\chi(t) = e^{ist}$, where s is a fixed real number; on the other hand, with the aid of Zermelo's axiom it is easy to construct discontinuous characters [12] (all of them are nonmeasurable functions).

Proof of Theorem 2. Each jump function $h(t) = \Sigma_{\lambda_k \leq t} h_k$ can be written in the form $h(t) = \Sigma\, h_k \epsilon(t-\lambda_k)$; therefore the functions $\epsilon(t-\lambda)$ generate H. Let M be a maximal ideal of the ring H and let $\chi(\lambda)$ be the number into which the homomorphism $H \to H/M$ carries the function $\epsilon(t-\lambda)$. Since $\|\epsilon(t-\lambda)\| = 1$, also $|\chi(\lambda)| \leq 1$. Evidently $\chi(0) = 1$, further, since $\epsilon(t-\lambda_1) * \epsilon(t-\lambda_2) = \epsilon(t-\lambda_1-\lambda_2)$, also $\chi(\lambda_1+\lambda_2) = \chi(\lambda_1)\chi(\lambda_2)$. Hence it follows directly that $|\chi(\lambda)| \equiv 1$, and, consequently, that $\chi(\lambda)$ is a character of the group Γ. Evidently the function $h(t) = \Sigma\, h_k \epsilon(t-\lambda_k)$ is carried by the mapping $H \to H/M$ to the number $\Sigma\, h_k \chi(\lambda_k)$. On the other hand, if a character $\chi(\lambda)$ of the group Γ is given, the assigning to the function $h(t) = \Sigma\, h_k \epsilon(t-\lambda_k)$ the number $\Sigma\, h_k \chi(\lambda_k)$ defines a homomorphism of the ring H into the field of complex numbers, so each character determines a maximal ideal. Evidently, distinct maximal ideals correspond to distinct characters, and conversely. This completes the proof of the first part of Theorem 2.

Now suppose that M_0 is an arbitrary maximal ideal of the ring H and that $U(M_0)$ is an arbitrary neighborhood of it; we show that there is a maximal ideal $M_{s_0} \in U(M_0)$. Denote by $\chi_\lambda(M)$ the value of the function $\epsilon(t-\lambda)$ at the maximal ideal M. Since the functions $\epsilon(t-\lambda)$ generate the ring H, by Theorem 14 of §5 the neighborhood $U(M_0)$ contains a neighborhood $U'(M_0)$ defined by the inequalities

$$|\chi_{\lambda_k}(M_0) - \chi_{\lambda_k}(M)| < \epsilon \quad (k = 1, 2, \cdots, n). \tag{1}$$

Let $\{\lambda_k\}$ be a basis for the group Γ, that is a set of numbers satisfying the following conditions:

a) Each real number is a finite linear combination with rational coefficients of numbers λ_a.

b) Each finite set of the numbers λ_a is linearly independent, that is, from the relation $\Sigma_{j=1}^m r_j \lambda_j = 0$, where the r_j are rational numbers, follow the relations $r_j = 0$ $(j = 1, 2, \cdots, m)$.

The construction of a basis can easily be carried out by transfinite induction [12].

Now consider the real numbers λ_k $(k = 1, 2, \cdots, n)$ appearing in the inequalities (1). By property a) it is possible to choose from the basis some finite set of numbers λ_{a_j} $(j = 1, 2, \cdots, m)$ such that each λ_k is a linear combination of the λ_{a_j} with rational coefficients. Let A be the least common denominator of all the coefficients of all these linear forms. λ_k can be expressed in terms of linear combinations with integral coefficients of the numbers $\mu_{a_j} = \frac{\lambda_{a_j}}{A}$. At the same time the numbers μ_{a_j} are, like the λ_{a_j}, linearly independent. Since the functions $\epsilon(t-\lambda_k)$ are products of the functions $\epsilon(t-\mu_{a_j})$, inside the neighborhood $U'(M_0)$ can be found a neighborhood $U''(M_0)$ of the form

$$|\chi_{\mu_{a_j}}(M_0)-\chi_{\mu_{a_j}}(M)| < \delta \quad (j = 1, 2, \cdots, m).$$

Since the numbers $\mu_{a_1}, \mu_{a_2}, \cdots, \mu_{a_m}$ are linearly independent, by a well known theorem of Kronecker there exists a real number s_0 such that the number $e^{i\mu_{a_j}s_0}$ differs from the number $\chi_{\mu_{a_j}}$ by less than δ for each $j = 1, 2, \cdots, m$. But this means that the maximal ideal M_{s_0} belongs to $U''(M_0)$, and therefore to $U(M_0)$. In view of the arbitraryness of $U(M_0)$ it follows that the set $\{M_{s_0}\}$ is everywhere dense in $\mathfrak{M}(H)$, which completes the proof of Theorem 2.

Corollary 1. *If $H(s)$ is an almost periodic function with absolutely convergent Fourier series $H(s) = \Sigma\, h_k e^{i\lambda_k s}$ and if $\inf_{-\infty<s<\infty} |H(s)| > 0$, then $\frac{1}{H(s)}$ is also an almost periodic function with absolutely convergent Fourier series.*

Since the multiplication of almost periodic functions corresponds to convolution of their Fourier coefficients, the proof follows simply from the fact that a continuous function whose lower bound on a dense set is positive can never vanish.

Corollary 2. *If $H(s)$ is an almost periodic function with absolutely convergent Fourier series and if $F(z)$ is an analytic function regular on the closure of the set of values of the function $H(s)$, then $F(H(s))$ is also an almost periodic function with absolutely convergent Fourier series.*

We remark that the topology induced on the line $-\infty < s < \infty$ by the space $\mathfrak{M}(H)$ differs from the ordinary topology of the line. We have determining neighborhoods of the form

$$\{\, |e^{i\lambda_k s} - e^{i\lambda_k s_0}| < \epsilon;\ k = 1, 2, \cdots, n\}.$$

Each of these inequalities defines a periodic system of intervals; the intersection of such systems gives an "almost periodic" system of intervals, consisting of infinitely many distinct intervals one of which occurs in every sufficiently long segment of the line.

We turn to the study of the maximal ideals in the ring $V^{(b)}$. The set of all maximal ideals of this ring is not yet determined, and therefore we can not give such complete results as in the preceding case.

Let $f(t) \in V^{(b)}$. The integral

$$F(s) = \int_{-\infty}^{\infty} e^{ist}\, df(t),$$

which evidently exists for all real values of s, is called the *Fourier-Stieltjes transform* of the function $f(t)$. Evidently, addition of functions in $V^{(b)}$ and multiplication of them by scalars correspond to just the same operations on their Fourier-Stieltjes transforms. To convolution of functions in $V^{(b)}$ corresponds multiplication of their Fourier-Stieltjes transforms. Indeed,

$$\int_{-\infty}^{\infty} e^{ist} d \int_{-\infty}^{\infty} f_1(t-u)\, df_2(u) = \int_{-\infty}^{\infty} \left(\int_{-\infty}^{\infty} e^{ist} df_1(t-u) \right) df_2(u) =$$

$$= \int_{-\infty}^{\infty} \left(\int_{-\infty}^{\infty} e^{is(t-u)} df_1(t-u) \right) e^{isu} df_2(u) = \int_{-\infty}^{\infty} e^{ist} df_1(t) \int_{-\infty}^{\infty} e^{isu} df_2(u).$$

In this way, making correspond to each function $f(t) \in V^{(b)}$ the value of its Fourier-Stieltjes transform at any fixed point s_0, we obtain a homomorphism of the ring $V^{(b)}$ into the complex field. We call the maximal ideal determined by this homomorphism M_{s_0}. Therefore

$$f(M_{s_0}) = \int_{-\infty}^{\infty} e^{is_0 t} df(t) = F(s_0).$$

If $s_1 \neq s_2$, then $M_{s_1} \neq M_{s_2}$, since there exists an absolutely continuous function $g(t) \in V^{(b)}$ such that $g(M_{s_1}) \neq g(M_{s_2})$.

Let us show that *each function* $f(t) \in V^{(b)}$ *is uniquely determined by its values on the maximal ideals* M_s, *i.e., by its Fourier-Stieltjes transform* $F(s)$. It is evident that this implies, in particular, that *the ring* $V^{(b)}$ *has no radical.*

Lemma 1. *The set G of all absolutely continuous functions $g(t) \in V^{(b)}$ is an ideal of the ring $V^{(b)}$.*

Proof. Let $g(t) \in G$ and let $f(t)$ be an arbitrary function from $V^{(b)}$ distinct from zero. We need to show that $f * g \in G$. Let $\epsilon > 0$ and take $\delta > 0$ such that

$$\Sigma \mid g(t_{k+1}) - g(t_k) < \frac{\epsilon}{\|f\|}$$

for each finite system of non-overlapping intervals (t_k, t_{k+1}) the sum of whose lengths is less than δ. Then

$$\Sigma \mid (f*g)(t_{k+1}) - (f*g)(t_k) \mid = \Sigma \left| \int_{-\infty}^{\infty} [g(t_{k+1}-\tau) - g(t_k-\tau)]\, df(\tau) \right| \leq$$

$$\leq \int_{-\infty}^{\infty} \Sigma \mid g(t_{k+1}-\tau) - g(t_k-\tau) \mid \mid df(\tau) \mid < \frac{\epsilon}{\|f\|} \operatorname{Var} f = \epsilon,$$

which proves the assertion of the lemma.

Now suppose that $f(t) \in V^{(b)}$ and that $f(M_s) \equiv F(s) \equiv 0$ for all s. Set

$$g_h(t) = \begin{cases} 0 & \text{for } t < -h, \\ 1 + \frac{t}{h} & \text{for } -h \leq t \leq 0, \\ 1 & \text{for } t > 0. \end{cases}$$

$g_h(t)$ is absolutely continuous. Therefore, by Lemma 1, the function

$$(f*g_h)(t) = \frac{1}{h}\int_0^h f(t+r)\,dr$$

is also absolutely continuous and hence belongs to the ring $V^{(a)}$. Also by assumption $(f*g_h)(M_s) = f(M_s)g_h(M_s) \equiv 0$. Applying the uniqueness theorem for Fourier transforms of absolutely integrable functions* we conclude from this that $(f*g_h)(t) \equiv 0$. But since $f(t)$ is continuous on the right $(f*g_h)(t) \to f(t)$ as $h \to 0$. Consequently $f(t) \equiv 0$, which was to be proved.

Before continuing the search for maximal ideals of the ring $V^{(b)}$ distinct from the M_s we note one corollary of the lemma proved above.

Corollary. *The only maximal ideals of the ring $V^{(b)}$ which do not contain the ideal G are the maximal ideals M_s.*

Indeed, suppose that M is a maximal ideal of the ring $V^{(b)}$ which does not contain the ideal G, i.e., there is an absolutely continuous function $g_0(t) \in V^{(b)}$ such that $g_0(M) \neq 0$. Matching to each element $\lambda\epsilon(t) + g(t)$ of the ring $V^{(a)}$ the number $\lambda + g(M)$ we determine a homomorphism of this ring into the field of complex numbers which is non-trivial since $g_0(t)$ is carried to a number different from zero. By Theorem 1 each absolutely continuous function $g(t) \in V^{(b)}$ is carried by this mapping to the number

$$\int_{-\infty}^{\infty} e^{is_0t}d_g(t) = \int_{-\infty}^{\infty} e^{is_0t}g'(t)\,dt.$$

Let $f(t)$ be an arbitrary function from $V^{(b)}$. By the lemma $g(t) = (g_0*f)(t)$ is an absolutely continuous function. Consequently,

$$(g_0*f)(M) = \int_{-\infty}^{\infty} e^{is_0t}d(g_0*f)(t) = g_0(M)f(M) = \int_{-\infty}^{\infty} e^{is_0t}dg_0(t)\cdot f(M).$$

But, on the other hand,

$$(g_0*f)(M) = \int_{-\infty}^{\infty} e^{is_0t}dg(t) = \int_{-\infty}^{\infty} e^{is_0t}dg_0(t)\int_{-\infty}^{\infty} e^{is_0t}df(t).$$

Since the first factor on the right-hand side is, by assumption, different from zero, then we get that

$$f(M) = \int_{-\infty}^{\infty} e^{is_0t}df(t)$$

for every $f(t) \in V^{(b)}$. But this says that $M = M_{s_0}$.

Therefore, *on all maximal ideals distinct from the ideals M_s all absolutely continuous functions of $V^{(b)}$ vanish.* We describe one class of such ideals.

Let us remark that *the set C of all continuous functions $c(t) \in V^{(b)}$ is an*

* This theorem will be proved for functions defined on an arbitrary commutative locally compact group in Appendix III (§20).

ideal in $V^{(b)}$. Indeed, suppose that $c(t) \in C$ and that $f(t)$ is an arbitrary function from $V^{(b)}$ different from zero. Let ϵ be an arbitrary fixed positive number. There exists an interval $[-A, A]$ such that the total variation of $c(t)$ outside this interval is less than $\frac{\epsilon}{2\|f\|}$. Take $\delta > 0$ such that $|c(t') - c(t)| < \frac{\epsilon}{2\|f\|}$ for all points t, t' of the interval $[-A, A]$ satisfying the inequality $|t - t'| < \delta$. Then $|c(t') - c(t)| < \frac{\epsilon}{\|f\|}$ for all points t, t' of the real line satisfying the given inequality. Therefore

$$|(c*f)(t') - (c*f)(t)| = \left| \int_{-\infty}^{\infty} [c(t'-\tau) - c(t-\tau)]\, df(\tau) \right| \le$$

$$\le \int_{-\infty}^{\infty} |c(t'-\tau) - c(t-\tau)|\, |df(\tau)| < \frac{\epsilon}{\|f\|} \operatorname{Var} f = \epsilon,$$

i.e., $(c*f)(t)$ is continuous, which proves our assertion. From this it follows that *the jump function of the convolution of two functions of* $V^{(b)}$ *is the convolution of the jump functions of these two functions*. Indeed, set $f_1(t) = c_1(t) + h_1(t)$, $f_2(t) = c_2(t) + h_2(t)$ $(c_1, c_2 \in C;\ h_1, h_2 \in H)$. Then

$$(f_1 * f_2)(t) = [(c_1 * c_2)(t) + (c_1 * h_2)(t) + (c_2 * h_1)(t)] + (h_1 * h_2)(t),$$

and the expression in the square brackets on the right-hand side is, as has just been shown, a continuous function, and the last term on the right-hand side is a jump function.

Now suppose that M is a maximal ideal of the ring H. By Theorem 2 it generates some character χ of the discrete additive group of real numbers. To each function $f(t) \in V^{(b)}$ let correspond the number $h(M) = \sum h_k \chi(\lambda_k)$, where $h(t)$ is the jump function of $f(t)$. From all that has been proved it follows that we have defined a homomorphism of the ring $V^{(b)}$ into the field of complex numbers. The maximal ideal determined by this homomorphism we denote by M_χ.

Evidently $c(M_\chi) = 0$ for each continuous function $c(t) \in V^{(b)}$ and each maximal ideal M_χ. It is not difficult to see that, conversely, if $c(M) = 0$ for each continuous function $c(t) \in V^{(b)}$, then $M = M_\chi$. Indeed, the maximal ideal M in the ring $V^{(b)}$ generates a maximal ideal in the ring H; consequently, by Theorem 2 there exists a character χ of the discrete additive group of all real numbers such that $h(M) = \sum h_k \chi(\lambda_k)$ for each function $h(t) = \sum_{\lambda_k \le t} h_k \in V^{(b)}$. By assumption for each $f(t) = c(t) + h(t) \in V^{(b)}$ we have $f(M) = c(M) + h(M) = h(M)$, so $M = M_\chi$.

For each maximal ideal M_s there is a continuous function $g(t)$ such that $g(M_s) \neq 0$. Hence, the maximal ideals M_χ are distinct from M_s.

The maximal ideals M_s and M_χ do not exhaust the maximal ideals of the ring $V^{(b)}$. We shall soon give other families of maximal ideals. To do this we need to discuss functions of bounded variation as set-functions.

Consider first a *non-decreasing* real function of bounded variation. Suppose that $f(t)$ is such a function and that F is an arbitrary bounded closed set on the line $-\infty < t < \infty$. We define

$$f(F) = \inf \{[f(t_1') - f(t_1)] + [f(t_2') - f(t_2)] + \cdots + [f(t_n') - f(t_n)]\},$$

where the lower bound is taken over all finite systems of non-overlapping intervals $(t_1, t_1'), (t_2, t_2'), \cdots, (t_n, t_n')$ covering F. For an arbitrary Borel set E of points of the line $-\infty < t < \infty$, we then set

$$f(E) = \sup f(F^E),$$

where the upper bound is taken over all bounded closed sets F^E contained in E. It can be shown that $f(E)$ is a completely additive function, that is, that for each finite or countable system of pairwise disjoint Borel sets $E_1, E_2, \cdots$, we have the equation

$$f(E_1 \cup E_2 \cup \cdots) = f(E_1) + f(E_2) + \cdots,$$

where $A \cup U$ denotes the union (sum) of the sets A and B. For the set $E_t = (-\infty, t)$ we have $f(E_t) = f(t+0) - f(-\infty)$.

We shall denote the total variation of the function $f(t)$ on the interval $(-\infty, t)$ (which is a function of t) by $\operatorname{Var} f(t)$. Let $f(t)$ be an arbitrary function from $V^{(b)}$. Then its real and imaginary parts, $\Re f(t)$ and $\Im f(t)$, are also functions in $V^{(b)}$. But each real function $\phi(t)$ in $V^{(b)}$ can be represented as a difference, $\phi(t) = \phi^+(t) - \phi^-(t)$, where $\phi^+(t)$ and $\phi^-(t)$ are non-decreasing functions of bounded variation determined from the function $\phi(t)$:

$$\phi^+(t) = \frac{1}{2}[\operatorname{Var} \phi(t) + \phi(t)], \quad \phi^-(t) = \frac{1}{2}[\operatorname{Var} \phi(t) - \phi(t)].$$

For each function $f(t) \in V^{(b)}$ we have now

$$f(E) = \Re^+ f(E) - \Re^- f(E) + i\Im^+ f(E) - i\Im^- f(E).$$

The equation $f(E_t) = f(t+0) - f(-\infty)$ shows that the function $f(t) \in V^{(b)}$ is uniquely determined by its set function $f(E)$. The function $f(t) \in V^{(b)}$ is continuous if and only if $f(E) = 0$ for each set E consisting of a single point. $f(t) \in V^{(b)}$ is absolutely continuous if and only if $f(E) = 0$ for each Borel set of Lebesgue measure zero. $f(t) \in V^{(b)}$ is singular if and only if it is continuous and there exists a Borel set E_0 of Lebesgue measure 0 such that $f(E) = 0$ for each Borel set E not intersecting E_0.

The convolution of functions of $V^{(b)}$ is carried to the following operation on the corresponding functions of sets:

$$(f_1 * f_2)(E) = \int_{-\infty}^{\infty} f_1(E-u)\, df_2(u),$$

where the integral is taken in the sense of Lebesgue-Stieltjes. This operation is closely related to the operation of arithmetic sum of sets.

By the *arithmetic sum* of sets A and B we mean the set consisting of all points of the form $a+b$, where $a \in A$, $b \in B$. For a symbol for the arithmetic sum of sets we shall use the sign $+$. Evidently, if A and B contain the point 0, then $A \subset A+B$ and $B \subset A+B$. It is not hard to verify that if A and B are bounded closed sets then $A+B$ is also bounded and closed. If A and B are analytic sets, then $A+B$ is also an analytic set [24]*. By nA we mean the sum $A+A+\cdots+A$, where A is repeated n times.

Lemma 2. *If A has positive measure, then $2A$ contains a segment.*

Proof. Denote by $f_A(u)$ the characteristic function of the set A and form the function

$$\phi(t) = \int_{-\infty}^{\infty} f_A(t-u) f_A(u)\, du.$$

This function is not identically zero, because

$$\int_{-\infty}^{\infty} \phi(t)\, dt = \left(\int_{-\infty}^{\infty} f_A(u)\, du\right)^2 = [m(A)]^2,$$

and $m(A)$, the measure of A, is positive by assumption. Since $f_A(u)$ is bounded, $\phi(t)$ is continuous (see page 169). Therefore there exists an interval in which $\phi(t)$ does not vanish. But $\phi(t)$, as follows immediately from its definition, can differ from zero only in points of the set $2A$. Consequently, $2A$ contains an interval, and the lemma is proved.

On the basis of this lemma we shall prove that there exists a perfect set F such that nF is a set of measure 0 for each n. One example of such a set is the set of all binary fractions (including 0) which have a terminating or nonterminating binary expansion in which the digit 1 appears only in the places with index a_n, where $\overline{\lim}_{n\to\infty} (a_{n+1}-a_n) = \infty$. To show this, observe that in order that a digit 1 shall reach the place numbered a_n+1 it must have come from digits in the places $a_{n+1}, a_{n+2}, \cdots$ and therefore can be represented as sums of fractions from F only if there are at least $a_{n+1}-a_n$ summands. Since $a_{n+1}-a_n$ can be arbitrarily large, it follows that the fraction whose a_Nth remainder is $\sum_{n=N}^{\infty} 2^{-a_n-1}$, can not be in any of the sets nF. But fractions of this form are everywhere dense on the line. Consequently, for no n does the set nF contain an interval. But if there were n such that nF had positive measure, the set $2nF$ would contain an interval, which is impossible. Hence all the nF have measure zero and our assertion is proved.

Now suppose that $\mathfrak{J}$ is a family of Borel sets $\mathfrak{F}_\sigma$ (i.e., finite or countable sums of closed sets) satisfying the following conditions: 1° If $\mathfrak{F}_\sigma \in \mathfrak{J}$ and if

* If the sets A and B are Lebesgue measurable but not analytic, then the set $A+B$ may be nonmeasurable; see [24].

$\mathfrak{F}'_\sigma$ is an arbitrary subset of the form $\mathfrak{F}_\sigma$ of the set $\mathfrak{F}_\sigma$, then $\mathfrak{F}'_\sigma \in \mathfrak{J}$; 2° If $\mathfrak{F}_{\sigma 1}$, $\mathfrak{F}_{\sigma 2}, \cdots$ are finitely or countably many sets in $\mathfrak{J}$, then $\mathfrak{F}_{\sigma 1} \cup \mathfrak{F}_{\sigma 2} \cup \cdots$ also belongs to $\mathfrak{J}$. 3° If $\mathfrak{F}_\sigma \in \mathfrak{J}$, then $\mathfrak{F}_\sigma - t$ also $\in \mathfrak{J}$ for all t, $-\infty < t < \infty$. 4° If $\mathfrak{F}_\sigma \in \mathfrak{J}$, then $2\mathfrak{F}_\sigma \in \mathfrak{J}$.

Evidently, if we take the system of all subsets of the type $\mathfrak{F}_\sigma$ of some set A, and apply to them, and to the sets resulting, the operations used in the above conditions 1°-4°, we get a family of sets $\mathfrak{J}_A$ satisfying the conditions enumerated.

We shall say that the function $f(E)$ is concentrated outside $\mathfrak{J}$ if $f(\mathfrak{F}_\sigma) = 0$ for every $\mathfrak{F}_\sigma \in \mathfrak{J}$; we shall say that the function f of $V^{(b)}$ is concentrated in $\mathfrak{J}$ if

$$\operatorname{Var} f(E) = \sup_{\substack{\mathfrak{F}_\sigma \subset E \\ \mathfrak{F}_\sigma \in \mathfrak{J}}} \operatorname{Var} f(\mathfrak{F}_\sigma)$$

for each Borel set E *. If f is concentrated in some set $\mathfrak{F}_\sigma \in \mathfrak{J}$, that is, $f(E) = f(E \cap \mathfrak{F}_\sigma)$ ** for each Borel set E, then f, evidently, is concentrated in $\mathfrak{J}$. But also conversely, *if f is concentrated in $\mathfrak{J}$, then it is concentrated in some $\mathfrak{F}_\sigma$ of $\mathfrak{J}$.*

For proof, if f is concentrated in $\mathfrak{J}$, then for each n there exists $\mathfrak{F}_{\sigma n} \in \mathfrak{J}$ such that $\operatorname{Var} f(\mathfrak{F}_{\sigma n}) > \operatorname{Var} f - \frac{1}{n}$. Then for $\mathfrak{F}_\sigma = \Sigma\, \mathfrak{F}_{\sigma n}$ we have $\operatorname{Var} f(\mathfrak{F}_\sigma) = \operatorname{Var} f$; also $\mathfrak{F}_\sigma \in \mathfrak{J}$ (property 2°). Evidently $f(E) = f(E \cap \mathfrak{F}_\sigma)$ for each Borel set E.

If f_1 is concentrated in $\mathfrak{J}$ and f_2 is concentrated outside of $\mathfrak{J}$, then $\operatorname{Var}(f_1 + f_2) = \operatorname{Var} f_1 + \operatorname{Var} f_2$. Indeed, f_1 is concentrated in some $\mathfrak{F}_\sigma \in \mathfrak{J}$; obviously f_2 is concentrated in $C\mathfrak{F}_\sigma$ ***. We have

$$\operatorname{Var}(f_1+f_2) = \operatorname{Var}[f_1(\mathfrak{F}_\sigma)+f_2(\mathfrak{F}_\sigma)] + \operatorname{Var}[f_1(C\mathfrak{F}_\sigma)+f_2(C\mathfrak{F}_\sigma)] =$$
$$= \operatorname{Var} f_1(\mathfrak{F}_\sigma) + \operatorname{Var} f_2(C\mathfrak{F}_\sigma) = \operatorname{Var} f_1 + \operatorname{Var} f_2.$$

If f_1 and f_2 are concentrated in $\mathfrak{J}$, then also every linear combination $\lambda_1 f_1 + \lambda_2 f_2$ is concentrated in $\mathfrak{J}$. Indeed, by the above results, f_1 is concentrated in some $\mathfrak{F}_{\sigma 1} \in \mathfrak{J}$ and f_2 in some $\mathfrak{F}_{\sigma 2} \in \mathfrak{J}$; therefore, both these functions, and consequently every linear combination of them, are concentrated in $\mathfrak{F}_\sigma = \mathfrak{F}_{\sigma 1} \cup \mathfrak{F}_{\sigma 2}$, and hence in $\mathfrak{J}$. Evidently, if *f_1 and f_2 are concentrated outside $\mathfrak{J}$, then every linear combination $\lambda_1 f_1 + \lambda_2 f_2$ is concentrated outside $\mathfrak{J}$.*

Each function f from $V^{(b)}$ can be decomposed in a unique way into a sum of two terms, also in $V^{(b)}$, one of which is concentrated in $\mathfrak{J}$ and the other outside $\mathfrak{J}$. By what is already proved, to show decomposability in the given form it suffices

* If $\mathfrak{J}$ is not empty (and this case we naturally exclude), then by properties 1° and 3° it contains all sets consisting of only one point; therefore sets $\mathfrak{F}_\sigma \in \mathfrak{J}$ and contained in E exist.

** $A \cap B$ denotes the intersection of the sets A and B.

*** CA denotes the complement of the set A.

to consider only nonnegative functions $f(E)$ from $V^{(b)}$, because each function $f(E)$ from $V^{(b)}$ is a linear combination of nonnegative functions. Set

$$f_{\mathfrak{J}}(E) = \sup_{\substack{\mathfrak{F}_\sigma \subset E \\ \mathfrak{F}_\sigma \in \mathfrak{J}}} f(\mathfrak{F}_\sigma), \quad f_{C\mathfrak{J}}(E) = f(E) - f_{\mathfrak{J}}(E).$$

It is easily verified by properties 1° and 2° that $f_{\mathfrak{J}}(E)$ (and, of course, $f_{C\mathfrak{J}}(E)$) are completely additive. Evidently $f(\mathfrak{F}_\sigma) = f_{\mathfrak{J}}(\mathfrak{F}_\sigma)$, and consequently $f_{C\mathfrak{J}}(\mathfrak{F}_\sigma) = 0$ for each $\mathfrak{F}_\sigma \in \mathfrak{J}$. Hence $f_{C\mathfrak{J}}$ is concentrated outside $\mathfrak{J}$. Moreover, $f_{\mathfrak{J}}(E) = \sup f(\mathfrak{F}_\sigma) = \sup f_{\mathfrak{J}}(\mathfrak{F}_\sigma)$, i.e., $f_{\mathfrak{J}}$ is concentrated in $\mathfrak{J}$. Suppose now that $f = f_1 + f_2 = f_3 + f_4$, where f_1 and f_3 are concentrated in $\mathfrak{J}$ and f_2 and f_4 outside $\mathfrak{J}$. Then

$$0 = (f_1 - f_3) + (f_2 - f_4),$$

and, since $f_1 - f_3$ is concentrated in $\mathfrak{J}$ and $f_2 - f_4$ is concentrated outside $\mathfrak{J}$, by an earlier result

$$0 = \operatorname{Var}(f_1 - f_3) + \operatorname{Var}(f_2 - f_4),$$

from which it follows that $f_3 = f_1$ and $f_4 = f_2$, i.e., the desired decomposition of the function f is unique.

The set $V_{C\mathfrak{J}}$ *of all functions from* $V^{(b)}$ *concentrated outside* $\mathfrak{J}$ *is an ideal of the ring* $V^{(b)}$. Indeed, as we have shown, $V_{C\mathfrak{J}}$ is a linear system. Suppose now that $f \in V_{C\mathfrak{J}}$ and that $\mathfrak{F}_\sigma$ is an arbitrary set of $\mathfrak{J}$; then $\mathfrak{F}_\sigma - t \in \mathfrak{J}$ for each t (property 3°); therefore, $f(\mathfrak{F}_\sigma - t) = 0$, and then $\int_{-\infty}^{\infty} f(\mathfrak{F}_\sigma - t) df_1(t) = 0$, that is, $f * f_1 \in V_{C\mathfrak{J}}$ for every $f_1 \in V^{(b)}$.

The set $F_{\mathfrak{J}}$ *of all functions of* $V^{(b)}$ *concentrated in* $\mathfrak{J}$ *is a subring of* $V^{(b)}$. Indeed, it has been shown that $V_{\mathfrak{J}}$ is a linear system. It remains to show that if f_1 and $f_2 \in V_{\mathfrak{J}}$, then also $f_1 * f_2 \in V_{\mathfrak{J}}$. For this it suffices to restrict oneself to the case where $f_1(E)$ and $f_2(E)$ are nonnegative. As we say in the proof of the linearity of the system $V_{\mathfrak{J}}$, the functions f_1 and f_2 are concentrated on some set $\mathfrak{F}_\sigma \in \mathfrak{J}$; moreover, we may assume that $\mathfrak{F}_\sigma$ contains the point 0 and therefore that $\lim_{n \to \infty} 2^n \mathfrak{F}_\sigma = X$ exists. X is the sum of countably many closed sets, and by properties 4° and 2° belongs to $\mathfrak{J}$. We show that the function $f_1 * f_2$ is concentrated in X. We have

$$\int_{-\infty}^{\infty} f_1(E - t) df_2(t) = \int_X f_1(E - t) df_2(t) =$$

$$= \int_X f_1((E \cap X) - t) df_2(t) + \int_X f_1((E \cap CX) - t) df_2(t).$$

But $X = 2X$; therefore if $t \in X$ the set $(E \cap CX) - t$ is entirely contained in CX, because otherwise the sets $E \cap CX \subset CX$ and $X + t \subset X$ would intersect, which is impossible. Consequently the last integral on the right-hand side equals zero, because in it $|f_1((E \cap CX) - t)| \leq \operatorname{Var} f_1(CX) = 0$ so we get:

$$\int_{-\infty}^{\infty} f_1(E-t)df_2(t) = \int_X f_1((E\cap X)-t)df_2(t) = \int_{-\infty}^{\infty} f_1((E\cap X)-t)df_2(t),$$

i.e., $f_1 * f_2$ is concentrated in X. Since $X \in \mathfrak{Z}$, $f_1 * f_2 \in V_{\mathfrak{Z}}$.

Suppose now that Γ is the discrete additive group of all real numbers and that $X_{\mathfrak{Z}}$ is the set of all characters $\chi(t)$ of the group measurable on each set $\mathfrak{F}_\sigma \in \mathfrak{Z}$. $X_{\mathfrak{Z}}$, in this case, contains all the continuous characters e^{ist}. Compute for each function $f \in V^{(b)}$ the integral $\int_{-\infty}^{\infty} \chi(t)df_{\mathfrak{Z}}(t)$, where $\chi(t)$ is an arbitrary fixed character from $X_{\mathfrak{Z}}$. This defines a homomorphism of the ring $V^{(b)}$ into the complex field. Indeed, $f_{1\mathfrak{Z}} + f_{2\mathfrak{Z}} \in V_{\mathfrak{Z}}$, $f_{1C\mathfrak{Z}} + f_{2C\mathfrak{Z}} \in V_{C\mathfrak{Z}}$, so $(f_1+f_2)_{\mathfrak{Z}} = f_{1\mathfrak{Z}} + f_{2\mathfrak{Z}}$, and therefore to the sum of functions corresponds the sum of the corresponding integrals. Further,

$$f_1 * f_2 = f_{1\mathfrak{Z}} * f_{2\mathfrak{Z}} + (f_{1\mathfrak{Z}} * f_{2C\mathfrak{Z}} + f_{1C\mathfrak{Z}} * f_{2\mathfrak{Z}} + f_{1C\mathfrak{Z}} * f_{2C\mathfrak{Z}});$$

the first term on the right-hand side is an element of $V_{\mathfrak{Z}}$, because $V_{\mathfrak{Z}}$ is a ring; the expression in parentheses is a function from $V_{C\mathfrak{Z}}$, because $V_{C\mathfrak{Z}}$ is an ideal of the ring $V^{(b)}$. Consequently, $(f_1 * f_2)_{\mathfrak{Z}} = f_{1\mathfrak{Z}} * f_{2\mathfrak{Z}}$. But for each character $\chi(t)$ from $X_{\mathfrak{Z}}$

$$\int_{-\infty}^{\infty} \chi(t)\, d(f_{1\mathfrak{Z}} * f_{2\mathfrak{Z}})(t) \text{ exists and } = \int_{-\infty}^{\infty} \chi(t)\, df_{1\mathfrak{Z}}(t) \int_{-\infty}^{\infty} \chi(t)\, df_{2\mathfrak{Z}}(t). \qquad (2)$$

Indeed, as we saw above in the proof that $V_{\mathfrak{Z}}$ is a ring, $f_{1\mathfrak{Z}}$ and $f_{2\mathfrak{Z}}$ are concentrated in some set X of $\mathfrak{Z}$ such that $2X = X$. Since the function $\chi(t)$ is measurable on X, the function $\chi(t+r) = \chi(t)\chi(r)$ is measurable in the topological square $X \times X$. Moreover

$$\iint_{X\times X} \chi(t+r)df_{1\mathfrak{Z}}(t)df_{2\mathfrak{Z}}(r) = \iint_{X\times X} \chi(t)\chi(r)df_{1\mathfrak{Z}}(t)df_{2\mathfrak{Z}}(r) =$$
$$= \int_X \chi(t)df_{1\mathfrak{Z}}(t) \int_X \chi(r)df_{2\mathfrak{Z}}(r) = \int_{-\infty}^{\infty} \chi(t)df_{1\mathfrak{Z}}(t) \int_{-\infty}^{\infty} \chi(r)df_{2\mathfrak{Z}}(r). \qquad (3)$$

But on the other hand applying the theorem of Fubini and noting the fact that $X \subset X - r$ if $r \in X$, we get:

$$\iint_{X\times X} \chi(t+r)df_{1\mathfrak{Z}}(t)df_{2\mathfrak{Z}}(r) = \int_X \Big(\int_X \chi(t+r)df_{1\mathfrak{Z}}(t)\Big)df_{2\mathfrak{Z}}(r) =$$
$$= \int_X \Big(\int_{X-r} \chi(t+r)df_{1\mathfrak{Z}}(t)\Big)\, df_{2\mathfrak{Z}}(r) = \int_X \Big(\int_X \chi(t)df_{1\mathfrak{Z}}(t-r)\Big)\, df_{2\mathfrak{Z}}(r) =$$
$$= \int_X \chi(t)d \int_X f_{1\mathfrak{Z}}(t-r)df_{2\mathfrak{Z}}(r) = \int_{-\infty}^{\infty} \chi(t)d \int_{-\infty}^{\infty} f_{1\mathfrak{Z}}(t-r)df_{2\mathfrak{Z}}(r). \qquad (4)$$

Combining (3) and (4) we get the desired equation (2). Thus we have shown that the result of convolution of two functions leads to the product of the corresponding integrals; with this our assertion is completely proved.

The maximal ideals generated in this way we denote by $M_{\chi\mathfrak{Z}}$. Thus

$$f(M_{\chi\mathfrak{J}}) = \int_{-\infty}^{\infty} \chi(t)\, df_{\mathfrak{J}}(t);$$

where it is assumed that the character $\chi(t)$ is measurable on all the sets $\mathfrak{F}_\sigma \in \mathfrak{J}$.

If the set $\mathfrak{J}$ contains even one set of positive measure, then this family coincides with the set of all $\mathfrak{F}_\sigma$. Indeed, suppose that in $\mathfrak{J}$ there is a set $\mathfrak{F}_\sigma$ of positive measure, then $2\mathfrak{F}_\sigma \in \mathfrak{J}$ (property 4°). But by Lemma 2 $2\mathfrak{F}_\sigma$ contains a whole interval; therefore $\mathfrak{J}$ contains an interval (property 1°). But then by properties 1°, 2°, and 3°, $\mathfrak{J}$ contains all the $\mathfrak{F}_\sigma$ sets.

But in this case $f_{\mathfrak{J}}(t)$ is simply $f(t)$; the only characters of the group Γ measurable on all the sets from $\mathfrak{J}$ are the continuous characters e^{ist}. In this way we find that here the maximal ideals $M_{\chi\mathfrak{J}}$ are the ideals M_s which are already known to us.

The opposite case, the smallest possible family $\mathfrak{J}$, consists of all countable sets (evidently this satisfies the conditions needed for a family $\mathfrak{J}$). For this family $f_{\mathfrak{J}}(t)$ is the jump function of $f(t)$; each character of the group Γ is measurable on every set of the family $\mathfrak{J}$. Consequently, we get here for the family of maximal ideals $M_{\chi\mathfrak{J}}$ the family M_χ found earlier.

Let us now consider a case when all the sets of the family $\mathfrak{J}$ have measure 0 but $\mathfrak{J}$ contains a set which is uncountable. Such a family would be, for example, the family $\mathfrak{J}_F$ constructed in the way described from an arbitrary perfect set F for which the sets nF have measure 0 for every n. We show that the maximal ideals $M_{\chi\mathfrak{J}}$ are all distinct from the ideals M_s and M_χ. Indeed, for each absolutely continuous function $g(t)$ we have, evidently, $g_{\mathfrak{J}}(E) \equiv 0$, because all the sets of $\mathfrak{J}$ have, by assumption, measure 0. Therefore, $g(M_{\chi\mathfrak{J}}) = 0$ for each $M_{\chi\mathfrak{J}}$ so $M_{\chi\mathfrak{J}}$ is distinct from all M_s, because for each M_s there is an absolutely continuous function $g(t)$ such that $g(M_s) \neq 0$. Suppose now that $\mathfrak{F}_\sigma$ is some uncountable set of $\mathfrak{J}$. In it is contained some perfect subset F. Suppose that $f(t)$ is any singular function distinct from zero all of whose growth is concentrated on the set F; the construction of such a function can be carried out in just the same way as the construction of the singular function that carries the Cantor set onto an interval. Since $f(E)$ is concentrated on F and $F \in \mathfrak{J}$, $f \in V_{\mathfrak{J}}$. By the uniqueness theorem for Fourier-Stieltjes integrals there exists an s_0 such that $\int_{-\infty}^{\infty} e^{is_0t} df(t) \neq 0$. Let $\chi(t)$ be an arbitrary character from $X_{\mathfrak{J}}$. Then also $e^{is_0t}\overline{\chi(t)} \in X_{\mathfrak{J}}$. Therefore $\phi(t) = \int_{-\infty}^{t} e^{is_0t}\overline{\chi(t)}\, df(t)$ exists. Evidently $\phi(t)$ is a function of bounded variation all of whose increase is on the set F. But

$$\int_{-\infty}^{\infty} \chi(t) d\phi_{\mathfrak{J}}(t) = \int_{-\infty}^{\infty} \chi(t) d\phi(t) = \int_{-\infty}^{\infty} \chi(t)\, d \int_{-\infty}^{t} e^{is_0\tau}\overline{\chi(\tau)}\, df(\tau) = \int_{-\infty}^{\infty} e^{is_0t} df(t) \neq 0.$$

Thus we have shown that for each maximal ideal $M_{\chi\mathfrak{J}}$ there exists a continuous function $\phi \in V^{(b)}$ such that $\phi(M_{\chi\mathfrak{J}}) \neq 0$. But from this it follows that $M_{\chi\mathfrak{J}}$ is distinct from all the M_χ, because $\phi(M_\chi) = 0$ for each continuous function $\phi \in V^{(b)}$.

Thus we have constructed, in addition to the well known maximal ideals M_s and M_χ, also some new maximal ideals*.

Theorem 3. *Suppose that* $f(t) = g(t) + h(t) + s(t)$ *is the decomposition of a function* $f(t) \in V^{(b)}$ *into an absolutely continuous part, a jump function, and a singular function. If*

1) $|F(s)| = \left|\int_{-\infty}^{\infty} e^{ist} df(t)\right| \geq c > 0$ *for all* s, $-\infty < s < \infty$, *and*

2) $\|s(t)\| < \inf_s \left|\int_{-\infty}^{\infty} e^{ist} dh(t)\right|$,

then $f(t)$ *has an inverse in the ring* $V^{(b)}$.

Proof. We have to show that

$$f(M) = g(M) + h(M) + s(M) \neq 0$$

for all maximal ideals M of the ring $V^{(b)}$. For $M = M_s$ this is true by condition 1). By the corollary to Lemma 1, for all other maximal ideals $g(M) = 0$, so that $f(M) = h(M) + s(M)$. The homomorphism $V^{(b)} \to V^{(b)}/M$ determines a homomorphism of the ring H into the field of complex numbers; by Theorem 2 under this homomorphism $h(t)$ is carried to a number in the closure of the set of values $h(M_s)$. Therefore, by condition 2) $|h(M)| > \|s(t)\| \geq |s(M)|$, and consequently $f(M) = h(M) + s(M) \neq 0$. Hence Theorem 3 is proved.

In the work [31] Wiener and Pitt raised the problem of constructing a function $f(t) \in V^{(b)}$ satisfying condition 1) but without an inverse. The author of the present work can not understand their proof and is therefore inclined to consider the problem open. If a function satisfying the given conditions exists, it follows that the set of maximal ideals of the type M_s is not a dense subset of the space $\mathfrak{M}(V^{(b)})$; indeed, were the set dense in $\mathfrak{M}(V^{(b)})$ then each function $f(t) \in V^{(b)}$ satisfying condition 1) would never vanish on any maximal ideal, and hence there would exist an inverse function.

By the uniqueness theorem for Fourier-Stieltjes transforms the ring $V^{(b)}$ is isomorphic to the ring $W^{(b)}$ of all functions $F(s)$ representable as Fourier-Steiltjes transforms of functions of bounded variation:

$$F(s) = \int_{-\infty}^{\infty} e^{ist} df(t),$$

with the ordinary algebraic operations and with norm

$$\|F(s)\| = \operatorname{Var} f.$$

* Since that time Yu. A. Šreider has given a general characterization of the set of maximal ideals of the ring $V^{(b)}$ and has constructed maximal ideals not included in any earlier construction. (Added in proof.)

We saw how complicated was the problem of maximal ideals in the space $W^{(b)}$. There is a much simpler structure for the maximal ideals of the quotient ring of $W^{(b)}$ by the ideal $I_{\alpha\beta}$ consisting of all functions $F(s)$ which vanish on the interval $\alpha \leq s \leq \beta$, or, what amounts to the same thing, the ring of all functions representable in the form

$$F(s) = \int_{-\infty}^{\infty} e^{ist}\,df(t) \quad (\alpha \leq s \leq \beta),$$

where $f(t)$ is some function of bounded variation. We introduce a norm in the ring by the formula (1) of § 2:

$$\| F(s) \| = \inf \operatorname{Var} f,$$

where the lower bound is taken over all functions $f(t)$ whose Fourier-Stieltjes transforms coincide with $F(s)$ on the interval $[\alpha, \beta]$.

Theorem 4. *The space of maximal ideals of the ring $W^{(b)}/I_{\alpha\beta}$ coincides with the segment $[\alpha, \beta]$.*

Proof. The maximal ideals of the ring $W^{(b)}/I_{\alpha\beta}$ are the images of those maximal ideals of the ring $W^{(b)}$ which contain the ideal $I_{\alpha\beta}$. Therefore it suffices to prove that $I_{\alpha\beta}$ is contained only in the maximal ideals M_s corresponding to points in the segment $\alpha \leq s \leq \beta$. Consider a function $Y(s)$ $(-\infty < s < \infty)$ equal to zero in $\alpha \leq s \leq \beta$ and nowhere else, equal to unity outside some large interval, and having a continuous second derivative. The functions $Y(s)$ and $X(s) = 1 - Y(s)$ are Fourier-Stieltjes transforms of some functions $y(t)$ and $x(t)$ in $V^{(b)}$, such that $x(t)$ is absolutely continuous. By the corollary of Lemma 1, $x(M) = 0$, so $y(M) = 1$ for all maximal ideals $M \neq M_s$. This same element y does not belong to any maximal ideal not one of the M_s, $\alpha \leq s \leq \beta$. Since $y \in I_{\alpha\beta}$, this shows that $I_{\alpha\beta}$ is contained only in the maximal ideals M_s, $\alpha \leq s \leq \beta$, and the theorem is proved.

Corollary. *Suppose that $F(s)$ is defined on an interval $\alpha \leq s \leq \beta$ and is on this interval a Fourier-Stieltjes transform of a function of bounded variation:*

$$F(s) = \int_{-\infty}^{\infty} e^{ist}\,df(t) \quad (\alpha \leq s \leq \beta).$$

If $F(s)$ vanishes nowhere in $[\alpha, \beta]$, then $\dfrac{1}{F(s)}$ is also on $[\alpha, \beta]$ the Fourier-Stieltjes transform of a function of bounded variation.

The following problem arises: Is it possible that every continuous function defined on the interval $[\alpha, \beta]$ be representable as a Fourier-Stieltjes transform of a function of bounded variation? The next theorem, which gives necessary and sufficient conditions for a function to belong to the ring $W^{(b)}/I_{\alpha\beta}$, gives a negative answer to this question.

Theorem 5. *In order that a function $F(s)$ defined on the interval $\alpha \leq s \leq \beta$*

be the Fourier-Stieltjes transform of a function of bounded variation:

$$F(s) = \int_{-\infty}^{\infty} e^{ist} df(t), \; f(t) \in V^{(b)} \; (\alpha \leq s \leq \beta), \tag{5}$$

it is necessary and sufficient that $F(s)$ be representable for each $l > \beta - \alpha$ as an absolutely convergent Fourier series

$$F(s) = \sum_{-\infty}^{\infty} c_n e^{i\frac{\pi}{l}ns}, \; \sum_{-\infty}^{\infty} |c_n| < \infty. \tag{6}$$

Proof. The sufficiency of condition (6) is evident, since the series (6) is a special case of the integral (5). We prove the necessity of condition (6). Without loss of generality we may assume $\alpha = 0$, $\beta = 1$, $l = \pi$. Consider the ring W of absolutely convergent trigonometric series $x = \sum_{-\infty}^{\infty} c_n e^{ins}$, $\sum_{-\infty}^{\infty} |c_n| < \infty$. As we have seen (see page 130), $\mathfrak{M}(W)$ consists of all points of the segment $-\pi \leq s \leq \pi$ with the end points identified. Consider the quotient ring W/I, where I is the closed ideal of those functions $x(s) \in W$ which vanish when $0 \leq s \leq 1$. By definition of the quotient ring, W/I consists of all functions which can be extended from $[0, 1]$ to $(-\pi, \pi)$ as functions $X^*(s) \in W$, and $\| X(s) \| = \inf \| X^*(s) \|$, where the infimum is taken over the set of all functions $X^*(s)$ which coincide with $X(s)$ on $[0, 1]$. Since the ring W contains all periodic functions with continuous derivative, among the functions $X(s) \in W/I$ can be found $X(s) = s$ and also e^{its} for every real t. We show that the norm $\| e^{its} \|$ is uniformly bounded for all t, $\|e^{its}\| \leq M$. Indeed, for $-1 \leq t \leq 1$ we have

$$\| e^{its} \| = \left\| \sum_0^{\infty} \frac{(its)^n}{n!} \right\| \leq \sum_0^{\infty} \frac{\|(its)^n\|}{n!} \leq \sum_0^{\infty} \frac{\|s\|^n}{n!} = e^{\|s\|} = M;$$

for integral t we have $\| e^{its} \| \leq 1$, because for $t = n$ (n integral) coinciding with the function e^{its} on $[0, 1]$ is the function e^{ins} which has norm 1; finally, for arbitrary t, $t = n + \vartheta$, where $|\vartheta| \leq 1$, we have

$$\| e^{its} \| = \| e^{ins} e^{i\vartheta s} \| \leq \| e^{ins} \| \, \| e^{i\vartheta s} \| \leq M.$$

Suppose now that $F(s)$, $0 \leq s \leq 1$, is any function representable by means of the integral (5). Replace the number s appearing in the integral (5) by the element s of the ring W/I; since, as we proved, $\| e^{ist} \|$ is bounded, the integral (5) converges in the norm of W/I and represents an element of this ring. This proves the theorem, since each element of W/I evidently satisfies the condition (6).

We can now, working on the basis of a result of Wiener [29, 30] * answer the question raised above.

Let $F(s)$ be a continuous function of period 2π not decomposable into an absolutely continuous Fourier series, that is, not belonging to the ring W. By the aforementioned theorem of Wiener there exists a point s_0 at which $F(s)$ is locally

* See, for example, Zygmund, *Trigonometric series*, page 140.

not in the ring W, that is, no function $\Phi(s)$ which coincides with $F(s)$ in a neighborhood of s_0 has an absolutely convergent Fourier series. Evidently, without loss of generality we can take $0 \leq s_0 \leq 1$. In this case $F(s)$ can not be represented for $0 \leq s \leq 1$ as a Fourier-Stieltjes transform; in the contrary case, by Theorem 5 there would be a function $\Phi(s)$ expandable in an absolutely convergent Fourier series and agreeing with $F(s)$ for $0 \leq s \leq 1$. Therefore there exist continuous functions not representable on an interval as Fourier-Stieltjes transforms.

APPENDIX III

Group rings of commutative locally compact groups and the theory of characters

In § 16 we considered the ring $V^{(a)}$ of all measurable absolutely integrable functions $x(t)$ of a real variable t with norm $\|x\| = \int_{-\infty}^{\infty} |x(t)|\, dt$ and multiplication ("convolution")

$$(x * y)(t) = \int_{-\infty}^{\infty} x(t-u)\, y(u)\, du,$$

with a formally adjoined unit. We saw that the maximal ideals of this ring, different from the maximal ideal M_∞ made up of the set L_1 of all measurable absolutely integrable functions $x(t)$, can be matched one-to-one with the points of the line $-\infty < s < \infty$; moreover, for the maximal ideal corresponding to the point s,

$$x(M_s) = \int_{-\infty}^{\infty} x(t)\, e^{ist}\, dt. \tag{1}$$

This shows also that the Fourier transform of an absolutely integrable function $x(t)$ can be regarded as a canonical mapping of it into a function $x(M)$ on the maximal ideals of the ring $V^{(a)}$.

When defining the ring $V^{(a)}$ we used only the fact that the real line $-\infty < t < \infty$ is a commutative group (under addition) on whose field of Borel sets is defined a completely additive measure invariant under translation (Lebesgue measure). As Haar [11]* showed, a measure satisfying these properties exists on every locally compact group which satisfies the second axiom of countability. For each such commutative group G there is an analog of the ring $V^{(a)}$, the ring $V^{(a)}(G)$ of all functions $x(g)$ $(g \in G)$ which are measurable and absolutely integrable with respect to Haar measure, with convolution

$$(x * y)(g) = \int x(g-h)\, y(h)\, dh \;**$$

* See also S. Banach, *On Haar's measure*, a note added to the book of S. Saks, *Theory of the integral*, Warsaw 1937. (Also translated into Russian in the second issue of Uspehi Mat. Nauk, pages 161-167.)

** We shall use the additive notation for the group operation in G. Henceforth, when no range of integration is written under an integral sign, it will be understood that the integration is extended over the whole group G.

for multiplication and with a unit adjoined whenever needed. The question arises: Is there a canonical representation of functions $x(g)$ as functions on the maximal ideals of the ring $V^{(a)}(G)$ which can be written as an integral of type (1), and what are the analogs of the functions e^{ist} here?

The proof that to the convolution of functions corresponds the product of their Fourier transforms depends only on the following property of the functions e^{ist}: $e^{is(u+v)} = e^{isu}e^{isv}$. Moreover, that the integral (1) exists for every $x(t) \in V^{(a)}$ depends on the boundedness of the functions e^{ist}, more precisely, $|e^{ist}| \equiv 1$. These properties, combined with continuity of the functions e^{ist}, show that these functions are the characters of the additive group of real numbers topologized in the usual manner.

In general, by a character of a commutative topological group G we mean a homomorphism of the group into the group K of all rotations of a circumference, topologized in the usual way. The group K can be represented analytically either as the additive group of real numbers modulo 2π or as the multiplicative group of complex numbers of absolute value 1. In the additive version of the group K a character χ of the group G is a continuous function $\chi(g)$ of arguments $g \in G$ taking real values modulo 2π and satisfying the equation

$$\chi(g+h) = \chi(g) + \chi(h) \pmod{2\pi}$$

(additive character). In the multiplicative version of the group K a character χ is given in the form of a continuous complex function $e^{i\chi(g)}$ which evidently satisfies the condition

$$e^{i\chi(g+h)} = e^{i\chi(g)}e^{i\chi(h)}, \quad |e^{i\chi(g)}| \equiv 1.$$

Thus we can expect that between the maximal ideals of the ring $V^{(a)}(G)$ (different from the set $L_1(G)$ of all absolutely integrable functions if it contains no unit under convolution) and the characters of the group G there is a one-to-one correspondence such that for the maximal ideal M_χ corresponding to the character χ

$$x(M_\chi) = \int x(g)\, e^{i\chi(g)}\, dg.$$

This allows us to regard the Fourier transform of an absolutely integrable function $x(g)$ as a canonical representation of it as a function on the maximal ideals of the ring $V^{(a)}(G)$. Here we need not assume beforehand than an arbitrary commutative locally compact group with the second axiom of countability has any nontrivial characters. We construct them from the maximal ideals of the ring $V^{(a)}(G)$ and show that there exist sufficiently many.

§ 18. Group Rings

Suppose that G is a locally compact commutative group with the second axiom of countability. On the field (B) of all Borel sets of the group G there is a *Haar measure*, that is, a completely additive, nonnegative function of sets, $m(E)$, finite on compact sets, different from zero on all open sets, and *invariant* under translation:

$$m(E+g) = m(E) \text{ for all } E \in (B) \text{ and } g \in G.$$

For commutative groups it follows also that the Haar measure is also *inverse invariant*:

$$m(-E) = m(E) \text{ for all } E \in (B).$$

We denote by $L_p(G)$ $(p \geq 1)$ the space of all complex functions $x(g)$ $(g \in G)$ measurable (B) and satisfying the condition

$$\| x \|^p = \int | x(g) |^p \, dg < \infty,$$

where the integral is with respect to Haar measure and the domain of integration is the whole group G. From the assumption that G satisfies the second axiom of countability it follows easily that $L_p(G)$ is separable.

From invariance of measure follows *invariance of the integral*:

$$\int x(g+h)\, dg = \int x(g)\, dg \text{ for all } x(g) \in L_1(G) \text{ and } h \in G.$$

In just the same way from inverse invariance of the measure follows *inverse invariance of the integral*:

$$\int x(-g)\, dg = \int x(g)\, dg \text{ for all } x(g) \in L_1(G).$$

To each element h of the group G corresponds the *translation operator* T_h:

$$T_h x(g) = x(g-h).$$

From the invariance of the integral it follows that if $x(g) \in L_p(G)$, then $T_h x(g) \in L_p(G)$ for all $h \in G$, and $\| T_h x \| = \| x \|$. Therefore the T_h make up a group of unitary operators in every $L_p(G)$. Haar measure also has the following important property: *In every $L_p(G)$ the operator T_h is a strongly continuous function of h, that is,*

$$\| T_h x - x \| \to 0 \text{ as } h \to 0; \tag{1}$$

in other words, if $x(g) \in L_p(G)$, then for each $\epsilon > 0$ there exists a neighborhood U of zero in the group G such that

$$\int | x(g-h) - x(g) |^p \, dg < \epsilon \text{ for all } h \in U. \tag{2}$$

A basic role in what follows will be played by the operation of *convolution*. We introduce this operation in the following theorem.

Theorem 1. *If $x(g)$ and $y(g) \in L_1(G)$, then the integral*

$$(x * y)(g) = \int x(g-h)\, y(h)\, dh \tag{3}$$

exists for almost all g and also belongs to $L_1(G)$; moreover

$$\| x * y \| \leq \| x \| \, \| y \|. \tag{4}$$

The convolution operation $*$ defined by formula (3) is associative,

$$(x * y) * z = x * (y * z),$$

and commutative,

$$x * y = y * x.$$

Proof. On the field of Borel sets of the topological square $G \times G$ of the group G, a completely additive measure μ (also a Haar measure) can be constructed with the aid of the Haar measure m given in G, uniquely determined by the following property: For each "rectangle" $X \times Y$, where $X, Y \in (B)$, we have the equation $\mu(X \times Y) = m(X)\,m(Y)$. It is easily verified that $x(g-h)\,y(h)$ as a function of the point $(g, h) \in G \times G$ is measurable. By Fubini's theorem* we get

$$\iint_{G\times G} x(g-h)\,y(h)\,d(g,h) = \int\left(\int x(g-h)\,y(h)\,dh\right)dg = \int y(h)\left(\int x(g-h)\,dg\right)dh,$$

and absolute convergence of one of these double or iterated integrals carries with it absolute convergence of the other two. But the last integral exists:

$$\int y(h)\left(\int x(g-h)\,dg\right)dh = \int y(h)\left(\int x(g)\,dg\right)dh = \int x(g)\,dg \int y(h)\,dh.$$

Therefore the middle integral exists; that is, $\int x(g-h)\,y(h)\,dh$ exists for almost all $g \in G$, is measurable as a function of g, and is absolutely integrable. Also

$$\int\left(\int x(g-h)\,y(h)\,dh\right)dg = \int x(g)\,dg \int y(h)\,dh.$$

The inequality (4) follows easily from this:

$$\|x * y\| = \int\left|\int x(g-h)\,y(h)\,dh\right| dg \le \int |x(g)|\,dg \int |y(h)|\,dh = \|x\|\,\|y\|.$$

Further, applying the theorem of Fubini and making the substitution $h \to h+k$ we get

$$\begin{aligned} x * (y * z) &= \int x(g-h)\left(\int y(h-k)\,z(k)\,dk\right)dh = \\ &= \int\left(\int x(g-h)\,y(h-k)\,dh\right)z(k)\,dk = \\ &= \int\left(\int x(g-k-h)\,y(h)\,dh\right)z(k)\,dk = (x * y) * z. \end{aligned}$$

Finally, making the substitutions $h \to g+h$ and $h \to -h$ we get

$$x * y = \int x(g-h)\,y(h)\,dh = \int x(-h)\,y(g+h)\,dh = \int y(g-h)\,x(h)\,dh = y * x.$$

This proves the theorem.

Theorem 1 shows that $L_1(G)$ is a commutative normed ring with convolution as multiplication. The question still remains as to whether this ring has an identity. There are two possible cases depending on whether the group G is discrete or not.

If G is discrete, then, as is easily verified, each of its elements has (one and the same) positive measure; denote it by c. In this case a unit under convolu-

* See S. Saks, *Theory of the integral*, ch. III, §9, Warsaw-Lwow, 1937.

tion exists; it is the function $x_0(g) = \frac{1}{c}$ if $g = 0$ and 0 if $g \neq 0$.

If G is not discrete, then it is easily verified that each of its elements has measure zero. Therefore here there is no function analogous to $x_0(g)$ in the preceding case which had norm 1 and also vanished for all (or almost all) $g \neq 0$. But we see that an identity can be only such a function.

Definition 1. A sequence of functions $x_n(g) \in L_1(G)$ will be called a *unit sequence shrinking to a point* $g_0 \in G$ if $x_n(g) \geq 0$, $\| x_n \| = 1$, and for each neighborhood U of zero there is a number $n(U)$ such that $x_n(g) = 0$ outside $U + g_0$ for all $n > n(U)$. We will say that a set of unit sequences *shrinks uniformly* to the corresponding elements if the number $n(U)$ can be chosen for each U and all sequences of this set.

An example of a unit sequence shrinking to the point zero of the group G is the sequence of functions $x_n(g) = \frac{1}{m(U_n)} f_{U_n}(g)$, where $\{ U_n \}$ is a basis of neighborhoods of zero and $f_{U_n}(g)$ is the characteristic function of the neighborhood U_n. Applying to all the $x_n(g)$ the operation T_{g_0} we get a unit sequence shrinking to g_0.

Lemma. *If* x_n *is a unit sequence shrinking to* g_0, *then* $x * x_n$ *converges in norm to* $T_{g_0}x$ *for every function* $x \in L_1(G)$:

$$\int | x(g-g_0) - \int x(g-h)\, x_n(h)\, dh \,|\, dg \to 0. \tag{5}$$

For a given function x *the limit in* (5) *is uniform for every family of unit sequences uniformly shrinking to* g_0 *and for each* $g_0 \in G$.

Proof. Fix $\epsilon > 0$. By (2) there is a neighborhood U such that $\int | x(g) - x(g-h) | \, dg < \epsilon$ for all $h \in U$. Take $n > n(U+g_0)$. Since $x(g-g_0) = \int x(g-g_0)\, x_n(h) dh$, making the substitutions $g \to g + g_0$ and $h \to h + g_0$ and applying the theorem of Fubini we get

$$\begin{aligned} \| T_{g_0}x - x * x_n \| &= \int | \int [x(g-g_0) - x(g-h)]\, x_n(h)\, dh \,|\, dg = \\ &= \int | \int [x(g) - x(g-h)]\, x_n(h+g_0)\, dh \,|\, dg \leq \\ &\leq \int (\int | x(g) - x(g-h) |\, dg)\, x_n(h+g_0)\, dh \leq \\ &\leq \sup_{h \in U} \int | x(g) - x(g-h) |\, dg < \epsilon. \end{aligned}$$

Since ϵ is arbitrary, the lemma is proved.

It is now easy to see that *if the group* G *is not discrete, then* $L_1(G)$ *contains no identity under convolution.* Indeed, suppose that $x_n(g)$ is a unit sequence shrinking to zero, and suppose, contrary to our assertion, that $L_1(G)$ contains a unit under convolution; denote it by $e(g)$. By the lemma $e * x_n$ converges in norm to e. But since e is an identity under convolution, $(e * x_n)(g) = x_n(g)$ for almost all g. Hence, $\int | x_n(g) - e(g) | \, dg \to 0$. But this means that $e(g) = 0$ for almost all $g \neq 0$, because outside each neighborhood of zero all but a finite number of the $x_n(g)$ vanish. Since by assumption G is not discrete, $\| e \| = 0$, and, consequently,

$e(g)$ can not be an identity of the ring $L_1(G)$.

Definition 2. By *the group ring* $V^{(a)}(G)$ of a locally compact group G with the second axiom of countability we mean the ring $L_1(G)$ if the group G is discrete and the ring $L_1(G)$ with a unit formally adjoined if G is not discrete; hence in this last case $V^{(a)}(G)$ consists of the elements $\mathfrak{z} = \lambda e + x(g)$, where $x(g) \in L_1(G)$, λ may take arbitrary complex values, e is the adjoined identity, and $\|\mathfrak{z}\| = |\lambda| + \|x\|$.

In clarification of the term "group ring" note the following:

The group ring of a finite group G is the ring of formal sums $\Sigma_{k=1}^{n} c_k g_k$, where g_k belong to G and c_k are arbitrary complex numbers. This definition can be easily generalized to a discrete group G. The group ring R_G is constructed in the following way: Its elements are the formal sums $x = \Sigma\, x(g)\, g$, where $x(g)$ are complex numbers and g runs over all elements of G, and

$$\|x\| = \Sigma\, |x(g)| < \infty \tag{6}$$

from which, in particular, it follows that for each $x \in R_G$ $x(g)$ is nonzero for at most a countable set of elements $g \in G$. The algebraic operations are carried out in the ordinary formal way, except that in multiplication one "collects similar terms", therefore, in particular,

$$xy = \Sigma\, x(h)\, h\, \Sigma\, y(h')\, h' = \Sigma\, \Sigma\, x(h)\, y(h')\, hh' = \Sigma\, z(g)\, g,$$

where

$$z(g) = \sum_{h+h'=g} x(h)\, y(h') = \sum_h x(h)\, y(g-h) = \sum_h x(g-h)\, y(h). \tag{7}$$

It is not hard to verify that R_G satisfies all the axioms of a normed ring. It contains G algebraically in the form of the group of "monomials" $1g = x_g$ ($x_g(h) = 1$ if $h = g$ and $= 0$ if $h \neq g$). The identity of the ring R_G is x_e, where e is the identity of G. However, for a non-discrete group G such a definition of the group ring is not suitable, for in its construction G is regarded as a discrete group, and the topology given in it is ignored. To give an element $x = \Sigma\, x(g)\, g$ of the ring R_G is equivalent to giving the function $x(g)$. Therefore R_G is isomorphic to the ring of functions $x(g)$ in which norm is defined by the formula (6), addition and multiplication by complex numbers are as usual, and multiplication of elements, corresponding to the formula (7), is defined by "convolution":

$$(x * y)(g) = \sum_h x(g-h)\, y(h). \tag{8}$$

As we have seen, in this form the group ring generalizes to locally compact groups satisfying the second axiom of countability if for the $x(g)$ we use measurable functions absolutely integrable with respect to Haar measure, and if instead of the sums (6) and (8) we use the corresponding integrals. Therefore, the ring $V^{(a)}(G)$ is truly a generalization of the group ring considered in the theory of finite groups.

§ 19. Relation between maximal ideals of a group ring and characters of the group

The relation between maximal ideals of the group ring and characters of the group reveals itself in most accessible form when G is a discrete commutative group.

Let M be a maximal ideal of the ring $V^{(a)}(G)$ or, what comes to the same thing, of the ring R_G. Then the function of g

$$M(g) = x_g(M) \tag{1}$$

is a multiplicative character of the group G.

In fact,

$$M(g+h) = x_{g+h}(M) = (x_g * x_h)(M) = x_g(M)\, x_h(M) = M(g)\, M(h). \tag{2}$$

Besides this

$$|M(g)| \le \|x_g\| = 1 \text{ and } M(0) = x_0(M) = 1,$$

from which by (2) it follows that $|M(g)| \equiv 1$.

Since each element x of the ring R_G can be represented in the form $x = \Sigma\, x(g)\, x_g$, by (1) we have for each $x \in R_G$

$$x(M) = \Sigma\, x(g)\, M(g) = \Sigma\, x(g) e^{i\chi_M(g)}, \tag{3}$$

where $\chi_M(g)$ is the additive character of the group G determined by the maximal ideal M.

Conversely, let χ be a character of the group G. Matching to each element $x \in R_G$ the number $\Sigma\, x(g)\, e^{i\chi(g)}$ gives a homomorphism of the ring R_G into the field of complex numbers. Indeed, we need only verify that the product of elements of the ring is carried into the product of the numbers corresponding to them. But

$$\sum_g (\sum_h x(g-h)\, y(h)) e^{i\chi(g)} = \sum_h (\sum_g x(g-h)\, e^{i\chi(g)})\, y(h) =$$

$$= \sum_h (\sum_g x(g-h)\, e^{i\chi(g-h)})\, e^{i\chi(h)}\, y(h) = \sum_g x(g)\, e^{i\chi(g)} \sum_h y(h)\, e^{i\chi(h)}.$$

This mapping is not trivial since x_0 is carried to 1. Denote the kernel of this homomorphism by M_χ. M_χ is a maximal ideal of the ring R_G. For it $x(M_\chi) = \Sigma\, x(g)\, e^{i\chi(g)}$ and, in particular,

$$x_g(M_\chi) = e^{i\chi(g)}. \tag{4}$$

The validity of formulas (1) and (4) shows that the maximal ideal M_χ determined by the character χ in turn determines this character: $\chi = \chi_{M_\chi}$.

From (3) it follows that distinct maximal ideals determine distinct characters. Indeed, if $\chi_{M_1}(g) = \chi_{M_2}(g)$, then by (3) $x(M_1) = x(M_2)$ for all $x \in R_G$, that is, M_1 and M_2 coincide.

Thus we have established a *one-to-one correspondence between the characters of a discrete commutative group G and the maximal ideals of its group ring R_G.*

Suppose now, as we shall henceforth, that the group G is not discrete. In this case $L_1(G)$ is a maximal ideal in the group ring $V^{(a)}(G)$. We denote it by M_∞; therefore

$$x(M_\infty) = 0 \text{ for all } x \in L_1(G).$$

We shall establish a one-to-one correspondence between characters of the group G and the maximal ideals distinct from M_∞ of the group ring $V^{(a)}(G)$.

The character of a discrete group G corresponding to a maximal ideal M of the group ring R_G is determined by the values taken at M by the "unit functions" $x_g(h)$. For non-discrete groups the role of these functions is played by unit sequences shrinking to elements of the group.

Theorem 2. *Let M be a maximal ideal $V^{(a)}(G)$ distinct from M_∞ and let x_n be a unit sequence shrinking to the element $g \in G$. Then*

$$\lim_{n\to\infty} x_n(M) = M(g) \tag{5}$$

exists and does not depend on the choice of the sequence x_n; also the approach is uniform for each uniformly shrinking family of unit sequences and all $g \in G$. $M(g)$ is a multiplicative character of the group G: $M(g) = e^{i\chi_M(g)}$.

Proof. Since $M \neq M_\infty$, there is a function $z(g) \in L_1(G)$ such that $z(M) \neq 0$. By the lemma of the preceding section

$$|(T_g z)(M) - z(M)\,x_n(M)| \leq \| T_g z - z * x_n \| \to 0, \tag{6}$$

that is,

$$x_n(M) \to \frac{(T_g z)(M)}{z(M)},$$

even uniformly for each uniformly shrinking to g family of unit sequences x_n and all $g \in G$. Fixing z we see that for all unit sequences x_n shrinking to g there exists a unique limit

$$M(g) = \frac{(T_g z)(M)}{z(M)}, \tag{7}$$

Fixing some sequence x_n we see that the limit does not depend on the choice of z.

$M(g)$ is a continuous function of g. Indeed, by (7) and relation (1) of § 18

$$|M(g') - M(g)| = \left| \frac{(T_{g'} z - T_g z)(M)}{z(M)} \right| \leq \frac{\|T_{g'} z - T_g z\|}{|z(M)|} \to 0 \quad \text{as} \quad g' \to g.$$

Let us show that $M(g)$ is a multiplicative character of the group G. First of all, $|x_n(M)| \leq \|x_n\| = 1$, so that

$$|M(g)| \leq 1. \tag{8}$$

In addition, it follows from (7) that

$$M(0) = 1. \tag{9}$$

It suffices now to prove

$$M(g+h) = M(g)\,M(h), \tag{10}$$

because then setting $h = -g$ in (10) and applying (8) and (9) we get that $|M(g)| = 1$. But if x_n is a unit sequence shrinking to g and if y_n is a unit sequence shrinking to h, then $x_n * y_n$, as is easily seen, is a unit sequence shrinking to $g+h$. Therefore

$$M(g+h) = \lim (x_n * y_n)(M) = \lim x_n(M) \lim y_n(M) = M(g)M(h).$$

This proves Theorem 2. We shall say that the maximal ideal M determines a character χ_M: $M(g) = e^{i\chi_M(g)}$

Remark. From equation (5) and relation (6), true for all $z(g) \in L_1(G)$, it follows that $(T_g x)(M) = M(g)x(M)$ for all functions $x(g) \in L_1(G)$.

Theorem 3. *If $M \neq M_\infty$, then for all $x \in L_1(G)$ we have the equation*

$$x(M) = \int x(g)\, e^{i\chi_M(g)}\, dg, \tag{11}$$

from which

$$\mathfrak{z}(M) = (\lambda e + x)(M) = \lambda + \int x(g)\, e^{i\chi_M(g)}\, dg. \tag{12}$$

Proof. It evidently suffices to prove (11) for nonnegative functions $x(g)$. By Theorem 2 for each $\epsilon > 0$ there is a neighborhood U of zero such that for each $g_0 \in G$ and each function $y(g)$ satisfying the conditions $y(g) \geq 0$, $\|y\| = 1$, and $y(g) = 0$ outside $U + g_0$ we have

$$|y(M) - e^{i\chi_M(g_0)}| < \epsilon. \tag{13}$$

Let D be the set on which $x(g) > 0$. Since G is a group with the second axiom of countability, as can be easily seen D can be decomposed into a countable family of pairwise disjoint sets $D_n \in (B)$, such that $g_n - g_n' \in U$ for all $g_n, g_n' \in D_n$. Since sets of measure 0 are negligeable we can suppose that all D_n have positive measure. Set $x_n(g) = x(g) f_{D_n}(g)$, where $f_{D_n}(g)$ is the characteristic function of the set D_n, so $x(g) = \Sigma\, x_n(g)$. Also, set $y_n(g) = \dfrac{x_n(g)}{\int x_n(g) dg}$. Since $y_n(g) \geq 0$, $\|y_n\| = 1$, and $y_n(g) = 0$ outside $U + g_n$ for each $g_n \in D_n$, applying (13) we get

$$\begin{aligned} |x_n(M) - \int x_n(g) e^{i\chi_M(g)} dg| &= \left| y_n(M) - \int y_n(g) e^{i\chi_M(g)} dg \right| \int x_n(g)\, dg = \\ &= \left| \int [y_n(M) - e^{i\chi_M(g)}] y_n(g) dg \right| \int x_n(g)\, dg \leq \\ &\leq \sup_{g_n \in D_n} |y_n(M) - e^{i\chi_M(g_n)}| \cdot \int x_n(g)\, dg \leq \epsilon \int x_n(g)\, dg, \end{aligned}$$

and consequently,

$$\begin{aligned} \left| x(M) - \int x(g)\, e^{i\chi_M(g)} dg \right| &= \left| \Sigma\, x_n(M) - \Sigma \int x_n(g) e^{i\chi_M(g)} dg \right| \leq \\ &\leq \epsilon\, \Sigma \int x_n(g)\, dg = \epsilon \int x(g)\, dg. \end{aligned}$$

Since ϵ is arbitrary this proves the equation (11).

Corollary. *Distinct maximal ideals determine distinct characters.*

Indeed, if $\chi_{M_1} = \chi_{M_2}$, then, by (12), $\mathfrak{z}(M_1) = \mathfrak{z}(M_2)$ for all $\mathfrak{z} \in V^{(a)}(G)$, so M_1 and M_2 coincide.

Theorem 4. *Let χ be a character of the group G. Then the mapping*

$$\lambda e + x(g) \to \lambda + \int x(g)\, e^{i\chi(g)}\, dg \tag{14}$$

is a homomorphism of the group ring $V^{(a)}(G)$ into the field of complex numbers. The maximal ideal M_χ which is the kernel of this mapping is distinct from M_∞.

Proof. The linearity of the mapping (14) is evident and it is only necessary to show it is multiplicative, that is, to show that to the product of elements of the group ring corresponds the product of the corresponding numbers. We introduce the symbol

$$\lambda + \int x(g)\, e^{i\chi(g)}\, dg = (\lambda e + x, \chi).$$

Applying the theorem of Fubini we get

$$(x * y, \chi) = \int\left(\int x(g-h)\, y(h)\, dh\right) e^{i\chi(g)}\, dg = \int\left(\int x(g-h)\, e^{i\chi(g)}\, dg\right) y(h)\, dh =$$

$$= \int\left(\int x(g-h)\, e^{i\chi(g-h)}\, dg\right) y(h)\, e^{i\chi(h)}\, dh =$$

$$= \int x(g)\, e^{i\chi(g)}\, dg \int y(h)\, e^{i\chi(h)}\, dh = (x, \chi)\,(y, \chi).$$

From this it follows without difficulty that

$$((\lambda e + x) * (\mu e + y), \chi) = (\lambda e + x, \chi)\,(\mu e + y, \chi)$$

for every $\lambda e + x,\ \mu e + y \in V^{(a)}(G)$. Hence the first assertion of the theorem if proved. For the proof of the other assertion take any function $x(g) \in L_1(G)$ for which $\int x(g)\, dg \neq 0$. Then $x(g)\, e^{-i\chi(g)}$ is not in M_χ, because

$$\int x(g)\, e^{-i\chi(g)} \cdot e^{i\chi(g)}\, dg = \int x(g)\, dg \neq 0.$$

Since, on the other hand, $x(g)\, e^{-i\chi(g)} \in M_\infty$, it follows that $M_\chi \neq M_\infty$:

Evidently, $(x, \chi) = x(M_\chi)$ for every function $x(g) \in L_1(G)$.

We will say that *the character χ determines the maximal ideal M_χ*. Theorem 3 shows that *a maximal ideal is determined by the character it determines, $M = M_{\chi_M}$*.

Theorem 5. *A character is determined by the maximal ideal it determines, $\chi = \chi_{M_\chi}$.*

Proof. By the preceding theorem in $L_1(G)$ there exists a function not contained in M_χ. Let $z(g)$ be such a function, i.e., $(z, \chi) \neq 0$. We have

$$(T_g z, \chi) = \int z(h-g)\, e^{i\chi(h)}\, dh = e^{i\chi(g)} \int z(h-g)\, e^{i\chi(h-g)}\, dh =$$

$$= e^{i\chi(g)} \int z(h)\, e^{i\chi(h)}\, dh = e^{i\chi(g)}\,(z, \chi),$$

whence

$$e^{i\chi(g)} = \frac{(T_g z, \chi)}{(z, \chi)} = \frac{(T_g z)(M_\chi)}{z(M_\chi)} = e \exp i\chi_{M_\chi}(g).$$

Remark. For the proof of Theorem 4 only the measurability of the character χ is needed. Therefore Theorem 5 proves that *from the measurability of a character χ follows its continuity.*

From Theorem 5 comes

Corollary. *Distinct characters determine distinct maximal ideals.*

Indeed, if M_{χ_1} coincides with M_{χ_2}, then also $\chi_1 = \chi_{M_{\chi_1}}$ coincides with $\chi_2 = \chi_{M_{\chi_2}}$.

Theorems 2-5 establish a *one-to-one correspondence between characters of the group G and maximal ideals of its group ring $V^{(a)}(G)$ distinct from M_∞*. In addition these theorems show that *the Fourier transform*

$$\int x(g)\, e^{i\chi(g)}\, dg$$

of an absolutely integrable function $x(g)$ can be considered as its canonical representation as a function on the maximal ideals of the group ring

$$\int x(g)\, e^{i\chi(g)}\, dg = x(M_\chi).$$

This new treatment of the Fourier transform can be used as a basis for constructing the harmonic analysis on commutative locally compact groups.

§20. Uniqueness theorem for the Fourier transform and sufficiency of the set of characters.

In view of the relation established above between characters of a commutative locally compact group satisfying the second axiom of countability and the maximal ideals of its group ring it is natural to expect that there is also a relation between the question of existence, for such a group, of a sufficiently large collection of characters and the question of existence for its group ring of a sufficiently large collection of maximal ideals.

Every commutative group has at least one character: $\chi(g) \equiv 0$. Say that a commutative group has *sufficiently many characters* if for each element g_0 different from zero there exists a character χ_0 such that $\chi_0(g_0) \neq 0$. Analogously, it can be said that a commutative normed ring R has sufficiently many maximal ideals if for each element $x_0 \in R$ different from zero there is a maximal ideal M_0 such that $x_0(M_0) \neq 0$. But this is nothing else than the assertion that there are no generalized nilpotent elements different from zero in R. In this section we prove this assertion for a group ring $V^{(a)}(G)$. In view of the results of the preceding section this will prove at the same time the theorem of uniqueness for Fourier transforms of absolutely integrable functions. From this we derive as a corollary that the groups under consideration have sufficiently many characters.

Theorem 6. *In the ring $V^{(a)}(G)$ there are no generalized nilpotent elements different from zero.*

Proof. From Theorem 3 it follows that *the ring $V^{(a)}(G)$ is symmetric*, and indeed that $\mathfrak{z}^* = \bar{\lambda} e + \overline{x(-g)}$ is an element conjugate to $\mathfrak{z} = \lambda e + x(g)$. Indeed, for $M = M_\infty$, we have $\mathfrak{z}^*(M_\infty) = \bar{\lambda} = \overline{\mathfrak{z}(M_\infty)}$, and for $M \neq M_\infty$, in view of Theorem 3, inverse invariance of the integral, and the equation $\chi(-g) = -\chi(g)$, we have

$$\mathfrak{z}^*(M) = \bar{\lambda} + \int \overline{x(-g)}\, e^{i\chi_M(g)} dg = \bar{\lambda} + \int \overline{x(g)\, e^{i\chi_M(g)}} dg = \overline{\mathfrak{z}(M)}.$$

The operation $\mathfrak{z} \to \mathfrak{z}^*$ satisfies all the conditions of Definition 10 of § 7. Therefore by Theorem 20 of § 7 our assertion will be proved if we show that on $V^{(a)}(G)$ there exists a sufficient set of positive functionals.

We show, first of all, a simple method of constructing such functionals. Let $\psi(h)$ be an arbitrary function of $L_2(G)$. Then also $\psi(g+h) \in L_2(G)$ for every $g \in G$. Therefore the "scalar product"

$$\phi(g) = \int \psi(g+h)\, \overline{\psi(h)}\, dh \tag{1}$$

exists for all $g \in G$. It is not difficult to see that $\overline{\phi(-g)} = \phi(g)$, $|\phi(g)| \leq \phi(0)$, and $\phi(g)$ is continuous. Indeed, making the substitution $h \to h + g$, we get

$$\overline{\phi(-g)} = \int \overline{\psi(-g+h)}\, \psi(h)\, dh = \int \overline{\psi(h)}\, \psi(g+h)\, dh = \phi(g);$$

further, by the Schwartz inequality,

$$|\phi(g)|^2 = \left| \int \psi(g+h)\, \overline{\psi(h)}\, dh \right|^2 \leq \int |\psi(g+h)|^2\, dh \int |\psi(h)|^2\, dh = \phi^2(0);$$

finally,

$$|\phi(g') - \phi(g)|^2 = \left| \int [\psi(g'+h) - \psi(g+h)]\, \overline{\psi(h)}\, dh \right|^2 \leq$$
$$\leq \int |\psi(h)|^2\, dh \int |\psi(g'+h) - \psi(g+h)|^2\, dh,$$

and the right hand side tends to zero as $g' \to g$. From the boundedness and continuity of $\phi(g)$ follows that the expression

$$f(\mathfrak{z}) = f(\lambda e + x(g)) = \lambda\phi(0) + \int \phi(g)\, x(g)\, dg \tag{2}$$

is defined for all $\mathfrak{z} \in V^{(a)}(G)$ and is a linear functional on $V^{(a)}(G)$. Let us show that this functional is positive, i.e., that

$$f(\mathfrak{z} * \mathfrak{z}^*) = |\lambda|^2\, \phi(0) + \lambda \int \phi(g)\, \overline{x(-g)}\, dg + \bar{\lambda} \int \phi(g)\, x(g)\, dg +$$
$$+ \int \phi(g)\, \left(\int x(g+h)\, \overline{x(h)}\, dh\right) dg \geq 0$$

for all $\mathfrak{z} = \lambda e + x(g) \in V^{(a)}(G)$. Since $\phi(0) \geq 0$ and

$$\int \phi(g)\, \overline{x(-g)}\, dg = \int \overline{\phi(-g)}\, \overline{x(-g)}\, dg = \overline{\int \phi(g)\, x(g)\, dg}, \tag{3}$$

for this it is sufficient to prove that

$$\left| \int \phi(g)\, x(g)\, dg \right|^2 \leq \phi(0) \int \phi(g)\, \left(\int x(g+h)\, \overline{x(h)}\, dh\right) dg. \tag{4}$$

But applying Fubini's theorem and Schwartz's inequality, we actually get

$$\left| \int \phi(g)\, x(g)\, dg \right|^2 = \left| \int \left(\int \psi(g+h)\, x(g)\, dg\right) \overline{\psi(h)}\, dh \right|^2 \leq$$
$$\leq \int |\psi(h)|^2\, dh \int \left| \int \psi(g+h)\, x(g)\, dg \right|^2 dh =$$
$$= \phi(0) \int \left[\int \left(\int \psi(g+h)\, \overline{\psi(g'+h)}\, dh\right) x(g)\, dg \right] \overline{x(g')}\, dg' =$$
$$= \phi(0) \int \left(\int \phi(g-g')\, x(g)\, dg\right) \overline{x(g')}\, dg' = \phi(0) \int \phi(g)\, \left(\int x(g+g')\, \overline{x(g')}\, dg'\right) dg.$$

We can now pass immediately to the proof of the theorem. If $\mathfrak{z} = \lambda e + x(g)$, where $\lambda \neq 0$, then for the positive functional M_∞ we have $M_\infty(\mathfrak{z}) = \lambda \neq 0$. Therefore it suffices to consider the case where $\mathfrak{z} = x(g)$. We show that if $\|x\| > 0$, then there can be found a positive linear functional f of the form considered such that $f(x) \neq 0$; in other words, there can be found a function $\psi(h) \in L_2(G)$ for which

$$\int(\int \psi(g+h)\overline{\psi(h)}\,dh)\,x(g)\,dg = \int\int x(g-h)\,\psi(g)\,\overline{\psi(h)}\,dg\,dh \neq 0. \tag{5}$$

Since each of the integrals

$$\int\int x(g-h)\,\xi(g)\,\eta(h)\,dg\,dh, \tag{6}$$

where $\xi(g)$, $\eta(g)$ are real and belong to $L_2(G)$, is a linear combination of integrals of the form (5):

$$\begin{aligned}\int\int x(g-h)\,\xi(g)\,\eta(h)\,dg\,dh = \frac{1}{4}\{ &\int\int x(g-h)[\xi(g)+\eta(g)][\xi(h)+\eta(h)]\,dg\,dh - \\ &\int\int x(g-h)[\xi(g)-\eta(g)][\xi(h)-\eta(h)]\,dg\,dg + \\ &+ i\int\int x(g-h)[\xi(g)+i\eta(g)][\xi(h)-i\eta(h)]\,dg\,dh - \\ &- i\int\int x(g-h)[\xi(g)-i\eta(g)][\xi(h)+i\eta(h)]\,dg\,dh\},\end{aligned}$$

it suffices to prove that there exists an integral of the form (6) different from zero.

Denote by $\Re x(g)$ and $\Im x(g)$, respectively, the real and imaginary parts of the function $x(g)$. Since, by assumption, $\|x\| > 0$, and $\epsilon > 0$ can be found such that some one of the sets $E\{\Re x(g) > \epsilon\}$, $E\{\Re x(g) < -\epsilon\}$, $E\{\Im x(g) > \epsilon\}$, $E\{\Im x(g) < -\epsilon\}$ has positive measure (obviously finite). Let $\xi(g)$ be the characteristic function of this set. Then the integral $\int x(g-h)\,\xi(g)\,dg$ is different from zero when $h = 0$. But it is easily seen, by the boundedness of the function $\xi(g)$, that this integral is a continuous function of h (see p. 169). Therefore, denoting by $\eta(h)$ the characteristic function of a sufficiently small neighborhood of the point $h = 0$, we have

$$\int\int x(g-h)\,\xi(g)\,\eta(h)\,dg\,dh \neq 0,$$

and the theorem is proved.

As we are already aware, in view of the results of § 19, Theorem 6 is nothing else but the

Theorem of uniqueness. *If* $x(g), y(g) \in L_1(G)$ *and*

$$\int x(g)\,e^{i\chi(g)}\,dg = \int y(g)\,e^{i\chi(g)}\,dg$$

for all characters χ *of the group* G, *then* $x(g)$ *and* $y(g)$ *coincide almost everywhere.*

Indeed, to say that in a commutative normed ring there are no generalized nilpotent elements different from zero is equivalent to saying that each element of the ring is completely determined by the values it takes on the maximal ideals.

Theorem 7. *A commutative locally compact group satisfying the second axiom of countability possesses sufficiently many characters.*

Proof. Let g_0 be an arbitrary element of the group G different from zero. It is necessary to show that there exists a character χ_0 such that $e^{i\chi_0(g_0)} \neq 1$. Since $g_0 \neq 0$ there exists a neighborhood of zero U sufficiently small that U and $U + g_0$ do not intersect. Let $f(g)$ be the characteristic function of the set U. Then $T_{g_0} f(g) = f(g - g_0)$ is the characteristic function of the set $U + g_0$ and $\|T_{g_0} f - f\| = 2\|f\| > 0$. Hence, by Theorem 6, $T_{g_0} f - f$ is not a generalized nilpotent element, that is, there exists a maximal ideal M_0 such that

$$(T_{g_0} f - f)(M_0) = (T_{g_0} f)(M_0) - f(M_0) \neq 0,$$

or

$$(T_{g_0} f)(M_0) \neq f(M_0). \tag{7}$$

Evidently, $M_0 \neq M_\infty$, because $(T_{g_0} f)(M_\infty) = f(M_\infty) = 0$. By the remark after Theorem 2 of § 19, the inequality (7) can be written in the form

$$M_0(g_0) f(M_0) \neq f(M_0).$$

Hence, first of all, it follows that $f(M_0) \neq 0$; dividing by $f(M_0)$ and denoting by χ_0 the character determined, according to Theorem 2 of § 19, by the maximal ideal M_0 we get

$$e^{i\chi_0(g_0)} = M_0(g_0) \neq 1,$$

and the theorem is proved.

§ 21. Positive definite functions

The functions with the aid of which we construct positive linear functionals on the ring $V^{(a)}(G)$ belong to the class called *positive definite functions*, that is, to the class of functions $\phi(g)$ characterized by the following structural property: For each finite set of elements of the group $g_1, g_2, \cdots, g_n$, and complex numbers $\xi_1, \xi_2, \cdots, \xi_n$, we have the equation

$$\sum_{k=1}^{n} \sum_{l=1}^{n} \phi(g_k - g_l)\, \xi_k \bar{\xi}_l \geq 0. \tag{1}$$

Indeed, if $\phi(g) = \int \psi(g+h)\overline{\psi(h)}\, dh$, where $\psi \in L_2(G)$, then substituting for $\phi(g)$ in the form (1), we get

$$\sum_{k=1}^{n} \sum_{l=1}^{n} \phi(g_k - g_l)\, \xi_k \bar{\xi}_l = \sum_{k=1}^{n} \sum_{l=1}^{n} \xi_k \bar{\xi}_l \int \psi(g_k - g_l + h)\overline{\psi(h)}\, dh =$$

$$= \sum_{k=1}^{n} \sum_{l=1}^{n} \xi_k \bar{\xi}_l \int \psi(g_k + h)\overline{\psi(g_l + h)}\, dh = \int \Big| \sum_{k=1}^{n} \xi_k \psi(g_k + h) \Big|^2 dh \geq 0.$$

The proof in § 20 of the positiveness of the functional determined by the function $\phi(g) = \int \psi(g+h)\overline{\psi(h)}\, dh$ used only properties of this function which followed from its continuity and positive definiteness, so *for each continuous, positive definite function $\phi(g)$ the expression*

$$f(\mathfrak{z}) = f(\lambda e + x(g)) = \lambda\phi(0) + \int \phi(g)\, x(g)\, dg \tag{2}$$

is itself a positive linear functional on the ring $V^{(a)}(G)$.

Indeed, from the inequality (1) if $n = 2$, $g_1 = 0$, $g_0 = g$, it easily follows that

$\overline{\phi(-g)} = \phi(g)$ and $|\phi(g)| \leq \phi(0)$. From the boundedness and continuity of $\phi(g)$ it follows that $f(\mathfrak{z})$ is a linear functional on $V^{(a)}(G)$. Further, from the inequality (1) and the continuity of $\phi(g)$, an analogous integral formula, derived by passage to the limit,

$$\iint \phi(g-h)\, x(g)\, \overline{x(h)}\, dg\, dh \geq 0,$$

is true for each $x(g) \in L_1(G)$. By Schwartz's inequality we have, therefore,

$$\left| \iint \phi(g-h)\, x(g)\, \overline{y(h)}\, dg\, dh \right|^2 \leq$$

$$\leq \iint \phi(g-h)\, x(g)\, \overline{x(h)}\, dg\, dh \iint \phi(g-h)\, y(g)\, \overline{y(h)}\, dg\, dh.$$

Choose here for $y(h)$ functions from a unit sequence shrinking to $h = 0$, to get, in the limit, the inequality (4) of § 20. But from this inequality combined with the equation (3) of § 20 we get the relation $\overline{\phi(-g)} = \phi(g)$ and with it positiveness of the functional (2).

Therefore, the absence of a radical from the group ring, the theorem of uniqueness for Fourier transforms of absolutely integrable functions, and the existence of sufficiently many characters are all corollaries of the existence on the group considered of sufficiently many positive definite functions.

Each multiplicative character $e^{i\chi(g)}$ is a positive definite function. Indeed

$$\sum_{k=1}^{n} \sum_{l=1}^{n} e^{i\chi(g_k - g_l)}\, \xi_k \bar{\xi}_l = \sum_{k=1}^{n} \sum_{l=1}^{n} e^{i\chi(g_k)}\, \overline{e^{i\chi(g_l)}}\, \xi_k \bar{\xi}_l = \left| \sum_{k=1}^{n} e^{i\chi(g_k)}\, \xi_k \right|^2 \geq 0.$$

We denote by X the set of all characters of the commutative locally compact group G which satisfies the second axiom of countability. Identifying characters with the corresponding maximal ideals, we can introduce in X the topology of the space $\mathfrak{M}$ of maximal ideals of the ring $V^{(a)}(G)$. Since $V^{(a)}(G)$ is separable (see p. 194) $\mathfrak{M}$ satisfies the second axiom of countability. Consequently, X, which is determined from $\mathfrak{M}$ by deleting the single point M_∞, is a locally compact space with the second axiom of countability. Suppose that $\Phi(P)$ is an arbitrary nonnegative, completely additive set function on the field of Borel sets of the space X. Then just as above one can convice oneself that the function

$$\phi(g) = \int_X e^{i\chi(g)} d\Phi(\chi) \tag{3}$$

is positive definite. Moreover, it is not difficult to see that $\phi(g)$ is continuous. Indeed, suppose that $g_n \to g$. Then $e^{i\chi(g_n)} \to e^{i\chi(g)}$ for all χ, and, by the theorem of Lebesgue on bounded convergence under the integral sign, we conclude that also $\phi(g_n) \to \phi(g)$.

It turns out that the converse proposition is also valid and is a generalization of a known theorem of Herglotz [13] and Bochner [2] on the representations of positive definite functions defined on the additive group of integers and on the additive group of real numbers.

Theorem 8. *Each continuous positive definite function defined on a commutative locally compact group* G *satisfying the second axiom of countability can be represented in just one way in the form* (3), *where* Φ *is a nonnegative, completely additive, regular* * *set function on the field of Borel sets of the space* X *of characters of* G.

Proof. It was proved above that the linear functional (2) determined by the continuous positive definite function $\phi(g)$ is positive. Therefore, by Theorem 19 of § 7, for each function $x(g) \in L_1(G)$ we have

$$f(x) = \int_{\mathfrak{M}} x(M)\, d\Phi(M),$$

where Φ is a uniquely determined, nonnegative, completely additive, regular set function on the field of Borel sets of the space $\mathfrak{M}$. Since $x(M_\infty) = 0$, we can suppose that the integral extends only over the space X:

$$f(x) = \int_X x(\chi)\, d\Phi(\chi).$$

Therefore, by Theorem 3 of § 19 we have

$$f(x) = \int_G x(g)\,\phi(g)\, dg = \int_X \left(\int_G x(g)\, e^{i\chi(g)} dg\right) d\Phi(\chi).$$

Applying Fubini's theorem to the right-hand side we get

$$\int x(g)\,\phi(g)\, dg = \int_G x(g) \left(\int_X e^{i\chi(g)} d\Phi(\chi)\right) dg.$$

Since $x(g)$ is an arbitrary function from $L_1(G)$, this equation shows that $\phi(g)$ coincides almost everywhere with $\int_X e^{i\chi(g)} d\Phi(\chi)$. But both functions are continuous, so

$$\phi(g) = \int_X e^{i\chi(g)} d\Phi(\chi),$$

and the theorem is proved.

If χ and ξ are two characters of the group G, then their sum $(\chi + \xi)(g) = \chi(g) + \xi(g) \pmod{2\pi}$ is also a character. Therefore, X is a group. It can be shown [23] that the group operation in X is continuous in the topology of X as a space of maximal ideals. Therefore, X is a locally compact group satisfying the second axiom of countability. Consequently, on X there is a Haar measure, and, in just the manner in which we constructed the character group X of G, we construct the character group G^* of the group X. On the other hand, each element g of the original group G determines a character of the group X, namely

$$g(\chi) = \chi(g),$$

where g is fixed and χ is variable. Here from the theorem about sufficiently many characters in X it follows that distinct elements of the group G determine distinct

* See the footnote on page 146.

characters of the group X. Evidently, the operations on elements of the group G are just those operations on them as characters of the group X. Therefore, G is algebraically contained in G^*. One of the important theorems of the Pontrjagin theory of characters is the law of duality, according to which G^* coincides with G both in elements and in topology. It turns out that this theorem can also be derived within the framework of our theory, in the final reckoning as a corollary of the existence on the group in question of a sufficiently large family of continuous positive definite functions [see 22 and 23]. For this one uses a generalization of a known theorem of Plancherel [14, see also 23] which is also attached to the theorem on representation of continuous positive definite functions.

APPENDIX IV

Extensions of maximal ideals

Each maximal ideal M of a commutative normed ring contains a maximal ideal of every subring with identity. Indeed, matching to each element x of the subring the number $x(M)$ we determine a homomorphism of the subring into the field of complex numbers which is non-trivial because the identity element is carried to the number 1; the kernel of this homomorphism, that is, the set of all elements of the subring contained in M, is a maximal ideal of the subring contained in M. The problem arises as to whether each maximal ideal of the subring is contained in a maximal ideal of the ring, that is, can every maximal ideal of the subring be "extended" to an ideal of the whole ring? For some classes of rings, for example, for symmetric and for regular, this possibility holds. But there exist many classes of rings where this extension, generally speaking, is impossible: in this case the problem becomes the determination and internal characterization of those maximal ideals in every larger ring. The present Appendix is devoted to an exposition of all these problems.

§ 22. Generalized zero divisors

In algebra a zero divisor is an element x such that the product of it with some $y \neq 0$ gives zero.

For example, in the ring C a function $x(t)$ which equals zero on some interval $\Delta \subset [0, 1]$ is a zero divisor, since the product of it with an arbitrary function $y(t) \in C$ which vanishes outside the interval but not identically equals zero.

The most natural generalization of this concept to normed rings is the following:

Definition 1. An element x of a normed ring R is called a *generalized zero divisor* if there exists a sequence $y_n \in R$ such that

$$1) \inf_n \|y_n\| > 0, \qquad 2) \lim_{n\to\infty} \|xy_n\| = 0.$$

Obviously an ordinary zero divisor is a generalized one. In finite-dimensional rings *, where from every bounded sequence can be chosen a convergent subsequence, the converse is also valid; each generalized zero divisor is a zero divisor in the ordinary sense. But, generally speaking, generalized zero divisors are not ordinary ones.

Let us consider in the ring C a function $x(t)$ which vanishes only at one point $t = 0$; it is not an (ordinary) zero divisor. Suppose, further, that $x_n(t)$ is a positive function vanishing in the segment $[\frac{1}{n}, 1]$ and having a maximum value in the interval $[0, \frac{1}{n}]$ such that $\|x_n(t)\| = 1$. Evidently $x(t)x_n(t) \to 0$; consequently $x(t)$ is a generalized zero divisor.

A generalized zero divisor can only be an element x without an inverse. In fact, suppose x^{-1} exists. If y_n is such that $\lim xy_n = 0$, then $\lim y_n = \lim x^{-1}xy_n = 0$, which means that conditions 1) and 2) of Definition 1 are inconsistent, that is, x is not a generalized zero divisor.

It is easy to verify that *each* element of the ring C which has no inverse is a generalized zero divisor. However, this property is not satisfied by all rings.

Consider, for example, the element $w(z) = z$ of the ring A (§ 1, Example 6), which, evidently, has no inverse. Suppose that the sequence $y_n(z)$ is such that $\|zy_n(z)\| \to 0$. On the boundary of the disk $|z| = 1$ and therefore $|y_n(z)| \to 0$. In view of the maximum modulus principle, $y_n(z)$ tends uniformly to zero in the whole disk, that is, $\|y_n(z)\| \to 0$. Consequently, $w(z) = z$ is not a generalized zero divisor of the ring A.

One can arrive at the concept of generalized zero divisor from the following considerations. Multiplication by the element x can be considered as a linear operation in the space R, giving a single-valued and continuous mapping of the space R into part of R: $R \to R$. The following cases are possible.

1. The mapping $R \to R$ is not one-to-one, that is, there exist distinct elements y and z such that $xy = xz$. In this case x is a zero divisor, since $x(y-z) = 0$.

2. The mapping $R \to R$ is one-to-one and the image fills up R. In this case x has an inverse, since there exists such a $y \in R$ that $xy = e$.

3. The mapping $R \to R$ is one-to-one but the image is not all of R but only some complete subspace $R' \subset R$. Then by the theorem of Banach the inverse mapping $R' \to R$ is also continuous, that is, from $xy_n \to 0$ it follows that $y_n \to 0$. This means that x is not a generalized zero divisor in R.

4. The mapping $R \to R$ is one-to-one but the image is some incomplete subspace $R' \subset R$. In this case the inverse mapping $R' \to R$ can not be continuous (for otherwise R' would be homeomorphic to R and consequently complete), that is, there exists a sequence $xy_n \to 0$ such that y_n does not $\to 0$. In this case the

* That is, rings whose elements make up a finite-dimensional space.

element x is a generalized zero divisor.

Theorem 1. *Let x be an arbitrary element of a commutative normed ring R and let λ_0 be a boundary point of the set S of values of the function $x(M)$; then $x-\lambda_0 e$ is a generalized zero divisor.*

Proof. Consider a sequence of numbers λ_n $(n = 1, 2, \cdots)$ not elements of S but converging to λ_0. In the ring R the elements $(x-\lambda_n e)^{-1}$ exist (§4, Theorem 9). Set $y_n = \frac{(x-\lambda_n e)^{-1}}{\|(x-\lambda_n e)^{-1}\|}$. Then $\|y_n\| = 1$. On the other hand, since

$$\|(x-\lambda_n e)^{-1}\| \geq \max_M |x(M)-\lambda_n|^{-1} \geq |\lambda_0-\lambda_n|^{-1} \to \infty,$$

then

$$(x-\lambda_0 e)\,y_n = (x-\lambda_n e)\,y_n + (\lambda_n-\lambda_0)\,y_n = \frac{e}{\|(x-\lambda_n e)^{-1}\|} + (\lambda_n-\lambda_0)\,y_n \to 0,$$

as we required.

From this follows easily

Theorem 2. *If 0 is the only generalized zero divisor of a commutative normed ring R, then R is isomorphic to the field of complex numbers.*

Proof. By Theorem 1 and the condition of Theorem 2 for each $x \in R$ there is a λ_0 such that $x-\lambda_0 e = 0$, that is, $x = \lambda_0 e$.

The following theorem of Mazur [17] is evidently contained in this result.

If $\|xy\| = \|x\|\,\|y\|$ for every $x, y \in R$, then R is the field of complex numbers.

Theorem 3. *Each element x of a commutative real normed ring R which has no inverse is a generalized zero divisor.*

Proof. If x has no inverse, then 0 is a boundary point of the set of values of the (nonnegative) function $x^2(M)$, so by Theorem 1 x^2 is a generalized zero divisor: that is, there exists a sequence $z_n \in R$ such that $\inf_n \|z_n\| > 0$ and $x^2 z_n = x(xz_n) \to 0$ $(n \to \infty)$. If also $\inf_n \|xz_n\| > 0$ then the theorem is proved; in the contrary case there exists a subsequence z_{n_k} for which $xz_{n_k} \to 0$, hence again x is a generalized zero divisor.

This property is possessed by every symmetric ring without radical. To prove this it is only necessary in the proof of Theorem 3 to replace x^2 by xx^* and make use of the remark at the end of § 11.

§23. Extension of maximal ideals of a symmetric ring

Theorem 4. *If a symmetric ring R_1 is a subring of any commutative normed ring R, then each maximal ideal $M_1 \subset R_1$ is contained in a maximal ideal $M \subset R$.*

The proof of this theorem is based on the following

Lemma. *If a real element $x \in R_1$ has in R an inverse element $x^{-1} = y$, then $y \in R_1$.*

Proof of the lemma. Suppose the contrary: suppose $y \overline{\in} R_1$ so that x has no inverse in R_1. Then 0 is a value of the function $x(M_1)$ $(M_1 \subset R_1)$; since the function $x(M_1)$ is real, all its values, even 0, are boundary points; by Theorem 1 x is a generalized zero divisor in R_1, and consequently in R; but in this case, as we proved at the beginning of §21, x can not have an inverse in R.

We remark that conjugate elements x and x^* in the ring R_1 are still conjugate in R, because the value of an element x on a maximal ideal M of the ring R coincides with the value of x on the maximal ideal M_1 of the ring R_1 which is the intersection of M and R_1.

Proof of Theorem 4. Let us suppose that some maximal ideal $M_1 \subset R_1$ is not contained in any maximal ideal of the ring R. This means that for each $M \subset R$ can be found an element $x \in M_1$ such that $x(M) \neq 0$. In this case $xx^*(M) > 0$. Since the function $xx^*(M)$ is continuous, there exists a neighborhood of the maximal ideal M in which $xx^*(M) > 0$. These neighborhoods form a covering of the space $\mathfrak{M}$ of maximal ideals of the ring R: by compactness of $\mathfrak{M}$, a finite subcovering can be chosen; suppose it determined by elements $x_1, x_2, \cdots, x_n$. Form the sum

$$x = x_1x_1^* + x_2x_2^* + \cdots + x_nx_n^* \in M_1.$$

The function $x(M)$ vanished nowhere in the space $\mathfrak{M}$. But then x has in R an inverse element x^{-1}, which by the lemma must belong to R_1; however, this contradicts the fact that x belongs to the maximal ideal $M_1 \subset R_1$ and therefore can have no inverse in R_1.

For non-symmetric rings Theorem 4 generally ceases to be true. For example, let $R_1 = A$ and let R be the ring of all functions $w(z)$ continuous on the circumference $|z| = 1$. The maximal ideal $M_1 \subset R_1$ containing the function $w(z) = z$ is not contained in any maximal ideal $M \subset R$, since in R this function already has an inverse.

As we saw in § 4, the set of all maximal ideals of the ring R_1 coincides with the set of all the points of the disk $|z| \leq 1$. Evidently only boundary points can be extended to maximal ideals of R; these are the maximal ideals $M_1^{\vartheta} \subset R_1$ consisting of functions vanishing at some point $z = e^{i\vartheta}$ $(0 \leq \vartheta < 2\pi)$. Let us show that these maximal ideals M_1^{ϑ} can be extended to maximal ideals not only of the special ring $R = C\{|z| = 1\}$, but to every ring $R \supset R_1$.

Indeed, by Theorem 1 the element $w(z) = z - e^{i\vartheta}$ is a generalized zero divisor and therefore has no inverse in every ring R containing R_1: therefore in every ring $R \supset R_1$ there is a maximal ideal $M \ni z - e^{i\vartheta}$. But we saw earlier that the intersection of R_1 with a maximal ideal of R is a maximal ideal of R_1; in this case it must be the ideal M_1^{ϑ} because $z - e^{i\vartheta}$ belongs only to M_1^{ϑ}.

§24. Extension of maximal ideals in the general case

The example considered in the ring of the preceding section naturally raises the following problem:

Let R_1 be a commutative normed ring. Do there exist maximal ideals $M_1 \subset R_1$ which can be extended to maximal ideals of every ring $R \supset R_1$? And if they exist, how can they be characterized internally?

The present section is devoted to a solution of this problem.

Let us emphasize in formulating the problem of existence that R is an *arbitrary* ring containing R_1. For each separate $R \supset R_1$ it is easy to see that there exist extendable maximal ideals of the ring R_1; it suffices to take an arbitrary maximal ideal $M \subset R$ and take the elements of it in R_1.

Definition 2. A closed set $F \subset \mathfrak{M}(R)$ is called *determining* if the absolute value of every function $x(M)$ attains its maximum on F.

It is evident that at least one determining set exists: $\mathfrak{M}$ itself.

Definition 3. The minimal determining set (that is, a determining set such that no proper subset of it is determining) is called the *boundary* of the set $\mathfrak{M}$ (or the *ring boundary*).

Proof of the existence of the boundary. If a determining set $F = F_1$ is not minimal, there exists a determining set $F_2 \subset F_1$; if F_2 is not minimal, there exists a determining set $F_3 \subset F_2$, and so on. Suppose that we have obtained a countable decreasing sequence of distinct determining sets $F_1 \supset F_2 \supset F_3 \supset \cdots$; their intersection F_ω is not empty by compactness of the space $\mathfrak{M}$, and is also determining: in fact, suppose that $m = \max_M |x(M)|$ and $M_n \in F_n$ is such that $|x(M_n)| = m$; since $|x(M)|$ is continuous, for each limit point M_ω of the sequence M_n we have $|x(M_\omega)| = m$; but $M_\omega \in F_\omega$ which means that F_ω is a detemining set. If F_ω is not minimal, we can extend the sequence of determining sets; in this way we get a transfinite sequence of decreasing closed sets; the intersection of all sets of this sequence is non-empty, by compactness of $\mathfrak{M}$, and is the desired minimal set.

Proof of uniqueness of the boundary. Suppose that in $\mathfrak{M}$ there exist two boundaries Γ_1 and Γ_2. Suppose that M_1 is some point of Γ_1; we shall show that in every neighborhood of it can be found a point of Γ_2. In this case, since Γ_2 is closed $M_1 \in \Gamma_2$, and by the arbitrariness of M_1, $\Gamma_1 \subset \Gamma_2$. But then $\Gamma_1 = \Gamma_2$, because Γ_2 is minimal.

Thus, suppose that U is some neighborhood of the point M_1; it is defined by n inequalities of the form $|x_i(M)| < \epsilon$ $(i = 1, 2, \cdots, n)$, where $x_1, x_2, \cdots, x_n$ are some elements of the ring which belong to M_1. Since Γ_1 is a minimal determining set there exists some function $y(M)$ $(y \in R)$ whose absolute value attains its maximum m on Γ_1 (equal by assumption to its maximum on all $\mathfrak{M}$) in the set U and outside this neighborhood is less than m on Γ_1; indeed, in the contrary case the closed set $\Gamma_1 - U \subset \Gamma_1$ would also be determining and Γ_1 would not be minimal. Without restriction of generality it can be assumed that $m = 1$; then, replacing y by one of its powers if necessary, it can be assumed that in the points of the set Γ_1 outside U, $|y(M) < \frac{\epsilon}{\max_i \|x_i\|}$. Then the products $x_i y$ $(i = 1, 2, \cdots, n)$ everywhere in Γ_1, and consequently in all of $\mathfrak{M}$, do not exceed ϵ in absolute value. But on Γ_2 there is a point M_2 for which $|y(M_2)| = 1$; since at that point $|(x_i y)(M_2)| < \epsilon$, then

$$|x_i(M_2)| < \epsilon \quad (i = 1, 2, \cdots, n).$$

But this asserts that $M_2 \in U$. With this the uniqueness of the boundary Γ of the set $\mathfrak{M}$ is established.

Theorem 5. *In order that a point $M_0 \in \mathfrak{M}$ belong to the boundary Γ it is necessary and sufficient that for each neighborhood $U(M_0)$ of this point there exist a function $y(M)$ $(y \in R)$ whose absolute value attains its maximum in $U(M_0)$ and is less than that outside $U(M_0)$.*

Proof. The necessity of the condition comes from the preceding argument, its sufficiency from the fact that when it holds every neighborhood of M_0 contains a point of the boundary.

Theorem 6. *For every ring $R \supset R_1$ each maximal ideal from the boundary Γ_1 of $\mathfrak{M}(R_1)$ can be extended to a maximal ideal of the ring R.*

Proof. First, note that if $x \in R_1$, then $\max_{M \subset R} |x(M)| = \max_{M_1 \subset R_1} |x(M_1)|$. This follows immediately from the formula

$$\max |x(M)| = \lim_{n\to\infty} \sqrt[n]{\|x^n\|},$$

if it is observed that all x^n are in R_1 and in R and that the norms computed in the two spaces coincide. Suppose now that some maximal ideal $M_1 \in \Gamma_1$ is contained in no maximal ideal of the ring R. This means that in the ring R there is no ideal containing all the elements of M_1, that is that the set of all sums of the form

$$\sum_{i=1}^{n} x_i z_i \quad (x_i \in M_1, \; z_i \in R)$$

coincides with the whole ring R; in particular, one such sum gives the identity element of the ring R:

$$e = \sum_{i=1}^{n} x_i z_i.$$

Without loss of generality we may assume here that $\max |x_i(M)| \leq 1$.

Take $\mu > \max_i \{\max_M |z_i(M)|\}$; consider the neighborhood of the point M_1 defined by the inequalities

$$|x_i(M)| < \frac{1}{2n\mu} \quad (i = 1, 2, \cdots, n; \; M \subset R_1),$$

and take a function $y(M)$ ($y \in R_1$) whose absolute value attains its maximum value 1 in this neighborhood and which outside the neighborhood does not exceed $\frac{1}{2n\mu}$ (see Theorem 5). By what has been done above the product $y \cdot \Sigma_{i=1}^{n} x_i z_i = ye = y$ attains its maximum absolute value 1 at a maximal ideal of the ring R. But, on the other hand,

$$\max_{M \subset R} \left| (y \cdot \sum_{i=1}^{n} x_i z_i)(M) \right| \leq \sum_{i=1}^{n} \max |y(M) x_i(M)| \max |z_i(M)| \leq$$

$$\leq \sum_{i=1}^{n} 1 \cdot \frac{1}{2n\mu} \cdot \mu = \frac{1}{2}.$$

We have achieved a contradiction and proved the validity of the theorem.

From this theorem we can again derive the theorem of § 23 on extension of the maximal ideals of a symmetric ring. For this it suffices to show that for a symmetric ring $\Gamma = \mathfrak{M}$. Let M_0 be an arbitrary point of $\mathfrak{M}$ and $U(M_0)$ a neighborhood of it. Consider a continuous function equal to zero outside $U(M_0)$ and 1 at M_0; by Theorem 16 of § 6 there is an element $x \in R$ such that $|f(M) - x(M)| < \frac{1}{3}$. In this case the maximum value of $|x(M)|$ must be taken in $U(M_0)$; this means that the condition of Theorem 5 is satisfied and $M_0 \in \Gamma$.

Since each regular* ring evidently satisfies the conditions of Theorem 5, then by Theorem 6 it follows that each maximal ideal of a regular ring R_1 can be extended to a maximal ideal of every ring $R \supset R_1$.

If in the ring R_1 $\|x\| = \max |x(M)|$ then it can be shown that there is a ring $R \supset R_1$ whose maximal ideals extend *only* those $M \subset R_1$ which belong to the boundary Γ of $\mathfrak{M}(R_1)$. An example of such a ring R is the ring of all continuous functions defined on Γ; it contains R_1 as a subring, and since Γ is a compact space, the set of all maximal ideals of the ring R coincides with Γ.

For such rings all the generalized zero divisors satisfy the following conditions**: *A function in R_1 is a generalized divisor of zero if and only if it vanishes somewhere in* Γ. In this case the boundary Γ can be characterized as the smallest closed set in which each generalized zero divisor vanishes.

Let us consider some examples.

1. Let R be a ring with one generator. Then $\mathfrak{M}(R)$, by Theorem 15 of § 5, can be regarded as a bounded closed subset of the complex plane. The boundary Γ in this case coincides with the ordinary topological boundary of the set $\mathfrak{M}(R)$. Indeed, by the maximum modulus principle, for interior (in the ordinary sense) points of $\mathfrak{M}$ the condition of Theorem 5 can not be satisfied, so they do not belong to Γ; if λ_0 is a boundary (usual sense) point of $\mathfrak{M}$, then there exists a sequence $\lambda_n \to \lambda_0$

* Recall that the ring R is called regular if for each maximal ideal M_0 and each neighborhood $U(M_0)$ there is an element $x \in R$ such that $x(M_0) \neq 0$ and $x(M) = 0$ outside $U(M_0)$.

** We shall not pause to give a proof of this here.

such that $\lambda_n \overline{\in} \mathfrak{M}$; the functions $(x-\lambda_n e)^{-1}$, where x is the generator, belong to the ring R and, beginning with some n, satisfy the condition of Theorem 5; hence $\lambda_0 \in \Gamma$.

2. Suppose that R is the ring of functions gotten from the set of all polynomials in two complex variables z_1 and z_2 by use of uniform convergence on the bicylinder $|z_1| \leq 1$, $|z_2| \leq 1$. It is easily seen that $\mathfrak{M}(R)$ coincides with the set of points of the bicylinder. The topological boundary of this set $\mathfrak{M}(R)$ consists of those points one of whose coordinates has absolute value one. The ring boundary consists of all points both of whose coordinates have absolute value one. Indeed, the set S of all those points is, as is easily seen, determining, so $\Gamma \subset S$; on the other hand, for each point $(e^{i\phi_1}, e^{i\phi_2})$ of the set S the function $(z_1 + e^{i\phi_1})$ $(z_2 + e^{i\phi_2})$ satisfies the conditions of Theorem 5, so $S \subset \Gamma$.

This example shows that the ring boundary Γ need not coincide with the topological boundary of the set $\mathfrak{M}$ (which is given by the natural imbedding of it in complex space (see Theorem 15 of § 5)).

3. Let R be the ring of functions obtained from the polynomials in two independent variables by use of uniform convergence of the bisquare $|z_1| + |z_2| \leq 1$. Here $\mathfrak{M}(R)$ is the set of points of the bisquare and the topological and ring boundaries coincide (the proof goes as in the next following example 4). *Every function $f(M)$ which has no inverse vanishes somewhere in the boundary.* Consider the case where $f(M) = f(z_1, z_2)$ is a polynomial; since it has no inverse, there is (z_1^0, z_2^0) such that $f(z_1^0, z_2^0) = 0$ and $|z_1^0| + |z_2^0| \leq 1$. Suppose $|z_1^0| + |z_2^0| < 1$. The polynomial $f(z_1^0, z_2)$ in the argument z_2 has the root z_2^0; if the parameter z_1^0 varies continuously, then the root z_2^0 also varies continuously; therefore, if we increase $|z_1^0|$ there comes a moment when $|z_1^0| + |z_2^0| = 1$. Hence for polynomials our assertion is proved. Suppose now that $f(M)$ is an arbitrary function from the ring which has no inverse, $f(M_0) = 0$. It can be approximated arbitrarily closely by polynomials, each vanishing at M_0 and, consequently, somewhere on the boundary. Therefore $\min |f(M)|$ on the boundary can not be positive.

Corollary. *Each element of the ring R with no inverse is a generalized zero divisor.*

4. Consider the set S_a of points $\{z\} = \{z_1, z_2, \cdots, z_n, \cdots\}$ of the infinite dimensional complex space satisfying the inequality $\Sigma_{i=1}^{\infty} |z_i| \leq a$. By R_a we denote the ring of limits of polynomials uniformly convergent on S_a, where each polynomial depends on only a finite number of variables. Let us determine the sets $\mathfrak{M}$ and $\Gamma(\mathfrak{M})$ for this ring. In S_a we introduce the usual Tyhonov topology, which makes it a compact space.

Let us determine $\mathfrak{M}(R_a)$. Just as in the ring with a finite number of generators each maximal ideal M determines a point $\{z(M)\} = \{z_1(M), z_2(M), \cdots, z_n(M), \cdots\}$

of the space considered. As we know, each point of S_a gives a maximal ideal of the ring R_a, that is, $S_a \subset \mathfrak{M}(R_a)$. Suppose now that M_0 is an arbitrary maximal ideal of the ring R_a and that $\rho_k e^{i\phi_k} = z_k(M_0)$. We consider the function $F(z) = \Sigma_{k=1}^{\infty} z_k e^{-i\phi_k}$. The series for this is absolutely convergent on S_a, so $F(z) \in R_a$. The norm of $F(z)$, that is, the maximum modulus of $F(z)$ on S_a, does not exceed a, and therefore $|F(M_0)| \leq a$. But $F(M_0) = \Sigma_{k=1}^{\infty} z_k(M_0) e^{-i\phi_k} = \Sigma_{k=1}^{\infty} \rho_k$; consequently, $\Sigma_{k=1}^{\infty} \rho_k \leq a$, $M_0 \in S_a$. That shows that $\mathfrak{M}(R_a)$ coincides with S_a.

We consider now a polynomial $P(z) = \Pi_{k=1}^{n} (z_k - z_k^0)$, where $z_k^0 \neq 0$ and the moduli $r_k = |z_k^0|$ have the property that the largest of their differences from their mean $c = \frac{\Sigma r_k}{n}$ is not greater than $\frac{a}{n}$. We shall show that the absolute value of this polynomial attains its maximum on S_a in a unique point. By compactness of S_a there is at least one point at which $|P(z)|$ attains its maximum; let it be $\zeta = \{\zeta_k\} = \{e^{i\phi_k}\rho_k\}$. Then $\phi_k = \pi + \arg z_k^0$, since in the contrary case we could translate ζ_k by a circle of radius ρ_k and increase $|P(z)|$ without going outside S_a. Therefore $|P(\zeta)| = \Pi_{k=1}^{n} (r_k + \rho_k)$. Thus, the point where this product attains its maximum must, evidently, satisfy the condition $\Sigma_{k=1}^{n} \rho_k = a$ or $\Sigma_{k=1}^{n} (r_k + \rho_k) = a + \Sigma_{k=1}^{n} r_k$. But these conditions on the factors $r_k + \rho_k$ for which the product is a maximum determine them uniquely; each of them is equal to

$$r_k + \rho_k = \frac{a + \sum_{k=1}^{n} r_k}{n} = \frac{a}{n} + c,$$

from which $\rho_k = \frac{a}{n} + c - r_k > 0$. Thus the point at which $|P(z)|$ attains its maximum is determined uniquely. Let us show that for every point $\{\zeta\} = \{\zeta_1, \zeta_2, \cdots, \zeta_n, 0, 0, \cdots\}$ with $\Sigma_{k=1}^{n} |\zeta_k| = a$ there is a polynomial $P(z)$ which attains its maximum modulus on S_a at just in this point $\{\zeta\}$. Let $\zeta_k = \rho_k e^{i\phi_k}$. Define the numbers r_k by the equations

$$r_k = c + \frac{a}{n} - \rho_k \quad (k = 1, 2, \cdots, n),$$

where c is any number greater than $\rho_k - \frac{a}{n}$ $(k = 1, 2, \cdots, n)$. It is easily seen that c is the arithmetic mean of the numbers r_k and that the largest difference between it and these numbers is not more than $\frac{a}{n}$. We set $z_k^0 = r_k e^{(\pi + \phi_k)i}$ and form the polynomial $P(z) = \Pi_{k=1}^{n} (z_k - z_k^0)$; by construction it attains its maximum in just one point, and that $\{\zeta\}$.

By Theorem 5 each point $\{\zeta\} = \{\zeta_1, \zeta_2, \cdots, \zeta_n, 0, 0, \cdots\}$ belongs to $\Gamma(\mathfrak{M})$. But these points make up a dense subset of S_a; since Γ is closed, $\Gamma = S_a = \mathfrak{M}$. Here we have an example of a nonsymmetric ring with norm determined by uniform convergence of functions $x(M)$ for which $\Gamma = \mathfrak{M}$. It is still an open question whether a ring with these properties can have a finite number of generators.

BIBLIOGRAPHICAL COMMENTS

The results presented in the section devoted to the general theory of commutative normed rings are essentially due to I. M. Gel′fand [5, 6]. Theorem 7 was first published by Mazur [17] (who derived it by quite different methods). Theorem 16′ is due to Stone [26, page 466 ff]; The results of § 7 are due to D. A. Raĭkov; one important particular case of Theorem 18 was published in [21]; the first rings with involution (both commutative and non-commutative) under some special assumptions were considered by I. M. Gel′fand and M. A. Naĭmark in a note included below in the supplementary list of literature under number 5.

The results of Appendix I are due to I. M. Gel′fand and G. E. Šilov [10].

The results of Appendix II are due essentially to I. M. Gel′fand [7, 8]. The proposition formulated as a corollary of Theorems 1 and 2 was found by Wiener and Pitt [31], Cameron [3], Pitt [19]. The construction of maximal ideals in the ring of functions of bounded variation presented in small type in § 17 is due to D. A. Raĭkov; it is first published here. Theorem 3 is due to Wiener and Pitt [31]. Theorems 4 and 5 are due to G. E. Šilov and represent his answer to a question raised by M. G. Kreĭn; first publication.

The results of Appendix III are due to I. M. Gel′fand and D. A. Raĭkov [9, 21, 23]. Theorem 8 was proved independently (with the aid of the Pontrjagin duality theory for commutative locally compact groups) by A. Ya. Povzner [20]*.

The results of Appendix IV are due to G. E. Šilov. The contents of §§ 22 and 23 were published in [25]; the result discussed in § 24 is here published for the first time.

BIBLIOGRAPHY

[1] S. Banach, *Théorie des opérations linéaires*, Warszawa, 1932.

[2] S. Bochner, *Vorlesungen über Fouriersche Integrale*, Akademische Verlagsgesellschaft, Leipzig, 1932.

[3] R. H. Cameron, *Analytic functions of absolutely convergent generalized trigonometric sums*, Duke Math. J. 3 (1937), 682-688.

[4] E. Čech, *On bicompact spaces*, Ann. of Math. (2) 38 (1937), 823-844.

[5] I. Gelfand, *On normed rings*, C. R. (Dokl.) Acad. Sci. URSS (N.S.) 23 (1939), 430-432.

[6] I. Gelfand, *Normierte Ringe*, Mat. Sb. N.S. 9(51) (1941), 3-24.

* And (also with the aid of the Pontrjagin theory) André Weil, published in "*L'intégration dans les groupes topologiques et ses applications*" Paris, 1940, but news of it reached Moscow in a review of it in Mathematical Reviews only in 1944. (Note added in proof.)

[7] I. Gelfand, *To the theory of normed rings, II, on absolutely convergent trigonometrical series and integrals*, C.R. (Dokl.) Acad. Sci. URSS (N.S.) 25 (1939), 570-572.

[8] I. Gelfand, *Über absolut konvergente trigonometrische Reihen und Integrale*, Mat. Sb. N.S. 9(51) (1941), 51-66.

[9] I. Gelfand and D. Raikov, *On the theory of characters of commutative topological groups*, C. R. (Dokl.) Acad. Sci. URSS (N.S.) 28 (1940), 195-198.

[10] I. Gelfand und G. Šilov, *Über verschiedene Methoden der Einführung der Topologie in die Menge der maximalen Ideale eines normierten Ringes*, Mat. Sb. N.S. 9(51) (1941), 25-39.

[11] A. Haar, *Der Massbegriff in der Theorie der kontinuierlichen Gruppen*, Ann. of Math. (2) 34 (1933), 147-169.

[12] G. Hamel, *Eine Basis aller Zahlen und die unstetigen Lösungen der Funktionalgleichung:* $f(x+y) = f(x) + f(y)$, Math. Ann. 60 (1905), 459-462.

[13] G. Herglotz, *Über Potenzreihen mit positivem, reellen Teil im Einheitskreis*, Ber. Verh. Sächs. Ges. Wiss. Leipzig. Math.-Phys. Kl. 63 (1911), 501-511.

[14] M. Krein, *Sur une généralisation du théorème de Plancherel au cas des intégrales de Fourier sur les groupes topologiques commutatifs*, C. R. (Dokl.) Acad. Sci. URSS (N.S.) 30 (1941), 484-488.

[15] P. Lévy, *Sur la convergence absolue des séries de Fourier*, Compositio Math. 1 (1934), 1-14.

[16] A. Markoff, *On mean values and exterior densities*, Mat. Sb. N.S. 4(46) (1938), 165-191.

[17] S. Mazur, *Sur les anneaux linéaires*, C. R. Acad. Sci. Paris 207 (1938), 1025-1027.

[18] R. E. A. C. Paley and N. Wiener, *Fourier transforms in the complex domain*, Amer. Math. Soc. Colloq. Publ., v. 19, New York, 1934.

[19] H. R. Pitt, *A theorem on absolutely convergent trigonometrical series*, J. Math. Phys. 16 (1938), 191-195.

[20] A. Powzner, *Über positive Funktionen auf einer Abelschen Gruppe*, C. R. (Dokl.) Acad. Sci. URSS (N.S.) 28 (1940), 294-295.

[21] D. Raikov, *Positive definite functions on commutative groups with an invariant measure*, C. R. (Dokl.) Acad. Sci. URSS (N.S.) 28 (1940), 296-300.

[22] D. Raikov, *Generalized duality theorem for commutative groups with an invariant measure*, C. R. (Dokl.) Acad. Sci. URSS (N.S.) 30 (1941), 589-591.

[23] D. A. Raikov, *Harmonic analysis on commutative groups with the Haar measure and the theory of characters*, Trudy Mat. Inst. Steklov, 14 (1945).

[24] W. Sierpiński, *Sur la question de la mesurabilité de la base de M. Hamel*, Fund. Math. 1 (1920), 105-111.

[25] G. Šilov, *On the extension of maximal ideals*, C. R. (Dokl.) Acad. Sci. URSS (N.S.) 29 (1940), 83-84.

[26] M. H. Stone, *Applications of the theory of Boolean rings to general topology*, Trans. Amer. Math. Soc. 41 (1937), 375-481.

[27] A. Tychonoff, *Über die topologische Erweiterung von Räumen*, Math. Ann. 102 (1929), 544-561.

[28] P. Urysohn, *Über die Metrisation der kompakten topologischen Räume*, Math. Ann. 92 (1924), 275-293.

[29] N. Wiener, *Tauberian theorems*, Ann. of Math. (2) 33 (1932), 1-100.

[30] N. Wiener, *The Fourier integral and certain of its applications*, Cambridge, 1933.

[31] N. Wiener and H. R. Pitt, *On absolutely convergent Fourier-Stieltjes transforms*, Duke Math. J. 4 (1938), 420-436.

Supplementary bibliography

[1] W. Ambrose, *Structure theorems for a special class of Banach algebras*, Trans. Amer. Math. Soc. 57 (1945), 364-386.

[2] S. Bochner, *On a theorem of Tannaka and Krein*, Ann. of Math. (2) 43 (1942), 56-58.

[3] S. Bochner and R. S. Phillips, *Absolutely convergent Fourier expansions for non-commutative normed rings*, Ann. of Math. (2) 43 (1942), 409-418.

[4] J. W. Calkin, *Two-sided ideals and congruences in the ring of bounded operators in Hilbert space*, Ann. of Math. (2) 42 (1941), 839-873.

[5] I. Gelfand and M. Neumark, *On the imbedding of normed rings into the ring of operators in Hilbert space*, Mat. Sb. N.S. 12(54) (1943), 197-213.

[6] N. Jacobson, *A topology for the set of primitive ideals in an arbitrary ring*, Proc. Nat. Acad. Sci. U. S. A. 31 (1945), 333-338.

[7] E. Hille, *On the theory of characters of groups and semi-groups in normed vector rings*, Proc. Nat. Acad. Sci. U. S. A. 30 (1944), 58-60.

[8] M. Krein, *A ring of functions on a topological group*, C. R. (Dokl.) Acad. Sci. URSS (N.S.) 29 (1940), 275-280.

[9] M. Krein, *On a special ring of functions*, C. R. (Dokl.) Acad. Sci. URSS (N.S.) 29 (1940), 355-359.

[10] M. Krein, *On almost periodic functions on a topological group*, C. R. (Dokl.) Acad. Sci. URSS (N.S.) 30 (1941), 5-8.

[11] M. Krein, *On positive functionals on almost periodic functions*, C. R. (Dokl.) Acad. Sci. URSS (N.S.) 30 (1940), 9-12.

[12] E. R. Lorch, *The spectrum of linear transformations*, Trans. Amer. Math. Soc. 52 (1942), 238-248.

[13] E. R. Lorch, *The theory of analytic functions in normed Abelian vector rings*, Trans. Amer. Math. Soc. 54 (1943), 414-425.

[14] E. R. Lorch, *The structure of normed Abelian rings*, Bull. Amer. Math. Soc. 50 (1944), 447-463.

[15] A. Powsner, *Sur les équations du type de Sturm-Liouville et les fonctions "positives"*, C. R. (Dokl.) Acad. Sci. URSS (N.S.) 43 (1944), 367-371.

[16] A. Povzner, *On equations of Sturm-Liouville type on a semi-axis*, Dissertation, Moscow, 1945.

[17] I. E. Segal, *The group ring of a locally compact group*, I, Proc. Nat. Acad. Sci. U. S. A. 27 (1941), 348-352.

[18] I. E. Segal, *Ring properties of certain classes of functions*, Dissertation, Yale Univ., 1940.

[19] G. E. Šilov, *On regular normed rings*, Trudy Mat. Inst. Steklov. 21 (1947).

[20] Yu. A. Šreĭder, *Investigation of maximal ideals of rings of functions of bounded variation*. (In preparation)

[21] M. H. Stone, *A general theory of spectra*, I, Proc. Nat. Acad. Sci. U. S. A. 26 (1940), 280-283.

[22] I. Vernikoff, S. Krein, and A. Tovbin, *Sur les anneaux semi-ordonnés*, C. R. (Dokl.) Acad. Sci URSS (N.S.) 30 (1941), 785-787

Translated by:
M. M. Day

15.

(with M. A. Najmark)

Normed rings with an involution and their representations

Izv. Akad. Nauk SSSR, Ser. Mat. **12** (1948) 445–480 [English translation, chapter VIII in: Commutative normed rings, I.M. Gelfand, D.A. Rajkov, and G.E. Shilov, pp. 240–274. Chelsea 1964]. Zbl. **31**:34

1. In this chapter[1] we study rings (mainly *non-commutative* rings) in which an operation of involution (∗-operation) is introduced axiomatically. The representation of such rings by operators in a Hilbert space are investigated by means of positive functionals. The results obtained are applied to the theory of representations of locally compact groups.

2. The theory of commutative normed rings, as developed in the papers [79] and [83] has proved a useful apparatus for the solution of various problems of analysis. It has also been applied with success in the theory of commutative topological groups.

For applications of analogous methods to non-commutative groups it has turned out to be necessary to develop a theory of non-commutative normed rings.

Since an operation of involution can be introduced in a group ring in a natural way, the problem that arises in the first place is the investigation of rings in which an operation of involution (∗-operation) is given.

One class of such rings (the so-called ∗-rings) was discussed by the authors in the paper [80]. In that paper it appeared that in the theory of rings with an involution an important role is played by positive functionals that is, functionals that satisfy the condition $f(x^*x) \geqq 0$.

Positive functionals on a group ring and the positive-definite functions on the group connected with them were used by Gelfand and Raikov in [82] in order to prove the existence and completeness of the system of continuous representations of a locally compact group.

---

[1] In the original Russian text, the present chapter was called an appendix.

[2] The present chapter is written by I. M. Gelfand and M. A. Naimark; it reproduces, with some improvements of an editorial nature, the paper that appeared in *Izv. AN. USSR., ser. mat.*, 12 (1948), 445-480. A section contained in the original concerned with the generalized Schur Lemma, which is rather far removed from the basic theme of the present chapter, is omitted here.

The present chapter is devoted to the general theory of rings with an involution and their representations in conjunction with the theory of positive functionals.

Certain methods to be explained here, particularly in § 50, are essentially an extension of the methods of the paper [82] to the case of an arbitrary ring with an involution.[3]

§ 46. Rings with an Involution and their Representations

1. A set R of elements $x, y, \ldots$ is called a *normed ring* if:

1. R is a ring, i.e., operations of addition and multiplication satisfying the usual algebraic conditions are defined in R. We also assume that R has a unit element e.

2. R is a linear vector space with multiplication by complex numbers, where this multiplication is permutable with the operation of multiplication of elements in R.

3. A norm is defined in R, i.e., every element x is associated with a number $|x|$ such that

$$|x+y| \leqq |x|+|y|, \qquad |xy| \leqq |x| \cdot |y|.$$

$$|x| \geqq 0 \quad \text{and is equal to zero only for} \quad x=0.$$

$$|\lambda x| = |\lambda|\,|x|, \qquad |e| = 1.$$

4. The ring is complete, i.e., from

$$\lim_{m,\, n \to \infty} |x_n - x_m| = 0$$

there follows the existence of an x such that

$$\lim_{n \to \infty} |x_n - x| = 0.$$

2. DEFINITION 1: A normed ring R is said to be a *ring with an involution* if an operation is defined in it that assigns to every element x an element x^* such that the following conditions are satisfied:

a) $(\lambda x + \mu y)^* = \bar{\lambda} x^* + \bar{\mu} y^*$; c) $(xy)^* = y^* x^*$;

b) $x^{**} = x$; d) $|x^*| = |x|$.

In the balance of this chapter it is to be understood, even without specific mention, that every ring under discussion is a ring with an involution.

An element x is called *Hermitian* if $x^* = x$.

[3] We should like to express our thanks here to D. A. Raikov, who has read the paper and made a number of valuable critical remarks.

Every element x can be represented in the form $x = x_1 + ix_2$, where x_1, x_2 are Hermitian elements. For it is sufficient to put

$$x_1 = [x + x^*]/2,$$
$$x_2 = [x - x^*]/2i.$$

The element x^*x is always Hermitian, because

$$(x^*x)^* = x^*x^{**} = x^*x.$$

In particular, since $e^* = e^*e$, we have $e = e^*$, i.e., e is a Hermitian element.

Some results of this chapter remain valid when only the algebraic conditions of our list of conditions are retained, namely, Axioms 1. and 2. and a), b), c).

A typical example of a ring with an involution is the ring K of all bounded linear operators in a Hilbert space. Here the $*$-operation is interpreted as the transition from an operator to its Hermitian conjugate.

In this context, it is natural to study homomorphic mappings of a ring with an involution into K that preserve the $*$-operation. Such a mapping will be called a *representation of the ring.* In other words, we introduce the following definition.

3. DEFINITION 2: We shall say that *a representation of a ring R is given* if every element $a \in R$ is associated with an operation $A \in K$ in a Hilbert space $\mathfrak{H}$ (we shall denote this for brevity as $a \to A$, or $A(a)$), provided the following conditions are satisfied:

1. If $a \to A$, $b \to B$, then $ab \to AB$ and $\lambda a + \mu b \to \lambda A + \mu B$;
2. If $a \to A$, then $a^* \to A^*$;
3. $e \to E$.

DEFINITION 3: The representation is called *cyclic* if the space contains a vector ξ_0 such that the vectors $A\xi_0$ (A are the operators corresponding to the elements of R) are everywhere dense in $\mathfrak{H}$. The vector ξ_0 is also called *cyclic*.

Suppose that two representations are given, the first of which assigns to the element a the operator $A(a)$ in a space $\mathfrak{H}$ and the other, the operator $A'(a)$ in a space $\mathfrak{H}'$. We shall say that these are *equivalent representations* if a one-to-one correspondence can be set up between $\mathfrak{H}$ and $\mathfrak{H}'$ in which the operator $A(a)$ corresponds to $A'(a)$.

A subspace $\mathfrak{H}_1 \subset \mathfrak{H}$ is called *invariant* if every vector of $\mathfrak{H}_1$ is carried by all the operators $A(a)$ into vectors of $\mathfrak{H}_1$.

By regarding all the operators of the representation only as operators in $\mathfrak{H}_1$ we obtain a representation of R in the space $\mathfrak{H}_1$. We shall call this representation a *part* of the original representation in $\mathfrak{H}$.

If $\mathfrak{H}_1$ is invariant, then its orthogonal complement is also invariant. For let ξ be orthogonal to $\mathfrak{H}_1$, i.e., $(\xi, \eta) = 0$ for all $\eta \in \mathfrak{H}_1$. Then

$$(A(a)\xi, \eta) = (\xi, A^*(a)\eta) = 0,$$

because $\mathfrak{H}_1$ is invariant with respect to the operators that are images of elements of R, and $A^*(a)$ is the image of a^*.

4. If a representation of a ring R in a space $\mathfrak{H}$ is given, then the space can be decomposed into the direct sum of invariant subspaces such that the representation is cyclic in each of them. For let $\xi_0 \neq 0$ be an arbitrary fixed vector of $\mathfrak{H}$. We consider the set of all vectors $A(a)\xi_0$, where a ranges over the entire ring R. The closure of this set forms an invariant subspace $\mathfrak{H}_1$ of $\mathfrak{H}$ in which the representation is cyclic. The orthogonal complement of this subspace is also invariant. In this space we proceed as before, etc. Using transfinite induction we arrive at the required decomposition.

DEFINITION 4: A representation is called *irreducible* if $\mathfrak{H}$ has no subspaces invariant with respect to all the operators $A(a)$, other than $\mathfrak{H}$ and 0.

If a representation is irreducible, then it is clear that every vector $\xi \neq 0$ is cyclic. Obviously, the converse is also true.

THEOREM 1: *A representation is irreducible if and only if every bounded operator B that is permutable with the operators $A(a)$ is a multiple of the unit element.*

Proof of the Necessity: Let B be permutable with all the $A(a)$. To begin with, we assume that B is Hermitian. Then every function of B is also permutable with $A(a)$. In particular, the projection operators $E(\lambda)$ that give the spectral decomposition of B are also permutable with $A(a)$. But this means that the subspaces corresponding to them are invariant with respect to $A(a)$. Since the representation is irreducible, this means that each of these subspaces is either the null space or the whole space. Thus, for every λ, $E(\lambda)$ is either zero or E. Since $(E(\lambda)\xi, \xi)$ increases monotonically with increasing λ, it follows from this that there exists a λ_0 such that $E(\lambda) = E$ for $\lambda > \lambda_0$ and $E(\lambda) = 0$ for $\lambda < \lambda_0$. Hence it follows that

$$B = \int_{-\infty}^{\infty} \lambda dE(\lambda) = \lambda_0 E.$$

If B is an arbitrary bounded operator, then B^* is also permutable with all the $A(a)$. For,

$$B^*A(a) = (A^*(a)B)^* = (BA^*(a))^* = A(a)B^*.$$

Therefore the Hermitian operators $(B+B^*)/2$ and $(B-B^*)/2i$ are multiples of the unit operators, and consequently so is B.

Proof of the Sufficiency: Let us assume the contrary, i.e., that the representation is reducible. Then there exists an invariant subspace $\mathfrak{H}_1$ different from 0 and from the whole space. We denote the projection operator corresponding to this subspace by E_1. Then E_1 is permutable with all the $A(a)$ (since $\mathfrak{H}_1$ is invariant with respect to $A(a)$). But E_1 is not a multiple of E, because $\mathfrak{H}_1$ is different from 0 and from the whole space. We have thus reached a contradiction. //

§ 47. Positive Functionals and their Connection with Representations of Rings

1. DEFINITION 5: *A positive linear functional* is a function $f(x)$ that assigns to every $x \in R$ a complex number $f(x)$ such that:

1. $f(\lambda x+\mu y)=\lambda f(x)+\mu f(y)$;
2. $f(x^*x) \geqq 0$ for every x.

The function $f(x)$ is called a *real linear functional* if:

1. $f(\lambda x+\mu y)=\lambda f(x)+\mu f(y)$;
2. $f(x)$ is continuous;
3. $f(x^*)=\overline{f(x)}$ for every x.

It is obvious that if $f(x)$ is a positive functional, then $f(e) \geqq 0$, because $e^*e=e$.

Every continuous linear functional $f(x)$ can be represented in the form $f=f_1+if_2$, where f_1 and f_2 are real functionals. For this it is sufficient to put

$$f_1(x)=[f(x)+\overline{f(x^*)}]/2 \quad \text{and} \quad f_2(x)=[f(x)-\overline{f(x^*)}]/2i.$$

It is easy to verify that f_1 and f_2 are real linear functionals.

Below, we shall show that every positive functional is real and hence that every linear combination of positive functionals with real coefficients is a real linear functional. The converse is not true in general. We shall later give a counter-example.

2. In what follows, we shall often make use of the following inequality: Let $f(x)$ be a positive functional. Then for arbitrary x and y we have

$$|f(y^*x)|^2 \leqq f(y^*y)f(x^*x). \tag{1}$$

The proof of this inequality is an exact replica of the usual proof of the Schwarz inequality.[4]

[4] In the original text the Cauchy-Bunyakovskiĭ Inequality.

THEOREM 1: *Every positive linear functional f is real and satisfies the inequality*

$$|f(x)| \leqq f(e)\,|x|. \tag{2}$$

Proof: Let us assume, to begin with, that $|x| < 1$ and $x^* = x$. We put

$$y = (e - x)^{\frac{1}{2}} =$$
$$= e - \frac{1}{2}x - \frac{1}{2!}\cdot\frac{1}{2}\cdot\frac{1}{2}x^2 - \frac{1}{3!}\frac{1}{2}\cdot\frac{1}{2}\cdot\frac{3}{2}x^3 - \ldots;$$

this series converges, because $|x| < 1$. In virtue of condition d) of Definition 1 of § 46, an involution is continuous; therefore $y^* = y$; moreover

$$yy^* = y^2 = e - x,$$

which is easy to prove by squaring the power series. Therefore .

$$f(e - x) = f(y^*y) \geqq 0,$$

i.e., $f(x)$ is real and $f(x) \leqq f(e)$.

We obtain $f(x) \geqq -f(e)$ similarly.

It is easy to get rid of the restriction $|x| < 1$. We then obtain: if $x^* = x$, then $f(x)$ is real and

$$|f(x)| \leqq f(e)\,|x|.$$

Hence it follows that $f(x^*) = \overline{f(x)}$ for every x. For,

$$f(x) = f\left(\frac{x + x^*}{2}\right) + if\left(\frac{x - x^*}{2i}\right),$$
$$f(x^*) = f\left(\frac{x + x^*}{2}\right) - if\left(\frac{x - x^*}{2i}\right);$$

$f((x + x^*)/2)$ and $f((x - x^*)/2i)$, by what has been proved above, are real; therefore $f(x^*) = \overline{f(x)}$. It now remains to prove that $f(x)$ is a continuous function of x. For this, it is sufficient to prove the inequality (2) for every $x \in R$. We have already proved this inequality for Hermitian elements x—in particular, for elements of the form x^*x. Thus,

$$f(x^*x) \leqq f(e)\,|x^*x|,$$

and therefore

$$f(x^*x) \leqq f(e)\,|x|^2. \tag{3}$$

On the other hand, by putting $y = e$ in (1), we obtain

$$|f(x)|^2 \leqq f(e)f(x^*x),$$

and therefore, by (3),

$$|f(x)|^2 \leqq f(e)^2 |x|^2.$$

Thus, the inequality (2) is proved for all elements $x \in R$. //

3. We shall now give an example of a ring and a real functional on it that is not representable as a linear combination of positive functionals.

Let the elements of the ring R be the set of complex functions $x(z)$ that are analytic for $|z| < 1$ and continuous in the circle $|z| \leqq 1$. We put $|x| = \max_{|z| \leqq 1} |x(z)|$. We define sums and products as the sums and products of the functions; x^* is defined by the equation $x^*(z) = \overline{x(\bar{z})}$. It will be shown in § 50 that every positive functional in this ring is of the form

$$f(x) = \int_{-1}^{+1} x(t)d\sigma(t),$$

where $\sigma(t)$ is a monotonic function given on the interval $[-1, +1]$ of the real t-axis.

We now consider the following real functional:

$$f_1(x) = [x(z_0) + \overline{x(\bar{z}_0)}]/2,$$

where z_0 is a fixed non-real number such that $|z_0| \leqq 1$. Then it is not difficult to verify that there does not exist a complex function of bounded variation $\sigma_1(t)$ such that

$$[x(z_0) + \overline{x(\bar{z}_0)}]/2 = \int_{-1}^{+1} x(t)d\sigma_1(t).$$

For let us suppose that there exists a $\sigma_1(t)$ such that

$$\int_{-1}^{+1} x(t)d\sigma_1(t) = [x(z_0) + \overline{x(\bar{z}_0)}]/2$$

for every function that is analytic in the unit circle. We substitute $x_n(z) = (1/n) \exp(inz)$ for $x(z)$ in this equation. Then the left-hand side of the equation tends to zero as $n \to \infty$, but the absolute value of the right-hand side tends to ∞; that is, we have reached a contradiction. This means that $f_1(x)$ is not representable as a linear combination of positive functionals.

Every representation of a ring R provides us with a set of positive functionals. For let ξ_0 be any vector of the space $\mathfrak{H}$. We put

$$f(a) = (A(a)\xi_0, \xi_0). \tag{4}$$

Then $f(a)$ is a positive functional. For,

$$f(a^*a) = (A(a^*a)\xi_0, \xi_0) =$$
$$= (A^*(a)A(a)\xi_0, \xi_0) = (A(a)\xi_0, A(a)\xi_0) \geqq 0.$$

THEOREM 2: *Every representation of a ring R with an involution is continuous. Furthermore,* $|A| \leqq |a|$.

Proof: Applying the inequality (3) to the positive functional

$$f(a) = (A(a)\xi_0, \xi_0),$$

we obtain

$$(A(a^*a)\xi_0, \xi_0) \leqq (\xi_0, \xi_0)\,|a|^2,$$

i.e.,

$$|A(a)\xi_0|^2 \leqq |a|^2\,|\xi_0|^2.$$

Since ξ_0 is an arbitrary vector, this inequality means that $|A| \leqq |a|$.

4. Our immediate object is to give a description of representations by means of positive functionals. It is better to do this first for cyclic representations.

Let two cyclic representations of a ring R be given, one in the space $\mathfrak{H}$ and one in the space $\mathfrak{H}'$. We denote the operators corresponding to an element a by $A(a)$ and $A'(a)$, respectively. Let ξ_0 and ξ_0' be cyclic vectors of the corresponding representations. We put

$$f(a) = (A(a)\xi_0, \xi_0) \quad \text{and} \quad f'(a) = (A'(a)\xi_0', \xi_0').$$

We shall show that if $f(a) = f'(a)$ *for every a,* then the representations are equivalent.

We set up a correspondence between the vectors of the spaces $\mathfrak{H}$ and $\mathfrak{H}'$ in the following way. Let $\xi = A\xi_0$. Then we associate with it the vector $\xi' = A'\xi_0'$. We shall show that this correspondence is isometric. It then follows that it is one to one. In order to prove the isometry, let us show that the scalar products of corresponding vectors coincide. Let

$$\xi_1 = A_1\xi_0, \quad \xi_1' = A_1'\xi_0',$$
$$\xi_2 = A_2\xi_0, \quad \xi_2' = A_2'\xi_0'.$$

Then

$$(\xi_1, \xi_2) = (A_1\xi_0, A_2\xi_0) = (A_2^*A_1\xi_0, \xi_0) = f(a_2^*a_1),$$
$$(\xi_1', \xi_2') = (A_1'\xi_0', A_2'\xi_0') = (A_2'^*A_1'\xi_0', \xi_0') = f'(a_2^*a_1).$$

Since $f(x) \equiv f'(x)$, we see that $(\xi_1, \xi_2) = (\xi_1', \xi_2')$, i.e., for elements of the form $A\xi_0$ (or of the form $A'\xi_0'$) the isometry is proved.

Since both representations are cyclic, the set of such elements is dense in $\mathfrak{H}$ and $\mathfrak{H}'$, respectively. We may therefore extend this correspondence by continuity to all of $\mathfrak{H}$ and $\mathfrak{H}'$.

We have thus seen that a cyclic representation is uniquely determined to within an equivalence by the positive functional (4). The problem now arises, Does there exist, for every positive functional, a representation in which this functional can be written in the form (4)? We shall show that the answer to this question is in the affirmative.

Suppose that a positive functional $f(x)$ in R is given. By making use of this functional, we introduce a scalar product in R in the following way. We put

$$(x, y) = f(y^*x).$$

We shall consider x to be equivalent to zero if

$$(x, x) = f(x^*x) = 0.$$

Two elements shall be called *equivalent* if their difference is equivalent to zero.

The set of elements that are equivalent to zero forms a left ideal in R. For suppose that $x \sim 0$ and that y is an arbitrary element. Then

$$f((yx)^*yx) = f(x^*y^*yx) = f(zx),$$

where $z = x^*y^*y$. By the inequality (1), we have

$$|f(zx)| \leq \sqrt{f(x^*x)}\sqrt{f(z^*z)},$$

and consequently $f(zx) = 0$, i.e., $yx \sim 0$. If $x_1 \sim 0$ and $x_2 \sim 0$, then $x_1 + x_2 \sim 0$, because

$$f((x_1 + x_2)^*(x_1 + x_2)) = \\ = f(x_1^*x_1) + f(x_2^*x_1) + f(x_1^*x_2) + f(x_2^*x_2).$$

The first and fourth terms on the right-hand side are equal to zero, by definition; and that the second and third terms are zero can again be deduced from (1).

Let us verify that the axioms for a scalar product are satisfied.

1. $(x, y) = \overline{(y, x)}$. For in virtue of the fact that a positive functional is real, we have

$$f(y^*x) = \overline{f((y^*x)^*)} = \overline{f(x^*y)}.$$

2. $(\lambda x + \mu y, z) = \lambda(x, z) + \mu(y, z)$. This is obvious.

3. $(x, x) \geqq 0$. This is obvious. The fact that $(x, x) = 0$ only for $x \sim 0$ follows from the definition of equivalence to zero.

We denote the space so obtained by $\tilde{\mathfrak{H}}$. In general, it is not complete. We denote its completion by $\mathfrak{H}$. We shall now construct a representation in this space in the following way: With every element a we associate an operator A in $\mathfrak{H}$ by the formula $Ax = ax$. We have only to verify that if $x_1 \sim x_2$, then $Ax_1 \sim Ax_2$. This is clear, because if $x_1 - x_2 \sim 0$, then $a(x_1 - x_2) \sim 0$, in virtue of the fact that the set of elements equivalent to zero forms an ideal.

Let us show that the operator A is bounded and that

$$| A | \leqq | a |. \tag{5}$$

By definition, $(Ax, Ax) = f(x^*a^*ax)$. We put

$$f_1(y) = f(x^*yx),$$

keeping x fixed. $f_1(y)$ is also a positive functional. For,

$$f_1(y^*y) = f(x^*y^*yx) = f((yx)^*yx) \geqq 0.$$

Therefore, by (2),

$$f_1(a^*a) \leqq f_1(e) \, | a^*a | \leqq | f_1(e) | \, | a |^2,$$

i.e.,

$$f(x^*a^*ax) \leqq f(x^*x) \, | a |^2,$$

or

$$(Ax, Ax) \leqq (x, x) \, | a |^2.$$

Thus,

$$| A |^2 = \sup_{(x, x) = 1} (Ax, Ax) \leqq | a |^2,$$

i.e., the inequality (5) is proved. We have shown that the operator A, defined on $\tilde{\mathfrak{H}}$, is bounded; therefore we can extend the definition to the closure $\mathfrak{H}$ of $\tilde{\mathfrak{H}}$. After this extension, the norm of A, as before, does not exceed $| a |$.

We shall now show that the mapping $a \to A$ is a representation of the ring R. First, it is obvious that if $a \to A$, $b \to B$, then $\lambda a + \mu b \to \lambda A + \mu B$ and $ab \to AB$.

Let us show that if $a \to A$, then a^* goes over into A^*—in other words, that $(ax, y) = (x, a^*y)$. But this is in fact the case, because $(ax, y) = f(y^*ax)$ and

$$(x, a^*y) = f((a^*y)^*x) = f(y^*ax) = (ax, y).$$

Let us show that the representation so obtained is cyclic. As our vector ξ_0, we choose the element $x=e$. Then the set of vectors $A\xi_0$ is in our case the set of all a, i.e., the entire space $\tilde{\mathfrak{H}}$, and it is therefore dense in $\mathfrak{H}$. This proves the cyclicity.

Next, it is obvious that for this choice of ξ_0 we have, by definition of the scalar product,

$$(A\xi_0, \xi_0) = (a, e) = f(e^*a) = f(a).$$

Thus, we have constructed a representation of the ring by means of the preassigned positive linear functional f. Our results can be combined in the form of a theorem:

Theorem 3: *To every cyclic representation of a ring R with a cyclic vector ξ_0 there corresponds a positive linear functional*

$$f(a) = (A(a)\xi_0, \xi_0), \tag{6}$$

where $A(a)$ is the operator corresponding to the element a. The representation is uniquely determined by the functional $f(x)$ to within equivalence.

Conversely, to every positive linear functional $f(a)$ there corresponds a cyclic representation such that $f(a)$ is defined by the formula (6).

5. If a representation $a \to A$ in a space $\mathfrak{H}$ is given, then to every element $\xi \in \mathfrak{H}$ there corresponds the positive functional $f(a) = (A(a)\xi, \xi)$.

When we replace the vector ξ by $\lambda\xi$, with $|\lambda| = 1$, then $f(a)$ remains unchanged.

In general, vectors that are not proportional may yield the same functional $f(a)$. However, we have the following theorem.

Theorem 4: *Let there be given an irreducible representation of a ring R. We put $(A\xi_1, \xi_1) = f_1(a)$ and $(A\xi_2, \xi_2) = f_2(a)$.*

If $f_1(a) \equiv f_2(a)$, then $\xi_1 = \lambda\xi_2$, where $|\lambda| = 1$.

Proof: Since the representation is irreducible and $\xi_1 \neq 0$, the set of vectors $A\xi_1$ is everywhere dense in $\mathfrak{H}$.

We define an operator U in the following way: if $\xi = A\xi_1$, then we put $U\xi = A\xi_2$. Let us show that the operator so constructed preserves the length of vectors. We have

$$(U\xi, U\xi) = (A\xi_2, A\xi_2) = (A^*A\xi_2, \xi_2) = \\ = (A^*A\xi_1, \xi_1) = (A\xi_1, A\xi_1) = (\xi, \xi),$$

and hence it follows that the operator U is uniquely determined. For if $A'\xi_1 = A''\xi_1$, then $\xi = (A' - A'')\xi_1 = 0$, and therefore $U\xi = 0$, i.e., $A'\xi_2 = A''\xi_2$. We extend the bounded operator U by continuity to all of $\mathfrak{H}$.

Let us show that the operator U is permutable with all the operators A

of the representation. For let A_0 be an operator of the representation and let the vector ξ have the form $\xi = A\xi_1$. Then $A_0\xi = A_0A\xi_1$, i.e., by definition of U, we have $UA_0\xi = A_0A\xi_2$. But

$$A_0U\xi = A_0UA\xi_1 = A_0A\xi_2.$$

Thus, for vectors of the form $\xi = A\xi_1$, we have $A_0U\xi = UA_0\xi$. In virtue of the fact that these elements are everywhere dense, we obtain $A_0U = UA_0$. Thus, U is permutable with all the operators of an irreducible representation and therefore, by Theorem 1 of § 46, $U = \lambda E$. But this means that

$$\xi_2 = U\xi_1 = \lambda\xi_1.$$

§ 48. Embedding of a Ring with an Involution in a Ring of Operators

1. Let a ring R with an involution be given. It may happen that the ring can be simplified, but simplified in such a way that the set of positive functionals—or, what is the same, of representations—remains the same.

For example, in § 47 we mentioned the ring of analytic functions given in a circle. Every positive functional in this ring is given by a formula

$$f(x) = \int_{-1}^{+1} x(t)d\sigma(t),$$

where $\sigma(t)$ is a monotonic function. But the same set of positive functionals belongs to the ring of all continuous functions on the interval $[-1, +1]$.

More accurately, we pose the problem in the following way.

To replace the norm $|x|$ in R by a norm $|x|_1$ as small as possible, but so that the set of positive functionals in R (i.e., the set of cyclic representations) remains the same. The new norm must be such that

$$\left.\begin{array}{lll} |e|_1 = 1, & |x+y|_1 \leqq |x|_1 + |y|_1, & |\lambda x|_1 = |\lambda|\,|x|_1, \\ |xy|_1 \leqq |x|_1 |y|_1, & |x^*|_1 = |x|_1, & |x|_1 \geqq 0. \end{array}\right\} \quad (1)$$

In the new norm there may be elements $x \neq 0$ for which $|x|_1 = 0$. In that case, we declare them to be equivalent to zero and thus go over to a new ring.

Furthermore, it may happen that in the new norm the ring R is not complete. Then we complete it. We also see that after introduction of the new norm the ring R becomes isomorphic to a ring of operators in a Hilbert space and the norm goes over into the operator norm in the Hilbert space.

2. LEMMA 1: *Let a set of norms* $|x|_\alpha$ *be introduced in R, each of them satisfying the condition* (1). *Let each of these norms be such that R can be mapped isomorphically, and with preservation of the involution and the norm, into a ring of operators in a Hilbert space. Lastly, for every x let* $\sup_\alpha |x|_\alpha < \infty$. *Then in the norm* $|x|_1 = \sup_\alpha |x|_\alpha$ *the ring R can be mapped isomorphically, and with preservation of the norm and the involution, into a ring of operators in a Hilbert space.*

Proof: Let us assume that for the norm $|x|_\alpha$ the ring is realized as a ring of operators in the space $\mathfrak{H}_\alpha$.

We consider the space $\mathfrak{H}$ that is the orthogonal direct sum of the spaces $\mathfrak{H}_\alpha$. If we denote the vectors in $\mathfrak{H}_\alpha$ by ξ^α, then the vectors in $\mathfrak{H}$ are $\xi = \{\xi^\alpha\}$, where $\{\xi^\alpha\}$ is different from zero for not more than a countable set of values α and

$$\sum_\alpha |\xi_\alpha|^2 = |\xi|^2 < \infty.$$

The scalar product is defined by the formula

$$(\xi, \eta) = \sum_\alpha (\xi_\alpha, \eta_\alpha).$$

To every a there corresponds, by definition, an operator $X_a{}^{(\alpha)}$ in $\mathfrak{H}_\alpha$, with $|X_a{}^{(\alpha)}| = |a|_\alpha$. In the space $\mathfrak{H}$, to the element a there corresponds the operator X_a, defined as follows:

$$X_a\{\xi_\alpha\} = \{X_a^{(\alpha)}\xi_\alpha\}.$$

The operator X_a is bounded. For

$$|X_a\xi|^2 = \sum_\alpha |X_a^{(\alpha)}\xi_\alpha|^2 \leqq \sum |X_a^{(\alpha)}|^2 |\xi_\alpha|^2 \leqq \sup_\alpha |X_a^{(\alpha)}|^2 \sum |\xi_\alpha|^2.$$

It is easy to see that $\sup_\alpha |X_a{}^{(\alpha)}| = |X_a| = |a|_1$. Thus we have realized the ring R with the norm $|a|_1$ in the form of a ring of operators in a Hilbert space. //

LEMMA 2: *In the ring R let a norm* $|x|_0$ *be introduced satisfying the condition* (1), *and let a representation* $a \to X_a$ *of R be given that is continuous in this norm.*[5] *Then* $|X_a| \leqq |a|_0$.

Proof: Since this representation is continuous, we can extend it to the completion R_0 of R. We then obtain a representation of the complete

[5] When we say that the representation $a \to X_a$ is continuous in the norm $|x|_0$ satisfying the conditions (1), then we also require that $|a|_0 = 0$ should imply $X_a = 0$, i.e., that elements equivalent to zero in this norm go over into 0.

ring R_0. For this, the inequality $|X_a| \leqq |a|_0$ holds, according to Theorem 2 of § 47. //

THEOREM 1: *In the ring R we can introduce a norm* $|x|_1$ *satisfying the following conditions:*

1. $|x|_1$ *satisfies* (1).
2. *Every representation of R is continuous in the norm* $|x|_1$.
3. *If any other norm* $|x|_2$ *also satisfies the conditions* 1. *and* 2., *then* $|x|_1 \leqq |x|_2$ *for every* x.
4. *The ring R with the norm* $|x|_1$ *can be mapped isomorphically, and with preservation of the involution and the norm, into a ring of operators in a Hilbert space.*

It is clear that the norm $|x|_1$ is uniquely determined by the conditions 1., 2., 3.

Proof: We consider the set of all cyclic representations of the given ring[6] and denote each cyclic representation by a symbol α. In this representation let the element a correspond to the operator $X_a^{(\alpha)}$. By Theorem 2 of § 47, $|X_a^{(\alpha)}| \leqq |a|$. We put

$$|a|_\alpha = |X_a^{(\alpha)}|;$$

$|a|_\alpha$ is a norm satisfying the conditions (1) of this section. Now we put $|a|_1 = \sup_\alpha |a|_\alpha$. Then we have $|a|_1 \leqq |a|$, since $|a|_\alpha \leqq |a|$.

By Lemma 1, the ring R with the norm $|a|_1$ satisfies the conditions (1) and can be realized as a ring of operators in a certain Hilbert space $\mathfrak{H}$.

Every cyclic representation of R is continuous in the norm $|a|_1$. For let a cyclic representation $a \to X_a^{(\alpha)}$ be given. Then

$$|X_a^{(\alpha)}| = |a|_\alpha \leqq \sup_\alpha |a|_\alpha = |a|_1.$$

Every representation splits into a direct sum of cyclic representations and is therefore also continuous in this norm.

Thus, we have shown that $|a|_1$ satisfies the conditions 1., 2., and 4. Let us show that the condition 3. is also satisfied. Let a norm $|x|_2$ satisfying the conditions (1) be given in which all the representations of R are continuous. Then we have, by Lemma 2,

$$|X_a^{(\alpha)}| \leqq |a|_2,$$

i.e., $|a|_\alpha \leqq |a|_2$. Therefore,

[6] We only take a cyclic representation, because every representation breaks up into cyclic ones.

$$|a|_1 = \sup_\alpha |a|_\alpha \leqq |a|_2.$$

This completes the proof of the theorem. //

THEOREM 2: *The equation*

$$|x|_1 = \sup \sqrt{f(x^*x)} \tag{2}$$

holds, where sup *is extended over all positive functionals f for which* $f(e) = 1$.

Proof: Every cyclic representation can be described (§ 47) by a positive linear functional f. Let us find the norm of the operator X_a corresponding to the element a in this representation. We have

$$(X_a x, X_a x) = f((ax)^*ax) = f(x^*a^*ax), \qquad (x, x) = f(x^*x).$$

We introduce the functional $f_x(y) = f(x^*yx)$, keeping x fixed. $f_x(y)$ is a positive functional and

$$f_x(e) = f(x^*x), \qquad |X_a|^2 = \sup (X_a x, X_a x),$$

where sup is taken over all x for which $(x, x) = 1$; thus,

$$|X_a|^2 = \sup f_x(a^*a),$$

where sup is taken over all x for which $f_x(e) = 1$.

Therefore $|X_a|^2 \leqq \sup f(a^*a)$, where sup is taken over all the positive functionals f for which $f(e) = 1$.

If we denote this representation by the subscript α, then

$$|a|_\alpha^2 \leqq \sup f(a^*a),$$

in virtue of the conditions that $f(e) = 1$ and f is a positive linear functional. Therefore

$$|a|_1^2 = \sup_\alpha |a|_\alpha^2 \leqq \sup_{f(e)=1} f(a^*a).$$

On the other hand,

$$f(a^*a) = (X_a e, X_a e).$$

Therefore, if $f(e) = 1$, then

$$f(a^*a) \leqq |X_a|^2 = |a|_\alpha^2 \leqq \sup_\alpha |a|_\alpha^2 = |a|_1^2,$$

and consequently

$$\sup_{f(e)=1} f(a^*a) \leqq |a|_1^2.$$

This completes the proof. //

3. *Note* 1: All the arguments of this section remain valid if there is no norm at all in the original ring, i.e., if it is defined only by the axioms 1., 2. (§ 46.1) and a), b), c), d) (§ 46.2); we need only impose the additional condition that for every x

$$\sup f(x^*x) < \infty,$$

where sup is taken over all positive f for which $f(e) = 1$.

Note 2: The set I of elements for which $|x|_1 = 0$ (we call them equivalent to zero) forms a two-sided ideal. These elements x can be characterized by the fact that they go over into 0 in every representation or, to put it differently, that every positive functional vanishes on them.

Let us show that I is in fact an ideal.

Let $|x|_1 = 0$ and let y be an arbitrary element. Then

$$|xy|_1 \leqq |x|_1 |y|_1 = 0,$$

and similarly, $|yx|_1 = 0$. Furthermore, if $|x|_1 = 0$ and $|y|_1 = 0$, then

$$|\lambda x + \mu y|_1 \leqq |\lambda| |x|_1 + |\mu| |y|_1 = 0.$$

Thus, we have shown that I is an ideal. In studying representations or positive functionals, we may replace R by the residue-class ring of this ideal. We denote this residue-class ring by R'. We shall call it a *reduced ring*.

Note 3: Every representation of R is continuous in the norm $|x|_1$ and is therefore a continuous representation of R'. But a continuous representation of R' can be extended to a representation of the completion of R'. We denote the completion of R' by $\bar{R}$.

Thus, *every representation of the ring R is also a representation of $\bar{R}$, and vice versa.*

Similarly, *every positive linear functional on R can be extended to $\bar{R}$, and vice versa.*

§ 49. Indecomposable Functionals and Irreducible Representations

1. In the finite-dimensional case, every representation splits into irreducible representations. In the general case, the existence of such representations is not clear a priori. Without touching on the problem of decomposing representations into irreducible ones, we shall show in this section that there exist irreducible representations. It is very convenient to do this in terms of positive functionals.

DEFINITION 1: We shall say that a positive functional f_1 is *subordinate to a functional* f ($f_1 \ll f$) if there exists a $\lambda > 0$ such that $\lambda f - f_1$ is a positive functional.

We construct a cyclic representation $a \to X_a$ (X_a are operators in a space $\mathfrak{H}$) corresponding to the functional f:

$$f(a) = (X_a \xi_0, \xi_0),$$

where ξ_0 is a cyclic vector. Let B be a bounded positive-definite operator in the Hilbert space $\mathfrak{H}$ that is permutable with all the operators of the representation. We put

$$f_1(a) = (X_a B \xi_0, \xi_0).$$

In particular, the functional f corresponds to the operator $B = E$. We claim that $f_1(a)$ *is a positive functional and is subordinate to* $f(a)$.

$f_1(a)$ is positive; for

$$f_1(a^* a) = (X_a^* X_a B \xi_0, \xi_0) = (X_a B \xi_0, X_a \xi_0) = (B X_a \xi_0, X_a \xi_0) \geqq 0,$$

since B is a positive-definite operator. Furthermore, f_1 is subordinate to f. For B is bounded. Therefore there exists a λ such that $(B\xi, \xi) \leqq \lambda(\xi, \xi)$, or

$$\lambda(\xi, \xi) - (B\xi, \xi) \geqq 0.$$

Putting $\xi = X_a \xi_0$, we obtain

$$\lambda (X_a^* X_a \xi_0, \xi_0) - (X_a^* X_a B \xi_0, \xi_0) \geqq 0,$$

i.e., $\lambda f - f_1$ is a positive functional; but this means that f_1 is subordinate to f.

Conversely, suppose that f_1 is a positive functional subordinate to f. We construct a cyclic representation by means of f (§ 47.4). *Then the functional* f_1 *corresponds to a positive-definite operator* B *that is permutable with all the operators of the representation.*

Let us prove this. We know that the space $\mathfrak{H}$ is obtained as the completion of the space $\tilde{\mathfrak{H}}$ formed from the classes of equivalent elements of R, and that the scalar product in $\tilde{\mathfrak{H}}$ is given by the formula

$$(x, y) = f(y^* x),$$

where the elements x for which $(x, x) = 0$, i.e., $f(x^* x) = 0$, are taken to be equivalent to zero.

In the space $\tilde{\mathfrak{H}}$ we consider the Hermitian form $f_1(y^* x)$. We shall show that f_1 is uniquely determined as a Hermitian bilinear form continuous in $\tilde{\mathfrak{H}}$. f_1 is subordinate to f. This means that there exists a λ such that $\lambda f - f_1$ is a positive functional. Thus, we have:

$$\lambda f(x^*x) - f_1(x^*x) \geqq 0,$$

and therefore

$$0 \leqq f_1(x^*x) \leqq \lambda f(x^*x).$$

This shows that $f(x^*x) = 0$ implies that $f_1(x^*x) = 0$. Hence we see, say by means of the inequality

$$|f_1(y^*x)|^2 \leqq f_1(x^*x) f_1(y^*y),$$

that the expression $f_1(y^*x)$ is zero.

We have thus shown that $f_1(y^*x)$ is uniquely determined, i.e., that $f_1(y^*x) = 0$ for $x \sim 0$. f_1 is a bounded bilinear form. In fact,

$$0 \leqq f_1(x^*x) \leqq \lambda f(x^*x).$$

Therefore

$$|f_1(y^*x)|^2 \leqq f_1(x^*x) f_1(y^*y) \leqq \lambda^2 f(x^*x) f(y^*y) = \lambda^2 (x, x)(y, y).$$

Being bounded, this bilinear form can be extended to the completion of $\tilde{\mathfrak{H}}$, i.e., to the space $\mathfrak{H}$. But to a bounded bilinear form in $\mathfrak{H}$ there corresponds a bounded operator B. Therefore there exists an operator B such that

$$f_1(y^*x) = (Bx, y).$$

Let us show that the operator B is permutable with all the operators X_a of the representation. For this purpose it is sufficient to show that

$$(BX_a x, y) = (Bx, X_a{}^* y).$$

But this is in fact so, because

$$(BX_a x, y) = (Bax, y) = f_1(y^*ax)$$

and

$$(Bx, X_a{}^* y) = (Bx, a^* y) = f_1((a^*y)^*x) = f_1(y^*ax).$$

Thus, we have proved the following theorem.

THEOREM 1: *Let $f(x)$ be a positive functional, $a \rightarrow X_a$ a cyclic representation corresponding to it, and ξ_0 a corresponding cyclic vector, i.e.,*

$$f(a) = (X_a \xi_0, \xi_0).$$

Then to every positive functional f_1 subordinate to f there corresponds a positive-definite operator B that is permutable with all the X_a, and

$$f_1(a) = (X_a B \xi_0, \xi_0).$$

Conversely, to every bounded positive-definite operator B that is permutable with all the operators X_a there corresponds a positive functional subordinate to f.

In particular, to $f(x)$ itself there corresponds the unit operator. Now let us consider linear combinations of positive functionals subordinate to $f(x)$ We call them functionals subordinate to the positive functional $f(x)$. They correspond to arbitrary bounded operators that are permutable with the operators X_a of the representation.

Indeed, every operator that permutes with the X_a can be represented as a linear combination of positive operators. Thus we have the following result:

In the set of functionals subordinate to the positive functional f we can introduce an operation of multiplication such that it becomes isomorphic to the ring of operators permutable with the operators X_a of the representation generated by the functional $f(x)$. Here f itself plays the role of the unit element of the ring.

We have turned the set C_f of functionals subordinate to f into a ring with an involution. Moreover:

1. To the functional f there corresponds the unit element of the ring.
2. The operation $*$ is defined as follows: $f^*(x) = \overline{f(x^*)}$.
3. Multiplication is connected with the $*$-operation by the usual condition:

$$(f_1 f_2)^* = f_2^* f_1^*.$$

2. Definition 2: A positive function f is called *indecomposable* if every functional f_1 subordinate to f is a multiple of f, i.e., $f_1(x) = \lambda f(x)$.

Theorem 2: *Let f be a positive functional. The representation corresponding to it is irreducible if and only if f is indecomposable.*

Proof: To every functional f_1 subordinate to f there corresponds an operator that is permutable with all the operators X_a of the representation. To f itself there corresponds the unit operator. The irreducibility of the representation is equivalent to the condition that every operator B that is permutable with the operators of the representation be a multiple of the unit (Theorem 1 of § 46), i.e., to the condition that every functional f_1 subordinate to f be a multiple of f. //

3. We shall now proceed to prove the existence of irreducible representations. By the theorem just proved, it is sufficient for this purpose to prove the existence of indecomposable positive functionals.

The set of positive functionals $f(x)$ such that $f(e) \leqq 1$ forms a convex set. For if $f_1(x)$ and $f_2(x)$ are positive and $f_1(e) \leqq 1$, $f_2(e) \leqq 1$, then

$$f(x) = \alpha f_1(x) + \beta f_2(x) \qquad (\alpha \geqq 0, \beta \geqq 0, \alpha + \beta = 1)$$

satisfies the same conditions. Therefore the existence of indecomposable positive functionals follows immediately from a theorem of Krein and Milman [85]. Furthermore, let x be an element of the ring such that $|x|_1 \neq 0$. Then there exists a positive functional f such that

$$f(x^*x) \neq 0, \qquad f(e) = 1.$$

On the other hand, by the same theorem of Krein and Milman the set of all positive functionals f satisfying the condition $f(e) \leqq 1$ is the least weakly closed convex set containing all the indecomposable positive functionals that satisfy the condition $f(e) \leqq 1$. Therefore there exists such an indecomposable functional f_0 satisfying the conditions

$$f_0(e) \leqq 1, \qquad f_0(x^*x) \neq 0.$$

Thus, we have proved the following theorem.

THEOREM 3: *Let x be an element of the ring $\overline{R}$ such that $|x|_1 \neq 0$. Then there exists an indecomposable positive functional f_0 satisfying the conditions*

$$f_0(e) \leqq 1, \qquad f_0(x^*x) \neq 0. \tag{1}$$

By Theorem 2 of the present section and Theorem 3 of § 47, this theorem can also be stated in the following way:

THEOREM 4: *Let x_0 be an element of a ring $\overline{R}$ such that $|x_0|_1 \neq 0$. Then there exists an irreducible representation $a \rightarrow X_a$ of the ring $\overline{R}$ such that the operator X_{x_0} corresponding to x_0 in this representation is different from zero.*

In fact, condition (1) can be rewritten in the form

$$|\xi_0| \leqq 1, \qquad |X_{x_0}\xi_0|^2 \neq 0.$$

The latter inequality means that $X_{x_0} \neq 0$.

§ 50. The Case of Commutative Rings

1. The whole picture becomes particularly simple when R is a commutative ring.

LEMMA 1: *If R is commutative, then the ring $\overline{R}$ is isomorphic to the ring of all continuous functions $x(M)$ on a compact space. Furthermore,*

$$x^*(M) = \overline{x(M)}.$$

Proof: $\overline{R}$ is a $*$-ring, i.e.,

$$|x^*x|_1 = |x|_1 \, |x^*|_1. \tag{1}$$

For $|x|_1 = \sup_f \sqrt{f(x^*x)}$, where f is a positive functional and $f(e) = 1$. By the inequality (2) of § 47, we have

$$\sup_f \sqrt{f(x^*x)} \leqq \sqrt{|xx^*|_1};$$

thus, $|x|_1^2 \leqq |xx^*|_1$. On the other hand, we always have

$$|xx^*|_1 \leqq |x|_1 \, |x^*|_1 = |x|_1^2,$$

so that

$$|xx^*|_1 = |x|_1^2 = |x|_1 \, |x^*|_1.$$

By Lemma 1 in [80], a commutative ring in which an involution and a norm are introduced satisfying the condition (1) is isomorphic to the ring of all continuous functions on a compact set $\mathfrak{M}_1$. This $\mathfrak{M}_1$ is the set of maximal ideals of $\bar{R}$.

2. DEFINITION: A maximal ideal M of a ring R is called *symmetric* if for every $x \in R$ we have $x^*(M) = \overline{x(M)}$. If M is a maximal ideal, then we denote by M^* the maximal ideal for which $x^*(M^*) = \overline{x(M)}$. It is easy to show that this M^* exists for every M. A symmetric maximal ideal is one for which $M^* = M$.

It can be shown that the set of symmetric maximal ideals of a ring R forms a closed subset in the set of all maximal ideals of R.

THEOREM 1: *$\bar{R}$ is isomorphic to the ring of all the continuous functions on the set $\mathfrak{M}_1$ of symmetric maximal ideals of R.*

Proof: To prove the theorem it is sufficient, by Lemma 1, to show that the set of maximal ideals of $\bar{R}$ is homeomorphic to the set of symmetric maximal ideals of R.

To every maximal ideal of $\bar{R}$ there corresponds a symmetric maximal ideal of R.

For let a homomorphism of $\bar{R}$ into the field of complex numbers be given. This homomorphism is at the same time a homomorphism of the ring R', which is part of $\bar{R}$. But R' is a residue-class ring of R. Therefore this homomorphism is at the same time a homomorphism of R itself into the field of complex numbers.

A maximal ideal of R is thus determined. This maximal ideal is symmetric, because all the maximal ideals of $\bar{R}$ are symmetric. The continuity of the correspondence between ideals follows immediately from the definition of the topology in the set of maximal ideals (see [79], § 7).

Conversely, let M be a symmetric maximal ideal of R. We consider the functional

$$f(x)=x(M).$$

This functional is positive. In fact,

$$f(x^*x)=x^*(M)x(M)=|x(M)|^2\geqq 0, \qquad f(e)=e(M)=1.$$

Therefore, it is equal to zero for elements for which $|x|_1=0$. Moreover, it is continuous in the norm $|x|_1$ and can therefore be extended to $\overline{R}$.

Thus, we can establish a one-to-one continuous correspondence between the symmetric maximal ideals M of R and the maximal ideals $\overline{M}$ of $\overline{R}$. If we denote an element of R and the corresponding element of $\overline{R}$ by the same letter x, then we have Theorem 2:

THEOREM 2: *Every positive linear functional $f(x)$ on a commutative ring R can be represented in a unique way in the form*

$$f(x)=\int x(M)d\sigma(\Delta), \tag{2}$$

where $\sigma(\Delta)$ is a positive completely additive set function on the set $\mathfrak{M}_1$ of symmetric maximal ideals of R.

Proof: Every positive functional on R can be extended to $\overline{R}$ (§ 48, Note 3). It then turns into a positive functional on the set of continuous functions on the compact set $\mathfrak{M}_1$. Such a functional is described by the formula (2), where $\sigma(\Delta)$ is a positive completely additive set function. The set function $\sigma(\Delta)$ is uniquely determined.

Conversely, if $\sigma(\Delta)$ is a positive set function on the set of symmetric maximal ideals, then $f(x)$, given by formula (2), is a positive functional. For

$$f(x^*x)=\int x^*(M)x(M)d\sigma(\Delta)=\int |x(M)|^2d\sigma(\Delta)\geqq 0.$$

This completes the proof. //

It follows from formula (2) that every indecomposable positive functional is of the form $f(x)=x(M_0)$, where M_0 is a fixed symmetric maximal ideal.

The theorem just proved means, strictly speaking, that every functional decomposes in a unique way into indecomposable positive functionals. We have required R to be commutative. This condition is essential, as is clear from the following theorem.

THEOREM 3: *Assume that every positive functional given on a ring R decomposes in a unique way into indecomposable positive functionals. Then the reduced ring $\overline{R}$ is commutative (i.e., $xy-yx$ is an element equivalent to zero in R).*

Proof: Let us consider an arbitrary continuous representation $a \to X_a$, where X_a is an operator in the space $\mathfrak{H}$. We shall show that $\mathfrak{H}$ is one-dimensional. Let us assume the contrary. Then $\mathfrak{H}$ contains at least two linearly independent vectors ξ_1 and ξ_2. We put

$$f_1(a) = (X_a\xi_1, \xi_1), \quad f_2(a) = (X_a\xi_2, \xi_2),$$

$$f_1'(a) = \frac{1}{2}(X_a(\xi_1 + \xi_2), \quad \xi_1 + \xi_2),$$

$$f_2'(a) = \frac{1}{2}(X_a(\xi_1 - \xi_2), \quad \xi_1 - \xi_2),$$

and assume that

$$\varphi(a) = f_1(a) + f_2(a) = f_1'(a) + f_2'(a).$$

$\varphi(a)$ is a positive functional. By Theorem 2 of § 49, the functionals $f_1(a)$, $f_2(a)$, $f_1'(a)$, $f_2'(a)$ are indecomposable. Therefore $f(a)$ decomposes in two ways into indecomposable functionals. These are distinct. For the vectors ξ_1, ξ_2, $\xi_1 + \xi_2$, $\xi_1 - \xi_2$ are not proportional; but on the other hand, by Theorem 4 of § 47, in the case of an irreducible representation by functionals of the form $(X_a\xi, \xi)$, the vector ξ is uniquely determined to within a factor.

Thus, every irreducible representation of R is one-dimensional and hence commutative. Since every element that is carried into zero by all continuous representations is equivalent to zero, $xy - yx$ is equivalent to zero, i.e., the reduced ring is commutative. //

3. *Example*: We denote by R_0' the set of functions given on the half-line $0 \leqq u < \infty$ such that

$$\| f \| = \int_0^\infty | f(u) | \sinh 2u du < \infty.$$

We define a multiplication $f = f_1 \times f_2$ in R_0', where $f(u)$ is given by the formula

$$f(u) = \int_0^\infty \int_{|u-t|}^{u+t} f_1(s) f_2(t) ds dt.$$

Further, we let R_0 denote the set of all elements of the form $\lambda e + f$, where e is a formally adjoined unit element and $f \in R_0'$.

It can be verified that this turns R_0 into a normed ring.

We also define an operation $*$ by putting

$$f^*(u) = \overline{f(u)}, \qquad (\lambda e + f)^* = \bar{\lambda} e + f^*.$$

It is easy to verify that this yields a commutative ring with an involution. Let us find the maximal ideals of this ring. The arguments here are exactly analogous to those given in [83].

Just as in the other case, we come to the conclusion that the maximal ideals are determined by homomorphisms

$$f \to \int_0^\infty f(s)\psi(s)\,ds,$$

where $\psi(s)$ is defined by the relation

$$\psi(s)\psi(t) = \int_{|t-s|}^{t+s} \psi(u)\,du.$$

The solutions of this equation are the functions

$$\psi(s) = [2 \sin \varrho s]/\varrho,$$

where ϱ is an arbitrary complex number. This homomorphism is defined for all elements of R_0 if and only if $\varrho = \varrho_1 + i\varrho_2$, where $|\varrho_2| \leqq 2$.

Thus, every maximal ideal of the ring is determined by a number $\varrho = \varrho_1 + i\varrho_2$, where $|\varrho_2| \leqq 2$, and with ϱ and $-\varrho$ determining the same maximal ideal.

Let M be the maximal ideal determined by ϱ. Then the maximal ideal M^* is determined by $\bar{\varrho}$. Therefore, symmetric maximal ideals are determined by the conditions $\varrho = \bar{\varrho}$ or $\varrho = -\bar{\varrho}$; thus, symmetric maximal ideals arise when ϱ is either real or pure imaginary. The corresponding homomorphisms are given by the formulas

$$f \to \int_0^\infty f(s)[2 \sin \varrho s]/\varrho\, d\varrho,$$

where ϱ is real, and by

$$f \to \int_0^\infty f(s)[2 \sinh \varrho s]/\varrho\, d\varrho,$$

where $0 \leqq \varrho \leqq 2$.

§ 51. Group Rings

1. As a special case of the rings studied earlier we shall now discuss the so-called group rings. This enables us to obtain certain results concerning representations of groups.

Let G be a locally compact group. For simplicity of the exposition we shall assume that the left and right invariant Haar measures on G coincide.

We examine the ring R' whose elements are the absolutely integrable functions $x(g)$ on the group.

Multiplication is given by the formula

$$x_1 \times x_2 = \int x_1(gg_1^{-1}) x_2(g_1) dg_1.$$

An involution is defined by the equation

$$x^*(g) = \overline{x(g^{-1})}.$$

The norm $|x|$ of the element x is taken to be

$$|x| = \int |x(g)| \, dg.$$

To this ring we adjoin a formal unit element (if the group is non-discrete), so that finally the elements of the ring can be expressed by the symbol $\lambda e + x$, where e is the formally adjoined unit element. The multiplication is naturally given by the formula

$$(\lambda_1 e + x_1)(\lambda_2 e + x_2) = \lambda_1 \lambda_2 e + (\lambda_1 x_2 + \lambda_2 x_1 + \lambda_1 \lambda_2 x_1 x_2),$$

and the involution and norm are extended as follows: We put

$$|\lambda e + x| = |\lambda| + |x|, \qquad (\lambda e + x)^* = \bar{\lambda} e + x^*.$$

We denote the ring so obtained by R and call it *the group ring of* G.

Theorem 1: *To every representation* $a + \lambda e \rightarrow X_a + \lambda E$ *of the group ring there corresponds a continuous unitary representation* $g \rightarrow T_g$ *of the group. Conversely, to every measurable unitary representation of the group there corresponds a representation* $a \rightarrow X_a$ *of the group ring. These representations are connected by the formula*

$$X_a = \int a(g) T_g dg.$$

Let us prove the theorem for a cyclic representation. Such a representation, as we know (§ 47), can be realized in the following way: The space $\mathfrak{H}$ is obtained by completing the space $\tilde{\mathfrak{H}}$ whose vectors are the elements $x, y, \ldots$ of R (considered to within equivalence). The scalar product is given by the formula

$$f(y^* x) = (x, y),$$

where f is a positive functional. To the element a there corresponds the operator X_a given by $X_a x = ax$. We have shown that here $|X_a| \leqq |a|$.

With the element g_0 of G we associate the operator $T_{g_0}x = y$, where $y(g) = x(g_0^{-1}g)$.

Let us show that this operator is unitary. Observe that the following equation holds for elements of R:

$$(T_{g_0}y)^*T_{g_0}x = y^*x.$$

For

$$y^*x = \int \overline{y(g_1g^{-1})}x(g_1)dg.$$

Therefore under the application of T_{g_0} scalar products do not change. Since, first of all, the operator T_g maps our set of functions into itself, it is clear that the operator T_{g_0} extended to the entire space $\mathfrak{H}$ is unitary. Let us now show that the representation $g \to T_g$ of G is continuous. For this purpose, we note that the functional f (like every positive functional) is continuous; on the other hand

$$|T_{g_0}x - x| = \int |T_{g_0}x - x| \, dg \to 0 \quad \text{as} \quad g_0 \to e,$$

and therefore

$$f((T_{g_0}x - x)^*(T_gx - x)) \to 0 \quad \text{as} \quad g_0 \to e.$$

This means that

$$(T_{g_0}x - x, T_{g_0}x - x) \to 0 \quad \text{as} \quad g_0 \to e.$$

This proves the continuity at the unit element of the group, and consequently at every other point.

It now remains to show that the representation of the ring generated by $g \to T_g$ is the one with which we started.

Let $a(g)$ be an absolutely integrable function. In the representation of R it corresponds to the operator $Ax = a \times x$. Our aim is to show that

$$Ax = \int a(g)T_gxdg.$$

or, in terms of scalar products,

$$(Ax, y) = \int a(g)(T_gx, y)dg,$$

i.e.,

$$f(y^*ax) = \int a(g)f(y^*T_gx)dg.$$

Since the functional f and the operation T_g are continuous, we can rewrite the right-hand side of this equation as follows:

$$f\left(y^* \int a(g)T_g x dg\right) = f(y^* ax).$$

This proves our statement.

We have shown that to every representation of the group ring there corresponds a representation of the group. The converse is easy to show. For suppose that a unitary representation of the group is given: $g \to T_g$.

Let us assume that the function T_g is weakly continuous, i.e., that $(T_g\xi, \eta)$ is a continuous function of g for every ξ and η. We put

$$A = \int a(g) T_g dg.$$

This operator exists, provided only that the function $a(g)$ is integrable. We thus obtain a representation of the group ring. It is easy to see that, conversely, to this operator there corresponds the representation T_g, provided only that T_g is cyclic.[7]

We make one further remark concerning the proof of the theorem. In constructing T_g, the representation we have used is not that of the extended ring R, but only that of the original ring R' without the adjoined unit element. But since the space $\mathfrak{H}$ was constructed by means of a completion of the space $\tilde{\mathfrak{H}}$ formed from elements of R, we have to show that the unextended ring R' leads to the same space $\mathfrak{H}$. For this purpose, we shall show that the element e is the limit, in the sense of our scalar product, of elements x of R'.

To every neighborhood V of the unit element of G we assign a function $e_V(g)$ satisfying the following conditions:

$$e_V(g) \geqq 0; \qquad e_V(g) = 0 \text{ if } g \notin V;$$
$$e_V(g^{-1}) = e_V(g); \qquad \int e_V(g) dg = 1.$$

We call such a system of functions a *unit system*. It is easy to show that a unit system has the following properties:

$$|x \times e_V - x| \to 0 \quad \text{as} \quad V \to e$$

(the limit is interpreted in the partially ordered system of neighborhoods given in the sense of their natural ordering). We have

$$f(e_V x) \to f(x) \quad \text{for every} \quad x \in R,$$

[7] In the case of a separable space $\mathfrak{H}$, this result can be strengthened. Specifically, it follows from the preceding arguments that if the function $(T_g\xi, \eta)$ is measurable for all $\xi, \eta \in \mathfrak{H}$, then the representation $g \to T_g$ coincides almost everywhere on G with a continuous representation.

i.e., $(e_V, x) \to (e, x)$ for every x. Moreover,

$$(e_V, e_V) = f(e_V^* e_V) \leqq f(e) | e_V^* e_V | = f(e),$$

—i.e., the lengths of the vectors e_V are bounded—so that e_V is weakly convergent. We denote this weak limit by ξ_0. Then we have

$$(\xi_0, x) = (e, x) \qquad \text{for every} \qquad x \in R.$$

In particular,

$$(e_V, \xi_0) = (e_V, e) = f(e_V),$$

and therefore $f(e_V)$ has a limit. Thus, the vector $\xi_0 - e$, which is orthogonal to all the elements x, splits off. It is not essential for the representation, because in this one-dimensional space every element x yields the null operator. By discarding this one-dimensional space we obtain the required result. For this 'parasitical' one-dimensional space to be absent it is necessary and sufficient that

$$\lim_{V \to e} f(e_V) = f(e). \; //$$

2. Applying Theorem 4 of § 49 to the group ring R and using Theorem 1 just proved, we obtain a fundamental result of Gelfand and Raikov [82] on the completeness of the system of irreducible representations of a locally compact group.

Using the expression (6) of § 47 for a positive functional and Theorem 1 of this section, we find that every positive functional in the group ring R is determined by the formula

$$f(\lambda e + a) = \lambda C + \int a(g)(T_g \xi, \xi) dg,$$

where $g \to T_g$ is a continuous unitary representation of G and the function $\varphi(g) = (T_g \xi, \xi)$ is a continuous positive-definite function on G (see [82]).

Conversely, every bounded measurable positive-definite function $\varphi(g)$ corresponds to a positive functional on the group ring defined by the formula

$$f(a) = \int a(g) \varphi(g) dg.$$

Hence it follows, in particular, that every bounded measurable positive-definite function coincides almost everywhere on G with a continuous positive-definite function (see [82]).

Positive-definite functions were the starting point in [82] for the construction of unitary representations of a locally compact group.

§ 52. Example of an Unsymmetric Group Ring

1. It is known that the group ring of a compact or a commutative group is symmetric (see [87]). However, it turns out that for a locally compact group this is in general not true.

We shall now give an example of a locally compact group whose group ring is unsymmetric.

2. Let G be the group of complex matrices of order 2 with determinant 1; let $\mathfrak{H}$ denote the subgroup of G that consists of unitary matrices.

According to a well-known definition, a double coset of $\mathfrak{H}$ in G is a set of elements of the form h_1gh_2, where g is fixed and h_1 and h_2 range over the whole subgroup $\mathfrak{H}$. Since $\mathfrak{H}$ is a compact subgroup, the set of elements of the form h_1gh_2 has finite measure whenever the set of elements g has finite measure.

We consider the set R_0' of functions $f(g)$ that are summable and constant on the double cosets of $\mathfrak{H}$ in G. The set R_0 of elements of the form $\lambda e + f$ $(f \in R_0')$ forms a subring of the group ring R. For assume that $f_1(g)$ and $f_2(g)$ belong to R_0'. We have to show that the function $f = f_1 \times f_2$ also belongs to R, i.e., is constant on the double cosets of $\mathfrak{H}$ in G. But

$$f(gh) = \int f_1(ghg_1^{-1})f_2(g_1)dg_1 =$$

$$= \int f_1(gg_2^{-1})f_2(g_2h)dg_2 = \int f_1(gg_2^{-1})f_2(g_2)dg_2 = f(g);$$

and similarly, $f(hg) = f(g)$ for all $h \in \mathfrak{H}$.

The ring R_0 is commutative. For the Haar measure on G is invariant under the transformation $g \rightarrow g^{-1}$ and therefore, when f_1 and $f_2 \in R_0'$, we have

$$\int f_1(hg^{-1})f_2(g)dg = \int f_1(g^{-1})f_2(gh)dg =$$

$$= \int f_1(g^{-1})f_2(hg)dg = \int f_2(hg^{-1})f_1(g)dg$$

for all $h \in \mathfrak{H}$.

3. In order to prove that the group ring is unsymmetric, we shall first show that R_0 is unsymmetric.

Let us therefore examine the ring R_0 in more detail. Every matrix g can be represented in the form $g = ha$, where h is unitary and a is a positive-definite Hermitian matrix. Every Hermitian matrix can be written in the form $a = h_1 \delta h_1^{-1}$, where h_1 is unitary and δ is a diagonal matrix. Since the group is unimodular, δ is of the form

$$\delta = \begin{pmatrix} \lambda & 0 \\ 0 & \lambda^{-1} \end{pmatrix}, \qquad \lambda > 0.$$

Thus, in every coset of $\mathfrak{H}$ in G there is a diagonal matrix $\begin{pmatrix} \lambda & 0 \\ 0 & \lambda^{-1} \end{pmatrix}$, and it is easy to see that every coset containing the matrix $\begin{pmatrix} \lambda & 0 \\ 0 & \lambda^{-1} \end{pmatrix}$ also contains $\begin{pmatrix} \lambda^{-1} & 0 \\ 0 & \lambda \end{pmatrix}$ and, apart from this, no further diagonal matrices.

Thus, every double coset is characterized by a number λ; and λ and λ^{-1} correspond to the same coset. It is convenient to consider $t = \log \lambda$ in place of λ. Then t and $-t$ determine the same coset.

Thus, we can regard the functions in R_0—that is, the functions that are constant on the double cosets, as even functions of t.

4. Now let us find the rule of multiplication and the norm in R_0. For this purpose, we observe that the space of left cosets of $\mathfrak{H}$ in G is the set of positive-definite matrices with determinant 1. The transformation of left cosets reduces to the transformation of the corresponding quadratic forms. It can be shown, furthermore, that this is the group of transformations of a three-dimensional Lobachevsky space, where the points of the Lobachevsky space are in one-to-one correspondence with the left cosets. A double coset is a collection of left cosets; therefore it corresponds to a set of points in the Lobachevsky space. In order to find this set, we note the following: The elements of our group can be regarded as transformations of the left cosets that consist in multiplying each coset on the right by a given element g of the group. An element g leaves the unit coset invariant if and only if it is an element of $\mathfrak{H}$.

Thus, the subgroup $\mathfrak{H}$ consists of the motions of a Lobachevsky space that leave a fixed point of the space in place.

Since a double coset is obtained from a left coset by multiplying it on the right by all the elements of $\mathfrak{H}$, this corresponds in the Lobachevsky space to a sphere with its center at a fixed point.

Thus, to a function on the group that is constant on double cosets there corresponds a function in the Lobachevsky space that is constant on spheres with a fixed center. The integral of a function over the group differs only by a constant factor from the integral of the corresponding function in the Lobachevsky space. Therefore the norm of such a function is equal to

$\int |f(\bar{g})|\, d\bar{g}$, where $d\bar{g}$ is the element of volume in the Lobachevsky space. If we denote the radius of the sphere by t, then the norm of such a function is equal to $\int_0^\infty |f(t)|\,\varphi(t)$, where $\varphi(t)$ is the area of the surface of the sphere.

5. We now proceed to compute the rule of multiplication for the functions $f(t)$ corresponding to the involution of functions on the group. Avoiding a direct computation, we use the following argument.

The maximal ideals of a commutative ring are given by homomorphisms

$$f \to \int_0^\infty f(t)\alpha_\rho(t)\varphi(t)dt,$$

where $\alpha_\rho(t)$ is a so-called spherical function of the given group. These functions are calculated in the paper [81]. They are equal to $[2 \sin \varrho t]/[\varrho \sinh 2t]$; every ϱ corresponds to its own homomorphism, i.e., its own maximal ideal. Under the homomorphism, products of functions go over into products of numbers. Therefore, if $f = f_1 \times f_2$, we have

$$\int f(u)\,\alpha_\rho(u)\,\varphi(u)\,du = \int f_1(u)\,\alpha_\rho(u)\,du \cdot \int f_2(u)\,\alpha_\rho(u)\,du.$$

In order to express f in terms of f_1 and f_2, we put

$$\left.\begin{aligned} \tilde{f}_1(u) = f_1(u)\cdot[\varphi(u)/\sinh 2u], \qquad \tilde{f}_2(u) = f_2(u)\cdot[\varphi(u)/\sinh 2u], \\ \tilde{f}(u) = f(u)\cdot[\varphi(u)/\sinh 2u]. \end{aligned}\right\} \tag{1}$$

Then the homomorphism is given by the formula

$$f \to \int_0^\infty \tilde{f}(u)\,\frac{2 \sin \rho u}{\rho}\,du.$$

We can satisfy condition (1) by putting

$$\tilde{f}(u) = \int_0^\infty \left(\int_{|t-u|}^{t+u} \tilde{f}_1(s)\tilde{f}_2(t)\,ds \right) dt.$$

For let us denote

$$\int_0^\infty \left(\int_{|t-u|}^{t+u} \tilde{f}_1(s)\tilde{f}_2(t)\,ds \right) dt$$

by $g(u)$. Then

$$\int_0^\infty g(u)\frac{2\sin\rho u}{\rho}\,du=$$

$$=\int_0^\infty\left(\int_0^\infty\left(\int_{|t-u|}^{t+u}\tilde{f}_1(s)\tilde{f}_2(t)\,ds\right)dt\right)\frac{2\sin\rho u}{\rho}\,du=$$

$$=\int_0^\infty\left(\int_0^\infty\tilde{f}_1(s)\left(\int_{|s-t|}^{s+t}\frac{2\sin\rho u}{\rho}\,du\right)ds\right)\tilde{f}_2(t)\,dt=$$

$$=\int_0^\infty\left(\int_0^\infty\tilde{f}_1(s)\frac{4\sin\rho s\sin\rho t}{\rho^2}\,ds\right)\tilde{f}_2(t)\,dt=$$

$$=\int_0^\infty\tilde{f}_1(s)\frac{2\sin\rho s}{\rho}\,ds\int_0^\infty\tilde{f}_2(t)\frac{2\sin\rho t}{\rho}\,dt,$$

i.e.,

$$\int_0^\infty g(u)\frac{2\sin\rho u}{\rho}\,du=\int_0^\infty\tilde{f}(u)\frac{2\sin\rho u}{\rho}\,du.$$

In virtue of the completeness of the system of functions $\sin\rho u$, it follows from this that $g(u)=\tilde{f}(u)$.

Now let us find all the maximal ideals of our ring. For this purpose, in what follows we shall denote the elements of the ring by $\tilde{f}(u)$. In terms of $\tilde{f}$, the norm is expressed by the formula

$$\int_0^\infty|\tilde{f}(u)|\sinh 2u\,du,$$

because

$$\int_0^\infty|f(u)|\,\varphi(u)\,du=\int_0^\infty[|f(u)|/\sinh 2u]\,\varphi(u)\cdot\sinh 2u\,du=$$

$$=\int_0^\infty|\tilde{f}(u)|\sinh 2u\,du.$$

Therefore the ring introduced here is isomorphic to the ring discussed in § 50. The maximal ideals of R_0 were examined in that same section. They were given by the formula

$$\int_0^\infty \tilde{f}(u)[2\sin \varrho u]\varrho^{-1}du,$$

where ϱ is the complex number $\varrho = \varrho_1 + i\varrho_2$, with $|\varrho_2| \leqq 2$.

Remark: This implies, among other things, that the remaining spherical functions corresponding to the given group are given by the formula $[2\sinh \varrho u]/[\varrho \sinh 2u]$ $(0 \leqq \varrho \leqq 2)$.

Since the transition from f to f^* corresponds to a replacement of $\tilde{f}(u)$ by $\overline{\tilde{f}}(u)$, it is clear that unsymmetric maximal ideals exist in the ring and consequently that the ring is unsymmetric.

6. Now let us show that the group ring R_0 is unsymmetric.

For this purpose, we note that if the element $f + \lambda e$ belongs to a subring of R_0 and has an inverse, then this inverse also belongs to R_0.

For let the element $\varphi + e$ of R be the inverse of the element $f + e \in R_0$. Let us show that $\varphi + e \in R_0$, i.e., that the function $\varphi(g)$ is constant on the double cosets of $\mathfrak{H}$. We have:

$$(f + e) \times (\varphi + e) = e.$$

Hence

$$\varphi = -f - f \times \varphi,$$

i.e.,

$$\varphi(g) = -f(g) - \int f(gg_1^{-1})\varphi(g_1)dg_1.$$

But then for $h \in \mathfrak{H}$

$$\varphi(hg) = -f(hg) - \int f(hgg_1^{-1})\varphi(g_1)dg_1 = \\ = -f(g) - \int f(gg_1^{-1})\varphi(g_1)dg_1 = \varphi(g),$$

because $f \in R_0'$. Thus, the function $\varphi(g)$ is invariant under left translation by $h \in \mathfrak{H}$. Similarly, using the equation

$$(\varphi + e) \times (f + e) = e,$$

we can show its invariance under a right translation. Therefore

$$\varphi \in R_0', \qquad \varphi + e \in R_0.$$

Assume now that $x \in R_0$ and that $(e + x^*x)^{-1}$ does not exist in R_0'. Such an element x can be found, because R_0 is unsymmetric. But then $(e + x^*x)^{-1}$ also does not exist in R. This completes the proof that the group ring R is unsymmetric.

7. In the paper [84] it is shown that Beurling's Theorem holds in commutative locally compact groups. The example we have given above shows that in non-commutative locally compact groups this theorem does not hold in general.

Let R be a normed ring with an involution. We shall say that the generalized Beurling Theorem holds in R if for every linear functional $f(x)$ in the ring R there exists an indecomposable positive functional that is a weak limit point of functionals $f(xa)$ $(a \in R)$.

THEOREM: *The generalized Beurling Theorem holds in a ring R if and only if R is a symmetric ring.*

Proof of the Necessity: Let R be an unsymmetric ring. Then there exists an element x_0 such that $(e + x_0^* x_0)^{-1}$ does not exist. Therefore, $e + x_0^* x_0$ belongs only to one maximal right ideal I_r of R. Since e does not belong to I_r, there exists a linear functional $f(x)$ such that $f(e) = 1$, $f(x) = 0$ for all $x \in I_r$. For $x \in I_r$, $a \in R$ we also have $xa \in I_r$, and therefore $f(xa) = 0$ for all $x \in I_r$. Hence every weak limit point $f_0(x)$ of the functionals $f_a(x) = f(xa)$ also vanishes on I_r. By assumption, Beurling's Theorem holds, i.e., among these limit points there is a positive normed indecomposable functional $f_0(x)$. Therefore, this functional also vanishes on I_r. In particular,

$$f_0(e + x_0^* x_0) = 0,$$

i.e.,

$$1 + f(x_0^* x_0) = 0;$$

but this is impossible, because $f(x_0^* x_0) \geqq 0$.

Proof of the Sufficiency: Let $f(x)$ be a linear functional. Let I_r denote the set of all elements $x \in R$ such that $f(xa) = 0$ for all $a \in R$. Obviously, I_r is a right ideal in R. According to [87] (see also [80]), there exists a positive functional $f_1(x)$ such that

$$f_1(e) = 1, \qquad f_1(xx^*) = 0 \quad \text{for} \quad x \in I_r. \tag{2}$$

The set of all functionals satisfying these conditions is a weakly closed bounded convex set in the space conjugate to R. By the theorem of Krein and Milman [85], this convex set contains at least one extreme point f_0; f_0 is an indecomposable positive functional satisfying the conditions (2). Hence it follows that $f_0(x) = 0$ for all $x \in I_r$, i.e., $f_0(x)$ is a weak limit point of the functionals $f_0(xa)$.

8. It is not difficult to show that if Beurling's Theorem in the formulation [84] holds in a group, then the Generalized Beurling Theorem holds in the group ring. It would be interesting to find out whether the converse is true.

From what has been shown above, it follows that *in the unimodular group of the second order, Beurling's Theorem does not hold.* Arguments similar to those of this section show that *Beurling's Theorem also fails to hold in every complex semi-simple Lie group.* This is connected with the existence of the so-called supplementary series of representations of these groups. (See the remark in § 52.5.)

Remark: Some of the results of this paper were obtained independently by I. E. Segal (see Bull. Amer. Math. Soc., Vol. 53 (1947), pp. 73-88).

Part III

Differential equations and mathematical physics

1.

Eigenfunction expansions for equations with periodic coefficients

Dokl. Akad. Nauk SSSR 73 (6) (1950) 1117–1120. Zbl. 37: 345

In this paper an elementary proof of eigenfunction Fourier integral expansion theorem for square-integrable functions will be given. So we shall get for second-order differential equations a new and very simple proof of the well-known theorem. For the case of a partial differential equation our is the first one as far as we know.

To make our arguments more concise we shall first explain the case of one independent variable. Extending the results to the case of several independent variables can be done automatically. Our proof is closely related to that of Plancherel theorem for locally compact abelian groups given by A. Weil [1].

1. All the subsequent arguments are based on the following simple lemma.

Lemma. *Every continuous function $f(x)$ which is identically zero outside some finite interval can be written in the form*

$$f(x)=\int_0^{2\pi} f_t(x)dt, \tag{1}$$

where the functions $f_t(x)$ satisfy the condition

$$f_t(x+1)=e^{it}f_t(x) \tag{2}$$

Moreover

$$\int_{-\infty}^{\infty} |f(x)|^2 dx=\frac{1}{2\pi}\int_0^{2\pi}\int_0^1 |f_t(x)|^2 dx dt. \tag{3}$$

This correspondence can be extended to all square-integrable functions.

Proof. Let $f(x)$ be a continuous function equal to zero outside some finite interval. Let

$$f_t(x)=\sum_{n=-\infty}^{\infty} f(x+n)e^{-\text{int}}. \tag{4}$$

In this series only a finite number of terms is not equal to zero and $f_t(x)$ is a continuous function of x and t. It is evident that $f_t(x)$ satisfies the condition (2).

Moreover

$$\frac{1}{2\pi}\int_0^{2\pi} f_t(x)dt=\frac{1}{2\pi}\int_0^{2\pi}\sum_n f(x+n)e^{-\text{int}}dt=f(x).$$

It remains only to prove the equality (3). Since for fixed x $f_t(x)=\sum_n f(x+n)e^{-int}$ is a trigonometric polynomial in t, we have

$$\frac{1}{2\pi}\int_0^1\int_0^{2\pi}|f_t(x)|^2\,dt\,dx=\int_0^1\sum_{n=-\infty}^{\infty}|f(x+n)|^2\,dx=\int_{-\infty}^{\infty}|f(x)|^2\,dx,$$

Let now $f_t(x)$ be a continuous function of variables x and t, $f_t(x+1)=e^{it}f_t(x)$ and $f(x)=\frac{1}{2\pi}\int_0^{2\pi}f_t(x)\,dt$; we have then

$$f(x+n)=\frac{1}{2\pi}\int_0^{2\pi}f_t(x+n)\,dt=\frac{1}{2\pi}\int_0^{2\pi}f_t(x)e^{int}\,dt$$

and therefore $f(x+n)$ are Fourier coefficients of the function $f_t(x)$, which is considered as a function of variable t for fixed x. Therefore

$$\frac{1}{2\pi}\int_0^1\left[\int_0^{2\pi}|f_t(x)|^2\,dt\right]dx=\int_0^1\sum|f(x+n)|^2\,dx=\int_{-\infty}^{\infty}|f(x)|^2\,dx.$$

2. Let us consider a selfadjoint differential operator of the order n which has coefficients that are periodic with respect to x, with period 1. Let us prove that every square integrable function can be expanded to the Fourier integral over all bounded eigenfunctions of this equation. Denote our operator by $L[y]$, consider it only on the interval $(0, 1)$ and set boundary conditions as follows:

$$y^{(k)}(1)=e^{-it}y^{(k)}(0)\qquad (k=0, 1, \ldots, n-1). \tag{5}$$

The operator $L[y]$ defined on functions given in the interval $0\leqq x\leqq 1$ and satisfying the boundary conditions (5), is a selfadjoint differential operator. This is easily seen for example from straightforward integration by parts. As an operator defined on a finite interval, it has a discrete spectrum. Let us denote normalized eigenfunctions of the operator $L[y]$ on the interval with the boundary conditions (5) by $y_{n,t}$ and corresponding eigenvalues by $\lambda_n(t)$. If we consider these eigenfunctions in the whole line then they satisfy the functional equation

$$y_{n,t}(x+1)=e^{-it}y_{n,t}(x). \tag{6}$$

Indeed, due to the boundary conditions (5) the functions $y_{n,t}(t)e^{-it}$ and $y_{n,t}(x+1)$ coincide for $x=0$ and so do their derivatives up to order $n-1$; therefore, the functions coincide everywere.

Let $f(x)$ be an arbitrary square-integrable function defined in the interval $(-\infty, +\infty)$. Then by the lemma, for almost all t in $[0, 2\pi]$, the function $f_t(x)$ from Sect. 1 is defined and $f_t(x)$ is a square integrable function of x, for almost all t. Let

$$a_n(t)=\int_0^1 f_t(x)y_{n,t}(x)\,dx.$$

The functions $a_n(t)$ are the Fourier coefficients of the expansion of the function $f_t(x)$ in the interval $(0, 1)$ over the functions $y_{n,t}(x)$. Suppose for the time being that $f(x)$ is a continuous function equal to zero outside a finite interval. Let us prove that $a_n(t)$ *are also the values of the Fourier transform of the function* $f(x)$ *for the infinite interval.*

Indeed

$$a_n(t)=\int_0^1 f_t(x)y_{n,t}(x)dx=\int_0^1 \sum_{k=-\infty}^{\infty} f(x+k)e^{-ikt}y_{n,t}(x)dx$$

$$=\int_0^1 \sum_{k=-\infty}^{\infty} f(x+k)y_{n,t}(x+k)dx=\int_{-\infty}^{\infty} f(x)y_{n,t}(x)dx. \tag{7}$$

Now we shall prove the Parseval equality on the interval $(-\infty, +\infty)$. Since $a_n(t)$ are the Fourier coefficients of the function $f_t(x)$ we conclude

$$\int_0^1 |f_t(x)|^2 dx=\sum_{n=-\infty}^{\infty} |a_n(t)|^2.$$

Therefore

$$\int_{-\infty}^{\infty} |f(x)|^2 dx=\frac{1}{2\pi}\int_0^1\int_0^{2\pi} |f_t(x)|^2 dx dt=\frac{1}{2\pi}\int_0^{2\pi}\sum_n |a_n(t)|^2 dt, \tag{8}$$

q.e.d.

Conversely $a_n(t)$ be a function of n and t such that

$$\frac{1}{2\pi}\int_0^{2\pi}\sum_n |a_n(t)|^2 dt<+\infty.$$

Let us prove that some square integrable function $f(x)$ corresponds to it. Indeed, the series $\sum_n |a_n(t)|^2$ is convergent for almost all t. For such t the series $\sum_n a_n(t)y_{n,t}(x)$ is convergent in mean with respect to x. Denote its sum by $f_t(x)$. The function $f_t(x)$ is a square integrable one and $\int_0^1 |f_t(x)|^2 dx=\sum_n |a_n(t)|^2$.

Since

$$\frac{1}{2\pi}\int_0^{2\pi}\sum_n |a_n(t)|^2 dt<+\infty, \tag{8'}$$

then according to lemma there also exists $f(x)$ and

$$\int_{-\infty}^{\infty} |f(x)|^2 dx=\frac{1}{2\pi}\int_0^{2\pi}\sum_n |a_n(t)|^2 dt.$$

The correspondence proved above for functions $f(x)$ equal to zero outside a finite interval can certainly be extended due to (8) to all square integrable functions; in such a way an isometric map of the space of all square integrable functions $f(x)$ on the set of all functions $a_n(t)$ with finite (8′) is established.

3. Now consider the case of partial differential equations. For the sake of concreteness we shall now denote by $L[u]$ an elliptic differential operator of the second order with coefficients that are continuous periodic functions of $x_1, \ldots, x_n$ with the period 1 with respect to each variable. Denote by $u_{n,t_1,\ldots,t_k}(x_1, \ldots, x_k)$ eigenfunctions of the operator $L[u]$ defined on the functions in the cube $0 \leqq x_i \leqq 1$ satisfying to the following boundary conditions:

$$e^{-i(x_1 t_1 + \ldots + x_k t_k)} u(x_1, \ldots, x_k)$$

is a periodic function. It is easy to prove that the operator with such boundary conditions is selfadjoint. We can repeat verbatim all arguments of Sect. 2 thus obtaining the Parseval equality for eigenfunctions of the differential operator defined on the functions in the whole space. It is also clear that $L[u]$ need not be a differential operator of the second order. The only impotant thing is that spectrum of the operator with the boundary conditions (9) be discrete. Discreteness of the spectrum of the operator $L[u]$ on the set of functions which are defined in the set $0 \leqq x_i \leqq 1$ and satisfy the boundary conditions (9) can be proved by a construction of the Green's function of the differential operator. This function can for example be constructed in such a way it was done by Hilbert in the case of the sphere by "parametrix" method.

Repeating verbatim the arguments given in Sect. 2 we obtain the equality

$$\frac{1}{(2\pi)^k} \sum_n \ldots \int_0^{2\pi} \int_0^{2\pi} |a_n(t_1, \ldots, t_k)|^2 \, dt_1 \ldots dt_k = \int_{-\infty}^{+\infty} \ldots \int_{-\infty}^{+\infty} |f(x_1, \ldots, x_k)|^2 \, dx_1 \ldots dx_k. \tag{8''}$$

Here $a_n(t_1, \ldots, t_k)$ is the Fourier transform of f, i.e.

$$a_n(t_1, \ldots, t_k) = \int_{-\infty}^{\infty} \ldots \int_{-\infty}^{\infty} f(x_1, \ldots, x_n) u_{n,t_1,\ldots,t_k}(x_1, \ldots, x_k) \, dx_1 \ldots dx_k.$$

References

1. Weil, A.: L'intégration dans les groupes topologiques et ses applications.

Received 13. VI. 1950

2.

(with B. M. Levitan)

On the determination of a differential equation from its spectral function

Izv. Akad. Nauk SSSR, Ser. Mat. **15** (1951) 309–361
[Transl., II. Ser., Am. Math. Soc. 1 (1955) 253–304]. Zbl. **44**:93

In this work there is given a method of reconstructing a differential equation of the second order from its spectral function $\rho(\lambda)$. This problem is reduced here to a certain linear integral equation. It is also explained how monotonic functions $\rho(\lambda)$ may serve as spectral functions of a differential equation of second order.

Introduction

We shall consider a differential equation of the second order given on the interval $(0, \infty)$,

$$y'' + (\lambda - q(x))\, y = 0 \tag{1}$$

with boundary conditions

$$y(0) = 1, \quad y'(0) = h. \tag{1'}$$

The function $q(x)$ is assumed to be continuous on any finite interval. It is known that an arbitrary function with integrable square may be expressed as a Fourier integral in the eigenfunctions of this equation for a given h. More precisely, there exists a monotonic function $\rho(\lambda)$, bounded on each finite interval, such that for any function $f(x)$ with integrable square the following equation holds:

$$\int_0^{+\infty} f^2(x)\, dx = \int_{-\infty}^{+\infty} E^2(\lambda)\, d\rho(\lambda). \tag{2}$$

Here $E(\lambda)$ is the Fourier transform of the function $f(x)$, i.e.,

$$E(\lambda) = \int_0^{\infty} f(x)\, \varphi(x, \lambda)\, dx, \; * \tag{3}$$

where $\phi(x, \lambda)$ is the solution of equation (1) with the initial conditions (1′).

We shall call the function $\rho(\lambda)$ the *spectral function* of equation (1) with conditions (1′). The meaning of the spectral function $\rho(\lambda)$ is particularly clear in the case of a discrete spectrum. In this case $\rho(\lambda)$ is the jump function with jumps at the spectral points $\lambda_1, \lambda_2, \cdots, \lambda_n, \cdots$. Moreover, the amount of the jumps equals

$$\frac{1}{\int_0^{\infty} \varphi^2(x, \lambda_n)\, dx}$$

and thus equation (2) becomes Parseval's equation for the Fourier series of the function $f(x)$.

Consequently, one may say that giving $\rho(\lambda)$ determines both the spectrum and the normalization of the eigenfunctions.

* For the sense in which this integral converges see §1.

This work is devoted to the solution of the following problem. Let a function $\rho(\lambda)$ be given. It is required to determine whether there exists an equation of the form (1) having the given $\rho(\lambda)$, and to give a method of actual computation of $q(x)$. The first problems of this type were treated by Ambarcumyan ([1]). The uniqueness of solution of the problem in this formulation was established by V. A. Marčenko ([2]). M. G. Kreĭn ([3]) then took up the question of the solvability of the inverse problem. By making use of his theory of the extension of positive definite functions, M. G. Kreĭn reduced to it the problem considered here and obtained a series of very interesting and important results. In particular, he completely solved by these methods the question of necessary and sufficient conditions for the existence of a cord with a distribution of mass having given spectra for two boundary conditions ([3a]).

In the present work, the above question is attacked with elementary means, namely, by reducing it to the solution of a certain integral equation. The idea of this work is very simple. Similarly to the way in which polynomials, orthogonal with respect to a given weight function, are constructed by orthogonalizing the powers of x, we construct from the spectral function $\rho(\lambda)$ the eigenfunctions $\phi(x, \lambda)$ by orthogonalizing the functions $\cos\sqrt{\lambda t}$ with respect to $\rho(\lambda)$.

We explain this in somewhat more detail. As is known, for the equation $y'' + \lambda y = 0$ and the boundary condition $y(0) = 1$, $y'(0) = 0$ we have $\rho(\lambda) = (2/\pi)\sqrt{\lambda}$ for $\lambda > 0$ and $\rho(\lambda) = 0$ for $\lambda < 0$.

In order to make the exposition stand out more we shall assume for a time that the function $\rho(\lambda)$ has the form

$$\rho(\lambda) = \begin{cases} \frac{2}{\pi}\sqrt{\lambda} + \sigma(\lambda) & \text{for} \quad \lambda > 0, \\ \sigma(\lambda) & \text{for} \quad \lambda < 0, \end{cases}$$

where the function $\sigma(\lambda)$ behaves itself sufficiently well at infinity; namely, we shall suppose that

$$\int_{-\infty}^{+\infty} |\lambda| \cdot |d\sigma(\lambda)| < +\infty.$$

We shall look for the eigenfunction $\phi(x, \lambda)$ in the form*

$$\varphi(x, \lambda) = \cos\sqrt{\lambda}\, x + \int_0^x K(x, t) \cos\sqrt{\lambda}\, t\, dt. \tag{4}$$

Here we introduce nonrigorous considerations which are justified for a wide class of functions $\rho(\lambda)$ in this work. We require that $\phi(x, \lambda)$ be an "orthonormal

* The fact that the eigenfunctions may be so represented in proved in ([4]) and ([5]). For a simpler proof of this see §1 of the present work

system'' with weight $\rho(\lambda)$, i.e., that

$$\int_{-\infty}^{+\infty} \varphi(x, \lambda)\varphi(y, \lambda)\, d\rho(\lambda) = \delta(x-y),$$

where $\delta(x)$ is Dirac's δ-function. Since, because of (4), for $y < x$, $\phi(y,\lambda)$ is a combination of $\cos\sqrt{\lambda t}$ for $t \leq y$, then, inverting equation (4), we may convince ourselves that conversely $\cos\sqrt{\lambda y}$ is a combination of $\phi(t,\lambda)$ for $t \leq y$. Thus, $\cos\sqrt{\lambda y}$ is orthogonal to $\phi(x,\lambda)$ if $y < x$:

$$\int_{-\infty}^{+\infty} \varphi(x, \lambda)\cos\sqrt{\lambda} y\, d\rho(\lambda) = 0, \quad \text{if} \quad y < x.$$

Let us substitute here in place of $\phi(x,\lambda)$ its value from equation (4). We obtain

$$\int_{-\infty}^{+\infty} \cos\sqrt{\lambda}x\cos\sqrt{\lambda}y\, d\rho(\lambda) + \int_{-\infty}^{+\infty} \cos\sqrt{\lambda}y\int_0^x K(x, t)\cos\sqrt{\lambda}t\, dt\, d\rho(\lambda) = 0. \quad (5)$$

In this form the integral we have written down diverges. However, the equation may be given a precise meaning; namely, we represent $\rho(\lambda)$ in the form

$$\rho(\lambda) = \frac{2}{\pi}\sqrt{\lambda} + \sigma(\lambda),$$

where $\lambda > 0$ and note that the following symbolic equation holds:

$$\frac{2}{\pi}\int_0^{+\infty} \cos\sqrt{\lambda}t\cos\sqrt{\lambda}y \cdot d\sqrt{\lambda} = \delta(y-t),$$

$$y > 0, \quad t > 0.$$

Then, changing the order of integration in (5) and making use of the symbolic equation

$$\int_0^{+\infty} \delta(y-t) f(t)\, dt = f(y),$$

we obtain (noting that $x \neq y$ and, consequently, that $\delta(x-y) = 0$)

$$\int_{-\infty}^{+\infty} \cos\sqrt{\lambda}x\cos\sqrt{\lambda}y\, d\sigma(\lambda) + K(x, y) +$$

$$+\int_0^x K(x, t)\left[\int_{-\infty}^{+\infty} \cos\sqrt{\lambda}t\cos\sqrt{\lambda}y\, d\sigma(\lambda)\right] dt = 0. \quad (6)$$

Since the function $\sigma(\lambda)$ is given, the function

$$f(x, y) = \int_{-\infty}^{+\infty} \cos\sqrt{\lambda}x\cos\sqrt{\lambda}y\, d\sigma(\lambda)$$

is also known.

Equation (6) for fixed x is a linear integral equation for the unknown function $K(x, y)$ and has the form*

$$f(x, y) + \int_0^x f(y, t) K(x, t)\, dt + K(x, y) = 0. \tag{7}$$

We show that equation (7) is solvable. In finding $K(x, t)$ we may, from formula (4), also find the eigenfunctions $\phi(x, \lambda)$ and consequently reconstruct equation (1).

In order to show that the constructed functions $\phi(x, \lambda)$ satisfy some differential equation, we proceed in the following manner.

First we prove (§5) that the constructed functions $\phi(x, \lambda)$ satisfy the Parseval equation defined by formulas (2) and (3) above. Making use of this and the integral equation (7) we then obtain in §7 a functional equation for the functions $\phi(x, \lambda)$:

$$\varphi(x, \lambda) \cos \sqrt{\lambda} t = \frac{1}{2}\left[\varphi(x+t, \lambda) + \varphi(x-t, \lambda)\right] + \int_{|x-t|}^{x+t} W(x, t, s) \varphi(s, \lambda)\, ds, \tag{8}$$

where the function $W(x, t, s)$ is constructed from the function $K(x, t)$. In order to obtain a differentiable equation we subtract $\phi(x, \lambda)$ from both sides of equation (8) and divide by t^2. We obtain

$$\varphi(x, \lambda) \frac{\cos \sqrt{\lambda} t - 1}{t^2} = \frac{\varphi(x+t, \lambda) - 2\varphi(x, \lambda) + \varphi(x-t, \lambda)}{2t^2} + \\ + \frac{1}{t^2} \int_{|x-t|}^{x+t} W(x, t, s) \varphi(s, \lambda)\, ds.$$

Passing to the limit as $t \to 0$, we obtain

$$-\frac{1}{2} \lambda \varphi(x, \lambda) = \frac{1}{2} \varphi''(x, \lambda) - \frac{1}{2} q(x) \varphi(x, \lambda),$$

where we have written

$$-\frac{1}{2} q(x) = \lim_{t \to 0} \frac{1}{t^2} \int_{x-t}^{x+t} W(x, t, s)\, ds.$$

The existence of all limits is proved in §7.

We remark that the function $q(x)$ in equation (1) may be immediately constructed from $K(x, t)$ by means of the following formula deduced in §1:

$$\frac{1}{2} q(x) = \frac{dK(x, x)}{dx}.$$

For an approximate solution of the stated problem, one may use the nonlinear Volterra integral equation derived in §2.

* A rigorous deduction of this equation and precise conditions on the function $\sigma(\lambda)$ are given in §§3, 4.

We now formulate the basic result of this work. In the work there are given, in a known sense, necessary and sufficient conditions for the existence of an equation with a given function $\rho(\lambda)$. In fact, these conditions are sufficient for the existence of an equation with a continuous function $q(x)$, and are necessary if $q(x)$ has a continuous derivative. The following theorem is proved.

Theorem. *Let a monotonic function $\rho(\lambda)$ be given on the whole axis. Let us represent it in the form*

$$\rho(\lambda)=\begin{cases}\frac{2}{\pi}\sqrt{\lambda}+\sigma(\lambda) & \text{for} \quad \lambda\geqslant 0,\\ \sigma(\lambda) & \text{for} \quad \lambda\leqslant 0.\end{cases}$$

Let us suppose that $\rho(\lambda)$ satisfies the following conditions.

1. *For each $x>0$ there exists the integral*

$$\int_{-\infty}^{0} e^{\sqrt{|\lambda|}\,x}\,d\rho(\lambda).$$

2. *The function*

$$a(x)=\int_{1}^{\infty}\frac{\cos\sqrt{\lambda}\,x}{\lambda}\,d\sigma(\lambda)$$

has a continuous fourth derivative.

If the function $\rho(\lambda)$ satisfies conditions 1 *and* 2 *then there exists a continuous function $q(x)$ and an h such that $\rho(\lambda)$ is the spectral function of equation* (1) *with boundary conditions* (1′). *Conversely, if the function $q(x)$ has a continuous derivative then the spectral function $\rho(\lambda)$ corresponding to it satisfies conditions* 1 *and* 2.

From the function $\sigma(\lambda)$ it is possible to determine the degree of smoothness of $q(x)$, namely, if the function $a(x)$ has derivatives of order $n+4$, then $q(x)$ has n derivatives. Conversely, one may show that if $q(x)$ has an $(n+1)$th continuous derivative, then $a(x)$ has continuous derivatives of order $n+4$.

Since all the conditions imposed upon the spectral function $\rho(\lambda)$ refer only to its behavior at $+\infty$, one may make the following interesting deduction. *The spectral function $\rho(\lambda)$ may be an arbitrary monotonic function on a finite interval.* Moreover, if on a finite interval one takes $\rho(\lambda)$ as an arbitrary monotonic function and extends it so that, beginning with some $\lambda>0$, it has the form $(2/\pi)\sqrt{\lambda}$, then the function $f(x,y)$, and thus also $K(x,y)$ and consequently $q(x)$, will be analytic functions. In this manner one may construct an equation with analytic coefficients and with a spectral function $\rho(\lambda)$ as badly behaved as one wishes upon a finite interval.

The proofs carried out in this article are sometimes cumbersone, but are simple in concept. All proofs may be significantly simplified if it is assumed that for $\lambda > 0$

$$\rho(\lambda) = \frac{2}{\pi}\sqrt{\lambda} + \sigma(\lambda),$$

where

$$\int_1^\infty \lambda\,|d\sigma(\lambda)| < +\infty,$$

for the case of a continuous spectrum (and the analogous inequality for the case of the pure point spectrum).

In §8 we examine the case when the spectral function is an orthogonal spectral function. This is analogous to the so-called determined case of the moment problem. We note that for the indeterminate case we have found all (and not only the orthogonal) spectral functions. We also note that for the case when the whole spectrum lies on the semi-axis (a, ∞) one always has the determined case (see §8).

In the last paragraph we consider the classical problem of eigenvalues for the equation

$$y'' + (\lambda - q(x))\,y = 0$$

on the finite interval (a, b) and with boundary conditions at a and b. If we designate by $\phi_n(x)$ the eigenfunctions satisfying the conditions

$$\varphi_n(a) = 1, \quad \varphi_n'(a) = 0,$$

and if λ_n are the eigenvalues and

$$\rho_n = \int_a^b \varphi_n^2(x)\,dx,$$

then the result obtained may be formulated as follows: *For every sequence of numbers λ_n and ρ_n satisfying the usual asymptotic equalities one may construct a function $q(x)$.* For more details see the corresponding paragraph.

§1. Some facts about differential equations of the second order

1. Let $q(x)$ $(0 \le x \le \infty)$ be continuous in each finite interval. Let us consider the differential equation

$$y'' + (\lambda - q(x))\,y = 0, \tag{1.1}$$

where λ is a real number. Let us designate by $\phi_1(x, \lambda)$ the solution of equation (1.1) satisfying the initial conditions

$$\varphi_1(0, \lambda) = 1, \quad \varphi_1'(0, \lambda) = 0,$$

and by $\phi_2(x, \lambda)$ the solution of the same equation satisfying the initial conditions

$$\varphi_2(0, \lambda) = 0, \quad \varphi_2'(0, \lambda) = 1.$$

Further, let us designate by h an arbitrary real number. Then the function

$$\varphi(x, \lambda) = \varphi_1(x, \lambda) + h\varphi_2(x, \lambda)$$

is also a solution of equation (1.1) and satisfies the boundary condition

$$\varphi'(0, \lambda) - h\varphi(0, \lambda) = 0. \tag{1.2}$$

If $h = 0$, then $\phi(x, \lambda) = \phi_1(x, \lambda)$. For $h = \infty$, we assume by definition that $\phi(x, \lambda) = \phi_2(x, \lambda)$.

It is known that for each function $q(x)$ and for each fixed h there exists at least one nondecreasing function $\rho(\lambda)$ $(-\infty \leq \lambda \leq +\infty)$ such that, for each function $f(x) \in L_2(0, \infty)$, the functions

$$E_n(\lambda) = \int_0^n f(x)\varphi(x, \lambda)\,dx$$

converge in the mean square with the measure $\rho(\lambda)$ to some function $E(\lambda)$, i.e.,

$$\lim_{n\to\infty} \int_{-\infty}^{+\infty} [E(\lambda) - E_n(\lambda)]^2\,d\rho(\lambda) = 0.$$

In addition, Parseval's equality holds:

$$\int_{-\infty}^{+\infty} E^2(\lambda)\,d\rho(\lambda) = \int_0^{+\infty} f^2(x)\,dx. \tag{1.3}$$

The function $\rho(\lambda)$ is called the *spectral function* of equation (1.1) for conditions (1.2).

2. In this paragraph we derive formulas expressing the eigenfunctions $\phi(x, \lambda)$ of equation (1.1) in terms of $\cos\sqrt{\lambda}\,x$, and analogous formulas expressing $\cos\sqrt{\lambda x}$ in terms of $\phi(x, \lambda)$. These formulas were considered earlier in ([4]) and ([5]). A new derivation of these formulas is given here.

We shall suppose that there exists the function $K(x, t)$ $(t \leq x)$ having continuous partial derivatives of first and second order and such that

$$\varphi(x, \lambda) = \cos\sqrt{\lambda}\,x + \int_0^x K(x, t)\cos\sqrt{\lambda}\,t\,dt. \tag{1.4}$$

We shall explain what conditions the function $K(x, t)$ must satisfy in order that $\phi(x, \lambda)$ be a solution of equation (1.1). Differentiating (1.4) twice with respect to x, we obtain:

$$\varphi''(x, \lambda) = -\lambda\cos\sqrt{\lambda}x + \frac{dK(x, x)}{dx}\cos\sqrt{\lambda}x - \sqrt{\lambda}K(x, x)\sin\sqrt{\lambda}x + \\ + \left.\frac{\partial K(x, t)}{\partial x}\right|_{t=x}\cos\sqrt{\lambda}x + \int_0^x \frac{\partial^2 K(x, t)}{\partial x^2}\cos\sqrt{\lambda}t\,dt. \tag{1.5}$$

Integrating the expression

$$\lambda\int_0^x K(x, t)\cos\sqrt{\lambda}\,t\cdot dt$$

twice by parts, we obtain:

$$\lambda \int_0^x K(x,t)\cos\sqrt{\lambda}t\,dt = \sqrt{\lambda}\,\sin\sqrt{\lambda}x\;K(x,x) + \frac{\partial K(x,t)}{\partial t}\Big|_{x=t}\cos\sqrt{\lambda}x -$$

$$-\frac{\partial K(x,t)}{\partial t}\Big|_{t=0} - \int_0^x \frac{\partial^2 K(x,t)}{\partial t^2}\cos\sqrt{\lambda}t\;dt.$$

Substituting this expression and (1.5) in equation (1.1), we obtain after reduction

$$\frac{dK(x,x)}{dx}\cdot\cos\sqrt{\lambda}x + \left(\frac{\partial K(x,t)}{\partial t} + \frac{\partial K(x,t)}{\partial x}\right)_{x=t}\cos\sqrt{\lambda}x + \frac{\partial K(x,t)}{\partial t}\Big|_{t=0} -$$

$$-q(x)\cos\sqrt{\lambda}x + \int_0^x\left[\frac{\partial^2 K}{\partial x^2} - \frac{\partial^2 K}{\partial t^2} - q(x)K(x,t)\right]\cos\sqrt{\lambda}t\,dt = 0.$$

In view of the uniqueness of the representation of a function as a Fourier-Stieltjes integral, there follows from the last equation the following partial differential equation:

$$\frac{\partial^2 K(x,t)}{\partial x^2} - q(x)K(x,t) = \frac{\partial^2 K(x,t)}{\partial t^2} \tag{1.6}$$

and the boundary conditions:

$$\frac{\partial K(x,t)}{\partial t}\Big|_{t=0} = 0, \tag{1.7}$$

$$\frac{dK(x,x)}{dx} = \frac{1}{2}q(x). \tag{1.8}$$

In order to find the function $K(x,x)$ from (1.8) one must know $K(0,0)$. If $\phi(x,\lambda)$ is represented in the form (1.4), then it is obvious that

$$\varphi(0,\lambda) = 1, \quad \varphi'(0,\lambda) = K(0,0). \tag{1.8}$$

Thus, in order that $\phi(x,\lambda)$ satisfy the boundary condition (1.2) one must set $K(0,0) = h$. Thus, from (1.8) follows

$$K(x,x) = h + \frac{1}{2}\int_0^x q(t)\,dt. \tag{1.9}$$

If the function $q(x)$ has a continuous derivative then, as is known, there exists a unique solution of equation (1.6) satisfying conditions (1.7) and (1.9). Hence there exists a function $K(x,t)$ satisfying (1.4). Solving equation (1.4) as a Volterra equation with unknown function $\cos\sqrt{\lambda t}$, we obtain

$$\cos\sqrt{\lambda}x = \varphi(x,\lambda) - \int_0^x K_1(x,t)\varphi(t,\lambda)\,dt. \tag{1.10}$$

Analogously we may show that $K_1(x,t)$ is the solution of the equation

$$\frac{\partial^2 K_1(x,t)}{\partial x^2} = \frac{\partial^2 K_1(x,t)}{\partial t^2} - q(t)K_1(x,t),$$

satisfying the conditions

$$\left(\frac{\partial K_1}{\partial t} - hK_1\right)_{t=0} = 0,$$

$$K_1(x, x) = h + \frac{1}{2}\int_0^x q(x)\,dx.$$

It is clear that formulas (1.4) and (1.10) also hold for $\lambda < 0$.

Furthermore, it is evident from formula (1.4) that, knowing $K(x, t)$, we also the eigenfunctions $\phi(x, \lambda)$ of equation (1.1), and then the equation itself since

$$\lambda - q(x) = -\frac{\varphi''(x, \lambda)}{\varphi(x, \lambda)}.$$

3. Let us now consider separately the case $h = \infty$. We set

$$\psi(x, \lambda) = \frac{\sin\sqrt{\lambda}x}{\sqrt{\lambda}} + \int_0^x L(x, t)\frac{\sin\sqrt{\lambda}t}{\sqrt{\lambda}}\,dt. \tag{1.4'}$$

Reasoning as above, we obtain for $L(x, t)$ the equation

$$\frac{\partial^2 L(x, t)}{\partial x^2} - q(x)L(x, t) = \frac{\partial^2 L(x, t)}{\partial t^2}$$

and the initial conditions

$$L(x, 0) = 0,$$

$$L(x, x) = \frac{1}{2}\int_0^x q(x)\,dx.$$

Solving equation (1.4′) for the unknown function $\frac{\sin\sqrt{\lambda t}}{\sqrt{\lambda}}$, we obtain

$$\frac{\sin\sqrt{\lambda}x}{\sqrt{\lambda}} = \psi(x, \lambda) - \int_0^x L_1(x, t)\psi(t, \lambda)\,dt. \tag{1.10'}$$

Thus, we have obtained formulas analogous to formulas (1.4) and (1.10). Also, the functions $L(x, t)$ and $L_1(x, t)$ have the same properties as the functions $K(x, t)$ and $K_1(x, t)$.

We assumed in paragraphs 2 and 3 that the function $q(x)$ had a continuous derivative. In this case, the functions $K(x, t)$ and $K_1(x, t)$ from paragraph 2 and the analogous functions from paragraph 3 have continuous second derivatives. If one assumes that $q(x)$ is simply continuous, one may proceed in the following manner. We approximate $q(x)$ by a sequence of functions $q_n(x)$ having a continuous derivative. The corresponding functions $K_n(x, t)$ have continuous second derivatives and converge to the function $K(x, t)$. The function $K(x, t)$ itself is known to have first derivatives but does not have second derivatives. However, the functions $\phi(x, \lambda)$ constructed from $K(x, t)$ have second derivatives and satisfy equation (1.1). Actually,

$$\varphi_n''(x, \lambda) + (\lambda - q_n(x))\varphi_n(x, \lambda) = 0.$$

Since $K_n \to K$, the functions ϕ_n converge uniformly in x to $\phi(x, \lambda)$, which means that the $\phi_n''(x, \lambda)$ converge uniformly also, i.e., $\phi(x, \lambda)$ has a continuous second derivative and satisfies equation (1.1).

§2. Derivation of the nonlinear integral equation

1. Let us consider first the case $h = \infty$. Integrating both sides of formula (1.10) from 0 to x and interchanging the order of integration, we obtain

$$\frac{\sin\sqrt{\lambda}\,x}{\sqrt{\lambda}} = \int_0^x \varphi(t, \lambda)\,dt - \int_0^x \varphi(t, \lambda)\,dt \int_t^x K_1(u, t)\,du =$$

$$= \int_0^x \varphi(t, \lambda)\left[1 - \int_t^x K_1(u, t)\,du\right] dt.$$

Let us fix the value of x in this formula. Then the above equation shows that* $\frac{\sin\sqrt{\lambda x}}{\sqrt{\lambda}}$ is a Fourier transform in the eigenfunctions of equation (1.1) of the following function of t

$$\Phi(x, t) = \begin{cases} 1 - \int_t^x K_1(u, t)\,du, & t \leqslant x, \\ 0, & t > x. \end{cases}$$

From Parseval's equation (1.3) follows

$$\int_{-\infty}^{+\infty} \frac{\sin\sqrt{\lambda}\,x \cdot \sin\sqrt{\lambda}\,y}{\lambda}\,d\rho(\lambda) = \int_0^y \Phi(x, t)\,\Phi(y, t)\,dt =$$

$$= \int_0^y dt - \int_0^y dt \int_t^x K_1(u, t)\,du - \int_0^y dt \int_t^y K_1(u, t)\,dt +$$

$$+ \int_0^y dt \int_t^x K_1(u,t)\,du \int_t^y K_1(v,t)\,dv.$$

If in the last formula one places $x = y$, one obtains (since the right side has meaning) that for each $x > 0$ there exists the integral

$$\int_{-\infty}^{0} \frac{\operatorname{sh}^2\sqrt{|\lambda|}\,x}{|\lambda|}\,d\rho(\lambda),$$

that is to say, the integral

$$\int_{-\infty}^{0} \operatorname{ch}\sqrt{|\lambda|}\,x\,d\rho(\lambda).$$

This fact has been mentioned earlier by V. A. Marčenko (2).

* For $\lambda < 0$ one understands $\frac{\sin\sqrt{\lambda x}}{\sqrt{\lambda}}$ to be $\frac{\operatorname{sh}\sqrt{|\lambda| x}}{\sqrt{|\lambda|}}$.

Let us now proceed to the derivation of the integral equation for $K_1(x, y)$. Since

$$\frac{\sin\sqrt{\lambda}\,x}{\sqrt{\lambda}} = \int_0^x \cos\sqrt{\lambda}\,t\,dt,$$

it follows from the Parseval formula for ordinary Fourier integrals $(y \leq x)$ that

$$\int_0^y dt = \frac{2}{\pi}\int_0^\infty \frac{\sin\sqrt{\lambda}\,x\sin\sqrt{\lambda}\,y}{\lambda}\,d(\sqrt{\lambda}).$$

Consequently $(y \leq x)$

$$F(x, y) = \int_{-\infty}^{+\infty} \frac{\sin\sqrt{\lambda}\,x\cdot\sin\sqrt{\lambda}\,y}{\lambda}\,d\sigma(\lambda) = -\int_0^y dt \int_t^x K_1(u, t)\,du - \\ - \int_0^y dt \int_t^y K_1(u, t)\,du + \int_0^y dt \int_t^x K_1(u, t)\,du \int_t^y K_1(v, t)\,dv, \qquad (2.1)$$

where

$$\sigma(\lambda) = \begin{cases} \rho(\lambda) - \frac{2}{\pi}\sqrt{\lambda}, & \lambda \geqslant 0, \\ \rho(\lambda), & \lambda < 0. \end{cases}$$

Since the right-hand side of the equation has a mixed derivative, $\partial^2 F/\partial x\partial y$ exists and, differentiating both sides of equation (2.1), we obtain for $K_1(x, y)$ the non-linear integral equation $(y \leq x)$:

$$\frac{\partial^2 F}{\partial x\,\partial y} = f(x, y) = -K_1(x, y) + \int_0^y K_1(x, t)\,K_1(y, t)\,dt. \qquad \text{(I)}$$

By $F(x, y)$ we understand

$$F(x, y) = \int_{-\infty}^{+\infty} \frac{\sin\sqrt{\lambda}\,x\cdot\sin\sqrt{\lambda}\,y}{\lambda}\,d\sigma(\lambda). \qquad (2.2)$$

In case $\sigma(\lambda)$ behaves sufficiently well at $+\infty$ (for example, var $[\sigma(\lambda)] < +\infty$), $f(x, y)$ may be given immediately by the formula

$$f(x, y) = \frac{\partial^2 F}{\partial x\partial y} = \int_{-\infty}^{+\infty} \cos\sqrt{\lambda}\,x.\cos\sqrt{\lambda}\,y\,d\sigma(\lambda). \qquad (2.2')$$

Since $K_1(x, x-0)$ is a continuous function and $f(x, y) = f(y, x)$, it follows from equation (I) that the function $f(x, y)$ is continuous for all values of the arguments.

Let ϵ be an arbitrary positive number. It follows from formula (2.1) that the following limit exists uniformly in each finite region:

$$\lim_{\varepsilon\to 0}\frac{F(x+\varepsilon,\ y+\varepsilon)-F(x-\varepsilon,\ y+\varepsilon)-F(x+\varepsilon,\ y-\varepsilon)+F(x-\varepsilon,\ y-\varepsilon)}{4\varepsilon^2}=$$

$$=\frac{\partial^2 F}{\partial x\,\partial y}=\lim_{\varepsilon\to 0}\int_{-\infty}^{+\infty}\left(\frac{\sin\varepsilon\sqrt{\lambda}}{\varepsilon\sqrt{\lambda}}\right)^2\cos\sqrt{\lambda}\,x\cos\sqrt{\lambda}\,y\,d\sigma(\lambda).$$

In particular, setting $x = y$ we obtain uniformly in each finite interval for $\epsilon\to 0$

$$\int_{-\infty}^{+\infty}\frac{\sin^2\varepsilon\sqrt{\lambda}}{\lambda}\cos^2\sqrt{\lambda}\,x\,d\sigma(\lambda)=o(\varepsilon).$$

For $x = 0$ we obtain

$$\int_{-\infty}^{+\infty}\frac{\sin^2\varepsilon\sqrt{\lambda}}{\lambda}\,d\sigma(\lambda)=o(\varepsilon).$$

From the last two equations it follows that uniformly in each finite interval

$$\int_{-\infty}^{+\infty}\frac{\sin^2\varepsilon\sqrt{\lambda}}{\lambda}\cos\sqrt{\lambda}\,x\,d\sigma(\lambda)=o(\varepsilon). \tag{2.3}$$

Condition (2.3) will play an important role hereafter.

2. Analogously one derives an integral equation for $h=\infty$. Integrating both sides of equation (1.9′) with the limits 0 and x, we obtain

$$\frac{1-\cos\sqrt{\lambda}\,x}{\lambda}=\int_0^x\psi(t,\lambda)\,\Phi(x,t)\,dt,$$

where

$$\Phi(x,t)=\begin{cases}1-\int_t^x L_1(u,t)\,du, & t\leqslant x,\\ 0, & t>x.\end{cases}$$

Consequently, it follows from Parseval's equation $(y\leq x)$:

$$\int_{-\infty}^{+\infty}\frac{(1-\cos\sqrt{\lambda}\,x)\,(1-\cos\sqrt{\lambda}\,y)}{\lambda^2}\,d\rho(\lambda)=\int_0^y dt-\int_0^y dt\int_t^x L_1(u,t)\,du-$$

$$-\int_0^y dt\int_t^y L_1(u,t)\,dt+\int_0^y dt\int_t^x L_1(u,t)\,du\int_t^y L_1(v,t)\,dv.$$

Since

$$\int_0^x\frac{\sin\sqrt{\lambda}\,t}{\sqrt{\lambda}}\,dt=\frac{1-\cos\sqrt{\lambda}\,x}{\lambda},$$

it follows from Parseval's formula for ordinary Fourier integrals that for $y\leq x$

$$\int_0^y dt=\frac{2}{3\pi}\int_0^\infty\frac{(1-\cos\sqrt{\lambda}\,x)\,(1-\cos\sqrt{\lambda}\,y)}{\lambda^2}\,d(\lambda^{3/2}).$$

Consequently,

$$F(x, y) = \int_{-\infty}^{+\infty} \frac{(1-\cos\sqrt{\lambda}x)(1-\cos\sqrt{\lambda}y)}{\lambda^2} d\sigma(\lambda) = -\int_0^y dt \int_t^x L_1(u, t)\, du -$$

$$-\int_0^y dt \int_t^y L_1(u, t)\, du + \int_0^y dt \int_t^x L_1(u, t)\, du \int_t^y L_1(v, t)\, dv,$$

where

$$\sigma(\lambda) = \begin{cases} \rho(\lambda) - \frac{2}{3\pi}\lambda^{3/2}, & \lambda \geqslant 0 \\ \rho(\lambda), & \lambda < 0. \end{cases}$$

Reasoning as in paragraph 1, we obtain for $L_1(x, y)$ the nonlinear integral equation $(y \leq x)$:

$$\frac{\partial^2 F(x, y)}{\partial x\, \partial y} = f(x, y) = -L_1(x, y) + \int_0^y L_1(x, t) L_1(y, t)\, dt.$$

In place of condition (2.3) we obtain the condition that for $\epsilon \to 0$

$$\int_{-\infty}^{+\infty} \frac{\sin^2 \varepsilon\sqrt{\lambda}}{\lambda} \cdot \frac{\sin^2 \sqrt{\lambda}x}{\lambda}\, d\sigma(\lambda) = o(\varepsilon) \tag{2.3'}$$

uniformly in each finite interval.

§3. Derivation of the linear integral equation

For each function $K(x, t)$ carrying $\cos\sqrt{\lambda x}$ into $\phi(x, \lambda)$ [formula (1.8)] we introduce a linear integral equation which is basic for the following work. It is easy to obtain it from the nonlinear equation. However, we shall derive it directly since further on (in the construction of $K(x, y)$ from $\rho(\lambda)$) we shall make use of the reasoning introduced below.

This equation has the following form:

$$f(x, y) + \int_0^x K(x, s) f(s, y)\, ds + K(x, y) = 0 \quad (y \leqslant x), \tag{II}$$

where the function $f(x, y)$ is defined in the preceding section.

Let us proceed to the derivation of this equation. We show first that for $b < y < a < x$ the functions

$$\int_a^x \varphi(t, \lambda)\, dt \quad \text{and} \quad \int_b^y \cos\sqrt{\lambda}\, t\, dt$$

are orthogonal with the weight $\rho(\lambda)$, i.e., that

$$I = \int_{-\infty}^{+\infty} \left[\int_a^x \varphi(t, \lambda)\, dt\right] \left[\int_b^y \cos\sqrt{\lambda}\, t\, dt\right] d\rho(\lambda) = 0. \tag{3.1}$$

In order to derive (3.1) we express $\cos\sqrt{\lambda t}$ in terms of $\phi(t,\lambda)$ according to formula (1.10):

$$h(b, y) = \int_b^y \cos\sqrt{\lambda}\, t\, dt =$$

$$= \int_b^y \varphi(t, \lambda)\, dt - \int_b^y dt \int_0^t K_1(t, s)\, \varphi(s, \lambda)\, ds =$$

$$= \int_b^y \varphi(t, \lambda)\, dt - \int_0^b \varphi(s, \lambda)\, ds \int_b^y K_1(t, s)\, dt - \int_b^y \varphi(s, \lambda)\, ds \int_s^y K_1(t, s)\, dt.$$

Consequently, the function $h(b, y)$ is a Fourier transform (in the functions $\phi(t,\lambda)$) of a function equal to 0 outside of the interval (b, y). Since the intervals (b, y) and (a, x) do not intersect, one has from Parseval's equation (1.3) $I = 0$.

In order to derive the integral equation (II) we now express in equation (3.1) $\phi(t,\lambda)$ in terms of $\cos\sqrt{\lambda t}$ [according to formula (1.4)]

$$\int_a^x \varphi(t, \lambda)\, dt = \int_a^x \cos\sqrt{\lambda}\, t\, dt + \int_0^a \cos\sqrt{\lambda}\, s\, ds \int_a^x K(t, s)\, dt +$$

$$+ \int_a^x \cos\sqrt{\lambda}\, s\, ds \int_s^x K(t, s)\, dt.$$

Thus,

$$I = \int_{-\infty}^{+\infty} \left[\int_a^x \cos\sqrt{\lambda}\, s\, ds\right]\left[\int_b^y \cos\sqrt{\lambda}\, s\, ds\right] d\rho(\lambda) +$$

$$+ \int_{-\infty}^{+\infty} \left[\int_0^a \cos\sqrt{\lambda}\, s. ds \int_a^x K(t, s)\, dt +\right.$$

$$\left. + \int_a^x \cos\sqrt{\lambda}\, s\, ds \int_s^x K(t, s)\, dt\right]\left[\int_b^y \cos\sqrt{\lambda}\, s\, ds\right] d\rho(\lambda) = 0. \qquad (3.1')$$

From Parseval's formula for ordinary Fourier integrals we have

$$\frac{2}{\pi}\int_0^\infty \left[\int_a^x \cos\sqrt{\lambda}\, s. ds\right]\left[\int_b^y \cos\sqrt{\lambda}\, s\, ds\right] d(\sqrt{\lambda}) = 0. \qquad (3.2)$$

Subtracting equation (3.2) from equation (3.1′), we obtain

$$\int_{-\infty}^{+\infty}\left[\int_a^x \cos\sqrt{\lambda}s\,ds\right]\left[\int_b^y \cos\sqrt{\lambda}s\,ds\right]d\sigma(\lambda)+\int_{-\infty}^{+\infty}\left[\int_0^a \cos\sqrt{\lambda}s\,ds\int_a^x K(t,s)\,dt+\right.$$

$$\left.+\int_a^x \cos\sqrt{\lambda}s\,ds\int_s^x K(t,s)\,dt\right]\left[\int_b^y \cos\sqrt{\lambda}s\,ds\right]d\sigma(\lambda)+$$

$$+\frac{2}{\pi}\int_0^\infty\left[\int_0^a \cos\sqrt{\lambda}s\,ds\int_a^x K(t,s)\,dt+\int_a^x \cos\sqrt{\lambda}s\,ds\int_s^x K(t,s)\,dt\right]\cdot$$

$$\cdot\left[\int_b^y \cos\sqrt{\lambda}s\,ds\right]d(\lambda^{1/2})=\int_{-\infty}^{+\infty}\frac{(\sin\sqrt{\lambda}x-\sin\sqrt{\lambda}a)\,(\sin\sqrt{\lambda}y-\sin\sqrt{\lambda}b)}{\lambda}\,d\sigma(\lambda)+$$

$$+\int_{-\infty}^{+\infty}\left[\int_0^a \cos\sqrt{\lambda}s\,ds\int_a^x K(t,s)\,dt+\int_a^x \cos\sqrt{\lambda}s\,ds\int_s^x K(t,s)\,dt\right]\cdot$$

$$\cdot\left[\int_b^y \cos\sqrt{\lambda}s\,ds\right]d\sigma(\lambda)+\int_b^y ds\int_a^x K(t,s)\,dt=0. \qquad (3.3)$$

For the definition of $\sigma(\lambda)$ see the prededing section. The summand

$$\int_b^y ds\int_a^x K(t,s)dt$$

is obtained from the last summand of the left side by applying Parseval's equation for ordinary Fourier integrals.

As before, we set

$$F(x,y)=\int_{-\infty}^{+\infty}\frac{\sin\sqrt{\lambda}x\,\sin\sqrt{\lambda}y}{\lambda}\,d\sigma(\lambda).$$

If the expression for $F(x,y)$ could be differentiated under the integral sign, then the integral equation (II) could be obtained directly from (3.3) by differentiation. In order not to impose superfluous conditions upon $\sigma(\lambda)$ we proceed in the following manner. Let us set

$$H(x,s)=\begin{cases}\int_a^x K(t,s)\,dt, & 0\leqslant s\leqslant a,\\ \int_s^x K(t,s)\,dt, & a\leqslant s\leqslant x,\\ 0, & s>x.\end{cases}$$

Equation (3.3) may be written in the form

$$F(x,y)-F(x,b)-F(a,y)+F(a,b)+$$
$$+\int_{-\infty}^{+\infty}\left[\int_0^x H(x,s)\cos\sqrt{\lambda}s\,ds\right]\left[\int_b^y \cos\sqrt{\lambda}s\,ds\right]d\sigma(\lambda)+\int_b^y ds\left[\int_a^x K(t,s)\,dt\right]=0. \qquad (3.4)$$

Since $H(x, x) = 0$ and $\partial H/\partial s$ is bounded, then, integrating by parts, we obtain

$$F(x, y) - F(x, b) - F(a, y) + F(a, b) -$$
$$-\int_{-\infty}^{+\infty}\left[\int_0^x \frac{\partial H}{\partial s}\cdot\frac{\sin\sqrt{\lambda}s}{\sqrt{\lambda}}\,ds\right]\left[\frac{\sin\sqrt{\lambda}y-\sin\sqrt{\lambda}b}{\sqrt{\lambda}}\right]d\sigma(\lambda)+\int_b^y ds\int_a^x K(t, s)\,dt = 0.$$

Changing the order of integration, we find

$$F(x, y) - F(x, b) - F(a, y) + F(a, b) -$$
$$-\int_0^x \frac{\partial H}{\partial s}\cdot[F(s, y) - F(s, b)]\,ds + \int_b^y ds\int_a^x K(t, s)\,dt = 0.$$

Integrating by parts, we obtain the equation

$$\int_0^x \frac{\partial H}{\partial s}[F(s, y) - F(s, b)]\,ds = -\int_0^x H(x, s)\left[\frac{\partial F(s, y)}{\partial s} - \frac{\partial F(s, b)}{\partial s}\right]ds.$$

Consequently, the following equation holds:

$$F(x, y) - F(x, b) - F(a, y) + F(a, b) +$$
$$+\int_0^x H(x, s)\left[\frac{\partial F(s, y)}{\partial s} - \frac{\partial F(s, b)}{\partial s}\right]ds + \int_b^y ds\int_a^x K(t, s)\,dt = 0. \qquad (3.5)$$

Changing the order of integration, it is easy to establish that

$$\int_0^x H(x, s)\left[\frac{\partial F(s, y)}{\partial s} - \frac{\partial F(s, b)}{\partial s}\right]ds = \int_a^x dt\int_0^t K(t, s)\left[\frac{\partial F(s, y)}{\partial s} - \frac{\partial F(s, b)}{\partial s}\right]ds.$$

Thus it follows from equation (3.5) that

$$F(x, y) - F(x, b) - F(a, y) + F(a, b) +$$
$$+\int_a^x dt\int_0^t K(t, s)\left[\frac{\partial F(s, y)}{\partial s} - \frac{\partial F(s, b)}{\partial s}\right]ds + \int_b^y ds\int_a^x K(t, s)\,dt = 0.$$

Differentiating the last equation with respect to x and y, we obtain

$$\frac{\partial^2 F}{\partial x\,\partial y} + \int_0^x K(x, s)\frac{\partial^2 F(s, y)}{\partial s\cdot\partial y}\,ds + K(x, y) = 0.$$

Setting $\partial^2 F/\partial x\,\partial y = f(x, y)$, we arrive at the integral equation:

$$f(x, y) + \int_0^x K(x, s) f(s, y)\,ds + K(x, y) = 0 \quad (y \leqslant x). \qquad \text{(II)}$$

For each fixed x equation (II) is a linear Fredholm equation with the symmetric kernel $f(x, y)$ and the unknown function $K(x, y)$ (x fixed).

§4. Investigation of the linear integral equation

1. Beginning with this section we shall take up the inverse problem, namely, the construction of the differential equation

$$y'' + (\lambda - q(x))\, y = 0$$

from the spectral function $\rho(\lambda)$. We shall obtain conditions on the function $\rho(\lambda)$ which are sufficient for there to exist a differential equation with a continuous function $q(x)$. These conditions are also necessary if one requires $q(x)$ to have a continuous derivative.

These conditions are as follows:

1. *For each x there exists the integral*

$$\int_{-\infty}^{0} e^{\sqrt{|\lambda|}\, x}\, d\rho(\lambda)\,.$$

2. *Let us place $\rho(\lambda) = \frac{2}{\pi}\sqrt{\lambda} + \sigma(\lambda)$ for $\lambda > 0$ and $\rho(\lambda) = \sigma(\lambda)$ for $\lambda < 0$. We require that the function*

$$a(x) = \int_{1}^{\infty} \frac{\cos\sqrt{\lambda}\, x}{\lambda}\, d\sigma(\lambda) \tag{4.1}$$

have a continuous fourth derivative.

It follows easily from conditions 1 and 2 that the function

$$F(x, y) = \int_{-\infty}^{+\infty} \frac{\sin\sqrt{\lambda}\, x \sin\sqrt{\lambda}\, y}{\lambda}\, d\sigma(\lambda) \tag{4.2}$$

has continuous fourth derivatives.

We note that, conversely, from the differentiability of (4.2) follows the differentiability of (4.1).

The necessity of condition 1 was proved earlier. If $q(x)$ has a continuous derivative, then $K(x, t)$ has a continuous second derivative. It follows from the nonlinear integral equation that $f(x, y)$ has a continuous second derivative, which means that the function $F(x, y)$ defined by the formula

$$F(x, y) = \int_{-\infty}^{+\infty} \frac{\sin\sqrt{\lambda}\, x \cdot \sin\sqrt{\lambda}\, y}{\lambda}\, d\sigma(\lambda)\,,$$

has continuous fourth derivatives.

In view of condition 1,

$$\int_{1}^{\infty} \frac{\sin\sqrt{\lambda}\, x \cdot \sin\sqrt{\lambda}\, y}{\lambda} \cdot d\sigma(\lambda)$$

has also fourth derivatives. It follows from this that also $a(x)$ has fourth derivatives.

Conversely, it is clear that it follows from conditions 1 and 2 that $F(x,y)$ has continuous fourth derivatives.

For the time being, we exclude from consideration the case when the spectrum is purely discrete, tending to $+\infty$. This case will be treated separately in §9.

Since $f(x,y)$ is a continuous function by hypothesis,

$$f(x,y) = \lim_{\varepsilon\to 0} \frac{F(x+\varepsilon, y+\varepsilon) - F(x-\varepsilon, y+\varepsilon) - F(x+\varepsilon, y-\varepsilon) + F(x-\varepsilon, y-\varepsilon)}{4\varepsilon^2} =$$

$$= \lim_{\varepsilon\to 0} \int_{-\infty}^{+\infty} \left(\frac{\sin\sqrt{\lambda}\varepsilon}{\varepsilon\sqrt{\lambda}}\right)^2 \cos\sqrt{\lambda}\,x \cos\sqrt{\lambda}\,y\, d\sigma(\lambda). \tag{4.3}$$

As was shown in §2, it follows from this that as $\epsilon \to 0$ one has uniformly in each finite interval of variation of x

$$\int_{-\infty}^{+\infty} \frac{\sin^2\sqrt{\lambda}\varepsilon}{\lambda} \cdot \cos\sqrt{\lambda}\,x\, d\sigma(\lambda) = o(\varepsilon). \tag{4.3'}$$

From the mean-value theorem, we have for $\lambda < 0$

$$\frac{\sin\sqrt{\lambda}\,\varepsilon}{\sqrt{\lambda}\,\varepsilon} = \operatorname{ch}\sqrt{|\lambda|}\,\theta\varepsilon \quad (0<\theta<1).$$

Thus

$$\int_{-\infty}^{0} \left(\frac{\sin\sqrt{\lambda}\,\varepsilon}{\varepsilon\sqrt{\lambda}}\right)^2 \cos\sqrt{\lambda x}\, d\rho(\lambda) = \int_{-\infty}^{0} \operatorname{ch}^2\sqrt{\lambda}\,\theta\,\varepsilon \cdot \operatorname{ch}\sqrt{|\lambda|}\,x\, d\rho(\lambda) < A < \infty$$

uniformly in each finite interval. From this we have uniformly in each finite interval as $\epsilon \to 0$

$$\int_{-\infty}^{0} \frac{\sin^2\sqrt{\lambda}\,\varepsilon}{\lambda} \cos\sqrt{\lambda}\,x\, d\rho(\lambda) = o(\varepsilon)$$

which means, in view of (4.3′), that as $\epsilon \to 0$ we have

$$\int_{0}^{\infty} \frac{\sin^2\sqrt{\lambda}\,\varepsilon}{\lambda} \cos\sqrt{\lambda}\,x\, d\sigma(\lambda) = o(\varepsilon) \tag{4.4}$$

uniformly in each finite interval. Condition (4.4) plays an essential role in the following.

We need in addition the following inequality

$$\int_{a}^{\infty} \frac{d\rho(\lambda)}{\lambda} < +\infty, \quad a > 0.$$

Let us prove it. Setting $x = y = 0$ in (4.3) we have:

$$\int_{0}^{\infty} \left(\frac{\sin\sqrt{\lambda}\,\varepsilon}{\sqrt{\lambda}\,\varepsilon}\right)^2 d\sigma(\lambda) = O(1).$$

From the ordinary Parseval equation we have

$$\frac{2}{\pi}\int_0^\infty \left(\frac{\sin\sqrt{\lambda}\,\varepsilon}{\sqrt{\lambda}\varepsilon}\right)^2 d\sqrt{\lambda}=\varepsilon^{-1}.$$

We thus obtain for $\rho(\lambda)$

$$\int_0^\infty \left(\frac{\sin\sqrt{\lambda}\varepsilon}{\sqrt{\lambda}\varepsilon}\right)^2 d\rho(\lambda)=\frac{1}{\varepsilon}+O(1).$$

Since $\frac{\sin\epsilon\sqrt{\lambda}}{\epsilon\sqrt{\lambda}}\geq\frac{2}{\pi}$ for $\epsilon\sqrt{\lambda}\leq\frac{\pi}{2}$ we have

$$\int_0^\infty \left(\frac{\sin\sqrt{\lambda}\varepsilon}{\sqrt{\lambda}\varepsilon}\right)^2 d\rho(\lambda)\geqslant\int_0^{\frac{1}{\varepsilon^2}} \left(\frac{\sin\sqrt{\lambda}\varepsilon}{\sqrt{\lambda}\varepsilon}\right)^2 d\rho(\lambda)\geqslant$$

$$\geqslant\frac{4}{\pi^2}\int_0^{1/\varepsilon^2} d\rho(\lambda)=\frac{4}{\pi^2}\left(\rho\left(\frac{1}{\varepsilon^2}\right)-\rho(0)\right).$$

Thus

$$\rho\left(\frac{1}{\varepsilon^2}\right)-\rho(0)\leqslant\frac{\pi^2}{4}\left(\frac{1}{\varepsilon}+O(1)\right)$$

which means

$$\rho(\lambda)\leqslant c\sqrt{\lambda}+O(1).$$

It follows easily from this that $\int_a^\infty\lambda^{-1}d\rho(\lambda)$ converges.

2. Let there be given a function $\rho(\lambda)$ satisfying conditions 1 and 2 and having an infinite number of points of increase on some finite interval. Let us construct with it the function

$$F(x,y)=\int_{-\infty}^{+\infty}\frac{\sin\sqrt{\lambda}x\cdot\sin\sqrt{\lambda}y}{\lambda}\,d\sigma(\lambda)\quad\text{and}\quad f(x,y)=\frac{\partial^2 F}{\partial x\,\partial y}.$$

We then have the following theorem.

Theorem. *For fixed x the integral equation*

$$f(x,y)+\int_0^x f(y,s)\,K(x,s)\,ds+K(x,y)=0 \tag{4.5}$$

for the unknown function $K(x,y)$ has a solution which is unique.

Proof. As is known, it is sufficient for the solvability of equation (4.5) that the homogeneous equation (x fixed)

$$\int_0^x f(s,y)\,h(s)\,ds+h(y)=0 \tag{4.6}$$

does not have a solution other than the trivial one. For this we show that the quadratic form

$$I(g)=\int_0^x\int_0^x f(s,y)\,g(s)\,g(y)\,ds\,dy+\int_0^x g^2(y)\,dy>0 \tag{4.7}$$

for an arbitrary function $g(y)$ is not 0.

Let us suppose to start with that $g(y)$ $(0 \leq y \leq x)$ is such that

1) $g(x) = 0$,

2) $g'(y)$ is continuous for $0 \leq y \leq x$.

Since $f(y, s) = \partial^2 F/\partial y\, \partial s$, we obtain, integrating the first summand in (4.7) by parts,

$$I(g) = \int_0^x g^2(y)\, dy + \int_0^x \int_0^x F(y, s)\, g'(y)\, g'(s)\, dy\, ds =$$

$$= \int_0^x g^2(y)\, dy + \int_0^x \int_0^x g'(y)\, g'(s)\, dy\, ds \int_{-\infty}^{+\infty} \frac{\sin \sqrt{\lambda} y}{\sqrt{\lambda}} \cdot \frac{\sin \sqrt{\lambda} s}{\sqrt{\lambda}}\, d\rho(\lambda) -$$

$$- \frac{2}{\pi} \int_0^x \int_0^x g'(y)\, g'(s)\, dy\, ds \int_0^\infty \frac{\sin \sqrt{\lambda} y}{\sqrt{\lambda}} \cdot \frac{\sin \sqrt{\lambda} s}{\sqrt{\lambda}}\, d(\sqrt{\lambda}) =$$

$$= \int_0^x g^2(y)\, dy + \int_{-\infty}^{+\infty} G^2(\lambda)\, d\rho(\lambda) - \frac{2}{\pi} \int_0^{+\infty} G^2(\lambda)\, d(\sqrt{\lambda}),$$

where

$$G(\lambda) = \int_0^x \frac{\sin \sqrt{\lambda} t}{\sqrt{\lambda}}\, g'(t)\, dt = \int_0^x \cos \sqrt{\lambda} t \cdot g(t)\, dt.$$

From Parseval's equation for ordinary Fourier integrals we have

$$\frac{2}{\pi} \int_0^\infty G^2(\lambda)\, d(\sqrt{\lambda}) = \int_0^x g^2(y)\, dy.$$

Thus

$$I(g) = \int_{-\infty}^{+\infty} G^2(\lambda) \cdot d\rho(\lambda). \tag{4.8}$$

We shall show that $g(t) = 0$ follows from $I(g) = 0$. Since

$$G(\lambda) = \int_0^x \cos \sqrt{\lambda t} \cdot g(t)\, dt,$$

$G(\lambda)$ is an entire analytic function of exponential growth in λ. Consequently, $G(\lambda)$ may have only isolated zeros with a limit point of ∞. If $\rho(\lambda)$ has an infinite number of points of increase on some finite interval, then it follows from the equation $I(g) = 0$, because of (4.8), that $G(\lambda) = 0$, i.e.,

$$\int_0^x \cos \sqrt{\lambda} t\, g(t)\, dt = 0$$

which means $g(t) = 0$.

Now let $g(y)$ be a continuous bounded function, not satisfying conditions 1) and 2). Let $g_n(y)$ converge in the mean square as $n \to \infty$ to $g(y)$ and satisfy conditions 1) and 2). It is easy to see that

$$\lim_{n\to\infty} I(g_n) = I(g).$$

Consequently,

$$I(g) = \lim_{n\to\infty} \int_{-\infty}^{+\infty} G_n^2(\lambda)\, d\rho(\lambda),$$

where

$$G_n(\lambda) = \int_0^x g_n(y) \cos\sqrt{\lambda}\, y\, dy.$$

Since $g_n(y)$ converges in the mean to $g(y)$, $G_n(\lambda)$ converges informally to $G(\lambda)$. If $I(g) = 0$, i.e.,

$$\int_{-\infty}^{+\infty} G_n^2(\lambda)\, d\rho(\lambda) \to 0,$$

then, since the function under the integral is nonnegative and approaches a limit, this is possible only for $G(\lambda) \equiv 0$. From this we obtain as before that $g(t) \equiv 0$.

Let us now show that the homogeneous equation (4.6) has only the trivial solution. Let $g(t)$ be a solution of equation (4.6). Multiply it by $g(y)$ and integrate from 0 to x. We obtain $I(g) = 0$, which means $g(y) \equiv 0$. Thus, we have shown that the linear equation (4.5) is solvable.

Since, by hypothesis, $f(x, y)$ is a continuous function, it follows from equation (4.5) that for each fixed x the function $K(x, y)$ is continuous in the variable y for $y \le x$. Furthermore, it follows from the same equation that if there exist the continuous derivatives $\partial^r f/\partial y^r$ $(r = 1, 2, \cdots, n)$, then there exist also the continuous derivatives $\partial^r K(x, y)/\partial y^r$.

3. The investigation of the behavior of the function $K(x, y)$ with respect to its first argument is somewhat more complicated. For this we make use of the following lemma:

Lemma. *Let there be given the integral equation*

$$g(x, a) = h(x, a) + \int_0^1 H(x, y; a)\, h(y, a)\, dy, \tag{4.9}$$

in which the kernel and the free term $g(x, a)$ are continuous functions of the parameter a and the independent variables. Then, if the homogeneous equation has only the trivial solution for $a = a_0$, the solution $h(x, a)$ is a continuous function of x and a in some neighborhood of the point $a = a_0$. If H and g have n continuous derivatives with respect to a, then the same holds for $h(x, a)$.

Proof. Let us place

$$H(x, y, a) = H(x, y, a_0) + H_1(x, y) = H_0 + H_1,$$

where $|H_1(x, y)| < \epsilon$ in case a is in some sufficiently small neighborhood of a_0. Equation (4.9) may be written symbolically in the form

$$g = h + Hh = h + H_0 h + H_1 h.$$

Applying the operator $(E + H_0)^{-1}$ to both sides of the last equation, we obtain

$$(E + H_0)^{-1} g = h + (E + H_0)^{-1} H_1 h. \tag{4.10}$$

Since the norm of the operator $(E + H_0)^{-1} H_1$ may be made arbitrarily small, equation (4.10) may be solved by the method of successive approximations. The lemma is proved.

Let us now apply the lemma to equation (II). We shall investigate the neighborhood of some point x_0. In equation (II) change s to sx and y to yx. We obtain the equation

$$f(y, yx) + x \int_0^1 K(x, sx) f(sx, yx)\, ds + K(y, yx) = 0,$$

i.e., the integral equation with the kernel $xf(sx, yx)$ and the free term $f(x, yx)$; here x is a parameter. Since $f(x, y)$ is a continuous function by hypothesis, the continuity of the function $K(x, y)$ in the whole set of variables follows from the lemma. Furthermore, it follows from the same lemma that $K(x, y)$ has continuous derivatives with respect to x of the same order as $f(x, y)$.

4. In order to prove the solubility of the equation for $h = \infty$ we set

$$F(x, y) = \int_{-\infty}^{+\infty} \frac{(1 - \cos\sqrt{\lambda}\, x)(1 - \cos\sqrt{\lambda}\, y)}{\lambda^2}\, d\sigma(\lambda),$$

where

$$\sigma(\lambda) = \begin{cases} \rho(\lambda) - \dfrac{3}{2\pi}\lambda^{3/2}, & \text{if} \quad \lambda \geqslant 0, \\ \rho(\lambda), & \text{if} \quad \lambda < 0 \end{cases}$$

and

$$f(x, y) = \frac{\partial^2 F}{\partial x \partial y}.$$

$L(x, y)$ satisfies the same equation as $K(x, y)$. In place of equation (4.4) we have

$$\int_0^\infty \frac{\sin^2 \varepsilon\sqrt{\lambda}}{\lambda} \cdot \frac{\sin^2 \varepsilon\sqrt{\lambda}\, x}{\lambda}\, d\sigma(\lambda) = o(\varepsilon) \tag{4.4'}$$

uniformly in each finite interval of variation of x. In place of $G(\lambda)$ we introduce

$$H(\lambda) = \int_0^x g(y) \frac{\sin\sqrt{\lambda}\, y}{\sqrt{\lambda}}\, dy.$$

The remainder of the proof goes over without change.

§5. Derivation of Parseval's equation

1. In the preceding section we showed that for a given monotonic function $\rho(\lambda)$ satisfying conditions 1 and 2 there exists a solution $K(x, t)$ of the linear integral equation (4.5). More precisely, for the solvability of this equation it is sufficient to replace condition 2 by a weaker one: the existence for the function $F(x, y)$ of a continuous derivative $f(x, y) = \partial^2 F/\partial x \partial y$. By means of the function $K(x, t)$ it is possible to construct the functions $\phi(x, \lambda)$ by the formula

$$\varphi(x, \lambda) = \cos \sqrt{\lambda}\, x + \int_0^x K(x, t) \cos \sqrt{\lambda}\, t\, dt.$$

In this paragraph we shall show that the functions $\phi(x, \lambda)$ are "orthonormal" functions with the weight $\rho(\lambda)$, i.e., more precisely, Parseval's equation holds for them in the form in which it was given in §1, paragraph 1. We shall again not make complete use of condition 2, but shall assume that $f(x, y) = \partial^2 F/\partial x \partial y$ has continuous derivatives of the first order.

Let us proceed to the proof of Parseval's equation.

We shall show first that the functions

$$h_1(\lambda) = \int_a^x \varphi(t, \lambda)\, dt \quad \text{and} \quad h_2(\lambda) = \int_b^y \varphi(t, \lambda)\, dt$$

in the case when the intervals (a, x) and (b, y) do not intersect are orthogonal with the weight $\rho(\lambda)$, i.e., we shall prove the equation

$$\int_{-\infty}^{+\infty} h_1(\lambda)\, h_2(\lambda)\, d\rho(\lambda) = 0. \tag{5.1}$$

Without restricting the generality, we shall assume that $b < y \leq a < x$. Reversing the order of the reasoning used in §3 for the linear integral equation (II), we obtain

$$\int_{-\infty}^{+\infty} \Big[\int_a^x \varphi(t, \lambda)\, dt\Big]\Big[\int_b^y \cos \sqrt{\lambda}\, t . dt\Big]\, d\rho(\lambda) = 0, \quad b < y \leqslant a < x, \tag{5.2}$$

where

$$\varphi(t, \lambda) = \cos \sqrt{\lambda}\, t + \int_0^t K(t, s) \cos \sqrt{\lambda}\, s\, ds. \tag{5.3}$$

We shall show that from formula (5.2) one may obtain the following equation of which we have need:

$$\int_{-\infty}^{+\infty} \Big[\int_a^x \varphi(t, \lambda)\, dt\Big]\Big[\int_b^y \varphi(t, \lambda)\, dt\Big]\, d\rho(\lambda) = 0. \tag{5.4}$$

Indeed, it follows from formula (5.3) that

$$\int_b^y \varphi(t,\lambda)\,dt = \int_b^y \cos\sqrt{\lambda}\,t\,dt + \int_0^b \cos\sqrt{\lambda}\,s\,ds \int_b^y K(t,s)\,dt +$$

$$+\int_b^y \cos\sqrt{\lambda}\,s\,ds \int_s^y K(t,s)\,dt = \int_b^y \cos\sqrt{\lambda}\,t\,dt + \int_0^y H(s,y)\cos\sqrt{\lambda}\,s\,ds,$$

where the function

$$H(s,y) = \begin{cases} \int_b^y K(t,s)\,dt, & 0 \leqslant s \leqslant b, \\ \int_s^y K(t,s)\,dt, & b \leqslant s \leqslant y \end{cases}$$

is continuous and has a derivative $\partial H/\partial s$.

Substituting the expression for $\int_b^y \phi(t,\lambda)dt$ in the left side of equation (5.4), and making use of equation (5.2), we obtain

$$\int_{-\infty}^{+\infty} \Big[\int_a^x \varphi(t,\lambda)\,dt\Big]\Big[\int_b^y \varphi(t,\lambda)\,dt\Big]\,d\rho(\lambda) =$$

$$= \int_{-\infty}^{+\infty} \Big[\int_a^x \varphi(t,\lambda)\,dt\Big]\Big[\int_0^y H(t,y)\cos\sqrt{\lambda}\,t\,dt\Big]\,d\rho(\lambda).$$

We shall now show that the last integral is equal to zero for each fixed $y < a$. For this we again make use of equation (5.2).

Let $P(s,y)$ $(s \leq y)$ be an arbitrary function. Replace b by s in formula (5.2), multiply both sides by $P(s,y)$, and integrate with respect to s between the limits 0 and y. We obtain (changing the order of integration)

$$\int_{-\infty}^{+\infty} \Big[\int_a^x \varphi(t,\lambda)\,dt\Big]\Big[\int_0^y P(s,y)\,ds \int_s^y \cos\sqrt{\lambda}\,t\,dt\Big]\,d\rho(\lambda) =$$

$$= \int_{-\infty}^{+\infty} \Big[\int_a^x \varphi(t,\lambda)\,dt\Big]\Big[\int_0^y \cos\sqrt{\lambda}\,t\,dt \int_0^t P(s,y)\,ds\Big]\,d\rho(\lambda) = 0.$$

Placing

$$H(t,y) = \int_0^t P(s,y)\,ds,$$

i.e.,

$$\frac{\partial H}{\partial t} = P(t,y),$$

we obtain (5.4).

Condition (5.4) is the orthogonality condition for the Fourier transform of the characteristic functions of nonintersecting intervals. In order to obtain Parseval's equation for step functions, one must still show that for an arbitrary interval (a, b) $(a < b)$ the following equation holds:

$$\int_{-\infty}^{+\infty} \left[\int_a^b \varphi(t,\lambda)\,dt\right]^2 d\rho(\lambda) = b - a, \tag{5.5}$$

which is Parseval's equation for the characteristic function of the interval (a, b).

Let us first show that a weaker condition holds: if $b - a \to 0$, then the equation

$$\int_{-\infty}^{+\infty} \left[\int_a^b \varphi(t,\lambda)\,dt\right]^2 d\rho(\lambda) = b - a + o(b-a) \tag{5.6}$$

is true; it is natural to call this the normalization condition. Let us place $b = \alpha + \epsilon$, $a = \alpha - \epsilon$. We shall first show that

$$\int_{-\infty}^{0} \left[\int_{\alpha-\varepsilon}^{\alpha+\varepsilon} \varphi(t,\lambda)\,dt\right]^2 d\rho(\lambda) = o(\varepsilon). \tag{5.7}$$

Since, by hypothesis, for $x \geq 0$

$$\int_{-\infty}^{0} \operatorname{ch}\sqrt{|\lambda|}\,x\,d\rho(\lambda) < \infty,$$

then because of formula (5.3) we have for all t

$$\int_{-\infty}^{0} \varphi^2(t,\lambda)\,d\rho(\lambda) < +\infty.$$

Thus, because of the Cauchy-Bunyakovskiĭ inequality,

$$\int_{-\infty}^{0} \left[\int_{\alpha-\varepsilon}^{\alpha+\varepsilon} \varphi(t,\lambda)\,dt\right]^2 d\rho(\lambda) \leqslant 2\varepsilon \int_{\alpha-\varepsilon}^{\alpha+\varepsilon} \left(\int_{-\infty}^{0} \varphi^2(t,\lambda)\,d\rho(\lambda)\right) dt = O(\varepsilon^2)$$

which holds uniformly in α for each finite interval. Thus, for the proof of (5.6) one must show that

$$\int_{0}^{\infty} \left[\int_{\alpha-\varepsilon}^{\alpha+\varepsilon} \varphi(t,\lambda)\,dt\right]^2 d\rho(\lambda) = 2\varepsilon + o(\varepsilon).$$

For $\lambda > 0$ we have

$$\int_{\alpha-\varepsilon}^{\alpha+\varepsilon} \varphi(t,\lambda)\,dt = \int_{\alpha-\varepsilon}^{\alpha+\varepsilon} \cos\sqrt{\lambda}\,t\,dt + \int_{\alpha-\varepsilon}^{\alpha+\varepsilon} dt \int_0^t K(t,s)\cos\sqrt{\lambda}\,s\,ds =$$

$$= \int_{\alpha-\varepsilon}^{\alpha+\varepsilon} \cos\sqrt{\lambda}\,t\,dt + \int_{\alpha-\varepsilon}^{\alpha+\varepsilon} \left[\frac{\sin\sqrt{\lambda}\,t}{\sqrt{\lambda}} K(t,t) - \int_0^t \frac{\sin\sqrt{\lambda}\,s}{\sqrt{\lambda}} \frac{\partial K(t,s)}{\partial s}\,ds\right] dt =$$

$$= \int_{\alpha-\varepsilon}^{\alpha+\varepsilon} \cos\sqrt{\lambda}\,t\,dt + \frac{O(\varepsilon)}{\sqrt{\lambda}}.$$

Then for $\lambda > 0$

$$\left[\int_{\alpha-\varepsilon}^{\alpha+\varepsilon} \varphi(t,\lambda)\,dt\right]^2 = \left[\int_{\alpha-\varepsilon}^{\alpha+\varepsilon} \cos\sqrt{\lambda}\,t.dt\right]^2 + 4\,\frac{\sin\sqrt{\lambda}\,\varepsilon}{\sqrt{\lambda}} \cdot \cos\sqrt{\lambda}\,\alpha\,\frac{O(\varepsilon)}{\sqrt{\lambda}} + \frac{O(\varepsilon^2)}{\lambda}.$$

Because of the boundedness of $\phi(t, \lambda)$ for arbitrary $\lambda > 0$,

$$\int_0^a \Big[\int_{\alpha-\varepsilon}^{\alpha+\varepsilon} \varphi(t, \lambda)\, dt\Big]^2 d\rho(\lambda) = O(\varepsilon^2). \tag{5.8}$$

Furthermore, we have

$$\int_a^\infty \Big[\int_{\alpha-\varepsilon}^{\alpha+\varepsilon} \varphi(t, \lambda)\, dt\Big]^2 d\rho(\lambda) = \int_a^\infty \Big[\int_{\alpha-\varepsilon}^{\alpha+\varepsilon} \cos\sqrt{\lambda}\, t\, dt\Big]^2 d\rho(\lambda) +$$

$$+ O(\varepsilon)\cdot \int_a^\infty \sin\sqrt{\lambda}\,\varepsilon\cdot\cos\sqrt{\lambda}\,\alpha\, \frac{d\rho(\lambda)}{\lambda} + O(\varepsilon^2)\int_a^\infty \frac{d\rho(\lambda)}{\lambda}. \tag{5.9}$$

Since (see §4)

$$\int_a^\infty \frac{d\rho(\lambda)}{\lambda} < \infty,$$

we have for $\epsilon \to 0$

$$\int_a^\infty \sin\sqrt{\lambda}\,\varepsilon\cdot\cos\sqrt{\lambda}\,\alpha\cdot \frac{d\rho(\lambda)}{\lambda} = o(1)$$

uniformly in a. From (5.8) and (5.9) and the estimate (4.4) follows

$$\int_0^\infty \Big[\int_{\alpha-\varepsilon}^{\alpha+\varepsilon} \varphi(t, \lambda)\, dt\Big]^2 d\rho(\lambda) = \int_a^\infty \Big[\int_{\alpha-\varepsilon}^{\alpha+\varepsilon} \cos\sqrt{\lambda}\, t\cdot dt\Big]^2 d\rho(\lambda) + o(\varepsilon) =$$

$$= \frac{2}{\pi}\int_0^\infty \Big[\int_{\alpha-\varepsilon}^{\alpha+\varepsilon} \cos\sqrt{\lambda}\, t\cdot dt\Big]^2 d\sqrt{\lambda} + 4\int_0^\infty \Big(\frac{\sin\sqrt{\lambda}\,\varepsilon}{\sqrt{\lambda}}\Big)^2 \cos^2\sqrt{\lambda}\alpha\, d\sigma(\lambda) + o(\varepsilon) =$$

$$= 2\varepsilon + o(\varepsilon),$$

i.e., condition (5.6) is proved.

We shall now show that from the orthogonality condition (5.4) and equation (5.6) follows equation (5.5), i.e., Parseval's equation for the characteristic function of the interval (a, b).

Let (a, b) be an arbitrary fixed finite interval. Set

$$a = a_0 < a_1 < a_2 < \cdots < a_{n-1} < a_n = b,$$

$$\max_{1 \leqslant i \leqslant n} (a_{i+1} - a_i) \to 0 \quad \text{as} \quad n \to \infty.$$

Because of (5.4) and (5.6),

$$\int_{-\infty}^{+\infty} \Big[\int_a^b \varphi(t, \lambda)\, dt\Big]^2 d\rho(\lambda) = \int_{-\infty}^{+\infty} \Big[\sum_{i=1}^n \int_{a_{i-1}}^{a_i} \varphi(t, \lambda)\, dt\Big]^2 d\rho(\lambda) =$$

$$= \sum_{i=1}^n \int_{-\infty}^{+\infty} \Big[\int_{a_{i-1}}^{a_i} \varphi(t; \lambda)\, dt\Big]^2 d\rho(\lambda) = \sum_{i=1}^n [(a_i - a_{i-1}) + o(a_i - a_{i-1})] \to b - a.$$

Since the left side in the last equation does not depend upon n, we obtain (5.5) by passing to the limit.

It is now easy to show Parseval's equation for an arbitrary step function. Let us consider a step function $f(x)$ of the form:

$$f(x) = d_i \text{ for } a_{i-1} \leqslant x < a_i \quad (i = 1, \ldots, n;\ a_0 = a,\ a_n = b),$$
$$f(x) = 0 \quad \text{outside the interval } (a, b).$$

From (5.4) and (5.6) follows

$$\int_{-\infty}^{+\infty} \left[\int_a^b f(x)\,\varphi(x, \lambda)\,dx\right]^2 d\rho(\lambda) = \sum_{i=1}^{n} d_i^2 (a_i - a_{i-1}) = \int_a^b f^2(x)\,dx.$$

We have thus obtained Parseval's equation for step functions. Since the set of step functions is dense in $L_2(0, \infty)$, Parseval's equation holds for all functions from $L_2(0, \infty)$.

We have proved the following theorem.

Theorem. *Let the function $\rho(\lambda)$ satisfy conditions 1 and 2 of §4* and let $K(x, y)$ be the solution of the integral equation (II). Let us place*

$$\varphi(x, \lambda) = \cos\sqrt{\lambda}\,x + \int_0^x K(x, t) \cos\sqrt{\lambda}\,t\,dt.$$

Then for each function $f(x) \in L_2(0, \infty)$ the functions

$$F_n(\lambda) = \int_0^n f(x)\,\varphi(x, \lambda)\,dx$$

as $n \to \infty$ converge in the mean square (with weight $\rho(\lambda)$) to some function $F(\lambda)$ (the generalized Fourier transform of the function $f(x)$), i.e.,

$$\lim_{n\to\infty} \int_{-\infty}^{+\infty} [F(\lambda) - F_n(\lambda)]^2\,d\rho(\lambda) = 0.$$

In addition, Parseval's formula holds:

$$\int_{-\infty}^{+\infty} F^2(\lambda)\,d\rho(\lambda) = \int_0^{\infty} f^2(x)\,dx.$$

If $g(x) \in L_2(0, \infty)$ is another function and $G(\lambda)$ is its Fourier transform, then

$$\int_{-\infty}^{+\infty} F(\lambda)\,G(\lambda)\,d\rho(\lambda) = \int_0^{\infty} f(x)\,g(x)\,dx.$$

Analogously, one may obtain Parseval's equation in the case $h = \infty$.

2. Parseval's equation is equivalent to mean-square convergence of the Fourier integral. In many cases it is necessary to have criteria of uniform convergence of a Fourier integral. We shall prove the following theorem:

Theorem. *Let $f(x)$ be a continuous function vanishing outside some finite interval $(0, a)$. Let us set*

* It is sufficient that the function $a(x)$ in condition 2 have three derivatives.

$$E(\lambda) = \int_0^a f(x)\,\varphi(x, \lambda)\,dx$$

and assume that for all $x \geq 0$

$$\int_{-\infty}^{+\infty} E(\lambda)\,\varphi(x, \lambda)\,d\rho(\lambda)$$

converges absolutely and uniformly in each finite interval, and consequently represents a continuous function. Under these hypotheses

$$f(x) = \int_{-\infty}^{+\infty} E(\lambda)\,\varphi(x, \lambda)\,d\rho(\lambda).$$

Proof. Let us set

$$h_n(x) = \int_{-n}^{n} E(\lambda)\,\varphi(x, \lambda)\,d\rho(\lambda).$$

Because of Parseval's equation,

$$\lim_{n\to\infty} \int_0^\infty [f(x) - h_n(x)]^2\,dx \to 0.$$

Consequently, and a fortiori for $N \geq a$,

$$\lim_{n\to\infty} \int_0^N [f(x) - h_n(x)]^2\,dx = 0. \tag{5.10}$$

By hypothesis, as $n \to \infty$, $h_n(x)$ converges uniformly in each finite interval to some function $h(x)$. Consequently, in (5.10) one may pass to the limit under the integral sign and obtain

$$\int_0^N [f(x) - h(x)]^2\,dx = 0.$$

Since $f(x)$ and $h(x)$ are continuous functions, there follows from the last equation

$$f(x) = h(x),$$

which was to be proved.

§6. Derivation of the differential equation for $\phi(x, \lambda)$

1. In the preceding two sections we have constructed for functions $\rho(\lambda)$, satisfying conditions 1 and 2 of §4, a system of functions $\phi(x, \lambda)$ for which Parseval's equation was proved. These functions were constructed from the kernel $K(x, t)$ by means of the formula

$$\varphi(x, \lambda) = \cos\sqrt{\lambda}\,x + \int_0^x K(x, s)\cos\sqrt{\lambda}\,s\,ds. \tag{6.1}$$

Along with equation (6.1) we have also need of its inversion:

$$\cos\sqrt{\lambda}\,x = \varphi(x,\lambda) - \int_0^x K_1(x,s)\,\varphi(s,\lambda)\,ds. \qquad (6.2)$$

We have shown that when condition 2 is fulfilled the function $K(x, t)$, and consequently also $K_1(x, t)$, has continuous second derivatives. In this paragraph we shall show that the functions $\phi(x, \lambda)$ are eigenfunctions of a differential equation of the form

$$y'' + (\lambda - q(x))\,y = 0, \qquad (6.3)$$

where the function $q(x)$ is continuous and satisfies the given boundary conditions at *0*. For this we have first obtained a certain functional equation for the function $\phi(x, \lambda)$ from which we can later obtain a differential equation (6.3) for $\phi(x, \lambda)$. Let us multiply both sides of equation (6.1) by $\cos\sqrt{\lambda t}$. We obtain

$$\varphi(x,\lambda)\cos\sqrt{\lambda}\,t = \frac{1}{2}[\cos\sqrt{\lambda}(x+t) + \cos\sqrt{\lambda}(x-t)] +$$
$$+ \frac{1}{2}\int_0^x K(x,s)\,[\cos\sqrt{\lambda}(s+t) + \cos\sqrt{\lambda}(s-t)]\,ds.$$

Substituting every place on the right-hand side in place of $\cos\sqrt{\lambda x}$ its value from formula (6.2), we obtain

$$\varphi(x,\lambda)\cos\sqrt{\lambda}\,t = \frac{1}{2}[\varphi(x+t,\lambda) + \varphi(x-t,\lambda)] + \int_0^{x+t} W(x,t,s)\,\varphi(s,\lambda)\,ds, \quad (6.4)$$

where $W(x, t, s)$ is a certain function constructed from the functions $K(x, t)$ and $K_1(x, t)$, of whose exact form we shall have no need. We only mention that, since $K(x, t)$ and $K_1(x, t)$ are continuous for $t \leq x$, $W(x, t, s)$ is continuous for $s \leq x+t$.

It is quite essential that in (6.4) the integration should actually be carried out not from *0* but from $|x-t|$. A rigorous proof of this fact is somewhat bulky. Consequently, we shall give some nonrigorous considerations from which the essence of the matter is clear. Let us multiply both sides of equation (6.4) by $\phi(s_0, \lambda)$, where $s_0 < x + t$, $s_0 \neq x - t$, and let us integrate both sides of the equation with $d\rho(\lambda)$. Since $\phi(x, \lambda)$ and $\phi(x_2, \lambda)$ for $x_1 \neq x_2$ are "orthogonal" with respect to $d\rho(\lambda)$, we obtain

$$W(x,t,s_0) = \int_{-\infty}^{+\infty} \varphi(x,\lambda)\cos\sqrt{\lambda}\,t\,\varphi(s_0,\lambda)\,d\rho(\lambda). \qquad (6.5)$$

From this it is evident that $W(x, t, s_0)$ is symmetric in the variables x and s_0. Since $W(x, t, s) = 0$ for $s_0 > x + t$, it follows from this symmetry that $W(x, t, s_0) = 0$ also in case $x > s_0 + t$, i.e., $s_0 < x - t$. This means that the integral (6.4) may be taken between the limits $x - t$ to $x + t$. We have reasoned unrigorously since the integral (6.5) diverges. The rigouous proof carried out below is based upon the same idea which is carried out above unrigorously.

Let us integrate both sides of equation (6.4) from x to $t(t \leq x)$. We obtain

$$\int_0^x \varphi(u,\lambda)\,du \int_0^y \cos\sqrt{\lambda}\,v\,dv = h(x,y;\lambda) = \frac{1}{2}\int_0^x\int_0^y \varphi(u+v,\lambda)\,du\,dv +$$

$$+\frac{1}{2}\int_0^x\int_0^y \varphi(u-v,\lambda)\,du\,dv + \frac{1}{2}\int_0^x\int_0^y du\,dv \int_0^{u+v} W(u,v,t)\,\varphi(t,\lambda)\,dt.$$

Changing the order of integration, we find

$$h(x,y;\lambda) = \int_0^{x+y} Z(x,y,t)\,\varphi(t,\lambda)\,dt, \tag{6.6}$$

where $Z(x, y, t)$ is a continuous function in its variables. Formula (6.6) shows that, for arbitrary x and y, $h(x, y; \lambda)$ is the Fourier transform of the continuous function $Z(x, y, t)$ vanishing outside the interval $(0, x+y)$. We shall show that for fixed x, y and t the integral

$$\int_{-\infty}^{+\infty} h(x,y;\lambda)\,\varphi(t,\lambda)\,d\rho(\lambda) \tag{6.7}$$

converges absolutely and uniformly with respect to t in each finite interval.

We first estimate the function $\phi(t, \lambda)$. For $\lambda < 0$ we have

$$|\varphi(t,\lambda)| \leqslant 1 + \int_0^t \left|K(t,s)\right| dt < a(t), \tag{6.8}$$

where $a(t)$ is bounded in each finite interval. Thus for $0 \leq \lambda < +\infty$ the function $\phi(t, \lambda)$ is uniformly bounded in t in each finite interval. For $\lambda < 0$ we have

$$|\varphi(t,\lambda)| \leqslant \operatorname{ch}\sqrt{|\lambda|}\,t\left(1 + \int_0^t |K(t,s)|\,ds\right) < a(t)\operatorname{ch}\sqrt{|\lambda|}\,t. \tag{6.9}$$

To obtain an estimate of the integral (6.7) we split it into the sum of two integrals: one with limits $-\infty$ to 0 and the other with the limits 0 to ∞. Let us first consider the integral between the limits 0 and ∞. Since the function

$$\psi(y,\lambda) = \int_0^y \cos\sqrt{\lambda}\,v\,dv = \int_0^y \varphi(v,\lambda)\,dv + \int_0^y \varphi(v,\lambda)\,dv \int_v^y K_1(t,v)\,dt$$

is the Fourier transform of the function

$$\Phi(y,v) = \begin{cases} 1 + \int_v^y K_1(t,v)\,dt, & v \leqslant y, \\ 0, & v > y, \end{cases}$$

and

$$\chi(x,\lambda) = \int_0^x \varphi(u,\lambda)\,du$$

is the Fourier transform of the function

$$e_x = \begin{cases} 1, & \text{if } u \leqslant x, \\ 0, & \text{if } u > x, \end{cases}$$

we have, by virtue of the Cauchy-Bunyakovskiĭ inequality, Parseval's equation, and the inequalities (6.8) (we recall that $h(x, y; \lambda) = \int_0^x \phi(u, \lambda)du \int_0^y \cos\sqrt{\lambda}\, v dv$)

$$\int_0^\infty |h(x, y; \lambda)| |\varphi(t, \lambda)| d\rho(\lambda) \leqslant a(t) \int_0^\infty |h(x, y, \lambda)| d\rho(\lambda) \leqslant$$

$$\leqslant a(t) \int_{-\infty}^{-\infty} |h| d\rho(\lambda) = a(t) \int_{-\infty}^{+\infty} |\chi(x, \lambda) \psi(y, \lambda)| d\rho(\lambda) \leqslant$$

$$\leqslant a(t) \sqrt{\int_{-\infty}^{+\infty} \chi^2(x, \lambda) d\rho(\lambda)}. \sqrt{\int_{-\infty}^{+\infty} \psi^2(y, \lambda) d\rho(\lambda)} = a(t) \sqrt{\int_0^y \Phi^2(u, v) dv}. \sqrt{x},$$

from which follows the absolute uniform convergence of the integral

$$\int_0^\infty h(x, y; \lambda) \varphi(t, \lambda) d\rho(\lambda).$$

We shall now consider the integral between the limits $-\infty$ and 0. From the the estimate (6.9) follows

$$|h(x, y; \lambda)| \leqslant A(x, y) \operatorname{ch} \sqrt{|\lambda|}(x + y),$$

where the function $A(x, y)$ does not depend upon λ. Thus

$$\int_{-\infty}^0 |h(x, y; \lambda)| |\varphi(t, \lambda)| d\rho(\lambda) \leqslant$$

$$\leqslant A(x, y) a(t) \int_{-\infty}^0 \operatorname{ch} \sqrt{|\lambda|}(x + y) \operatorname{ch} \sqrt{|\lambda|}\, t . d\rho(\lambda).$$

Because of the conditions laid down upon $\rho(\lambda)$ the last integral converges. Thus (since because of (6.6) $h(x, y; \lambda)$ is the Fourier transform of the function Z) it follows from the theorem of paragraph 2, §5 that

$$Z(x, y, t = \begin{cases} \int_{-\infty}^{+\infty} h(x, y; \lambda) \varphi(t, \lambda) d\rho(\lambda), & t \leqslant x + y, \\ 0, & t > x + y. \end{cases}$$

Integrating $Z(x, y, s)$ with respect to s between the limits 0 and t, we obtain

$$Z_1(x, y, t) = \int_0^t Z(x, y, s) ds =$$

$$= \int_{-\infty}^{+\infty} \left(\int_0^x \varphi(u, \lambda) du \right) \left(\int_0^y \cos \sqrt{\lambda}\, v dv \right) \left(\int_0^t \varphi(s, \lambda) ds \right) d\rho(\lambda).$$

From the foregoing we obtain for $t > x + y$

$$\frac{\partial Z_1}{\partial t} = Z(x, y, t) = 0.$$

Since the function $Z_1(x, y, t)$ is symmetric in the variables x and t,

$$\frac{\partial Z_1}{\partial x} = 0$$

for $x > t + y$, i.e., for $t < x - y$. Consequently, if t is outside the interval $(x - y, x + y)$, then

$$\frac{\partial Z}{\partial x} = \frac{\partial}{\partial x}\left(\frac{\partial Z_1}{\partial t}\right) = \frac{\partial}{\partial t}\left(\frac{\partial Z_1}{\partial x}\right) = 0.$$

Differentiating the equation (6.6) with respect to x, we obtain

$$\varphi(x, \lambda)\int_0^t \cos\sqrt{\lambda}v\, dv = \int_{x-t}^{x+t} \frac{\partial Z}{\partial x}\cdot\varphi(s, \lambda)\, ds.$$

Differentiating the last equation with respect to t, we find

$$\varphi(x, \lambda)\cos\sqrt{\lambda}t = \frac{1}{2}[\varphi(x+t, \lambda) + \varphi(x-t, \lambda)] + \int_{x-t}^{x+t} W(x,t,s)\,\varphi(s,\lambda)\, ds. \quad (6.10)$$

Formula (6.10) allows one to obtain without particular difficulty the differential equation for $\phi(x, \lambda)$. Subtract $\phi(x, \lambda)$ from both sides of formula (6.10) and divide by t^2. We obtain

$$\varphi(x, \lambda)\frac{\cos\sqrt{\lambda}t - 1}{t^2} = \frac{\varphi(x+t,\lambda) - 2\varphi(x,\lambda) + \varphi(x-t,\lambda)}{2t^2} + \frac{1}{t^2}\int_{x-t}^{x+t} W(x, t, s)\,\varphi(s, \lambda)\, ds.$$

As $t \to 0$ the left side of this equation approaches $\frac{\lambda}{2}\phi(x, \lambda)$, the first term of the right side approaches $\frac{1}{2}\phi''(x, \lambda)$ (the differentiability of $\phi(x, \lambda)$ follows from the differentiability of K); consequently, the limit

$$\lim_{t\to 0}\frac{1}{t^2}\int_{x-t}^{x+t} W(x, t, s)\,\varphi(s, \lambda)\, ds$$

exists. Let us compute this limit.

Expanding the function $\phi(x, \lambda)$ in a Taylor series, we obtain

$$\varphi(s, \lambda) = \varphi(x, \lambda) + (s - x)\varphi'(x, \lambda) + O(t^2).$$

Thus

$$\int_{x-t}^{x+t} W(x, t, s)\,\varphi(s, \lambda)\, ds =$$

$$= \varphi(x,\lambda)\int_{x-t}^{x+t} W(x, t, s)\, ds + \varphi'(x, \lambda)\int_{x-t}^{x+t}(s - x)\, W(x, t, s)\, ds + O(t^3).$$

Since the function $W(x, t, s)$ is continuous for $x - t \leq s \leq x + t$, then for $t \to 0$

$$W(x, t, s) = W(x, t, x) + o(1).$$

Thus *

* We note that $$\int_{x-t}^{x+t}(s - x)\, W(x, t, x)\, ds = 0.$$

$$\int_{x-t}^{x+t} W(x,t,s)\,\varphi(s,\lambda)\,ds = \varphi(x,\lambda)\int_{x-t}^{x+t} W(x,t,s)\,ds +$$

$$+\varphi'(x,\lambda)\int_{x-t}^{x+t}(s-x)\left[W(x,t,x)+o(1)\right]ds + o(t^2) =$$

$$=\varphi(x,\lambda)\int_{x-t}^{x+t} W(x,t,s)\,ds + o(t^2).$$

Then,

$$\frac{1}{t^2}\int_{x-t}^{x+t} W(x,t,s)\,\varphi(s,\lambda)\,ds = \varphi(x,\lambda)\cdot\frac{1}{t^2}\int_{x-t}^{x+t} W(x,t,s)\,ds + o(1).$$

Since, for fixed x, $\phi(x,\lambda)$ cannot vanish identically in λ and since in the last equation the limit on the left exists, the following limit exists

$$\lim_{t\to 0}\frac{1}{t^2}\int_{x-t}^{x+t} W(x,t,s)\,ds = -\frac{1}{2}\,q(x).$$

Consequently, $\phi(x,\lambda)$ satisfies the differential equation

$$-\lambda\varphi(x,\lambda) = \varphi''(x,\lambda) - q(x)\,\varphi(x,\lambda)$$

or

$$\varphi''(x,\lambda) + (\lambda - q(x))\,\varphi(x,\lambda) = 0.$$

From the last equation follows

$$q(x) - \lambda = -\frac{\varphi''(x,\lambda)}{\varphi(x,\lambda)}.$$

Since $\phi(x,\lambda)$ cannot vanish identically for fixed x, $q(x)$ is a continuous function. From formula (6.1) follows

$$\varphi(0,\lambda) = 1, \quad \varphi_x'(0,\lambda) = K(0,0) = h.$$

Thus $\rho(\lambda)$ determines not only the differential equation for $\phi(x,\lambda)$, but also the initial conditions.

2. Let us now consider the case $h = \infty$. Let

$$\psi(x,\lambda) = \frac{\sin\sqrt{\lambda}x}{\sqrt{\lambda}} + \int_0^x L(x,t)\,\frac{\sin\sqrt{\lambda}t}{\sqrt{\lambda}}\,dt, \tag{6.11}$$

$$\frac{\sin\sqrt{\lambda}x}{\sqrt{\lambda}} = \psi(x,\lambda) - \int_0^x L_1(x,t)\,\psi(t,\lambda)\,dt. \tag{6.12}$$

It follows from (6.11) for $y < x$ that

$$\psi(x,\lambda)\,\frac{\sin\sqrt{\lambda}y}{\sqrt{\lambda}} = \frac{\sin\sqrt{\lambda}x}{\sqrt{\lambda}}\cdot\frac{\sin\sqrt{\lambda}y}{\sqrt{\lambda}} + \int_0^x L(x,t)\,\frac{\sin\sqrt{\lambda}t}{\sqrt{\lambda}}\cdot\frac{\sin\sqrt{\lambda}y}{\sqrt{\lambda}}\,dt =$$

$$= \frac{1}{2}\int_{x-y}^{x+y}\frac{\sin\sqrt{\lambda}t}{\sqrt{\lambda}}\,dt + \int_0^x L(x,t)\left[\int_{|t-y|}^{t+y}\frac{\sin\sqrt{\lambda}u}{\sqrt{\lambda}}\,du\right]dt.$$

Reasoning exactly as in the preceding case, we obtain

$$\psi(x,\lambda)\frac{\sin\sqrt{\lambda}y}{\sqrt{\lambda}}=\frac{1}{2}\int_{x-y}^{x+y}\psi(t,\lambda)\,dt+\int_{x-y}^{x+y}W_1(x,y,t)\psi(t,\lambda)\,dt, \qquad (6.13)$$

where the function $W_1(x, y, t)$ is continuous for all of its arguments and, in particular, $W_1(x, y, t \pm x) = 0$ and $\partial W_1/\partial y$ is continuous for $x - y \leq t \leq x + y$. Differentiating (6.13) with respect to y, we obtain

$$\psi(x,\lambda)\cos\sqrt{\lambda}y=\frac{1}{2}[\psi(x+y,\lambda)+\psi(x-y,\lambda)]+\int_{x-y}^{x+y}\frac{\partial W_1(x,y,t)}{\partial y}\psi(t,\lambda)\,dt.$$

This equation is analogous to (6.4) and thus the following differential equation for ψ holds:

$$\psi''(x,\lambda)+(\lambda-q(x))\psi(x,\lambda)=0.$$

§7. Examples

1. Let us suppose that the function $\rho(\lambda) = \frac{2}{\pi}\sqrt{\lambda} + \sigma(\lambda)$ for $\lambda \geq 0$ and $\rho(\lambda)=0$ for $\lambda < 0$ where the function $\sigma(\lambda)$ satisfies the following conditions:

1) $\rho(\lambda)$ is nondecreasing;

2) $\int_0^\infty \lambda^k |d\sigma(\lambda)| < +\infty, \quad k = 0, 1;$

3) $\rho(\lambda)$ has on some finite interval an infinite number of points of increase.

In the case under consideration

$$F(x,y)=\int_0^\infty\frac{\sin\sqrt{\lambda}x\cdot\sin\sqrt{\lambda}y}{\lambda}\,d\sigma(\lambda),$$

$$f(x,y)=\int_0^\infty\cos\sqrt{\lambda}x\cos\sqrt{\lambda}y.d\sigma(\lambda),$$

$$\frac{\partial f(x,y)}{\partial x}=-\int_0^\infty\sqrt{\lambda}\cos\sqrt{\lambda}x\cos\sqrt{\lambda}y.d\sigma(\lambda),$$

$$\frac{\partial^2 f}{\partial x^2}=\int_0^\infty\lambda\cos\sqrt{\lambda}x\cos\sqrt{\lambda}y\,d\sigma\,\lambda).$$

In view of condition 2) all these functions are continuous. Thus all of our results are applicable and, consequently, there exists a differential operator of the second order $L(y) = y'' - q(x)y$ with a continuous function $q(x)$ for which Parseval's equation can be written with the given function $\rho(\lambda)$. In particular, there follows from this the following interesting result.

There exists an equation of the form

$$y''+(\lambda-q(x))y=0, \quad 0\leqslant x<\infty,$$

in which the function $\rho(\lambda)$ will be on a finite interval an arbitrary monotonic bounded function given in advance.

2. In a series of particular cases we can actually find the function $q(x)$ for a spectral function $\rho(\lambda)$. Let us take one very simple example. Let

$$\rho(\lambda) = \frac{2}{\pi}\sqrt{\lambda} + \alpha \cdot e(\lambda - \lambda_0) \quad \text{for} \quad \lambda > 0,$$
$$\rho(\lambda) = \alpha e(\lambda - \lambda_0) \quad \text{for} \quad \lambda < 0,$$

where $e(\lambda - \lambda_0) = 0$ if $\lambda < \lambda_0$, and $e(\lambda - \lambda_0) = 1$ if $\lambda > \lambda_0$, and α is a positive number.

The differential equation constructed with this $\rho(\lambda)$ must have a continuous spectrum for $\lambda > 0$ and a point of the discrete spectrum for $\lambda = \lambda_0$. In this case the function $\sigma(\lambda)$ has the form

$$\sigma(\lambda) = \alpha e(\lambda - \lambda_0)$$

and consequently

$$f(x, y) = \int_{-\infty}^{+\infty} \cos\sqrt{\lambda}\, x \cdot \cos\sqrt{\lambda}\, y\, d\sigma(\lambda) = \alpha \cos\sqrt{\lambda_0}\, x \cdot \cos\sqrt{\lambda_0}\, y.$$

The integral equation (II) will be an equation with the degenerate kernel $f(x, y)$ and has the form

$$\alpha \cos\sqrt{\lambda_0}\, x \cos\sqrt{\lambda_0}\, y + \alpha \int_0^x K(x, s) \cos\sqrt{\lambda_0}\, s . \cos\sqrt{\lambda_0}\, y\, ds + K(x, y) = 0.$$

From this we have

$$K(x, y) = \frac{-\alpha \cos\sqrt{\lambda_0}\, x \cos\sqrt{\lambda_0}\, y}{1 + \alpha \int_0^x \cos^2\sqrt{\lambda_0}\, s\, ds}.$$

Let us now find the equation and the boundary conditions. From formula (1.8′) we have

$$h = K(0, 0) = -\alpha.$$

Furthermore, from formula (1.8) we have

$$\frac{1}{2} q(x) = \frac{dK(x, x)}{dx} = \frac{d}{dx}\left(\frac{-\alpha \cos^2\sqrt{\lambda_0}\, x}{1 + \alpha \int_0^x \cos^2\sqrt{\lambda_0}\, s\, ds}\right).$$

The corresponding eigenfunctions are computed by the formula

$$\varphi(x, \lambda) = \cos\sqrt{\lambda}\, x + \int_0^x K(x, t) \cos\sqrt{\lambda}\, t\, dt,$$

i.e.,

$$\varphi(x, \lambda) = \cos\sqrt{\lambda}\, x - \frac{\alpha \cos\sqrt{\lambda_0}\, x}{1 + \alpha \int_0^x \cos^2\sqrt{\lambda_0}\, s ds} \cdot \int_0^x \cos\sqrt{\lambda_0}\, t \cos\sqrt{\lambda}\, t dt.$$

As is evident from the definition of the function $\rho(\lambda)$, the spectrum consists of

the numbers $\lambda \geq 0$ corresponding to the continuous spectrum and the point λ_0 of the discrete spectrum. The eigenfunction $\phi(x, \lambda_0)$ of the discrete spectrum has in this case the form

$$\varphi(x, \lambda_0) = \frac{\cos\sqrt{\lambda_0}\, x}{1+\alpha\int\limits_0^x \cos^2\sqrt{\lambda_0}\, s\, ds}.$$

If $\lambda_0 < 0$, then instead of $\cos\sqrt{\lambda_0}\, x$ one must write $\cosh\sqrt{|\lambda|}\, x$.

Analogously one may find an equation and eigenfunctions for the case

$$\rho(\lambda) = \begin{cases} \frac{2}{\pi}\sqrt{\lambda} + \sigma(\lambda) & \text{for } \lambda \geqslant 0, \\ \sigma(\lambda) & \text{for } \lambda < 0, \end{cases}$$

where $\sigma(\lambda)$ is a step function with a finite number of jumps. In this case the integral equation also turns out to be degenerate.

In the beginning of the paper, we compared an arbitrary equation with the simple equation $y'' + \lambda y = 0$. Indeed, for this reason, we express the eigenfunctions by means of $\cos\sqrt{\lambda}\, t$. Without difficulty we may take some other equation instead of the "selected" equation. In particular, one may write formulas for the eigenfunctions of an equation obtained from the given one if to its spectral function $\rho(\lambda)$ one adds a step function with a finite number of jumps.

§8. The condition of orthogonality of an expansion in a Fourier integral

Let $\phi(x, \lambda)$ be the solution of the equation

$$y'' + [\lambda - q(x)]\, y = 0,$$

satisfying boundary conditions

$$y(0) = 1, \quad y'(0) = h,$$

and let $\rho(\lambda)$ be a spectral function. The latter means, as we have said, that for each function $f(x) \in L_2(0, \infty)$ there exists the Fourier transform

$$E(\lambda) = \underset{n\to\infty}{\text{l.i.m.}} \int\limits_0^n f(x)\, \varphi(x, \lambda)\, dx$$

and the Parseval equation holds:

$$\int\limits_0^\infty f^2(x)\, dx = \int\limits_{-\infty}^{+\infty} E^2(\lambda)\, d\rho(\lambda).$$

Let $\Delta = (\lambda, \lambda + \Delta)$ be an arbitrary finite interval. Let us set

$$E_\Delta(x) = \int\limits_\Delta \varphi(x, \lambda)\, d\rho(\lambda).$$

The spectral function $\rho(\lambda)$ is called *orthogonal* if, for an arbitrary interval Δ, $E_\Delta(x) \in L_2(0, \infty)$ and for any two finite intervals Δ and Δ'

$$\int_0^\infty E_\Delta(x) E_{\Delta'}(x)\,dx = \int_{\Delta\Delta'} d\rho(\lambda). \tag{8.1}$$

In the general case the differential equation may have nonorthogonal spectral functions. In the present paragraph there is given a sufficient condition in order that the function $\rho(\lambda)$ satisfying the conditions of the preceding paragraph be an orthogonal spectral function. In particular, from this condition follows that $\rho(\lambda)$ is known to be orthogonal if it is concentrated on the interval $(c, +\infty)$, $c > -\infty$.

Let us denote by $L_2\{\rho\}$ the real Hilbert space of functions $g(\lambda)$ satisfying the condition

$$\int_{-\infty}^{+\infty} g^2(\lambda)\,d\rho(\lambda) < +\infty.$$

Lemma. *In order that the spectral function $\rho(\lambda)$ be orthogonal it it sufficient that the set of functions of the form*

$$F_a(\lambda) = \int_0^a f(x)\cos\sqrt{\lambda}\,x\,dx,$$

be dense in $L_2\{\rho\}$, where a is an arbitrary positive number and $f(x)$ is a continuous function having a continuous first derivative.

Proof. First of all let us show that the functions $F_a(\lambda)$ belong to $L_2\{\rho\}$. The existence of the integral

$$\int_{-\infty}^{0} F_a^2(\lambda)\,d\rho(\lambda)$$

follows immediately from condition 1, §4. Furthermore, integrating by parts, we obtain

$$|F_a(\lambda)| = O\left(\frac{1}{\sqrt{\lambda}}\right).$$

Consequently,

$$|F_a(\lambda)|^2 = O\left(\frac{1}{\lambda}\right).$$

Thus it follows from the result of §4 that

$$\int_0^\infty |F_a(\lambda)|^2\,d\rho(\lambda) < +\infty.$$

Therefore, $F_a(\lambda) \in L_2\{\rho\}$.

Using the formula

$$\cos\sqrt{\lambda}\,x = \varphi(x,\lambda) - \int_0^x K_1(x,t)\,\varphi(t,\lambda)\,dt,$$

we obtain

$$F_a(\lambda) = \int_0^a f(x)\left[\varphi(x,\lambda) - \int_0^x K_1(x,t)\,\varphi(t,\lambda)\,dt\right]dx =$$

$$= \int_0^a \left[f(x) - \int_x^a K_1(t,x)\,f(t)\,dt\right]\varphi(x,\lambda)\,dx = \int_0^a g(x)\,\varphi(x,\lambda)\,dx,$$

where

$$g(x) = f(x) - \int_x^a K_1(t, x) f(t)\, dt. \tag{8.2}$$

Equation (8.2) is a Volterra equation. Thus $g(x)$ may be considered as an arbitrary function with a continuous derivative. It then remains to show that if the set of functions

$$\int_0^a g(x) \varphi(x, \lambda)\, dx$$

is dense in $L_2\{\rho\}$, then the spectral function $\rho(\lambda)$ is orthogonal. Let us denote by $F(\lambda)$ an arbitrary function from $L_2\{\rho\}$ and let the functions

$$F_n(\lambda) = \int_0^n f_n(x) \varphi(x, \lambda)\, dx$$

converge in the mean to $F(\lambda)$ as $n \to \infty$. It follows from Parseval's equation $(n > m)$ that

$$\int_0^n |f_n(x) - f_m(x)|^2\, dx = \int_{-\infty}^{+\infty} |F_n(\lambda) - F_m(\lambda)|^2\, d\rho(\lambda).$$

Thus the functions $f_n(x)$ converge in the mean square to some limit $f(x)$, and $F(\lambda)$ is the Fourier transform for $f(x)$, i.e., we have shown that an arbitrary function $F(\lambda) \in L_2\{\rho\}$ is a Fourier transform of some function $f(x)$.

Let

$$F_\Delta(\lambda) = \begin{cases} 1, & \text{if } \lambda \in \Delta, \\ 0, & \text{if } \lambda \overline{\in} \Delta. \end{cases}$$

The function $F_\Delta(\lambda) \in L_2\{\rho\}$. Thus, from the above, there exists a function $f_\Delta(x) \in L_2(0, \infty)$ whose Fourier transform is $F_\Delta(\lambda)$. We shall show that

$$f_\Delta(x) = \int_\Delta \varphi(x, \lambda)\, d\rho(\lambda). \tag{8.3}$$

From the last equation and Parseval's equation it follows that

$$\int_\Delta^\infty f_\Delta(x) f_{\Delta'}(x)\, dx = \int_{\Delta\Delta'} d\rho(\lambda),$$

i.e., the orthogonality of the spectral function $\rho(\lambda)$.

There now remains to prove equation (8.3). Let us denote by $h(x)$ $(x \geq 0)$ an arbitrary continuous function vanishing outside of a finite interval and let

$$H(\lambda) = \int_0^\infty h(x) \varphi(x, \lambda)\, dx.$$

Because of Parseval's equation

$$\int_0^\infty h(x) f_\Delta(x)\, dx = \int_{-\infty}^\infty H(\lambda) \cdot F_\Delta(\lambda)\, d\rho(\lambda) = \int_\Delta H(\lambda)\, d\rho(\lambda) =$$

$$= \int_\Delta d\rho(\lambda) \int_0^\infty h(x) \varphi(x, \lambda)\, dx = \int_0^\infty h(x) \int_\Delta \varphi(x, \lambda)\, d\rho(\lambda).$$

Since $h(x)$ is an arbitrary function, then it follows from the last equation that almost everywhere

$$f_\Delta(x) = \int_\Delta \varphi(x, \lambda)\, d\rho(\lambda).$$

The last equation, because of the continuity of the right-hand side, holds everywhere.

Let us point out a criterion of completeness for the set of functions $F_a(\lambda)$ in $L_2\{\rho\}$.

Theorem. *If there exists a number* $\alpha < 2$ *such that for sufficiently large* $x > 0$

$$\int_{-\infty}^{0} e^{\sqrt{|\lambda|}\,x} d\rho(\lambda) < e^{x^\alpha}, \tag{8.4}$$

then the set of functions $F_a(\lambda)$ *is dense in* $L_2\{\rho\}$.

Proof. The set of functions $F_a(\lambda)$ generates in $L_2\{\rho\}$ a linear subspace. After taking the closure, we obtain the closed subspace $\mathfrak{M}\{\rho\}$. One must now show that if condition (8.4) is satisfied then $\mathfrak{M}\{\rho\} = L_2\{\rho\}$. Let us assume the contrary. Then in $L_2\{\rho\}$ one may find an element $g(\lambda)$ orthogonal to $\mathfrak{M}\{\rho\}$. In particular, for an arbitrary function

$$F_a(\lambda) = \int_0^a f(t) \cos\sqrt{\lambda}\, t\, dt$$

we have

$$\int_{-\infty}^{\infty} g(\lambda) F_a(\lambda)\, d\rho(\lambda) = 0. \tag{8.5}$$

Let us assume that for fixed $x > 0$ ($f(t) = 0$ for $t > a$)

$$f_x(t) = \frac{1}{2}\{f(x+t) + f(x-t)\}.$$

Let us introduce the function

$$F_a(\lambda;\ x) = \int_0^\infty f_x(t) \cos\sqrt{\lambda}\, t\, dt.$$

It is easy to see that

$$F_a(\lambda;\ x) = \cos\sqrt{\lambda}\, x F_a(\lambda).$$

Then, since $F_a(\lambda; x)$ belongs to $\mathfrak{M}\{\rho\}$, the following equation holds for each x:

$$\int_{-\infty}^{\infty} \cos\sqrt{\lambda}\, x\, g(\lambda) F_a(\lambda)\, d\rho(\lambda) = 0. \tag{8.6}$$

For $\lambda < 0$

$$|F_a(\lambda)| \leqslant C \cdot e^{\sqrt{|\lambda|}\, a}. \tag{8.7}$$

For sufficiently large x the following estimate follows from the estimates (8.4) and (8.7) and the Cauchy-Bunyakovskiĭ inequality:

$$\left| \int_{-\infty}^{0} \cos \sqrt{\lambda} x g(\lambda) F_a(\lambda) d\rho(\lambda) \right| \leqslant C \int_{-\infty}^{0} \cos \sqrt{\lambda} x e^{\sqrt{|\lambda|} a} |g(\lambda)| d\rho(\lambda) \leqslant$$

$$\leqslant C \left(\int_{-\infty}^{0} \operatorname{ch}^2 \sqrt{|\lambda|} x e^{2\sqrt{|\lambda|} a} d\rho(\lambda) \right)^{1/2} \left(\int_{-\infty}^{0} |g(\lambda)|^2 d\rho(\lambda) \right)^{1/2} \leqslant C_1 \exp(x^{\alpha}). \quad (8.7')$$

The following theorem ([7]) holds:

For each $x > 0$ *let*

$$\int_{-\infty}^{\infty} \cos \sqrt{\lambda} x \, d\sigma(\lambda) = 0,$$

where this integral converges absolutely for each x. *If there exists* $\alpha < 2$ *such that for all sufficiently large* x

$$\int_{-\infty}^{0} \cos \sqrt{\lambda} x |d\sigma(\lambda)| < e^{x^{\alpha}},$$

then $\sigma(\lambda) \equiv \text{const.}$

From (8.6), (8.7′) and this result, it follows that for an arbitrary finite interval Δ

$$\int_{\Delta} g(\lambda) F_a(\lambda) d\rho(\lambda) = 0. \quad (8.8)$$

For $\lambda > 0$ the set of functions $F_a(\lambda)$ is dense among all continuous functions. Thus it follows from (8.8) that for each interval Δ with positive end-points

$$\int_{\Delta} g(\lambda) d\rho(\lambda) = 0. \quad (8.9)$$

Let N be an arbitrary positive number. From (8.9) follows:

$$\int_0^N g^2(\lambda) d\rho(\lambda) = \int_0^N g(\lambda) d \int_0^{\lambda} g(\mu) d\rho(\mu) = 0.$$

Letting $N \to \infty$, we obtain

$$\int_0^{\infty} g^2(\lambda) d\rho(\lambda) = 0.$$

Furthermore, from (8.9) and (8.5) follows

$$\int_{-\infty}^{0} g(\lambda) F_a(\lambda) d\rho(\lambda) = 0. \quad (8.10)$$

From condition 1, §4 it follows that

$$\int_{-\infty}^{0} d\rho(\lambda) < \infty.$$

Then, applying the Cauchy-Bunyakovskiĭ inequality, we obtain

$$\int_{-\infty}^{0} |g(\lambda)| d\rho(\lambda) \leqslant \left(\int_{-\infty}^{0} g^2(\lambda) d\rho(\lambda) \right)^{1/2} \left(\int_{-\infty}^{0} d\rho(\lambda) \right)^{1/2} < \infty.$$

Changing the order of integration in (8.10), we obtain

$$\int_0^a f(x)\,dx \int_{-\infty}^0 g(\lambda)\cos\sqrt{\lambda}\,x\,d\rho(\lambda) = 0.$$

Because of the arbitrariness of the functions $f(x)$ it follows from the last equation that

$$\int_{-\infty}^0 g(\lambda)\cos\sqrt{\lambda}\,x\,d\rho(\lambda) = 0.$$

Applying again the result indicated above we obtain

$$\int_{\Delta} g(\lambda)\,d\rho(\lambda) = 0.$$

Thus

$$\int_{-\infty}^0 g^2(\lambda)\,d\rho(\lambda) = 0,$$

which together with the equation obtained earlier gives

$$\int_{-\infty}^{\infty} g^2(\lambda)\,d\rho(\lambda) = 0,$$

and the theorem is proved.

Remark I. If $h = \infty$, then one may place

$$F_a(\lambda) = \int_0^a f(x)\frac{\sin\sqrt{\lambda}\,x}{\sqrt{\lambda}}\,dx.$$

In place of condition (8.4) one may require that for all sufficiently large x the following estimate is satisfied:

$$\int_{-\infty}^0 \frac{\sin\sqrt{\lambda}\,x}{\sqrt{\lambda}}\,d\rho(\lambda) < \exp(x^{\alpha}),$$

where $a < 2$. The rest of the proof is analogous.

Remark II. If $\rho(\lambda) = \text{const.}$ for $\lambda < c$, $c > -\infty$, then for sufficiently large x

$$\int_{-\infty}^0 \cos\sqrt{\lambda}\,x\,d\rho(\lambda) = \int_c^0 \cos\sqrt{\lambda}\,x\,d\rho(\lambda) \leqslant Ae^{\sqrt{|c|}\,x}.$$

Thus the estimate (8.4) is satisfied and consequently the spectral measure is orthogonal.

§9. The case of a single limit point at infinity of the spectrum

1. Up to now we have excluded the case when the spectrum has a single limit point at infinity. To this group of boundary problems belongs the classical Sturm-Liouville problem. In the present section we shall consider the case of a discrete spectrum on a half-line. In the following two sections we shall consider the clas-

sical Sturm-Liouville problem. We first introduce a concept. Let E be some set of real numbers. Let us designate by $N(E)$ the number of numbers of the set E included in the interval $(-N, N)$. Furthermore, let us set

$$\mu = \overline{\lim_{N\to\infty}} \frac{N(E)}{N}.$$

The number μ will hereafter be called the density of the set E.

Let us consider the Sturm-Liouville problem on the finite interval $(0, l)$:

$$y'' + \{\lambda - q(x)\} y = 0, \tag{9.1}$$

$$y'(0) - hy(0) = 0, \tag{9.2}$$

$$y'(l) + Hy(l) = 0, \tag{9.3}$$

where h and H are real numbers. Let us designate by $\lambda_1(l), \lambda_2(l), \cdots, \lambda_n(l), \cdots$ the eigenvalues of the boundary problem (9.1)–(9.3). As is known,

$$\sqrt{\lambda_n(l)} = \frac{\pi}{l} n + O\left(\frac{1}{n}\right). \tag{9.4}$$

Let $E(l)$ be the set of numbers $\sqrt{|\lambda_n(l)|}$ $(n = 1, 2, \cdots)$. From formula (9.4) follows that in this case

$$\mu = \frac{l}{\pi}.$$

Now let $g(x)$ be continuous in each finite interval. Let us consider equation (9.1) together with the boundary condition (9.2) on the interval $(0, \infty)$. It is known that if $q(x) \geq 0$ and tends monotonically to ∞, the spectrum is discrete, positive and has a unique limit point at infinity. In addition, the nth eigenfunction has n zeros. Let us denote the points in the spectrum in this last case by $\lambda_1, \lambda_2, \cdots, \lambda_n, \cdots$, and let E be the set of points $\sqrt{\lambda_n}$ $(n = 1, 2, \cdots)$. We shall show that the density of the set E equals infinity.

Let us consider the boundary problem (9.1)–(9.3). Since the density of the set $E(l)$ equals l/π, one may select a number N sufficiently large that

$$\frac{N[E(l)]}{N} > \frac{l}{2\pi}.$$

We shall denote by $\phi(x, \lambda)$ the solution of equation (9.1) satisfying the first boundary condition (9.2). It is clear that the number of zeros of the function $\phi(x, \lambda)$ in the interval $(0, \infty)$ is not less than the number of zeros of the same function in the interval $(0, l)$. Thus it follows from Sturm's theorem that the number of eigenvalues of the boundary problem (9.1)–(9.3) in the interval $(0, N)$ is not less than the number of eigenvalues of the boundary problem (9.1)–(9.3) on the infinite interval. Thus,

$$\frac{N(E)}{N} \geqslant \frac{N[E(l)]}{N} > \frac{l}{2\pi}.$$

Since l is chosen arbitrarily, then

$$\overline{\lim_{N\to\infty}}\ \frac{N(E)}{N} = \infty, \tag{9.5}$$

which is what we wish to prove.

2. Let us now consider the inverse problem. Let the function $\rho(\lambda)$ satisfy conditions 1 and 2 of §4, and be a pure step function with a single limit point at infinity of the jumps. Let us denote the points of increase of the function $\rho(\lambda)$ by $\lambda_1, \lambda_2, \cdots, \lambda_n, \cdots$. Furthermore, let us denote by E the set of points

$$\sqrt{|\lambda_n|}\ (n = 1, 2, \cdots).$$

We shall assume that the density of the set E is infinite. We shall now show that under these conditions the inverse problem is solvable.

The only place in which we have made use of the presence of limit points of increase of the function $\rho(\lambda)$ within a finite distance was in the solvability of equation (II):

$$f(x, y) + \int_0^x f(y, s) K(x, s)\, ds + K(x, y) = 0. \tag{II}$$

Thus we must show that if the density of the set E is infinite then equation (II) is solvable. In §4 we showed that equation (II) is known to be solvable if it follows from equation

$$\int_{-\infty}^{\infty} G^2(\lambda)\, d\rho(\lambda) = 0, \tag{9.6}$$

where

$$G(\lambda) = \int_0^x g(t) \cos\sqrt{\lambda}\, t\, dt,$$

and $g(t)$ is a square-integrable function, that $G(\lambda) = 0$ and consequently that $g(t)$ vanishes almost everywhere.

Equation (9.6) can hold only if the jumps of the function $\rho(\lambda)$ coincide with the zeros of $G(\lambda)$. We shall show that this is impossible. Let us denote by $\lambda_{1,x}$, $\lambda_{2,x}, \cdots, \lambda_{n,x}, \cdots$ the set of all real zeros of the function $G(\lambda)$ and by E_x the set of real numbers $\sqrt{|\lambda_{n,x}|}$ $(n = 1, 2, \cdots)$. Furthermore, let us designate by μ_x the density of the set E_x. It is known ([8]) that

$$\mu_x \leqslant \frac{x}{\pi}.$$

Since by hypothesis $\mu = \infty$, it follows from equation (9.6) that, for $x < \infty$, $G(\lambda) \equiv 0$; thus equation (II) is solvable.

If the points of increase of the function $\rho(\lambda)$ are bounded from below, then, in view of the results of the preceding paragraph, the eigenfunctions are orthogonal.

§10. The case of a finite interval

Let us consider the equation

$$y'' + \{\lambda - q(x)\}\, y = 0 \tag{10.1}$$

on the interval $(0, l)$; we shall suppose that the function $q(x)$ is continuous in each interval $(0, l')$ where $l' < l < \infty$. Together with equation (10.1) we shall consider the boundary condition at the point $x = 0$:

$$y(0) = 1, \quad y'(0) = h. \tag{10.2}$$

Let $\phi(x, \lambda)$ be the solution of equation (10.1) satisfying conditions (10.2). Just as in the first section, one may show that if the function $K(x, t)$ is a solution of the partial differential equation

$$\frac{\partial^2 K}{\partial x^2} - q(x) K = \frac{\partial^2 K}{\partial t^2}$$

and satisfies the conditions

$$K(x, x) = h + \int_0^x q(t)\, dt \quad (x < l), \quad \left.\frac{\partial K}{\partial t}\right|_{t=0} = 0,$$

then

$$\varphi(x, \lambda) = \cos \sqrt{\lambda}\, x + \int_0^x K(x, t) \cos \sqrt{\lambda}\, t\, dt \quad (x < l). \tag{10.3}$$

Transforming the last equation, we obtain

$$\cos \sqrt{\lambda}\, x = \varphi(x, \lambda) - \int_0^x K_1(x, t)\, \varphi(t, \lambda)\, dt \quad (x < l).$$

Let $\rho(\lambda)$ be a spectral function for equation (10.1), i.e., for any function $f(x) \in L_2(0, l)$ there exists the Fourier transform

$$E(\lambda) = \underset{l' \to l}{\text{l. i. m.}} \int_0^{l'} f(x)\, \varphi(x, \lambda)\, d\lambda$$

and Parseval's equation holds:

$$\int_0^l f^2(x)\, dx = \int_{-\infty}^{\infty} E^2(\lambda)\, d\rho(\lambda).$$

If $\rho(\lambda)$ is an orthogonal spectral function, then the set of functions of the form

$$\int_0^{l'} f^2(x)\, \varphi(x, \lambda)\, dx \quad (l' < l),$$

where $f(x)$ is an arbitrary continuous function, is dense in $L_2\{\rho\}$. Let us set

$$\sigma(\lambda) = \begin{cases} \rho(\lambda) - \frac{2}{\pi} \sqrt{\lambda}, & \lambda \geqslant 0, \\ \rho(\lambda), & \lambda < 0, \end{cases} \tag{10.4}$$

$$F(x, y) = \int_{-\infty}^{\infty} \frac{\sin \sqrt{\lambda}\, x \cdot \sin \sqrt{\lambda}\, y}{\lambda}\, d\sigma(\lambda) \quad (x, y < l). \tag{10.5}$$

Just as in §§2 and 3 one may show that there exist continuous derivatives $(x, y < l)$

$$\frac{\partial^2 F}{\partial x \partial y} = f(x, y), \quad \frac{\partial f}{\partial x}, \quad \frac{\partial f}{\partial y}$$

and the following integral equations hold:

$$f(x, y) = K_1(x, y) - \int_0^y K_1(x, t) K_1(y, t)\, dt \quad (y \leqslant x,\ x < l), \tag{I}$$

$$f(x, y) + \int_0^x f(y, t) K(x, t)\, dt + K(x, y) = 0 \quad (y \leqslant x,\ x < l). \tag{II}$$

2. Let us consider the inverse problem. Let $\rho(\lambda)$ be a monotonic function bounded in each finite interval and satisfying the following two conditions:

1. *For each $x < 2l$ there exists the integral*

$$\int_{-\infty}^{\infty} \operatorname{ch} \sqrt{|\lambda|}\, x\, d\rho(\lambda).$$

2. *The function*

$$a(x) = \int_1^{\infty} \frac{\cos \sqrt{\lambda}\, x}{\lambda}\, d\sigma(\lambda)$$

has a continuous fourth derivative if $0 \leq x < 2l$. **In this equation $\sigma(\lambda)$ is determined from (10.4).**

It follows from condition 2 that the function $F(x, y)$, determined by equation (10.5), has the continuous partial derivatives

$$\frac{\partial^2 F}{\partial x \partial y} = f(x, y), \quad \frac{\partial^2 f}{\partial x^2}, \quad \frac{\partial^2 f}{\partial y^2}.$$

Furthermore, just as in §4, one may show that if $\rho(\lambda)$ has an infinite set of points of increase in any finite interval, then equation (II) is solvable for $x < l$. Equation (II) is also solvable in the case which was considered in the preceding section. Repeting the reasoning of §§5, 6, and 7, one may prove Parseval's equation and reconstruct the equation.

Thus, conditions 1 and 2 are sufficient in order that the function $\rho(\lambda)$ be a spectral function of equation (1) for the interval $(0, l)$.

Let us now see under what supplementary condition the function $\rho(\lambda)$ becomes an orthogonal spectral function. Just as in §8 one may show that in order that the function $\rho(\lambda)$ be an orthogonal spectral function, it is sufficient that (in addition to conditions 1 and 2) the set of functions

$$F_{l'}(\lambda) = \int_0^{l'} f(x) \cos \sqrt{\lambda}\, x\, dx \quad (l' < l),$$

be dense in $L_2\{\rho\}$, where $f(x)$ is an arbitrary continuous differentiable function. The last situation occurs in that case when it follows from the orthogonality of the function $g(\lambda) \in L_2\{\rho\}$ to all the functions $F_{l'}(\lambda)$, i.e., from the equation

$$\int_{-\infty}^{\infty} g(\lambda) F_{l'}(\lambda)\, d\rho(\lambda) = 0$$

that

$$\int_{-\infty}^{\infty} g^2(\lambda)\, d\rho(\lambda) = 0.$$

Let us denote by N an arbitrary positive number. Changing the order of integration, we obtain

$$\int_{-\infty}^{N} g(\lambda) F_{l'}(\lambda) d\rho(\lambda) = \int_{0}^{l'} f(x) \int_{-\infty}^{N} g(\lambda) \cos \sqrt{\lambda} x \, d\rho(\lambda). \tag{10.6}$$

Let us suppose that as $N \longrightarrow \infty$ the functions

$$\int_{-\infty}^{N} g(\lambda) \cos \sqrt{\lambda} x \, d\rho(\lambda)$$

converge in the mean square on the interval $(0, l)$ to some function* $G(x)$. Passing to the limit in equation (10.6), we obtain

$$\int_{-\infty}^{\infty} g(\lambda) F_{l'}(\lambda) d\rho(\lambda) = \int_{0}^{l'} f(x) G(x) dx = 0.$$

Since $f(x)$ is an arbitrary function, then it follows from the last equation that $G(x)$ vanishes almost everywhere on the interval $(0, l)$. Thus we have obtained the following theorem.

Theorem. *Let $g(\lambda)$ be an arbitrary function belonging to $L_2\{\rho\}$. If it follows that*

$$\int_{-\infty}^{\infty} g^2(\lambda) d\rho(\lambda) = 0,$$

whenever, as $N \longrightarrow \infty$, the functions

$$\int_{-\infty}^{N} g(\lambda) \cos \sqrt{\lambda} x \, d\rho(\lambda)$$

converge to 0 in the mean square on the interval $(0, l)$ then $\rho(\lambda)$ is an orthogonal spectral function.

Remark. The case $h = \infty$ may be treated analogously.

§11. The classical Sturm-Liouville problem

1. The results of the preceding sections allow one to characterize relatively easily the spectral functions corresponding to the classical Sturm-Liouville problem. Let the function $q(x)$ be continuous in the closed interval $(0, l)$. Let us denote by $\phi(x, \lambda)$ a solution of the equation

$$y'' + \{\lambda - q(x)\} y = 0, \tag{11.1}$$

satisfying the conditions

$$y(0) = 1, \quad y'(0) = h, \tag{11.2}$$

where h is an arbitrary real number. Let H be another real number. We are going to consider at the point $x = l$ the second boundary condition

$$(y' + Hy)_{x=l} = 0. \tag{11.3}$$

* For orthogonal spectral functions $\rho(\lambda)$ this always holds.

Let $\lambda_1, \lambda_2, \cdots, \lambda_n, \cdots$ be the eigenvalues of the boundary-value problem (11.1)–(11.3). As is known, the spectrum of the problem under consideration is bounded below. Let λ_0 be the smallest eigenvalue. Equation (11.1) may be written in the form

$$y'' + \{(\lambda - \lambda_0) - (q(x) - \lambda_0)\} y = 0.$$

Setting $\lambda - \lambda_0 = \mu$, we obtain a boundary-value problem with a non-negative spectrum. Thus, without destroying the generality of our reasoning, we may hereafter suppose that the spectrum is non-negative. The eigenfunctions $\phi(x, \lambda_n)$ corresponding to the different eigenvalues are mutually orthogonal, i.e.,

$$\int_0^l \varphi(x, \lambda_n)\varphi(x, \lambda_m)\,dx = 0 \quad (n \neq m).$$

Further, let us set

$$\rho_n = \int_0^l \varphi^2(x, \lambda_n)\,dx$$

(we recall that the eigenfunctions are selected so that they satisfy conditions (11.2)). For each function $f(x) \in L_2(0, l)$ Parseval's equation holds:

$$\int_0^l f^2(x)\,dx = \sum_{n=0}^{\infty} \frac{1}{\rho_n}\left\{\int_0^l f(x)\varphi(x, \lambda_n)\,dx\right\}^2.$$

If $H \neq \infty$, then the following asymptotic formula holds for the eigenvalues λ_n:

$$\sqrt{\lambda_n} = \frac{\pi}{l} n + O\left(\frac{1}{n}\right). \tag{11.4}$$

For the eigenfunctions one has the asymptotic formula:

$$\varphi(x, \lambda_n) = \cos n \frac{\pi}{l} x + O\left(\frac{1}{n}\right).$$

Consequently,

$$\rho_n = \frac{l}{2} + O\left(\frac{1}{n}\right). \tag{11.5}$$

In the case under consideration, the functions

$$K(x, t), \quad K_1(x, t), \quad f(x, t), \quad \frac{\partial f}{\partial x}, \quad \frac{\partial f}{\partial y} \quad (t \leqslant x)$$

are continuous up to the point $x = l$. Thus the integral equations (I) and (II) hold up to the points $y = l$ and $x = l$.

2. Let us now pass to the solution of the inverse problem. Let $\lambda_0, \lambda_1, \cdots, \lambda_n, \cdots$ be different non-negative numbers satisfying the asymptotic formula (11.4), and at the points $\sqrt{\lambda_n}$ let there be given positive jumps ρ_n satisfying the asymptotic formula (11.5). We shall prove the following lemma.

Lemma I. *The system of functions* $\cos\sqrt{\lambda_n}\,x$ *is complete on the interval* $(0, l)$ *in the space* L_2.

Proof. Changing the numbers λ_n to $\mu_n = l^2\lambda_n/\pi^2$, we reduce the interval $(0, l)$ to the interval $(0, \pi)$. Let us denote $\sqrt{\mu_n}$ by τ_n. Furthermore, using the following theorem of Levinson* [(9), p. 6]:

Let us denote by $\Lambda(u)$ the number of numbers r_n which are $\leq u$. If

$$\int_1^v \frac{\Lambda(u)}{u}\, du > v - \frac{1}{8}\ln v - c, \tag{11.6}$$

where c is a constant, then it follows from the equations

$$\int_0^\pi \cos\sqrt{\lambda_n}\, x \cdot f(x)\, dx = 0 \quad (n = 0, 1, \ldots, \quad f(x) \in L_2)$$

that $f(x) = 0$ almost everywhere, i.e., the system $\cos\sqrt{\mu_n}x$ is complete in L_2.

In our case $r_n = \sqrt{\mu_n} = n + O(1/n)$. Thus, for sufficiently large u, $\Lambda(u) = [u] + 1$ which means $\Lambda(u)/u \geq 1$. Thus, equation (11.6) is satisfied. Changing x to $\pi x/l$ we see that the system $\cos\sqrt{\lambda_n}x$ is complete in $L_2(0, l)$.

Lemma II.* *If the series* $\sum_{n=0}^{\infty} c_n \cos\sqrt{\lambda_n}x$ *converges to* 0 *in the mean square on the interval* $(0, l)$, *then for each* n *we have* $c_n = 0$.

Proof. In the proof of the preceding lemma we have already mentioned that we may restrict ourselves to the interval $(0, \pi)$. Let us define linear operator A by the formula

$$A \cos nx = \cos\sqrt{\lambda_n}\, x.$$

We shall show that the operator A may be represented in the form $A = E + P$, where P is a completely continuous operator. In fact, let us select as a basis the functions $\cos nx$. Let the operator A correspond to the matrix a_{mn}. Consequently, by definition,

$$a_{mn} = \frac{1}{\pi}\int_0^\pi \cos\sqrt{\lambda_n}\, x \cos mx\, dx =$$

$$= \frac{1}{\pi}\int_0^\pi \left(\cos nx + \frac{\alpha_n(x)}{n}\right)\cos mx\, dx = \delta_{mn} + \frac{1}{n}\frac{1}{\pi}\int_0^\pi \alpha_n(x)\cos mx\, dx = \delta_{mn} + p_{mn}.$$

Furthermore, we have as a result of the ordinary Parseval equation

$$\sum_{mn} p_{mn}^2 = \sum_n \frac{1}{n^2}\alpha_n{}^2(x) < +\infty.$$

Consequently, the operator P with the matrix p_{mn} is completely continuous.

Because of Lemma I, the operator A takes the whole space $L_2(0, \pi)$ into the whole space $L_2(0, \pi)$. Consequently, because of the Fredholm theory, there exists a bounded inverse operator $B = A^{-1}$, where $B = E + Q$, and Q is a completely continuous operator. Consequently,

$$B\cos\sqrt{\lambda_n}\, x = \cos nx.$$

Let us suppose that the series $\sum_{n=0}^{\infty} c_n \cos\sqrt{\lambda_n}\, x$ converges to 0 in the mean

* Levinson considers the interval $(-\pi, \pi)$ and the functions $e^{i\tau_n x}$, $-\infty < \tau_n < +\infty$. Assuming $\tau_{-n} = -\tau_n$ and the function $f(x)$ to be even, we obtain the result used.

square. Applying the operator B, we find that the series $\sum_{n=0}^{\infty} c_n \cos nx$ converges to 0 in the mean square, which means, in view of the orthogonality of the functions $\cos nx$, that $c_n = 0$ for $n = 0, 1, 2, \cdots$, which completes the proof.

On the basis of Lemmas I and II, we shall prove the following theorem:

Theorem I. *Let the positive numbers* $\lambda_0, \lambda_1, \cdots, \lambda_n, \cdots$ *obey the asymptotic formula* (11.4) *and let the positive numbers* $\rho_0, \rho_1, \cdots, \rho_n, \cdots$ *obey the asymptotic formula* (11.5). *Let*

$$\rho(\lambda) = \sum_{\lambda_n \leqslant \lambda} \frac{1}{\rho_n}.$$

If the function $\rho(\lambda)$ *satisfies condition 2 of the preceding section, then there exists a function* $K(x, t)$ *having continuous partial derivatives of the first and second orders for* $0 \leq t \leq x \leq l$; *here the functions*

$$\varphi(x, \lambda_n) = \cos \sqrt{\lambda_n}\, x + \int_0^x K(x, t) \cos \sqrt{\lambda_n}\, t\, dt \tag{11.7}$$

are solutions of the equation

$$y'' + (\lambda_n - q(x)\, y) = 0, \tag{11.8}$$

where

$$q(x) = \frac{1}{2} \frac{dK(x, x)}{dx}, \quad (0 \leqslant x \leqslant l).$$

The functions $\phi(x, \lambda_n)$ *form a complete orthogonal system on the interval* $(0, l)$.

Proof. First we shall prove that the linear integral equation (II) is solvable for $x \leq l$. On the basis of the result of §4 it is sufficient to prove that it follows from the equation

$$\int_0^\infty G^2(\lambda)\, d\rho(\lambda) = 0, \tag{11.9}$$

where $G(\lambda) = \int_0^x g(t) \cos \sqrt{\lambda}\, t\, dt$ $(0 \leq x \leq l)$ that $g(t) = 0$ almost everywhere. In view of the definition of the function $\rho(\lambda)$ equation (11.9) is possible only when for all $n = 0, 1, 2, \ldots$

$$G(\lambda_n) = \int_0^x g(t) \cos \sqrt{\lambda_n}\, t\, dt = 0 \quad (0 \leqslant x \leqslant l).$$

Because of Lemma I the last equation is possible only if $g(t) = 0$ almost everywhere. Thus equation (II) is solvable for all x up to $x = l$. Repeating the reasoning of §§5 and 6, we prove the completeness of the system of functions (11.7) and obtain for them the differential equation (11.8).

Let us show that the functions $\phi(x, \lambda_n)$ are orthogonal. On the basis of the theorem proved in the preceding section it is sufficient to show that from the convergence to 0 of the series $\sum_0^\infty \rho_n^{-1} a_n \cos \sqrt{\lambda_n} x$ follows that $a_n = 0$ for all n. But this is the import of Lemma II. Thus the theorem is completely proved.

3. Let us now show how to reconstruct the boundary condition at the end $x = l$. It follows from formula (11.7) that

$$\varphi(0, \lambda_n) = 1, \quad \varphi_x'(0, \lambda_n) = K(0, 0) = h. \tag{11.10}$$

Furthermore, we have

$$\varphi''(x, \lambda_n) + (\lambda_n - q(x))\varphi(x, \lambda_n) = 0,$$
$$\varphi''(x, \lambda_m) + (\lambda_m - q(x))\varphi(x, \lambda_m) = 0.$$

Multiply the first equation by $\phi(x, \lambda_m)$ and the second by $\phi(x, \lambda_n)$ and subtract one from the other. We obtain the identity

$$\varphi''(x, \lambda_n)\varphi(x, \lambda_m) - \varphi''(x, \lambda_m)\varphi(x, \lambda_n) = (\lambda_m - \lambda_n)\varphi(x, \lambda_n)\varphi(x, \lambda_m).$$

Noting that the first part of this identity is

$$[\varphi'(x, \lambda_n)\varphi(x, \lambda_m) - \varphi'(x, \lambda_m)\varphi(x, \lambda_n)]',$$

we obtain, integrating between the limits 0 and l and using orthogonality and the conditions (11.10), $\phi'(l, \lambda_n)\phi(l, \lambda_m) - \phi'(l, \lambda_m)\phi(l, \lambda_n) = 0$.

From the last equation follows

$$\frac{\varphi'(l, \lambda_n)}{\varphi(l, \lambda_n)} = \frac{\varphi'(l, \lambda_m)}{\varphi(l, \lambda_m)}.$$

Thus the ratio $\phi'(l, \lambda_n)/\phi(l, \lambda_n)$ is constant, which gives the second boundary condition.

4. If in the derivation of the integral equations (see §3) the function

$$f_x(t) = \begin{cases} 1 & \text{for} \quad t \leqslant x \\ 0 & \text{for} \quad t > x \end{cases}$$

is not expanded in a Fourier integral but in a Fourier series in $\cos n\frac{\pi}{l}x$, then for the function $F(x, y)$ one obtains the formula

$$F(x, y) = \sum_{n=1}^{\infty}\left[\frac{1}{\rho_n}\frac{\sin\sqrt{\lambda_n}\,x \cdot \sin\sqrt{\lambda_m}\,y}{\lambda_n} - \frac{2l}{\pi^2}\frac{\sin n\frac{\pi}{l}x \cdot \sin n\frac{\pi}{l}y}{n^2}\right] +$$
$$+ \frac{1}{\rho_0}\frac{\sin\sqrt{\lambda_0}\,x\,\sin\sqrt{\lambda_0}\,y}{\lambda_0} - \frac{xy}{l}.$$

Thus condition 2 of the preceding section may be reformulated as follows:

The function

$$a(x) = \sum_{n=1}^{\infty}\left[\frac{1}{\rho_n}\frac{\cos\sqrt{\lambda_n}x}{\rho_n} - \frac{2l}{\pi^2}\frac{\cos n\frac{\pi}{l}x}{n}\right]$$

has a continuous fourth derivative for $0 \leq x \leq 2l$ (in the point $x = 2l$ it has a left-hand derivative and in the point $x = 0$ a right-hand one).

The following theorem holds.

Theorem 2. *If*

$$\sqrt{\lambda_n} = \frac{\pi}{l}n + \frac{b_1}{n} + \frac{b_3}{n^3} + O\left(\frac{1}{n^4}\right), \tag{11.11}$$

$$\rho_n = \frac{2}{l} + \frac{a_1}{n^2} + O\left(\frac{1}{n^4}\right), \tag{11.12}$$

where a_1, b_1, b_3 *are constants, then there exists a function* $q(x)$ *with given* λ_n

and ρ_n.

Proof. It is sufficient to show, in view of Theorem 1, that the function $a(x)$ has a continuous fourth derivative for $0 \leq x \leq 2l$. If the asymptotic formulas (11.11) and (11.12) are substituted in the expression for $a(x)$, then, on the one hand one may single out series which may be differentiated four times such that the four-times differentiated series converges uniformly. On the other hand, we may single out series which may be obtained by formal integration of the series $\sum \frac{\sin n\pi x/l}{n\pi/l}$. Since

$$\sum_{n=1}^{\infty} \frac{\sin n\frac{\pi}{l}x}{n\frac{\pi}{l}} = \frac{l-x}{2} \quad (0 < x < 2l),$$

then for $0 \leq x \leq 2l$ to the formal derivatives of this series correspond continuous functions; and consequently, the theorem is proved.

Remark 1. If h or $H = \infty$, then for λ_n another asymptotic formula holds:

$$\sqrt{\lambda_n} = \frac{\pi}{l}\left(n + \frac{1}{2}\right) + O\left(\frac{1}{n}\right).$$

If also h and $H = \infty$, then

$$\sqrt{\lambda_n} = \frac{\pi}{l}(n+1) + O\left(\frac{1}{n}\right).$$

In both cases ρ_n obeys the asymptotic formula

$$\rho_n = \frac{1}{n^2\frac{\pi^2}{l^2}}\left[1 + O\left(\frac{1}{n}\right)\right].$$

In the derivation of the integral equations the function $f_x(t)$ is then expanded in a series in $\sum \frac{\sin n\pi x/l}{n\pi/l}$. Then

$$F(x, y) = \sum_{n=0}^{\infty}\left[\frac{1}{\rho_n}\frac{(1-\cos\sqrt{\lambda_n}x)(1-\cos\sqrt{\lambda_n}y)}{\lambda_n^2} - \right.$$
$$\left. -\frac{l}{2}\left(1-\cos n\frac{\pi}{l}x\right)\left(1-\cos n\frac{\pi}{l}x\right)\right].$$

Remark 2. The differentiability of $q(x)$ is connected with the number of terms of the asymptotic expansion for ρ_n and λ_n. For example, for infinite differentiability of $q(x)$ it is necessary and sufficient that the classical infinite asymptotic expansions hold for ρ_n and λ_n.

BIBLIOGRAPHY

[1] V. A. Ambarzumian. *Über eine Frage der Eigenwerttheorie.* Z. Physik 53, 690–695 (1929).

[2] V. A. Marčenko. *Certain questions of the theory of a second-order differential operator.* Dokl. Akad. Nauk SSSR (N.S.) 72, 457–460 (1950).

[3] M. G. Kreĭn. *Solution of the inverse Sturm-Liouville problem.* Dokl. Akad. Nauk SSSR (N.S.) 76, 21–24 (1951).

[3a] M. G. Kreĭn. *Determination of the density of a nonhomogeneous symmetric cord from its frequency spectrum.* Dokl. Akad. Nauk SSSR (N.S.) 76, 345–348 (1951).

[4] B. M. Levitan. *Application of generalized displacement operators to second-order linear differential equations.* Uspehi Mat. Nauk (N.S.) 4, no. 1 (29), 3–112 (1949); Amer. Math. Soc. Transl. no. 59 (1951).

[5] A. Ya. Povzner. *On differential equations of Sturm-Liouville type on a half-axis.* Mat. Sb. N.S. 23(65), 3–52 (1948); Amer Math. Soc. Transl. no. 5 (1950).

[6] G. Borg. *Eine Umkehrung der Sturm-Liouvilleschen Eigenwertaufgabe. Bestimmung der Differentialgleichung durch die Eigenwerte.* Acta Math. 78, 1–96 (1946).

[7] B. M. Levitan. *On a uniqueness theorem.* Dokl. Akad. Nauk SSSR (N.S.) 76, 485–488 (1951).

[8] E. C. Titchmarsh. *The zeros of certain integral functions.* Proc. London Math. Soc. (2) 25, 283–302 (1926).

[9] N. Levinson. *Gap and density theorems.* Colloq. Publ. vol. 26. Amer. Math. Soc., New York, 1940.

[10] L. A. Čudov. *The inverse Sturm-Liouville problem.* Mat. Sb. N.S. 25(67), 451–456 (1949).

3.

(with B.M. Levitan)

On a simple identity for eigenvalues of second order differential operator

Dokl. Akad. Nauk SSSR **88** (4) (1953) 593–596. Zbl. **53**: 60

It is well known that the sum of eigenvalues of a matrix and the trace of an integral operator can be easily calculated.

Let λ_n be the eigenvalues of second order differential equation

$$-y''+[q(x)-\lambda]\,y=0 \tag{1}$$

with boundary conditions

$$y'(0)-h\,y(0)=0, \qquad y'(\pi)+H\,y(\pi)=0. \tag{2}$$

The series $\lambda_1+\lambda_2+\ldots$ diverges, however we will introduce the analog to the notion of trace for that class of differential operators and then we will give the expression for this trace.

First we notice that

$$\lambda_n=n^2+C+O(1/n^2), \qquad C=\frac{1}{\pi}\left[\frac{1}{2}\int_0^\pi q(t)\,dt+h+H\right] \tag{3}$$

if $q(x)$ is a differentiable function.

Denote by μ_n eigenvalues of the equation

$$-y''-\mu y=0 \tag{4}$$

with boundary conditions (2). Without loss of generality we suppose $\int_0^\pi q(t)\,dt=0$ otherwise we have to replace λ by $\lambda-\frac{1}{\pi}\int_0^\pi q(t)\,dt$. It follows from (3) that the series $(\lambda_1-\mu_1)+(\lambda_2-\mu_2)+\ldots$ converges. The last sum can be considered as a regularized trace for the differential operator (1), (2). The goal of this paper is to calculate this sum. The following theorem gives the result, which is very simple.

Theorem. *Let λ_n be eigenvalues of the equation* (1) *with boundary conditions* (2) *and let μ_n be eigenvalue of* (4) *with the same boundary conditions* (2). *Assume that $q(x)$ is a differentiable function and $\int_0^\pi q(x)\,dx=0$. Then*

$$\sum_{n=1}^{\infty}(\lambda_n-\mu_n)=\tfrac{1}{4}[q(0)+q(\pi)].$$

First we prove some lemmas.

Lemma 1. *Under the assumptions of the theorem we have*

$$\sum_{n=1}^{\infty}(\lambda_n-\mu_n)=\lim_{\zeta\to\infty}\zeta^2\sum_{n=1}^{\infty}[(\zeta+\mu_n)^{-1}-(\zeta+\lambda_n)^{-1}]. \tag{5}$$

Proof. It is clear that $\zeta^2[(\zeta+\mu_n)^{-1}-(\zeta+\lambda_n)^{-1}]\to\lambda_n-\mu_n$ as $\zeta\to+\infty$. It follows from (3) that the series $\sum(\lambda_n-\mu_n)$ converges absolutely. That enables us to change "lim" and "$\sum$" in the right hand side of (5).

Lemma 2. *Let $G(x, y, \zeta)$ be the Green's function of the equation $-y''+[q(x)+\zeta]y$ with boundary conditions (2) and let $G_0(x, y, \zeta)$ be the Green's function of the same equation with $q(x)\equiv 0$. Then*

$$\sum_{n=1}^{\infty}(\lambda_n-\mu_n)=\lim_{\zeta\to+\infty}\zeta^2\left\{\int_0^{\pi}(G_0(x,x,\zeta)dx-\int_0^{\pi}G(x,x,\zeta)dx\right\}. \tag{6}$$

Proof. The operator $-y''+[q(x)y+\zeta]y\equiv Ly$ with boundary conditions (2) has eigenvalues $\lambda_n+\zeta$. The inverse operator $L^{-1}y=\int_0^{X}G(x,t,\zeta)y(t)dt$, has eigenvalues $(\lambda_n+\zeta)^{-1}$. The us

$$\sum_{n=1}^{\infty}(\lambda_n+\zeta)^{-1}=\int_0^{\pi}G(x,x,\zeta)dx,\qquad \sum_{n=1}^{\infty}(\mu_n+\zeta)^{-1}=\int_0^{\pi}G_0(x,x,\zeta)dx.$$

Now (6) follows from Lemma 1.

Thus the calculation of the sum $\sum(\lambda_n-\mu_n)$ which is the subject of this paper, reduces to the study of asymptotic behavior of the kernel $G(x, y, \zeta)$ as $\zeta\to+\infty$. To simplify further calculations we consider the case $h=H=0$. If $G(x, y, \zeta)$ is the Green's function of the equation (1) with these boundary conditions, then

$$G(x,y,\zeta)=\begin{cases}\dfrac{1}{W(\zeta)}u(x,\zeta)v(y,\zeta), & x<y\\[2mm] \dfrac{1}{W(\zeta)}u(y,\zeta)v(x,\zeta), & x>y\end{cases}$$

where $W(\zeta)=-u(x,\zeta)v'(x,\zeta)+u'(x,\zeta)v(x,\zeta)$ and $u(x,\zeta)$, $v(x,\zeta)$ are the solutions of (1) satisfying the conditions $u(0,\zeta)=1, u'(0,\zeta)=0$ and $v(\pi,\zeta)=1$, $v'(\pi,\zeta)=0$ respectively. Therefore

$$\int_0^{\pi}G(x,x,\zeta)dx=\frac{1}{W(\zeta)}\int_0^{\pi}u(x,\zeta)v(x,\zeta)dx.$$

To calculate the right hand side of the last equality we put $x=\pi$ in the expression for $W(\zeta)$. Then we get $W(\zeta)=u'_x(\pi,\zeta)$. The integral in the right hand side can be calculated by the usual method*: Finally we get

$$\int_0^{\pi}G(x,x,\zeta)dx=\frac{1}{u'_x(\pi,\zeta)}\frac{d}{d\zeta}u'_x(\pi,\zeta)=\frac{d}{d\zeta}\ln u'_x(\pi,\zeta).$$

* First we calculate the integral $\int_0^{\pi}u(x,\zeta)v(x,\zeta_1)dx$ by the method which is used to prove orthogonality of eigenfunctions. Then we get the result putting $\zeta_1\to\zeta$.

Thus from Lemma 2 we get

Lemma 3. *The following equality is valid:*

$$\sum_{n=1}^{\infty}(\lambda_n-\mu_n)=\lim_{\zeta\to\infty}\cdot\zeta^2\left[\frac{d}{d\zeta}\ln u'_{0x}(\pi,\zeta)-\frac{d}{d\zeta}\ln u'_x(\pi,\zeta)\right],$$

here $u(x,\zeta)$ is the solution of the equation $-u''+[q(x)+\zeta]u=0$ such that $u(0)=1, u'(0)=0$, and $u_0(x,\zeta)$ is the similar solution of the equation with $q(x)\equiv 0$ i.e. $u_0(x,\zeta)=\operatorname{ch}\sqrt{\zeta}x)$.

Now we find the asymptotic behavior of $u'_x(\pi,\zeta)$ as $\zeta\to+\infty$. It is known that the following representation holds for $u(x,\zeta)$:

$$u(x,\zeta)=\operatorname{ch}\sqrt{\zeta}x+\int_0^x K(x,t)\operatorname{ch}\sqrt{\zeta}t\,dt. \tag{7}$$

The function $K(x,t)$ in (7) satisfies the following equation and boundary conditions

$$\begin{aligned}&K''_{xx}(x,t)-q(x)K(x,t)=K''_{tt}(x,t);\\&K'_t(x,0)=0,\quad K(x,x)=\tfrac{1}{2}\int_0^x q(t)dt.\end{aligned} \tag{8}$$

Let us differentiale (7) with respect to x and put $x=\pi$. Then

$$u'_x(\pi,\zeta)=\sqrt{\zeta}\operatorname{sh}\pi\sqrt{\zeta}+\int_0^\pi K'_x(\pi,t)\operatorname{ch}\sqrt{\zeta}t\,dt+K(\pi,\pi)\operatorname{ch}\sqrt{\zeta}\pi.$$

The last term in the right hand side of this equality vanishes, since we have $K(\pi,\pi)=0$ according to the assumptions of theorem and to (8). Therefore

$$-\ln\frac{u'_x(\pi,\zeta)}{u'_{0x}(\pi,\zeta)}=-\ln\left(1+\frac{1}{\sqrt{\zeta}\operatorname{sh}\pi\sqrt{\zeta}}\int_0^\pi K'_x(\pi,t)\operatorname{ch}\sqrt{\zeta}t\,dt\right).$$

Denote $\zeta=z^2$. We have to study the behavior of the right hand side as $\zeta\to+\infty$. Picking out the terms of order $O(e^{-\pi z})$ we get

$$-\ln\frac{u'_x(\pi,\zeta)}{u'_{0x}(\pi,\zeta)}=-\ln\left[1+\frac{1}{z}\int_0^\pi e^{z(t-\pi)}K'_x(\pi,t)dt+O(e^{-\pi z})\right]$$

hence

$$\begin{aligned}-\frac{d}{d\zeta}\ln\frac{u'_x(\pi,\zeta)}{u'_{0x}(\pi,\zeta)}&=-\frac{1}{2z}\frac{d}{dz}\ln\left[1+\frac{1}{z}\int_0^\pi e^{z(t-\pi)}K'_x(\pi,t)dt+O(e^{-\pi z})\right]\\&=\frac{1}{2z^3}\int_0^\pi K'_x(\pi,t)e^{z(t-\pi)}dt-\frac{1}{2z^2}\int_0^\pi K'_x(\pi,t)e^{z(t-\pi)}(t-\pi)\,dt\\&\quad+\text{terms of lower order.}\end{aligned}$$

After replacing the function $K'_x(\pi, t)$ by $K'_x(\pi, \pi)(1+O(t-\pi))$ we can calculate the integrals and finally get

$$-\frac{d}{d\zeta}\ln\frac{u'_x(\pi, \zeta)}{u'_{0x}(\pi, \zeta)}=\frac{1}{z^4}K'_x(\pi, \pi)+O\left(\frac{1}{z^5}\right), \qquad \zeta=z^2.$$

Then we obtain from Lemma 3.

Lemma 4. *The following equality is true*

$$\sum_{n=1}^{\infty}(\lambda_n-\mu_n)=-\lim_{\zeta\to\infty}\zeta^2\frac{d}{d\zeta}\ln\frac{u'_x(\pi, \zeta)}{u'_{0x}(\pi, \zeta)}=K'_x(\pi, \pi).$$

To complete the proof of theorem it remains only to compute $K'_x(\pi, \pi)$. Let us consider the equation and boundary conditions (8). Suppose $x=\xi+\eta$, $t=\xi-\eta$ and denote $\tilde{K}(\xi, \eta)=K(x, t)=K(\xi+\eta, \xi-\eta)$. Then the equation and boundary conditions (8) are transformed to

$$\tilde{K}''_{\xi\eta}(\xi, \eta)=q(\xi+\eta)\tilde{K}(\xi, \eta), \tag{9}$$

$$\tilde{K}'_\xi(\xi, \xi)-\tilde{K}'_\eta(\xi, \xi)=0, \qquad \tilde{K}(\xi, 0)=\tfrac{1}{2}\int_0^\xi q(t)dt. \tag{10}$$

We have also $K'_x(\pi, \pi)=\frac{1}{2}[\tilde{K}'_\xi(\pi, 0)+\tilde{K}'_\eta(\pi, 0)]$. It follows from (10) that

$$\tilde{K}'_\xi(\pi, 0)=q(\pi)/2.$$

Now put in (9) $\eta=0$ and integrate this equation with respect to ξ from 0 to π. This gives us

$$\tilde{K}'_\eta(\pi, 0)-\tilde{K}'_\eta(0, 0)=\int_0^\pi q(\tau)\tilde{K}(\tau, 0)d\tau.$$

Using (10), we get

$$0=\tilde{K}'_\xi(0, 0)-\tilde{K}'_\eta(0, 0)=\tfrac{1}{2}q(0)-\tilde{K}'_\eta(0, 0).$$

Thus

$$\tilde{K}'_\eta(\pi, 0)=\tfrac{1}{2}q(0)+\int_0^\pi \tilde{K}(\tau,0)q(\tau)d\tau=\tfrac{1}{2}q(0)+2\int_0^\pi \tilde{K}(\tau, 0)d\tilde{K}(\tau, 0)$$
$$=\tfrac{1}{2}q(0)+[\tilde{K}(\pi, 0)]^2=\tfrac{1}{2}q(0)$$

since $\tilde{K}(\pi, 0)=\tilde{K}(\pi, \pi)=\int_0^\pi q(t)dt=0.$

Finally we have

$$K'_x(\pi, \pi)=\tfrac{1}{2}[\tilde{K}'_\xi(\pi, 0)+\tilde{K}'_\eta(\pi, 0)]=\tfrac{1}{4}[q(\pi)+q(0)].$$

This equality and Lemma 4 prove the theorem for $y'(0)=y'(\pi)=0$. The proof of theorem with general boundary conditions may be obtained in the same way

References

1. Gelfand, I.M., Levitan, B.M.: Izv. Akad. Nauk SSSR, Ser. Mat. **15** (1951) 309.

4.
(with R. A. Minlos)

Solution of quantum field equations

Dokl. Akad. Nauk SSSR **97** (2) (1954) 209–212. Zbl. **58**: 232

In this note we present the solution of quantum field equations in a closed form. In order to obtain it a new and apparently unusual mathematical formalism is worked out. It seems to us that application of this or some analogous formalism cannot be avoided in modern quantum electrodynamics. On the other hand the formalism we invent here is essentially very simple one and it creates no problems when manipulated. In this (first) note we shall describe this method in a most primitive manner as to avoid unnecessary mathematical casualities.

Consider the simplest case: the so-called Thirring equation. The generalisation of the formalism to other fields will be explained in the end of the note.

The Thirring equations is

$$\Box\,\psi(x)-\kappa^2\psi(x)=\lambda\psi^2(x)+J(x).$$

where $J(x)$ is an external current, $\psi(x)$ is field operator for some particle, $x=(x_0, x_1, x_2, x_3)$ is a point in four-dimensional space-time.

It is essential for us to write equations not for operators but for functionals of the external current $J(x)$ in Schwinger's form [1]. What we shall deal with is however not Schwinger equations for Green function, operators, mass etc. themselves (one can prove that this system is not complete and has infinitely many solutions depending on an arbitrary function) but some other system of equations. Schwinger equations can be obtained as a corollary of our system. Our system has another valuable advantage: it is linear.

Following Schwinger we shall write $\langle f\rangle$ for the vacuum expectation (normalised) of an operator f [1]. Then the upper equation gives rise to

$$(\Box-\kappa^2)\langle\psi(x)\rangle=\lambda\langle\psi^2(x)\rangle+J(x). \tag{1}$$

Here the expectation $\langle\psi(x)\rangle$ is the functional of $J(x)$ still to be found.

Now we want to linearise these equations. This can be done in the following manner. Let L be the Lagrange function (action) for our field. It has the form $L=L_1+\int J\psi\, d^4x$, where L_1 includes terms of the Lagrange function free of $J(x)$. The variation δL of L generated by the variation δJ while ψ remains constant is $\delta L=\int\psi\,\delta J\, d^4x$; so $\dfrac{\delta L}{\delta J(x)}=\psi(x)$.

We have also (see [1])

$$i\frac{\delta\langle f\rangle}{\delta J}=\left\langle f\frac{\delta L}{\delta J}\right\rangle_+ -\langle f\rangle\left\langle\frac{\delta L}{\delta J}\right\rangle \tag{2}$$

where f is some operator that does not depend on $J(x)$ explicitly and $\left\langle f\frac{\delta L}{\delta J}\right\rangle_+$ stands for the expectation of the chronologically ordered product (the so-called T-product). If an operator depends on J explicitly we must add the derivative in respect to J.

Now consider the function*

$$z=\langle e^{-iL}\rangle^{-1}.$$

Next using (2) and noting that e^{-iL} depends on J explicitly and $\frac{\delta L}{\delta J(x)}=\langle\psi(x)\rangle$ we can obtain $-i\frac{\delta z}{\delta J(x)}=z\langle\psi(x)\rangle$. From (2) we have also

$$\begin{aligned}-\frac{\delta^2 z}{(\delta J(x))^2}&=-\frac{i\delta}{\delta J(x)}[\langle\psi(x)\rangle z]\\&=z(\langle\psi^2(x)\rangle-\langle\psi(x)\rangle^2+\langle\psi(x)\rangle^2)=z\langle\psi^2(x)\rangle.\end{aligned} \tag{3}$$

So equation (1) when multiplied by z gives finally

$$-i(\Box-\kappa^2)\frac{\delta z}{\delta J(x)}+\lambda\frac{\delta^2 z}{[\delta J(x)]^2}=J(x)z \tag{4}$$

which is already linear.

In order to solve this equation we shall consider $J(x)$ as l-periodic any coordinative function defined on some lattice. The period of the lattice will tend to zero and the period l of the function to infinity. So we have a finite set of numbers $J_0, J_1, \ldots, J_n$ (where $n+1$ equals the number of vertices of our lattice in some cube with the side of the length l) instead of the function $J(x)$. The functional z is now represented by a function $z(J_0, \ldots, J_n)$ of $n+1$ variable. Instead of the variational derivative $\frac{\delta z}{\delta J(x)}$ we shall obtain the partial derivative $\frac{1}{\omega}\frac{\partial z}{\partial J(x)}$, where ω equals the four-dimensional volume of the cell in our lattice and k is the number of the vertex where x is situated. The differential operator $\Box-\kappa^2$ should be replaced by the difference operator, whose matrix we shall denote by $\|R_{ik}\|$. Then (4) transforms to

$$-i\sum_i R_{ik}\frac{\partial z}{\omega\partial J_k}+\lambda\frac{\partial^2 z}{\omega^2(\partial J_i)^2}=J_i z. \tag{4'}$$

* As E.S. Fradkin and Acad. L.D. Landau kindly informed the authors, this function can be interpreted as a vacuum expectation of the S-matrix that corresponds to the external current $J(x)$ appearing at the moment $-\infty$ and vanishing at $+\infty$ (while the interaction does not vanish, so that the vacuum is considered with respect to this interaction).

Now we shall make the Fourier transform: $z=\int \exp[i\sum s_k J_k]\xi d^4 s$. Then (4′) may be written as $\frac{1}{\omega}\sum_k R_{ik}s_k\xi-\frac{\lambda}{\omega^2}S^2\xi=i\frac{\partial\xi}{\partial s_k}$. Resolving this equation, one can find z in the form

$$z=c\int \exp\left[i\sum s_k J_k+\frac{i\lambda}{\omega^2}\sum\frac{s_k^3}{3}-\frac{i}{2\omega}\sum_{i,k}R_{ik}s_i s_k\right]ds_1\dots ds_n.$$

If we put $is_k=\omega t_k$ this will become

$$z=c'\int \exp\left[\left(\sum t_k J_k-\lambda\sum\frac{t_k^3}{3}+\frac{i}{2}\sum R_{ik}t_i t_k\right)\omega\right]dt_1\dots dt_n. \quad (5)$$

In order to fix the value of this integral we have to choose the contour of integration in every complex plane t_i. Since the convergence depends on the t_k^3 term for large $|t_k|$ our contour should tend to infinity in any plane inside one of three sectors where $\operatorname{Re} t_k^3>0$, that is S_1 where $|\arg t_k|<\pi/6$, S_2 where $\pi/2<\arg t_k<5\pi/6$ and S_3 where $7\pi/6<\arg t_k<3\pi/2$. There are three nontrivial contours that give nonzero value to our integral. They are: the contour going from S_1 to S_2, from S_2 to S_3 and from S_3 to S_1. We can limit ourselves by two of them since the integral along the third one is equal to the sum of integrals along two chosen contours. Choosing these contours independently for every plane we shall obtain 2^{n+1} linear independent solutions of (4′). One can prove that any solution of (4′) is equal to some linear combinations of mentioned above. The integral analogous to (5) in one dimension case is the Airy integral (the Bessel function of order 1/3), so it seems natural to call (5) the Airy integral. On needs also some initial values for our differential equations. We choose the following one: let $\langle\psi_k\rangle=0$ for $J_k=0$ $(k=0,1,\dots,n+1)$ so that $\partial z/\partial J_k=0$ for $J_k=0$.

As we have mentioned already there are 2^{n+1} contours generating different solutions. Now we shall point some considerations that can wipe out all the arbitrariness from the choice of contours. Let us recollect, that we choose our contours in all the planes t_k where k stands for the number of the x-point of the lattice in 4-dimensional space. Since "tails" of the contour may lie only in three definite sectors, the demand that the contour should depend continuously on the choice of the x-point implies that only two fundamentally different surfaces of integration in $(n+1)$-dimensional space can exist. The first one is obtained when all the contours in t_k-planes lead from S_1 to S_2, and the second if they lead from S_2 to S_3. So we can see that the only possible solution is a linear combination of the two mentioned above. The proper choice of constants should imply $\partial z/\partial J_1=0$ for $J_k=0$, then we shall also have $\partial z/\partial J_k=0$.

Now we shall exhibit the extension of computations above to the case of arbitrary field. Let us consider for example the interaction of an arbitrary boson with electromagnetic field. The general relativistic invariant equation of the first order for the case is

$$L^{\mu}_{\alpha\beta}\frac{\partial\psi_\beta}{\partial x_k}-ieA_\mu L^{\mu}_{\alpha\beta}\psi_\beta+i\kappa\psi_\alpha=\eta_\alpha;\qquad \Box A_\mu=s_\mu+J_\mu.$$

with the current s_μ given by the formula

$$s_\mu = \tfrac{1}{2}(\psi_\alpha^+ L^\mu_{\alpha\beta}\psi_\beta + \psi_\alpha L^\mu_{\alpha\beta}\psi_\beta^+),$$

η_α are "sources" and J_μ are external currents. Let L be the Lagrange function (action) for the system consisting of particles and electromagnetic field. We introduce a function $z = \langle e^{-iL}\rangle^{-1}$ and after calculations described above obtain the system of linear equations for this case:

$$L^\mu_{\alpha\beta}\frac{\partial}{\partial x_\mu}\frac{\delta z}{\delta\eta_\beta(x)} - ieL^\mu_{\alpha\beta}\frac{\delta^2 z}{\delta\eta_\beta(x)\delta J_\mu(x)} + ix\frac{\delta z}{\delta\eta_\alpha(x)} = \eta_\alpha(x)z,$$

$$\Box\frac{\delta z}{\delta J_\mu(x)} = J_\mu(x)z + \tfrac{1}{2}L^\mu_{\alpha\beta}\left(\frac{\delta^2 z}{\delta\eta_\alpha^+\delta\eta_\beta(x)} + \frac{\delta^2 z}{\delta\eta_\alpha(x)\delta\eta_\beta^+(x)}\right).$$

Here Z is of course the function of $\eta_\alpha(x)$ and $J_\mu(x)$. The solution of the problem can still obtained in the form of a contour integral that is analogous to the preceding one, that is

$$z = \int \exp\left[i\sum s_{\alpha,k}\eta^+_{\alpha,k} + s^+_{\alpha,k}\eta_{\alpha,k} + J_{\mu,k}\;{}_{\mu,k} + L\right]\prod ds_{\alpha,k}dt_{\mu,k};$$

where $L = L(s,t)$ is the Lagrange function (action) with $s_{\alpha,k}, t_{\mu,k}$ standing for $\psi_\alpha(x)$, $A_\mu(x)$ respectively, and k is the number of the point in the lattice; $\partial/\partial x_\mu$ is replaced by the difference operator. The choice of the single solution is again based on the continuity of the contour as a function of x and on initial conditions when all currents are absent.

Some final remarks

1. A matrix element of any effect can be obtained as the fraction of some derivative of z with respect to external fields over z when all the external fields vanish. The number of differentiations with respect to every field equals to the number of external lines for this kind of particles.

2. When considering fermions one must assume that external sources are anticommuting, as Schwinger has noted in [1].

When this work was reported the authors were told that the final formulae have much in common with those in the work by Feynman [2].

Authors are grateful to academician L.D. Landau and to E.S. Fradkin for valuable discussions on the topic.

References

1. Schwinger, J.: Proc. Nat. Acad. (7) **37** (1951) 452.
2. Feynman, R.P.: Phys. Rev. (3) **80** (1950) 440.

Received 28. IV. 1954

5.

(with V. B. Lidskij)

On the structure of the regions of stability of linear canonical systems of differential equations with periodic coefficients

Usp. Mat. Nauk **10** (1) (1955) 3–40 [Transl., II. Ser., Am. Math. Soc. **8** (1958) 143–181]
Zbl. **64**:89

Introduction

We consider a linear canonical system of $2k$ differential equations with periodic coefficients

$$\frac{dy_s}{dt} = \frac{\partial \mathrm{H}}{\partial y_{k+s}}, \qquad \frac{dy_{k+s}}{dt} = -\frac{\partial \mathrm{H}}{\partial y_s} \qquad (s = 1, 2, \ldots, k),$$

where $\mathrm{H} = \sum_{i,j=1}^{2k} h_{ij}(t) y_i y_j$, $h_{ij}(t) = h_{ji}(t)$ being piecewise continuous real functions with the common period ω.

This system arises in investigating the stability of a periodic mechanical process characterised by a non-linear Hamilton–Jacobi equation. To it are reducible numerous problems of physics and technology: questions of parametric resonance, automatic regulation, and so on.

This system may be written in matrix form as follows:

$$\frac{d}{dt} Y = IH(t) Y, \tag{1}$$

where $H(t)$ is a real symmetric matrix whose elements are the functions $h_{ij}(t)$, $Y(t)$ is a matrix of linearly independent solutions, and the matrix I has the form

$$I = \begin{pmatrix} 0 & E_k \\ -E_k & 0 \end{pmatrix} \tag{2}$$

where E_k is the unit matrix of order k. For brevity we shall sometimes term the system (1) canonical, and the matrix $Y(t)$ the matrix of solutions, or simply the solution.

The system (1) is termed *stable* if all its solutions are bounded as $t \to \infty$.

In the theory of stability and in practice great interest attaches to *strongly stable* canonical systems of equations. Such we shall term stable systems all of whose solutions remain bounded also when the system receives an arbitrary sufficiently small perturbation, its form being preserved.

The following is a strict definition of a strongly stable canonical system:

A canonical system

$$\frac{d}{dt} Y = IH(t) Y$$

is termed strongly stable if it is stable and if there exists an $\epsilon > 0$ *such that for any real symmetric matrix* $\widetilde{H}(t)$, *with piecewise continuous elements, satisfying the* *condition**

$$\tilde{H}(t+\omega) = \tilde{H}(t) \text{ and } \|\tilde{H}(t) - H(t)\| < \epsilon,$$

all solutions of the system $\frac{d}{dt} Y = I\tilde{H}(t) Y$ *are bounded as* $t \to \infty$.

It follows from the definition of strongly stable systems that they form an open set in the set of all canonical systems.

In this article we consider the conditions under which two strongly stable canonical systems

$$\frac{d}{dt} Y_1 = IH_1(t) Y_1 \tag{1'}$$

and

$$\frac{d}{dt} Y_2 = IH_2(t) Y_2 \tag{1''}$$

are such that one can be deformed into the other by continuous variation of the coefficients, without violating the strong stability. In a strict form the problem may be put thus: to find the conditions on the systems (1′) and (1″) under which there exists a piecewise continuous in t and continuous in ν symmetric matrix $H(t, \nu)$, (subject to $H(t+\omega, \nu) = H(t, \nu)$, $0 \le \nu \le 1$), such that $H(t, 0) = H_1(t)$, $H(t, 1) = H_2(t)$ and such that the system

$$\frac{d}{dt} Y = IH(t, \nu) Y$$

is strongly stable for all $0 \le \nu \le 1$.

If some two strongly stable systems (1′) and (1″) can be deformed one into the other in this way, then we say that the systems belong to *one stability-region*, and in the contrary event to *distinct stability-regions*.

In this paper will be given necessary and sufficient conditions, to be satisfied by the *matrices of solutions of canonical systems*, in order that the systems corresponding to them should belong to one stability-region. With these conditions stands revealed the structure of the stability-regions of canonical systems, which turns out to be unexpectedly simple.

The formulation of the basic result we shall give in the third section, prefacing this with various information about stable and strongly stable canonical systems of equations.

In particular, the second section is in the main devoted to an exposition of fundamental results of M. G. Kreĭn, who obtained sufficient conditions for the strong stability of a canonical system.

We observe that the structure of the stability-regions of a canonical system of two equations ($k = 1$) was previously investigated by V. A. Yakubovič [1].

* As the norm of a symmetric matrix $A(t)$ will be understood $\max|\alpha_s(t)|$, where $\alpha_s(t)$ ($s = 1, \cdots, 2k$) are the characteristic values of $A(t)$ for given t.

§1. Conditions for the stability of a canonical system

Let us consider the solution-matrix $Y(t)$ of the canonical system $\frac{d}{dt}Y = IH(t)Y$, normalised at zero by the condition

$$Y(0) = E_{2k}, \tag{3}$$

where E_{2k} is the unit matrix of order $2k$. Since all the coefficients of the system (1) are periodic functions with the common period ω, it follows that $Y(t+\omega)$ is, just as $Y(t)$, a solution-matrix of this system. There holds moreover

$$Y(t+\omega) = Y(t) \cdot Y(\omega). \tag{4}$$

For the left and right sides of (4) are solution-matrices of the system (1) which, in view of (3), coincide for $t = 0$. By the uniqueness theorem they coincide for all t.

Applying the equation (4) n times, we arrive at the relation

$$Y(t+n\omega) = Y(t) \cdot Y^n(\omega). \tag{5}$$

We remark that the elements of the matrix $Y(t)$ are bounded on the interval $0 \le t \le \omega$. In order that they should remain bounded for increasing values of the argument it is necessary and sufficient, as follows from (5), that the elements of the matrix $Y^n(\omega)$ should remain bounded as $n \to \infty$.

Thus the stability or instability of the system (1), as indeed any system with periodic coefficients, depends on the behaviour of the matrix $Y^n(\omega)$ as $n \to \infty$.

It is known that the behaviour of the elements of a matrix A^n as $n \to \infty$ is determined by the Jordan normal form of this matrix. In fact in order that the elements of the matrix should remain bounded as $n \to \infty$, it is necessary and sufficient that all roots of the characteristic polynomial of the matrix A should be not greater than unity in absolute value, and that those which have modulus unity should correspond to linear elementary divisors.

Passing to the formulation of stability-conditions for the system (1), we mention the theorem of Lyapunov–Poincaré ([2], p. 266), according to which in the case of the system (1) together with every root ρ of the characteristic polynomial of the matrix $Y(\omega)$ there occurs a reciprocal root ρ^{-1}*.

In view of this for the stability of the system (1) *it is necessary and sufficient that all roots of the characteristic polynomial of the matrix* $Y(\omega)$ *should have unit modulus, and that all elementary divisors of the matrix* $Y(\omega)$ *should be linear; the latter is, as is known, equivalent to the requirement that the matrix* $Y(\omega)$ *should be reducible to diagonal form.*

The conditions just given for the stability of the system are however, insufficient for the strong stability of the system, i.e., in order that all solutions should remain bounded for any sufficiently small perturbation of the system.

* We give a proof of this fact in §4.

Conditions for strong stability are given in the next section.

In order to make the subsequent reasoning more graphic let us consider a canonical system of four equations ($k = 2$). In this case the matrix $Y(\omega)$ is of the fourth order. Assume to begin with that all roots of the corresponding characteristic polynomial are distinct and distributed on the unit circle of the complex plane in the manner represented in Fig. 1.

It is easy to see that the system in question is stable. Let us show that it is strongly stable. In fact for a sufficiently small variation of the coefficient-matrix $H(t)$ of the system all roots of the characteristic polynomial of the matrix $Y(\omega)$ must by continuity remain distinct, while elementary divisors of higher than the first order cannot arise during the process. If under this deformation one of the roots, ρ_1 for example, became of modulus greater than unity and left the unit circle, this would in view of the reality of the matrix $Y(\omega)$ entail a displacement of the complex-conjugate root into a point symmetrically placed in regard to the real axis. We should then arrive at a contradiction with the Lyapunov–Poincaré theorem, since the root ρ_1 would not then have a counterpart in the shape of a root reciprocal in value.

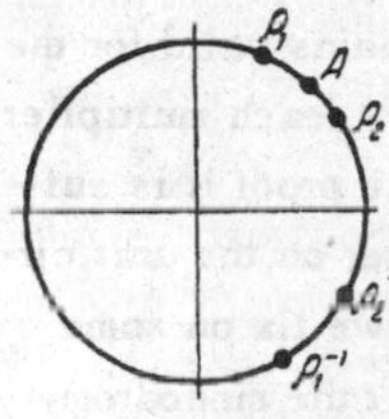

Fig. 1.

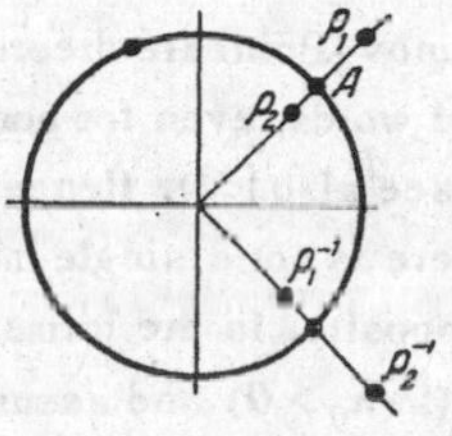

Fig. 2.

If, however, the monodromy matrix has multiple roots, located at some point A of the unit circle, then in general they can leave the circle* to take up a position represented in Fig. 2 (we emphasise that if under a perturbation of a canonical system, its form being preserved, the roots leave the unit circle, then they must necessarily leave it in both directions on account of the Lyapunov–Poincaré theorem). As was shown by M. G. Kreĭn (see [3]), the departure of multiple roots from the unit circle is not always possible and depends on the properties of roots which have come together.

We examine these questions in more detail.

* This circumstance renders essentially more complicated the investigation of the question of stability of a system of many equations, in comparison with the analogous problem for a system of two equations of the form (1) or for a single second-order equation $y'' + p(t)y = 0$ with periodic $p(t)$, reducible to a system of two equations. The point is that in the case of a stable system of two equations there are on the unit circle only two complex-conjugate roots, which can come together and leave it only through the points $+1$ or -1, while in the general case any point of the circle can be a point of departure.

It is precisely in connection with this that the classical results of Lyapunov concerning a single equation $y'' + p(t)y = 0$ were generalised to systems of any number of equations only very recently.

§2. Conditions for the strong stability of a system

Following M. G. Kreĭn, we shall in future term the matrix $Y(\omega)$ the monodromy matrix, and the roots of its characteristic equation multipliers. We consider the system

$$\frac{d}{dt} Y = IH(t)\,Y + \lambda IQ(t)\,Y, \tag{6}$$

where $H(t)$ and $Q(t)$ are real symmetric matrices with the common period ω, $Q(t)$ being a positive definite matrix and λ a certain parameter.

For $\lambda = 0$ (and for real λ) the system (6) is canonical.

Considering the behaviour of the multipliers of the system for complex values of the parameter λ, M. G. Kreĭn established that if one displaces *the parameter* λ *from the real axis into the upper half-plane, then from the general number* $2k$ *of multipliers of the system* k *multipliers will appear inside the unit circle, and* k *multipliers outside it, not one multiplier remaining on the unit circle.* The first k multipliers are called multipliers of the first kind, the second k multipliers of the second kind.*

Let us pause to prove this fact.

As will be shown in §4, the Lyapunov–Poincaré theorem remains valid for the system (6) even for non-real λ; in other words, even for non-real λ each multiplier occurs in company with its reciprocal (see also [3]). Hence for the proof it is sufficient to establish that for $\operatorname{Im}\lambda \neq 0$ there is not a single multiplier on the unit circle. The proof we carry out from the opposite. In the formula (6) we fix on some non-real value of the parameter $\lambda = \lambda_0$ ($\operatorname{Im}\lambda_0 > 0$) and assume that the monodromy matrix has a proper value ρ_0 of unit modulus. We denote by f_0 the corresponding proper vector of the monodromy matrix. We consider also the vector

$$\gamma_0(t) = Y(t)\,f_0, \tag{7}$$

where $Y(t)$ is a matrix of solutions of the system (6), normalised at zero by the condition $Y(0) = E_{2k}$. The vector $\gamma_0(t)$ is obviously a solution of the system (6), in other words there holds the relation

$$\frac{d}{dt}\gamma_0 = IH(t)\,\gamma_0 + \lambda_0 IQ(t)\,\gamma_0. \tag{8'}$$

We mention also that when the parameter t undergoes a shift of amount ω, the vector $\gamma_0(t)$ is multiplied by ρ_0. For

$$\gamma_0(t+\omega) = Y(t+\omega)\,f_0 = Y(t)\cdot Y(\omega)\,f_0 = \rho_0 Y(t)\,f_0 = \rho_0 \cdot \gamma_0(t).$$

In particular for $t = \omega$ we have:

$$\gamma_0(\omega) = \rho_0 \cdot \gamma_0(0). \tag{8''}$$

* In §5 we shall give a direct discussion of the two kinds of multipliers, without recourse to consideration of the system (6).

In view of (8′) and (8″) we may consider the vector $y_0(t)$ as the solution of a boundary problem for the system (8′) with the boundary condition (8″), and λ_0 as the eigen-value corresponding to this solution. As is easily verified, if $|\rho_0| = 1$ this boundary problem is self-adjoint, which leads to a contradiction, since for a self-adjoint boundary problem all eigen-values are real.

Let us carry out the proof in more detail. We apply to both sides of the equation $dy/dt = IH(t)y + \lambda IQ(t)y$ the matrix I^{-1}, and then multiply both sides scalarly* by the vector $y(t)$. From the resulting equation, $(I^{-1}dy/dt, y) = (Hy, y) + \lambda(Qy, y)$, we subtract the complex-conjugate. Since the expressions (Hy, y) and (Qy, y) are real, $H(t)$ and $Q(t)$ being symmetric matrices, we have as a result:

$$\left(I^{-1}\frac{dy}{dt}, y\right) - \left(y, I^{-1}\frac{dy}{dt}\right) = 2i \operatorname{Im}\lambda\,(Qy, y). \tag{9}$$

We now transform the expression on the left of (9). For this we observe that the matrix I (see (2)) is orthogonal and skew-symmetric:

$$I^{-1} = I^* = -I, \tag{10}$$

I^* denoting the adjoint matrix to I. Hence

$$\left(I^{-1}\frac{dy}{dt}, y\right) - \left(y, I^{-1}\frac{dy}{dt}\right) = -\left(I\frac{dy}{dt}, y\right) - \left(Iy, \frac{dy}{dt}\right) = -\frac{d}{dt}(Iy, y). \tag{11}$$

Using (11), we integrate the equation (9) with respect to t from 0 to ω, getting as a result:

$$(Iy(0), y(0)) - (Iy(\omega), y(\omega)) = 2i \operatorname{Im}\lambda \int_0^\omega (Qy, y)\,dt. \tag{12}$$

But by (8″) $(Iy(\omega), y(\omega)) = |\rho|^2 (I(y(0), y(0))$, so that if $|\rho| = 1$ the left-hand side of (12) vanishes, which is impossible for $\operatorname{Im}\lambda \neq 0$, since $Q(t)$ is a positive-definite matrix.

We now explain how two cases can arise when a canonical system has repeated multipliers on the unit circle. As λ shifts into a non-real region, the multipliers either leave the point in question of the complex plane in one and the same sense from the circumference, or leave it in different senses.** In the first case we shall speak of repeated multipliers of like type, in the second case of repeated multipliers of unlike type.

Using the concept of multipliers of the first and second type, M. G. Kreĭn established the following important fact:

If a stable canonical system does not have repeated multipliers of unlike type, then it is strongly stable.

For the proof, obviously, it is necessary to show that if a stable system has

* As the scalar product (g, h) of two vectors g and h is to be understood $\Sigma_s g_s \overline{h}_s$, where g_s and h_s are the components of the vectors.

** This of course does not contradict the remark made at the end of the previous section, since if λ is complex the system (6) is no longer canonical.

repeated multipliers of like type, then for any sufficiently small deformation of the matrix of coefficients the multipliers remain on the unit circle, it being at the same time impossible for non-linear elementary divisors at arise.

The first part of this assertion may be proved as follows. We assume that at some point A of the unit circle there are repeated multipliers, for definiteness of of the first type of the system (6) for $\lambda = 0$. Assume further that for some small deformation of the matrix $H(t)$ the multipliers are displaced from the unit circle and assume a position indicated in Fig. 2. We show that this is impossible. To see this let us first give the parameter λ a complex value from the upper half-plane, sufficiently small in modulus. The multipliers will thereby move inside the unit circle. After this we effect the above continuous deformation of the matrix $H(t)$, and then move the parameter λ from the upper half-plane to zero. It is clear that as a result the multipliers will not be able to take up a final position on different sides of the circumference, as represented in Fig. 2, since during the whole of the deformation they were inside the unit circle.

The fact that elementary divisors of above the first order cannot form in the case of repeated multipliers of like type needs a more complicated proof. In a supplement (see sub-section 7) we give a formula from which it follows that if the monodromy matrix of the system (6) has for $\lambda = 0$ a Jordan box with multipliers to which corresponds a point A of the complex plane, then as the parameter λ moves into the non-real region the multipliers move from the point A in different senses from the circumference, which refutes the assumption of the possibility of the formation of a Jordan box in the case of multipliers of like type. In the supplement we prove an assertion converse to the theorem of M. G. Kreĭn, to wit that *if a stable system* (1) *has repeated multipliers of unlike type, then they can be displaced from the unit circle by an arbitrarily small deformation of the matrix* $H(t)$. Taking this fact into account, we may formulate a necessary and sufficient condition for the strong stability of the system (1) in the following form: *a canonical system is strongly stable if and only if it is stable and if among the multipliers there are none which are repeated and of unlike type.*

§3. Formulation of the basic result

In view of the above conditions for strong stability it is possible to indicate at once a whole series of canonical systems which belong to distinct stability regions. We make to begin with one observation regarding the distribution of the multipliers of a strongly stable system.

As will be shown in §6 (see Lemma 2), the monodromy matrix of a strongly stable system has all its multipliers non-real, the multipliers of the first and second kinds being distributed on the unit circle symmetrically in regard to the real axis (if ρ is a multiplier of the first kind, then $\bar{\rho}$ is a multiplier of the second kind) *.

Let us now consider two strongly stable systems with the coefficient-matrices $H_1(t)$ and $H_2(t)$. To the matrix $H_1(t)$ let there correspond a system with multipliers distributed as in Fig. 3, where the signs ⊕, ⊖ stand for multipliers of the first and second kinds, respectively. Likewise to the matrix $H_2(t)$ let there correspond a system with multipliers distributed as in Fig. 4.

It is easy to see that the systems in question belong to distinct stability regions. For as the multipliers move continuously along the unit circle they do not change their particular type, in view of which a continuous deformation of the matrix $H_1(t)$ into $H_2(t)$ would inevitably produce on the unit circle a coincidence of multipliers of unlike type, in contravention of the conditions for strong stability. We thus arrive at a simple necessary condition for two systems to belong to one stability region. The necessary condition is in fact that the two systems should enjoy a *conformal* distribution of multipliers of the first and second kinds on the unit circle. More precisely, that the ordering of the signs ⊕ and ⊖ in an anti-clockwise transit of the upper semi-circle should coincide as between the two systems. Here in the event of repeated multipliers each ⊕ and ⊖ is to be counted a number of times equal to the multiplicity of the corresponding multiplier (we recall once more that repeated multipliers of unlike type do not occur for strongly stable systems). One calculates without difficulty that for a system of $2k$ equations there are 2^k possible types of distribution of multipliers of the first and second kinds on the unit circle.

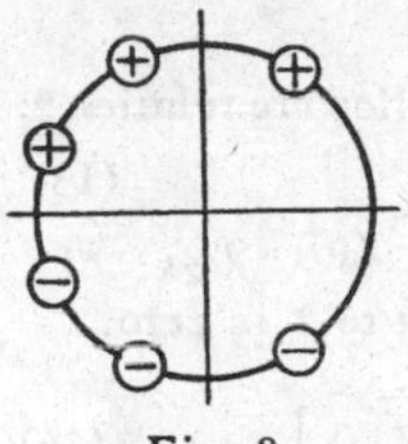

Fig. 3.

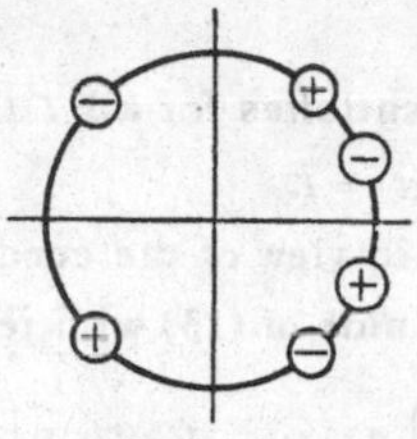

Fig. 4.

The basic result of the present article is the following

Theorem 1. *Each stability region of a canonical system of $2k$ equations of the form (1) is determined by one of the 2^k possible types of relative distribution of multipliers on the unit circle and, in addition, by an integer n, the number*[1)] *of the stability region* ($-\infty < n < \infty$). The number n we will sometimes term the index of rotation of the system.

Theorem 1 asserts that the question of whether some strongly stable system belongs to this or that stability region is determined by the type of distribution of multipliers of the first and second kinds on the unit circle and, in addition, by one

* These facts may be easily proved also by shifting the parameter λ in the system (6) into the complex region.

1) Translator's note: "Number" in the sense of serial number.

(!) integer, to be associated in a certain way with the given system. Hence if for two strongly stable systems with a conformal distribution of multipliers on the unit circle these numbers coincide, then the coefficient-matrices of these systems can be continuously deformed one into the other within the corresponding stability region.

We shall establish the nature of this integer n, associated with a given strongly stable system, and in fact the whole of Theorem 1, on the basis of certain simple algebraic and topological properties of the solution-matrices of canonical systems of differential equations.

We are now to consider these properties.

§4. Curves in the group of real symplectic matrices

We consider the solution-matrix $Y(t)$ of a canonical system, normalised at the origin by the condition $Y(0) = E_{2k}$.

We shall prove that the matrix $Y(t)$ satisfies for all t the following relation*:

$$Y^*(t)IY(t) = I. \tag{13}$$

That (13) holds for $t = 0$ is obvious in view of the condition $Y(0) = E_{2k}$. We prove that the derivative of the left-hand side of (13) with respect to t is zero. For

$$\frac{d}{dt}[Y^*(t)IY(t)] = \left[\frac{d}{dt}Y^*(t)\right]IY(t) + Y^*(t)I\left[\frac{d}{dt}Y(t)\right]. \tag{14}$$

We now substitute on the right of (14) expressions for the derivatives of the solution-matrices on the basis of the identities $d/dt\,Y = IH(t)Y$ and $d/dt\,Y^* = Y^*H^*I^*$, and use the facts that $H^*(t) = H(t)$, $I^* = I^{-1}$ and $I^2 = -E_{2k}$ getting:

$$\frac{d}{dt}[Y^*(t)IY(t)] = Y^*(t)H(t)Y(t) - Y^*(t)H(t)Y(t) = 0.$$

This proves (13).

The relation (13) will play a basic role in what follows. We remark that from it there follows easily the Lyapunov–Poincaré theorem, which we cited in the derivation of the stability conditions. For putting $t = \omega$ in (13) we get:

$$Y^*(\omega)IY(\omega) = I \tag{15}$$

or

$$Y^*(\omega) = IY^{-1}(\omega)I^{-1}. \tag{16}$$

In view of the formula (16), the matrix $Y^*(\omega)$ is similar to the matrix $Y^{-1}(\omega)$, which means that the characteristic polynomials of these matrices coincide. Since on the other hand the roots of the characteristic polynomial of the matrix $Y(\omega)$ are equal to the roots of the characteristic polynomial of the matrix $Y^*(\omega)$* and are the reciprocals of the roots of the inverse matrix $Y^{-1}(\omega)$, we reach the conclusion

* Y^* is the adjoint matrix of Y, that is to say the matrix obtained by transposing Y and replacing all elements by their complex-conjugates. In the present case the condition of reality implies that Y^* is merely the transpose of Y.

that each root of the characteristic polynomial of the matrix $Y(\omega)$ occurs in company with its reciprocal.

We remark that the identity (13), when written in the form $Y'(t)IY(t) = I$, where $Y'(t)$ denotes the transposed matrix, is valid for the solution-matrix of the system (6) also for non-real values of the parameter λ. For in deducing (13) we used only the symmetry of the coefficient-matrix and not its reality. Hence the Lyapunov–Poincaré theorem holds in this case too, which fact we used in the last section.

Matrices which satisfy the condition (13) ($Y\ IY = I$) are termed symplectic. As is not hard to verify, they form a group under multiplication. Hence the solution-matrix of the system (1), normalised by the condition $Y(0) = E_{2k}$, constitutes in the accepted terminology a curve in the group of real symplectic matrices $\{Y\}$.

The following simple proposition, a converse to its predecessor, is very important for what follows.

Lemma 1. *Every curve $Y(t)$ in the group of real symplectic matrices ($0 \le t \le \omega$, $Y(0) = E_{2k}$), possessing a piece-wise continuous derivative, satisfies some matrix equation of the form (1) on the interval $[0, \omega]$.*

Proof. We differentiate the identity $Y^*(t)IY(t) = I$, which the matrix $Y(t)$ satisfies by hypothesis. We multiply the resulting relation

$$\left[\frac{d}{dt} Y^*(t)\right] IY(t) + Y^*(t)I\left[\frac{d}{dt}Y(t)\right] = 0$$

on the left by $[Y^*(t)]^{-1}$ and on the right by $Y^{-1}(t)$. Using the facts that $I = (I^{-1})^*$ and $-I = I^{-1}$, we write this identity in the following form:

$$(Y^{-1}(t))^*\left(\frac{d}{dt}Y(t)\right)^*(I^{-1})^* = I^{-1}\left(\frac{d}{dt}Y(t)\right)\cdot Y^{-1}(t). \tag{17}$$

The formula (17) shows that the matrix $I^{-1}[d/dt\, Y(t)] \cdot Y^{-1}(t)$ is symmetric for all t. We write

$$I^{-1}\left[\frac{d}{dt}Y(t)\right]Y^{-1}(t) = H(t),$$

and then $dY/dt = IH(t)Y(t)$, and Lemma 1 is proved. We have established a one-to-one correspondence between equations of the form $d/dt\, Y = IH(t)Y$ and curves $Y(t)$ ($Y(0) = E_{2k}$, $0 \le t \le \omega$) in the group of real symplectic matrices $\{Y\}$. This makes it possible to reduce the question which interests us to the solution of the following problem.

In the group of real symplectic matrices $\{Y\}$ are given two curves $Y_1(t)$ and $Y_2(t)$ with a common starting-point $Y_1(0) = Y_2(0) = E_{2k}$. Both curves have piece-wise continuous derivatives and correspond to strongly stable systems of equations of the form (1).

It is required to determine under what conditions the curve $Y_1(t)$ can be deformed into the curve $Y_2(t)$ in such a way that during the process of deformation it remains the solution of a strongly stable system.

More rigorously: the problem is posed of the conditions for the existence in the group $\{Y\}$ of a deformation $Y(t, \nu)$ $(0 \leq t \leq \omega, 0 \leq \nu \leq 1)$ such that $Y(t, 0) = Y_1(t)$, $Y(t, 1) = Y_2(t)$, $Y(0, \nu) = E_{2k}$, the derivative $d/dt\, Y(t, \nu)$ exists, is piece-wise continuous in t and continuous in ν, and the curve $Y(t, \nu)$ for each fixed ν corresponds to a strongly stable system.

It is easy to see that the systems of differential equations to which the curves $Y_1(t)$ and $Y_2(t)$ correspond belong to one and the same stability-region if and only if there exists a deformation $Y(t, \nu)$ as described above. In this connection we will for brevity sometimes speak of curves belonging to the same stability-region, meaning that the correspondingly strongly stable systems have this property.

We now indicate in general outline the solution of the problem we have formulated.

Hitherto we have distinguished the kinds of multiplier according to the behaviour of the multiplier when the parameter λ moves into the real non-region. Below we show that the kind of multiplier can be completely determined if we only know the matrix $Y(\omega)$ (the end of the curve $Y(t)$). This makes it possible to determine only from the end of the curve whether the curve belongs to a strongly stable system or not (see §5). After this we effect the solution of the basic problem posed above, regarding the deformation of the two curves $Y_1(t)$ and $Y_2(t)$ with a fixed starting-point E_{2k}, in two stages.

In §6 we investigate when the ends of these curves, the matrices $Y_1(\omega)$ and $Y_2(\omega)$, can be continuously deformed one into the other, retaining the condition of strong stability. As will be explained in §6, for this purpose it is necessary and sufficient that the two matrices $Y_1(\omega)$ and $Y_2(\omega)$ should have a conformal* distribution of multipliers of the first and second kinds on the unit circle.

After the ends of the curves have been deformed one into the other, it remains to bring together the curves themselves, leaving their common starting-point E_{2k} and common end undisturbed. In §§7 and 8 it is established that such a deformation is not always possible. To bring it about it is necessary and sufficient that the curves $Y_1(t)$ and $Y_2(t)$ under consideration should have the same rotation indices, these being integers associated in a certain way with the curves.

In §8 we derive a formula for the calculation of the rotation index of a curve $Y(t)$ $(0 \leq t \leq \omega)$ (the number of the stability-region to which the curve belongs). Let us here remark that, in carrying out the deformation, we shall not concern ourselves with the differentiability of the curves. The point is that if a continuous deformation

* The definition of conformal distribution was given in the previous section.

$Y(t, \nu)$ is possible, then it can always be smoothed by an inessential modification and replaced by one which is differentiable.

§5. Symplectic matrices of stable type

Let us agree to term the real symplectic matrix Y a *matrix of stable type* if it is the monodromy matrix of some strongly stable system of differential equations of the form (1).

Let us consider the conditions which characterise matrices of stable type*. As was said in §1, a matrix of stable type must be reducible to diagonal form and all its proper values must have unit modulus. The whole question consists in how to determine the type of multiplier from the inner properties of the matrix.

Let us consider some real symplectic matrix Y_0 which is reducible to diagonal form and all of whose proper values have unit modulus. For convenience of exposition we shall suppose that the matrix in question, Y_0, is the monodromy matrix of the system (6) for $\lambda = 0$. This assumption does not affect the generality of the investigation to be given since, as follows from Lemma 1 and the connectedness of the group $\{Y\}$ (see Lemma 3, §7), any real symplectic matrix can be considered as the monodromy matrix of some canonical system.

Let $f_1, f_2, \cdots, f_{2k}$ *be a system of linearly independent proper vectors of* Y_0. *We show that proper vectors* f_{s_1} *and* f_{s_2} *corresponding to distinct proper values are I-orthogonal, i.e.,*

$$(If_{s_1}, f_{s_2}) = 0. \tag{18}$$

Proof. For the matrix I in the expression (If_{s_1}, f_{s_2}) we substitute the matrix $Y_0^* I Y_0$, which is equal to it by hypothesis; we get as a result:

$$(If_{s_1}, f_{s_2}) = (Y_0^* I Y_0 f_{s_1}, f_{s_2}) = (IY_0 f_{s_1}, Y_0 f_{s_2}) = (I\rho_{s_1} f_{s_1}, \rho_{s_2} f_{s_2}) = \rho_{s_1}\bar{\rho}_{s_2}(If_{s_1}, f_{s_2}). \tag{19}$$

Since the proper values ρ_{s_1} and ρ_{s_2} have unit moduli and are distinct, it follows that $\rho_{s_1}\bar{\rho}_{s_2} \neq 1$, so that it follows from (19) that $(If_{s_1}, f_{s_2}) = 0$, as was to be proved.

From the fact just proved it follows that if ρ_0 is a non-repeated proper value of the symplectic matrix Y_0 and f_0 is a corresponding proper vector, then the form

$$(If_0, f_0) \neq 0. \tag{20}$$

For in the contrary event the vector If_0 would be orthogonal in the ordinary sense to all proper vectors of the matrix Y_0, which form a basis in the whole $2k$-dimensional space, and so would therefore vanish. But from the relation $If_0 = 0$ it would follow that $f_0 = 0$, since the matrix I is non-singular. This contradiction establishes (20).

* When this article had already been written, we became acquainted with the manuscript of a work of M. G. Kreĭn, written for the collection of articles in memory of A. A. Andronov. In this work M. G. Kreĭn finds independently analogous conditions characterising matrices of stable type.

We observe further that $(If_0 f_0)$ is purely imaginary, since I is a skew-symmetric matrix.

It will be convenient to introduce the Hermitean matrix J by the formula

$$J = -iI. \tag{21}$$

In the supplement to this article (see sub-section 4) we show that if $(Jf, f) = -i(If, f) > 0$, then as the parameter λ of the system (6) moves into the upper half-plane, the multiplier ρ_0 moves inside the unit circle of the complex plane, and is therefore a multiplier of the first kind in the sense of the definition given in the second section.

In an analogous way it will be shown that if $(Jf, f) < 0$, then ρ_0 is a multiplier of the second kind.

Thus *if ρ is a non-repeated multiplier, then its type is determined by the sign of the quadratic form (Jf, f), where f is the proper vector corresponding to the proper value ρ of the monodromy matrix in question.*

We now consider the general case in which ρ_0 is a proper value of Y_0 with multiplicity r. We choose a basis $f_1, f_2, \cdots, f_r$ in the corresponding proper subspace and consider the rth order Hermitean matrix (the "Gram's matrix")

$$L_{\beta_0} = \begin{pmatrix} (Jf_1, f_1)\,(Jf_2, f_1)\ldots(Jf_r, f_1) \\ (Jf_1, f_2)\,(Jf_2, f_2)\ldots(Jf_r, f_2) \\ \cdots\cdots\cdots\cdots \\ (Jf_1, f_r)\,(Jf_2, f_r)\ldots(Jf_r, f_r) \end{pmatrix}. \tag{22}$$

Matrices of the form (22) will occur fairly often in the course of the work; let us for brevity denote them as follows:

$$L_{\beta_0} = ((Jf_\nu, f_\mu)) \quad (\nu, \mu = 1, 2, \cdots, r).$$

We show that the matrix L_{β_0} is non-singular. For in the contrary event there would exist a linear dependence between its columns, and numbers $\alpha_1, \alpha_2, \cdots, \alpha_r$, not all zero, could be found such that

$$\alpha_1(Jf_1, f_j) + \alpha_2(Jf_2, f_j) + \cdots + \alpha_r(Jf_r, f_j) = 0 \quad (j = 1, 2, \cdots, r). \tag{23}$$

Let us denote by h the vector $\alpha_1 f_1 + \alpha_2 f_2 + \cdots + \alpha_r f_r$. It follows from (23) that $(Jh, f_j) = 0$ $(j = 1, 2, \cdots, r)$. In other words the vector h is I-orthogonal to all the vectors f_j, which form a basis in the proper subspace corresponding to the proper value ρ_0. Since the vector h itself lies in this subspace, it follows from the formula (18) that it is I-orthogonal to all the remaining proper vectors of the matrix Y_0, so that $Ih = 0$. Hence $h = 0$ and we arrive at a contradiction, since it was assumed that among the numbers $\alpha_1, \alpha_2, \cdots, \alpha_r$ there were some different from zero.

Thus the martix L_{β_0} is non-singular. Since L_{β_0} is a Hermitean matrix, it follows that all its proper values are real. They are all different from zero since L_{β_0} is non-singular.

Let the number of positive proper values of L_{β_0} be r', and the number of negative ones r'', $(r' + r'' = r)$*. In the supplement to the present work (see sub-section 5) we show that in this case if the parameter λ of the system (6) moves into the upper half-plane, then from the point ρ of the unit circle r' multipliers leave for the interior of the unit circle while r'' multipliers leave the circle in the other direction. This means in the sense of the definition given in the second section, that ρ is a point of "confluence" of r' multipliers of the first kind and r'' multipliers of the second kind. Hence in order the matrix Y_0 should not have repeated multipliers of unlike type, it is necessary and sufficient that all matrices of the type L_β should be *sign-definite*, i.e., that each of them should have proper values of one sign only. Here *multipliers of the first kind correspond to positive-definite matrices* L_β, *and multipliers of the second kind to negative-definite ones.*

As is known, the matrix L_β constructed for the vectors $f_1, f_2, \cdots, f_r$ is sign-definite if and only if the quadratic form (Jh, h) is definite** on the subspace spanned by these vectors.

Bringing together the above considerations, we can formulate conditions characterising matrices of stable type in the following form:

The real symplectic matrix Y is a matrix of stable type if and only if it is reducible to diagonal form, all its proper values are of unit modulus, and the quadratic form (Jh, h) is definite on every proper subspace of the matrix Y.

§6. Confluence of the ends of the curves

Following the general plan outlined at the end of the fifth section of this work, we pass to the sutdy of conditions under which two matrices of stable type can be deformed one into the other through matrices of stable type.

It follows from the conditions which characterise matrices of stable type that for this defomation it is *necessary* that the two matrices should have a *conformal** distribution of multipliers of the first and second kind on the unit circle, since otherwise in the course of the continuous deformation there would inevitably arise a confluence of multipliers of unlike type.

Below we show that this condition is *sufficient*. In other words, if two matrices of stable type have a conformal distribution of multipliers on the unit circle, then they can be continuously deformed one into the other through matrices of stable typel Since there exist altogether 2^k distinct types of distribution of multipliers, we shall thereby establish that the set of matrices of stable type consists of 2^k

* We recall that by the law of inertia the number of positive and negative proper values of the matrix L_{β_0} does not depend on the choice of the basis in the proper subspace in question.

** I.e., either $(Jh, h) > 0$ $(h \neq 0)$ for all vectors os the subspace, or else $(Jh, h) < 0$.

*** See §3.

connected components. We shall easily derive the sufficiency of the condition from the following lemma.

Lemma 2. *The symplectic matrix Y is a matrix of stable type if and only if it can be represented in the form*

$$Y = GRG^{-1}, \tag{24}$$

where G and R are real symplectic matrices. Here the matrix R has the special form

$$R = \begin{pmatrix} \cos\theta & \sin\theta \\ -\sin\theta & \cos\theta \end{pmatrix}, \tag{25}$$

where θ is a real diagonal matrix of order k, whose diagonal elements $\theta_1, \theta_2, \cdots, \theta_k$ satisfy the conditions $|\theta_s| < \pi$ $(s = 1, 2, \cdots, k)$ and

$$\theta_{s'} \neq \theta_{s''} \quad (1 \leq s', s'' \leq k). \tag{26}$$

Remark. In the course of the proof it will be established that the numbers $\theta_1, \theta_2, \cdots, \theta_k$ are the principal values of the arguments of the multipliers of the first kind.

Proof. 1) We show to begin with that a matrix Y which can be represented in the form (24) is a matrix of stable type. From the formula $Y = GRG^{-1}$ and the orthogonality of the matrix R it follows that the matrix Y is reducible to diagonal form, all its proper values being of modulus unity.

It thus remains to show that the Hermitean quadratic form (Jh, h) is definite on the proper subspaces of the matrix Y.

We remark that since the matrix G is symplectic applying it to the vector h does not change the value of the quadratic form (Jh, h). For in fact

$$(JGh, Gh) = G^* JGh, h) = (Jh, h).$$

On the other hand it follows from the formula $Y = GRG^{-1}$ that the proper subspaces of the matrices Y and R, corresponding to the same proper value, are connected by a linear transformation determined by the matrix G. Hence it is sufficient to show that the form (Jh, h) is definite on every proper subspace of the matrix R. This fact may be established by direct calculation. We introduce a notation for the unit vectors,

$$e_1 = (1, 0, 0, \cdots, 0),\ e_2 = (0, 1, \cdots, 0), \cdots, e_{2k} = (0, 0, \cdots, 1),$$

and consider the vectors

$$m_s = \frac{e_s + ie_{s+k}}{\sqrt{2}} \text{ and } m_{-s} = \frac{e_s - ie_{s+k}}{\sqrt{2}} \qquad (s = 1, 2, \ldots, k). \tag{27}$$

The vectors m_s, m_{-s} $(s = 1, \cdots, k)$ form an orthogonal system of proper vectors of the matrix R with the proper values

$$\rho_s = e^{i\theta_s} \text{ and } \rho_{-s} = e^{-i\theta_s} \quad (s = 1, 2, \ldots, k).$$

In other words there hold the equalities

$$Rm_s = e^{i\theta_s} m_s, \quad Rm_{-s} = e^{-i\theta_s} m_{-s}; \tag{28}$$
$$(m_s, m_s) = 1, (m_{-s}, m_{-s}) = 1 \ (s = 1, 2, \ldots, k), (m_{s'}, m_{s''}) = 0,$$
$$s' \neq s'' \ (-k \leqslant s', \ s'' \leqslant k).$$

We observe incidentally that the matrix I, given by (2), is a matrix of the form (25) for $\theta_s = \frac{1}{2}\pi$, $(s = 1, 2, \cdots, k)$. Putting in the formulae (28) $\theta_s = \frac{1}{2}\pi$, we get:

$$Im_s = im_s \text{ and } Im_{-s} = -im_{-s}. \tag{29}$$

We now consider a proper subspace of the matrix R corresponding to some proper value ρ. We denote by $m_{s_1}, m_{s_2}, \cdots, m_{s_r}$ vectors of the system (27) forming a basis in this subspace. We remark that all the indices $s_1, s_2, \cdots, s_r$ must have the same sign. For in view of the formulae (28) and the condition (26) of the theorem we have that proper vectors of R, with indices of different signs, correspond to unequal proper values, and therefore belong to different proper subspaces. Let us assume for definiteness that $s_1, s_2, \cdots, s_r > 0$, and consider a vector $h = \Sigma_{j=1}^r \alpha_j m_{s_j}$ not equal to zero. Using the left of the formulae (29) we get: $Ih = \Sigma_{j=1}^r \alpha_j Im_{s_j} = ih$, whence it follows that

$$(Jh, h) = -i(Ih, h) = (h, h) > 0.$$

It is therewith proved that the form (Jh, h) is definite, in fact positive-definite, on a proper subspace to which correspond proper vectors with positive indices. The latter implies that the numbers $e^{i\theta_s}$ $(s = 1, 2, \cdots, k)$ are multipliers of the first kind.

Considering a subspace determined by vectors with negative indices we obviously establish that the form (Jh, h) is negative-definite. With this we complete the proof of the first part of the lemma.

2) We shall prove the converse assertion. Let us assume that Y is a matrix of stable type. In particular this means that the matrix Y is reducible to diagonal form and that all its proper values lie on the unit circle. We observe that none of the proper values of the matrix Y can be real. For in the contrary event in view of the reality of the matrix one could find a real proper vector $f \neq 0$, corresponding to this proper value. But for a real proper vector the form $(Jf, f) = -i(If, f) = 0$, which contradicts the condition of definiteness.

Let ρ be a proper value of the matrix Y of multiplicity r and let, for definiteness, the form (Jh, h) be positive-definite on the corresponding proper subspace.

We consider in this subspace a basis of r vectors $f_{s_1}, f_{s_2}, \cdots, f_{s_r}$, chosen and normalised so as to fulfill the equations

$$(Jf_{s_\nu}, f_{s_\mu}) = 0, \ \nu \neq \mu; \ (Jf_{s_\nu}, f_{s_\mu}) = 1, \ \nu = \mu \ (0 \leqslant \nu, \ \mu \leqslant r). \tag{30}$$

As is known, it is always possible to select such a basis in virtue of the Hermitean and positive-definite character of the quadratic form (Jh, h) on the subspace in question.

We observe that since the matrix Y is real, the number $\bar{\rho}$ too is a proper value of the matrix Y. The vectors $f_{-s_1} = \bar{f}_{s_1}, f_{-s_2} = \bar{f}_{s_2}, \cdots, f_{-s_r} = \bar{f}_{s_r}$, obviously form a basis in the corresponding subspace; furthermore, as is easily verified, using the reality of the matrix I, the formulae (30) imply the equations

$$(Jf_{-s_\nu}, f_{-s_\mu}) = 0, \quad \nu \neq \mu, \quad (Jf_{-s_\nu}, f_{-s_\mu}) = -1, \ \nu = \mu \ (0 \leqslant \nu, \mu \leqslant r). \tag{31}$$

Since $\rho \neq \bar{\rho}$, the proper vectors $f_{s_1}, f_{s_2}, \cdots, f_{s_r}$, and $f_{-s_1}, f_{-s_2}, \cdots, f_{-s_r}$ are linearly independent. It is furthermore easy to see that they are all I-orthogonal. For vectors denoted by indices of the one sign this property follows from (30) and (31). Vectors with indices of different signs are I-orthogonal in virtue of (18), since they correspond to distinct proper values.

Choosing and normalising in a similar way proper vectors of the matrix Y corresponding to pairs of complex-conjugate proper values, we arrive as a result at a system of $2k$ linearly independent proper vectors of Y

$$f_1, f_2, \cdots, f_k, f_{-1}, f_{-2}, \cdots, f_{-k}, \tag{32}$$

satisfying the conditions

$$(Jf_{s'}, f_{s''}) = 0, \ s' \neq s'' \ (-k \leqslant s', s'' \leqslant k); \tag{33}$$

$$(Jf_s, f_s) = 1, \ (Jf_{-s}, f_{-s}) = -1 \ (s = 1, 2, \ldots, k). \tag{34}$$

Proper vectors whos suffixes differ only in sign are complex-conjugate.

We denote by $e^{i\theta_1}, e^{i\theta_2}, \cdots, e^{i\theta_k}, e^{-i\theta_k}, e^{-i\theta_1}, \cdots, e^{-i\theta_k}$ the proper values corresponding to the proper vectors (32), each being written a number of times equal to its multiplicity. Obviously the first k of these are multipliers of the first kind, the second k multipliers of the second kind. We choose the arguments θ_s so that $|\theta_s| < \pi$. It is then easy to see that the condition $\theta_{s'} \neq \theta_{s''}$ $(s', s'' = 1, 2, \cdots, k)$ will hold. In the contrary event there would be two vectors of the system (32) with indices of different signs belonging to the same proper subspace of the matrix Y, which would contravene the definiteness condition on the form (Jh, h).

We now consider a system of real vectors given by

$$g_s = \frac{f_s + f_{-s}}{\sqrt{2}}, \quad g_{s+k} = \frac{f_s - f_{-s}}{i\sqrt{2}}. \tag{35}$$

Apart from a constant factor, the vectors g_s and g_{s+k} obviously coincide with the real and imaginary parts of f_s $(s = 1, 2, \cdots, k)$:

$$f_s = \frac{g_s + ig_{s+h}}{\sqrt{2}}, \quad f_{-s} = \frac{g_s - ig_{s+h}}{\sqrt{2}}. \tag{35'}$$

Taking real and imaginary parts in the equation $Yf_s = e^{i\theta_s} f_s$ we get the formulae

$$\left.\begin{aligned} Yg_s &= \cos\theta_s g_s - \sin\theta_s \cdot g_{s+k}, \\ Yg_{s+k} &= \sin\theta_s \cdot g_s + \cos\theta_s g_{s+k}. \end{aligned}\right\} \tag{36}$$

We denote by G a matrix whose columns are the vectors $g_1, g_2, \cdots, g_{2k}$. With the help of the matrix G the formulae (36) may be written as follows: $YG = GR$, where R is a matrix of the form (25). Since the vectors $g_1, g_2, \cdots, g_{2k}$ are linearly independent, the matrix G is non-singular. Multiplying the equation $YG = GR$ on the right by G^{-1} we get:

$$Y = GRG^{-1}.$$

We now show that G is a symplectic matrix, Writing the vectors $g_1, g_2, \cdots, g_{2k}$ we the form (35) and using (33) and (34) it is not difficult to verify the following equations:

$$(Ig_{s'}, g_{s''}) = 0, \quad |s' - s''| \neq k, \quad s', s'' = 1, 2, \ldots, 2k,$$
$$(g_s, Ig_{s+k}) = 1 \text{ and } (g_{s+k}, Ig_s) = -1.$$

These equations are equivalent to the matrix relation $G^* IG = I$, which proves that G is symplectic. This completes the proof of the second part of the lemma.

From Lemma 2 there follows almost automatically the sufficiency of the condition of a conformal distribution of multipliers. For let the matrices Y_1 and Y_2 have this property. We show that these matrices can be continuously deformed one into the other through matrices of stable type. For this purpose we write Y_1 and Y_2 in the form (25):

$$Y_1 = G_1 R_1 G_1^{-1} \text{ and } Y_2 = G_2 R_2 G_2^{-1}.$$

We first deform the matrix R_1 into R_2 in such a way that on the unit circle no confluence takes place of multipliers of unlike type. Such a deformation is obviously possible in view of the conformal distribution of multipliers. After this we deform the matrix G_1 into G_2, generally speaking, along any curve which connects these matrices in the group of real symplectic matrices. As follows from Lemma 2, the deformed matrix remains a matrix of stable type throughout the above process of deformation.

§7. Deformation of curves with united end-points

The possibility of bringing together the end-points of two curves via matrices of stable type still does not mean that the systems of differential equations, corresponding to these curves, belong to one stability-region. The point is that in the group of real symplectic matrices not every pair of curves with common end-points can be continuously deformed into coincidence. We take up this question in the present section, a basic role being played by the following:

Lemma 3. *The group of real symplectic matrices is homeomorphic to the topological product of the circumference of a circle and a simply-connected* topological space.*

For the proof we represent a real symplectic matrix Y in the polar form

$$Y = S \cdot U, \tag{37}$$

where S is a positive-definite symmetric matrix, and U is an orthogonal matrix. As is known, such a representation is always possible for a non-singular matrix,

* A topological space is called simply-connected if it is connected and if an arbitrary closed curve in the space can be continuously shrunk up to a point.

and is unique. It is not hard to show that the matrices S and U depend continuously on Y. The subsequent argument extends through four sub-sections, whose content we now briefly indicate. First we show (sub-section 1) that the matrices S and U are symplectic. Then (sub-section 2) will be shown that the set of $2k$th order matrices which are positive-definite, symmetric and also symplectic is homeomorphic to a $k(k+1)$-dimensional Euclidean space. After this we show in sub-section 3 that the group of $2k$th order matrices which are both symplectic and orthogonal is homeomorphic (and moreover isomorphic in the sense of matrix multiplication) to the group of complex unitary matrices of order k. Finally, in sub-section 4, it will be shown that the group of unitary matrices can be decomposed into the topological product of the circumference of a circle and the group of unimodular unitary matrices*. Since the group of unitary unimodular matrices is simply-connected (see for example [2]), we arrive on collecting these facts at the proof of the lemma.

We now set out the proofs.

1. We show that *the matrices S and U, figuring in the polar representation* (37) *are symplectic.* For this we write the symplectic condition $Y^*IY = I$ in the following equivalent form**:

$$Y = I^{-1}(Y^*)^{-1}I. \tag{38}$$

Substituting the matrix SU for Y we get:

$$Y = SU = I^{-1}(S^*)^{-1}(U^*)^{-1}I.$$

We write this equation in the following form:

$$Y = I^{-1}(S^*)^{-1}I \cdot I^{-1}(U^*)^{-1}I. \tag{39}$$

Since S is a symmetric positive-definite matrix, and I is an orthogonal matrix, it follows that the matrix $I^{-1}(S^*)^{-1}I = I^*S^{-1}I$ is also symmetric positive-definite. In addition it is obvious that the matrix $I^{-1}(U^*)^{-1}I$ is orthogonal. Hence the right-hand side of (39) constitutes a polar representation of the matrix Y. By the uniqueness of the latter we have $S = I^{-1}(S^*)^{-1}I$ and $U = I^{-1}(U^*)^{-1}I$. According to (38) these relations show that the matrices S and U are symplectic.

2. We show that *the set* $\{S\}$ *of symmetric, positive-definite and symplectic matrices of order* $2k$ *is homeomorphic to a* $k(k+1)$*-dimensional Euclidean space.*

We use the fact that every symmetric positive-definite matrix S can be uniquely represented in the form

$$S = e^A \tag{40}$$

where A is a real symmetric matrix (the logarithm of the matrix S) and the symbol e^A is understood to mean the series $e^A = E_{2k} + A + A^2/2! + A^3/3! + \cdots$.

The formula (40) establishes a homeomorphic correspondence between posi-

* The results of sub-sections 1, 2, 4 also follow from the general theory of semi-simple Lie groups.

** This relation is to be obtained by multiplying both sides of the equation $Y^*IY = I$ on the left first by $(Y^*)^{-1}$, and then by I^{-1}.

tive-definite symmetric matrices of order $2k$ and all symmetric matrices of the same order (for the proof see [4],* p. 26). The symplectic condition, satisfied by the matrices $\{S\}$, naturally imposes a certain restriction on the logarithms of these matrices. To determine the nature of this restriction we substitute the matrix e^A in the symplectic condition, written in the form (38). This leads us to the formula

$$e^A = I^{-1}(e^A)^{-1}I. \tag{41}$$

Using the fact that $(e^A)^{-1} = e^{-A}$, we transform the right of (41) to the following form:

$$I^{-1}(e^A)^{-1}I = I^{-1}e^{-A}I = e^{-I^{-1}AI}. \tag{42}$$

The validity of the last equation may easily be checked by expanding both members in series. Comparing (41) and (42) we conclude that

$$e^A = e^{-I^{-1}AI}.$$

Since the matrix $-I^{-1}AI$ is real and symmetric together with A, it follows in view of the uniqueness of the representation (40) that

$$A = -I^{-1}AI. \tag{43}$$

These real symmetric matrices $\{A\}$ which are logarithms of matrices of the set $\{S\}$, satisfy the condition (43). Let us now show that, on the other hand, if the real symmetric matrix A satisfies the condition (43), then the matrix e^A belongs to the set $\{S\}$. The proof of this is not difficult. For by hypothesis we have $e^A = e^{-I^{-1}AI}$; reading (42) from right to left we get $e^A = I^{-1}(e^A)^{-1}I$, whence since the matrix $e^A = S$ is symmetric there follows the formula $S = I^{-1}(S^*)^{-1}I$. According to (38) this means that S is a symplectic matrix.

Thus the set of matrices $\{S\}$ is homeomorphic to the set of all real symmetric matrices $\{A\}$ which satisfy the condition (43).

We now show that the set $\{A\}$ in its turn is homeomorphic to a $k(k+1)$-dimensional Euclidean space; our assertion will therewith be proved. We represent the matrix A in the form

$$A = \begin{pmatrix} a_1 & a_2 \\ a_3 & a_4 \end{pmatrix},$$

where a_1, a_2, a_3, a_4 are square matrices of order k. Carrying out the multiplication on the right-hand side of (43) and comparing similarly placed matrices of order k we get: $a_1 = -a_4$, $a_2 = a_3$.

On the other hand, since the matrix A is symmetric there hold the equations $a_1^* = a_1$, and $a_2^* = a_3$, whence it follows that the symmetric matrices $\{A\}$ under consideration have the form

$$A = \begin{pmatrix} a_1 & a_2 \\ a_2 & -a_1 \end{pmatrix}, \tag{44}$$

where a_1 and a_2 are real symmetric matrices of order k. It is easy to verify that

* Translator's note: P. 14 of the original edition (Princeton, 1946).

the set of matrices of the form (44) is homeomorphic to a $k(k+1)$-dimensional Euclidean space. To establish the correspondence it is sufficient to associate with each matrix A a point whose coordinates are an ordered system of elements of the matrices a_1 and a_2 situated on and above the principal diagonals. With this we complete the proof of the assertion formulated at the beginning of sub-section 2.

3. We prove that *the group* $\{U\}$ *of real symplectic orthogonal matrices of order* $2k$ *is homeomorphic (and in fact isomorphic in the sense of matrix multiplication) to the group* $\{w\}$ *of complex unitary matrices of order* k.

Proof. By hypothesis there hold for each matrix U the relations $U^*IU = I$ and $UU^* = E_{2k}$. Multiplying the first of these equations by U and using the second we arrive at the relation

$$UI = IU. \tag{45}$$

We now put the matrix U in the form $U = \begin{pmatrix} u_1 & u_2 \\ u_3 & u_4 \end{pmatrix}$ and carry out the multiplication on both sides of (45). Comparing similarly placed matrices of order k we get:

$$u_1 = u_4 \quad \text{and} \quad u_2 = -u_3. \tag{46}$$

Thus every symplectic orthogonal matrix U has the form

$$U = \begin{pmatrix} u_1 & u_2 \\ -u_2 & u_1 \end{pmatrix}. \tag{47}$$

From the relation $UU^* = E_{2k}$ we now get by direct multiplication:

$$u_1u_1^* + u_2u_2^* = E_{2k};\ u_1u_2^* - u_2u_1^* = 0. \tag{48}$$

As is easily seen, the equations (48) are equivalent to:

$$(u_1 + iu_2)(u_1 + iu_2)^* = E_k.$$

Thus we have shown that the matrices u_1 and u_2 completely determine the matrix U and that the matrix of order k

$$w = u_1 + iu_2 \tag{49}$$

is unitary.

On the other hand, if w is some kth order unitary matrix and $\tilde{u}_1$ and $\tilde{u}_2$ are its real and imaginary parts, then for them there hold the equations (48). Hence the matrix $\tilde{U}$, constructed according to (47), is orthogonal. Matrices of the form (47) commute with I. Multiplying the relation $I\tilde{U} = \tilde{U}I$ on the right by $\tilde{U}^*$, we see that $\tilde{U}$ is a symplectic matrix. Associating with every symplectic orthogonal matrix U a unitary matrix w according to (49), we obtain the required homeomorphism.

Let us show that the relation obtained

$$U \longleftrightarrow w \tag{50}$$

is an isomorphism. For let $U^{(1)} \longleftrightarrow w^{(1)}$ and $U^{(2)} \longleftrightarrow w^{(2)}$. Representing $U^{(1)}$ and $U^{(2)}$ in the form (47) and carrying out the multiplication we get:

$$U^{(1)}U^{(2)} = \begin{pmatrix} u_1^{(1)}u_1^{(2)} - u_2^{(1)}u_2^{(2)} & u_1^{(1)}u_2^{(2)} + u_2^{(1)}u_1^{(2)} \\ -u_1^{(1)}u_2^{(2)} - u_2^{(1)}u_1^{(2)} & u_1^{(1)}u_1^{(2)} - u_2^{(1)}u_2^{(2)} \end{pmatrix}$$

and, on the other hand:

$$w^{(1)}w^{(2)}=(u_1^{(1)}+iu_2^{(1)})(u_1^{(2)}+iu_2^{(2)})=u_1^{(1)}u_1^{(2)}-u_2^{(1)}u_2^{(2)}+i(u_1^{(1)}u_2^{(2)}+u_2^{(1)}u_1^{(2)}).$$

Hence in this correspondence $w^{(1)}w^{(2)}$ corresponds to the matrix $u^{(1)}u^{(2)}$. Since furthermore the unity of the group $\{w\}$, the matrix E_k, obviously corresponds here to the unity E_{2k} of the group $\{U\}$, the isomorphism of the groups $\{U\}$ and $\{w\}$ is proved.

4. We now consider the proof of the fact that the group of complex unitary matrices can be decomposed into the topological product of the circumference of a circle and the group of unitary unimodular matrices.

For let w be some unitary matrix and $\operatorname{Det} w = e^{i\psi}$ *. We consider the unitary matrix

$$v=\begin{pmatrix} e^{i\psi} & & & & 0 \\ & 1 & & & \\ & & 1 & & \\ & & & \ddots & \\ 0 & & & & 1 \end{pmatrix} \tag{51}$$

and represent w in the following form:

$$w = v(v^{-1}w).$$

The group of matrices $\{v\}$ is homeomorphic to a circumference (to establish the correspondence it is sufficient to associate with every matrix v the point $e^{i\psi}$ of the unit circle in the complex plane). The matrices of the form $v^{-1}w$ are obviously unimodular.

Let us now finish the proof of Lemma 3.

We have already indicated that the unitary unimodular matrices form a simply connected topological group. A simple proof of this fact is given in the book of H. Weyl ([5], p. 360)**. Collecting the above facts, we see that the group of real symplectic matrices of order $2k$ is homeomorphic to the topological product of a circumference (the group of matrices $\{v\}$) and the group of unitary unimodular matrices of order k and a $k(k+1)$-dimensional Euclidean space. Since the topological product of two simply-connected spaces is simply-connected, Lemma 3 is proved.

It follows from Lemma 3 that not every two curves $Y_1(t)$ and $Y_2(t)$ $(0 \le t \le \omega)$ with common ends $Y_1(0) = Y_2(0) = E_{2k}$ and $Y_1(\omega) = Y_2(\omega)$ are homotopic, i.e., can be continuously deformed one into the other, their end-points remaining fixed.

For by the lemma to each real symplectic matrix Y there corresponds a fully determined point on the unit circle on the complex plane (the projection of the matrix Y on the circle), whose location, as we saw, is determined by the determinant of the matrix w. The projections on the circle of the matrix curves $Y_1(t)$ and $Y_2(t)$

* As is known, the determinant of a unitary matrix has modulus equal to unity.

** Translator's note: P. 268 of the original.

here form two continuous curves with common ends, executing in general a different number of circuits of the circle. It is obvious that if the projections of the curves $Y_1(t)$ and $Y_2(t)$ on the circle cannot be continuously deformed one into the other, then the matrix curves themselves are non-homotopic. If on the other hand the projections of the matrix curves on the circle are homotopic, then to set up the continuous deformation it remains in view of the lemma to bring together their projections on the simply-connected topological space. But in the case of a simply-connected space any two curves are homotopic (see [6], p. 234).

Thus in order that the curves $Y_1(t)$ and $Y_2(t)$ can be continuously deformed one into the other, their ends remaining fixed, it is necessary and sufficient that their projections on the circle should have this property.

Let us for brevity denote by $\operatorname{Arg}\zeta(t)|_a^b$ the change in the argument of some complex-valued function $\zeta(t)$, due to the variation of the real parameter t on the interval $[a, b]$. If $w_1(t)$ and $w_2(t)$ are curves in the group of unitary matrices, corresponding to the curves $Y_1(t)$ and $Y_2(t)$ according to the formulae (37) and (49), then the necessary and sufficient condition for the curves to be homotopic takes the form

$$\operatorname{Arg}\operatorname{Det} w_1(t)|_0^\omega = \operatorname{Arg}\operatorname{Det} w_2(t)|_0^\omega. \tag{52}$$

For brevity we shall in the sequel write $\operatorname{Arg} Y(t)|_0^\omega$ for the quantity $\operatorname{Arg}\operatorname{Det} w(t)|_0^\omega$. Thus by definition

$$\operatorname{Arg} Y(t)|_0^\omega = \operatorname{Arg}\operatorname{Det} w(t)|_0^\omega,$$

where $w(t)$ is a curve in the group of unitary matrices, corresponding to the curve $Y(t)$. The condition for the curves $Y_1(t)$, $Y_2(t)$ to be homotopic may then be briefly written in the form:

$$\operatorname{Arg} Y_1(t)|_0^\omega = \operatorname{Arg} Y_2(t)|_0^\omega. \tag{53}$$

To conclude this section we indicate a simple means of determining the quantity $\operatorname{Arg} Y(t)|_0^\omega$.

We represent the curve $Y(t)$ in the polar form $Y(t) = S(t)\,U(t)$ and introduce the unitary matrix M, whose columns are the vectors m_s and m_{-s} (see (27)):

$$M = \frac{1}{\sqrt{2}}\begin{pmatrix} E_k & E_k \\ iE_k & -iE_k \end{pmatrix}.$$

Taking into account the general form possessed by orthogonal symplectic matrices U, it is easy to verify the formula

$$M^*UM = \begin{pmatrix} w & 0 \\ 0 & \bar{w} \end{pmatrix}, \text{ where } w = u_1 + iu_2. \tag{54}$$

We now transform the equation $Y(t) = S(t)\,U(t)$ as follows:

$$M^*Y(t)M = [M^*S(t)M]\,[M^*U(t)M]. \tag{55}$$

We denote by $z(t)$ the upper diagonal minor of order k of the matrix $M^*Y(t)M$. It is obvious that

$$z(t) = ((Y(t)m_\mu, m_\nu)) \quad (1 \le \nu, \mu \le k). \tag{56}$$

We now prove the formula

$$\operatorname{Arg} Y(t)|_0^{\omega} = \operatorname{Arg} \operatorname{Det} z(t)|_0^{\omega}. \tag{57}$$

For this we use (55). Since the matrix $M^*U(t)M$ has the box-diagonal form (54), there follows immediately from (55) the equation

$$z(t) = s_k(t) \cdot w(t), \tag{58}$$

where $s_k(t)$ denotes the upper diagonal minor of the positive-definite Hermitean matrix $M^*S(t)M$. The determinant of $s_k(t)$ $(0 \le t \le w)$ is real and non-zero. Hence from (58) it follows at once that $\operatorname{Arg} \operatorname{Det} z(t)|_0^{\omega} = \operatorname{Arg} \operatorname{Det} w(t)|_0^{\omega}$, which proves the formula (57).

§8. Determination of the number* of a stability-region for a given curve

To complete the proof of Theorem 1 we have to solve a subsidiary problem, consisting in the following.

We assume that the two curves $Y_1(t)$ and $Y_2(t)$ $(0 \le t \le \omega,\ Y_1(0) = Y_2(0) = E_{2k})$ have the property that their ends $Y_1(\omega)$ and $Y_2(\omega)$ lie in one and the same connected component of the matrices of stable type, i.e., have a conformal distribution of multipliers. We deform the curve $Y_1(t)$ in such a way that its end, without leaving the component in question, is brought into coincidence with the end of the curve $Y_2(t)$.

Let us assume that as a result of this deformation we obtain two curves for which the condition of homotopicity is not satisfied.

Does this mean that the curves in question belong to distinct stability regions? Could it not happen that if we had brought the ends together, via matrices of stable type, *in another way* we should have arrived at two homotopic curves?

We shall presently give a negative answer to the latter question.

For if such a statement held, then obviously within some component of the matrices of stable type, there could be found two non-homotopic curves joining the matrices $Y_1(\omega)$ and $Y_2(\omega)$.

The proof of the fact that such curves do not exist we give as a separate lemma, which we shall use again in one case.

Lemma 4. *Two continuous curves $Y_1(r)$ and $Y_2(r)$ $(0 \le r \le 1)$ with common end-points $Y_1(0) = Y_2(0)$, $Y_1(1) = Y_2(1)$, lying entirely in some component of the matrices of stable type, can always be continuously deformed one into the other.*

Remark. It will be clear from the proof that the assertion of Lemma 4 remains valid in the case when the end-point of the curves $Y_1(r)$ and $Y_2(r)$ is the unit matrix $(Y_1(1) = Y_2(1) = E_{2k})$.

Proof. By Lemma 2 for every fixed value of the parameter r the matrices $Y_1(r)$ and $Y_2(r)$ can be written in the form

* Translator's note: Serial number.

$$Y_1(\tau) = G_1(\tau)R_1(\tau)G_1^{-1}(\tau) \text{ and } Y_2(\tau)\ \omega\ G_2(\tau)R_2(\tau)G_2^{-1}(\tau), \tag{59}$$

where $R_1(\tau)$ and $R_2(\tau)$ are symplectic orthogonal matrices of the form (25), and $G_1(\tau)$ and $G_2(\tau)$ are certain symplectic matrices. Examining the proof of Lemma 2, it is not hard to see that the matrices $R(\tau)$ and $G(\tau)$ can always be chosen so that they depend continuously on the parameter and furthermore satisfy the conditions

$$R_1(0) = R_2(0),\ \ R_1(1) = R_2(1); \tag{60}$$

$$G_1(0) = G_2(0),\ \ G_1(1) = G_2(1). \tag{61}$$

Let us first show that the thus determined curves $R_1(\tau)$ and $R_2(\tau)$ are homotopic. For the multipliers of the matrices $Y_1(\tau)$ and $Y_2(\tau)$ in the interval $0 \le \tau \le 1$ describe on the unit circle in the complex plane paths which do not go from one half-plane to the other. Hence the paths described by the multipliers, which coincide pairwise for $\tau = 0$ and $\tau = 1$, can be continuously deformed one into the other. As a result of this deformation the curves $R_1(\tau)$ and $R_2(\tau)$ obviously coincide. After this deformation we seek to deform the curve $G_1(\tau)$ into $G_2(\tau)$. We use the fact that the matrices $G(\tau)$ in the representations (24), (59) are determined to within right multiplication by a matrix commuting with $R(\tau)$.

Hence the condition for the curves $G_1(\tau)$ and $G_2(\tau)$ to be homotopic

$$\operatorname{Arg}\operatorname{Det} w_1(\tau)\big|_0^1 = \operatorname{Arg}\operatorname{Det} w_2(\tau)\big|_0^1 \tag{62}$$

can always be satisfied, if necessary multiplying one of the curves by a matrix which commutes with $R(\tau)$.

For let us assume that in the original choice of $G_1(\tau)$ and $G_2(\tau)$ the left of (62) turned out to exceed the right by an amount Δ. We observe that by the condition (61) we have $\Delta = 2\pi m$, where m is an integer.

We represent $G_2(\tau)$ in the polar form

$$G_2(\tau) = S_2(\tau)U_2(\tau)$$

and introduce an orthogonal symplectic matrix $N(\tau)$ of the form (25)

$$N(\tau) = \begin{Bmatrix} \cos\tau 2\pi m & 0 & \dots & 0 & \sin\tau 2\pi m & 0 & \dots & 0 \\ 0 & 1 & \dots & 0 & 0 & 0 & \dots & 0 \\ \vdots & \vdots & & \vdots & \vdots & \vdots & & \vdots \\ 0 & 0 & \dots & 1 & 0 & 0 & \dots & 0 \\ -\sin\tau 2\pi m & 0 & \dots & 0 & \cos\tau 2\pi m & 0 & \dots & 0 \\ 0 & 0 & \dots & 0 & 0 & 1 & \dots & 0 \\ \vdots & \vdots & & \vdots & \vdots & \vdots & & \vdots \\ 0 & 0 & \dots & 0 & 0 & 0 & \dots & 1 \end{Bmatrix}$$

It is easy to verify that $N(\tau)$ commutes with $R_2(\tau)$. We multiply $G_2(\tau)$ on the right by $N(\tau)$. Since $N(0) = N(1) = E_{2k}$ the conditions (61) are not thereby broken. On the other hand the condition (62) for the curves $G_1(\tau)$ and $\widetilde{G}_2(\tau) = G_2(\tau)N(\tau)$ to be homotopic will now be fulfilled; for the orthogonal part of $\widetilde{G}_2(\tau)$ in the polar representation (37) is obviously equal to $U_2(\tau)N(\tau)$; to it by the isomorphism (50) corresponds a unitary matrix

$$\widetilde{w}_2(\tau) = w_2(\tau)\, w_N(\tau),$$

where $w_N(\tau)$ is a matrix of the form (51) ($\psi = \tau 2\pi m$).

Since $\operatorname{Det} w_N(\tau) = e^{i\tau 2\pi m}$, it follows that $\operatorname{Arg}\operatorname{Det}\widetilde{w}_2(\tau)|_0^1 = \operatorname{Arg}\operatorname{Det} w_2(\tau)|_0^1 + 2\pi m$, as asserted.

To complete the proof of the lemma it remains to observe that under the continuous deformation of $G_1(\tau)$ into $G_2(\tau)$ the curve $G_1^{-1}(\tau)$ is continuously deformed into $G_2^{-1}(\tau)$. This fact follows immediately from the formula

$$G_1^{-1}(\tau) = I^{-1} G_1^*(\tau) I,$$

which in turn follows from the symplectic condition. The proof of Lemma 4 is complete.

We now consider a set of curves $[Y_\alpha(t)]$ ($0 \le t \le \omega$, $Y_\alpha(0) = E_{2k}$) in the group $\{Y\}$, which have piecewise continuous derivatives and satisfy the following conditions: a) the ends of all the curves lie in one component of the matrices of stable type; b) after the ends of any two curves* have been moved into coincidence there hold the homotopic conditions (53).

The set $[Y_\alpha(t)]$ obviously constitutes a certain stability-region. We show that apart from the type of distribution of the multipliers, the region in question can be characterised by one integer, the number of the stability-region; at the same time we indicate means for calculating it.

We take any curve $Y(t)$ belonging to the stability-region $[Y_\alpha(t)]$, and represent its end-point $Y(\omega)$ in the form (24):

$$Y(\omega) = GRG^{-1}. \tag{63}$$

Together with the curve $Y(t)$ we consider the curve $Y_g(t)$ defined by

$$Y_g(t) = G^{-1} Y(t)\, G. \tag{64}$$

It is not hard to verify that the curve $Y_g(t)$ also belongs to the set $[Y_\alpha(t)]$.

For in the first place $Y_g(0) = E_{2k}$; in the second place $Y_g(\omega) = G^{-1}Y(\omega)G = R$ and therefore the end of the curve $Y_g(t)$ is a matrix of stable type with the same type of distribution of the multipliers as the ends of the curves $[Y_\alpha(t)]$. Finally, if in (64) we deform the matrix G to the unit matrix, the curve $Y_g(t)$ as a result of this deformation will move into coincidence with the curve $Y(t)$. Here the end-point of the curve which undergoes deformation will not leave the component of the matrices of stable type, which proves the the curve $Y_g(t)$ belongs to the set $[Y_\alpha(t)]$.

* Wherever mention is made of the moving together of end-points, a deformation is to be understood which does not take the end-points outside the component of the matrices of stable type.

We join the end of the curve $Y_g(t)$, the matrix R^*, to the beginning, the matrix E_{2k}, by a continuous curve which we determine by the formula

$$R(\tau)=\begin{pmatrix} \cos(1-\tau)\theta & \sin(1-\tau)\theta \\ -\sin(1-\tau)\theta & \cos(1-\tau)\theta \end{pmatrix} \qquad (0\leqslant\tau\leqslant 1). \tag{65}$$

We emphasise that in view of the condition (26), which the diagonal elements of the matrix θ must satisfy, *all points of the curve $R(\tau)$, excepting the end-point, belong to the component of the matrices of stable type.*

When we join the end and the beginning of the curve $Y_g(t)$ by means of the matrix $R(\tau)$, we obtain a certain closed curve. The change in the argument of the determinant of the unitary matrix w, corresponding to this curve according to the formulae (37) and (49), is obviously compounded of $\operatorname{Arg} Y_g(t)|_0^\omega$ and $\operatorname{Arg} R(\tau)|_0^1$, the sum of these quantities being an integral multiple of 2π in view of the fact that the curve in question is closed. We put:

$$(2\pi)^{-1}[\operatorname{Arg} Y_g(t)|_0^\omega + \operatorname{Arg} R(\tau)|_0^1] = n. \tag{66}$$

We shall associate the integer n with the given stability-region $[Y_\alpha(t)]$. In view of Lemma 4 n does not depend on the choice of the curve $Y(t)$ from the set $[Y_\alpha(t)]$

Let us consider, on the other hand, some curve

$$\tilde{Y}(t)\ (0\leq t\leq\omega,\ \tilde{Y}(0)=E_{2k}),$$

whose end, the matrix $\tilde{Y}(\omega)$, lies in the same component as the ends of the curves $[Y_\alpha(t)]$. If for the curve $\tilde{Y}(t)$ there holds (66) with the same value of the integer n, then after moving the end-point of $\tilde{Y}(t)$ into coincidence with the end-point of some curve of the set $[Y_\alpha(t)]$ the condition of homotopocity will be fulfilled (supposing the contrary, we immediately arrive at a contradiction with Lemma 4). This means that the curve $\tilde{Y}(t)$ belongs to the stability-region $[Y_\alpha(t)]$. Thus the integer n, calculated according to (66), together with the type of distribution of the multipliers, completely determines the stability-region $[Y_\alpha(t)]$, which proves Theorem 1.

We now transform (66) to a form more convenient for calculations.

1. Let us first expose the meaning of the expression $\operatorname{Arg} R(\tau)|_0^1$. To the matrix $R(\tau)$ (see (65)) there corresponds according to (49) the unitary matrix

$$\cos(1-\tau)\theta + i\sin(1-\tau)\theta = e^{i(1-\tau)\theta},$$

where θ is a diagonal matrix, along whose diagonal stand the principal values of the first kind of the matrix $Y(\omega)$: $\theta_1, \theta_2, \cdots, \theta_k$. Hence

$$\operatorname{Arg} R(\tau)|_0^1 = \operatorname{Arg} \operatorname{Det} e^{i(1-\tau)\theta}|_0^1 = \operatorname{Arg} \prod_{s=1}^{k} e^{i\theta_s(1-\tau)}|_0^1 = -\sum_{s=1}^{k}\theta_s. \tag{67}$$

2. We turn now to the first term on the left of (66). According to the formulae (57) and (56) $\operatorname{Arg} Y_g(t)|_0^\omega = \operatorname{Arg} \operatorname{Det} z(t)|_0^\omega$, where

$$z(t) = ((Y_g(t)m_\mu, m_\nu)) \quad (1\leq\nu,\mu\leq k).$$

* We recall that R is an orthogonal symplectic matrix of special form (see (25)).

By definition $Y_g(t) = G^{-1}Y(t)G$. Since the matrix G is symplectic there holds the equation $G^{-1} = I^{-1}G^*I$. Hence

$$(Y_g(t)m_\mu, m_\nu) = (G^{-1}Y(t)Gm_\mu, m_\nu) = (I^{-1}G^*IYGm_\mu, m_\nu) = (IYGm_\mu, GIm_\nu). \quad (68)$$

Since $Im_\nu = im_\nu$ (see the left-hand formula (29)), the equation (68) may be written as follows:

$$(Y_g(t)m_\mu m_\nu) = (-iIY(t)Gm_\mu, m_\nu) = (JY(t)Gm_\mu, m_\nu).$$

The vectors $Gm_s = f_s$ $(s = 1, \cdots, k)$ are the proper vectors of the matrix $Y(\omega)$, corresponding to the multipliers of the first kind. In fact

$$Y(\omega)f_s = GRG^{-1}Gm_s = GRm_s,$$

but in view of (28) $Rm_s = e^{i\theta_s}m_s$ and therefore $Y(\omega)f_s = e^{i\theta_s}Gm_s = e^{i\theta_s}f_s$. However this fact follows directly from the formulae (27) and (35′).

Thus the matrix $z(t)$ assumes the following form:

$$z(t) = ((JY(t)f_\mu, f_\nu)) \quad (1 \le \nu, \mu \le k),$$

where f_s $(s = 1, \cdots, k)$ are proper vectors of the matrix $Y(\omega)$, the end of the curve $Y(t)$, corresponding to multipliers of the first kind.

We note that $z(t)$ is the matrix of a certain bilinear form. Hence if in the subspace spanned by the vectors f_s $(s = 1, \cdots, k)$, we pass to another basis $h_j = \sum_{s=1}^k c_{js}f_s$ $(j = 1, 2, \cdots, k)$, where (c_{js}) is some non-singular matrix, then $z(t)$ is transformed to the form $cz(t)c^*$; its determinant will thereby be multiplied by a certain constant, which of course will not affect the quantity $\operatorname{Arg}\operatorname{Det} z(t)|_0^\omega$.

In consequence of these transformations we reach the following result, which we formulate as a rule for ascertaining the number of a stability-region for a given curve:

Let there be given the solution matrix $Y(t)$ $(0 \le t \le \omega)$ of a strongly stable system, normalised by a condition at zero $Y(0) = E_{2k}$. We choose an arbitrary basis $h_1, h_2, \cdots, h_k$ in the subspace spanned by the proper vectors of the monodromy matrix $Y(\omega)$ which correspond to multipliers of the first kind, and set up the matrix

$$z(t) = ((JY(t)h_\mu, h_\nu)) \quad (1 \le \nu, \mu \le k).$$

Then the number of the region, to which the given system belongs, is to be found from the formula

$$n = (2\pi)^{-1}\Big(\operatorname{Arg}\operatorname{Det} z(t)\,|_0^\omega - \sum_{s=1}^k \theta_s\Big), \quad (69)$$

where the θ_s are the principal values of the arguments of the multipliers of the first kind.

We note in conclusion that it is not hard to adduce examples of curves which belong to stability-regions determined by an arbitrary prescribed distribution of multipliers and an arbitrary number n. For this purpose it is convenient to use the curves $R(t)$ of the form (25).

§9. The case of a system of two equations

We illustrate the above results by the example $k = 1$ (a system of two equations), the case investigated in the paper [1].

It is easy to verify directly that for a real second-order matrix Y the symplectic condition $Y^* IY = I$ coincides with the unimodular condition, i.e., with the condition $\mathrm{Det}\, Y = 1$.

We show that the group of second-order symplectic matrices is homeomorphic with the interior of a torus.

For this we represent the matrix Y in the polar form $Y = SU$, or, in more detail,

$$Y = \begin{pmatrix} s_{11} & s_{12} \\ s_{21} & s_{22} \end{pmatrix} \begin{pmatrix} \cos\psi & \sin\psi \\ -\sin\psi & \cos\psi \end{pmatrix} \tag{70}$$

Starting from the conditions $s_{11}s_{22} - s_{12}^2 = 1,\ s_{11} > 0$ we parametrise the elements of the symmetric positive-definite matrix S as follows:

$$\left.\begin{aligned} s_{11} &= \cosh r + \sinh r \cdot \cos\sigma, \\ s_{22} &= \cosh r - \sinh r \cdot \cos\sigma, \\ s_{12} = s_{21} &= \sinh r \cdot \sin\sigma \end{aligned}\right\} \tag{71}$$

$$(0 \le \sigma \le 2\pi,\ \ 0 \le r \le \infty).$$

We introduce the usual toral system of coordinates. (Fig. 5).

With each matrix Y, determined by the parameters ψ, σ and r, we shall associate the point of the torus with coordinates

$$\phi = \psi,\ \ \theta = \sigma,\ \ r = \tanh^2 r. \tag{72}$$

It is not hard to verify that this correspondence is (1,1) and bicontinuous.

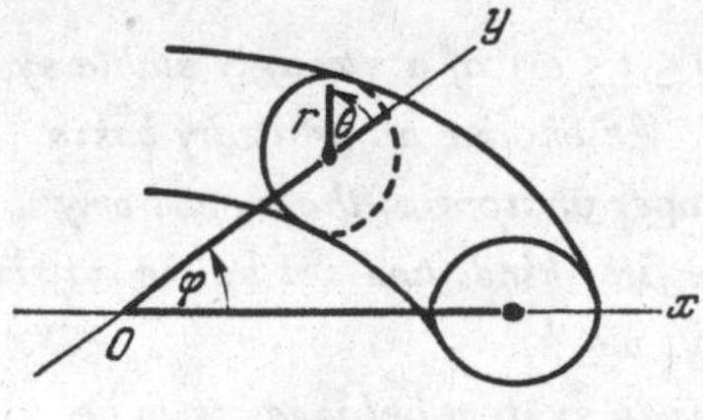

Fig. 5.

Let us find on our model the regions which correspond to the matrices of stable type. In the present case the matrices of stable type are those whose characteristic polynomials have non-real roots. For being complex-conjugate and reciprocal in value, both roots are necessarily situated on the unit circle and are simple. The condition that the roots of a real unimodular matrix Y should be non-real reduces, as is easily verified, to the requirement that the trace of Y should have modulus less than 2.

By direct calculation from the formula (70) we get: $\mathrm{Tr}\, Y = \cos\psi \cdot \mathrm{Tr}\, S$, or in view of (71) $\mathrm{Tr}\, Y = 2\cos\psi \cdot \cosh r$.

Thus the matrices of stable type are distinguished by the condition

$$2|\cos\psi|\cosh r < 2.$$

To find these regions on the torus, we set up the equation for its boundary, putting $|\cos\psi| \cosh r = 1$. The corresponding dependence between the coordinates of points of the torus, derived with the help of (72), may be expressed as follows:

$$r = \sin^2 \phi \quad (0 \le \phi \le 2\pi). \tag{73}$$

This is in fact the required equation for the boundary. The corresponding surface in the torus is shaded in Fig. 6.

Points of the torus lying inside the shaded surface correspond to matrices of stable type. The set of these points falls into two connected components, since the boundary at $\phi = 0$ and $\phi = \pi$ shrinks up to a point, by (73). This corresponds

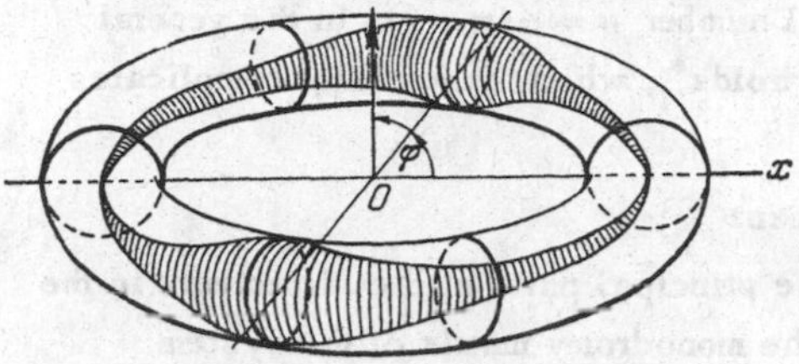

Fig. 6.

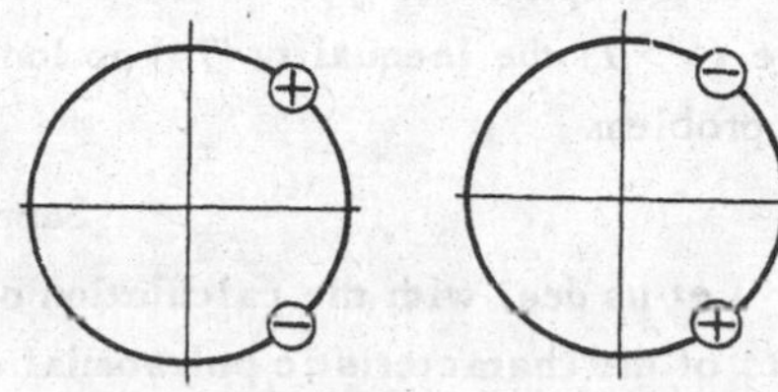

Fig. 7.

to the two distinct types in regard to the distribution of multipliers (see Fig. 7).

The matrix E_2 corresponds on the torus to the point with the coordinates $\phi = 0$, $r = 0$. The curve $Y(t)$ $(0 \le t \le \omega, Y(0) = E_2)$ corresponds to a strongly stable system if its end falls into one of the components of the matrices of stable type inside the shaded region. The quantity $\operatorname{Arg} Y(t)|_0^\omega$ here is equal to the angle ϕ, swept out on the horizontal plane by the projection of the vector, whose end describes the curve $Y(t)$ on the torus (see (70) and (72)). A stability-region is formed by the curves $[Y_\alpha(t)]$ $(0 \le t \le \omega)$, whose end-points lie in one component of the matrices of stable type and which, in addition, make the identical number of circuits of the torus in the same direction.

To determine the number of a region, according to the way this was done in §8, it is necessary to take some curve $Y(t)$ belonging to the region in question, and to join its end $Y(\omega)$ with the unit matrix by a continuous curve, all interior points of which are matrices of stable type*. As a result we get a certain closed curve. The number of complete circuits made by this curve, taken with a plus or minus sign according to the sense of the circuit with increasing t, is equal to the serial number of the stability-region.

We note that in the particular case in question $(k = 1)$, the numeration of the stability-regions may be carried out otherwise.

Taking into account the way in which the components of the matrices of stable type are disposed (see (73) and Fig. 5), it is easy to make the deduction that for all curves of some stability-region $[Y_\alpha(t)]$ there holds, in this case, an inequality

$$m\pi < \operatorname{Arg} Y_\alpha(t)|_0^\omega < (m+1)\pi, \tag{74}$$

* In §8 we first of all went from the curve $Y(t)$ to the curve $Y_g(t)$. This was done in order to obtain an effective formula for the number of the stability-region.

m being an integer. The equality sign is here excluded, since the cross-section of the torus for $\phi = 0$ and $\phi = \pi$ does not contain matrices of stable type. This inequality makes it possible to associate with each stability-region an integer $m = 0, \pm 1, \pm 2, \cdots$ (see [1]). With such a numeration to regions with even m there corresponds a distribution of multipliers as represented in Fig. 7 on the left (type a), and the serial number $n = \frac{1}{2}m$, while to regions with odd m corresponds a distribution of multipliers of type b and the serial number $n = \frac{1}{2}(m+1)$. In the general case $(k > 1)$ the inequality (74) no longer holds*, which essentially complicates the problem.

Supplement

Let us deal with the calculation of the principal parts of the increments in the roots of the characteristic polynomial of the monodromy matrix of the system

$$\frac{d}{dt} Y = IH(t)\,Y + \lambda IQ(t)\,Y \tag{75}$$

as functions of the parameter λ.

The formulae which we obtain make it possible to follow up the behaviour of the multipliers as the parameter λ varies in the neighbourhood of the origin. From them will be deduced the properties of the multipliers which we indicated in the first sections of this paper.

1°. Let us consider the monodromy matrix $Y(\omega, \lambda)$ of the system (75). In virtue of known theorems concerning the dependence upon a parameter of solutions of differential equations, the elements of this matrix are entire analytic functions of λ. Hence the roots $\rho_1(\lambda), \rho_2(\lambda), \cdots, \rho_{2k}(\lambda)$ of the characteristic equation

$$\operatorname{Det}\,[Y(\omega, \lambda) - \rho E_{2k}] = 0 \tag{76}$$

can be represented in the nieghbourhood of $\lambda = 0$ by convergent series

$$\rho_s(\lambda) = \rho_s + \sum_{\nu=1}^{\infty} a_\nu^s \lambda^{\frac{\nu}{n_s}} \qquad (s = 1, 2, \ldots, 2k), \tag{77}$$

where the ρ_s are the multipliers of $Y(\omega, 0)$, n_s is a positive integer $\leq 2k$, and the a_ν^s are constant coefficients.

In other words, the equation (76) determines $2k$ functions $\rho_s(\lambda)$, which can be represented in the nieghbourhood of $\lambda = 0$ by the series (77), and which are such that there holds the identity

$$\operatorname{Det}\,[Y(\omega, \lambda) - \rho E_{2k}] = \prod_{s=1}^{2k} [\rho - \rho_s(\lambda)]. \tag{78}$$

Below we calculate the first terms in the expansions (77).

* We note in passing that one can show that in the general case $(k > 1)$ for curves belonging to one stability-region there holds the bound

$$\left| \operatorname{Arg} Y_{\alpha'}(t)\big|_0^{\omega} - \operatorname{Arg} Y_{\alpha''}(t)\big|_0^{\omega} \right| < k\pi,$$

which cannot be improved.

2°. Let us first calculate the derivative of the monodromy matrix $Y(\omega, \lambda)$ with respect to the parameter λ. For this we substitute in the equation (75) the solution-matrix $Y(t, \lambda)$, normalised by the condition

$$Y(0, \lambda) = E_{2k}, \tag{79}$$

and differentiate with respect to λ both sides of the resulting identity. Putting $\lambda = 0$ we get:

$$\frac{d}{dt} Y'_\lambda = IH(t) Y'_\lambda + IQ(t) Y. \tag{80}$$

Here we have used Y'_λ as an abbreviation for $\partial = Y(t, \lambda)/\partial\lambda|_{\lambda=0}$. It follows from (80) that the matrix Y'_λ satisfies in respect of t a linear non-homogeneous equation.

By a known formula for the solution of a non-homogeneous equation (see [7], p. 384)

$$Y'_\lambda = Y(t, 0) \cdot C + Y(t, 0) \cdot \int_0^t Y^{-1}(r, 0) IQ(r) Y(r, 0) dr, \tag{81}$$

where C is a constant matrix. We determine C from the initial conditions. We put in (81) $t = 0$. Since, in view of the condition (79)

$$\frac{\partial}{\partial\lambda} Y(0, \lambda)|_{\lambda=0} = 0 \text{ and } Y(0, 0) = E_{2k},$$

it follows from (81) that $C = 0$. Hence

$$Y'_\lambda = Y(t, 0) \int_0^t Y^{-1}(r, 0) IQ(r) Y(r, 0) dr. \tag{82}$$

We transform the right-hand side of (82) using the equation $Y^{-1}(r, 0) I = IY^*(r, 0)$, which follows from the symplectic condition. Putting then in (82) $t = \omega$, we get:

$$\frac{\partial}{\partial\lambda} \cdot Y(\omega, \lambda)|_{\lambda=0} = Y(\omega) \cdot I \cdot \int_0^\omega Y^*(\tau) Q(\tau) Y(\tau) d\tau. \tag{83}$$

On the right of (83) we have omitted for brevity indication of the value of the parameter λ.

3°. Since it is analytic the matrix $Y(\omega, \lambda)$ can be expanded in the neighbourhood of $\lambda = 0$ in a convergent series of powers of $\Delta\lambda$, the increment, with matrix coefficients*

$$Y(\omega, \lambda) = Y(\omega) + \sum_{\nu=1}^{\infty} B_\nu (\Delta\lambda)^\nu \tag{84}$$

$$(Y(\omega) = Y(\omega, 0)).$$

Here the matrix B_1, which stands before the first power of $\Delta\lambda$, is equal to $\partial = Y(t, \lambda)/\partial\lambda|_{\lambda=0}$, and therefore has the form (83).

We denote by D the Jordan form of the matrix $Y(\omega)$, and by F the matrix which brings $Y(\omega)$ to the Jordan form $(D = F^{-1} Y(\omega) F)$. Transforming $Y(\omega, \lambda)$ by means of the matrix F, we reduce the characteristic equation (76) to the form

$$\text{Det}(D + \Delta\lambda \widetilde{B} + \cdots - \rho E_{2k}) = 0. \tag{85}$$

* $\Delta\lambda = \lambda - \lambda_0$; since $\lambda_0 = 0$, we have $\Delta\lambda = \lambda$.

In this formula the dots stand for terms containing $\Delta\lambda$ to a power higher than the first, while the matrix $\widetilde{B} = F^{-1}B_1F$. To facilitate subsequent calculations we transform $\widetilde{B}$ as follows:

$$\widetilde{B} = F^{-1}B_1F = (F^{-1}Y(\omega)F)(F^{-1}IF^{*-1})\left(F^* \cdot \int_0^\omega Y^*(\tau)Q(\tau)Y(\tau)\,d\tau \cdot F\right). \quad (86)$$

Introducing the notations

$$F^*JF = L \text{ and } F^* \cdot \int_0^\omega Y^*(\tau)Q(\tau)Y(\tau)\,d\tau \cdot F = \widetilde{Q}, \quad (87)$$

we write $\widetilde{B}$ in the form

$$\widetilde{B} = iDL^{-1}\widetilde{Q}. \quad (88)$$

Let us assume that the system (75) is stable for $\lambda = 0$. In this case the matrix D is diagonal, and all the multipliers have unit modulus. We note that under these conditions the Hermitean matrix L is box-diagonal. For the columns of F are the proper vectors of $Y(\omega)$, and the proper vectors corresponding to unequal proper values are, as we know, I-orthogonal. Thus L has the form

$$L = \begin{pmatrix} L_{\rho'} & 0 & & \\ 0 & L_{\rho''} & & \\ & & \cdot & \\ & & & \cdot \end{pmatrix}, \quad (89)$$

where the L_ρ are Hermitean matrices of the form (22), of order equal to the multiplicity of the corresponding proper value ρ.

As to the matrix $\widetilde{Q}$, its elements $\widetilde{q}_{ij}$ obviously have the form

$$\widetilde{q}_{ij} = \int_0^\omega (Y^*(\tau)Q(\tau)Y(\tau)f_j, f_i)\,d\tau, \quad (90)$$

where f_i and f_j are columns of F.

4°. We pass to the calculation of the first terms of the expansion (77). We consider first of all the case when ρ_0 is a simple proper values of the matrix $Y(\omega)$. Without loss of generality one may assume that ρ_0 is the upper diagonal element of D; with this condition the proper vector of $Y(\omega)$ which corresponds to the proper value ρ_0 will form the first column of F.

We substitute in the equation (85) the function $\rho(\lambda)$, which coincides for $\lambda = 0$ with ρ_0. In the resulting identity we remove the factor $\Delta\lambda$ from the first row of the determinant outside the determinantal sign and cancel it. Expanding further the determinant according to elements of the first row and collecting terms containing $\Delta\lambda$ we reduce the identity to the form

$$(\Pi(\rho_s - \rho(\lambda)) + O(|\Delta\lambda|))\left(\frac{\Delta\rho}{\Delta\lambda} - \sigma\right) + O(|\Delta\lambda|) = 0. \quad (91)$$

In this formula Δ_ρ denotes the difference $\rho(\lambda) - \rho_0$, σ the upper diagonal element of the matrix $\widetilde{B}$, and the product extends over all multipliers ρ_s of the matrix $Y(\omega)$ distinct from ρ_0. Making $\Delta\lambda \to 0$ in (91) we conclude that the limit of the ratio $\Delta\rho/\Delta\lambda$ exists and equals σ. Using the simple structure of the matrix L, we find easily from (88) that

$$\sigma = i\rho_0 \frac{\int\limits_0^\omega (Y^*(\tau)\, Q(\tau)\, Y(\tau)\, f_0,\ f_0)\, d\tau}{(Jf_0,\ f_0)}, \tag{92}$$

where f_0 is the proper vector of $Y(\omega)$ which corresponds to the proper value ρ_0. Thus in the case of a simple multiplier there holds the formula

$$\Delta\rho = \sigma \cdot \Delta\lambda + \cdots, \tag{93}$$

where σ has the form (92).

The formulae (92) and (93) establish a direct link between the nature of the displacement of a multiplier from the unit circle under the shift $\Delta\lambda$ of the parameter into the non-real region, and the sign of the formula (Jf_0, f_0).

For if $Q(t)$ is a positive-definite matrix, then in the numerator of the ratio (92) the integral is positive, and under a shift $\Delta\lambda$ into the upper half-plane the multiplier $\rho(\lambda)$ obviously moves inside the unit circle for $(Jf_0, f_0) > 0$, and in the opposite sense from the circle for $(Jf_0, f_0) < 0$.

It also follows from (93) that as the parameter λ increases along the real axis a multiplier of the first kind moves along the unit circle in the counter-clockwise sense, and a multiplier of the second kind in the opposite direction. This fact, first established in the paper [3], plays an important part in the derivation of stability-criteria for the system (1) (see [3] and [8]).

5°. We consider now the case when ρ_0 is a proper value of $Y(\omega)$ with multiplicity r. In this case r roots of the equation (76) $\rho_{s_1}(\lambda), \rho_{s_2}(\lambda), \cdots, \rho_{s_r}(\lambda)$ coincide for $\lambda = 0$ with ρ_0. Without loss of generality we may assume that the top r diagonal elements of the matrix D are equal to ρ_0. Under these conditions the first r columns of the matrix F constitute a basis in the proper subspace of $Y(\omega)$ corresponding to the proper value ρ_0. We substitute in the equation (85) one of the functions $\rho_{s_j}(\lambda)$ $(1 \le j \le r)$. After this from each of the first r rows of the determinant we extract the factor $\Delta\lambda$, taking these factors outside the determinantal sign and dividing the identity by $(\Delta\lambda)^r$. Expanding the determinant according to the first r rows and collecting terms containing the factor $\Delta\lambda$, we give the identity the following form:

$$(\Pi(\rho_s - \rho_{s_j}(\lambda)) + O(|\Delta\lambda|))\ \mathrm{Det}\left(\frac{\Delta\rho_{s_j}}{\Delta\lambda} E_r - \widetilde{B}_{(r)}\right) + O(|\Delta\lambda|) = 0. \tag{94}$$

In this formula $\Delta\rho_{s_j} = \rho_{s_j}(\lambda) - \rho_0$, $\widetilde{B}_{(r)}$ is the upper diagonal minor of order r of the matrix $\widetilde{B}$, and the product is taken over all multipliers ρ_s of the matrix $Y(\omega)$ distinct from ρ_0. The identity (94) may be considered as a polynomial in $\Delta\rho_{s_j}/\Delta\lambda$ with coefficients depending continuously on the parameter $\Delta\lambda$. Making $\Delta\lambda \to 0$ and using the theorem on the continuous dependence of the roots of a polynomial upon the coefficients, we reach the conclusion that the limit of $\Delta\rho_{s_j}/\Delta\lambda$ as $\Delta\lambda \to 0$ exists and coincides with one of the roots of the equation

$$\mathrm{Det}(\sigma E_r - \widetilde{B}_{(r)}) = 0. \tag{95}$$

The equation (95) is sometimes termed *secular*. Thus in the present case there hold formulae

$$\rho_{s_j}(\lambda) = \rho_0 + \sigma_j \cdot \Delta\lambda + \cdots \quad (j = 1, 2, \cdots, r), \tag{96}$$

where the σ_j are the roots of the equation (95). In the expansions (96) there figure all r roots of the secular equation.

For let us assume that $\widetilde{\sigma}$ is a root of the equation (95) which is distinct from all the σ_j. We consider the function $\widetilde{\rho} = \rho_0 + \widetilde{\sigma}\Delta\lambda$. We substitute $\widetilde{\rho}$ on the left of (94) in place of $\rho_{s_j}(\lambda)$. It is easy to see that then all terms of zero order in $\Delta\lambda$ cancel out. This means that the substitution of $\widetilde{\rho}$ in the left-hand side of the characteristic equation (76) leads to the cancelling of all terms up to the order $|\Delta\lambda|^r$ inclusive (we recall that when obtaining (94) we carried out a division by $(\Delta\lambda)^r$). We remark on the other hand that if we replace $\widetilde{\rho}$ on the right of (78) by ρ, then, as is easy to verify, the term of order $(\Delta\lambda)^r$ is preserved. We arrive at a contradiction.

Thus the behaviour of the multipliers of a stable system in the case, of primary importance to us, of multiple roots is determined by the character of the roots of the secular equation (95).

Let us turn to the investigation of the matrix $\widetilde{B}_{(r)}$. Since L is a box-diagonal matrix, we obtain easily from (88) that:

$$\widetilde{B}_{(r)} = i\rho_0 L_{\rho 0}^{-1} \widetilde{Q}_{(r)}, \tag{97}$$

where $L_{\rho 0}$ is a Hermitean matrix of the form (22), and $\widetilde{Q}_{(r)}$ is the upper diagonal minor of order r of the Hermitean matrix $\widetilde{Q}$, determined by the formula (90). We denote by $\pi_1, \pi_2, \cdots, \pi_\nu$ the roots of the characteristic polynomial of the matrix $L_{\rho 0}^{-1}\widetilde{Q}_{(r)}$. By (97) there hold the equations

$$\sigma_j = i\rho_0 \pi_j \quad (j = 1, 2, \cdots, r). \tag{98}$$

In view of this the formula (96) may be written as follows:

$$\Delta\rho_{s_j} = i\rho_0 \cdot \pi_j \Delta\lambda + \cdots \quad (j = 1, 2, \cdots, r). \tag{99}$$

We will for the present assume that the symmetric matrix $Q(t)$ appearing in the equation (75) is positive-definite for all t. It then follows from (87) that the Hermitean matrix $\widetilde{Q}$ is also, by Sylvester's theorem, its minor, the Hermitean matrix $\widetilde{Q}_{(r)}$, will have this property. Let us assume that among the proper values of the Hermitean matrix $L_{\rho 0}^{-1}$ there are r' and r'' negative $(r' + r'' = r)$. In this case among the numbers $\pi_1, \pi_2, \cdots, \pi_r$ too there will be just r' positive and r'' negative.

This fact follows from the following general proposition: when a Hermitean matrix A is multiplied by a positive-definite Hermitean matrix B all proper values of AB remain real, and the distribution signs of the proper values of the matrix A

is preserved. For the proof we make a similarity transformation of AB, multiplying it on the left by the matrix $B^{+\frac{1}{2}}$ and on the right by the matrix $B^{-\frac{1}{2}}$. As a result we arrive at the Hermitean matrix $B^{\frac{1}{2}}AB^{\frac{1}{2}}$, which by the law of inertia retains the distribution of signs of the proper values of the matrix A.

Turning now the the formulae (99), we establish easily that as $\Delta\lambda$ is displaced into the upper half-plane, r' multipliers leave the point ρ_0 of the unit circle for the interior of the unit circle, while r'' multipliers are displaced in the other sense from the circle. The latter, as we have already noted, means that ρ_0 is a point of confluence of r' multipliers of the first kind and r'' multipliers of the second kind.

6°. The formulae (99) make it possible to trace the nature of the movement of the multipliers for a shift $\Delta\lambda$ along the real axis. From them it follows in particular that if among the numbers $\pi_1, \pi_2, \cdots, \pi_r$ there are some non-real, then, having given $\Delta\lambda$ a real value, sufficiently small in absolute value, it is possible to displace the multipliers from the circle. We shall soon show that in the case when the matrix L_ρ^{-1} is not definite $(r' \cdot r'' \neq 0)$, it is always possible to choose the matrix $Q(t)$ in the system (75) so that the equation

$$\operatorname{Det}(\pi E_r - L_{\rho 0}^{-1}\widetilde{Q}_{(r)}) = 0 \tag{100}$$

has complex roots ($Q(t)$ for this purpose must of course be taken to be indefinite).

It will therewith be proved that a stable system (1) *having repeated multipliers of unlike type is not strongly stable.* For the proof we note to begin with that, since the symmetric matrix $Q(t)$ can be chosen arbitrarily, the matrix $\int_0^\omega Y^* QY\,dr$, appearing on the right of (87), can be taken to be any real symmetric matrix of order $2k$.

For putting $Q(t) = \omega^{-1} \cdot Y^{*-1}(t)AY^{-1}(t)$ it is obvious that we get $\int_0^\omega Y^* QY\,dr = A$.

Thus the matrix $\widetilde{Q}$ defined by (87) has the form

$$\widetilde{Q} = F^* AF, \tag{101}$$

where F is a matrix which reduces $Y(\omega)$ to the Jordan form, and A is any real symmetric matrix.

Let us assume to begin with that ρ_0 is a non-real multiplier. In this case since $Y(\omega)$ is real $r \leq k$. So far we have not laid any restrictions on the basis in the proper subspace of $Y(\omega)$ corresponding to ρ_0. We now choose it so that the matrix $L_{\rho 0}^{-1}$ should take the form

$$L_{\rho 0}^{-1} = \begin{pmatrix} 1 & 0 & \\ 0 & -1 & 0 \\ & & * \; * \\ 0 & & * \; * \end{pmatrix}$$

We denote the vectors of the new basis by $f_1, f_2, \cdots, f_r$. Since the matrix A may be chosen at will, we try to choose it now so that the matrix $\widetilde{Q}_{(r)}$ should assume the following form:

$$\tilde{Q}_{(r)}=\begin{pmatrix} \begin{matrix}0 & 1\\ 1 & 0\end{matrix} & 0\\ 0 & \begin{matrix}* & * & *\\ * & * & *\end{matrix}\end{pmatrix}.$$

If this can be done then, as is not hard to check, the equation (100) will have non-real roots, and our aim will be achieved. Since $\widetilde{Q}_{(r)}$ is the upper diagonal minor of $\widetilde{Q}$, and the vectors $f_1, f_2, \cdots, f_r$ are by hypothesis the left-hand columns of the matrix F, it follows in view of (101) that the matter reduces to the choice of a real symmetric matrix A, for which there hold the equations

$$\begin{aligned}&(f_1,\ Af_1)=0,\quad (f_1,\ Af_2)=1,\quad (f_1Af_j)=0;\\ &(f_2,\ Af_1)=1,\quad (f_2,\ Af_2)=0,\quad (f_2Af_j)=0\quad (3\leqslant j\leqslant r).\end{aligned}\tag{102}$$

We use the circumstance that the vectors $f_1, f_2, \cdots, f_r, \bar{f}_1, \bar{f}_2, \cdots, \bar{f}_r$ are linearly independent. The latter is related to the fact that the vectors $\bar{f}_1, \bar{f}_2, \cdots, \bar{f}_r$ are proper vectors of $Y(\omega)$ with the proper value $\bar{\rho}_0$, which by hypothesis is distinct from ρ_0.

We deonote by f_j' and f_j'' respectively the real and imaginary parts of the vector f_j, and extend the system of linearly independent real vectors $f_1', f_1'', f_2', f_2'', \cdots, f_r', f_r''$ to a complete basis, adjoining to it the real vectors $f_{r+1}', f_{r+1}'', \cdots, f_k', f_k''$. We denote by T a matrix whose columns are the vectors f_j', f_j'' and introduce f for convenience a new notation for the unit vectors:
$e_1'=(1,\ 0,\ \ldots,\ 0)$, $e_1''=(0,\ 1,\ \ldots,\ 0), \ldots,$ $e_k'=(0,\ \ldots,\ 1,\ 0)$, $e_k''=(0,\ \ldots,1)$.
Furthermore we put $e_j=e_j'+ie_j''$.

There hold the obvious equations

$$f_j = Te_j.$$

We now introduce the real symmetric matrix of order $2k$

$$C=(c_{ij})=\begin{bmatrix}\begin{matrix}0&0&1&0\\0&0&0&0\\1&0&0&0\\0&0&0&0\end{matrix} & 0\\ 0 & \begin{matrix}*&*&*\\ *&*&*\\ *&*&*\end{matrix}\end{bmatrix}$$

and consider the symmetric matrix $T^{*-1}CT^{-1}$, denoting it by A. A is in fact the desired matrix. For let us verify that the conditions (102) are satisfied (we confine ourselves to checking the first row):

$$\begin{aligned}(f_1,\ Af_j)&=(f_1,\ T^{*-1}CT^{-1}f_j)=(Te_1,\ T^{*-1}CT^{-1}Te_j)=(e_1,\ Ce_j)=\\ &=(e_1',\ Ce_j')+i(e_1'',\ Ce_j')-i(e_1',\ Ce_j'')+(e_1'',\ Ce_j'')=\\ &=c_{1,\,2j-1}+ic_{2,\,2j-1}-ic_{1,\,2j}+c_{2,\,2j}\quad (f=1,\ 2,\ \ldots,\ k),\end{aligned}$$

and the validity of (102) follows directly from the form of the matrix C.

Thus in the case when ρ_0 is a nonreal multiplier the assertion is proved. If ρ_0 is a real number, then it can happen that r is greater than k, but in view of the reality of $Y(\omega)$ in this case the basis in the proper subspace can be chosen so as to be real. Then the matrices $L^{-1}_{\rho_0}$ and $\widetilde{Q}_{(r)}$ are real (the former up to a scalar factor), and the proof proceeds without difficulty.

We make incidentally one important remark. If the proper subspace of $Y(\omega)$ corresponding to the multiplier ρ_0 is such that the form (Jh, h) is definite on it, the matrix $L^{-1}_{\rho_0}$ being therefore definite, then the equation (100) for any Hermitean matrix $\widetilde{Q}_{(r)}$ has only real roots. As follows from (99), in this case the differentials of the multipliers $\rho_{s_i}(\lambda)$ for real $\Delta\lambda$ lie on a line touching the circle at the point ρ_0. This fact is a confirmation (and in essence a proof) of the fact that for repeated multipliers of like type the system (1) is strongly stable.

7°. In conclusion we give the formula for the first terms of the expansion (77) in the case when the monodromy matrix has a Jordan box of order m:

$$\begin{pmatrix} \rho_0 & 1 & & 0 \\ & \rho_0 & 1 & \\ & & \ddots & 1 \\ 0 & & & \rho_0 \end{pmatrix}.$$

The formula in question has the form

$$\rho_{s_j}(\lambda) = \rho_0 + \rho_0 \cdot \sqrt[m]{\sigma \cdot \Delta\lambda} + O(|\Delta\lambda|^{\frac{2}{m}}) \quad (j = 1, 2, \ldots, m). \tag{103}$$

By σ_ω we have denoted the expression

$$\sigma = \frac{i \int_0^\omega (Y^* Q Y f_1, f_1)\, d\tau}{\rho_0^{m-1} (J f_m, f_1)},$$

in which f_1 is a proper vector of $Y(\omega)$, and f_m is the last associated vector:

$$Y(\omega) f_1 = \rho_0 f_1, \; Y(\omega) f_2 = \rho_0 f_2 + f_1, \; \ldots, \; Y(\omega) f_{m-1} = \rho_0 f_{m-1} + f_m.$$

In the formula (103) to each function $\rho_{s_j}(\lambda)$ corresponds one of the m values of the expression $\sqrt[m]{\sigma\Delta\lambda_0}$, where it may be shown that for even m σ must be real, σ being purely imaginary if m is odd.

To derive the formula (103) would be a good exercise for the reader.

From (103) it is immediately to be seen that under a shift $\Delta\lambda$ into the non-real region, part of the multipliers leave the point ρ_0 of the unit circle for the interior of the circle, part leaving the circle in the other sense, whence it follows that under the confluence of multipliers of like type on the unit circle there is no possibility of a Jordan box arising.

REFERENCES

[1] V. A. Yakubovič, *Criteria of stability for systems of two differential equations of canonical type with periodic coefficients*, Dokl. Akad. Nauk SSSR (N.S.) 78 (1951), 221-224. (Russian)

[2] A. M. Lyapunov, *Problème général de la stabilité du mouvement*, 2nd ed., ONT I, Leningrad-Moscow, 1935. (Russian) [French translation of pp. 5-278, 344-348 published by Princeton Univ. Press, 1947]

[3] M. G. Kreĭn, *Generalisation of certain investigations of A. M. Lyapunov on linear differential equations with periodic coefficients*, Dokl. Akad. Nauk SSSR (N.S.) 73 (1950), 445-448. (Russian)

[4] C. Chevalley, *Theory of Lie groups, I*, Princeton, 1946.

[5] H. Weyl, *The classical groups, their invariants and representations*, Princeton, 1939.

[6] L. S. Pontryagin, *Continuous groups*, Gostehizdat, Moscow-Leningrad, 1938. (Russian) [English translation published by Princeton Univ. Press, 1939]

[7] F. R. Gantmaher, *Theory of matrices*, Gostehizdat, Moscow, 1953. (Russian)

[8] M. G. Neĭgauz and V. B. Lidskiĭ, *On the boundedness of the solutions of linear systems of differential equations with periodic coefficients*, Dokl. Akad. Nauk SSSR (N.S.) 77 (1951), 189-192. (Russian)

Translated by:
Dr. F. V. Atkinson

6.

(with A.G. Kostyuchenko)

Eigenfunction expansions for differential and other operators

Dokl. Akad. Nauk SSSR **103** (3) (1955) 349–352. Zbl. **65**: 104

1. The well known general spectral theorem enables one to represent a selfadjoint operator A in the form $A=\int A\,dE_\lambda$ where E_λ is the spectral family of the operator A. However this theorem does not lead to eigen-expansions for differential operators. Some results on eigen-expansions for special classes of selfadjoint differential operators were obtained in [1]–[5] and each case was the subject of a new piece of research.*

In this paper we will prove a general theorem on eigen-expansions. This result covers practically all essential theorems on spectral eigen-expansions for selfadjoint differential operators. These eigen-expansions involve ordinary eigenfunctions, or generalized eigenfunctions when there are no ordinary eigenfunctions.

2. Let Φ be a linear space of smooth functions with the topology determined by a countable family of norms. Examples are: 1) the space S of functions with all derivatives decreasing faster than $(1+x^2)^{-n}$ for all n; 2) the space S_M, consisting of smooth functions such that $|\varphi^{(p)}(x)|\leq c_p M(x)$, where $M(x)$ is some positive square integrable and monotonous for sufficiently large x function; with the system of norms $\|\varphi\|_n=\sup\limits_{x,\,p\leq n}|\varphi^{(p)}(x)|\cdot M^{-1}(x)$. Further for different classes of differential operators we will choose different spaces Φ (see examples at the end of the paper).

Denote by Φ' the space of continuous linear functionals on Φ. We will consider these functionals as generalized functions. Let (φ,ψ) be a continuous positive bilinear form in Φ; we will consider it as the inner product. Denote by $\bar{\Phi}$ the completion of the space Φ with respect to the inner product (φ,ψ). Then $\bar{\Phi}$ is a (complete) Hilbert space. Each element $\bar{\varphi}\in\bar{\Phi}$ determines continuous linear functional $T(\varphi)=(\bar{\varphi},\varphi)$, hence $\bar{\Phi}\subset\Phi'$ and we have the inclusions $\Phi\subset\bar{\Phi}\subset\Phi'$.

Let A be an operator defined for all $\varphi\in\Phi$ such that $A\varphi\in\bar{\Phi}$ if $\varphi\in\Phi$. We assume that A is a symmetric operator, i.e. $(A\varphi,\psi)=(\varphi,A\psi)$; assume also A that is continuous as an operator from Φ to $\bar{\Phi}$ and can be extended to a selfadjoint operator**.

---

* M.G. Krein obtained his results on eigenexpansions using the method of directed functionals [7].

Mautner [5] proved an eigenexpansion theorem in a more general situation, namely, for a selfadjoint operator whose resolvent is an integral operator with Carleman's kernel. We would like also to draw attention to the interesting paper [11].

** This is true for each real symmetric differential operator as it has equal deficiency indices.

Further we assume that $A\varphi\in\Phi$ for $\varphi\in\Phi$. In this case the adjoint operator A' maps Φ' into itself. The function $\psi\in\Phi'$ is said to be a generalized eigenfunction if $A'\psi=\lambda\psi$. In the general case (when $A\Phi\not\subset\Phi$) we, unfortunately, have to define the generalized eigenfunction $\psi\in\Phi'$ by differentiation of the spectral function (see Theorem 1).

Theorem 1. *Let Φ be a linear space and A be a symmetric operator as defined above. Then A has a complete system of generalized eigenfunctions.*

We sketch the proof. Denote by E_λ the spectral family of selfadjoint operator $\bar{A}$. Let $g(\alpha)$ be the system of generating vectors, i.e. let the subspaces, generated by elements $E_\lambda g^{(\alpha)}$, be mutually orthogonal for different α and their sum be the whole space. For each λ we may consider $E_\lambda g^{(\alpha)}$ not as an element of Hilbert space $\bar{\Phi}$ but as an element of the space Φ', since $\bar{\Phi}\subset\Phi'$. We will prove that $E_\lambda g^{(\alpha)}$, considered as a function of parameter λ with values in Φ' has Radon-Nikodým derivative with respect to the measure $\sigma^{(\alpha)}(\lambda)=(E_\lambda g^{(\alpha)}, g^{(\alpha)})$. This derivative is the required generalized eigenfunction. The following lemma shows the existence of the derivative.

Lemma. *Let Φ be a linear space (S or S_M), g_λ be a functional in Φ' depending on λ such that for each $\varphi\in\Phi$ the function $g_\lambda(\varphi)$ is of bounded variation. Then g_λ has the Radon-Nikodým derivative with respect to the measure $\sigma(\lambda)$*.*

The proof of the lemma depends on the following fact: there exist an r, such that $g_\lambda\in\Phi_r'$ for all λ where Φ_r' is the adjoint to the normed space** Φ_r (so far as $g_\lambda(\varphi)$ is of bounded variation). One may show also that

$$\sup(\|g_{\lambda_2}-g_{\lambda_1}\|_r+\ldots+\|g_{\lambda_n}-g_{\lambda_{n-1}}\|_r)\leqq k \tag{1}$$

for each sequence λ_j. Thus $g_\lambda\in\Phi_r'$ for all λ, g_λ satisfies (1) and Φ_r' is adjoint to the normed separable space Φ_r. In this situation the existence of the weak derivative has been proved (see [6] or [7]). This implies the existence of Radon-Nikodým derivative of $E_\lambda g^{(\alpha)}$ with respect to the measure $\sigma^{(\alpha)}(\lambda)$. Weak and strong convergence in the space Φ' being equivalent, we have also the strong derivative. Besides, if the function $g_\lambda(\varphi)$ is absolutely continuous with respect to $\sigma(\lambda)$, then we have the representation $g_\lambda=\int_{-\infty}^{\lambda} f_\lambda d\sigma(\lambda)$.

Now we apply lemma to $g_\lambda^{(\alpha)}=E_\lambda g^{(\alpha)}$; then we obtain generalized eigenfunctions as the derivatives of $g_\lambda^{(\alpha)}$. The completeness of the system of generalized functions follows from the completeness of the system $E_\lambda g^{(\alpha)}$.

* If $\Phi=S_M$ then the derivative is the functional in S_{M_1} where $M_1(x)$ is of order $M(x)|x|^r$ for some fixed r.

** Φ_r is the completion of Φ with respect to the norm $\|\cdot\|_r$.

Theorem 2. *Let $f_\lambda^{(\alpha)}$ be the system of generalized eigenfunctions constructed in Theorem 1. Then for each $\varphi \in \Phi$ there exists the Fourier transform*

$$c^\alpha(\lambda) = (f_\lambda^{(\alpha)}, \varphi) \quad \text{and} \quad (\varphi, \varphi) = \sum_\alpha \int |c^\alpha(\lambda)|^2 d\sigma^{(\alpha)}(\lambda). \tag{2}$$

This formula also gives isometry of the Hilbert space $\bar{\Phi}$ and the completion of the space of all functions $c^{(\alpha)}(\lambda)$ with respect to inner product defined in (2).

Proof of Theorem 2. From the general theory of selfadjoint operators one can deduce that to each $f \in \bar{\Phi}$ there corresponds a function $c^{(\alpha)}(\lambda)$ *with*

$$f = \sum_\alpha \int c^{(\alpha)}(\lambda) dE_\lambda g^{(\alpha)} \quad \text{and} \quad (f_1, f_2) = \sum_\alpha \int c_1^{(\alpha)}(\lambda) c_2^{(\alpha)}(\lambda) d\sigma^{(\alpha)}(\lambda). \tag{3}$$

We have to prove that $c^{(\alpha)}(\lambda) = (f_\lambda^{(\alpha)}, \varphi)$ for $\varphi \in \Phi$. Using (3) and Theorem 1, we obtain

$$(f_1, \varphi) = \sum_\alpha \int c_1^{(\alpha)}(\lambda) dE_\lambda (g^\alpha, \varphi) = \sum_\alpha \int c_1^{(\alpha)}(\lambda) \overline{(f_\lambda^{(\alpha)}, \varphi)} d\sigma^{(\alpha)}(\lambda).$$

Let us compare the last equality with (3) (we put in (3) $f_2 = \varphi$). As f_1 is an arbitrary function, we have $c^{(\alpha)}(\lambda) = (f_\lambda^{(\alpha)}, \varphi)$. This proves Theorem 2.

Now we give some examples.

In all our examples the differential operator maps into itself, if its coefficients are smooth functions. Otherwise one has to consider generalized eigenfunctions that are, as we said earlier, derivatives of $E_\lambda g$.

Example 1. Let us consider ordinary differential equation of order $2n$:

$$Ly = y^{(2n)} + q_1 y^{(2n-1)} + \ldots + q_{2n} y = \lambda y \tag{4}$$

where L is selfadjoint operator and $-\infty < x < +\infty$.

Let the coefficients q_i of equation (4) satisfy

$$|q_i(x)| \leq c(1 + |x|^\beta).$$

Consider the space S_β of smooth functions φ, such that $|\varphi^{(p)}(x)| \leq c_{p\varepsilon}(1 + |x|^{\beta + 1/2 + \varepsilon})^{-1}$, $\varepsilon > 0$ and suppose that operator L acts in this space S_β. Then according to Theorem 1 the operator L has a complete system of eigenfunctions. They are functionals on the space S_β. Thus we gave

Theorem 3. *If the coefficients q_i of the equation (4) satisfy inequalities $|q_i(x)| \leq c(1 + |x|^\beta)$, then there exists a complete system of eigenfunctions $y_i(x, \lambda)$, $i = 1, \ldots, n$, and an integer r such that, when integrated r times, $y_i(x, \lambda)$ grows not faster than $o(1 + |x|^{\beta + 3/2 + \varepsilon})$.*

If the functions $y_i(x, \lambda)$ do not oscillate too much (for example, if q_i are bounded functions) then the functions $y_i(x, \lambda)$ grow not faster than $c(1$

$+|x|^{\beta+3/2+\varepsilon}$). For the second order equations similar result was proved by E.E. Shnol [8].

Let us consider now the selfadjoint partial differential operator A defined by the expression

$$\sum a_{k_1\ldots k_n}(x)\frac{\partial^{m_i}}{\partial x_1^{k_1}\ldots\partial x_n^{k_n}}, \quad x=(x_1,\ldots,x_n), \quad -\infty\leqq x_i\leqq+\infty. \tag{5}$$

Let the coefficients $a_{k_1\ldots k_n}$ satisfy to the inequality

$$|a_{k_1\ldots k_n}(x)|\leqq c(1+|x|^\beta) \quad \text{where} \quad |x|=\sqrt{x_1^2+\ldots+x_n^2}. \tag{6}$$

We define the operator A on functions $\varphi\in S_{\beta,n}$ where $S_{\beta,n}$ is the linear space of functions such that $|\varphi^{(p)}(x)|\leqq c_{p\varepsilon}(1+|x|^{\beta+n/2+\varepsilon})^{-1}$ where ε is any positive number.

One may show that every generalized eigenfunction of an elliptic equation is an ordinary solution of this equation. Thus we have

Theorem 4. *Let A be selfadjoint elliptic operator* (5) *and its coefficients satisfy* (6). *Then the operator A has the complete system of eigenfunctions $u(x,\lambda)$, such that for some integer r its r times integral grows not faster than $|x|^{\beta+n/2+\varepsilon}$.*

If we would like to get sharper estimates in the infinity for eigenfunctions we have to introduce spaces of the type S_L [10].

Example 2. Let L be an arbitrary selfadjoint differential operator (not necessarily elliptic one) with continuous coefficients. Let the function $M^{-1}(x)$ grows faster, then all coefficients and S_M be the space described before.

Then Theorem 1 implies

Theorem 5. *Every selfadjoint differential operator has the complete system of generalized eigenfunctions.*

Evidently, according to Theorem 2, generalized eigenfunctions satisfy the Parserval equality.

Example 3. Let A and B be selfadjoint differential operators, and B is a positive operator. Choose the function $M^{-1}(x)$ growing faster than all the coefficients of A and B. Let us complete the space $\Phi=S_M$ with respect to the inner product $(B\varphi,\psi)$*. Then we have the following result.

Theorem 6. *Let A and B be differential operators as before. Then the equation $A+\lambda B=0$ has the complete system of generalized eigenfunctions, considered as functionals on Φ.*

* If the equation is determined in the bounded domain, then the space is constructed on the same way.

In particular, such the system of eigenfunctions exists for the equation $\frac{\partial^2 u}{\partial x^2} - \lambda \frac{\partial^2 \Delta u}{\partial z^2} = 0$. This equation was considered by S.L. Sobolev and R.A. Aleksandrjan [9].

Example 4. Let us consider, on a smooth compact manifold $\mathfrak{M}$, the dynamic system with invariant measure of the form $x_i' = X_i(x_1, \ldots, x_n)$, $i = 1, \ldots, n$. Let Φ be the space of smooth functions on $\mathfrak{M}$. The generator of the system $L = i \sum X_i \frac{\partial}{\partial x_i}$ is a selfadjoint operator in Hilbert space H consisting of functions on $\mathfrak{M}$ that are square integrable with respect to invariant measure σ.

Suppose that the operator L is defined for all functions $\varphi \in \Phi$. Then Theorem 1 implies the following

Theorem 7. *The dynamic system described before, has a complete system of generalized eigenfunctions.*

References

1. Weil, H.: Götting. Nachr. **37** (1909)
2. Krein, M.G.: Dokl. Akad. Nauk SSSR **74** (1950) 9
3. Kodaira, K.: Am. J. Math. **72** (1950) 502
4. Povzner, A.Ya.: Mat. Sb. **32** (1) (1953)
5. Mautner, F.: Proc. Nat. Acad. Sci. **39** (1953) 49
6. Gelfand, I.M.: Mat. Sb. **4** (1938) 46
7. Krein, M.G.: Dokl. Akad. Nauk SSSR **53** (1) (1946)
8. Shnol, E.E.: Usp. Mat. Nauk **9** (4) (1954)
9. Aleksandrjan, R.A.: Dokl. Akad. Nauk SSSR **73** (4) (1950)
10. Kostyuchenko, A.G.: Dokl. Akad. Nauk SSSR **103** (1) (1955)
11. Gårding, L.: Comptes Rendus du Douzième Congrès des Math. Scandinaves **44** (1953)

Received April 16, 1955

7.

On identities for eigenvalues of a second order differential operator

Usp. Mat. Nauk **11** (1) (1956) 191–198. Zbl. **70**: 83

The paper [1] contains a formula that enables one to compute the sum $\sum_{n=1}^{\infty}(\lambda_n - n^2 - a_0)$, where λ_0 are eigenvalues of the equation

$$-y'' + p(x)y = \lambda y, \qquad y(0) - y(\pi) = 0$$

and $a_0 = \int_0^\pi p(x)dx$. For a simple proof of this formula see [2].

This formula is an analogue of the trace formula for differential operators*: it generalizes to differential operators an assertion that the sum of eigenvalues of a matrix is equal to its trace, i.e. to the sum of its diagonal elements. However for a matrix we can compute not only the sum of eigenvalues, but the sum of powers of eigenvalues as well. In this paper we get the similar result for a second order differential operator. Corresponding results were obtained by L.A. Dikij and by myself approximately at the same time; however the proofs are different. We think that both methods are of some interest, so we have decided to publish both proofs. In essence each proof depends on some summation method, or as one usually says in physics, on some method of regularization of the sum

$$\sum_{n=1}^{\infty} \lambda_n^k. \tag{1}$$

The regularization method accepted in this paper may be called "the Hadamard regularization". This method is reduced to the subtraction of some terms from λ_n^k so that the series (1) turns into a convergent one. (Hadamard used similar regularization for the computation of divergent integrals, see [4]). The regularization method used by Dikij, is similar to M. Riesz's method [5] for the computation of divergent integrals; it is reduced to the analytic continuation of $\sum \lambda_n^{-s}$ as a function on s.

1. Let us consider a differential equation

$$-y'' + p(x)y = \lambda y \tag{2}$$

* One should mention here an interesting paper by M.G. Krein [3], where he obtained some relations from traces in a class of operators in the Hilbert space. It would be interesting to study the relation of our results with those of M.G. Krein.

with boundary conditions $y(0)=y(\pi)=0$. The function $p(x)$ is supposed to be smooth. To make the formulae simpler we assume that in some neighbourhoods of the points $x=0$ and $x=\pi$ the function $p(x)$ is identically zero (one could assume instead that $p(x)$ and the boundary conditions are periodic). In the general case the final formulae are somewhat more complicated. The symbolic form of the equation (2) is

$$(-D^2+p)y=\lambda y$$

where D is the differentiation operator and p is the operator of multiplication by $p(x)$. Let us consider now the operator $(-D^2+p+\zeta E)^{-1}$. This is an integral operator whose kernel is the Green's function of the operator $-y''+p(x)y+\zeta y$, $y(0)=y(\pi)=0$ and eigenvalues are $(\lambda_n+\zeta)^{-1}$, λ_n being eigenvalues of the equation (2). The trace of any integral operator equals the sum of its eigenvalues, so

$$\mathrm{Sp}(-D^2+p+\zeta E)^{-1}=\sum_{n=1}^{\infty}\frac{1}{\lambda_n+\zeta}. \tag{3}$$

We will find asymptotic expansion of the right-hand side and left-hand side of this equality in powers of ζ^{-1}. These series contain integral and half-integral powers of ζ^{-1}. Half-integral powers are related to asymptotics of eigenvalues, and integral powers give the desired identities.

2. Let us consider first the right hand side of (3). It is known that eigenvalues λ_n of (2) have the following asymptotics:

$$\lambda_n=n^2+c_0+\frac{c_2}{n^2}+\dots. \tag{4}$$

Let us write down $(\lambda_n+\zeta)^{-1}$ as an asymptotic series in ζ

$$(\lambda_n+\zeta)^{-1}=\sum_{k=0}^{\infty}(-1)^k\frac{\lambda_n^k}{\zeta^{k+1}}. \tag{5}$$

With this formula we cannot yet write down asymptotic expansion of $\sum_{n=1}^{\infty}(\lambda_n+\zeta)^{-1}$ in powers of ζ^{-1}, because the series $\sum_{n=1}^{\infty}\lambda_n^k$ is divergent. It appears however that the final expression for the asymptotics of $\sum_{n=1}^{\infty}(\lambda_n+\zeta)^{-1}$ is similar to the one obtained by formal summation of equalities (5) for all n. Namely, integral powers of ζ^{-1} in the asymptotic series for $\sum_{n=1}^{\infty}(\lambda_n+\zeta)^{-1}$ are multiplied by coefficients that are equal to regularized sums $\sum_{n=1}^{\infty}\lambda_n^k$ (see below). The final expression contains also half-integral powers $\zeta^{-n+1/2}$; one can see that these powers are closely related to the regularization. Let us give precise statements.

The regularized sum $\sum_{n=1}^{\infty}\lambda_n^k$ (k is fixed) with λ_n having asymptotics (4), is the sum of terms that are obtained from λ_n^k by the subtraction of the divergent terms of asymptotics. For simplicity we will substract also half of the constant term of the asymptotic expansion. The regularized sum will be denoted by $\sum_{n=1}^{\infty}{}' \lambda_n^k$. So, for example,

$$\sum_{n=1}^{\infty}{}' \lambda_n = \sum_{n=1}^{\infty} (\lambda_n - n^2 - c_0) - \tfrac{1}{2} c_0,$$

$$\sum_{n=1}^{\infty}{}' \lambda_n^2 = \sum_{n=1}^{\infty} (\lambda_n^2 - n^4 - 2n^2 c_0 - 2c_1 - c_0^2) - \frac{2c_2 + c_0^2}{2}.$$

To describe half-integral coefficients of the expansion we will consider the inversion of the asymptotic series (4). Namely, if $\lambda_n = n^2 + c_0 + \frac{c_2}{n^2} + \ldots \equiv \varphi(n^2)$, where $\varphi(n^2)$ denotes this asymptotic series, then its inversion gives

$$n^2 = \psi(\lambda) \equiv \lambda + b_0 + \frac{b_1}{\lambda} + \ldots. \tag{6}$$

Theorem 1. *Let λ_n have asymptotic expansion* (4). *Then the sum* $\sum_{n=1}^{\infty} (\lambda_n + \zeta)^{-1}$ *has the following asymptotic expansion in powers of* ζ^{-1}:

$$\sum_{n=1}^{\infty} (\lambda_n + \zeta)^{-1} = \sum_{k=0}^{\infty} (-1)^k \frac{A_k}{\zeta^{k+1}} + \pi \frac{d}{d\zeta} \sqrt{-\psi(-\zeta)} \tag{7}$$

where the asymptotic power $\psi(\zeta)$ *is defined by* (6), *and coefficients* A_k *are given by*

$$A_k = \sum_{n=1}^{\infty}{}' \lambda_n^k.$$

Proof. Let us first substitute λ_n into its approximate asymptotic expression:

$$\tilde{\lambda}_n = \tilde{\varphi}(n^2) = n^2 + c_0 + \frac{c_2}{n^2} + \ldots + \frac{c_{2m}}{n^{2m}}.$$

Later m will tend to infinity. It is clear that when m increases, the inverse asymptotic series $\tilde{\psi}(\lambda)$ coincides with the asymptotic series $\psi(\lambda)$ up to terms of higher and higher order. Let us decompose the rational function in n^2

$$\frac{1}{n^2 + c_0 + \frac{c_2}{n^2} + \ldots + \frac{c_{2m}}{n^{2m}} + \zeta}$$

into elementary fractions:

$$\frac{1}{\tilde{\lambda}_n + \zeta} = \sum_{i=0}^{m} \frac{a_i(\zeta)}{n^2 - \alpha_i(\zeta)} \tag{8}$$

where $\alpha_i(\zeta)$ are roots of the equation

$$x^{m+1} + (c_0 + \zeta) x^m + c_2 x^{m-1} + \ldots + c_{2m} = 0. \tag{9}$$

One can easily see that the roots of the equation (9) have the following behavior for $\zeta \to \infty$: one of the roots, say $\alpha_0(\zeta)$ tends to infinity and has the asymptotic behavior

$$\alpha_0(\zeta) = -\zeta + d_0 + \frac{d_1}{\zeta} + \dots$$

and all other roots tend to zero, and their asymptotic expansions contain only negative (not necessary integral) powers of ζ. It is clear that $\alpha_0(\zeta) = \tilde{\psi}(-\zeta)$. Let us note also that $a_0(\zeta)$ is the residue of the function

$$\frac{1}{x + c_0 + \frac{c_2}{x} + \dots + \frac{c_{2m}}{x^m} + \zeta}$$

in the point $x = \tilde{\psi}(-\zeta)$, i.e.

$$a_0(\zeta) = \frac{1}{\tilde{\varphi}'[\tilde{\psi}(-\zeta)]} = \tilde{\psi}'(-\zeta).$$

Therefore

$$\frac{1}{\lambda_n + \zeta} - \frac{\tilde{\psi}'(-\zeta)}{n^2 - \tilde{\psi}(-\zeta)} = \sum_{i=1}^{m} \frac{a_i(\zeta)}{n^2 - \alpha_i(\zeta)}. \tag{10}$$

As $\alpha_i(\zeta) \to 0$ for $\zeta \to \infty$ $(i = 1, 2, \dots, m)$, to obtain the asymptotic of the sum in the right hand side of (10), one has to write

$$\frac{1}{n^2 - \alpha_i(\zeta)} = \frac{1}{n^2} + \frac{\alpha_i(\zeta)}{n^4} + \frac{\alpha_i^2(\zeta)}{n^6} + \dots$$

so

$$\sum_{n=1}^{\infty} \frac{a_i(\zeta)}{n^2 - \alpha_i(\zeta)} = a_i(\zeta) \sum_{n=1}^{\infty} \frac{1}{n^2} + \alpha_i(\zeta) a_i(\zeta) \sum_{n=1}^{\infty} \frac{1}{n^4} + \dots. \tag{11}$$

To get an asymptotic expansion of the expression

$$\sum_{n=1}^{\infty} \sum_{i=1}^{\infty} \frac{a_i(\zeta)}{n^2 - \alpha_i(\zeta)}$$

in powers of ζ, one has to sum up m series of the form (11) and collect terms with given powers of ζ. As $\alpha_i(\zeta)$ begins from a negative power of ζ, any given power of ζ appears in only a finite number of terms in the series (11). However, it would be difficult to get asymptotics from this formula. The only statement we need is this:

If we perform the asymptotic expansion in powers of ζ (for $\zeta \to \infty$) of the right hand side (so also of the left hand side) of (10), and sum up over n all coefficients at a fixed power ζ^{-k}, then the corresponding series would be convergent.

Let us obtain now the asymptotic expansion of the left hand side of (10). Let us expand each term in series on ζ^{-1} (bearing in mind that $\psi(-\zeta) = \zeta + \dots$ for large ζ):

$$\frac{1}{\lambda_n+\zeta}=\frac{1}{\zeta}-\frac{\lambda_n}{\zeta^2}+\frac{\lambda_n^2}{\zeta^3}-\dots$$

$$\frac{\psi'(-\zeta)}{n^2-\psi(-\zeta)}=-\frac{\psi'(-\zeta)}{\psi(-\zeta)}+n^2\frac{\psi(-\zeta)}{\psi^2(-\zeta)}+\dots$$

Combining coefficients in ζ^{-n}, we get

$$\frac{1}{\lambda_n+\zeta}-\frac{\psi'(-\zeta)}{n^2-\psi(-\zeta)}=\sum_{k=0}^{\infty}\frac{(-1)^k}{\zeta^{k+1}}(\lambda_n^k-\dots),$$

where points denote combinations of nonnegative powers of n^2. We have proved already that the sum over n of the coefficients at $\frac{1}{\zeta^{k+1}}$ converges. So the series

$$\sum_{n=1}^{\infty}(\lambda_n^k-\dots) \tag{12}$$

converges. Therefore the combinations of nonnegative powers of n^2, denoted earlier by points, are initial terms in the asymptotic expansion of λ_n^k; in fact only after their substraction does the series (12) become convergent. So

$$\sum_{n=1}^{\infty}\left(\frac{1}{\lambda_n+\zeta}-\frac{\psi'(-\zeta)}{n^2-\psi(-\zeta)}\right)\sim\sum_{k=0}^{\infty}\frac{(-1)^k}{\zeta^{k+1}}\left[\sum_{n=1}^{\infty}(\lambda_n^k-\dots)\right],$$

where in $\sum_{n=1}^{\infty}(\lambda_n^k-\dots)$ each summand is λ_n^k minus the divergent asymptotic terms. To find asymptotic expansion of

$$\sum_{n=1}^{\infty}\frac{\psi'(-\zeta)}{n^2-\psi(-\zeta)}$$

we will use the formula

$$\sum_{n=1}^{\infty}\frac{1}{z^2+4\pi^2n^2}=-\frac{1}{2z^2}+\frac{1}{4z}\frac{e^z+1}{e^z-1}.$$

Then

$$\sum_{n=1}^{\infty}\frac{\tilde{\psi}'(-\zeta)}{n^2+[-\tilde{\psi}(-\zeta)]}=\frac{\pi\tilde{\psi}'(-\zeta)}{2\sqrt{-\tilde{\psi}(-\zeta)}}\cdot\frac{e^{2\pi\sqrt{-\tilde{\psi}(-3)}}+1}{e^{2\pi\sqrt{-\tilde{\psi}(-\zeta)}}-1}$$
$$+\frac{\tilde{\psi}'(-\zeta)}{2\psi(-\zeta)}\approx\frac{\pi\tilde{\psi}'(-\zeta)}{2\sqrt{-\tilde{\psi}(-\zeta)}}+\frac{\tilde{\psi}'(-\zeta)}{2\tilde{\psi}(-\zeta)}.$$

The last equality is an asymptotic one up to exponentially decreasing terms of the form $e^{-\zeta}$. Then

$$\sum_{n=1}^{\infty}\frac{1}{\tilde{\lambda}_n+\zeta}\approx\sum_{k}(-1)^k\frac{(\lambda_n^k-\dots)}{\zeta^{k+1}}+\frac{\tilde{\psi}'(-\zeta)}{2\tilde{\psi}(-\zeta)}+\frac{\pi\tilde{\psi}'(-\zeta)}{2\sqrt{-\tilde{\psi}(-\zeta)}}.$$

Expanding $\frac{\tilde{\psi}'(-\zeta)}{2\tilde{\psi}\cdot(-\zeta)}$ in powers of ζ we get as coefficients free terms of asymptotics of $\tilde{\lambda}_n^k$ divided by two; these are terms that have to be substracted from the sum as was mentioned before. So

$$\sum_{n=1}^{\infty} \frac{1}{\tilde{\lambda}_n+\zeta} = \sum_{k=0}^{\infty} (-1)^k \frac{\sum_{n=1}^{\infty}{}' \lambda_n^k}{\zeta^{k+1}} + \pi \frac{d}{d\zeta} \sqrt{-\tilde{\psi}(-\zeta)}$$

To replace $\tilde{\lambda}_n$ to λ_n we use the fact that the series $\sum_n (\lambda_n^k - \tilde{\lambda}_n^k)$ converges for $k<m$ and $\sum_n (\lambda_n^k - \tilde{\lambda}_n^k) = \sum_n{}' \lambda_n^k - \sum_n{}' \tilde{\lambda}_n^k$

$$\sum_{n=1}^{\infty} \frac{1}{\lambda_n+\zeta} = \sum_{n=1}^{\infty} \left(\frac{1}{\lambda_n+\zeta} - \frac{1}{\tilde{\lambda}+\zeta}\right) + \sum_{n=1}^{\infty} \frac{1}{\tilde{\lambda}_n+\zeta}$$

$$= \sum_{k=0}^{m-1} (-1)^k \frac{\sum_n (\lambda_n^k - \tilde{\lambda}_n^k)}{\zeta^{k+1}} + O\left(\frac{1}{\zeta^{m+1}}\right) + \sum_{k=0}^{m-1} (-1)^k \frac{\sum_n{}' \tilde{\lambda}_n^k}{\zeta^{k+1}}$$

$$+ \pi \frac{d}{d\zeta} \sqrt{-\tilde{\psi}(-\lambda)} = \sum_{k=0}^{m-1} (-1)^k \frac{\sum_n{}' \lambda_n^k}{\zeta^{k+1}} + \pi \frac{d}{d\zeta} \sqrt{-\tilde{\psi}(-\zeta)} + O\left(\frac{1}{\zeta^{m+1}}\right).$$

As m is arbitrary, we can take $m \to \infty$ thus proving our assertion.

3. Let us express now the left-hand side of the formula (3) in terms of $p(x)$ and its derivatives. Let us assume for the simplicity that $p(x)=0$ in some neighborhoods of the end points of the interval $[0, \pi]$. We have

$$\begin{aligned}(-D^2+p+\zeta E)^{-1} &= (-D^2+\zeta E)^{-1}[E+p(-D^2+\zeta E)^{-1}]^{-1} \\ &= (-D^2+\zeta E)^{-1} - (-D^2+\zeta E)^{-1} p(-D^2+\zeta E)^{-1} \\ &\quad + (-D^2+\zeta E)^{-1} p(-D^2+\zeta E)^{-1} p(-D^2+\zeta E)^{-1} + \dots .\end{aligned}$$

This is an asymptotic series in ζ (and even convergent in norm topology). We have to find out the asymptotic expansion of each summand. The operator $(-D^2+\zeta E)$ is an integral operator in the interval $(0, \pi)$. Its kernel is the Green's function of the differential operator $-y''+\zeta y$, $y(0)=y(\pi)=0$. It has the form

$$G(x_1, x_2) = \begin{cases} \dfrac{1}{\sqrt{\zeta}\,\mathrm{sh}\,\pi\sqrt{\zeta}}\, \mathrm{ch}\sqrt{\zeta}x_1 \cdot \mathrm{ch}\sqrt{\zeta}(\pi - x_2), & (x_1 < x_2) \\[2ex] \dfrac{1}{\sqrt{\zeta}\,\mathrm{ch}\,\pi\sqrt{\zeta}}\, \mathrm{ch}\sqrt{\zeta}(\pi - x_1) \cdot \mathrm{ch}\sqrt{\zeta}x_2, & (x_1 > x_2). \end{cases}$$

Therefore this kernel is the sum of $\frac{1}{2\sqrt{\zeta}} e^{-\sqrt{\xi}|x_1-x_2|}$ and of some function that decreases exponentially as $\zeta \to \infty$ when x_1 and x_2 lie inside the interval $(0, \pi)$. As we have to integrate the product of the kernel with the function $p(x)$, equal to 0 near endpoints of the interval, the second summand is not essential for the asymptotic behavior. So

$$\operatorname{Sp}(-D^2+p+\zeta E)^{-1}=\operatorname{Sp}(-D^2+\zeta E)^{-1}$$
$$-\operatorname{Sp}\{(-D^2+\zeta E)^{-1}p(-D^2+\zeta E)^{-1}\}+\ldots,$$

$$\operatorname{Sp}(-D^2+\zeta E)^{-4}=\sum_{n=1}^{\infty}\frac{1}{n^2+\zeta}\approx\frac{\pi}{2\sqrt{\zeta}}+\frac{1}{2\zeta},$$

$$\operatorname{Sp}\{(-D^2+\zeta E)^{-1}p(-D^2+\zeta E)^{-1}\}=\frac{1}{4\zeta}\int_0^\pi\int_0^\pi e^{-\sqrt{\zeta}|x_2-x_1|}e^{-\sqrt{\zeta}|x_1-x_2|}p(x_2)dx_1dx_2$$
$$=\frac{1}{4\zeta}\int_0^\pi p(x_2)\int_0^\pi e^{-2\sqrt{\zeta}|x_2-x_1|}dx_1dx_2$$
$$=\frac{1}{4\zeta\sqrt{\zeta}}\int_0^\pi p(x)dx.$$

All other traces can be expanded in the same way

$$\operatorname{Sp}\{(-D^2+\zeta E)^{-1}p(-D^2+\zeta E)^{-1}p\ldots p(-D^2+\zeta E)^{-1}\}$$
$$=\frac{1}{2^n(\sqrt{\zeta})^n}\int_0^\pi\ldots\int_0^\pi e^{-\sqrt{\zeta}(|x_1-x_2|+|x_2-x_3|+\ldots+|x_n-x_1|)}p(x_2)p(x_3)\ldots p(x_n)dx_1\ldots dx_n.$$

For the asymptotic behavior only the neighbourhood of the point $x_1=\ldots=x_n$ is crucial. So we expand $p(x_{i+1})$ in powers of $x_{i+1}-x_1=\xi_i$:

$$p(x_{i+1})=\sum\frac{D^k}{k!}p(x_1)\xi_i^k.$$

Let us change the variables $\sqrt{\zeta}\,\xi_i=\eta_i$. Then the coefficients are

$$\frac{1}{(\sqrt{\zeta})^{k_1+k_2+\ldots+k_n+2n-1}}\int\ldots\int e^{-(|\eta_1|+|\eta_1-\eta_2|+\ldots+|\eta_{n-1}|)}d\eta_1\ldots d\eta_{n-1}.\tag{13}$$

This implies that non-zero coefficients may appear only for odd powers of $\sqrt{\zeta}$, because an integral changing sign for $\eta_i\to-\eta_i$ should be equal to zero.

Theorem 2. *Let us consider a differential operator*

$$(-D^2+p(x)+\zeta E)y\qquad\left(D=\frac{d}{dx}\right)$$

with boundary conditions $y(0)=y(\pi)=0$. Let us assume that $p(x)$ is smooth and equal to zero near points $x=0$ and $x=\pi$. Then we have the following asymptotic equality

$$\operatorname{Sp}(-D^2+p+\zeta E)^{-1}=\frac{\pi}{2\sqrt{\zeta}}+\frac{1}{2\zeta}+\sum_{k=1}^{\infty}\frac{c_k}{\zeta^{k+1/2}}.\tag{14}$$

The coefficients c_k are linear combinations of expressions of the form

$$\int_0^x p^{(k_1)}(x)\, p^{(k_2)}(x) \dots p^{(k_n)}(x)\, dx,$$

and

$$k_1 + k_2 + \dots + k_n + n = 2k.$$

Theorem 1 gives the asymptotic expansion of the left hand side of (3), and Theorem 2 gives that of the right hand side of (3). Comparing coefficients for integral powers of ζ, we get the following theorem.

Theorem 3. *Let the function $p(x)$ in the equation $-y'' + p(x)y = \lambda y$, $y(0) = y(\pi) = 0$, be smooth and equal to zero near $x = 0$ and $x = \pi$; then*

$$\sum_{n=1}^{\infty}{}' \lambda_n^k = 0 \tag{15}$$

here λ_n are eigenvalues and $\sum'$ is the regularized sum of eigenvalues defined on the page 512.

The above method can be easily generalized to the case when $p(x)$ has singularities in some points, or is not equal to zero near endpoints. The expression in the right hand side of (15) in this case will, in general be nonzero.

References

1. Gelfand, I.M. and Levitan, B.M.: On a simple identity for eigenvalues of second order differential operator. Dokl. Akad. Nauk SSSR **88** (4) (1953) 593–596.
2. Dikij, L.A.: On a formula of Gelfand and Levitan, Usp. Mat. Nauk **8** (2) (1953).
3. Krein, M.G.: On trace formulas in perturbation theory. Mat. Sb. **33** (3) (1953) 597–626.
4. Hadamard, J.: Le problème de Cauchy et les èquations aux dérivées partielles linéaires hyperboliques, Paris, 1932.
5. Riesz, M.: Intégrale de Riemann-Liouville et problèmes de Cauchy, Acta Math. **81** (1949) 1–223.

Received March 24, 1955

8.

Some problems in the theory of quasilinear equations

Usp. Mat. Nauk **14** (2) (1959) 87–158 [Transl., II. Ser., Am. Math. Soc. **29** (1963) 295–381]
Zbl. **96**:66

CONTENTS

Introduction 295
§1. Evolution systems of equations. 296
§2. System of quasilinear equations of first order. Discontinuous solutions. 298
§3. Fundamental relations of thermodynamics. 305
§4. Equations of hydrodynamics of an ideal liquid. 309
§5. Equations of motion in the presence of viscosity. 315
§6. System of Lagrangian coordinates. 317
§7. Characteristics. 320
§8. Stability of discontinuous solution. 325
§9. Decay of arbitrary discontinuity. 328
§10. Existence and uniqueness theorems. 334
§11. Two-dimensional evolution systems of equations. 335
§12. Linear equations with discontinuous coefficients. 339
§13. Equations of magnetohydrodynamics. 347
§14. Characteristic cones in the equations of magnetohydrodynamics. 351
§15. Problem of thermal self-ignition. 355
§16. Problem of establishment of a chemical process. 361
§17. Normal flame propagation. 370
Supplement 375
Bibliography 380

Introduction

The present article is based on a course of lectures delivered by the author in 1957–1958 at the Moscow State University. The notes of these lectures were written up by K. V. Brušlinskiĭ and K. V. D'jačenko; without their participation, this article could not have been published.

Since there is no general theory of quasilinear equations at present, the author did not take it upon himself to develop in this course, in all possible rigor, those few initial data available on this subject. The Cauchy problem was well studied, essentially to completion, for one quasilinear equation. A rigorous exposition of this material was given by O. A. Oleĭnik in an exhaustive article published in Uspehi Mat. Nauk 12 (1957), no. 3 (75), 3–73. The present elementary course contains many considerations, proved frequently at the "physical level of rigor," and is essentially a more fully developed formulation of several problems in the theory of quasilinear equations. It is precisely because of the incompleteness of this theory and the abundance of interesting problems, which still await their solution, that we found it advantageous to publish these notes in so concise a form. It would be very pleasant were these notes to contribute to a solution of at least some of the problems raised here.

The author is indebted to K. I. Babenko, S. K. Godunov, and V. F. D'jačenko for discussion of several problems. §15 was written by G. I. Barenblatt, who also made several comments on §§16 and 17. The author also profited from the interesting article by P. Lax, published in Comm. Pure Appl. Math. 10(1957), 537–566.

There are three appendices to this article, one of which is included in the article and the others, written by O. A. Oleĭnik, is placed after the present article.

§1. Evolution systems of equations. In the study of processes that vary with time, we frequently deal with differential equations which contain an independent variable t (time), for which usually the Cauchy problem is formulated. We shall call a system of such equations an *evolution* system, if the Cauchy problem is correct for it, i.e., if small changes in the initial data at $t = 0$ correspond to small changes in the solution of the system of equations for $t > 0$ (in a limited range of variation of t).

An important role will be played later by two examples of systems of linear equations with constant coefficients

$$\frac{\partial u_i}{\partial t} = \sum_{k=1}^{n} A_{ik} \frac{\partial u_k}{\partial x} \quad (i = 1, 2, \cdots, n), \tag{1.1}$$

and

$$\frac{\partial u_i}{\partial t} = \sum_{k=1}^{n} A_{ik} \frac{\partial u_k}{\partial x} + \sum_{k=1}^{n} B_{ik} \frac{\partial^2 u_k}{\partial x^2} \quad (i = 1, 2, \cdots, n) \tag{1.2}$$

or, as we shall frequently write in vector form,

$$\frac{\partial u}{\partial t} = A \frac{\partial u}{\partial x}, \tag{1.1}$$

$$\frac{\partial u}{\partial t} = A \frac{\partial u}{\partial x} + B \frac{\partial^2 u}{\partial x^2}. \tag{1.2}$$

Let us verify the correctness of the Cauchy problem for the system (1.1) with particular solutions of the form

$$u(x, t) = \xi e^{i(\lambda t + \mu x)}. \tag{1. 3}$$

Substituting this expression in the system (1. 1), we obtain $\lambda\xi = \mu A\xi$.

In order for the last system of equations for ξ to have a non-trivial solution, it is necessary that λ/μ be an eigenvalue of the matrix A. Since we consider the Cauchy problem on the entire straight line $-\infty < x < +\infty$, we should consider μ real. Let $\lambda/\mu = a + ib$; then

$$e^{i(\lambda t + \mu x)} = e^{i\mu(x + at)} \cdot e^{-\mu bt}.$$

From this it is seen that if the initial data are taken in the form of rapidly oscillating waves $e^{i\mu x}$, with large μ, then no matter how small the amplitude of the wave, by the instant of time $t = 1$ the amplitude increases by a factor $e^{-\mu b}$, and this factor, for b different from 0, can be as large as desired, by choosing a large μ (positive or negative, depending on the sign of b).

Thus, *for the evolution system* (1. 1) *it is necessary that the matrix A have all eigenvalues real.*

If it is assumed that all the eigenvalues of the matrix are different, then, by using the Fourier transform of the solutions of the system, it is possible to show that this condition is also sufficient.

Let us turn now to the system (1. 2) and consider again solutions of the form (1. 3). Substituting (1. 3) in the system (1. 2), we obtain

$$i\lambda\xi = i\mu A\xi - \mu^2 B\xi, \tag{1. 4}$$

from which it follows that $i\lambda$ should be an eigenvalue of the matrix $i\mu A - \mu^2 B$ (μ real). In this case incorrectness takes place, as in the investigation of the system (1. 1), if $\operatorname{Im}\lambda \to -\infty$ as $\mu \to \infty$ or $\mu \to -\infty$. It is natural to expect that when $\mu \to \pm\infty$ the asymptotic behavior of λ will be affected not by the entire matrix $i\mu A - \mu^2 B$, but only by the principal term $-\mu^2 B$, i. e., for large μ, instead of the system (1. 4) it is enough to consider the system

$$i\lambda\xi = -\mu^2 B\xi.$$

From this it is seen that as $\mu \to \pm\infty$, $-i\lambda/\mu^2$ should tend to the eigenvalue of the matrix B, which we denote by $a + ib$. If $a = 0$, then the principal term in $\operatorname{Im}\lambda = a\mu^2$ vanishes, and to determine the sign of $\operatorname{Im}\lambda$ it is necessary to have a more detailed investigation, now without discarding the matrix $i\mu A$. If $a \neq 0$, then $\operatorname{Im}\lambda \sim a\mu^2$ and *for the evolution system* (1. 2) *it is necessary that the eigenvalues of the matrix B have a positive real part.* As in the case of the system (1. 1), it is possible to demonstrate also the sufficiency of this condition.

§ 2. **System of quasilinear equations of first order. Discontinuous solutions.** In this article we shall be interested in the qualitative features of quasilinear equations, which distinguish them from linear ones. The principal feature is the existence in the solution of discontinuities, which can appear even for sufficiently smooth initial data.* This fact is well known in hydrodynamics.

Linear equations do not have such a singularity. In their investigation one frequently uses as a model equations with constant coefficients, since (when the coefficients are continuous functions) it is possible to neglect their variations in small neighborhoods. To investigate this continuous solution such a model is no longer suitable, since on the discontinuity lines the solutions become discontinuous along with the coefficients that depend on the discontinuity. We shall therefore leave the dependence of the coefficients of the equation on the functions sought, and we shall consider a quasilinear system of the form

$$\frac{\partial u}{\partial t} + A(u)\frac{\partial u}{\partial x} = 0. \qquad (2.1)$$

The matrix $A(u)$ will be assumed to satisfy the condition of the evolution for any fixed u.

We shall impose one essential requirement on the discontinuous solutions of this system. By introducing in this system (2.1) additional, formally small "smoothing" terms, one can get rid of the discontinuous solutions, and "smear out" the discontinuities. It is natural to require that the discontinuous solutions be stable with respect to a certain class of such disturbances, i.e., *that they be obtained as limits of continuous solutions of perturbed systems* and, in addition, that these limiting solutions be independent of the specific type of the perturbation.

For the system under consideration we introduce the perturbation by adding to the right-hand side of (2.1) the term $\epsilon B\, \partial^2 u/\partial x^2$ $(\epsilon > 0)$, where B is an arbitrary constant matrix, without disturbing the evolution nature of the system. In analogy with hydrodynamics, this addition will be called viscosity.

Thus, we are interested in the conditions under which a discontinuous solution of the system (2.1) can be obtained as the limit, when $\epsilon \to 0$, of continuous solutions of the system

---

*Let us consider, for example, the equation $\partial u/\partial t + u\partial u/\partial x = 0$ with initial condition $u(0, x) = \sin x$. The general solution of this equation has the form $u = \phi(x - ut)$; it is easy to verify that under our initial conditions it would become indeterminate at a certain instant t_0. As we shall see below, it is precisely at this instant that it should be replaced by a suitably chosen discontinuous solution, thereby insuring the solution of the Cauchy problem for any instants of time t.

$$\frac{\partial u}{\partial t} + A(u)\frac{\partial u}{\partial x} = \epsilon B \frac{\partial^2 u}{\partial x^2} \tag{2.2}$$

and is furthermore independent of the matrix B.

Let us consider a small vicinity of the point lying on the line of discontinuity.

In this vicinity one can assume in the first approximation that the solution of the system (2.1) depends only on $x - \omega t$, where $x - \omega t =$ const determines the tangent of the line of discontinuity at this point. It is natural to expect that the continuous solutions of the system (2.2), which approximate this discontinuous solution, will also be of the same type. Since it is clear that everything will be determined by the behavior of the solutions in the vicinity of the discontinuity, it is enough to consider the solutions of the type $u = u(\xi)$, $\xi = x - \omega t$. Substituting this expression in (2.2), we obtain

$$-\omega\frac{du}{d\xi} + A(u)\frac{du}{d\xi} = \epsilon B\frac{d^2u}{d\xi^2}, \tag{2.3}$$

i.e., a system of ordinary differential equations.

It is easy to verify directly that the solution of this system satisfies the equation

$$u(\xi, \epsilon) = u\left(\frac{\xi}{\epsilon}, 1\right).$$

Thus, as ϵ changes, the solution, as a contour in the space u, goes into itself, and only the parametrization changes.

Let us assume that this system has a solution satisfying the following boundary conditions:*

$$u(\xi, \epsilon) \to u^+ \text{ for } \xi \to +\infty,$$
$$u(\xi, \epsilon) \to u^- \text{ for } \xi \to -\infty.$$

It is then clear that the limiting solution (when $\epsilon \to 0$) is discontinuous and has the form

$$u(\xi) = u^+ \text{ for } \xi > 0,$$
$$u(\xi) = u^- \text{ for } \xi < 0.$$

It remains to clarify the conditions of the independence of this limiting solution

* The determination of the conditions under which such a solution exists is an interesting problem in the theory of systems of ordinary equations.

on the specific type of the matrix B. Integrating the system (2. 3) with respect to ξ from $-\infty$ to $+\infty$ we obtain*

$$-\omega(u^+ - u^-) + \int_{-\infty}^{+\infty} A(u)\frac{du}{d\xi}\,d\xi = 0$$

or

$$-\omega(u^+ - u^-) + \int_\Gamma A(u)\,du = 0, \tag{2. 4}$$

where the second term is a curvilinear integral in the space u. The contour Γ is specified parametrically in the form $u = u(\xi)$, where $u(\xi)$ is the solution of the system (2. 3), i. e., it depends on the matrix B. The limiting solution, which, as we have seen, is made up of u^- and u^+, will not depend on B, if condition (2. 4), which these u^+ and u^- should satisfy, is also independent of B; this will occur if

$$A(u)\,du = df(u), \tag{2. 5}$$

i. e., if $A(u)\,du$ is a total differential of a certain vector function $f(u)$. In this case (2. 4) assumes the form

$$-\omega(u^+ - u^-) + f(u^+) - f(u^-) = 0 \tag{2. 6}$$

and yields the conditions that relate the values of u on both sides of the discontinuity. The system of equation (2. 1) is written in this case in the form

$$\frac{\partial u}{\partial t} + \frac{\partial f(u)}{\partial x} = 0, \tag{2. 7}$$

which we shall call *divergent*.

Thus, if the system of equations has a divergent form (2. 7), then its discontinuous solutions can be obtained as the limits (as $\epsilon \to 0$), independent of the matrix B, of the continuous solutions of the system

$$\frac{\partial u}{\partial t} + \frac{\partial f(u)}{\partial x} = \epsilon B\frac{\partial^2 u}{\partial x^2}. \tag{2. 8}$$

These solutions should satisfy, in the region of smoothness, the system of equation (2. 7), and should satisfy conditions (2. 6) on the discontinuities. It must be noted, however, that not all functions, which satisfy formally (2. 7) and (2. 6), will be limits of the solutions of equation (2. 8). This will be discussed later in §8.

---

*We assume, naturally, that when $u(\xi, \epsilon)$ tends on to u^+ or u^-, we have $du/d\xi \to 0$.

It is interesting to note the following fact. Let us assume that one and the same system of equations can be written in several divergent forms. Then these forms of notation, being equivalent from the point of view of the smooth solutions, will not be equivalent for the discontinuous solutions.

Let us illustrate this with the following example. We take one equation

$$\frac{\partial u}{\partial t} + u\frac{\partial u}{\partial x} = 0, \tag{2.9}$$

and write it in two divergent forms:

$$\frac{\partial u}{\partial t} + \frac{\partial \frac{u^2}{2}}{\partial x} = 0, \tag{2.9'}$$

$$\frac{\partial \frac{u^2}{2}}{\partial t} + \frac{\partial \frac{u^3}{3}}{\partial x} = 0. \tag{2.9''}$$

Writing for each of these equations the condition (2.6), it is easy to verify that their discontinuous solutions are not equivalent. This remark will be used later on.

So far we have considered only the class of linear viscosities, i. e., the case of constant matrices B. Let us show that it is impossible to broaden this class arbitrarily. For this purpose we use the example of equation (2.9) just analyzed. If the viscosity is taken to be linear, $\epsilon\partial^2 u/\partial x^2$, then the conditions on this discontinuity (2.6) will have the form

$$-\omega(u^+ - u^-) + \frac{(u^+)^2 - (u^-)^2}{2} = 0.$$

If the viscosity is taken, however, in the form $\epsilon(1/u)\,(\partial^2 u/\partial x^2)$, then it is obvious that this corresponds to the introduction of a linear viscosity for the equation written in the form (2.9) and (2.9″), and, consequently, the conditions on the discontinuity (2.6) will have the form

$$-\omega\,\frac{(u^+)^2 - (u^-)^2}{2} + \frac{(u^+)^3 - (u^-)^3}{3} = 0,$$

i. e., the limiting discontinuous solution of our equation depends on the form of the viscosity.

It would be interesting to indicate a class of viscosities, broader than the class of linear viscosities, in which there is no dependence of the limiting solution on the specific form of viscosity. Apparently this class will represent divergent viscosities, i. e., viscosities of the form

$$\epsilon \frac{\partial b}{\partial x}\left\{u, \frac{\partial u}{\partial x}\right\},$$

which naturally do not disturb the evolution nature of the system.

Another interesting problem is that of the character of the "smeared" discontinuity, obtained when viscosity of some type or another is introduced, i. e., the process of the character of the continuous solutions of the disturbed equation, approximating the discontinuous solution. We shall consider by way of an example the case of one equation

$$\frac{\partial u}{\partial t} + \frac{\partial f(u)}{\partial x} = \epsilon \frac{\partial b}{\partial x}, \tag{2. 10}$$

where b has the form

$$b = \begin{cases} \left(\dfrac{\partial u}{\partial x}\right)^{\alpha} \quad \alpha > 0 & \text{for } \dfrac{\partial u}{\partial x} > 0, \\ 0 & \text{for } \dfrac{\partial u}{\partial x} < 0. \end{cases}$$

The solution, as before, will be sought in the form $u = u(\xi)$, where $\xi = x - \omega t$. Then instead of (2. 10) we obtain the equation

$$-\omega \frac{du}{d\xi} + \frac{df(u)}{d\xi} = \epsilon \frac{db}{d\xi}, \tag{2. 10$'$}$$

for which there exists an integral

$$\epsilon b\left(\frac{du}{d\xi}\right) + \omega u - f(u) = C. \tag{2. 11}$$

If on any interval of variation of ξ there exists a solution different from a constant, then on this interval, by virtue of the definition $b(\partial u/\partial\xi)$, we have from (2. 11)

$$\frac{du}{d\xi} = \epsilon^{-\frac{1}{\alpha}} \Phi(u)^{\frac{1}{\alpha}},$$

where

$$\Phi(u) = C - \omega(u) + f(u).$$

Hence

$$d\xi = \epsilon^{\frac{1}{\alpha}} \Phi^{-\frac{1}{\alpha}}(u)\, du.$$

Integrating the latter relations in the limits from u^- to u^+, we find that the region along the ξ axis, where the variation in u takes place, will be finite and equal to

$$\Delta \xi = \epsilon^{\frac{1}{\alpha}} \int_{u^-}^{u^+} \Phi^{-\frac{1}{\alpha}}(u)\, du,$$

if the written integral converges (it is easy to verify that this takes place when $\alpha > 1$). In this case the solution $u(\xi)$ will have the form shown in Figure 1. If the interval diverges ($0 < \alpha \leq 1$), then the zone of "smearing" will be infinite, i. e., the variation of $u(\xi)$ takes place over the entire ξ axis, and it will tend to the values u^- and u^+ only when $\xi \to \infty$.

However, the limiting solution (as $\xi \to 0$) consists in either case of two constants u^- and u^+, satisfying the relation on the discontinuity (2. 16).

We have obtained discontinuous solutions of our system as limits of continuous solutions of equations with "viscosity," and from those considerations we derive the conditions on the discontinuity. These can be obtained also by other "internal" means. For this purpose we give another, more universal definition of the solution.

Let $u(x, t)$ be a smooth solution of the system of equations

$$\frac{\partial u}{\partial t} + \frac{\partial f(u)}{\partial x} = 0, \tag{2. 7}$$

and $\phi(x, t)$ any smooth finite (i. e., vanishing outside a certain finite region) vector function.

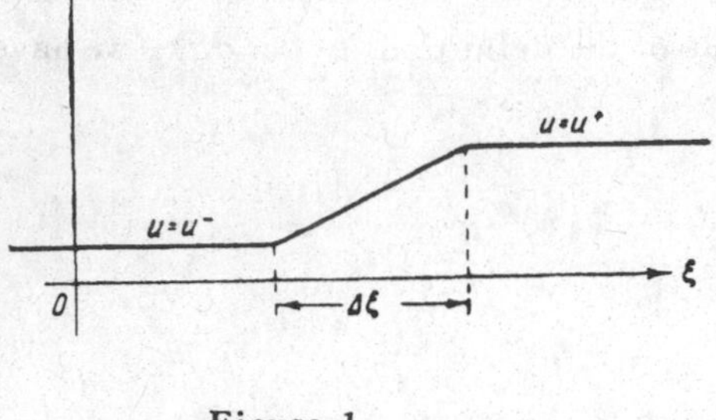

Figure 1

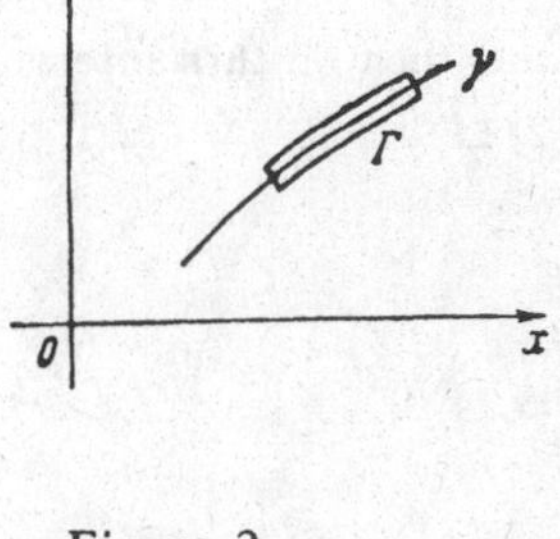

Figure 2

Since $u(x, t)$ is a solution, we have

$$\iint \phi \cdot \left[\frac{\partial u}{\partial t} + \frac{\partial f(u)}{\partial x}\right] dx\, dt = 0.$$

Integrating by parts, using the finite nature of ϕ, we obtain

$$\iint \left[u \frac{\partial \phi}{\partial t} + f(u) \frac{\partial \phi}{\partial x}\right] dx\, dt = 0. \tag{2.12}$$

To the contrary, from the latter equation, which is satisfied for all smooth finite ϕ, follows (2.7) for smooth u. But (2.12) can be verified also in the case when u is discontinuous. Therefore it is natural to call the functions u, for which (2.12) is satisfied for all smooth finite ϕ, the generalized solution of the system (2.7). This definition can sometimes be given in a different form. Namely, it is specified that over any closed contour

$$\oint_{\Gamma} (-u\, dt + f(u)\, dx) = 0. \tag{2.13}$$

The latter equation can be obtained by applying Green's formula to the integral

$$\iint \left[\frac{\partial u}{\partial t} + \frac{\partial f(u)}{\partial x}\right] dx\, dt,$$

taken over the region bounded by the contour Γ.

From condition (2.13) it is possible to obtain the conditions on discontinuity (2.6), without using the transition to the limit in the system of equation (2.8). Let the solution experience a discontinuity along a certain curve γ in the plane x, t. We take the contour γ in the form of a narrow curvilinear rectangle near a small portion of this curve (Figure 2), and let the width of the rectangle be so small compared to the length, so that the integral over those portions of the contour Γ, which are transverse to γ, can be neglected. Then

$$\int_{\Gamma} (u\, dx - f\, dt) = (u^{+}\omega - f^{+})\,\Delta t - (u^{-}\omega - f^{-})\,\Delta t + o(\Delta t),$$

where $\omega = dx/dt$ corresponds to the slope of the tangent to γ, Δt is the length of the projection of the contour on the t axis, and u^{+} and u^{-} are the values of u to the right and to the left on γ. As $\Delta t \to 0$, the foregoing approximate equation goes into

$$\omega(u^{+} - u^{-}) = f(u^{+}) - f(u^{-}). \tag{2.6}$$

To conclude this section, we give a simple example of the quasilinear system, the one-dimensional problem of nonlinear theory of elasticity. Let $u(x, t)$ be the displacement of the point x at the instant of time t, $\epsilon = \partial u/\partial x$ the strain at this point, and σ the stress causing this strain, $\sigma = \phi(\epsilon)$ is a specified function. The equation of motion is written in the form

$$\frac{\partial^2 u}{\partial t^2} = \frac{\partial \sigma}{\partial x}.$$

If one goes to the pair of functions ϵ and $v = \partial u/\partial t$, we obtain for these the quasilinear system

$$\frac{\partial v}{\partial t} = \frac{\partial \phi(\epsilon)}{\partial x},$$

$$\frac{\partial \epsilon}{\partial t} = \frac{\partial v}{\partial x}.$$

In linear theory of elasticity ϵ is assumed small, and then $\phi(\epsilon)$ can be considered linear.

Another very important example of a quasilinear system, hydrodynamics, will be expounded in detail in the subsequent sections.

§ 3. Fundamental relations of thermodynamics. In order to go over to the problems connected with hydromechanics, we will find it useful to dwell, although briefly, on the formal foundations of thermodynamics, i. e., to explain the physical quantities dealt with and the relations that these quantities obey. For this purpose we shall use the formalism of the Legendre transformations. Let us review these.

Let $\eta = f(\xi)$ be a function relating ξ and η. The Legendre transformation is the transition from ξ and η to the following pair of quantities:

$$p = f'(\xi), \qquad H(p) = -f(\xi) + p\xi. \tag{3.1}$$

The function $H(p)$ has the meaning of a vertical distance between the tangent to the curve of the function $f(\xi)$ at the point and to a line $\eta = p\xi$, passing through the origin (Figure 3) parallel to this tangent.

It was found that the inverse transformation is realized in accordance with this same rule. In fact

$$dH(p) = -f(\xi)\,d\xi + p\,d\xi + \xi\,dp = \xi\,dp,$$

i. e.,

$$\xi = H'(p), \qquad f(\xi) = -H(p) + \xi p. \tag{3.2}$$

For the Legendre transformations to be mutually unique, it is necessary that $f'(\xi)$ be monotonic, i.e., that $f(\xi)$ be convex. If this convexity takes place, then $H(p)$ will also be convex, since

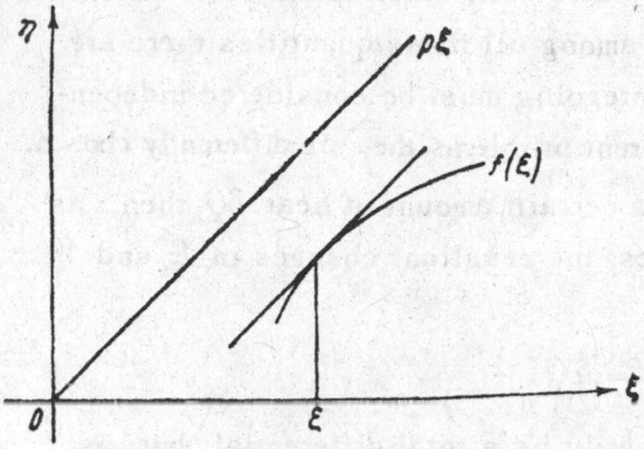

Figure 3

$$\frac{d^2 H}{dp^2} = \frac{d\xi}{dp} = \frac{1}{\frac{dp}{d\xi}} = \frac{1}{\frac{d^2 f}{d\xi^2}}.$$

It must also be noted that if the curve $f(\xi)$ has a kink (discontinuity in the derivative), then the curve of $H(p)$ contains a corresponding linear segment, and vice versa.

The Legendre transformation of a function of two variables $a = f(\xi_1, \xi_2)$ is called the transition from ξ_1, ξ_2, η to p_1, p_2, H

$$p_1 = \frac{\partial f}{\partial \xi_1}, \; p_2 = \frac{\partial f}{\partial \xi_2}, \; H(p_1, p_2) = -f(\xi_1, \xi_2) + \xi_1 p_1 + \xi_2 p_2. \qquad (3.3)$$

In vector form, the Legendre transformation is written the same as for one variable:

$$p = f'(\xi),$$
$$H(p) = -f(\xi) + \xi p$$

(by ξp one must understand here the scalar product of the vectors ξ and p; $f'(\xi)$ is a vector). The inverse transformation is carried out by means of the same formulas.

The uniqueness of the Legendre transformation takes place when

$$\frac{\partial(p_1, p_2)}{\partial(\xi_1, \xi_2)} = \begin{vmatrix} f''_{\xi_1 \xi_1} & f''_{\xi_1 \xi_2} \\ f''_{\xi_2 \xi_1} & f''_{\xi_2 \xi_2} \end{vmatrix} \neq 0. \qquad (3.4)$$

We shall henceforth assume that the matrix is positive definite (then (3.4) is known to be satisfied). From the fact that $f''(\xi)$ is positive definite it follows that $H''(p)$ is positive definite, for, as is readily verified,

$$f''(\xi) \cdot H''(p) = I,$$

where I is the unit matrix.

Let us turn now to thermodynamics.

A gas in the state of statistical equilibrium is characterized by the following quantities: pressure p, density ρ or a specific volume $V = 1/\rho$, temperature T, and internal energy E. Sometimes one considers still other quantities, which are functions of the foregoing ones. However, among all these quantities there are only two independent ones. Which of the foregoing must be considered independent is more or less insignificant. In different problems they are differently chosen.

If one transfers to a unit mass of gas a certain amount of heat δQ, then, as follows from the first law of thermodynamics, the resultant changes in E and V should be related as follows:

$$dE + pdV = \delta Q.$$

The expression $dE + pdV$ need not necessarily be a total differential, but, as follows from the well-known theorems of the theory of differential equations, it must admit of an integrating factor. (We recall that, as already noted, among E, p and V there are only two independent quantities.) We denote this integrating factor by $1/r$. We then have

$$dE + pdV = rdS = \delta Q.$$

As is proved in thermodynamics, from the axioms that make up the contents of the second law of thermodynamics, it follows that among the different integrating factors $1/r$, there is one which depends only on the temperature T. This is usually simply assumed to be T.

In this case we arrive at the following basic thermodynamic identity:

$$dE = TdS - pdV. \tag{3.5}$$

The function of state S is called the *entropy*. We choose V and S as the independent variables. The function $E(V, S)$ for each gas is different, and characterizes the gas, and is called its equation of state. As we shall see below, the condition for the correctness for the equations of hydrodynamics with heat conduction imposes on $E(V, S)$ the condition that the matrix $E''(V, S)$ of the second derivatives be positive definite.

It should be noted that for real gases and liquids, this condition is not always satisfied. It is found, however, that the solutions of the equations of hydrodynamics, even for such equations of state for which there is a range of variable V and S such that $E''(V, S)$ is not positive definite, are so constructed that the values of V and S in those places where the solution is smooth do not fall in this region. The transition through these regions in the solutions of the equations is always jump-like, discontinuous.

In order not to complicate the picture, we shall assume for the time being that for all equations of state with which we deal, the condition of positive definitness of the matrix $E''(V, S)$ is satisfied. In addition, it is necessaey to stipulate that the pressure and the temperature, i. e., E_V and E_S, be positive. In all other respects the function $E(V, S)$ can be arbitrary. From (3. 5) we obtain directly

$$T = \frac{\partial E}{\partial S}, \quad p = -\frac{\partial E}{\partial V}, \tag{3.6}$$

i. e., by assuming V and S to be the independent variables, we can obtain with the aid of the equation of state an expression for all the remaining quantities that characterize the state of the gas. One can choose other pairs of independent variables. For this purpose it is convenient to use the Legendre transformation. If it is performed over the function $E(V, S)$ with respect to the two variables V and S, we arrive at the independent variables T and p (see (3. 3) and (3. 6)) and their function*

$$\Phi = E - TS + pV, \tag{3.7}$$

which is called the *thermodynamic potential.* Thus, the equation of state should specify Φ as the function of T and p. Then S and V are determined as its partial derivatives

$$\Phi = \Phi(T, p), \quad \frac{\partial \Phi}{\partial T} = -S, \quad \frac{\partial \Phi}{\partial p} = V. \tag{3.8}$$

If the Legendre transformation of the function $E(V, S)$ is taken only with respect to the variable V, we arrive at the independent variables p and S and their function*

$$W = E + pV. \tag{3.9}$$

This function is called *the heat contents, the heat function*, or *the enthalpy.* The derivatives of these functions $W = W(S, p)$ have the form

$$\frac{\partial W}{\partial S} = T, \quad \frac{\partial W}{\partial p} = V. \tag{3.10}$$

Finally, we can carry out the Legendre transformation of the function $E(V, S)$ only with respect to the variable S. Then the independent variables will be V and T, and the equation of state must be written for the so-called *free energy* F:*

---

* Formula (3. 3) for the Legendre transformation yields $-\Phi$, $-W$, and $-F$, but we have reversed the sign so as to deal with positive quantities.

$$F = E - TS. \tag{3.11}$$

The derivatives of the function $F = F(V, T)$ yield

$$S = -\frac{\partial F}{\partial T}, \quad p = -\frac{\partial F}{\partial V}.$$

This is the basic list of thermodynamic quantities and the relations between them. We establish that all thermodynamic characteristics of the gas are fully determined by specifying its equation of state, i. e., one convex function of two variables, for example, $E = E(V, S)$.

§4. **Equations of hydrodynamics of an ideal liquid.** Hydrodynamics is the study of the motion of liquids or gases, and it therefore deals, in addition to thermodynamic quantities, also with the velocity of particles $u = (u_1, u_2, u_3)$, a vector function in three-dimensional space. All the foregoing quantities are functions of the time t and of the coordinates of the point x_1, x_2, x_3. They obey the equations of hydrodynamics, which reflect the physical laws of conservation of mass, momentum, and energy.

Inasmuch as these laws take place for any liquid and for any gas, the equations of hydrodynamics are completely independent of the specific medium with which we deal.

Specific liquids or gases are characterized by their equation of state (the function $E(V, S)$), which we must bear in mind, since without it the entire system of equations will contain more unknowns than necessary, i.e., will be unclosed.

The mathematical contents of hydrodynamics represents indeed a study of the foregoing equations: the laws of conservation with allowance for the equations of state, which are physically clear and a most completely worked out example of a system of quasilinear equations.

Let us proceed to derive the equations.* Here we shall deal (throughout this section) with the so-called ideal liquid, i. e., such in which there is no viscosity and heat conduction.

1. *The law of conservation of mass (continuity equation).* Let D be a fixed volume of space, and Γ its boundary. The change in mass in this volume, referred to a unit time, is (ρ is the density)

$$\frac{\partial}{\partial t} \iiint_D \rho \, dx_1 \, dx_2 \, dx_3 = \iiint_D \frac{\partial \rho}{\partial t} dx_1 \, dx_2 \, dx_3 .$$

---

* For more details see, for example, [9].

This change in mass should be equal to the amount of matter which flows into our volume per unit time through the boundary Γ. This amount can be calculated with the aid of the surface integral

$$-\iint_{\Gamma} \rho u_n \, d\sigma,$$

where u_n is the projection of the velocity on the outward normal to Γ, and $d\sigma$ is the element of the surface. By the Gauss-Ostrogradskiĭ formula

$$\iint_{\Gamma} \rho u_n \, d\sigma = \iiint_{D} \operatorname{div}(\rho \mathbf{u}) \, dx_1 \, dx_2 \, dx_3 .$$

Thus, we see that

$$\iiint_{D} \left[\frac{\partial \rho}{\partial t} + \operatorname{div}(\rho \mathbf{u}) \right] dx_1 \, dx_2 \, dx_3 = 0$$

for any volume D, which is possible only if the integrand vanishes. Thus, we arrive at the equation

$$\frac{\partial \rho}{\partial t} + \operatorname{div}(\rho \mathbf{u}) = 0,$$

which in coordinate form is written

$$\frac{\partial \rho}{\partial t} + \frac{\partial \rho u_k}{\partial x_k} = 0. \tag{4. 1}$$

Here and throughout we assume summations over repeated indices from one to three.

2. *Law of conversation of momentum and Euler's equation.* Let us write down Newton's second law for the element of volume of the substance dV. The mass included in this volume is $\rho \, dV$. The acceleration experienced by particles of this volume is $d\mathbf{u}/dt$. In differentiation we have in mind not the rate of variation of $\mathbf{u}$ at a given point in space, which could have been calculated from the formula for $\partial \mathbf{u}/\partial t$, but the rate of variation $\mathbf{u}$ of a fixed particle of the liquid. Since over the time dt the displacement of the particle is described by a vector with components $dx_1 = u_1 \, dt$, $dx_2 = u_2 \, dt$, $dx_3 = u_3 \, dt$, the acceleration $d\mathbf{u}/dt$ is written in the form

$$\frac{d\mathbf{u}}{dt} = \frac{\partial \mathbf{u}}{\partial t} + u_k \frac{\partial \mathbf{u}}{\partial x_k} . \tag{4. 2}$$

The particles can acquire an acceleration only under the action of the forces applied to them. Since we do not consider viscosity, i. e., friction forces in the liquid or in a gas, the only forces that act on our volume will be the presure forces.

The resultant of the pressure forces applied to a certain volume is calculated with the aid of the integral

$$-\iint p \cdot \mathbf{n}\, d\sigma = -\iiint \operatorname{grad} p \cdot dV,$$

where the double integral is taken over the surface bounding the volume, and the triple integral is taken over the entire volume ($\mathbf{n}$ is a unit vector outwardly normal to the surface).

In our case of an infinitesimally small volume, the resultant of the forces of pressure can be written simply as $dV \operatorname{grad} p$. After calculating the products of the mass included in the volume and the acceleration it acquires, and after equating this product to the force acting on the volume, we obtain

$$\rho \frac{\partial \mathbf{u}}{\partial t} + \rho u_k \frac{\partial \mathbf{u}}{\partial x_k} = -\operatorname{grad} p, \tag{4.3}$$

which was called Euler's equation. We note that this equation is not divergent, i.e., it does not represent a conservation law. In order to obtain divergent equations, we must multiply equation (4.1) by $\mathbf{u}$ and add the result to (4.3). The resultant equations

$$\left.\begin{aligned} &\frac{\partial \rho u_i}{\partial t} + \frac{\partial \Pi_{ik}}{\partial x_k} = 0 \qquad (i = 1, 2, 3), \\ &\Pi_{ik} = p\delta_{ik} + \rho u_i u_k \end{aligned}\right. \tag{4.4}$$

will now be divergent. It is found that they represent the law of conservation of momentum. In fact, the equivalence of equations (4.4) are the integral equations

$$\frac{\partial}{\partial t} \iiint_D \rho u_i\, dx_1\, dx_2\, dx_3 + \iint_\Gamma p n_i\, d\sigma + \iint_\Gamma \rho u_i (\mathbf{u} \cdot \mathbf{n})\, d\sigma = 0 \tag{4.4'}$$

which has the following meaning.

The derivative of

$$\iiint_D \rho u_i\, dx_1\, dx_2\, dx_3$$

represents the change per unit time of the ith component of the momentum, included in a fixed volume D,

$$\iint_\Gamma \rho u_i (\mathbf{u}\mathbf{n})\, d\sigma$$

is the amount of the ith component momentum which has flowed out during the same time through Γ together with the liquid leaving the volume D, and $\iint_\Gamma p n_i \, d\sigma$ is the change in the momentum due to the acceleration under the influence of the pressure p of the particles in the volume D. $\Pi_{ik} = \rho u_i u_k$ is called *the tensor of the momentum flux.*

3. *The law of conservation of energy.* The energy per unit volume of the matter consists of the internal energy ρE, and the kinetic energy $\rho |\mathrm{u}| 2/2$. The change in the energy of the volume D per unit time is

$$\frac{\partial}{\partial t} \iiint_D \left(\rho E + \frac{\rho u^2}{2} \right) dx_1 \, dx_2 \, dx_3 = \iiint_D \frac{\partial}{\partial t} \left(\rho E + \frac{\rho u^2}{2} \right) dx_1 \, dx_2 \, dx_3 .$$

This change is made up of two parts. The first consists of the influx of energy, due to the transfer of energy of the liquid flowing into the volume. The influx is written in the form of the integral

$$- \iint_{\mathrm{I}} \left(\rho E + \frac{\rho u^2}{2} \right) u_n \, d\sigma .$$

The second part consists of the increase in the energy due to the work performed on the volume D by the pressure forces. This work is written as

$$- \iint p u_n \, a\sigma .$$

We arrive at the equation

$$\frac{\partial}{\partial t} \iiint_D \left(\rho E + \frac{\rho u^2}{2} \right) dx_1 \, dx_2 \, dx_3 + \iint_\Gamma \left(p + \rho E + \frac{\rho u^2}{2} \right) u_n \, d\sigma = 0.$$

Transforming the surface integral, as in the derivation of the equation of continuity, by means of the Gauss-Ostrogradskiĭ formula and using the arbitrariness of the volume D, we obtain the following energy equation for an ideal liquid:

$$\frac{\partial}{\partial t} \left(\rho E + \frac{\rho u^2}{2} \right) + \frac{\partial}{\partial x_k} \left[\rho u_k \left(\frac{p}{\rho} + E + \frac{u^2}{2} \right) \right] = 0. \qquad (4.5)$$

In the general case of a viscous and heat conducting liquid, the energy equation will be more complicated, although it can be derived by the same method. It is merely necessary to take into account the influx of heat due to the heat conduction and the change in the energy due to the work performed by the forces of internal friction.

The resultant system

$$\frac{\partial \rho}{\partial t} + \frac{\partial \rho u_k}{\partial x_k} = 0,$$

$$\frac{\partial \rho u_i}{\partial t} + \frac{\partial (p\sigma_{ik} + \rho u_i u_k)}{\partial x_k} = 0, \tag{4.6}$$

$$\frac{\partial \left[\rho E + \rho \frac{u_i u_i}{2}\right]}{\partial t} + \frac{\partial}{\partial x_k}\left[\rho u_k \left[E + \frac{p}{\rho} + \frac{u_i u_i}{2}\right]\right] = 0$$

together with the equation of state $E = E(V, S)$ and the thermodynamic relation $p(V, S) = -E_V(V, S)$ (we recall that $V = 1/r$) represents a complete system of five equations with five unknown functions, which can be chosen to be, for example, u_1, u_2, u_3, V and S.

Let us note the following fact. If we consider the smooth solutions of the system (4.6), it is possible, as a consequence of this system, to obtain an equation for the entropy

$$\frac{\partial \rho S}{\partial t} + \frac{\partial \rho u_k S}{\partial x_k} = 0, \tag{4.7}$$

and to consider instead of the system (4.6) the system of equations (4.1), (4.4), and (4.7); we shall call this system (4.6′). However, as we noted in § 2, different divergent forms of notation for one and the same system may not be equivalent from the point of view of discontinuous solutions. This takes place here, too. Therefore, if we are interested not only in smooth solutions, we must make a choice between (4.6) and (4.6′). As examples of quasilinear systems, they are quite equivalent, but whereas (4.6) describes correctly a real physical process, the system (4.6′) (equation (4.7)) leads to the law of conservation of entropy of the particles of the liquid, which is not satisfied on shock waves.

Let us go on to consider the question of the evolution nature of the system of equations of hydrodynamics (4.6). Here we shall confine ourselves to the least cumbersome case of plane-parallel motion, when all the quantities are independent of the coordinates x_2 and x_3, and the velocity is directed along the axis x_1, which will be simply denoted by x.

We shall investigate smooth solutions of the equations of hydrodynamics. We shall therefore consider not only the divergent forms of the equations. We make up a system of hydrodynamic equations consisting of the equation of continuity, the Euler equation, and the equation for entropy

$$\frac{\partial \rho}{\partial t} + u\frac{\partial \rho}{\partial x} + \rho\frac{\partial u}{\partial x} = 0,$$
$$\frac{\partial u}{\partial t} + u\frac{\partial u}{\partial x} + \frac{1}{\rho}\frac{\partial p}{\partial x} = 0, \tag{4. 8}$$
$$\frac{\partial S}{\partial t} + u\frac{\partial S}{\partial x} = 0.$$

We change over to $V = 1/\rho$ and make use of the fact that $p = -E_V(V, S)$. Then $\partial p/\partial x = -E_{VV}\partial V/\partial x - E_{VS}\,\partial S/\partial x$, and the equations (4. 7) are rewritten

$$\frac{\partial V}{\partial t} = -u\frac{\partial V}{\partial x} + V\frac{\partial u}{\partial x},$$
$$\frac{\partial u}{\partial t} = VE_{VV}\frac{\partial V}{\partial x} - u\frac{\partial u}{\partial x} + VE_{VS}\frac{\partial S}{\partial x},$$
$$\frac{\partial S}{\partial t} = -u\frac{\partial S}{\partial x}.$$

From the characteristic equation

$$(\lambda + u)[(\lambda + u)^2 - V^2 E_{VV}] = 0$$

it is seen that this will take place when $E_{VV} > 0$. This is the only requirement that is imposed in the case of an ideal liquid on the equation of state by the equations of hydrodynamics. They cannot impose any limitations whatever in principle on the derivatives of $E(V, S)$ with respect to S, as can be seen from the following remark. Equation (4. 8) will not change if the entropy S is replaced by any function of S (all that changes is the temperature T, which does not enter into the equation), and we can thus make S, and consequently the dependence of E on S, sufficiently arbitrary.

However, the requirement that the entire matrix $E''(V, S)$ be positive definite nevertheless follows from the equation of hydrodynamics, considered even with allowance of heat conduction.

To conclude this section, we call attention to one simple case, namely, the equation of isothermal motion. In this case a constant temperature T is maintained in the gas. For this purpose it is necessary to have additional energy to remove this energy, and therefore the equation of energy (4. 5) does not hold. The equation of conservation of mass (4. 1) and of momentum (4. 4) remain unchanged and the quantities p and S contained in these equations are connected by the final equation $T \equiv \text{const}$.

Thus, the isothermal motion is completely determined by the equations

$$\frac{\partial\rho}{\partial t} + \frac{\partial\rho\, u_k}{\partial x_k} = 0, \tag{4. 1}$$

$$\frac{\partial\rho u_i}{\partial t} + \frac{\partial\rho u_i u_k}{\partial x_k} + \frac{\partial p}{\partial x_i} = 0 \tag{4. 4''}$$

and the condition that the temperature be constant. This example of a quasilinear system will be used later on.

§ 5. Equations of motion in the presence of viscosity. Let us consider now a viscous liquid. This means that in the derivation of the equations for this liquid we should take into account the forces of internal friction. It is found that from the most general considerations regarding viscosity, dictated by the physical aspect of the matter, we can obtain a specific form for its influence on the motion. Obviously, the equation of continuity

$$\frac{\partial\rho}{\partial t} + \frac{\partial\rho u_k}{\partial x_k} = 0 \tag{5. 1}$$

will take place also for a viscous liquid, since its derivation is not based on the presence of any forces whatever.

The equations of the law of conservation of momentum for an ideal liquid were obtained in the following form:

$$\frac{\partial\rho u_i}{\partial t} + \frac{\partial\Pi_{ik}}{\partial x_k} = 0.$$

The tensor of the density of the momentum flux is made up of the transfer of the momentum proper by the moving medium ($\rho u_i u_k$) and the pressure forces ($p\,\delta_{ik}$). It is clear that also in the case of a viscous liquid, the general form of the equation should remain the same, except that in the expression for Π_{ik} it is necessary to add friction forces, i. e., to write

$$\Pi_{ik} = \rho u_i u_k + p\delta_{ik} - \nu_{ik},$$

where ν_{ik} is a tensor, the meaning of which consists of the following: ν_{ik} is the ith component of the friction force of the viscosity, acting in a given point on a unit area, perpendicular to the axis x_k. We obtain the form of this tensor from the following considerations. Firstly, the viscosity must not change when the entire moving liquid is subject to translational motion. In addition, the viscosity must vanish if the liquid moves in translation as a solid body. It follows therefore that ν_{ik} is independent of the velocity u and is only a function of the derivatives of the velocity components. At small velocity gradients one can assume that ν_{ik} is a function of only its first derivatives and is furthermore linear. This assumption is well confirmed experimentally. We now consider motion that consists of rotation with unit angular velocity performed by the entire liquid as a whole about the axis x_3. This motion is described by the velocities

$$u_1 = -x_2,$$
$$u_2 = x_1,$$
$$u_3 = 0.$$

It is natural to expect that in this case no viscosity forces will act in the liquid, i.e., $\nu_{ik} = 0$. Let

$$\nu_{ik} = \sum_{\alpha,\beta=1}^{3} A_{\alpha\beta}\frac{\partial u_\alpha}{\partial x_\beta}. \quad (5.2)$$

Substituting in this formula values of the velocities, and equating ν_{ik} to 0, we find that $A_{12} = A_{21}$. Analogously, considering rotations about the axes x_2 and x_1, we arrive at the equations $A_{32} = A_{23}$ and $A_{31} = A_{13}$. We thus see that the components of the viscosity tensor depend only on symmetrical combinations of the derivatives of the form $\partial u_i/\partial x_i$ or $a_{ik} = \partial u_i/\partial x_k + \partial u_k/\partial x_i$. Further limitations are imposed on the form of ν_{ik} by the fact that ν_{ik} is a tensor, i.e., by the law of transformation of the components upon going to a new system of coordinates. It is also necessary to take into account that the liquid is isotropic, i.e., all the directions in the liquid are of equal value. It follows therefore that ν_{ik} is expressed in terms of $\partial u_i/\partial x_i$ and in terms of a_{ik} in a similar manner in any rectangular system of coordinates.

It can be verified that a general form of the tensor, satisfying these conditions, is

$$\nu_{ik} = \alpha a_{ik} + \beta a_{ll}\delta_{ik}.$$

It is customary to write the expression for the viscosity in a somewhat different form:*

$$\nu_{ik} = \eta\left[a_{ik} - \frac{1}{3}a_{ll}\delta_{ik}\right] + \xi a_{ll}\delta_{ik},$$

$$\nu_{ik} = \eta\left\{\frac{\partial u_i}{\partial x_k} + \frac{\partial u_k}{\partial x_i} - \frac{2}{3}\frac{\partial u_l}{\partial x_l}\delta_{ik}\right\} + \xi\frac{\partial u_l}{\partial x_l}\delta_{ik}. \quad (5.3)$$

---

* Such a notation is dictated by the condition that the trace of the matrix, which is the coefficient of η, vanish.

The coefficients η and ξ that enter into (5.3) are called the coefficients first and second viscosity, respectively.

It now remains for us to write the equation of the law of conservation of energy for the viscous liquid. For an ideal liquid this law is written in the form

$$\frac{\partial}{\partial t}\iiint_D \left[\rho E + \frac{\rho u^2}{2}\right] dx_1\, dx_2\, dx_3 + \iint_\Gamma \left[p + \rho E + \frac{\rho u^2}{2}\right] u_n\, d\sigma = 0.$$

For a viscous liquid it is necessary to add in the second (surface) integral the work performed by the friction forces, i. e., to add the integral

$$-\iint_\Gamma u_k \nu_{ik} n_i\, d\sigma.$$

We then arrive at the equality

$$\frac{\partial}{\partial t}\iiint \left[\rho E + \frac{\rho u^2}{2}\right] dx_1\, dx_2\, dx_3 + \iint \left[\left[p + \rho E + \frac{\rho u^2}{2}\right] u_k - u_i \nu_{ki} n_k\right] d\sigma = 0,$$

from which follows the law of conservation of energy in the following divergent differential form

$$\frac{\partial}{\partial t}\left[\rho E + \frac{\rho u^2}{2}\right] + \frac{\partial}{\partial x_k}\left[\rho u_k\left[E + \frac{p}{\rho} + \frac{u^2}{2}\right] - u_i \nu_{ik}\right] = 0.$$

§6. System of Lagrangian coordinates. Thus far we have considered hydrodynamics in the system of the so-called Euler coordinates, taking the independent variables to be the time t and the cartesian* coordinates of stationary, fixed points in space.

It is sometimes convenient to use other coordinates, namely, those connected with fixed (although moving) particles of the considered liquid or gas. Here the position of the particle in space, i. e., its Eulerian coordinates, will be determined by the functions of time and of the coordinates of the particle. Such a system of coordinates is usually called a system of Lagrangian coordinates. It is frequently employed to solve one-dimensional problems in hydrodynamics, and we shall therefore dwell on the case of plane-parallel motion along the x axis.

The Eulerian coordinates will be t and x. It is natural to choose for the Lagrangian coordinates again the time, which we shall denote, to avert confusion, by t' and the quantity

---

*Other coordinate systems (cylindrical, spherical) can be used, depending on the specific problem.

$$q = \int_0^x \rho(x, t)\, dx. \tag{6. 1}$$

If the origin $x = 0$ is chosen such that the region $x \leq 0$ does not intersect the region of solution of the problem, then (6. 1) determines the massive liquid, which is located to the left of the point x, i. e., actually q is connected with the particles of the liquid.

Inasmuch as the path covered by the particle during the time dt' is $u\,dt'$, we have for fixed q

$$dx = u\, dt',$$

and for fixed t from (6. 1) it follows that

$$dq = \rho\, dx.$$

Hence, recalling that $t' = t$, we obtain the differential relations

$$dx = u\, dt' + \frac{1}{\rho}\, dq, \tag{6. 2}$$

$$dq = -\rho u\, dt + \rho\, dx, \tag{6. 3}$$

which will give us all the partial derivatives of the new coordinates with respect to the old ones and will allow us to perform coordinate transformation in the equations

$$\frac{\partial}{\partial t} = \frac{\partial}{\partial t'} - \rho u \frac{\partial}{\partial q}, \qquad \frac{\partial}{\partial x} = \rho \frac{\partial}{\partial q}. \tag{6. 4}$$

In going over to Lagrangian coordinates, the continuity equation vanishes, since the continuity of the mass is contained in the very choice of the coordinate q. The equations of momentum and energy are transformed in accordance to formulas (6. 4). By way of an additional equation, there appears the condition of integrability of equation (6. 2) for x, i. e.,

$$\frac{\partial u}{\partial q} = \frac{\partial}{\partial t}\left(\frac{1}{\rho}\right). \tag{6. 5}$$

In addition, if it is necessary in the problem to know the coordinate x, it is obtained from (6. 2), i. e., for many of the equations

$$\frac{\partial x}{\partial t} = u, \quad \frac{\partial x}{\partial q} = \frac{1}{\rho}, \tag{6. 6}$$

which are compatible, owing to (6. 5).

Thus, the system of equations of plane motion in Langrangian coordinates has the form

$$\frac{\partial V}{\partial t} - \frac{\partial u}{\partial q} = 0,$$

$$\frac{\partial u}{\partial t} + \frac{\partial p}{\partial q} = 0, \tag{6.7}$$

$$\frac{\partial \left(E + \frac{u^2}{2}\right)}{\partial t} + \frac{\partial (pu)}{\partial q} = 0.$$

At the end of § 4 we determine the isothermal motion. In this case its equations are particularly simple

$$\frac{\partial V}{\partial t} - \frac{\partial u}{\partial q} = 0, \qquad \frac{\partial u}{\partial t} + \frac{\partial p}{\partial q} = 0, \tag{6. 8}$$

where p is a function of V, specified with the aid of the relation $T(p, V) = \text{const}$. It is easy to establish that the system (6. 8) is an evolution system, *if* $p'(V) \leq 0$. This requirement is always assumed to be satisfied and can furthermore be derived directly from the thermodynamic relations. The quantity $\sqrt{-p'(V)}$ is, as will be explained below, the velocity of propagation of sound perturbation in a gas, and is therefore called the *isothermal velocity of sound.*

One can approach the transformation in Lagrangian coordinates from still another point of view, which will explain the close connection between the transformation and the system of equations itself and which will make it possible to carry out analogous constructions in the case of any system in divergent form.

The continuity equation

$$\frac{\partial \rho}{\partial t} + \frac{\partial \rho u}{\partial x} = 0$$

expresses in other words the fact that

$$dq = \rho\, dx - \rho u\, dt \tag{6. 9}$$

is a total differential of a certain function q. It can be taken to be a new coordinate. With this, the continuity equation will be satisfied automatically, and in its place will appear the condition that

$$dx = u\, dt + \frac{1}{\rho}\, dq$$

be a total differential.

In our case (6.9) coincides with (6.3), i.e., we obtain the same coordinate as before, from perfectly general considerations.

Let us carry over this consideration to the case of the derivative of the system

$$\frac{\partial u_i}{\partial t} + \frac{\partial f_i(u)}{\partial x} = 0 \qquad (i = 1, \cdots, n). \tag{2.7'}$$

Each of its equations generates a transformation of coordinates, which we shall call, by analogy with the preceding, "Lagrangian."

Let us fix the number of the equation i. In accordance with (2.7'), the expression

$$dq = u_i\,dx - f_i(u)\,dt \tag{6.10}$$

is a total differential and can be taken to be the differential of the coordinate q. Equation (6.10) makes it possible to carry out transformations of coordinates and the remaining $n-1$ equations, in analogy with (6.4):

$$\begin{aligned} \frac{\partial}{\partial t} &= \frac{\partial}{\partial t'} - f_i(u)\frac{\partial}{\partial q}, \\ \frac{\partial}{\partial x} &= u_i\frac{\partial}{\partial q}, \end{aligned} \tag{6.11}$$

and the ith equation itself is replaced by the condition that the right-hand side of the equation

$$dx = \frac{f_i(u)}{u_i}\,dt + \frac{1}{u_i}\,dq \tag{6.12}$$

be a total differential, i.e.,

$$\frac{\partial}{\partial t}\left[\frac{1}{u_i}\right] - \frac{\partial}{\partial q}\left[\frac{f_i(u)}{u_i}\right] = 0. \tag{6.13}$$

§7. Characteristics. In the study of partial differential equations, and also in the numerical solution of such equations, a very important role is played by certain special surfaces, which are called characteristic surfaces or simply characteristics. We shall consider them for the case of the most general quasilinear system of equations of first order

$$A\frac{\partial u}{\partial t} + B\frac{\partial u}{\partial x} = f, \tag{7.1}$$

where u is the unknown vector function, f the known vector function, and A and B are specified matrices, which, like f, can depend on t and x and all the unknown u_k.

The characteristics of a system of equations are called such surfaces, on which the Cauchy problem is either insolvable or solvable in a non-unique manner. In other words, if in such a surface there is specified an analytic function, then we cannot use the Kovalevskaja method to find uniquely all its partial derivatives, using the differential equations. In the case considered here, when there are only two independent variables, the characteristics will be lines. Let u be given along this certain line. Then along this line we know the differential

$$du = \frac{\partial u}{\partial t}\, dt + \frac{\partial u}{\partial x}\, dx. \tag{7.2}$$

Together with the system of equation (7.1), this relation should be considered as a system of algebraic equations for $\partial u_i / \partial t$ and $\partial u_i / \partial x$, namely:

$$\begin{aligned} A \frac{\partial u}{\partial t} + B \frac{\partial u}{\partial x} &= f, \\ I\, dt \frac{\partial u}{\partial t} + I\, dx \frac{\partial u}{\partial x} &= du, \end{aligned} \tag{7.3}$$

where I is a unit matrix. We thus have a system of $2n$ equations for the determination of $2n$ partial derivatives $\partial u_i /\partial t$, $\partial u_i /\partial x$ with the aid of the specified values u of the line under consideration.

The foregoing definition of the characteristic can now be formulated algebraically: a line is the characteristic if the determinant of this system vanishes.

The matrix of the system (of order $2n$) consists of four matrices of nth order, and the determinant is conveniently written in the following manner:

$$\begin{vmatrix} A & B \\ I\, dt & I\, dx \end{vmatrix} = 0. \tag{7.4}$$

We shall use the following lemma, which takes place for all matrices of nth order A, B, C, D: if $AD = DA$ and $BC = CB$, then

$$\begin{vmatrix} A & B \\ C & D \end{vmatrix} = |\, AD - BC \,|.$$

Obviously, it is also applicable to our case, and therefore equation (7.4) can now be written with the aid of the nth order determinant

$$|A\,dx - B\,dt| = 0. \tag{7.5}$$

This is the condition that dt and dx define the characteristic direction.

In particular, if the system (7.1) has a simpler form

$$\frac{\partial u}{\partial t} + A\frac{\partial u}{\partial x} = 0, \tag{7.6}$$

then the characteristics are found from the equation

$$|I\,dx - A\,dt| = 0,$$

i. e.,

$$\left|I\cdot\frac{dx}{dt} - A\right| = 0.$$

In other words, the slope of the characteristic dx/dt is the root of the characteristic equation of the matrix A. In § 1 we established that the system (7.6) is an evolution system, if the eigenvalues of the matrix A are real. Thus in our case it follows from the evolution nature of the system that its characteristics are real.

Let us return again to system (7.1).

For any real solution, the function u and its derivative should satisfy (7.1) and (7.2), i. e., the system of algebraic equations (7.3) should be compatible. For this purpose it is necessary and sufficient that the rank of the expanded matrix of the system

$$\begin{pmatrix} A & B & f \\ I\,dt & I\,dx & du \end{pmatrix} \tag{7.7}$$

coincide with the rank of the matrix of the coefficients of the unknowns. The latter is assumed to be $2n-1$, which corresponds to the assumption that all the characteristic directions are different.

If the rank (7.7) is also equal to $2n-1$, then the last column of matrix is a linear combination of the remaining ones, i. e.,

$$f_i = \sum_{k=1}^{n} a_{ik}\xi_k + \sum_{k=1}^{n} b_{ik}\eta_k,$$

$$du_i = \xi_i\,dt + \eta_i\,dx.$$

Eliminating ξ_k from the first relations with the aid of the lower ones, we obtain

$$f_i\,dt - \sum_{k=1}^{n} a_{ik}\,du_k = \sum_{k=1}^{n} b_{ik}\,\eta_k\,dt - \sum_{k=1}^{n} a_{ik}\,\eta_k\,dx$$

or, in vector form,

$$f\,dt - A\,du = (B\,dt - A\,dx)\,\eta. \tag{7.8}$$

This means that the vector $f\,dt - A\,du$ is a linear combination of the columns of the matrix $B\,dt - A\,dx$, i.e., the rank of the matrix

$$(B\,dt - A\,dx,\ f\,dt - A\,du) \tag{7.9}$$

is $n-1$. Equating to zero the determinant made up of any $n-1$ columns $B\,dt - A\,dx$ and the vector $f\,dt - A\,du$, we obtain the desired differential relation on the characteristic. It is obvious that the vanishing of any other determinant of nth order of the matrix (7.9), other than (7.5), will be the result of the relation obtained, since the columns of the matrix $B\,dt - A\,dx$ are linearly dependent.*

Thus, we obtain at each point (x, t) n characteristic directions and n relations between the differentials du_k on them. This makes it possible to solve the Cauchy problem for an evolution system with the aid of characteristics. From each point it is possible to draw characteristics until they meet the line of the initial data (Figure 4), write out n relations for the increments of the functions u_k, and by solving simultaneously to obtain the values at a given point.

Essentially, these considerations serve as the basis for the approximate method of solving equations with the aid of characteristics.

In the case of the system

$$\frac{\partial u}{\partial t} + A\frac{\partial u}{\partial x} = 0$$

the relation on the characteristic (7.8) has the form

$$-\,du = (A\,dt - I\,dx)\,\eta,$$

and since, as we have seen above, $dx/dt = \lambda$, where λ are the eigenvalues of the matrix A,

$$du = (\lambda I - A)\,\eta\,dt.$$

It is seen from this that du belongs to the subspace, into which all space is transformed by the operator with degenerate matrix $\lambda I - A$.

---

* Analogously, we can obtain from (7.7), instead of (7.8),

$$f\,dx - B\,du = (A\,dx - B\,dt)\,\xi,$$

i.e., the relations on the characteristics followed from the fact that the matrix $(A\,dx - B\,dt,\ f\,dx - B\,du)$ has a rank $n-1$. This fact should be used, for example, if A is a degenerate matrix.

This is orthogonal to the eigenvector of the matrix A, corresponding to the given λ, and is streched over the remaining $n-1$ eigenvectors. Such is the geometrical meaning, in this case, of the relations on the characteristics, if one considers du as a vector in the space $(u_1, \cdots, u_n)$.

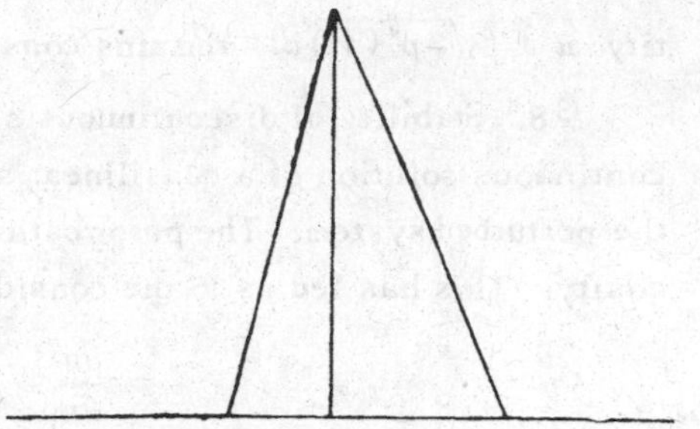

Figure 4

It is interesting to indicate that the characteristics are found to be lines along which small perturbations of the solution of the system propagate.

For this purpose we consider the system (7.5); let its solution be $u+v$, where u is the certain solution, which we shall consider known, and v is a small perturbation. Substituting this in (7.5), using the fact that u is a solution, and discarding terms above the first order of smallness in v, we obtain a system of linear equations

$$\frac{\partial v}{\partial t} + A(u)\frac{\partial v}{\partial x} = -\sum_{k=1}^{n} \frac{\partial A(u)}{\partial u_k} v_k \frac{\partial u}{\partial x}, \tag{7.10}$$

Its coefficients in front of the derivatives coincide with the coefficients of (7.5), from which it follows that it has the same system of characteristics. On the other hand, in the case of a linear system it is known that if the initial data are specified at a point, then from there on this perturbation will propagate along the characteristics. If the initial data are specified in a finite region, then the boundary of this perturbed region will also move with time along the characteristic.

In hydrodynamics, small perturbations are called sonic or acoustical, and therefore, in accordance with the foregoing, the slope of the characteristic determines the velocity of sound.

In conclusion, let us consider an example, a system of equations of hydrodynamics (isothermal case) in Lagrangian coordinates with one spatial variable

$$\frac{\partial u}{\partial t} + \frac{\partial p(V)}{\partial x} = 0,$$

$$\frac{\partial V}{\partial t} - \frac{\partial u}{\partial x} = 0.$$

The eigenvalues of the matrix of this system (velocity of sound) are $\pm\sqrt{-p'(V)}$, and consequently the equations of the characteristics have the form

$dx/dt = \pm\sqrt{-p'(V)}$. It is easy to show that along the characteristics the quantity $u \pm \int\sqrt{-p'(V)}\,dt$ remains constant (it is called the Riemann invariant).

§8. Stability of discontinuous solution. In §2 we have required that a discontinuous solution of a quasilinear system be a limit of continuous solutions of the perturbed system. The perturbation was introduced in the form of linear viscosity. This has led us to the consideration of only divergent systems

$$\frac{\partial u}{\partial t} + \frac{\partial f(u)}{\partial x} = 0, \tag{2.7}$$

and also to the conditions on the lines of discontinuity

$$-\omega(u^+ - u^-) + f(u^+) - f(u^-) = 0, \tag{2.6}$$

where u^-, u^+ are the values of u on the discontinuity, and $\omega = dx/dt$ is the slope of the line of discontinuity.

It is easy to show that not every discontinuous solution of the system (2.7), which satisfies the conditions (2.6) on the discontinuity lines, is an *admissible* solution, i.e., it can be obtained as a limit of the continuous solutions (as $\epsilon \to 0$) of the perturbed system

$$\frac{\partial u}{\partial t} + \frac{\partial f(u)}{\partial x} = \epsilon\frac{\partial^2 u}{dx^2}. \tag{8.1}$$

Let us attempt to find the supplementary conditions that insure the admissibility of the discontinuous solution. For this purpose, as before, we confine ourselves to solutions of the form $u = u(\xi)$, where $\xi = x - \omega t$. Then the system of equation (8.1) assumes the form

$$-\omega\frac{du}{d\epsilon} + \frac{df(u)}{d\xi} = \epsilon\frac{d^2 u}{d\xi^2} \tag{8.2}$$

and the problem of the admissibility reduces to the following:

Let ω, u^-, and u^+ satisfy relations (2.6). *Under what conditions does there exist for the system of equation* (8.2) *a continuous solution, satisfying the following boundary conditions:*

$$\begin{aligned} u(\xi) &\to u^+ \quad \text{for} \quad \xi \to +\infty, \\ u(\xi) &\to u^- \quad \text{for} \quad \xi \to -\infty. \end{aligned} \tag{8.3}$$

Actually, if such a solution exists, then as $\epsilon \to 0$, it goes into the discontinuous solution of the unperturbed system (2.7), which has the form

$$\begin{aligned} u(\xi) &= u^+ \quad \text{for} \quad \xi > 0, \\ u(\xi) &= u^- \quad \text{for} \quad \xi < 0. \end{aligned}$$

Therefore the conditions of the existence of the continuous solution of the system (8. 2) are conditions of the admissibility of the discontinuous solution u^- and u^+.

The formulation of the problem of finding conditions of admissibility can be modified. For this purpose we integrate the system of equation (8. 2) once with respect to ξ; we obtain

$$\epsilon \frac{du}{d\xi} = -\omega u + f(u) + C.$$

We choose the constant C such that the right-hand side vanishes when $u = u^-$ and $u = u^+$. This is possible since ω, u^-, and u^+ satisfy the conditions (2. 6). We denote

$$-\omega u + f(u) + C = \Phi(u), \tag{8. 4}$$

so that $\Phi(u^-) = \Phi(u^+) = 0$. We arrive at the following problem of qualitative theory of ordinary equations: *under what conditions does there exist for the system of equations*

$$\epsilon \frac{du}{d\xi} = \Phi(u) \tag{8. 5}$$

an integral curve, joining the two singular points of this system; or, *more accurately, does there exist a solution satisfying the boundary conditions*

$$\begin{aligned} u(\xi) &\to u^+ \quad \text{for} \quad \xi \to +\infty, \\ u(\xi) &\to u^- \quad \text{for} \quad \xi \to -\infty, \end{aligned} \tag{8. 3}$$

where u^- and u^+ satisfy the conditions $\Phi(u^-) = \Phi(u^+) = 0$, and $\Phi(u)$ is determined from equation (8. 4).

In the case when (8. 5) is not a system, but a single equation, the problem is solved in the following manner. We consider the problem of the behavior of the integral curves in the plane ξ, u. Let there be between the two zeroes u^- and u^+ of the function $\Phi(u)$ at least one zero u^*, $\Phi(u^*) = 0$. Then $u(\xi) = u^*$ is a solution of the equation, which separates u^- from u^+. Consequently, there is no solution that joins u^- and u^+. If, however, u^- and u^+ are neighboring zeroes of the function $\Phi(u)$, then for the existence of the necessary solution it is essential that in the band between $u = u^-$ and $u = u^+$ the sign of $du/d\xi$, i. e., the sign of $\Phi(u)$, coincide with the sign of the difference $u^+ - u^-$. This requirement is equivalent to the conditions $\Phi'(u^+) < 0$ and $\Phi'(u^-) > 0$, or, using the expression for $\Phi(u)$, (8. 4), to the condition $f'(u^+) < \omega < f'(u^-)$.

Thus, for the case of one equation one can formulate the following two conditions of admissibility: *First*: for the admissibility of the discontinuity u^-, u^+

it is essential that u^- and u^+ be *neighboring* zeroes of the function $\Phi(u)$. *Second*: the discontinuity u^-, u^+ is admissible if the inequality

$$f'(u^+) < \omega < f'(u^-) \tag{8.6}$$

is satisfied. The fulfillment of both conditions insures the admissibility of the discontinuity.

To find the conditions of admissibility in the case of a system of equations one can attempt to generalize the foregoing pair of conditions. In a correct generalization of the first condition lies indeed the entire difficulty of the problem formulated above. The second condition is relatively simple to generalize, as will be shown now.

The point is that this condition (8.6) can be obtained in an entirely different manner, without resorting to the addition of viscosity. Namely, we stipulate that the discontinuous solution of the equation

$$\frac{\partial u}{\partial t} + \frac{\partial f(u)}{\partial x} = 0, \tag{2.7}$$

satisfying on the discontinuity the condition

$$-\omega(u^+ - u^-) + f(u^+) - f(u^-) = 0, \tag{2.6}$$

be stable with respect to small perturbations of the solution itself. To find the conditions of stability we use the same discontinuous solution

$$\begin{aligned} u &= u^+ \quad \text{for} \quad x - \omega t > 0, \\ u &= u^- \quad \text{for} \quad x - \omega t < 0. \end{aligned} \tag{8.7}$$

We add to it a small perturbation δu and substitute $u + \delta u$ into equation (2.7). Discarding terms of order $(\delta u)^2$ and higher, we obtain an equation for δu

$$\frac{\partial \delta u}{\partial t} + f' \frac{\partial \delta u}{\partial x} = 0 \tag{8.8}$$

with a piecewise-constant coefficient f'

$$\begin{aligned} f' &= f'(u^+) \quad \text{for } x - \omega t > 0, \\ f' &= f'(u^-) \quad \text{for } x - \omega t < 0. \end{aligned} \tag{8.8'}$$

The discontinuous solution (8.7) will be stable, if the perturbation δu along the discontinuity line $x - \omega t = 0$ will tend to 0 as $t \to \infty$.

Let us specify a perturbation at the initial instant, and let this perturbation tend to zero as $x \to \pm\infty$. The solution of equation (8.8) is $\delta u = \text{const}$ along the characteristic $dx/dt = f'$. Therefore the stability will be insured, if the values of δu to the left and to the right on the line of discontinuity can be determined from the initial data. For this purpose it is necessary that the characteristics $dx/dt = f'$ be "incoming" on this discontinuity from both sides of the discontinuity line. Comparing the slopes of the characteristics $f'(u^+)$ and $f'(u^-)$ with the slope of the discontinuity line ω, we obtain the condition

$$f'(u^+) < \omega < f'(u^-). \tag{8.6}$$

Therefore, for the purpose of generalizing the second condition of admissibility (8.6), we require stability of the discontinuity solution of the *system* of equation (2.7) with respect to small perturbations.

Proceeding as in the case of a single equation, we obtain the system of equations (8.8) and (8.8′), where δu is now a vector, and f' is a matrix of order n.

Let us vary, furthermore, the relations on the discontinuity (2.6), and obtain

$$\omega(\delta u^+ - \delta u^-) + f'(u^+)\delta u^+ - f'(u^-)\delta u^- - \delta\omega(u^+ - u^-) = 0. \tag{8.9}$$

The stability again reduces to the possibility of unique determination of δu^+ and δu^- on the discontinuity. For their determination we have n relations (8.9) which include still another unknown $\delta\omega$, i.e., the total number of unknowns is $2n + 1$, and there are only n equations available for their determination. To obtain the lacking $n + 1$ relations, we proceed as follows. The constant matrix f' in the system (8.8), has, by virtue of its evolution nature, n real eigenvalues; we assume these to be, furthermore, different. Then f' is reduced to the diagonal form $k \cdot f' \cdot k^{-1} = \Lambda = (\lambda_i \delta_{ij})$; multiplying (8.8) to the left by the matrix k, we obtain

$$\frac{\partial}{\partial t}(k\delta u) + \Lambda \frac{\partial}{\partial x}(k\delta u) = 0 \tag{8.10}$$

with a diagonal matrix Λ. From this we find that along the characteristic $dx/dt = \lambda_i$, the ith component of the vector $k\delta u$ is conserved. Consequently, the system of equations brings on the discontinuity as many relations as $\delta u^{\pm}$, as there are "incoming" characteristics on the discontinuity line. It follows therefore that *the discontinuous solution is stable, if the number of characteristics "incoming" on the discontinuity from the left and from the right is equal to* $n + 1$. Comparing again the slopes of the characteristics and the slope of the discontinuity line, one can write this condition briefly in the form of the inequalities

$$\lambda_{k-1}(u^-),\ \lambda_k(u^+) < \omega < \lambda_k(u^-),\ \lambda_{k+1}(u^+), \tag{8.11}$$

which are satisfied for any k. With this, the λ_i are assumed numbered in increasing order of magnitude. When $n = 1$, (8.11) obviously goes into (8.6).

§9. Decay of arbitrary discontinuity. In the theory of quasilinear systems, a particular role is played by the so-called problem of the decay of an arbitrary discontinuity. As will be shown below, the solution of this problem is very important for the clarification of problems of existence in the uniqueness of the Cauchy problem, for the determination of the asymptotic values of the solution $t \to \infty$. In addition, this problem is of interest in itself. It consists of the following.

It is required to construct a solution* of the evolution system

$$\frac{\partial u}{\partial t} + \frac{\partial f(u)}{\partial x} = 0, \tag{2.7}$$

satisfying the discontinuous initial condition

$$u(x, 0) = \begin{cases} u^- & \text{for} \quad x < 0, \\ u^+ & \text{for} \quad x > 0, \end{cases} \tag{9.1}$$

where u^- and u^+ are arbitrary constants.

The system of equations and the initial conditions are subject to groups of similar transformations of the plane x, t, namely, $x \to \alpha x$ and $t \to \alpha t$. It is therefore natural to seek a solution of the problem in the form $u = u(\xi)$, where $\xi = x/t$. Then, instead of (2.7) and (9.1) we have a system of ordinary equations

$$-\xi \frac{du}{d\xi} + \frac{df(u)}{d\xi} = 0 \tag{9.2}$$

with boundary conditions

$$u(\xi) \to u^- \quad \text{for} \quad \xi \to -\infty,$$
$$u(\xi) \to u^+ \quad \text{for} \quad \xi \to +\infty.$$

We note that $u(\xi) \equiv \text{const}$ is the solution of the system. To describe the non-trivial solutions, we rewrite the system (9.2) in the form

$$(f'(u) - \xi I)\frac{du}{d\xi} = 0, \tag{9.3}$$

where $f'(u)$ is the matrix of the derivatives, I is the unit matrix. It is seen from this that if $du/d\xi \neq 0$, then $du/d\xi$ should be an eigenvector of the matrix $f'(u)$, and ξ should be the corresponding eigenvalue. By virtue of the evolution nature of the system, all the eigenvalues $\lambda_k(u)$ are real. The corresponding eigenvectors $l_k(u)$ form, in the space u, n vector fields, which are linearly independent if all the $\lambda(u)$ are different. This means that we can draw n families of trajectories of the systems of the equations $du/ds = l_k(u)$. We shall call them *the characteristic trajectories*. Naturally, by way of a solution it is

* By solution we mean here an admissible solution (see §8).

possible to use only those portions of the trajectories on which the corresponding $\lambda_k(u)$ change monotonically, since $\lambda_k = \xi$, and in the opposite case $u(\xi)$ will not be a single-valued function.

In addition, *discontinuous transitions* are possible, i. e., a transition from the point u_1 to the point u_2, satisfying

$$-\omega(u_2 - u_1) + f(u_2) - f(u_1) = 0$$

and completed when $\xi = \omega$. Since for the stability of such a transition u must satisfy the inequality $\lambda_k(u_2) < \omega < \lambda_k(u_1)$ for a certain k, then the discontinuous transitions can also be broken up into m families, depending on the number k. The one-dimensional set of points u satisfying $-\omega(u - u_0) + f(u) - f(u_0) = 0$, i. e., points in which it is possible to go from the point u_0 via a discontinuous transition of the kth family, we shall call, using the language of hydrodynamics, the *shock adabat* of the kth family. Using the same language, we shall call the continuous solution (non-trivial) a *rarefaction wave*, and the discontinuous transition a *shock wave.*

Thus, the solution of the problem of the decay of an arbitrary discontinuity is made up of solutions of three types, constants, rarefaction waves, and shock waves, and the latter two are subdivided into n families.

Let us consider the problem of the case of a discontinuity for a *single equation.* In this case the problem always has a solution, which is furthermore unique. Instead of proving this fact, we shall consider an example, which is sufficiently arbitrary, which demonstrates well the method of constructing the solution for any equation. Thus, we have one equation (2.7), where the function $f(u)$, which enters into the equation, has the form shown in Figure 5. Let u^- (left-hand value of the initial data) have the value indicated in Figure 5, and in addition, the points u_1, u_2, and u_3 are noted, defined in the following manner. The point u_3 is the one in which a ray drawn through $f(u^-)$ is tangent to the curve $f(u)$. u_1 is the point of intersection of this ray with $f(u)$, and u_2 is the point of inflection of the curve $f(u)$.

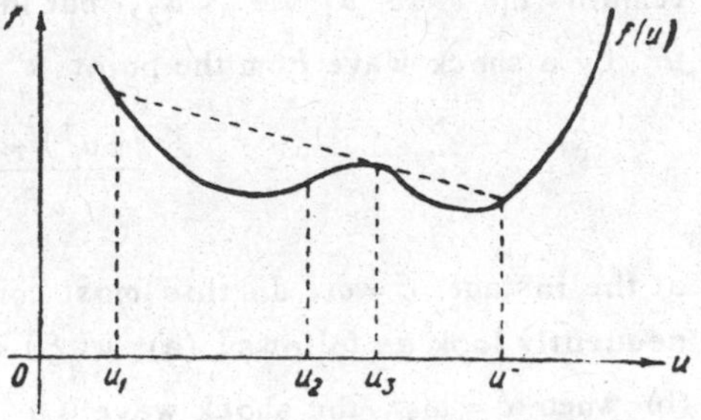

Figure 5

Our problem is to go from u^- into any point u^+ on the u axis. We first find the aggregate of points $\Omega(u^-)$, into which it is possible to go from u^- by a shock wave. We know that such points should satisfy relation (2.6),

$$f(u) - f(u_0) = \omega(u - u_0)$$

and in addition, the two conditions of admissibility formulated in the preceding section. The first of these is the condition of the "neighborness" of the roots $\Phi(u) = 0$, where in the given case

$$\Phi(u) = f(u) - \omega u - (f(u^-) - \omega u^-).$$

From this we obtain that u should lie outside the integral (u_1, u_3). The second condition is

$$f'(u) < \omega < f'(u^-).$$

It forbids transitions at the point $u > u^-$. Thus, $\Omega(u^-)$ consists of the following points: $u \leq u_1$, and $u_3 \leq u < u^-$. We have replaced here the sign $<$ by $\leq$, but if one recalls the method of obtaining these conditions (§8) it is easy to verify that this is legitimate.

We next find the aggregate of points $L(u^-)$, to which it is possible to go from u^- by a rarefaction wave. Here we should move along the characteristic trajectory, but by virtue of the one-dimensionality of the space u, this trajectory will be the u axis. It is necessary to see to it only that $\xi = f(u)$ increase as one moves from the point u^-. Satisfying this condition are the points $u > u^-$, which indeed form $L(u^-)$. Thus, if $u^+ \leq u_1$ or $u^+ \geq u_3$, then the transition from u^- to u^+ is defined. We are left with the case $u_1 < u^+ < u_3$. To determine the solution in this case, we note that a discontinuous transition from u^- into u_3 occurs at

$$\xi = \omega = \frac{f(u_3) - f(u^-)}{u_3 - u^-} = f'(u_3).$$

Consequently, nothing prevents us from continuing the solution from u_3 by a rarefaction wave in a direction of increasing $f' = \xi$, i. e., to the point u_2. There remains the case $u_1 < u^+ < u_2$, but in this case it is possible to go to the point u^+ by a shock wave from the point u^* of the interval (u_2, u_3), satisfying

$$\frac{f(u^+) - f(u^*)}{u^+ - u^*} = \omega = f'(u^*),$$

at the instant $\xi = \omega$. In this most complicated case the entire solution will consequently look as follows: (a) $u(\xi) \equiv u^-$ when $\xi < \omega_3 = (f(u_3) - f(u^-))/(u_3 - u^-)$, (b) when $\xi = \omega_3$, the shock wave (u^-, u_3), (c) rarefaction wave $u = u(\xi)$ on the section $\omega_3 < \xi < \omega_1$, where $u(\xi)$ is determined from $f'(u) = \xi$, (d) shock wave u^*, u^+, when $\xi = \omega_1 = f'(u^*) = (f(u^+) - f(u^*))/(u^+ - u^*)$ and, finally (e) $u(\xi) \equiv u^+$ when $\xi > \omega_1$. It is easy to visualize the solution in the remaining cases of location of u^+.

From the foregoing example the principle of the construction of the solution for the case of any function $f(u)$ and any pair u^-, u^+ becomes clear, as does the uniqueness of this solution.

In the case of a *system* of equations, we do not have effective conditions for the admissibility of the discontinuous solution (see §8), nor do we have a proof of the uniqueness of the admissible solution of the problem of the decay of the discontinuity. To obtain an idea of a method of constructing a solution in this case, let us consider by way of an example a system of gas dynamic equations in Lagrangian coordinates (isothermal case):

$$\begin{aligned} \frac{\partial u}{\partial t} + \frac{\partial p(v)}{\partial x} &= 0, \\ \frac{\partial v}{\partial t} - \frac{\partial u}{\partial x} &= 0, \end{aligned} \tag{9.4}$$

where $p(v) = 1/v$, and let us construct a solution satisfying the discontinuous initial conditions: u^-, v^- when $x < 0$, u^+, v^+, when $x > 0$ ($v^-, v^+ > 0$). As was noted above, the solution should be made up of constants, rarefaction waves, and shock waves. We begin with the latter.

The set of points u, v, satisfying

$$-w(u - u^-) + \left(\frac{1}{v} - \frac{1}{v^-}\right) = 0,$$

$$-\omega(v - v^-) - (u - u^-) = 0,$$

$$\lambda_1(v) = -\frac{1}{v} < \omega < -\frac{1}{v^-},$$

defines the adabat of the first family $\Omega_1(u^-, v^-)$, corresponding to the point u^-, v^-, i.e., the aggregate of points in which it is possible to go from u^-, v^- by a shock wave of the first family. We already know the process by which to obtain the characteristic trajectory of the first family $L_1(u^-, v^-)$, passing through the point u^-, v^-, i.e., the aggregate of points, in which it is possible to go from the point u^-, v^- by a rarefaction wave of the first family

$$v = v^- e^{u - u^-}, \quad -\frac{1}{v} > -\frac{1}{v^-}.$$

$\Omega_1(u^-, v^-)$ and $L_1(u^-, v^-)$ are shown in Figure 6. At the point u^-, v^- itself, which separates Ω_1 and L_1, these curves have not only a common tangent, but also a common curvature (this fact is true for any system). In the foregoing example of a singular equation, we could supplement the obtained aggregate of points Ω_1 and L_1 by using the rarefaction waves and shock waves of the same (first) family. Here, however, as can be readily verified, this cannot be done (it will be made clear later that there is no need for it). This is explained by the fact that the given system has the property of "convexity," namely, on moving along the characteristic trajectory $L_1(u^-, v^-)$, from the point u^-, v^-, we shall have at all times a monotonic increase in the eigenvalue λ_1, equal to $-1/v$. On the other hand, on the set $\Omega_1(u^-, v^-)$ there is no point for which the corresponding ω would equal the eigenvalue λ_1 at this point. In the example of a single equation this was not the case, because the function $f(u)$ was not convex (in the usual sense) and this made it possible to carry out additional constructions. We shall discuss the "convexity" of the system of functions later (§ 10).

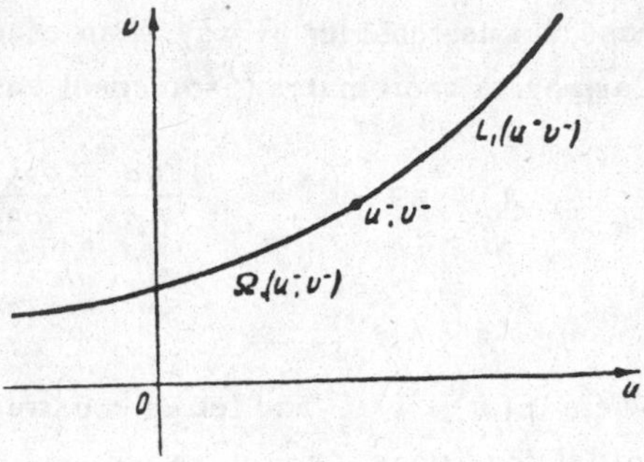

Figure 6

Let us turn to the construction of the solution. For each point u, v of the curve $\Omega_1 + L_1$ we construct the adabat of the second family $\Omega_2(u, v)$ and the characteristic trajectory of the second family $L_2(u, v)$. Then $\Omega_2(\Omega_1 + L_1)$ together with $L_2(\Omega_1 + L_1)$ cover the entire half-plane $v > 0$. It is easy to see that any point u^+, v^+ of the upper half-plane is reached in the same manner, and furthermore uniquely.

Thus, the solution of the problem of the decay of the discontinuity for the system (9.4) in the half-plane $v > 0$ always has a solution, which is furthermore unique.

To conclude this section, let us dwell on the following problem. It is required to determine the asymptote (as $t \to \infty$) of the solution of the system of equations with the viscosity

$$\frac{\partial u}{\partial t} + \frac{\partial f(u)}{\partial x} = B\frac{\partial^2 u}{\partial x^2}$$

for specified initial data in the form

$$u(0, x) = \begin{cases} u^- & \text{for } x < x^-, \\ u_0(x) & \text{for } x^- < x < x^+, \\ u^+ & \text{for } x^+ < x, \end{cases}$$

where $u_0(x)$ is an arbitrary function ("smeared" discontinuity u^-, u^+). It would be interesting to prove that the principal term of the asymptote coincides with the solution of the problem of the decay of the discontinuity (u^-, u^+) for the same system without viscosity.

The question of the asymptote of the solution of such a problem for the case of a single equation with a convex function $f(u)$ was considered in [2].

§10. Existence and uniqueness theorems. In the theory of evolution systems of quasilinear equations, the question of existence of uniqueness of the solution of the Cauchy problem remains open.*

However, simple considerations make it possible to separate the basic difficulties in this problem. Namely, we know that at each point of continuity of the solution of the system, there are brought along n characteristics, from the region of smaller values of t, n linearly-independent relations between the differentials of the sought functions. By solving these, we can continue uniquely the solution, but we cannot do so if the characteristics touch each other. It would be interesting to investigate this case and to show that from the point of tangency of the characteristics it is possible to draw a stable shock wave. If this can be done, then further solution will be continued with the aid of n relations on the discontinuity line: the shock wave and $n+1$ relations, which are brought along the characteristics. The basic difficulty arises at the instant when two shock waves meet or when the characteristics intersect. In this case the point of intersection is a discontinuity point in the solution, and the values to the left and to the right are independent at this point. Therefore, in order to construct the solution in the vicinity of such a point, it is necessary to solve the problem of the decay of an arbitrary discontinuity. Thus, the basic difficulty is the proof of the existence and uniqueness of the solution of the problem of the decay of the discontinuity.

In §2 we have defined the solution as the limit of continuous solutions of the perturbed system. Apparently such an approach to the solution insures its existence and uniqueness. If this is really so, one can justify still another approach towards solving this problem. As we have noted in §8, the condition of admissibility of the continuous solution reduces to the generalized condition of the "neighborliness" of the zeroes of a certain function $\Phi(u)$ and to the condition of stability of the discontinuity (8.2). In the case of one equation the first condition is known to be satisfied, if the function $f(u)$ is convex (i.e., $f''(u) \neq 0$), for in this case $\Phi(u)$ will have a total of two zeroes. Therefore it is possible to attempt to generalize to the system of functions $f(u)$ the concept of convexity,

* See [3]–[7], devoted to this problem.

and thereby lift the first condition. If we succeed in doing so, then *for the uniqueness in the class of convex systems it is enough to specify merely the stability of the discontinuous solution.* Along this path, several definitions of convexity were proposed. We shall cite two of these. First: let $\lambda(u)$ be the eigenvalue of the matrix $f'(u)$, $l(u)$ the corresponding eigenvector, and then the system of the function $f(u)$ is called convex, if $(l\cdot \operatorname{grad} \lambda) \neq 0$.* Second: let the pair u_1 and u_2 satisfy the condition $f(u_2) - f(u_1) = \omega(u_2 - u_1)$ for a certain ω; then the system of functions $f(u)$ is convex if for any such pair ω differs from each eigenvalue $\lambda(u_1)$, $\lambda(u_2)$. It is easy to see that these definitions are generated in the process of constructing a solution for the problem of the decay of the discontinuity, with the tendency towards ensuring maximum simplicity of this solution.

§ **11. Two-dimensional evolution systems of equations.** Thus far we have considered problems connected with the evolution systems of quasilinear equations for functions of two independent variables t and x. In hydrodynamics such equations describe one-dimensional processes, and we therefore called the corresponding problems one-dimensional.

Problems that deal with two or more spatial coordinates we shall call multi-dimensional. In this solution one encounters certain qualitatively new problems, and they arise principally upon going from one measurement to two, i. e., different multi-dimensional problems differ much less from two-dimensional ones, than two-dimensional ones differ from one-dimensional ones. This causes us to concentrate all our attention on equations with three independent variables t, x, and y, as being the simplest.

For what follows we shall find it convenient to establish certain information concerning systems of linear equations with three independent variables.

$$\frac{\partial u}{\partial t} + A\frac{\partial u}{\partial x} + B\frac{\partial u}{\partial y} = 0, \tag{11. 1}$$

where A and B are known matrices of order n (for simplicity, constant).

The first question concerns the conditions that must be satisfied by these matrices in order for the system (11. 1) to be an evolution one. Again, as in § 1, we shall seek its solution in the form

$$u(t, x, y) = u \cdot e^{i(\lambda t + \mu x + \nu y)}, \tag{11. 2}$$

where u is a constant vector, and μ and ν are any real numbers. The latter

* This condition was first encountered in [7].

is a result of the fact that arbitrary initial data $u(0, x, y)$ can be expanded in plane waves of the indicated type. Substituting (11. 2) in (11. 1) we obtain equations for u:

$$\lambda u + \mu A u + \nu B u = 0$$

or

$$(\lambda I + \mu A + \nu B)\, u = 0.$$

This system has non-zero solutions under the condition

$$\mathrm{Det}\, |\lambda I + \mu A + \nu B| = 0, \tag{11. 3}$$

which is an algebraic equation for the determination of λ. It is homogeneous in λ, μ, ν, i. e., its solutions λ are proportional to μ and ν. If λ is not real, then we can make its imaginary part negative by a suitable choice of μ and ν, and make this part as large as desired in absolute magnitude. This means, as in § 1, that we obtain a solution which increases within a finite time by a factor as large as desired, i. e., the Cauchy problem for the system is not correct.

Thus, the system (11. 1) is an evolution system if all the solutions of equation (11. 3) or, in other words, all the eigenvalues of the matrix $\mu A + \nu B$ for all real μ and ν, are real.

The same result could be obtained directly from the fact (see § 1) that the system

$$\frac{\partial u}{\partial t} + A \frac{\partial u}{\partial x} = 0$$

is an evolution system if A has real eigenvalues. Let us consider solutions which do not depend on y; then the matrix A should satisfy this requirement. We then rotate the axes x and y, or, what is the same, consider solutions which depend only on $\mu x + \nu y$; then the systems will be one-dimensional and the matrix $\mu A + \nu B$ will play the role of A.

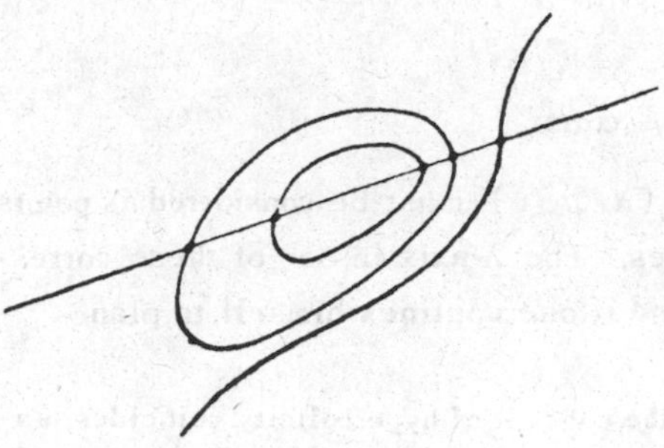

Figure 7

In order not to complicate the arguments, we shall again assume that all the eigenvalues are furthermore different. This means that to each pair of real numbers μ, ν, there corresponds n different real numbers λ, solutions of the nth order of equation (11. 3), i. e., in the space (λ, μ, ν) there are located over each point in the plane (μ, ν) n points λ, which form thereby n sheets of the surface, which is described by equation (11.3). The latter, as indicated above, is homogeneous, meaning that this surface is conic. We shall denote it *the characteristic cone* of the system (11.1).

One can consider (11.3) the same as an equation of a curve in projective coordinates on a plane. This curve is, for example, the section of the cone by the plane $\lambda = 1$. Since the cone consists of n sheets, the curve consists of n ovals. Such a construction is used to determine in general form a hyperbolic system of equations: the system is called hyperbolic if in the indicated projective plane there exists a point such that each line passing through it intersects the section of the characteristic cone in exactly n points (Figure 7).* Such a point, for example, is the intersection of the plane with the axis λ. In other words, each plane in the space (λ, μ, ν) passing through the λ-axis intersects the characteristic cone on exactly n generatrices.

We note that any line in our projective plane crosses the ovals in not more than n points, since the points of intersection are solutions of an algebraic equation of nth degree.

Let us consider a pencil of lines passing through the origin, and let us shift its center such that it remains inside the innermost oval. Since we do not leave this oval, not one of the lines can lose its points of intersection with the curve. It follows therefore that the internal oval is always concave, or otherwise certain lines should acquire new points of intersection, i. e., acquire more than n such points. We have also established that if the system is hyperbolic with respect to the λ-axis, then the same can be said also concerning any other straight line inside the innermost oval. In other words, we can turn the λ-axis at will, provided it does not leave the confines of this oval, and this will not influence the hyperbolic nature of the system.

Let us explain now what this means in the coordinate space (t, x, y). The coordinates λ, μ, and ν appeared here in the consideration of plane waves (11.2), i. e., to each triplet (λ, μ, ν) there corresponds in coordinate space a plane

$$\lambda t + \mu x + \nu y = \text{const.}$$

This means that the points (t, x, y) and (λ, μ, ν) must be considered as points of conjugate three-dimensional affine spaces. The λ-axis in one of these corresponds to the plane $t = \text{const}$ in the other, and if one confines himself to planes

* In the case of first-order system (11.1), the concept of hyperbolicity coincides, as we have seen, with the concept of evolution, although generally speaking the latter is broader than the former.

passing through the origin, this will be merely the plane (x, y), on which we specify usually the initial data. The replacement of the λ-axis by any line passing through the point $(\lambda_0, \mu_0, \nu_0)$ corresponds to our specifying the initial data on the plane

$$\lambda_0 t + \mu_0 x + \nu_0 y = 0. \qquad (11.\ 4)$$

The Cauchy problem remains in this case correct, provided only the point $(\lambda_0, \mu_0, \nu_0)$ does not leave the innermost oval.

We shall say that the plane (11.4) is "orthogonal" to a vector $(\lambda_0, \mu_0, \nu_0)$.*

If the vector $(\lambda_0, \mu_0, \nu_0)$ will run over a convex internal cone, then the plane orthogonal to it will in coordinate space also envelope a certain convex dual cone. If the plane of the initial data is located outside this cone, i. e., it is a reference plane for it, then, as we have established, the Cauchy problem for equation (11. 1) will always be correct. If, however, the plane intersects the cone, then the Cauchy problem will generally speaking not be correct.

Corresponding to the remaining skirts of the characteristic cone are, in the same dual manner, certain other cones which now lie inside the cone just constructed. It is seen therefore, in particular, that this entire dual cone, which we shall call *the cone of characteristics*, intersects the plane $t =$ const always on bounded curves. This follows from the fact that the λ-axis, by the hyperbolic definition, is always inside the skirt of the characteristic cone, i. e., it is not its generatrix.

If some oval of the characteristic cone goes to infinity (in the plane $\lambda = 1$), then the corresponding oval of the characteristic cone in the plane $t = 1$ passes through the origin.

In conclusion, let us tie in the contents of this section with the characteristics introduced earlier in §7. For equations with three independent variables, the characteristic directions at each point will be the two-dimensional planes

$$\lambda t + \mu x + \nu y = \text{const.}$$

One can show here that the coefficients λ, μ, and ν should satisfy the equation of the characteristic cone in the conjugate space, and consequently, the plane itself is tangent to the cone of the characteristics in coordinate space.

As already noted, a perturbation specified in any point of space will propagate along the characteristics. This means that the external skirt of the characteristic cone corresponds to the largest velocity of propagation, and is therefore of greatest interest. Outside this skirt the solution remains unperturbed. Readers

*This is an affine term, since we speak of conjugate spaces.

interested in more details of the role of other parts of the cone of characteristics are refered to [14].

§ 12. Linear equations with discontinuous coefficients. In certain problems a quasilinear system of equations can be modeled in the vicinity of the discontinuity of its solution by means of a linear system with discontinuous coefficients. Outside the discontinuity these coefficients can be considered constant. Equations with discontinuous coefficients were considered in [8], where energy integrals were constructed for it and a uniqueness theorem was proved for different boundary conditions.

In the present section we shall deal with the system

$$\frac{\partial u}{\partial t} + A\frac{\partial u}{\partial x} + B\frac{\partial u}{\partial y} = 0, \tag{12. 1}$$

where the matrices A and B have constant but different values on both sides of the discontinuity. The latter will be set for simplicity at $x = 0$, i. e., we have

$$A = \begin{cases} A^- & \text{for} \quad x < 0, \\ A^+ & \text{for} \quad x > 0, \end{cases} \quad B = \begin{cases} B^- & \text{for} \quad x < 0, \\ B^+ & \text{for} \quad x > 0. \end{cases} \tag{12. 2}$$

The constant matrices A^-, A^+, B^-, and B^+ should insure hyperbolicity of the system (12. 1), each in its own half-space, and can be arbitrary in all other respects.

To obtain a solution of such a system at any point of space, we should specify the boundary condition at $x = 0$, which relates the solutions of the two different systems in the different half-spaces.

For example, one of the simplest conditions is the requirement of continuity of the solution on the line of discontinuity of the coefficients, i. e., the condition

$$u^- = u^+ \quad \text{for } x = 0, \tag{12. 3}$$

where u^- and u^+ are the solutions of the system (12. 1) for $x < 0$ and $x > 0$, respectively.

The equations themselves suggest another boundary condition, which is in a certain sense more natural, and which we shall soon derive. We proceed in a way analogous to that of § 2, where we introduce the concept of the generalized solution. We take the scalar product of equation (12. 1) and an arbitrary smooth finite function $\phi(t, x, y)$ and integrate the resultant expression (which by virtue of the equations vanishes) by parts. We obtain

$$0 = \iiint \left[\frac{\partial u}{\partial t} + A \frac{\partial u}{\partial x} + B \frac{\partial u}{\partial y} \right] \phi \, dt \, dx \, dy =$$

$$= - \iiint u \left[\frac{\partial \phi}{\partial t} + A' \frac{\partial \phi}{\partial x} + B' \frac{\partial \phi}{\partial y} \right] dt \, dx \, dy + \iint_{x=0} (A^- u^- - A^+ u^+) \phi \, dt \, dy ,$$

where A' and B' are the transpose matrices of A and B. We shall, as before, call the function u, which causes the first integral in the left-hand side of this equation to vanish, *the generalized solution of the system* (12.1). It is then necessary to stipulate also that the second integral over the discontinuity plane vanish for any ϕ, i.e., that the following boundary condition be satisfied

$$A^- u^- = A^+ u^+ \quad \text{for} \quad x = 0, \tag{12.4}$$

which we shall call *the natural boundary condition.*

It can be obtained not only for the discontinuity on the line $x = 0$, but for any line in the plane (x, y). Let the tangent at the point of discontinuity considered be defined by the expression

$$\mu x + \nu y = c.$$

We introduce for the time being new spatial coordinates

$$x_1 = \mu x + \nu y - c,$$
$$y_1 = y,$$

in which the system (12.1) assumes the form

$$\frac{\partial u}{\partial t} + (\mu A + \nu B) \frac{\partial u}{\partial x_1} + B \frac{\partial u}{\partial y_1} = 0,$$

and the tangent to the discontinuity line will be, as before, $x_1 = 0$. This means, following (12.4), that the natural boundary condition in the given case will have the form

$$(\mu A^- + \nu B^-) u^- = (\mu A^+ + \nu B^+) u^+. \tag{12.5}$$

We shall deal soon with discontinuities only for $x = 0$ and we shall use boundary conditions (12.3) or (12.4).

The main problem which will interest us now is the following: *under what conditions will the Cauchy problem be correct in the entire space for the system of equations*

$$\frac{\partial u}{\partial t} + A \frac{\partial u}{\partial x} + B \frac{\partial u}{\partial y} = 0$$

with discontinuous coefficients (12.2) *and with one of the foregoing boundary conditions* (12.3) *and* (12.4).

We shall first indicate its solution for two variables, which will be used by us later. In analogy with the investigation of the stability of discontinuity in §8, it is easy to establish that at each point of discontinuity there should arrive from the left and from the right exactly n characteristics (and not $n+1$ as in §8, since it is not necessary to determine here the speed of the discontinuity), and that the relations that are brought by these characteristics should be linearly independent.

We now proceed to three independent variables.

Since the coefficients of the system A and B depend only on x and not on t or y, our problem is invariant with respect to shifts in t and y. This brings to mind a Fourier transformation with respect to these variables. We must recall, however, that the problem is solved with respect to time t only in the half-space $t > 0$, and we are interested in the behavior of the solutions with increasing t, i. e., positive and negative directions of the t-axis are not of equal weight.

We shall therefore confine ourselves to the Fourier transformations only with respect to y. This dictates the following dependence of the solution on y:

$$u(t, x, y) = e^{i\nu y} u(t, x), \tag{12. 6}$$

where ν is a real number, the same for both half-spaces (for otherwise the boundary conditions would depend on y). Substituting (12. 6) in (12. 1), we obtain the following system with two independent variables:

$$\frac{\partial u}{\partial t} + A \frac{\partial u}{\partial x} + i\nu Bu = 0. \tag{12. 7}$$

We next assume that our Cauchy problem is correct and (12. 1), and consequently also (12. 7) has solution bounded in time. This allows us to go to a Laplace transformation in t. We introduce

$$v(x) = \int_{\infty}^{0} u(x, t) e^{-i\lambda t} dt. \tag{12. 8}$$

The integral should exist when $\operatorname{Im} \lambda < 0$. Since under these assumptions

$$\int_0^\infty \frac{\partial u}{\partial t} e^{-i\lambda t} dt = -u(0, x) + i\lambda v(x),$$

we obtain, by multiplying (12. 8) by $e^{-i\lambda t}$ and integrating from 0 to ∞, the following equations for $v(x)$:

$$i\lambda v + A \frac{dv}{dx} + i\nu Bv = u(0, x). \tag{12. 9}$$

Here A and B are discontinuous matrices (12. 2), and $\bar{\bar{u}}(0, x)$ is a known vector function, making up the initial data of the problem. The solution should exist and should be unique for $\operatorname{Im} \lambda < 0$ and any real ν, and should be bounded for all x and satisfy one of the boundary conditions for $x = 0$:

$$v^- = v^+, \tag{12. 3'}$$

$$A^- v^- = A^+ v^+. \tag{12. 4'}$$

The latter are obtained by applying the same Laplace transformation to the boundary conditions (12. 3) and (12. 4) for the functions $u(t, x)$.

In other words, the differential operator

$$i\lambda I + i\nu B + A \frac{d}{dx}$$

with boundary conditions (12. 3′) or (12. 4′) should have an inverse one for real ν and $\operatorname{Im} \lambda < 0$. The inverse operator should be integral and should be completely determined by its kernel, the matrix Green's function, which is constructed, as usual, from solutions of the homogeneous equation

$$i\lambda v + i\nu B v + A \frac{dv}{dx} = 0. \tag{12. 9}$$

Since A and B are constant to the right and to the left of the discontinuity, then on each of the half-axes $x < 0$ and $x > 0$, the solutions of equation (12. 9) are constructed in the well-known manner. Let $\mu_k^-(\lambda, \nu)$ be the roots of the characteristic equation

$$\det | \lambda I + B^- \nu + A^- \mu | = 0.$$

Then for $x < 0$ there exist n linearly independent solutions

$$v = v_-^{(k)} e^{i\mu_k^- x} \quad (k = 1, \cdots, n). \tag{12.10}$$

Analogously, if $\mu_k^+(\lambda, \nu)$ are the roots of the analogous equation to the right, then for $x > 0$ the solutions will be

$$v = v_+^{(k)} e^{i\mu_k^+ x} \quad (k = 1, \cdots, n), \tag{12.10'}$$

where $v_-^{(k)}$ and $v_+^{(k)}$ are constant vectors.

We are interested in the solutions of the problem, which are bounded for all values of x, and therefore in the construction of the Green's function there should participate only the bounded expotentials from (12. 10). We denote by r^- the number of solutions to the left, bounded for $x \to -\infty$, and by r^+ the number of solutions to the right, bounded for $x \to +\infty$. In other words, r^- is the number of the μ_k^- such that $\operatorname{Im} \mu_k^- < 0$, and r^+ is the number of μ_k^+ such that $\operatorname{Im} \mu_k^+ > 0$. The pair of numbers ($r^-$, r^+) will be called the index of the system (12.9) on the discontinuity $x = 0$.

Each of the solutions (12.10) is determined accurate to an arbitrary constant. These constants can be determined only from the boundary condition (12.3′) or (12.4′), which contains n relations. Since the Green's function should be determined uniquely, and from n relations on the discontinuity we can find only n constants, there should participate in the construction of the Green's function exactly n functions (12.10). This means that there should be n of the needed bounded solutions among (12.10), or, in other words, the index of the system should satisfy the condition

$$r^- + r^+ = n. \tag{12.11}$$

If $r^- + r^+ > n$, then, as indicated above, the Green's function will contain more constants than we can determine, i.e., the problem will be solved in a non-unique manner, and if $r^- + r^+ < n$, then we shall not have enough parameters to satisfy the boundary conditions, and the problem will generally speaking not have any solution.

In addition, it is necessary that the system of algebraic equations for the indicated constants not be degenerate. In other words, the n solutions (12.10) selected by us should be linearly independent at $x = 0$, i.e., the vectors $v_-^{(k)}$ and $v_+^{(k)}$ should be linearly independent.

The conditions we obtained for the regularity of the index is for the time being difficult to verify, since the calculation of the index in this form as we have defined it is quite cumbersome. We shall indicate here one of its important properties, which will simplify the matters considerably, We shall prove that the index is determined by the system in a unique manner, i.e., it is independent of λ or ν. Actually, r^- is the number of roots μ_k^- of the equation

$$\operatorname{Det} | \lambda I + \mu A^- + \nu B^- | = 0 \tag{12.12}$$

for $\operatorname{Im} \lambda < 0$, $\operatorname{Im} \nu = 0$, such that $\operatorname{Im} \mu_k^- < 0$. We know that under real λ, μ, and ν this is the equation of the characteristic cone. Therefore, if λ and ν are varied and one of the roots μ_k^- goes from the lower half-plane to the upper one or vice versa, than it should intersect the real axis on the discontinuity. But in this case μ and ν will be real, and consequently, as indicated above, λ should also become real in spite of the condition $\operatorname{Im} \lambda < 0$. This means that the number r^- is the same for all real ν and for all λ from the lower half-plane. One can prove analogously that r^+ is independent of λ or ν.

This leads to an important practical corollary. In the determination of the index of the system one can obtain $\nu = 0$. By virtue of (12.6) this means that we consider solutions which are independent of y, i. e., solutions of the system of equations

$$\frac{\partial u}{\partial t} + A\frac{\partial u}{\partial x} = 0. \tag{12. 13}$$

This means that the problem of the conditions on the boundary can be solved in terms of the one-dimensional problem, in which, as we know, there should arrive at the discontinuity n characteristics, bearing n linearly independent relations. It is obvious that the characteristics of the one-dimensional system (12. 13) form in the plane (x, y, t) planes parallel to the y-axis and tangent to the characteristic cone of the initial system (12. 1).

Thus, *for the correctness of the Cauchy problem it is necessary that there arrive on the plane of the discontinuity* $x = 0$ *from the left and from the right exactly* n *characteristic planes, parallel to the* y*-axis.* These determine the index: to the left there are $n - r^-$, and to the right $n - r^+$. In addition, the relations formulated above, for example, in terms of the vectors v_k^- and v_+^k should be independent of these.

If the latter requirement is not satisfied, then the Green's function cannot be constructed uniquely, i. e., we cannot solve the Cauchy problem for the specified initial conditions. On the contrary, the corresponding homogeneous problem, which is obtained after employing the Laplace transform, will have a non-trivial solution. This corresponds to the following.

We shall seek the solution of the equations

$$\frac{\partial u}{\partial t} + A\frac{\partial u}{\partial x} + B\frac{\partial u}{\partial y} = 0$$

in the form

$$u = e^{i(\lambda t + \nu y)} f(x).$$

Then for $f(x)$ we obtain the equations

$$i\lambda f + i\nu B f + A\frac{df}{dx} = 0,$$

which are identical with (12.9′). Specifying again that the solution be bounded as $x \to \pm\infty$, we find, as above, that the general solution has the form

$$f^-(x) = \sum_{k=1}^{r^-} c_k^- v_-^{(k)} e^{i\mu_k^- x} \qquad \text{for } x < 0,$$
$$f^+(x) = \sum_{k=1}^{r^+} c_k^+ v_+^{(k)} e^{i\mu_k^+ x} \qquad \text{for } x > 0. \tag{12.14}$$

We substitute this in the boundary conditions for $x = 0$, for example, $u^- = u^+$. We then obtain

$$\sum_{k=1}^{r^-} c_k^- v_-^{(k)} = \sum_{k=1}^{r^+} c_k^+ v_+^{(k)}.$$

This vector equation consists of n linear homogeneous equations for n constants c_k^-, c_k^+ (since $r^+ + r^- = n$). Equating the determinant of this system to 0, we obtain the additional equation relating μ_k, λ, and ν. Together with the characteristic equation (12. 2) it determines, accurate to a factor, certain exclusive values of λ, μ_k, and ν at which the solution under consideration holds. We shall call such solutions *the natural oscillations* of the system of equations. They can be of two types:

(1) Exclusive values μ_k are real. Then the solutions $f(x)$ (see (12. 14)) are oscillating, undamped for all x. This corresponds to the natural oscillations of the entire space.

(2) All the μ_k are not real. Then the waves (12. 14) are exponentially attenuated as $|x| \to \infty$, i. e., as the distance from the discontinuity increases, and represent a purely edge effect. Such oscillations are called *surface waves* or *Rayleigh waves.*

So far we have dealt with conditions imposed on the coefficients of the equations, if boundary conditions (12. 3) and (12. 4) are specified. We now formulate the inverse problem.

Assume that certain arbitrary coefficients (12. 2) *are specified and satisfy the condition of hyperbolicity in each half-plane separately. What should be the boundary conditions on the discontinuity, so that the Cauchy problem for the system* (12. 1) *is correct?* The foregoing discussions make it possible to give the following answer: *it is necessary to calculate the index of the system* (r^-, r^+) *and to specify exactly* $r^- + r^+$ *conditions at* $x = 0$.

For example, in the one-dimensional problem with one equation

$$\frac{\partial u}{\partial t} + a\frac{\partial u}{\partial x} = 0,$$

where a has a discontinuity at $x = 0$, one must have 0, 1, or 2 conditions on the discontinuity, depending on the picture of the characteristics in the vicinity of the discontinuity (Figure 8 (a), (b), and (c), respectively). Here the index is (0, 0), (0, 1), and (1, 1).

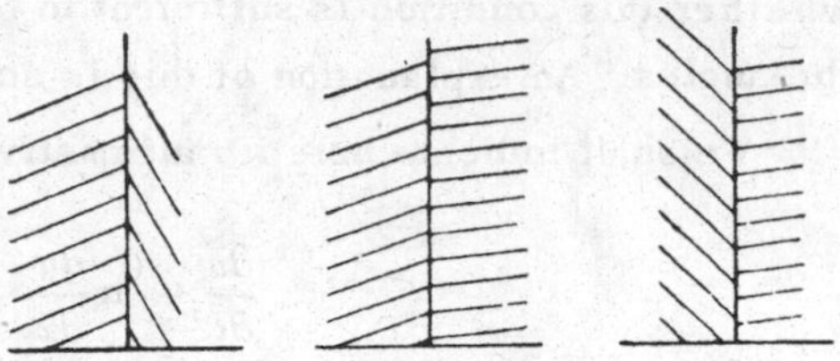

(a) (b) (c)

Figure 8

In general, the behavior of the index and its use in similar problems makes it possible to determine correctly the transfer of conditions from point to point. If these conditions correspond to the sense of the problem, then the problem, as a rule, can be solved and furthermore uniquely. If there is no discontinuity, the the transfer of conditions is automatically correctly carried out with the aid of smooth, or more accurately, differentiable coefficients, and therefore all the theorems of existence and uniqueness usually require differentiability. Without it, the theorems are not true, as can be seen from the example of this very same single equation, where $a(x)$ is a continuous but not differentiable function. When

$$a(x) = \begin{cases} -\sqrt{|x|}, & \text{if } x < 0, \\ \sqrt{x}, & \text{if } x > 0, \end{cases}$$

the problem has not a unique solution, and when

$$a(x) = \begin{cases} \sqrt{|x|}, & \text{if } x < 0, \\ -\sqrt{x}, & \text{if } x > 0 \end{cases}$$

there exists no solution at all, as can be seen from the picture of the characteristics in Figure 9.

This example shows that the index can be introduced for the same purpose in the study of equations with variable coefficients, having singular points, since we are again faced with the problem of the transfer of the conditions through such points.

To conclude this section we shall raise the question of the conditions of the correctness of the system (4.1) in the case of three regions with different values of matrices A and B. These regions are separated by three straight line rays, leaving a single point 0, as shown in Figure 10. On these rays are specified certain boundary conditions. It is obvious that on each ray it is necessary to the correctness of the index, i.e., conditions of type (12.11) but it is not clear whether this condition is sufficient to reconcile all the data at the vertices of the angles. An explanation of this is an interesting unsolved problem.

We shall indicate here its affirmative solution for the case of one equation

$$\frac{\partial u}{\partial t} + a\frac{\partial u}{\partial x} + b\frac{\partial u}{\partial y} = 0,$$

where a, b are constants which assume different values in regions I, II and III (Figure 10).

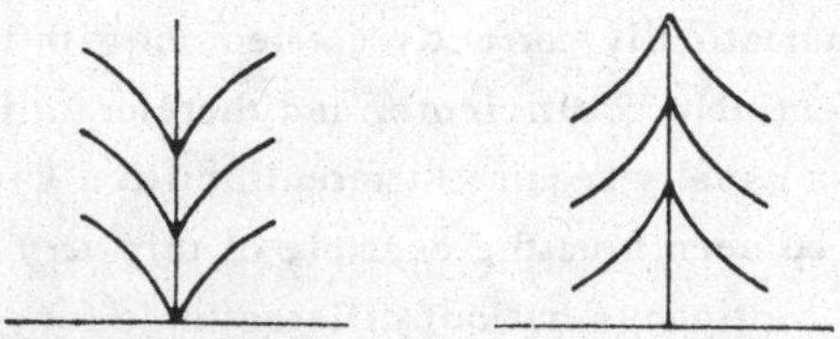

Figure 9

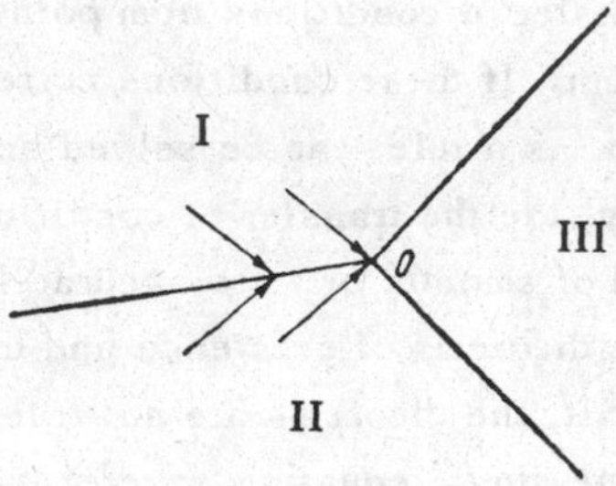

Figure 10

By way of boundary conditions on the rays, we select, for example, the continuity of u. The general solution of this equation is given by an arbitrary function of two variables

$$u = \phi(x - at,\ y - bt),$$

i.e., the solution is constant along the lines $x - at = c_1$, $y - bt = c_2$. The projections on the plane (x, y) of these lines are parallel to the vector (a, b), which is separately defined for each region. In the case when data are brought on the t-axis (at the point 0 in Figure 10) along two lines (for example, as indicated in the diagram by the arrows), the same situation will take place for the entire ray (which separates I and II in the case shown), which contradicts the assumption of the regularity of the index on each ray. This proves that the conditions of regularity of the index on the boundaries of each of two regions are sufficient to make the problem solvable as a whole.

§ 13. Equations of magnetohydrodynamics. This section and the following one will be devoted to an exposition of certain facts, pertaining to magnetohydrodynamics, which can serve as a good illustration of the theory of quasilinear equations in spite of the fact that in itself it does not attract the attention of the mathematicians. This relatively new branch of physics is devoted to the study of electrically-conducting liquid or gas in a magnetic field. The decisive significance in this case attaches to different problems in astrophysics, which have caused the rapid development of magnetohydrodynamics during recent times. It is found that in the investigation of large-scale processes in cosmic space the gas masses contained there should be considered, in spite of their strong rarefaction, as continuous media. Furthermore, they are good conductors.

Mathematically one has to deal with rather complicated equations, which describe the intersection of the moving medium with the magnetic field. The charge particles of the liquids or of the gas move under the action of the field, and in moving they themselves influence the changes in the field. Since the equations of ordinary hydrodynamics should take into account electromagnetic forces, Maxwell's equations for the electromagnetic field must be written with allowance for the currents that are induced as a result of the motion of particles in the field. We shall not stop for a detailed justification and derivation of the equations, and we shall confine ourselves to remarks on their origin and cite the equations themselves, indicating in each case the degree of idealization of the physical phenomena. Readers interested in the subject in greater detail are referred to the book [10] or the survey article [11], where an extensive bibliography is also given.

The basic concepts and equations of hydrodynamics were developed in §§ 4 and 5. Here we shall note corrections necessitated by the magnetic field, and we shall now engage in the field equations. The field is characterized at each point by a certain electric and magnetic field E and H, and these values mean not the fields of the individual particles, but an average over the volumes which are sufficiently large compared with the characteristic dimensions, connected with the particles. E and H should satisfy the following Maxwell's equations:

$$\operatorname{div} H = 0,$$

$$\operatorname{rot} H = \frac{4\pi}{c} j, \tag{13.1}$$

$$\operatorname{rot} E = -\frac{1}{c}\frac{\partial H}{\partial t}.$$

Here j is the vector of the electric current, which by Ohm's law has a value

$$j = \sigma E', \tag{13.2}$$

where σ is the conductivity, assumed constant, and E' is the electric vector in the coordinate system moving together with the conductor, of value

$$E' = E + \frac{1}{c}[vH]; \tag{13.3}$$

v denotes the velocity of motion of the medium, and the constant quantity c denotes the velocity of light.

The first of equation (3. 1) is connected with the absence of sources of magnetic field, while the second and third describe the interaction of the electric and magnetic fields.

To reduce the number of equations, one can eliminate from (13. 1) the vector E. Expressing it with the aid of the second equation (with allowance for (13. 2) and (13. 3)) in terms of H and substituting in the third, we obtain

$$\frac{\partial H}{\partial t} = \operatorname{rot}\,[vH] + \frac{c^2}{4\pi\sigma}\,\Delta H, \tag{13. 4}$$

where $\Delta = \dfrac{\partial^2}{\partial x_1^2} + \dfrac{\partial^2}{\partial x_2^2} + \dfrac{\partial^2}{\partial x_3^2}$ is the Laplace operator. Equation (13. 4) describes the diffusion of the magnetic field. The role of the coefficient of diffusion is played by the quantity $1/\sigma$, which is the reciprocal of the conductance.

Equation (13. 4) in conjunction with Maxwell's first equation must be understood as a solution of the Cauchy problem in the class of functions satisfying the condition $\operatorname{div} H = 0$.

Thus, we have obtained an equation for the magnetic field, containing, true, in addition to H, an unknown function v, which does not enable us to solve it independently of the equations of hydrodynamics. We have established in § 4 that the equations of hydrodynamics correspond to the laws of conservation of mass, momentum, and energy for arbitrary volumes of liquid or gas. Obviously, these laws of conversation should be satisfied also in this case, except that it is necessary to take into account the additional force and energy, due to the presence of the magnetic field. Therefore the continuity equation (4. 1) should be rewritten in general without modification

$$\frac{\partial \rho}{\partial t} + \frac{\partial \rho v_k}{\partial x_k} = 0. \tag{13. 5}$$

The equations corresponding to the change in the velocity will be taken in the form of the conservation law (4. 4):

$$\frac{\partial \rho v_i}{\partial t} + \frac{\partial \Pi_{ik}}{\partial x_k} = 0 \qquad (i = 1, 2, 3), \tag{13. 6}$$

except that in the expression for Π_{ik}, in addition to the momentum transfer, the friction forces, and the viscosity (see (5. 1), (5. 2), and (5. 3)) one must add terms corresponding to the forces with which the field acts on the moving liquid. The force acting on a unit volume of the liquid is

$$f = \frac{1}{c}[jH].$$

Using (13. 1) we can readily obtain from this the expression

$$f = \frac{1}{4\pi}\ [\text{rot}\ H, H],$$

i. e.,

$$4\pi f_1 = -H_2\left[\frac{\partial H_2}{\partial x_1} - \frac{\partial H_1}{\partial x_2}\right] + H_3\left[\frac{\partial H_1}{\partial x_3} - \frac{\partial H_3}{\partial x_1}\right] =$$

$$= H_k\frac{\partial H_i}{\partial x_k} - \frac{1}{2}\frac{\partial}{\partial x_1}H^2 = \frac{\partial}{\partial x_k}\left[H_1 H_k - \frac{1}{2}H^2\delta_{1k}\right];$$

f_2 and f_3 are obtained analogously, and we have

$$f_i = \frac{1}{4\pi}\frac{\partial}{\partial x_k}\left[H_i H_k - \frac{1}{2}H^2\delta_{ik}\right]\quad (i = 1, 2, 3).$$

From this it follows that one must add in Π_{ik} the term

$$-\frac{1}{4\pi}\left[H_i H_k - \frac{1}{2}H^2\delta_{ik}\right],$$

which is called the tensor of Maxwell stresses.

In order not to complicate the picture, we shall assume the liquid ideal, i. e., the coefficients of viscosity and heat conduction are set equal to zero. We then have in (13. 6)

$$\Pi_{ik} = \rho v_i v_k + p\,\delta_{ik} - \frac{1}{4\pi}\left[H_i H_k - \frac{1}{2}H^2\delta_{ik}\right]. \tag{13.7}$$

The law of conservation of energy for an ideal liquid was written in the form (4. 5). We must now add to the energy density $\rho(\epsilon + v^2/2)$ the density of magnetic energy $H^2/8\pi$ (the energy of the electric field can be disregarded, since it is small compared with the magnetic), and the density of the flux of electromagnetic energy is given by the Poynting vector

$$\frac{c}{4\pi}[EH] = \frac{1}{4\pi}[H[vH]] - \frac{c^2}{16\pi^2\sigma}[H\ \text{rot}\ H]$$

(here again we express E in terms of H as in (13. 1)). Thus, the energy equation has the form

$$\frac{\partial}{\partial t}\left[\rho\epsilon + \frac{\rho v^2}{2} + \frac{H^2}{8\pi}\right] + \frac{\partial q_k}{\partial x_k} = 0, \tag{13. 8}$$

where

$$q = \rho v\left[\epsilon + \frac{p}{\rho} + \frac{v^2}{2}\right] + \frac{1}{4\pi}[H[vH]] - \frac{c^2}{16\pi^2\sigma}[H \ \mathrm{rot}\ H] \tag{13. 9}$$

is the vector of the energy flux density.

Finally, we make still another simplification: we assume the conductivity σ to be infinite. As indicated above, this means that we neglect magnetic diffusion, which is characterized by the quantity $1/\sigma$. Such an assumption is natural, once we have decided to neglect the viscosity and heat conduction.

After all the above we write down again the resultant system of equations

$$\begin{gathered} \mathrm{div}\, H = 0, \\ \frac{\partial H}{\partial t} - \mathrm{rot}\,[vH] = 0, \\ \frac{\partial \rho}{\partial t} + \frac{\partial \rho v_k}{\partial x_k} = 0, \\ \frac{\partial \rho v_i}{\partial t} + \frac{\partial \Pi_{ik}}{\partial x_k} = 0, \\ \frac{\partial}{\partial t}\left[\rho\epsilon + \rho\frac{v^2}{2} + \frac{H^2}{8\pi}\right] + \frac{\partial q_k}{\partial x_k} = 0, \end{gathered} \tag{13. 10}$$

where

$$\Pi_{ik} = \rho v_i v_k + p\delta_{ik} - \frac{1}{4\pi}\left[H_i H_k - \frac{1}{2}H^2\delta_{ik}\right],$$

$$q = \rho v\left[\epsilon + \frac{p}{\rho} + \frac{v^2}{2}\right] + \frac{1}{4\pi}[H[vH]].$$

In addition, the thermodynamic quantities ρ, p, and ϵ are related by the equation of state.

§ 14. **Characteristic cones in the equations of magnetohydrodynamics.** In this section we shall clarify the problem of the propagation of small perturbations, i. e., on the characteristics and magnetic hydrodynamics. For this purpose we should, as is said, linearize equation (3. 10), i. e., write linear equations for small perturbations of a certain solution, discarding terms of first and higher orders of smallness, and obtain the corresponding characteristic cone as was done in § 11. By way of the basic, unperturbed solution we take the constants p_0, ρ_0, v_0, and H_0, where v_0 is set equal to 0. (It is obvious that any constant satisfies equation (13. 10).) This corresponds to having a medium with constant density ρ_0 and pressure p_0 at rest, and a constant magnetic field H is applied to this medium.

The last of the equation (13. 10) signifies, as in ordinary hydrodynamics of an ideal medium, that the entropy is constant, and after the perturbation gives us the trivial entropy characteristics. We shall therefore assume the entropy perturbation to be equal to zero and henceforth will not be interested in the equation of energy.

We denote small perturbations of the solution by the same letters, but without the index "zero": ρ, p, v and H, and put $\rho_0 + \rho$, $p_0 + p$, v, $H_0 + H$ in equation (3. 10), leaving only first-order terms. In addition, by varying in the same manner the equation of state and using the constancy of the entropy, we have

$$p = a^2 \rho,$$

where

$$a = \sqrt{\left[\frac{\partial p_0}{\partial \rho_0}\right]_s}$$

is the hydrodynamic velocity of sound. Using this, we obtain the following linear equations for the perturbations of ρ, v and H:

$$\begin{gathered} \operatorname{div} H = 0, \\ \frac{\partial H}{\partial t} = \operatorname{rot}\,[vH_0], \\ \frac{\partial \rho}{\partial t} + \rho_0 \frac{\partial v_k}{\partial x_k} = 0, \\ \frac{\partial v}{\partial t} + \frac{a^2}{\rho_0} \operatorname{grad} \rho = -\frac{1}{4\pi}[H_0 \operatorname{rot} H]. \end{gathered} \tag{14. 1}$$

Furthermore, repeating the arguments of § 11, we shall seek the solution (14. 1) in the form of a plane monochromatic wave

$$\rho = \rho e^{i(kr - \omega t)}, \quad v = v e^{i(kr - \omega t)}, \quad H = H e^{i(kr - \omega t)}. \tag{14. 2}$$

Here now ρ, v, and H denote constant quantities, $r = (x_1, x_2, x_3)$ is the radius vector of the point, $k = (k_1, k_2$ and $k_3)$ the wave vector, which specifies the direction of the wave. At fixed t, the surfaces of the level of our functions will be the planes

$$kr = k_1 x_1 + k_2 x_2 + k_3 x_3 = \text{const.}$$

The velocity of the monochromatic wave (the so-called "phase velocity," unlike the "group velocity" of propagation of the wave packet) is, as can be readily seen, $\omega / |k|$. It must be determined together with the constants ρ, v, and H. Substituting (14. 2) in (14. 1) we obtain the following equations:

$$\begin{aligned} &kH = 0, \\ &-\omega H = [k[vH_0]], \\ &-\omega\rho + \rho_0 kv = 0, \\ &-\omega v + \frac{a^2}{\rho_0}\rho k = -\frac{1}{4\pi\rho_0}[H_0[kH]]. \end{aligned} \tag{14. 3}$$

It follows from the second equation that H is perpendicular to k, as to one of the factors of the vector product, and therefore the first equation is automatically satisfied, and we can disregard it. The remaining equations are linear and homogeneous for ρ, v, and H. The vanishing of the determinant yields an equation for ω.

For convenience in calculations we choose in a special manner a system of spatial coordinates (x_1, x_2, and x_3), and adopt it to the "selected" directions contained in the problem, which are the vectors H_0 and k. Normally, we direct the x_1-axis along k, the x_2-axis is taken in the plane of k and H_0. In these coordinates, as can be readily calculated,

$$[k[vH_0]] = (0, |k|(v_2 H_{01} - v_1 H_{02}), |k| v_3 H_{01}),$$
$$[H_0[kH]] = (|k| H_{02} H_2 - |k| H_{01} H_2, -|k| H_{01} H_3),$$

i. e., equation (14. 3) assumes the form

$$\begin{aligned} &-\omega H_1 = 0, \\ &-\omega H_2 = |k|(H_{01} v_2 - H_{02} v_1), \\ &-\omega H_3 = |k| H_{01} v_3, \\ &-\omega(\rho/\rho_0) + |k| v_1 = 0, \\ &-\omega v_1 + a^2(\rho/\rho_0)|k| = -(1/4\pi\rho_0)|k| H_{02} H_2, \\ &-\omega v_2 = (|k|/4\pi\rho_0) H_{01} H_2, \\ &-\omega v_3 = (|k|/4\pi\rho_0) H_{01} H_3. \end{aligned} \tag{14. 4}$$

Hence $H_1 = 0$, the oscillations of H are transverse, as in a magnetic field in vacuum. The remaining equations break up into two independent systems, and this corresponds to two independent types of oscillations, which can be considered separately. The first system consists of the third and seventh equations and contains only v_3 and H_3. Setting the determinant of these equal to zero, we obtain

$$\omega^2 - \frac{|k|^2 H_{01}^2}{4\pi\rho_0} = 0,$$

i. e., the velocity of the wave $\omega/|k| = H_{01}/\sqrt{4\pi\rho_0}$. Such a solution describes purely electromagnetic oscillations: these do not contain the density of matter, and the velocity of the waves is independent of the hydrodynamic velocity of sound. Obviously, the expression for ω has an invariant meaning

$$\omega^2 = \frac{(kH_0)^2}{4\pi\rho_0}, \tag{14.5}$$

no longer with our special system of coordinates.

The second system consists of the remaining four equations for ρ, v_1, v_2, and H. Its determinant is

$$-(a^2|k|^2 - \omega^2)\left(\omega^2 - \frac{|k|^2 H_{01}^2}{4\pi\rho_0}\right) + \omega^2 \frac{|k|^2 H_{02}^2}{4\pi\rho_0}.$$

Inasmuch as

$$|k|^2 H_{02}^2 = |k|^2 (H_{01}^2 + H_{02}^2) - |k|^2 H_{01}^2 = k^2 H_0^2 - (kH_0)^2 = |[kH_0]|^2,$$

this yields the following equation for ω in invariant form:

$$(a^2 k^2 - \omega^2)\left(\omega^2 - \frac{(kH_0)^2}{4\pi\rho_0}\right) = \omega^2 \frac{|[kH_0]|^2}{4\pi\rho_0}. \tag{14.6}$$

Thus, the characteristic cone in magnetic electrodynamics is a sixth order surface in three-dimensional projective space $(k_1 : k_2 : k_3 : \omega)$, described by equations (14.5) and (14.6). It breaks up into two parallel planes (14.5) and a fourth order surface (14.6). In addition, it can be noted that it is a surface of revolution with axis H_0.

Figure 11 shows its section with a certain axial plane in the space $\omega = 1$, while Figure 12 shows the section, for $t = 1$, of the corresponding dual cone of the characteristics in the space (x_1, x_2, x_3, t). Corresponding to the internal circle on the first drawing is the external circle on the second.

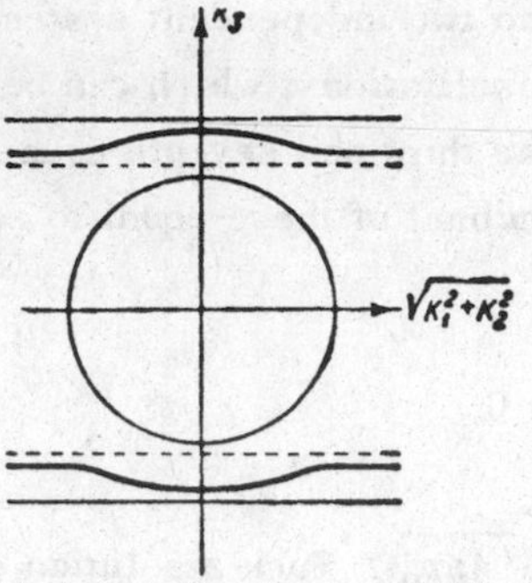

Figure 11

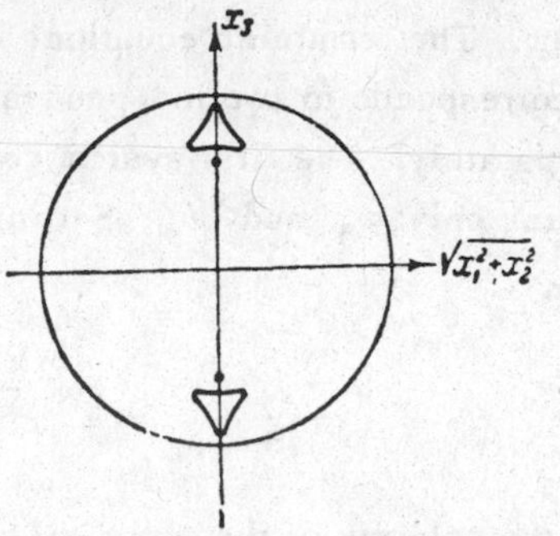

Figure 12

The curves which go into infinity correspond to the curvilinear triangles: in general the points of inflection always correspond to the point of return in the dual figure. The indicated curves have three points of inflection each: two finite and one at ∞. Finally, the two planes correspond simply to two points.

§ 15. Problem of thermal self-ignition. Many interesting problems connected with parabolic quasilinear equations are raised in the theory of ignition and the theory of flame propagation.

Let us consider first the problem of thermal self-ignition of a chemically active mixture of gases in a vessel. Let a certain vessel of arbitrary shape be filled by such a mixture. During the course of the reaction, which proceeds in the vessel, certain components of the mixture are destroyed and heat is consequently liberated.

Under the simplest assumptions, when the rate of the reaction depends only on the temperature T of the mixture and on the concentration n of one of the components of the mixture, which is destroyed during the course of the reaction, the course of the reaction is described by a system of equations of heat conduction and diffusion.

$$\frac{\partial T}{\partial t} = k\Delta T + qf(T, n), \tag{15.1}$$

$$\frac{\partial n}{\partial t} = D\Delta n - \epsilon f(T, n), \tag{15.2}$$

so that $\epsilon f(T, n)$ is the rate of the reaction (ϵ is a dimensional factor, and the function $f(T, n)$ is assumed dimensionless), the thermal effect of the reaction is proportional to q/ϵ, and the coefficients of heat conduction and diffusion are assumed constants. The boundary conditions on the walls of the vessel are taken in the following form: $T = T_0 = \text{const}$ and the normal derivative of the concentration n is 0 (the walls are impermeable).

In principle the reaction can proceed in two ways. In the first one the heat, formed as a result of the reaction, does not have time to be carried away through the walls of the vessel, the mixture is heated, and the rate of the reaction increases, until a shortage of the combustible component of the mixture begins to manifest itself. In the second type of reaction, equilibrium between the amount of heat liberated during the course of the reaction and the amount of heat carried away through the walls of the vessel is rapidly established. In the former case one speaks of self-ignition of the mixtures; the problem is formulated as follows: under what conditions is the second type of reaction established? The formulation of this problem belongs to Ja. B. Zel′dovič and D. A. Frank-Kameneckiĭ (see [15], where further bibliography is available).

In the second type of reaction, in the case of sufficiently small ϵ, one can on a certain interval $(0, 2t_0)$ neglect the burning up of the combustible component, i. e., one can assume $n \equiv n_0 = \text{const}$, where n_0 is the initial concentration of the combustible component. Substituting this relation in (15. 1), we find that on the interval of time $(0, t_0)$ the temperature T will be determined from the equation

$$\frac{\partial T}{\partial t} = k\Delta T + qf(T, n_0). \tag{15.3}$$

At sufficiently large k, the solution of this latter equation, starting with $t = t_0$, will be as close as desired to the solution of the equation

$$k\Delta T + qf(T, n_0) = 0 \tag{15.4}$$

and, consequently, the distribution of the temperature, determined from the system (15. 1) and (15. 2) at $t_0 < 2t < 2t_0$, will be close to this solution, if of course the latter exists.

Thus, if the assumptions made on the coefficients ϵ and k are satisfied (and these hold for a broad class of chemical reactions), the definition of the type of the reaction reduces to the question of the existence of the solution of the Dirichlet problem for equation (15. 4) subject to the condition $T = T_0$ on Γ.

Let us note the "intermediate" characteristic of the asymptote, provided by the solution of this Dirichlet problem: this asymptote is correct not for a very large value of the time, since for such values of the time the burning up of the combustible component of the mixture is essential, and not at too small values of the time, since for such values of the time the initial distribution of the temperature, which can differ considerably from stationary, is essential.

Thus, as ϵ is decreased we obtain a succession of functions, which approximate the limiting function over an ever increasing interval of variation of t.

Obviously, the problem includes two parameters, which have a dimensionality of a length: the parameter $l = \sqrt{k/q\epsilon}$ and the parameter R, the characteristic dimension of the vessel, which will be taken to be its largest radius. We denote their ratio by λ, and, going to dimensionless variables, we reduce the problem to the following.

Determine under what λ there exists a solution of the equation

$$\Delta u + \phi(u) = 0, \tag{15.5}$$

satisfying the condition $u = 0$ on a surface, the equation of which is given in the form $\mathbf{r} = \lambda \mathbf{r}_0$ $(\max|\mathbf{r}_0| = 1$; $\mathbf{r}$ is the radius vector of the point of the surface).

It would be interesting to prove that from the existence of the solution for a certain vessel follows the existence of the solution for vessels that fit into it (and having the same maximum radius).

We shall develop, following D. A. Frank-Kameneckiĭ, the solution of the problem for the case of symmetrical vessels: a vessel with flat walls, a cylindrical vessel, and a spherical vessel. Here we shall take for the function $\phi(u)$ the expression $\phi(u) = 2e^u$, which is obtained from the basic laws of chemical kinetics and certain additional considerations. It would be quite interesting to carry out the general investigation, at least for the case of the function $\phi(u)$ which is qualitatively similar to the function e^u.

In this case equation (15.5) assumes the form

$$\frac{1}{\xi^\nu} \frac{d}{d\xi} \xi^\nu \frac{du}{d\xi} + 2e^u = 0 \qquad \left(\xi = \frac{r}{l}\right) \tag{15.6}$$

($\nu = 0$ for a vessel with plane walls, $\nu = 1$ for a cylindrical vessel, and $\nu = 2$ for a spherical vessel). The condition on the boundary reduces to the form

$$u = 0 \quad \text{for} \quad \xi = \lambda = \frac{R}{l}. \tag{15.7}$$

From the symmetry of the vessel we also have the condition

$$\frac{du}{d\xi} = 0 \quad \text{for} \quad \xi = 0. \tag{15.8}$$

We should ascertain under what values of λ does this problem have a solution and if solutions exist for certain λ, how many are there.

To solve the problem formulated in this manner we note that the equation and the boundary condition (15.8) are invariant with respect to the group of transformations

$$u(\xi, \alpha) = \alpha + u_0(\xi e^{\frac{\alpha}{2}}), \tag{15.9}$$

i.e., if $u_0(\xi)$ satisfies equation (15.6) and the condition (15.8), then for any α it also satisfies equation (15.6) and the condition (15.8). This can be readily verified directly.

Let us assume, for the sake of being definite, that $u_0(\xi)$ satisfies also the condition

$$u_0(0) = 0. \tag{15.10}$$

Then the family of functions (15.9), as can be readily seen, represents all the solutions of equation (15.6), satisfying the condition (15.8), since by choosing α in a suitable manner one can make the function $u(\xi, \alpha)$ assume any specified value at $\xi = 0$.

Let us show now that the family of the functions (15.9) fills only part of the plane $u\xi$. For this purpose we find the envelopes of this family. Differentiating (15.9) with respect to α, we have

$$0 = 1 + u_0(\xi e^{\frac{u}{2}}) \frac{1}{2} \xi e^{\frac{u}{2}},$$

so that the equations of the envelopes assume the form

$$u = u_0 S_0 + 2 \ln \frac{S_0}{\xi}, \tag{15.11}$$

where S_0 is the root of the equation

$$2 + u_0'(S_0) S_0 = 0. \tag{15.12}$$

Thus, to find the envelopes it is necessary to know the function $u_0(\xi)$.

It is now necessary to carry out an analysis separately for plane, cylindrical, and spherical vessels.

In the case of a plane vessel, equation (15.6) has the form

$$\frac{d^2 u}{d \xi^2} + 2e^u = 0; \tag{15.13}$$

it is readily integrated, and we obtain

The first of equation (3.1) is connected with the absence of sources of magnetic field, while the second and third describe the interaction of the electric and magnetic fields.

To reduce the number of equations, one can eliminate from (13.1) the vector E. Expressing it with the aid of the second equation (with allowance for (13.2) and (13.3)) in terms of H and substituting in the third, we obtain

$$\frac{\partial H}{\partial t} = \operatorname{rot}[vH] + \frac{c^2}{4\pi\sigma}\Delta H, \tag{13.4}$$

where $\Delta = \frac{\partial^2}{\partial x_1^2} + \frac{\partial^2}{\partial x_2^2} + \frac{\partial^2}{\partial x_3^2}$ is the Laplace operator. Equation (13.4) describes the diffusion of the magnetic field. The role of the coefficient of diffusion is played by the quantity $1/\sigma$, which is the reciprocal of the conductance.

Equation (13.4) in conjunction with Maxwell's first equation must be understood as a solution of the Cauchy problem in the class of functions satisfying the condition $\operatorname{div} H = 0$.

Thus, we have obtained an equation for the magnetic field, containing, true, in addition to H, an unknown function v, which does not enable us to solve it independently of the equations of hydrodynamics. We have established in §4 that the equations of hydrodynamics correspond to the laws of conservation of mass, momentum, and energy for arbitrary volumes of liquid or gas. Obviously, these laws of conversation should be satisfied also in this case, except that it is necessary to take into account the additional force and energy, due to the presence of the magnetic field. Therefore the continuity equation (4.1) should be rewritten in general without modification

$$\frac{\partial \rho}{\partial t} + \frac{\partial \rho v_k}{\partial x_k} = 0. \tag{13.5}$$

The equations corresponding to the change in the velocity will be taken in the form of the conservation law (4.4):

$$\frac{\partial \rho v_i}{\partial t} + \frac{\partial \Pi_{ik}}{\partial x_k} = 0 \qquad (i = 1, 2, 3), \tag{13.6}$$

except that in the expression for Π_{ik}, in addition to the momentum transfer, the friction forces, and the viscosity (see (5.1), (5.2), and (5.3)) one must add terms corresponding to the forces with which the field acts on the moving liquid. The force acting on a unit volume of the liquid is

$$\frac{1}{\xi^2}\frac{d}{d\xi}\xi^2\frac{du}{d\xi} + 2e^u = 0. \tag{15.16}$$

Such an equation, in connection with the problem of isothermal equilibrium of gas spheres, was considered by Èmden (1907). In particular Èmden calculated by numerical integration the function $u_0(\xi)$.

Putting $u = v - 2\ln\xi$, $\eta = \ln\xi$, we obtain

$$\frac{d^2v}{d\eta^2} + \frac{dv}{d\eta} - 2 + 2e^v = 0, \tag{15.17}$$

which does not contain the independent variable. Putting furthermore $\psi = dv/d\eta$ and assuming $e^v = z$ as the independent variable, equation (15.17) is reduced to the form

$$\frac{d\psi}{dz} = \frac{2 - \psi - 2z}{\psi z}. \tag{15.18}$$

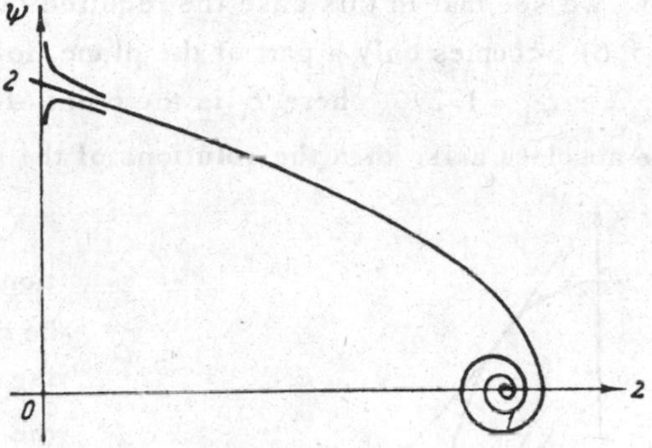

Figure 14

An investigation of this equation in the right half of the plane z (corresponding to the physically meaningful real values of v) yields the following picture of the integral curves (Figure 14). Leaving the singular point (0, 2) (of the saddle-point type), the separating integral curve corresponds to the curves for which $(du/d\xi)_{\xi=0} = 0$. In fact, we have

$$\frac{du}{d\xi} = \frac{du}{d\eta}\cdot\frac{1}{\xi} = \left[\frac{dv}{d\eta} - 2\right]\frac{1}{\xi} = \frac{\psi - 2}{\xi}. \tag{15.19}$$

On the other hand, we have $\psi = dv/d\eta = dz\cdot\xi/z\cdot d\xi$, i.e., we have $(1/\psi)\,dz/z \sim d\xi/\xi$ meaning that when $z \sim 0$ and $\psi \sim 2$ we have $\xi \sim c\sqrt{z}$, since the point (0, 2) of the integral curve corresponds to $\xi = 0$ and $du/d\xi = 0$.

The separating curve crosses the abscissa axis many times near the focus (1, 0). At each point of intersection we have $\psi = dv/d\eta = 0$, from which we obtain

$$\frac{du}{d\eta} - 2 = 0, \text{ i.e., } \xi\frac{du}{d\xi} + 2 = 0, \tag{15.20}$$

so that for each point of intersection equation (15.12) is satisfied. Let $S_0^{(i)}$ be the abscissa corresponding to the ith point of intersection. (The values of $S_0^{(i)}$ can be obtained by using the Èmden calculations.) The family of curves (15.9) has in this case an infinite number of parallel envelopes

$$u = 2\ln\frac{S_0^{(i)}}{\xi} + u_0(S_0^{(i)}),$$

which converges to the curve $u = \ln(1/\xi)$ and which look as shown in Figure 15. We see that in this case the required family of integral curves of equation (15.6) occupies only a part of the plane, located above the first envelope. Thus, if $\lambda > \bar{\xi}_1 = 1.29$, where ξ is the point of intersection of the first envelope with the abscissa axis, then the solutions of the problem do not exist.

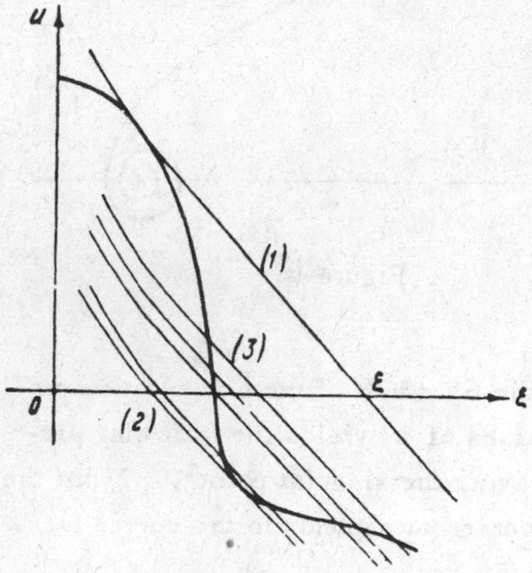

Figure 15

However, if $\lambda \leq \bar{\xi}_1$, then the construction of the set of stationary solutions will be more complicated than in the preceding cases. Namely, if $\lambda = \lambda_0 = \bar{\xi}$, then the solution will be correct if $\bar{\xi}_3 < \lambda < \lambda_0$, where $\bar{\xi}_3$ is the point of intersection of the third envelope with the abscissa axis, then there will be two solutions, and at $\lambda = \xi_3$ there will be three solutions, etc; finally, when $\lambda = \bar{\xi}_\infty = 1$ there will be an infinite number of solutions. The number of solutions then begins to decrease, so that when $\xi_{2n+2} > \lambda > \bar{\xi}_{2n}$ there are $2n + 1$ solutions, and when $\lambda = \xi_{2n}$ there are only $2n$. Finally, when $0 < \lambda < \bar{\xi}_2$, there is only one solution.

§16. **Problem of establishment of a chemical process.** Another formulation of the problem is of interest, when the reaction proceeds with formation of an intermediate product, an active center.

In this problem, along with the initial matter, one considers also two intermediate products: one of them, the "active centers," is capable of diffusing and its diffusion causes the propagation of the reaction over the volume. With this, the active centers are capable of multiplying in the presence of another intermediate substance – the catalyst.

Probably, such relations are possible in a mixture of H_2 and O_2 in a certain range of temperatures:*

$$H + O_2 = HO_2,$$
$$H + HO_2 + 2H_2 = 3H + H_2O + OH.$$

Actually, the first process takes place only in a triple collision

$$(H + O_2 + M = HO_2 + M)$$

and the third particle absorbs the energy.

The second process can be represented as the result of several sucessive reactions. We do not consider the further fate of OH. The diffusion of HO_2 and O_2 is considered slow compared with the diffusion of the H atoms. The theory developed here is similar in structure to the relations between the centers and the temperature in a large number of reactions, if one replaces the word catalyst by the word heat.

The problem is formulated as follows: is the introduced portion of active centers sufficient to cause a reaction over the entire volume of chemically-active mixture, or will the beginning of the reaction have a local character and will not cover the entire volume of the mixture. We develop below the solution of this problem, belonging to Ja. B. Zel'dovič and the author. We denote the concentration of the atomic hydrogen, oxygen, and the intermediate product HO_2 by a, g, and h, respectively. The equations that describe their variations are obtained from the foregoing equations of chemical reactions in the following manner.

The rate of combustion of the oxygen, in the first-order assumption of the reaction, is proportional to the product ag; we shall assume it equal to $-lag$, where l is a constant. The weight of creation of the autocatalyst HO_2 represents the algebraic sum of the velocities of the production of the autocatalyst as a result of the first reaction (equal to lag) and its annihilation in the course of the second reaction, equal to $-kah$, under the assumption that this reaction is also of first order. Finally, the rate of variation of the concentration of the atoms of hydrogen is made up of the velocity $D\Delta a(\Delta = \partial^2/\partial x^2 + \partial^2/\partial y^2 + \partial^2/\partial z^2)$ and their accumulation due to the diffusion, the velocity $-lag$ of their annihilation during the course of the first reaction, and the velocity $2kah$ of production of new atoms of hydrogen due to the reaction with the autocatalyst (from the equation of the second reaction it follows that the speed of formation of H is twice as large as the speed of consumption of HO_2: hence the coefficient $2k$).

Thus, we have the following differential equations for a, g, and h:

*See, for example, N. N. Semenov [17].

$$\frac{\partial a}{\partial t} = D\Delta a - lag + 2kah,$$
$$\frac{\partial g}{\partial t} = - lag, \qquad (16.1)$$
$$\frac{\partial h}{\partial t} = lag - kah,$$

where D, l, and k are constant coefficients.

During the initial instant of time there is a certain concentration of oxygen molecules, which we assume to be unity, and a certain initial distribution of concentration of atoms of hydrogen, which is natural to specify by means of a certain finite (i. e., non-vanishing only in a certain finite region) function $n_0(x, y, z)$. This function characterizes the initial induction of the reaction. There are no molecules of the autocatalyst HO_2 at the initial instant, i. e., the initial data for the equation (16. 1) are

$$\text{for } t = 0, \quad a = a_0(x, y, z), \quad g = 1, \quad h = 0. \qquad (16.2)$$

In general, if the initial distribution of the concentration of the active substance is a finite function, then it is natural to assume that the process stops, if at $t \to \infty$ the solution at infinitely distant points remains equal to the initial data. To the contrary, if the solution at infinity has limits as $t \to \infty$, different from the initial values, we shall say that the process has gone to a stationary mode. Here it is necessary to specify in addition that the solution with the initial data $g = 1$, $a = 0$, and $h = 0$ be stationary and a stable solution of the equations, i. e., the solution with slightly perturbed initial data, should differ little from the unperturbed solution.

It is obvious that the initial data (16. 2) satisfy this condition. Actually, the constants $a = 0$, $g = 1$, $h = 0$ (and in general any constant, provided $a = 0$) satisfy (16. 1) and the stability of such a solution is relatively simple to prove.

We must now find the conditions that make it possible to determine, from the function $a_0(x, y, z)$, whether our reactions will start or will be damped out. This is conveniently done in the following manner. Assume that damping takes place, i. e., that as a result of the initial induction a certain amount of matter has reacted, and the process has stopped. We seek the resultant stationary distribution of the concentration, i. e., the form of the combusted "crater" (although the term is not very applicable to gases). The conditions under which this problem has a solution will be indeed the conditions of the attenuation of the process.

We can show that if induction takes place, then at each point (x, y, z) of the function $g(x, y, z, t)$ tends to 0 as $t \to \infty$ (i. e., the oxygen will burn out completely) and, thus, if at least at one point $\lim g(x, y, z, t) \neq 0$, the process damps out.

From the second equation (16.1) subject to condition (16.2) we have

$$g = e^{-l \int_0^t a(x, y, z, t) dt}. \tag{16.3}$$

From this we obtain for the attenuation case of interest to us

$$u(x, y, z) = \int_0^\infty a(x, y, z, t)\, dt < \infty \tag{16.4}$$

at least in one point of space. This function determines the total number of hydrogen atoms participating in the reaction at a given point, and $e^{-lu} = g(x, y, z, \infty)$ is a concentration of the oxygen which remains after the reaction, i.e., characterizes the sought form of the "crater."

Using (16.3), we can integrate the third equation of (16.1). With allowance of (16.2) we obtain

$$h = \frac{1}{\kappa - 1} (g - g^\kappa), \tag{16.5}$$

where $\kappa = k/l$. The first equation of (16.1) will be rewritten, using the remaining ones, in the following manner:

$$\frac{\partial a}{\partial t} = D \Delta a - 2 \frac{\partial h}{\partial t} - \frac{\partial g}{\partial t}. \tag{16.6}$$

We denote

$$p(x, y, z, t) = \int_0^t a(x, y, z, t)\, dt.$$

Integrating (16.6) from 0 to t and using (16.2), (16.3), and (16.5), we obtain

$$a(x, y, z, t) - a_0(x, y, z) = D\Delta p - \frac{\kappa+1}{\kappa-1} e^{-lp} + \frac{2}{\kappa-1} e^{-kp} + 1. \tag{16.7}$$

We have seen that during damping $p(x, y, z, t)$ tends to a finite function $u(x, y, z)$ as $t \to \infty$, and as a result of the conversion of the integral (16.4), $a(x, y, z, t)$ tends to 0. Going to the limit as $t \to \infty$ in (16.7) we obtain the basic equation for $u(x, y, z)$:

$$D\Delta u - \frac{\kappa+1}{\kappa-1} e^{-lu} + \frac{2}{\kappa-1} e^{-ku} + a_0(x, y, z) + 1 = 0.$$

As mentioned above, during damping, $u(x, y, z)$ should vanish at infinitely remote points. This is indeed the boundary condition.

We go over to dimensionless variables

$$x\sqrt{\frac{l}{D}} \to x, \quad y\sqrt{\frac{l}{D}} \to y, \quad z\sqrt{\frac{l}{D}} \to z, \quad lu \to v.$$

Then $v(x, y, z)$ will be the solution of the equation

$$\Delta v - \frac{\kappa+1}{\kappa-1} e^{-v} + \frac{2}{\kappa-1} e^{-\kappa v} + a_0(x, y, z) + 1 = 0 \tag{16. 8}$$

with the following boundary condition: $v(x, y, z) \to 0$ as $r = \sqrt{x^2 + y^2 + z^2} \to \infty$. The problem of interest to us now assumes the following form: for what functions $a_0(x, y, z)$ does this problem have a solution, and when does it not.

We shall indicate the solution in only two simplest cases.

Let $a_0(x, y, z) = \text{const}$. Then, obviously, the solution should not depend on x, y, z, i. e., it is also a constant, and is determined by the equation*

$$f(v) = a_0, \tag{16. 9}$$

where

$$f(v) = \frac{\kappa+1}{\kappa-1} e^{-v} - \frac{2}{\kappa-1} e^{-\kappa v} - 1. \tag{16. 10}$$

An elementary investigation of the function $f(v)$ leads to the following results: $f(0) = 0$, $f'(0) = 1$, $\lim\limits_{v \to \infty} f(v) = 1$; there exists a $v_0 > 0$ such that $f(v) > 0$ when $v < v_0$, $f(v) < 0$ when $v > v_0$, and finally $f(v)$ has a unique positive maximum $f(v_1)$ at the point $v_1 < v_0$ (Figure 16).

It is seen from this that equation (16. 9) has two solutions if $a_0 < f(v_1) = \max\limits_{v \geq 0} f(v)$, one solution if $a_0 = f(v_1)$, and no solution if $a_0 > f(v_1)$, i. e., $a_0 = f(v_1)$ is the minimal value for a_0, starting with which the reaction will proceed further and go to the end. At smaller a_0 it will damp out. It can be shown that of the two solutions obtained from the intersection of the curve $f(v)$ with a horizontal straight line of height a_0, the smaller one would always have a physical meaning, i. e., that lying on the rising part of $f(v)$, shown by means of a heavy line in Figure 16. This follows at least from the fact that from the meaning of the problem, the solution tends to zero as $a \to 0$.

The second case which we shall consider is a one-dimensional problem, so that $a_0(x, y, z)$ and $v(x, y, z)$ depend only on one variable x. Then equation (16. 8) with allowance for (16. 10) becomes

---

* The same situation will obtain if we neglect diffusion.

$$\frac{d^2 v}{dx^2} - f(v) + a_0(x) = 0. \tag{16.11}$$

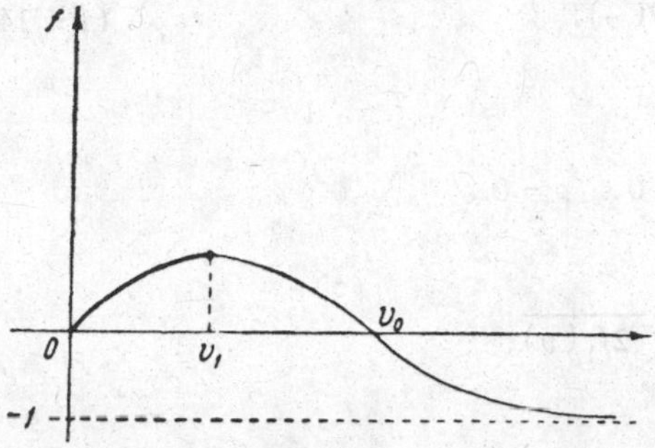

Figure 16

We shall consider the function $a_0(x)$ not to be merely finite, but concentrated in a single point $x = 0$, i. e., we put

$$a_0(x) = a_0 \delta(x),$$

where $\delta(x)$ is the Dirac δ-function and a_0 is a constant.

With this, the principal terms of (16.11) near the point $x = 0$ will be

$$\frac{d^2 v}{dx^2} + a_0 \delta(x) = 0.$$

Integrating this equation with respect to x, we find that the derivative dv/dx experiences at $x = 0$ a discontinuity, equal to $-a_0$.

This denotes that equation (16.11) can be solved without $a_0(x)$ separately for $x > 0$ and for $x < 0$, and the resultant solutions can be "glued together" at $x = 0$ with the aid of the condition

$$v'(+0) - v'(-0) = -a_0.$$

We notice further that by virtue of the evenness of the function $a_0(x)$ selected by us and of the invariance of equation (16.11) under substitution of $-x$ for x, the solution should also be an even function of x, i. e., the process is symmetrical with respect to $x = 0$. Hence

$$v'(+0) = -v'(-0).$$

This enables us to solve the problem only on a half-line, for example $x > 0$, where one needs the equation

$$\frac{d^2 v}{dx^2} - f(v) = 0, \tag{16.12}$$

with boundary conditions

$$v'(0) = -\frac{a_0}{2}, \quad \lim_{x \to \infty} v(x) = 0. \tag{16.13}$$

Going over from the variables x, v to the variables v, $p = dv/dx$, we reduce the order of the equation, namely, we have instead of (16. 12)

$$p \frac{dp}{dv} = f(v). \qquad (16.14)$$

The boundary condition is taken to be:*

$$\text{for } v = 0, \quad p = 0.$$

Integrating, we obtain

$$p = \pm\sqrt{2f_1(v)},$$

where

$$f_1(v) = \int_0^v f(v)\,dv.$$

From the properties of $f(v)$ it follows that $f_1(v)$ has a single positive zero, v_2, and $f_1(v) > 0$ when $0 < v < v_2$, $f_1(v) < 0$ when $v > v_2$. When $v = 0$, $f_1(v)$ has a double zero. We are therefore interested in the real solutions of the problem, and v should change only in the segment $0 \le v \le v_2$.

Integrating

$$\frac{dv}{dx} = \pm\sqrt{2f_1(v)},$$

we obtain**

$$x = -\int_{v_2}^{v} \frac{dv}{\sqrt{2f_1(v)}} + C.$$

When $v = 0$ we should have $x = +\infty$ and therefore we have taken the minus sign in front of the root (since $v < v_2$).

From this we can readily see the character of the dependence of x on v, which, when inverted, yields the function $v(x)$ sought. Figure 17 shows its curve for $C = 0$. The presence of an arbitrary constant C in the latter equation leads us, to obtain the necessary solution, to shift the curve parellel to the x-axis. The final position is fixed in this case by the boundary condition $v'(0) = -a_0/2$.

*Since there are no grounds to consider in the problem the function $v(x)$ to be oscillating at infinity, we assume that as $x \to \infty$, v tends to 0 together with the derivative $p = dv/dx$.

**We note that this integral converges at the point $v = v_2$ and diverges as $v \to 0$.

The curve on Figure 17 has a point of inflection, corresponding to $v = v_0$ (see Figure 16), and each value of the derivative $v'(x)$ between the null and the value at the point of the inflection is assumed by the function twice. The consideration already employed, that as $a \to 0$, $v(x)$ should tend to 0 at each point, brings to mind that the part of the curve below the point of inflection has a physical meaning: its points should coincide with the v-axis in accordance with the boundary condition. This part of the curve corresponds to the section of the increasing function $f_1(v)$ and to the section where $f(v)$ is positive, i.e., to values $v < v_0$, as can be seen from Figures 16 and 17.

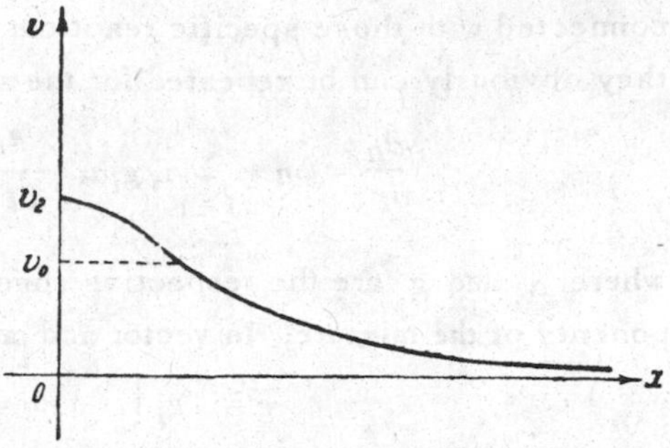

Figure 17

It follows from the foregoing that the problem posed has a solution if $a_0/2$ does not exceed $\max|v'(x)|$, i.e., since

$$\frac{dv}{dx} = -\sqrt{2f_1(v)},$$

if

$$\frac{a_0}{2} \leq \max\sqrt{2f_1(v)} = \sqrt{2\int_0^{v_0} f(v)\,dv}.$$

In this case the process will attenuate. If a_0 is greater than the foregoing limit, a reaction will be induced.

In a similar manner it is possible to solve the problem in the case of a spherical symmetry of the function $a_0(x, y, z)$, i.e., when it depends only on $z = \sqrt{x^2 + y^2 + z^2}$; it is merely necessary to go over to spherical coordinates, and the corresponding differential equation must be solved numerically.

Thus, we have explained by means of example that there exists every time a certain critical function $a_0(x, y, z)$ such that if the initial portion of the atomic hydrogen does not exceed it, the solution of the problem of the form of the burned-out "crater" exists, and if it exceeds it, the solution does not exist, and then the reaction is induced. The problem with the critical function a_0 has one solution, and with a_0 less than critical it has two solutions. It would be interesting to ascertain the meaning of the second, discarded parasitic solution.

The foregoing considerations have quite general character, i. e., they are not connected with those specific reactions with which we just dealt. For example, they obviously can be repeated for the system of equations

$$\frac{\partial a}{\partial t} = \Delta a + \sum_{i=1}^{m} \alpha_i g_i a, \quad \frac{\partial g_i}{\partial t} = \sum_{k=1}^{m} a_{ik} g_k a \qquad (i = 1, \cdots, m),$$

where a and g are the respective concentrations of the active and inactive components of the mixture. In vector and matrix notation

$$g = \{g_i\}, \qquad \alpha = \{\alpha_i\}, \qquad A = (a_{ik})$$

we have

$$\frac{\partial u}{\partial t} = \Delta a + (\alpha, g)\, a,$$

$$\frac{\partial g}{\partial t} = aAg.$$

Hence

$$g = e^{\int_0^t a\,dt A} \cdot g(0), \qquad ag = A^{-1} \frac{\partial g}{\partial t},$$

$$\frac{\partial a}{\partial t} = \Delta a + \left\{\alpha, A^{-1} \frac{\partial g}{\partial t}\right\}.$$

For the form of the burned-up "crater"

$$v(x, y, z) = \int_0^\infty a(x, y, z, t)\, dt$$

we obtain an equation

$$\Delta v + (\alpha, A^{-1} g) - (\alpha, A^{-1} g(0)) + a_0 (x, y, z) = 0,$$

where

$$g = e^{vA} g(0).$$

If a linear transformation is made on g such that the matrix A becomes diagonal with eigenvalues λ_i, the latter equation assumes the form

$$\Delta v + \sum_{i=1}^{m} \frac{\beta_i}{\lambda_i} (e^{\lambda_i v} - 1)\, g(0) + a_0 (x, y, z) = 0.$$

It would be interesting to see to what more general problems this method can be extended. For example, what can be said concerning the solution of the equation

$$\Delta v - f(v) + a_0(x, y, z) = 0$$

with more or less arbitrary function $f(v)$? Which of the equations of this form correspond to certain chemical reactions?

§17. Normal flame propagation. Imagine an infinitely long tube, filled with a combustible mixture, on one of the points of which is initiated the combustion process. The flame can propagate along a tube, and within a certain time the rate of propagation of the flame becomes constant. We consider this phenomenon with a specific example, where the speed of combustion is determined by the temperature T of the mixture and the concentration a of the matter that becomes annihilated during the course of the reaction

$$\begin{aligned} \rho \frac{\partial a}{\partial t} &= \frac{\partial}{\partial x}\left[D_\rho \frac{\partial a}{\partial x}\right] - \Phi(a, T), \\ \rho c \frac{\partial T}{\partial t} &= \frac{\partial}{\partial x}\left[\lambda \frac{\partial T}{\partial x}\right] + h\,\Phi(a, T). \end{aligned} \tag{17.1}$$

Here D, λ, and c are the coefficients of diffusion, heat conduction, and heat capacity, respectively, h is the thermal effect of the reaction, ρ the density of the mixture, $\Phi(a, T)$ is the speed of the reaction. The quantities D, λ, c, k, and ρ will be considered constant; the function $\Phi(a, T)$ will be considered specified. The initial data have the form shown in Figure 18, i. e., to the left a zero temperature and a maximum concentration $\bar{a}$, to the right after combustion of the entire substance, a maximum temperature (taken to be unity) and a zero concentration of the combustible component of the mixture.

The system (17.1) is still difficult to investigate. However, for the case when the combustible component of the mixture and the product of the reaction have nearly equal molecular weights, the following relation holds:

$$D\rho c = \lambda.$$

From this, obviously, it follows that the enthalpy $H = ha + cT$ satisfies the ordinary equation of heat conduction

$$\frac{\partial H}{\partial t} = \frac{\partial}{\partial x}\left[D\frac{\partial H}{\partial x}\right]. \tag{17.2}$$

For simplicity we choose initial conditions such that when $t = 0$ we have $H = \text{const}$. Then by virtue of (17.2), H will be equal to that constant at any t, i. e.,

$$ha + cT = \text{const} \tag{17.3}$$

and a is found to be a linear known function of T. Substituting (17. 3) in (17. 1), we obtain one equation for one unknown function T

$$\frac{\partial T}{\partial t} = \frac{\lambda}{\rho c} \frac{\partial^2 T}{\partial x^2} + F(T),$$

where $F(T) = (h/\rho c)\Phi(a(T), T)$. Choosing, in addition $x\sqrt{\rho c/\lambda}$ as the new spatial variable and denoting it again by x, we obtain finally

$$\frac{\partial T}{\partial t} = \frac{\partial^2 T}{\partial x^2} + F(T). \tag{17. 4}$$

The values $T = 1$ and $T = 0$ (right and left values of T in the initial data) should be solutions of the equation. Consequently, $F(T)$ should vanish at the points 0 and 1. In the remaining points of the segment (0, 1), $F(T)$ should be non-negative from the physical meaning of the problem. In addition, $T = 1$ and $T = 0$ should be stable solutions. Let us find the conditions under which this requirement will be satisfied. We impose on the solution $T = 1$ a perturbation $\tau(t)$.

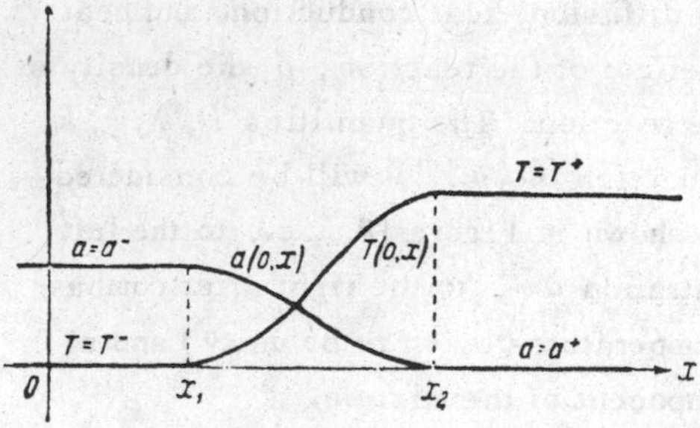

Figure 18

Substituting $1 + \tau(t)$ in (17.4), we obtain, leaving the principal terms,

$$\frac{d\tau}{dt} = F'(1)\tau.$$

This means that in order for the perturbation not to increase with time, it is necessary that the condition $F'(1) \leq 0$ be satisfied. This condition is obviously satisfied if $F(T)$ is differentiable, since $F(1) = 0$, $F(T)$ is non-negative. In order for the stability requirement to be satisfied also for $T = 0$, it is necessary to put $F(T) = 0$ on the small interval $0 \leq T \leq \epsilon$. Thus, the function $F(T)$ should have the form shown in Figure 19.

Let us show that there exists a unique stationary solution(i. e., a wave traveling with a constant velocity).* We shall seek it in the form of a wave traveling to the left with a velocity ω,

$$T(x, t) = T(x + \omega t).$$

Substituting this in (17. 4), we obtain

$$\omega \frac{dT}{d\xi} = \frac{d^2 T}{d\xi^2} + F(T),$$

* We do so following Ja. B. Zel'dovič (see [16]).

where $\xi = x + \omega t$. We introduce $p = dT/d\xi$ as the unknown function of T. Then the order of the equation is reduced,

$$p\frac{dp}{dT} - \omega p + F(T) = 0. \tag{17. 5}$$

The solution of this equation should satisfy the condition

$$p = 0 \quad \text{for } T = 0 \quad \text{and for } T = 1. \tag{17. 6}$$

Since equation (17. 5) is of first order, and there are two conditions, there exists in general no solution, with the exception perhaps of certain exclusive values of the parameter ω which are of interest to us.

The integral curves of equation (17. 5) behave, as shown in Figure 20. From the singular point (the saddle (1, 0)) there emerge two solutions of the problem,

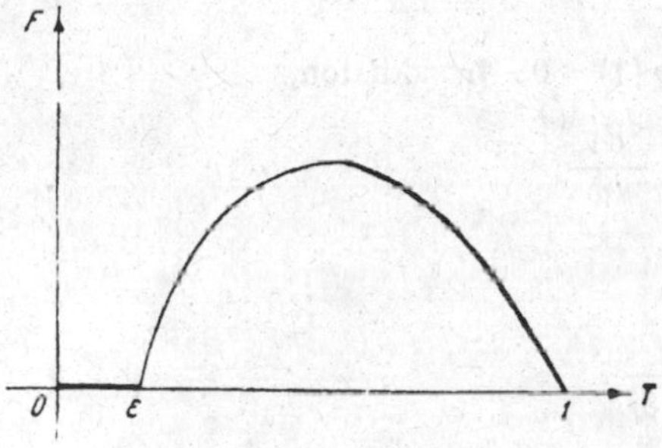

Figure 19

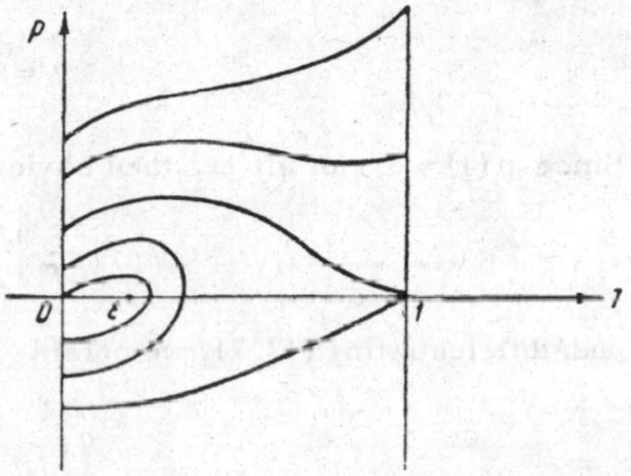

Figure 20

one in the upper half-plane and the other in the lower. The slopes of these curves at the singular point can be readily obtained, for example, by expanding the solution in a Taylor series in the vicinity of $T = 1$. These are the roots of the quadratic equation

$$p_1^2 - \omega p_1 + F_1 = 0, \quad \text{where } F_1 = \left.\frac{dF}{dT}\right|_{T=1}, \tag{17. 7}$$

i. e.,

$$p_1 = \frac{\omega}{2} \pm \sqrt{\left(\frac{\omega}{2}\right)^2 - F_1}.$$

Inasmuch as $F_1 \leq 0$ (see Figure 19), it follows that

$$p_1^{(1)} \geq \omega, \quad p_1^{(2)} \leq 0. \tag{17. 8}$$

The first root corresponds to the lower solution, the second to the upper one.

When $\omega = 0$, equation (17.5) can be readily integrated, and we obtain

$$p^2 = -2\int_0^T F(T)\,dT.$$

Extracting the square root, we obtain the two solutions just indicated. In particular,

$$p(0) = \pm\sqrt{2\int_0^1 F(T)\,dT} \neq 0. \qquad (17.9)$$

We now show that as ω increases both of the obtained values of $p(0)$ diminish without limit. It will follow from this that the lower solution will never pass through the point (0, 0), and the upper one will pass at a certain value $\omega > 0$. We introduce the derivative of our solution (so far it does not matter whether it is the upper one or the lower one) with respect to ω:

$$q = \frac{\partial p(T, \omega)}{\partial \omega}.$$

Since $p(1) = 0$, for all ω, then obviously $q(1) = 0$. In addition,

$$q_1 = \left.\frac{dq}{dT}\right|_{T=1} = \frac{dp_1}{d\omega},$$

and, differentiating (17.7), we obtain

$$q_1 = \frac{p_1}{2p_1 - \omega}.$$

As a result of (17.8), $q_1 > 0$ for both solutions, i.e., the curves $q(T)$ fall from the point (1, 0) into the lower half-plane, and $q(T)$ is negative near this point. It cannot pass through zero for the following reasons. The equation for q is obtained by differentiating (17.5)

$$\frac{dq}{dT} = \frac{F(T)}{p^2}\,q + 1.$$

If somewhere $q = 0$, then at this point $dq/dT = 1$, which is impossible upon going from the lower half-plane to the upper one in moving from right to left. This means that $q(T)$ is always negative.

Finally, when $0 \leq T \leq \epsilon$ (see Figure 19), $dq/dT = 1$, since $F(T) = 0$. It follows from this that $q(T)$ descends downwards, when T approaches 0, at least on ϵ, i.e.,

$$q(0) \leq -\epsilon \qquad (17.10)$$

for both solutions. But $q(0) = dp(0)/d\omega$, and we have from (17.9) and (17.10)

$$p^{(1)}(0, \omega) \leq -\sqrt{2\int_0^1 F(T)\,dT - \epsilon\omega}, \tag{17.11}$$

$$p^{(2)}(0, \omega) \leq \sqrt{2\int_0^1 F(T)\,dT - \epsilon\omega}. \tag{17.12}$$

Since $\epsilon > 0$ is fixed, and independent of ω, then $p(0, \omega)$ diminishes without limit, and, as already noted above, there follows from this the existence of a stationary solution of equation (17.5). In addition, it follows from (17.10) that the solution is unique: actually, $p(0, \omega)$ changes monotonically and therefore cannot pass through zero twice.

That the solution goes into a stationary mode for a somewhat different formulation of the problem has been demonstrated in [12]. Undoubedly this fact holds also in our case.

We give here one simple and elegant consideration of the stability of the stationary solution $T(x + \omega t)$ under small perturbations (see [13]).

Let the perturbed solution have the form

$$T = T_0(\xi) + \tau(x, t),$$

where $T_0(\xi)$ is the unperturbed solution, $\xi = x + \omega t$, and $\tau(x, t)$ is a small perturbation. Substituting this in (17.6) and retaining only linear terms, we obtain

$$\frac{\partial \tau}{\partial t} = \frac{\partial^2 \tau}{\partial x^2} + F'(T_0)\,\tau.$$

If we go over from the variables (x, t) to (ξ, t), we obtain the equation

$$\frac{\partial \tau}{\partial t} = \frac{\partial^2 \tau}{\partial \xi^2} - \omega \frac{\partial \tau}{\partial \xi} + F'(T_0)\,\tau, \tag{17.13}$$

the coefficients of which depend only on ξ. This makes it possible to employ the Fourier method, i. e., to seek a solution in the form

$$\tau(\xi, t) = e^{\lambda t}\,\tau(\xi). \tag{17.14}$$

Then $\tau(\xi)$ should be a solution of the problem on the eigenvalues for the equation

$$\frac{d^2\tau}{d\xi^2} - \omega\frac{d\tau}{d\xi} + F'(T_0)\,\tau = \lambda\tau$$

with boundary conditions $\tau \to 0$ as $\xi \to \pm\infty$. It is necessary to show that the problem does not have positive eigenvalues, since they would correspond to increasing $\tau(\xi; t)$.

It is known that the maximum eigenvalue corresponds to the eigenfunction that has no zeroes. This function can be found in the following manner. We have already noted that the solutions of the equations are, in addition to $T_0(\xi)$, its shifts. This means that a small shift is a particular case of the perturbed solution, and their difference is one of the solutions τ of the linear problem (17.13). This difference is independent of t, i.e., (17.14) takes place with $\lambda = 0$, and has no zeroes, since $T_0(\xi)$ is monotonic ($p = dT_0/d\xi > 0$; see Figure 20). Thus, the difference in the shifts $T_0(\xi)$ is indeed the eigenfunction without zeroes, and $\lambda = 0$ is the maximum eigenvalue, which proves the stability.

In conclusion we note an interesting problem, the solution of which would include the particular cases treated in the present section, as well as the question raised in the end of §9. Namely: *ascertain the behavior of the solutions of the Cauchy problem for the system of equations*

$$\frac{\partial u}{\partial t} = k\frac{\partial^2 u}{\partial x^2} + \frac{\partial F_1(u)}{\partial x} + F_2(u)$$

with initial data of the type shown in Figure 18, *as* $t \to \infty$.

Supplement

In this supplement we give several remarks belonging to K. I. Babenko and the author and published in greater detail in [8].

We turn to the problem of the correctness and uniqueness of the Cauchy problem for the system of linear hyperbolic equations. Without analyzing this question in detail (the simplest and most convincing proof belongs to L. Gårding) we wish to analyze here the algebraic basis of these problems: the construction of the so-called energy integrals for hyperbolic systems of equations. The energy integrals were first constructed in general form by I. G. Petrovskiĭ, I. Leray, and L. G. Gårding. We consider a hyperbolic system of equations

$$\frac{\partial u}{\partial t} = A\left[\frac{\partial}{\partial x_1}, \cdots, \frac{\partial}{\partial x_m}\right]u,$$

where u is the vector with coordinates $(u_1, \cdots, u_n)$ and the elements $a_{ik}(\partial/\partial x_1, \cdots, \partial/\partial x_n)$ of the matrix A are polynomials of $\partial/\partial x_1, \cdots, \partial/\partial x_n$. More briefly, we shall write the system in the form

$$\frac{\partial u}{\partial t} = A(p)u, \tag{1}$$

where

$$p = (p_1, \cdots, p_m) \equiv \left(\frac{\partial}{\partial x_1}, \cdots, \frac{\partial}{\partial x_m} \right).$$

It is assumed that the matrix $A(i\xi)$ for any vector $\xi = (\xi_1, \cdots, \xi_n)$ has purely imaginary eigenvalues, and the eigenvectors form a basis. Our purpose is to construct a matrix $B(p)$ of differential operators, satisfying the following conditions:

(1) $B(i\xi)$ is a positive definite Hermitian matrix.

(2) $B(i\xi)\,A(i\xi) + [B(i\xi)\,A(i\xi)]^* = 0.$

Such a matrix $B(p)$ will be called the energy integral. This name is justified by the fact that, as will be proved,

$$\int_{-\infty}^{+\infty} (B(p)u, u)\,dx \geq 0$$

for any function u and for any solution u

$$\int_{-\infty}^{+\infty} (Bu, u)\,dx = \text{const.}$$

By virtue of condition (1), the matrix $B(p)$ will be a positive definite Hermitian operator in the space of the vector functions u, the coordinates of which are infinitely differentiable finite functions

$$B(p)u = \int_{-\infty}^{\infty} B(i\xi)\tilde{u}(\xi)e^{i\xi x}\,d\xi, \quad \text{where } \tilde{u}(\xi) = \frac{1}{(2\pi)^m}\int_{-\infty}^{\infty} u(x)e^{i\xi x}\,dx. \tag{2}$$

In order to construct the matrix $B(i\xi)$, satisfying all the foregoing conditions, it is enough to note that from condition (2) it follows that the matrix $B(i\xi)$ converts the vectors $A(i\xi)$ in the eigenvectors of the matrix $A^*(i\xi)$. Therefore, if $\eta_1, \eta_2, \cdots, \eta_n$ are the eigenvectors of the matrix $A^*(i\xi)$, $\eta_e = (\eta_{e1}, \eta_{e2}, \ldots, \eta_{en})$ and $\eta = (\bar{\eta}_{ke})$, then the matrix $B(i\xi)$ should have the form $B(i\xi) = \eta^*\eta$. It follows therefore directly that condition (1) is satisfied. Furthermore, since

$$A^*(i\xi)\;B(i\xi) = \left(\sum_{e=1}^{n} \bar{\lambda}_e \eta_{ei}\bar{\eta}_{ek}\right), \quad BA = \left(\sum_{e=1}^{n} \lambda_e \eta_{ei}\bar{\eta}_{ek}\right),$$

where λ_e are the eigenvalues of the matrix $A(i\xi)$, then condition (2) follows from the fact that λ_e are pure imaginary. We choose now a normalization of the eigenvectors η_k in such a way, that the elements of the matrix $B(i\xi)$ will be integral rational functions of ξ.

In fact, if we use as the coordinates of the vector η_e the cofactors of the elements of a certain row of the matrix $A^*(i\xi) = \overline{\lambda}_e E$, we obtain

$$\eta_{ek} = \sum_{r=0}^{n-1} \overline{\lambda}^r{}_e \alpha_{rk}(\xi) \quad (k = 1, 2, \cdots, n), \tag{3}$$

where $\alpha_{rk}(\xi)$ are integral rational functions of ξ, since the elements of the matrix $A^*(i\xi)$ depend rationally on ξ. We note that not all the coordinates of the vector η_e vanish, since the rank of the matrix $A^*(i\xi) - \overline{\lambda}_e E$ is $n-1$. Therefore, on the basis of (3), the element b_{ik} of the matrix $B(i\xi)$ will be

$$b_{ik} = \sum_{r=1}^{n} \sum_{r=0}^{n-1} \overline{\lambda}_e \alpha_{ri}(\xi) \sum_{s=0}^{n-1} \overline{\lambda_e^s \alpha_{sk} \xi} = \sum_{r,\, s=0}^{n-1} (-1)^s \alpha_{ri} \overline{\alpha_{sk}} \sum_{e=1}^{n} \lambda_e^{-r+s}.$$

But by Newton's formula $\sum_{e=1}^{n} \overline{\lambda}_e^{r+s}$ is rationally expressed in terms of the elements of the matrix $A^*(i\xi)$. Therefore b_{ik} is an integral rational function of ξ. Thus, the energy integral sought has been constructed. By virtue of the construction of the matrix $B(i\xi)$, if μ is a solution of the system (1), then

$$\frac{d}{dt} \int_{-\infty}^{\infty} (Bu, u)\, dx = \int_{-\infty}^{\infty} (BAu, u) dx + \int_{-\infty}^{\infty} (Bu, Au) dx = 0.$$

Thus, *if the system* (1) *is hyperbolic, then there exists such a positive definite Hermitian matrix B, the elements of which are differential operators and for which the relation*

$$\frac{d}{dt} \int_{-\infty}^{\infty} (Bu, u)\, dx = 0 \tag{4}$$

holds for any solution of equation (1).

For hyperbolic systems with variable coefficients one can construct an energy integral and its construction will be the basis of the energy integral obtained above.

We note that if there exists a hyperbolic system with variable coefficients, and also with junior terms, then there exists an energy integral, for which the following relation holds:

$$\frac{d}{dt} \int_{-\infty}^{\infty} (Bu, u)\, dx \le C \int_{-\infty}^{\infty} (Bu, u) dx, \tag{5}$$

if the coefficients of the systems are sufficiently smooth.

It would be interesting to construct energy integrals for equations with discontinuous coefficients. Unfortunately, we can do this only for two variables. It goes without saying that the condition of the possibility of constructing such an integral is closely related with the concept, introduced in §12, of the index of a system of equations. Assume that we have a hyperbolic system

$$\frac{\partial u}{\partial t} + \frac{\partial (Au)}{\partial x} = \sigma, \tag{6}$$

where $A = A(x, t)$ experiences a discontinuity along a certain smooth curve $x = \phi(t)$. We shall denote the right-hand and left-hand value on the discontinuity with the functions by u^- and u^+, respectively. The condition on the discontinuity will be taken in the form

$$A^+u^+ - A^-u^- = D(u^+ - u^-), \tag{7}$$

where $D = d\phi/dt$. Let us consider first the case when the discontinuity is on the line $t = 0$, and the matrix A is constant on the left and on the right of the line $t = 0$. We put $A = A^-$ when $x < 0$ and $A = A^+$ when $x > 0$. The eigenvalues of the matrices A^- and A^+ will be denoted by λ^- and λ^+. The characteristics corresponding to the eigenvalues $\lambda^- > 0$ and $\lambda^+ < 0$ will be called the incoming ones, and let there be r incoming characteristics to the left of the discontinuity and s to the right of the discontinuity. The index of the discontinuity line will be called, as earlier in §12, the two numbers (r, s). It is clear that for uniqueness of the solution of the Cauchy problem it is necessary that the relation $r + s \geq n$ be satisfied. Considering that this relation is satisfied, we construct the energy integral of the system (6). The energy integral is specified by the formula $\int_{-\infty}^{\infty} (Bu, u)dx$, where B is a positive definite Hermitian matrix, for which the following relation holds:

$$A' B = BA. \tag{8}$$

Here the prime denotes the transition to the transposed matrix. If the matrix A is discontinuous, the matrix B will also be discontinuous. The construction of the matrix B will be carried out for each of the regions $x > 0$ and $x < 0$. We consider the region $x > 0$. Let $\lambda_1^+, \lambda_2^+, \cdots, \lambda_n^+$ and $\eta_1^+, \eta_2^+, \cdots, \eta_n^+$ be the eigenvalues and the eigenvectors of the matrix $(A^+)'$. We put

$$B^+ = \left(\sum_{e=1}^{n} \eta_{ei}^+ \eta_{ek}^+ \right), \tag{9}$$

where η_{ei} are the coordinates of the eigenvector η_e. By virtue of the definition, relation (8) holds and $(Bu, u) \geq 0$. The matrix B^- is defined analogously. We consider

$$E(u; t) = \int_{-\infty}^{\infty} (Bu, u)dx = \int_{-\infty}^{0} (B^-u, u)dx + \int_{0}^{\infty} (B^+u, u)\,dx.$$

If the vector u is the solution of the system (6), then

$$\frac{dE(u;\,t)}{dt} = -\int_{-\infty}^{\infty}\left(B\frac{\partial(Au)}{\partial x}\,u\right)dx - \int_{-\infty}^{\infty}\left(Bu,\frac{\partial(Au)}{\partial x}\right)dx.$$

Integrating by parts under the sign of the second integral and using (8), we obtain

$$\frac{dE(u;\,t)}{dt} = (B^-u^-,\,A^-u^-) - (B^+u^+,\,A^+u^+). \tag{10}$$

But from relation (7), considering that $D = 0$, we have $A^+,\,u^+ = A^-,\,u^- = f$. Therefore (10) can be rewritten

$$\frac{dE(u;\,t)}{dt} = -(B^+(A^-)^{-1}f,\,f) + (B^+(A^+)^{-1}f,\,f). \tag{11}$$

But by virtue of (9), $(BA^{-1}f,f) = \sum_{e=1}^{n}\frac{1}{\lambda_e}\left(\sum_{i=1}^{n}\eta_{ei}f_i\right)^2$, where f_i are the coordinates of the vector f, and B and A can assume values B^+, A^+, and B^-, A^-.

Taking into consideration the last relations, we obtain

$$\frac{dE(u;\,t)}{dt} = -\sum_{e=1}^{n}\frac{1}{\lambda_e^-}\left(\sum_{i=1}^{n}\eta_{ei}^-f_i\right)^2 + \sum_{e=1}^{n}\frac{1}{\lambda_e^+}\left(\sum_{i=1}^{n}\eta_{ei}^+f_i\right)^2. \tag{12}$$

We assume that the eigenvectors, corresponding to the incoming characteristics, form a basis. We break up all the terms of the right-hand side of (12) into two groups. In the first group we include all the terms corresponding to the incoming characteristics, and in the second group we include all the terms corresponding to the outgoing characteristics. Then the components of the first group, by virtue of the assumed linear independence, give a non-degenerate quadratic form. Therefore the normalization of the eigenvectors should make the right-hand side of relation (12) non-positive. Thus

$$\frac{dE(u;\,t)}{dt} \le 0. \tag{13}$$

We have thus shown that *if the eigenvectors corresponding to the incoming characteristics form a basis, then for equation* (1) *there exists an energy integral* $E(u,t)$ *such that* $E(u,t) \le E(u;0)$.

If the matrix A is not constant on both sides of the discontinuity, then by using the same construction, we obtain instead of inequality (13) (assuming that the elements of the matrix A have bounded derivatives on both sides of the discontinuity) the following inequality: $dE(u;t)/dt \le CE(u;t)$. The case of an arbitrary discontinuity $x = \phi(t)$ reduces to the foregoing simple change of variables.

BIBLIOGRAPHY

[1] O. A. Oleĭnik, *Discontinuous solutions of non-linear differential equations,* Uspehi Mat. Nauk 12 (1957), no. 3(75), 3–73. (Russian)

[2] A. M. Il'in and O. A. Oleĭnik, *Behavior of solutions of the Cauchy problem for certain quasilinear equations for unbounded increase of the time,* Dokl. Akad. Nauk SSSR 120 (1958), 25–28. (Russian)

[3] A. N. Tihonov and A. A. Samarskiĭ, *On discontinuous solutions of a quasilinear equation of first order,* Dokl. Akad. Nauk SSSR 99 (1954), 27–30. (Russian)

[4] S. K. Godunov, *On uniqueness of the solution of the hydrodynamic equation,* Mat. Sb. (N. S.) 40 (82) (1956), 467–478. (Russian)

[5] O. A. Oleĭnik, *On the uniqueness of the generalized solution of the Cauchy problem for a non-linear system of equations occurring in mechanics,* Uspehi Mat. Nauk 12 (1957), no. 6 (78), 169–176. (Russian)

[6] O. A. Ladyženskaja, *On the construction of discontinuous solutions of quasilinear hyperbolic equations as limits of solutions of the corresponding parabolic equations when the "coefficient of viscosity" tends towards zero,* Dokl. Akad. Nauk SSSR 111 (1956), 291–294. (Russian)

[7] P. D. Lax, *Hyperbolic systems of conservation laws.* II, Comm. Pure Appl. Math. 10 (1957), 537–566.

[8] K. I. Babenko and I. M. Gel'fand, *Notes on hyperbolic systems,* Nauč. Dokl. Vysš. Školy. Fiz. - Mat. Nauk 1958, no. 1, 12–18. (Russian)

[9] L. D. Landau and E. M. Lifshitz, *The mechanics of a continuous media,* GITTL, Moscow, 1953. (Russian)

[10] ———, *Electrodynamics of continuous media*, GITTL, Moscow, 1957. (Russian)

[11] S. I. Syrovatskiĭ, *Magnetic hydrodynamics,* Uspehi Fiz. Nauk 62 (1957), 247. (Russian)

[12] A. M. Kolmogorov, I. G. Petrovskiĭ, and N. S. Piskunov, *Investigation of the equation of diffusion, connected with the increase in the amount of matter, and its application to one biological problem,* Bull. Moskov. Gos. Univ. A 1 (1937), no. 6, 1–26. (Russian)

[13] G. I. Barenblatt and Ja. B. Zel'dovič, *On stability of flame-propagation,* Prikl. Mat. Meh. 21 (1957), 856–859. (Russian)

[14] V. A. Borovikov, *The fundamental solution of a linear partial differential equation with constant coefficients*, Dokl. Akad. Nauk SSSR **119** (1958), 407–410. (Russian)

[15] D. A. Frank-Kamanckiĭ, *Diffusion and heat transfer in chemical kinetics*, Moscow, 1947. (Russian)

[16] Ja. B. Zel'dovič, *On the theory of flame propagation*, J. Physical Chem. 22 (1948), 27–48.

[17] N. N. Semenov, *Some problems of chemical kinetics and reactivity*, Izdat Akad. Nauk SSSR, Moscow, 1958 (Russian); English transl., Princeton Univ. Press, Princeton, N. J., 1959.

Translated by:
Office of Technical Services
U. S. Department of Commerce

9.

(with I. I. Piatetski-Shapiro)

On a theorem of Poincaré

Dokl. Akad. Nauk SSSR **127** (3) (1959) 490–493. Zbl. **107**: 171

The following theorem was proved by A. Poincaré. Let g be a one-to-one continuous mapping of the unit circle onto itself. Write it as $g(\varphi)=\varphi+F(\varphi)$, where $F(\varphi)$ is a continuous periodic function of φ, $0\leqq\varphi\leqq 2\pi$. Then the limit

$$\lim_{n\to\infty}\frac{F(\varphi)+F(g(\varphi))+\ldots+F(g^{n-1}(\varphi))}{n}=2\pi\mu$$

exists and does not depend on φ. This limit is called "the mean rotation number".

This theorem was found by Poincaré in connection with his study of differential equations on a torus. Namely, let C denote a meridian of a torus. The curve beginning in an arbitrary point of the circle C meets this circle in the first time at some another point. Thus, to each dynamical system on a torus corresponds a continuous one-to-one mapping of the circle onto itself. The number μ defined above shows that, on average, to one rotation along the latitude there correspond μ rotations along the longitude.

In this note we give a generalization of Poincaré's theorem to the case of dynamical systems on arbitrary compact smooth manifolds.

Let M be a compact differentiable manifold on which a dynamical system is defined, i.e. an one-parameter group $x\to x_t$ with an invariant measure μ. To this dynamical system is associated a vector field called the field of directions: $l(x)=dx_t/dt$.

To formulate the condition on $l(x)$ in a more convenient way we assume that M is supplied with a Riemannian metric. This does not restrict the generality because a Riemannian metric may be defined on any differentiable manifold.

We shall assume that the following conditions hold.

1) The field of directions is integrable, i.e., the vector $l(x)$ exists for a.e. $x\in M$, and

$$\int_M |l(x)|\,d\mu<\infty; \tag{1}$$

here $|l(x)|$ is the length of a vector with respect to the Riemannian metric on M.

2) Our dynamical system is ergodic, i.e., the measure of any invariant set is equal to 0 or 1.

We shall define an analogue of the "Poincaré mean rotation numbers". Denote by $\gamma_1, \ldots, \gamma_p$ a basis of the (first) integral homology group of M. Let x be an arbitrary point of M; we suppose that the trajectory beginning at x comes back the infinite number of times into a fixed simply connected neighbourhood U of x.

For such a trajectory we can define an analogue of the "mean rotation numbers" as follows. Let $t_1 < t_2 < \ldots (t_k \to \infty)$ be a sequence of values of the parameter t such that x_t belongs to U. We connect x and x_{t_k} by an arbitrary path contained in U. Clearly, the homology class of the resulting cycle $\gamma(t_k)$ does not depend on the choice of a connecting path.

We decompose $\gamma(t_k)$ with respect to the basis of the group of integral cycles:

$$\gamma(t_k) = m_1(t_k)\gamma_1 + \ldots + m_p(t_k)\gamma_p.$$

We call the limit

$$\lim_{k \to \infty} \frac{m_i(t_k)}{t_k} = \lambda_i \tag{2}$$

(if it exists) the mean rotation frequency with respect to γ_i. We shall prove that the limit (2) exists and equals the same number for a.e. trajectories.

Let $\omega_i = \sum a_{ik} dx^k$ denote the closed one-dimensional differential form such that

$$\int_{\gamma_i} \omega_i = 1, \qquad \int_{\gamma_j} \omega_i = 0 \quad (i \neq j).$$

The existence of such a form follows from the De Rham's theorem. We have

$$m_i(t_k) = \int_x^{x(t_k)} \omega_i + O(1)$$

where the integration is along the trajectory x_t, $0 \leqq t \leqq t_k$.

To prove that the limit (2) exists we introduce the auxiliary function

$$\varphi_i(x) = \lim_{t \to 0} \frac{1}{t} \int_x^{x_t} \omega_i. \tag{3}$$

By condition (1), $\varphi_i(x)$ is integrable.

According to the Birkhoff's ergodic theorem, the limit

$$\lim_{t \to \infty} \frac{1}{t} \int_x^{x_t} \omega_i = \lim_{t \to \infty} \frac{1}{t} \int_0^t \varphi_i(x_t) dt = \int_M \varphi_i(x) d\mu \tag{4}$$

exists for almost all $x \in M$. It follows that the limits (2) exist.

Thus, the following theorem is proved:

Theorem. *Let $\gamma_1, \ldots, \gamma_p$ be a basis of the integral homology group of a compact manifold M, which is equipped with ergodic dynamical system. Then for almost all trajectories the mean rotation frequencies λ_i $(i = 1, \ldots, p)$ exist and are computed by the formula* (4). *The numbers λ_i up to integral unimodular transformation are invariants under diffeomorphisms of M.*

Now we give some examples.

1. M is an n-dimensional torus. Consider the dynamical system on M defined by the differential equations $dx_i/dt=\alpha_i$, $i=1, \ldots, n$. We assume that our torus is obtained by the identification of opposite sides of the hypercube $0 \leqq x_i \leqq 1$, $i=1, \ldots, n$. We choose a basis of one-dimensional cycles as follows: γ_k is the circle $0 \leqq x_k \leqq 1$, $1 \leqq k \leqq n$. It is easy to verify that $\lambda_i=\alpha_i$ $(i=1, \ldots, n)$.

2. Let G be the group of real matrices order two, and Γ a discrete subgroup of G without elements of finite order, such that the quotient space $M=G/\Gamma$ is compact. It is known that M is a homogeneous space under the action of G; we write down the action of G on M as $x \to xg$. With every one-parameter subgroup of G is associated a dynamical system on M: $x_t-xg(t)$. In such a form this dynamical system was first studied in the paper by I.M. Gelfand and S.V. Fomin [2]. One can consider M as the space of linear elements of a certain surface F of constant negative curvature (cf. [2]). The first homology group of M has dimension $2p$, where p is the genus of F.

There are two kinds of non-conjugate one-parameter subgroups in G:

$$g_1(t)=\begin{pmatrix} e^t & 0 \\ 0 & e^{-t} \end{pmatrix}, \quad g_2(t)=\begin{pmatrix} 1 & t \\ 0 & 1 \end{pmatrix}$$

We shall compute the "frequencies" λ_i for the dynamical system corresponding to these subgroups. We let

$$g=\begin{pmatrix} g_{11} & g_{12} \\ g_{21} & g_{22} \end{pmatrix}$$

We introduce the parameters (cf. [2]) $\omega_1=g_{11}-ig_{12}$, $\omega_2=g_{21}-ig_{22}$. We set $\tau=\omega_1/\omega_2$, $\theta=\arg\omega_2$. Clearly, $\operatorname{Im}\tau=1/|\omega_2|^2>0$, and θ is defined modulo 2π. The parameters τ and θ uniquely determine $g \in G$.

The manifold M is obtained by the identification of the points (cf. [2]) (τ, θ) and $\left(\frac{\gamma_{11}\tau+\gamma_{12}}{\gamma_{21}\tau+\gamma_{22}}, \theta+\arg(\gamma_{21}\tau+\gamma_{22})\right)$, where $\gamma=\begin{pmatrix} \gamma_{11} & \gamma_{12} \\ \gamma_{21} & \gamma_{22} \end{pmatrix} \in \Gamma$.

We can choose a basis of closed one-dimensional differential forms in the form

$$f_1(\tau)\,d\tau, \ldots, f_p(\tau)\,d\tau, \quad \overline{f_1(\tau)}\,d\bar{\tau}, \ldots, \overline{f_p(\tau)}\,d\bar{\tau},$$

where p is the genus of F (the $f_k(\tau)$ are analytic functions of τ). It is easy to verify that the $\varphi_i(\tau, \theta)$ for the first dynamical system have the form

$$\begin{aligned} \varphi_i(\tau, \theta) &= -2if_i(\tau)\,y e^{-2\theta i}, \quad 1 \leqq i \leqq p \\ \varphi_i(\tau, \theta) &= 2i\overline{f_i(\tau)}\,y e^{2\theta i}, \quad p+1 \leqq i \leqq 2p. \end{aligned} \tag{5}$$

The formulae (4) and (5) immediately imply that all frequencies $\lambda_i=0$. Similar expressions for the φ_i can be also obtained for the second dynamical system. They again imply that all $\lambda_i=0$.

These results can be made more precise, namely one can show that $m_i(t, g) = O(t^{2/3})$ for almost every trajectory of the first dynamical system, and $m_i(t, g) = O(\ln t)$ for every trajectory of the second system.

We note that in the case of the first dynamical system it is impossible to obtain non-trivial estimates for all trajectories simultaneously. Indeed, choose an initial point $g_0 \in M$ so that $g_0 g(t_0) g_0^{-1} = \gamma_0 \in \Gamma$. Then the trajectory beginning at g_0 is closed, and it is clear that it is homologically non-trivial for a suitable choice of γ_0.

This example shows the complex behaviour of the function $m_i(t, g)$. Note that a similar situation occurs in the problem of estimating of trigonometrical sums, and so, in this problem we have $m_i(t) \sim ct$ $(c \neq 0)$.

The case of the second dynamical system is simpler. Namely, one can show that the trajectory x_t $(0 \leqq t \leqq T)$ is homological to a path of length $O(\ln T)$. which implies our result.

The estimate $m_i(t, g) = O(t^{2/3})$ (for a.e. g) is easily derived from the following asymptotic formula:

$$\int_M |m_i(t, g)|^2 \, dg \sim ct, \tag{6}$$

where g is the initial point of the trajectory, and c is a constant.

We shall give a brief derivation of (6). Let H denote the Hilbert space of all square integrable functions $f(g)$ on M. We have the unitary representation of G in H, namely, to each $g_0 \in G$ corresponds the operator $T_{g_0}(T_{g_0} f(g) = f(g g_0))$. One can show [3] that H is a countable sum of the mutually orthogonal subspaces H_k, $k = 1, 2, \ldots$ such that the operators T_g induce on each H_k an irreducible representation of G.

It turns out that for fixed i the function $\varphi_i(g)$ belongs to an irreducible representation which is isomorphic to the discrete series representation with signature 2 [2]. This representation has the following canonical realization. Denote by A the set of all holomorphic square integrable functions on the unit disk $|z| < 1$. Define the operators U_g of the representation as follows:

$$U_g f(z) = f(gz) g'(z),$$

where* $gz = (\alpha z + \beta)(\bar{\beta} z + \bar{\alpha})^{-1}$.

Under the natural isomorphism with the canonical realization the function $\varphi_i(g)$ turns into a constant, which implies at once that $m_i(t, g)$ is the image of the function

$$\int_0^T U_{g_1(t)} f(z) \, dt = \frac{\tanh t}{1 - z \tanh t} \qquad (f(z) = \text{const}).$$

The correspondence is unitary, and therefore, we have the identity

$$\int_M |m(t, g)|^2 \, dg = \int_{|z|<1} \left| \frac{\tanh t}{1 - z \tanh t} \right|^2 dx\,dy \sim ct \quad \text{as } t \to \infty.$$

---

* The group of real matrices is realized as the group of fractional linear transformations of the unit disk.

The formula (6) is proved. Evidently, it shows that the measure of the set of all g such that $|m(t,g)| > t^{1/2+\varepsilon}$ $(\varepsilon > 0)$ (t is fixed) tends to zero as $t \to \infty$.

Added in proof. After the submission of the present paper the authors become aware of the article [4] where a theorem close to our theorem on mean rotation frequencies is proved. However, the most interesting result of our note (see Example 2) has no counterpart in [4].

References

1. A. Poincaré.
2. Gelfand, I.M., Fomin, S.V.: Usp. Mat. Nauk **7** (1) (1952) 118.
3. Gelfand, I.M., Piatetski-Shapiro, I.I.: Usp. Mat. Nauk **14** (2) (1959).
4. Schwarzman: Ann. Math. **66** (2) (1957) 270.

Received May 14, 1959

10.

(with L. A. Dikij)

Fractional powers of operators and Hamiltonian systems

Funkts. Anal. Prilozh. **10** (4) (1976) 13–29. MR **55**:6484. Zbl. **346**:35085
[Funct. Anal. Appl. **10** (1976) 259–273]

In [1] it was discovered that the nonlinear Korteweg–de Vries (KdV) equation admits in some sense of an exact integration procedure. The principal bases of this procedure were clarified in [2]: It was connected with the problem of seeking differential operators of any order for the Sturm–Liouville operator $L = -d^2/dx^2 + u(x)$, which would commute in the maximally possible way with L. Usually this is called the construction of Lax's L,A-pairs.* In [3, 4] it was shown that the KdV equation is a Hamiltonian system having a complete collection of first integrals in involutions. Already in [2] it was noticed that instead of a second-order operator L we can choose higher-order operators and look for those which pair with them. An algorithm for this was proposed in [5]. This was done in the modern way in [6] on the basis of a development of the technique in [7]. As a result, a system of equations was constructed generalizing the KdV equation and the complete integrability was proved of the corresponding stationary time-independent equations (also see survey [8]). As far as we know, the Hamiltonian mechanics of these systems, analogous to that for the case of second-order operators L, has not been constructed anywhere.

In the present article we shall show that by a sequential application of the technique suggested in the authors' previous articles [9, 10] we can construct a theory of generalized systems of the KdV type, including the Hamiltonian structure.

1. Ring of Polynomials of $u_k(x)$, $u_k'(x)$, ... By A we denote the ring of polynomials of several functions $u_k(x)$ and their derivatives of any order. The algebra and the variational calculus in such a ring of one function were presented in detail in [9]. Here we list briefly the information needed for the case of any finite number of functions. Differentiations or "vector fields" $\sum b_{k,i}\frac{\partial}{\partial u_k^{(i)}}$, $b_{k,i}\in A$ act in the ring. The collection of differentiations is named TA. There is one preferred differentiation $\frac{d}{dx}=\sum u_k^{(i+1)}\frac{\partial}{\partial u_k^{(i)}}$. Let dA/dx be the set of elements of A, representable in the form df/dx, $f\in A$. We set $\tilde{A} = A/(dA/dx)$. The elements of $\tilde{A}$ are called functionals [if we examine some boundary conditions on $u_k(x)$, making it possible to talk about the integrals $\int f\,dx$, e.g., the condition of dying out at $\pm\infty$ or of periodicity, etc., then a one-to-one correspondence exists between the equivalence

*In what follows we shall call them P,L-pairs since, firstly, the letter A will be firmly occupied, and, secondly, it honors P. Lax.

Moscow State University. Translated from Funktsional'nyi Analiz i Ego Prilozheniya, Vol. 10, No. 4, pp. 13-29, October-December, 1976. Original article submitted June 22, 1976.

classes with respect to dA/dx and the integrals]. The mapping which associates with each $f \in A$ its class $\tilde{f}$ in $\tilde{A}$ is denoted $\int dx$: $\tilde{f} = \int f\,dx$. Here the integral is defined purely algebraically without any convergence conditions. Let us determine the operators of the variational derivatives $\frac{\delta}{\delta u_k} = \sum \left(-\frac{d}{dx}\right)^i \frac{\partial}{\partial u_k^{(i)}}$. It can be proved that for $f \in dA/dx$, it is necessary and sufficient that all $\delta f/\delta u_k = 0$. For this reason $\delta/\delta u_k$ can be transferred from A to $\tilde{A}$ (but they take values as before in A). Obviously, $\frac{\delta f}{\delta u_k} = \frac{\delta}{\delta u_k}\int f\,dx$. Differential operators commuting with d/dx have the form $\sum b_k^{(i)} \frac{\partial}{\partial u_k^{(i)}}$, where $b_k^{(i)} = \left(\frac{d}{dx}\right)^i b_k$. These operators can be taken as acting in $\tilde{A}$. If we apply the obvious formula for integration by parts to $\int f'g\,dx = -\int fg'\,dx$, then we obtain

$$\left(\sum b_k^{(i)} \frac{\partial}{\partial u_k^{(i)}}\right)\tilde{f} = \int \sum b_k^{(i)} \frac{\partial f}{\partial u_k^{(i)}}\,dx = \sum \int b_k \left(-\frac{d}{dx}\right)^i \frac{\partial f}{\partial u_k^{(i)}}\,dx = \sum_k b_k \frac{\delta}{\delta u_k} f\,dx.$$

Later on we examine the module of the differential forms $TA^* = \{\sum a_{ij\ldots}^{rs\ldots} \delta u_i^{(r)} \wedge \delta u_j^{(s)} \wedge \ldots\}$ over A. In it acts the operator δ, i.e., the exterior differential, and d/dx (acts both on the coefficients as well as on $\delta u_i^{(r)}$). We set $\widetilde{TA}^* = TA^* \big/ \frac{d}{dx} TA^*$. The operators δ and d/dx commute; therefore, δ can be examined in $\widetilde{TA}^*$ as well; here $\delta \int \omega\,dx = \int \delta\omega\,dx$.

There holds

THEOREM 1. If $f \in A$, then δf can be uniquely represented in the form $\sum R_k \delta u_k + \frac{d}{dx}\omega$, where $R_k \in A$, and ω is some 1-form. Here the coefficients R_k equal $\delta f/\delta u_k$. Another formulation: $\delta \int f\,dx$ can be uniquely written as $\int \sum R_k \delta u_k\,dx$, and here $R_k = \frac{\delta}{\delta u_k}\int f\,dx$.

2. Differential Operator L and Its Resolvent. We examine the operator

$$L\left(-i\frac{d}{dx}\right) = \sum_{k=0}^{n} u_k(x)\left(-i\frac{d}{dx}\right)^k; \quad u_n \equiv 1, \quad u_{n-1} \equiv 0 \tag{1}$$

The coefficients $u_k(x)$ are arbitrary functions.* The symbol of this operator is $L(\xi) = \sum_k u_k \xi^k$. By $\circ$ we denote the operation of multiplication of symbols:

$$\sigma_1 \circ \sigma_2 = \sum_\nu \frac{1}{\nu!}\left(\frac{\partial}{\partial \xi}\right)^\nu \sigma_1 \cdot \left(-i\frac{d}{dx}\right)^\nu \sigma_2.$$

Let b be the symbol inverse to $L(\xi) - z$:

$$b \circ (L(\xi) - z) = (L(\xi) - z) \circ b = 1 \tag{2}$$

(i.e., the symbol of the resolvent). We seek b in the form†

$$b(\xi, x; z) = \sum_{l,m} B_{l,m} (-1)^{\frac{l+m}{n}} \xi^m (\xi^n - z)^{-1-\frac{l+m}{n}}.$$

Only those nonnegative l and m for which $(l + m)/n$ is an integer are present in the sum. Equation (2) yields the recurrence relations

*Almost all the results — the construction of P,L-pairs, the Hamiltonian formalism — are preserved when the u_k are matrices. Here for simplicity we restrict ourselves to the scalar case, but we hope to return to the more general case in another article wherein we shall apply another technique which is more natural for the matrix case.
†Such expansions were analyzed in [11] for a second-order operator and in [12] for any elliptic pseudodifferential operator. For our purposes the technique of symbols is especially convenient since it enables us to carry out purely local analyses without using boundary conditions or spectra.

$$B_{0,0} = 1, \quad B_{l,m} = 0, \text{ if } l < 0 \text{ or } m < 0,$$

$$B_{l+n,\,m} = \sum_{k=0}^{n} \sum_{\nu=0}^{k}{}' \binom{k}{\nu} u_k \left(-i\frac{d}{dx}\right)^{\nu} B_{l+(k-\nu),\, m-(k-\nu)} \tag{4}$$

(the prime on the summation sign denotes that the values $k = n$ and $\nu = 0$ are omitted). From the recurrence formula it is easy to get that the $B_{l,m}$ are polynomials of u_k and their derivatives.

3. Fractional Power. For $|\xi| \geqslant 1/2$ we examine

$$\tilde{a}(\xi, x; s) = \frac{1}{2\pi i} \int_{\Gamma} z^s b(\xi, x; z)\, dz. \tag{5}$$

The contour Γ is shown in Fig. 1. The integral converges for $\operatorname{Re} s < -1$. By z^s we mean $|z|^s e^{s \arg z}$, where $-\pi/2 \leqslant \arg z \leqslant 3\pi/2$. Let $\chi(\xi)$ be a smooth function, $\chi(\xi) = 1$ when $|\xi| \geq 1$ and $\chi(\xi) = 0$ when $|\xi| \leq 1/2$. We introduce the symbol

$$a(\xi, x; s) = \chi(\xi)\, \tilde{a}(\xi, x; s) \tag{6}$$

[generally speaking, the class of functions of ξ and x, distinguished in the finite range of ξ, is named the symbol; when speaking of a symbol $a(\xi, x; s)$ we shall have in mind a written concrete representative of an equivalence class. As a matter of fact, those important characteristics which we shall need do not depend upon the choice of the representative of the class, for instance, upon the smoothing function $\chi(\xi)$]. It is not difficult to prove the formulas

$$\frac{b(\xi, x; z_1) - b(\xi, x; z_2)}{z_1 - z_2} = b(\xi, x; z_1) \circ b(\xi, x; z_2) \tag{7}$$

(the functional equation of the resolvent) and

$$a(\xi, x; s_1) \circ a(\xi, x; s_2) = a(\xi, x; s_1 + s_2). \tag{8}$$

From Eq. (7) it follows that

$$\frac{\partial b(\xi, x; z)}{\partial z} = b(\xi, x; z) \circ b(\xi, x; z). \tag{9}$$

Substituting expansion (3) into Eq. (5), we have

$$a(\xi, x; s) = \frac{1}{2\pi i} \int_{\Gamma} \chi(\xi) \sum_{l,m} B_{l,m} (-1)^{\frac{l+m}{n}} \xi^m (\xi^n - z)^{-1-\frac{l+m}{n}} z^s dz = \sum_{l,m} B_{l,m} (\xi^n)^{s-\frac{l+m}{n}} \xi^m \cdot \binom{s}{\frac{l+m}{n}}.$$

Introducing the notation

$$A_l(s) = \sum_m B_{l,m} \binom{s}{\frac{l+m}{n}}, \tag{10}$$

we obtain

$$a(\xi, x; s) = \chi(\xi) \sum_{l=0}^{\infty} A_l(s)\, (\xi^n)^s\, \xi^{-l}, \tag{11}$$

where, according to the choice of branch, $\xi^n = 0$ for $\xi^n > 0$ and $\arg \xi^n = \pi$ for $\xi^n < 0$.

4. Diagonal of the Kernel. If $\sigma(\xi, x)$ is some symbol or, more precisely, a representative of the class of the symbol, we set

$$\overline{\sigma}(x) = \frac{1}{2\pi} \int_{-\infty}^{\infty} \sigma(\xi, x)\, d\xi,$$

if this integral converges. $\overline{\sigma}(x)$ depends upon the choice of the representative of the class. [$\overline{\sigma}(x)$ is the diagonal of the kernel of the operator which can be constructed from the symbol;

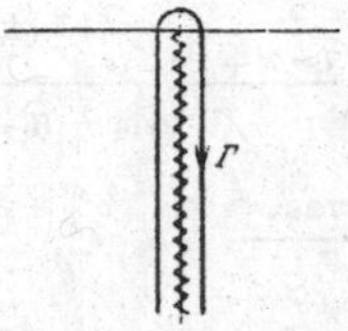

Fig. 1

such an operator is defined ambiguously by the symbol, i.e., to within a smoothing operator.] We consider

$$\bar{a}_k(x; s) = \frac{1}{2\pi} \int_{-\infty}^{\infty} \xi^k a(\xi, x; s)\, d\xi,$$

where k is any integer. Later on we shall be interested in the analytic continuation of this function onto the whole plane of the complex variable s, or, more precisely, in the residues of this function. It is easy to see that they are determined by the symbol and do not depend upon the choice of the smoothing function $\chi(\xi)$. They coincide with the residues of the integral

$$J(s) = \frac{1}{2\pi} \int_{|\xi| \geq 1} \sum_l A_l(s) (\xi^n)^s \xi^{-l+k}\, d\xi.$$

We have

$$J(s) = -\frac{1}{\pi} \sum_l A_l(s) \frac{1}{2(ns - l + k + 1)} [(-1)^{l+k} + 1]$$

for even n and

$$J(s) = -\frac{1}{\pi} \sum_l A_l(s) \frac{1}{2(ns - l + k + 1)} [e^{\pi i[s+l+k]} + 1]$$

for odd n. The residue at $s = (l - k - 1)/n$ equals

$$-\frac{1}{2\pi n} A_l\left(\frac{l-k-1}{n}\right) \varepsilon_{l,k},$$

where $\varepsilon_{l,k} = (-1)^{l+k} + 1$ for even n and $\varepsilon_{l,k} = e^{\pi i((l-k-1/n)+l+k)} + 1$ for odd n.

5. Asymptotics of the Diagonal of the Resolvent's Kernel. We rewrite Eq. (5) as

$$\bar{a}(\xi, x; s) = -\frac{1}{2\pi} (e^{\frac{3}{2}\pi i s} - e^{-\frac{\pi}{2} i s}) \int_0^{\infty} z^s b(\xi, x; -iz)\, dz = -i \frac{e^{\frac{\pi}{2} i s} \sin \pi s}{\pi} \int_0^{\infty} z^s b(\xi, x; -iz)\, dz.$$

The integral converges for $-1 < \operatorname{Re} s < -1/2$. Having set $b_k = \xi^k b$, we have

$$\bar{a}_k(x; s) = -\frac{i}{\pi} e^{\frac{\pi}{2} i s} \sin \pi s \int_0^{\infty} z^s \overline{\chi(\xi) b_k(\xi, x; -iz)}\, dz.$$

By the Mellin inversion formula (see [13]) we obtain

$$\bar{b}_k(x; -iz) = \frac{1}{2\pi} \int_{\gamma - i\infty}^{\gamma + i\infty} z^{-s-1} e^{-\frac{\pi}{2} i s} \frac{\pi \bar{a}_k(x; s)}{\sin \pi s}\, ds.$$

Hence follows the asymptotic behavior

$$\bar{b}_k(x; -iz) = \frac{i}{2n} \sum_l A_l\left(\frac{l-k+1}{n}\right) \frac{e^{-\frac{\pi}{2} i \frac{l-k-1}{n}}}{\sin \frac{\pi}{n}(l-k-1)} z^{\frac{-l+k+1}{n}-1} \varepsilon_{l,k}. \tag{12}$$

6. Theorem on Variational Derivatives.

THEOREM 2.

$$\frac{\delta}{\delta u_k} \bar{b}(x; z) = -\sum_{\nu=0}^{k} \binom{k}{\nu} \frac{\partial}{\partial z}\left(-i\frac{d}{dx}\right)^{\nu} \bar{b}_{k-\nu}(x; z). \tag{13}$$

To prove Theorem 2 we compute the differential δb and we write it as $R_k \delta u_k + d/dx(\ )$; then $\delta b/\delta u_k = R_k$ (see Paragraph 1). We apply operator δ to $b \circ [L(\xi) - z] = 1$:

$$\delta b \circ [L(\xi) - z] + \sum_k b \circ \delta u_k \circ \xi^k = 0.$$

We multiply from the right by b and we apply the operator $\bar{\ }$:

$$\delta \bar{b} = -\sum_k \overline{b \circ \delta u_k \circ \xi^k \circ b}.$$

For any σ_1 and σ_2 there holds $\overline{\sigma_1 \circ \sigma_2} = \overline{\sigma_2 \circ \sigma_1} + \frac{d}{dx}(\ )$, which follows easily from the definition of the multiplication $\sigma_1 \circ \sigma_2$. We have

$$\delta \bar{b} = -\sum_k \overline{\delta u_k \circ \xi^k \circ b \circ b} + \frac{d}{dx}(\ ).$$

Now Eq. (9) yields $\frac{\delta \bar{b}}{\delta u_k} = -\overline{\xi^k \circ \frac{\partial b}{\partial z}}$. It remains to compute the product of symbols:

$$\frac{\delta \bar{b}}{\delta u_k} = -\sum_{\nu} \frac{1}{\nu!} \overline{\left[\left(\frac{\partial}{\partial \xi}\right)^{\nu} \xi^k\right] \cdot \left(-i\frac{d}{dx}\right)^{\nu} \frac{\partial b}{\partial z}} = -\sum_{\nu=0}^{\infty} \binom{k}{\nu} \overline{\xi^{k-\nu}\left(-i\frac{d}{dx}\right)^{\nu} \frac{\partial b}{\partial z}} = -\sum_{\nu=0}^{\infty} \binom{k}{\nu}\left(-i\frac{d}{dx}\right)^{\nu} \frac{\partial \bar{b}_{k-\nu}}{\partial z}.$$

QED.

COROLLARY 1.

$$\frac{\delta}{\delta u_k} A_l\left(\frac{l-1}{n}\right) = \frac{l-1}{n} \sum_{\nu=0}^{\infty} \binom{k}{\nu}\left(-i\frac{d}{dx}\right)^{\nu} A_{l-n+k-\nu}\left(\frac{l-n-1}{n}\right). \tag{14}$$

The proof of Corollary 1 can be obtained without difficulty from the connection of $\bar{b}_k$ and A_l [see (12)].

7. Lax's P,L-Pairs. Let us consider Eq. (10) with $s = N/n$, where N is an integer not divisible by n. We indicate by

$$P_1(\xi) = \sum_{l=0}^{N} A_l\left(\frac{N}{n}\right) (\xi^n)^{N/n} \xi^{-l}$$

the part of the symbol of a nonnegative power of ξ. Let

$$P(\xi, x) = \sum_{l=0}^{N} A_l\left(\frac{N}{n}\right) \xi^{N-l}.$$

Then, when n is odd $P_1 \equiv P$, and when n is even $P_1 \equiv (\operatorname{sign} \xi)^N P$. $P(\xi, x)$ is the symbol of the differential operator

$$P\left(-i\frac{d}{dx}, x\right) = \sum_{l=0}^{N} A_l\left(\frac{N}{n}\right)\left(-i\frac{d}{dx}\right)^{N-l}. \tag{15}$$

We compute the commutator $[P(\xi, x), L(\xi)]$, using the fact that $a(\xi, x; s)$ commutes with $L(\xi)$ [since $L(\xi) = a(\xi, x; 1)$ and in accord with Eq. (8)]. We have

$$[P_1(\xi, x), L(\xi)] = \left[L(\xi), \sum_{l=N+1}^{\infty} A_l\left(\frac{N}{n}\right)\xi^{N-l}\right]\cdot \varepsilon =$$

$$= \sum_{\nu=0}^{\infty}\sum_{k=0}^{n}\sum_{l=N+1}^{\infty} (-i)^{\nu}\left[\binom{k}{\nu} u_k \xi^{k-\nu} A_l^{(\nu)}\left(\frac{N}{n}\right)\xi^{N-l} - \binom{N-l}{\nu} A_l\left(\frac{N}{n}\right)\xi^{N-l-\nu} u_k^{(\nu)}\xi^k\right]\cdot \varepsilon,$$

where $\varepsilon = 1$ for odd n and $\varepsilon = (\operatorname{sign} \xi)^N$ for even n. Further,

$$[P(\xi, x), L(\xi)] = \sum_{l=N+1}^{\infty}\sum_{\nu=0}^{\infty}\sum_{k=0}^{n}(-i)^{\nu}\left[\binom{k}{\nu} A_l^{(\nu)}\left(\frac{N}{n}\right) u_k - \binom{N-l}{\nu} A_l\left(\frac{N}{n}\right) u_k^{(\nu)}\right]\xi^{N-l+k-\nu}. \tag{16}$$

There are no negative powers of ξ in the left-hand side of Eq. (16); therefore, they are mutually annulled in the right-hand side; the maximal power of ξ is $n - 2$.

Thus, for every integer N not divisible by n, we have found a differential operator $P(-id/dx, x)$ such that the commutator $[P, L]$ is a differential operator of order $n - 2$. This is a direct generalization of Lax's theorem for $n = 2$.

8. Systems of KdV Type. Let the functions u_r depend upon one more variable t. We write the operator equation

$$\frac{d}{dt} L(\xi) = [P, L]. \tag{17}$$

This is equivalent to a system for the functions u_r

$$\frac{du_r}{dt} = \sum_{\substack{N-l+k-\nu=r\\(l>N)}} (-i)^{\nu}\left[\binom{k}{\nu} A_l^{(\nu)}\left(\frac{N}{n}\right) u_k - \binom{N-l}{\nu} A_l\left(\frac{N}{n}\right) u_k^{(\nu)}\right] \tag{18}$$

$$(r = 0, 1, \ldots, n-2).$$

We call this system a system of KdV type.

THEOREM 3. System (18) can be written as

$$\frac{d\mathbf{u}}{dt} = l\,\frac{\delta}{\delta \mathbf{u}} A_{N+n+1}\left(\frac{N+n}{n}\right)\cdot\frac{n}{N+n}, \tag{19}$$

where $\mathbf{u} = (u_0, u_1, \ldots, u_{n-2})$, $\frac{\delta}{\delta \mathbf{u}} = \left(\frac{\delta}{\delta u_0}, \ldots, \frac{\delta}{\delta u_{n-2}}\right)$, l is a matrix consisting of the differential operators

$$l_{rs} = \sum_{\gamma=0}^{n-1-r-s}\left[\binom{\gamma+r}{r} u_{r+s+\gamma+1}\left(-i\frac{d}{dx}\right)^{\gamma} - \binom{\gamma+s}{s}\left(i\frac{d}{dx}\right)^{\gamma} u_{r+s+\gamma+1}\right]. \tag{20}$$

Proof. We can invert Eq. (14):

$$A_{l-n+k}\left(\frac{l-n-1}{n}\right) = \frac{n}{l-1}\sum_{\nu=0}^{k}\binom{k}{\nu}\left(i\frac{d}{dx}\right)^{\nu}\frac{\delta}{\delta u_{k-\nu}} A_l\left(\frac{l-1}{n}\right). \tag{21}$$

We substitute these expressions into the right-hand side of Eq. (18). The first one of the two summands yields

$$\frac{n}{N+n}\sum_{\mu,\nu,s}(-i)^{\nu} i^{\mu}\binom{r+s+\mu+\nu+1}{\nu}\binom{s+\mu}{s} u_{r+s+\mu+\nu+1}\left(\frac{d}{dx}\right)^{\mu+\nu}\frac{\delta}{\delta u_s} A_{N+n+1}\left(\frac{N+n}{n}\right).$$

We set $\mu + \nu = \gamma$ and we make use of the identity

$$\sum_{\nu=0}^{\gamma}(-1)^{\nu}\binom{r+s+\gamma+1}{\nu}\binom{s+\gamma-\nu}{s} = (-1)^{\gamma}\binom{\gamma+r}{r}, \tag{22}$$

easily provable by induction over s. We have

$$\frac{n}{N+n}\sum_{s=0}^{n-2}\sum_{\gamma=0}^{n-1-r-s}\binom{\gamma+r}{r}u_{r+s+\gamma+1}\left(-i\frac{d}{dx}\right)^{\gamma}\frac{\delta}{\delta u_s}A_{N+n+1}\left(\frac{N+n}{n}\right).$$

The second summand in the right-hand side of Eq. (18) yields

$$-\frac{n}{N+n}\sum_{\mu,\nu,s}(-i)^{\nu}\binom{-1-\mu-s}{\nu}i^{\mu}\binom{s+\mu}{\mu}u^{(\nu)}_{r+s+\mu+\nu+1}\left(\frac{d}{dx}\right)^{\mu}\frac{\delta}{\delta u_s}A_{N+n+1}\left(\frac{N+n}{n}\right).$$

Setting $\mu + \nu = \gamma$, we have

$$-\frac{n}{N+n}\sum_{s=0}^{n-2}\sum_{\gamma=0}^{n-1-r-s}\binom{\gamma+s}{s}i^{\gamma}\sum_{\nu=0}^{\gamma}\binom{\gamma}{\nu}u^{(\nu)}_{r+s+\gamma+1}\left(\frac{d}{dx}\right)^{\gamma-\nu}\frac{\delta}{\delta u_s}A_{N+n+1}\left(\frac{N+n}{n}\right)=$$

$$=-\frac{n}{N+1}\sum_{s=0}^{n-2}\sum_{\gamma=0}^{n-1-r-s}\binom{\gamma+s}{s}\left(i\frac{d}{dx}\right)^{\gamma}u_{r+s+\gamma+1}\frac{\delta}{\delta u_s}A_{N+n+1}\left(\frac{N+n}{n}\right).$$

The sum of the two terms yields what we require.

9. First Integrals.

THEOREM 4. For any p, $\int A_p\left(\frac{p-1}{n}\right)dx$ is a first integral of Eq. (19).

(We recall, see Paragraph 1, that the integral is defined purely algebraically as an equivalence class with respect to dA/dx; i.e., we need not speak about any convergence for the integral. If we restrict ourselves to the class of functions for which $\int f\,dx$ exists in the analytic sense and $\int \frac{d}{dx} f\,dx = 0$, then the integral in Theorem 1 can be understood in such a sense.)

Proof of Theorem 4. Let u satisfy system (19). We differentiate the equations $b \circ [L(\xi) - z] = 1$ with respect to t:

$$b_t \circ [L(\xi) - z] + b \circ L_t(\xi) = 0.$$

But $L_t = P \circ L - L \circ P$ [since this is equivalent to (19)]. We substitute this into the equation and we apply the operation $\bar{\ }$:

$$\bar{b}_t + \overline{b \circ P \circ L \circ b} - \overline{b \circ L \circ P \circ b} = 0.$$

Taking into account that $b \circ L = L \circ b = 1 + zb$, we obtain $\bar{b}_t + \overline{b \circ P} - \overline{P \circ b} = 0$. We remember that $\overline{\sigma_1 \circ \sigma_2} = \overline{\sigma_2 \circ \sigma_1} + d/dx(\ )$. Hence, $\bar{b}_t = d/dx(\ )$. All the coefficients in the expansion $\bar{b}_t$ in powers of z turn out to be expressions of derivative type, $(\bar{b}_p)_t = d/dx(\ )$. It remains to note that $\bar{b}_p$ is proportional to $A_p[(p-1)/n]$.

THEOREM 5. For any p and q (p − 1 and q − 1 are not divisible by n) there exists $J_{q,p}$, a polynomial of u_r and their derivatives, such that

$$\sum_{r,s=0}^{n-2}\left[l_{rs}\left(\frac{\delta}{\delta u_s}A_q\left(\frac{q-1}{n}\right)\right)\right]\cdot\frac{\delta}{\delta u_r}A_p\left(\frac{p-1}{n}\right)=\frac{d}{dx}J_{q,p}. \tag{23}$$

This is an immediate corollary of the preceding theorem:

$$0=\frac{d}{dt}\int A_p\left(\frac{p-1}{n}\right)dx=\int\sum\frac{\partial}{\partial u_r^{(k)}}A_p\left(\frac{p-1}{n}\right)\cdot(u_r^{(k)})_t\,dx=$$

$$=\int\sum\left(-i\frac{d}{dx}\right)^{k}\frac{\partial}{\partial u_r^{(k)}}A_p\left(\frac{p-1}{n}\right)\cdot(u_r)_t\,dx=\int\sum\frac{\delta}{\delta u_r}A_p\left(\frac{p-1}{n}\right)\cdot l_{rs}\frac{\delta}{\delta u_s}A_{N+n+1}\left(\frac{N+n}{n}\right)dx.$$

Setting $q = N + n + 1$ and allowing for the arbitrariness of N, we obtain the theorem's assertion.

10. Space A^{n-1}. Lattices of a Lie Algebra. The space A^{n-1} consists of the collections $f = (f_0, \ldots, f_{n-2})$, where $f_k \in A$. In this space we now introduce a lattice of a Lie

algebra. Namely, with each **f** we associate a differential operator $\partial_f = \sum f_k^{(i)} \frac{\partial}{\partial u_k^{(i)}}$ commuting with d/dx. The space A^{n-1} turns out to be in a one-to-one correspondence with the space of such differential operators. The Lie algebra lattice, existing in the latter and introducible by the usual commutation $[\partial_f, \partial_g] = \partial_f\partial_g - \partial_g\partial_f$, is carried over to A^{n-1}. Namely,

$$[\mathbf{f}, \mathbf{g}]_0 = \partial_f \mathbf{g} - \partial_g \mathbf{f} \tag{24}$$

(∂_f and ∂_g act componentwise on **g** and **f**). We have marked the commutator with a subscript 0 since we shall be introducing another commutator right away.

THEOREM 6. The matrix-valued differential operator l maps A^{n-1} onto a Lie subalgebra relative to the commutator $[\ ]_0$, i.e., for any **f** and **g** there exists an element **h** such that

$$[l\mathbf{f}, l\mathbf{g}]_0 = l\mathbf{h}. \tag{25}$$

Here **h** is given by the formula

$$\mathbf{h} = \partial_{l\mathbf{g}}\mathbf{g} - \partial_{l\mathbf{g}}\mathbf{f} + \mathbf{h}_1, \tag{26}$$

where

$$(h_1)_k = \sum_{\substack{r,\, s=0 \\ r+s<k}}^{n-2} \left[\binom{\gamma+r}{r}(-i)^\gamma f_s^{(\gamma)} g_r - \binom{\gamma+s}{s}(-i)^\gamma f_s g_r^{(\gamma)}\right],$$

$$\gamma = k - r - s - 1. \tag{27}$$

Theorem 6 can be proved by direct computation, using formulas of type (22). We cannot present these calculations here in view of their awkwardness.

If we exclude constants from ring A and, correspondingly, from A^{n-1}, then, as is easy to verify, the mapping l is a monomorphism. Then **h**, constructed from **f** and **g**, is a commutator induced from $[\ ]_0$ by mapping l. We shall denote it by $[\ ]_1$, i.e., $[l\mathbf{f}, l\mathbf{g}]_0 = l[\mathbf{f}, \mathbf{g}]_1$.

11. Poisson Brackets. Let $\tilde{F}, \tilde{G} \in \tilde{A}$ be two functionals. The following function:

$$\{\tilde{F}, \tilde{G}\} = -\int \sum_{r,\, s}\left[l_{rs}\left(\frac{\delta \tilde{G}}{\delta u_s}\right)\right] \cdot \frac{\delta \tilde{F}}{\delta u_r}\, dx \tag{28}$$

is called the Poisson bracket of these functionals. From Eq. (20) for l_{rs} it is obvious that $l_{rs}^* = -l_{sr}$ (the asterisk denotes the formally adjoint differential operator). Hence follows the skew-symmetry of the Poisson bracket. The Jacobi identity is not obvious; it will follow from the next theorem below. We note that when n = 2 the vectors are turned into scalars and $l = -2id/dx$; we arrive at the Gardner–Zakharov–Faddeev brackets.

THEOREM 7. The variational gradient operation $\delta/\delta u$, mapping $\tilde{A}$ into A^{n-1}, leads the Poisson bracket { } into the commutator $[\ ]_1$:

$$\frac{\delta}{\delta u}\{\tilde{F}, \tilde{G}\} = \left[\frac{\delta}{\delta u}\tilde{F}, \frac{\delta}{\delta u}\tilde{G}\right]_1. \tag{29}$$

We prove this theorem at once by the method applied in [14] to prove this same fact but in a somewhat different situation.

LEMMA 1. Let $\tilde{R} \in \tilde{A}$ be an arbitrary functional,

$$m_{rs} = \sum_k \frac{\partial}{\partial u_s^{(k)}}\left(\frac{\delta \tilde{R}}{\delta u_r}\right) \cdot \left(\frac{d}{dx}\right)^k.$$

Then the matrix-valued differential operator m is formally self-adjoint, m* = m, or

$$\sum_k \left(-\frac{d}{dx}\right)^k \frac{\partial}{\partial u_s^{(k)}}\left(\frac{\delta \tilde{R}}{\delta u_r}\right) = \sum_k \frac{\partial}{\partial u_r^{(k)}}\left(\frac{\delta \tilde{R}}{\delta u_s}\right)\left(\frac{d}{dx}\right)^k. \tag{30}$$

To prove the equality of the two operators it is sufficient to prove that they act alike on the vector-valued form $\delta u = (\delta u_0, \ldots, \delta u_{n-2})$ (since the forms $\delta u_k^{(i)}$ are linearly independent over A). Let us verify this action for the left- and right-hand sides of the equality to be proved. We have

266

$$\sum_{r,k}\left(-\frac{d}{dx}\right)^k \frac{\partial}{\partial u_s^{(k)}}\left(\frac{\delta R}{\delta u_r}\right)\delta u_r = \sum_{k}\left(-\frac{d}{dx}\right)^k \frac{\partial}{\partial u_s^{(k)}}\sum_r \frac{\delta R}{\delta u_r}\delta u_r = \frac{\delta}{\delta u_s}\delta R,$$

$$\sum_{r,k}\frac{\partial}{\partial u_r^{(k)}}\left(\frac{\delta R}{\delta u_s}\right)\left(\frac{d}{dx}\right)^k \delta u_r = \sum_{r,k}\frac{\partial}{\partial u_r^{(k)}}\left(\frac{\delta R}{\partial u_s}\right)\delta u_r^{(k)} = \delta\frac{\delta R}{\delta u_s} = \frac{\delta}{\delta u_s}\delta R.$$

Lemma 1 has been proved.

We go on to prove Theorem 7. We compute $\delta\{\tilde{F}, \tilde{G}\}$ and we represent it in the form $\int\sum R_k \delta u_k dx$, then (see Paragraph 1) $R_k = \frac{\delta}{\delta u_k}\{F, \tilde{G}\}$. We have

$$\delta\{F,\tilde{G}\} = -\int\delta\sum_{r,s}\left[l_{rs}\left(\frac{\delta\tilde{G}}{\delta u_s}\right)\right]\cdot\frac{\delta\tilde{F}}{\delta u_r}dx =$$

$$= -\int\sum_{r,s}\left[l_{rs}\left(\delta\frac{\delta\tilde{G}}{\delta u_s}\right)\right]\cdot\frac{\delta\tilde{F}}{\delta u_r}dx - \int\sum_{r,s}\left[l_{rs}\left(\frac{\delta\tilde{G}}{\delta u_s}\right)\right]\delta\frac{\delta\tilde{F}}{\delta u_r}dx - \int\sum_{r,s}\left[(\delta l_{rs})\left(\frac{\delta\tilde{G}}{\delta u_s}\right)\right]\cdot\frac{\delta\tilde{F}}{\delta u_r}dx.$$

The first two terms equal

$$+\int\sum_{r,s,k,i}\left[\frac{\partial}{\partial u_k^{(i)}}\left(\frac{\delta\tilde{G}}{\delta u_s}\right)\left(\frac{d}{dx}\right)^i\delta u_k\right]\cdot\left[l_{sr}\left(\frac{\delta\tilde{F}}{\delta u_r}\right)\right]dx - \int\sum_{r,s,k,i}\left[\frac{\partial}{\partial u_k^{(i)}}\left(\frac{\delta\tilde{F}}{\delta u_r}\right)\left(\frac{d}{dx}\right)^i\delta u_k\right]\cdot\left[l_{rs}\left(\frac{\delta\tilde{G}}{\delta u_s}\right)\right]dx.$$

By Lemma 1 this expression equals

$$\int\sum_{r,s,k,i}\left[l_{sr}\left(\frac{\delta\tilde{F}}{\delta u_r}\right)\right]^{(i)}\frac{\partial}{\partial u_s^{(i)}}\left(\frac{\delta G}{\delta u_k}\right)\delta u_k dx - \dots$$

(the dots denote terms differing by a commutation of $\tilde{F}$ and $\tilde{G}$). We obtain

$$\int\sum_{k}\partial_{l\frac{\delta F}{\delta u}}\frac{\delta\tilde{G}}{\delta u_k}\delta u_k dx - \dots$$

The third term equals

$$-\int\sum_{r,s}\sum_{\gamma=0}^{n-1-r-s}\left[\binom{\gamma+r}{r}\delta u_{r+s+\gamma+1}\left(-i\frac{d}{dx}\right)^\gamma - \binom{\gamma+s}{s}\left(i\frac{d}{dx}\right)^\gamma\delta u_{r+s+\gamma+1}\right]\left(\frac{\delta\tilde{G}}{\delta u_s}\right)\frac{\delta\tilde{F}}{\delta u_r}dx =$$

$$= -\int\sum_{r,s=0}^{n-2}\left[\binom{\gamma+r}{r}(-i)^\gamma\left(\frac{\delta\tilde{G}}{\delta u_s}\right)^{(\gamma)}\frac{\delta\tilde{F}}{\delta u_r} - \binom{\gamma+s}{s}\frac{\delta\tilde{G}}{\delta u_s}\left(\frac{\delta\tilde{F}}{\delta u_r}\right)^{(\gamma)}\right]\delta u_k dx.$$

Collecting all the terms and keeping Eqs. (26) and (27) in mind, we obtain the required equality

$$\delta\{F,\tilde{G}\} = \sum_k\int\left(\left[\frac{\delta}{\delta u}F, \frac{\delta}{\delta u}\tilde{G}\right]_{1;k}\delta u_k dx.\right.$$

COROLLARY 2. If the Poisson bracket of two functionals equals zero, then the differential operators $\partial_{l\frac{\delta\tilde{F}}{\delta u}}$, $\partial_{l\frac{\delta\tilde{G}}{\delta u}}$ commute.

COROLLARY 3. For any p and q the operators $\partial_{l\frac{\delta}{\delta u}\mathcal{L}}$, $\partial_{l\frac{\delta}{\delta u}\mathcal{M}}$, where $\mathcal{L} = A_p\left(\frac{p-1}{n}\right)$, $\mathcal{M} = A_q\left(\frac{q-1}{n}\right)$, commute.

Remark 1. We can examine an equation, somewhat more general than (19),

$$\frac{du}{dt} = l\frac{\delta}{\delta u}\mathcal{L}, \tag{31}$$

267

where the Hamiltonian $\mathcal{L}$ is an arbitrary linear combination $\sum_{p=1}^{p_0} c_p A_p\left(\frac{n-1}{n}\right)$. Then, as before, for every q we can find an element J_q such that for $\mathcal{H}_q = A_q\left(\frac{q-1}{n}\right)$

$$\sum_{r,\,s=0}^{n-2}\left[l_{r,s}\left(\frac{\delta}{\delta u_s}\cdot \mathcal{H}_q\right)\right]\cdot \frac{\delta}{\delta u_r}\mathcal{L} = \frac{d}{dx} J_q$$

and the differential operators $\partial_{l\delta\,\delta u\,\mathcal{L}}$, $\partial_{l\delta|\delta u\cdot\mathcal{H}_q}$ commute.

12. Stationary Equations. Now let u be independent of t and satisfy the stationary equation corresponding to Eq. (31)

$$\frac{\delta}{\delta u_r}\mathcal{L} = 0 \qquad (r = 0, \ldots, n-2). \tag{33}$$

Equation (32) shows that the quantities J_q are first integrals of a stationary system. Our next problem is to show that a stationary system can be represented in Hamiltonian form and to compute the Poisson brackets of the first integrals indicated. The Poisson brackets turn out to be equal to zero; i.e., the first integrals turn out to be in involutions.

By $I_{\mathcal{L}}$ we denote an ideal in ring A, generated by the left-hand sides of Eqs. (32) and all their derivatives with respect to x. In other words, to this ideal belong the polynomials of u_i, u_i', . . ., which vanish by virtue of system (33). We set $A_{\mathcal{L}} = A/I_{\mathcal{L}}$.

We now restrict somewhat the generality of the analysis by introducing additional requirements whose meaning reduces to the possibility of solving Eqs. (32) relative to the highest derivatives. In ring A there is the following graduation: The number n − i + k is called the weight of factor $u_i^{(k)}$, while the sum of the weights of the factors is called the sum of the monomial. [This graduation arises naturally from the very origin of the u_i as the coefficients of operator (1); besides, this is not important just now.] The collection of terms of the highest weight is called the leading part of $\mathcal{L}$. It is easy to see that all terms of the polynomials $A_p(s)$ have one and the same weight p; therefore, the leading part of $\mathcal{L}$ is $A_{p_0}[(p_0 - 1)/n]$. In each variational derivative $\delta/\delta u_r\mathcal{L}$ we pick out the linear part of highest weight

$$\frac{\delta}{\delta u_0}\mathcal{L} = \sum_{j=0}^{n-2} k_{0j} u_j^{(p_0-2n+j)} + \ldots$$

$$\ldots\ldots\ldots\ldots\ldots\ldots$$

$$\frac{\delta}{\delta u_i}\mathcal{L} = \sum_{j=0}^{n-2} k_{ij} u_j^{(p_0-2n+i+j)} + \ldots$$

$$\ldots\ldots\ldots\ldots\ldots\ldots$$

The terms not written out either are of lesser weight or are nonlinear. In this and other cases they contain derivatives of the functions u_j, of orders lower than in the terms written out. It is easy to see that $k_{ij} = (-1)^{p_0-2n+i+j} k_{ji}$. From now on we shall examine only those Lagrangians for which

$$\Delta_0 \equiv k_{00} \neq 0,\quad \Delta_1 \equiv \begin{vmatrix} k_{00} & k_{01}\\ k_{10} & k_{11}\end{vmatrix} \neq 0, \ldots, \Delta_{n-2} \equiv \begin{vmatrix} k_{00} & \cdots & k_{0\,n-2}\\ \cdots & \cdots & \cdots\\ \cdots & \cdots & \cdots\\ \cdots & \cdots & \cdots\\ k_{n-2\,0} & \cdots & k_{n-2\,n-2}\end{vmatrix} \neq 0. \tag{34}$$

Hence, it follows already that p_0 must be even and that $p_0 \geqslant 2n$. We set $p_0 - 2n = 2\mu$. We shall analyze only this case (for systems of KdV type this signifies that n + N + 1 is even).

LEMMA 2. As independent generators in ring A we can take the system

$$\begin{aligned} &\{u_i^{(s)}\} \quad (i = 0, \ldots, n-2;\ s \leqslant 2\mu + 2i - 1), \text{ (a)}\\ &\left\{\left(\frac{d}{dx}\right)^r \frac{\delta\mathcal{L}}{\delta u_i}\right\} \quad (i = 0, \ldots, n-2;\ r = 0, 1, 2, \ldots). \text{ (b)} \end{aligned} \tag{35}$$

268

The functions (35a) are called the principal derivatives. We prove Lemma 2 by induction over the weight of the derivatives $u_i^{(s)}$, which we must express in terms of the quantities (35). Suppose that this has been done for the weight $2\mu + n + m$. We write the system

$$\left(\frac{d}{dx}\right)^m \frac{\delta \mathcal{L}}{\delta u_0} = \sum_{j=0}^{m} k_{0j} u_j^{(2\mu+m+j)} + \sum_{j=m+1}^{n-2} k_{0j} u_j^{(2\mu+m+j)} + \ldots$$

$$\cdots\cdots\cdots\cdots\cdots\cdots\cdots\cdots\cdots$$

$$\frac{\delta \mathcal{L}}{\delta u_m} = \sum_{j=0}^{m} k_{mj} u_j^{(2\mu+m+j)} + \sum_{j=m+1}^{n-2} k_{mj} u_j^{(2\mu+m+j)} + \ldots,$$

whence we express the derivatives $u_0^{(2\mu+m)}, \ldots, u_m^{(2\mu+2m)}$, since the system's determinant is nonzero.

LEMMA 3. If

$$\sum_{r,i} a_i^r \left(\frac{d}{dx}\right)^r \frac{\delta \mathcal{L}}{\delta u_i} = 0, \tag{36}$$

where $a_i^r \in A$, then $a_i^r \in I_{\mathcal{L}}$.

Proof. We express a_i^r in terms of the generators (35). If in these expressions there were terms containing only the generators (35a) and not (35b), then the left-hand side of (36) would contain terms linear with respect to (35b), which would not mutually annihilate anything, and that contradicts the independence of the generators.

Definition 1. A vector field (differentiation) ξ is called tangent if it contains the ideal $I_{\mathcal{L}}$: $\xi I_{\mathcal{L}} \subset I_{\mathcal{L}}$. Two tangent fields are equivalent if $(\xi - \eta) A \subset I_{\mathcal{L}}$. The set of equivalence classes is called $TA_{\mathcal{L}}$.

LEMMA 3. A tangent field $\xi \in TA_{\mathcal{L}}$ is uniquely defined by the principal coordinates ξ_i^s $(s \leqslant 2\mu + 2i - 1)$, which can be taken arbitrarily.

The proof is carried out similarly to the proof of Lemma 2. The nonprincipal coordinates are determined successively from the systems

$$\xi \left(\frac{\delta \mathcal{L}}{\delta u_0}\right)^{(m)} = \sum_{j=0}^{m} k_{0j} \xi_j^{2\mu+j+m} + \sum_{j=m+1}^{n-2} k_{0j} \xi_j^{2\mu+j+m} + \ldots$$

$$\cdots\cdots\cdots\cdots\cdots\cdots\cdots\cdots\cdots$$

$$\xi \frac{\delta \mathcal{L}}{\delta u_0} = \sum_{j=0}^{m} k_{mj} \xi_j^{2\mu+j+m} + \sum_{j=m+1}^{n-2} k_{mj} \xi_j^{2\mu+j+m} + \ldots$$

The left-hand sides belong to the ideal. The coordinates are uniquely determined in $TA_{\mathcal{L}}$

13. Characteristics of the Ideal's Elements. We construct the mapping $I_{\mathcal{L}} \big/ \frac{d}{dx} I_{\mathcal{L}} \to A_{\mathcal{L}}^{n-1}$, which we shall call a characteristic of the element $I_{\mathcal{L}} \big/ \frac{d}{dx} I_{\mathcal{L}}$. Let $f \in I_{\mathcal{L}}$ be any representative of the class. As an element of the ideal it can be written (not uniquely) as

$$f = \sum_{i,s} a_i^s \left(\frac{\delta \mathcal{L}}{\delta u_i}\right)^{(s)} \qquad (a_i^s \in A).$$

With it we associate $a_i = \left(\sum_s \left(-\frac{d}{dx}\right)^s a_i^s\right)_{\mathcal{L}}$ (the notation $(\ )_{\mathcal{L}}$ signifies the natural projection $A \to A_{\mathcal{L}}$). It is necessary to show that this is indeed a single-valued mapping of the class $I_{\mathcal{L}} \big/ \frac{d}{dx} I_{\mathcal{L}}$ onto the class $A_{\mathcal{L}}^{n-1}$. Let $f = \sum_{i,s} b_i^s \left(\frac{\delta \mathcal{L}}{\delta u_i}\right)^{(s)}$ be another way of writing the same f in terms of the generators of the ideal. By Lemma 3, 2, $a_i^s - b_i^s \in I_{\mathcal{L}}$ and, therefore, a_i =

269

b_i. Now let g be another representative of this same class, $g=\sum_{i,s} b_i^s\left(\frac{\delta\mathscr{L}}{\delta u_i}\right)^{(s)}$. Then $f-g \in \frac{d}{dx} I_{\mathscr{L}}$, i.e.,

$$\sum_{i,s}(a_i^s-b_i^s)\left(\frac{\delta\mathscr{L}}{\delta u_i}\right)^{(s)}=\frac{d}{dx}\sum c_i^s\left(\frac{\delta\mathscr{L}}{\delta u_i}\right)^{(s)},$$

whence

$$a_i^s-b_i^s-(c_i^s)'-c_i^{s-1}\in I_{\mathscr{L}},\quad \sum_s\left(-\frac{d}{dx}\right)^s a_i^s-\sum_s\left(-\frac{d}{dx}\right)^s b_i^s\in I_{\mathscr{L}}.$$

Definition 2. $F\in A$ is called a first integral if $\frac{dF}{dx}\in I_{\mathscr{L}}$. Its class $(F)_{\mathscr{L}}$ also will be called a first integral.

Obviously, the class of dF/dx in $I_{\mathscr{L}}\Big/\frac{d}{dx}I_{\mathscr{L}}$ is uniquely determined by class $(F)_{\mathscr{L}}$.

Definition 3. The characteristic of dF/dx in $I_{\mathscr{L}}\Big/\frac{d}{dx}I_{\mathscr{L}}$ is called the characteristic of the first integral.

14. Hamiltonian Lattice. As we know (Paragraph 1), a certain form $\Omega^{(1)}$ exists such that

$$\delta\mathscr{L}=\sum_k\frac{\delta\mathscr{L}}{\delta u_k}\delta u_k+\frac{d}{dx}\Omega^{(1)}.$$

Let $\Omega^{(2)}=\delta\Omega^{(1)}$, i.e.,

$$\frac{d}{dx}\Omega^{(2)}=-\sum_k\delta\frac{\delta\mathscr{L}}{\delta u_k}\wedge\delta u_k. \tag{37}$$

We shall treat $\Omega^{(2)}$ as a form over $TA_{\mathscr{L}}$ with values in $A_{\mathscr{L}}$. As the differential of $\Omega^{(1)}$, this form is closed.

THEOREM 8. $\Omega^{(2)}$ is a nondegenerate form in $TA_{\mathscr{L}}$.

Proof. In $d\Omega^{(2)}/dx$ we pick out the terms highest with respect to the total weight of the differentials

$$\frac{d}{dx}\Omega^{(2)}=-\sum k_{ij}\delta u_j^{(2\mu+i+j)}\wedge\delta u_i+\dots$$

Hence we find that

$$\Omega^{(2)}=-\sum\tilde{k}_{ij}(\delta u_j^{(2\mu+i+j-1)}\wedge\delta u_i-\delta u_j^{(2\mu+i+j-2)}\wedge\delta u_i'+\dots$$
$$\dots+(-1)^{i+j-1}\delta u_j\wedge\delta u_i^{(2\mu+i+j-1)})+\dots,$$

where

$$\tilde{k}_{ij}=\begin{cases}k_{ij}, & i\neq j,\\ k_{ij}/2, & i=j,\end{cases}$$

while the dots denote terms which contain $\delta u_j^{(s)}\wedge\delta u_i^{(r)}$ with $s+r<2\mu+i+j-1$. Later on it is necessary to express $\Omega^{(2)}$ in terms of the differentials of the principal derivatives [and of the differentials of variables (35b), but the latter yield a form zero in $TA_{\mathscr{L}}$].

We do not present the simple but lengthy calculations, but write out at once the resulting formula. If we introduce the bordered determinant

$$\Delta_{m;i,l} = \begin{vmatrix} k_{00} & \dots & k_{0m} & k_{0l} \\ \dots & \dots & \dots & \dots \\ \dots & \dots & \dots & \dots \\ \dots & \dots & \dots & \dots \\ k_{m0} & \dots & k_{mm} & k_{ml} \\ k_{i0} & \dots & k_{im} & k_{il} \end{vmatrix}, \qquad \frac{\Delta_{-1;i,l}}{\Delta_{-1}} = k_{il},$$

then

$$\Omega^{(2)} = \Omega^{(2)}_{*} + \Omega^{(2)}_{\mathscr{L}},$$

where $\Omega^{(2)}_{\mathscr{L}} = \sum a \cdot \delta \left(\frac{\delta L}{\delta u_i}\right)^{(r)} \wedge \delta q_j$, while the q_j are any variables.

$$\Omega^{(2)}_{*} = \sum_{i \geqslant l} \sum_{r=i-l}^{i-1} (-1)^{r-1} \frac{\Delta_{i-1-r;\,i,\,l}}{\Delta_{i-1-r}} \delta u_l^{(s)} \wedge \delta u_i^{(r)} + \sum_{i>l} \sum_{s=0}^{l-1} (-1)^{s} \frac{\Delta_{l-1-s;\,l,\,i}}{\Delta_{l-1-s}} \delta u_l^{(s)} \wedge \delta u_i^{(r)} +$$
$$+ \sum_{i \geqslant l} \sum_{r=i}^{2\mu+i-1} (-1)^{r-1} \bar{k}_{il} \delta u_l^{(s)} \wedge \delta u_i^{(r)} + \dots \qquad (r+s=2\mu+i+l-1). \tag{38}$$

The nondegeneracy of the form signifies that whatever be the 1-form ω, we can find a vector ξ such that $\omega = -i(\xi)\,\Omega^{(2)}$ in $TA_{\mathscr{L}}$. Obviously, we can take it right away that ω contains only the differentials of the principal variables, $\omega = \sum_{s \leqslant 2\mu+2i-1} \omega_i^s \delta u_i^{(s)}$, and as $\Omega^{(2)}$ we can take $\Omega^{(2)}_{*}$.

We obtain the sequence of systems

$$\left\{ \omega_{n-2}^{2\mu+2n-5} = -\frac{\Delta_{n-3;\,n-2,\,n-2}}{\Delta_{n-3}} \xi_{n-2}^{0}, \right.$$

$$\left\{ \begin{aligned} \omega_{n-3}^{2\mu+2n-7} &= -\frac{\Delta_{n-4;\,n-3,\,n-3}}{\Delta_{n-4}} \xi_{n-3}^{0} + \frac{\Delta_{n-4;\,n-2,\,n-3}}{\Delta_{n-4}} \xi_{n-2}^{1} + \dots, \\ \omega_{n-2}^{2\mu+2n-6} &= -\frac{\Delta_{n-4;\,n-3,\,n-2}}{\Delta_{n-4}} \xi_{n-3}^{0} + \frac{\Delta_{n-4;\,n-2,\,n-2}}{\Delta_{n-4}} \xi_{n-2}^{1} + \dots, \end{aligned} \right.$$

$$\dots\dots\dots\dots\dots\dots\dots\dots\dots\dots$$

$$\left\{ \begin{aligned} \omega_0^{(2\mu-1)} &= -\frac{\Delta_{-1;\,0,\,0}}{\Delta_{-1}} \xi_0^0 + \dots + (-1)^{n-1} \frac{\Delta_{-1;\,n-2,\,0}}{\Delta_{-1}} \xi_{n-2}^{n-2} + \dots \\ &\dots\dots\dots\dots\dots\dots\dots\dots \\ \omega_{n-2}^{(2\mu+n-3)} &= -\frac{\Delta_{-1;\,0,\,n-2}}{\Delta_{-1}} \xi_0^0 + \dots + (-1)^{n-1} \frac{\Delta_{-1;\,n-2,\,n-2}}{\Delta_{-1}} \xi_{n-2}^{n-2} + \dots \end{aligned} \right.$$

Besides the ones written down there are more terms containing the coordinates of ξ, determined from the preceding systems. The determinants of all these systems are nonzero by Sylvester's theorem, $\det(\Delta_{n-3-m;\,i,k}) = \Delta^m_{n-3-m} \cdot \Delta_{n-2}$. These are still not all the systems needed. We need to finish writing several more systems for the determination of the missing coordinates. They all have one and the same matrix, just as in the last of the systems written out. Thus, all the coordinates ξ_i^s $(s \leqslant 2\mu + 2i - 1)$ are determined in succession.

Remark 2. As we saw from the proof, if as ω and $\Omega^{(2)}$ we take forms containing only the differentials of the fundamental variables, then the equation $\omega = -i(\xi)\,\Omega^{(2)}$ can be solved exactly, i.e., in A and not just in $A_{\mathscr{L}}$.

15. Construction of the Vector Field Corresponding to a First Integral. After the symplectic form $\Omega^{(2)}$ has been constructed from a given $\mathscr{L}$, we can develop the usual concepts of Hamiltonian mechanics (we refer to [9] for details). $\mathscr{L}$ is a Langrangian, d/dx is a vector field corresponding to the equation, ξ_F are the Hamiltonian vector fields corresponding to $F \in A_{\mathscr{L}}$, i.e., $\delta F = -i(\xi_F)\,\Omega^{(2)}$. If $F, G \in A_{\mathscr{L}}$, then their Poisson brackets are $\xi_F G = \xi_G F$. The vector field corresponding to the Poisson brackets of F and G is the commutator of the vector fields ξ_F and ξ_G; therefore, the Poisson brackets equal zero if and only if the corresponding vector fields commute. If $F \in A_{\mathscr{L}}$ is a first integral, i.e., dF/dx = 0 in $A_{\mathscr{L}}$, then field ξ_F commutes with vector field d/dx.

When speaking about forms we shall distinguish three kinds of equalities:

1) the identity equality $\omega_1 = \omega_2$, i.e., the coincidence of all the coefficients in A; 2) equality over $A_{\mathcal{L}}$, $\omega_1 = \omega_2 (A_{\mathcal{L}})$, when the values of the forms, treated as elements of $A_{\mathcal{L}}$, coincide on the tangent vector fields; 3) equivalence, $\omega_1 \simeq \omega_2$, when the values of the forms, treated as elements of $A_{\mathcal{L}}$, coincide on all the vector fields (i.e., the coefficients for the forms differ by the elements of ideal $I_{\mathcal{L}}$).

LEMMA 5. If $\delta F \simeq 0$, then $F \in I_{\mathcal{L}}$ and its characteristic equals zero.

Proof. We shall use variables (35). The relation $\delta F \simeq 0$ signifies that $\xi F \in I_{\mathcal{L}}$ for every vector field ξ. Let $F = F_1 + F_2$, where F_1 depends only on variables (35a) and $F_2 \in I_{\mathcal{L}}$. If as ξ we take the partial derivatives with respect to the variables (35a), then we obtain $\partial F_1 / \partial u_i^{(s)} = 0, F_1 = 0$. Hence $F = \sum a_i^r \left(\frac{\delta \mathcal{L}}{\delta u_i}\right)^{(r)}$. But $\partial F/\partial \left(\frac{\delta \mathcal{L}}{\delta u_i}\right)^{(r)} \in I_{\mathcal{L}}$. Then all $a_i^r \in I_{\mathcal{L}}$, and the characteristic equals zero.

THEOREM 9. If F is a first integral and $\{f_i\}$ is its characteristic, then the vector field corresponding to this first integral is

$$\xi_F = -\sum_{i,s} f_i^{(s)} \frac{\partial}{\partial u_i^{(s)}} = -\partial_{\bar{f}}. \tag{39}$$

Proof. Let $\Omega^{(2)}$ be defined by the exact equality

$$\frac{d\Omega^{(2)}}{dx} = -\sum \delta \frac{\delta \mathcal{L}}{\delta u_i} \wedge \delta u_i.$$

As before we represent $\Omega^{(2)}$ as $\Omega_*^{(2)} + \Omega_{\mathcal{L}}^{(2)}$, where $\Omega_*^{(2)}$ depends only on the fundamental variables (35a) and $\Omega_{\mathcal{L}}^{(2)}$ is a form of type $\sum a\delta \left(\frac{\delta \mathcal{L}}{\delta u_i}\right)^{(r)} \wedge \delta q_j$, where the q_j are any coordinates. F can be reckoned as depending only on the fundamental variables. By Remark 2 to Theorem 8 in Paragraph 15 there exists a vector field ξ such that the exact equality $\delta F = -i(\xi)\,\Omega_*^{(2)}$ holds. Then

$$\delta F = -i(\xi)\,\Omega^{(2)} + i(\xi)\Omega_{\mathcal{L}}^{(2)}.$$

We apply d/dx to both sides. We note that $\frac{d}{dx} i(\xi)\,\Omega^{(2)} = i(\xi) \frac{d}{dx} \Omega^{(2)} + i\left(\left[\frac{d}{dx}, \xi\right]\right)\Omega^{(2)}$. But d/dx and ξ commute in $A_{\mathcal{L}}$, i.e., in the vector field [d/dx, ξ] all coordinates belong to $I_{\mathcal{L}}$. Its convolution with any form yields a form all of whose coefficients belong to $I_{\mathcal{L}}$, i.e., is equivalent to zero. We have

$$\delta \frac{dF}{dx} \simeq -i(\xi) \frac{d\Omega^{(2)}}{dx} + \frac{d}{dx} i(\xi)\,\Omega_{\mathcal{L}}^{(2)}$$

or

$$\delta \sum f_i^r \left(\frac{\delta \mathcal{L}}{\delta u_i}\right)^{(r)} \simeq i(\xi) \sum \delta \frac{\delta \mathcal{L}}{\delta u_i} \wedge \delta u_i + \frac{d}{dx} i(\xi) \sum a\delta \left(\frac{\delta \mathcal{L}}{\delta u_i}\right)^{(r)} \wedge \delta q_j.$$

Taking into account that $\xi \left(\frac{\delta \mathcal{L}}{\delta u_i}\right)^{(r)} \in I_{\mathcal{L}}$, since $\xi \in TA_{\mathcal{L}}$, and discarding the forms equivalent to zero, we obtain

$$\delta \sum f_i^r \left(\frac{\delta \mathcal{L}}{\delta u_i}\right)^{(r)} \simeq -\sum \xi_i^0 \delta \frac{\delta \mathcal{L}}{\delta u_i} - \frac{d}{dx} \sum (\xi q_j)\, a\delta \left(\frac{\delta \mathcal{L}}{\delta u_i}\right)^{(r)} \simeq -\delta \left[\sum \xi_i^0 \frac{\delta \mathcal{L}}{\delta u_i} + \frac{d}{dx} \sum (\xi q_j)\, a \left(\frac{\delta \mathcal{L}}{\delta u_i}\right)^{(r)}\right],$$

i.e.,

$$\delta \left[\sum f_i^r \left(\frac{\delta \mathcal{L}}{\delta u_i}\right)^{(r)} + \sum \xi_i^0 \frac{\delta \mathcal{L}}{\delta u_i} - \frac{d}{dx} \sum (\xi q_j)\, a \left(\frac{\delta \mathcal{L}}{\delta u_i}\right)^{(r)}\right] \simeq 0.$$

By Lemma 5 the expression within the brackets has a zero characteristic. But the last term is an arbitrary element of the ideal and so its characteristic is zero. Therefore, the characteristic of $\sum f_i^{(r)} \left(\frac{\delta \mathcal{L}}{\delta u_i}\right)^{(r)}$ equals the characteristic of $-\xi_i^0 \frac{\delta \mathcal{L}}{\delta u_i}$, i.e.,

272

$$\xi_i^0 = -f_i = -\sum_r \left(-\frac{d}{dx}\right)^r f_i^{(r)}.$$

Now Eq. (39) follows from the fact that ξ_F commutes in $A_{\mathcal{L}}$ with d/dx; hence, $\xi_i^r = \left(\frac{d}{dx}\right)^r \xi_i^0$.

The preceding analysis was of a more or less general nature. We now turn to the case $\mathcal{L} = \sum_{p=1}^{p_0} c_p A_p \left(\frac{n-1}{n}\right)$. Equation (32) shows that the quantities J_q are the first integrals of system (33), with characteristics $f_r = \sum_{s=0}^{n-2} l_{rs}\left(\frac{\delta}{\delta u_s} \cdot \mathcal{M}_q\right)$ or $f = l\frac{\delta}{\delta u} \cdot \mathcal{M}_q$.

THEOREM 10. The first integrals J_q of system (33) are in involutions among themselves for any q.

Proof. By what was said at the end of Paragraph 10 the operators $\partial_{l\frac{\delta}{\delta u}\cdot\mathcal{M}_{q_1}}$ and $\partial_{l\frac{\delta}{\delta u}\cdot\mathcal{M}_{q_2}}$ commute.

LITERATURE CITED

1. C. S. Gardner, J. M. Green, M. D. Kruskal, and R. M. Miura, "Method for solving the KdV equation," Phys. Rev. Lett., 19, No. 19, 1095-1097 (1967).
2. P. Lax, "Integrals of nonlinear equations of evolution and solitary waves," Commun. Pure. Appl. Math., 21, No. 5, 467-490 (1968).
3. C. S. Gardner, "Korteweg—de Vries equation and generalizations. IV," J. Math. Phys., 12, No. 8, 1548-1551 (1971).
4. V. E. Zakharov and L. D. Faddeev, "The Korteweg—de Vries equation is a fully integrable Hamiltonian system," Funkts. Anal. Prilozhen., 5, No. 4, 18-27 (1971).
5. V. E. Zakharov and A. B. Shabat, "A scheme for the integration of the nonlinear equations of mathematical physics by the method of the inverse scattering problem," Funkts. Anal. Prilozhen., 8, No. 3, 43-53 (1974).
6. I. M. Krichever, "Reflection-free potentials in a background of finitely zoned ones," Funkts. Anal. Prilozhen., 9, No. 2, 77-78 (1975).
7. S. P. Novikov, "A periodic problem for the Korteweg—de Vries equation. I,"Funkts'. Anal. Prilozhen., 8, No. 3, 54-66 (1974).
8. B. A. Dubrovin, V. B. Matveev, and S. P. Novikov, "Nonlinear equations of Korteweg—de Vries type, finitely zoned linear operators, and Abelian manifolds," Usp. Mat. Nauk, 30, No. 1, 55-136 (1975).
9. I. M. Gel'fand and L. A. Dikii, "Asymptotics of the resolvent of the Sturm—Liouville equations and the algebra of the Korteweg—de Vries equations," Usp. Mat. Nauk, 30, No. 5, 67-100 (1975).
10. I. M. Gel'fand and L. A. Dikii, "Lattice of a Lie algebra in formal variational calculus," Funkts. Anal. Prilozhen., 10, No. 1, 18-25 (1976).
11. L. A. Dikii, "The zeta-function of an ordinary differential equation on a finite segment," Izv. Akad. Nauk SSSR, Ser. Matem., 19, 187-200 (1955).
12. R. T. Seeley, "The powers A^s of an elliptic operator," Matematika, 12, No. 1, 96-112 (1968).
13. E. Titchmarsh, Introduction to the Theory of the Fourier Integral, Clarendon Press, Oxford (1937).
14. I. M. Gel'fand, Yu. I. Manin, and M. A. Shubin, "Poisson brackets and the kernel of a variational derivative in formal variational calculus," Funkts. Anal. Prilozhen., 10, No. 4, 30-34 (1976).

11.

(with L. A. Dikij)

A family of Hamiltonian structures related to nonlinear integrable differential equations

Prepr. Inst. Appl. Mat. **136** (1978). MR **81**: 58027

1. Introduction

1.1. In [1–3] a construction of nonlinear differential equations which admit of the so-called Lax commutation presentation $L_t=[P,L]$ was given. Here P and L are differential operators with respect to the variable x; the coefficients of P are differential polynomials in the coefficients of the operator L. It was also proved that these equations are Hamiltonian in a certain sympletic structure. The complicated construction of this structure (the Poisson bracket and the symplectic form) was considerably simplified and clarified by Lebedjev and Manin [4] and Adler [5], who showed that these structures can be obtained by a standard Kirillov-Kostant construction for an infinite Lie algebra. Earlier this fact was proved for the discrete case by Kostant.

It is well-known that for a special case of Lax systems, namely for the KdV equation it is possible to introduce two Hamiltonian structures and even a one-parameter family of Hamiltonian structures. Arbitrary linear combinations of two Poisson brackets turn out to be a Poisson bracket (i.e. satisfy the Jacobi identity). It is natural to assume that this situation holds for more general Lax equations. In [5] a possible construction of a new Poisson bracket was suggested but the necessary property (the Jacobi identity) was not proved. Here we shall prove this assertion.

1.2. It is well-known that the theory of the KdV equation can be constructed on the basis of the third-order equation

$$-R'''+2uR+4(u+\zeta)R'=0 \tag{1.2.1}$$

for the diagonal of the kernel of a resolvent which is the inverse of the differential operator $L+\zeta=-d^2/dx^2+u+\zeta$. In the case of a general nth order operator $L=\sum_{k=0}^{n} u_k(d/dx)^k$ the Volterra integral operator T with kernel $T(x,y)$ which satisfies the equation $(L_x+\zeta)\,T(x,y)=0$ with respect to the first argument and the equation $(L_y^*+\zeta)\,T(x,y)=0$ with respect to the second one is called the resolvent. Here L^* is the adjoint of L. Then $\Delta^+(T)=(L+\zeta)\,T$ and $\Delta^-(T)=T(L+\zeta)$ are differential operators of order $\leq n-1$. In [3] these operators are the basis of the whole theory. It is obvious that

$$(L+\zeta)\Delta^-(T)-\Delta^+(T)(L+\zeta)=0. \tag{1.2.2}$$

Adler [5] suggested considering more general operators $\Delta^+(X)=[(L+\zeta)X]_+$ and $\Delta^-(X)=[X(L+\zeta)]_+$. The subscript + distinguishes the differential part of an operator which, generally, can be a sum of an integral (Volterra) and of a differential operators.

The equation

$$(L+\zeta)\Delta^-(X)-\Delta^+(X)(L+\zeta)=0 \tag{1.2.3}$$

is the determining equation for the resolvents, this will be proved later on (see the theorem in 3.7). Thus Eq. (1.2.3) is an analog of (1.2.1) for general operators L.

Another analog was introduced earlier in our papers [2] (Eq. (38)) and [3] (Eq. (3.10)). The new equation (1.2.3) suggested by Adler is more suitable; it contains fewer variables.

1.3. The Hamiltonian structures relate to the third-order equation (in the simplest case of the second-order operator L) and to Eq. (1.2.3) (in the general case). Let $f(\{u_k^{(i)}\})$ be a polynomial in the coefficients u_k of the operator L and their derivatives of all orders; denote it as $f[u]$. Let $\tilde{f}$ be a functional of the form $\tilde{f}=\int f[u]\,dx$. This functional generates a Volterra operator X_f whose kernel has the following property: $(\partial/\partial x)^k X_f(x,y)|_{y=x}=\delta\tilde{f}/\delta u_k$ where $\delta\tilde{f}/\delta u_k$ is the variational derivative of the functional. Then the Poisson bracket depending on a parameter ζ is defined as

$$\{\tilde{f},\tilde{g}\}_\zeta=\operatorname{Tr}\{[(L+\zeta)\Delta^-(X_f)-\Delta^+(X_f)(L+\zeta)]\,X_g\} \tag{1.3.1}$$

(The trace of an operator which is a sum of a differential operator and an integral operator is the trace of the integral part).

We prove here that the Jacobi identity holds for this bracket.

1.4. It is easy to write the expression for the bracket (1.3.1) in explicit form, in terms of variational derivatives. In [3] the expressions

$$\begin{aligned}\Delta^-(X)&=\sum_{\alpha,\beta=0}^{n-1}(\tilde{l}^*_{\alpha,\beta}X_{\alpha,0})\cdot(d/dx)^\beta\\ \Delta^+(X)&=\sum_{\alpha,\beta=0}^{n-1}(\tilde{l}_{\beta\alpha}X_{\alpha,0})\cdot(d/dx)^\beta\end{aligned} \tag{1.4.1}$$

were given; here

$$X_{\alpha,\beta}=(\partial|\partial x)^\alpha(\partial/\partial y)^\beta X(x,y)|_{y=x}$$

and

$$\tilde{l}_{\alpha\beta}=\sum_{\gamma=0}^{n-1-\alpha-\beta}\binom{\gamma+\alpha}{\alpha}u_{\alpha+\beta+\gamma+1}(d/dx)^\gamma,\quad \alpha+\beta\leqq n-1;\ \tilde{l}_{\alpha\beta}=0,\ \alpha+\beta\geqq n \tag{1.4.2}$$

and $\tilde{l}^*_{\alpha\beta}$ is the adjoint operator of $\tilde{l}_{\alpha\beta}$. Then

$$(L+\zeta)\Delta^-(X)-\Delta^+(X)(L+\zeta)=\sum_{\alpha,\beta=0}^{n-1}[(m_{\alpha\beta}+\zeta l_{\alpha\beta})\,X_{\beta,0}(d/dx)^\alpha$$

where $l_{\alpha\beta}=\tilde{l}_{\alpha\beta}-\tilde{l}^*_{\beta\alpha}$,

$$m_{\alpha\beta}=\sum_{k,\eta\geqq 0}\sum_{\nu=k-\beta}^{\eta}(-1)^{\nu+\eta}\binom{k}{\nu}\binom{\eta+\alpha-\nu}{\alpha}u_k(d/dx)^{\eta}\circ u_{\alpha+\beta+\eta-k+1}$$
$$-\sum_{\substack{k,\eta\\(k\leqq\beta)}}\binom{\eta+\beta-k}{\eta}u_{\alpha+\beta+\eta-k+1}\circ(d/dx)^{\eta}u_k. \tag{1.4.3}$$

As can be proved, the matrix of these coefficients has the property $m^*_{\alpha\beta}=-m_{\beta\alpha}$. We prove also

Theorem. *The bracket* $\{\tilde{f},\tilde{g}\}_\zeta$ *satisfies the Jacobi identity.*

1.5. From the very beginning we deal with the class of Volterra operators X for which $X_{\alpha\beta}$ are formal series in $\zeta^{-1/n}$, $X_{\alpha,\beta}=\sum_{r=r_0}^{\infty}X_{\alpha,\beta,r}\zeta^{-r/n}$. Equation (1.2.3) an analog of the third-order equation (1.2.1), can be written in terms of $m_{\alpha\beta}$ and $l_{\alpha\beta}$ as

$$\sum_{\beta=0}^{n-1}m_{\alpha\beta}X_{\beta,0,r}+\sum_{\beta=0}^{n-1}l_{\alpha\beta}X_{\beta,0,r+n}=0,\qquad \alpha=0,\dots,n-1. \tag{1.5.1}$$

1.6 The paper also contains our old results proved in a different way. Let $T^{[1]}$, $T^{[2]}$ be two resolvents. We construct functionals

$$\tilde{F}^{[i]}_r=\int T^{[i]}_{0,0,r}\,dx,\qquad i=1,2.$$

Then for all r, s, i and j the identity $\{\tilde{F}^{[i]}_r,\tilde{F}^{[j]}_s\}=0$ holds. Thus an infinite set of functionals is presented which commute with respect to a Hamiltonian structure introduced with the aid of the Poisson bracket. If one of the functionals is taken as a Hamiltonian, the others are first integrals in involution.

1.7. The plan of this article is the following. The subsections 2, 3 are preparatory. They in fact repeat the material of the corresponding parts of [3]. The formal algebraic definition of a ring of differential and integral (Volterra) operators as well as the definition of the resolvents is given. In [1]-[3], from one article to another, we changed our technique. So in [1] we used fractional powers of operators and in [2], [3] we passed on to resolvents exeptionally. We think the theory became more consistent. In [1], [2] we constructed the ring of operators as a ring of symbols of the pseudodifferential operators. In [3] we dealt with kernels of integral operators defined formally. Each approach seems to have its advantages and disadvantages. (In [4], [5] the symbols of pseudodifferential operators and the fractional powers were used.) In Subsection 4 the basic results are obtained, the symplectic form and the Poisson bracket are introduced, and the theorem formultated earlier is proved. In subsection 5 it is shown how the constructed Hamiltonian systems can be written in the Lax form (although in our approach we can do without this form).

2. The ring of formal operators

The material of this subsection corresponds to the two first subsections of [3]; we refer the reader to this article for more detail.

2.1. We consider a differential ring A with differentiation d/dx. It is assumed that the kernel of the operator d/dx is isomorphic to the field $\mathbb{C}$ and $A=\mathbb{C}+A_0$ where A_0 is an ideal. The elements of A_0 are called "elements without constants". An example of a ring A is the ring of the polynomials in symbols $w_k^{(j)}$, $k=1,\dots,K$; $j=0,1,\dots$ where $dw_k^{(j)}/dx=w_k^{(j+1)}$. The symbols $w_k^{(0)}=w_k$ are the free differential generators of the differential ring.

2.2. The image of the operator d/dx is denoted dA/dx. The quotient $\tilde{A}=A/dA/dx$ is called the space of functionals. The natural mapping $A\to\tilde{A}$ is denoted as $f\mapsto\tilde{f}=\int f\,dx$. Thus the elements $\tilde{f}\in\tilde{A}$ are f to within the derivatives. Accordingly, $\int f'g\,dx=-\int fg'\,dx$ where $f'=df/dx$, $g'=dg/dx$.

2.3. $A((z^{-1}))$ will denote the differential ring of the formal Laurent series

$$f(z)=\sum_{r=r_0}^{\infty} f_r z^{-r}\,,\qquad f_r\in A$$

where z is a parameter. The operations are natural. The invertible elements of this ring are those for which $f_{r_0}\in\mathbb{C}$. The operation $\int dx$ can be extended to the elements of $A((z^{-1}))$.

2.4. Let $A((\xi-x))$ be the ring of the formal Taylor series

$$f(\xi)=\sum_{p=0}^{\infty}\frac{(\xi-x)^p}{p!}f^{(p)},\qquad f^{(p)}\in A.$$

It is assumed that $df_p/dx=f_{p+1}$, i.e. $f_p=f_0^{(p)}$. This ring is isomorphic to the ring A through the mapping

$$f\in A\mapsto\sum_{p=0}^{\infty}\frac{(\xi-x)^p}{p!}f^{(p)}.$$

The ring $A((\xi-x,z^{-1}))$ can be constructed in the same way from the formal series

$$f(\xi,z)=\sum_{p=0}^{\infty}\frac{(\xi-x)^p}{p!}f_p(z),\qquad f_p(z)\in A((z^{-1})),$$

where $f_{p+1}(z)=df_p(z)/dx$. This ring is isomorphic to $A((z^{-1}))$.

2.5. Let $A((\xi-x,\eta-x))$ be the ring of the double series

$$X(\xi,\eta)=\sum_{p,q=0}^{\infty}\frac{(\xi-x)^p(\eta-x)^q}{p!\,q!}X_{p,q},\qquad X_{p,q}\in A$$

where

$$\frac{d}{dx} X_{p,q} = X_{p+1,q} + X_{p,q+1}. \tag{2.5.1}$$

The ring $A((\xi - x, \eta - x, z^{-1}))$ is defined similarly; here $X_{p,q} \in A((z^{-1}))$ i.e.

$$X_{p,q} = \sum_{r=r_0(p,q)}^{\infty} X_{p,q,r} z^{-r}. \tag{2.5.2}$$

Besides the ordinary multiplication of formal series we introduce a new associative operation, the convolution:

$$X(\xi, \eta) \circ Y(\xi, \eta) = \int_{\eta}^{\xi} X(\xi, \zeta) Y(\zeta, \eta) d\zeta.$$

Here X and Y must be replaced by the corresponding series and integrated termwise. (This is the usual rule for calculating of the kernel of the product of two Volterra operators with the kernels X and Y.)

$A((\xi - x, \eta - x, z^{-1}))$ is a ring with respect to this convolution. We call it the ring of (formal) Volterra operators and denote as R_-. Further we shall drop the circle in the notation of the convolution since no other multiplication will be used.

2.6. The ring R_+ of differential operators consists of expressions

$$L = \sum_{k=0}^{K} f_k (d/dx)^k, \qquad f_k \in A((z^{-1})).$$

The operations are defined in the usual way: when two terms multiply one must transpose the operator d/dx to the end of the expression using the commutation rule $(d/dx) \circ f = f(d/dx) + f'$. The ring R_+ is isomorphic to the ring of operators

$$L = \sum_{k=0}^{K} f_k(\xi)(d/d\xi)^k, \qquad f_k(\xi) \in A((\xi - x, z^{-1}))$$

If $f_k(\xi) \in A((\xi - x))$ for all k we may say that operator does not depend on z.

2.7. The ring of operators R is defined as the direct sum of R_+ and R_-: $R = R_+ + R_-$. It only remains to introduce the multiplication of the elements of R_+ and R_-. This is done in the usual manner when differential and Volterra operators are multiplied:

$$\sum_{k=0}^{m} f_k(\xi)(d/d\xi)^k \circ X(\xi, \eta) = M_1 + F_1, \qquad M_1 \in R_+, \ F_1 \in R_-,$$

where

$$M_1 = \sum_{k=1}^{m} \sum_{\alpha=0}^{k-1} f_k(\xi)(d/d\xi)^{k-1-\alpha} \circ [(\partial/\partial\xi)^{\alpha} X(\xi, \eta)]_{\eta=\xi}$$

$$F_1 = \sum_{k=0}^{m} f_k(\xi)(\partial/\partial\xi)^k X(\xi, \eta)$$

and

$$X(\xi,\eta)\circ\sum_{k=0}^{m} f_k(\xi)(d/d\xi)^k = M_2+F_2, \qquad M_2\in R_+,\ F_2\in R_-$$

where

$$M_2=\sum_{k=1}^{m}\sum_{\alpha=0}^{k-1}(-1)^\alpha(\partial/\partial\eta)^\alpha f_k(\eta)X(\xi,\eta)]_{\eta=\xi}\cdot(d/d\xi)^{k-1-\alpha}$$

$$F_2=\sum_{k=0}^{m}(-1)^k(\partial/\partial\eta)^k f_k(\eta)X(\xi,\eta).$$

The subscripts $+$ and $-$ symbolize the projection of R onto R_+ and R_-.

2.8. It easily follows from (2.5.1) that

$$(d/dx)X(x,x)=(\partial/\partial\xi)X(\xi,\eta)|_{\xi=\eta=x}+(\partial/\partial\eta)X(\xi,\eta)|_{\xi=\eta=x},$$

and

$$(\partial/\partial\xi)X(\xi,\eta)|_{\xi=\eta=x}=(d/dx-\partial/\partial\eta)X(\xi,\eta)|_{\xi=\eta=x}.$$

A more general identity also holds:

$$(\partial/\partial\xi)^p X(\xi,\eta)|_{\xi=\eta=x}=(d/dx-\partial/\partial\eta)^p X(\xi,\eta)|_{\xi=\eta=x}. \tag{2.8.1}$$

2.9. An element $X\in R_-$ is said to be bounded if there exists r_0 such that all $r_0(p,q)>r_0$ in (2.5.2).

An element $X\in R_-$ is said to belong to the exponent $\varepsilon\in\mathbb{C}$ if $X=\exp(\varepsilon(\xi-\eta)z)\bar{X}$ where $\bar{X}$ is a bounded element of R_-, $\exp(\varepsilon(\xi-\eta)z)$ is an element of R_- defined as

$$\exp(\varepsilon z(\xi-\eta))=\sum_{p,q\geqq 0}\frac{(\xi-x)^p(\eta-x)^q}{p!\,q!}(-1)^q(\varepsilon z)^{p+q}.$$

Direct calculation shows that the product of two elements belonging to ε_1 and ε_2 is a sum of two elements belonging to ε_1 and ε_2.

If an operator is represented as a sum of operators belonging to some exponents, this representation is called its canonical representation. If it exists, it is unique.

2.10. If $F\in R$, we define

$$\operatorname{Tr}F=\operatorname{Tr}F_-=\int(F_-)_{00}\,dx.$$

Theorem. a) *If* $F,G\in R$ *then* $\operatorname{Tr}FG=\operatorname{Tr}GF$,
b) *If* $F,G\in R_-$ *then* $\operatorname{Tr}FG=0$,
c) *If* $F\in R_+$ *then* $\operatorname{Tr}F=0$.

Proof. c) is obvious. b) follows from the fact that $\operatorname{Tr}X=\int X(x,x)dx$, $X\in R_-$ and from the fact that for $X=FG$ we have $X(x,x)=0$ (see 2.5). It remains to check a) for $F\in R_+$ and $G\in R_-$ or vice versa. It is sufficient to consider the following two cases: F is the operator of multiplication by a function $f(\xi)$ and

$F=d/d\xi$. In the first case

$$[f(\xi)G(\xi,\eta)]_{\xi=\eta=x}=f_0G_{00}=G_{00}f_0=[G(\xi,\eta)f(\eta)]_{\xi=\eta=x}.$$

In the second case

$$\begin{aligned}\operatorname{Tr}(FG-GF)&=\int[\partial/\partial\xi\, G(\xi,\eta)+\partial/\partial\eta\, G(\xi,\eta)]_{\xi=\eta=x}\cdot dx\\&=\int(G_{10}+G_{01})\,dx=\int G'_{00}\,dx=0\end{aligned}$$

(see 2.5).

Theorem 2. *If* $F=\sum_{k=0}^{n}\exp(\varepsilon_k(\xi-\eta)z)\bar{F}^{[k]}$, $G=\sum_{k=0}^{n}\exp(\varepsilon_k(\xi-\eta)z)\bar{G}^{[k]}$ *are the canonical representations of* $F, G\in R_-$ *and*

$$FG=\sum_{k=0}^{n}\exp(\varepsilon_k(\xi-\eta)z)(\overline{FG})^{[k]},\quad GF=\sum_{k=0}^{n}\exp(\varepsilon_k(\xi-\eta)z)(\overline{GF})^{[k]}$$

are canonical representations of the products, then

$$\operatorname{Tr}(\overline{FG})^{[k]}=\operatorname{Tr}(\overline{GF})^{[k]}.$$

The proof can be carried out by a direct calculation:

$$\begin{aligned}(FG)^k_{00}=\sum_{l\neq k}\sum_{q,r=0}^{\infty}&[(-1)^{q+r}F^{[k]}_{0,q}G^{[l]}_{r,0}\\&+F^{[l]}_{0,q}G^{[k]}_{r,0}]\binom{q+r}{q}[z(\varepsilon_k-\varepsilon_l)]^{-q-r-1}.\end{aligned}$$

(In [3] there was a mistake in this formula and we correct it here.) Thus

$$\begin{aligned}(FG-GF)^{[k]}_{0,0}=\frac{d}{dx}\sum_{l\neq k}\sum_{q,r=0}^{\infty}\sum_{\beta=0}^{q}&\binom{q}{\beta}[(-1)^{q+r}F^{(q-\beta)}_{\beta,0}G_{r,0}\\&-F_{r,0}G^{(q-\beta)}_{\beta,0}]\binom{q+r}{q}[z(\varepsilon_k-\varepsilon_l)]^{-q-r-1}-(F\leftrightarrow G).\end{aligned}$$

(The symbol $(F\leftrightarrow G)$ denotes a similar term with transposed F and G.) This proves the theorem.

2.11. We introduce a scalar product in R:

$$(F,G)=\operatorname{Tr}FG.$$

This product is indefinite. For $F\in R_+$, $G\in R_+$ or $F\in R_-$, $G\in R_-$ we have $(F,G)=0$. However for every $F\in R_+$ (or $F\in R_-$) an element $G\in R_-$ (or $G\in R_+$) can be found such that $(F,G)\neq 0$, the scalar product is nondegenerate.

3. Resolvents

3.1. We consider a differential operator

$$L=\sum_{k=0}^{n} u_k(\xi)(d/d\xi)^k \tag{3.1.1}$$

independent of z. Let $u_n=1$. By the resolvent we mean an operator $T\in R^-$ such that $(L-z^n)T\in R_+$ and $T(L-z^n)\in R_+$. (In the Introduction z^n was denoted as $-\zeta$.) The exact resolvent is an operator such that

$$(L-z^n)T=T(L-z^n)=1.$$

Theorem. *The exact resolvent exists, is unique, and admits of a canonical representation,*

$$T=\sum_{k=1}^{n} T^{[k]}$$

where $T^{[k]}$ belong to the exponents ε_k which are the nth roots of unity. $T^{[k]}$ are resolvents.

The proof of this theorem is in [3]. It is based on the Neumann series

$$T=T_0-T_0L_1T_0+T_0L_1T_0L_1T_0-\ldots$$

where T_0 is the exact resolvent of the simplest operator $L_0=(d/d\xi)^n$ and $L_1=L-L_0$. The resolvent T_0 can be written exactly in the form of the canonical representation:

$$T_0(\xi,\eta)=\sum_{k=1}^{n} \exp(-\varepsilon_k(\eta-\xi)z)/[n(\varepsilon_k z)^{n-1}], \qquad \varepsilon_k^n=1.$$

The coefficients of the expansion of this kernel in $\xi-x$, $n-x$ are $(T_0^{[k]})_{p,q}=(-1)^q(\varepsilon_k z)^{p+q-n+1}$.

From the Neumann expansion above the formula

$$T_{p,q}^{[k]}=(-1)^q(\varepsilon_k z)^{p+q-n+1}+t_{p,q}^{[k]}$$

can be obtained. Here $t_{p,q}^{[k]}$ is the sum of terms of smaller degrees in z. Moreover, all $t_{p,q,r}^{[k]}\in A_0$ i.e. do not contain constants.

3.2. Proposition (see [3]). *Let L be an operator of form* (3.1.1). *Let T be one of its resolvents such that the elements $T_{j,0,r}$, $j\leqq n-1$, $r=0,1,\ldots$ do not contain constants. Then $T=0$.*

Thus resolvents are completely determined by their constants in the coefficients $T_{j,0,r}$, $j\leqq n-1$.

Theorem. *The set of all resolvents of the operator L is an n-dimensional space over the field* $\mathbb{C}((z^{-1}))$ *of series with constant coefficients. The canonical components of the exact resolvents can be taken as a basis.*

For the proof it suffices to note that for the canonical component $T^{[k]}$ the constants in $T^{[k]}_{j,0,r}$ are ε_k^{j-n+1} when $r=n-1-j$ and there are no other constants. We construct the resolvents

$$Q^{[j]}=\frac{1}{n}\sum_k (\varepsilon_k z)^{n-1-j}T^{[k]}, \quad j\leqq n-1.$$

It is easy to see that each of these resolvents has only one nonzero constant (in $Q^{[j]}_{j,0,0}$) and this constant is equal to 1. Linear combinations of these constants with arbitrary coefficients belonging to $\mathbb{C}(z)$ make up the whole resolvent space.

3.3. We write an equation, suggested for the first time by Adler [5] and we demonstrate that this is the equation for the resolvents. Let L be the operator (3.1.1), $\hat{L}=L-z^n$. We write the equation

$$\hat{L}(X\hat{L})_+-(\hat{L}X)_+\hat{L}=0 \tag{3.3.1}$$

for $X\in R_-$. The equivalent forms for this equation are

$$\hat{L}(XL)_+-(LX)_+\hat{L}=0,$$
$$L(XL)_+-(LX)_+L=z^n[X,L]_+, \tag{3.3.2}$$

and

$$\hat{L}(X\hat{L})_--(\hat{L}X)_-\hat{L}=0 \tag{3.3.3}$$

($[X,L]=XL-LX$). The equivalence of (3.3.3) and (3.3.1) follows from the fact that the sum of the left-hand sides of these equations is equal to zero.

The equation (3.3.1) imposes constraints not on the whole operator X, i.e. not on the whole set of the coefficients $X_{p,q}$ but only on those for which $p+q\leqq n-1$.

Two operators X and Y are said to be *m*-equivalent if $X_{p,q}=Y_{p,q}$ for $p+q\leqq m$. Thus the equation (3.3.1) determines a solution to within $n-1$-equivalence.

3.4. It is also useful to write Eq. (3.3.1) in terms of $X_{p,q}$. According to 2.7, we have

$$(LX)_+=\sum_{k-1}^{n}\sum_{\alpha=0}^{k-1}u_k(\xi)(d/d\xi)^{k-1-\alpha}\circ[(\partial/\partial\xi)^\alpha X(\xi,\eta)]_{\eta=\xi}.$$

We shall write this expression as a differential operator with respect to x instead of ξ (see 2.6):

$$\begin{aligned}(LX)_+ &= \sum_{k=1}^{n}\sum_{\alpha=0}^{k-1} u_k(d/dx)^{k-1-\alpha}\circ[(\partial/\partial\xi)^\alpha X(\xi,\eta)]_{\eta=\xi=x}\\ &= \sum_{k=1}^{n}\sum_{\alpha=0}^{k-1} u_k(d/dx)^{k-1-\alpha}X_{\alpha,0}\\ &= \sum_{k=1}^{n}\sum_{\alpha+\beta\leqq n-1} u_k\binom{k-1-\alpha}{\beta}(d/dx)^{k-1-\alpha-\beta}X_{\alpha,0}](d/dx)^\beta\\ &= \sum_{\alpha+\beta\leqq n-1}(\tilde{l}_{\beta\alpha}X_{\alpha,0})(d/dx)^\beta \end{aligned}\tag{3.4.1}$$

where

$$\tilde{l}_{\beta\alpha}=\sum_{\gamma=0}^{n-1-\alpha-\beta}\binom{\gamma+\beta}{\beta}u_{\alpha+\beta+\gamma+1}(d/dx)^\gamma.\tag{3.4.2}$$

Similarly

$$(XL)_+=\sum_{k=1}^{n}\sum_{\alpha=0}^{k-1}(-1)^\alpha(\partial/\partial\eta)^\alpha u_k(\eta)X(\xi,\eta)]_{\eta=\xi=x}(d/dx)^{k-1-\alpha}.$$

Taking into account (2.8.1) we have

$$\begin{aligned}(XL)_+ &= \sum_{k=1}^{n}\sum_{\alpha=0}^{k-1}(-1)^\alpha\left(\frac{d}{dx}-\frac{\partial}{\partial\xi}\right)u_k(\eta)X(\xi,\eta)]_{\eta=\xi=x}(d/dx)^{k-1-\alpha}\\ &= \sum_{k=1}^{n}\sum_{\alpha=0}^{k-1}\sum_{\beta=0}^{\alpha}(-1)^{\alpha+\beta}\binom{\alpha}{\beta}(d/dx)^{\alpha-\beta}(u_kX_{\beta,0})](d/dx)^{k-1-\alpha}\\ &= \sum_{k=1}^{n}\sum_{\alpha+\beta\leqq k-1}\binom{k-1-\alpha}{\beta}(-d/dx)^{k-1-\alpha-\beta}(u_kX_{\beta,0})(d/dx)^\alpha\\ &= \sum_{\alpha+\beta\leqq n-1}(\tilde{l}^*_{\alpha\beta}X_{\alpha,0})(d/dx)^\beta, \end{aligned}\tag{3.4.3}$$

where $\tilde{l}^*_{\alpha\beta}$ is the adjoint operator of $\tilde{l}_{\alpha\beta}$.

Thus the right-hand side of Eq. (3.3.2) can be written as

$$-z^n\sum_{\alpha+\beta\leqq n-1}(l_{\beta\alpha}X_{\alpha,0})(d/dx)^\beta$$

where $l_{\alpha\beta}=\tilde{l}_{\alpha\beta}-\tilde{l}^*_{\beta\alpha}$ is a differential operator acting on $X_{\alpha,0}$. The left-hand side of Eq. (3.3.2) involves some more complicated operator arising from the multiplication of L by $(XL)_+$ and also $(LX)_+$ by L. It is not at all obvious that the resulting operator is of order less than or equal to $n-1$. This becomes clear if the left-hand side of Eq. (3.3.1) is replaced by the left-hand side of the Eq. (3.3.3), equal to the former to within the sign. Denote

$$\begin{aligned}&\sum_k u_k(d/dx)^k\circ\sum_{\alpha+\beta\leqq n-1}(\tilde{l}^*_{\alpha\beta}X_{\alpha,0})(d/dx)^\beta\\ &\quad-\sum_{\alpha+\beta\leqq n-1}(\tilde{l}_{\beta\alpha}X_{\alpha,0})(d/dx)^\beta\circ\sum_k u_k(d/dx)^k=-\sum_{\alpha,\beta\leqq n-1}(m_{\beta\alpha}X_{\alpha,0})(d/dx)^\beta.\end{aligned}$$

Then Eq. (3.3.2) takes the form

$$\sum_{\alpha,\beta\leqq n-1}(m_{\beta\alpha}X_{\alpha,0})(d/dx)^{\beta}=z^{n}\sum_{\alpha+\beta\leqq n-1}(l_{\beta\alpha}X_{\alpha,0})(d/dx)^{\beta}$$

i.e.

$$\sum_{\alpha=0}^{n-1-\beta}\mathrm{m}_{\beta\alpha}X_{\alpha,0}=z^{n}\sum_{\alpha=0}^{n-1-\beta}l_{\beta\alpha}X_{\alpha,0},\qquad \beta=0,1,\ldots,n-1.$$

The matrices $m_{\alpha\beta}$ and $l_{\alpha\beta}$ have the property $m^{*}_{\alpha\beta}=-m_{\beta\alpha}$ and $l^{*}_{\alpha\beta}=-l_{\beta\alpha}$. The expression for $m_{\alpha\beta}$ was given in Introduction.

3.5. If the series $X_{\alpha\beta}=\sum_{r=r_0(\alpha)}^{\infty}X_{\alpha,0,r}z^{-r}$ is substituted into (3.4.5) we obtain

$$\sum_{\alpha=0}^{n-1-\beta}m_{\alpha\beta}X_{\alpha,0,r}=\sum_{\alpha=0}^{n-1-\beta}l_{\alpha\beta}X_{\alpha,0,r+n},\qquad \beta=0,1,\ldots,n-1 \tag{3.5.1}$$

which yields a recurrence formula for $X_{\alpha,0,r+n}$. However the equation does not display exact recurrence. The matrix $l_{\alpha\beta}$ has the structure

$$l=n\begin{pmatrix} * & * & * & \frac{d}{dx} & 0\\ * & & \cdot^{\cdot^{\cdot}} & & \\ * & \cdot^{\cdot^{\cdot}} & & & \\ \frac{d}{dx} & & & 0 & \\ 0 & & & & \end{pmatrix} \tag{3.5.2}$$

Hence the right-hand side of Eq. (3.5.1) contains $X_{\alpha,0,r+n}$ only for $\alpha\leqq n-2$. The derivatives of $X_{\alpha,0,r+n}$ can be determined in succession from this equation. $X_{n-1,0,r}$ must be expressed in a different way. For $\beta=n-1$ the right-hand side is equal to zero and hence so is the left-hand side. In other words, $\sum m_{\alpha,n-1}X_{\alpha,0,r}=0$. If we make sure that $m_{n-1,n-1}\neq 0$, $X_{n-1,0,r}$ can be expressed in terms of $X_{\alpha,0,r}$ with smaller α. This equation for the elimination of $X_{n-1,0}$ can be represented in another form. If we come back to the Eq. (3.3.2) we see that the right-hand side involves a differential operator of the $(n-1)$th order. Rewriting the left-hand side as $-L(XL)_{-}+(LX)_{-}L$ and equating to zero the term with the nth derivative we obtain

$$([X,L]_{-})_{00}=0 \tag{3.5.3}$$

This equation is used to eliminate $X_{n-1,0}$. It can be written in another form. From

$$\left[\sum_{k=0}^{n}(-\partial/\partial\eta)^{k}u_{k}(\eta)X(\xi,\eta)-\sum_{k=0}^{n}u_{k}(\xi)(\partial/\partial\xi)^{k}X(\xi,\eta)\right]_{\xi=\eta=x}=0,$$

we obtain in succession

$$\left[\sum_{k=0}^{n}\left(\frac{-d}{dx}+\frac{\partial}{\partial\xi}\right)^{k} u_k(\eta)X(\xi,\eta)-\sum_{k=0}^{n} u_k(\xi)\left(\frac{\partial}{\partial\xi}\right)^{k} X(\xi,\eta)\right]_{\xi=\eta=x}=0,$$

and

$$\sum_{k=0}^{n} u_k\left\{\sum_{\alpha=0}^{k}\binom{k}{\alpha}(-d/dx)^{\alpha} X_{k-\alpha,0}-X_{k,0}\right\}=0,$$

i.e.

$$nX_{n-1,0}=-\sum_{k=0}^{n-2} u_k\sum_{\alpha=1}^{k}\binom{k}{\alpha}(-d/dx)^{\alpha-1}X_{k-\alpha,0}+C, \tag{3.5.4}$$

3.6. Proposition. *A solution of Eq.* (3.3.1) *is determined to within* $n-1$-*equivalence by the set of constants in* $X_{k,0,\bar{r}}$, $k\leqq n-1$.

Proof. Let X be a solution without constants in its terms. Let $\bar{r}$ be the smallest number for which at least one of $X_{k,0,\bar{r}}$, $k\leqq n-1$ does not vanish. The recurrence formulae (3.5.1) for $r=\bar{r}-n$ give $\sum l_{\alpha\beta}X_{\alpha,0,\bar{r}}=0$. Taking into account (3.5.2) we obtain $dX_{n-2,0,\bar{r}}/dx=0$ and $X_{n-2,0,\bar{r}}=C$. By assumption all the constants are zero, and hence $X_{n-2,0,\bar{r}}=0$. We find in succession $X_{n-3,0,\bar{r}}=X_{n-4,0,\bar{r}}=\ldots=X_{0,0,\bar{r}}=0$. Then (3.5.4) yields $X_{n-1,0,\bar{r}}=0$. Thus all $X_{k,0,\bar{r}}$, $k\leqq n-1$ vanish, which contradicts our assumption.

3.7. Theorem. *The resolvents and only the resolvents satisfy equation* (3.3.1).

Proof. The fact that the resolvents are solutions of the equation is obvious, namely, according to the definition of a resolvent $\hat{L}X$ and $X\hat{L}\in R_+$ in this case. Thus the subscripts + in the equation can be omitted, and the equation is satisfied trivially. The converse follows from the fact that, as was already seen (in 3.2), resolvents can have an arbitrary set of constants in the coefficients $X_{j,0,r}$, $j\leqq n-1$ and these constants determine the solution uniquely.

4. Hamiltonian structures

4.1. We extend the ring A with the help of the independent differential generators $v_0, v_1, \ldots$, i.e. we consider the differential polynomials in v_k with coefficients belonging to A. The resulting differential ring is denoted as $A(v_0, v_1, \ldots)$, or simply $A(v)$. Accordingly, $\tilde{A}(v)=A(v)/(dA(v)/dx)$. The elements of $\tilde{A}(v)$ are denoted $\tilde{f}(v)=\int f(v)dx$ where $f(v)\in A(v)$. Each $\tilde{f}(v)\in\tilde{A}(v)$ generates a mapping $R_+\to\tilde{A}$, namely

$$L=\sum u_k(d/dx)^k\mapsto\tilde{f}(L)=\int f(u)dx\in\tilde{A}$$

where $f(u)$ is the result of the substitution of the coefficients $u_k\in A$ of the operator L for the symbols v_k. Therefore the elements of $A(v)$ are called functions on R_+. Let us consider a formal expression $M(v)=\sum g_k(v)(d/dx)^k$ (with a finite number of terms). This formal differential operator gives rise to a mapping $R_+ \to R_+$. Namely,

$$L \xrightarrow{M(L)} \sum_k g_k(u)(d/dx)^k$$

where $g_k(u)$ is the result of the substitution of the coefficients $u_k \in A$ of the operator $L \in R_+$ for the symbols v_k. The mapping $M(L)$ is called a vector field on R_+.

4.2. Now we define the action of vector fields in R_+ on the functions on R_+. Let

$$M(L)\colon L=\sum_{k=0}^{n} u_k(d/dx)^k \mapsto \sum_{k=0}^{m} g_k(u)(d/dx)^k, \quad m \leqq n,$$

and

$$\tilde{f}(L)=\int f(u)\,dx.$$

Put

$$M(L)\widetilde{f(L)}=\int \sum_{k,i} \partial f(u)/\partial u_k^{(i)} \cdot g_k^{(i)}(u)\,dx. \tag{4.2.1}$$

Integrating by parts we derive

$$M(L)\widetilde{f(L)}=\int \sum_k \delta f(u)/\delta u_k \cdot g_k(u)\,dx \tag{4.2.2}$$

where

$$\delta f(u)/\delta u_k=\sum_i (-d/dx)^i\, \partial f/\partial u_k^{(i)}$$

is the variational derivative.

Not that in a special case when $\widetilde{f(L)}$ depends on L linearly (i.e. $f(u)$ depends linearly on $u_k^{(i)}$) formula (4.2.1) implies that $M(L)\widetilde{f(L)}=f(M(L))$, i.e. the coefficients of the operator $M(L)=\sum g_k(u)(d/dx)^k$ should be substituted into $f(L)$ for the coefficients of the operator L.

If $\widetilde{f(L)}$ depends on L quadratically, i.e. $f(u)=f_1(u,u)$ where f_1 is a bilinear form which can be written as $\widetilde{f(L)}=\widetilde{f_1(L,L)}$, then

$$M(L)\widetilde{f_1(L,L)}=\widetilde{F_1(M(L),L)}+\widetilde{f_1(L,M(L))}$$

etc.

We present also an alternative form of (4.2.2). Let X_f be a Volterra operator for which

$$(X_f)_{p,0}=\delta f/\delta u_p, \qquad p \leqq n.$$

Then

$$M(L)\widetilde{f(L)}=(M(L),X_p)=\operatorname{Tr}(M(L)X_f).$$

4.3. Now we define a mapping of R_- into the space of vector fields on R_+ thus:

$$\begin{aligned} X \mapsto M_X(L) &= (\hat{L}(X\hat{L})_+ - (\hat{L}X)_+\hat{L})\cdot z^{-n} \\ &= z^{-n}(L(XL)_+ - (LX)_+L) - [X,L]_+. \end{aligned} \tag{4.3.1}$$

As is already known if L is an operator (3.1.1), the kernel of the mapping (4.3.1) is formed of resolvents (to within the $n-1$-equivalence). Note that there is a limiting case of this mapping when $z\to\infty$; we denote it $M_X^\infty(L)$. We have

$$M_X^\infty(L)=[L,X]_+. \tag{4.3.2}$$

The kernel of this mapping consists of first terms of the expansion of the resolvents in z^{-1}. (The right-hand side of Eq. (4.3.2) contains only the coefficients $X_{p,q}$, $p+q\leqq n-1$, and they must be expanded in series in z^{-1}). It is sufficient to take only the basic resolvents $Q^{[j]}$, $j\leqq n-1$ (see 3.2). Thus the kernel of the mapping (4.3.2) is n-dimensional.

4.4. We calculate the commutator of the vector fields $M_X(L)$ and $M_Y(L)$ which is denoted as $[[M_X(L), M_Y(L)]]$. The commutator is a vector field $M(L)$ such that for any $\tilde{f}(L)$

$$M(L)\tilde{f}(L)=M_X(L)M_Y(L)\tilde{f}(L)-M_Y(L)M_X(L)\tilde{f}(L).$$

It the fields depend on L linearly, the rule for calculating the commutator is

$$[[M_X(L), M_Y(L)]]=M_Y(M_X(L))-M_X(M_Y(L)).$$

(This means that in the first term $M_X(L)$ must be substituted for the argument L of the field $M_Y(L)$ and the same in the second term.) If the fields depend on L polylinearly then one of the fields must be substituted into the other for all the arguments in succession.

The field (4.3.1) depends on L quadratically. We find

$$\begin{aligned}[[M_X(L), M_Y(L)]]=z^{-2n}\{&M_X(L)(Y\hat{L})_+ +\hat{L}(YM_X(L))_+\\ &-(M_X(L)Y)_+\hat{L}-(\hat{L}Y)_+M_X(L)\}-(X\leftrightarrow Y).\end{aligned}$$

(The symbol $(X\leftrightarrow Y)$ has the same meaning as above.) Now

$$\begin{aligned}[[M_X(L), M_Y(L)]]=z^{-2n}\{&(\hat{L}(X\hat{L})_+-(\hat{L}X)_+\hat{L})(Y\hat{L})_+ +\hat{L}(Y(\hat{L}(X\hat{L})_+\\ &-(\hat{L}X)_+\hat{L}))_+ -((\hat{L}(X\hat{L})_+ -(\hat{L}X)_+\hat{L})Y)_+\hat{L}\\ &-(\hat{L}Y)_+(\hat{L}(X\hat{L})_+-(\hat{L}X)_+\hat{L})\}-(X\leftrightarrow Y).\end{aligned}$$

The two terms $(\hat{L}X)_+\hat{L}(Y\hat{L})_+$ and $(\hat{L}Y)_+\hat{L}(X\hat{L})_+$ mutually cancel (if the terms meant by the symbol $(X\leftrightarrow Y)$ are taken into account), and we have

$$\begin{aligned}z^{-2n}\{&\hat{L}((X\hat{L})_+(Y\hat{L})_+ +(Y\hat{L}(X\hat{L})_+)_+ -(Y(\hat{L}X)_+\hat{L})_+)+(-(\hat{L}(X\hat{L})_+Y)_+\\ &+((\hat{L}X)_+\hat{L}Y)_+ +(\hat{L}Y)_+(\hat{L}X)_+)\hat{L}-(X\leftrightarrow Y).\end{aligned}$$

Owing to $(X\leftrightarrow Y)$ it is possible to interchange X and Y in each of the terms supplying it with the opposite sign. So we transform

$$\begin{aligned}(X\hat{L})_+(Y\hat{L})_+ +(Y\hat{L}(X\hat{L})_+)_+ &=((X\hat{L})_+(Y\hat{L})_+)_+ -(X\hat{L}(Y\hat{L})_+)_+\\ &=-((XL)_-(YL)_+)_+ =-((XL)_-YL)_+\end{aligned}$$

etc. We obtain

$$\begin{aligned}[[M_X(L), M_Y(L)]] &= z^{-2n}\{\hat{L}(-(X\hat{L})_- Y\hat{L} - Y(\hat{L}X)_+\hat{L})_+ \\ &\quad + (-\hat{L}(X\hat{L})_+ Y + \hat{L}X(\hat{L}Y)_-)_+\hat{L}\} - (X \leftrightarrow Y) \\ &= z^{-2n}\{\hat{L}((X(\hat{L}Y)_+ - (X\hat{L})_- Y)\hat{L})_+ \\ &\quad + (\hat{L}(X(\hat{L}Y)_- - (X\hat{L})_+ Y))_+\hat{L}\} - (X \leftrightarrow Y).\end{aligned}$$

Further, $X(\hat{L}Y)_- - (X\hat{L})_+ Y = -X(\hat{L}Y)_+ + (X\hat{L})_- Y$. Moreover, the expression $Z = X(\hat{L}Y)_+ - (X\hat{L})_- Y$ can be equipped with the subscript "$-$" since if we substitute this expression with the opposite subscript "$+$" into the formula we shall obtain zero:

$$\hat{L}(Z_+\hat{L})_+ - (\hat{L}Z_+)_+\hat{L} = \hat{L}Z_+\hat{L} - \hat{L}Z_+\hat{L} = 0.$$

If all this is taken into account, we have

Theorem. *The commutator of two vector fields $M_X(L)$, and $M_Y(L)$ is a vector field of the same type:*

$$[[M_X(L), M_Y(L)]] = -M_{[X,Y]_L}(L) \tag{4.4.1}$$

where $[X, Y]_L$ is a Volterra operator:

$$[X, Y]_L = z^{-n}(X(\hat{L}Y)_+ - (X\hat{L})_- Y)_- - (X \leftrightarrow Y). \tag{4.4.2}$$

4.5. Remarks. 1) The Volterra operator $[X, Y]_L$ depends on the point $L \in R_+$ where the commutator of vector fields is calculated.

2) The set of vector fields $M_X(L)$ taken for all $L \in R_+$ and all $X \in R_-$ form an involutive distribution in the Frobenius sense. In a finite-dimensional situation this means that the space is stratified into submanifolds to which the fields $M_X(L)$ are tangent. We do not discuss here in what sense this is correct in our infinite-dimensional case. In fact we do not need these submanifolds. (In the limitting case $z \to \infty$, i.e. for the vector fields $M_X^\infty(L) = [L, X]_+$, these submanifolds are the orbits of the coadjoint action of the group of the Volterra operators with the augmented unity, see [4], [5]).

3) On an involutive distribution in a tangent bundle differential forms can be defined. The Lie differential of a form also make sense:

$$\begin{aligned}d\omega(\xi_1, \xi_2, \ldots) &= \sum_i (-1)^{i-1}\xi_i\omega(\xi_1, \ldots, \hat{\xi}_i, \ldots) \\ &\quad + \sum_{i<j} (-1)^{i+j}\omega([\xi_i, \xi_j], \xi_1, \ldots, \hat{\xi}_i, \ldots, \hat{\xi}_j, \ldots).\end{aligned}$$

In the infinite-dimensional case this formula determines the differential of a form defined on our integrable distribution.

4) When $z \to \infty$, we obtain $[X, Y]_L \to [X, Y]$.

5) It can be proved in the usual way (see e.g. [6]) that the definition of d is local. This means that $d\omega(\xi_1, \xi_2, \ldots)$ depends on the fields $\xi_1, \xi_2, \ldots$ at only one

point, although formally the definition involves the fields in the vicinity of the point.

4.6. Let

$$\omega(M_X(L), M_Y(L)) = \tfrac{1}{2}(L, [X, Y]_L) \tag{4.6.1}$$

be a form on the fields of the type $M_X(L)$ at the point L. The form is well defined (this means here that the definition depends on $M_X(L)$ and not on X), which follows from the

Lemma. *The form* (4.6.1) *can be written as*

$$\omega(M_X(L), M_X(L)) = (M_X(L), Y) = -(M_Y(L), X). \tag{4.6.2}$$

The proof can be carried out using a simple calculation:

$$\begin{aligned}
&\omega(M_X(L), M_Y(L))\\
&\quad = -\tfrac{1}{2} z^{-n} \operatorname{Tr}\{(\hat{L}(X(\hat{L}Y)_+ - (X\hat{L})_- Y)_-)\} - (X \leftrightarrow Y)\\
&\quad = \tfrac{1}{2} z^{-n} \operatorname{Tr}\{(\hat{L}X)_- \hat{L}Y - \hat{L}(X\hat{L})_- Y\} - \tfrac{1}{2} z^{-n} \operatorname{Tr}\{\hat{L}Y(\hat{L}X)_+ - Y\hat{L}(X\hat{L})_+\}\\
&\quad = \tfrac{1}{2} z^{-n}((\hat{L}X)_- \hat{L} - \hat{L}(X\hat{L})_-, Y) - \tfrac{1}{2} z^{-n}((\hat{L}X)_+ \hat{L} - \hat{L}(X\hat{L})_+, Y)\\
&\quad = (M_X(L), Y).
\end{aligned}$$

4.7. Theorem. *The form ω is closed.*

Proof. We have

$$\begin{aligned}
d\omega(M_X(L), M_Y(L), M_Z(L)) &= M_X(L)\omega(M_Y(L), M_Z(L))\\
&\quad - \omega([[M_X(L), M_Y(L)]], M_Z(L)) + (\text{cycle}).
\end{aligned}$$

The symbol (cycle) denotes the sum of terms which are obtained from the foregoing expression by the cyclic permutations. For brevity we do not write the argument L.

$$\begin{aligned}
M_X\,\omega(M_Y, M_Z) &= z^{-n} M_X(M_Y, Z) = z^{-n} M_X \operatorname{Tr}(\hat{L}(Y\hat{L})_+ - (\hat{L}Y)_+ \hat{L})Z\\
&= z^{-n} \operatorname{Tr}(M_X(Y\hat{L})_+ + \hat{L}(YM_X)_+ - (M_XY)_+ \hat{L} - (\hat{L}Y)_+ M_X)Z\\
&= z^{-n} \operatorname{Tr} M_X[(Y\hat{L})_+ Z + (Z\hat{L})_- Y - Y(\hat{L}Z)_- - Z(\hat{L}Y)_+]\\
&= z^{-n} \operatorname{Tr} M_X[-(Y\hat{L})_- Z + (Z\hat{L})_- Y + Y(\hat{L}Z)_+ - Z(\hat{L}Y)_+]\\
&= z^{-n} \operatorname{Tr} M_X[Y(\hat{L}Z)_+ - (Y\hat{L})_- Z] - (Z \leftrightarrow Y) = (M_X, [Y, Z]_L).
\end{aligned}$$

Taking into account the cyclic permutations we see that the two terms in the expression for $d\omega$ are equal. Now we must prove that

$$(M_X, [Y, Z]_L) + (\text{cycle}) = 0,$$

that is

$$(M_X, Y(LZ)_+ - (YL)_- Z) + (\text{permut}) = 0.$$

The symbol (permut) designates the sum of all the expressions which differ from the foregoing one by permutations of X, Y, and Z. The sign must be taken depending on the signature of the permutation. We find

$$\begin{aligned}(M_X, Y(\hat{L}Z)_+ -(Y\hat{L})_- Z)&+(\text{permut})\\ = -(\hat{L}(X\hat{L})_- -(\hat{L}X)_- \hat{L}, &Y(\hat{L}Z)_+)-(\hat{L}(X\hat{L})_+ \\ -(\hat{L}X)_+ \hat{L}, &(Y\hat{L})_- Z)+(\text{permut}).\end{aligned}$$

Two of these terms cancel out:

$$\begin{aligned}&\operatorname{Tr}(-\hat{L}(X\hat{L})_- Y(\hat{L}Z)_+ +(\hat{L}X)_+ \hat{L}(Y\hat{L})_- Z+(\text{permut})\\ &=\operatorname{Tr}(-(\hat{L}Z)_+ \hat{L}(X\hat{L})_- Y+(\hat{L}X)_+ \hat{L}(Y\hat{L})_- Z)+(\text{permut})=0.\end{aligned}$$

The remaining terms are

$$\begin{aligned}&-\operatorname{Tr}\{(\hat{L}X)_- \hat{L}Y(\hat{L}Z)_+ -(Z\hat{L})(X\hat{L})_+(Y\hat{L})_-\}+(\text{permut})\\ =\;&\operatorname{Tr}\{\hat{L}X(\hat{L}Y)_+(\hat{L}Z)_- -(X\hat{L})(Y\hat{L})_+(Z\hat{L})_-\}+(\text{permut}).\end{aligned}$$

Now we prove

Lemma. *For any operators A, B, and C*

$$\operatorname{Tr}(AB_+ C_-)+(\text{cycle})=\operatorname{Tr}(ABC). \tag{4.7.1}$$

Proof.

$$\begin{aligned}&\operatorname{Tr}(AB_+ C_-)+(\text{cycle})=\tfrac{2}{3}\operatorname{Tr}(AB_+ C_-)+\tfrac{1}{3}\operatorname{Tr}(A_+ B_+ C_-)\\ &\quad+\tfrac{1}{3}\operatorname{Tr}(A_- B_+ C_-)+(\text{cycle})=\tfrac{2}{3}\operatorname{Tr}(AB_+ C_-)+\tfrac{1}{3}\operatorname{Tr}(A_+ B_+ C)\\ &\quad+\tfrac{1}{3}\operatorname{Tr}(A_- BC_-)+(\text{cycle})=\tfrac{1}{3}\operatorname{Tr}(2AB_+ C_- +A_+ B_+ C+A_- BC_-)+(\text{cycle})\\ &=\tfrac{1}{3}\operatorname{Tr}(AB_+ C+ABC_-)+(\text{cycle})=\tfrac{1}{3}\operatorname{Tr}(AB_+ C+AB_- C)+(\text{cycle})\\ &=\tfrac{1}{3}\operatorname{Tr}(ABC)+(\text{cycle})=\operatorname{Tr}(ABC).\end{aligned}$$

Applying this lemma to the expression calculated above we obtain

$$\operatorname{Tr}(\hat{L}X\hat{L}Y\hat{L}Z-X\hat{L}Y\hat{L}Z\hat{L})=0.$$

Thus $d\omega(M_X, M_Y, M_Z)=0$, for any X, Y, and Z, i.e. $d\omega=0$. This completes the proof of the theorem.

A special case of the form ω for $z\to\infty$ is

$$\omega^\infty([L,X]_+,[L,Y]_+)=([L,X]_+,Y)=-(L,[X,Y])$$

This form is also closed.

4.8. The formula (4.2.4) for the case when $M(L)=M_Y(L)$ for some $Y\in R_-$ is written

$$\begin{aligned}M_Y(L)\widetilde{f(L)}=(M_Y(L),X_{\tilde{f}})&=\omega(M_Y(L),M_{X_{\tilde{f}}}(L))\\ &=-(M_{X_{\tilde{f}}}(L),M_Y(L)).\end{aligned} \tag{4.8.1}$$

According to the general definition of the correspondence between functions and vector fields which can be established due to a symplectic form, we say that the field $M_{X_f}(L)$ corresponds to the function $\widetilde{f(L)}$:

$$\tilde{f} \mapsto M_{X_{\tilde{f}}}. \tag{4.8.2}$$

The Volterra operator $X_{\tilde{f}}$ does not depend on z and hence it cannot be a resolvent of L; whence it follows that the field $M_{X_{\tilde{f}}}$ which depends on z^n linearly cannot vanish identically although it is not excluded that for some particular values of z, finite or infinite, it can vanish.

4.9. The Poisson bracket of two functions $\widetilde{f(L)}$ and $\widetilde{g(L)}$ is determined by

$$\{\widetilde{f(L)}, \widetilde{g(L)}\}_z = M_{X_{\tilde{f}}}\widetilde{g(L)} = -M_{X_{\tilde{g}}}(L) = \omega(M_{X_{\tilde{f}}}(L), M_{X_{\tilde{g}}}(L)). \tag{4.9.1}$$

Theorem. *The mapping* (4.8.2) *possesses the property*

$$M_{X_{\{\tilde{f}, \tilde{g}\}_z}} = [[M_{X_{\tilde{f}}}, M_{X_{\tilde{g}}}]]. \tag{4.9.2}$$

Proof. We proved that the form ω is closed calculating the values of the form $d\omega$ on the three fields $M_X(L)$, $M_Y(L)$, and $M_Z(L)$ where X, Y, and Z were independent of L. Now use the fact that $d\omega$ is equal to zero on the three fields $M_{X_{\tilde{f}}}(L)$, $M_{X_{\tilde{g}}}(L)$, and $M_Z(L)$ where $X_{\tilde{f}}$, and $X_{\tilde{g}}$, depend on L. The possibility of this application of the Theorem 4.7 is based on the locality of the definition of $d\omega$ (see the end of 4.5)*.

Using (4.8.1), (4.9.1), and (4.6.2) we have

$$\begin{aligned}
0 &= d\omega(M_{X_{\tilde{f}}}, M_{X_{\tilde{g}}}, M_Z) = M_{X_{\tilde{f}}}\omega(M_{X_{\tilde{g}}}, M_Z) - M_{X_{\tilde{g}}}\omega(M_{X_{\tilde{f}}}, M_Z) \\
&\quad + M_Z\omega(M_{X_{\tilde{f}}}, M_{X_{\tilde{g}}}) - \omega([[M_{X_{\tilde{f}}}, M_{X_{\tilde{g}}}]], M_Z) \\
&\quad + \omega([[M_{X_{\tilde{f}}}, M_Z]], M_{X_{\tilde{g}}}) - \omega([[M_{X_{\tilde{g}}}, M_Z]], M_{X_{\tilde{f}}}) \\
&= -M_{X_{\tilde{f}}}M_Z\tilde{g} + M_{X_{\tilde{g}}}M_Z\tilde{f} + M_Z\{\tilde{f}, \tilde{g}\}_z - ([[M_{X_{\tilde{f}}}, M_{X_{\tilde{g}}}]], Z) + [[M_{X_{\tilde{f}}}, M_Z]]\tilde{g} \\
&\quad - [[M_{X_{\tilde{g}}}, M_Z]]\tilde{f} = -M_Z M_{X_{\tilde{f}}}\tilde{g} + M_Z M_{X_{\tilde{g}}}\tilde{f} + M_Z\{\tilde{f}, \tilde{g}\}_z - ([[M_{X_{\tilde{f}}}, M_{X_{\tilde{g}}}]], Z) \\
&= -M_Z\{\tilde{f}, \tilde{g}\}_z - ([[M_{X_{\tilde{f}}}, M_{X_{\tilde{g}}}]], Z) \\
&= (M_{X_{\{\tilde{f}, \tilde{g}\}_z}} - [[M_{X_{\tilde{f}}}, M_{X_{\tilde{g}}}]], Z).
\end{aligned}$$

This yields (4.9.2) by virtue of the arbitrariness of Z.

4.10. Theorem. *For the Poisson bracket* $\{\tilde{f}, \tilde{g}\}_z$ *the Jacobi identity holds:*

$$\{\{\tilde{f}, \tilde{g}\}_z, \tilde{h}\}_z + \{\{\tilde{g}, \tilde{h}\}_z, \tilde{f}\}_z + \{\{\tilde{h}, \tilde{f}\}_z, \tilde{g}\}_z = 0. \tag{4.10.1}$$

Proof. Let us find the image of the left-hand side of this equality under the mapping (4.8.2). If the left-hand side is nonzero, then the image is nonzero at least for one value of z. On the other hand according to Theorem 4.9, the

* What has been said provides an important improvement of the technique suggested in [4], [5] in comparison with our previous works [1], [3] where calculations were more cumbersome.

image of the Poisson bracket is the commutator of vector fields, and for the commutator the Jacobi identity is satisfied. Hence the image of the lefthand side of (4.10.1) must be zero. This contradiction proves the theorem.

4.11. Now the term "the Poisson bracket" is justified. We have obtained a one-parameter family of Poisson brackets linearly dependent on the parameter z^n. The limiting case for z is the following

$$\{f,g\}_\infty = ([L, X_{\tilde{f}}]_+, X_{\tilde{g}}) = \int \sum_{i+j\leqq n-1} (l_{ij}\delta\tilde{f}/\delta u_j)\delta\tilde{g}/\delta u_i dx$$

(l_{ij} as in 3.4).

4.12. The equation (4.8.1) can be written as $d\widetilde{f(L)} = -i(M_{X_{\tilde{f}}})\omega$ where i denotes the operator of substitution of a vector field into the form. This implies that the field $M_{X_{\tilde{f}}}$ preserves the form (the Lie derivative of the form along the field is equal to zero). In other words, $M_{X_{\tilde{f}}}$ is a Hamiltonian field.

4.13. Let $T^{[i]}$ be the canonical components of the exact resolvent. Let

$$\tilde{F}_r^{[i]} = \int T^{[i]}_{0,0,r} dx, \quad i=1,\ldots,n;\ r=n-1, n, \ldots \tag{4.13.1}$$

Theorem. *The identity holds*

$$\{F_r^{[i]}, F_s^{[j]}\}_z = 0.$$

Proof. We use the fact (see e.g. [3]) that

$$\delta T^{[i]}_{0,0}/\delta u_k = \partial/\partial(z^n)(T^{[i]}_{k,0})$$

i.e.

$$T^{[i]}_{0,0,r} = [(-r+n)/n]\, T^{[i]}_{k,0,r-n} \tag{4.13.2}$$

Eq. (3.4.3) for a resolvent gives

$$\sum_{\alpha=0}^{n-1} m_{\alpha\beta} T^{[i]}_{\alpha,0,r-n} = \sum_{\alpha=0}^{n-1-\beta} l_{\alpha\beta} T^{[i]}_{\alpha,0,r}, \quad \beta = 0, 1, \ldots, n-1$$

and

$$\sum_{\alpha=0}^{n-1} m_{\alpha\beta} T^{[j]}_{\alpha,0,s-n} = \sum_{\alpha=0}^{n-1-\beta} l_{\alpha\beta} T^{[j]}_{\alpha,0,s}, \quad \beta = 0, 1, \ldots, n-1.$$

Multiply the first equation by $T^{[j]}_{\beta,0,s-n}$ and the second by $T^{[i]}_{\beta,0,r-n}$, sum up with respect to β, and add together the two equations. Due to the skew-symmetry of the matrix $m_{\alpha,\beta}(m^*_{\alpha\beta} = -m_{\beta\alpha})$ the left-hand side of this equation is an exact derivative, and hence

$$\sum_{\alpha+\beta\leqq n-1} [(l_{\alpha\beta} T^{[i]}_{\alpha,0,r}) T^{[j]}_{\beta,0,s-n} - (l_{\alpha\beta} T^{[i]}_{\alpha,0,r-n}) T^{[j]}_{\beta,0,s}] \in dA/dx$$

Eq. (4.13.2) yields

$$\int \sum_{\alpha+\beta\leqq n-1}\left(l_{\alpha\beta}\frac{\delta T^{[i]}_{0,0,r+n}}{\delta u_\alpha}\right)\frac{\delta T^{[j]}_{0,0,s}}{\delta u_\beta}dx$$
$$=\frac{(r+n)s}{r(s+n)}\int \sum_{\alpha+\beta\leqq n-1}\left(l_{\alpha\beta}\frac{\delta T^{[i]}_{0,0,r}}{\delta u_\alpha}\right)\frac{\delta T^{[j]}_{0,0,s+n}}{\delta u_\beta}dx$$

which implies

$$\{F^{[i]}_{r+n}, F^{[j]}_s\}_\infty=\frac{(r+n)s}{r(s+n)}\{F^{[i]}_r, F^{[j]}_{s+n}\}_\infty.$$

The formula obtained can be iterated with the number s increasing and the number r decreasing until r becomes negative. Then $T^{[i]}_{0,0,r}=0$. This proves that (4.13.1) is valid for $z=\infty$. Then Eq. (4.13.3) implies that it is valid $z=0$ and for an arbitrary z since the general Poisson bracket is a linear combination of the Poisson brackets for $z=\infty$ and for $z=0$. This proves the theorem.

4.14. If one of the $F^{[i]}_r$ is taken as a Hamiltonian we obtain a Hamiltonian equation having an infinite set of first integrals $F^{[j]}_s$ in involution.

4.15. Note. Consider a submanifold of the operators (3.1.1) in R_+ for which $u_{n-1}=0$. In order that $M_X(L)$ be tangent to this submanifold at the point L it is necessary and sufficient that X satisfy the condition (3.5.3) (or its equivalent form (3.5.4)), which is the elimination condition for $X_{n-1,0}$.

The construction of vector fields corresponding to functions $f(L)$ is performed in some other way: the operator $X_{\tilde f}$ satisfies the condition $(X_{\tilde f})_{\alpha,0}=\delta\tilde f/\delta u_\alpha$ only for $\alpha\leqq n-2$, and $(X_{\tilde f})_{n-1,0}$ must be expressed according to the formula (3.5.4). The rest remains unchanged. Note that, generally, in this case the Poisson bracket depends on u_k not quadratically, but cubically, owing to u_k in Eq. (3.5.4).

5. Lax *P*, *L* pairs

5.1. We have constructed Hamiltonian systems having infinite sets of first integrals in involution without using Lax pairs. Nevertheless, for the sake of completeness, we demonstrate how they can be written in the Lax form.

This form is written

$$L_t=[P,L], \tag{5.1.1}$$

where t is a parameter on which the coefficients u_k of the operator L depend. It is assumed that the coefficients of the differential operator P are differential polynomials in u_k. Now we must rewrite the Hamiltonian systems constructed in 4.14 in the form (5.1.1).

We use the simplest mapping $X\mapsto M_X(L)$ i.e. $X\mapsto[L,X]_+$. The Hamiltonian vector field corresponding to the Hamiltonian $\tilde F^{[i]}_{r+n}$ is $[L, X_{\tilde F^{[i]}_{r+n}}]_+$. Remembering (4.13.2) we see that

$$[L, X_{\tilde{F}^{[i]}_{r+n}}]_+ = ([L, T^{[i]}]_+)_r \cdot \frac{-r}{n}$$

where the subscript r designates the coefficient of z^{-r} in the series in z^{-1}. Thus, to within an unimportant constant $-r/n$, the Hamiltonian equation is the following:

$$L_t = [L, T^{[i]}_r] \tag{5.1.2}$$

(the operator L does not depend on z). The form of this equation is close to the Lax form. The only distinction is that $T^{[i]}_r$ is a Volterra integral and not a differential operator.

5.2. All the coefficients $T^{[i]}_{p,q}$ are one-way infinite series in z^{-1}

$$T^{[i]}_{p,q} = \sum_{r=r_0(p,q)} T^{[i]}_{p,q,r} z^{-r}.$$

However the set of all $r_0(p,q)$ is unbounded (the operator $T^{[i]}$ is unbounded, see (2.9)). Thus $T^{[i]}$ can be regarded as a two-way infinite series in z^{-1}.

According to the definition of a resolvent the expression $\Delta^{[i]} = (L+\zeta)T^{[i]}$ is a differential operator ($\zeta = -z^n$). It can be expanded into a one-way infinite series. We also consider the expression $(L+\zeta)^{-1}$; by definition,

$$(L+\zeta)^{-1} = \sum_{k=0} (-L)^k \zeta^{-k-1}.$$

Two series $\Delta^{[i]}$ and $(L+\zeta)^{-1}$ can be multiplied:

$$P = (L+\zeta)^{-1} \Delta^{[i]}. \tag{5.2.1}$$

It would be incorrect to conclude that

$$P = (L+\zeta)^{-1}((L+\zeta)T^{[i]}) = ((L+\zeta)^{-1}(L+\zeta))T^{[i]} = T^{[i]}$$

because the multiplication of series is not associative when one of them is a two-way series (even if it is assumed that all the multiplications make sense). In fact, the left-hand side of this "equality" is a differential operator, whereas the right-hand side is the integral one.

Lemma. *The operator* $K = P - T^{[i]}$ *commutes with* L.

Proof. Calculate

$$(L+\zeta)K = (L+\zeta)((L+\zeta)^{-1}\Delta^{[i]}) - (L+\zeta)T^{[i]}$$
$$= ((L+\zeta)(L+\zeta)^{-1})\Delta^{[i]} - \Delta^{[i]} = 0.$$

(The associative law holds for three operators $(L+\zeta)$, $(L+\zeta)^{-1}$, and $\Delta^{[i]}$ because they are one-way infinite series in z^{-1}.) Further,

$$((L+\zeta)T^{[i]})(L+\zeta) = (L+\zeta)(T^{[i]}(L+\zeta))$$

because two terms here are finite series. This implies that

$$\begin{aligned}K(L+\zeta)&=((L+\zeta)^{-1}\Delta^{[i]})(L+\zeta)-T^{[i]}(L+\zeta)\\&=(L+\zeta)^{-1}(((L+\zeta)T^{[i]})(L+\zeta))-T^{[i]}(L+\zeta)\\&=(L+\zeta)^{-1}((L+\zeta)T^{[i]}(L+\zeta)))-T^{[i]}(L+\zeta)=0.\end{aligned}$$

The operator K commutes with $L+\zeta$ and hence with L as well.

Corollary. $[L,P]=[L,T^{[i]}]$, *and Eq.* (5.1.2) *can be written as*

$$L_t=[L,P_r] \tag{5.2.2}$$

Now the equation has the Lax form as desired.

Note that the lemma expresses a remarkable fact: the P-operator and the resolvent are differential and integral parts of an element of the ring R commutating with the operator L (i.e. the element of the centralizer of L).

The operator P_r is obtained as the coefficient of z^{-r} in the expansion of

$$P=(L+\zeta)^{-1}\cdot\Delta^{+}(T^{[k]}),\qquad \zeta=-z^n. \tag{5.2.3}$$

The generating function P generalizes Dubrovin's function which was constructed for $n=2$, see [8].

References

1. Gelfand, I.M., Dikij, L.A.: Fractional powers of operators and Hamiltonian systems, Funkt. Anal. **10** (4) (1976) 13–39.
2. Gelfand, I.M., Dikij, L.A.: Resolvents and Hamiltonian systems, Funkt. Anal. **11** (2) (1977) 11–27.
3. Gelfand, I.M., Dikij, L.A.: Jet calculus and nonlinear Hamiltonian system, Funkt. Anal. **12** (2) (1978) 8–23.
4. Lebedjev, D.R., Manin, Yu.I.: The Gelfand-Dickey Hamiltonian operator and the coadjoint representation of the Volterra group. Funkt. Anal. **13** (4) 40–46.
5. Adler, M.: On a trace functional for formal pseudodifferential operators and the symplectic structure of the Korteweg-de Vries type equation, Invent. Math. **50** (3) (1979) 219–248.
6. Bishop, R.L., Crittenden, R.J.: Geometry of manifolds. Academic Press, 1964.
7. Gelfand, I.M., Dikij, L.A.: A Lie algebra structure in the formal variational calculus, Funkt. Anal. **10** (1) (1976) 18–25.
8. Dubrovin, B.A.: The periodic problem for the Korteweg-de Vries equation in the class of finite-gap potentials, Funkt. Anal. **9** (3) (1975) 41–52.

12.

(with L. A. Dikij)

Asymptotic behaviour of the resolvent of Sturm-Liouville equations and the algebra of the Korteweg-de Vries equations

Usp. Mat. Nauk **30** (5) (1975) 67–100. MR **58**:22746, Zbl. **461**:35072
[Russ. Math. Surv. **30** (5) (1975) 77–113]

This paper is concerned with a group of problems associated with recent results on non-linear equations of Korteweg–de Vries type. The second chapter is basically of a survey character and investigates connections between these equations and trace formulae. In the first chapter we develop a new algebraic formalism of the calculus of variations.

Contents

Introduction 77
Chapter 1. The algebra of polynomials in u, u', u'' . . . and the formal calculus of variations 78
Chapter 2. The asymptotic expansion of the resolvent of a Sturm–Liouville equation 94
Chapter 3. The Korteweg–de Vries equation 102
Appendix 1. A generating function for the polynomials $R_l[u]$ . . 107
Appendix 2. Another recurrence method for obtaining the coefficients of the asymptotic expansion of the resolvent 109
References 112

Introduction

Work done in the last 20 years [1]–[3] (generalizing [4]) has provided an asymptotic expansion of the kernel of the resolvent of the equation

$$(1) \qquad -\frac{d^2\varphi}{dx^2}+[u(x)+\zeta]\,\varphi=0$$

in powers of ζ^{-1}. The motivation for this problem came from the desire to give a meaning to traces of positive powers of differential operators (see also [6]–[8]). But the impetus for the study of the asymptotic behaviour of the resolvent came from different directions, and this study has been pursued by many authors (see [5], [9] and others) and continues to the present day [10]. Recently these problems have acquired a completely new significance owing to a close connection with the theory, dating from the mid-60's, of the integrability of the non-linear Korteweg–de Vries equation

[11]–[13]. The connection between the Korteweg–de Vries equation and the theory of traces was pointed out in [13]. Novikov and Dubrovin in their papers [14] and [15] made substantial new progress in the theory of the Korteweg–de Vries equation. Although their papers are formally concerned only with the case of periodic boundary conditions for (1), they contain many important algebraic relations of a local character that are independent of the boundary conditions. This aspect of these papers to some extent stimulated us to write this survey. Finally, we refer to the very recent paper by Marchenko [16], which also deals with the periodic problem; our account has some points of contact with this paper, which we shall note below, and we also mention the latest paper by Lax [17].

Perhaps the most surprising facet, which we did not find mentioned in any of the older papers and which became clear to us only after Novikov's work, is the fact that the coefficients of the asymptotic expansion of the kernel of the resolvent, which we had obtained, can be taken as Lagrange functions, after which we obtain a fully integrable Hamiltonian system. This fact forms the core of the present paper. We try to explain its algebraic nature, independent of the boundary conditions for the Sturm–Liouville equations and of the corresponding spectral methods. In this context we find it advantageous to develop a special algebra of polynomials in a function $u(x)$ and its derivatives u', u'', ..., which includes elements of a formal calculus of variations and of formal Hamiltonian mechanics in the ring of such polynomials. Such an algebra arose in connection with what is now called formal geometry ([18], [19]). We hope to return to this problem in connection with work on the Gauss–Bonnet theorems for characteristic classes and on eigenvalues of partial differential operators (see also [19]).[1]

CHAPTER 1

The algebra of polynomials in u, u', u'' ... and the formal calculus of variations

1. Polynomials in u, u', u'' ... We consider polynomials in several symbols u, u', u'', ..., $u^{(k)}$, ... with coefficients in a field of characteristic zero. These polynomials form an algebra over this field, which we denote by A. By A_N we denote the set of all homogeneous polynomials of degree N. Then

$$A = \sum A_N. \tag{1}$$

[1] NOTE IN PROOF. After this paper was sent to the printers, we became acquainted with two extremely interesting and useful preprints: P. D. Lax, Periodic solutions of the K–de V equation; H. P. McKean and P. van Moerbeck, The spectrum of Hill's equation. In his paper Lax gives explicit formulae similar to ours. We also point out the very interesting papers of M. Kac.

We define the differential order of the monomial $a_{k_1 \dots k_r} u^{(k_1)} \dots u^{(k_r)}$ to be $k = k_1 + \dots + k_r$ (where $u^{(0)}$ is defined to be u). For brevity $[u]$ is used to denote the set of arguments $u, u', u'', \dots$ of the polynomials.

In our algebra it makes sense to talk of linear differential operators, or vector fields,

$$\xi = \sum_{l=0}^{\infty} \xi_l [u] \frac{\partial}{\partial u^{(l)}}, \tag{2}$$

with coefficients $\xi_1 \in A$. The set of such vector fields forms a module TA over A. Among the vector fields we distinguish a canonical one, which we denote by d/dx; it is defined as follows:

$$\frac{d}{dx} = \sum_{l=0}^{\infty} u^{(l+1)} \frac{\partial}{\partial u^{(l)}}. \tag{3}$$

From this definition it follows that

$$\frac{d}{dx} u^{(l)} = u^{(l+1)}, \quad u^{(l)} = \frac{d}{dx} \dots \frac{d}{dx} u = \left(\frac{d}{dx}\right)^l u. \tag{4}$$

The commutator of two vector fields ξ and η is

$$[\xi, \eta] = \sum_{l=0}^{\infty} \sum_{k=0}^{\infty} \left[\xi_k \left(\frac{\partial}{\partial u^{(k)}} \eta_l \right) - \eta_k \left(\frac{\partial}{\partial u^{(k)}} \xi_l \right) \right] \frac{\partial}{\partial u^{(l)}}.$$

Therefore, the commutator of an arbitrary field with the canonical field $\frac{d}{dx}$ is

$$\left[\xi, \frac{d}{dx} \right] = \sum_{l=0}^{\infty} \left(\xi_{l+1} - \frac{d}{dx} (\xi_l) \right) \frac{\partial}{\partial u^{(l)}}. \tag{5}$$

Fields that commute with the canonical field form a submodule. These fields are characterized by the property $\xi_l \, [u] = (d/dx)^l \, \xi_0 [u]$, where $\xi_0 [u]$ can be any element of A. Between A and the module of vector fields commuting with d/dx there is an isomorphism (of A-modules). But, in addition, the vector fields form a Lie algebra, and this enables us to transfer this structure to A. The commutator of two elements of A is given by

$$[f[u], g[u]] = \sum_{k=0}^{\infty} \left[\left(\left(\frac{d}{dx} \right)^k f \right) \frac{\partial}{\partial u^{(k)}} g - \left(\left(\frac{d}{dx} \right)^k g \right) \frac{\partial}{\partial u^{(k)}} f \right].$$

We define a differential linear form as a formal finite sum $\omega = \Sigma \omega_l [u] \; du^{(l)}$, where $\omega_l \in A$ and $du^{(l)}$ are certain new symbols. The value of the form on the vector field ξ is given by $\omega(\xi) = \Sigma \omega_l \xi_l$. Differential forms of higher orders are defined similarly. Exterior products, the differential of a form, and the convolution $i(\xi)\omega$ of a vector field with a form are defined in the usual way. Then all the usual formulae are satisfied, for example, the Lie formula for a differential

(6) $$d\omega(\xi, \eta, \zeta, \ldots) = \xi\omega(\eta, \zeta, \ldots) - \eta\omega(\xi, \zeta, \ldots) + \ldots \ldots - \omega([\xi, \eta], \zeta, \ldots) + \omega([\xi, \zeta], \eta, \ldots) - \ldots$$

The fact that the symbol $du^{(l)}$ is equal to the differential of $u^{(l)}$ is easily verified.

The set of all r-forms

$$\omega = \sum_{k_1 \ldots k_r} \omega_{k_1 \ldots k_r}[u]\, du^{(k_1)} \wedge \ldots \wedge du^{(k_r)},$$

with coefficients in A_N, is denoted by $A_N^{(r)}$. We set $A^{(r)} = \Sigma A_N^{(r)}$. We define the differential order of the form

$$cu^{(l_1)} \ldots u^{(l_t)}\, du^{(k_1)} \wedge \ldots \wedge du^{(k_r)}$$

to be $k = l_1 + \ldots + l_t + k_1 + \ldots + k_r$. Further, we define the action of d/dx on forms by:

(7) $$\frac{d}{dx} \sum \omega_{k_1 k_2 \ldots k_r} du^{(k_1)} \wedge du^{(k_2)} \wedge \ldots \wedge du^{(k_r)} =$$
$$= \sum \left(\frac{d}{dx} \omega_{k_1 k_2 \ldots k_r}\right) du^{(k_1)} \wedge du^{(k_2)} \ldots \wedge du^{(k_r)} +$$
$$+ \sum \omega_{k_1 k_2 \ldots k_r} du^{(k_1+1)} \wedge du^{(k_2)} \ldots \wedge du^{(k_r)} + \ldots$$
$$\ldots + \sum \omega_{k_1 \ldots k_r} du^{(k_1)} \wedge du^{(k_2)} \wedge \ldots \wedge du^{(k_r+1)},$$

that is, all the $u^{(k)}$ are differentiated both where they occur in coefficients and also where they stand under the differential sign. The operator d/dx commutes with d: $\left[\frac{d}{dx}, d\right] = 0$.

We define the action of the operator $\partial/\partial u^{(i)}$ on forms as

(8) $$\frac{\partial}{\partial u^{(i)}} \sum \omega_{k_1 k_2 \ldots k_r} du^{(k_1)} \wedge du^{(k_2)} \wedge \ldots \wedge du^{(k_r)} = \sum \frac{\partial \omega_{k_1 k_2 \ldots k_r}}{\partial u^{(i)}} du^{(k_1)} \wedge du^{(k_2)} \wedge \ldots \wedge du^{(k_r)}.$$

WARNING. Observe that $\frac{d}{dx}\, \omega \neq \Sigma u^{(l+1)} \left(\frac{\partial}{\partial u^{(l)}}\, \omega\right)$. Indeed, the right-hand side is only the first term in (7). The operator d/dx is defined by (3) only in its action on functions, that is, 0-forms. We mention the commutation relations

(9) $$\left[\frac{\partial}{\partial u^{(i)}}, \frac{d}{dx}\right] = \begin{cases} \dfrac{\partial}{\partial u^{(i-1)}}, & i \neq 0, \\ 0, & i = 0, \end{cases} \qquad \left[\frac{\partial}{\partial u^{(i)}}, d\right] = 0.$$

The following operator is important: it is the variational derivative

(10) $$\frac{\delta}{\delta u} = \sum_{k=0}^{\infty} (-1)^k \left(\frac{d}{dx}\right)^k \frac{\partial}{\partial u^{(k)}}.$$

Now $\delta/\delta u$ commutes with d, since d/dx and $\partial/\partial u^{(k)}$ do so. The operator

d/dx maps

$$A_N^{(r)} \to A_N^{(r)}.$$

The operator $\delta/\delta u$ maps

$$A_N^{(r)} \to A_{N-1}^{(r)},$$

and d maps:

$$A_N^{(r)} \to A_{N-1}^{(r+1)}.$$

THEOREM. *The sequence of mappings*

$$0 \to A_N^{(r)} \xrightarrow{d/dx} A_N^{(r)} \xrightarrow{\delta/\delta x} A_{N-1}^{(r)}, \quad N \neq 0$$

is exact.

In other words, a form ω can be written as $\frac{d}{dx}\,\omega_1$ if and only if $\frac{\delta}{\delta u}\,\omega = 0$ (the exactness of the sequence at the first term is trivial).

In one direction the theorem is very easy to prove:

$$\frac{\delta}{\delta u}\frac{d}{dx} = \sum_{k=0}^{\infty}(-1)^k\left(\frac{d}{dx}\right)^k \frac{\partial}{\partial u^{(k)}}\frac{d}{dx} =$$

$$= \sum_{k=0}^{\infty}(-1)^k\left(\frac{d}{dx}\right)^{k+1}\frac{\partial}{\partial u^{(k)}} + \sum_{k-1}^{\infty}(-1)^k\left(\frac{d}{dx}\right)^k\frac{\partial}{\partial u^{(k-1)}} = 0.$$

To prove the converse, we shall develop in the next section an auxiliary apparatus (of independent interest) that will enable us not only to prove the existence of the form, but also to construct it explicitly. For another algorithm, which is very useful in practice, see the footnote on p. 87.

NOTE ON THE DEFINITIONS. The operators d/dx and $\partial/\partial u^{(i)}$ in (7) and (8) are just the Lie derivatives of ω with respect to d/dx and $\partial/\partial u^{(i)}$ and are usually denoted by $L_{d/dx}$ and $L_{\partial/\partial u^{(i)}}$.

The Lie derivative with respect to the vector field ξ is defined as $L_\xi = i(\xi)\,d + di\,(\xi)$. From this it follows that all Lie derivatives commute with the differential: $dL_\xi = L_\xi d = di\,(\xi)\,d$.

We claim that $L_{d/dx}\,\omega$ is, in fact, equal to the right-hand side of (7). To show this we first note that $L_{d/dx}du^{(i)} = dL_{d/dx}u^{(i)} = du^{(i+1)}$. The equation $L_\xi\omega_1 \wedge \omega_2 = (L_\xi\omega_1) \wedge \omega_2 + \omega_1 \wedge (L_\xi\omega_2)$ completes the proof. Similarly we find that $L_{\partial/\partial u^{(i)}}\,du^{(j)} = dL_{\partial/\partial u^{(i)}}\,u^{(j)} = d\delta_{ij} = 0$. Hence $L_{\partial/\partial^{(i)}}$ is equal to the right-hand side of (8).

Our reason for simply writing d/dx and $\partial/\partial u^{(i)}$ rather than $L_{d/dx}$ and $L_{\partial/\partial u^{(i)}}$ is that a Lagrangian function L will play a fundamental part in what follows, and formulae containing both $L_{d/dx}$ and L are ugly in appearance.[1]

1) In his excellent book on mechanics, Godbillon [21] avoids this difficulty because the Lagrangian appears only at the very end of the last chapter.

2. The symbolic notation for polynomials in u, u', u'', ... Let $\widetilde{A}_N^{(r)}$ be the set of polynomials $\widetilde{F}(\xi; z)$ (where $\xi = (\xi_1, \xi_2, \ldots, \xi_N)$ and $z = (z_1, z_2, \ldots, z_r)$) that are symmetric in ξ and antisymmetric in z. With each such polynomial we associate a form in $A_N^{(r)}$ by the following rule: for a monomial we set

$$a\xi_1^{k_1} \ldots \xi_N^{k_N} z_1^{l_1} \ldots z_r^{l_r} \to au^{(k_1)} \ldots u^{(k_N)} du^{(l_1)} \wedge \ldots \wedge du^{(l_r)},$$

to a sum of monomials there corresponds the sum of the forms. It is easy to see that every form is associated with one and only one polynomial in ξ and z, which arises as follows. With the monomial $au^{(k_1)} \ldots u^{(k_N)} du^{(l_1)} \wedge \ldots \wedge du^{(l_r)}$ we associate $S_{N,r} a \xi_1^{k_1} \ldots \xi_N^{k_N} z_1^{l_1} \ldots z_r^{l_r}$, where $S_{N,r}$ is the operator of symmetrization with respect to ξ and of anti-symmetrization with respect to the z, that is, $S_{N,r} = \dfrac{1}{N!r!} \sum_{\sigma_N, \sigma_r} (-1)^\nu \sigma_N \sigma_r$, where σ_N ranges over all permutations of $1, \ldots, N$, and σ_r over the permutations of $1, \ldots, r$; $\nu = 0$ if σ_r is even and $\nu = 1$ otherwise. The action of the operators σ_N and σ_r on a function $F(\xi_1, \ldots, \xi_N; z_1, \ldots, z_r)$ is clear. For example,

$$uu'u''\, du \wedge du'' \leftrightarrow \frac{1}{12} (\xi_2\xi_3^2 + \xi_3\xi_2^2 + \xi_3\xi_1^2 + \xi_1\xi_3^2 + \xi_1\xi_2^2 + \xi_2\xi_1^2)(z_2^2 - z_1^2).$$

We denote the set of these symbolic polynomials by $\widetilde{A}_N^{(r)}$. The operators $\dfrac{d}{dx}$, $\dfrac{\delta}{\delta u}$, d, $\dfrac{\partial}{\partial u^{(i)}}$ on forms go over into certain operators $\dfrac{\widetilde{d}}{dx}$, $\dfrac{\widetilde{\delta}}{\delta u}$, $\widetilde{d}$, $\dfrac{\widetilde{\partial}}{\partial u^{(i)}}$ on symbolic polynomials. We find that

$$\text{(11)} \quad \left(\frac{\widetilde{d}}{dx} \widetilde{F}\right)(\xi_1, \ldots, \xi_N; z_1, \ldots, z_r) = \\ = (\xi_1 + \ldots + \xi_N + z_1 + \ldots + z_r)\widetilde{F}(\xi_1, \ldots, \xi_N; z_1, \ldots, z_r).$$

This property is obvious.

$$\text{(12)} \quad \left(\frac{\widetilde{\partial}}{\partial u} \widetilde{F}\right)(\xi_1, \ldots, \xi_{N-1}; z_1, \ldots, z_r) = N\widetilde{F}(\xi_1, \ldots, \xi_{N-1}, 0; z_1, \ldots, z_r).$$

PROOF. We note that

$$\frac{\partial}{\partial u} u^{(k_1)} \ldots u^{(k_N)} = \sum_{i=1}^{N} u^{(k_1)} \ldots \frac{\partial u^{(k_i)}}{\partial u} \ldots = \sum_{i=1}^{N} \delta_{k_i 0} u^{(k_1)} \ldots \hat{u}^{(k_i)} \ldots;$$

here $\delta_{k_i 0}$ is the Kronecker symbol, and the symbol ^ indicates the omission of the corresponding term. Hence

$$\frac{\tilde{\partial}}{\partial u} S_{N,r}\xi_1^{k_1}\dots\xi_N^{k_N} z_1^{l_1}\dots z_r^{l_r} = S_{N,r}\sum_{i=1}^{N}\delta_{k_i,0}\xi_1^{k_1}\dots\hat{\xi}_i^{k_i}\dots\xi_N^{k_N} z_1^{l_1}\dots z_r^{l_r} =$$

$$= \sum_{i=1}^{N}\tilde{F}(\xi_1,\dots,\xi_{i-1},0,\xi_{i+1},\dots,\xi_N; z_1,\dots,z_r) =$$

$$= N\tilde{F}(\xi_1,\dots,\xi_{N-1},0; z_1,\dots,z_r).$$

Similarly we find

(13) $$\left(\frac{\tilde{\partial}}{\partial u^{(k)}}\tilde{F}\right)(\xi_1,\dots,\xi_{N-1},z_1,\dots,z_r) = N\cdot\frac{1}{k!}\frac{\partial^k}{\partial\xi_N^k}\tilde{F}(\xi_1,\dots,\xi_{N-1},0;z_1,\dots,z_r),$$

(14) $$\left(\frac{\tilde{\delta}}{\delta u}\tilde{F}\right)(\xi_1,\dots,\xi_{N-1};z_1,\dots,z_r) = \sum_{k=0}^{\infty}(-1)^k(\xi_1+\dots+\xi_{N-1}+z_1,+\dots+z_r)^k\times$$

$$\times N\cdot\frac{1}{k!}\frac{\partial^k}{\partial\xi_N^k}\tilde{F}(\xi_1,\dots,\xi_{N-1},0;z_1,\dots,z_r) =$$

$$= N\tilde{F}(\xi_1,\dots,\xi_{N-1},-\xi_1-\dots-\xi_{N-1}-z_1-\dots-z_r;z_1,\dots,z_r).$$

If $\tilde{F}_1(\xi_1,\dots,\xi_{N_1};z_1,\dots,z_{r_1})$ and $\tilde{F}_2(\xi_1,\dots,\xi_{N_2};z_1,\dots,z_{r_2})$ correspond to the forms ω_1 and ω_2, then the form corresponding to $\omega_1\wedge\omega_2$ is

(15) $$S_{N_1+N_2,\,r_1+r_2}[\tilde{F}_1(\xi_1,\dots,\xi_{N_1};z_1,\dots,z_{r_1})\times \tilde{F}_2(\xi_{N_1+1},\dots,\xi_{N_1+N_2};z_{r_1+1},\dots,z_{r_1+r_2})],$$

(16) $$(\tilde{d}\tilde{F})(\xi_1,\dots,\xi_{N-1};z_1,\dots,z_{r+1}) = S_{N-1,\,r+1}\tilde{F}(\xi_1,\dots,\xi_{N-1},z_{r+1};z_1,\dots,z_r).$$

The properties (15) and (16) are obvious.

COMPLETION OF THE PROOF OF THE THEOREM OF THE PRECEDING SECTION. By means of the symbolic calculus which we have developed, we can now prove the theorem in both directions. From (11) and (14) it is immediately clear that $\frac{\tilde{\delta}}{\delta u}\,\frac{\tilde{d}}{dx} = 0$. For by substituting $\xi_N = -\xi_1 - \dots - \xi_{N-1} - z_1 - \dots - z_r$ in the expression $\xi_1 + \dots + \xi_N + z_1 + \dots + z_r$ we obtain zero. Suppose now that the $\frac{\delta}{\delta u}\,\omega = 0$. Then for the corresponding function $\tilde{F}$ we have $\frac{\tilde{\delta}}{\delta u}\,\tilde{F} = 0$. This means that the polynomial $\tilde{F}$ vanishes on substituting $\xi_N = -\xi_1 - \dots - \xi_{N-1} - z_1 - \dots - z_r$. Hence, $\tilde{F}$ is divisible by

$\xi_N - (-\xi_1 - \ldots - \xi_{N-1} - z_1 - \ldots - z_r)$, and the quotient is again a polynomial symmetric in ξ and antisymmetric in z, say $\widetilde{F} = \frac{d}{dx}\widetilde{G}$. If ω_1 is the form corresponding to $\widetilde{G}$, then $\omega = \frac{d}{dx}\omega_1$, as required.

If $\omega = \frac{d}{dx}\omega_1$, then we also write $\omega_1 = \left(\frac{d}{dx}\right)^{-1}\omega$.

NOTE. $(d/dx)^{-1}$ is a well-defined operator: there are no constants of integration. We consider d/dx separately acting on each $A_N^{(r)}$, where there are no constants.

3. The Lagrangian, the Hamiltonian, 1- and 2-forms. We fix an element $L[u] \in A$ and call it the Lagrangian. We denote by M_L the ideal in A generated by the elements $\frac{\delta}{\delta u}L$, $\frac{d}{dx}\frac{\delta}{\delta u}L$, $\left(\frac{d}{dx}\right)^2\frac{\delta}{\delta u}L, \ldots$, and by A_L the quotient algebra A/M_L.

THEOREM 1. *In* A *we have*

$$\frac{\delta}{\delta u}u'\frac{\delta}{\delta u}=0 \tag{17}$$

(but not in $A^{(r)}$, $r \neq 0$*).*

We prove the theorem by the symbolic calculus of §2. Let $F \in A_N$. Then $u'\frac{\delta}{\delta u}F$ corresponds to the function

$$\begin{aligned}\xi_N\widetilde{F}(\xi_1, \ldots, \xi_{N-1}, -\xi_1 - \ldots - \xi_{N-1}) + \\ + \xi_1\widetilde{F}(\xi_2, \ldots, \xi_N, -\xi_2 - \ldots - \xi_N) + \\ + \xi_2\widetilde{F}(\xi_3, \ldots, \xi_N, \xi_1, -\xi_3 - \ldots - \xi_N - \xi_1) + \ldots,\end{aligned}$$

and $\frac{\delta}{\delta u}u'\frac{\delta}{\delta u}F$ to

$$\begin{aligned}(-\xi_1 - \xi_2 - \ldots - \xi_{N-1})\widetilde{F}(\xi_1, \ldots, \xi_{N-1}, -\xi_1 - \ldots - \xi_{N-1}) + \\ + \xi_1\widetilde{F}(\xi_2, \ldots, -\xi_1 - \ldots - \xi_{N-1}, \xi_1) + \\ + \xi_2\widetilde{F}(\xi_3, \ldots, -\xi_1 - \ldots - \xi_{N-1}, \xi_1, \xi_2) + \ldots = 0,\end{aligned}$$

as required.

COROLLARY. *The operator*

$$H = -\left(\frac{d}{dx}\right)^{-1}u'\frac{\delta}{\delta u}. \tag{18}$$

is well-defined.

We call (18) the Hamiltonian operator and $H[u] = HL[u]$ the Hamiltonian corresponding to the Lagrangian $L[u]$. We consider the operator $d - du\cdot\frac{\delta}{\delta u}$ from A to $A^{(1)}$:

$$\left(d - du\cdot\frac{\delta}{\delta u}\right)F[u] = dF - \left(\frac{\delta}{\delta u}F\right)du.$$

Then the following theorem holds.

THEOREM 2.

$$\frac{\delta}{\delta u}\left(d-du\frac{\delta}{\delta u}\right)=0. \tag{19}$$

PROOF. Let $F \in A_N$. Then

$$\frac{\tilde{\delta}}{\delta u}\, d\tilde{F}=\frac{\tilde{\delta}}{\delta u} N\tilde{F}(\xi_1, \xi_2, \ldots, \xi_{N-1}, z)=$$
$$=N(N-1)\tilde{F}(\xi_1, \xi_2, \ldots, -\xi_1-\xi_2-\ldots-\xi_{N-2}-z, z),$$
$$\frac{\tilde{\delta}}{\delta u}\tilde{F}=N\tilde{F}(\xi_1, \xi_2, \ldots, \xi_{N-1}, -\xi_1-\ldots-\xi_{N-1}),$$
$$\widetilde{\frac{\delta}{\delta u}F\,du}=N\tilde{F}(\xi_1, \xi_2, \ldots, \xi_{N-1}, -\xi_1-\ldots-\xi_{N-1}),$$
$$\frac{\tilde{\delta}}{\delta u}\widetilde{\frac{\delta}{\delta u}F\,du}=N(N-1)\tilde{F}(\xi_1, \xi_2, \ldots, \xi_{N-2}, -\xi_1-\ldots-\xi_{N-2}-z, z),$$

that is, $\frac{\tilde{\delta}}{\delta u}\left(\widetilde{\frac{\delta}{\delta u}F\,du-dF}\right) = 0$. This proves the theorem.

COROLLARY. *The operator*

$$\Omega^{(1)}=\left(\frac{d}{dx}\right)^{-1}\left(d-du\frac{\delta}{\delta u}\right) \tag{20}$$

is well defined.

We call it the 1-form operator and

$$\Omega^{(2)}=d\left(\frac{d}{dx}\right)^{-1}\left(d-du\frac{\delta}{\delta u}\right)=\left(\frac{d}{dx}\right)^{-1} du\wedge d\frac{\delta}{\delta u} \tag{21}$$

the 2-form operator. We call the results of applying these operators to a given Lagrangian the 1- and 2-forms of this Lagrangian,

$$\Omega^{(2)}=\left(\frac{d}{dx}\right)^{-1} du\wedge d\left(\frac{\delta}{\delta u}L\right).$$

The form $\Omega^{(2)}$ is closed, being the differential of the form $\Omega^{(1)}$.

NOTE. The form $du \wedge d\left(\frac{\delta}{\delta u}L\right)$ is highly degenerate. The operator $\left(\frac{d}{dx}\right)^{-1}$ carries it into the $\Omega^{(2)}$– form, which, as we see below, is non-degenerate, at least in the most important cases.

4. Momenta. We introduce the momentum operators in A

$$p_i=\sum_{k=0}^{\infty}(-1)^k\left(\frac{d}{dx}\right)^k\frac{\partial}{\partial u^{(k+i+1)}} \qquad (i=0, 1, 2, \ldots)$$

(when $i = -1$, the right-hand side is just $\frac{\delta}{\delta u}$). Clearly,[1]

---

1) Another relation $p_0\frac{\delta}{\delta u} = 0$ can easily be verified by means of the symbolic calculus of § 2.

$$\frac{d}{dx}\, p_i = -p_{i-1} + \frac{\partial}{\partial u^{(i)}}.$$

We define the moments $p_i = p_i L$ as the result of the actions of the p_i on the Lagrangian.

THEOREM. $H = \sum_{i=0}^{\infty} p_i u^{(i+1)} - 1 \quad \Omega^{(i)} = \sum_{i}^{\infty} p_i du^{(i)}$.

PROOF.

$$\frac{d}{dx}\left(\sum_{i=0}^{\infty} p_i u^{(i+1)} - 1\right) = \sum_{i=1}^{\infty} u^{(i+1)} \sum_{k=0}^{\infty} (-1)^k \left(\frac{d}{dx}\right)^{k+1} \frac{\partial}{\partial u^{(k+i+1)}} +$$

$$+ \sum_{i=0}^{\infty} u^{(i+2)} \sum_{k=0}^{\infty} (-1)^k \left(\frac{d}{dx}\right)^k \frac{\partial}{\partial u^{(k+i+1)}} - \sum_{i=0}^{\infty} u^{(i+1)} \frac{\partial}{\partial u^{(i)}} =$$

$$= -u' \sum_{k=0}^{\infty} (-1)^{k+1} \left(\frac{d}{dx}\right)^{k+1} \frac{\partial}{\partial u^{(k+1)}} + \sum_{i=1}^{\infty} u^{(i+1)} \frac{\partial}{\partial u^{(i)}} - \sum_{i=0}^{\infty} u^{(i+1)} \frac{\partial}{\partial u^{(i)}} =$$

$$= -u' \frac{\delta}{\delta u} = \frac{d}{dx} \boldsymbol{H}.$$

The equality of these derivatives implies that of the operators themselves. The second part of the theorem is proved in exactly the same way.

COROLLARY. $\Omega^{(2)} = \sum_{i=0}^{\infty} dp_i \wedge du^{(i)}$.

Let ξ and η be two vector fields commuting with $\frac{d}{dx}$. According to (5) we have $\xi_i = \left(\frac{d}{dx}\right)^i \xi_0$, $\eta_i = \left(\frac{d}{dx}\right)^i \eta_0$. We calculate

$$\frac{d}{dx}\Omega^{(2)}(\xi, \eta) = \left(\frac{d}{dx}\Omega^{(2)}\right)(\xi, \eta) = du \wedge d\left(\frac{\delta}{\delta u} L\right)(\xi, \eta) =$$

$$= \xi_0 \eta\left(\frac{\delta}{\delta u} L\right) - \eta_0 \xi\left(\frac{\delta}{\delta u} L\right);$$

and obtain the useful identity

$$(22) \quad \xi_0 \sum_{i=0}^{\infty} \left[\left(\frac{d}{dx}\right)^i \eta_0\right] \frac{\partial}{\partial u^{(i)}} \frac{\delta}{\delta u} L - \eta_0 \sum_{i=0}^{\infty} \left[\left(\frac{d}{dx}\right)^i \xi_0\right] \frac{\partial}{\partial u^{(i)}} \frac{\delta}{\delta u} L = $$

$$= \frac{d}{dx}\left[\Omega^{(2)}(\xi, \eta)\right].$$

We say that two Lagrangians are equivalent if they differ by a function of form $\frac{d}{dx} F$. For equivalent Lagrangians the elements $\frac{\delta}{\delta u} L$ coincide and so do also the ideal M_L, the manifold A_L, the Hamiltonian operators, the 2-forms and so on. From the point of view of our present theory we may always replace a Lagrangian by an equivalent one. Among all the Lagrangians equivalent to a given one we can choose one for which the leading derivative occurs raised to the smallest power. It is easy to see that the term of this Lagrangian containing the leading derivative to the highest power must have the form $a(u^{(k_1)})^{l_1} (u^{k_2})^{l_2}, \ldots, k_1 > k_2 > \ldots$, where

$l_1 \geqslant 2$, that is, the leading derivative must occur to a power higher than the first (otherwise it would be possible to lower it by adding an expression of the form $\frac{d}{dx}F$).[1] A Lagrangian equivalent to the given one with the leading derivative of lowest order is called reduced. The terms containing the leading derivative to the highest power are uniquely determined in the reduced Hamiltonian.

5. Vector fields and forms in A_L. We now define the tangent space TA_L to the manifold A_L. We consider the set of vector fields $\xi \in TA$ for which $\xi F \in M_L$ whenever $F \in M_L$. This is a linear subspace of TA, which we denote by TA_L^*. We distinguish in this space the set of vectors ξ such that $\xi F \in M_L$ for every $F \in A$ (in the coordinate representation of such vector fields, all the $\xi_i \in M_L$). These fields ξ form a subspace $TA_L^{(0)}$. We set $TA_L = TA_L^*/TA_L^0$ and call it the tangent space to A_L.

To each $\xi \in TA_L$ there corresponds the sequence of coordinates $\{\xi_i\}$, $\xi_i \in A_L$.

For an arbitrary Lagrangian L the vector field d/dx belongs to TA_L^*; we use d/dx also to denote the corresponding element of TA_L.

Let $\omega = A^{(r)}$ be a given form. We restrict it to the vector fields $\xi \in TA_L^*$. Then the composition of the mappings $TA_L^* \times TA_L^* \times \ldots \times TA_L^* \to A \to A_L$ generates a mapping, which is a form on TA_L^* with values in A_L. It is easy to see that this form vanishes if any one of its arguments belongs to TA_L^0. Hence it can be regarded as a form on $TA_L = TA_L^*/TA_L^0$. Thus, any form $\omega \in A^{(r)}$ induces a form in $A_L^{(r)}$ the space of forms on TA_L with values in A_L. Thus, there is a mapping $A^{(r)} \to A_L^{(r)}$ and $\sum_r A^{(r)} \to \sum_r A_L^{(r)}$. It is easy to see that the kernel of this mapping is invariant under the action of d. Indeed, this is clear for 0-forms, that is, functions. If $F \in M_L$, then $dF(\xi) = \xi F \in M_L$ for $\xi \in TA_L^*$, that is, dF regarded as a form in $A_L^{(1)}$ vanishes. For forms of higher dimension we must use the formula (6) for the differential, together with the fact that the commutator of two fields in TA_L^* again belongs to TA_L^*. From this we conclude that the operator d can be carried over to $A_L^{(r)}$.

---

1) Thus we have the following algorithm for reduction of the Hamiltonian. We consider the terms with the leading derivative. If this derivative occurs to the first power, then the term is transformed according to the models

$$(u')^2 u''' = -2u'(u'')^2 + \frac{d}{dx}[(u')^2 u''] \quad \text{or} \quad u'u''u''' = -\frac{1}{2}(u'')^3 + \frac{d}{dx}\left[\frac{1}{2}u'(u'')^2\right].$$

This process is continued until the leading derivative occurs to a power higher than the first. In the special case when $L = \frac{d}{dx}F$, the leading derivative cannot occur to a power higher than the first. In this case the process of reducing the Lagrangian continues until it becomes zero, that is, the reduced Lagrangian in this case is zero. We note that incidentally we have also obtained another algorithm for determining the primitive $\left(\frac{d}{dx}\right)^{-1} L$.

We define the action of $\xi\ \omega$ for $\omega \in A_L^{(r)}$ and $\xi \in TA_L$ in exactly the same way.

The spaces TA_L and $A_L^{(r)}$ are modules over A_L.

NOTE. We have restricted ourselves to vector fields and forms induced by fields and forms in A. A more general definition of a vector field would be as a derivation in A_L and of a form as a linear functional on vector fields. We have no need for this more general definition in what follows.

6. A correspondance between functions and vector fields. A vector field $\xi \in TA_L$ and a function $F \in A_L$ are said to be associated if $dF = -i\,(\xi)\,\Omega^{(2)}$ (the form $\Omega^{(2)}$ must, of course, be understood as an element of $A_L^{(2)}$).

LEMMA 1. *The vector field* $\frac{d}{dx}$ *and the Hamiltonian* H *are associated.*

PROOF. Considering H and $\frac{d}{dx}$ as elements of A and TA, we establish the formula

$$dH = -i\left(\frac{d}{dx}\right)\Omega^{(2)} - \left(\frac{\delta}{\delta u} L\right) du,$$

from which the lemma follows immediately by projecting onto $A_L^{(1)}$. To prove this it is sufficient to show that the result of applying d/dx to the two sides of (22) coincide. We have

$$\frac{d}{dx} dH = d\frac{d}{dx} H = -d\left(u' \frac{\delta}{\delta u} L\right) = -\left(\frac{\delta}{\delta u} L\right) du' - u'd\left(\frac{\delta}{\delta u} L\right),$$

$$\frac{d}{dx}\left[-i\left(\frac{d}{dx}\right)\Omega^{(2)} - \left(\frac{\delta}{\delta u} L\right) du\right] =$$
$$= i\left(\frac{d}{dx}\right)\left(-du \wedge d\left(\frac{\delta}{\delta u} L\right)\right) - \frac{d}{dx}\left(\left(\frac{\delta}{\delta u} L\right) du\right) =$$
$$= -u'd\left(\frac{\delta}{\delta u} L\right) + \frac{d}{dx}\left(\frac{\delta}{\delta u} L\right) du - \frac{d}{dx}\left(\frac{\delta}{\delta u} L\right) du - \left(\frac{\delta}{\delta u} L\right) du' =$$
$$= -u'd\left(\frac{\delta}{\delta u} L\right) - \left(\frac{\delta}{\delta u} L\right) du'.$$

This proves the formula, and with it the lemma.

NOTE. We do not clarify in general form the conditions for the existence of a vector field associated with a given element $F \in A_L$, nor the uniqueness of the association. We introduce below a fairly important class of Lagrangians, to which, in particular, the examples in the following chapters belong. For this class, one and only one vector field $\xi \in TA_L$ is associated with each $F \in A_L$, and this field is associated with only one element $F \in A_L$.

LEMMA 2. *If* ξ_1 *and* ξ_2 *are two vector fields in* TA_L, *then the vector field* $[\xi_1, \xi_2]$ *and the function* $\Omega^{(2)}(\xi_1, \xi_2) \in A_L$ *are associated.*

PROOF. Let η be an arbitrary field in TA_L. Using the fact that the form $\Omega^{(2)}$ is closed we obtain by means of (6)

$$0 = (d\Omega^{(2)})(\xi_1, \xi_2, \eta) = \xi_1\Omega^{(2)}(\xi_2, \eta) - \xi_2\Omega^{(2)}(\xi, \eta) + \eta\Omega^{(2)}(\xi_1, \xi_2) -$$
$$-\Omega^{(2)}([\xi_1, \xi_2], \eta) + \Omega^{(2)}([\xi_1, \eta], \xi_2) - \Omega^2([\xi_2, \eta], \xi_1) =$$
$$=[-d(\Omega^{(2)}(\xi_1, \xi_2)) - i([\xi_1, \xi_2])\Omega^{(2)}](\eta),$$

that is,

$$d(\Omega^{(2)}(\xi_1, \xi_2)) = i([\xi_1, \xi_2])\Omega^{(2)},$$

which establishes the lemma.

When the vector field ξ associated with an element $F \in A_L$ is unique, then the function $\Omega^{(2)}(\xi_1, \xi_2) \in A_L$, where ξ_1 and ξ_2 are fields associated with F_1 and F_2, is called the Poisson bracket of F_1 and F_2 and is denoted by (F_1, F_2).

7. **Invariants.** A form $\omega \in A_L^{(r)}$ (in particular, a function) is said to be invariant if

$$\frac{d}{dx}\omega = 0. \tag{23}$$

A vector field $\xi \in TA_L$ is said to be invariant if

$$\left[\xi, \frac{d}{dx}\right] = 0. \tag{24}$$

The Hamiltonian H and the 2-form $\Omega^{(2)}$ are invariant. In fact, for $H \in A$ we have $\dfrac{dH}{dx} = -u' \dfrac{\delta}{\delta u} L \in M_L$, that is, regarding H as an element of A_L, $\dfrac{dH}{dx} = 0$, as required. Moreover, for all $\xi, \eta \in TA_L^*$, $\left(\frac{d}{dx}\Omega^{(2)}\right)(\xi, \eta) = -(\xi u)\left(\eta \frac{\delta}{\delta u} L\right) + (\eta u)\left(\xi \frac{\delta}{\delta u} L\right) \in M_L$. Thus, in TA_L $\left(\frac{d}{dx}\Omega^{(2)}\right)(\xi, \eta) = 0$, that is, $\dfrac{d}{dx}\Omega^{(2)} = 0$. From (24) and from the formula (5) for the commutator it follows that for an invariant vector field $\xi_i = \dfrac{d}{dx}\xi_{i-1}$, or

$$\xi_i = \left(\frac{d}{dx}\right)^i \xi_0. \tag{25}$$

We consider the invariants of A in greater detail. An element $F \in A$ is said to be an invariant or a first integral of the equation $\dfrac{\delta}{\delta u} L = 0$ if $\dfrac{d}{dx} F \in M_L$, that is,

$$\frac{d}{dx} F = a_0 \frac{\delta}{\delta u} L + a_1 \frac{d}{dx}\frac{\delta}{\delta u} L + a_2 \left(\frac{d}{dx}\right)^2 \frac{\delta}{\delta u} L + \ldots + a_r \left(\frac{d}{dx}\right)^r \frac{\delta}{\delta u} L. \tag{26}$$

Two first integrals F_1 and F_2 are said to be equivalent, $F_1 \sim F_2$, if $F_1 - F_2 \in M_L$. In particular, if $G \in M_L$ is a first integral, then it is equivalent to zero. Thus, an equivalence class of first integrals determines an invariant function in A_L.

LEMMA. *Let F be a first integral. Then there exists an equivalent first integral F_1 for which*

$$\frac{d}{dx} F_1 = a \frac{\delta}{\delta u} L.$$

Moreover, if (26) *holds for F, then*

$$a = \sum_{i=0}^{r} (-1)^i \left(\frac{d}{dx}\right)^i a_i. \tag{27}$$

PROOF. We show that if the r in (26) is non-zero, then we can find a first integral equivalent to F for which r is smaller by 1 and the sum on the right-hand side of (27) is the same as for F. From this the lemma follows directly. We have

$$\frac{d}{dx} F = \sum_{i=0}^{r-1} a_i \left(\frac{d}{dx}\right)^i \frac{\delta}{\delta u} L + \frac{d}{dx}\left[a_r \left(\frac{d}{dx}\right)^{r-1} \frac{\delta}{\delta u} L\right] - \left(\frac{d}{dx} a_r\right)\left(\frac{d}{dx}\right)^{r-1} \frac{\delta}{\delta u} L,$$

that is, for $F_1 = F - a_r \left(\frac{d}{dx}\right)^{r-1} \frac{\delta}{\delta u} L$,

$$\frac{d}{dx} F_1 = \sum_{i=0}^{r-2} a_i \left(\frac{d}{dx}\right)^i \frac{\delta}{\delta u} L + \left(a_{r-1} - \left(\frac{d}{dx} a_r\right)\right)\left(\frac{d}{dx}\right)^{r-1} \frac{\delta}{\delta u} L.$$

That the sum (27) is unaltered is clear. This proves the lemma.

The quantity $a[u]$ given by (27) is called the characteristic of the first integral F. To justify this term we must prove that the characteristic is uniquely determined by the first integral. Non-uniqueness may occur because of non-uniqueness of the representation (26). We say that two characteristics are equivalent if $a_1 - a_2 \in M_L$. Then the following theorem holds.

THEOREM 1. *If $a_1[u]$ is a characteristic of the first integral $F_1[u]$ and $a_2[u]$ a characteristic of $F_2[u]$ and if $F_1 \sim F_2$, then $a_1 \sim a_2$.*

We prove this theorem here under an additional hypothesis on the Lagrangian, which we call non-degeneracy. We require that the relation

$$b_0 \frac{\delta}{\delta u} L + b_1 \frac{d}{dx} \frac{\delta}{\delta u} L + b_2 \left(\frac{d}{dx}\right)^2 \frac{\delta}{\delta u} L + \ldots = 0$$

implies that all the $b_i \in M_L$. We note that the class of Lagrangians to be introduced in the next section ("normal" Lagrangians) are non-degenerate. Given two expansions of form (26) for the same $F \in A$, with coefficients a_i' and a_i'', then the condition of non-degeneracy implies that $a_i' - a_i'' \in M_L$, hence $a' \sim a$ for the characteristics.

Now let F_1 and F_2 be two equivalent first integrals. Their characteristics differ by a characteristic of the first integral $F = F_1 - F_2$, which is equivalent to zero. But $F \in M_L$, that is, $F = \sum_{i=0}^{\infty} b_i \left(\frac{d}{dx}\right)^i \frac{\delta}{\delta u} L$. So

$$\frac{d}{dx}F=\sum_{i=0}^{\infty}\left(\frac{d}{dx}b_i\right)\left(\frac{d}{dx}\right)^i\frac{\delta}{\delta u}L+\sum_{i=0}^{\infty}b_i\left(\frac{d}{dx}\right)^{i+1}\frac{\delta}{\delta u}L=$$

$$=\left(\frac{d}{dx}b_0\right)\frac{\delta}{\delta u}L+\sum_{i=1}^{\infty}\left(\frac{d}{dx}b_i+b_{i-1}\right)\left(\frac{d}{dx}\right)^i\frac{\delta}{\delta u}L.$$

The characteristic can be evaluated without difficulty:

$$\frac{d}{dx}b_0+\sum_{i=1}^{\infty}(-1)^i\left(\frac{d}{dx}\right)^i\left(\frac{d}{dx}b_i+b_{i-1}\right)=0.$$

This proves the theorem.

The importance of the characteristic of an invariant lies in the fact that under certain conditions satisfied by the Lagrangian the following theorem holds.

THEOREM 2. *The vector field constructed from the characteristic $a[u]$ of a first integral $F[u]$ by the formula*

(28) $$\xi=\sum_{i=0}^{\infty}\left[\left(\frac{d}{dx}\right)^i a\right]\frac{\partial}{\partial u^{(i)}},$$

is tangent and is associated with the invariant $F[u]$.

We do not know the most general conditions under which this theorem is true. It holds for the "normal" Lagrangians to be introduced in the next section, to which, in particular, the Lagrangians in the second part of this paper belong.

Let $a[u]\in A$ be an arbitrary element. We ask when there exist a Lagrangian $L\in A$ and a first integral $F\in A$ of which $a[u]$ is a characteristic; in other words, when do there exist $F, L\in A$ such that $\frac{d}{dx}F=a[u]\frac{\delta}{\delta u}L$ or

(29) $$\frac{\delta}{\delta u}a[u]\frac{\delta}{\delta u}L=0.$$

Everything reduces to a study of the operator $\frac{\delta}{\delta u}\circ a[u]\circ\frac{\delta}{\delta u}: A\to A$. The kernel of this operator consists of all Lagrangians L for which $a[u]$ is a characteristic of a first integral, and the first integral F is given by the formula

(30) $$F=\left(\frac{d}{dx}\right)^{-1}a[u]\frac{\delta}{\delta u}L.$$

8. Normal Lagrangians. We say that a Lagrangian is normal if

(31) $$\frac{\partial}{\partial u^{(2n)}}\left(\frac{\delta}{\delta u}L\right)=1,$$

where $u^{(2n)}$ is the leading derivative occurring in $\frac{\delta}{\delta u}L$. This means that

$\frac{\delta}{\delta u} L = u^{(2n)} + f(u, u', \ldots, u^{(2n-1)})$, where f is a polynomial. By repeated differentiation of this relation we deduce that all the derivatives of u beginning with $u^{(2n)}$ can be expressed as polynomials in $u, u' \ldots u^{(2n-1)}$ and in $\frac{\delta}{\delta u} L, \frac{d}{dx} \frac{\delta}{\delta u} L, \left(\frac{d}{dx}\right)^2 L, \ldots$

Passing to A_L we can reformulate this as follows: all the leading derivatives $u^{(2n)}, u^{(2n+1)}, \ldots$, as elements of A_L, are polynomials in the lower derivatives. Hence, every polynomial in $u, u', u'', \ldots$ can be expressed as a polynomial in $u, u', \ldots, u^{(2n-1)}$. We establish that normal Lagrangians are non-degenerate, in the sense of the preceding section. Let

$$\sum_{k=0}^{\infty} a_k [u] \left(\frac{d}{dx}\right)^k \frac{\delta}{\delta u} L = 0,$$

where $a_k [u]$ are elements of A. We rewrite each a_k as a polynomial in the variables $u, u', \ldots, u^{(2n-1)}, \frac{\delta}{\delta u} L, \frac{d}{dx} \frac{\delta}{\delta u} L, \left(\frac{d}{dx}\right)^2 \frac{\delta}{\delta u} L, \ldots$. Since these variables are independent, the terms of the form $\varphi(u, \ldots, u^{(2n-1)}) \left(\frac{d}{dx}\right)^k \frac{\delta}{\delta u} L$ in the sum must vanish separately. Hence the coefficients a_k cannot contain terms independent of $\frac{\delta}{\delta u} L, \frac{d}{dx} \frac{\delta}{\delta u} L, \ldots$, that is, must belong to M_L, as required.

The condition (26) is equivalent to the fact that the Lagrangian L, in an abbreviated form, has $u^{(n)}$ as a leading derivative, and $u^{(n)}$ occurs as a separate term $\frac{1}{2} (-1)^n [u^{(n)}]^2$, while the remaining terms contain $u^{(n)}$ in powers no higher than the first. From the formulae for the Hamiltonian and for the form $\Omega^{(2)}$

$$H = \sum_{i=0}^{n-1} p_i u^{(i+1)} - L, \qquad \Omega^{(2)} = \sum_{i=0}^{n-1} dp_i \wedge du^{(i)}$$

it follows that H depends only on $u, u', \ldots, u^{(2n-1)}$, and

$$\Omega^{(2)} = \sum_{k,\, l=0}^{2n-1} \Omega^{(2)}_{k,\, l}\, du^{(k)} \wedge du^{(l)},$$

where the matrix $\Omega^{(2)}_{k,l}$ has the form

$$\begin{pmatrix} * & \cdot & \cdot\cdot & & * & 1 & \\ \cdot & \cdot\cdot & \cdot\cdot & & -1 & \cdots & \\ \cdot & \cdot\cdot & \cdot\cdot & & \cdot & \cdots & \\ * & 1 & \cdot\cdot & & \cdot & & \\ -1 & & & & & 0 & \end{pmatrix}.$$

Only elements above the diagonal can be non-zero, and they depend on $u, u', \ldots, u^{(2n-1)}$.

What does TA_L represent in the case of a normal Lagrangian? We claim that the first $2n$ coordinates of a vector field $\xi \in TA_L^*$ can be arbitrary elements of A_L, while the remaining ones are uniquely determined by the former. Indeed, for a vector field $\xi \in TA$ to belong to TA_L^* it is necessary and sufficient that $\xi \left(\frac{d}{dx}\right)^i \frac{\delta}{\delta u} L \in M_L$ $(i = 0, 1, 2, \ldots)$, that is,

$$\xi_0 \frac{\partial}{\partial u}\left(\frac{\delta}{\delta u} L\right) + \ldots + \xi_{2n-1} \frac{\partial}{\partial u^{(2n-1)}}\left(\frac{\delta}{\delta u} L\right) + \xi_{2n} \in M_L,$$

$$\xi_0 \frac{\partial}{\partial u}\left(\frac{d}{dx}\frac{\delta}{\delta u} L\right) + \ldots \qquad\qquad + \xi_{2n+1} \in M_L,$$

$$\ldots\ldots\ldots\ldots\ldots\ldots\ldots\ldots\ldots\ldots$$

or, to within terms belonging to M_L $\xi_{2n}, \xi_{2n+1}, \ldots$, are linear combinations of the first $2n$ coordinates. As elements of A_L, the coordinates $\xi_{2n}, \xi_{2n+1} \ldots$ are linear combinations of the first $2n$ coordinates, which are completely arbitrary. Thus, TA_L is a $2n$-dimensional module over A_L.

LEMMA 1. *If the Lagrangian is normal, then with each $F \in A_L$ there is associated one and only one vector field $\xi \in TA_L$. Every vector field is associated with at most one $F \in A_L$ (to within a constant.)*

PROOF. Let $F \in A_L$ be an arbitrary element of A_L. It can be regarded as a polynomial in $u, u', \ldots, u^{(2n-1)}$. Then the equality $-\, i(\xi)\, \Omega^{(2)} = dF$ means that

$$\sum_{j=0}^{2n-1} \Omega_{ji}^{(0)} \xi_j = -\frac{\partial F}{\partial u^{(i)}} \qquad (i = 0, \ldots, 2n-1).$$

These equations can be solved successively if we recall the form of the matrix $\Omega_{ji}^{(2)}$. After the first $2n$ coordinates have been found (as polynomials in $u, u', \ldots, u^{(2n-1)}$), the remaining ones are uniquely determined, as explained above.

There remains the last statement of the lemma. Suppose that one field $\xi \in TA_L$ is associated with two functions $F_1, F_2 \in A_L$. Then the difference $F = F_1 - F_2$ satisfies $dF = 0$. If we take for $F \in A$ the representative of the residue class $F \in A_L$ that depends on $u, u', \ldots, u^{(2n-1)}$, then from $\partial F/\partial u^{(i)} \in M_L$ we obtain $\partial F/\partial u^{(i)} = 0$, that is, $F =$ const, as required.

We consider now invariant functions and vector fields for normal Lagrangians.

First we prove that if $F \in A_L$ is an invariant, then its associated vector field $\xi \in TA_L$ is invariant. For by Lemma 6 in §6, $\left[\xi, \frac{d}{dx}\right]$ is the vector field corresponding to $\Omega^{(2)}\left(\xi, \frac{d}{dx}\right) = [i(\xi)\, \Omega^{(2)}\left(\frac{d}{dx}\right) = -\frac{d}{dx} F = 0$. Hence

$\left[\xi, \frac{d}{dx}\right] = 0$, as required.

Now we show that Theorem 2 of the preceding section holds for normal Lagrangians.

Let $F \in A$ be a first integral. We may regard it as depending only on $u, u', \ldots u^{(2n-1)}$. Then $\frac{d}{dx} F$ contains derivatives up to $u^{(2n)}$, and the leading derivative occurs linearly. In addition, $\frac{d}{dx} F \in M_L$. Hence $\frac{d}{dx} F = a[u] \frac{\delta}{\delta u} L$, where a depends on $u, u', \ldots, u^{(2n-1)}$. Clearly, $a = \partial F / \partial u^{(2n-1)}$. Now from $-i(\xi)\, \Omega^{(2)} = dF$, recalling the form of $\Omega^{(2)}_{ji}$, we obtain what is required:

$$-\sum_{j=0}^{2n-1} \Omega^{(2)}_{j,\, 2n-1} \xi_j = \frac{\partial F}{\partial u^{(2n-1)}}, \quad -\xi_0 = \frac{\partial F}{\partial u^{(2n-1)}} = a, \quad \xi_i = \left(\frac{d}{dx}\right)^i \xi_0 = -\left(\frac{d}{dx}\right)^i a.$$

CHAPTER 2

The asymptotic expansion of the resolvent of a Sturm–Liouville equation

1. The asymptotic series for the resolvent. We consider the second order linear differential equation

$$-\varphi'' + [u(x) + \zeta]\varphi = 0. \tag{1}$$

A kernel of the resolvent is defined to be a function $R(x, y; \zeta)$ that

a) is continuous in the pair of variables x, y;

b) is symmetric: $R(x, y; \zeta) = R(y, x; \zeta)$;

c) satisfies (1) as a function of either of the variables x, y when the other is held fixed and $x \neq y$.

d) is such that for fixed y, the derivative R'_x has a jump discontinuity of height 1 at $x = y$. By symmetry, this is equivalent to

$$\lim_{x \to y} (R'_x - R'_y) = 1; \tag{2}$$

e) $R(x, y; \zeta)$ converges to zero as $\zeta \to +\infty$ through positive values faster than any power of ζ.

The notion of resolvent defined here includes as a particular case the Green's function for any fixed self-adjoint boundary conditions (and even for a broad class of conditions depending on ζ).

It is clear that a linear combination of two resolvents such that the sum of the coefficients is 1 is again a resolvent. The difference of two resolvents is a smooth solution of (1) in each of the variables; from this we can deduce that this difference decreases exponentially as $\zeta \to \infty$ not only off the diagonal, but also on it. We denote the restriction of $R(x, y; \zeta)$ to the diagonal by $R(x; \zeta)$.

We now construct an asymptotic expansion

$$(3)\qquad R(x;\zeta)=\sum_{l=0}^{\infty}\frac{R_l[u]}{\zeta^{l+\frac{1}{2}}}$$

as $\zeta\to\infty$. The coefficients $R_l[u]$ are polynomials in u and its derivatives $u, u', \ldots$ Two different resolvents have one and the same asymptotic series, since they differ by an exponentially small quantity. In what follows, other asymptotic expansions occur, not just for $R(x;\zeta)$. We always understand the equality of two functions to mean equality of their asymptotic expansions as $\zeta\to+\infty$ (that is, as equality of formal power series in powers of $\zeta^{-1/2}$).

The method of obtaining the asymptotic expansion of $R(x;\zeta)$, which we now describe briefly, was developed more fully in [1]. It is based on the expansion of the operator $\left(-\frac{d^2}{dx^2}+u+\zeta\right)^{-1}$ in powers of $\left(-\frac{d^2}{dx^2}+\zeta\right)^{-1}$:

$$\left(-\frac{d^2}{dx^2}+u+\zeta\right)^{-1}=\left(-\frac{d^2}{dx^2}+\zeta\right)^{-1}\left[1+u\left(-\frac{d^2}{dx^2}+\zeta\right)^{-1}\right]^{-1}=$$
$$=\left(-\frac{d^2}{dx^2}+\zeta\right)^{-1}-\left(-\frac{d^2}{dx^2}+\zeta\right)^{-1}u\left(-\frac{d^2}{dx^2}+\zeta\right)^{-1}+\ldots$$

For the kernel of the resolvent we obtain accordingly the asymptotic series

$$R-R_0-R_0\circ u\circ R_0+R_0\circ u\circ R_0\circ u\circ R_0-\ldots,$$

where R_0 is the kernel of the resolvent of (1) for $u=0$, while $R_0\circ u\circ R_0\ldots$ is the superposition

$$\int\ldots\int R_0(x,x_1;\zeta)\,u(x_1)\,R_0(x_1,x_2;\zeta)\,u(x_2)\ldots dx_1\ldots dx_N.$$

The limits of integration are arbitrary, except that the point x must be within the interval of integration, since altering the limits only adds an exponential term. The equation is to be understood as one between formal (asymptotic) power series; it is meaningful, since for each fixed power of ζ only finitely many non-zero terms contribute to the right-hand side. As a representative of the resolvent $R_0(x,y;\zeta)$ we can take $e^{-\sqrt{\zeta}|x-y|}/2\sqrt{\zeta}$.

For the general term of the series we obtain the expression

$$\frac{(-1)^N}{2^{N+1}\zeta^{\frac{N+1}{2}}}\int\ldots\int e^{-\sqrt{\zeta}\,(|x-x_1|+|x_1-x_2|+\ldots+|x_N-x|)}u(x_1)\ldots u(x_N)\,dx_1\ldots dx_N.$$

To find an asymptotic expansion of this expression, we represent $u(x_i)$ as a series

$$u(x_i)=\sum\frac{(x_i-x)^{k_i}}{k_i!}\,u^{(k_i)}(x).$$

We obtain the asymptotic form as

$$\sum_{N=0}^{\infty}\sum_{(k)}(-1)^N\frac{M_{k_1\ldots k_N}u^{(k_1)}\ldots u^{(k_N)}}{2^{N+1}(\sqrt{\zeta})^{k_1+\ldots+k_N+2N+1}},$$

where

(4) $$M_{k_1\ldots k_N}=\frac{1}{k!\ldots k_N!}\times$$

$$\times\int_{-\infty}^{\infty}\ldots\int_{-\infty}^{\infty}e^{-\{|\eta_1|+|\eta_1-\eta_2|+\ldots+|\eta_N|\}}\eta_1^{k_1}\ldots\eta_N^{k_N}\,d\eta_1\ldots d\eta_N.$$

The coefficients are non-zero only if $k_1+\ldots+k_N$ is even, that is, the expansion is in odd powers of $\sqrt{\zeta}$. So we have obtained an expansion of the form (3), with

(5) $$R_l[u]=\sum_{N=l}^{\infty}\sum_{k_1+\ldots+k_N=2l-2N}M_{k_1\ldots k_N}u^{(k_1)}\ldots u^{(k_N)}.$$

This is, in fact, a polynomial in u, u' ..., and is homogeneous in the grading $k_1+\ldots+k_N+2N$, that is, in the sum of twice the degree N in the variables u, u', ... and of the differential order $k_1+k_2+\ldots+k_N$. The resulting expression for the coefficients in the form of integrals is complicated. These formulae acquire a more lucid and practically useful form if the symbolic calculus of polynomials in u, u', u'' ... developed in Chapter I, § 2 is used. This is done in Appendix I. In Appendix II we describe another method, leading to recurrence formulae for $R_l[u]$, which was developed in [2], [3]. Although these recurrence formulae are more complicated than those occurring in the subsequent sections, they have the advantage that the method used to obtain them is suitable not only for ordinary differential equations, but also in more general cases (see [9]).

2. A third order linear equation for the resolvent and the Riccati equation. We prove that $R(x;\zeta)$ satisfies the equation

(6) $$-2RR''+(R')^2+4(u+\zeta)R^2=1.$$

(By our convention, this equation is to be interpreted as meaning that the left- and right-hand sides differ by a term that is exponentially small as $\zeta\to\infty$). The dashes always denote $\frac{d}{dx}$. We find

$$R'(x,\zeta)=\left[\frac{\partial}{\partial x}R(x,y;\zeta)+\frac{\partial}{\partial y}R(x,y;\zeta)\right]_{y=x-0},$$

$$R''(x,\zeta)=\left[\frac{\partial^2}{\partial x^2}R(x,y;\zeta)+2\frac{\partial^2}{\partial x\,\partial y}R(x,y;\zeta)+\frac{\partial^2}{\partial y^2}R(x;y;\zeta)\right]_{y=x-0}=$$
$$=[(u(x)+\zeta)R(x,y;\zeta)+(u(y)+\zeta)R(x,y;\zeta)+2R_{xy}(x,y;\zeta)]_{y=x-0}.$$

We now observe that $RR_{xy}=R_xR_y+c(\zeta)$, where $c(\zeta)$ does not depend on x, y and tends exponentially to zero as $\zeta\to\infty$. For both R_x and R satisfy (1) with respect to y. Consequently, the Wronskian $R(R_x)_y-R_y(R_x)$ does not depend on y. By symmetry, it does not depend on x either. Off the diagonal $R(x, y; \zeta)$ is exponentially small, hence so is $c(\zeta)$. Furthermore, we have

$$-2RR''+(R')^2+4(u(x)+\zeta)R^2=$$
$$=-2R[(u(x)+\zeta)R(x,y;\zeta)+(u(y)+\zeta)R(x,y;\zeta)+2R_{xy}(x,y;\zeta)]_{y=x-0}+$$
$$+[R_x(x,y;\zeta)+R_y(x,y;\zeta)]^2_{y=x-0}+4(u(x)+\zeta)R^2(x,x;\zeta)=$$
$$=[-4R_xR_y+R_x^2+2R_xR_y+R_y^2]_{y=x-0}-4c(\zeta)=(R_x-R_y)^2_{y=x-0}-4c(\zeta)=$$
$$=1-4c(\zeta),$$

as was claimed.

As a consequence of (6) we see that $R(x; \zeta)$ satisfies the third order linear equation

$$-R'''+4(u(x)+\zeta)R'+2u'(x)R=0. \tag{7}$$

This can be verified by differentiating (6) and dividing by R. (But (7) could have obtained more simply. If φ_1, φ_2 is a fundamental system of solutions of (1), then $R(x{:}\ \zeta)$ is always a linear combination $\varphi_1^2, \varphi_2^2, \varphi_1\varphi_2$. Each of these expressions satisfies (7), as we can see by direct verification. Now (6) contains rather more information than (7) (the constant of integration depends on the normalization of R). Equation (7) is also used in [16].[1]

The equations (6) and (7) yield a recurrence method for determining the expansion coefficients R_l. Substituting (3) in (6) we obtain

$$R_0=\frac{1}{2},\quad R_1=-\frac{1}{4}u,\quad R_{l+1}=2\sum_{h=0}^{l-1}R_hR''_{l-h}-\sum_{k=1}^{l-1}R'_kR'_{l-k}-$$
$$-4u\sum_{k=0}^{l}R_kR_{l-k}-4\sum_{k=1}^{l}R_kR_{l-k+1}\qquad (l=1,2,\dots). \tag{8}$$

But (7) gives simpler formulae than these. Substituting (3) in (7) we obtain

$$R'_{l+1}=\frac{1}{4}R'''_l-uR'_l-\frac{1}{2}u'R_l, \tag{9}$$

that is, $R'_{l+1} = \frac{1}{4}R''_1 - u\,R_l + \frac{1}{2}\left(\frac{d}{dx}\right)^{-1} u'R_l$. The simplicity of this relation is due to the linearity of (7). However, here we do not obtain R_{l+1} but only its derivative. In Chapter 1 we have described an algorithm

[1] Apparently (7) plays an important part in Sturm–Liouville theory. It arises, for example, in the following situation (see [20]). Consider the equation $y'' + q(x)\,y = 0$. Under the change of variables $x = F(\xi)$, $y = \sqrt{F'(\xi)}\,Y(\xi)$ it goes over into $Y'' + Q(\xi)\,Y = 0$, $Q = (F')^2 q - \frac{3}{4}(F'')^2(F')^2\frac{1}{2}F'''/F$. Under an infinitesimal deformation of the variables $Y(\xi) = y/\sqrt{1+\ \varepsilon f'}$ where ε is a small parameter we have $Q = q + (4qf' + 2q'f + f''')\,2\,\varepsilon\ + 0(\varepsilon^2)$. If f satisfies $f''' + 2q'f + 4qf' = 0$, which is the same as (7), then $Q = q + 0(\varepsilon)^2$, that is, the equation is invariant to within terms of smaller than the highest order. From this it follows that there is a one-parameter group of transformations $x = F_t(\xi)$ (with a corresponding transformation of y) under which the equation is invariant. This group is given by the equation $d\xi/dt = R(\xi)$ with the condition $\xi|_{t=0} = x$. R is one solution of (7). There are three such independent groups corresponding to the three independent solutions of (7).

for calculating $\left(\frac{d}{dx}\right)^{-1}$.

We exhibit the first few coefficients R_l:

$$(10)\quad \begin{cases} R_0=\frac{1}{2}, \\ R_1=-\frac{1}{4}u, \\ R_2=\frac{1}{16}(3u^2-u''), \\ R_3=-\frac{1}{64}(10u^3-10uu''-5(u')^2+u^{\mathrm{IV}}), \\ R_4=\frac{1}{256}(35u^4-70u(u')^2-70u^2u''+21(u'')^2+ \\ \qquad\qquad +28u'u'''+14uu^{\mathrm{IV}}-u^{\mathrm{VI}}), \\ R_5=-\frac{1}{1024}(126u^5-630u^2(u')^2+504uu'u'''+462(u')^2u''+ \\ \qquad +378u(u'')^2-54u'u^{\mathrm{V}}-114u''u^{\mathrm{IV}}-69(u''')^2- \\ \qquad\qquad -420u^3u''+126u^2u^{\mathrm{IV}}-18uu^{\mathrm{VI}}+u^{\mathrm{VIII}}. \end{cases}$$

We now write (6) in another form and show how it is connected with the Riccati equation. We introduce two functions:

$$(11)\qquad \chi_R=\frac{1}{2iR}, \qquad \chi_I=-\frac{\chi_R'}{2\chi_R}.$$

Then (6) transforms into the system

$$(12)\qquad \begin{cases} \chi_R'+2\chi_R\chi_I=0, \\ \chi_I'+\chi_R^2-\chi_I^2+u+\zeta=0, \end{cases}$$

which can be combined into a single equation by setting $\chi = \chi_R = + i\varkappa_I$, so we obtain

$$(13)\qquad i\chi'+\chi^2+u+\zeta=0,$$

that is, a Riccati equation. If we expand χ by means of (13) as an asymptotic series in powers of $\sqrt{\zeta}$, then, as is easily seen from (12), all the odd powers refer to χ_R and the even ones to χ_I. For the coefficients of the expansion

$$(14)\qquad \chi=\sum_{l=-1}^{\infty}\frac{\chi_l}{(\sqrt{\zeta})^l}=\sum_{l=-1}^{\infty}\frac{\chi_l^{(R)}}{\zeta^{l+\frac{1}{2}}}+i\sum_{l=0}^{\infty}\frac{\chi_l^{(I)}}{\xi^l}$$

we obtain the recurrence formulae

$$(15)\qquad \begin{cases} \chi_{-1}=i, \quad \chi_0=0, \quad \chi_1=iu/2, \\ \chi_l=-\frac{1}{2}\chi_{l-1}'+\frac{i}{2}\sum_{r=1}^{l-2}\chi_r\chi_{l-1-r}. \end{cases}$$

3. Variational relations between the coefficients of the asymptotic form.

FIRST RELATION.

(16) $$\frac{\delta}{\delta u} R(x;\zeta) = \frac{\partial}{\partial \zeta} R(x;\zeta),$$

or in terms of the coefficients R_l;

(16′) $$\frac{\delta}{\delta u} R_l[u] = -\left(l - \frac{1}{2}\right) R_{l-1}[u].$$

A purely algebraic proof of this relation, based on the formulae obtained previously for the coefficients $R_l[u]$, is given in Appendix 1. Here we give a very simple proof, using the spectral representation of the resolvent.

Let $R(x, y; \zeta)$ be a resolvent for some fixed self-adjoint boundary conditions at the ends of an arbitrary interval containing x. The coefficients $R_l[u]$, as we know, do not depend on the specific form of the resolvent. If λ_n are the eigenvalues of the boundary problem and φ_n are the eigenfunctions, then

$$R(x;\zeta) = \sum \frac{\varphi_n^2(x)}{\lambda_n + \zeta}.$$

On variation we obtain

$$\frac{\delta}{\delta u} \int R(x;\zeta)\, dx = \frac{\delta}{\delta u} \sum \frac{1}{\lambda_n + \zeta} = -\sum \frac{\delta \lambda_n}{\delta u} \cdot \frac{1}{(\lambda_n + \zeta)^2}.$$

By perturbation theory,

$$\Delta \lambda_n = \int \varphi_n^2(x)\, \delta u(x)\, dx + O(\delta u^2),$$

from which, by definition of the variational derivative $\dfrac{\delta \lambda_n}{\delta u} = \varphi_n^2(x)$, we obtain

$$\frac{\delta}{\delta u} \int R(x;\zeta)\, dx = -\sum \varphi_n^2(x) \cdot \frac{1}{(\lambda_n + \zeta)^2} = \frac{\partial}{\partial \zeta} \sum \frac{\varphi_n^2(x)}{(\lambda_n + \zeta)} = \frac{\partial R(x,\zeta)}{\partial \zeta}.$$

In accordance with the accepted usage we replace $\dfrac{\delta}{\delta u}\int \ldots dx$ by $\dfrac{\delta}{\delta u} R$, which proves the relation.

SECOND RELATION.

(17) $$\frac{\partial}{\partial u} R_l[u] = -\left(l - \frac{1}{2}\right) R_{l-1}[u].$$

This relation can be established without difficulty by induction, using the recurrence formulae (8).[1] Symbolically it can be written as

(17′) $$\frac{\partial}{\partial u} R(x;\zeta) = \frac{\partial}{\partial \zeta} R(x;\zeta).$$

(We say symbolically because this equation is meaningful only for asymptotic series; otherwise it is not clear how to define $\partial/\partial u$-differentiation

---

1) (16′) and (17) show that $\frac{\delta}{\delta u} R_l = \frac{\partial}{\partial u} R_l$ or $p_0 R_l = 0$, where p_0 is the first momentum operator (Chapter 1.) The note on page 85 makes it clear that this is a consequence of the fact that R_l is the variational derivative.

with respect to u for fixed $u', u'', \ldots$). Intuitively, the equation is obvious; it is a consequence of the fact that $R(x, \zeta)$ depends on u and ζ not separately but only through their sum.

EXAMPLE. $R_2 = \frac{1}{16}(3u^2 - u'')$, $R_3 = -\frac{1}{64}(10u^3 - 10uu'' - 5(u')^2 + u^{IV})$.

Here the relation can be verified without difficulty.

THIRD RELATION

(18) $$\frac{\delta}{\delta u}\chi(x;\zeta) = \frac{\partial}{\partial\zeta}\chi(x;\zeta) + \frac{d}{dx}(\ ),$$

where $\frac{d}{dx}(\ )$ denotes the derivative of a function that is expanded as an asymptotic series whose coefficients are polynomials in $u, u'' \ldots$ In terms of the coefficients (18) means that

(18′) $$\frac{\delta}{\delta u}\chi_l[u] = -\frac{l-2}{2}\chi_{l-2}[u] + \frac{d}{dx}(\ ).$$

PROOF. First we show that

(19) $$\frac{\partial}{\partial u}\chi_l[u] = -\frac{l-2}{2}\chi_{l-2}[u],$$

or, symbolically,

(19′) $$\frac{\partial}{\partial u}\chi(x;\zeta) = \frac{\partial}{\partial\zeta}\chi(x;\zeta).$$

(19) is proved by induction, using the recurrence formula (15). The intuitive meaning is the same as that of (17) and (17′).

It remains for us to note that the partial derivative $\partial\chi_l/\partial u$ coincides with the variational derivative to within terms of the form $\frac{d}{dx}(\quad)$.

The relation obtained for χ_l is weaker than the relation for R_l by the presence of an additional undetermined term of the type of a derivative.

FOURTH RELATION ([15]).

(20) $$\frac{\delta}{\delta u}\chi_R(x;\zeta) = -\frac{1}{2\chi_R(x;\zeta)}.$$

We first prove a weaker identity having an additional term of the type of a derivative.

The function χ_R satisfies the equation obtained by eliminating χ_I from the system (12):

(21) $$\frac{\chi_R''}{2\chi_R} - \frac{3}{4}\frac{(\chi_R')^2}{\chi_R^2} + \chi_R^2 + u + \zeta = 0.$$

We differentiate this equation with respect to ζ and, for simplicity, write $\chi_R = f$, $\partial\chi_R/\partial\zeta = g$:

$$\frac{g''}{2f} - \frac{f''g}{2f^2} - \frac{3}{2}\frac{f'g'}{f^2} + \frac{3}{2}\frac{f'^2g}{f^3} + 2fg + 1 = 0,$$

$$g = -\frac{1}{2f} - \frac{g''f^2 - ff''g - 3ff'g' + 3f'^2g}{4f^4} = -\frac{1}{2f} - \frac{1}{4}\left(\frac{g'f - gf'}{f^3}\right)',$$

that is,

$$\frac{\partial \chi_R}{\partial \zeta}=-\frac{1}{2\chi_R}+\frac{d}{dx}(\ ).$$

Now we take the variational derivative, using the connection (11) between χ_R and R and the first variational relation:

$$\frac{\partial}{\partial \zeta}\frac{\delta}{\delta u}\chi_R=-\frac{\delta}{\delta u}\frac{1}{2\chi_R}=-i\frac{\delta}{\delta u}R=-i\frac{\partial R}{\partial \zeta}=-\frac{1}{2}\frac{\partial}{\partial \zeta}\frac{1}{\chi_R},$$

that is,

$$(22)\qquad \frac{\partial}{\partial \zeta}\left(\frac{\delta}{\delta u}\chi_R+\frac{1}{2\chi_R}\right)=0,$$

from which (21) follows, since the expression in brackets tends to zero as $\zeta\to\infty$.

The relation (21) can be rewritten as

$$(23)\qquad \frac{\delta}{\delta u}\chi_R=-iR$$

or, in terms of the coefficients,

$$(24)\qquad \frac{\delta}{\delta u}\chi_l^{(R)}=-iR_l.$$

Comparing this with (16) and (17), we see that the R_l are obtained as the variational derivatives of $\chi_l^{(R)}$ and R_{l+1}. Compared with $\chi(x;\zeta)$, the function $R(x;\zeta)$ has the advantage that it is self-reproducing under the action of the operator of variational derivative. As we shall see below, what is important in fact, are the variational derivatives of the coefficients R_l (or χ_l), that is, the coefficients R_l.

4. The basic property of the coefficients R_l. The main application of the coefficients R_l to the Korteweg–de Vries equation rests on the following result.

THEOREM. *The quantities R_kR_l' are derivatives, that is, there exist polynomials $P_{k,l}[u]$ in u, u' ... such that*

$$(25)\qquad R_kR_l'=\frac{d}{dx}P_{k,l}.$$

PROOF. By means of the recurrence relation (9) it can easily be verified that

$$R_kR_{l+1}'-R_{k+1}'R_l=\left(\frac{1}{4}R_kR_l''+\frac{1}{4}R_k''R_l-R_k'R_l'-uR_kR_l\right)'.$$

Moreover, it is clear that $R_kR_l'+R_k'R_l=(R_kR_l)'$. Therefore, the quantity R_kR_l' differs from $R_{k+1}R_{l-1}'$ by the sign and by a term of the type of a derivative. The chain of these equations can be continued until the second index becomes zero. But $R_0'=0$.

COROLLARY. $\frac{\delta}{\delta u}(R_kR_l')=0.$

Since the quantities $P_{k,l}$ play an important role in what follows, some formulae for them will be useful. We can write down a generating function

(26) $$P(x;\zeta_1,\zeta_2)=\sum_{k,\,l=0}^{\infty}\frac{P_{k,\,l}[u]}{\zeta_1^{k+\frac{1}{2}}\zeta_2^{l+\frac{1}{2}}}.$$

In terms of this generating function, (25) indicates that

(27) $$R(x;\zeta_1)\,R'(x;\zeta_2)=\frac{d}{dx}P(x;\zeta_1,\zeta_2).$$

We now use the third order equation (7):

$$-R'''(x;\zeta_1)+4(u+\zeta_1)R'(x;\zeta_1)+2u'R(x;\zeta_1)=0,$$
$$-R'''(x;\zeta_2)+4(u+\zeta_2)R'(x;\zeta_2)+2u'R(x;\zeta_2)=0.$$

Multiplying the first equation by $R(x;\zeta_2)$, the second by $R(x;\zeta_1)$, and adding, we obtain

(28) $$2(\zeta_1-\zeta_2)[P(x;\zeta_1,\zeta_2)-P(x;\zeta_2,\zeta_1)]=$$
$$=-R''(x;\zeta_1)R(x;\zeta_2)-R(x;\zeta_1)R''(x;\zeta_2)+R'(x;\zeta_1)R'(x;\zeta_2)+$$
$$+4uR(x;\zeta_1)R(x;\zeta_2)+2(\zeta_1+\zeta_2)R(x;\zeta_1)R(x;\zeta_2)+c(\zeta_1,\zeta_2).$$

Here c is a quantity independent of x. It is completely determined by the condition that the right-hand side is divisible by $\zeta_1-\zeta_2$. To evaluate it, we must replace R on the right-hand side by the first term of the expansion $1/(2\sqrt{\zeta})$ and require that the result should vanish. We compute without trouble that $c(\zeta_1,\zeta_2)=-(\zeta_1+\zeta_2)/2\sqrt{\zeta_1\zeta_2}$. Now (28) defines the antisymmetric part of the generating function $P(x;\zeta_1,\zeta_2)$. The symmetric part can be obtained more simply:

(29) $$P(x;\zeta_1,\zeta_2)+P(x;\zeta_1,\zeta_2)=R(x;\zeta_1)R(x;\zeta_2).$$

CHAPTER 3

The Korteweg–de Vries equation

1. The Novikov equations (the stationary Korteweg-de Vries equation). We define a Novikov equation (or stationary Korteweg-de Vries equation) as an ordinary differential equation of the form

(1) $$\sum_{k=0}^{n+1}c_kR_k[u]=0$$

for the function u, where the c_k are arbitrary fixed constants and $c_{n+1}=1$. The order of this equation is $2n$. By (17) in Chapter 2 this equation is equivalent to

(2) $$\frac{\delta}{\delta u}\sum_{k=0}^{n+2}d_kR_k[u]=0,\qquad d_k=-c_{k-1}\Big/\left(k-\frac{1}{2}\right),$$

which is the Euler–Lagrange variational equation for the Lagrangian $L(u,u',\ldots,u^{(2n+2)})=\sum_{k=0}^{n+2}d_kR_k$. Hence we can apply here the theory of

Chapter 1. First we construct a Hamiltonian. This satisfies $\frac{dH}{dx} = -u' \frac{\delta}{\delta u} L$. This is the general formula. But in our case, when the Lagrangian is a linear combination of the R_k, we can go further and write down explicitly not only the derivative of the Hamiltonian, but also the Hamiltonian itself. We observe that the correspondance between Lagrangians and Hamiltonians is linear. Therefore, it is sufficient to consider the case when the Lagrangian is equal to a single R_k. We can write down a generating function for the Hamiltonians H_k corresponding to the Lagrangians R_k:

$$H(x; \zeta) = \sum_{k=0}^{\infty} \frac{H_k[u]}{\zeta^{k+\frac{1}{2}}}.$$

Then

$$\text{(3)} \qquad \frac{dH}{dx} = -u' \frac{\delta}{\delta u} R = -u' \frac{\partial R}{\partial \zeta}.$$

Recalling the third order equation (7) of Chapter 2, we obtain $H' = -\frac{\partial}{\partial \zeta}\left[\frac{1}{2} R''' - 2(uR)' - 2\zeta R'\right]$; hence

$$H = \frac{\partial}{\partial \zeta}\left(\frac{1}{2} R'' - 2uR - 2\zeta R\right),$$

and this means that

$$\text{(4)} \qquad H_k = (2k-1)\left(R_k - \frac{1}{4} R''_{k-1} + uR_{k-1}\right).$$

This formula expresses H_k in terms of R_k. From (3) we obtain

$$\text{(5)} \qquad H'_k = \left(k - \frac{1}{2}\right) u' R_{k-1},$$

which enables us to express R_k in terms of H_k.

2. First integrals of the Novikov equations. The system (1) or (2) has n independent first integrals. For multiplying (1) by R'_l and using (25) of Chapter 2 we have

$$\frac{d}{dx} \sum_{k=0}^{n+1} c_k P_{k,l}[u] = 0,$$

that is, the quantities

$$\text{(4)} \qquad I_l = \sum_{k=0}^{n+1} c_k P_{k,l}[u] \qquad (l = 1, 2, \ldots, n)$$

are invariant.

The independence of these integrals is established as follows: the l-th integral contains the term $\pm[u^{(n+l-1)}]^2 \cdot 2^{-2(n+l+1)}$. The remaining terms of this integral and also all the preceding integrals can contain $u^{(n+l-1)}$ only in products by lower derivatives. If we now put all the variables equal to zero except for $u^{(n+l-1)}$, we see that I_l is independent of all the preceding integrals. The independence of all the integrals in the set follows from this. We note, that in (6) we can take l greater than n and still obtain first integrals, but they are no longer independent of the preceding ones.

The vector fields $\xi_{I_l} = \xi_l$ corresponding to the integrals I_l according to Theorem 3 in Ch.1 §7, can be found as follows. We must find a characteristic of each invariant, that is, an element a such that $\frac{dI_l}{dx} = a \frac{\delta}{\delta u} L$. In our case $a = R_l'$. Then the vector field is $\xi_l = -\sum_i R_l^{(i+1)} \frac{\partial}{\partial u^{(i)}}$. This equation must hold in A_L, that is, for solutions of the equation. Thus, the vector field corresponding to the invariant I_l is

$$\xi_l = -\sum_{i=0}^{\infty} R_l^{(i+1)} \frac{\partial}{\partial u^{(i)}}, \tag{7}$$

where the derivatives higher than $u^{(2n-1)}$ in the coefficients must be eliminated by means of the equation. It is remarkable that the vector field ξ_1 in this form does not depend on the equation, only the method of eliminating the higher derivatives depends on the equation.

The commutator of two such vector fields is equal to

$$[\xi_l, \xi_m] = \sum_i \sum_k \left(R_l^{(k+1)} \frac{\partial R_m^{(i+1)}}{\partial u^{(k)}} - R_m^{(k+1)} \frac{\partial R_l^{(i+1)}}{\partial u^{(k)}} \right) \frac{\partial}{\partial u^{(i)}}. \tag{8}$$

In [14] Novikov states that integrals that are polynomials in $u, u', u'', \ldots$, must be in involution, that is, their Poisson brackets vanish, or the corresponding vector fields commute. This proposition reduces to the fact that the right-hand sides of (8) vanish, except for the leading derivatives with the help of any of the equations for all n. Hence these right-hand sides must be identically equal to zero, with the exception of the leading derivatives:

$\sum_k \left(R_l^{(k+1)} \frac{\partial R_m^{(i+1)}}{\partial u^{(k)}} - R_m^{(k+1)} \frac{\partial R_l^{(i+1)}}{\partial u^{(k)}} \right) = 0 \ (i = 0, 1, \ldots)$. This is equivalent to the single identity

$$\sum_{k=0}^{\infty} \left(R_l^{(k+1)} \frac{\partial R_m}{\partial u^{(k)}} - R_m^{(k+1)} \frac{\partial R_l}{\partial u^{(k)}} \right) = 0. \tag{9}$$

A general proof of this identity will be published separately. It is trivial to verify for $l = 1$, when $I_l = H$. For $R_1 = -\frac{1}{4} u$ and

$$\sum_{k=0}^{\infty} \left(u^{(k+1)} \frac{\partial R_m}{\partial u^{(k)}} - R_m^{(k+1)} \frac{\partial u}{\partial u^{(k)}} \right) = R_m' - R_m' = 0.$$

It can also easily be verified for $l = 2$ and arbitrary m.

In conclusion we give a table of invariants for the first few Novikov equations.

1) The Lagrangian $L = R_3$. The equation is $3u^2 - u'' = 0$. The invariant is $I_1 = H = u^3 - \frac{1}{2}(u')^2$.

2) The Lagrangian R_4. The equation is

$10u^3 - 10uu'' - 5(u')^2 + u^{\mathrm{IV}} = 0$. The invariants are:

$$I_1 = H = \frac{5}{2}u^4 - 5u(u')^2 + u'''u' - \frac{1}{2}(u'')^2$$

and

$$I_2 = 12u^5 - 10u^3u'' - 15u^2(u')^2 + 6uu'u''' - (u')^2u'' + 2u(u'')^2 - \frac{1}{2}(u''')^2.$$

3) The Lagrangian R_5. The equation is

$$35u^4 - 70uu'^2 - 70u^2u'' + 21(u'')^2 + 28u'u''' + 14uu^{\mathrm{IV}} - u^{\mathrm{VI}} = 0.$$

The invariants are

$$I_1 = H = 7u^5 - 35u^2(u')^2 + 14(u')^2u'' + 14uu'u''' - 7u(u'')^2 - u'u^{\mathrm{V}} + u''u^{\mathrm{IV}} - \frac{1}{2}(u''')^2,$$

$$\begin{aligned} I_2 = 35u^6 - 35u^4u'' - 140u^3(u')^2 + 84u^2u'u''' - \\ - 7u^2(u'')^2 + 70u(u')^2u'' - \frac{35}{2}(u')^4 - 6uu'u^{\mathrm{V}} + 6(u')^2u^{\mathrm{IV}} + \\ + 6uu''u^{\mathrm{IV}} - 10u(u''')^2 - 18u'u''u''' + 6(u'')^3 + u'''u^{\mathrm{V}} - \frac{1}{2}(u^{\mathrm{IV}})^2, \end{aligned}$$

$$\begin{aligned} I_3 = 150u^7 - 350u^5u'' - 525u^4(u')^2 + 280u^3u'u''' + \\ + 35u^4u^{\mathrm{IV}} + 210u^3(u'')^2 + 700u^2(u')^2u'' - 30u^2u'u^{\mathrm{V}} - \\ - 10u(u')^2u^{\mathrm{IV}} - 40u^2u''u^{\mathrm{IV}} + 10(u')^3u''' - 180uu'u''u''' - \\ - 50u^2(u''')^2 - 205(u')^2(u'')^2 - 10u(u'')^3 + 20u'u''u^{\mathrm{V}} + \\ + 10uu'''u^{\mathrm{V}} + (u'')^2u^{\mathrm{IV}} - 2u'u'''u^{\mathrm{IV}} + 2u(u^{\mathrm{IV}})^2 - \frac{1}{2}(u^{\mathrm{V}})^2. \end{aligned}$$

3. The Korteweg–de Vries equation. We assume that the function u depends additionally on a parameter t. As before, we use dashes to denote the derivatives with respect to x. (Generalized) Korteweg–de Vries equations are of the form

$$\frac{\partial u}{\partial t} = \frac{\partial}{\partial x}\sum_{k=0}^{n+1} c_k R_k[u]. \tag{10}$$

Thus, Novikov equations are Korteweg–de Vries equations for stationary functions u (independent of t). The original (non-generalized) Korteweg–de Vries equation is the particular case of (10)

$$\frac{\partial u}{\partial t} = \frac{\partial}{\partial x} R_2[u].$$

In contrast to the preceding, we must now choose a class of functions u. In fact, the integral $\int \ldots dx$ must have a meaning and an integral of a function of the type of a derivative must vanish. For example, the functions may be damped at infinity, or periodic, or almost periodic, and so on.

THEOREM. *The quantities*

$$I_l = \int R_l[u]\,dx \tag{11}$$

are invariants of the solutions of (10).

PROOF.

$$\frac{dI_l}{dt}=\int\frac{\partial}{\partial t}R_l[u]\,dx=\int\sum_{k=0}^{\infty}\frac{\partial R_l}{\partial u^{(k)}}u_t^{(k)}\,dx=\int\Big(\sum_{k=0}^{\infty}(-1)^k\Big(\frac{d}{dx}\Big)^k\frac{\partial R_l}{\partial u^{(k)}}\Big)\cdot u_t\,dt=$$

$$=\int\frac{\delta}{\delta u}R_l[u]\cdot\sum_{k=0}^{n+1}c_kR'_k[u]\,dx=-\Big(l-\frac{1}{2}\Big)\sum_{k=0}^{n+1}c_k\int R_{l-1}[u]\,R'_k[u]\,dx=0,$$

since $R_{l-1}R'_k$ is an expression of the type of a derivative.[1]

Since the paper of Lax [12] it has been usual to base the theory of the Korteweg–de Vries equation on the construction of so-called $(\mathcal{L}, A)$-pairs. We have avoided this above, but for completeness we now describe this method. Let $\mathcal{L}$ be the operator

$$\mathcal{L}=-\frac{d^2}{dx^2}+u(x).$$

Let A be a differential operator of arbitrary order whose coefficients are polynomials in u, u', ..., whose commutator with $\mathcal{L}$ is the operator of multiplication by some function $\widetilde{A}[u]$. Then we call $u_t = \widetilde{A}[u]$ a generalized Korteweg–de Vries equation. It is equivalent to the operator equation $\mathcal{L}_t = [A, \mathcal{L}]$. Hence it follows that if u is a solution of $u_t = \widetilde{A}[u]$, then under a change of the parameter t the operator $\mathcal{L}$ is transformed into a similar operator $\mathcal{L}(t) = T^{-1}(t)\,\mathcal{L}(0)\,T(t)$, and its spectrum is invariant. Therefore, the quantities (11) must be invariants of the equation, because they are the coefficients of the asymptotic expansion of the trace of the resolvent and therefore expressible in terms of the spectrum.

How can such an operator A be constructed? One of the possible ways is given in [15]. These operators appear in the expansion of ζ,

$$\mathcal{A}=\Big(\frac{1}{2}R\frac{d}{dx}-\frac{1}{4}R'\Big)(\mathcal{L}+\zeta)^{-1}=\sum_{k=0}^{\infty}\frac{A_k}{\zeta^{k+\frac{1}{2}}}$$

in powers of $\sqrt{\zeta}$. The fact that the resulting operators have the required property follows from the formula $[\mathcal{L}, \mathcal{A}] = R'(x;\zeta)$, which is most

---

1) In an infinite-dimensional space we can introduce a symplectic structure by means of the 2-form (see [13]):

$$\omega(\delta u,\delta v)=\int_{-\infty}^{\infty}\int_{-\infty}^{x}[\delta u(x)\,\delta v(x_1)-\delta u(x_1)\,\delta v(x)]\,dx_1\,dx.$$

Then for any functional $H = \int H[u]\,dx$ as Hamiltonian, we can construct the Hamiltonian equation $u_t = \frac{\partial}{\partial x}\frac{\delta}{\delta u}H$. The equation (10) corresponds to the Hamiltonian $H = \sum_{k=0}^{n+1}\frac{c_k}{-(k+\frac{1}{2})}R_{k+1}$. The fact that $R_lR'_k$ is an expression of the type of a derivative is in this context equivalent to the fact that the Poisson brackets of the functionals (11) vanish. Therefore, the quantities I_l are integrals in involution of the Korteweg–de Vries equation.

easily verified using the third order equation (7) in Chapter 2:

$$[\mathcal{L}, \mathcal{A}] = \frac{1}{2}\left[\mathcal{L}, R\frac{d}{dx} - \frac{1}{2}R'\right](\mathcal{L}+\zeta)^{-1} = \frac{1}{4}\left(R''' - 4R'\frac{d^2}{dx^2} - 2Ru'\right)(\mathcal{L}+\zeta)^{-1} =$$

$$= \frac{1}{4}\left(4uR' - 4R'\frac{d^2}{dx^2}\right)(\mathcal{L}+\zeta)^{-1} = R'(\mathcal{L}+\zeta)(\mathcal{L}+\zeta)^{-1} = R' = \frac{d}{dx}\sum_{l=0}^{\infty}\frac{R_l}{\zeta^{l+\frac{1}{2}}}.$$

So we arrive at the equation $\frac{\partial}{\partial t}u = \frac{\partial}{\partial x}R_l$, that is, the Korteweg–de Vries equation.

APPENDIX 1

A generating function for the polynomials $R_l[u]$

The expressions (4) in Chapter 2 for the coefficients of the polynomials $R_l[u]$ are complicated and difficult to apply in practice. We now write the polynomials $R_l[u]$ in symbolic form and find for them a generating function of a completely transparent form.

We denote by $R_l^{(N)}[u]$ the homogeneous part of $R_l[u]$ of degree N in the variables $u, u', u'' \dots$ In agreement with the symbolic notation of Chapter 1, §2, the function $R_l^N[u]$ is associated with a polynomial $\widetilde{R}^{(N)}(\xi_1, \dots, \xi_N)$. Specifically,

(1) $$\widetilde{R}_l^{(N)}(\xi_1, \xi_2, \dots, \xi_N) = \frac{(-1)^N}{2^{N+1}}S_N\left(\sum_{k_1+\dots+k_N=2l-2N} M_{k_1\dots k_N}\xi_1^{k_1}\dots\xi_N^{k_N}\right).$$

We consider the series

(2) $$\widetilde{R}^{(N)}(\xi_1, \xi_2, \dots, \xi_N; \zeta) = \sum_{l=N}^{\infty}\frac{R_l^{(N)}(\xi_1, \xi_2, \dots, \xi_N)}{\zeta^{l+\frac{1}{2}}},$$

whose coefficients in the expansion in powers of $\sqrt{\zeta}$ are given by (1). We transform the expressions (4) for $M_{k_1\dots k_N}$ in Chapter 2 using the Fourier transform

$$\frac{1}{2\sqrt{\zeta}}e^{-\sqrt{\zeta}|\eta|} = \frac{1}{2\pi}\int\frac{e^{i\alpha\eta}}{\alpha^2+\zeta}d\alpha.$$

Then we have

(3) $$\widetilde{R}^{(N)}(\xi_1, \xi_2, \dots, \xi_N; \zeta) = \frac{(-1)^N}{(2\pi)^{N+1}}S_N\left(\int\dots\int\frac{e^{i\alpha_1\eta_1+i\alpha_2(\eta_1-\eta_2)+\dots+i\alpha_{N+1}\eta_N}}{(\alpha_1^2+\zeta)\dots(\alpha_{N+1}^2+\zeta)}\times\right.$$
$$\times e^{\xi_1\eta_1+\dots+\xi_N\eta_N}d\alpha_1\dots d\alpha_{N+1}d\eta_1\dots d\eta_N.$$

We now evaluate these integrals. Integrating first with respect to $\eta_1, \ldots, \eta_N$, and then with respect to $\alpha_1, \ldots, \alpha_{N+1}$, we obtain

$$\widetilde{R}^{(N)}(\xi_1, \ldots, \xi_N; \zeta) = \frac{(-1)^N}{2\pi} \times$$

$$\times S_N\left(\int \frac{\delta(i\xi_1+\alpha_1-\alpha_2)\,\delta(i\xi_2+\alpha_2-\alpha_3)\ldots\delta(i\xi_N+\alpha_N-\alpha_{N+1})}{(\alpha_1^2+\zeta)\ldots(\alpha_{N+1}^2+\zeta)}\,d\alpha_1\ldots d\alpha_{N+1}\right) = \frac{(-1)^N}{2\pi}\times$$

$$\times S_N\left(\int \frac{d\alpha_{N+1}}{[-(\xi_1+\ldots+\xi_N+i\alpha_{N+1})^2+\zeta][-(\xi_2+\ldots+\xi_N+i\alpha_{N+1})^2+\zeta]\ldots[-(\xi_N+i\alpha_{N+1})^2+\zeta]\cdot[-(i\alpha_{N+1})^2+\zeta]}\right).$$

Finally

$$(4) \qquad \widetilde{R}^{(N)}(\xi_1, \ldots, \xi_N; \zeta) = \frac{(-1)^N}{2\sqrt{\zeta}} \times$$

$$\times S_N\left(\sum_{r=1}^{N+1} \frac{1}{[\zeta-(\sqrt{\zeta}-\xi_1-\ldots-\xi_{r-1})^2]\ldots[\zeta-(\sqrt{\zeta}-\xi_{r-1})]^2[\zeta-(\sqrt{\zeta}+\xi_r)^2]\ldots[\zeta-(\sqrt{\zeta}+\xi_r+\ldots+\xi_N)^2]}\right).$$

The first term is absent for $r = 1$, and the second for $r = N + 1$. We now give another form of this expression, as an expansion in elementary fractions:

$$(5) \qquad \widetilde{R}^{(N)}(\xi_1, \xi_2, \ldots, \xi_N; \zeta) =$$

$$= \frac{1}{2\sqrt{\zeta}} \sum_{r=1}^{N+1} \sum_{\substack{k=1 \\ (r\neq k)}}^{N+1} (-1)^{N+r+k} b_r \cdot b_k \cdot \begin{cases} \dfrac{(\xi_k+\ldots+\xi_{r-1})^2}{4\zeta-(\xi_k+\ldots+\xi_{r-1})^2}, & k<r, \\[2mm] \dfrac{(\xi_r+\ldots+\xi_{k-1})^2}{4\zeta-(\xi_r+\ldots+\xi_{k-1})^2}, & k>r, \end{cases}$$

$$b_r = \prod_{j=1}^{r-1} (\xi_j+\ldots+\xi_{r-1})^{-1} \cdot \prod_{j=r}^{N} (\xi_r+\ldots+\xi_j)^{-1}.$$

In the expression for b_r the first term is absent for $r = 1$ and the second for $r = N + 1$.

We show how the generating function $\widetilde{R}^{(N)}(\xi_1, \ldots, \xi_N)$ is used to prove the basic variational relation (16) or (17) of Chapter 2. It is sufficient to prove this relation separately for each homogeneous part of $R_l[u]$, that is, for $R_l^{(N)}[u]$. We claim that

$$\frac{\widetilde{\delta}}{\delta u} \widetilde{R}_l^{(N)} = -\left(l-\frac{1}{2}\right) \widetilde{R}_{l-1}^{(N-1)} \quad \text{or} \quad \frac{\widetilde{\delta}}{\delta u} \widetilde{R}^{(N)} = -\frac{\partial}{\partial\zeta} \widetilde{R}^{(N-1)}.$$

To prove this we choose the generating function in the form (3), without integrating. Then

$$\frac{\widetilde{\delta}}{\delta u} \widetilde{R}^{(N)}(\xi_1, \ldots, \xi_{N-1}; \zeta) = \frac{(-1)^N N}{(2\pi)^{N+1}} \cdot \frac{1}{N!} \times$$

$$\times \sum_{(k)} \int\ldots\int \frac{e^{-i\alpha_1\eta_1+i\alpha_2(\eta_1-\eta_2)+\ldots+i\alpha_{N+1}\eta_N}}{(\alpha_1^2+\zeta)\ldots(\alpha_{N+1}^2+\zeta)} e^{\xi_1\eta_{k_1}+\xi_2\eta_{k_2}+\ldots+\xi_{N-1}\eta_{k_{N-1}}+(-\xi_1-\ldots-\xi_{N-1})\eta_{k_N}} \times$$

$$\times d\alpha_1\ldots d\alpha_{N+1}\,d\eta_1\ldots d\eta_N.$$

Here $(k) = (k_1, k_2, \ldots, k_N)$ comprises all permutations of the indices $1, \ldots, N$. We divide all the terms of the sum into three types, for $k_N = 1$, $k_N = N$ and $k_N \neq 1, N$. In the case $k_N = 1$ we set $\eta_k - \eta_2 = \tilde{\eta}_{ki}$ $(i = 1, \ldots, N-1)$. Then we have

$$\frac{(-1)^N}{(2\pi)^{N+1}} \cdot \frac{1}{(N-1)!} \times$$

$$\times \sum_{(k_1, \ldots, k_N)} \int \cdots \int \frac{e^{-i\alpha_1\eta_1 + i\alpha_2(-\tilde{\eta}_2) + i\alpha_3(\tilde{\eta}_2 - \tilde{\eta}_3) + \ldots + i\alpha_{N+1}\tilde{\eta}_N + i\alpha_{N+1}\eta_1}}{(\alpha_1^2 + \zeta) \ldots (\alpha_{N+1}^2 + \zeta)} \times$$

$$\times e^{\xi_1 \tilde{\eta}_{k_1} + \ldots + \xi_{N-1}\tilde{\eta}_{k_{N-1}}} \, d\alpha_1 \ldots d\alpha_{N+1} \, d\eta_1 \, d\tilde{\eta}_2 \ldots d\tilde{\eta}_N.$$

Integrating with respect to η_1, α_1 and relabelling $\alpha_2, \ldots, \alpha_{N+1} \to \alpha_1, \ldots, \alpha_N$; $\tilde{\eta}_2, \ldots, \tilde{\eta}_N \to \eta_1, \ldots, \eta_{N-1}$, we obtain

$$-\frac{(-1)^{N-1}}{(2\pi)^N} \cdot \frac{1}{(N-1)!} \sum_{(k_1, \ldots, k_{N-1})} \int \cdots \int \frac{e^{-i\alpha_1\eta_1 + i\alpha_2(\eta_1 - \eta_2) + \ldots + i\alpha_N\eta_{N-1}}}{(\alpha_1^2 + \zeta) \ldots (\alpha_{N-1}^2 + \zeta)(\alpha_N^2 + \zeta)^2} \times$$

$$\times e^{\xi_1\eta_{k_1} + \ldots + \xi_{N-1}\eta_{k_{N-1}}} \, d\alpha_1 \ldots d\alpha_N \, d\eta_1 \ldots d\eta_{N-1}.$$

This is equal to the term in the derivative $\dfrac{\partial}{\partial\zeta} \tilde{R}^{(N-1)}(\xi_1, \ldots, \xi_{N-1}; \zeta)$ arising from the differentiation with respect to ζ of the last bracket of the denominator. The other two cases $k_N = N$ and $k_N \neq 1, N$ are dealt with in exactly the same way. We obtain the terms in the derivative corresponding, respectively, to the first and the remaining brackets of the denominator. We omit the details, which differ little from the case above. This proves the formula.

APPENDIX 2

Another recurrence method for obtaining the coefficients of the asymptotic expansion of the resolvent

This method was proposed in [2], [3]. In [9] it was developed in full generality for arbitrary elliptic pseudo-differential operators on manifolds of any dimension.

The method is also based on an expansion of the kernel of the resolvent as a series in the iterations of the kernel of the simplest resolvent, for the equation $-\dfrac{d^2\varphi}{dx^2} + \zeta\varphi = 0$. But instead of the series used in Chapter 2 §1, now write down a series in which the factors of each term occur in a different order: on the right the iterations of the kernel R_0, and on the left the operators of multiplication by a function. Explicitly, we write

$$(1)\qquad R(x, y; \zeta)=\sum_{l=0}^{\infty}\sum_{m=0}^{l}(-1)^{\frac{l+m}{2}}B_{l,m}[u]\,D^m R_0^{\left(1+\frac{l+m}{2}\right)}(x, y; \zeta),$$

where

$$D=i\frac{d}{dx},\qquad R_0^{(n)}=R_0\circ R_0\circ\ldots\circ R_0.$$

The sum is over pairs l, m such that $l+m$ is even, the $B_{l,m}[u]$ are polynomials in $u, u', \ldots$, which are chosen so that the series (1) is asymptotic: if the series is truncated at $l=T$, then the difference $R-R_T$ between the left- and the right-hand sides is $O(\zeta^{-T/2})$. This can be done as follows. Both sides of (6) are multiplied on the left by $D^2+u+\zeta$. On the right-hand side, using the commutation relations for the operators of differentiation and of multiplication by a function, the operators are permuted, so that they occur in the same order as in (1), that is, that on the left there is a multiplication by a function, then a differentiation, then the iterations of the kernel R_0. Here we must bear in mind

$$(D^2+\zeta)R_0(x, y; \zeta)=\delta(x-y),\ (D^2+u+\zeta)R(x, y; \zeta)=\delta(x-y),$$
$$(D^2+\zeta)R_0^{(n)}(x, y; \zeta)=R_0^{n-1}(x, y; \zeta).$$

Then the $B_{l,m}$ are chosen so that all the terms are mutually annihilated. This leads to recurrence relations to determine the coefficients:

$$(2)\qquad \begin{cases} B_{0,0}\equiv 1,\ B_{l,m}\equiv 0 \quad\text{for}\quad m<0 \quad\text{or}\quad l<0,\\ B_{l,m}=-B''_{l-2,m}+uB_{l-2,m}+2iB'_{l-1,m-1}.\end{cases}$$

Each term of (1) can be evaluated directly, by means of the Fourier transform

$$D^m(D^2+\zeta)^{-n}\frac{1}{2\pi}\int e^{i\xi(x-y)}\,d\xi=\frac{1}{2\pi}\int(-\xi)^m(\xi^2+\zeta)^{-n}e^{i\xi(x-y)}\,d\xi.$$

For $x=y$ these expressions are non-zero only when m is even, and then they are equal to

$$\frac{1}{2\pi}\,\frac{\Gamma\left(\frac{m+1}{2}\right)\Gamma\left(n-\frac{m+1}{2}\right)}{\Gamma(n)}\,\zeta^{-n+\frac{m+1}{2}}$$

([22], 3.241). It is now straightforward to show that the series (1) has the required property of being asymptotic. We find, ultimately,

$$R(x; \zeta)=\sum_{l=0}^{\infty}\sum_{m=0}^{\infty}B_{l,m}[u]\cdot\frac{1}{2}\begin{pmatrix}\frac{l-1}{2}\\ \frac{l+m}{2}\end{pmatrix}(-1)^{\frac{l}{2}}\zeta^{-\frac{1}{2}-\frac{l}{2}}.$$

Here l and m are even. We write the result as follows:

$$R_l[u]=\frac{1}{2}A_{2l,\,l-\frac{1}{2}}[u]\cdot(-1)^l,$$

where

$$A_{l,s} = \sum_{m=0}^{l} \binom{s}{\frac{l+m}{2}} B_{l,m}[u],$$

and $B_{l,m}$ is determined by the recurrence relation (2).[1]

The technique just developed throws a new light on the procedure for finding Lax pairs ($\mathscr{L}$, A) (see Ch.3 §3). To the asymptotic expansion of the resolvent (1) there corresponds the asymptotic expansion of the operators

$$(D^2+u+\zeta)^{-1} = \sum_{l=0}^{\infty} \sum_{m=0}^{l} (-1)^{\frac{l+m}{2}} B_{l,m}[u]\, D^m (D^2+\zeta)^{-1-\frac{l+m}{2}},$$

where the operators $D^2 + u + \zeta$ and $D^2 + \zeta$ must, of course, be taken under certain boundary conditions. The results are purely local and depend on the function $u(x)$ only in a neighbourhood of the point x in question, and not on the choice of boundary conditions. It is simplest to assume zero boundary conditions at the ends of an interval containing x and $u(x)$ to vanish in a neighbourhood of the end-points. The asymptotic character of the series is to be understood in the sense that if the series is truncated at the l-th stage, the remainder is $O(D^{-2-l})$. The latter means that if it is multiplied by D^{2+l} the remainder is a bounded operator (for greater detail see [2]). Now if both sides of the equation are multiplied by $(-\zeta)^s$, and then integrated with respect to $z = -\zeta$ over a contour enclosing the whole spectrum of $D^2 + u$ and D^2 (but not $z = 0$; we may assume that zero is not a point of the spectrum), we obtain

$$(D^2+u)^s = \sum_{l=0}^{\infty} A_{l,s}[u]\, D^l (D^2)^{s-l},$$

where the $A_{l,s}[u]$ are the polynomials in u, u', ... defined above; the series is asymptotic in the sense explained.

Now let $s = n + \frac{1}{2}$, where n is an integer. We denote by Q the part of the series corresponding to negative powers of D:

$$Q = \sum_{l=0}^{2n+1} A_{l,n+\frac{1}{2}}[u]\, D^{2n+1-l}.$$

The operator $(D^2 + u)^{n+\frac{1}{2}}$ commutes with $D^2 + u$, but

$$(D^2+u)^{n+\frac{1}{2}} = Q + A_{2n+2,\, n+\frac{1}{2}} D (D^2)^{-1} + O(D^{-2}).$$

---

[1] For comparison with Seeley's paper [9] it should be noted that there the coefficients $\gamma_{j,k}$ are connected with $B_{l,m}$ by the relations $\gamma_{k,j} = (-1)^{j-k} B_{j,\,2k-j}\, \xi^{2k-j}$. Seeley's recurrence relations are somewhat more complex: the difference occurs because these recurrence formulae arise from multiplying (1) by $D^2 + u + \zeta$ not on the left, as in our case, but on the right.

Hence

$$[Q, D^2+u] = -[A_{2n+2,\, n+\frac{1}{2}} D(D^2)^{-1}, D^2+u] + O(D^{-1})$$

or

$$[Q, D^2+u] = 2iA'_{2n+2,\, n+\frac{1}{2}} + O(D^{-1}).$$

The remaining terms must vanish, because the commutator of two differential operators can only be a differential operator. Thus, the $(2n+1)$-th order operator Q has the required property – its commutator with D^2+u is the operator of multiplication by a function,

$$[Q, D^2+u] = 2iA'_{2n+2,\, n+\frac{1}{2}} = 4i\frac{d}{dx}R_{n+1}[u].$$

As might be expected, the commutator is the right-hand side of the Korteweg–de Vries equation.

References

[1] I. M. Gel'fand, On identities for eigenvalues of a differential operator of the second order, Uspekhi Mat. Nauk **11**:1 (1956), 191–198. MR **18**–129.

[2] L. A. Dikii, The zeta-function of an ordinary differential equation on a finite interval, Izv. Akad. Nauk SSSR Ser. Mat. **19** (1955), 187–200. MR **17**–619.

[3] L. A. Dikii, Trace formulae for Sturm–Liouville operators, Uspekhi Mat. Nauk **13**:3 (1958), 111–143. MR **20**–6655.

[4] I. M. Gel'fand and B. M. Levitan, On a simple identity for the eigenvalues of a differential operator of the second order, Dokl. Akad. Nauk SSSR **88** (1953), 593–596. MR **13**–558.

[5] S. Minakshisundaram, On a generalization of Epstein's zeta-function, Canadian J. Math. **1** (1949), 320–327. MR **11**–357.

[6] M. G. Krein, On a trace formula in perturbation theory, Mat. Sb. **33** (1953), 597–626. MR **15**–720.

[7] L. D. Fadeev, An expression for the trace of the difference between two singular differential operators of Sturm–Liouville type, Dokl. Akad. Nauk SSSR **115** (1957), 878–880. MR **20**–1029.

[8] V. S. Buslaev and L. D. Faddeev, On trace formulae for a singular differential Sturm–Liouville operator, Dokl. Akad. Nauk SSSR **132** (1960), 13–16. = Soviet Math. Dokl. **1** (1960), 451–454.

[9] R. T. Seeley, The index of elliptic systems of singular integral operators, J. Math. Anal. Appl. **7** (1963), 289–309. MR **28** #2464. = Matematika **12** (1968), 96–112.

[10] V. S. Buslaev, On the asymptotic behaviour of spectral characteristics of exterior problems for the Schrödinger operator, Izv. Akad. Nauk SSSR Ser. Mat. **39** (1975), 149–235.

[11] C. S. Gardiner, J. M. Green, M. D. Kruskal and R. M. Miura, A method for solving the Korteweg–de Vries equation, Phys. Rev. Letters **19** (1967), 1095–1097.

[12] P. Lax, Integrals of non-linear equations of evolution and solitary waves, Comm. Pure Appl. Math. 21 (1968), 467–490. MR 38–3620.

[13] V. E. Zakharov and L. D. Faddeev, The Korteweg–de Vries equation as a completely integrable Hamiltonian system, Funktsional. Anal. i Prilozhen. 5:4 (1971), 18–27. = Functional Anal. Appl. **5** (1971), 280–287.

[14] S. P. Novikov, The periodic problem for the Korteweg–de Vries equation. I, Funktsional. Anal. i Prilozhen. 8:3 (1974), 54–66. = Functional Anal. Appl. **8** (1974), 236–246.

[15] B. A. Dubrovin, The periodic problem for the Korteweg–de Vries equation in a class of finite zone potentials, Funktsional. Anal. i Prilozhen. 9:3 (1975), 41–52. = Functional Anal. Appl. **9** (1975), 215–223.

[16] V. A. Marchenko, The periodic Korteweg–de Vries problem, Mat. Sb. **95** (1974), 331–356.

[17] P. Lax, Periodic solutions of the Korteweg–de Vries equations, Lectures in Appl. Math. **15** (1974), 85–96.

[18] I. M. Gel'fand, The cohomology of infinite-dimensional Lie algebras; some questions of integral geometry, Actes Congrès Internat. Mathématiciens, Nice 1970, Vol.1, 35–111 (in English).

[19] A. M. Gabrielov, I. M. Gel'fand and M. V. Losik, Combinatorial calculation of characteristic classes, Funktsional. Anal. i Prilozhen. 9:2 (1975), 12–28. = Functional Anal. Appl. **9** (1975), 103–115.

[20] V. F. Lazutkin and T. V. Pankratova, Normal forms and versal deformations for the Hill equation, Funktsional. Anal. i Prilozhen. 9:4 (1975), 41–47. = Functional Anal. Appl. **9** (1975), 306–311.

[21] C. Godbillon, Géométrie différentielle et mécanique analytique, Hermann, Paris 1969. MR 39 # 3416.
Translation: *Differentsial'naya geometriya i analiticheskaya mekhanika*, Mir, Moscow 1973.

[22] I. S. Gradshtein and I. M. Ryzhik, *Tablitsy integralov, sum, ryadov i proizvedenii*, (Tables of integrals, sums, series and products), Fizmatgiz, Moscow 1963.

Received by the Editors, 26 May 1975

Translated by G. and R. Hudson

13.

(with L. A. Dikij)

The resolvent and Hamiltonian systems

Funkts. Anal. Prilozh. **11** (2) (1977) 11–27. MR **56**:1359, Zbl. **357**:58005.
[Funct. Anal. Appl. **11** (1977) 93–105]

In [1] and in a number of papers extending it, devoted to the Korteweg–de Vries (KdV) equation, the system of first equations of this equation was constructed and a method was proposed for solving this equation, using the theory of linear first-order equations. Lax [2] explained the connection of this to the so-called LA pairs (we call them PL pairs). Gardner [3] and Zakharov and Faddeev [4] constructed the Hamiltonian form of the KdV equation, relative to which the first integrals proved to be involution. The methods in [1, 2] were generalized to many other systems. Each of them uses the construction of PL pairs. We know three general algorithms for the construction of PL pairs. The first one of them, due to Zakharov and Shabat [5], is the method of "clothes" which gives the possibility of constructing a whole series of interesting new systems. This method arises from an analysis of inverse problems. The second method is based on the asymptotic behavior of eigenfunctions and was proposed for the case when L is a matrix differential operator of second order by Dubrovin [6] and for both an n-th-order scalar operator as well as for a general n-th-order matrix operator by Krichever [7, 8] (also see survey [9]). The solutions of the stationary problem have been written out in explicit form in terms of θ functions in [8]; Its and Matveev [10] also obtained an analogous solution. The ideas expressed by example of a second-order operator by Novikov [11], and also in [23], were developed in [6–8]. Here we do not specially touch upon the periodic aspect of the problem (in this regard see also Marchenko [12], Lax [13], and McKean and V. Moerbeke [14]) and also upon other important sides of the problem (see [24–27]). The third method is connected with the study of the asymptotic behavior of the resolvent. In [15] the authors revealed that the expressions, constructed by them earlier in [16–18], for the diagonal of the resolvent and for the powers of an operator, essentially are expressions for the first integrals which were discovered in the KdV theory. It was shown that PL pairs can be expressed by the terms of the asymptotics of the resolvent. After the appearance of Krichever's paper the authors posed themselves the problem of ascertaining whether PL pairs can be constructed also by the methods of the resolvent and the symbolic calculus of differential operators. This was done in [19] for an n-th-order scalar operator with the aid of fractional powers of an operator and in the present article directly with the aid of the resolvent for the general matrix case. Besides the method for constructing PL pairs the main problem of this paper is the construction of a Hamiltonian structure for the equations obtained. We remark that not only the construction of the Hamiltonian but also the construction of the symplectic form and of the Poisson brackets are nontrivial here. For the case of scalar operators this was done in [19] and for the general case, here. The techniques in this article and in the preceding ones differ. In [19], as we have already said, we investigated the fractional powers of operator L; in that same paper, for the jet of the diagonal of the resolvent we introduced an analog of the so-called "third-order equation" which indeed is a third-order equation in the scalar case with n = 2 [see formula (3) later]. We should stress that in the problems being examined there are two Hamiltonian formalisms, viz., for nonstationary and for stationary equations. The connection between these for a second-order scalar operator was established in [20, 21]. In this and in the preceding papers by the authors, this connection was established in the general case and the system of first integrals of the stationary problem was constructed, generalizing those which were obtained in [13, 15] for n = 2. In conclusion, we make two remarks. 1) Although the method proposed in [15] makes it possible to construct PL pairs, the presentation of the fundamental results was made without them, relying only on a number of identities for the resolvent. We tried to carry out this program here too. Unfortunately, we did not succeed in this in one important section, although we assume that this is possible. 2) In essence, the giving of the Poisson brackets is the construction of an infinite-dimensional Lie algebra; thus, here we have constructed, what seems to us, important nontrivial examples of infinite-dimensional Lie algebras.

Moscow State University. Translated from Funktsional'nyi Analiz i Ego Prilozheniya, Vol. 11, No. 2, pp. 11–27, April–June, 1977. Original article submitted December 22, 1976.

The plan of the article is the following. We have the differential operator

$$L\left(-i\frac{d}{dx}\right)=\sum_{k=0}^{n}U_k(x)\left(-i\frac{d}{dx}\right)^k, \tag{1}$$

where $U_k(x)$ are m-th-order matrices. Let $u_{k,\alpha\beta}(x)$, $k = 0, 1, \ldots, n$, $\alpha, \beta = 1, \ldots, m$, be the elements of these matrices. By $f[u_{k,\alpha\beta}]$ we denote a polynomial of the functions $u_{k,\alpha\beta}(x)$ and their derivatives of any orders. Let us examine functionals of the form $\int f[u_{k,\alpha\beta}]dx$ (where the sense of the integral $\int dx$ will be defined more exactly later in Sec. 1). In the linear space $\tilde{A}$ of such functionals we introduce a commutator, viz., the Poisson brackets, so that $\tilde{A}$ becomes a Lie algebra (Sec. 2). Now the problem arises of finding a system of functionals such that all paired Poisson brackets equal zero, i.e., an involutive system. To construct such a system we extend to the general case the method proposed in [15] for the case n = 2 and m = 1, based on the expansion of the diagonal of the nucleus R(x, y; z) of the resolvent of operator (1) in powers of the spectral parameter z. There holds the asymptotic expansion

$$R(x, x; z)=\sum_{p=0}^{\infty}R_p z^{-\frac{p+1}{n}-1}, \tag{2}$$

where the elements of the matrices R_p are polynomials of the elements $u_{k,\alpha\beta}$ and their derivatives (Theorem 3). Then $\mathcal{H}_p=\int \mathrm{Sp}\, R_p\, dx$ and form the needed system of functionals in involutions. The proof of this fact can, in principle, be carried out with the aid of an equation which is satisfied by the function R(x, x; z). In case n = 2 and m = 1 this equation is

$$-R'''+4(u-z)R'+2u'R=0. \tag{3}$$

For the expansion coefficients R_p this equation yields a recurrence system from which they can be successively determined, and the involutiveness of $\mathcal{H}_p$ can be proved by induction. Such a way was taken in [15]. Such an equation for R can be written out also in the general case (Theorem 6). Now this will be a system of equations in which together with R(x, x; z) there will also occur $(d/dy)R(x, y; z)|_{y=x}$, $(d^2/dy^2)R(x, y; z)|_{y=x}$, i.e., the jet of the nucleus of the resolvent on the diagonal. This too yields a recurrence system for the determination of R_p. But the proof of the involutiveness of $\mathcal{H}_p$ meets with technical difficulties which we did not overcome here, and we were compelled to use the additional techniques in the form of PL pairs. If we take one of the $\mathcal{H}_p$ as the Hamiltonian and construct the corresponding Hamiltonian equation relative to the Poisson brackets introduced, then this will be a generalization of the KdV equation. The involutiveness of the system $\{\mathcal{H}_p\}$ is equivalent to the fact that all the $\mathcal{H}_p$ are the first integrals of the equation constructed with respect to one of them. It turns out that any of the equations of such type can be written in the form $L_t = [P, L]$, where P is some matrix differential operator (for n = 2 and m = 1 such a way of writing was suggested by Lax [2]). From Lax's way of writing the equations being examined it is now easy to obtain the involutiveness of $\mathcal{H}_p$. In the last part of the article we discuss stationary equations. For these equations we have constructed the Hamiltonian formalism. We mention that in contrast to a nonstationary system, which is infinite-dimensional, a stationary system is a finite-dimensional Hamiltonian system. The variable x plays the role of t; this is a finite system of ordinary differential equations. It turns out that each first integral of a nonstationary system generates a first integral of a stationary one and, moreover, involutive first integrals generate involutive ones.

1. Certain Definitions. We present in brief some of the concepts introduced in [15, 21]. We consider a ring A whose elements are polynomials of the letters $u_1, \ldots, u_N$, $u_1', \ldots, u_N'$, $u_1^{(j)}, \ldots, u_N^{(j)}, \ldots$. Let there be given $a_{k,j} \in A$, $k = 1, \ldots, N$; $j = 0, 1, \ldots$. Then the formula $\xi f=\sum_{k,j} a_{k,j}\,\partial f/\partial u_k^{(j)}$, $f \in A$, gives a differentiation ξ in ring A. A singular role is played by the differentiation d/dx defined by the formula $df/dx=\sum u_k^{(j+1)}\cdot \partial f/\partial u_k^{(j)}$. It is easy to see that $u_k^{(i)} = (d/dx)^i u_k$. By dA/dx we denote the set of elements $f \in A$, representable in the form $f = dg/dx$, $g \in A$. We say that f_1 is equivalent to f_2 if $f_1 - f_2 \in dA/dx$. The class of equivalent elements containing f is denoted $\tilde{f}$. The space of equivalence classes is denoted $\tilde{A}$, i.e., by definition, $\tilde{A} = A/(dA/dx)$. The analyst, instead of speaking about classes of equivalent elements, says that the functions have been defined to within total derivatives. The map $A \to \tilde{A}$, associating with each $f \in A$ its class $\tilde{f} \in \tilde{A}$, is denoted $\int dx\colon f \mapsto \tilde{f}=\int f\,dx \in \tilde{A}$. From the definitions it follows that for any $f_1, f_2 \in A$ there holds the formula for integration by parts

$$\int f_1\cdot\frac{df_2}{dx}\,dx=-\int\frac{df_1}{dx}\cdot f_2\,dx.$$

The elements of $\tilde{A}$ are called functionals. This name is justified by the two examples presented later of the contensive theory covered by this scheme. 1) Let $u_k(x)$, $k = 1, \ldots, N$, be arbitrary smooth periodic functions on the segment $[0, a]$. Then for each $f \in A$ by $f[u_k]$ we mean the function that is obtained if in the polynomial $f(u_1, \ldots, u_N, \ldots, u_k^{(j)}, \ldots)$ instead of the letters $u_k^{(j)}$, $k = 1, \ldots, N$; $j = 0, 1, \ldots$, we substitute the function $u_k^{(j)}(x)$, the j-th derivative of the function $u_k(x)$. It is easy to see that $f_1 \sim f_2$ if and only if for any smooth periodic functions $u_k(x)$ we have $\int_0^a f_1[u_k]\,dx = \int_0^a f_2[u_k]\,dx$ (Lax's lemma [13]). Therefore, the elements $\bar{f} \in \tilde{A}$ can be identified with the functionals $J(u_1, \ldots, u_N) = \int_0^a f[u_k]\,dx$. 2) The second example is constructed analogously, but instead of periodic functions we examine functions $u_k(x)$ on $(-\infty, \infty)$, tending to zero sufficiently rapidly as $|x| \to \infty$, for example, of class $\mathcal{S}$. Then $\bar{f} \in \tilde{A}$ can be identified with the functional $J \cdot (u_1, \ldots, u_N) = \int_{-\infty}^{\infty} f[u_k]\,dx$, where f is some representative of class $\bar{f}$.

In a ring A we can introduce the operator $\delta/\delta u_k$: $A \to A$, called the variational derivative, by the formula $\delta f/\delta u_k = \sum(-d/dx)^j \cdot \partial f/\partial u_k^{(j)}$. The following is valid.

LEMMA 1. $f \in dA/dx$ if and only if $\delta f/\delta u_k = 0$ for all k.

The operator $\delta/\delta u_k$ of variational derivative can be defined also in $\tilde{A}$ in the following manner. If $f \in A$ is some representative of class $\bar{f} \in \tilde{A}$, then we set $\delta\bar{f}/\delta u_k = \delta f/\delta u_k$. According to Lemma 1, this definition is independent of the choice of f. Thus, by definition,

$$\frac{\delta \bar{f}}{\delta u_k} = \frac{\delta}{\delta u_k}\int f\,dx = \frac{\delta f}{\delta u_k}.$$

We pass on to the definition of differential forms. The finite sum

$$\omega = \sum a_{k_1 \ldots k_p}^{j_1 \ldots j_p}\, \delta u_{k_1}^{(j_1)} \wedge \delta u_{k_2}^{(j_2)} \wedge \ldots \wedge \delta u_{k_p}^{(j_p)}, \tag{4}$$

where $a_{k_1 \ldots k_p}^{j_1 \ldots j_p} \in A$ and $\delta u_k^{(j)}$ are new independent variables, is called a differential form. The sum of two forms and the multiplication of a form by an element from A are defined in the obvious way. Further, as usual, we can define outer multiplication and exterior differential δ. If ξ is a differentiation, defined earlier, and ω is a p-form, then by $i(\xi)\omega$ we mean the $(p-1)$-form obtainable, as usual, by substituting the vector field ξ into form ω; here $i(\partial/\partial u_k^{(j)})\,\delta u_r^{(l)} = 1$ if $k = r$ and $j = l$ and $i(\partial/\partial u_k^{j})\,\delta u_r^{(l)} = 0$ otherwise. By $\xi\omega$ we mean $\delta i(\xi)\omega + i(\xi)\delta\omega$ (i.e., ξ acts on the form as a Lie derivative). It is not difficult to find what $\xi\omega$ equals if $\xi = d/dx$:

$$\frac{d}{dx}\omega = \sum\left(\frac{d}{dx} a_{k_1 \ldots k_p}^{j_1 \ldots j_p}\right)\delta u_{k_1}^{(j_1)} \wedge \delta u_{k_2}^{(j_2)} \wedge \ldots + \sum a_{k_1 \ldots k_p}^{j_1 \ldots j_p}\,\delta u_{k_1}^{(j_1+1)} \wedge \delta u_{k_2}^{(j_2)} \wedge \ldots + \sum a_{k_1 \ldots k_p}^{j_1 \ldots j_p}\,\delta u_{k_1}^{(j_1)} \wedge \delta u_{k_2}^{(j_2+1)} \wedge \ldots + \ldots.$$

There holds

THEOREM. If $f \in A$, then δf is uniquely representable in the form $\delta f = \sum a_k \delta u_k + (d\omega/dx)$, where $a_k \in A$ and ω is some 1-form. Here, $a_k = \delta f/\delta u_k$.

Among all differentiations ξ, particularly important are those which commute with d/dx. It is easy to show that they have the form

$$\xi = \sum_{k,j}\left[\left(\frac{d}{dx}\right)^j f_k\right]\frac{\partial}{\partial u_k^{(j)}}, \quad f_k \in A. \tag{5}$$

We denote the collection of such differentiations by $T\tilde{A}$. Their value is that we can define their action in $\tilde{A}$ since if $f_1 \sim f_2$, then $\xi f_1 \sim \xi f_2$ for such ξ.

We denote the collection $(f_1, f_2, \ldots, f_N)$ by $\mathfrak{f}$. Such collections form a space A^N. Equation (5) shows that A^N and $T\tilde{A}$ are in a one-to-one correspondence. A differentiation ξ constructed with respect to f is denoted $\partial_{\mathfrak{f}}$. As was said, $\partial_{\mathfrak{f}}$ can be understood either as the operator $A \to A$ or as the operator $\tilde{A} \to \tilde{A}$. Sometimes we denote $(d/dx)^j f$ as $f^{(j)}$.

2. Poisson Brackets. The generators of the element $u_k^{(j)}$ of ring A, beginning as of now, will be grouped into matrices. Let there be given a diagonal matrix with the elements $a = (a_\alpha \delta_{\alpha\beta})$, $\alpha, \beta = 1, \ldots, m$, $a_\alpha \neq 0$. We construct a ring A_α in the following way. We consider the matrices

$$U_k = (u_{k,\alpha\beta}), \quad k = 0, 1, \ldots, n-1, \quad \alpha, \beta = 1, \ldots, m.$$

Concerning the matrix U_{n-1} we take it that $u_{n-1,\alpha\alpha} = 0$. In addition, if $a_\alpha = a_\beta$, then $u_{n-1,\alpha\beta} = u_{n-1,\beta\alpha} = 0$. All the nonzero elements $u_{k,\alpha\beta}$, as well as the symbols $u^{(j)}_{k,\alpha\beta}$, $j = 1, 2, \ldots$, we take as the generators of ring A_a.

Remarks. 1) As a matter of fact the ring A_a depends not on a but only on which a_α are equal to one another. In the notation A_a we shall drop the subscript a wherever this does not cause misunderstandings. 2) We can replace the diagonal matrix a by any constant matrix having an inverse. We can replace the condition on U_{n-1} here by this: that U_{n-1} is representable in the form of the commutator $[a, V]$.

For convenience of notation, in what follows we introduce further the matrix $U_n = a$.

If all the diagonal elements a_α are distinct, then N now equals $nm^2 - m$; otherwise, N is correspondingly smaller. By $g = \partial f/\partial U_k^{(j)}$ we mean the matrix with elements $g_{\alpha\beta} = \partial f/\partial u^{(j)}_{k,\alpha\beta}$ and by $\partial f/\partial U_k^{*(j)}$ the transposed matrix. We define $\delta f/\delta U_k$ and $\delta f/\delta U_k^*$ analogously.

Now we define the space A^N in the following manner. Its elements are collections of matrices $\mathbf{F} = (F_1, F_2, \ldots, F_{n-1})$ whose elements $f_{k,\alpha\beta} \in A$. Here the same requirements concerning the zero element imposed on U_{n-1} are imposed on matrix F_{n-1}; in particular, the diagonal elements equal zero.

In A^N we define the scalar product

$$(\mathbf{F}, \mathbf{G}) = \sum_{k=0}^{n-1} \mathrm{Sp}(F_k^* G_k). \tag{6}$$

We introduce the operators $\delta/\delta\mathrm{U}$, $\delta/\delta\mathrm{U}^*$: $\tilde{A} \to A^N$ as

$$\frac{\delta\tilde{f}}{\delta\mathrm{U}} = \left(\frac{\delta\tilde{f}}{\delta U_0}, \ldots, \frac{\delta\tilde{f}}{\delta U_{n-1}}\right), \quad \frac{\delta\tilde{f}}{\delta\mathrm{U}^*} = \left(\frac{\delta\tilde{f}}{\delta U_0^*}, \ldots, \frac{\delta\tilde{f}}{\delta U_{n-1}^*}\right).$$

Analogously, we define $\partial f/\partial\mathbf{U}$.

We define the operator l: $A^N \to A^N$, playing a fundamental role later on:

$$(l\mathbf{F})_r = \sum_{s=0}^{n-1} l_{rs} F_s, \tag{7}$$

where l_{rs} are differential operators acting by the formulas

$$l_{rs}F_s = \sum_{\gamma=0}^{n-1-r-s} \left\{\binom{\gamma+r}{r} U_{r+s+\gamma+1}\left(-i\frac{d}{dx}\right)^{\gamma} F_s - \binom{\gamma+s}{s}\left(i\frac{d}{dx}\right)^{\gamma}(F_s U_{r+s+\gamma+1})\right\} \quad (l_{rs} = 0 \ \text{ for } \ r+s > n-1). \tag{8}$$

The expression

$$\{\tilde{f}, \tilde{g}\} = -\int\left(\frac{\delta\tilde{f}}{\delta\mathrm{U}}, l\frac{\delta\tilde{g}}{\delta\mathrm{U}^*}\right)dx \tag{9}$$

is called the Poisson bracket of the two functionals $\tilde{f}, \tilde{g} \in \tilde{A}$. In order that the definition be well posed it is necessary that the space $\tilde{A}$ rigged with these brackets be turned into a Lie algebra. The bilinearity is obvious; the skew symmetry follows elementarily from the form of the expression for l_{rs}. It is entirely difficult to verify the Jacobi identity. This requires the introduction of certain supplementary structures, to which we pass right away.

In the preceding section we introduced the correspondence between A^N and $T\tilde{A}$, which in the present notation has the form

$$\mathbf{F} \mapsto \partial_{\mathbf{F}} = \sum f^{(j)}_{k,\alpha\beta}\frac{\partial}{\partial u^{(j)}_{k,\alpha\beta}} = \sum_j\left(\mathbf{F}^{(j)}, \frac{\partial}{\partial\mathbf{U}^{(j)}}\right).$$

In $T\tilde{A}$ there exists the natural structure of a Lie algebra with the commutator $[\partial_{\mathbf{F}}, \partial_{\mathbf{G}}] = \partial_{\mathbf{F}}\partial_{\mathbf{G}} - \partial_{\mathbf{G}}\partial_{\mathbf{F}}$. The correspondence $\mathbf{F} \to \partial_{\mathbf{F}}$ carries this commutator into A^N where it will have the form $[\mathbf{F}, \mathbf{G}] = \partial_{\mathbf{F}}\mathbf{G} - \partial_{\mathbf{G}}\mathbf{F}$. The operator $\partial_{\mathbf{F}}$ acts on $\mathbf{G}$ componentwise.

THEOREM 1. For any $\mathbf{F}, \mathbf{G} \in A^N$ there exists $\mathbf{H} \in A^N$ such that

$$[l\mathbf{F}, l\mathbf{G}] = l\mathbf{H}. \tag{10}$$

Here

$$\mathbf{H} = \partial_{l\mathbf{F}}\mathbf{G} - \partial_{l\mathbf{G}}\mathbf{F} + \mathbf{H}_1,$$

$$(H_1)_k = \sum_{\substack{r,s \\ r+s\leqslant k-1}} \left[\binom{k-s-1}{r}(-i)^{k-r-s-1}F_s^{(k-r-s-1)}G_r - \binom{k-r-1}{s}(-i)^{k-r-s-1}G_r^{(k-r-s-1)}F_s\right]. \tag{11}$$

From the assertion of this theorem it follows, in particular, that operator l maps A^N onto a Lie subalgebra.

Theorem 1 is proved by direct computation. There are no principal difficulties but the proof is cumbersome and will be presented in the Supplement (Sec. 12).

Constants can be excluded from ring A and from space A^N; after this has been done we denote them by A_0 and A_0^N. From $\tilde{A}$ we exclude the functionals $\tilde{f}$ such that $\delta\tilde{f}/\delta U$ are constants, i.e., the functionals $\int f\,dx$, where f are linear functions. We denote the space left by $\tilde{A}_0$. It is easy to see that l effects a one-to-one map of A_0^N onto the subalgebra A^N. We can transfer the structure of a Lie algebra from the image to the preimage, having by the same token turned A_0^N into an algebra with a commutator $[\ ,\ ]_l$ defined by the equality

$$l\,[\mathbf{F}, \mathbf{G}]_l = [l\mathbf{F}, l\mathbf{G}] \tag{12}$$

[i.e., $[\mathbf{F}, \mathbf{G}]_l$ is the $\mathbf{H}$ from (10)].

THEOREM 2. For any $\tilde{f}$, $\tilde{g} \in \tilde{A}_0$ we have

$$\frac{\delta}{\delta U^*}\{\tilde{f}, \tilde{g}\} = \left[\frac{\delta\tilde{f}}{\delta U^*}, \frac{\delta\tilde{g}}{\delta U^*}\right]_l. \tag{13}$$

This theorem establishes the isomorphism of the Lie algebra A_0^N (with a commutator $[\ ,\ ]_l$) and of the Lie algebra $\tilde{A}_0$ (with the Poisson brackets $\{\ ,\ \}$ as the commutator). From it, in particular, follows the Jacobi identity for the Poisson brackets (by the same token the Jacobi identity is proved not only in $\tilde{A}_0$ but in the whole $\tilde{A}$; the Poisson brackets vanish for exceptional functionals).

We prove Theorem 2 by using the ideas in [22]. For any $\tilde{f} \in \tilde{A}$ we introduce the differential operator $D(\tilde{f})$: $A^N \to A^N$,

$$(D(\tilde{f})\mathbf{H})_{k,\alpha\beta} = \sum_{r,\gamma\varepsilon} \frac{\partial}{\partial u_{k,\alpha\beta}^{(j)}}\left(\frac{\delta f}{\delta u_{r,\gamma\varepsilon}}\right) h_{r,\gamma\varepsilon}^{(j)}.$$

LEMMA 2. The differential operator is formally self-adjoint, i.e., for any $\mathbf{H}_1$, $\mathbf{H}_2 \in A^N$

$$(D(\tilde{f})\mathbf{H}_1, \mathbf{H}_2) - (D(\tilde{f})\mathbf{H}_2, \mathbf{H}_1) \in \frac{dA}{dx}$$

or, in coordinates,

$$\sum \frac{\partial}{\partial u_{k,\alpha\beta}^{(j)}}\left(\frac{\delta\tilde{f}}{\delta u_{r,\gamma\varepsilon}}\right)[(h_1)_{r,\gamma\varepsilon}^{(j)}(h_2)_{k,\alpha\beta} - (h_2)_{r,\gamma\varepsilon}^{(j)}(h_1)_{k,\alpha\beta}] \in \frac{dA}{dx}.$$

In the simplest case of n = 2 and m = 1, Lemma 2 was proved in [15, 13] and in the needed generality, in [22] (this proof was reproduced in [19]).

Proof of Theorem 2. We rewrite formula (9) in coordinates:

$$\{\tilde{f}, \tilde{g}\} = -\int \sum_{\substack{r,s,\gamma,\\ \alpha,\beta,\varepsilon}} \frac{\delta\tilde{f}}{\delta u_{r,\beta\alpha}}\binom{\gamma+r}{r} u_{r+s+\gamma+1,\beta\varepsilon}\left(-i\frac{d}{dx}\right)^{\gamma}\frac{\delta\tilde{g}}{\delta u_{s,\alpha\varepsilon}}\,dx + (\tilde{f} \rightleftarrows \tilde{g}), \tag{14}$$

where $(\tilde{f} \rightleftarrows \tilde{g})$ denotes the term differing from the one written by a permutation of $\tilde{f}$ and $\tilde{g}$. In order that we do not have to repeat the sign Σ in what follows, we adopt the following completely customary convention: if any letter is encountered twice in the indices (such as r, s, α, β, ε in the last formula) or is encountered in the indices and as a power exponent (such as γ), then a summation is made over these letters; the sign Σ is deleted. Let us take the differential δ on both sides of Eq. (14). Then $\delta\{\tilde{f}, \tilde{g}\}$ will be represented as a sum of three summands in correspondence with the number of factors under the integral sign in the right-hand side. We have

$$\delta\{\tilde{f}, \tilde{g}\}|_{\mathrm{I}} = -\int \frac{\partial}{\partial u_{k,\eta\zeta}^{(i)}}\left(\frac{\delta\tilde{f}}{\delta u_{r,\beta\alpha}}\right)\delta u_{k,\eta\zeta}^{(i)}\binom{\gamma+r}{r} u_{r+s+\gamma+1,\beta\varepsilon}\left(-i\frac{d}{dx}\right)^{\gamma}\frac{\delta\tilde{g}}{\delta u_{s,\alpha\varepsilon}}\,dx + (\tilde{f} \rightleftarrows \tilde{g}).$$

According to Lemma 2 this equals

$$-\int \frac{\partial}{\partial u_{r,\beta\alpha}^{(i)}}\left(\frac{\delta\tilde{f}}{\delta u_{k,\eta\zeta}}\right)\delta u_{k,\eta\zeta}\left(\frac{d}{dx}\right)^{i}\binom{\gamma+r}{r} u_{r+s+\gamma+1,\beta\varepsilon}\left(-i\frac{d}{dx}\right)^{\gamma}\frac{\delta\tilde{g}}{\delta u_{s,\alpha\varepsilon}}\,dx + (\tilde{f} \rightleftarrows \tilde{g}).$$

Further,

$$\begin{aligned}\delta\{\bar f, \bar g\}|_{II} &= -\int \frac{\partial}{\partial u^{(i)}_{k,\eta\zeta}}\left(\frac{\delta \bar g}{\delta u_{s,\alpha\varepsilon}}\right)\delta u^{(i)}_{k,\eta\zeta}\left(i\frac{d}{dx}\right)^{\gamma}\binom{\gamma+r}{r}\frac{\delta \bar f}{\delta u_{r,\beta\alpha}}u_{r+s+\gamma+1,\beta\varepsilon}\,dx + (\bar f \rightleftarrows \bar g)\\ &= -\int \frac{\partial}{\partial u^{(i)}_{s,\alpha\varepsilon}}\left(\frac{\delta \bar g}{\delta u_{k,\eta\zeta}}\right)\delta u_{k,\eta\zeta}\left(\frac{d}{dx}\right)^{i}\left(i\frac{d}{dx}\right)^{\gamma}\binom{\gamma+r}{r}\frac{\delta \bar f}{\delta u_{r,\beta\alpha}}u_{r+s+\gamma+1,\beta\varepsilon}\,dx + (\bar f \rightleftarrows \bar g)\\ &= \int \frac{\partial}{\partial u^{(i)}_{r,\alpha\varepsilon}}\left(\frac{\delta \bar f}{\delta u_{k,\eta\zeta}}\right)\delta u_{k,\eta\zeta}\left(\frac{d}{dx}\right)^{i}\left(i\frac{d}{dx}\right)^{\gamma}\binom{\gamma+s}{s}\frac{\delta \bar g}{\delta u_{s,\beta\alpha}}u_{r+s+\gamma+1,\beta\varepsilon}\,dx - (\bar f \rightleftarrows \bar g).\end{aligned}$$

Therefore,

$$\delta\{\bar f, \bar g\}|_{I} + \delta\{\bar f, \bar g\}|_{II} = -\operatorname{Sp}\int \partial_{l\frac{\delta \bar g}{\delta U^*}}\left(\frac{\delta \bar f}{\delta U^*_k}\right)\cdot \delta U_k\,dx + (\bar f \rightleftarrows \bar g).$$

The third term is:

$$\begin{aligned}\delta\{\bar f, \bar g\}|_{III} &= -\int \frac{\delta \bar f}{\delta u_{r,\beta\alpha}}\binom{\gamma+r}{r}\delta u_{r+s+\gamma+1,\beta\varepsilon}\left(-i\frac{d}{dx}\right)^{\gamma}\frac{\delta \bar g}{\delta u_{s,\alpha\varepsilon}}\,dx + (\bar f \rightleftarrows \bar g)\\ &= -\operatorname{Sp}\int \binom{k-s-1}{r}(-i)^{k-r-s-1}\left(\frac{\delta \bar g}{\delta U^*_s}\right)^{(k-r-s-1)}\left(\frac{\delta \bar f}{\delta U^*_r}\right)\delta U_k\,dx + (\bar f \rightleftarrows \bar g).\end{aligned}$$

The sum of the three summands, with due regard to (11), yields the required equality

$$\delta\{\bar f, \bar g\} = \int \operatorname{Sp}\left(\left[\frac{\delta f}{\delta U^*}, \frac{\delta \bar g}{\delta U^*}\right]_l\right)_k \delta U_k\,dx.$$

The theorem is proved.

Remark. To prove the Jacobi identity for the Poisson brackets we used not only the fact that l maps A^N onto a Lie subalgebra, i.e., the existence of the H in Eq. (10), but in an explicit form of this H. It is interesting to explain how essential this is. The second interesting question arising here is: do the iterations of l, i.e., l^2, l^3, . . . , map the Lie algebra A^N onto its subalgebras, i.e., does there not arise an infinite series of commutators. This is exactly so for the special case $n = 2$ and $m = 1$.

We now pass on to the second part of the paper, viz., to the construction of functionals $\mathscr{H}_p$ found in involutions relative to the Poisson brackets constructed.

3. Diagonal of the Resolvent's Nucleus. In the preceding section a Lie algebra $\tilde A$ was constructed from the matrices U_k. In order to find a system of functionals in involution, we construct from the matrices U_k the differential operator

$$L\left(-i\frac{d}{dx}\right) = \sum_{k=0}^{n} U_k(x)\left(-i\frac{d}{dx}\right)^k. \tag{15}$$

The functionals we need will be constructed with the aid of the diagonal of the resolvent's nucleus, i.e., the operator $[L(-id/dx) - z^n E]^{-1}$. It is significantly more convenient to use, instead of the operators, their symbols, since the definition of the operator is nonlocal and requires boundary conditions which do not exist here. The algebra of symbols is very well known, but we present certain definitions; it is convenient for us to introduce the spectral parameter z in the very definition of a symbol as a variable having equal rights with ξ: $\sigma_m(x, \xi, z)$ is called a symbol of order m if for every finite segment of the measurement of x and for integers α, β, and γ we can find a constant C such that

$$\left|\left(\frac{\partial}{\partial \xi}\right)^{\alpha}\left(\frac{\partial}{\partial z}\right)^{\beta}\left(\frac{\partial}{\partial x}\right)^{\gamma}\sigma_m\right| < C(|\xi| + |z| + 1)^{m-\alpha-\beta}. \tag{16}$$

Let us consider the series

$$\sigma(x, \xi, z) = \sum_{k=0}^{\infty} \sigma_{m_k}(x, \xi, z), \tag{17}$$

where σ_{m_k} are symbols of orders m_k; m_k tends monotonically to $-\infty$. The series will be understood to be asymptotic in the following sense. We set $\sigma|_K = \sum_0^K \sigma_{m_k}$. Two series are said to be equivalent if $\sigma^{(1)}|_K - \sigma^{(2)}|_K$ is a symbol whose order m tends to $-\infty$ as $K \to \infty$. We will not distinguish equivalent series. The class of equivalent series will be called the total symbol. We retain the same notation $\sigma(x, \xi, z)$ for the total symbol as for one series from the equivalence class.

The operation

$$\sigma_{(1)} \circ \sigma_{(2)} = \sum_{\alpha=0}^{\infty} \frac{(-i)^\alpha}{\alpha!} \left(\frac{\partial}{\partial \xi}\right)^\alpha \sigma_{(1)} \left(\frac{\partial}{\partial x}\right)^\alpha \sigma_{(2)} \tag{18}$$

is called the product of total symbols.

We introduce further the operation of the integration of a symbol with respect to ξ. If $\sigma_m(x, \xi, z)$ is a symbol of order $m < 1 - \varepsilon$, then we set

$$\bar{\sigma}_m(x, z) = \frac{1}{2\pi} \int \sigma_m(x, \xi, z)\, d\xi. \tag{19}$$

The sense of $\bar{\sigma}_m(x, z)$ is the following. From the symbol $\sigma_m(x, \xi, z)$ we construct the so-called canonic operator $\mathrm{Op}(\sigma_m)$,

$$\mathrm{Op}(\sigma_m)\, u(x) = \frac{1}{2\pi} \int e^{i\xi x} \sigma_m(x, \xi, z)\, \hat{u}(\xi)\, d\xi,$$

where $\hat{u}(\xi) = \int e^{-i\xi y} u(y)\, dy$. Then $\frac{1}{2\pi} \int e^{i\xi(x-y)} \sigma_m(x, \xi, z)\, d\xi$ is the nucleus of this operator and $\bar{\sigma}_m(x, z)$ is the diagonal of the nucleus. Thus, the operation of integration with respect to ξ is the transition from a symbol to the diagonal of the nucleus of the canonic operator corresponding to this symbol. Integration with respect to ξ can be carried out, in general, over any complex contour. From (16) it is easy to obtain the bound

$$|\bar{\sigma}_m(x, z)| < C\, |z|^{m+1}. \tag{20}$$

For a total symbol $\sigma(x, \xi, z)$ we define $\bar{\sigma}(x, z)$ as the series

$$\bar{\sigma}(x, z) = \sum_{k=0}^{\infty} \bar{\sigma}_{m_k}(x, z). \tag{21}$$

This series will be taken as asymptotic with respect to the degree of decrease in z in absolutely the same sense as series (17) with respect to the degree of decrease in the pair ξ, z. It is easy to see that by virtue of bound (20), under a term-by-term integration with respect to ξ of two equivalent series (17) we obtain equivalent series (21).

Later on we encounter only such symbols for which the corresponding $\bar{\sigma}_m(x, z)$ are distinguished in the asymptotic series in powers of z. By substituting these series into the right-hand side of (21), we obtain asymptotic series for $\bar{\sigma}(x, z)$, corresponding to total symbols. By virtue of bound (20), to obtain the coefficient of any power of z it is necessary to sum only a finite number of terms in the right-hand side of relation (21).

Let us return to operator (15). To it corresponds the symbol $L(\xi) = \sum_{k=0}^{n} U_k(x)\xi^k$. By $b(x, \xi, z)$ we denote the symbol of the resolvent, defined by the equalities

$$b \circ [L(\xi) - z^n E] = [L(\xi) - z^n E] \circ b = E. \tag{22}$$

In order to solve this equation we write $L(\xi)$ in the form $U_n\xi^n + L_1(\xi)$, i.e., we pick out the leading part of the symbol. Then the solution of Eq. (22) can be written as the series

$$b(x, \xi, z) = \sum_{k=1}^{\infty} (-1)^{k-1} (U_n\xi^n - z^n E)^{-1} \circ L_1(\xi) \circ (U_n\xi^n - z^n E)^{-1} \circ \ldots \circ L_1(\xi) \circ (U_n\xi^n - z^n E)^{-1}. \tag{23}$$

The quantity $\bar{b}(x, z)$, i.e., the diagonal of the nucleus of the resolvent, constructed from $b(x, \xi, z)$ by an integration with respect to ξ, is given the special notation $R(x, z)$ in view of the importance of this quantity. The contour of integration with respect to ξ must not pass through the roots of $\det(U_n\xi^n - z^n E)$. The function $R(x, z)$ depends upon the contour. In all we have $nm - 1$ independent functions.

<u>THEOREM 3.</u> $R(x, z)$ can be expanded into an asymptotic series in powers of z

$$R(x, z) = \sum_{p=0}^{\infty} R_p[u_{k,\alpha\beta}]\, z^{-p-n+1}, \tag{24}$$

where the elements of matrix R_p are polynomials of $u^{(j)}_{k,\alpha\beta}$, i.e., the elements of matrices U_k and their derivatives; in other words, they belong to ring A.

Proof. It is sufficient to show that the integral with respect to ξ of each term of series (23) can be expanded into an asymptotic series. If we make use of Eq. (18) and change the integration variable $\xi = z\eta$, then we obtain a sum of terms of the form

$$z^{-nk+\sum i_r-\sum j_r+1}\int c_1(\eta)\, U_{i_1}^{(j_1)} c_2(\eta)\ldots U_{i_{k-1}}^{(j_{k-1})} c_k(\eta)\, d\eta,$$

where the $c_i(\eta)$ are certain matrices independent of x. The integral is a matrix whose elements belong to A. QED.

Remark. The number $n-k+j$ is called the weight of the element $u_{k,\alpha\beta}^{(j)}$. The weight of a monomial is the sum of the weights of its factors. From the preceding formula it is clear that all monomials in R_p have the weight p.

Now we set

$$\mathcal{H}(z) = \int \mathrm{Sp}\, R(x, z)\, dx. \tag{25}$$

Here by $\mathcal{H}(z)$ we mean the asymptotic series $\mathcal{H}(z) = \sum_0^\infty \mathcal{H}_p z^{-p-n+1}$ obtained by a termwise integration of the asymptotic series (24), i.e.,

$$\mathcal{H}_p = \int \mathrm{Sp}\, R_p\,[u_{k,\alpha\beta}]\, dx. \tag{26}$$

4. The Jet of a Nucleus of the Resolvent on the Diagonal. In order to compute the variational derivatives of $\mathcal{H}_p$, to which we pass shortly, as well as for other purposes, we require quantities more general than R(x, z). More precisely, we set

$$S_k(x, z) = \int \xi^k b(x, \xi, z)\, d\xi. \tag{27}$$

(In particular, $S_0 = R$.) In exactly the same way that Theorem 3 was proved, we can show that $S_k(x, z)$ expands into an asymptotic power series

$$S_k(x, z) = \sum_{p=-k}^{\infty} S_{k,p} z^{-p-n+1}, \tag{28}$$

where the elements of matrices S_{kp} belong to A.

[If $\sigma_m(x, \xi, z)$ is a symbol of order m, then for $k < -m-1$ we can compute $s_k(x, z) = \int \xi^k \sigma_m(x, \xi, z)\, d\xi$; if r(x, y) is the nucleus of $\mathrm{Op}(\sigma_m)$, then $s_k(x, z) = \left(i\frac{d}{dy}\right)^k r(x, y)\big|_{y=x}$.] The collection of functions S_k is called the jet of the nucleus of the resolvent on the diagonal. We recall that, depending on the choice of the contour, we have defined $nm-1$ sequences of functions S_k.

5. Theorem on the Variational Derivative. THEOREM 4.

$$\frac{\delta}{\delta U_k}\mathcal{H} = -\sum_{\alpha=0}^{k}\binom{k}{\alpha}\left(-i\frac{d}{dx}\right)^{\alpha}\frac{\partial S_{k-\alpha}^*}{\partial(z^n)}, \tag{29}$$

where $\mathcal{H}$ has been defined by Eq. (25); as before, the asterisk denotes a transposed matrix.

Proof. On both sides of the identity $b\circ(L(\xi)-z^n E) = E$ we take the differential δ and we multiply the equality from the right by b; we obtain

$$\delta b + b\circ\sum_{k=0}^{n}\delta U_k \xi^k \circ b = 0.$$

We take the matrix trace and we integrate with respect to ξ. From formula (18) we see that for any symbols, $\mathrm{Sp}(\sigma_{(1)}\circ\sigma_{(2)}) = \mathrm{Sp}(\sigma_{(2)}\circ\sigma_{(1)}) + \frac{\sigma}{\partial x}(\ ) + \frac{\sigma}{\partial\xi}(\ )$. We have

$$\delta\,\mathrm{Sp}\, R + \frac{1}{2\pi}\int \mathrm{Sp}(\delta U_k \xi^k \circ b\circ b)\, d\xi = \frac{d}{dx}(\ ). \tag{30}$$

If we take into account the easily provable identity $b\circ b = \partial b/\partial z^n$ and compute $\xi^k\circ\partial b^*/\partial z^n$ by Eq. (18), then we obtain just what we needed.

COROLLARY.

$$\frac{\partial S_k^*}{\partial(z^n)} = -\sum_{\alpha=0}^{k} \binom{k}{\alpha}\left(i\frac{d}{dx}\right)^{\alpha} \frac{\delta}{\delta U_{k-\alpha}} \mathcal{H}. \tag{31}$$

This inversion of Eq. (29) is obtained without difficulty if we take advantage of the formula for binomial coefficients

$$\sum_{\alpha=0}^{\mu} (-1)^{\alpha} \binom{k}{\alpha}\binom{k-\alpha}{k-\mu} = \begin{cases} 0, & \mu \neq 0, \\ 1, & \mu = 0. \end{cases} \tag{32}$$

We remark, further, that Eq. (29) for the coefficients of the asymptotic expansions can be written as

$$\frac{\delta}{\delta U_k} \mathcal{H}_p = \frac{p-1}{n} \sum_{\alpha=0}^{k} \binom{k}{\alpha}\left(-i\frac{d}{dx}\right)^{\alpha} S^*_{k-\alpha,\, p-n}. \tag{33}$$

6. Hamiltonian Equations. We take one of the $\mathcal{H}_p$ as the Hamiltonian in $\tilde{A}$ and we construct the Hamiltonian equations. This means that for any functional $\tilde{f} = \int f\,dx$ we require that

$$\frac{d\tilde{f}}{dt} = \{\mathcal{H}_p, \tilde{f}\} \tag{34}$$

or

$$\int \mathrm{Sp}\left(\frac{\partial f}{\partial U_r^{(j)}} \dot{U}_r^{(j)}\right) dx = \int \mathrm{Sp}\left(\frac{\delta \tilde{f}}{\delta U_r^*} \dot{U}_r\right) dx = \int \mathrm{Sp}\left(\frac{\delta \tilde{f}}{\delta U_r^*}\, l_{rs}\, \frac{\delta \mathcal{H}_p}{\delta U_s^*}\right) dx.$$

In order that this hold for any $\tilde{f} \in \tilde{A}$, it is necessary and sufficient that

$$\frac{\partial}{\partial t} U_r = \sum_{s=0}^{n-1} l_{rs} \frac{\delta \mathcal{H}_p}{\delta U_s^*}, \quad r = 0, \ldots, n-1. \tag{35}$$

This is the system of equations obtained.

We can examine a more general system if as the Hamiltonian we take not one of the functionals $\mathcal{H}_p$ but a linear combination $\sum c_p \mathcal{H}_p$.

We write system (35) in still another form, making use of Eq. (33) and of expression (8) for l_{rs}:

$$\frac{\partial}{\partial t} U_k = \sum_{r=k+1}^{n} \sum_{\alpha=0}^{r-k-1} \left[i^{\alpha} \binom{r-k-1}{\alpha} S_{r-k-\alpha-1,\, p-n} U_r^{(\alpha)} - (-i)^{\alpha} \binom{r}{\alpha} U_r S^{(\alpha)}_{r-k-\alpha-1,\, p-n} \right] \frac{p-1}{n}. \tag{36}$$

7. Involutiveness of $\{\mathcal{H}_p\}$. First Integrals. We have to prove the following fundamental theorem.

THEOREM 5. For all p and q

$$\{\mathcal{H}_p, \mathcal{H}_q\} = 0. \tag{37}$$

The theorem's assertion is equivalent to this: that all $\mathcal{H}_q$ are the first integrals of system (35) [or (36)], since this system is equivalent to (34) for any $\tilde{f} \in \tilde{A}$.

In principle, the theorem can be proved by direct computation with the use of equations which are satisfied by the functions S_k and which will be introduced in the next section. This was done in [15] for the case n = 2 and m = 1. But in the general case this is technically difficult and has not been done. The theorem will be proved with the aid of PL pairs.

Remark. If as the Hamiltonian we take the linear combination $\sum c_p \mathcal{H}_p$, then, as before, all the $\mathcal{H}_q$ remain as first integrals:

8. System of Equations for the Jet of the Resolvent. THEOREM 6. The functions S_k satisfy the equations

$$\begin{aligned} &\sum_{k=0}^{n} \sum_{\alpha=0}^{\infty} (-i)^{\alpha} \binom{k}{\alpha} U_k S^{(\alpha)}_{k+l-\alpha} = z^n S_l, \\ &\sum_{k=0}^{n} \sum_{\alpha=0}^{\infty} i^{\alpha} \binom{k+l}{\alpha} S_{k+l-\alpha} U_k^{(\alpha)} = z^n S_l, \quad l = 0, \pm 1, \pm 2, \ldots \end{aligned} \tag{38}$$

Before we prove this theorem we remark that if we restrict ourselves to $l = 0, \ldots, n-1$, then we obtain a closed system of equations for the quantities $S_0, S_1, \ldots, S_{2n-1}$.

If we substitute the asymptotic expansions for S_k into this system, we obtain a recurrence system for determining the S_{kp}. This is the simplest method for computing the S_{kp}. From the recurrence system we find not the S_{kp} themselves but their derivatives with respect to x; a completely nontrivial fact stemming from Theorem 3 is that from the dS_{kp}/dx we can find the matrices S_{kp} whose elements are polynomials of $u^{(j)}_{l,\alpha\beta}$.

Let us present some more special cases of Theorem 6. If n = 2 and m = 1, then the closed system of equations consists of four equations for the quantities S_0, S_1, S_2, and S_3. From it we can eliminate three quantities without difficulty, while for $S_0 = R$ we obtain the equation $-R''' + 4(u - z^2)R' + 2u'R = 0$. This is a very well-known equation for the diagonal of the nucleus of the resolvent of a second-order operator, which is most easily obtained by treating this diagonal as the product of two solutions of the second-order equation $L\varphi = z^2\varphi$.

A second special case is n = 1. Here from the system of two equations we eliminate one unknown and we are left with one equation for $S_0 = R$: $-i(U_1R)' = [U_1R, (U_0 - zE)U_1^{-1}]$. This equation is easy to obtain also by taking as R the matrix $\|\varphi_i(x, z)\psi_j(x, z)\|$, where $\{\varphi_i\} = \varphi$ is the solution of the equation $L\varphi = z\varphi$ and ψ is the solution of the adjoint equation.

Proof of Theorem 6. We multiply identity (22) by ξ^l, $l = 0, \pm 1, \pm 2, \ldots$. We uncover the operation $\circ$ by Eq. (18) and we integrate the identity along any closed contour in the plane of ξ. We obtain the theorem's assertion by recalling the definition of the S_k.

9. PL Pairs. THEOREM 7. We set

$$P = -\sum_{k=0}^{\infty} S_{-k-1}\xi^k. \tag{39}$$

Then there holds the identity

$$P - \sum_{l=1}^{n}\sum_{r=l}^{n}\sum_{\alpha=0}^{r-l} i^\alpha \binom{r-l}{\alpha} S_{r-l-\alpha}U_r^{(\alpha)}\xi^{l-1} \circ (L(\xi) - z^n E)^{-1} = (L(\xi) - z^n E)^{-1} \circ \sum_{l=1}^{n}\sum_{r=l}^{n}\sum_{\alpha=0}^{r-l} (-i)^\alpha \binom{r}{\alpha} U_r S^{(\alpha)}_{r-l-\alpha}\xi^{l-1}. \tag{40}$$

Proof. Let us prove the first equality:

$$\sum_{k=0}^{\infty} S_{-k-1}\xi^k \circ (L(\xi) - z^n E) = \sum_{k=0}^{\infty}\sum_{\alpha=0}^{\infty} (-i)^\alpha \binom{k}{\alpha} S_{-k-1}\xi^k * \sum_{r=0}^{n} U_r^{(\alpha)}\xi^r - z^n \sum_{k=0}^{\infty} S_{-k-1}\xi^k$$

$$= \sum_{k=0}^{\infty}\sum_{\alpha=0}^{k} i^\alpha \binom{-k+\alpha-1}{\alpha} S_{-k-1} \sum_{r=0}^{n} U_r^{(\alpha)}\xi^{k-\alpha+r} - z^n \sum_{k=0}^{\infty} S_{-k-1}\xi^k$$

$$= \sum_{l=-\infty}^{-1} \left\{ \sum_{r=0}^{\min(n,-1-l)} \sum_{\alpha=0}^{\infty} i^\alpha \binom{r+l}{\alpha} S_{r+l-\alpha}U_r^{(\alpha)} - z^n S_l \right\} \xi^{-l-1}.$$

Keeping in mind Eq. (38) for the jet of the resolvent, we have that this expression is equal to

$$-\sum_{l=-n}^{-1}\sum_{r=-l}^{n}\sum_{\alpha=0}^{r+l} i^\alpha \binom{r+l}{\alpha} S_{r+l-\alpha}U_r^{(\alpha)}\xi^{-l-1} = \sum_{l=1}^{n}\sum_{r=l}^{n}\sum_{\alpha=0}^{r-l} i^\alpha \binom{r-l}{\alpha} S_{r-l-\alpha}U_r^{(\alpha)}\xi^{l-1},$$

QED. The second equality is proved in exactly the same way.

CORROLLARY.

$$[P, L(\xi) - z^n E] = [P, L(\xi)] = \sum_{l=1}^{n}\sum_{r=l}^{n}\sum_{\alpha=0}^{r-l} \left[i^\alpha \binom{r-l}{\alpha} S_{r-l-\alpha}U_r^{(\alpha)} - (-i)^\alpha \binom{r}{\alpha} U_r S^{(\alpha)}_{r-l-\alpha} \right] \xi^{l-1}.$$

If we expand P into an asymptotic series in z,

$$P = \sum_{p=1}^{\infty} P_p z^{-p-n+1}$$

(the P_p are the symbols of differential operators of increasing order as p grows), then we have

$$[P_p, L] = \sum_{l=1}^{n}\sum_{r=l}^{n}\sum_{\alpha=0}^{r-l} \left[i^\alpha \binom{r-l}{\alpha} S_{r-l-\alpha,\,p}U_r^{(\alpha)} - (-i)^\alpha \binom{r}{\alpha} U_r S^{(\alpha)}_{r-l-1,\,p} \right] \xi^{l-1}. \tag{41}$$

Thus, there exist differential operators P_p of arbitrarily high order such that their commutators with L are operators of order n − 1 (even n − 2 in the scalar case).

10. Proof of Theorem 5. Comparing Eq. (41) with Eq. (36), we see that Eq. (36) can be written as

$$L_t = [P_p, L] \tag{42}$$

[here we did not write down the unessential constant multiplier $(p-1)/n$; we reckon that the change of variable $t \to nt/(p-1)$ has been made]. Using this so-called "Lax" way of writing the equation we can easily prove the invariance of the quantities $\mathcal{H}_p$. As a matter of fact, we again recall the equation $b \circ [L(\xi) - z^n E] = E$. Let the U_k occurring here satisfy Eq. (36) or Eq. (42). We differentiate the equation written down with respect to t: $b_t \circ [L(\xi) - z^n E] + b \circ L_t(\xi) = 0$. We replace L_t from (42), we multiply this equality from the right by b, we take the matrix trace, and we integrate with respect to ξ:

$$\int \operatorname{Sp} b_t d\xi + \int \operatorname{Sp}(b \circ P_p \circ L \circ b)\, d\xi - \int \operatorname{Sp}(b \circ L \circ P_p \circ b)\, d\xi = 0.$$

From $b \circ L = L \circ b = E + zb$ we get that

$$\int \operatorname{Sp} b_t d\xi + \int \operatorname{Sp}(b \circ P_p)\, d\xi - \int \operatorname{Sp}(P_p \circ b)\, d\xi = 0.$$

But, as we noted, $\int \operatorname{Sp}(\sigma_1 \circ \sigma_2)\, d\xi - \int \operatorname{Sp}(\sigma_2 \circ \sigma_1)\, d\xi = \frac{d}{dx}(\ )$. Therefore, $\frac{\partial}{\partial t} \int \operatorname{Sp} b\, d\xi = \frac{d}{dx}(\ )$ or $\operatorname{Sp} R_t = \frac{d}{dx}(\ )$. From $\mathcal{H} = \int \operatorname{Sp} R\, dx$ we obtain $\mathcal{H}_t = 0$, which signifies the invariance of $\mathcal{H}$ or of all of its coefficients $\mathcal{H}_q$ in the expansion in powers of z; $d\mathcal{H}_q/dt = 0$. As was said in Sec. 7, this is equivalent to the involutiveness of the system $\{\mathcal{H}_p\}$, i.e., to the assertion of Theorem 5.

11. Stationary Equations. For U_k, $k = 0, \ldots, n-1$, not depending on parameter t, system (35) can be turned into the system

$$\frac{\delta \mathcal{H}_q}{\delta U_s} = 0, \quad s = 0, 1, \ldots, n-1. \tag{43}$$

This is a so-called stationary system. This system of ordinary differential equations is written in Lagrange form; $\mathcal{H}_q$ plays the role of the Lagrangian. The finite-dimensional manifold of solutions of such a system is invariant for any nonstationary equation of form (35), with any $\mathcal{H}_p$, in view of the involuteness of $\mathcal{H}_p$ and $\mathcal{H}_q$. Further, any $\mathcal{H}_p$ generates a first integral of system (43). This can be seen as follows. In view of (9), relation $\{\mathcal{H}_p, \mathcal{H}_q\} = 0$ signifies the existence of $J_{p,q} \in A$ such that

$$\operatorname{Sp} \sum_{r,s} \left(\frac{\delta \mathcal{H}_p}{\delta U_r}\right)^* l_{rs} \left(\frac{\delta \mathcal{H}_q}{\delta U_s}\right)^* = \frac{d}{dx} J_{p,q}. \tag{44}$$

Hence we see that $J_{p,q}$ is a first integral of system (43), since $(d/dx)J_{p,q}$ vanishes for the solutions of Eqs. (43).

Since system (42) has a variational form, it can be put into a Hamiltonian form. We will not set forth this theory here. This was done in [15, 21] for n = 2 and m = 1 and in [19] for any n and m = 1. The matrix case is not materially different from the scalar case. We note only the most essential result which can be proved here. From the involutiveness of $\mathcal{H}_{p_1}$ and $\mathcal{H}_{p_2}$ in the nonstationary theory follows the involutiveness of $J_{p_1,q}$ and $J_{p_2,q}$ relative to the stationary Hamiltonian formalism.

12. Supplement. Proof of Theorem 1. Let us compute the left-hand side of relation (10): $\partial_{lF} lG - \partial_{lG} lF = l(\partial_{lF} G - \partial_{lG} F) + [(\partial_{lF} l)G - (\partial_{lG} l)F]$. We need to prove that the summand in the brackets equals lH_1, where H_1 is given by Eq. (11). We have

$$[(\partial_{lF} l)\, G - (\partial_{lG} l)\, F]_r = \sum_{s=0}^{n-2-r} \sum_{\varepsilon=0}^{n-2-r-s} \left[\binom{\varepsilon + r}{r} (lG)_{r+s+\varepsilon+1} \left(-i\frac{d}{dx}\right)^\varepsilon G_s - \binom{\varepsilon + s}{s} \left(i\frac{d}{dx}\right)^\varepsilon G_s (lF)_{r+s+\varepsilon+1}\right] - (F \leftrightarrows G)$$

[as before, by $(F \leftrightarrows G)$ we have denoted the term differing from the one written by a permutation of F and G]. Let us show that $\sum_{i=1}^{6} A_i = 0$, where

$$A_1 = \sum_{\substack{s, \varepsilon, m, \gamma \\ (s+\varepsilon+m+\gamma \leq n-2-r)}} \binom{\varepsilon + r}{r} \binom{\gamma + r + s + \varepsilon + 1}{\gamma} U_{r+s+\varepsilon+m+\gamma+2} \left[\left(-i\frac{d}{dx}\right)^\gamma F_m\right] (-i)^\varepsilon G_s^{(\varepsilon)} - (F \leftrightarrows G),$$

$$A_2 = -\sum_{s, \varepsilon, m, \gamma} \binom{\varepsilon + r}{r} \binom{\gamma + m}{m} \left[\left(i\frac{d}{dx}\right)^\gamma (F_m U_{r+s+\varepsilon+m+\gamma+2})\right] (-i)^\varepsilon G_s^{(\varepsilon)} + (F \leftrightarrows G),$$

$$A_3 = -\sum_{s, \varepsilon, m, \gamma} \binom{\varepsilon + s}{s} \binom{\gamma + r + s + \varepsilon + 1}{\gamma} \left(i\frac{d}{dx}\right)^\varepsilon [G_s U_{r+s+\varepsilon+m+\gamma+2} (-i)^\gamma F_m^{(\gamma)}] + (F \leftrightarrows G),$$

$$A_4 = \sum_{s,\varepsilon,m,\gamma} \binom{\varepsilon+s}{s}\binom{\gamma+m}{\gamma}\left(i\frac{d}{dx}\right)^{\varepsilon}\left[G_s\left(i\frac{d}{dx}\right)^{\gamma}(F_m U_{r+s+\varepsilon+m+\gamma+2})\right] - (F \leftrightarrows G),$$

$$A_5 = -\sum_{s,\varepsilon,m,\gamma} \binom{\varepsilon+r}{r}\binom{\gamma+s}{s} U_{r+s+\varepsilon+m+\gamma+2}\left(-i\frac{d}{dx}\right)^{\varepsilon}[(-i)^{\gamma}F_m^{(\gamma)}G_s] + (F \leftrightarrows G),$$

$$A_6 = \sum_{s,\varepsilon,m,\gamma} \binom{\varepsilon+\gamma+m+s+1}{\varepsilon}\binom{\gamma+s}{s}\left(i\frac{d}{dx}\right)^{\varepsilon}[(-i)^{\gamma}F_m^{(\gamma)}G_s U_{r+s+\varepsilon+m+\gamma+2}] - (F \leftrightarrows G).$$

It turns out that $A_1 + A_5 = 0$, $A_2 + A_3 = 0$, and $A_4 + A_6 = 0$. The verification of these equalities is similar. For example, we transform

$$A_5 = -\sum_{s,\varepsilon,m,\gamma} \binom{\varepsilon+r}{r}\binom{\gamma+s}{s} U_{r+s+\varepsilon+m+\gamma+2}\,(-i)^{\varepsilon+\gamma}\sum_{\varepsilon_1=0}^{\varepsilon}\binom{\varepsilon}{\varepsilon_1}F_m^{(\gamma+\varepsilon-\varepsilon_1)}G_s^{(\varepsilon_1)} + (F \leftrightarrows G).$$

We set $\gamma_1 = \gamma + \varepsilon - \varepsilon_1$, then

$$A_5 = -\sum_{s,\varepsilon_1,m,\gamma_1}\sum_{\varepsilon=\varepsilon_1}^{\gamma_1+\varepsilon_1}\binom{\varepsilon+r}{r}\binom{\gamma_1-\varepsilon+\varepsilon_1+s}{s}\binom{\varepsilon}{\varepsilon_1} U_{r+s+\varepsilon_1+m+\gamma_1+2}\,(-i)^{\varepsilon_1+\gamma_1}F_m^{(\gamma_1)}G_s^{(\varepsilon_1)} + (F \leftrightarrows G).$$

We have

$$\sum_{\varepsilon=\varepsilon_1}^{\gamma_1+\varepsilon_1}\binom{\varepsilon+r}{r}\binom{\gamma_1-\varepsilon+\varepsilon_1+s}{s}\binom{\varepsilon}{\varepsilon_1} = \binom{\varepsilon_1+r}{r}\sum_{\varepsilon=\varepsilon_1}^{\gamma_1+\varepsilon_1}\binom{\varepsilon+r}{\varepsilon_1+r}\binom{\gamma_1+\varepsilon_1+s-\varepsilon}{\gamma_1+\varepsilon_1+s-(\gamma_1+s_1)}.$$

By virtue of the summation formula for binomial coefficients,

$$\sum_{\gamma=n}^{m}\binom{k-\gamma}{k-m}\binom{\gamma-l}{n-l} = \binom{k-l+1}{m-n} \quad (k \geqslant m \geqslant n \geqslant l),$$

we obtain

$$A_5 = -\sum_{s,\varepsilon_1,m,\gamma_1}\binom{\varepsilon_1+r}{r}\binom{\gamma_1+r+s+\varepsilon_1+1}{\gamma_1} U_{r+s+\varepsilon_1+m+\gamma_1+2}(-i)^{\varepsilon_1+\gamma_1}F_m^{(\gamma_1)}G_s^{(\varepsilon_1)} + (F \leftrightarrows G),$$

which differs from A_1 only in sign. The remaining two identities are proved in the same way. We should not forget the terms not written out, hidden under the symbol $(F \leftrightarrows G)$. As a matter of fact, the terms from A_2 written out are mutually annihilated by the terms from A_3 not written out, and conversely. Precisely the same is true of the pair A_4 and A_6.

LITERATURE CITED

1. C. S. Gardner, J. M. Green, M. D. Kruskal, and R. M. Miura, "Method for solving the KdV equation," Phys. Rev. Lett., 19, No. 19, 1095-1097 (1967).
2. P. Lax, "Integrals of nonlinear equations of evolution and solitary waves," Commun. Pure Appl. Math., 21, No. 5, 467-490 (1968).
3. C. S. Gardner, "Korteweg–de Vries equation and generalizations. IV," J. Math. Phys., 12, No. 8, 1548-1551 (1971).
4. V. E. Zakharov and L. D. Faddeev, "The Korteweg–de Vries equation is a fully integrable Hamiltonian system," Funkts. Anal. Prilozhen., 5, No. 4, 18-27 (1971).
5. V. E. Zakharov and A. B. Shabat, "A scheme for the integration of the nonlinear equations of mathematical physics by the method of the inverse problem in scattering theory," Funkts. Anal. Prilozhen., 8, No. 3, 54-66 (1974).
6. B. A. Dubrovin, "Finitely zoned linear differential operators and Abelian manifolds," Usp. Mat. Nauk, 31, No. 4, 259-260 (1976).
7. I. M. Krichever, "Algebraic-geometric construction of the Zakharov–Shabat equations and their periodic solutions," Dokl. Akad. Nauk SSSR, 227, No. 2, 291-294 (1976).
8. I. M. Krichever, "Integration of nonlinear equations by the methods of algebraic geometry," Funkts. Anal. Prilozhen., 11, No. 1, 15-31 (1977).
9. B. A. Dubrovin, V. B. Matveev, and S. P. Novikov, "Nonlinear equations of Korteweg–de Vries type, finitely zoned linear operators, and Abelian manifolds," Usp. Mat. Nauk, 31, No. 1, 55-136 (1976).
10. A. R. Its and V. B. Matveev, "On one class of solutions of the KdV equation," in: Problems of Mathematical Physics [in Russian], No. 8, Izd. Leningrad State Univ., Leningrad (1976), pp. 70-92.
11. S. P. Novikov, "The periodic Korteweg–de Vries problem. I," Funkts. Anal. Prilozhen., 8, No. 3, 54-66 (1974).

12. V. A. Marchenko, "The periodic Korteweg–de Vries problem," Mat. Sb., 95, 331-356 (1974).
13. P. Lax, "Periodic solutions of the KdV equations," Lect. Appl. Math., 15, 85-96 (1974).
14. N. P. McKean and P. van Moerbeke, "Sur le spèctre de quelque operateur et le variété de Jacobi," Sém. Bourbaki (1975-1976), p. 474.
15. I. M. Gel'fand and L. A. Dikii, "Asymptotic behavior of the resolvent of Sturm–Liouville equations and the algebra of Korteweg–de Vries equations," Usp. Mat. Nauk, 30, No. 5, 67-100 (1975).
16. I. M. Gel'fand and B. M. Levitan, "On one simple identity for the eigenvalues of the Sturm–Liouville operator," Dokl. Akad. Nauk SSSR, 88, No. 4, 593-596 (1953).
17. I. M. Gel'fand, "On identities for the eigenvalues of a second-order differential operator," Usp. Mat. Nauk, 11, No. 1, 191-198 (1956).
18. L. A. Dikii, "Zeta-functions of an ordinary differential equation on a finite segment," Izv. Akad. Nauk SSSR, Ser. Mat., 19, 187-200 (1955).
19. I. M. Gel'fand and L. A. Dikii, "Fractional powers of operators and Hamiltonian systems," Funkts. Anal. Prilozhen., 10, No. 4, 13-29 (1976).
20. O. I. Bogoyavlenskii and S. P. Novikov, "On the relation between the Hamiltonian formalisms of stationary and nonstationary problems," Funkts. Anal. Prilozhen., 10, No. 1, 9-13 (1976).
21. I. M. Gel'fand and L. A. Dikii, "Structure of a Lie algebra in the formal calculus of variations," Funkts. Anal. Prilozhen., 10, No. 1, 18-25 (1976).
22. I. M. Gel'fand, Yu. I. Manin, and M. A. Shubin, "Poisson brackets and the nucleus of a variational derivative in the formal calculus of variations," Funkts. Anal. Prilozhen., 10, No. 4, 30-34 (1976).
23. B. A. Dubrovin and S. P. Novikov, "Periodic problem for the Korteweg–de Vries and the Sturm–Liouville equations. Their connection with algebraic geometry," Dokl. Akad. Nauk SSSR, 219, No. 3, 19-22 (1974).
24. N. P. McKean and P. v. Moerbeke, "The spectrum of Hill's equation," Invent. Math., 30, No. 3, 217-274 (1975).
25. N. P. McKean and E. Trubovitz, "Hill's operator and hyperelliptic function theory in the presence of infinitely many branch points," Commun. Pure Appl. Math., 29, No. 2, 143-226 (1976).
26. M. Kac and P. v. Moerbeke, "On an explicitly soluble system of nonlinear differential equations related to certain Toda lattices," Adv. Math., 16, No. 2, 160-169 (1975).
27. M. Kac and P. v. Moerbeke, "A complete solution of the periodic Toda problem," Proc. Nat. Acad. Sci. USA, 72, No. 8, 2879-2880 (1975).

14.
(with L. A. Dikij)

Integrable nonlinear equations and the Liouville theorem

Funkts. Anal. Prilozh. **13** (1) (1979) 8–20. MR **80i**: 58027, Zbl. **423**: 34003
[Funct. Anal. Appl. **13** (1979) 6–15]

The stationary higher-order Korteweg–de Vries equations (Novikov equations) are a remarkable example of nonlinear ordinary differential equations admitting, as is shown in [1-4], an explicit solution in terms of a theta-function. On the other hand, it was shown in [5] that these equations have a Hamiltonian structure and a complete set of first integrals in involution. The existence of these first integrals was obtained independently by Lax [6]. There is a classical theorem of Liouville which says that the existence of n first integrals in involution for a Hamiltonian system of order 2n assures integrability in quadratures. We will show in this article that all the explicit formulas of the solution, including the theta-functions, can be obtained by successive Liouville integration. We would like to note here that algebrogeometric procedures such as the Abel transform arise here of their own accord as a natural consequence of the classical Liouville theorem.

The article is based on concepts and theorems of our paper [5], but everything needed is stated, with the reader referred to [5] only for proofs. The article is in fact elementary and requires no knowledge beyond the rudiments of analysis and algebra.

1. Construction of the Higher Order KdV Equations

We denote by A the ring of differential polynomials in one variable u, i.e., polynomials in u, u', u", . . ., $u^{(i)}$, . . ., in which the operator d/dx acting by the rule $du^{(i)}/dx = u^{(i+1)}$ $(u^{(0)} = u)$ appears (u can be understood to be an arbitrary function u(x), $u^{(i)}$ its derivatives, and d/dx ordinary differentiation). Let $A[[z^{-1}]]$ be the ring of formal Laurent power series, i.e., series of the form $\sum_{k=k_0}^{\infty} f_k z^{-k}$, $f_k \in A$, k_0 an arbitrary integer. The invertible elements of the ring $A[[z^{-1}]]$ are the ones for which f_{k_0} = constant. The operator d/dx can be carried over to the ring $A[[z^{-1}]]$, where it acts term by term.

The following third-order equation is fundamental in the whole theory:

$$-R''' + 4(u + \zeta)R' + 2u'R = 0, \quad \zeta = z^2. \tag{1.1}$$

We will seek a solution $R \in A[[z^{-1}]]$; $R = \sum_{k=k_0}^{\infty} R_k z^{-k}$, where the R_k are differential polynomials, $R_k \in A$. Equation (1.1) appeared practically simultaneously in various papers. In [5], it

Moscow State University. Translated from Funktsional'nyi Analiz i Ego Prilozheniya, Vol. 13, No. 1, pp. 8-20, January-March, 1979. Original article submitted September 21, 1978.

appears as the equation for the diagonal of the resolvent of the equation $-\varphi'' + (u+\zeta)\varphi = 0$. If Eq. (1.1) is multiplied by R and integrated, we obtain

$$-2RR'' + (R')^2 + 4(u+\zeta)R^2 = c(\zeta), \tag{1.2}$$

where $c(\zeta) = \Sigma c_k z^{-k}$ is an element of $A[[z^{-1}]]$ with constant coefficients. Conversely, Eq. (1.2) with arbitrary c(ζ) reduces by differentiation to (1.1). Equations (1.1) and (1.2) give recursion formulas for the coefficients R_k:

$$-R_k''' + 4uR_k' + 4R_{k+2}' + 2u'R_k = 0; \tag{1.3}$$

$$-2\sum_{l=k_0}^{k-k_0} R_l R_{k-l}'' + \sum_{l=k_0}^{k-k_0} R_l' R_{k-l}' + 4u\sum_{l=k_0}^{k-k_0} R_l R_{k-l} + 4\sum_{l=k_0}^{k-k_0+2} R_l R_{k+2-l} = c_k, \quad k > 2k_0 - 2, \tag{1.4}$$

$$4R_{k_0}^2 = c_{2k_0-2}.$$

Relations (1.4) determine R uniquely (up to sign) in terms of c(ζ). Here and throughout the sequel we denote by $\bar{R} = \sum_{k=1}^{\infty} \bar{R}_k z^{-k}$ the solution of Eq. (1.2) for $c(\zeta) \equiv 1$ and call this solution standard. It is easy to see from the recursion formulas that all $\bar{R}_{2k} = 0$. The first $\bar{R}_k$ are given by:

$$\bar{R}_1 = \frac{1}{2}, \quad \bar{R}_3 = -\frac{1}{4}u, \quad \bar{R}_5 = \frac{1}{16}(3u^2 - u''). \tag{1.5}$$

The highest derivative of u in the differential polynomial $\bar{R}_k$ is $u^{(k-3)}$; it appears only in the linear term. (For purposes of comparison with the notation of [5], we point out that $\bar{R}_{2k+1}$ was denoted there by R_k.) It is clear that $d(\zeta) = \sqrt{c(\zeta)}$ is also an element of $A[[z^{-1}]]$ with constant coefficients. The solution of Eq. (1.2) corresponding to some c(ζ) has the form $R = d(\zeta)\bar{R}$. Thus, all the solutions of Eq. (1.2) differ from the standard solution by a factor in $A[[z^{-1}]]$ with constant coefficients.

The solution of Eq. (1.1) is a formal Laurent series whose coefficients are differential polynomials in u. If in place of the letter u we substitute the function u(x), we obtain a Laurent series with functional coefficients. We determine for which functions u(x) this series is a polynomial.

Proposition 1.1 (Its, Matveev). In order for Eq. (1.1) to have a solution which is a polynomial in z, it is necessary and sufficient that the function u(x) satisfy for some n the differential equation

$$\sum_{l=0}^{n+1} d_{-2l+1}\bar{R}_{2l+1} = 0, \tag{1.6}$$

where the d_k are certain numbers. (Since $\bar{R}_k$ is a polynomial in $u, u', \ldots, u^{(k-3)}$, (1.6) is a differential equation of order 2n.)

Proof. Let Eq. (1.1) have a solution which is a polynomial in z. Since z appears in an equation of the form $\zeta = z^2$, the even and odd parts of the solution are also solutions. Since the equation can be multiplied by any power of z, we may assume that it has the form $R = \sum_{k=-n}^{0} R_{2k}\zeta^{-k}$. It is clear from Eq. (1.2) that the corresponding c(ζ) is a polynomial in ζ of degree $2n+1$, $c(\zeta) = \sum_{k=-2n-1}^{0} c_{2k}\zeta^{-k}$. Then $d(\zeta) = \sqrt{c(\zeta)} = \sum_{k=-n}^{\infty} d_{2k-1}\zeta^{-k+(1/2)}$. We have $R = d(\zeta)\bar{R}$, whence

$$R_{2k} = \sum_{l=0}^{k+n} d_{2k-2l-1}\bar{R}_{2l+1}, \quad k = -n, -n+1, \ldots \tag{1.7}$$

Recall that $R_2 = R_4 = \ldots = 0$. In particular,

$$R_2 = \sum_{l=0}^{n+1} d_{1-2l}\bar{R}_{2l+1} = 0,$$

i.e., Eq. (1.6) is proved.

Conversely, let u(x) satisfy Eq. (1.6) for some coefficients $d_1, d_{-1}, \ldots, d_{-2n-1}$. We construct R_{2k}, $k = -n, \ldots, 0$ in terms of Eqs. (1.7). We show that

7

$$R = \sum_{k=-n}^{0} R_{2k}\zeta^{-k} = \sum_{k=0}^{n} d_{-2k+1} \sum_{l=0}^{k} \bar{R}_{2l+1}\zeta^{k-l} \tag{1.8}$$

is a solution of Eq. (1.1). In fact,

$$-R''' + 4uR' + 4\zeta R' + 2u'R =$$

$$= \sum_{k=0}^{n} d_{-2k-1} \sum_{l=0}^{k} [\zeta^{k-l}(-\bar{R}'''_{2l+1} + 4u\bar{R}'_{2l+1} + 2u'\bar{R}_{2l+1}) + 4\zeta^{k-l+1}\bar{R}'_{2l+1}] =$$

$$= \sum_{k=0}^{n} d_{-2k-1} \sum_{l=0}^{k} \zeta^{k-l}(-\bar{R}'''_{2l+1} + 4u\bar{R}'_{2l+1} + 4\bar{R}'_{2l+3} + 2u'\bar{R}_{2l+1}) - \sum_{k=0}^{n} d_{-2k-1}\cdot 4\bar{R}'_{2k+3} = -4\sum_{k=1}^{n+1} d_{-2k+1}\bar{R}'_{2k+1},$$

and hence by (1.6) $-R''' + 4(u+\zeta)R' + 2u'R = 0$, as required.

Equation (1.6) will be called the stationary higher-order KdV equation, or the Novikov equation. The ordinary stationary KdV equation is the special case when $\bar{R}_5 + d\bar{R}_1 = 0$ or $3u^2 - u'' = d/2$.

2. First Integrals

In this section we find the first integrals of Eq. (1.6). Consider the element $\hat{R} \in A[[\zeta^{-1}]]$, defined by

$$\hat{R} = \sum_{k=-n}^{0} R_{2k}\zeta^{-k}, \tag{2.1}$$

where $R_{2k} = \sum_{l=0}^{k+n} d_{2k-2l-1}\bar{R}_{2l+1}$ and the $\bar{R}_{2l+1}$ are the standard differential polynomials. We remark that (2.1) differs from (1.8) in that the R_{2k} here are differential polynomials, while in (1.8) a function of x (viz., solution of Eq. (1.6)) is substituted for the letter u. Since $\hat{R}$ is not a solution of (1.1), we do not obtain a polynomial with constant coefficients upon substituting $\hat{R}$ into the left hand side of (1.2), but rather some polynomial

$$P(\zeta) = \sum_{l=0}^{2n+1} J_l\zeta^l, \tag{2.2}$$

whose coefficients J_l are differential polynomials. If we now substitute the function u(x) for u in $\hat{R}$, where u(x) is a solution of Eq. (1.6), then as was shown in the proof of Proposition 1.1, $\hat{R}$ becomes a solution (1.8) of Eq. (1.1) and the coefficients J_l of the polynomial $P(\zeta)$ become constants. Thus, the J_l are polynomials in $u^{(i)}$ which are constants by Eq. (1.6), i.e., are the first integrals of this equation. Thus, the following proposition is proved.

Proposition 2.1. The differential polynomials J_l are the first integrals of Eq. (1.6).

Remark. These first integrals contains some trivial ones. Indeed, the J_l for l = 2n + 1, . . ., n are identically constant and do not depend on u. These constants are uniquely determined by the coefficients d_k in Eq. (1.6). In fact, (2.1) can be rewritten in the form

$$\hat{R} = \sum_{k=0}^{n} d_{-2k+1}\zeta^{k+1/2}\cdot \sum_{l=0}^{\infty} \bar{R}_{2l+1}\zeta^{-l-1/2} - \sum_{k=0}^{n} d_{-2k-1}\zeta^{k}\cdot \sum_{l=k+1}^{\infty} \bar{R}_{2l+1}\zeta^{-l} = \bar{R}\cdot \sum_{k=0}^{n} d_{-2k+1}\zeta^{k+1/2} + O(\zeta^{-1}),$$

where $O(\zeta^{-1})$ denotes an element in $A[[z^{-1}]]$, containing ζ to negative powers only. Substituting the expression obtained into the left-hand side of (1.2), and taking into account its bilinearity and the definition of $\bar{R}$, we obtain

$$P(\zeta) = \zeta\Big(\sum_{k=0}^{n} d_{-2k+1}\zeta^{k}\Big)^2 + O(\zeta^{n-1}).$$

The leading n + 2 coefficients of the polynomial $P(\zeta)$ are contained only in the first summand and are therefore constants.

The following proposition gives a connection between the first integrals constructed above and the ones proposed in [5, 6]. We will not use this proposition in the sequel.

Proposition 2.2.

$$\frac{d}{dx}J_l = -8\sum_{k=0}^{n-l} d_{-2k-2l-1}\bar{R}_{2k+1}\cdot \sum_{k=1}^{n+1} d_{-2k+1}R'_{2k+1}.$$

For the proof, it is in fact necessary to substitute (2.1) into the left-hand side of (1.2), differentiate with respect to x, and make use of the recursion relations (1.3). We omit this simple calculation.

In [5, 6] the first integrals were defined as those elements $F_l \in A$ such that $dF_l/dx = \bar{R}_l \sum d_{-2k+1}\bar{R}'_{2k+1}$. The proposition proved above shows that the first integrals J_l are linear combinations of the F_l and conversely (the transition matrix is triangular and nonsingular). The independence of these first integrals is established very simply in [5].

3. Some Concepts of Variational Calculus in the Ring A

The content of this section is discussed in more detail in [5]. A differential form is a formal sum

$$\sum f_{i_1\ldots i_k} du^{(i_1)} \wedge \ldots \wedge du^{(i_k)},$$

where the $f_{i_1\ldots i_k} \in A$ (or $A[[z^{-1}]]$). $du^{(0)}$ will be denoted by du. There is the relation $du^{(i)} \wedge du^{(k)} = -du^{(k)} \wedge du^{(i)}$. The operator d, the differential of a form, is defined in the usual way by

$$df = \sum (\partial f/\partial u^{(i)})\, du^{(i)}, \quad d\sum f_i du^{(i)} = \sum (\partial f_i/\partial u^{(j)})\, du^{(j)} \wedge du^{(i)} \text{ etc.}$$

The operator d/dx acts on forms according to the rule

$$\frac{d}{dx}\sum f_{i_1\ldots i_k} du^{(i_1)} \wedge \ldots \wedge du^{(i_k)} = \sum f'_{i_1\ldots i_k} du^{(i_1)} \wedge \ldots \wedge du^{(i_k)} +$$
$$+\sum f_{i_1\ldots i_k} du^{(i_1+1)} \wedge \ldots \wedge du^{(i_k)} + \ldots + \sum f_{i_1\ldots i_k} du^{(i_1)} \wedge \ldots \wedge du^{(i_k+1)}.$$

The operators d and d/dx commute. We state three lemmas proved in [5].

LEMMA 3.1. If $f \in A$ (or $f \in A[[z^{-1}]]$), then df can be expressed uniquely in the form

$$df = g\, du + \frac{d}{dx}\omega^{(1)}, \tag{3.1}$$

where $g \in A$ (respectively, $A[[z^{-1}]]$) and $\omega^{(1)}$ is some 1-form. Moreover, g is equal to

$$g = \sum_{k=0}^{\infty}\left(-\frac{d}{dx}\right)^k \frac{\partial}{\partial u^{(k)}} f; \tag{3.2}$$

g is called the variational derivative of f and denoted by $\delta f/\delta u$.

LEMMA 3.2. If $f \in A$ then there exists a $g \in A$ such that

$$\frac{dg}{dx} = -u'\frac{\delta f}{\delta u}. \tag{3.3}$$

Any element $\mathcal{L} \in A$ such that $\delta\mathcal{L}/\delta u$ contains the highest derivative only in the linear term, $cu^{(2n)}$, $c \neq 0$, will be called a normal Lagrangian. Let $\mathcal{L}$ be a normal Lagrangian. Consider the differential equation $\delta\mathcal{L}/\delta u = 0$. This equation is solvable for the highest derivative $u^{(2n)}$. The phase of this equation is the space of $(u, u', \ldots, u^{(2n-1)})$. We decompose $d\mathcal{L}$ by Eq. (3.1),

$$d\mathcal{L} = \frac{\delta\mathcal{L}}{\delta u}du + \frac{d}{dx}\omega^{(1)} \tag{3.4}$$

and put

$$\omega^{(2)} = d\omega^{(1)}, \tag{3.5}$$

so that $\omega^{(2)}$ is a closed 2-form.

LEMMA 3.3. If $\mathcal{L}$ is a normal Lagrangian, then $\omega^{(2)}$ is a nondegenerate form.

Thus, $\omega^{(2)}$ is a symplectic form. It follows from (3.4) that

$$\frac{d}{dx}\omega^{(2)} = -d\left(\frac{\delta \mathcal{L}}{\delta u}\right) \wedge du. \tag{3.6}$$

It follows, in particular, that if u(x) is a solution of the equation $\delta\mathcal{L}/\delta u = 0$, then $d\omega^{(2)}/dx = 0$, i.e., $\omega^{(2)}$ is conserved by virtue of the equation. Thus the following theorem is proved.

THEOREM 3.4. If $\mathcal{L}$ is a normal Lagrangian, then the equation $\delta\mathcal{L}/\delta u = 0$ is Hamiltonian with respect to the symplectic form $\omega^{(2)}$ defined by Eqs. (3.4), (3.5).

Remark. It is proved in [5] that the Hamiltonian corresponding to the equation $\delta\mathcal{L}/\delta u = 0$ with respect to the form $\omega^{(2)}$ is defined by the equality $H = -(d/dx)^{-1}u'\cdot\delta\mathcal{L}/\delta u$ (the existence of such a differential polynomial H follows from Lemma 3.2). We will not need the Hamiltonian in what follows.

4. Hamiltonian Nature of the Stationary KdV Equations

THEOREM 4.1 (see [5]). $\delta\bar{R}/\delta u = \partial\bar{R}/\partial\zeta$ holds for $\bar{R}$ defined as in Sec. 1, or for the coefficients $\bar{R}_k$

$$\frac{\delta\bar{R}_{2k+1}}{\delta u} = -\left(k - \frac{1}{2}\right)\bar{R}_{2k-1}. \tag{4.1}$$

We give the proof of this theorem which is due to Yusin [7]. $\bar{R}$ satisfies the equation

$$-2\bar{R}\bar{R}'' + (\bar{R}')^2 + 4(u+\zeta)\bar{R}^2 = 1. \tag{4.2}$$

Applying the operator d to this equation gives

$$-\bar{R}d\bar{R}'' - \bar{R}''d\bar{R} + \bar{R}'d\bar{R} + 4(u+\zeta)\bar{R}\,d\bar{R} + 2\bar{R}^2du = 0.$$

We multiply by $\bar{R}_\zeta/2\bar{R}^2$, where $\bar{R}_\zeta = \partial\bar{R}/\partial\zeta$. By means of simple transformations we get

$$\left[\left(-\frac{\bar{R}_\zeta}{2\bar{R}}\right)'' - \left(\frac{\bar{R}'\bar{R}_\zeta}{2\bar{R}^2}\right)' - \frac{\bar{R}''\bar{R}_\zeta}{2\bar{R}^2} + \frac{2(u+\zeta)\bar{R}_\zeta}{\bar{R}}\right]d\bar{R} + \bar{R}_\zeta du = \frac{d}{dx}\left[-\left(\frac{\bar{R}_\zeta}{2\bar{R}}\right)'d\bar{R} + \frac{\bar{R}_\zeta}{2\bar{R}}d\bar{R}' - \frac{\bar{R}'\bar{R}_\zeta}{2\bar{R}^2}d\bar{R}\right].$$

It is easy to see that by the same Eq. (4.2) (differentiated with respect to ζ), the expression in the square brackets in the left-hand side is equal to -1. Hence

$$d\bar{R} = \bar{R}_\zeta du + \frac{d}{dx}\left[-\frac{\bar{R}_\zeta}{2\bar{R}}d\bar{R}' + \frac{\bar{R}_\zeta'}{2\bar{R}}d\bar{R}\right]. \tag{4.3}$$

It follows from Lemma 3.1 that $\bar{R}_\zeta = \delta\bar{R}/\delta u$. This proves the theorem.

We remark that the expression in square brackets in Eq. (4.3) is not written out in [7]. It plays no role in the calculation of the variational derivative. But this expression is important in another respect. In accordance with the rationale of Lemma 3.1, the form appearing inside the sign d/dx is the form $\omega^{(1)}$ of the equation $\delta\bar{R}/\delta u = 0$. However, in order for this to acquire a precise meaning, it is necessary to write down both sides of Eq. (4.3) in the form of series in powers of ζ^{-1} and to take the coefficient of some power. We obtain the next proposition.

Proposition 4.2. To the equation $\bar{R}_{2n+3} = 0$, i.e.,

$$\delta\left(-\frac{2}{2n+3}\bar{R}_{2n+5}\right)/\delta u = 0$$

there corresponds a form $\omega^{(1)}$ of the form

$$\omega^{(1)} = -\frac{2}{2n+3}\left(-\frac{\bar{R}_\zeta}{2\bar{R}}d\bar{R}' + \frac{\bar{R}_\zeta'}{2\bar{R}}d\bar{R}\right)_{n+5/2}, \tag{4.4}$$

where the index n + 5/2 indicates the coefficient of $\zeta^{-n-5/2}$ in the expansion.

Remark. It is not the form $\omega^{(1)}$ itself which is of importance to us, but rather the symplectic form $\omega^{(2)} = d\omega^{(1)}$. This means that any form of the type df can be added to $\omega^{(1)}$ i.e., it is not the form that is important but the cohomology class of $\omega^{(1)}$. In place of (4.4), we take another form which differs by $d(\bar{R}_\zeta\bar{R}'/2\bar{R})$:

$$\omega^{(1)} = -\frac{2}{2n+3}\left[\frac{\partial}{\partial\zeta}\left(\frac{\bar{R}'}{2\bar{R}}d\bar{R}\right)\right]_{n+5/2} = (\bar{w}\,d\bar{R})_{n+3/2};\quad \bar{w} = \frac{\bar{R}'}{2\bar{R}}. \tag{4.5}$$

10

COROLLARY.

$$\omega^{(2)} = (d\bar{w} \wedge d\bar{R})_{n+3/2} = \sum_{k=1}^{n} d\bar{w}_{2(n-k)+2} \wedge d\bar{R}_{2k+1}. \tag{4.6}$$

This expression for the symplectic form was first found by another method by Al'ber [8]. We also indicate the form of the Hamiltonian, although it is not required,

$$H = \frac{1}{2}(\bar{R}''_{2n+3} - 4u\bar{R}_{2n+3} + 4\bar{R}_{2n+5}).$$

Only the phase variables $u, u', \ldots, u^{(2n-1)}$ appear in the right-hand side of this expression: the variables $u^{(2n)}$, $u^{(2n+1)}$, $u^{(2n+2)}$ appearing in the individual terms of the right-hand side cancel each other.

The equation $\bar{R}_{2n+3} = 0$ has order 2n; the symplectic form (4.6) contains n summands. Thus, the symplectic form is represented in canonical form and $\bar{R}_{2k+1}$ and $w_{2(n-k)+2}$, $k = 1, \ldots, n$, are canonically conjugate variables.

We turn to the general equations (1.6).

THEOREM 4.3. The symplectic form for Eq. (1.6) is

$$\omega^{(2)} = (d\hat{w} \wedge d\hat{R})_{\mathrm{I}}, \text{ where } \hat{w} = \sum_{1}^{\infty} w_{2k}\zeta^{-k} = \hat{R}'/2\hat{R}. \tag{4.7}$$

The index 1 denotes the coefficient of ζ^{-1} in the expansion, and $\hat{R}$ was defined previously in Eq. (2.1). In other words,

$$\omega^{(2)} = \sum_{k=1}^{n} dw_{2k} \wedge dR_{-2k+2}.$$

Proof. If the left-hand side of the equation is a linear combination $\delta\mathscr{L}/\delta u = \Sigma c_k \delta\mathscr{L}_k/\delta u$, then it is obvious that $\mathscr{L} = \Sigma c_k \mathscr{L}$ and $\omega^{(2)} = \Sigma c_k \omega_k^{(2)}$. Thus, we have for Eq. (1.6)

$$\omega^{(2)} = \sum_{l=1}^{n} d_{-2l-1} \sum_{k=1}^{n} d\bar{w}_{2(-l-k)+2} \wedge d\bar{R}_{2l+1} = -\sum_{l=1}^{n}\sum_{k=0}^{l-1} d_{-2l-1} \cdot d\bar{R}_{2(l-k)+1} \wedge d\bar{w}_{2k+2}$$

$$= \sum_{k=0}^{n-1} d\bar{w}_{2k+2} \wedge d\Big(\sum_{l=k+1}^{n} d_{-2l-1}\bar{R}_{2(l-k)+1}\Big) = \sum_{k=0}^{n-1} d\bar{w}_{2k+2} \wedge dR_{-2k} = (d\bar{w} \wedge d\hat{R})_{\mathrm{I}}.$$

Compared with the equation which it is required to prove, we have here the difference that w_{2k} appears in place of $\bar{w}_{2k}$, i.e., in place of the coefficients of the expansion $\hat{w} = \hat{R}/2\hat{R}$ there appear the coefficients of the expansion $\bar{w} = \bar{R}'/2\bar{R}$. We show that the expansions of $\hat{w}$ and $\bar{w}$ coincide up to sufficiently high powers of ζ^{-1} which no longer have any importance for us. In Sec. 2 we have already written the equation

$$\hat{R} = \sum_{k=-n}^{0} d_{2k-1}\zeta^{-k+1/2} \cdot R + O(\zeta^{-1}).$$

It follows without difficulty from this that $\hat{w} = \bar{w} + O(\zeta^{-n-1})$. Since $d\hat{R} = O(\zeta^{n-1})$, the replacement of $\bar{w}$ by $\hat{w}$ in the expression for $\omega^{(2)}$ does not change the coefficient of ζ^{-1}. The theorem is proved.

In the formula for $\omega^{(2)}$ it is important for us to replace $\bar{w}$ by $\hat{w}$ because $\bar{w}$ is a formal series in the powers of ζ^{-1} and $\hat{w}$ is a rational function of ζ; this will be required in what follows.

5. Liouville Integration

Assume we are given a coordinate neighborhood $\mathcal{O}$ of some 2n-dimensional manifold $\mathcal{M}$, and and that a symplectic form $\omega^{(2)}$ is defined in this neighborhood. To each function f defined on $\mathcal{O}$, there corresponds a vector field ξ_f such that $df = -i(\xi_f)\,\omega^{(2)}$. This field is called the Hamiltonian field corresponding to the Hamiltonian f. The differential equation $\dot{x} = \xi_f$ $(x \in \mathcal{O})$ is a Hamiltonian equation. The Poisson brackets of the two functions f, g are given by $\{f, g\} = -\xi_g f$. The function J is a first integral of the Hamiltonian equation $\dot{x} = \xi_H$ if and only if $\{J, H\} = 0$. Two first integrals J_1, J_2 are said to be in involution if $\{J_1, J_2\} = 0$. A collection of n independent functions whose pairwise Poisson brackets are equal to zero is called a complete involutive system of functions (we will say "involutive system" for

11

short). Two involutive systems $p_1, \ldots, p_n$ and $q_1, \ldots, q_n$ are canonically conjugate if $\{p_i,q_j\} = \delta_{ij}$. Liouville's theorem asserts that a system canonically conjugate to a given system can be constructed by means of quadratures.

Let us prove this theorem. Assume given a system $q_1, \ldots, q_n$. We construct a system of coordinate functions $s_1, \ldots, s_n, q_1, \ldots, q_n$, where the choice of the functions $s_1, \ldots, s_n$ is a matter of indifference. As before, let $\omega^{(1)}$ be a form such that $\omega^{(2)} = d\omega^{(1)}$. We write it in coordinates, $\omega^{(1)} = \Sigma\alpha_i ds_i + \Sigma\beta_i dq_i$. Consider the manifold of points $x \in \mathcal{O}$, for which $q_i = c_i$, $i = 1, \ldots, n$, where $c = (c_1, \ldots, c_n)$ is an arbitrary n-tuple of numbers. We denote this manifold by $\mathcal{N}_c$. It is easy to see that the involutiveness of the system $q_1, \ldots, q_{\mathcal{N}}$ means that $\mathcal{N}_c$ is a Lagrangian manifold, i.e., the restriction of the form $\omega^{(2)}$ to $\mathcal{N}_c$ is zero. Then the restriction of the form $\omega^{(1)}$ to $\mathcal{N}_c$ is locally the differential of some function: $\omega^{(1)} \mid \mathcal{N}_c = d_s V(s, q)$, or. $\alpha_i = \partial V/\partial s_i$. This function can be found by means of quadratures. The form $\omega^{(1)}$ on all of $\mathcal{O}$ has the form

$$\omega^{(1)} = \sum_1^n \frac{\partial V}{\partial s_i} ds_i + \sum_1^n \beta_i dq_i = dV(s, q) + \sum_1^n \left(\beta_i - \frac{\partial V(s, q)}{\partial q_i}\right) dq_i. \quad (5.1)$$

We introduce the quantities $p_i = \beta_i - (\partial V/\partial q_i)$. Then $\omega^{(1)} = \Sigma p_i dq_i$ (up to an exact differential of no importance) and $\omega^{(2)} = \sum_1^n dp_i \wedge dq_i$. The form $\omega^{(2)}$ is nondegenerate by hypothesis, so that all the $p_1, \ldots, p_n, q_1, \ldots, q_n$ are jointly independent. The quantities $p_1, \ldots, p_n$ form the desired involutive system canonically conjugate to the given system.

Assume now that the system $q_1, \ldots, q_n$ is a system of first integrals of some Hamiltonian equation, i.e., $q_i = q_i^0 = \text{const}$. Then it follows from the first of the equations

$$\dot{q}_i = \frac{\partial H}{\partial p_i}, \quad \dot{p}_i = -\frac{\partial H}{\partial q_i}$$

that H does not depend on $p_1, \ldots, p_n$. Hence, $\dot{p}_i = -\partial H/\partial q_i = \text{const}$, i.e., $p_i = -(\partial H/\partial q_i)x + p_i^0$, $p_i^0 = \text{const}$. In these coordinates the system is integrated without difficulty. It remains to return to the original variables.

6. Integration of the Stationary Higher Order KdV Equations

We return to the integration of Eq. (1.6). We have a Hamiltonian equation in the space of variables $u, u', \ldots, u^{(2n-1)}$ with symplectic form (4.7) and first integrals $J_l, l = 0,1,\ldots, n-1$. In [9] the involutiveness of the first integrals was proved; however, we will obtain this independently. In accordance with the procedure of Liouville integration, it will be necessary to pass to the coordinates $J_0, \ldots, J_{n-1}, s_1, \ldots, s_n$, where the $s_1, \ldots, s_n$ are certain functions. It is convenient to make this change of coordinates in stages: First we pass from the coordinates $u, u', \ldots, u^{(2n-1)}$ to $R_{-2n+2}, \ldots, R_0, R'_{-2n+2}, \ldots, R'_0$ (we denote this coordinate system by (R, R')), and then pass to $R_{-2n+2}, \ldots, R_0, w_2, \ldots, w_{2n}$ (coordinates (R, w)), and, finally, to the coordinates (J, s). The coordinates (R, R') are expressed polynomially in terms of $u, \ldots, u^{(2n-1)}$. Since the highest derivative $u^{(2n-2k-2)}$ appearing in R_{-2k} is present only in the linear term (and analogously for the derivative $u^{(2n-2k-1)}$ in R'_{-2k}), the inverse transformation of coordinates is also polynomial. The same holds for the next transformation from coordinates (R, R') to (R, w). The coordinate system (R, w) is convenient in that the symplectic form is expressed canonically (4.7). Equation (1.6) can also be written down in this coordinate system, although this will not be needed below:

$$\frac{dR_{2k}}{dx} = (2\hat{R}\hat{w})_k, \qquad k = -(n-1), \ldots, 0, \qquad \frac{dw_{2k}}{dx} = \left(-\hat{w}^2 - \frac{P(\zeta)}{4\hat{R}^2}\right)_k, \quad k = 1, \ldots, n.$$

Here the index k denotes the coefficient of ζ^{-k} in the expansion. We remark that only the first 2n + 2 terms of the polynomial $P(\zeta)$ are required in the second formula, which, as we see, are determined by the coefficients d_k of Eq. (1.6).

Passing to the coordinates (J, s), we remark that although the $s_1, \ldots, s_n$ are arbitrary, in practice their choice is limited by considerations of convenience: The form $\omega^{(1)}$ must have a simple expression in these coordinates. A good choice of such coordinates is to use the n roots ζ_i of the polynomial $\hat{R} = \sum_{k=-(n+1)}^{0} R_{2k}\zeta^{-k}$. We show how the coordinates (R, R') and (J, ζ) are related. The R_{2k} and ζ_i are related as the coefficients and roots of the polynomial

$$\hat{R} = \frac{d_{-2n-1}}{2}\prod_{j=1}^{n}(\zeta-\zeta_j), \tag{6.1}$$

i.e., the R_{2k} are the elementary symmetric functions in the ζ_i. As for R'_{2k}, we have the next lemma.

LEMMA 6.1.

$$\hat{R}'|_{\zeta=\zeta_l} = \sqrt{P(\zeta_l)}, \quad l=1,\ldots,n. \tag{6.2}$$

The proof is obtained by substituting $\zeta=\zeta_l$ in

$$-2\hat{R}\hat{R}'' + (\hat{R}')^2 + 4(u-\zeta)\hat{R}^2 = P(\zeta).$$

Equation (6.2) gives a linear system for the determination of R'_{2k}, $k=-n+1,\ldots,0$, in terms of ζ_l and J_k, the coefficients of the polynomial $P(\zeta)$.

Thus, the R_{2k} are expressed only in terms of the ζ_l and the R'_{2k} in terms of ζ_l and J_l.

Assume now that $u^{(i)}$ are the derivatives of some function $u = u(x)$. (That is, in phase space there is a curve $u^{(i)} = u^{(i)}(x)$ such that the differential equations $du^{(i)}/dx = u^{(i+1)}(x)$ are satisfied.) Then $\hat{R}$ becomes a function of x and $d\hat{R}/dx = \hat{R}'$. The roots ζ_l of the polynomial $\hat{R}$ also depend on x, $\zeta_l(x)$. We differentiate (6.1) with respect to x for fixed ζ and put $\zeta=\zeta_l$. We have $\hat{R}'|_{\zeta=\zeta_l} = -\frac{d_{-2n-1}}{2}\prod_{j\neq l}(\zeta_l-\zeta_j)\zeta_l(x)$. Substituting in (6.2), we obtain the following result.

LEMMA 6.2.

$$\frac{d}{dx}\zeta_l(x) = \frac{2}{d_{-2n-1}}\cdot\frac{\sqrt{P(\zeta_l)}}{\prod_{j\neq l}(\zeta_j-\zeta_l)}. \tag{6.3}$$

COROLLARY. If $u = u(x)$ is a solution of Eq. (1.6), then the coefficients J_k of the polynomial $P(\zeta)$ are constants and Eq. (6.3) is the expression for Eq. (1.6) on the invariant surface $J_k = c_k = \text{const}$ in the coordinates (ζ, J).

System (6.3) was first obtained by Dubrovin [2].

We now write out the form $\omega^{(1)}$ in these coordinates. In the coordinates (R, R'), it has the form $\omega^{(1)} = (\hat{R}'d\hat{R}/2\hat{R})_1$. This expression is equal to the sum of the residues of the rational function $\hat{R}'d\hat{R}/2\hat{R}$,

$$\sum_{k=1}^{n}\frac{\hat{R}'(\zeta_k)\,d\hat{R}(\zeta_k)}{2\hat{R}_\zeta(\zeta_k)} = \sum_{k=1}^{n}\frac{\sqrt{P(\zeta_k)}\prod_{j\neq k}(\zeta_k-\zeta_j)\,d\zeta_k}{2\prod_{j\neq k}(\zeta_k-\zeta_j)},$$

or

$$\omega^{(1)} = \frac{1}{2}\sum_{k=1}^{n}\sqrt{P(\zeta_k)}\,d\zeta_k. \tag{6.4}$$

We have thus proved the following proposition.

Proposition 6.3. In coordinates (J, ζ), system (1.6) has form (6.3) together with the equations $dJ_k/dx = 0$, and the form $\omega^{(1)}$ is given by (6.4).

We remark that the form in (6.4) contains only the differentials $d\zeta_h$ but not dJ_k, the variable J_k appearing only in the coefficients. If the form $\omega^{(1)}$ is bounded on the manifold $\mathcal{N}_c$: $J_k = c_k$, then it will be integrable since the variables are separated. Therefore, $\mathcal{N}_c$ is a Lagrangian manifold, i.e., the J_k are in involution. (By the way, we note that the variables ζ_l are also in involution.)

According to the procedure for Liouville integration, we find

$$V = \frac{1}{2}\sum_{k=1}^{n}\int_{\infty}^{\zeta_k}\sqrt{P(\zeta_k)}\,d\zeta_k$$

13

and finally, we find the system of variables

$$\varphi_l = \frac{\partial V}{\partial J_l} = \frac{1}{4}\sum_{k=1}^{n}\int_{\infty}^{\zeta_k}\frac{\zeta_k^l}{\sqrt{P(\zeta_k)}}d\zeta_k, \quad l = 0, \ldots, n-1 \tag{6.5}$$

conjugate to the J_l. In these variables the system is integrated, J_k = const, and the φ_k are linear functions of x. The last fact can also be verified directly:

$$\frac{d\varphi_l}{dx} = \frac{1}{4}\sum_{k=1}^{n}\frac{\zeta_k^l\cdot\zeta_k'(x)}{\sqrt{P(\zeta_k)}} = \frac{1}{4}\sum_{k=1}^{n}\frac{\zeta_k^l}{\prod\limits_{j\neq k}(\zeta_j-\zeta_k)}\cdot\frac{2}{d_{-2n-1}} = \begin{cases}0, & l<n-1\\ 2/d_{-2n-1}, & l=n-1.\end{cases}$$

In other words,

$$J_l = c_l,\ l = 0, \ldots, n-1;\ \varphi_l = b_l,\ l = 1, \ldots, n-1, \varphi_{n-1} = 2x/d_{-2n-1} + b_{n-1}. \tag{6.6}$$

7. Explicit Form of the Solution of a Stationary KdV Equation

It now remains to express solution (6.6) in the old coordinates $u, u', \ldots, u^{(2n-1)}$. For the sake of completeness, we describe this change of variables in this section, although this transformation is already known from [1-4]. The system has been integrated in the coordinates $J_0, \ldots, J_{n-1}, \varphi_1, \ldots, \varphi_n$, or (J, φ). Equation (6.5) gives the transition from (J, ζ) to (J, φ). Since the multivalued function $\sqrt{P(\zeta)}$ is being integrated, the points ζ_k must be considered as points on the Riemann surface of this function. Thus, the set of coordinates ζ_k is a point of a symmetric Riemann surface of degree n. The new coordinates φ are n numbers which are not uniquely defined, since there are noncontractible closed paths on the Riemann surface over which the integral is not zero. The values of the sum of the integrals (6.5) over these paths are called periods. Thus, the φ_k are defined modulo a discrete group of periods. This group has 2n generators, since the genus of the Riemann surface is n (all the notions needed from algebraic function theory can be found in more detail in, e.g., [10]). Thus, the point $\varphi_1, \ldots, \varphi_n$ belongs to the torus equal to the quotient space of C^n by the group of periods. This torus is called the Jacobi variety of the Riemann surface. Equations (6.5) give a mapping of a Riemann surface of degree n onto the Jacobi variety of the Riemann surface. This is the so-called Abel transform, which in this context could be called the Liouville–Abel transform. The Abel transform will be expressed in a more standard form if we replace the φ_k by certain of their linear combinations. Indeed, we choose on the Riemann surface a canonical basis system of closed paths $\alpha_1, \ldots, \alpha_n, \beta_1, \ldots, \beta_n$, determined by the roots $E_1, E_2, \ldots, E_{2n+1}$ of the polynomial $P(\zeta)$. Among all these paths, the only pairs that intersect are α_i and B_i, i = 1, . . ., n. The Abel differentials of the first kind are the differentials $dU = \sum_{j=0}^{n-1} c_j\zeta^j d\zeta/\sqrt{P(\zeta)}$. They are holomorphic on the Riemann surface. It is possible to find n differentials of the first kind $dU_i(\zeta) = \sum_{j=0}^{n-1} c_{ij}\zeta^j d\zeta/\sqrt{P(\zeta)}$, $i = 1, \ldots, n$, such that $\oint_{a_k} dU_i = \delta_{ik}$. We put $\oint_{b_k} dU_i = B_{ik}$. It can be proved (see [10]) that $B_{ik} = B_{ki}$ and that the matrix $\tau_{ik} = \mathrm{Im}B_{ik}$ is positive definite. We now introduce in place of the φ_i the system of variables

$$\psi_i = \sum_{k=1}^{n}\int_{\infty}^{\zeta_k} dU_i(\zeta_k) = \sum_{j=0}^{n-1} c_{ij}\varphi_j, \quad i = 1, \ldots, n. \tag{7.1}$$

We remark that the solution of our Hamiltonian system can be written in these coordinates as

$$\psi_i(x) = 2c_{in-1}\cdot x/d_{-2n-1} + h_i,\ i = 1, \ldots, n, \tag{7.2}$$

where the h_i are arbitrary constants.

Thus, we now need to return to the coordinates (J, ζ) from (J, ψ), i.e., invert the Abel transform (7.1). This is done with the aid of a theta function defined as

$$\theta(\mathbf{p}) = \sum_{\mathbf{k}}\exp\{\pi i(B\mathbf{k}, \mathbf{k}) + 2\pi i(\mathbf{p}, \mathbf{k})\}, \tag{7.3}$$

where $B = (B_{ik})$, $\mathbf{p}\in C^n$, $\mathbf{k}$ is a vector with integer coefficients and (,) is the ordinary inner product. Series (7.3) converges absolutely. The θ-function has the properties

$$\theta(p+\delta_k)=\theta(p),\quad \theta(p+B_k)=\theta(p)\,e^{-\pi i(B_{kk}+2p_k)},$$

where δ_k is a vector with the coordinates δ_{ik}, $i=1,\ldots,n$, and the B_k are vectors with coordinates B_{ik}. The following theorem holds.

Riemann's THEOREM. There exist constants K_i determined by the coefficients of the polynomial $P(\zeta)$ (the Riemann constants) such that $\zeta_1,\ldots,\zeta_n$ are solutions of the equation

$$\sum_{k=1}^{n}\int_{\infty}^{\zeta_k} dU_i(\zeta_k)\,d\zeta_k=l_i-K_i,\quad i=1,\ldots,n,$$

with certain l_i if and only if they are zeros of the function

$$\tilde\theta(\zeta)=\theta(U_1(\zeta)-l_1,\ \ldots,\ U_n(\zeta)-l_n),$$

where $U_i(\zeta)=\int_{\infty}^{\zeta} dU_i(\zeta)$ (it can be shown that this function has exactly n zeros).

Thus, in order to invert the Abel transform, it is necessary to find the zeros of the function $\tilde\theta(\zeta)$. The elementary symmetric functions of the zeros can be calculated by contour integration. Indeed, let us cut the Riemann surface along the contours α_i, β_i. Let γ be the contour consisting of all the lines of all the contours. Then replacing $(1/2\pi i)\oint_{\gamma}\zeta\,d\ln\tilde\theta(\zeta)$ by the sum of the residues at the points ζ_k and ∞, we obtain that $\zeta_1+\ldots+\zeta_n=-d^2/dx^2\ln\theta\times(l_1,l_2,\ldots,l_n)+C_1$, where $C_1=\sum_{k=1}^{n}\oint_{\alpha_k}\zeta\,dU_k(\zeta)$. The remaining elementary symmetric functions can be found in the same way, although just the first elementary function is important for us. In the variables (R, w), this function gives $R_{-2n+2}/R_{-2n}=-\Sigma\zeta_i$. In the original variables $u,u',\ldots,u^{(2n-1)}$ we have $R_{-2n+2}/R_{-2n}=d_{-2n+1}/d_{-2n-1}-u/2$, and therefore $u=2d_{-2n+1}+2\Sigma\zeta_i$. From this we obtain the main result of [3]. The solution of Eq. (1.6) has the form

$$u=-2\frac{d^2}{dx^2}\ln\theta(l_1,l_2,\ldots,l_n)+C, \tag{7.4}$$

where $l_i=2c_{in-1}\cdot x/d_{-2n-1}+h_i$, $i=1,\ldots,n$, and $C=2C_1+2d_{-2n+1}/d_{-2n-1}$.

LITERATURE CITED

1. S. P. Novikov, "Periodic problem for the Korteweg–de Vries equation," Funkts. Anal. Prilozhen., 8, No. 3, 54-66 (1974).
2. B. A. Dubrovin, "The periodic Korteweg–de Vries problem in a class of finite zone potentials," Funkts. Anal. Prilozhen., 9, No. 3, 41-52 (1975).
3. A. R. Its and V. B. Matveev, "Schrödinger operators with finite-zone spectrum and N-soliton solutions of the Korteweg–de Vries equations," Teor. Mat. Fiz., 23, No. 1, 51-68 (1975).
4. N. F. McKean and F. V. Moerbeke, "Sur le spèctre de quelques opérateurs et la variété de Jacobi," Sém. Bourbaki (1975-1976), p. 474.
5. I. M. Gel'fand and L. A. Dikii, "Asymptotics of the resolvent of Sturm–Liouville equations and the algebra of the Korteweg–de Vries equations," Usp. Mat. Nauk, 20, No. 5, 67-100 (1975).
6. P. Lax, "Periodic solutions of the KdV equations," Lect. Appl. Math., 15, 85-96 (1974).
7. B. V. Yusin, "Proof of a variational relation among the coefficients of the asymptotics of a Sturm–Liouville equation," Usp. Mat. Nauk, 33, No. 1, 233-234 (1978).
8. S. I. Al'ber, "Study of equations of the Korteweg–de Vries type by the method of recursion relations," Preprint (1976), pp. 1-13.
9. I. M. Gel'fand and L. A. Dikii, "Lie algebra structure in the formal calculus of variations," Funkts. Anal. Prilozhen., 10, No. 1, 18-25 (1976).
10. N. G. Chebotarev, The Theory of Algebraic Functions [in Russian], Gostekhizdat, Moscow (1948).

15.

(with I.Ya. Dorfman)

Hamiltonian operators and algebraic structures related to them

Funkts. Anal. Prilozh. **13** (4) (1979) 13–31. MR **81c**: 58035, Zbl. **428**: 58009
[Funct. Anal. Appl. **13** (1980) 248–262]

1. Algebraic Approach to the Concept of Symplectic Structure

The basic objects of consideration will be a Lie algebra $\mathfrak{A}$ and a left $\mathfrak{A}$-module M. This means that a bilinear operation is given which assigns to each pair $a \in \mathfrak{A}$, $m \in M$ an element $am \subset M$ such that $a_1a_2m - a_2a_1m = [a_1, a_2]m$.

Examples. 1. Let X be a smooth, finite-dimensional manifold. The elements of the Lie algebra $\mathfrak{A}$ are vector fields a on X; M is the space of C^∞-functions on X. The operation $[a_1, a_2]$ is the commutator of vectors, and am is the result of the action of the field a on the function $m(x) \in M$.

2. A second important example of this situation arises in the formal calculus of variation [1]. The next section will be devoted to it. We note only that in the second example M has no ring structure; actually, the ring structure of M is not used in Example 1 either.

Our objective is to construct a Hamiltonian structure on the pair $(\mathfrak{A}, M)$ using so far as possible only the structures present in $\mathfrak{A}$, and M. There are thus two basic operations: the commutator $[a_1, a_2]$ and the action am. We shall attempt to express all operations in terms of these basic operations.

We first define forms. A q-form is a multilinear, skew-symmetric function $\omega(a_1, \ldots, a_q)$ on $\mathfrak{A}$ with values in M. A 0-form is by definition a fixed element $m \in M$. The differential d is given by the usual formula

$$d\omega(a_1, \ldots, a_{q+1}) = \sum_i (-1)^{i+1} a_i\omega(a_1, \ldots \hat{a}_i \ldots, a_{q+1}) + \sum_{i<j} (-1)^{i+j} \omega([a_i, a_j], a_1, \ldots \hat{a}_i \ldots \hat{a}_j \ldots, a_{q+1}).$$

In particular, dm is a linear form on $\mathfrak{A}$ with values in M; the value of the 1-form dm on an element $a \in \mathfrak{A}$ is given by the formula $(dm)(a) = am$. If ξ is a 1-form, i.e., a linear mapping $\xi : \mathfrak{A} \to M$, then $(d\xi)(a_1, a_2) = a_1\xi(a_2) - a_2\xi(a_1) - \xi([a_1, a_2])$.

It is easy to verify that $d^2 = 0$. A form ω is called closed if $d\omega = 0$.

We denote by C^q the space of q-forms (it is sometimes also called the space of q-cochains on $\mathfrak{A}$ with values in M). We observe that C^q is also a left $\mathfrak{A}$-module; the action of $\mathfrak{A}$ is given by

$$L_am = am,\ m \in C^0,\ (L_a\xi)(x) = a\xi(x) - \xi([a, x]),\ \xi \in C^1, \text{ etc.}$$

The general formula is

$$L_a = i_ad + di_a,$$

where $(i_a\omega)(x_1, \ldots, x_{q-1}) = \omega(a, x_1, \ldots, x_{q-1})$ for any $\omega \in C^q$. Here C^q is actually an $\mathfrak{A}$-module, since it can be shown that $L_{a_1}L_{a_2} - L_{a_2}L_{a_1} = L_{[a_1, a_2]}$.

Let $\omega \in C^2$ be a fixed 2-form.

In $\mathfrak{A} \oplus M$ we consider the subspace of pairs (a, m) such that $\omega(a, x) = -xm$ for all $x \in \mathfrak{A}$. Then, let

$$\mathfrak{H} = \{(a, m) : \omega(a, x) = -xm \quad \text{for all} \quad x \in \mathfrak{A}\}.$$

On $\mathfrak{A} \oplus M$ we introduce the following operations:

$$[(a_1, m_1), (a_2, m_2)]_1 = ([a_1, a_2], \tfrac{1}{2}(a_1m_2 - a_2m_1)), \tag{1.1}$$

Institute of Applied Mathematics, Academy of Sciences of the USSR. Institute of Chemical Physics, Academy of Sciences of the USSR. Translated from Funktsional'nyi Analiz i Ego Prilozheniya, Vol. 13, No. 4, pp. 13-30, October-December, 1979. Original article submitted April 16, 1979.

$$[(a_1, m_1), (a_2, m_2)]_2 = ([a_1, a_2], \omega(a_1, a_2)). \tag{1.1'}$$

We note that on $\mathfrak{H}$ these operations coincide, and we therefore henceforth denote the restriction of each of the operations to $\mathfrak{H}$ by the single symbol [,].

THEOREM 1.1. If $d\omega = 0$, then the subspace $\mathfrak{H}$ is closed relative to the operation [,] and is hence a Lie algebra with respect to it.

Proof. For any $a_1, a_2, x \in \mathfrak{A}$ such that $(a_1, m_1), (a_2, m_2) \in \mathfrak{H}$, we have

$$\begin{aligned} 0 = d\omega(a_1, a_2, x) &= a_1\omega(a_2, x) + a_2\omega(x, a_1) + x\omega(a_1, a_2) - \\ &- \omega([a_1, a_2], x) - \omega([a_2, x], a_1) - \omega([x, a_1], a_2) = \\ &= x\omega(a_1, a_2) - \omega([a_1, a_2], x) - a_1 x m_2 + a_2 x m_1 - [a_2, x]\, m_1 - \\ &- [x, a_1]\, m_2 = x\omega(a_1, a_2) - \omega([a_1, a_2], x) + x(a_2 m_1 - a_1 m_2) = \\ &= x\omega(a_1, a_2) - \omega([a_1, a_2], x) - 2x\omega(a_1, a_2) = -x\omega(a_1, a_2) - \omega([a_1, a_2], x). \end{aligned}$$

We have thus proved that $([a_1, a_2], \omega(a_1, a_2)) \in \mathfrak{H}$. We shall prove the Jacobi identity on $\mathfrak{H}$.* The expression (cycle) here and below denotes the sum over cyclic permutations. We have

$$\begin{aligned} &[[(a_1, m_1), (a_2, m_2)], (a_3, m_3)] + (\text{cycle}) = [([a_1, a_2], \omega(a_1, a_2)), (a_3, m_3)] + (\text{cycle}) = \\ &= ([[a_1, a_2], a_3], \omega([a_1, a_2], a_3)) + (\text{cycle}) = \Big([[a_1, a_2], a_3], \tfrac{1}{2}(\omega([a_1, a_2], a_3) - \\ &- a_3\omega(a_1, a_2))\Big) + (\text{cycle}) = \Big(0, \tfrac{1}{2} d\omega(a_1, a_2, a_3)\Big) = (0, 0), \end{aligned}$$

as required.

We denote by N the projection of $\mathfrak{H}$ onto M. Let $m_1, m_2 \in N$. We choose $a_1, a_2 \in \mathfrak{A}$ such that $(a_1, m_1) \in \mathfrak{H}$ $(a_2, m_2) \in \mathfrak{H}$. We set

$$\{m_1, m_2\} = \omega(a_1, a_2). \tag{1.2}$$

It is not hard to see that this operation does not depend on the choice of a_1, a_2. Indeed, if a_1, a_1' are such that $(a_1, m_1) \in \mathfrak{H}$, $(a_1', m_1) \in \mathfrak{H}$, then $(a_1 - a_1', 0) \in \mathfrak{H}$, whence $\omega(a_1 - a_1', a_2) = -a_2 \cdot 0 = 0$.

COROLLARY 1.2. If $d\omega = 0$, then the operation { , } satisfies the Jacobi identity and hence N becomes a Lie algebra.

Proof. For any $(a_1, m_1) \in \mathfrak{H}$, $(a_2, m_2) \in \mathfrak{H}$ we have

$$[(a_1, m_1), (a_2, m_2)] = ([a_1, a_2], \{m_1, m_2\}).$$

It follows from Theorem 1.1 that the Jacobi identity is satisfied on $\mathfrak{H}$ and hence for the brackets { , }.

Definition. The closed 2-form ω is called a symplectic structure on $(\mathfrak{A}, M)$. The operation { , } given on N by formula (1.2) is called the Poisson bracket corresponding to this structure.

We suppose that in the space of 1-forms a subspace $\Omega^1 \subset C^1$ is fixed which contains the differentials of all 0-forms (i.e., of elements of M). In concrete situations considered below Ω^1 will be specially described.

Further, let $H : \Omega^1 \to \mathfrak{A}$ be a linear operator. We call H skew-symmetric if for any $\xi_1, \xi_2 \in \Omega^1$

$$\xi_1(H\xi_2) = -\xi_2(H\xi_1). \tag{1.3}$$

With any skew-symmetric operator $H : \Omega^1 \to \mathfrak{A}$ we connect a 2-form ω_H defined on the image Im H by

$$\omega_H(a_1, a_2) = (H^{-1}a_2)(a_1), \qquad a_1, a_2 \in \operatorname{Im} H. \tag{1.4}$$

We shall show that the 2-form is well defined. Suppose a_1, a_2 belong to the image Im H of the operator H, i.e., they have the form $a_i = H\xi_i$. Then $(H^{-1}a_2)(a_1)$ by definition is $\xi_2(H\xi_1)$ and from the skew-symmetry of the operator H it follows that the value of the 2-form ω_H depends only on a_1, a_2.

Definition. We say that a skew-symmetric operator $H : \Omega^1 \to \mathfrak{A}$ is Hamiltonian if a) the image of the operator Im H is a subalgebra of the Lie algebra $\mathfrak{A}$ and b) $d\omega_H = 0$ on this subalgebra.

---

*We emphasize that the Jacobi identity for operations (1.1)-(1.1') is satisfied only on $\mathfrak{H}$, in contrast to the operation $[(a_1, m_1), (a_2, m_2)]_3 = ([a_1, a_2], a_1 m_2 - a_2 m_1)$ which gives a Lie-algebra structure on all of $\mathfrak{A} \oplus M$. However, $\mathfrak{H}$ is not closed with respect to $[,]_3$.

Thus, if a Hamiltonian operator $H : \Omega^1 \to \mathfrak{A}$ is given, then there is a symplectic structure ω_H on (Im H, M) defined by formula (1.4).

We shall describe the set $\mathfrak{H}$ corresponding to this structure; by definition

$$\mathfrak{H} = \{(a, m) : (H^{-1}a)(x) = xm \quad \text{for all} \quad x \in \text{Im } H\} = \{(a, m) : a = Hdm\}.$$

We see that in our case the set N coincides with M, and the bilinear operation $\{,\}_H$ which we call the Poisson bracket is thus defined on M. According to formula (1.2), it has the explicit form

$$\{m_1, m_2\}_H = dm_2(Hdm_1) = (Hdm_1)\, m_2, \qquad m_1, m_2 \in M. \tag{1.5}$$

The following assertion holds on the basis of Corollary 1.2.

<u>THEOREM 1.3.</u> If the operator $H : \Omega^1 \to \mathfrak{A}$ is Hamiltonian, then M is a Lie algebra relative to the Poisson bracket $\{,\}_H$. The mapping $Hd : M \to \mathfrak{A}$ is a morphism of this Lie algebra into the original Lie algebra $\mathfrak{A}$.

<u>Definition.</u> We say that two skew-symmetric operators $H : \Omega^1 \to \mathfrak{A}$ and $K : \Omega^1 \to \mathfrak{A}$ form a **Hamiltonian pair** if for any constants λ, μ the operator $(\lambda H + \mu K) : \Omega^1 \to \mathfrak{A}$ is Hamiltonian.

We shall see below that Hamiltonain pairs are closely related to integrable systems (finite-dimensional and infinite-dimensional), viz., by means of Hamiltonian pairs it is possible to construct systems of equations and give sets of integrals of them.

2. Symplectic Structures in the Formal Variational Calculus

Let A be an algebra of polynomials of symbols $u_\alpha^{(i)}$, where α runs through some set of indices I, i = 0, 1, 2, ... (it is intended that each polynomial depend only on a finite number of $u_\alpha^{(i)}$).* In the formal variational calculus [1-3] "differentiation with respect to x" is defined by

$$\frac{d}{dx} = \sum_{i,\alpha} u_\alpha^{(i+1)} \frac{\partial}{\partial u_\alpha^{(i)}}.$$

There is an equivalence relation in A: $f \sim 0$ if $f = (d/dx)g$. An equivalence class is called a functional, and the set of functionals is denoted by $\tilde{A}$. The mapping assigning to each $f \in A$ its equivalence class $\tilde{f} \in \tilde{A}$ is written as follows: $\tilde{f} = \int f dx$. In the nonformal theory where the u_α are not symbols but rather functions of x (e.g., periodic or rapidly decreasing functions) the $u_\alpha^{(i)}$ are their derivatives, and the notation d/dx and $\int \cdot\, dx$ can be interpreted in the usual sense. We denote by $\bar{A}$ the set of all $\bar{f} = \{f_\alpha\}$, where the index α runs through the set I.

It can be shown that any differentiation of the algebra A has the form

$$\partial = \sum_{\alpha, i} h_{\alpha, i} \frac{\partial}{\partial u_\alpha^{(i)}}, \qquad h_{\alpha, i} \in A. \tag{2.1}$$

The set of all differentiations forms a Lie algebra relative to the commutator $[\partial_1, \partial_2] = \partial_1\partial_2 - \partial_2\partial_1$. Differentiations commuting with d/dx form a subalgebra of this Lie algebra. The condition $[\partial, d/dx]$, as is evident from formula (2.1), is equivalent to $h_{\alpha,i} = (d/dx)^i h_\alpha$, $h_\alpha \in A$. Thus, differentiations commuting with d/dx are in one-to-one correspondence with elements $\bar{h} = \{h_\alpha\} \in \bar{A}$; viz., to any $\bar{h} \in \bar{A}$ there corresponds the differentiation

$$\partial_{\bar{h}} = \sum_{\alpha, i} h_\alpha^{(i)} \frac{\partial}{\partial u_\alpha^{(i)}}. \tag{2.2}$$

Here $h_\alpha^{(i)}$ denotes $(d/dx)^i h_\alpha$. In particular, the differentiation d/dx is the differentiation given by the element $\bar{h}$ with components $h_\alpha = u_\alpha^{(1)}$.

Any differentiation of form (2.2) we call a vector field and identify it with the element $\bar{h} = \{h_\alpha\} \in \bar{A}$. The commutator of vector fields $\bar{h}$ and $\bar{g}$ has the form $[\bar{h}, \bar{g}] = \bar{k}$, where

* To be specific we consider an algebra of polynomials, but it is possible to take A to be an algebra of smooth functions, an algebra of rational functions, and other algebras.

$$k_\alpha = [\bar{h}, \bar{g}]_\alpha = \sum_{i,\beta}\left(h_\beta^{(i)}\frac{\partial g_\alpha}{\partial u_\beta^{(i)}} - g_\beta^{(i)}\frac{\partial h_\alpha}{\partial u_\beta^{(i)}}\right). \tag{2.3}$$

The commutation operation which was introduced in [1] will play a basic role below.

We thus have a Lie algebra, the space $\bar{A}$, equipped with the commutation operation (2.3). We shall now construct a left $\mathfrak{A}$ -module M. To this end we introduce the partial variational derivatives $\delta/\delta u_\alpha: A \to A$ by the formula

$$\frac{\delta}{\delta u_\alpha} = \sum_{i=0}^{\infty}\left(-\frac{d}{dx}\right)^i \frac{\partial}{\partial u_\alpha^{(i)}}.$$

We denote by $\delta/\delta\bar{u}$ the mapping assigning to each $f \in A$ the element $\delta f/\delta\bar{u}$ with coordinates $(\delta/\delta u_\alpha)f$ (it is obvious that there are only a finite number of nonzero coordinates). We have the equality

$$\frac{\delta}{\delta \bar{u}} \circ \frac{d}{dx} = 0,$$

which implies that $\delta/\delta\bar{u}$ is well defined on the set $\tilde{A}$ of functionals. The mapping $\delta/\delta\bar{u}: \tilde{A} \to \bar{A}$ is thus defined.

We introduce the action of a vector field $\bar{h} \in \bar{A}$ on a functional $\tilde{f} \in \tilde{A}$ as follows:

$$\bar{h}\tilde{f} = \int\left(\sum_\alpha \frac{\delta \tilde{f}}{\delta u_\alpha} h_\alpha\right) dx. \tag{2.4}$$

It is not hard to verify that this action defines on $\tilde{A}$ the structure of a left $\bar{A}$-module. This will be of our $\mathfrak{A}$-module M.

According to the definition adopted in Sec. 1, the space C^1 consists of all possible linear mappings $\xi: \bar{A} \to \tilde{A}$. In C^1 we consider the subspace Ω^1 of 1-forms ξ with values on vector fields $\bar{h} \in \bar{A}$ given by

$$\xi(\bar{h}) = \int\left(\sum_\alpha \xi_\alpha h_\alpha\right) dx, \tag{2.5}$$

where $\xi_\alpha \in A$, $\alpha \in I$ and only a finite number of the ξ_α are nonzero.

Below we shall with no special mention account with the fact that the sequences $\{\xi_\alpha\}$ are in one-to-one correspondence with elements of Ω^1, and for the 1-form ξ we shall use the notation $\xi = \{\xi_\alpha\}$. We agree also to denote by $(\xi, \bar{h})$ the value of the 1-form ξ on the vector field $\bar{h}$.

We note that Ω^1 contains the differentials of all functionals. Indeed, if $\tilde{f} \in \tilde{A}$ is a functional, then by definition

$$d\tilde{f}(\bar{h}) = \bar{h}\tilde{f} = \int\left(\sum_\alpha \frac{\delta \tilde{f}}{\delta u_\alpha} h_\alpha\right) dx,$$

so that on setting $\xi_\alpha = (\delta/\delta u_\alpha)\tilde{f}$, we obtain a representation of the 1-form $\xi = d\tilde{f}$ in the form (2.5).

We thus have the following objects:

1) a Lie algebra $\mathfrak{A}$ which is the space $\bar{A}$ of vector fields equipped with the commutator by formula (2.3);

2) a module M which is the space $\tilde{A}$ of functionals; the action of $\bar{A}$ on $\tilde{A}$ is given by formula (2.4);

3) a space Ω^1 consisting of 1-forms with values on vector fields given by formula (2.5).

In accordance with 1, we now consider linear operations acting from Ω^1 to $\bar{A}$ and among them seek Hamiltonian operators.

We call a linear operator $H: \Omega^1 \to \bar{A}$ a matrix differential operator or, more briefly, a differential operator, if it acts in the following manner:

$$(H\xi)_\alpha = \sum_\beta H_{\alpha\beta}\xi_\beta, \quad \alpha, \beta \in I, \quad \xi \in \Omega^1, \tag{2.6}$$

where all the $H_{\alpha\beta}$ have the form

$$H_{\alpha\beta}=\sum_{i=0}^{n(\alpha,\beta)} a_{\alpha\beta i}\left(\frac{d}{dx}\right)^i,\quad a_{\alpha\beta i}\in A. \tag{2.7}$$

We do not require here that the orders n(α, β) be bounded.

The adjoint of an operator H of the form (2.6) is the operator H^*: $\Omega^1\to\bar{A}$ for which

$$(H^*)_{\alpha\beta}=\sum_{i=0}^{n(\beta,\alpha)}\left(-\frac{d}{dx}\right)^i\circ a_{\beta\alpha i}.$$

It is obvious that operators which are skew-symmetric in the sense of Sec. 1 are operators satisfying the condition $H^*=-H$.

Below in Sec. 5 we shall establish necessary and sufficient conditions that operators H: $\Omega^1\to\bar{A}$ be Hamiltonian (both for differential operators and for operators of general form). In Sec. 6 we shall then construct examples of Hamiltonian operators (only differential operators will be encountered in the examples).

To conclude this section we reformulate the general result (Theorem 1.3) in terms of the formal variational calculus. We recalled that according to formula (1.5) the Poisson bracket connected with any Hamiltonian operator H: $\Omega^1\to\bar{A}$ has the form

$$\{\tilde{f},\tilde{g}\}_H=d\tilde{g}(Hd\tilde{f})=\int\left(\sum_{\alpha,\beta}H_{\alpha\beta}\frac{\delta\tilde{f}}{\delta u_\beta}\right)\frac{\delta\tilde{g}}{\delta u_\alpha}dx.$$

THEOREM 2.1. Let H: $\Omega^1\to\bar{A}$ be a Hamiltonian operator. Then $\tilde{A}$ is a Lie algebra relative to the Poisson bracket $\{,\}_H$. The mapping $H(\delta/\delta\bar{u})$: $\tilde{A}\to\bar{A}$ is a morphism of Lie algebras, i.e.,

$$H\frac{\delta}{\delta\bar{u}}\{\tilde{f},\tilde{g}\}_H=\left[H\frac{\delta}{\delta\bar{u}}\tilde{f},H\frac{\delta}{\delta\bar{u}}\tilde{g}\right].$$

3. Hamiltonian Operators and Hamiltonian Pairs in the Finite-Dimensional Theory*

Let X be a finite-dimensional smooth manifold to each point $x\in X$ of which there corresponds a linear operator $H(x)$: $T_X^*X\to T_XX$ which depends smoothly on the point x and is skew-symmetric. (Skew-symmetry here means that $(H(x)\xi,\eta)=-(\xi,H(x)\eta)$ for any $\xi,\eta\in T_X^*X$; the bracket (,) here and below denotes the canonical pairing of a space with its dual.) We shall assume for simplicity that X is even-dimensional and all the operators H(x) are invertible, although with appropriate modifications all that follows could be carried out in the general case.

As already mentioned, we are in the framework of the general situation described in Sec. 1; there are the following objects:

1) a Lie algebra $\mathfrak{A}$ which is the Lie algebra of vector fields on X;

2) an $\mathfrak{A}$ -module M which is the space of C^∞-functions on X;

3) for Ω^1 we choose the space of differential 1-forms on X.

The set of operators $\{H(x),x\in X\}$ obviously defines an operator H: $\Omega^1\to\mathfrak{A}$. We shall now give a necessary and sufficient condition that the operator H be Hamiltonian as the condition that a certain trivalent tensor on X vanish. This idea was put forth in the work [9].

We introduce local coordinates $x_1,\ldots,x_{2n}$ on X; induced coordinates then arise on TX and T*X so that the operators H(x) can be represented by matrices $(H_{\alpha\beta}(x))$ (by the skew-symmetry $H_{\alpha\beta}(x)=-H_{\beta\alpha}(x)$).

Let $H,K:\Omega^1\to\mathfrak{A}$ be two operators of the type indicated. We construct a trivalent tensor [H, K] on X as follows:

$$[H,K]_{ijk}=\sum_\beta(H_{ij,\beta}K_{\beta k}+K_{ij,\beta}H_{\beta k})+\text{(cycle)}. \tag{3.1}$$

---

*The authors are indebted to D. V. Alekseevskii for discussion of the results presented here.

Here and below $H_{ij,\beta} = \partial H_{ij}/\partial x_\beta$; the word (cycle), as before, denotes the sum over cyclic permutation of the indices i, j, k. The tensor [H, K] is called the Schouten–Nijenhuis bracket of the tensors H and K (see [9]).

THEOREM 3.1. The operator $H: \Omega^1 \to \mathfrak{A}$ is Hamiltonian if and only if

$$[H, H] = 0. \tag{3.2}$$

COROLLARY 3.2. The operators $H, K: \Omega^1 \to \mathfrak{A}$ form a Hamiltonian pair if and only if the following three conditions are satisfied:

$$[H, H] = 0, \quad [H, K] = 0, \quad [K, K] = 0. \tag{3.3}$$

We shall now indicate how Hamiltonian pairs are related to integrable dynamical systems on X. For this we require the following result.

THEOREM 3.3. Let H, K be a Hamiltonian pair of operators. Let $\varphi, \psi, \chi \in \Omega^1$ be 1-forms such that

$$K\psi = H\varphi, \tag{3.4}$$

$$K\chi = H\psi. \tag{3.5}$$

Then the following formula holds*:

$$Kd\chi K = Hd\psi K + Kd\psi H - Hd\varphi H. \tag{3.6}$$

Proof. We choose a coordinate atlas such that the operator K has the form

$$K = \begin{pmatrix} 0 & -E \\ E & 0 \end{pmatrix}.$$

This is possible by the theorem of Darboux. Differentiating relations (3.4) and (3.5) with respect to x_j, we obtain, respectively,

$$\sum_\alpha H_{i\alpha,\,j}\varphi_\alpha + \sum_\alpha H_{i\alpha}\varphi_{\alpha,\,j} - \sum_\alpha K_{i\alpha}\psi_{\alpha,\,j} = 0,$$

$$\sum_\alpha H_{i\alpha,\,j}\psi_\alpha + \sum_\alpha H_{i\alpha}\psi_{\alpha,\,j} - \sum_\alpha K_{i\alpha}\chi_{\alpha,\,j} = 0.$$

We multiply the first of these equalities by $H_{j\beta}$, the second by $K_{j\beta}$, subtract the second from the first, and sum on j. We obtain

$$\sum_{j,\,\alpha} K_{i\alpha}\chi_{\alpha,\,j}K_{j\beta} = \sum_{j,\,\alpha} H_{i\alpha}\psi_{\alpha,\,j}K_{j\beta} + \sum_{j,\,\alpha} K_{i\alpha}\psi_{\alpha,\,j}H_{j\beta} - \sum_{j,\,\alpha} H_{i\alpha}\varphi_{\alpha,\,j}H_{j\beta} + \sum_{j,\,\alpha} H_{i\alpha,\,j}\psi_\alpha K_{j\beta} + \sum_{j,\,\alpha} H_{\alpha i,\,j}\varphi_\alpha H_{j\beta}. \tag{3.7}$$

Since the pair H, K is Hamiltonian, it follows that [H, K] = 0; hence

$$\sum_{j,\,\alpha} H_{i\beta,\,j}K_{j\alpha}\psi_\alpha + \sum_{j,\,\alpha} H_{\alpha i,\,j}K_{j\beta}\psi_\alpha + \sum_{j,\,\alpha} H_{\beta\alpha,\,j}K_{ji}\psi_\alpha = 0$$

or according to (3.4)

$$\sum_{j,\,\alpha} H_{i\beta,\,j}H_{j\alpha}\varphi_\alpha = \sum_{j,\,\alpha} H_{i\alpha,\,j}\psi_\alpha K_{j\beta} - \sum_{j,\,\alpha} H_{\beta\alpha,\,j}\psi_\alpha K_{ji}. \tag{3.8}$$

From the condition [H, H] = 0 we have

$$\sum_\alpha \Big(\sum_j (H_{i\beta,\,j}H_{j\alpha} + H_{\alpha i,\,j}H_{j\beta} + H_{\beta\alpha,\,j}H_{ji})\Big)\varphi_\alpha = 0. \tag{3.9}$$

Comparing (3.8) and (3.9), we obtain

$$\sum_{j,\,\alpha} [(H_{i\alpha,\,j}\psi_\alpha K_{j\beta} - H_{\beta\alpha,\,j}\psi_\alpha K_{ji}) + (H_{\alpha i,\,j}\varphi_\alpha H_{j\beta} - H_{\alpha\beta,\,j}\varphi_\alpha H_{ji})] = 0. \tag{3.10}$$

From (3.7) and (3.10) we obtain finally

---

*We point out that in formula (3.6) the 2-forms $d\varphi$, $d\psi$, $d\chi$ are understood as operators acting from $\mathfrak{A}$ to Ω^1; the operator $d\varphi$ according to this agreement acts as follows: $(d\varphi(h_1), h_2) = d\varphi(h_1, h_2)$, $h_1, h_2 \in \mathfrak{A}$.

$$\sum_{j,\alpha} K_{i\alpha}(\chi_{\alpha,j}-\chi_{j,\alpha})K_{j\beta}=\sum_{j,\alpha} H_{i\alpha}(\psi_{\alpha,j}-\psi_{j,\alpha})K_{j\beta}+\sum_{j,\alpha} K_{i\alpha}(\psi_{\alpha,j}-\psi_{j,\alpha})H_{j\beta}-\sum_{j,\alpha} H_{i\alpha}(\varphi_{\alpha,j}-\varphi_{j,\alpha})H_{j\beta}.$$

This is the coordinate representation of formula (3.6).

We now proceed to the construction of Hamiltonian systems on a manifold X which are connected with a given Hamiltonian pair H, K.

THEOREM 3.4. Suppose that the first cohomology group of the manifold X with constant coefficients is trivial. Let $H, K: \Omega^1 \to \mathfrak{A}$ be a Hamiltonian pair of operators, and let f_0, f_1 be functions on X satisfying the relation $Kdf_1 = Hdf_0$. Then

1) there exists a sequence $f_0, f_1, f_2, \ldots$ of functions on X satisfying the system of equalities

$$Kdf_{i+1}=Hdf_i, \quad i=0,1,2,\ldots; \tag{3.11}$$

b) for fixed j on X there is the Hamiltonian dynamical system

$$\frac{dx}{dt}=Kdf_j, \tag{3.12}$$

and all the f_i, $i = 0, 1, 2, \ldots$ are integrals of it*;

c) the integrals f_i of the system (3.12) are in involution both with respect to the Poisson bracket $\{\ ,\ \}_H$ and with respect to $\{\ ,\ \}_K$.

Proof. We construct the sequence $f_2, f_3, \ldots$ inductively. Suppose that functions $f_2, f_3, \ldots, f_n$ satisfying relations (3.11) have already been constructed. We construct the 1-forms $\varphi = df_{n-1}$, $\psi = df_n$, $\chi = K^{-1}H\psi$. Since $d\varphi = d\psi = 0$, it follows from Theorem 3.3 that $d\chi = 0$. Since the first cohomology group is trivial, there exists a function (we take this to be f_{n+1}) such that $\chi = df_{n+1}$. Continuing the construction, it is now possible to define f_{n+2}, etc.

We shall show that the system $f_0, f_1, f_2, \ldots$ is involutive. For any pair of indices i, j, $i > j$, we have

$$\{f_i,f_j\}_K=(Kdf_i,\ df_j)=-(df_{i-1},\ Hdf_j)=-(df_{i-1},\ Kdf_{j+1})=\{f_{i-1},f_{j+1}\}_K.$$

Continuing this chain of equalities, we arrive at either $\{f_s, f_s\}_K = 0$ or $\{f_{s+1}, f_s\}_K = (Kdf_{s+1}, df_s) = (Hdf_s, df_s) = 0$ depending on the parity of $i - j$; this is what is required. A similar argument goes through for the bracket $\{,\}_H$.

Finally, we shall show that all the f_i are integrals of system (3.12). This assertion follows from the equalities

$$\frac{d}{dt}(f_i(x(t)))=\left(df_i,\ \frac{dx}{dt}\right)=(df_i,\ Kdf_j)=-\{f_i,f_j\}_K=0.$$

Remark. Another system of integrals of Eq. (3.12) is the system of eigenvalues of the operator $A = KH^{-1}$.

4. Regular Operators

Suppose there are two operators $H, K: \Omega^1 \to \mathfrak{A}$. If operator K is Hamiltonian, then on X there is the symplectic structure

$$\langle h_1,h_2\rangle=\omega_K(h_1,h_2)=(h_1,\ K^{-1}h_2),\quad h_i\in\mathfrak{A}. \tag{4.1}$$

Conversely, if there is a symplectic structure $\langle , \rangle$ on X, then it is possible to recover K by this formula.

We shall now assume that the symplectic structure $\langle , \rangle$ is fixed and express in terms of this structure the fact that the operator H forms together with K a Hamiltonian pair. We first prove the following result.

Proposition 4.1. The operator H forms with the operator K a Hamiltonian pair if and only if the 2-forms $\omega_H(h_1, h_2) = (h_1, H^{-1}h_2)$ and $\omega_{K,H}(h_1, h_2) = (h_1, K^{-1}HK^{-1}h_2)$ are simultaneously closed.

Proof. Suppose that the operators H, K form a Hamiltonian pair. By definition, this means that for any λ the following 2-form is closed:

$$\eta_\lambda(h_1,h_2)=(h_1,\ (\lambda H+K)^{-1}h_2). \tag{4.2}$$

*In symplectic (for K) coordinates $(p_1, \ldots, p_n, q_1, \ldots, q_n)$ the system (3.12) can be represented in the standard fashion: $dp_i/dt = -\partial f_j/\partial q_i$, $dq_i/dt = \partial f_j/\partial p_i$.

Expanding this in powers of λ, we obtain

$$\eta_\lambda(h_1, h_2) = (h_1, K^{-1}h_2) - \lambda (h_1, K^{-1}HK^{-1}h_2) + \dots,$$

whence it follows, in particular, that $\omega_{K,H}$ is closed.

We prove the converse. Suppose that the forms ω_H and $\omega_{K,H}$ are closed. We choose coordinates in which K has the form

$$K = \begin{pmatrix} 0 & -E \\ E & 0 \end{pmatrix}.$$

In these coordinates $(K^{-1})_{ij} = -J_{ji}$, and therefore $(K^{-1}HK^{-1})_{ik} = \sum_{\beta,\gamma} K_{i\beta}H_{\beta\gamma}K_{\gamma k}$. The condition that the form $\omega_{K,H}$ be closed is

$$\sum_{\beta,\gamma} (K_{i\beta}H_{\beta\gamma,j}K_{\gamma k} + K_{j\beta}H_{\beta\gamma,k}K_{\gamma i} + K_{k\beta}H_{\beta\gamma,i}K_{\gamma j}) = 0.$$

Multiplying the equality obtained by $K_{i\lambda}K_{k\mu}K_{j\nu}$ and summing on i, k, j, we obtain (using the fact that $-\sum_j K_{ij}K_{jl} = \delta_{i,l}$)

$$\sum_j (H_{\lambda\mu,j}K_{\nu j} + H_{\nu\lambda,j}K_{\mu j} + H_{\mu\nu,j}K_{\lambda j}) = 0.$$

We have thus obtained the relation $[K, H] = 0$. The relation $[H, H] = 0$ follows from the fact that ω_H is closed. The operators H and K thus form a Hamiltonian pair. The proof of the proposition is complete.

We observe that the form ω_H can be given the form

$$\omega_H (h_1, h_2) = (h_1, H^{-1}h_2) = \langle h_1, Ah_2\rangle. \quad (4.3)$$

Here $\langle\ ,\ \rangle$ is the symplectic structure, and $A = KH^{-1}$ is an operator acting in the space $\mathfrak{A}$ of vector fields. It is obvious that A is self-adjoint with respect to $\langle\cdot\rangle$, i.e.,

$$\langle Ah_1, h_2\rangle = \langle h_1, Ah_2\rangle, \quad h_1, h_2 \in \mathfrak{A}.$$

We can represent the form $\omega_{K,H}$

$$\omega_{K,H} (h_1, h_2) = (h_1, K^{-1}HK^{-1}h_2) = \langle h_1, A^{-1}h_2\rangle. \quad (4.4)$$

Proposition 4.1 can be now formulated as follows: the operator H forms with the operator K a Hamiltonian pair if and only if the forms $\langle h_1, Ah_2\rangle$ and $\langle h_1, A^{-1}h_2\rangle$ are simultaneously closed.

Definition. We shall say that a regular operator is given on the manifold X if for any $x \in X$ there is given an invertible operator $A(x)\colon T_XX \to T_XX$ which is self-adjoint with respect to the existing symplectic structure and the forms $\langle h_1, Ah_2\rangle$ and $\langle h_1, A^{-1}h_2\rangle$ are closed.

Any Hamiltonian pair $H, K\colon \Omega^1 \to \mathfrak{A}$ defines a regular operator $A = KH^{-1}$, and, conversely, if there is a regular operator A it is possible to recover a Hamiltonian pair: it is necessary to take as K the operator corresponding to the symplectic structure according to (4.1) and then set $H = A^{-1}K$ (H is hereby skew-symmetric).

THEOREM 4.2. Let A be a regular operator on X. Then for any vector fields $h_1, h_2 \in \mathfrak{A}$ there is the formula†

$$[Ah_1, Ah_2] - A\,[Ah_1, h_2] - A\,[h_1, Ah_2] + A^2[h_1, h_2] = 0. \quad (4.5)$$

Proof. On the basis of the operator A we construct the Hamiltonian pair $H, K\colon \Omega^1 \to \mathfrak{A}$, $A = KH^{-1}$. The adjoint operator A* to A acts in the space of forms: $A^* = H^{-1}K$. We rewrite formula (3.6) in the following way:

$$d\chi (K\xi_1, K\xi_2) - d\psi (K\xi_1, H\xi_2) - d\psi (H\xi_1, K\xi_2) + d\varphi (H\xi_1, H\xi_2) = 0.$$

We set $h_i = H\xi_i$. Since $\psi = (A^*)^{-1}\varphi$, $\chi = (A^*)^{-2}\varphi$, we have

$$d (A^{*-2}\varphi) (Ah_1, Ah_2) - d (A^{*-1}\varphi) (Ah_1, h_2) - d (A^{*-1}\varphi) (h_1, Ah_2) + d\varphi (h_1, h_2) = 0$$

or, using the definition of d,

† This condition means that A is constant in some system of coordinates.

$$\varphi(A^{-2}[Ah_1, Ah_2]) - \varphi(A^{-1}[Ah_1, h_2]) - \varphi(A^{-1}[h_1, Ah_2]) + \varphi([h_1, h_2]) = 0.$$

Since φ is an arbitrary 1-form, this implies (4.5). The proof of the theorem is complete.

It turns out that a function of a regular operator is also a regular operator under certain conditions. The corresponding result is the following.

THEOREM 4.3. Let A be a regular operator on X. We suppose that the closure of the set $\Lambda(A) = \bigcup_{x \in X} \{\lambda \in C : \det(\lambda E - A(x)) = 0\}$ is compact in the complex plane. Let $\Phi(\lambda)$ be a function which is analytic and does not vanish in a neighborhood of $\Lambda(A)$. Then $\Phi(A) = \{\Phi(A(x)) : T_X X \to T_X X\}$ is a regular operator.

Proof. We choose an open set containing $\Lambda(A)$ in which the function $\Phi(\lambda)$ is analytic and has no zeros and which contains a contour Γ enclosing $\Lambda(A)$. We consider the following family of operators depending on the parameter $\lambda \in \Gamma$:

$$R(\lambda) = (\lambda E - A)^{-1}.$$

We shall show that all the $R(\lambda)$, $\lambda \in \Gamma$ are regular. Since A is a regular operator, it follows that A^{-1} is also regular. We construct the Hamiltonian pair $H, K: \Omega^1 \to \mathfrak{A}$ corresponding to the operator A^{-1} (we recall that $A^{-1} = KH^{-1}$). The operators K, $H_1 = \lambda K - H$ also form a Hamiltonian pair. The following operator is therefore regular:

$$KH_1^{-1} = K(\lambda K - H)^{-1} = (\lambda E - HK^{-1})^{-1} = (\lambda E - A)^{-1} = R(\lambda).$$

By the formulas of the operational calculus

$$\Phi(A) = \frac{1}{2\pi i} \oint_\Gamma \Phi(\lambda) R(\lambda) d\lambda, \tag{4.6}$$

$$(\Phi(A))^{-1} = \frac{1}{2\pi i} \oint_\Gamma (\Phi(\lambda))^{-1} R(\lambda) d\lambda. \tag{4.7}$$

Since the form $\langle h_1, R(\lambda)h_2\rangle$ is closed for all $\lambda \in \Gamma$, it follows from (4.6) and (4.7) that the forms $\langle h_1, \Phi(A)h_2\rangle$ and $\langle h_1, (\Phi(A))^{-1}h_2\rangle$ are also closed. Hence the operator $\Phi(A)$ is regular, and the theorem is proved.

5. Conditions for the Hamiltonian Property in the Formal Calculus of Variations

In order to indicate critiera that an operator be Hamiltonian analogous to the finite-dimensional formula (3.2), we must obviously reformulate the condition that the form ω_H defined by formula (1.4) be closed directly in terms of the form generating the operator H.

We suppose to begin that the operator H is a differential operator, i.e., it is defined by formulas (2.6), (2.7). For any $h \in A$ and any $\alpha, \beta, \gamma \in I$ we construct the following differential operator:

$$D^{(\beta)}_{H_{\alpha\gamma}} h = \sum_{i,j} h^{(i)} \frac{\partial a_{\alpha\gamma i}}{\partial u^{(j)}_\beta} \left(\frac{d}{dx}\right)^j.$$

For any form $\xi \in \Omega^1$ we further construct the operator $(D_H\xi): \bar{A} \to \bar{A}$ by giving its matrix

$$(D_H\xi)_{\alpha\beta} = \sum_\gamma D^{(\beta)}_{H_{\alpha\gamma}} \xi_\gamma. \tag{5.1}$$

We shall call a q-tensor (in contrast to q-forms defined earlier) a q-linear operation on Ω^1 with values in $\tilde{A}$.

Let $H, K: \Omega^1 \to \bar{A}$ be two differential operators.

Definition. The Schouten–Nijenhuis bracket of operators H and K is the three-tensor [H, K] the value of which on 1-forms ξ_1, ξ_2, ξ_3 is

$$[H, K](\xi_1, \xi_2, \xi_3) = ((D_H\xi_1) K\xi_2, \xi_3) + ((D_K\xi_1) H\xi_2, \xi_3) + (\text{cycle}). \tag{5.2}$$

THEOREM 5.1. In order that a skew-symmetric operator H: $\Omega^1 \to \bar{A}$ be Hamiltonian it is necessary and sufficient that*

$$[H, H] = 0. \tag{5.3}$$

Remark 1. The theorem asserts, in particular, that if condition (5.3) is satisfied the image of the operator H in $\bar{A}$ is closed with respect to the commutator (2.3), i.e., for any $\xi, \eta \in \Omega^1$ there is a 1-form $\zeta \in \Omega^1$ such that $[H\xi, H\eta] = H\zeta$. Without presenting the proof of the theorem here, we indicate the explicit expression for ζ:

$$\zeta = \sum_{i,\beta}\left(\frac{\partial\eta}{\partial u_\beta^{(i)}}(H\xi)_\beta^{(i)} - \frac{\partial\xi}{\partial u_\beta^{(i)}}(H\eta)_\beta^{(i)}\right) + (D_H\xi)^*\eta. \tag{5.4}$$

Remark 2. It is evident from condition (5.3) that any skew-symmetric operator with constant coefficients is Hamiltonian. This result was established in [4].

COROLLARY 5.2. In order that skew-symmetric operators H, K: $\Omega^1 \to \bar{A}$ form a Hamiltonian pair, it is necessary and sufficient that the following set of conditions be satisfied:

$$[H, H] = 0, \quad [H, K] = 0, \quad [K, K] = 0. \tag{5.5}$$

We have introduced the Hamiltonian concept in a general fashion not only for differential operators. It is found that condition (5.3) as well is meaningful for arbitrary operators, and Theorem 5.1 holds. For this we require the concept of the Fréchet differential (for vector fields, 1-forms, and also for operators).

The value of the Fréchet differential of the vector field $\bar{h} \in \bar{A}$ on a vector field $\bar{g} \in \bar{A}$ is the vector field $\bar{h}'(g)$ defined by the formula

$$\bar{h}'(\bar{g}) = \sum_{i,\beta}\frac{\partial\bar{h}}{\partial u_\beta^{(i)}} g_\beta^{(i)}.$$

The informal meaning of this concept is clear: $\bar{h}'(\bar{g})$ is the principal linear part of the increment in $\bar{h}$ when the functions $u_\alpha(x)$ are given an increment of $g_\alpha(x)$.

In similar fashion we define the value of the Fréchet differential of a 1-form $\xi \in \Omega^1$ on a vector field $\bar{g} \in \bar{A}$:

$$\xi'(\bar{g}) = \sum_{i,\beta}\frac{\partial\xi}{\partial u_\beta^{(i)}} g_\beta^{(i)}.$$

Suppose further that H: $\Omega^1 \to \bar{A}$ is a linear operator. The value of the Fréchet differential of the operator H on the vector field $\bar{g} \in \bar{A}$ is the linear operator $H'(\bar{g})$ with value on an arbitrary element $\xi \in \Omega^1$ given by

$$(H'(\bar{g}))(\xi) = (H\xi)'(\bar{g}) - H(\xi'(\bar{g})).$$

Clarification: usually the differential of an operator accounts with the dependence of H on $u_\alpha^{(i)}$. In the sense of our constructions, however, the element ξ to which the operator is applied also depends on $u_\alpha^{(i)}$; the definition accounts with this.

We introduce one further piece of notation. For fixed $\xi \in \Omega^1$ the vector field $(H'(\bar{g}))(\xi)$ depends linearly on $\bar{g}$ and thus defines a linear mapping acting in $\bar{A}$. We shall denote this mapping by $D_H\xi$. Thus $D_H\xi: \bar{A} \to \bar{A}$,

$$(D_H\xi)\,\bar{g} = (H'(\bar{g}))(\xi).$$

It can be verified that this notation is consistent with formula (5.1), i.e., for differential operators H, $D_H\xi$ can be computed using this formula.

We have thus given meaning to the Schouten–Nijenhuis bracket in the case of arbitrary operators and not only in the case of differential operators. Theorem 5.1 hereby remains valid. Formula (5.4) also holds. We can write it more compactly with the help of Fréchet differentials:

*If in the finite-dimensional theory the operator $D_H\xi$ is introduced so that $(D_H\xi)_{\alpha\beta} = \sum_\gamma H_{\alpha\gamma,\beta}\xi_\gamma$, then formula (5.2) gives an invariant expression for the finite-dimensional Schouten–Nijenhuis bracket (3.1). Theorem 5.1 is thus an infinite-dimensional analogue of Theorem 3.1.

$$\zeta = \eta'(H\xi) - \xi'(H\eta) + (D_H\xi)^*\eta. \tag{5.4'}$$

In computing it is convenient to use various forms of condition (5.3). We shall here present three conditions equivalent to (5.3) (we recall that H is a skew-symmetric operator; this is used in proving equivalence). Thus, the following conditions are equivalent to one another and to condition (5.3):

$$(D_H\xi_1) H\xi_2 - (D_H\xi_2) H\xi_1 = H(D_H\xi_2)^*\xi_1 \quad \text{for all} \quad \xi_1, \xi_2 \in \Omega^1, \tag{5.6}$$

$$(D_H\xi_1) H\xi_2 + \frac{1}{2} H(D_H\xi_1)^*\xi_2 \text{ symmetrically in } \xi_1, \xi_2 \in \Omega^1, \quad H = K - K^*, \tag{5.7}$$

where K satisfies the relation

$$\sum_{\sigma\in S_3} \operatorname{sgn}\sigma\,((D_K\xi_{\sigma(1)}) K\xi_{\sigma(2)} - K(D_K\xi_{\sigma(2)})^*\xi_{\sigma(1)}, \xi_{\sigma(3)}) = 0 \tag{5.8}$$

for all $\xi_1, \xi_2, \xi_3 \in \Omega^1$ (here S_3 is the group of permutations).

Using the form (5.7) of the Hamiltonian condition, it is possible to write this condition as a system of equations for the coefficients $a_{\alpha\beta i}$ of the differential operator H. To simplify notation we introduce for any $h_1, h_2 \in A$ and any $\lambda, \alpha, \gamma \in I$ the following notation:

$$T_{\lambda\alpha\gamma}(h_1, h_2) = \sum_\beta ((D^{(\beta)}_{H_{\lambda\gamma}} h_1) H_{\beta\alpha} h_2 + \frac{1}{2} H_{\lambda\beta}(D^{(\beta)}_{H_{\alpha\gamma}} h_1)^* h_2).$$

Condition (5.7) then means that for any $h_1, h_2 \in A$ and any $\lambda, \alpha, \gamma \in I$

$$T_{\lambda\alpha\gamma}(h_1, h_2) = T_{\lambda\gamma\alpha}(h_2, h_1). \tag{5.9}$$

According to our construction,

$$T_{\lambda\alpha\gamma}(h_1, h_2) = \sum_{i,j} t_{\lambda\alpha\gamma ij} h_1^{(i)} h_2^{(j)},$$

where each coefficient $t_{\lambda\alpha\gamma ij} \in A$ can be explicitly expressed in terms of the $a_{\alpha\beta i}$. This implies that the desired system of equations for the coefficients $a_{\alpha\beta i}$ of the operator H which is equivalent to the Hamiltonian property has the following form:

$$t_{\lambda\alpha\gamma ij} = t_{\lambda\gamma\alpha ji}. \tag{5.10}$$

In some special cases it is possible to solve system (5.10) in general form or at least find particular solutions of it. These special cases are described in the next section.

6. Examples of Hamiltonian Operators in the Formal Calculus of Variations

6.1. We shall formulate conditions for the Hamiltonian property for an operator H with

$$H_{ij} = \sum_k a_{ijk} u_k^{(0)}, \quad a_{ijk} \in \mathbf{C}, \quad a_{ijk} + a_{jik} = 0. \tag{6.1}$$

We shall interpret the constants a_{ijk} as a composition rule $\times$ in the linear space $L = L(e_0, e_1, \ldots)$ with basis $e_0, e_1, \ldots$:

$$e_i \times e_j = \sum_k a_{ijk} e_k.$$

System (5.10) is equivalent to the following conditions on the operation $\times$:

$$a \times b = -b \times a, \quad (a \times b) \times c + (b \times c) \times a + (c \times a) \times b = 0.$$

Thus, prescribing a Hamiltonian operator of form (6.1) means prescribing on L the structure of a Lie algebra.

6.2. We shall formulate conditions for the Hamiltonian property of an operator H with

$$H_{ij} = \sum_k \left(c_{ijk} u_k^{(1)} + d_{ijk} u_k^{(0)} \frac{d}{dx}\right), \quad c_{ijk} \in \mathbf{C}, \quad d_{ijk} = c_{ijk} + c_{jik}. \tag{6.2}$$

As in the preceding case, we shall interpret the collection c_{ijk} as a composition rule $\circ$ in $L = L(e_0, e_1, \ldots)$. System (5.10) gives the following conditions on the operation $\circ$:

$$(a \circ b) \circ c = (a \circ c) \circ b, \quad (a \circ b) \circ c + c \circ (a \circ b) = (c \circ b) \circ a + a \circ (c \circ b). \tag{6.3}$$

There is thus a one-to-one correspondence between Hamiltonian operators of the form (6.2) and algebras with the axioms (6.3).

The following remarks are due to S. I. Gel'fand.

Remark 1. Any algebra with the axioms (6.3) is a Lie algebra relative to the operation $a \times b = a \circ b - b \circ a$.

Remark 2. It is possible to construct algebras with the property (6.3) according to the following recipe: take an associative commutative algebra with a fixed differentiation ∂ and introduce a new operation $\circ$ according to the formula $a \circ b = a\, \partial b$.

Examples. In the algebra Q[x] of polynomials in x we prescribe a differentiation by the condition

$$\partial x = \sum_{s=0}^{n} \lambda_s x^s, \quad \lambda_0, \ldots, \lambda_n \in \mathbb{C}.$$

We construct the operation $\circ$ according to the recipe of S. I. Gel'fand; it gives a collection of constants $c_{ijk} = j\lambda_{k-i-j+1}$ (in particular, the standard differentiation $\partial x = 1$, gives $c_{ijk} = j\delta_{i+j-1,k}$). A Hamiltonian operator corresponding to the standard differentiation has the following form:

$$H_{ij} = ju^{(1)}_{i+j-1} + (i + j)\, u^{(0)}_{i+j-1} \frac{d}{dx}.$$

This operator first arose in [5] in connection with the system of Benney equations; we have thus obtained still another proof of the Hamiltonian property of this operator. Hamiltonian operators corresponding to an arbitrary collection $\lambda_0, \ldots, \lambda_n$ have so far apparently not been indicated anywhere.

6.3. The following example generalizes the considerations of parts 6.1 and 6.2:

$$H_{ij} = \sum_k \left(a_{ijk} u^{(0)}_k + c_{ijk} u^{(1)}_k + d_{ijk} u^{(0)}_k \frac{d}{dx}\right), \quad a_{ijk} = -a_{jik}, \quad d_{ijk} = c_{ijk} + c_{jik}. \tag{6.4}$$

As before, we shall interpret the collections of constants a_{ijk} and c_{ijk} as two rules of composition $\times$ and $\circ$ in $L = L(e_0, e_1, \ldots)$. System (5.10) gives the following conditions on $\times$ and $\circ$:

a) the operation $\times$ gives a Lie algebra structure on L;

b) the operation $\circ$ satisfies conditions (6.3);

c) the operations $\times$ and $\circ$ are related as follows:

$$(c \circ a) \times b - (c \circ b) \times a + (c \times a) \circ b - (c \times b) \circ a - c \circ (a \times b) = 0. \tag{6.5}$$

As in the preceding parts, the question arises of the existence of such a pair of operations.

Remark. If in ana algebra with axioms (6.3) the operation $\times$ is introduced by the rule $a \times b = a \circ b - b \circ a$, then it is guaranteed that condition (6.5) is satisfied. Thus, if the constants c_{ijk} define an operation satisfying (6.3), then the operator

$$H_{ij} = \sum_k \left((c_{ijk} - c_{jik})\, u^{(0)}_k + c_{ijk} u^{(1)}_k + (c_{ijk} + c_{jik})\, u^{(0)}_k \frac{d}{dx}\right)$$

is Hamiltonian.

6.4. We consider the operator $H = K - K^*$, where K is defined by

$$K_{ij} = \sum_{k,l} c_{ijkl} u^{(0)}_k \left(\frac{d}{dx}\right)^l, \quad c_{ijkl} \in \mathbb{C}. \tag{6.6}$$

As in the preceding cases, we introduce in $L = L(e_0, e_1, \ldots)$ the multiplication rule (depending, however, on the numerical parameter λ):

$$e_i \overset{\lambda}{\circ} e_j = \sum_{k,l} c_{ijkl} e_k \lambda^l .$$

Using the form (5.8) of the Hamiltonian condition, it is possible to show that in order that the operator H be Hamiltonian it suffices that for any i_1, i_2, i_3

$$\sum_{\sigma \in S_3} \operatorname{sgn} \sigma \left[\left(e_{i_{\sigma(3)}} \overset{\lambda_{\sigma(1)}}{\circ} e_{i_{\sigma(1)}}\right) \overset{\lambda_{\sigma(2)}}{\circ} e_{i_{\sigma(2)}} - e_{i_{\sigma(3)}} \overset{\lambda_{\sigma(1)}+\lambda_{\sigma(2)}}{\circ} \left(e_{i_{\sigma(1)}} \overset{\lambda_{\sigma(2)}}{\circ} e_{i_{\sigma(2)}}\right)\right] = 0$$

hold identically in $\lambda_1, \lambda_2, \lambda_3 \in C$. For this it suffices in turn that the condition

$$\left(e_{i_1} \overset{\lambda_1}{\circ} e_{i_2}\right) \overset{\lambda_2}{\circ} e_{i_3} = e_{i_1} \overset{\lambda_1+\lambda_2}{\circ} \left(e_{i_2} \overset{\lambda_2}{\circ} e_{i_3}\right) \tag{6.7}$$

be satisfied identically in λ_1, λ_2 for any i_1, i_2, i_3. We set $e_\xi = \sum_i e_i \xi^i$. Condition (6.7) is then equivalent to the condition that the equality

$$\left(e_\xi \overset{\lambda_1}{\circ} e_\eta\right) \overset{\lambda_2}{\circ} e_\zeta = e_\xi \overset{\lambda_1+\lambda_2}{\circ} \left(e_\eta \overset{\lambda_2}{\circ} e_\zeta\right) \tag{6.8}$$

hold identically in $\lambda_1, \lambda_2, \xi, \eta, \zeta \in C$.

Example. The operator introduced in [3] has the form (6.6) with the choice of constants $c_{ijkl} = \binom{l+i}{i} \delta_{i+j+l+1,\,k}$. We shall write out the operation $\overset{\lambda}{\circ}$ corresponding to this choice of constants. It is not hard to verify that it has the following simple form:

$$e_\xi \overset{\lambda}{\circ} e_\eta = \frac{1}{\lambda + \xi - \eta} (e_{\lambda+\xi} - e_\eta).$$

From this formula it follows easily that condition (6.8) is satisfied. We thus obtain the assertion that the operator is Hamiltonian (it was first proved in the work [3], and a shorter proof was then given in [6–8]).

6.5. We now consider operators with coefficients depending on a function u and its derivatives. Let

$$H = \sum_{i=0}^{n} a_i \left(\frac{d}{dx}\right)^i$$

be a differential operator of order n with coefficients a_i belonging to an algebra A of smooth functions of u, u', u", ... H may be considered a special case of an operator of the form (2.6)–(2.7) when the set of indices I consists of a single element.

By solving the system (5.10) for $a_0, \ldots, a_n$, it is possible in principle to describe all Hamiltonian operators of the given order n (only odd n are considered due to the skew-symmetry). However, the computations become complicated with increasing n. We have carried them through for n = 1, n = 3. We present the result.

THEOREM 6.1. a) The general form of a differential operator of first order which is Hamiltonian is

$$H = g \circ \frac{d}{dx} \circ g,$$

where

$$g = g(u, u', u'') = \left(v_s\left(u, \frac{u'^2}{2}\right) + v_t\left(u, \frac{u'^2}{2}\right) u''\right)^{-1},$$

and v = v(s, t) is an arbitrary function of two variables.

b) The general form of a differential operator of third order which is Hamiltonian is

$$H = h(u) \circ \left[\sqrt{C_1 + C_2 \int_0^u \frac{d\xi}{h(\xi)}} \circ \frac{d}{dx} \circ \sqrt{C_1 + C_2 \int_0^u \frac{d\xi}{h(\xi)}} + \left(\frac{d}{dx}\right)^3\right] \circ h(u),$$

where $h(\xi)$ is an arbitrary function of a single variable and C_1 and C_2 are arbitrary constants.

7. Hamiltonian Pairs and the Lenhart Scheme

We now formulate a result which is an analogue of Theorem 3.4 (the modifications in the formulation are occasioned by the fact that even in simple cases the operators of the formal calculus of variations are not invertible).

THEOREM 7.1. Let H, K: $\Omega^1 \to \bar{A}$ be a Hamiltonian pair of operators and suppose that the condition KTK = 0, T: $\bar{A} \to \Omega^1$ implies that T = 0. Suppose further that there is a sequence $\xi_0, \xi_1, \ldots \in \Omega^1$ satisfying the system of relations

$$K\xi_{i+1} = H\xi_i, \tag{7.1}$$

and $\xi_0, \xi_1 \in (\delta/\delta\bar{u})\,\bar{A}$. Then

a) there exists a sequence of functions $\tilde{f}_0, \tilde{f}_1, \ldots \in \tilde{A}$ such that

$$\xi_i = \frac{\delta}{\delta\bar{u}}\tilde{f}_i; \tag{7.2}$$

b) for fixed j the functions $\tilde{f}_j$, i = 0, 1, ..., are integrals (conservation laws) of the system

$$\frac{\partial\bar{u}}{\partial t} = K\frac{\delta}{\delta\bar{u}}\tilde{f}_j \tag{7.3}$$

(here $\bar{u} = \{u_\alpha\} \in \bar{A}$);

c) the integrals $\tilde{f}_i$ are in involution both with respect to the Poisson bracekt $\{\ ,\ \}_H$ and with respect to $\{\ ,\ \}_K$.

We shall not present the proof of the theorem here. It can be carried out in analogy with the proof of Theorem 3.4 using the Fréchet differentials introduced in Sec. 5 in place of partial derivatives (for the case K = d/dx a similar argument is given in [11]).

Formulas (7.1)-(7.2) can be written in the form

$$K\frac{\delta}{\delta\bar{u}}\tilde{f}_{i+1} = H\frac{\delta}{\delta\bar{u}}\tilde{f}_i \tag{7.4}$$

and (7.4) may be interpreted as a recurrence procedure for finding $\tilde{f}_i$. This procedure is often called the Lenhart scheme; the authors learned of it from a report of P. Lax in Moscow in 1976 (see [10]) where the first example of a Lenhart scheme was described. This example is the following:

$$K = \frac{d}{dx},\quad H = \left(\frac{d}{dx}\right)^3 + \frac{2}{3}u\frac{d}{dx} + \frac{1}{3}u',\quad \tilde{f}_0 = 3\bar{u}.$$

Equations (7.3) which arise are the higher Korteweg–de Vries equations, and the $\tilde{f}_i$ are their integrals in involution. Versions of the Lenhart scheme and some examples (in the nonformal theory) are given in [12-15]; the role of Hamiltonian pairs was apparently first indicated in the work of Magri [14]. A formalization of the Lax method of investigating the Korteweg–de Vries equations is given in [11].

The following example of a Hamiltonian pair is investigated in [16]:

$$K = 2(C_1 + C_2u)\frac{d}{dx} + C_2u',\qquad H = \left(\frac{d}{dx}\right)^3 + 2(D_1 + D_2u)\frac{d}{dx} + D_2u'$$

(here C_i, D_i are constants). The Lenhart scheme with $f_0 = (C_1 + C_2u)^{1/2}$ gives equations which are, in a certain sense, intermediate between the Korteweg–de Vries equation and the H. Dym equation $u_t = (u^{-1/2})_{xxx}$. They all have an infinite collection of conservation laws and higher analogues.

An important example of a Hamiltonian pair was constructed in the work [8] by considering the resolvent of the operator $\sum_{k=0}^{n} u_k (d/dx)^k$. This pair consists of an operator K with coefficients depending linearly on $u_\alpha^{(i)}$ (we have considered just this operator in an example in Sec. 6.4 of the present paper) and another operator H of more complex nature with coefficients depending on $u_\alpha^{(i)}$ quadratically.

LITERATURE CITED

1. I. M. Gel'fand and L. A. Dikii, "Asymptotics of the resolvent of Sturm–Liouville equations and the algebra of Korteweg–de Vries equations," Usp. Mat. Nauk, 30, No. 5, 67-100 (1975).
2. I. M. Gel'fand and L. A. Dikii, "The structure of a Lie algebra in the formal calculus of variations," Funkts. Anal. Prilozhen., 10, No. 1, 18-25 (1976).

3. I. M. Gel'fand and L. A. Dikii, "Fractional powers of operators and Hamiltonian systems," Funkts. Anal. Prilozhen., 10, No. 4, 13-29 (1976).
4. I. M. Gel'fand, Yu. I. Manin, and M. A. Shubin, "Poisson brackets and the kernel of the variational derivatives in the formal calculus of variations," Funks. Anal. Prilozhen., 10, No. 4, 30-34 (1976).
5. B. A. Kuperschmidt and Yu. I. Manin, "Equations of long waves with a free surface. I. Conservation laws and solutions," Funks. Anal. Prilozhen., 11, No. 3 (1977); 31-42; "II. Hamiltonian structure and higher equations," Funks. Anal. Prilozhen., 12, No. 1, 25-32 (1978).
6. M. Adler, "On a trace functional for formal pseudodifferential operators and the symplectic structure of the Korteweg–de Vries equation," Invent. Math., 50, No. 3, 219-248 (1979).
7. D. R. Lebedev and Yu. I. Manin, "A Hamiltonian operator of Gel'fand–Dikii and the coadjoint representation of the Volterra group," Preprint ITÉF, No. 155 (1978).
8. I. M. Gel'fand and L. A. Dikii, "A family of Hamiltonian structures connected with integrable, nonlinear differential equations," Preprint Inst. Prikl. Mat. Akad. Nauk SSSR. No. 136 (1978).
9. M. Flato, A. Lichnerowicz, and D. Sternheimer, "Algebras de Lie attachées a une variété canonique," Preprint (1975).
10. P. D. Lax, "Almost periodic solutions of the KdV equation," SIAM Rev., 18, No. 3, 351-375 (1976).
11. I. Ya. Dorfman, "On the formal variational calculus in an algebra of smooth cylinder functions." Funkts. Anal. Prilozhen., 12, No. 2, 32-39 (1978).
12. P. J. Olver, "Evolution equations possessing infinitely many symmetries," J. Math. Phys., 18, No. 6, 1212-1215 (1977).
13. M. Adler, "Some algebraic relations common to a set of integrable partial and ordinary differential equations," Preprint (1978).
14. F. Magri, "A simple model of the integrable Hamiltonian equation," J. Math. Phys., 19, No. 5, 1156-1162 (1978).
15. P. P. Kulish and A. G. Reiman, "The hierarchy of symplectic forms for the Schrödinger and Dirac equations on the line," Zap. Nauchn. Sem. Leningr. Otd. Mat. Inst., 77, 134-147 (1978).
16. I. M. Gel'fand and I. Ya. Dorfman, "Integral equations of KdV-H. Dym type," in: Modern Problems of Computational Mathematics and Mathematical Physics [in Russian], Moscow (1979).

16.

(with I. Ya. Dorfman)

The Schouten bracket and Hamiltonian operators

Funkts. Anal. Prilozh. **14** (3) (1980) 71–74. MR **82e**:58039. Zbl. **444**:58010
[Funct. Anal. Appl. **14** (1980) 223–226]

1. Main Objects. Let $\mathfrak{A}$ be an arbitrary Lie algebra and M be a left $\mathfrak{A}$-module, i.e., the following relations hold: $a_1a_2\mathrm{m} - a_2a_1\mathrm{m} = [a_1, a_2]\mathrm{m}$. Such a situation arises in a formal variational calculation in the following way: we examine the algebra of polynomials in $u_\alpha^{(i)}$, where α varies over the set I, $i = 0, 1, \ldots$; the operator $d/dx = \sum_{i,\alpha} u_\alpha^{(i+1)}\partial/\partial u_\alpha^{(i)}$ plays in A the role of differentiation with respect to x; as Lie algebra $\mathfrak{A}$ we take the space $\bar{A}$ of sequences $h = \{h_\alpha\}$, $h_\alpha \in A$, equipped with the commutator

$$[h, g]_\alpha = \sum_{i,\beta}\left(h_\beta^{(i)} \frac{\partial g_\alpha}{\partial u_\beta^{(i)}} - g_\beta^{(i)} \frac{\partial h_\alpha}{\partial u_\beta^{(i)}} \right); \tag{1}$$

as the module M we take the space $\tilde{A} = A/(d/dx)A$; the elements of $\tilde{A}$ are called functionals; an action of $\bar{A}$ in $\tilde{A}$ is given by the formula $h\tilde{f} = \int \sum_\alpha (\delta f/\delta u_\alpha)\, h_\alpha\, dx$. The commutator (1), introduced in [1-2], is of great importance in this theory. The other example is the following one: $\mathfrak{A}$ is a Lie algebra of the vector fields on a manifold X, $M = C^\infty(X)$.

Let C^q be the space of q-linear skew-symmetric functions $\omega(a_1, \ldots, a_q)$ on $\mathfrak{A}$ with range in M, $C^0 \equiv M$. The space $\oplus_q C^q$ becomes a complex if we introduce the differential d: $C^q \to C^{q+1}$ in the usual manner (see the formula in [3]); if $m \in M$ is a 0-form this formula gives $dm(a) = am$; if ξ: $\mathfrak{A} \to M$ is a 1-form, then $d\xi(a_1, a_2) = a_1\xi(a_2) - a_2\xi(a_1) - \xi([a_1, a_2])$. For $a \in \mathfrak{A}$ we define the operator i_a: $C^q \to C^{q-1}$ of the inner product on a as follows: $(i_a\omega)(a_1, \ldots, a_{q-1}) = \omega(a, a_1, \ldots, a_{q-1})$, $i_a \equiv 0$ on C^0. We introduce the operator L_a of the Lie derivative in the following way: $L_a = i_a d + d i_a$.

Taking into consideration that, generally speaking, there is no ring structure in M (e.g., it is impossible to multiply functionals) we cannot pick out the local forms using the so-called $\mathcal{F}$-linearity condition; so, we introduce the following additional objects: let a space $\Omega^q \subset C^q$ be fixed in each C^q so that the following axioms hold: 1) $d\Omega^q \subset \Omega^{q+1}$, 2) $i_a\Omega^q \subset \Omega^{q-1}$. Hereafter for $\xi \in \Omega^1$, $a \in \mathfrak{A}$ we denote the value $\xi(a) \in M$ by (ξ, a) or (a, ξ). We will assume that $(\xi, a) = 0$ for every $\xi \in \Omega^1$ implies $a = 0$.

2. Schouten Bracket. Let H: $\Omega^1 \to \mathfrak{A}$ be a linear operator. We call it skew-symmetric if $(H\xi_1, \xi_2) = -(\xi_1, H\xi_2)$ for all $\xi_1, \xi_2 \in \Omega^1$. For a pair of skew-symmetric operators H, K: $\Omega^1 \to \mathfrak{A}$ we construct a new object: their Schouten bracket [H, K]. This is a trilinear mapping from Ω^1 to M defined by the formula

$$[H, K](\xi_1, \xi_2, \xi_3) = (KL_{H\xi_1}\xi_2, \xi_3) + (HL_{K\xi_1}\xi_2, \xi_3) + (\text{cycle}), \tag{2}$$

where the term (cycle) here and throughout this paper indicates the summation over all cyclical permutations of the indices (1, 2, 3). We say that a skew-symmetric operator is called Hamiltonian if and only if [H, H] = 0.*

Proposition 1. If H: $\Omega^1 \to \mathfrak{A}$ is a Hamiltonian operator, then M is a Lie algebra with respect to the bracket $\{m_1, m_2\}_H = (Hdm_1, dm_2)$. The mapping $Hd: M \to \mathfrak{A}$ is a morphism of Lie algebras.

Proof. For arbitrary $m_1, m_2, m_3 \in M$ we have

$$0 = {}^1/_2\,[H, H](dm_1, dm_2, dm_3) = (Hdi_{Hdm_1}dm_2, dm_3) + (\text{cycle}) = (Hd\{m_1, m_2\}_H, dm_3) + (\text{cycle}) = \{\{m_1, m_2\}_H, m_3\}_H + (\text{cycle}).$$

*In Sec. 1 of [3] the definition of the Hamiltonian nature was given in terms of symplectic structure. It is not difficult to show that the definition accepted in this paper is the generalization of the definition of [3]; the difference is that here we do not require Im H to be a subalgebra of the Lie algebra $\mathfrak{A}$. Also, we can verify that the definition of Schouten bracket accepted in [3] [Eq. (5.2)] within the framework of a formal variational calculation is exactly the same as the one introduced here.

Institute of Applied Mathematics, Academy of Sciences of the USSR. Chemical Physics Institute, Academy of Sciences of the USSR. Translated from Funktsional'nyi Analiz i Ego Prilozheniya, Vol. 14, No. 3, pp. 71-74, July-September, 1980. Original article submitted March 19, 1980.

so the Jacobian identity holds. Further,

$$(Hd\{m_1, m_2\}_H, \xi) = (L_{Hdm_2}\xi, Hdm_1) + (L_{H\xi}dm_1, \quad Hdm_2) = ([Hdm_1, \quad Hdm_2], \quad \xi),$$

and since ξ is an arbitrary element of Ω^1 this means that $Hd\{m_1, m_2\}_H = [Hdm_1, Hdm_2]$. The proposition is proved.

Thus, it is natural to call an element $Hdm \in \mathfrak{A}$ a Hamiltonian field with a Hamiltonian $m \in M$.

3. Hamiltonian Pairs. We say that two skew-symmetric operators $H, K: \Omega^1 \to \mathfrak{A}$ form a Hamiltonian pair if any linear combination of them is a Hamiltonian operator or, what is the same, if the following equalities hold: $[H, H] = [H, K] = [K, K] = 0$.

THEOREM 2. Let the operators $H, K: \Omega^1 \to \mathfrak{A}$ form a Hamiltonian pair. Let 1-form $\varphi, \psi, \chi \in \Omega^1$ satisfy the equalities

$$K\psi = H\varphi, \qquad K\chi = H\psi. \tag{3}$$

Then for any $\xi, \eta \in \Omega^1$ the following equation holds:

$$d\chi\,(K\xi, K\eta) = d\psi\,(K\xi, H\eta) + d\psi\,(H\xi, K\eta) - d\varphi\,(H\xi, H\eta). \tag{4}$$

Proof. We have

$$\begin{aligned} d\chi\,(K\xi, K\eta) &= (L_{K\xi}\chi, K\eta) - (K\eta)(\psi, H\xi) = \\ &= -(L_{K\chi}\eta, K\xi) - (L_{K\eta}\xi, K\chi) - (K\eta)(\psi, H\xi) = -(L_{H\psi}\eta, K\xi) - (L_{K\eta}\xi, H\psi) - \\ &- (K\eta)(\psi, H\xi) - (L_{H\eta}\xi, K\psi) + (L_{H\xi}\psi, K\eta) + (L_{K\xi}\psi, H\eta) + (L_{K\psi}\eta, H\xi) - \\ &- (K\eta)(\psi, H\xi) = -(L_{H\xi}\varphi, H\eta) + (L_{H\xi}\psi, K\eta) + (L_{K\xi}\psi, H\eta) - (K\eta)(\psi, H\xi) = \\ &= -d\varphi\,(H\xi, H\eta) - (H\eta)(\varphi, H\xi) + d\psi\,(H\xi, K\eta) + (K\eta)(\psi, H\xi) + d\psi\,(K\xi, H\eta) + \\ &+ (H\eta)(\psi, K\xi) - (K\eta)(\psi, H\xi) = d\psi\,(K\xi, H\eta) + d\psi\,(H\xi, K\eta) - d\varphi\,(H\xi, H\eta). \end{aligned}$$

Here we have used successively the conditions $[K, K] = 0$, $[H, K] = 0$, and $[H, H] = 0$, everywhere taking into account the equalities (3).

The following result is the algebraic representation of the so-called Lenart scheme (see [3, Secs. 3, 7]). Let us introduce the following.

Definition. If $m \in M$, $a \in \mathfrak{A}$, and $am = 0$ then m is called the conservation law for a.

Let $\mathcal{H}^1_K(\mathfrak{A}, M, \Omega^1)$ denote the factor-space $\{\chi \in \Omega^1 : d\chi\,(K\xi, K\eta) = 0 \ \forall \xi, \eta \in \Omega^1\}/dM$

THEOREM 3. Let $\mathcal{H}^1_K(\mathfrak{A}, M, \Omega^1) = 0$ and let the Hamiltonian pair of the operators $H, K: \Omega^1 \to \mathfrak{A}$ be given. If a sequence $\xi_0, \xi_1, \xi_2, \ldots \in \Omega^1$ satisfies the equalities $K\xi_{i+1} = H\xi_i$, $\xi_0, \xi_1 \in dM$, then:

a) there exist $m_0, m_1, m_2, \ldots \in M$ such that $\xi_i = dm_i$, $i = 0, 1, \ldots$,

b) for fixed j, the elements m_j are the conservation laws for each Hamiltonian field $a_i = Kdm_i$,

c) the conservation laws m_i are in involution both with respect to the bracket $\{\ ,\ \}_H$ and the bracket $\{\ ,\ \}_K$.

The proof of this theorem is analogous to that of Theorem 3.4 of [3].

4. Nyehuis Bracket. Let there exist a relation between elements of $\mathfrak{A}$, i.e., in $\mathfrak{A} \times \mathfrak{A}$ a subset $\mathcal{A} \subset \mathfrak{A} \times \mathfrak{A}$ is fixed. Using $\mathcal{A}$ we construct the relation between elements of Ω^1, i.e., a subset $\mathcal{A}^* \subset \Omega^1 \times \Omega^1$, defined in such a way: $(\eta_1, \eta_2) \in \mathcal{A}^* \Leftrightarrow \eta_2(a_1) - \eta_1(a_2) = 0$ for all $(a_1, a_2) \in \mathcal{A}$ (it is natural to call $\mathcal{A}^*$ adjoint to $\mathcal{A}$). Let us define in addition the relation between triplets $\eta_1, \eta_2, \eta_3 \in \Omega^1$, i.e., a subset $\mathcal{A}_1^* \subset \Omega^1 \times \Omega^1 \times \Omega^1$, in the following manner: $(\eta_1, \eta_2, \eta_3) \in \mathcal{A}_1^* \Leftrightarrow (\eta_1, \eta_2) \in \mathcal{A}^* \,\&\, (\eta_2, \eta_3) \in \mathcal{A}^*$. The mapping $[\mathcal{A}, \mathcal{A}]: \mathcal{A} \times \mathcal{A} \times \mathcal{A}_1^* \to M$, which acts on elements $\bar{a} = (a_1, a_2) \in \mathcal{A}$, $\bar{b} = (b_1, b_2) \in \mathcal{A}$, $\bar{\eta} = (\eta_1, \eta_2, \eta_3) \in \mathcal{A}_1^*$ according to the formula

$$[\mathcal{A}, \mathcal{A}]\,(\bar{a}, \bar{b}, \bar{\eta}) = \eta_1([a_2, b_2]) - \eta_2([a_2, b_1]) + [a_1, b_2]) + \eta_3([a_1, b_1]), \tag{5}$$

is called the Nyenhuis bracket of $\mathcal{A}$ (with itself). Using polarization we can also define the bracket of two relations $[\mathcal{A}, \mathcal{B}]$. Note, as it is not difficult to obtain from Eq. (5), that if $[\mathcal{A}, \mathcal{A}] = 0$, then $[\mathcal{A}^{-1}, \mathcal{A}^{-1}] = 0$, where $\mathcal{A}^{-1}$ denotes the inverse relation defined in such a way: $(a, b) \in \mathcal{A}^{-1} \Leftrightarrow (b, a) \in \mathcal{A}$.

Let us explain the reason for using the term "Nyenhuis bracket." If $\mathcal{A}$ is the graph of a linear operator $A: \mathfrak{A} \to \mathfrak{A}$, then $\bar{a} = (a_1, Aa_1)$, $\bar{b} = (b_1, Ab_1)$ and Eq. (5) becomes $\eta_1([Aa_1, Ab_1] - A[Aa_1, b_1] - A[a_1, Ab_1] + A^2[a_1, b_1])$. It is exactly in such a way that the Nyenhuis bracket of an operator A with itself is defined in differential geometry (see, for example, [4]).

Proposition 4. Let $H, K\colon \Omega^1 \to \mathfrak{A}$ be a Hamiltonian pair. Then the relation $\mathcal{A}_{HK} = \{(H\xi,\ K\xi),\ \xi \in \Omega^1\} \subset \mathfrak{A} \times \mathfrak{A}$ satisfies the condition: $[\mathcal{A}_{HK}, \mathcal{A}_{HK}] = 0$.

The proof is obtained directly from the definitions. This result is an algebraic representation of Theorem 4.2 of [3] which asserts that the "quotient" of two Hamiltonian operators (if it has meaning) satisfies the condition that the Nyenhuis bracket vanishes.

5. Realization of $\bar{A}$. Let us consider the ring of polynomials in $u^{(i)}_{k\alpha\beta}$, $1 \le \alpha, \beta \le l$, $0 \le k \le n-1$, $i = 0, 1, \ldots$. As was already noted in Paragraph 1, the Lie algebra $\mathfrak{A}$ consists of all possible collections $h = \{h_{k\alpha\beta}\}$, $h_{k\alpha\beta} \in A$. We need also to define Ω^1. An element $\xi \in \Omega^1$ will be defined by the set $\{\xi_{k\alpha\beta}\}$, $\xi_{k\alpha\beta} \in A$; the value of the 1-form ξ on the element $h \in \bar{A}$ is defined by the equation $(\xi, h) = \int \Big(\sum\limits_{k,\alpha,\beta} \xi_{k\alpha\beta} h_{k\alpha\beta} \Big) dx \in \bar{A}$.

Let us introduce the ring of formal integrodifferential operators. This ring was independently used in this theory by Manin [5] and Gel'fand and Dikii [6]. So, a formal series of the form $\sum\limits_{-\infty}^{N} a_k (d/dx)^k$ is called here the formal integrodifferential operator, where a_k are matrices of order $l \times l$ consisting of elements of A. If we introduce the operation of multiplication with the aid of the relation $(d/dx)^{-1}a = \sum\limits_{j=0}^{\infty} (-1)^j a^{(j)} \times (d/dx)^{-j-1}$, then the set R of all formal integrodifferential operators becomes an associative ring. R is a direct sum of the subring R_+ of differential operators and the subring R_- of integral operators. Further we shall denote by B_+ (B_-) the differential (integral) part of an operator $B \in R$. On the space R we introduce the trace Sp according to the equation $\mathrm{Sp}\Big(\sum\limits_{-\infty}^{N} a_k (d/dx)^k\Big) = \int (\mathrm{tr}\, a_{-1})\, dx \in \bar{A}$, where tr is the usual matrix trace. The trace Sp has an important property: Sp (AB) = Sp (BA).

Let us determine a relationship between 1-forms and elements of R_-. For every 1-form $\xi = \{\xi_{k\alpha\beta}\} \in \Omega^1$ we pick $X_\xi \in R_-$ according to the equation $X_\xi = \sum\limits_{k=0}^{n-1} (d/dx)^{-k-1} \xi_k^t$, where ξ_k^t is a matrix which is the transpose of the matrix with the elements $(\xi_{k\alpha\beta})$. In addition we associate with every element $h \in \mathfrak{A} = \bar{A}$ the differential operator $F_h \in R_+$ according to the formula $F_h = \sum\limits_{k=0}^{n-1} h_k (d/dx)^k$, where h_k is a matrix with elements $(h_{k\alpha\beta})$. Thus, R_- is the realization of Ω^1 and R_+ is the realization of $\bar{A}$. Moreover $(\xi, h) = \mathrm{Sp}(X_\xi F_h)$.

The equation $hf = \sum\limits_{k,\alpha,\beta,i} (\partial f/\partial u^{(i)}_{k\alpha\beta})\, h^{(i)}_{k\alpha\beta}$, $h \in \bar{A}$, $f \in A$, defines the action of $\bar{A}$ on A; it is obvious that h(fg) = (hf)g + f(hg), so $\bar{A}$ is realized in the Lie algebra of differentiations on the ring A. For $B = \sum\limits_{-\infty}^{N} b_k (d/dx)^k \in R$ we set $hB = \sum\limits_{-\infty}^{N} (hb_k)(d/dx)^k$, where h acts on matrices b_k componentwise. Then, as is not difficult to prove, h(BC) = (hB)C + B(hC), so $\bar{A}$ is realized in the Lie algebra of differentiations on the ring R. Equation (1) is transformed into the relation $F_{[a,b]} = aF_b - bF_a$, whence we obtain $(L_a\xi, h) = \mathrm{Sp}\,((aX_\xi)F_h + X_\xi(hF_a))$.

6. Example: the Second Hamiltonian Structure of the Lax Equation. Let us describe the so-called second Hamiltonian structure of the Lax equation. The hypothesis concerning its Hamiltonian nature was propounded by Adler [7]; the proof of this Hamiltonian nature was given in [8]. Here we follow the idea of [8].

Let us fix an operator L of order n and for a given $\xi \in \Omega^1$ construct the differential operator

$$F_h = L\,(X_\xi L)_+ - (LX_\xi)_+ L \equiv (LX_\xi)_- L - L\,(X_\xi L)_-. \tag{6}$$

Thus, the element $h = H\xi \in \mathfrak{A} = \bar{A}$ is associated with every $\xi \in \Omega^1$. It is easy to verify that the constructed operator $H\colon \Omega^1 \to \mathfrak{A}$ is skew-symmetric.

THEOREM 5. Let $L = \sum\limits_{k=0}^{n-1} u_k (d/dx)^k + C$, where u_k are matrices with elements $(u_{k\alpha\beta})$ and C is a differential operator with constant coefficients of order not greater than n. Then the operator H which corresponds to L is a Hamiltonian operator.

Proof. For brevity, let us set $X_i = X_{\xi_i}$. We have

$$-(HL_{H\xi_1}\xi_3, \xi_2) + (\text{cycle}) = \mathrm{Sp}\,(((H\xi_1)\,X_3)\,F_{H\xi_2} + X_3\,(H\xi_2)\,F_{H\xi_1}) + (\text{cycle}) =$$

$$= \mathrm{Sp}(((H\xi_2)X_1)F_{H\xi_3} + ((H\xi_2)F_{H\xi_1})X_3) + (\text{cycle}) =$$
$$= \mathrm{Sp}([F_{H\xi_2}(X_1L)_+ + L(X_1F_{H\xi_2})_+ - (F_{H\xi_2}X_1)_+L - (LX_1)_+F_{H\xi_2}]X_3) + (\text{cycle}) =$$
$$= \mathrm{Sp}(F_{H\xi_2}[(X_1L)_+X_3 + (X_3L)_-X_1 - X_1(LX_3)_- - X_3(LX_1)_+]) + (\text{cycle}) =$$
$$= \sum_{\sigma\in S_3} \mathrm{sgn}\,\sigma\, \mathrm{Sp}(F_{H\xi_{\sigma(2)}}[X_{\sigma(1)}(LX_{\sigma(3)})_+ - (X_{\sigma(1)}L)_-X_{\sigma(3)}]) =$$
$$= \sum_{\sigma\in S_3} \mathrm{sgn}\,\sigma\, \mathrm{Sp}\,[(LX_{\sigma(2)})_-LX_{\sigma(1)}(LX_{\sigma(3)})_+ - (X_{\sigma(1)}L)_-X_{\sigma(3)}L(X_{\sigma(2)}L)_+] = 0.$$

In these transformations we have used the formulas of Paragraph 5, the equality $hL = F_h$, the commutative property of the trace, and in the last equality the following fact: $\sum_{\sigma\in S_3} \mathrm{sgn}\,\sigma\,\mathrm{Sp}\,(P_{\sigma(1)})_-P_{\sigma(2)}(P_{\sigma(3)})_+ = 0$ for all P_1, P_2, $P_3 \in R$.

COROLLARY. Let an operator L satisfy the conditions of Theorem 5; H is the operator which corresponds to L, according to (6); $A_1, \ldots, A_s$ are constant $l \times l$ matrices, $K_1, \ldots, K_s \colon \Omega^1 \to \mathfrak{A}$ are operators defined by the relations $F_{K_i\xi} = A_i(X_\xi L)_+ - (LX_\xi)_+A_i$. Then the operators $H, K_1, \ldots, K_s$ form a Hamiltonian family, i.e., every two of them form a Hamiltonian pair.

The proof of the corollary is obtained by replacing L by $L - \sum_1 \zeta_i A_i$. In the special case where $C = (d/dx)^n$, $s = 1$, $A_1 = E$, K and H are the first and second Hamiltonian structures of the Lax equation.

LITERATURE CITED

1. I. M. Gel'fand and L. A. Dikii, Usp. Mat. Nauk, 30, No. 5, 67-100 (1975).
2. I. M. Gel'fand and L. A. Dikii, Funkts. Anal. Prilozhen., 10, No. 1, 18-25 (1976).
3. I. M. Gel'fand and I. Ya. Dorfman, Funkts. Anal. Prilozhen., 13, No. 4, 13-30 (1979).
4. A. P. Stone, Can. J. Math., 25, No. 5, 903-907 (1973).
5. Yu. I. Manin, Sovrem. Probl. Mat., 11, 5-152 (1978).
6. I. M. Gel'fand and L. A. Dikii, Funkts. Anal. Prilozhen., 10, No. 4, 13-29 (1976).
7. M. Adler, "On a trace functional for formal pseudodifferential operators and the symplectic structure of the Korteweg–de Vries equation," Preprint (1979).
8. I. M. Gel'fand and L. A. Dikii, "The family of Hamiltonian structures connected with integrable nonlinear equations," Preprint Inst. Prikl. Mat. Akad. Nauk SSSR, No. 136 (1978).

226

17.

(with I.Ya. Dorfman)

Hamiltonian operators and the classical Yang-Baxter equation

Funkts. Anal. Prilozh. **16** (4) (1982) 1–9 [Funct. Anal. Appl. **16** (1982) 241–248]
Zbl. **527**: 58018

In classical differential geometry and in mechanics one can construct an algebraization of the basic concepts in the following way: One takes as basic object a ring of functions, in terms of which one defines all other basic objects -- vector fields (as differentiations in the ring of functions), differential forms as forms multilinear with respect to this ring, etc. However, in recent years, in connection with integrable systems it became necessary to construct a Hamiltonian formalism in the calculus of variations (cf. [1-7]). The authors' papers [8-10] arose from the observation that a construction based on the use of a ring of functions is not well compatible with the calculus of variations. In fact, the collection of functionals in the calculus of variations does not form a ring.* Hence, the authors chose another path for the introduction of basic differential geometric objects. It seemed to us that the system of definitions given below (and also in [8-10]) is essential not only in the calculus of variations, but also in many other cases, some of which are cited later. Roughly speaking, this approach consists of the fact that in place of the ring of functions, for constructing such a scheme one is given in advance a Lie algebra $\mathfrak{a}$ and a complex of $\mathfrak{a}$-modules (Ω, d) with a differential compatible with the action of the Lie algebra (for the precise axioms, cf. below). All the examples given later differ from one another in the choice of Lie algebra and the complex. It seems to us that the possibilities of such an approach are far from exhausted.

The goal of the paper is to show that in the scheme indicated there is contained a series of meaningful and important examples. These are: Hamiltonian formalism of the calculus of variations [3, 4, 8]; classical Yang–Baxter equation; Hamiltonian formalism on finite-dimensional manifolds; cohomology of Lie algebras, including the Lie algebra of vector fields [11, 12]; and deformations of Lie algebras (cf., e.g., [13, 18]). The goal of the paper is also the definitive establishment of a connection between Hamiltonian operators and symplectic structures. Here the concept of generalized nondegenerate 2-cocycle (which only coincides with the concept of nondegenerate 2-cocycle on a subalgebra of the Lie algebra for finite-dimensional Lie algebras) is important.

1. Classical Yang–Baxter Equation. Since this equation will be considered below as an example, we give it first in the form in which it occurs in the mathematical literature. However, it seems to us that the algebraic treatment of the classical Yang–Baxter equation which is given in Sec. 4, being equivalent to the one given here, is more natural, not requiring, in particular, the inclusion of the Lie algebra in an associative ring. One can get the classical Yang–Baxter equation (cf. [14, 15]) in the following way: Let $\mathfrak{g}$ be a finite-dimensional Lie algebra, U be an arbitrary associative algebra with unit 1 containing it (one has in mind that $[a, b] = ab - ba$, $a, b \in \mathfrak{g}$; one can assume that U is the universal enveloping algebra for $\mathfrak{g}$). We denote by $\varphi^{ij}: \mathfrak{g}\otimes\mathfrak{g} \to U\otimes U\otimes U$, $1 \leqslant i, j \leqslant 3$, $i \neq j$, the mappings, acting on an element $a\otimes b \in \mathfrak{g}\otimes\mathfrak{g}$ as follows: a is placed in the i-th place, b in the j-th place, 1 in the remaining free place (e.g., $\varphi^{13}(a\otimes b) = a\otimes 1\otimes b$). Suppose further that r is a function of two complex variables with values in $\mathfrak{g}\otimes\mathfrak{g}$. The relation

$$[r^{12}(u_1, u_2), r^{13}(u_1, u_3)] + [r^{12}(u_1, u_2), r^{23}(u_2, u_3)] + [r^{13}(u_1, u_3), r^{23}(u_2, u_3)] = 0, \qquad (1)$$

where $r^{ij} = \varphi^{ij}\circ r$, is called the classical Yang–Baxter equation. It is frequently considered simultaneously with the so-called unitary condition

*Of course, we can construct a ring, by considering, along with the functionals $\int f\,dx$, the sums of their products, but such a construction is no more interesting than the imbedding of a linear space in a ring, due to the consideration of the sums of products of elements.

Institute of Applied Mathematics, Academy of Sciences of the USSR. Institute of Chemical Physics, Academy of Sciences of the USSR. Translated from Funktsional'nyi Analiz i Ego Prilozheniya, Vol. 16, No. 4, pp. 1-9, October-December, 1982. Original article submitted June 1, 1982.

$$r^{12}(u_1, u_2) = -r^{21}(u_2, u_1). \tag{2}$$

In the present paper, in particular, it will be shown that the system (1)-(2) is the Hamiltonian condition in the sense of the authors' papers [8-10] (for suitable choice of complex) for the operator which corresponds canonically to the function r.

2. Introduction of the Basic Differential Geometric Objects. Let $\mathfrak{a}$ be some (possibly infinite-dimensional) Lie algebra, (Ω, d) be a complex of linear or topological linear spaces, i.e., $\Omega = \Omega^0 \oplus \Omega^1 \oplus \ldots \oplus \Omega^q \oplus \ldots$, $d: \Omega \to \Omega$ is a linear map such that $d\Omega^q \subset \Omega^{q+1}$, $d^2 = 0$. If the spaces Ω^q are topological, then the operator d and the operators introduced below in terms of it are assumed to be continuous.

We shall say that (Ω, d) is a complex over the Lie algebra $\mathfrak{a}$, if with any $a \in \mathfrak{a}$ there is associated a linear map $i_a: \Omega \to \Omega$, $i_a\Omega^q \subset \Omega^{q-1}$, $i_a\Omega^0 = \{0\}$, satisfying the conditions

$$i_a i_b + i_b i_a = 0, \quad [i_a d + d i_a, i_b] = i_{[a, b]}.$$

The operator $L_a = i_a d + d i_a$ (carrying Ω^q into Ω^q for all q) is called the Lie derivative in the direction $a \in \mathfrak{a}$. The operators L_a define a representation of the Lie algebra $\mathfrak{a}$ in Ω^q, since from the formulas given for them it follows easily that $L_{[a,b]} = [L_a, L_b]$.

In what follows we shall use the following notation: For $a \in \mathfrak{a}$, $m \in \Omega^0$ we denote by am the element $i_a dm \in \Omega^0$; further, for $\xi \in \Omega^1$ by (ξ, a) or (a, ξ) we denote the element $i_a\xi \in \Omega^0$. We shall assume that from $(\xi, a) = 0$ for all $\xi \in \Omega^1$ it follows that $a = 0$.

Examples of complexes over Lie algebras will be given below.

3. Schouten Brackets and Hamiltonian Operators. The goal of this section is to define a symplectic structure in the realms of introduced concepts. The connection with the usual definition of a symplectic structure will be given in Sec. 6.

We call an operator $H: \Omega^1 \to \mathfrak{a}$ skew-symmetric, if $(H\xi, \eta) = -(\xi, H\eta)$ for any $\xi, \eta \in \Omega^1$. The Schouten bracket of two skew-symmetric operators $H, K: \Omega^1 \to \mathfrak{a}$ means the trilinear map $[H, K]: \Omega^1 \times \Omega^1 \times \Omega^1 \to \Omega^0$, defined by the formula

$$[H, K](\xi_1, \xi_2, \xi_3) = (KL_{H\xi_1}\xi_2, \xi_3) + (HL_{K\xi_1}\xi_2, \xi_3) + (\text{cycle}), \tag{3}$$

where the word (cycle) here and later means the sum over the cyclic permutations of the indices 1, 2, 3. The operator $H: \Omega^1 \to \mathfrak{a}$ will be called Hamiltonian if it is skew-symmetric and $[H, H] = 0$.

Any Hamiltonian operator $H: \Omega^1 \to \mathfrak{a}$ allows us to construct the Hamiltonian formalism connected with it. The Poisson bracket is introduced by the formula $\{m_1, m_2\} = (Hdm_1, dm_2)$, $m_1, m_2 \in \Omega^0$; on Ω^0 it satisfies the Jacobi identity. The analog of the Hamiltonian correspondence between functions and vector fields in the present theory is the map $Hd: \Omega^0 \to \mathfrak{a}$; one can prove that this map is a morphism of Lie algebras. A detailed account of all these facts is contained in [8-10], so we shall not dwell on them and shall proceed to the definition of a symplectic structure.

If there is a skew-symmetric operator $H: \Omega^1 \to \mathfrak{a}$ then it is easy to show that on its image $L = \operatorname{Im} H \subset \mathfrak{a}$ the pairing

$$\omega_H(a, b) = (a, H^{-1}b) \in \Omega^0, \tag{4}$$

is well-defined, i.e., if $a, b \in L = \operatorname{Im} H$, then $\omega_H(a, b)$ is uniquely defined. More generally, (4) makes sense for all pairs $(a, b) \in L_1 \times L$, where $L = \operatorname{Im} H$, $L_1 = (\operatorname{Ker} H)^\perp$ is the subspace in $\mathfrak{a}$, consisting of all a such that $(a, \xi) = 0$ for all $\xi \in \operatorname{Ker} H$. It is obvious that $L \subset L_1$. We now characterize Hamiltonian operators in terms of the pairing (4). We introduce some definitions.

Suppose that fixed in the Lie algebra $\mathfrak{a}$ are two linear subspaces $L \subset L_1 \subset \mathfrak{a}$ such that $[L, L] \subset L_1$ (we stress that neither L nor L_1 is necessarily a subalgebra of the Lie algebra $\mathfrak{a}$).

Definition. A bilinear mapping $\omega: L_1 \times L \to \Omega^0$ will be called a generalized 2-cocycle on $(\mathfrak{a}, \Omega)$ if:

(a) ω is skew-symmetric on $L \times L$,

(b) for any $a_1, a_2, a_3 \in L$ one has

$$a_1\omega(a_2, a_3) - \omega([a_1, a_2], a_3) + (\text{cycle}) = 0. \tag{5}$$

It is obvious in the case when L is a subalgebra of the Lie algebra $\mathfrak{a}$ that the present definition turns into the ordinary definition of a 2-cocycle on this subalgebra (with values in Ω^0).

We shall now consider elements $\xi \in \Omega^1$ only on the subspace L_1. Thus, two elements ξ_1, $\xi_2 \in \Omega^1$ will be identified if $(\xi_1, a) = (\xi_2, a)$ for all $a \in L_1$. We denote the corresponding quotient-space by $\Omega^1_{L_1}$.

Definition. We shall call a generalized 2-cocycle ω nondegenerate, or a generalized symplectic structure on $(\mathfrak{a}, \Omega)$, if there exists an isomorphism $i: L \to \Omega^1_{L_1}$ such that

$$(a, ib) = \omega(a, b), \ a \in L_1, \quad b \in L. \tag{6}$$

THEOREM. Any Hamiltonian operator $H: \Omega^1 \to \mathfrak{a}$ defines a generalized symplectic structure ω_H on $(\mathfrak{a}, \Omega)$ if one sets $L = \operatorname{Im} H$, $L_1 = (\operatorname{Ker} H)^\perp$, $\omega_H(a, b) = (a, H^{-1}b)$.

Conversely, for any generalized symplectic structure one can construct a Hamiltonian operator.

Proof. Let $H: \Omega^1 \to \mathfrak{a}$ be a Hamiltonian operator. We verify that $[L, L] \subset L_1$, i.e., that for any $a_1, a_2 \in \operatorname{Im} H$ we have $[a_1, a_2] \in (\operatorname{Ker} H)^\perp$. Let $a_i = H\xi_i$, $\xi \in \operatorname{Ker} H$. Then, according to (3) we have

$$0 = -\frac{1}{2}[H, H](\xi_1, \xi_2, \xi) = (L_{H\xi}\xi_2, H\xi) + (L_{H\xi_2}\xi, H\xi_1) + (L_{H\xi}\xi_1, H\xi_2) = (L_{H\xi_2}\xi, H\xi_1) = -(\xi, [H\xi_2, H\xi_1]) = (\xi, [a_1, a_2]),$$

which is what was required.

The skew-symmetry of ω_H on $L \times L$ obviously follows from the skew-symmetry of the operator H. Further, for $a_i = H\xi_i$, $i = 1, 2, 3$, we have

$$a_1\omega_H(a_2, a_3) - \omega_H([a_1, a_2], a_3) + (\text{cycle}) = (H\xi_1)(H\xi_2, \xi_3) - ([H\xi_1, H\xi_2], \xi_3) + \\ + (\text{cycle}) = -(H\xi_3, L_{H\xi_1}\xi_2) + (\text{cycle}) = \frac{1}{2}[H, H](\xi_1, \xi_2, \xi_3) = 0,$$

i.e., (5) holds.

We verify the nondegeneracy of ω_H. First we construct the map i. With any $b \in L = \operatorname{Im} H$ we associate the equivalence class of the element $H^{-1}b \in \Omega^1$ in the quotient-space $\Omega^1_{L_1}$ (since any two preimages of the element b differ by an element of Ker H, and $L_1 = (\operatorname{Ker} H)^\perp$, this association is well-defined). We denote this element by $\overline{H^{-1}b} \in \Omega^1_{L_1}$. We get a map $i: L \to \Omega^1_{L_1}$, defined by the formula $ib = \overline{H^{-1}b}$. It is an isomorphism. In fact, it is obvious that it is a map "onto." Further, if $b \in \operatorname{Ker} i$, then by the definition of $\Omega^1_{L_1}$ we have $(a, H^{-1}b) = 0$ for all $a \in L_1 = (\operatorname{Ker} H)^\perp$ and, in particular, for all a having the form $a = H\xi$, $\xi \in \Omega^1$. Whence, $(\xi, b) = 0$ for any $\xi \in \Omega^1$, and hence, $b = 0$. Thus, any Hamiltonian operator defines a generalized symplectic structure.

Conversely, suppose we are given a pair $L \subset L_1 \subset \mathfrak{a}$ and a nondegenerate generalized 2-cocycle ω on $(\mathfrak{a}, \Omega)$. For any element $\xi \in \Omega^1$ we consider the corresponding equivalence class $\bar{\xi} \in \Omega^1_{L_1}$. By virtue of the nondegeneracy, there exists a uniquely defined element $b = i^{-1}\bar{\xi} \in L$ such that $(a, \xi) = \omega(a, b)$, $a \in L_1$. Associating with the element $\xi \in \Omega^1$ the element $b = i^{-1}\bar{\xi} \in L$, we get a linear operator $H: \Omega^1 \to L \subset \mathfrak{a}$ such that $(a, \xi) = \omega(a, H\xi)$, $\xi \in \Omega^1$, $a \in L_1$. The skew-symmetry of the operator H follows easily from the skew-symmetry of ω on $L \times L$. The vanishing of the Schouten bracket $[H, H]$ follows directly from (5). Thus, the operator H is Hamiltonian and the theorem is proved.

Remark. For a skew-symmetric operator H, the Hamiltonian condition $[H, H] = 0$ can be reformulated in the following equivalent way: For any $\xi, \eta \in \Omega^1$ one has

$$[H\xi, H\eta] = H(i_{H\xi}d\eta - i_{H\eta}d\xi + di_{H\xi}\eta).$$

In particular, the image Im H of a Hamiltonian operator is closed with respect to the commutator in $\mathfrak{a}$.* From this fact and the theorem proved it follows that a nondegenerate generalized 2-cocycle defines a Lie subalgebra L of the Lie algebra $\mathfrak{a}$ and an ordinary 2-cocycle on it.

4. Interpretation of the Classical Yang–Baxter Equation. Let the Lie algebra $\mathfrak{a}$ be a topological Lie algebra. We construct a special form of a complex Ω over $\mathfrak{a}$, introducing the spaces Ω^q, $q \geqslant 0$, and also the operators d and i_a, $a \in \mathfrak{a}$, in the following way. We set $\Omega^0 = \mathbf{R}$ (or $\mathbf{C}$); further, let Ω^q be the space of multilinear, skew-symmetric forms $\omega(a_1, \ldots, a_q)$, continuous with respect to their arguments, assuming values in $\mathbf{R}$ (or $\mathbf{C}$). We set

$$d\omega(a_1, \ldots, a_{q+1}) = \sum_{i<j} (-1)^{i+j} \omega([a_i, a_j], a_1, \ldots, \hat{a}_i, \ldots, \hat{a}_j, \ldots, a_{q+1}),$$

$$(i_a\omega)(a_1, \ldots, a_{q-1}) = \omega(a, a_1, \ldots, a_{q-1}).$$

As is easy to verify, all the necessary requirements (cf. Sec. 2) hold here and thus there arises a complex Ω over the Lie algebra $\mathfrak{a}$.

Definition. In the present case we shall call the Hamiltonian operator H: $\Omega^1 \to \mathfrak{a}$ the Yang–Baxter operator, and the vanishing condition for the Schouten bracket [H, H], which in this situation assumes the form

$$(\xi_1, [H\xi_2, H\xi_3]) + (\text{cycle}) = 0, \quad \xi_1, \xi_2, \xi_3 \in \Omega^1, \tag{7}$$

is called the *classical Yang–Baxter equation*.

We give the most needed explanations. Since $\Omega^1 = \mathfrak{a}^*$, the operator H is an element of the space Hom $(\mathfrak{a}^*, \mathfrak{a})$ In the case of finite-dimensional or nuclear spaces there is a natural correspondence between Hom $(\mathfrak{a}^*, \mathfrak{a})$ and $\mathfrak{a} \otimes \mathfrak{a}$. Hence, the unknown function appearing in the classical Yang–Baxter equation is usually considered as an element of $\mathfrak{a} \otimes \mathfrak{a}$. In our view the unnaturality of this interpretation manifests itself in that to define the classical Yang–Baxter equation (cf. Sec 1) one must imbed the Lie algebra in an associative algebra with unit, and then prove that it does not depend on the method of imbedding. The definition of the classical Yang–Baxter equation that we have given does not require any supplementary structures.

The traditional form of the classical Yang–Baxter equation (Sec. 1) can be obtained from our definition in the following way: As Lie algebra $\mathfrak{a}$ we consider the algebra of currents on $\mathbf{C}$ with values in the finite-dimensional Lie algebra $\mathfrak{g}$. We shall assume that the topologies in $\mathfrak{a}$ and $\mathfrak{a}^*$ are chosen so that the elements Hom $(\mathfrak{a}^*, \mathfrak{a})$ are defined by kernels $r(u_1, u_2)$, assuming values in $\mathfrak{g} \otimes \mathfrak{g}$; skew-symmetric operators H: $\Omega^1 \to \mathfrak{a}$ are then defined by the kernels which satisfy the condition

$$r(u_1, u_2) = -\sigma^{21} r(u_2, u_1), \tag{8}$$

where σ^{21}: $\mathfrak{g} \otimes \mathfrak{g} \to \mathfrak{g} \otimes \mathfrak{g}$ is the permutation operator, acting according to the rule $\sigma^{21}(a \otimes b) = b \otimes a$ (cf. (8) with the unitary condition (2)). We fix a basis $e_1, \ldots, e_n$ in the Lie algebra $\mathfrak{g}$. Then, as is easy to verify, (7) can be rewritten in terms of the kernel r as follows†:

$$\sum_{\alpha,\beta} (r^{j\alpha}(u_2, u_1) r^{k\beta}(u_3, u_1) [e_\alpha, e_\beta]^i + r^{k\alpha}(u_3, u_2) r^{i\beta}(u_1, u_2) [e_\alpha, e_\beta]^j + \\ + r^{i\alpha}(u_1, u_3) r^{j\beta}(u_2, u_3) [e_\alpha, e_\beta]^k) = 0, \quad i, j, k = 1, \ldots, n. \tag{9}$$

It only remains to see that the system of equations obtained is (taking (8) into account) the coordinate description of (1). We thus see that the system (1)-(2) is the description of the Hamiltonian condition for the operator H. Hence, by virtue of the theorem of Sec. 3 we can characterize the solutions of this system as generalized nondegenerate 2-cocycles on $(\mathfrak{a}, \Omega)$.

In the case when the Lie algebra $\mathfrak{a}$ is finite-dimensional, a generalized nondegenerate 2-cocycle on $(\mathfrak{a}; \Omega)$ is an R- or C-valued 2-cocycle on a subalgebra of the Lie algebra $\mathfrak{a}$ (in fact, a simple argument using dimensional considerations shows that in the case of a finite-dimensional Lie algebra $\mathfrak{a}$ there follows from the nondegeneracy condition the coincidence of the subspaces L and L_1). Now, in the case when is an infinite-dimensional Lie algebra, in gen-

*This was also noted independently by Yu. L. Daletskii.

†One should not confuse the coordinates of the tensor r^{ij} with the notations r^{12}, r^{13}, and r^{23}, used traditionally in the present theory.

eral there also exist solutions of the classical Yang—Baxter equation (7) for which the inclusion $L \subset L_1$ is strict.

We note that the fact that the Yang—Baxter equation occurs in the development of our scheme was understood by V. G. Drinfel'd. For the case of finite-dimensional Lie algebras, for which the Lie algebra can be integrated to a group, he also formulated (for $\Omega^0 = \mathbf{C}$, $\Omega^1 = \mathfrak{g}^*$) the corresponding condition in group terms [16].

5. Differential Geometric Objects of the Formal Calculus of Variation. We consider the ring A of polynomials in the letters $u_\alpha^{(i)}$, where α runs through some set I, finite or infinite, i = 0, 1, 2, ... Before introducing the Lie algebra and the complex over it, connected with the formal calculus of variations, we describe an auxiliary object, the de Rham complex of the ring A. We consider the Lie algebra $\mathfrak{a}$ of all differentiations ∂ of the ring A and we introduce the space Ω^q, consisting of all skew-symmetric forms $\omega(\partial_1, \ldots, \partial_q)$, multilinear over A, assuming values in the ring A. We define operators d: $\Omega^q \to \Omega^{q+1}$ and i_∂: $\Omega^q \to \Omega^{q-1}$, $\partial \in \mathfrak{a}$, respectively, by the formulas

$$d\omega(\partial_1, \ldots, \partial_{q+1}) = \sum_i (-1)^{i+1} \partial_i \omega(\partial_1, \ldots, \hat{\partial}_i, \ldots, \partial_{q+1}) + \sum_{i<j} (-1)^{i+j} \omega([\partial_i, \partial_j], \partial_1, \ldots, \hat{\partial}_i, \ldots, \hat{\partial}_j, \ldots, \partial_{q+1}),$$

$$(i_\partial \omega)(\partial_1, \ldots, \partial_{q-1}) = \omega(\partial, \partial_1, \ldots, \partial_{q-1}).$$

There arises a complex $\Omega(A)$ over the Lie algebra $\mathfrak{a}$ of differentiations of the ring A, called *the de Rham complex* of the ring A. In the Lie algebra $\mathfrak{a}$ we single out the differentiation d/dx, defined by the formula $(d/dx)f = \sum_{i,\alpha} u_\alpha^{(i+1)} \partial f/\partial u_\alpha^{(i)}$. We consider the subalgebra $\mathfrak{a}_1$ of the Lie algebra $\mathfrak{a}$, consisting of differentiations, commuting with d/dx.

We introduce the basic complex $(\tilde{\Omega}, d)$ in the following way: We shall consider $\omega \in \Omega^q$ equivalent to 0 if ω is the Lie derivative $L_{d/dx}\omega_1$ of some form $\omega_1 \in \Omega^q$. The collection of q forms equivalent to 0 we denote by Ω_0^q and we set $\tilde{\Omega}^q = \Omega^q/\Omega_0^q$. In the space $\tilde{\Omega} = \bigoplus \tilde{\Omega}^q$ there are well-defined mappings d: $\tilde{\Omega}^q \to \tilde{\Omega}^{q+1}$ and i_∂: $\tilde{\Omega}^q \to \tilde{\Omega}^{q-1}$ for any $\partial \in \mathfrak{a}_1$. It is easy to verify that for d and i_∂ the requirements of Sec. 2 hold and thus there arises a complex $(\tilde{\Omega}, d)$ over the Lie algebra $\mathfrak{a}_1$.

The pair consisting of the Lie algebra of differentiations ∂ of the ring A such that $[\partial, d/dx] = 0$, and the complex $(\tilde{\Omega}, d)$ constructed above is the basic object of the formal calculus of variations.

The Lie algebra $\mathfrak{a}_1$, and also the spaces $\tilde{\Omega}_0$ and $\tilde{\Omega}_1$, can be described in a somewhat more interesting way. As is easy to see, any differentiation from $\mathfrak{a}_1$ has the form $\partial = \sum_{i,\alpha} h_\alpha^{(i)} \partial/\partial u_\alpha^{(i)}$, where $h_\alpha^{(i)} \overset{\text{def}}{=} (d/dx)^i h_\alpha$. We denote by A^I the collection of all sequences $h = \{h_\alpha\}$, $h_\alpha \in A$, $\alpha \in I$. Any such sequence we shall call a *vector field*. Since any differentiation $\partial \in \mathfrak{a}_1$ is uniquely determined by the collection $h_\alpha = \partial u_\alpha$, one can interpret the Lie algebra $\mathfrak{a}_1$ as the space of vector fields A^I, provided with the commutator k = [h, g], where

$$k_\beta = \sum_{\alpha, i} \left(h_\alpha^{(i)} \frac{\partial g_\beta}{\partial u_\alpha^{(i)}} - g_\alpha^{(i)} \frac{\partial h_\beta}{\partial u_\alpha^{(i)}} \right), \quad \beta \in I. \tag{10}$$

The commutator in the space of vector fields was first introduced by (10) in [4]. Further, the space $\tilde{\Omega}^0$ obviously coincides with the quotient-space $\tilde{A} = A/(d/dx)A$ and thus consists of elements of the form $\int f\,dx$, $f \in A$; such elements are called *functionals* in the formal calculus of variations. Finally, one can verify that the space $\tilde{\Omega}^1$ can be identified with the space $\bar{A}$ of sequences $\xi = \{\xi_\alpha\}$, $\xi_\alpha \in A$, $\alpha \in I$ such that $\xi_\alpha \neq 0$ only for a finite collection of indices α. The pairing $(\xi, h) \in \tilde{A}$ between a 1-form $\xi \in \bar{A}$ and a vector field $h \in A^I$ is defined by the formula $(\xi, h) = \int \sum_\alpha \xi_\alpha h_\alpha\, dx \in \tilde{A}$. There is a detailed account of these facts in [10].

We consider Hamiltonian operators. Let H: $\tilde{\Omega}^1 \to \mathfrak{a}_1$ be a Hamiltonian operator. According to our conventions, one can assume that H acts from the space $\bar{A}$ of 1-forms into the space A^I of vector fields, and thus one can define a finite matrix $H = (H_{\alpha\beta})$, $H_{\alpha\beta}$: $A \to A$, $\alpha, \beta \in I$. Usually in the calculus of variations one assumes the operators $H_{\alpha\beta}$ to be differentiations, i.e., to have the form*

*The fact that $H_{\alpha\beta}$ has this form gives the locality of the Poisson brackets, i.e., in different notation $\{u_\alpha(x), u_\beta(y)\} = \sum a_{\alpha\beta i} \delta^{(i)}(x - y)$.

$$H_{\alpha\beta}=\sum_{i=0}^{n(\alpha,\beta)} a_{\alpha\beta i}\left(\frac{d}{dx}\right)^i, \quad a_{\alpha\beta i}\in A.$$

The skew-symmetry of such an operator H reduces to the relations

$$\sum_{i=0}^{n(\sigma,\beta)} a_{\alpha\beta i}\left(\frac{d}{dx}\right)^i=-\sum_{i=0}^{n(\beta,\alpha)}\left(-\frac{d}{dx}\right)^i \circ a_{\beta\alpha i}. \tag{11}$$

The condition of vanishing of the Schouten bracket [H, H] can, as shown in [8], be rewritten in the following form:

$$((D_H\xi_1)H\xi_2, \xi_3)+(\text{cycle})=0, \quad \xi_1, \xi_2, \xi_3\in \bar{A}, \tag{12}$$

where by $D_H\xi$, for fixed $\xi\in\bar{A}$, one denotes the operator $(D_H\xi)$: $\bar{A}\to\bar{A}$ with matrix elements

$$(D_H\xi)_{\alpha\beta}=\sum_{i,j,\gamma}\frac{\partial a_{\alpha\gamma i}}{\partial u_\beta^{(j)}}\xi_\gamma^{(?)}\left(\frac{d}{dx}\right)^j.$$

On the basis of (11)-(12) one can write down the conditions for the operator H to be Hamiltonian, in the form of a system of equations on the coefficients $a_{\alpha\beta i}\in A$, solving which, in the end, we get examples of Hamiltonian operators of the calculus of variations. A detailed description of the Hamiltonian formalism in the calculus of variations, and also an analysis of examples, can be found in [8-10].

6. Finite-Dimensional Hamiltonian Manifolds. Let M be a smooth m-dimensional manifold; $\mathfrak{a}$ be the Lie algebra of vector fields on M; and (Ω, d) be the complex of differential forms on M, i.e., the de Rham complex of the ring C^∞ (M) of smooth functions on M (cf. Sec. 5). The pair consisting of the indicated Lie algebra $\mathfrak{a}$ and complex (Ω, d) over it is the basic object of consideration of this section.

Let us assume that with any point $x\in M$ there is associated a linear operator H(x): $T_x^*M\to T_xM$, depending smoothly on x and skew-symmetric for any $x\in M$, i.e., such that $(H(x)\xi, \eta)=-(\xi, H(x)\eta)$, $\xi, \eta\in T_x^*M$. In other words, we assume that on M there is given a bivector field $H(x)=\{H^{ij}(x)\}$. Such a field obviously determines some skew symmetric operator H: $\Omega^1\to\mathfrak{a}$. The condition that the operator H be Hamiltonian, i.e., the condition that the Schouten bracket [H, H] vanish, as is easy to see, can be written in coordinates as follows:

$$\sum_\alpha\left(H^{i\alpha}\frac{\partial}{\partial x^\alpha}H^{jk}+H^{j\alpha}\frac{\partial}{\partial x^\alpha}H^{ki}+H^{k\alpha}\frac{\partial}{\partial x^\alpha}H^{ij}\right)=0, \quad i, j, k=1,\ldots,m. \tag{13}$$

Definition. We shall call a manifold M, equipped with a bivector field H(x) and satisfying (13) a *Hamiltonian manifold*.

On any Hamiltonian manifold, according to the general scheme given above, one can construct a Hamiltonian formalism, i.e., there are introduced the Poisson brackets of functions

$$\{f, g\}=\sum_{i,j}H^{ij}(x)\frac{\partial f}{\partial x^j}\frac{\partial g}{\partial x^i}, \quad f, g\in C^\infty(M),$$

satisfying the Jacobi identity, there is given a correspondence between functions on M (Hamiltonians) and Hamiltonian vector fields, etc. A Hamiltonian manifold is a rather natural generalization of a symplectic manifold. In fact, according to the theorem of Sec. 3, with any Hamiltonian manifold M one can associate a pair of spaces $L\subset L_1\subset\mathfrak{a}$ and a nondegenerate generalized 2-cocycle ω, $L_1\times L$. If the operator H(x) for any $x\in M$ effects an isomorphism of T_x^*M with T_xM (which is possible only for even m), then, as is easy to see, the space L coincides with the entire Lie algebra $\mathfrak{a}$, and the given 2-cocycle ω is a closed nondegenerate bilinear form on M, i.e., a symplectic structure.

A case, intermediate between Hamiltonian and symplectic manifolds, was investigated by A. Lichnerowicz [17]. These are the so-called *Poisson manifolds*, i.e., manifolds equipped with a bivector field of constant rank $2n<m$, satisfying (13). As shown in [17], any Poisson manifold admits a fibration of codimension $m-2n$ by a symplectic manifold. The fibers are tangent to the subspaces $\operatorname{Im}H(x)\subset T_xM$, and the symplectic structure on them is induced by the

bivector field. This result can be obtained on the basis of the property of Hamiltonian operators mentioned above, which is that the image of such an operator is closed with respect to the commutator in $\mathfrak{a}$. In fact, if one considers the distribution $\{\operatorname{Im} H(x)\}$ on the manifold M, then by the Frobenius theorem it is integrable and thus defines the indicated fibration. The general case of a Hamiltonian manifold when the rank of the operator H(x) can vary from point to point is more complicated. A series of structures arising here is studied in A. A. Kirillov [19].

We note that the complex (Ω, d) considered does not exhaust the interesting examples of complexes over the Lie algebra of vector fields. For example, the complex $(\bigoplus_q \Lambda^q \mathfrak{a}^*, \partial)$, which arises upon considering the cohomology of the Lie algebra $\mathfrak{a}$ with constant coefficients (cf. [11-13]) is important. Another interesting complex over $\mathfrak{a}$ is obtained if as Ω^q one takes the collection of all q-linear mappings of $\mathfrak{a}$ into itself; the action of the Lie algebra $\mathfrak{a}$ on Ω^q is induced by the adjoint action on $\mathfrak{a}$. This complex is closely connected with deformations of Lie algebras (cf. [13, 18]).

Complexes over Lie algebras arising in connection with fiber spaces also deserve attention. Suppose one has a vector bundle $E \xrightarrow{\pi} M$. We denote by G the group of automorphisms of it. Let (Ω, d) be a complex of differential forms on E, satisfying the natural requirements of compatibility with the structure of the bundle (or, in our terms, being a complex over the Lie algebra of the group G). According to the general scheme, here there can be defined Hamiltonian operators. Since the group G contains as a subgroup the group of currents on the manifold M with values in GL(n), and also (in the case of a trivial bundle) the group of automorphisms of the base M, the Hamiltonian condition here generalizes the classical Yang–Baxter equation, on the one hand, and the Hamiltonian condition (13), on the other. The corresponding formulas will be published separately.

The infinite-dimensional analog of this situation, which one gets if the Lie algebra of vector fields on the base M is replaced by the Lie algebra of vector fields of the formal calculus of variations, is interesting.

LITERATURE CITED

1. V. E. Zakharov and L. D. Faddeev, "The Korteweg–de Vries equation is a completely integrable Hamiltonian system," Funkts. Anal., 5, No. 4, 18-27 (1971).
2. C. S. Gardner, "Korteweg–de Vries equation and generalizations, IV," J. Math. Phys., 12, No. 8, 1548-1551 (1971).
3. I. M. Gel'fand and L. A. Dikii, "Asymptotic resolvents of Sturm–Liouville equations and the algebra of Korteweg–de Vries equations," Usp. Mat. Nauk, 30, No. 5, 67-100 (1975).
4. I. M. Gel'fand and L. A. Dikii, "Fractional powers of operators and Hamiltonian systems," Funkts. Anal., 10, No. 4, 13-29 (1976).
5. I. M. Gel'fand and L. A. Dikii, "Resolvents and Hamiltonian systems," Funkts. Anal., 11, No. 2, 11-27 (1977).
6. O. I. Bogoyavlenskii and S. P. Novikov, "Connection of the Hamiltonian formalisms of stationary and nonstationary problems," Funkts. Anal., 10, No. 1, 9-13 (1976).
7. F. Magri, "A simple model of the integrable Hamiltonian equation," J. Math. Phys., 19, No. 5, 1156-1162 (1978).
8. I. M. Gel'fand and I. Ya. Dorfman, "Hamiltonian operators and algebraic structures connected with them," Funkts. Anal., 13, No. 4, 13-30 (1979).
9. I. M. Gel'fand and I. Ya. Dorfman, "Schouten brackets and Hamiltonian operators," Funkts. Anal., 14, No. 3, 71-74 (1980).
10. I. M. Gel'fand and I. Ya. Dorfman, "Hamiltonian operators and infinite-dimensional Lie algebras," Funkts. Anal., 15, No. 3, 23-40 (1981).
11. I. M. Gel'fand and D. B. Fuks, "Cohomology of the Lie algebra of formal vector fields," Izv. Akad. Nauk SSSR, Ser. Mat., 34, No. 2, 322-337 (1970).
12. I. M. Gel'fand and D. B. Fuks, "Cohomology of the Lie algebra of vector fields on a circle," Funkts. Anal., 2, No. 4, 92-93 (1968).
13. I. M. Gel'fand, B. L. Feigin, and D. B. Fuks, "Cohomology of the Lie algebra of formal vector fields with coefficients in its dual space and variations of characteristic classes of bundles," Funkts. Anal., 8, No. 2, 13-29 (1974).
14. P. P. Kulish and E. K. Sklyanin, "Solutions of the Yang–Baxter equation," in: Differential Geometry, Lie Groups and Mechanics, III (Zap. Nauchn. Semin. LOMI, Vol. 95) Nauka, Leningrad (1980), pp. 129-160.

15. A. A. Belavin and V. G. Drinfel'd, "Solutions of the classical Yang—Baxter equation for simple Lie algebras," Funkts. Anal., 16, No. 3, 1-29 (1982).
16. V. G. Drinfel'd, "Hamiltonian structures on Lie groups, Lie bialgebras, and the geometric meaning of the classical Yang—Baxter equations," Dokl. Akad. Nauk SSSR, 266, (1982).
17. A. Lichnerowicz, "Les variétés de Poisson et leurs algebres de Lie associées," J. Diff. Geom., No. 12, 253-300 (1977).
18. M. Flato and D. Sternheimer, "Deformations of Poisson brackets, separate and joint analyticity in group representations, nonlinear group representations and physical applications," in: Harmonic Analysis and Representations of Semisimple Lie Groups, Lect. NATO Adv. Study Inst., Liege 1977, Dordrecht (1980), pp. 385-448.
19. A. A. Kirillov, "Local Lie algebras," Usp. Mat. Nauk, 31, No. 4, 57-76 (1976).

Izrail Moiseevich Gelfand
(On his fiftieth birthday)

Usp. Mat. Nauk **19** (3) (1964) 187–206
[Russ. Math. Surv. **19** (4) (1964) 163–180]

On August 20 1963 was the fiftieth birthday of Izrail' Moiseevich Gel'fand, Professor at the University of Moscow and Corresponding Member of the USSR Academy of Sciences, whose name enjoys wide popular esteem both at home and abroad.

Gel'fand was born at Krasnye Okny in the province of Odessa on August 20 (7) 1913. After an incomplete secondary education, he came to Moscow in 1930 and at first took casual work (for example, as door-keeper at the Lenin Library). At the same time, he began to teach mathematics, (at first elementary and then higher mathematics as well), for various courses and at evening institutes. He began to attend mathematics lectures and seminars at the University of Moscow; as he himself has said, the first mathematics school in his life was M.A. Lavrent'ev's seminar on the theory of functions of a complex variable. In 1932, Gel'fand was admitted as a research student; his supervisor was A.N. Kolmogorov, who set him to work in the field of functional analysis. At that time, this was a subject in which only a very small group of mathematicians in Moscow was interested. The choice of topics in Gel'fand's first papers was also influenced by conversations with L.A. Lyusternik and A.I. Plesner.

Even in his first papers, Gel'fand obtained results which have since passed into the 'golden treasury' of functional analysis. For example, in [2] he proved the theorem: in a complete normed space any closed convex centrally symmetric set that contains an interval on every ray starting from the origin contains a whole sphere. Nowadays, a similar property is used as the definition of an important class of linear topological spaces ('barrelled' spaces).

Gel'fand's thesis 'Abstract functions and linear operators' (1935) contains a number of theorems on the general form of linear operators in normed spaces. But perhaps even more important than these theorems is the method that Gel'fand developed for their proof. By applying to a function $x(t)$ with values in a normed space E any linear functional f, we obtain an ordinary function that can be studied by the methods of classical analysis; and since by virtue of the Hahn-Banach theorem linear functionals in E exist, this provides adequate information about the abstract function $x(t)$ itself. These observations now seem self-evident, but they were first exploited in Gel'fand's thesis. His paper [9] on one-parameter groups of operators should also be mentioned.

Gel'fand's most important achievement in the pre-war years, however, was the theory he created of commutative normed rings and this was the subject of his doctoral thesis (1938). Although mathematicians before

Gel'fand had studied normed rings (Riesz, Nagumo, Mazur, Ditkin), it was he who brought to light the fundamental concept of a maximal ideal which made it possible to unite previously uncoordinated facts and to create an interesting new theory. Gel'fand's theory of normed rings revealed close connexions between Banach's general functional analysis and classical analysis. For example, Wiener's well known theorem: if a function $x(t)$ has an absolutely convergent Fourier expansion and does not vanish, then $\frac{1}{x(t)}$ has the same properties, turned out to be essentially algebraic. Gel'fand's proof of this by the method of normed rings, which took five lines, demonstrated the power of the new theory and at the same time brought it to the attention of the mathematical world. The fundamental theorem of the theory concerning the mapping of any commutative normed ring into the ring of continuous functions on a certain compactum remains to this day one of the most outstanding achievements of functional analysis. An important component part of this theorem should be mentioned: the full ring R of all continuous functions on a compactum is distinguished from all others in having an involution $x(t) \to \overline{x(t)}$ which is an antilinear automorphism $x \to x^*$ of the ring R with the property that $|x^*x| = |x^*| \cdot |x|$ for any $x \in R$.

The closest noncommutative analogue of the ring of all continuous functions on a compactum is the ring of linear operators in Hilbert space, where there is also an involution, the transition to the adjoint operator. So, in the next brilliant paper [16], Gel'fand proved, jointly with M.A. Naimark (1942), that every (noncommutative) normed ring with an involution can be realized as a certain ring of linear operators in a Hilbert space with its natural involution. This paper provided a link between the theory of normed rings and the theory of infinite-dimensional representations of groups, which was developed by Gel'fand and Naimark in the post-war years.

Within the scope of a short article it would be difficult to throw light on the whole of Izrail' Moiseevich's creative work, which began more than 30 years ago and has included not only fundamental developments in functional analysis, but also the theory of differential equations, problems of computational mathematics and more recently physiology and biocybernetics as well. We shall, therefore, dwell only on some of the principal trends in his work.

Work on the theory of representations and on automorphic functions

Algebraic questions in analysis have always occupied a prominent place in Gel'fand's scientific interests. In particular, in the early 1940's, the theory of the representations of continuous groups attracted his attention and he guessed that here was a field which would most strikingly combine algebraic and analytic aspects.

The theory of finite-dimensional representations, mainly in connexion with finite groups, had been constructed by Frobenius and Schur. Fundamental investigations on finite-dimensional representations of continuous groups are due to E. Cartan and H. Weyl. In particular, for the representations of compact groups, an exhaustive theory was built up in the well known paper by Peter and Weyl. For non-compact groups, however, the situation was far more complicated. On the one hand, it had been shown that such groups cannot in general have non-trivial unitary finite-dimensional representations, and on the other hand, on examining the infinite-dimensional representations of such groups, substantial complications of a set-theoretic nature were revealed. Thus, even the very formulations of the basic problems were not clear. It was Gel'fand who succeeded in finding the correct approach. He noticed that unitary representations were of fundamental importance and evolved a deep and important theory of infinite-dimensional unitary representations of locally compact groups.

In 1943, I.M. Gel'fand and D.A. Raikov [17] showed that for any locally compact group there exist irreducible unitary representations in a Hilbert space. The next problem after this was to describe these representations for the most important groups. It was not at all clear whether a sufficiently explicit description could be given, even for groups such as the group of complex second order matrices.

From 1944 to 1948, Gel'fand and Naimark constructed a theory of infinite-dimensional representations of the classical complex Lie groups. They established that irreducible unitary representations of these groups can be given by simple explicit formulae. We shall indicate first the representation formulae they obtained for the case of the group G of complex second order matrices with determinant 1. This group has two series of irreducible unitary representations, a fundamental and a supplementary series. The representations of the fundamental series are constructed in the space of functions $f(z)$, $z = x + iy$, of integrable square modulus. The operator $T(g)$ corresponding to the matrix $g = \begin{pmatrix}\alpha & \beta \\ \gamma & \delta\end{pmatrix}$ is defined in the following way.

$$T(g) f(z) = f\left(\frac{\alpha z+\gamma}{\beta z+\delta}\right) |\beta z+\delta|^{is+n-2} (\beta z+\delta)^{-n}, \qquad (1)$$

where n is an integer and s a real number.

The representations of the supplementary series are constructed in the space of functions $f(z)$ with a different scalar product; the operators of the representation are defined by formula (1), where $n = 0$ and $s = i\rho$ is an imaginary number with $-2 \leqslant \rho \leqslant 2$.

It is worth noting that the group of complex second order matrices is not only interesting as an object of pure mathematics. This group is locally isomorphic to the Lorentz group, and for this reason Gel'fand and Naimark's results made an important contribution to theoretical physics. (Dirac had earlier succeeded in finding isolated representations of this group).

An important achievement of Gel'fand and Naimark was their discovery of manifolds on which irreducible representations of the classical complex Lie groups are realized in the most natural way. These manifolds are most simply described in the case of the group G of all linear transformations of a complex n-dimensional space with determinant 1. In this case, the points of the manifold are the 'flags', that is, the sequences of linear subspaces $H_1 \subset H_2 \subset \ldots \subset H_{n-1}$ (H_k denotes a k-dimensional subspace for

$k = 1, \ldots, n - 1$). The elements g of G determine in the natural way the transformations $z \to zg$ in the manifold of all flags. The operators of the irreducible representations are defined in the space of functions $f(z)$ by the formula

$$T(g) f(z) = f(zg)\,\alpha(z, g),$$

where $\alpha(z, g)$ is a certain function which can be simply described and depends on $2n - 2$ parameters (the parameters of the representation). If a flag is interpreted as a chain of subspaces with gaps, for example the chain $H_1 \subset H_{n-1}$, then on the resulting manifolds the so-called degenerate representations of G are constructed in an analogous way.

Gel'fand and Naimark showed that for the irreducible unitary representations of the classical groups it was possible to define the trace of the operator $T(g)$ as a generalized function on the group and they obtained an explicit formula for this function. They proved that the representation is uniquely defined, to within equivalence, by its trace.

After these results the problem naturally arose of describing the unitary representations of the real semisimple Lie groups. Even today this problem is not completely solved despite the efforts of many mathematicians. However, in papers by I.M. Gel'fand and M.I. Graev, important results in this direction have been obtained. It has been established, in particular, that there exist as many different series of representations as there are non-isomorphic maximal abelian subgroups in the group and that each such series is realized in a space of functions that are analytic in some of the variables. In the most important cases formulae for the characters of the unitary representations were found. Although in the main these results were obtained for groups of real n-th order matrices, they shed light on the situation in the general case.

We may mention also the elegant work of I.M. Gel'fand and M.L. Tsetlin on finite-dimensional representations. In these papers all finite-dimensional representations for the unimodular and the orthogonal groups were described explicitly.

Questions concerning harmonic analysis on classical Lie groups have been prominent in Gel'fand's papers. If $f(g)$ is a function on the group G, it is natural to call the operator function $T_\pi(f)$ defined on the set of irreducible representations $T_\pi(g)$ of G, the Fourier transform of f. (In the case when G is the group of straight line motions, this definition coincides with that of the ordinary Fourier transform.) The problem arises of finding a formula for the inverse Fourier transform. This problem was solved by Gel'fand and Naimark for the classical complex Lie groups.

Another problem of harmonic analysis on these same groups solved by Gel'fand is the description of the Fourier transform of 'sufficiently well-behaved' functions on the group (the analogue of the Paley-Wiener theorem). The importance of this problem is that its solution sheds light on the structure of the space of all representations of the group. Gel'fand discovered, in fact, that the Fourier transform of a function $f(g)$, apart from some natural conditions on growth and analyticity, satisfies certain algebraic relations. These relations occur at particular points of the representation space and are associated with the existence in G of degenerate (in particular, finite-dimensional) representations. One of the most interesting parts of harmonic analysis is the theory of zonal harmonics. These are defined as follows. Let G be a group, U a subgroup of G, and T_g an irreducible unitary representation of G in a Hilbert space H. If there exists in H a vector ξ, invariant with respect to the operators T_u and contained in U, then the function $\varphi(g) = (\xi, T_g\xi)$ is called a zonal harmonic. The classical zonal harmonics are obtained when G is the group of rotations of a sphere and U is a subgroup of rotations about a fixed axis. With a different choice of the groups G and U many other special functions can be obtained.

Gel'fand adapted the methods of the theory of normed rings to the study of zonal harmonics. His idea was to examine the ring of those functions on the group that are constant on the double cosets with respect to U. This ring is commutative in the case of a symmetric space G/U. From this it follows immediately that in the space of any irreducible representation there is not more than one vector invariant with respect to the operators T_u, $u \in U$. This fact is the foundation of the whole theory.[1] It can be proved that the zonal harmonics $\varphi(g)$ give homomorphisms of the ring R into the field of complex numbers:

$$M_{\varphi}(f) = \int f(g)\,\varphi(g)\,dg.$$

In this connexion Gel'fand introduced the so-called Laplace operators on groups, that is, differential operators, that commute with the motions and have spherical functions as their eigenfunctions.

One of the strongest results in the theory of spherical functions is the explicit formula obtained by Gel'fand and Naimark for zonal harmonics on the noncompact symmetric spaces associated with the complex Lie groups. These investigations were continued for the real Lie groups by Bhanu Murti, F.I. Karpelevich and S.G. Gindikin.

Various applications of the theory of representations are generally linked with harmonic analysis on homogeneous spaces. It applies for example to the study of automorphic functions, which can be carried out within the framework of harmonic analysis on a homogeneous space with a discrete stationary group.

In this field, Gel'fand and Graev proposed the very effective horosphere method. This method allowed them to construct a harmonic analysis for a number of important spaces. The essence of the horosphere method is as follows. Let X be a homogeneous space on which the complex semisimple Lie group G operates. The trajectories of the maximal nilpotent subgroup of G are called horospheres in X (for the ordinary Lobachevskii space this definition is equivalent to the usual definition of a horosphere). If the function $f(x)$, $x \in X$, is integrated with respect to all possible horospheres, we obtain a function $\varphi(\omega)$ in the space of all horospheres. In general, this horosphere space has a simpler structure than the original space X, and the resolution of $\varphi(\omega)$ into irreducible

[1] For compact symmetric spaces this fact was established by E. Cartan.

representations can be effected in an elementary way.

The problem that remains to be solved is one of integral geometry: to recover from $\varphi(\omega)$ the original function $f(x)$.

In recent years Gel'fand and Graev have been studying the representations of groups over arbitrary fields. They have succeeded in obtaining strong results on the representations of the Chevalley-Dickson groups over finite fields. (As is well known, these groups are represented by groups of matrices, analogous to the complex semisimple groups.)

This classical field is one in which several mathematicians have worked, from Frobenius onwards, and it has been the unexpected possibility of applying to these problems the methods of infinite-dimensional representations which has led to this progress.

For the group of second order matrices over an arbitrary continuous locally compact field K, Gel'fand and Graev obtained a common description of the irreducible unitary representations for all fields. It may be noted that in the course of this research several interesting classes of special functions on a locally compact field were discovered.

Gel'fand's interests have always included the classical as well as the new branches of mathematics. One of them is the theory of automorphic functions. Gel'fand made the remarkable observation that the theory of automorphic functions is essentially a part of representation theory. More precisely, almost all problems of the theory of automorphic functions can be formulated within the framework of the following problem of the theory of representations. A representation of a semisimple Lie group G in the space of functions $f(x)$, $x \in X$ is given, where X is a homogeneous space with a discrete stationary subgroup. The problem is to resolve this into irreducible representations.

In a paper in 1952 on geodesic flows on manifolds of constant negative curvature, I.M. Gel'fand and S.V. Fomin showed that the dimension of the space of automorphic forms was equal to the multiplicity with which the representation of the so-called discrete series enters into the given representation. In recent years Gel'fand and I.I. Pyatetskii-Shapiro began a systematic study of the spectrum of representations of the group G in the spaces G/Γ, where G is an arbitrary semisimple Lie group and Γ a discrete subgroup of G. In this investigation the horosphere method, of which we spoke above, was applied with success.

With the help of this method, the space of functions $f(x)$, $x \in G/\Gamma$, is mapped into the space of functions $\varphi(\omega)$ defined on the set of compact horospheres. By the same token the study of the spectrum of the representation falls into two parts; a study of the spectrum in the space of functions $\varphi(\omega)$, and in the kernel of the mapping $f(x) \to \varphi(\omega)$. It was shown that the kernel of this mapping, that is, the space of functions on G/Γ whose integrals over any compact horosphere vanish, always has a discrete spectrum (see Trudy Mosk. Matem. Obshch., 12). For studying the space of functions $\varphi(\omega)$ the methods of the theory of perturbations were employed. The analogues of the S-function of quantum mechanics which arise in this connexion are important number-theoretic functions of Riemann ζ-function type.

The combination of algebraic methods and extensive use of analytic technique is characteristic of the theory of infinite-dimensional representations. This theory in its turn greatly influenced analysis and was applied to a number of problems. A typical example was the description of all the relativistic invariant equations, given by I.M. Gel'fand and A.M. Yaglom [31].

Work on differential equations

Over the years, various questions connected with differential equations have attracted Izrail' Moiseevich's attention. First and foremost among these come the researches of I.M. Gel'fand and B.M. Levitan on the inverse Sturm-Liouville problem, which have received wide acclaim.

Let us examine the equation

$$y'' + (\lambda - q(x))\, y = 0 \tag{1}$$

on the semi-axis $[0, +\infty)$. The eigenfunctions $\varphi(x, \lambda)$ of this equation are usually normalized by boundary conditions at the origin

$$\varphi(0, \lambda) = 1, \quad \varphi'(0, \lambda) = h. \tag{1'}$$

The resolution of an arbitrary function $f(x)$ as a Fourier integral with respect to the eigenfunctions of (1) takes the form

$$f(x) = \int_{-\infty}^{+\infty} F(\lambda)\, \varphi(x, \lambda)\, d\rho(\lambda),$$

where $\rho(\lambda)$ is a monotonic function with growth points on the spectrum. The function $\rho(\lambda)$ is called the spectral function of the Sturm-Liouville problem. The problem known as the inverse Sturm-Liouville problem is the task of discovering the function $q(x)$ in (1) according to some specified spectral characteristics of the equation.

In various formulations the inverse problem has interested V.A. Ambartsumyan, G. Borg, N. Levinson, V.A. Marchenko, M.G. Krein and several physicists. In particular, M.G. Krein solved in 1951 the inverse problem for two given spectra of (1) which corresponded to various boundary conditions at the ends of a finite interval.

The most general and fruitful form of the inverse problem turned out to be the determination of $q(x)$ in (1) for a given function $\rho(\lambda)$.

V.A. Marchenko (1950) and the Swedish mathematician G. Borg first tackled the problem in this formulation and proved that not more than one function $q(x)$ can correspond to a given $\rho(\lambda)$.

Gel'fand and Levitan found a linear integral equation connecting the functions $q(x)$ and $\rho(\lambda)$. This equation not only settled the question of the existence of $q(x)$, but also indicated a constructive procedure for its determination.

At the heart of this solution lay the idea of orthogonalization of the system of functions $\cos\sqrt{\lambda x}$ $(0 \leqslant \lambda < \infty)$ by the weight $\rho(\lambda)$, suggested by analogy with the problem of moments. In the problem of moments multiplication by λ generates on the basis of the orthogonal polynomials the Jacobi matrix (the finite-difference analogue of the Sturm-Liouville operator), and in just the same way on the basis of the orthogonalized cosines the Sturm-Liouville operator emerges.

This work aroused great interest among mathematicians and theoretical physicists and stimulated a number of important investigations (M.G. Krein, V.A. Marchenko, N. Levinson, L.D. Fadeev, Jost and Kohn, Newton, Kay and Moses; etc.)

The physicists' interest lay in the scattering problem of quantum mechanics, where the Gel'fand-Levitan equation made the determination of the potential of the field possible according to the phase of the scattering and yielded a complete solution of the problem in the radially symmetric case.

Another important result for the Sturm-Liouville operator, obtained by Gel'fand in collaboration with B.M. Levitan and L.A. Dikii, was the set of formulae for the traces of such an operator, considered on a finite interval [54]. Let $\lambda_1, \lambda_2, \ldots, \lambda_n$ be the eigenvalues of (1) under the boundary conditions $y(0) = y(\pi) = 0$, and let $\int_0^\pi q(x)dx = 0$. Then

$$\sum_{n=1}^{\infty} (\lambda_n - n^2) = -\frac{q(0)+q(\pi)}{4}. \tag{2}$$

The convergence of the series (2) follows from the classical theory for λ_n, but the possibility of finding its sum in an explicit form was a surprise. Similar formulae, containing integer powers of λ_n, were derived by Gel'fand and Dikii on the assumption that $q(x)$ was sufficiently smooth. The formulae for the traces can be used for the numerical evaluation of the first eigenvalues of a problem. This work was continued by various other writers. In particular, L.D. Fadeev obtained the analogue of formula (2) for the continuous spectrum.

In the theory of hamiltonian systems with coefficients periodic with respect to time, the paper of I.M. Gel'fand and V.B. Lidskii [60] is well known. In this, a complete description is given of the topological structure of the stability regions of linear hamiltonian systems. The successful development of this topic was due to the algebraization of the problem. It was shown in the paper that to each linear hamiltonian system there corresponds a certain function $y(t)$ $(0 < t < \omega)$ in the group G of real symplectic matrices, and conversely, to each such curve there corresponds a hamiltonian system.

Gel'fand has proposed a number of problems in the theory of partial differential equations which have considerably influenced the development of various parts of this theory during the last 20 years or so. In the seminar which he conducted at the University of Moscow in 1945/6, he raised the question of describing the domain of definition of the closure of differential operators under appropriate boundary conditions and that of finding well-posed (or correct) boundary value problems, for example, for all elliptic differential equations. Subsequently these problems were satisfactorily solved by M.I. Vishik and O.A. Ladyzhenskaya, participants in this seminar, and also in papers by L. Hörmander and various other mathematicians.

In his paper [103], Gel'fand posed the question of the homotopy classification of all systems of differential equations that are elliptic in the sense of I.G. Petrovskii and the boundary problems for them. He had spoken about this problem also in the 1945-6 seminar. As is well known, elliptic boundary value problems are normally soluble in a finite region, that is, the corresponding homogeneous problem has a finite number k of linearly independent solutions and the inhomogeneous problem is soluble only when l additional conditions on the right hand sides are satisfied. The number $\varkappa = k - l$ is called the index of the boundary value problem and is its main homotopy invariant. Gel'fand's problem was solved for a whole range of cases in papers by A.I. Vol'pert, A.S. Dynin, M.S. Agranovich and others. Its general solution was given recently by Atiyah and Singer.

Paper [90] is different from the usual mathematical articles. In it are many fruitful ideas, not always rigorously justified. We may mention the important concept of the state of evolution of a system which greatly influenced investigations into the structure and stability of shock waves in ordinary and magneto-hydrodynamics. The mathematical study of the equations of magnetohydrodynamics, outlined in this paper, was one of the first and had a powerful influence on all later work on this question. The article also included a formulation by I.M. Gel'fand and Ya. B. Zel'dovich and the solution of a problem on the establishment of a given chemical reaction, which proceeds with the formation of an intermediate product of the active centres.

The theory of distributions

I.M. Gel'fand was one of the first Soviet mathematicians to appreciate the future prospects and the importance of the work of S.L. Sobolev, and later L. Schwartz, on the theory of generalized functions. In the subsequent development of this theory, Izrail' Moiseevich's papers and those of his students and collaborators played a leading part. In the paper [58] by I.M. Gel'fand and G.E. Shilov, the idea that generalized functions could and must be constructed on various spaces of fundamental functions had already taken shape, the most suitable class of spaces being chosen for each group of problems. This idea transformed the theory of distributions into a flexible instrument, which found applications in the theories of partial differential equations, representations, stochastic processes, integral geometry and so on. About 10 years ago, Gel'fand planned a series of monographs devoted to the application of the ideas and methods of the theory of distributions to various branches of functional analysis and related fields. This series *'Obobshchennye funktsii'*, 'Generalized functions', in which five books have already appeared, has now achieved international recognition and fame; the results obtained by Gel'fand and his school have been the foundation of the series. Let us look briefly at the contents of these books. The first three volumes, written by Gel'fand and Shilov, are devoted to the development of the techniques of generalized functions, that is, a description of various classes of fundamental spaces and the corresponding spaces of linear functionals, and the construction of a theory of Fourier transforms for generalized functions. These classes of spaces were the basis for the investigations into the correctness and uniqueness of Cauchy's problem for systems of partial differential equations.

The first volume also gives an account of the results of I.M. Gel'fand and Z.Ya. Shapiro concerning generalized functions of several variables, and develops in particular the theory of homogeneous generalized functions and functions of n variables that are concentrated on manifolds of fewer than n dimensions.

Gel'fand repeatedly expressed his view that the general theorem on the spectral decomposition of a self-adjoint operator could not be regarded as the definitive solution of the problem of spectral analysis and that it was important to construct side by side with the corresponding decompositions the units and specific eigenfunctions (generally speaking, generalized functions) corresponding to individual points of the spectrum. This idea was realized in a paper by I.M. Gel'fand and A.G. Kostyuchenko.

The most detailed presentation of these questions is given in the third volume of the series.

The range of questions examined in the last two volumes of '*Obobshchennye funktsii*' differs appreciably from the topics of the first three, which were largely associated with questions in the theory of differential equations. The fourth volume, written by Gel'fand in collaboration with N.Ya. Vilenkin, is devoted mainly to the theory of nuclear spaces, introduced by A. Grothendiek. Gel'fand gave another formulation of this concept and explained its important rôle in various problems, for example in studying measures in linear spaces. In connexion with the above task of finding generalized eigenfunctions of self-adjoint (and unitary) operators, Gel'fand introduced the concept of the so-called equipped Hilbert space. This can be explained as follows.

Let us examine a countable Hilbert nuclear space Φ in which an additional scalar product has been introduced, apart from the scalar products which define the topology of the space. When Φ is enlarged by means of this new scalar product, we obtain a Hilbert space H in which Φ is an everywhere dense set. In addition to Φ and H, we consider the space Φ' of linear functionals on Φ. The triplet Φ, H, Φ' is called an equipped Hilbert space. This concept leads to results that are elegant and complete, such, for example, as the following: every self-adjoint operator in an equipped Hilbert space possesses a complete system of generalized eigenvectors corresponding to the eigenvalues.

In addition the fourth volume contains the theory of positive definite distributions (Chapter 2), the theory of generalized stochastic processes which Gel'fand had constructed in 1955 and which gave a precise mathematical basis to such notions (popular with physicists) as 'white noise', and also the theory of measure in linear topological spaces.

The fifth volume, written by Gel'fand jointly with M.I. Graev and N.Ya. Vilenkin, discusses harmonic analysis on the Lorentz group and homogeneous spaces associated with it (in particular, on the three-dimensional Lobachevskii space). Integral geometry provided the foundation for these investigations. Integral geometry, in the sense in which it is understood in the book, represents a new trend in functional analysis, linking it with classical ideas in geometry. Essentially, it consists in the transition from functions defined on a set of certain geometric entities to functions defined on a set of different entities. For example, if a function $f(x)$, defined on the hyperboloid $x_0^2 - x_1^2 - x_2^2 - x_3^2 = 1$, is integrated along all its various generators ω, a function $\varphi(\omega)$ is obtained which is defined on the set of generators. This gives rise to the following problem of integral geometry: to recover from $\varphi(\pi)$ the initial function $f(x)$. Gel'fand observed that several problems of the theory of representations could easily be reduced to the solution of similar problems of integral geometry.

The solutions of many interesting problems of integral geometry are given in the book. Some of these are not directly connected with the theory of representations. Thus, the book contains a study of the simplest transformation of integral geometry that associates with a function in an affine space its integrals over the hyperplanes. This transformation is closely connected with the ordinary Fourier transform; it is, however, more geometric and this gives it a certain advantage.

Work on computational mathematics and physiology

Gel'fand has made a very substantial contribution to the development of computational mathematics. He must be given great credit for his discovery of general methods for the numerical solution of equations of mathematical physics and also for the solution of particular applied problems.

Together with his collaborators, Gel'fand made advances in the spectral theory of the stability of difference operators. An account of some of the results of these papers is given by S.K. Godunov and V.S. Ryaben'kii in their book *'Vvedenie v teoriyu raznostnykh skhem'* ('Introduction to the theory of difference schemes'); see also [121]. In particular, I.M. Gel'fand and K.I. Babenko were the first to analyze the influence of the boundary on the spectrum of a difference operator. The requirements of stability lead as a rule to the necessity of using implicit difference schemes. The question therefore arose of solving a system consisting of a large number of algebraic equations. With this aim in view, a simple stable algorithm, which has been widely introduced into computational practice under the name of 'double sweep', was applied. This algorithm allows the implicit scheme to be solved in the case of both univariate and multivariate problems.

From the first Gel'fand understood the fundamental importance for computation of methods for investigating systems involving a large number of variables, where the customary methods of analysis proved ineffective. The paper [74] in which a straightforward calculation of the path (or continual) integral was undertaken for the first time attracted considerable attention.

Computation by the Monte Carlo method of mean trajectories of particles in space was subsequently applied to the solution of problems in transfer physics [89], [97]. These papers stimulated a great deal of interest in the creation of quadrature formulae for integrals of high multiplicity. The methods for determining the minima of functions of a large number of variables [112] were a further development in the study of complicated systems. Gel'fand proposed a non-local search method for minima (the so-called 'ravine' method). The method was used for the solution of problems of phase-shift analysis of nucleon scattering [108], [109] and for deciphering the structure of crystalline amino-acids [131], [132].

In recent years Gel'fand has taken a deep interest in the study of complicated physiological systems. His various important ideas in this field (continual control systems and excitable media, the tactics of movements [110], the principle of least interaction [120] and others) have encouraged a talented group of young physiologists to gather round him.

Teaching activities

When speaking of Gel'fand's creative work as a scholar, one cannot help but mention also his teaching activities. He began lecturing at the Moscow State University 30 years ago, when he was a young man of 20, and he has continued to work there right up to the present time. One of the characteristic features of all Izrail' Moiseevich's activities has been the extremely close bond between his research work and his teaching. The formulation of new problems and unexpected questions, a tendency to look at even well known things from a new point of view characterizes Gel'fand as a teacher, regardless of whether at a given moment he is holding a conversation with schoolchildren or with his own colleagues.

The first of Gel'fand's students was G.E. Shilov, who came to him as a research student 25 years ago. Since that time he has supervised dozens of students, several of whom have already become prominent scholars in their own right. Besides the 'first class' students whom he has trained in research, many mathematicians have in one way or another come under his influence in lectures, seminars and private discussions. His seminar on functional analysis is well known; it celebrated its twentieth anniversary on the very same day that its instructor celebrated his fiftieth birthday. This seminar has long been one of the major centres of development in functional analysis and of mathematical training for young people.

A distinctive feature of Gel'fand's creative work has been his skill in organizing purposeful, concerted work in a team. A large number of Gel'fand's papers have been written in collaboration with his colleagues and students, often quite young ones, for whom such combined work has been an exceedingly valuable experience. Thus it is practically impossible to separate his own research work proper from his teaching and supervising activities.

Gel'fands' lectures have always enjoyed a wide popularity. Often resembling a lively conversation, they have always demanded of the audience attention and participation and stimulated their imagination by providing food for independent meditation.

The textbooks written by Gel'fand on linear algebra, calculus of variations (jointly with S.V. Fomin) and the theory of group representations (jointly with Z.Ya. Shapiro and R.A. Minlos) are widely known.

Soviet mathematicians warmly congratulate Izrail' Moiseevich and wish him many years of health and outstanding new successes in science.

M.I. Vishik, A.N. Kolmogorov, S.V. Fomin, G.E. Shilov.

LIST OF I.M. GEL'FAND'S PUBLICATIONS

DAN = Doklady Akademii Nauk. UMN = Uspekhi Matematicheskikh Nauk.
RMS = Russian Mathematical Surveys.

1936

1. Sur une lemme de la théorie des espaces linéaires. Izv. Nauchno-issled. Inst. Matem. Khar'kov Univ., seriya 4, 13, 35-40.

1937

2. On the theory of abstract functions. DAN 17, 237-240.
3. Operators and abstract functions. DAN 17, 241-244.

1938

4. Abstrakte Funktionen und lineare Operatoren. Mat. Sb. 4 (46), 235-286.

1939

5. On rings of continuous functions. DAN 22, 11-15 (with A.N. Kolmogorov).
6. On normed rings. DAN 23, 430-432.
7. On absolutely convergent trigonometric series and integrals. DAN 25, 570-572.
8. On the ring of almost periodic functions. DAN 25, 573-574.
9. On one-parameter groups of operators in a normed space. DAN 25, 711-716.

1940

10. On the theory of characters of commutative topological groups. DAN 28, 195-198 (with D.A. Raikov).

1941

11. Normed rings. Mat. Sb. 9 (51), 3-24.
12. One some methods of introducing a topology into the set of maximal ideals of a normed ring. Mat. Sb. 9 (51), 25-40. (with G.E. Shilov).
13. Ideals and prime ideals in normed rings. Mat. Sb. 9 (51), 41-48.
14. On the theory of characters of abelian topological groups. Mat. Sb. 9 (51), 49-50.
15. On absolutely convergent trigonometric series and integrals. Mat. Sb. 9 (51), 51-56.

1943

16. On the embedding of a normed ring into the ring of operators in a Hilbert space. Mat. Sb. 12 (54), 197-217 (with M.A. Naimark).
17. Irreducible unitary representations of locally compact groups. Mat. Sb. 13 (55), 301-316 (with D.A. Raikov).
= Amer. Math. Soc. Transl. (2), 36 (1964), 1-15.

1944

18. Irreducible unitary representations of locally compact groups. DAN 42, 203-205 (with D.A. Raikov).

1946

19. Unitary representations of the Lorentz group. Journ. of Phys. 10, 93-94 (with M.A. Naimark) (in English).
20. On unitary representations of the complex unimodular group. DAN 54, 194-198 (with M.A. Naimark).
21. Commutative normed rings. UMN 1, 2, 48-146 (with D.A. Raikov and G.E. Shilov). = Amer. Math. Soc. Transl. (2), 5 (1957), 115-220.
22. Unitary representations of the group of linear transformations of the straight line. DAN 55, 571-574 (with M.A. Naimark).
23. The fundamental series of irreducible representations of the complex unimodular group. DAN 56, 3-4 (with M.A. Naimark).
24. Unitary representations of the Lorentz group. Izvest. Akad. Nauk 2, 411-504 (with M.A. Naimark).
25. Unitary representations of semisimple Lie groups I. Mat. Sb. 21 (63), 405-434 (with M.A. Naimark).
    = Amer. Math. Soc. Transl. (1) (reprinted), 9 (1962), 1-41.
26. Supplementary and degenerate series of unitary representations of the unimodular group. DAN 58, 1577-1580 (with M.A. Naimark).

1948

27. *Lektsii po lineinoi algebre.* (Lectures on linear algebra) Moscow-Leningrad Gostekhizdat.
28. Normed rings with an involution. Izvest. Akad. Nauk (seriya matem.) 12, 445-480 (with M.A. Naimark).
29. General relativistically invariant equations and infinite-dimensional representations of the Lorentz group. DAN 59, 655-659 (with A.M. Yaglom).
30. Integral equations. Article in the Great Soviet Encyclopedia.
31. General relativistically invariant equations and infinite-dimensional representations of the Lorentz group. Zhurnal Eksper. Teoret. Fiz. 18, 703-733 (with A.M. Yaglom).
32. The trace in the fundamental and supplementary series of representations of the complex unimodular group. DAN 61, 9-11 (with M.A. Naimark).
33. Relativistically invariant equations corresponding to a definite charge and a definite energy. DAN 63, 371-374 (with A.M. Yaglom).
34. Pauli's theorem for general relativistically invariant equations. Zhurnal Eksper. Teoret. Fiz. 18, 1096-1104 (with A.M. Yaglom).
35. Charge conjugacy for general relativistically invariant equations. Zhurnal Eksper. Teoret. Fiz. 18, 1105-1111 (with A.M. Yaglom).
36. On the connexion between the representations of a semisimple Lie group and those of its maximal compact subgroup. DAN 63, 225-228 (with M.A. Naimark).
37. The analogue of Plancherel's formula for the complex unimodular group. DAN 63, 609-612 (with M.A. Naimark).

1950

38. The centre of the infinitesimal group ring. Mat. Sb. 26 (68), 103-112.
39. The connexion between the unitary representations of the complex unimodular group and those of its unitary subgroup. Izvest. Akad. Nauk 14, 239-260 (with M.A. Naimark).
40. Spherical functions in symmetric Riemann spaces. DAN 70, 5-8.
    = Amer. Math. Soc. Transl. (2), 37 (1964), 39-43.
41. Finite-dimensional representations of the group of unimodular matrices. DAN 71, 825-828 (with M.L. Tsetlin).
42. Finite-dimensional representations of the group of orthogonal matrices. DAN 71, 1017-1020 (with M.L. Tsetlin).
43. Eigenfunction expansions of equations with periodic coefficients. DAN 73, 1117-1120.

44. Unitary representations of the classical groups. Trudy Matem. Inst. V.A. Steklov. 36, 1-288 (with M.A. Naimark).

1951

45. Unitary representations of Lie groups and geodesic flows on surfaces of constant negative curvature. DAN 76, 771-774 (with S.V. Fomin).
46. *Lektsii po lineinoi algebre*. 2nd Edition, revised and enlarged. Moscow-Leningrad, Gostekhizdat.
= Translation: Lectures on linear algebra, New York, Interscience 1961.
47. On the determination of a differential equation by its spectral function. Izvest. Akad. Nauk (seriya matem.) 15, 309-361 (with B.M. Levitan).
48. Remark on N.K. Bari's paper 'Biorthogonal systems and bases in a Hilbert space'. Uchen. Zap. MGU, 140 (matem.), 4, 224-225.

1952

49. Representations of the group of rotations of 3-dimensional space and their applications. UMN 7, 1, 3-117 (with Z.Ya. Shapiro).
50. Geodesic flows on surfaces of constant negative curvature. UMN 7, 1, 118-137 (with S.V. Fomin).
51. Unitary representations of the unimodular group that contain an identity representation of the unitary subgroup. Trudy Mosk. Matem. Obshch. 1, 423-473 (with M.A. Naimark).
52. Unitary representations of real simple Lie groups. DAN 86, 461-463 (with M.I. Graev).
53. On the spectrum of non-self-adjoint differential operators. UMN 7, 6, 183-184.

1953

54. On a simple identity for the eigenvalues of a differential equation. DAN 88, 593-596 (with B.M. Levitan).
55. The fundamental series of representations of the real unimodular group. Izvest. Akad. Nauk (seriya matem.) 17, 189-249 (with M.I. Graev).
56. On a general method of decomposition of the regular representation of a Lie group into irreducible representations. DAN 92, 221-224 (with M.I. Graev).
57. The analogue of Plancherel's theorem for real unimodular groups. DAN 92, 461-464 (with M.I. Graev).
58. Fourier transforms of rapidly increasing functions and questions of the uniqueness of the solution of Cauchy's problem. UMN 8, 6, 3-54 (with G.E. Shilov).
= Amer. Math. Soc. Transl. (2), 5 (1957), 221-274.

1954

59. Solution of the quantum field equations. DAN 97, 209-212 (with R.A. Minlos).

1955

60. On the structure of the regions of stability of linear canonical systems of differential equations with periodic coefficients. UMN 10, 1, 3-40 (with V.B. Lidskii).
= Amer. Math. Soc. Transl. (2), 8 (1958), 143-181.
61. Generalized random processes. DAN 100, 853-856.
62. The traces of unitary representations of the real unimodular group. DAN 100, 1037-1040 (with M.I. Graev).
63. The analogue of Plancherel's formula for the classical groups. Trudy Mosk. Matem. Obshch. 4, 375-404 (with M.I. Graev).
= Amer. Math. Soc. Transl. (2), 9 (1958), 123-154.
64. Homogeneous functions and their applications. UMN 10, 3, 3-70 (with Z.Ya. Shapiro).
= Amer. Math. Soc. Transl. (2), 8 (1958), 21-85.
65. On a new method in theorems concerning the uniqueness of the solution of Cauchy's problem for systems of linear partial differential equations. DAN 102, 1065-1068 (with G.E. Shilov).
66. Eigenfunction expansions of differential and other operators. DAN 103, 349-352 (with A.G. Kostyuchenko).

1956
67. Integration in function spaces and its applications in quantum physics. UMN 11, 1, 77-114 (with A.M. Yaglom).
68. On identities for the eigenvalues of a differential operator of the second order, UMN 11, 1, 191-198.
69. Some observations on the theory of spherical functions on symmetric Riemann manifolds. Trudy Mosk. Matem. Obshch. 5, 311-351 (with F.A. Berezin). = Amer. Math. Soc. Transl. (2), 21 (1962), 193-238.
70. Quelques applications de la théorie des fonctions généralisées. Journ. de Math. pures et appl. 38, 383-413 (with G.E. Shilov).
71. On the general definition of the amount of information. DAN 111, 745-748 (with A.N. Kolmogorov and A.M. Yaglom).
72. Some problems of functional analysis. UMN 11, 6, 3-12. = Amer. Math. Soc. Transl. (2), 16 (1960), 315-324.
73. Representations of groups. UMN 11, 6, 13-40 (with F.A. Berezin, M.I. Graev and M.A. Naimark). = Amer. Math. Soc. Transl. (2), 16 (1960), 325-353.
74. On the numerical evaluation of continuous integrals. Zhurnal Eksper. Teoret. Fiz. 31, 1106-1107 (with N.N. Chentsov).
75. On quantities with anomalous parity. Zhurnal Eksper. Teoret. Fiz. 31, 1107-1109 (with M.L. Tsetlin).
1957
76. On the computation of the amount of information about a stochastic function contained in another such function. UMN 12, 1, 3-52 (with A.M. Yaglom).
77. On subrings of the ring of continuous functions. UMN 12, 1, 247-251. = Amer. Math. Soc. Transl. (2) 16 (1960), 477-479.
78. Some aspects of functional analysis and algebra. Proc. International Congress of Math. 1954, Amsterdam, 1, 253-276 (in English).
1958
79. *Obobshchennye funktsii i deistviya nad nimi.* (Generalized functions and operations on them), Moscow. Fizmatgiz. (with G.E. Shilov). = Translation: Verallgemeiuerte Funktionen und das Rechnen mit ihnen, Deutscher Verlag der Wissenschaften, Berlin 1960.
80. *Prostranstva osnovnykh i obobshchennykh funktsii.* (Spaces of fundamental and generalized functions), Moscow, Fizmatgiz. (with G.E. Shilov). = Translation: Räume von Grundfunktionen und verallgemeinerten Funktionen, Deutscher Verlag der Wissenschaften, Berlin 1960.
81. *Nekotorye voprosy teorii differentsial'nykh uravnenii.* (Some questions in the theory of differential equations.), Moscow. Fizmatgiz. (with G.E. Shilov).
82. Some observations on hyperbolic systems. Nauchn. dokl. vyssh. shkoly, no. 1, 12-18 (with K.I. Babenko).
83. *Predstavleniya gruppy vrashchenii i gruppy Lorentsa*, Moscow, Fizmatgiz. (with R.A. Minlos and Z.Ya. Shapiro). = Translation: Representations of the rotation and Lorentz groups, London, Pergamon Press 1963.
84. Path integrals. Proceedings of the Third All-Union Mathematics Congress 3, 521-531 (with A.M. Yaglom and R.A. Minlos).
85. Representations of Lie groups. Proceedings of the Third All-Union Mathematics Congress 3, 246- 254 (with F.A. Berezin, M.I. Graev and M.A. Naimark).
86. The amount of information and entropy for continuous distributions. Proceedings of the Third All-Union Mathematics Congress 3, 300-320 (with A.N. Kolmogorov and A.M. Yaglom).
87. The theory of systems of partial differential equations. Proceedings of the Third All-Union Mathematics Congress 3, 65-72 (with I.G. Petrovskii and G.E. Shilov).
88. The theory of the compression and pulsation of a plasma column under a powerful pulse discharge. Fizika plazmy i problema upravlyaemykh termoyadernykh reaktsii 4, 201-222 (with C.I. Braginskii and R.P. Fedorenko).

89. The computation of continuous integrals by the Monte Carlo method. Izv. Vyssh. Uchebn. Zaved., ser. matem., 32-45 (with N.N. Chentsov and A.S. Frolov).

1959

90. Some problems of the theory of quasi-linear equations. UMN 14, 2, 87-158. = Amer. Math. Soc. Transl. (2), 29 (1963), 295-381.

91. The theory of representations and the theory of automorphic functions. UMN 14, 2, 171-194 (with I.I. Pyatetskii-Shapiro).

92. Some questions in analysis and differential equations. UMN 14, 3, 3-19. = Amer. Math. Soc. Transl. (2), 26 (1963), 201-219.

93. The geometry of homogeneous spaces, the representations of groups in homogeneous spaces and related questions in integral geometry. Trudy Mosk. Matem. Obshch. 8, 321-390 (with M.I. Graev). = Amer. Math. Soc. Transl. (2), 37 (1964), 351-429.

94. The resolution into irreducible components of representations of the Lorentz group in the spaces of functions defined on symmetric spaces. DAN 127, 250-253 (with M.I. Graev).

95. On a theorem of Poincaré, DAN 127, 490-493 (with I.I. Pyatetskii-Shapiro).

96. On the structure of a ring of rapidly decreasing functions on a Lie group. DAN 124, 19-21 (with M.I. Graev).

1960

97. The application of the Monte Carlo method to the solution of a kinetic equation. Proceedings of the Second International Conference on the Peaceful Use of Atomic Energy. (Geneva, 1958), 2, 628-683 (with N.N. Chentsov, S.M. Feinberg and A.S. Frolov).

98. On positive definite generalized functions. UMN 15, 1, 185-190 (with Sya Do-Shin).

99. Integral geometry and its relation to the theory of representations. UMN 15, 2, 155-164 = RMS 15, 2, 143-151.

100. Fourier transforms of rapidly decreasing functions on complex semisimple groups. DAN 131, 496-499 (with M.I. Graev).

101. *Nekotorye primeneniya garmonicheskogo analiza. Osnashchennye gil'bertovy prostranstva.* Some applications of harmonic analysis. Equipped Hilbert spaces. Moscow. Fizmatgiz. (with N.Ya. Vilenkin).

102. On continuous models of control systems. DAN 131, 1242-1245 (with M.L. Tsetlin).

103. On elliptic equations. UMN 15, 3, 121-132 = RMS 15, 3, 113-123.

104. On a paper by K. Hoffman and I.M. Singer. UMN 15, 3, 239-240.

105. *Kommutativnye normirovannye kol'tsa*. Moscow. Fizmatgiz. (with D.A. Raikov and G.E. Shilov). = Translation: Commutative normed rings, New York, Chelsea 1964.

106. Integrals over hyperplanes of fundamental and generalized functions. DAN 135, 1307-1310 (with M.I. Graev). = Soviet Mathematics, 1 (1960), 1369-1372.

1961

107. *Variatsionnoe ischislenie*. Moscow. Fizmatgiz. (with S.V. Fomin). = Translation: Calculus of variations, Englewood Cliffs (N.J.), Prentice Hall 1963).

108. Phase-shift analysis of *pp*-scattering at an energy of 95 *Mev*. Zhurnal Eksper. Teoret. Fiz. 40, 1106-1111 (with V.A. Borovikov et al).

109. Phase-shift analysis of *pp*-scattering at an energy of 150 *Mev*. Zhurnal Eksper. Teoret. Fiz. 40, no 5, 1338-1342 (with A.F. Grashin and L.N. Ivanova).

110. Some considerations on the tactics of making movements. DAN 139, 1250-1253 (with M.L. Tsetlin and V.S. Gurfinkel).

111. Magnetic surfaces of the 3-path helical magnetic field excited by a crimped field. Zhurnal Tekh. Fiz. 31, 1164-1169 (with M.I. Graev et al).

1962

112. On some methods for the control of complex systems. UMN 17, 1, 3-25 (with M.L. Tsetlin) = RMS 17, 1, 95-117.

113. An example of the theoretical determination of a magnetic field that does not have magnetic surfaces. DAN 143, 81-83 (with N.M. Zueva, A.I. Morozov et al).
114. An application of the horosphere method to the spectral analysis of functions in real and imaginary Lobachevskii space. Trudy Mosk. Matem. Obshch. 11, 243-308 (with M.I. Graev).
115. *Integral'naya geometriya i svyazannye s nei voprosy teorii predstavlenii.* (Integral geometry and related questions of the theory of representations), Moscow. Fizmatgiz. (with M.I. Graev and N.Ya. Vilenkin).
116. The category of group representations and the problem of the classification of irreducible representations. DAN 146, 757-760 (with M.I. Graev).
= Soviet Mathematics, 3 (1962), 1378-1381.
117. Unitary representations in homogeneous spaces with discrete stationary groups. DAN 147, 17-20 (with I.I. Pyatetskii-Shapiro).
= Soviet Mathematics, 3 (1962), 1528-1531.
118. Unitary representations in the G/Γ space, where G is a group of real nth order matrices and Γ a subgroup of integral matrices. DAN 147, 275-278 (with I.I. Pyatetskii-Shapiro).
= Soviet Mathematics, 3 (1962), 1574-1577.
119. The construction of irreducible representations of simple algebraic groups over a finite field. DAN 147, 529-532 (with M.I. Graev).
= Soviet Mathematics, 3 (1962), 1646-1649.
120. On the techniques of control of complex systems in relation to physiology. Symposium 'Biological aspects of cybernetics'. Moscow, Publ. Akad. Nauk SSSR, 66-73 (with M.L. Tsetlin and V.S. Gurfinkel').
121. On difference-schemes for the solution of the heat conduction equation. The 'double sweep' method for the solution of difference equations. Supplements I and II to *'Vvedenie v teoriyu raznostnykh skhem'*. 'Introduction to the theory of difference-schemes', by S.K. Godunov and V.S. Ryaben'kii. Moscow. Fizmatgiz. (with O.V. Lokushchievskii).

1963

122. Intracellular irritation of the various compartments of a frog's heart. DAN 148, 973-976 (with S.A. Kovalev and L.I. Chailakhyan).
123. On the structure of a toroidal magnetic field that does not have magnetic surfaces. DAN 148, 1286-1289 (with M.I. Graev, A.I. Morozov et al).
124. Irreducible unitary representations of the group of unimodular matrices of the second order with elements from a locally compact field. DAN 149, 499-502 (with M.I. Graev).
= Soviet Mathematics, 4 (1963), 397-400.
125. Automorphic functions and the theory of representations. Trudy Mosk. Matem. Obshch. 12, 389-412 (with I.I. Pyatetskii-Shapiro).
= Proc. Moscow Math. Soc., 12 (1965).
126. Automorphic functions and the theory of representations. Proc. International Congress of Math. Stockholm, 1962, 74-85 (in English).
127. Plancherel's formula for the group of unimodular matrices of the second order with elements from a locally compact field. DAN 151, 262-264 (with M.I. Graev).
= Soviet Mathematics, 4 (1963), 397-400.
128. Representations of a group of second order matrices with elements from a locally compact field and special functions on locally compact fields. UMN 18, 4, 29-99 (with M.I. Graev)
= RMS 18, 4, 29-100.
129. On the possible mechanism of change of immunological tolerance. Uspekhi sov. biologii 55, 428-439 (with A.Ya. Fridenshtein).
130. On the synchronization of motor units and some related ideas. Biofizika 8, 475-488 (with V.S. Gurfinkel', Ya.M. Kots, M.L. Tsetlin and M.L. Shik).
131. The determination of crystal structure using the method of non-local search. DAN 152, 1045-1048 (with I.I. Pyatetskii-Shapiro and Yu.G. Fedorov).
= Soviet Mathematics, 4 (1963), 1487-1490.
132. The determination of crystal structures by the method of R-factor minimalization. DAN 153, 93-96 (with B.K. Vainshtein, R.A. Kayushina and Yu.G. Fedorov).

Translated by R.F. Wheeler

Izrail Moiseevich Gelfand
(On his sixtieth birthday)

Usp. Mat. Nauk SSSR **29** (1) (1974) 193–246
[Russ. Math. Surv. **29** (1) (1974) 3–61]

On 2 September 1973 Izrail' Moiseevich Gel'fand, Corresponding Member of the USSR Academy of Sciences, mathematician of worldwide renown, celebrated his sixtieth birthday.

His scientific work began 40 years ago, when he was a student of Kolmogorov. His very first articles, on the theory of abstract functions, contain a number of fundamental results, which are often mentioned in text-books on functional analysis.

In the late 1930's Gel'fand created the theory of commutative normed rings (Banach algebras), which was to become one of the most important branches of modern functional analysis. The combination of analytic and algebraic methods, which is characteristic of this work, is typical of a number of Gel'fand's later researches. In the early 1940's Gel'fand, together with Naimark, investigated (non-commutative) normed rings with an involution and established that any such ring can be realized as a ring of linear operators in a Hilbert space.

A natural continuation of this work was the extensive cycle of investigations of Gel'fand, together with Raikov and Naimark, on the theory of infinite-dimensional representations of locally compact groups. His interest in this subject, largely created by his own papers and those of his collaborators and students, and more recently of students of his students, lasted throughout the next three decades. The results and methods of the theory of representations developed by Gel'fand and his school have found wide applications to problems on the physics of elementary particles and quantum field theory.

Research in the theory of representations provided the original stimulus for a cycle of papers by Gel'fand and his collaborators on integral geometry. In turn, these papers proved to be closely connected with the theory of generalized functions, based on the work of Schwartz and Sobolev.

The series of monographs on the theory of generalized functions created by Gel'fand and his students is very well known in the mathematical world.

He has always regarded mathematics as a single science, and so his work on very abstract questions is naturally combined with intensive and fruitful activity in applied problems. For more than twenty years he was Head of a section of the Institute of Applied Mathematics of the USSR Academy of Sciences. The general methods of approximate integration of partial differential equations developed by Gel'fand and his collaborators have formed an essential stage in the development of the whole of computational mathematics and have led to the solution of a number of specific applied problems.

The work of Gel'fand and his school played a significant part in the creation of the theory of inverse Sturm-Liouville problems, which arise in the theory of oscillations and the quantum theory of dispersion.

About 15 years ago, Gel'fand began his research in biology. On the basis of actual biological results, he developed important general principles of the organization of control in complex multi-cell systems. These ideas, apart from their biological significance, served as a starting point for the creation of new methods for finding an extremum, which were successfully applied to problems of X-ray structural analysis, problems of recognition, and so on. The research teams established by Gel'fand at the Moscow State University and the USSR Academy of Sciences introduced many new ideas into contemporary thinking on mechanisms for the control of movements in man and animals and mechanisms for regulating the behaviour of cells in a multi-cell organism.

Gel'fand's biological work is characterized by the same clarity in posing problems, the ability to find non-trivial new approaches, and the combination of concreteness and breadth of general concepts that distinguish his mathematical research.

As founder of an extensive scientific school, Gel'fand showed great powers of organizational, public and pedagogical work. For many years he was a member of the editorial board of the "Uspekhi Matematicheskikh Nauk", chief editor of the journal "Functional Analysis", and director of the Inter-Faculty Laboratory of Mathematical Methods in Biology at the Moscow State University. From 1968 to 1970, he was President of the Moscow Mathematical Society and is now an Honorary Member of it. For about 10 years a correspondence school in mathematics for schoolchildren living in the country and workers' settlements, which was founded and is directed by him, has been running successfully.

Gelfand's scientific achievements have received wide international recognition: he is an Honorary Member of the American National Academy of Sciences, the American Academy of Sciences and Arts, the Royal Irish Academy, the American Mathematical Society and the London Mathematical Society. On the eve of his sixtieth birthday he was awarded an Honorary

Doctorate of the University of Oxford.

For his services to the development of science and the training of specialists, he has been awarded the Order of Lenin three times, also the Order of the Workers' Red Banner and the Order of the Badge of Honour. He has won the Lenin Prize and two State Prizes.

His many-sided activity is extraordinarily intensive and fruitful. In the last decade alone he has published more than 100 papers covering various questions of mathematics and biology. We wish Izrail' Moiseevich Gel'fand many years of further creative activity.

THE WORK OF I. M. GEL'FAND ON FUNCTIONAL ANALYSIS, ALGEBRA, AND TOPOLOGY

S. G. Gindikin, A. A. Kirillov, and D. B. Fuks

This survey is timed to coincide with the sixtieth birthday of I. M. Gel'fand. The authors have confined themselves to those branches of mathematics in which he has been engaged during the last decade, and in the various branches the chronology of the articles covered by the survey is different.

Gel'fand's research in the theory of group representations, which has lasted for thirty years, falls into several cycles; the majority of his results are widely known and were dealt with in the survey[1] [1*] on the occasion of his fiftieth birthday. For this reason we deal here only with the results of the last ten years.

His first articles on integral geometry appeared more than ten years ago, but this branch of mathematics is still in a formative phase. Because of this we include in the survey an outline of Gel'fand's basic research in integral geometry, not excluding some that is comparatively old.

Topology is a new branch of his scientific activity; all the work in this field was carried out between 1968 and 1973.

In the preparation of the survey we had the help of I. N. Bernshtein, M. I. Graev, and D. A. Kazhdan.

Contents

§1. Theory of representations 5
§2. Integral geometry 17
§3. Cohomology of infinite-dimensional Lie algebras 26
References 33

§1. Theory of representations

Gel'fand's papers over 40–50 years on the theory of infinite-dimensional group representations (jointly with Raikov, Naimark, Graev, and Berezin)

1) References marked with an asterisk relate to the bibliography at the end of this article, those without an asterisk to the bibliography of articles by Gel'fand (since 1963), which is also published in the present issue of the journal.

essentially opened a new stage of representation theory and largely determined its content. These papers are very well known; a short survey of them is contained in [1*], written on the occasion of his fiftieth birthday.

Here we shall speak about some of his new results obtained during the last ten years.

1.1. Representations of matrix groups over local fields. The first objects of research in the theory of infinite-dimensional representations were the classical Lie groups over the field of complex numbers. Of course, this is no accident. Firstly, the classical groups arise naturally in very different branches of mathematics, mechanics, and physics. Secondly, the classical groups and their representations form one of the most beautiful mathematical theories, which gives a profound aesthetic delight to anyone familiar with it. Finally, the choice of the complex field greatly facilitates the research, since the structure of the groups and their representations is significantly simpler in this case than for other fields.

However, recently, under the influence of modern algebraic geometry and number theory, there has been a tendency in many branches of mathematics, first algebra and then analysis, to move from the complex field to more general fields.

For functional analysis one is interested primarily in fields for which one can construct a meaningful analogue to classical harmonic analysis. As an example, we have the so-called local fields, among them the fields of real, of complex, and of $\mathfrak{p}$-adic numbers, also fields of formal power series over a finite field. The name "local field" derives from one of the ways of constructing these fields – by means of a localization, that is, completion with respect to a certain "local" norm of one of the so-called "global" fields (for example, the field of algebraic numbers or the field of functions on an algebraic curve over a finite field). More detailed information on local and global fields can be found in the book by Weil [2*].

One of the basic properties of a local field K is the fact that its additive group K^+ is locally compact and self-dual in the sense of Pontryagin. We note that this property characterizes local fields among all infinite topological fields. For a local field we can define analogues to the classical elementary and special functions (taking complex values): exponential, gamma and beta functions, Bessel functions, and so on. There is a natural concept of Fourier and Mellin transforms, and also spaces of test and of generalized functions.

The question of the structure of algebraic groups over a local field and their representations is very interesting and difficult. Even for the classical groups over the real field the question of classifying the representations is unsolved; in particular, there is no classification yet of the unitary irreducible representations of these groups.

The simplest of the classical groups is the group $SL(2, K)$ of unimodular matrices of order 2. In the case $K = \mathbf{C}$ this is the well-known Lorentz group, and in the case $K = \mathbf{R}$ it is isomorphic to the subgroup of the Lorentz group that preserves one spatially similar vector. The theory of the unitary representations of these groups was constructed in 1947 in the well known articles of Gel'fand and Naimark, and of Bargman. For an arbitrary local field K (of characteristic $\neq 2$) a complete list of all the irreducible unitary representations was first obtained by Gel'fand and Graev.

Apart from the representations of the fundamental and complementary series (which are constructed exactly as in the case of the complex and the real field), the group $SL(2, K)$ has three discrete series of representations and one singular representation. The representations of the discrete series are enumerated by the characters of the groups of units of quadratic extensions of K. The construction of these representations due to Gel'fand and Graev uses an analogue to the concept of "a function close to being analytic in the upper half-plane" for the local field K. Namely, this is a natural term for the Fourier transforms of functions on K concentrated on the "half-line" $N(L) \subset K$ that consists of norms of elements of a quadratic extension L of K. In the case $K = \mathbf{R}$, $L = \mathbf{C}$, we have $N(\mathbf{C}) = \{x \in \mathbf{R},\ x \geqslant 0\} = \mathbf{R}_+$. It is known that Fourier transforms of functions with support in $\mathbf{R}_+$ can be continued analytically into the upper half-plane.

The existence of the singular representation is connected with a specific non-Archimedean field. This representation has no analogue in the classical fields.

A detailed account of the result described here can be found in the second chapter of [142].

We mention that in the third chapter of this book there is an account of a group-theoretical approach, due to Gel'fand, to the theory of automorphic functions and related problems of number theory. This approach makes it possible to give a simpler and more natural proof of the remarkable results of A. Selberg (the so-called trace formula). The fruitfulness of this approach was later confirmed by numerous articles at home and abroad.

In the last decade the theory of representations of the groups $GL(n, K)$, where K is a local field, has been the subject of intensive research. Papers by Langlands, Weil, Harish-Chandra, Godement, Jacquet and Deligne have revealed many new laws and interesting connections. In particular, we mention a very deep conjecture of Weil and Langlands on the connection between irreducible infinite-dimensional representations of the group $GL(n, K)$ and the n-dimensional representations of the Shafarevich-Weyl group $W(K)$. For $n = 1$ this conjecture is essentially equivalent to a fundamental theorem in class field theory. In the case $n = 2$ and odd p, one of the local variants of the conjecture is proved in an article by Jacquet and Langlands [3*].

Recently Gel'fand and Kazhdan have suggested a refinement of the Weil–Langlands conjecture, based on a comparison of two special functions. One of them, the so-called L-function of Artin, is constructed by means of an n-dimensional representation of the Shafarevich – Weyl group $W(K)$. The other was introduced by Gel'fand and Kazhdan and is constructed by means of a pair of irreducible representations of the groups $GL(n, K)$ and $GL(n-1, K)$. For a detailed account of this conjecture and some results confirming it, see [4*].

Here we merely mention a precise formulation of the Gel'fand – Kazhdan conjecture (see [231]).

Let K be a local field, $W(K)$ the Shafarevich – Weyl group and I the inertia subgroup. The definition of these groups (see [2*], Appendix I) is clear from the diagram:

$$\begin{array}{ccccccccc} 1 \to & I & \to & \mathrm{Gal}(\bar{K}/K) & \to & \mathrm{Gal}(\bar{k}/k) & \to 1 \\ & \| & & \uparrow & & \uparrow & \\ 1 \to & I & \longrightarrow & W(K) & \longrightarrow & \mathbf{Z} & \to 1, \\ & & & \uparrow & & \uparrow & \\ & & & 1 & & 1 & \end{array}$$

where $\bar{K}$ is the algebraic closure of K, k is the residue class field of the ring of integers in K with respect to its maximal ideal, $\bar{k}$ is its algebraic closure, and $\mathbf{Z}$ is generated by the Frobenius automorphism F.

If σ is a finite-dimensional representation of $W(K)$ in a space V, then we denote by V^I the subspace of V consisting of the invariants of the group I. The Artin L-function of σ is defined by the formula

$$L(s, \sigma) = \det(1 - q^s\sigma(F)|_{V^I})^{-1}.$$

Here s is a complex variable and q the number of elements of k.

Now let $G_n = GL(n, K)$, P_n the subgroup of G_n consisting of matrices with the last row $(0, \ldots, 0, 1)$ and $Z_n \subset P_n$ the subgroup of upper unitriangular matrices. We identify G_{n-1} with the subgroup $P_n \cap P_n' \subset G_n$ (the prime denotes transposition).

In a sense, P_n contains the "whole non-commutativity" of G_n. For the classical cases $K = \mathbf{R}$ or $\mathbf{C}$ this is expressed, for example, by the fact that the irreducible unitary representations of G_n remain irreducible on restriction to P_n or by the fact that the Lie skew field of G_n (see below, § 1.2) is a central extension of the Lie skew field of P_n.

For a disconnected field K the analogous property of P_n is expressed in a more complicated way.

We recall that a representation π of G_n in a linear complex space V is called cuspidal if for any horospherical subgroup $N \subset G_n$ (that is, a group of block-triangular matrices with unit blocks along the main diagonal) the vectors of the form $\pi(n)v - v$, $n \in N$, $v \in V$, generate V.

It turns out that all the irreducible cuspidal representations of G_n on restriction to P_n go over into one and the same irreducible representation ρ of P_n. Namely, ρ is the right regular representation in a space V_θ of functions on P_n having the properties:

1) $f(zp) = \theta(z)f(p)$ for $z \in Z_n$, $p \in P_n$;

2) $f(p)$ is locally constant on P_n;
3) $f(p)$ has compact support mod Z_n.

Here $\theta(z)$ denotes the unitary character of Z_n given by the formula $\theta(z) = \chi\left(\sum_{k=1}^{n-1} z_{k,k+1}\right)$, where χ is a non-trivial additive character of K.

In other words, ρ is the representation of P_n induced by the 1-dimensional representation of Z_n. We note that by virtue of 1) each function $f \in V_\theta$ is completely determined by its restriction to the subgroup $G_{n-1} \subset P_n$.

We denote by s_n the element of G_n with coefficients

$$(s_n)_{ij} = \begin{cases} 0 & \text{if } i + j \neq n + 1, \\ (-1)^{i+1} & \text{if } i + j = n + 1. \end{cases}$$

It is easy to verify that G_n is generated by s_n and P_n. Thus, to define the representation π of G_n in V_θ completely it is sufficient to specify the operator $\pi(s_n)$.

It is convenient to introduce an operator C_π in V_θ by the formula

$$(C_\pi f)(g) = [\pi(s_n s_{n-1} g) f](1) \text{ for } g \in G_{n-1}.$$

It is easy to verify that C_π, like $\pi(s_n)$, uniquely determines the representation π, and, moreover, that it commutes with displacements by elements $g \in G_{n-1}$.

Now let τ be an irreducible representation of G_{n-1} that occurs in the decomposition of the space V'_θ dual to V_θ. It is known that in this case the multiplicity of τ is at most 1. Those representations for which it is equal to 1 are called non-degenerate. In particular, all cuspidal representations are of this kind.

The operator C'_π, being permutable with the displacements by elements of G_{n-1}, is a multiple of the unit operator on each irreducible component of V'_θ. So we obtain a complex number depending on the pair of representations. This is the Gel'fand–Kazhdan Γ-function.

In the definition of this function we have fixed a certain non-trivial character χ of K. Therefore, we denote the function by $\Gamma_\chi(\pi, \tau)$.

Now let $\widehat{W(K)}_n$ be the set of equivalence classes of n-dimensional representations of $W(K)$ and $\Pi_n(K)$ the set of equivalence classes of irreducible admissible[1] non-degenerate representations of $GL(n, K)$. The Weil–Langlands conjecture is that there exists a one-to-one mapping $\varkappa_n$: $\widehat{W(K)}_n \to \Pi_n(K)$.

Suppose that such a mapping $\varkappa_n$ exists and consider the function

$$\varepsilon_\chi(\pi, \tau) = \Gamma_\chi(\varkappa_n(\pi), \varkappa_{n-1}(\tau)) \frac{L(1/2, \tilde{\pi} \otimes \tau)}{L(1/2, \pi \otimes \tilde{\tau})},$$

[1] A representation π is called admissible if the stabiliser of each vector is open and the space of vectors with given stabiliser is finite-dimensional.

where the tilde sign denotes the contragredient representation. The Gel'fand – Kazhdan conjecture is that for a suitable choice of $\varkappa_n$ and χ (for all natural numbers n and all local fields L containing K) the function $\varepsilon_\chi(\pi, \tau)$ has simple functorial properties with respect to the representations π, τ and and the field L.

Namely, let us choose $\chi_L(x) = \chi_K(\mathrm{tr}_K^L x)$ as the additive character of L, where χ is a fixed character of K. One assertion of the conjecture is that $\varepsilon_\chi(\pi, \tau)$ depends only on the tensor product $\pi \otimes \tilde{\tau} = \sigma$. Therefore, in what follows we write $\varepsilon_L(\sigma)$ instead of $\varepsilon_\chi(\pi, \tau)$ bearing in mind the equality

$$\varepsilon_{\chi_L}(\pi, \tau) = \varepsilon_L(\pi \otimes \tilde{\tau}).$$

The first functorial property of $\varepsilon_L(\sigma)$ is stated as follows. If the representation σ is reducible and has a subrepresentation σ' and factor-representation σ'', then

$$\varepsilon_L(\sigma) = \varepsilon_L(\sigma')\, \varepsilon_L(\sigma'').$$

This enables us to extend the function $\varepsilon_L(\sigma)$ to the Grothendieck group $\Gamma(L)$ of the category of representations of $W(L)$. (In other words, for any formal integral combination of the form $\sigma = \sum_k n_k \sigma_k$, $\sigma_k \in \widehat{W(L)}_{m_k}$ we can define $\varepsilon_L(\sigma)$ by $\varepsilon_L(\sigma) = \prod_k \varepsilon_L(\sigma_k)^{n_k}$.

We can now state the second functorial property of $\varepsilon_L(\sigma)$.

If $\sigma \in \Gamma(L)$ and $\deg \sigma = 0$ (that is, $= \sum_k n_k \sigma_k$, $\sigma_k \in \widehat{W(L)}_{m_k}$ and $\sum_k n_k \deg \sigma_k = \sum_k n_k m_k = 0$), then

$$\varepsilon_M(\mathrm{Ind}_{W(L)}^{W(M)} \sigma) = \varepsilon_L(\sigma)$$

for any subfield M of L (containing K).

The functorial properties described above determine $\varepsilon_L(\sigma)$ uniquely (though implicitly) if it is known for all one-dimensional representations. For this case the function $\varepsilon_L(\sigma)$ is defined as follows. By a fundamental theorem in local class field theory there is a canonical isomorphism between the group $GL(1, L) = L^\times$ and the factor group of $W(L)$ with respect to the closure of its derived group. We choose the dual mapping for our $\varkappa_1\colon \widehat{W(L)}_1 \to \Pi_1(L)$. Then, to each one-dimensional representation σ of $W(L)$ there corresponds a multiplicative character $\varkappa_1(\sigma)$ of L.

We put $\varepsilon_L(\sigma) = 1$ if the character σ is unramified, that is, trivial on the inertia subgroup $I \subset W(L)$, and

---

[1] A representation π is called admissible if the stabiliser of each vector is open and the space of vectors with given stabiliser is finite-dimensional.

$$\varepsilon_L(\sigma) = \int_L [\varkappa_1(\sigma)](x)\, \chi_L(x)\, d^*x$$

if σ is ramified.[1]

The statement of the conjecture is now complete.

1.2. Non-commutative algebraic geometry. The basis of modern algebraic geometry is commutative algebra – the advanced theory of commutative rings. However, many questions in analysis naturally give rise to non-commutative rings, which deserve equally careful study. By analogy with the commutative case we can pose the question of "birational classification" of these rings, that is, the classification of the corresponding skew fields of fractions. It ought to be said that in the non-commutative case the very definition of the skew field of fractions is not obvious and not trivial. For a wide class of rings it can be constructed in the following natural way. Let R be a ring without divisors of zero, so that from $xy = 0$ if follows that $x = 0$ or $y = 0$. Consider all possible "left fractions" of the form xy^{-1} and "right fractions" $y^{-1}x$, where $y \neq 0$. We say that a left fraction xy^{-1} is equivalent to a right fraction $a^{-1}b$ if $ax = by$ (this is formally equivalent to $xy^{-1} = a^{-1}b$). We regard two left (right) fractions as equivalent if they are equivalent to one and the same right (left) fraction.

A ring R is called a right Ore ring if any two non-zero elements x, y of R have a common non-zero right multiple a, that is, $a = xx_1 = yy_1$ for some x_1, y_1 in R. Left and two-sided Ore rings are defined in a similar way. The latter will simply be called Ore rings.

It is easy to verify that in a right Ore ring every left fraction is equivalent to some right fraction, and any two left fractions have equivalent right fractions with a common denominator. This enables us to define all four arithmetical operations in the set of classes of equivalent fractions for any Ore ring R without divisors of zero, by putting

$$xa^{-1} \pm ya^{-1} = (x \pm y)a^{-1},\ (xy^{-1})^{-1} = yx^{-1},$$
$$xa^{-1} \cdot ay^{-1} = xy^{-1}.$$

The skew field D obtained in this way is called the skew field of fractions of R.

A fairly large supply of Ore rings is ensured by the following simple lemmas (see [144]).

LEMMA 1. *Any left (right) Noetherian ring without divisors of zero is a left (right) Ore ring.*

LEMMA 2. *If a ring R has a filtration $R = \cup R_k$ such that the adjoint ring $\mathrm{gr}\, R = \bigoplus_k R_k/R_{k-1}$ is left (right) Noetherian and has no divisors of zero, then R itself has the same properties.*

An important example of a non-commutative Ore ring is the ring of

---

1) This integral diverges, but it can be regularized by means of a standard technique (see 142], Chap. 2). Its value is sometimes called the Γ-function of L and is denoted by $\Gamma_L(\kappa(\sigma_1))$.

differential operators with polynomial coefficients in n variables. A slightly more general object, a so-called Weyl algebra $A_n(K)$, is defined as the algebra over a field K with generators $p_1, \ldots, p_n, q_1, \ldots, q_n$ and relations

$$(1.1.) \qquad p_i q_j - q_j p_i = \delta_{ij}, \quad p_i p_j = p_j p_i, \quad q_i q_j = q_j q_i$$

(canonical commutation relations). We denote the skew field of fractions of $A_n(K)$ by $D_n(K)$. In the particular case when K is the field of rational functions of k variables $z_1, \ldots, z_k$, we call $D_n(K)$ a standard skew field and denote it by $D_{n,k}$. It can be shown that standard skew fields are pairwise non-isomorphic for different values of n and k (see [144]).

Among the Ore rings are the enveloping algebras (in another terminology, associative envelopes) of Lie algebras.[1] If $\mathfrak{g}$ is a Lie algebra, $X_1, \ldots, X_N$ a basis of it, and [,] the operation of commutation, then the enveloping algebra $U(\mathfrak{g})$ is the algebra with the generators $X_1, \ldots, X_N$ and the relations

$$(1.2) \qquad X_i X_j - X_j X_i = [X_i, X_j], \qquad 1 \leqslant i, \; j \leqslant N.$$

We denote the skew field of fractions of $U(\mathfrak{g})$ by $D(\mathfrak{g})$ and call it the Lie skew field of $\mathfrak{g}$.

The first investigations of the algebraic structure of Lie skew fields were begun on Gel'fand's initiative. In [143] the conjecture was advanced that for any algebraic Lie algebra $\mathfrak{g}$ over an algebraically closed field of characteristic zero, the Lie skew field $D(\mathfrak{g})$ is isomorphic to one of the standard skew fields $D_{n,k}$. This was proved in [144] for all nilpotent Lie algebras and for a full matrix algebra. In [175] a weak form of this conjecture is proved for all semisimple Lie algebras, namely, that a certain algebraic extension $\widetilde{D(\mathfrak{g})}$ of $D(\mathfrak{g})$ is isomorphic to a standard skew field. It is interesting that $\widetilde{D(\mathfrak{g})}$ is a Galois extension, and the role of the Galois group is played by the Weyl group of $\mathfrak{g}$.

The proof of this fact makes use of a special realization of the algebra $U(\mathfrak{g})$ in the form of an algebra of differential operators. Namely, consider an algebraic variety A that is the factor space of a group G (regarded as an algebraic group) with respect to the maximal unipotent subgroup N. This variety is well known in representation theory. In papers by Gel'fand and Graev on integral geometry and the method of horospheres it is called the fundamental affine space of G.

The action of G on A enables us to realize the elements of the Lie algebra $\mathfrak{g}$ in the form of vector fields on A, and the elements of $U(\mathfrak{g})$ as differential operators on A.

Apart from the structure of a right G-space, the variety A (see §§1.2,

---

1) For the definition of Lie algebras, see § 3 of this article.

1.3) is also equipped with the structure of a left H-space, where H is the Cartan subgroup of G that normalizes N. Let R be the algebra of all differential operators on A that commute with the action of H. The skew field of fractions of R is the extension $\widetilde{D(\mathfrak{g})}$ mentioned above. Further study of R promises to be very interesting (see [192], [219], [5*]).

Very recently, information has come in to the effect that the Gel'fand–Kirillov conjecture has been verified for soluble algebraic Lie algebras.

We now show how the results described here are connected with the theory of unitary representations of Lie groups. The isomorphism $D(\mathfrak{g}) \approx D_{n,k}$ means that by a suitable change of variables the commutation relations (1.2) reduce to the canonical relations (1.1). If the change to the new variables could be performed by means of polynomials, and not rational functions, then all the problems of the theory of unitary representations would be solved. For by the Stone – von Neumann theorem, the commutation relations (1.1) have (to within equivalence) a unique irreducible realization in a Hilbert space. Namely, in the space $L^2(\mathbf{R}^n)$, to an element p_j there corresponds the differential operator $\partial/\partial x_j$, and to the element q_j the operator of multiplication by x_j. Giving arbitrary numerical values to the remaining generators $z_1, \ldots, z_k$ we would obtain a k-parameter family of irreducible representations of the original Lie algebra in the space of functions of n variables.

Owing to the presence of denominators in the transition formulae, this "naive" approach enables us to construct only the so called **representations** of general position. To study the general case we need more complete information on the structure of $U(\mathfrak{g})$ and $D(\mathfrak{g})$. In particular, it is useful to know the structure of the two-sided ideals in $U(\mathfrak{g})$ and to verify the truth of the fundamental conjecture for the skew fields of fractions of the corresponding factor algebras.

1.3. Representations of infinite-dimensional groups. Infinite-dimensional groups have been encountered for a long time in mathematics and its applications. Thus, the configuration space in fluid mechanics is a group of diffeomorphisms (for an incompressable fluid, the group of diffeomorphisms preserving volume) of a certain domain; in quantum field theory an important role is played by the so-called group of flows, which can be realized as a group of functions on a certain manifold with values in the Heisenberg group and pointwise multiplication; in functional analysis infinite-dimensional groups arise naturally as groups of invertible (unitary, isometric, multiplicative, and so on) operators in linear topological (in particular, Hilbert and Banach) spaces.

The theory of representations of infinite-dimensional groups is not yet worked out. Apparently, the situation here is very different from that for finite-dimensional groups.

We mention here a very interesting example of a representation of a group of flows, which was constructed in recent papers by Gel'fand (with Graev and Vershik).

The group in question consists of measurable bounded functions on a space X with measure μ that take values in the Lie group $SL(2, \mathbf{R})$. Two functions that coincide almost everywhere are identified. The group law is given by pointwise multiplication.

In other words, this group G is $SL(2, R(X))$, where $R(X)$ is the ring of equivalence classes of real bounded measurable functions on X. Clearly, this example allows various modifications and generalizations.

In particular, a simpler group G is obtained if instead of $SL(2, \mathbf{R})$ we take simply $\mathbf{R}$, and instead of the ring $R(X)$ of classes of measurable functions on a measure space we consider the ring $D(X)$ of smooth functions on a manifold. In this case the classification of irreducible unitary representations of G reduces to a full description of the generalized functions on X. For G is commutative, all its irreducible unitary representations are one-dimensional, and, as is easily seen, have the form

$$f \longmapsto e^{i\langle F, f\rangle}$$

where F is a certain linear functional on $D(X)$.

Thus, the theory of representations of groups of this type can be regarded as the non-commutative analogue to the theory of generalized functions. In an article by Gel'fand and Graev [171], irreducible representations are constructed for the case of a group of functions with values in $SU(2)$, which are analogues to generalized functions with finite support, that is, delta-functions and their derivatives.[1] However, they did not succeed (and apparently, for the case $SU(2)$ it is impossible) in defining an analogue to the integral, that is, an irreducible representation that depends essentially on the value of the function f at all points of X.

Unexpectedly it turned out that for the group of functions with values in $SL(2, \mathbf{R})$ an analogue to the integral exists. For a detailed account of the six different approaches to this construction, see the survey [242]. Here we give just a brief statement of the result.

THEOREM. *There is a one-parameter family of faithful irreducible unitary representations of the group $SL(2, R(X))$ that go into an equivalent representation under all automorphisms of the group generated by automorphisms of the measure space X.*

1.4. Representation theory and problems of linear algebra. After the problem of classifying the irreducible infinite-dimensional representations of the complex classical groups had been solved, it was natural to turn to a classification of arbitrary representations. In the unitary case, this problem reduces to the classification of irreducible representations. Namely, in the theory of operator algebras it is proved that every unitary representation of a separable

[1] This result has an intuitive interpretation with the help of the method of orbits: see [6*].

locally compact group can be written uniquely as a direct integral of so-called primary representations.

For groups of type I in von Neumann's sense (in particular, for all semisimple Lie groups) every primary representation is a multiple of an irreducible one.

In the non-unitary case the situation is more complicated. Firstly, in this case there are reducible, but indecomposable representations. This means that the representation operators have a common invariant subspace, which, however, does not have an invariant complement. Moreover, the decomposition of an arbitrary representation into indecomposable ones is not always possible nor unique. The same difficulties arise in an infinitesimal approach to the problem, that is, by considering arbitrary infinite-dimensional $\mathfrak{g}$-modules, where $\mathfrak{g}$ is a semisimple Lie algebra.

Gel'fand suggested (see [191]) singling out a special subcategory in the category of $\mathfrak{g}$-modules. Namely, let $\mathfrak{g}$ be a semisimple Lie algebra and $\mathfrak{h}$ a subalgebra of it. We denote by $U(\mathfrak{g})$ the enveloping algebra of $\mathfrak{g}$. We call a $\mathfrak{g}$-module $(\mathfrak{g}, \mathfrak{h})$-finite if it is finitely generated as a $U(\mathfrak{g})$-module and is semisimple with a spectrum of finite multiplicity as a $U(\mathfrak{h})$-module. The study of the category of $(\mathfrak{g}, \mathfrak{h})$-finite modules is of the greatest interest in the following two cases:

1) $\mathfrak{g}$ is a semisimple Lie algebra, and $\mathfrak{h}$ is the subalgebra corresponding to the maximal compact subgroup of the Lie group adjoint to $\mathfrak{g}$.

2) $\mathfrak{g}$ is a semisimple Lie algebra and $\mathfrak{h}$ is a reductive subalgebra of the same rank (in particular, a Cartan subalgebra).

In the first case the $(\mathfrak{g}, \mathfrak{h})$-finite modules are also called Harish-Chandra modules, since it follows from the results of Harish-Chandra that precisely such $\mathfrak{g}$-modules arise from the irreducible representations of the corresponding Lie group.

A complete classification of the Harish-Chandra modules is known only for the simplest case when $\mathfrak{g}$ is the Lie algebra of the Lorentz group. Each indecomposable Harish-Chandra module is characterized by two complex invariants λ_1 and λ_2 (the eigenvalues of the Laplace operators on the group).

This leads to the problem of describing the category $C(\lambda_1, \lambda_2)$ of Harish-Chandra modules with given values of the invariants. It turns out that this problem reduces to certain finite-dimensional problems of linear algebra. For the majority of values of λ_1 and λ_2 the corresponding problem of linear algebra is very simple and consists in the reduction of canonical form of a single nilpotent matrix. At "singular" points there arises the more difficult and interesting problem of reducing to canonical form a pair of nilpotent matrices A and B with $AB = BA = 0$. In contrast to the first problem, the answer here depends not only on discrete, but also on continuous parameters. It can be shown that the $\mathfrak{g}$-modules so obtained correspond to certain representations of the Lorentz group. Thus, the indecomposable representations of the Lorentz group many admit a non-trivial deformation.

The reduction to problems of linear algebra is possible also in a more general situation. For any semisimple Lie algebra $\mathfrak{g}$ the category of indecomposable Harish-Chandra modules with fixed eigenvalues of the Laplace operators is isomorphic to the category of indecomposable representations of a certain finite-dimensional algebra. However, the latter problem often turns out to be "insoluble" in the sense that part of it requires the reduction to canonical form of a pair of arbitrary matrices. (As was shown in [180], this last problem in turn requires a solution of the problem of classifying k arbitrary matrices for any k; it is natural to regard this problem as the standard of "insolubility" and to regard as "insoluble" all problems that reduce to it.)

It is possible, however, that the whole complexity of the problem of classifying Harish-Chandra modules reduces to the one above. It would be very interesting to state this assertion precisely and prove it.

Another interesting category is obtained if we take for $\mathfrak{h}$ the Cartan subalgebra of $\mathfrak{g}$. Let $\mathfrak{b}$ be the Borel subalgebra containing $\mathfrak{h}$ and $\mathfrak{n}$, its unipotent radical. In the category of $(\mathfrak{g}, \mathfrak{h})$-finite modules we pick out the subcategory $\mathcal{O}$ consisting of those modules M in which all the elements of $U(\mathfrak{n})$ are finite (that is, $\dim U(\mathfrak{n})\xi < \infty$ for all $\xi \in M$).

In this category it is very natural to construct a theory of dominant Cartan weight in the infinite-dimensional case. Although the study of this category was only begun recently and is far from complete, the results obtained are very interesting and have already found applications. For a detailed account of these results, see [203], [221].

Here we just quote one of them.

Let χ be a linear functional on $\mathfrak{h}$ and let M_χ be the module in the category $\mathcal{O}$ that is obtained by factoring out in $U(\mathfrak{g})$ the left ideal generated by $\mathfrak{n}$ and the elements of the form $X - \langle \chi - \rho, X \rangle$, where $X \in \mathfrak{h}$ and ρ is the half-sum of the positive roots of $\mathfrak{g}$ with respect to $\mathfrak{b}$. We represent the Weyl group W of $\mathfrak{g}$ as the union of subsets W_i so that $w \in W_i$ takes exactly i positive roots into negative roots.

Let V be the finite-dimensional $\mathfrak{g}$-module with dominant weight λ. Then there is an exact sequence of $\mathfrak{g}$-modules:

$$0 \leftarrow V \leftarrow C_0 \leftarrow C_1 \leftarrow \ldots \leftarrow C_s \leftarrow 0,$$

where $C_i = \bigotimes_{w \in W_i} M_{w\lambda}$.

This theorem has as corollaries the well-known Borel – Weil – Bott theorem, the Weyl formula for characters, the Kostant formula for the weight multiplicity and the Harish-Chandra theorem on the structure of the ideal corresponding to the module V.

To conclude this section, we mention that a complete description of the category $\mathcal{O}$ would be very interesting for representation theory, but so far it has not been achieved. Even the structure of submodules of a given module M_χ is still unknown (examples show that among these submodules there may be some that do not coincide with one of the M_χ).

1.5. Representation theory and algebraic topology. The connection between these branches of mathematics is almost as old as these branches themselves. Already in papers of E. Cartan and H. Weyl, representation theory was used to find the Betti numbers of the classical groups. The well-known Borel – Weil – Bott theorem is an example of the reverse influence. It gives a descriptive realization of the irreducible representations of a compact semisimple Lie group G in terms of algebraic topology: the representation space is a cohomology space with coefficients in a certain linear bundle over the homogeneous space $X = G/H$, where H is the Cartan subgroup of G.

The space $X = G/H$ and its generalizations play an important role in many questions of the theory of representations of semisimple groups. An essentially new step in the study of these spaces was taken in a recent article [239]. Readers of the Russian Mathematical Surveys can find an account of this work in the survey [241]; therefore we confine ourselves here to a brief description of the result.

The cohomology groups of the space $X = G/H$ can be calculated by two different methods.

a) The geometrical method consists in indicating an explicit decomposition of X into cells of even dimension, indexed by the elements of the Weyl group W of G. (In particular, it follows immediately from this that the Euler characteristic of X is equal to the order of W.)

b) The analytical method reduces the problem to a description of the G-invariant differential forms on G/H by means of de Rham's theorem, and this in turn reduces to a description of the W-invariant polynomials on the Cartan subalgebra.

Although the two methods are both natural, they lead to different bases in the cohomology groups, and explicit formulae for changing from one basis to the other were not known at the time of publication of [239].

We mention that as a corollary of these explicit formulae the authors obtain a description of the action of the group W (which is obviously the automorphism group of the G-space X) on the cohomology of X.

§2. Integral geometry

The early work of Gel'fand and his collaborators on integral geometry arose in connection with problems in the theory of group representations. We begin, however, with a more elementary question on the Radon integral transform, where many essential features of integral geometry problems are well illustrated. This range of questions goes back to the problem which was posed by Radon in 1917 [7*] and solved by John [8*]. A systematic study of the Radon transform was undertaken in connection with the preparation of the book [9*].

2.1. The Radon transform. The *Radon transform* of a function $f(x)$ on $\mathbf{R}^n$ is the function on the set of hyperplanes in $\mathbf{R}^n$ defined by

$$\hat{f}(\xi, p) = \int_{(\xi, x)=p} f(x)\, d\omega, \qquad dx = dp\, d\omega.$$

This transform is closely connected with the many-dimensional Fourier transform $\tilde{f}(\xi)$ by means of the one-dimensional Fourier transform:

$$\tilde{f}(\alpha\xi) = \int_{-\infty}^{\infty} \hat{f}(\xi, p)\, e^{i\alpha p}\, dp, \qquad \alpha \in \mathbf{R}^1.$$

The relation between $\hat{f}$ and $\tilde{f}$ enables one to obtain the properties of the Radon transform comparatively easily, starting from the corresponding properties of the Fourier transform; however, in a number of problems, the use of the Radon transform is preferable to that of the Fourier transform. The method of plane waves in the theory of differential equations with constant coefficients, based on the Radon transform, makes it possible to solve a number of important problems, for example, the investigation of a fundamental solution of a strictly hyperbolic operator (the Herglotz – Petrovskii formula).

In the applications of the Radon transform an important role is played by the inversion formula. The form of this formula depends essentially on the parity of n (the dimension of the space). If n is odd, then to obtain $f(x)$ it is sufficient to average $\partial^{n-1}\hat{f}(\xi, p)/\partial p^{n-1}$ over the set of hyperplanes passing through x (with respect to a certain canonical measure); if n is even, then instead of the operator of differentiation with respect to p in the inversion formula we have the Riemann – Liouville operator (fractional differentiation). For this reason, the inversion formula is local in the odd-dimensional case, that is, to reconstruct $f(x)$ it is sufficient to know the values of $\hat{f}$ on hyperplanes passing through a small neighbourhood of x. In the even-dimensional case, in the same situation it is necessary to know the integrals over all hyperplanes. This distinction is closely connected with the fact that the Huygens principle holds in an odd-dimensional but not in an even-dimensional space.

Problems connected with the Radon transform are distinguished by their originality and at the same time their proximity to classical analysis. Gel'fand was always glad to pose problems of this kind to students. Here is one example (see "Mathematical Education", No. 2, 1957, p. 274). Suppose that $n = 3$ and that f is the characteristic function of some solid T. Average over $\xi \in S^2$ (S^2 is a sphere) the quantity $u(\xi) = \int_{-\infty}^{\infty} \left| \frac{\partial \hat{f}(\xi, p)}{\partial p} \right|^2 dp.$

We obtain a geometrical characteristic $\overline{w}(T)$, which has the dimension of a volume; what is its geometrical meaning?[1]

It is natural to present the theory of the Radon transform in the language of generalized functions. This was actually done in the first edition of "Generalized Functions" (although the term "Radon transform" was not used there). A systematic account of the Radon transform in the language of generalized functions is contained in the fifth volume of this series [9*]. First of all, there is the question of how to define the Radon transform $\hat{f}$ by a generalized function,[2], say $f \in \mathscr{S}'(\mathbf{R}^n)$. As in the case of the Fourier transform, an important role is played here by the analogue to the Plancherel theorem for the Radon transform. As a result, the problem of constructing $\hat{f}$ reduces to finding a functional F such that $(f, \phi) = (F, \hat{\phi}(\xi, p))$ for all $\phi \in \mathscr{S}'(\mathbf{R}^n)$. It is immediately obvious that in a certain sense the choice of the generalized function F is not unique. Thus, if $(f, \phi) = \int_{\mathbf{R}^n} \phi(x)dx$, then $(f, \phi) = (F, \phi) = \int_{-\infty}^{\infty} \hat{\phi}(\xi, p)dp$ for every ξ.

The reason is that the functions $\hat{\phi}$ for $\phi \in S$ obviously satisfy conditions of homogeneity, smoothness and rapid decrease with respect to p. But apart from these there is another condition, namely that the moments of $\hat{\phi}$ with respect to p are polynomials in ξ. Gel'fand sometimes calls this the "Cavalieri condition", because in the simplest case it expresses the fact that $\int \hat{\phi}(\xi, p)dp$ is dependent on ξ. It is precisely equivalent to the smoothness of the Fourier transform $\tilde{f}$ at zero. It is natural to look for F in the class of functionals on the test functions ψ that satisfy all these conditions on ϕ except the Cavalieri condition. However, F is defined to within functionals that vanish on the subspace of test functions satisfying the Cavalieri condition. Examples of such "inessential" generalized functions are the functionals $\int_{-\infty}^{\infty} [\psi(\xi', p) - \psi(\xi'', p)]\, dp$. All the "inessential" generalized functions can be described; in particular, they turn out to be polynomials in p. Here is an interesting corollary:[3] if $\mathscr{L}$ is a set of hyperplanes which allow us to reconstruct $\int_{\mathbf{R}^n} f(x)dx$, provided that we know the integrals of $f \in \mathscr{S}(\mathbf{R}^n)$ over them, then, for some ξ, $\mathscr{L}$ contains almost all hyperplanes $(\xi, x) = p$.

The explicit calculation of the Radon transforms of the characteristic

1) The "secret" of this problem is revealed by means of the Plancherel formula for the Radon transform [9*], which shows immediately that $\overline{w}(T)$ is proportional to a volume.

2) $\mathscr{S}'$ is the Schwartz space of generalized functions of moderate growth, adjoint to the space $\mathscr{S}$ of functions that decrease rapidly, together with all their derivatives.

3) This corollary can be regarded in a certain sense as a statement about the uniqueness of the classical Cavalieri principle.

functions of certain unbounded domains [9*] is instructive. A result of this is a method of determining, for example, the regularized values of the areas of all conic sections.

Finally, the analogue of the Radon transform in $\mathbf{C}^n$ is investigated in [9*]. The situation in $\mathbf{C}^n$ recalls that in $\mathbf{R}^{2k+1}$, and the inversion formula for the complex Radon transform is local.

An examination of this very elementary part of integral geometry shows what an important role is played in it by the discussion of beautiful individual problems and by carrying the results to the point of explicit formulae. We shall see later that a whole series of features that come to light for the Radon transform are also characteristic of other problems of integral geometry.

2.2. Integrals of a homogeneous polynomial along level curves and Plancherel's formula. In the book by Gel'fand and Naimark [10], which students and collaborators of Gel'fand used to call the "blue" book, probably the most difficult part is that devoted to the derivation of the analogue to Plancherel's formula for the complex classical Lie groups. Later, Gel'fand and his collaborators repeatedly thought over these questions again and, as we shall see, these speculations gave a significant stimulus to the development of integral geometry. This activity is closely connected with work on generalized functions. In fact, generalized functions are implicity present in the "blue" book. Subsequently these connections were thoroughly thought out and extended.

A basic step in the derivation of Plancherel's formula is the calculation of the values at the unit element of a function on a group if its integrals over the conjugacy classes are known. This can be done by a simple limiting process when the group, and hence the conjugacy classes, are compact. In the non-compact case the limit does not exist and some regularizing procedure is needed.

As a model we consider the problem of reconstructing $f(0)$ for $f(x) \in \mathscr{S}(\mathbf{R}^n)$ in terms of $I_P(f; c)$, the integrals of f along the level curves $L_P(c) = \{x : P(x) = c\}$ of a homogeneous polynomial P. If the level curves are compact (for example, $P = x_1^2 + x_2^2$), then for a suitable normalization of the measure $f(0) = \lim_{c \to 0} I_P(f; c)$. But if the L_P are hyperbolas ($P = x_1^2 - x_2^2$), then $f(0)$ coincides, to within a factor, with $\lim_{c \to 0} I'_P(f; c)$.

This problem turned out to be connected with work on homogeneous generalized functions. We recall that in his lecture at the Amsterdam Congress of Mathematicians Gel'fand posed the problem of investigating the analytic continuation with respect to λ of a generalized function (P^λ, ϕ). Examples show that for some P there is a pole, at which the residue is proportional to $\delta(x)$. Then standard devices of the theory of generalized functions [11*] enable us to find a formula for reconstructing

$\phi(0)$ in terms of $I_P(\phi; c)$. It turns out that in this way the result on hyperbolas mentioned above can be obtained and, more generally, the results of the celebrated paper of M. Riesz on Riemann – Liouville integrals for the wave equation enable us to solve the problem for the case when $P(x_1, \ldots, x_n, x_{n+1}) = x_1^2 + \ldots + x_n^2 - x_{n+1}^2$. Gel'fand and Graev [9*], [12*] generalized the results of Riesz to arbitrary quadratic forms. This result enabled them to solve the problem of reconstructing a function from the integrals over the conjugacy classes for the classical complex groups. We note that although Gel'fand's problem for P^λ is completely solved [13*], more precise information on the residues of P^λ, which is necessary for the solution of our problem of integral geometry (the existence of a pole at which the residue is proportional to $\delta(x)$), had to be obtained in several cases.

Gel'fand and Graev also considered [14*] the case of real groups. A characteristic example is the group $SL(n; \mathbf{R})$ of real matrices of order n and determinant 1. Here the situation is complicated by the fact that the conjugacy classes are of essentially different types. We shall explain this in the simplest example of the group $SL(2; \mathbf{R})$, restating the problem without using the language of groups. Let f be a rapidly decreasing infinitely differentiable function on the hyperboloid $x_1^2 + x_2^2 - x_3^2 - x_4^2 = 1$. We consider its integrals over the sections by three-dimensional planes. It turns out that f can be reconstructed if we know the integrals over sections that are two-sheet hyperboloids, but it cannot be reconstructed if we only know the integrals over sections that are one-sheet hyperboloids. We note that in this problem it is essential that we only assume f to decrease more rapidly than any power; if f is of compact support or decreases at least exponentially, then sections that are one-sheet hyperboloids are sufficient. This corresponds to the fact that for functions of the first type the Fourier transform is only infinitely differentiable, but for functions of the second type it is analytic.

2.3. The method of horospheres. A principal role in determining the content of this cycle of problems, which Gel'fand attributed to integral geometry, is played by the article [15*]. The term "integral geometry" goes back to Blaschke. However, the problems of integral geometry in Gel'fand's sense are essentially different from the old meaning of this term. The definition of the class of questions within the sphere of integral geometry has been revised and extended many times. At the time of writing [15*] and vol. 5 of "Generalized functions" [9*] the general situation was as follows. There are two homogeneous spaces with the same group of motions G: $X_1 = G/G_1$, $X_2 = G/G_2$. Then under certain conditions we can define an integral transform of functions on X_1 into functions on X_2 by integrating the first functions over the trajectories of G_2 on X_1. Integral geometry should investigate this kind of transform of functions on homogeneous manifolds.

Speaking more concretely, the starting point of [15*] is the recognition of the basic role of the maximal soluble subgroup K and the maximal nilpotent group Z in the constructions of the "blue" book [10*]. To fix our ideas, let $G = SL(n; \mathbf{C})$ be the group of complex unimodular matrices of order n with determinant 1, Z the subgroup of upper unitriangular matrices and $X = G/Z$. We consider the regular representation[1] of G in $L^2(X)$ (with respect to invariant measure): $T_g f(x) = f(g(x))$. We may suppose that the f are those functions on G for which $f(zg) = f(g)$, $z \in Z$, and then $T_{g_0} f(g) = f(gg_0)$. Let H be the subgroup of diagonal matrices (the maximal Cartan subgroup). Then on $L^2(X)$, apart from the operators T_g, the "left displacement" operators $\Lambda_h f(g) = f(hg)$ act on the elements $h \in H$: (see §1.2, p. 12–13). It can be verified immediately that because H is contained in the normalizer of Z ($h^{-1}Zh = Z$, $h \in H$), Λ_h preserves the condition $f(zg) = f(g)$.

Using harmonic analysis on commutative groups, we can decompose $L^2(X)$ into a continuous sum of Hilbert spaces invariant under Λ_h: under the action of Λ_h, $h \in H$, the functions of each space are multiplied by some character of H. It is remarkable that this is simultaneously a decomposition of the representation T_g into irreducible representations of G, where representations are equivalent if and only if the characters of H corresponding to them are obtained from one another under the action of elements of the Weyl group (in the present case, the group of permutations of the diagonal elements of the matrix h).

It is also remarkable that the set of unitary representations so obtained is sufficient to decompose the regular representation in $L^2(G)$: $T_{g_0} f(g) = f(gg_0)$. The situation is as follows. We consider on G the trajectories of Z and its conjugates. They are called *horospheres.* Horospheres have the form $\Omega(g_1, g_2) = \{g_1 Z g_2\}$; in the general situation they are defined by a pair of cosets $\{g_1 Z\}, \{Z g_2\}$ that is, by a pair $x_1, x_2 \in X = G/Z$. We set up a correspondence between $f \in L^2(G)$ and its integrals over the horospheres $\hat{f}(x_1, x_2)$. To a right displacement on G there corresponds the natural transformation $\hat{f}(x_1, x_2) \mapsto \hat{f}(x_1, g(x_2))$, where g acts on the second argument x_2 as on a point of the homogeneous space $X = G/Z$. In this way, the decomposition of the resulting representation in $\hat{f}$ into irreducible ones is automatically obtained from the decomposition of the quasiregular representation in $L^2(X)$.

In turn, the decomposition of the regular representation requires a clarification of the properties of the mapping $f(g) \mapsto \hat{f}(x_1, x_2)$. Firstly, one has to determine whether this mapping has a kernel, and to obtain the decomposition in explicit form one must have the inversion formula $\hat{f} \mapsto f$. Part of the results of the "blue" book can be interpreted as a proof that the mapping $f \mapsto \hat{f}$ has no kernel and a derivation of the explicit

1) It is sometimes called quasiregular.

inversion formula.

In its structure this formula recalls the inversion formula for the Radon transform. It is natural to define the concept of "parallel" horospheres, the trajectories of the same horospherical subgroup gZg^{-1} (g is fixed); they are parametrized by elements of H. A canonical operator L is defined on H, which in exponential coordinates can be written as a differential operator with constant coefficients. This operator is applied to $\hat{f}$ (along each set of parallel horospheres), and then $L\hat{f}$ is averaged over the set of horospheres passing through g. The function so obtained differs from $f(g)$ only by a constant factor.

Suppose that $f(g)$ is invariant under left displacements by elements of a subgroup G_0, that is, f can be interpreted as a function on the homogeneous space G/G_0. If G_0 is compact, then $L^2(G/G_0)$ can be embedded in $L^2(G)$, and the previous scheme enables us to decompose the regular representation in $L^2(G/G_0)$. The case $G_0 = U$, the maximal compact subgroup, is worth special mention.[1] Then $S = G/U$ is a symmetric Riemannian space of non-negative curvature. In this case $\hat{f}$ does not depend on x_1 and the horospheres on S are defined geometrically; in the case of the Lobachevskii space ($SL(2; \mathrm{C})/SU(2)$) they coincide with the horospheres for this space in the accepted sense. The resulting problem of integral geometry on a Lobachevskii space turns out to be more appropriate than another possible analogue to the Radon problem when we consider integrals of a function over Lobachevskii planes.

As for the case of a non-compact subgroup G_0, in principle we can always operate according to the "method of horospheres". However, questions on the investigation of the kernel of the mapping $f \mapsto \hat{f}$ and the decomposition of the resulting representation into functions $\hat{f}$ need special consideration. The latter question can lead to difficulties associated with the separation of the set of horospheres on G/G_0 into orbits under the action of G. In [15*] the important case $G_0 = H$ is analysed. From this result there follows the decomposition of the Kronecker product of two representations of the fundamental series into irreducible ones; for the Lorentz group this decomposition was obtained earlier by Naimark.

The method of horospheres proved very fruitful in the investigation of the case when G_0 is a discrete subgroup whose factor space G/G_0 has finite invariant volume [142]. In this case the mapping $f \mapsto \hat{f}$ has a kernel, and in favourable cases this kernel separates the discrete spectrum in the decomposition of G/G_0 into irreducible ones, and the orbits in the set of horospheres correspond to the "exits" of the fundamental domain on the boundary.

The study of the case when G is a real semisimple Lie group is far from complete. In this case, of course, a quasiregular representation can be

1) In this example we may take U to be the subgroup of unitary matrices.

decomposed into irreducible ones just as in the case of complex groups. However, the resulting representations are insufficient for the decomposition of the regular representation on G. There are still the so-called representations of the discrete series (and mixed series). This is apparent from the fact that the mapping $f \mapsto \hat{f}$ of $L^2(G)$ has a kernel. It turns out that the kernel does not contain elements invariant under left displacements by elements of the maximal compact subgroup U. We can therefore pose the problem of reconstructing $f \in L^2(S)$, where S is the symmetric space G/U, if we know the integrals over horospheres. It turns out [16*] that in this case also there exists an operator L on H (acting along a set of parallel horospheres) such that $f(g)$ is proportional to the average of $L\hat{f}$ over the set of horospheres passing through g. However, in contrast to the case of complex groups, the operator L is not local, in general. This complicates the discovery of its explicit form. For a number of symmetric spaces the operator L is calculated in [16*].

2.4. Complexes of planes in $\mathbf{C}^n$. A natural development of the Radon problem is the problem of reconstructing a function if its integrals over planes of dimension $k < n - 1$ are known. Let $H = H_{n,k}$ be the manifold of such planes in $\mathbf{C}^n$. For $k < n - 1$ the dimension of H is greater than n, and so the problem of reconstructing the function using the integrals over all planes of H is overdetermined. It is natural to limit ourselves to certain subsets of H and, in particular, to the n-dimensional submanifolds of H (complexes).

Work on this problem was stimulated by the fact that the problem of integral geometry for $SL(2; \mathbf{C})$ (or $SL(2; \mathbf{R})$) (see §2.3) is equivalent to the problem of reconstructing a function in $\mathbf{C}^3$ (or $\mathbf{R}^3$) if its integrals along all complex (real) lines intersecting the hyperbola $\{t, t^{-1}, 0\}$, where $t \in \mathbf{C}$ (or $t \in \mathbf{R}$), are known. This set of lines is three-dimensional. In the complex case we can obtain the inversion formula, but in the real case there is no unique solution of the reconstruction problem.

Suppose that $f(x) \in \mathscr{S}(\mathbf{R}^n)$, $\hat{f}(h) = \int_h f$, $h \in H_{n,k}$. Let $G = G_{n,k}$ be the corresponding Grassman manifold (of k-dimensional subspaces of $\mathbf{C}^n$) and $\pi\colon H_{n,k} \to G_{n,k}$ the canonical bundle. For each $x \in \mathbf{C}^n$ consider the set H_x of planes $h \in H_{n,k}$ passing through x; H_x is a section of the bundle $H \to G$. By means of π we identify H_x with the base. Suppose that $M \subset H$ is invariant under displacements in $\mathbf{C}^n$. Then M is completely determined by its projection $\pi M \subset G$. It is not difficult to show that in order to reconstruct f from $\hat{f}|_M$ the following condition is necessary and sufficient:

(∗) *almost all hyperplanes of* $\mathbf{C}^n$ *contain planes of* M.

If (∗) is satisfied, then to reconstruct f it is sufficient to integrate $\hat{f}$ so as to obtain integrals over almost all hyperplanes, and then to use the inversion formula for the Radon transform.

It is shown in [153] that if πM is a cycle ($\dim_{\mathbf{R}} \pi M = 2k$), then the reconstruction can be effected in the following way. We write out explicitly the operator $\varkappa$ which, for each $x \in \mathbf{C}^n$ assigns to $\phi \in \mathscr{S}(H)$ a differential form $\varkappa_x \phi$ on $H_x \cong G_{n,k}$ of type (k, k); $\varkappa$ is a differential operator with constant coefficients, which acts along the fibres of the bundle $H \to G$. For $k < n - 1$ it turns out that $\phi \in \mathscr{S}(H)$ lies in the image of $f \mapsto \hat{f}$ if and only if the forms $\varkappa_x \phi$ are closed for all x. Under this condition

$$\int_{\pi M} \varkappa_x \hat{f} = c_M f(x),$$

and $c_M \neq 0$ if and only if (*) is satisfied.

Thus, in the case of complexes invariant under displacements we obtain a universal inversion formula.

2.5. Admissible complexes. The local character of the inversion formula enables us to weaken substantially the requirement that M is invariant under displacements. Let M be an analytic manifold in $H_{n,k}$ such that for almost all $x \in \mathbf{C}^n$ the sets $M_x = M \cap H_x$ are cycles in $G_{n,k}$ on which the form $\varkappa_x \phi$ is calculated over $\phi|_M$. If M satisfies (*), then the formula makes it possible to reconstruct f from $\hat{f}|_M$. Such manifolds are called *admissible complexes.*

We now return to the problem of integral geometry on $G = SL(p; \mathbf{C})$ (§2.3). We embed G in $\mathbf{C}^n$, $n = p^2 - 1$. Then the horospheres $\Omega(x_1, x_2)$ can be interpreted as planes of dimension $k = p(p-1)/2$ in this space. It turns out that the complex of these planes is admissible and the inversion formula of §2.3 provides an inversion formula for the problem of integral geometry on $SL(p; \mathbf{C})$ [16*].

In the question of describing the admissible complexes the essential results concern only the case $k = 1$ (line complexes). All admissible line complexes in general position have been classified [164]. It turns out that for $n = 3$ each admissible complex consists either of lines intersecting a fixed curve or of lines touching a fixed surface.

In subsequent papers on integral geometry a number of results were obtained that are connected with the application of differential forms to integral transforms [176]. This article discusses a general (non-homogeneous) formulation of problems of integral geometry that arise if there is an incidence relation between points of a pair of manifolds. An example of such a problem, connected with a pair of Grassman manifolds, is considered in [185]. The article [184] is concerned with integral geometry in a projective space. The lecture [191] outlines a method by which, using the results of this article, one can obtain the Paley – Wiener theorem on $G = SL(p; \mathbf{C})$. We recall that it is a question of conditions on $\hat{f}(x_1, x_2)$, $x_1, x_2 \in X = G/Z$, that ensure that f belongs to the space $\mathscr{S}(G)$. If we embed G in $\mathbf{C}P^n$, $n = p^2 - 1$, then the latter condition signifies the possibility of extending f to an infinitely differentiable function on $\mathbf{C}P^n$

that vanishes at infinity. An analogue to the Paley – Wiener theorem is obtained if we carry over these conditions to $\hat{f}(x_1, x_2)$ by means of the inversion formula in projective space.

§3. Cohomology of infinite-dimensional Lie algebras

A long series of papers by Gel'fand, written in 1968–1972, the majority jointly with Fuks, is devoted to an investigation of the cohomology of infinite-dimensional Lie algebras.

A formal definition of the cohomology of a Lie algebra was given by Chevalley and Eilenberg in 1948 (see [17*]); in fact, this concept was already known to E. Cartan. During the last two decades the cohomological theory of Lie algebras has been intensively developed, and without going into details it can be said that the cohomology of real and complex finite-dimensional Lie algebras are now as well known as the cohomology of the corresponding Lie groups; the reader can find the details in the proceedings of the Séminaire "Sophus Lie" [18*]. Nonetheless, until Gel'fand's articles appeared, no attempt had been made to find out anything about the cohomology of infinite-dimensional Lie algebras, although examples of these algebras had been known, so to speak, from time immemorial.

Why was this so? Firstly, there was no reason to expect that a cohomology defined by means of infinite-dimensional objects would turn out to be finite-dimensional and therefore meaningful from the point of view of homology theory. Secondly, those branches of topology and analysis whose algebraic roots spring from the cohomology of infinite-dimensional Lie algebras were frozen for the same reason (in retrospect we can say that Gel'fand's work provided a stimulus for the development of these branches). In any event, nothing was known about the cohomology of infinite-dimensional Lie algebras up to 1968 and the first, very partial, results of Gel'fand were completely unexpected.

3.1 General Definitions (Source: Chapter 13 of "Homological algebra" by Cartan and Eilenberg). A *Lie algebra* over a field k is a vector space $\mathfrak{g}$ (over k) with a multiplication $\mathfrak{g} \otimes_{k} \mathfrak{g} \to \mathfrak{g}$ (the product of two elements $\xi, \eta \in \mathfrak{g}$ is called their *commutator* and is denoted by $[\xi, \eta]$) with the two properties: (i) $[\xi, \xi] = 0$ for any $\xi \in \mathfrak{g}$; (ii) $[\xi, [\eta, \zeta]] + [\eta, [\zeta, \xi]] + [\zeta, [\xi, \eta]] = 0$ for any $\xi, \eta, \zeta \in \mathfrak{g}$. A vector space M is called a $\mathfrak{g}$-module if there is a homomorphism $\mathfrak{g} \otimes M \to M$ (the image of an element $\xi \otimes x$ under this homomorphism is usually denoted by ξx) such that $\xi(\eta x) - \eta(\xi x) = [\xi, \eta]x$ for any $\xi, \eta \in \mathfrak{g}, = \in M$.

A *q-dimensional cochain* of the algebra $\mathfrak{g}$ with values (coefficients) in M is defined to be a q-linear skew-symmetric function on $\mathfrak{g}$ with values in M; the q-dimensional cochains form a vector space, which is denoted by $C^q(\mathfrak{g}; M)$. The formula

$$(3.1)\quad dL(\xi_1, \ldots, \xi_{q+1}) = \sum_{1\leqslant s<t\leqslant q+1} (-1)^{s+t-1} L([\xi_s, \xi_t], \xi_1, \ldots, \hat{\xi}_s, \ldots, \hat{\xi}_t, \ldots, \xi_{q+1}) + \sum_{1\leqslant s\leqslant q+1} (-1)^s \xi_s L(\xi_1, \ldots, \hat{\xi}_s, \ldots, \xi_{q+1})$$

defines a homomorphism $d = d^q(\mathfrak{g})$: $C^q(\mathfrak{g}; M) \to C^{q+1}(\mathfrak{g}; |M)$, which is called a derivation. The composition $d^q(\mathfrak{g}) \circ d^{q-1}(\mathfrak{g})$ is trivial, because of $\operatorname{Im} d^{q-1}(\mathfrak{g}) \subset \operatorname{Ker} d^q(\mathfrak{g})$. The factor space $\operatorname{Ker} d^q(\mathfrak{g})/\operatorname{Im} d^{q-1}(\mathfrak{g})$ is called the qth *homology space of* $\mathfrak{g}$ *with coefficients in* M and is denoted by $H^q(\mathfrak{g}; M)$.

A module M is called *trivial* if $\xi x = 0$ for any $\xi \in \mathfrak{g}$, $x \in M$; any vector space can be provided with the structure of a trivial $\mathfrak{g}$-module, in particular, the field **k** itself. In the case of a trivial module the second term on the right-hand side of (3.1) is equal to zero and can be discarded.

If M is a commutative and associative **k**-algebra and the operators of $\mathfrak{g}$ are derivations, in particular, if $M = \mathbf{k}$, then the formula

$$(3.2)\quad LL'(\xi_1, \ldots, \xi_{q+r}) = \sum_{1\leqslant i_1<\ldots<i_q\leqslant q+r} (-1)^{i_1+\ldots+i_q-q(q+1)/2} L(\xi_{i_1}, \ldots, \xi_{i_q}) \cdot L'(\xi_1, \ldots, \hat{\xi}_{i_1}, \ldots, \hat{\xi}_{i_q}, \ldots, \xi_{q+r})$$

defines a multiplication $C^q(\mathfrak{g}; M) \otimes C^r(\mathfrak{g}; M) \to C^{q+r}(\mathfrak{g}; M)$ which turns $C^*(\mathfrak{g}; M) = \oplus_q C^q(\mathfrak{g}; M)$ into an associative skew-commutative graded **k**-algebra. This multiplication is connected with d by the Leibniz formula and therefore it carries over to the cohomology. Thus, in this case, $H^*(\mathfrak{g}; M) = \oplus_q H^q(\mathfrak{g}; M)$ is also an associative skew-commutative **k**-algebra.

If $\mathfrak{h}$ is a subalgebra of a Lie algebra $\mathfrak{g}$, then the cochains $F \in C^q(\mathfrak{g}; M)$, satisfying the conditions

$$L(\xi_1, \ldots, \xi_q) = 0 \text{ for } \xi_1 \in \mathfrak{h},$$
$$dL(\xi_1, \ldots, \xi_{q+1}) = 0 \text{ for } \xi_1 \in \mathfrak{h},$$

form a subspace of $C^q(\mathfrak{g}; M)$ which is denoted by $C^q(\mathfrak{g}, \mathfrak{h}; M)$. Obviously $d[C^q(\mathfrak{g}, \mathfrak{h}; M)] \subset C^{q+1}(\mathfrak{g}, \mathfrak{h}; M)$ which enables us to define the cohomology, starting out from the spaces $C^q(\mathfrak{g}, \mathfrak{h}; M)$. The spaces $H^q(\mathfrak{g}, \mathfrak{h}; M)$ so obtained are called cohomology spaces of $\mathfrak{g}$ *modulo* $\mathfrak{h}$. All we have said about multiplication in cochains and cohomology carries over to this "relative" case.

As a rule, the Lie algebras and modules over them have an additional structure, a topology. *In such a case, without saying so explicitly, we assume that all cochains are continuous functions of their arguments and we understand cohomology in the corresponding sense.* We must warn the reader that this terminological licence is not generally accepted: what we call cohomology below is usually called continuous cohomology.

In conclusion, we comment briefly on these definitions. If $\mathfrak{g}$ is the Lie algebra of a finite-dimensional connected real Lie group G, then the q-dimensional cochains of $\mathfrak{g}$ with coefficients in the trivial $\mathfrak{g}$-module $\mathbf{R}$ are naturally identified with the left-invariant exterior differential forms of degree q on G. Under this identification the derivation d goes into the usual exterior differential, so that the cohomology of $\mathfrak{g}$ is naturally identified with the de Rham cohomology of G. If G is compact, then the latter homomorphism is an isomorphism and $H^*(\mathfrak{g}; \mathbf{R}) = H^*(G; \mathbf{R})$. If G is semi-simple, then it is easy to see that $H^*(\mathfrak{g}; \mathbf{R}) = H^*(G^*; \mathbf{R})$, where G^* is the compact form of G (see [18*]). Finally, if $\mathfrak{g}$ is the Lie algebra of a compact connected Lie group G and $\mathfrak{h}$ the Lie algebra of a subgroup H of G, then

$$H^*(\mathfrak{g}, \mathfrak{h}; \mathbf{R}) = H^*(G/H; \mathbf{R}).$$

Cohomology with coefficients in non-trivial modules can also be explained in the language of Lie groups.

3.2. Lie algebras of smooth vector fields. The case of a circle. Smoothness is understood below as the property of belonging to the class C^∞.

It is standard knowledge that smooth vector fields on a smooth manifold M form a Lie algebra with respect to the Poisson bracket; this algebra, equipped with the C^∞-topology, is denoted by $\mathfrak{a}(M)$. The problem arises of calculating the cohomology of this algebra both with trivial coefficients and with coefficients in natural $\mathfrak{a}(M)$-modules, such as the space $C^\infty(M)$ of smooth functions on M, the space $\Omega^r(M)$ of smooth exterior differential forms of degree r on M, all possible spaces of smooth tensor fields on M, and so on. One of the main results of Gel'fand and Fuks is the following theorem [178]:

For any smooth manifold M and any integer q the space $H^q(\mathfrak{a}(M); \mathbf{R})$ is finite-dimensional.

A similar statement holds also for cohomology with coefficients in many non-trivial modules, in particular, in all modules listed above (see [19*]). The article [178] and later publications by various authors contain certain information on the structure of the groups $H^q(\mathfrak{a}(M); \mathbf{R})$; nonetheless, so far these groups have not been completely calculated. Postponing to §4 an account of the progress made here, we mention the results of an earlier article [169], which contains a complete calculation of the spaces $H^q(\mathfrak{a}(M);\mathbf{R})$ when M is a circle S^1.

We note that the smooth vector fields on S^1 are naturally identified with smooth functions, and the commutation operation is defined by the formula

$$[f, g] = fg' - f'g.$$

Here is a complete description of the cohomology $H^q(\mathfrak{a}(S^1); \mathbf{R})$: the space $H^q(\mathfrak{a}(S^1);\mathbf{R})$ is trivial for $q = 1$ and one-dimensional for $q \geqslant 2$; the ring $H^*(\mathfrak{a}(S^1);\mathbf{R})$ is the tensor product (over $\mathbf{R}$) of the ring of polynomials in one two-dimensional generator a and the exterior algebra in one three-dimensional generator b, and the generators a and b are represented by cochains $\alpha \in C^2(\mathfrak{a}(S^1);\ \mathbf{R})$, $\beta \in C^3(\mathfrak{a}(S^1);\ \mathbf{R})$ defined by the formulae

$$\alpha(f_1, f_2) = \int_{S^1} \begin{vmatrix} f_1' & f_2' \\ f_1'' & f_2'' \end{vmatrix} d\theta,$$

$$\beta(f_1, f_2, f_3) = \int_{S^1} \begin{vmatrix} f_1 & f_2 & f_3 \\ f_1' & f_2' & f_3' \\ f_1'' & f_2'' & f_3'' \end{vmatrix} d\theta.$$

3.3. Lie algebras of formal vector fields. The subsequent results on the cohomology of algebras $\mathfrak{a}(M)$ are based on another group of theorems: on the cohomology of Lie algebras of formal vector fields ([188], [190], [195], [196]). We ought to mention that these theorems, which arose in their time as lemmas to theorems on the cohomology of Lie algebras of smooth vector fields, now have an important independent significance, owing to their connection with the recently constructed theory of characteristic classes of foliations; this is dealt with in §3.5.

We denote by W_n the naturally topologized Lie algebra of formal vector fields in $\mathbf{R}^n$; we recall that a formal vector field is an expression of the form $\sum_{i=1}^{n} a_i(x_1, \ldots, x_n) \frac{\partial}{\partial x_i}$ where the a_i are formal power series, and that the commutator of two formal vector fields

$$\xi = \sum_{i=1}^{n} a_i(x_1, \ldots, x_n) \frac{\partial}{\partial x_i}, \qquad \eta = \sum_{i=1}^{n} b_i(x_1, \ldots, x_n) \frac{\partial}{\partial x_i}$$

is defined as the formal vector field

$$[\xi, \eta] = \sum_{i=1}^{n} \sum_{j=1}^{n} \left(a_j \frac{\partial b_i}{\partial x_j} - b_j \frac{\partial a_i}{\partial x_j} \right) \frac{\partial}{\partial x_i}.$$

The formal vector fields of the form

$$\sum_{i,j} a_{ij} x_i \frac{\partial}{\partial x_j} \tag{3.3}$$

with $a_{ij} \in \mathbf{R}$ constitute, as is easy to see, a subalgebra of W_n isomorphic to the Lie algebra $\mathfrak{gl}(n, \mathbf{R})$ of matrices of order n. This subalgebra is also denoted by $\mathfrak{gl}(n, \mathbf{R})$ and the part of it composed of fields of the form (3.3) with $a_{ij} = -a_{ji}$ by $\mathfrak{o}(n)$. W_n-modules are very common: $\mathbf{R}$; the space S_n of formal power series in $\mathbf{R}^n$; the space Ω_n^r of formal exterior differential forms of degree r in $\mathbf{R}^n$; other spaces of formal tensor fields in $\mathbf{R}^n$.

We denote by V_n the $2n^{\text{th}}$ skeleton of the standard complex of the classifying space of the unitary group $U(n)$, by X_n the inverse image of V_n in the total space of the universal $U(n)$-bundle (so that X_n is a $U(n)$-space and $X_n/U(n) = V_n$) and by Y_n the factor space $X_n/SO(n)$. It is obvious, in particular, that $X_1 = Y_1 = S^3$. We state the main result of [188] (in a somewhat modified form; see [32*], [20*].

There are ring isomorphisms $H^*(W_n;\ \mathbf{R}) = H^*(X_n, \mathbf{R})$, $H^*(W_n, \mathfrak{o}(n);\ \mathbf{R}) =$

$= H^*(Y_n; \mathbf{R})$, $H^*(W_n, \mathfrak{gl}(n, \mathbf{R}); \mathbf{R}) = H^*(V_n; \mathbf{R})$ *such that the diagram*

$$\begin{array}{ccc} H^*(W_n; \mathbf{R}) & = & H^*(X_n; \mathbf{R}) \\ \uparrow & & \uparrow \\ H^*(W_n, \mathfrak{o}(n); \mathbf{R}) & = & H^*(Y_n; \mathbf{R}) \\ \uparrow & & \uparrow \\ H^*(W_n, \mathfrak{gl}(n, \mathbf{R}); \mathbf{R}) & = & H^*(V_n; \mathbf{R}), \end{array}$$

in which the left-hand series of arrows is induced by the embedding of complexes, and the right-hand series of arrows by natural projections, is commutative.

From this theorem it follows, in particular, that multiplication in the ring $H^*(W_n; \mathbf{R})$ is trivial (that is, the product of any two elements of positive dimension is zero).

The cohomology of W_n with coefficients in Ω_n^r are described as follows:

$$H^q(W_n; \Omega_n^r) = \begin{cases} 0, & \text{if } q < r, \\ H^r(W_n; \Omega_n^r) \otimes H^{q-r}(U(n); \mathbf{R}), & \text{if } q \geqslant r; \end{cases}$$

the dimension of the space $H^r(W_n; \Omega_n^r)$ is 1 for $r = 0$ and is the number of partitions of r if $r > 0$.

A somewhat improved version of the proof of these theorems was presented at a recent lecture by Godbillon at the Bourbaki seminar (see [20*]).

The fact that the spaces $H^*(W_n; \Omega_n^r)$ are finite-dimensional, which follows from Theorem 3, is a consequence of the general theorem on finite dimensionality of the cohomology of a wide class of infinite-dimensional Lie algebras with coefficients in a wide class of modules (see [196]). The cohomology of some algebras of this class have been calculated by Rozenfel'd [21*]. However, as before, little is known about the cohomology of Lie algebras of formal vector fields not satisfying the conditions of the general theorems of [196]. This cohomology continues to be studied by Gel'fand and his students. Of the progress in this direction we should mention the results on Hamiltonian vector fields [230] and a recent article by Goncharova [22*], who calculates the cohomology of the algebra W_1 with coefficients in a wide class of W_1-modules, in particular, with coefficients in an arbitrary space of formal tensor fields on a line.

3.4. The Lie algebra of smooth vector fields. The diagonal complex. As we said in §3.2, the cohomology of the Lie algebra $\mathfrak{a}(M)$ of smooth vector fields on a smooth manifold M is completely known only in the case $M = S^1$. However, there is significant (we can now say, exhaustive) information on the homology of an important subcomplex of the cochain complex of $\mathfrak{a}(M)$, which is singled out in [178] and called *diagonal*.

A cochain $L \in H^q(\mathfrak{a}(M); \mathbf{R})$ is called diagonal if for any vector fields $\xi_1, \ldots, \xi_q$ with $\bigcap_{i=1}^q \operatorname{supp} \xi_i = \emptyset$, where supp is the support,

$L(\xi_1, \ldots, \xi_q) = 0$. The space of q-dimensional diagonal cochains is denoted by $C^q_\Delta(\mathfrak{a}(M); \mathbf{R})$ and it is easy to verify that $d[C^{q-1}_\Delta(\mathfrak{a}(M); \mathbf{R})] \subset \subset C^q_\Delta(\mathfrak{a}(M); \mathbf{R})$. The space

$$\{(\operatorname{Ker} d) \cap [C^q_\Delta(\mathfrak{a}(M); \mathbf{R})]\}/d\,[C^{q-1}_\Delta(\mathfrak{a}(M); \mathbf{R})]$$

is called the qth space of diagonal cohomology of $\mathfrak{a}(M)$ and is denoted by $H^q_\Delta(\mathfrak{a}(M); \mathbf{R})$.

In [178] a spectral sequence is constructed with

$$E_2^{pq} = \begin{cases} 0, & \text{if } q = 0, \\ H^{p+n}(M; \mathbf{R}) \otimes H^q(W_n; \mathbf{R}) & \text{if } q \neq 0, \end{cases}$$

where $n = \dim M$, which converges to $H^*_\Delta(\mathfrak{a}(M), \mathbf{R})$. The question of the derivations of this spectral sequence was open for a long time; various special cases were investigated in articles by Gel'fand, Kazhdan, and Fuks [189], [215], [220] and Losik [23*], until finally Losik [24*], and independently Guillemin [25*], succeeded in solving the problem completely.

Here is this solution (for simplicity we confine ourselves to the orientable case). Let M be an oriented manifold, and $(X(M), p, M)$ the bundle with structure group $SO(n)$ and standard fibre X_n, associated with the tangent bundle of M. If $\{F_r^{pq}\}$ is the real cohomological spectral sequence of the bundle $(X(M), p, M)$, then there are isomorphisms that commute with derivations:

$$E_r^{pq} \to F_r^{p+n,q} \qquad (r \geqslant 2,\ q \neq 0).$$

In particular, for $q > n$ the space $H^q_\Delta(\mathfrak{a}(M); \mathbf{R})$ is isomorphic to $H^{q-n}(X(M); \mathbf{R})$ (for $q \leqslant n$ the space $H^q_\Delta(\mathfrak{a}(M); \mathbf{R})$ is trivial).

As for the "off-diagonal" cohomology of $\mathfrak{a}(M)$ its calculation is reduced in [178] to the calculation of a sequence of spectral sequences, analogous to that of Theorem 4.1, but more complicated. With the help of these spectral sequences the theorem on finite dimensionality was also proved in [178] (see §3.2).

In the problem of calculating the cohomology of $\mathfrak{a}(M)$ with non-trivial coefficients the position is very similar to that of the problem of calculating $H^*(\mathfrak{a}(M); \mathbf{R})$. Here also there is the diagonal complex, (see [26*], [190]), and for the calculation of its cohomology there is a spectral sequence, constructed by Gel'fand and Fuks [190] and calculated by Losik [27*]. However, the second term of the spectral sequence of [190] is expressed in terms of the cohomology of the algebra W_n with non-trivial coefficients, which is not always known (see §3.2).

3.5. Application: characteristic classes of foliations. To conclude this section, we mention the effect which the results by Gel'fand and his collaborators had in a classical domain at the junction of topology, algebra, and differential geometry: the theory of foliations.

In the June issue of the Comptes Rendus for 1971 there appeared a note by two young French mathematicians, Godbillon and Vey [28*], in which they associate with each oriented foliation of codimension 1 on a smooth manifold M a certain element of the space $H^3(M; \mathbf{R})$ having a number of curious properties. The construction of Godbillon and Vey is quite elementary and no heavy algebra is used, but nevertheless it is stated in their note that the class so constructed originates from a non-trivial element of $H^3(W_1; \mathbf{R})$ (see §3.3). This observation indicated the direction in which to look for multi-dimensional analogues of the Godbillon – Vey classes, and they were found very quickly. Generalized Godbillon – Vey classes were discovered independently by two pupils of Gel'fand, Bernshtein and Rozenfel'd [29*], by Malgrange (though he did not publish an article of his own) and by Haefliger and Bott [30*], [31*]. Referring the interested reader for the details to the surveys [32*], [33*], we briefly describe the results of the proposed constructions.

We say that on a smooth n-dimensional manifold M a foliation of codimension q is defined, where $0 \leqslant q \leqslant n$, if M is decomposed into disjoint connected subsets F_α with the following property: for each point $x \in M$ there exist a neighbourhood U of it and a submersion ϕ: $U \to \mathbf{R}^q$ such that the connected components of $F_\alpha \cap U$ are mapped by ϕ into separate points. A smooth mapping f of a smooth manifold N into a smooth manifold M with foliation $\mathscr{F}$ is called transversal to $\mathscr{F}$ if the composition of f with each of the submersions ϕ is a submersion. If this condition is satisfied, then the connected components of the sets $f^{-1}(F_\alpha)$ form a foliation on N of the same codimension as $\mathscr{F}$; this is called the foliation of $\mathscr{F}$ induced by f and is denoted by $f^*\mathscr{F}$. Two foliations $\mathscr{F}_0$ and $\mathscr{F}_1$ on M are called homotopic if there is a foliation $\mathscr{F}$ on $M \times \mathbf{R}$ such that $i_0^*\mathscr{F} = \mathscr{F}_0$, $i_1^*\mathscr{F} = \mathscr{F}_1$ where i_0 and i_1 are the embeddings $M \to M \times \mathbf{R}$ defined by the formulae $i_0(x) = (x, 0)$, $i_1(x) = (x, 1)$.

By the definition of a foliation, the kernels of the differentials of the submersions ϕ form a certain subbundle $\tau(\mathscr{F})$ of the tangent bundle $\tau(M)$ of the manifold M; this subbundle is called tangential to $\mathscr{F}$, and the factor-bundle $\nu(\mathscr{F}) = \tau(M)/\tau(\mathscr{F})$ is called normal to $\mathscr{F}$. The foliation $\mathscr{F}$ is called orientable (oriented) if $\nu(\mathscr{F})$ is orientable (oriented). Clearly, if f: $N \to M$ is a smooth mapping, transversal to $\mathscr{F}$, then

$$\nu\ (f^*(\mathscr{F})) = f^*(\nu(\mathscr{F})).$$

We say that *an r-dimensional lower characteristic class α of oriented foliations of codimension q* is defined if to each oriented foliation $\mathscr{F}$ of codimension q on any smooth manifold M (of any dimension) there corresponds an element $\alpha(\mathscr{F})$ of the space $H^r(M; \mathbf{R})$, and $\alpha(f^*\mathscr{F}) = f^*\alpha(\mathscr{F})$ for any smooth mapping f: $N \to M$ transversal to $\mathscr{F}$. We say that *an r-dimensional upper characteristic class β of oriented foliations of codimension*

q is defined if to each oriented foliation $\mathscr{F}$ of codimension q on any smooth manifold M there corresponds an element $\beta(\mathscr{F})$ of the space $H^r(\mathscr{N}(\mathscr{F}); \mathbf{R})$ where $\mathscr{N}(\mathscr{F})$ is the total space $SO(q)$ of the bundle associated with $\nu(\mathscr{F})$ and for any smooth mapping $f\colon N \to M$ transversal to $\mathscr{F}$ the equality $\tilde{f}^*\beta(\mathscr{F}) = \beta(f^*\mathscr{F})$ holds, where $\tilde{f}$ is the mapping $\mathscr{N}(f^*\mathscr{F}) \to \mathscr{N}(\mathscr{F})$ induced by f.

It is easy to show that both lower and upper characteristic classes are homotopy invariants.

The set of all lower (upper) r-dimensional characteristic classes of oriented foliations of codimension q forms a vector space denoted by $\mathrm{Char}_r(q)$ ($\mathrm{Char}^r(q)$); we emphasize that this notation is not generally accepted.

The result of the Godbillon – Vey – Bernshtein – Rozenfel'd – Malgrange – Haefliger – Bott construction are homomorphisms

$$H^r(W_q; \mathbf{R}) \to \mathrm{Char}^r(q),$$

$$H^r(W_q, \mathfrak{o}(q); \mathbf{R}) \to \mathrm{Char}_r(q),$$

which interpret elements of the spaces $H^r(W_q; \mathbf{R})$, $H^r(W_q, \mathfrak{o}(q); \mathbf{R})$, described by the theorem of 3.3 as characteristic classes of foliations. In particular, the image of the natural generator of the space $H^3(W_1; \mathbf{R})$ under the first homomorphism (which coincides with the second if $q = 1$) is the Godbillon – Vey class.

There are examples that confirm that each of these homomorphisms is non-trivial. These examples enable us to establish for the first time that an isomorphism of the normal bundles of two foliations does not imply that these foliations are homotopic. Moreover, as Thurston proved [34*], there are foliations on a three-dimensional sphere whose Godbillon – Vey class is a preassigned element of the group $H^3(S^3; \mathbf{R})$; thus, *on a three-dimensional sphere there is a continuum of pairwise non-homotopic foliations of codimension* 1.

It just remains to add that the theory of characteristic classes of foliations, stimulated by the work of Gel'fand, is undergoing a period of rapid development. Results on the cohomology of the algebras $\mathfrak{a}(M)$ also find their place in this theory (see [30*]). We do not know what further results will be obtained from it, but already we can say that here, too, Gel'fand has performed his usual role of originator.

References

[1] M. I. Vishik, A. N. Kolmogorov, S. V. Fomin, and G. E. Shilov, Izrail' Moiseevich Gel'fand (on his fiftieth birthday), Uspekhi Mat. Nauk **19**:3 (1964), 187–204. MR **31** #20.
= Russian Math. Surveys **19**:3 (1964), 163–180.

[2] A. Weil, Basic number theory, Springer–Verlag, Berlin–Heidelberg–New York 1967. MR **38** #3244.
Translation: *Oznovy teorii chisel*, Izdat. Mir, Moscow 1972.

[3] H. Jacquet and R. Langlands, Automorphic forms on GL_2, Lecture Notes in Mathematics 114, Springer–Verlag, Berlin–Heidelberg–New York 1973.
Translation: *Avtomorfnye formy na* GL_2, Izdat. Mir, Moscow 1973.

[4] I. M. Gel'fand and D. A. Kazhdan, Representations of $GL(n, K)$, Proc. Summer School Representation Theory, Hungary 1971 (in English).

[5] N. N. Shapovalov, A conjecture by Gel'fand and Kirillov, Funktsional. Anal. i Prilozhen. 7:2 (1973), 93–94.
= Functional. Anal. Appl. **7** (1973), 165–166.

[6] A. A. Kirillov, Representations of certain infinite-dimensional Lie groups, Vestnik Moskov. Univ. Ser. I Mat. Mekh. no. 1 (1974).

[7] J. Radon, Über die Bestimmung von Funktionen durch ihre Integralwerte längs gewisser Mannigfaltigkeiten, Ber. Verh. Sächs. Akad. **69** (1917), 262–277.

[8] F. John, Bestimmung einer Funktion aus ihren Integralen über gewisse Mannigfaltigkeiten, Math. Ann. **109** (1934), 488–520.

[9] I. M. Gel'fand, M. I. Graev, and N. Ya. Vilenkin, *Integral'naya geometriya i svyazannye s nei voprosy teorii predstavlenii*, Gosudarstv. Izdat. Fiz.-Mat. Lit., Moscow 1962. MR **28** # 3324.
Translation: Integral geometry and representation theory, Academic Press, New York–London 1966.

[10] I. M. Gel'fand and M. A. Naimark, Unitary representations of the classical groups, Trudy Mat. Inst. Steklov **36** (1950). MR **13**–722.
Translation: Unitäre Darstellungen der Klassischen Gruppen, Akademic–Verlag, Berlin 1957.

[11] I. M. Gel'fand and G. E. Shilov, *Obobshchennye funktsii i deistviya nad nimi*, Gosudarstv. Izdat. Fiz.-Mat. Lit., Moscow 1958. MR **20** # 4182.
Translation: Generalized functions, properties and operations, Academic Press, New York–London 1964.

[12] I. M. Gel'fand and M. I. Graev, Analogues of the Plancherel formula for the classical groups, Trudy Moskov. Mat. Obshch. **4** (1955), 375–404. MR **17**–173.
= Amer. Math. Soc. Transl. (2), **9**, 123–154. MR **19**–1181.

[13] I. N. Bernshtein and S. I. Gel'fand, Meromorphy of the function P^λ, Funktsional. Anal. i Prilozhen. **3**:1 (1969), 84–85. MR **40** # 723.
= Functional. Anal. Appl. **3** (1969), 68–69.

[14] I. M. Gel'fand and M. I. Graev, An analogue of Plancherel's theorem for the real unimodular groups, Dokl. Akad. Nauk SSSR **92** (1953), 461–464. MR **15**–683.

[15] I. M. Gel'fand and M. I. Graev, The geometry of homogeneous spaces, representation of groups in homogeneous spaces and related questions of integral geometry, Trudy Moskov. Mat. Obshch. **8** (1959), 321–390. MR **23** # A4013.

[16] S. G. Gindikin and F. I. Karpelevich, A problem of integral geometry, In memoriam N. G. Chebotarev, 30–43. Izdat. Kazan. Univ., Kazan 1964. MR **33** # 4875.

[17] C. Chevalley and S. Eilenberg, Cohomology theory of Lie groups and Lie algebras, Trans. Amer. Math. Soc. **63** (1948), 85–124. MR **9**–567.

[18] V. N. Reshetnikov, Cohomology of Lie algebras of smooth vector fields with non-trivial coefficients, Dokl. Akad. Nauk SSSR **208** (1973), 1041–1043.
= Soviet Math., Dokl. **14** (1973), 230–240.

[19] N. Bourbaki, Eléments de mathématique, Vol. XXVI. Groupes et algèbres de Lie. Hermann & Cie, Paris 1960. MR **24** # A2641.

Translation : *Teoriya algebra Li. Topologiya grupp Li.* Izdat. Inostr. Lit., Moscow 1965.

[20] K. Godbillon, Cohomologies d'algèbres de champs de vecteur formels, Sém. Bourbaki, 25-e année 1972/73, no. 421.
= Uspekhi Mat. Nauk **28**:4 (1973), 139–151.

[21] B. I. Rozenfel'd, Cohomology of some infinite-dimensional Lie algebras, Funktsional. Anal. i Prilozhen. **5**:4 (1971), 84–85. MR **46** #8267.
= Functional Anal. Appl. **5** (1971), 340–344.

[22] L. V. Goncharov, Cohomology of Lie algebras of formal vector fields on a line, Funktsional. Anal. i Prilozhen. **7**:2 (1973), 6–14; **7**:3 (1973), 33–44.
= Functional Anal. Appl. **5** (1973), 91–97; 194–203.

[23] M. V. Losik, Cohomology of the Lie algebra of vector fields with coefficients in the trivial identity representation, Funktsional. Anal. i Prilozhen. **6**:1 (1972), 24–36. MR **46** #912.
= Functional. Anal. Appl. **6** (1972), 21–30.

[24] M. V. Losik, A topological interpretation of the homology of the diagonal complex of the Lie algebra of vector fields with coefficients in the trivial identity representation, Funktsional. Anal. i Prilozhen. **6**:3 (1972), 79–80. MR **47** #1074.
= Functional Anal. Appl. **6**, 242–243.

[25] V. W. Guillemin, Remarks on some results of Gel'fand and Fuks, Bull. Amer. Math. Soc. 78 (1972), 539–540. MR **45** #7739.

[26] M. V. Losik, Cohomology of infinite-dimensional Lie algebras of vector fields, Funktsional. Anal. i Prilozhen. **4**:2 (1970), 43–53. MR **43** #5544.
= Functional Anal. Appl. **4** (1970), 127–135.

[27] M. V. Losik, Cohomology of Lie algebras of vector fields with non-trivial coefficients Funktsional. Anal. i Prilozhen. **6**:4 (1972), 44–46. MR **47** #1075.
= Functional Anal. Appl. **6** (1972), 289–291.

[28] C. Godbillon and J. Vey, Un invariant des feutilletages de codimension un, C. R. Acad. Sci. (Paris) **273** (1971), A92–A95. MR **44** #1046.

[29] I. N. Bernshtein and B. I. Rozenfel'd, Characteristic classes of foliations, Funktsional. Anal. i Prilozhen. **6**:1 (1972), 68–69, MR **45** #6026.

[30] A. Haefliger, Sur les classes charactéristiques des feuilletages, Sém. Bourbaki, 24-e année 1971/1972, no. 412.

[31] R. Bott and A. Haefliger, On characteristic classes of Γ-foliations, Bull. Amer. Math. Soc. 78 (1972), 1039–1044. MR **46** #6370.

[32] D. B. Fuks, Characteristic classes of foliations, Uspekhi Mat. Nauk **28**:2 (1973), 3–18.
= Russian Math. Surveys **28**:2 (1973), 1–16.

[33] I. N. Bernshtein and B. I. Rozenfel'd, Homogeneous spaces of infinite-dimensional Lie algebras and characteristic classes of foliations, Uspekhi Mat. Nauk **28**:4 (1973), 103–138.
= Russian Math. Surveys 28:4 (1973), 107–142.

[34] W. Thurston, Non-cobordant foliations of S^3, Bull Amer, Math. Soc. **78** (1972), 511–514. MR **45** #7741.

THE WORK OF I. M. GEL'FAND IN APPLIED AND COMPUTATIONAL MATHEMATICS

O. V. Lokutsievskii and N. N. Chentsov

Twenty years ago computational mathematics was more of an art than a science. Today it is an extensive scientific discipline, which not only uses the ideas and results of very abstract branches of mathematics, but also, through its problems, stimulates their development. The ideas and results of I. M. Gel'fand, his students and collaborators have played a prominent role in this transformation. Computations of record difficulty, successfully carried out under his direction, have indeed demonstrated the perspective of the new approaches in computational practice.

The central problem facing computational mathematics in the late 1940's and early 1950's was the solution of non-stationary problems of a continuous medium, in the first place quasilinear equations of gas dynamics, initially in one spatial variable and then in several. Gel'fand was one of the first to realize that with the appearance of the EVM computer a more perspicuous method of solving such problems was a difference method, and he concentrated his efforts on this. The standard that difference methods had reached at that time did not allow them to be used directly for the solution of any complicated problems. Difference methods needed substantial development, and Gel'fand's great service in carrying out this development so brilliantly is indisputable.

Although, as we have said, his work in this field is connected with the solution of various concrete problems, the concepts originated by him have a general character and are widely applied in modern computational mathematics.

Gel'fand was the first Soviet mathematician to remark that an approximation of a differential operator L by differences does not need to be based on an approximation of derivatives converging to L by the corresponding difference relations. This led him to an apparently very fruitful method of constructing a difference scheme, which gives an optimal approximation with respect to order for the solutions of the original differential equation. A number of convenient and often unexpected difference schemes were obtained in this way. The understanding of an approximation described above, which was wider than that existing previously, greatly extended the range of difference schemes that could be considered. As we have already said, many of them turned out to be very convenient. However, for reasons not at first understood, some did not stand up to practical test. An investigation into the reasons for this led Gel'fand to a substantial refinement of the idea of approximating a differential operator by differences: he demanded that necessary bounds should exist for the *independent* convergence of the steps of the mesh to zero (Gel'fand called schemes that satisfy this condition flexible).

An approximation of a differential operator by differences does not ensure the convergence of the solution of the difference problem to the solution of the original differential problem under mesh refinement. Convergence is guaranteed only for those schemes which, apart from the approximation properties, also have stability, a property analogous to the original problem being well-posed, but which does not follow by approximation from being well-posed.

Thus, there is an actual question of the stability of various difference schemes. In the case of the Cauchy problem it is solved (with reasonable bounds of validity) comparatively easily: for fixed coefficients, a spectral investigation of stability can be carried out, as a rule, by Fourier's method. The problem of spectral investigation of a difference scheme with a condition on one end (for a half-line) is more difficult, but soluble. Essentially more difficult, and for non-self-adjoint problems practically unrealizable, is the direct spectral investigation of stability of a difference problem with two boundaries (for a segment).

Gel'fand and Babenko suggested a very effective procedure for investigating the stability of such problems. This consists in first investigating the stability of the Cauchy problem and then that of each of the two boundary problems that arise when one of the boundaries is removed to infinity. It turns out that the spectrum of the original problem can be described in this way. (In view of the presence of two independent parameters in the original problem, space and time, the very idea of its spectrum needs certain refinements, which we cannot go into here.) As a consequence, the argument of Gel'fand and Babenko described above provided a further development and overflowed into the theory of spectra of families of difference operators of Godunov and Ryaben'kii.

Difference schemes intended for the solution of evolutionary problems are divided into two sharply contrasting classes: explicit and implicit. Explicit schemes, that is, those describable by triangular matrices, are stable only for certain ratios of the steps of the different variables. Namely, if h is the space step, and τ the time step, then for hyperbolic systems there must be an inequality of the form $\tau/h \leqslant C$, and for the heat conduction equation $\tau/h^2 \leqslant C$. Such a restriction, which inevitably arises in view of the necessity of effectively considering the sphere of influence of a differential operator in its difference realization, leads to enormously laborious computations in the use of difference schemes. Therefore the question of the effective use of implicit schemes in the solution of evolutionary problems of mathematical physics is very real.

The inconvenience of the latter consists in the fact that, because the matrices for changing to the next time layer are not triangular, there arises the problem of solving an algebraic system with a large number of unknowns. In a paper by Gel'fand and Lokutsievskii [121] it was shown that iterative methods of solving such systems cannot be regarded as satisfactory: in view of the stability requirement, they cannot significantly reduce the volume of work in comparison with explicit schemes. Therefore, in the development there arises the need for a convenient precise method of solution of such systems. Algorithms that existed at that time (for example, the so-called "range") seemed perfectly acceptable at first glance. However, in their numerical realization, to preserve accuracy they needed a practically unattainable number of spare symbols (proportional to the order of the system).

In the paper by Gel'fand and Lokutsievskii [121] a precise algorithm is suggested that does not have this deficiency. This algorithm, based on the use of the specific character of matrices of such systems, needs for its realization a number of operations proportional to the number of equations and only a small number of symbols. This algorithm has wide application in practical computation.

In the middle 1950's Gel'fand stood out as one of the pioneers of the solution of multi-dimensional stationary and non-stationary problems of gas dynamics. In this work his collaborators Babenko, D"yachenko, Lokutsievskii, Rusanov, Fedorenko, Chentsov and others took part. Fundamental facts of the theory of multi-dimensional schemes were explained, and effective methods were developed for computing spatial flows, using schemes of explicit-implicit type (that is, explicit in one spatial variable and implicit in the other). In addition, a matrix method suggested by Keldysh was investigated. In subsequent years work in this direction was widely expanded by a group under the direction of Babenko, which became an independent section and the leading scientific association in this field. The work carried out by Babenko and his collaborators was honoured by a State Prize in 1966.

In the computation of problems in the physics of a continuous medium, it is convenient to calculate the isolated singularities of the solution separately, making use of any analytical or semianalytical methods. In the early 1950's Gusarov, D"yachenko, Zhukov and Lokutsievskii, under the direction of Gel'fand and with his collaboration, constructed the solution of an automodel problem on the collapse of a spherical cavity, and also a problem on a convergent spherical shock wave in an ideal gas. In both cases the solution of the automodel problem reduces to the integration of a system of ordinary differential equations in the independent variable $\xi = tr^{-k}$, where t is the time and r the distance from the centre. Usually the automodel exponent k is determined from dimensional considerations. Here dimensional considerations were inapplicable, and it was necessary to find a value of k for which the integral curve $(x(\xi), y(\xi))$ joining two fixed points (x_0, y_0) and (x_1, y_1) passes through a given singular point (here we can draw an analogy with the problem of eigenvalues for linear equations)[1]. Gel'fand was the first to discover that in a certain interval of values of the

[1] In the problem of a convergent shock wave this was shown by Guderley, Landau and Stanyukovich, who posed and solved this problem in the 1940's.

adiabatic exponent of gas the automodel exponent k of the solution is not uniquely determined by the stated conditions. He conjectured that only one automodel solution determined by them is stable, and therefore physically realizable. In the course of later research, continued by Brushlinskii, Kazhdan and others,[1] investigation of problems of this type made great progress. However, Gel'fand's conjecture has not been verified so far, although it has been confirmed by difference integration of the equations of gas dynamics.

For the quasilinear equations of a continuous medium, in particular, the equations of motion of a compressible gas, no complete general theory has yet been constructed. It has not even been possible to give a sufficiently general definition of what is meant by a solution of a quasilinear system of partial differential equations. It is known that a smooth initial motion of a gas can change in the course of time: discontinuities (shock waves) can arise in it, discontinuities of derivatives (weak discontinuities), and so on. Moreover, examples are known, admittedly "non-physical" ones, where after a definite moment of time the solution generally does not exist. Also, it is not known when a generalized discontinuous solution is unique. Therefore, in the course of computing the characteristic field of a gas, if we encounter an incomprehensible behaviour of the computed values, we can attribute this behaviour to some unexpected singularity of the flow itself. (Incidentally, in this way a number of important effects of the flow of a gas and a plasma were first found in numerical computation and only later discovered in physical experiments.) But one might think that this incomprehensible behaviour of numbers is caused by deficiencies in the difference scheme, in other words, it is a "computing effect". (Unfortunately, examples of a similar kind occur very frequently.) Moreover, we cannot exclude the possibility that there is no solution of the problem in differential equations itself, and the difference scheme "does not know what to do next". Finally, the difference scheme can "jump" from one solution to another, in view of their non-uniqueness. Because the physical intuition of specialists in gas dynamics is quite well developed, the last two possibilities in the solution of problems put to them are unlikely. We note that intuitively obtained stable conservative schemes having difference analogues of physical laws of conservation behave "correctly".

What is the "physical" class of evolutionary systems of quasilinear equations in which we can establish the existence and uniqueness of generalized solutions of natural problems? This question, in some form, was solved in the mid-1950's for an equation of the first order. In the academic year 1957/58, Gel'fand gave a course of lectures at the Moscow

[1] See the paper by Brushlinskii and Kazhdan; Automodel solutions of certain problems in gas dynamics, Uspekhi Mat. Nauk **18**:2 (1963), 3–36. MR **30** # 2796.
= Russian Math. Surveys **18**:2 (1963), 3–22.

State University on the theory of quasilinear partial differential equations, the fundamental aspects of which are reproduced in [90]. The ideas advanced by him noticeably stimulated interest in this theory, and in the succeeding years his audience (chiefly Godunov) established many new results in this field. However, we are still a long way from creating a general theory for the system

$$\frac{\partial u}{\partial t}+\frac{\partial \varphi(v)}{\partial x}=0; \qquad \frac{\partial v}{\partial t}+\frac{\partial \psi(u)}{\partial x}=0.$$

We only remark that the laws of conservation and the conditions of stability, as Babenko and Gel'fand explained earlier, prohibit the appearance of certain types of discontinuities in the solutions of hyperbolic systems of quasilinear equations [82].

The accumulated experience of computing complicated gas dynamic flows put Gel'fand and his collaborators among the first to undertake a numerical solution of problems in magnetohydrodynamics. In the late 1950's, under the direction of Artsimovich and Leontovich, experimental procedures were developed in which compression of a plasma cord occurred under the action of a powerful electric discharge. The magnetic field in such a "pinch", which compresses the cord, leads to the rise of a cylindrical shock wave, which collapses onto the axis. At the instant of collapse close to the axis, for a sufficiently high pressure and temperature, a reaction of thermo-nuclear synthesis can begin. The mathematical theory of "pinch", given by Leontovich, Osoviets and Braginskii, leads to the Cauchy problem for a system of quasilinear equations of magnetohydrodynamics with boundary conditions closed by the electrotechnical equation. The numerical computations carried out by Gel'fand and Fedorenko [88] were among the first in the world on nonstationary magnetohydrodynamic flow. In this joint work with physicists there developed the now customary interaction of physical experiment on the apparatus and mathematical computation on the EVM. The good agreement of experimental and computed values strengthened confidence in the adequacy of the mathematical model and the accuracy of the numerical method. At the same time, the computations gave a significantly more detailed and complete picture of the flow than the results of direct measurements and a deeper understanding of the essence of the phenomenon under investigation. The last remarks could apply as a whole to the later work of Gel'fand and his collaborators on the computation of the structure of toroidal magnetic fields. This problem arose in connection with the construction of stellarators, where such a magnetic field must maintain a hot plasma. Close to the axis of a stellarator, each line of force in general position winds ergodically on a torus-shaped magnetic surface, so that a set of coaxal surfaces is formed, which cover the magnetic axis of the system. Such a magnetic configuration retains charged particles well. It was initially expected that when the field is disturbed, the magnetic surfaces would only be slightly deformed. However, direct numerical integration of the equations of the lines of force drawn by Zuev gave a very different picture. Namely, even under weak perturbations new families of embedded tori appeared about the axis, and the domain of closed surfaces is appreciably reduced [111], [113], [123]. And under large perturbations the magnetic surfaces begin to decompose with the formation of a Birkhoff fibre structure. A similar field ceases to retain plasma. After the computations had been carried out and understood, the fibre structure was observed in experiments.

The scientific interests of Gel'fand in the 1960's were also connected with research into complex systems described by a large number of variables, and a study of their organization and behaviour. He was the first to understand the importance of creating computational methods for such systems, for which the usual local methods of analysis prove ineffective.

A significant number of problems on large systems reduce to the discovery of the absolute minimum of a function of very many variables on the corresponding range.

Universal methods for solving such problems (of the type of excess of all values on an ε-net) prove completely ineffective. Hence each method of solving experimental problems can work effectively only for a definite class of functions. Gel'fand made an essential contribution to the methods for finding an extremum, which began to be developed intensively in the 1950's. He and Tsetlin introduced the idea of "well-organized" functions [244], [112]. For the analysis of a whole series of mathematical, physical and biological problems they remarked: in "reasonably" posed problems on the minimum of the contours of a function, as a rule, the situation is like that of mountainous country, with steep slopes of the mountain ranges and narrow valleys gently sloping towards the minimum points. Hence gradient methods, which are very convenient for a quick descent into a hollow, fail for motion along the bottom of a weakly curving valley, where the gradients in many directions are small and only a relatively large step leads to a steep slope. The "ravine method", proposed by Gel'fand and Tsetlin in 1959, is roughly as follows. Suppose that we know two not very close points on the bottom of a valley. They determine a direction. Then if we take a large step in this direction we obtain a point, which is, as a rule, on the slope. From there, by a local method (with a small step), we descend into the valley. The descent continues until the gradient becomes small. The process is then repeated. Adaptational algorithms have been developed for choosing the gradient and ravine steps, the tactics of motion, when the ravine is not one-dimensional, and so on, see [112], [151].

The ravine method was used to solve phase analysis problems of proton-proton scattering [108], [109], and then for the X-ray-structure analysis of complex compounds. The methods of determining molecular crystal structures, developed by Gel'fand and his collaborators Vul, Ginzburg, Ivanova, Neigauz and Fedorov and described in [151], rank favourably with other methods of solving this problem. Many structures have been deciphered with its help, including some quite complex ones. The same method was successfully used in neutron analysis.[1] The ravine method has proved useful in many other questions – from the computation of heterogeneous defences to the computation of bubble chambers.

The ravine method proved suitable also in a general problem of recognition, when Gel'fand moved from the deciphering of structures to a wider range of problems. In the first place he was interested in a problem of medical diagnosis and prognosis of the outcome of a disease. They attracted him particularly, because the process of adoption of the solution by the doctor is rather complicated, and they developed in him general principles of thinking. On the other hand, a concrete formulation enables one to restrict this activity to narrow professional limits and to formulate precise criteria for the quality of the solution of a problem.

Work on diagnostics has proceeded and still proceeds in close collaboration with physicians. For each of the problems a questionnaire is developed, information gathered from archives, and then an educational programme builds up a decisive principle. In this work, as well as purely mathematical questions, one has to encounter the following features. Statistics of diseases in the archives are usually not numerous. To a certain degree the doctor's record of the outcome of a disease is affected by the treatment of the patient's symptoms. Furthermore, the characteristics which the doctor usually names or which can be found in the literature frequently do not correspond to the ideas by which an experienced doctor actually acts at the patient's bedside. And for a formal algorithm the latter criteria are valuable. Therefore a clear and reasonable statement of the problem and an adequate questionnaire arise only in the course of the work and are whole investigations in themselves. The final test is carried out in the clinic, when in the processing of the results of the investigation the outcome of the disease is still largely unknown. At this stage it is also determined whether an operationally presented prognosis would be useful to the doctor or whether its contents would be emasculated in the course of work on the problem.

At the present time, for nearly three years the Institute of Neurology of the Academy of Medical Sciences (director K. V. Shmidt) has issued a machine prognosis of the outcome of a haemorrhage with a complete definition of the evidence for surgical treatment [197]. The machine predicts whether the patient will live for 10 days: a) under conservative methods of treatment and b) after a surgical operation for removal of a haematoma. The prognosis is issued on the basis of information on the patient gathered in the first six hours of his stay in the clinic. Preliminary

---

1) R. A. Alikhanov, E. B. Vul, and J. G. Fedorov, Structure and magnetism of solid oxygen, Acta Cristallographica **21**:7 (1966), A–92.

analysis shows that in 90% of cases the prognosis is proved correct.

With Gel'shtein (Institute of Cardiovascular Surgery of the Academy of Medical Sciences) he created a method of determining the degree of pulmonary hypertonia in one of the congenital diseases from a given electrocardiogram and phonocardiogram without directly probing the heart [210], [216]; an algorithm was constructed for a prognosis under infarction of the myocard (with Martynov of the General Military Hospital) and so on. It is still difficult to formulate principles for solving diagnostic problems in general form. However, in all the work that has been carried out, the result depends to a greater degree on the adequacy of the medical information than on the method of treatment, if the latter is reasonable enough.

Another large group of practical problems of recognition, to the solution of which Gel'fand attracted his collaborators (Guberman, Izvekov, Rotvain and others) is connected with questions of geological and seismic prognosis. They constructed algorithms and established prognostic maps for ore deposits, developed methods of recognizing oil-bearing strata in multilayer deposits from mining-geophysical data, and so on. Here they used both algorithms that require preliminary training on examples and also algorithms based on the ravine method, which use the a priori intrinsic structure of a system of prognosticable objects. The practical value of this work proved to be very great.

In the joint research of the Institute of Applied Mathematics of the USSR Academy of Sciences and the Institute of Earth Physics of the USSR Academy of Sciences (director M. A. Sadovskii) Gel'fand, Guberman, Keilis-Borok and Rantsman discovered a very general regularity in the origin of powerful earthquakes [227], [235].

According to established ideas, destructive earthquakes are confined to zones of active breaking of the earth's core and particularly to their intersection. These general considerations are insufficient for a concrete prognosis, all the more, because some of the breaks can be certainly distinguished only on photographs of the cosmos. Hence a system of symptoms was developed, and the zones of breaking were ordered according to rank. Then the breaks of the three highest ranks were represented. They all have a large extent and cut the earth's core to a depth of 15 km and more. It appears that the epicentres of destructive earthquakes with energy $\geqslant 10^{22}$ ergs are situated close to the intersections of breaks of the first three ranks. (Significant destruction in the epicentre itself can occur in weaker earthquakes of 10^{18} ergs, but these were not considered.) Then an instructional algorithm was constructed to recognize those intersections of breaks where earthquakes are possible, even though they are not known to have occurred. A number of regions were considered (Pamir and Tyan-Shan, Anatolia and the Armenian upland, the Balkans, Nevada and California) on the basis of which a criterion was constructed and a concrete prognosis given. For decision rules, only very rough and easily accessible geological data were used. It turned out that the criteria for the different regions were similar

[232]. To test the reliability of the criteria on the data up to 1912, sites of possible powerful earthquakes were predicted in Pamir and Tyan-Shan. And in the course of 60 years powerful earthquakes did, in fact, occur only in the predicted sites. The same was true for other regions. The results obtained are of great practical interest.

Gel'fand was the first to understand clearly the significance and importance of the models proposed by Tsetlin, of the appropriate behaviour of automata in random surroundings for the study of multilevel systems with non-individualized control, in the first place, biological systems. In contrast to the traditional formal-logical treatment of automata models of biological systems, Gel'fand, Gurfinkel and Tsetlin first formulated [120] a "behavioural" approach to the construction of such models. Under this approach, a system is regarded as consisting of separate subsystems having a high degree of autonomy and "special interests". Then the family of remaining subsystems and the outside world form an external medium for each subsystem in which it must act in the best way, that is, so as to minimize the unfavourable influence of the external medium. The behavioural approach to automata models turns out to be fruitful not only in physiological movements, where it was originally formulated in [120] (the application of these ideas to physiology is dealt with below, in the survey of Gel'fand's biological work), but also in etology, cell physiology, sociology and the control of large systems of mass service.[1]

The paper [120] laid the foundations of the theory of collective behaviour of stochastic automata. In particular, it formulated (in the language of physiology) a variational principle of least interaction for a system working appropriately in an external medium. A general formulation of the principles of organization of behaviour of a family of autonomic objects, as games, was given by Gel'fand, Pyatetskii-Shapiro and Tsetlin in [133]. There they formally defined certain classes of automata games which are very interesting for the simulation of the behaviour of complex systems. This work determined for many years the development of research into models of collective behaviour.

A whole series of trends in Gel'fand's activity was devoted to the creation of computational methods and the solution of concrete applied problems. We mention here the computations of problems of the physics of transfer [97], the computations of electrostatic oscillations of a rarefied plasma [161], the computations of the behaviour of collections of automata [139]. Gel'fand was the initiator of this work and took a large part in the posing of the problems and the choice of methods of computation. Finally, we recall one of his recent papers [230] where, by computations on the EVM, he, Fuks and Kalinin discovered new non-trivial cohomology classes of Lie algebras of Hamiltonian formal vector fields in R^2.

1) M. L. Tsetlin, A. V. Butrimenko and S. L. Ginzburg, An algorithm of communication network control. Problemy kibernetiki, no. 20 (1968). MR 44 # 3775.

THE WORK OF I. M. GEL'FAND IN BIOLOGY

M. B. Berkinblit, Yu. M. Vasil'ev and M. L. Shik

I

I. M. Gel'fand began to give his attention to biology in 1958. Up to 1960–1961 this work was of a theoretical character: Gel'fand conducted a small seminar in which actual investigations of the physiology of movements and the physiology of the heart were examined. The first work of physiology carried out by Gel'fand and Tsetlin ended this period. It was devoted to an analysis of the propagation of activity in nerve fibres and the heart muscle, which were considered as continuous media. The theory developed in this article was used as a further stimulus for a whole series of investigations of the propagation of activity in various excitable structures. It must be observed, however, that this was Gel'fand's only work in physiology, as this term was accepted at that time: well chosen experimental material was suitable for the existing mathematical apparatus. Gel'fand's work in biology had only just begun.

Very quickly Gel'fand remarked that physiologists, although experimentalists, are often inclined to a formal approach to their work. Although they deal daily with living organisms or at least living organs and tissue, physiologists are essentially ignorant of their "living" properties. For a physiologist, cells have dimensions, they are membranes with definite electrical properties, contractility or other mechanical properties, but are almost entirely devoid of a whole series of other qualities, which are studied separately by other scientific disciplines: biochemistry, immunology, embryology, and so on.

A period in the world of conventional ideas does not generally harm the work of experimentalists, since in the course of experiments, on the direct investigation of a living object, necessary corrections are introduced – "surprises" always "earth" the world of ideas. However, in obtaining a description of the properties of a living object at second hand – by studying the publications of other investigators – there is a real danger of substantially distorting the picture.

For this reason Gel'fand, together with Tsetlin, Gurfinkel and Fomin, organized a laboratory in which physiologists, physicists and mathematicians could interact at all stages of the research. Thinking out principles, planning an experiment, carrying it out, analysing and interpreting the results of measurements, developing a non-contradictory description of the results (a model) and generalizing it – the whole edifice of scientific work is erected by joint labour.

This laboratory was initially (1960–1967) in the Institute of Biological Physics of the USSR Academy of Sciences, and then became part of the Institute for Problems of Information Transmission of the USSR Academy of Sciences. In this laboratory, a number of projects were carried out on the control of movements and the physiology of the cerebellum. Later, Gel'fand's main interest in biology was the behaviour of cells. This theme was the basis for the organization, with Yu. M. Vasil'ev, of the Interfaculty Laboratory of Mathematical Methods in Biology at the Moscow State University.

II

As we have already said, the first work of Gel'fand and Tsetlin was to propose a formal model of excitable tissues, regarded as continuous media [149].

In this model, points of the medium have a state of rest and excitation, absolute refraction (where a point cannot excite any action) and relative refraction (when, to excite a point, greater action is necessary than at rest). The fact that the speed of propagation of excitation in the medium depends on the phase of relative refraction was taken into account. A variant of the model was also considered, where the points of the medium have spontaneous activity, that is, in a definite time after the last excitation each point is excited again, even if the external stimulus is missing. Thus, this model medium has the fundamental properties inherent in real excitable tissues.

A theoretical analysis carried out by Gel'fand and Tsetlin showed that the propagation of excitation in such media has a number of interesting properties. For example, the propagation of impulses in a homogeneous ring of active tissue is self-synchronizing: independently of the initial phases and the initial arrangement of impulses in the ring, a condition is set up so that the impulses are arranged equidistantly in the ring and are propagated with constant speed.

Considering a two-dimensional excitable tissue whose points are spontaneously active, Gel'fand the Tsetlin showed that such a tissue has a property analogous to memory – the medium "remembers" the phase and point of application of the external stimulus [149].

If the points of an active medium have different periods of spontaneous activity, in the condition that is set up the period of excitation of any point of the medium is equal to the minimal period of spontaneous activity – the medium is synchronized by the most active point. The possibility that the leading centre of the medium is synchronized automatically by the active cell that works with greatest frequency was verified experimentally in the work of Gel'fand and others.

The work of Gel'fand and Tsetlin provided a great affiliated direction of research – simulation of pathological conditions in heart tissue by means of continuous excitable media. It is interesting to note that certain theoretical results on the circulation and interaction of waves of excitation, obtained for the heart muscle, proved to be applicable to the propagation of depression in the cortex of the large hemispheres, as the well-known Czech physiologist Ya. Buresh showed.

In the domain of neurophysiology Gel'fand jointly with Tsetlin began by considering the operational control of movements. At this time they were engaged in the problem of looking for an extremum of a function of several variables.

But the construction (realization) of a movement can be regarded as the minimization of the deviations of a curve which, for definiteness, represents the movement of an extremity in a phase space whose coordinates are angles in the joints of this extremity and the speeds of their variation from an ideal curve in that space, the standard. For the general case, irrespective of the physiology of movements, Gel'fand and Tsetlin proposed the so-called "ravine method", which gives a greater reduction of excess than a better "organized" function. A "well organized" function is understood to be one for which the variables on which it depends can be split into two groups: a small number of essential ones, on which it depends weakly, and the inessential ones, on which it depends strongly and of which there may be many. This idea of organization is thus connected with the "effective dimensionality" of a function. We recall that in 1935 N. A. Bernstein pointed out that the coordination of movements was none other than a lowering of the number of degrees of freedom of the motive apparatus. Gel'fand and Tsetlin suggested that the nervous system uses the "ravine method" in the construction of movements [149]. This conjecture proved to be heuristically useful for five to seven years in the work of the newly created laboratory of the physiology of movements. It stimulated experimental work in the study of mechanisms for preserving vertical attitude, physiological tremor, the simplest arbitrary movements of man, and finally the control of locomotion (walking and running) of living things, the result of which was the establishment of a number of new facts, and above all the formation of the idea of synergy [150], [183] (the term itself is borrowed from classical neurology, which had almost "seen" it, strange as it may seem, in a number of pathological situations). Synergy is defined as a system of joints (muscles) which in a given class of movements change (work) in a connected manner, although generally speaking (for example, for movements of another type) they may be independent. Thus, synergy is a functional concept. Clearly, the formation of synergy sharply reduces the number of degrees of freedom of motive apparatus and makes a movement "well organized".

However, the question arises, how synergies are formed. Gel'fand and Tsetlin were then working on the theory of automata games. They had in mind a system of elements having simple rules of individual behaviour and a variable matrix of interaction. The matrix itself can be thought of as realized on the same automata as form the second level of the system, and the variation of this matrix as realizable automata that form the third level of the system, and so on. A remarkable property of the system so constructed is that the adaptivity of its behaviour in a changing external medium can be achieved by means of non-individualized control by the activity of elements of the underlying level. It is sufficient to change the system of their interaction, which is a far simpler problem for the overlying level than the generation of a programme of specialized control by each of the elements of the underlying level.

Something close to such a representation of physiology is meant when describing tonic influences of the brain on the activity of automatisms of the spinal cord. But in the Gel'fand-Tsetlin theory this idea is formulated much more sharply, which enables one to see a number of fascinating advantages of such a system in comparison with individualized control over non-interacting elements. This idea stimulated experimental investigations of a system of control by arbitrary movements of man and locomotion of animals.

It was shown that a cat with the large hemispheres and interstitial cord removed, unable to stand and walk either at will or under the influence of natural irritation, could be made to move by electric irritation of a definite "locomotor" region of the middle brain. Changing the strength of the irritation can control the speed of movement and walk of the cat, changing it from a walk to a trot and a gallop. This excited locomotion lacks the elegance of the natural walk of a cat with a whole brain, but all the basic kinematic and electromyographic characteristics are preserved.

The development of a specimen with controllable locomotion made it possible to apply the modern technique of studying the impulsive activity of single nervous cells to an animal in motion, and during motive activity that imitates natural activity. This led to progress in the solution of a number of problems of neurophysiology and the physiology of movements.

The ideas about a multilevel system of non-individualized control developed by Gel'fand and Tsetlin, in which the upper levels are not concerned with the development of programmes for the separate control elements, naturally led to the idea that signals from below enter the uppermost levels not from the control elements but from the intermediate levels. In particular, for the control by movements the upper stages of the nervous system can obtain signals not from muscles or tendons but, for example, from the apparatus of the spinal cord. However, these general considerations do not give any indication which actual nerve paths can transmit such signals.

The Swedish neurophysiologists Lundberg and Oscarsson, on the basis of an analytical study of ascending paths of the spinal cord of motionless animals, advanced the hypothesis that some of these paths do not have the usual afferent tracts that transmit signals upwards from muscles, skin, and so on, and communicate information about the state of the spinal cord through the overlying levels of the nervous system. To verify this hypothesis it was necessary to investigate the behaviour of these actual steps during the motion of an animal.

Such an investigation was carried out by Gel'fand and others on a specimen with controllable locomotion [224]–[226]. In this series of papers signals were studied which enter the cerebellum by ascending routes during the cat's walk. It was shown that in agreement with ideas on non-individualized control and the Lindberg – Oscarsson hypothesis such routes in the spinal cord exist, in which signals about the processes occurring in the spinal cord during locomotion are passed upwards into the cerebellum. The character of these signals changes only a little even after all the sensitive fibres going into the cat's spinal cord, muscles, sinews and so on have been completely cut, provided that after such an operation the animal continues to walk [226]. In this series of papers other paths of the spinal cord were studied which during locomotion pass signals upwards about the work of the various muscles, that is, about the behaviour of the control apparatus [224], [236].

In the papers mentioned above, Gel'fand studied the difference between signals entering the cerebellum by different routes; in another series of papers [159], [181], [194], [198] he showed that in the cortex of the cerebellum itself signals entering by different routes are processed by different methods. In particular, he showed that signals entering the cortex of the cerebellum through reticular formation regulate the intensity of signals entering by other routes [198], [208], [218].

Of course, we cannot explain here the content of all the work on physiology carried out by Gel'fand. Still less can we explain in one article the results obtained in work in which Gel'fand did not formally take part, but which was carried out under the influence of personal conversations that various people had with him or under the influence of participants in the physiological seminar which he led for several years. For all physiologists, whatever their future fate as experimentalists, contact with Gel'fand was and still is a most important school of thought.

III

During the last ten years Gel'fand has been carrying out work on cell biology. This research has been done with a group of collaborators from the Laboratory of Mathematical Methods in Biology at the Moscow State University and the Institute of Experimental and Clinical Oncology of the USSR Academy of Medical Sciences. The basic trend of this work was the study of the reactions of a cell on the medium surrounding it, including other cells. Interest in this problem lies first of all in the fact that disturbance of the interactions with the external medium is apparently the main difference between a tumour cell and a normal one. Despite all the successes of biology, up to now we know how to distinguish tumour cells from normal ones only by their behaviour on the tissue level, only by their changing interrelations with other cells. It is still not clear which cellular changes determine the disturbed behaviour of a tumour cell in a multicell system. In the first stage of the work it was important to understand which components of a cell could be responsible for the interaction of this cell with the medium. Analysis of the data led to the conclusion that such a structure is a cellular surface – the membrane of the cell and the formations connected with it. Ideas were expressed and justified to the effect that steady hereditary changes of the cellular surface are the basis of abnormal behaviour of a tumour cell, including irregular reproduction and disturbed contact interactions. In recent years views of this kind have become almost conventional; the study of the surface of a tumour cell has now become one of the main directions of the biology of tumours.

The characteristic experimental work of Gel'fand and his colleagues began with a study of a peculiar and previously uninvestigated object: cell complexes in an ascitic tumour of liver (hepatoma) muscle [136], [141], [145]. It is a question of tumour cells that preserve the ability inherent in their normal ancestors (epithelial liver cells) of forming the simplest multicell formations: cell complexes of 2, 3 or more rarely, 4–5 cells. These small groups prove to be a very convenient system for studying interactions of cells. Despite the fact that the cells of an ascitic tumour multiply actively, the ratio of the fraction of individual cells and the fraction of cells forming different complexes remains constant. Analysis of the statistical distribution of dividing cells in complexes of different types made it possible to establish those periods of the life cycle of a cell in which the formation and destruction of complexes take place. It turned out that complexes are formed as a result of the non-divergence of daughter cells after division. When two contacting cells of a complex enter synchronously in the next division, then the complex usually breaks down. Investigation of ascitic hepatomata disclosed two new types of contact intracellular interaction:

1) Synchronization of mitotic cycles of contacting cells. This is expressed by the fact that the majority of cells in the complexes synchronously go through a phase of synthesis of DNA and synchronously enter mitosis.

2) Contact acceleration of multiplication. In contacting cells in complexes the phase G_1 of a mitotic cycle is reduced to 10–11 hours compared with 17–18 hours for individual cells.

The discovery of these phenomena is evidence of the fact that even in tumourous tissue multiplication can be regulated by the distinctive contact interactions with other cells. It is possible that these interactions have a pathological character here: each cell accelerates the multiplication of its neighbours, and the synchronization of mitoses can promote the collapse of intercellular connections and consequently the collapse of the tissue structure.

After their work on the ascitic hepatoma, Gel'fand and his collaborators began research on normal and tumour cells in another experimental system – cultures of fibroblasts. The advantage of this system is the possibility of various direct and prolonged observations of cells, and also of comparing the behaviour of normal cells with the behaviour of many different lines of tumour cells that arise from these normal cells.

Normal fibroblasts form in the culture the likeness of organized tissue: they stretch parallel to one another and the structures of the epigastrium on which they grow. One of the directions of the work of Gel'fand and his collaborators has been the study of the processes that determine the stretched form and regular arrangement of fibroblasts. They disclosed two groups of such morphogenetic processes. The first group is formed by processes connected with the attachment of a cell to the epigastrium on which it grows. In the culture of such an epigastrium there is most often a surface of glass or plate. When a cell is spread on the epigastrium, a peculiar thin plate structure is formed on its periphery, whose surface is firmly attached to the epigastrium This structure was called "lamellar cytoplasm" [222].

The second basic group of morphogenetic processes consists of the processes of polarization of a cell that stretch it in a definite direction. An important result of the work of Gel'fand and his collaborators was the discovery of a peculiar intracellular process that leads to the stable polarization of a cell. This process, called "stabilization of the cell surface" is connected with the formation in the cell of peculiar structures – microtubes. It was shown that the substance that selectively destroys microtubes (colcemid) breaks down the polarization of the cell and makes the movements of the surface of the cell disordered. In a cell treated with colcemid the division of the surface into fixed and movable parts disappears; all parts of the surface become movable. Thus, a process of stabilization of the cell surface sensitive to colcemid determines not only the form of the cell, but also the capacity for ordered directed movements. The stabilizing mechanism discovered in experiments on fibroblasts undoubtedly plays an important role in the polarization of many other types of cell. It is apparently a question of one of the central mechanisms that determine the form and ordered arrangement of cells in the course of embryonic development.

Processes of this kind, which determine the form of a cell as a whole, also determine the oriented arrangement of the organoids in such a cell and in the first place the arrangement of the nucleus. The direction of the long axis of the nucleus can apparently serve as an index of the inner orientation of the structures in the cell [209]. For example, an analysis, recently carried out by Gel'fand and his collaborators, of microcinefilms in which the cell cultures were removed, showed that from the position of the long axis of the nucleus of an interphase cell one can quite accurately predict the direction of fissure of its subsequent division. This work was the prerequisite for a new direction of research: the study of the inner polarized organization of a cell.

As we have said, in cultures of tumour cells the ordered arrangement of these cells is abruptly disturbed. Until recently, the view was generally accepted in the literature that these disturbances are the result of disturbance of the so-called contact inhibition of movements; it was assumed that tumour cells, in contrast to normal ones, can move freely on the surface of other cells. However, microcinematographical analysis of collisions between cells carried out by Gel'fand and his collaborators showed that this is not true: in tumour cells the ability to inhibit motion is completely preserved by contact.

The discovery of the ordered arrangement of tumour cells in a culture turned out to be connected with another of their peculiarities, namely the process mentioned above of attachment of cells to the epigastrium and the formation of lamellar cytoplasm [222], [237]. Investigation of seven lines of transformed fibroblasts by means of a screen electron microscope showed that in all the lines the formation of lamellar cytoplasm was disturbed to some degree. Inferior spreading of the cells on the epigastrium affects the character of the intracellular collisions; precisely this apparently determines the changed morphology of the culture. If a mixed culture is grown from tumour and normal fibroblasts, then the less attached tumour cells are pushed aside by the better attached normal cells [240]. This experimental system (mixed culture) apparently reproduces in a simplified form the interrelations between tumour and normal cells in an organism. It is possible that tumour cells, being badly attached to the natural epigastria (collagenic fibres, basal membranes, and so on), can be thrown out from the plates of normal cells connected with such epigastria; after this, the tumour cells can spread out over the surfaces of such plates. Thus the peculiarity of tumour cells, discovered in the culture, may be the basis of one of the most characteristic features of the behaviour of malignant cells: their invasiveness, that is, their ability to penetrate between normal structures.

Another fundamental direction of research into normal and tumour fibroblasts in cultures was the study of the regularity of multiplication in this system. The growth of a normal culture stops after the cells increase to a definite density of population, measured by the number of cells on one surface of the epigastrium. Mechanisms for regulating the influence of population density on multiplication are implicit. In the first papers [148], [213] devoted to this question it was shown that such a regulation has local character: the intensity of multiplication of a cell is determined by the number of cells in its immediate neighbourhood only. This conclusion was reached as a result of experiments in which a new method of "injuring the culture" was used: it turned out that if a small part of the cell plate in a thick culture is removed, then the cells migrate from the remaining part of the plate to the freed part of the epigastrium. For several hours such cells begin to synthesise DNK, and then divide. Multiplication of the cells situated alongside, which remain in the thick part of the culture, remains slowed down.

The method of "injury" gives the possibility of comparing cells in identical conditions of nutrition but in a different local environment; such a method is useful in the study of a wide circle of problems.

Theoretical arguments on the regulation of cell multiplication by changing the surface, developed by Vasil'ev and Gel'fand [166], [170], gave a basis for supposing that the synthesis of DNA and the division of cells in a thick culture can be stimulated not only by injury, but also by many other actions that reversibly change the state of the surface.

These propositions were confirmed experimentally [174], [182], [193]. It was shown that the synthesis of DNA in thick cultures of normal fibroblasts can be stimulated by agents of a different character: ferments (hyaluronidase, ribonuclease) detergents (digitonin) and agents that destroy microtubes (colcemid, vinblastin). The discovery of the stimulating action of these various agents leads to the possibility of searching for changes, common to all these factors, that reduce to activation of multiplication.

It is impossible to stimulate multiplication of thick cultures of tumour cells by many of those actions that activate normal cultures.

It is obvious from what we have said that the work of Gel'fand and his collaborators on cell biology, which extends to the present time, has already given a number of significant results.

In this work they have explained and analyzed a number of processes that regulate the form of a cell, its capability for directed movements and also the intensity of multiplication. They have significantly changed and refined ideas on which cellular reactions are disturbed by tumour conversion. This work is still going on.

For a number of years, Gel'fand has been conducting a seminar on cells, in which biologists working in various foundations in Moscow have taken part. The unique feature of this seminar is that work relating to many different branches of biology, biochemistry, cytology, immunology and so on, is discussed at a very high level. For many of those taking part the seminar has been a real school of general biological thinking.

LIST OF THE PRINTED PAPERS BY I. M. GEL'FAND[1]

DAN = Doklady Akademii Nauk; translation SMD = Soviet Mathematics Doklady.
UMN = Uspekhi Matematicheskikh Nauk; translation RMS = Russian Mathematical Surveys.
FAB = Funktsional'nye Analiz i Prilozheniya; translation FAA = Functional Analysis and its Applications.

1) The beginning of the list was published in UMN **19**:3 (1964), 199–205. = RMS **19**:3 (1964), 175–180.

1963

[133] Some classes of games and automata games, DAN **152**, 845–848 (with I. I. Pyatetskii-Shapiro and M. L. Tsetlin). MR **28** # 1068.

[134] The structure of the ring of finite functions on the group of second-order unimodular matrices with elements from a disconnected locally compact field, DAN **153**, 512–515 (with M. I. Graev). MR **33** # 4183.
= SMD **4**, 1697–1700.

[135] Categories of finite-dimensional spaces, Vestnik Moskov. Univ. Ser. I.Mat. Mekh. no. 4, 27–48 (with G. E. Shilov). MR **28** # 1223.

1964

[136] Characteristics of cell complexes of ascitic mouse hepatoma 22, DAN **156**, 168–170 (with Yu. M. Vasil'ev, V. I. Gel'shtein and A. G. Malenkov).

[137] Representations of adèle groups, DAN **156**, 487–490 (with M. I. Graev and I. I. Pyatetskii-Shapiro). MR **29** # 2237.
= SMD **5**, 657–661.

[138] Investigation of recognition activity, Biofizika **9**, 710–717 (with V. S. Gurfinkel).

[139] Homogeneous automata games and their simulation on digital computers, Avtomat. i Telemekh. **25**, 1572–1580 (with V. I. Bryzgalov, V. S. Gurfinkel and M. L. Tsetlin). MR **30** # 1897.

1965

[140] Finite-dimensional irreducible representations of the unitary and complete linear group and special functions associated with them, Izv. Akad. Nauk SSSR, Ser. Mat. **29**, 1329–1356 (with M. I. Graev). MR **34** # 1450.

[141] Cell complexes in ascitic hepatomata of mice and rats, in the collection "*Cell differentiation and induction mechanisms*", Izdat. Nauka, Moscow, 220–232 (with Yu. M. Vasil'ev, V. I. Gel'shtein and A. G. Malenkov).

1966

[142] *Teoriya predstavlenii i avtomorfnye funktsii*, Izdat. Nauka, Moscow (with M. I. Graev and I. I. Pyatetskii-Shapiro). MR **36** # 3725.
Translation: Theory of representations and automorphic functions, Saunders, Philadelphia – London – Toronto 1969.

[143] Skew fields connected with the enveloping algebras of Lie algebras, DAN **167**, 503–505 (with A. A. Kirillov). MR **33** # 4108.
= SMD 7, 407–409.

[144] Sur les corps liés aux enveloppantes des algèbres de Lie. Inst. Hautes Études Sci. Publ. Math. no. 31, 509–523 (with A. A. Kirillov), MR **33** # 7731.

[145] Local interactions of cells in cell complexes of ascitic hepatoma 22 DAN **167**, 437–439 (with Yu. M. Vasil'ev, V. I. Gel'shtein and A. G. Malenkov).

[146] Interrelationships of contacting cells in the cell complexes of mouse ascites hepatoma, Internat. J. Cancer I, 451–462 (in English) (with Yu. M. Vasil'ev, V. I. Gel'shtein and A. G. Malenkov).

[147] Integral geometry on a manifold of k-dimensional planes, DAN **168**, 1236–1238 (with M. I. Graev and E. Ya. Shapiro). MR **33** # 6565.
= SMD 7, 801–804.

[148] The behaviour of fibroblasts of a cell culture on removal of part of the mono-layer,

DAN **171**, 721–724 (with Yu. M. Vasil'ev and L. V. Erofeeva).

[149] Mathematical simulation of mechanisms of the central nervous system , in the collection "*Models of structuro-functional organization of some biological systems*", Izdat. Nauka, Moscow, 9–26 (with M. L. Tsetlin).

[150] Some questions in the investigation of movement, in the collection "*Models of structuro-functional organisation of certain biological systems*", Izdat. Nauka, Moscow, 264–276 (with V. S. Gurfinkel and M. L. Tsetlin).

[151] *Metod ovragov v zadachakh rentgenostrukturnogo analiza* (The ravine method in problems of X-ray structure analysis), Izdat. Nauka, 1–77 (with E. B. Vul, S. L. Ginzburg and Yu. G. Fedorov).

[152] *Lektsii po lineinoi algebre* (Lectures on linear algebra), Izdat. Nauka, Moscow, 1–280. MR **34** # 4274.

1967

[153] Integral geometry on k-dimensional planes, FAP **1**: 1, 15–31 (with M. I. Graev and Z. Ya. Shapiro). MR **35** # 3620.
= FAA **1**, 14–27.

[154] Categories of Harish-Chandra models over the Lie algebra of the Lorentz group, DAN **176**, 243–246 (with V. A. Ponomarev). MR **36** # 6552.
= SMD **8**, 1065–1068.

[155] Classification of indecomposable infinitesimal representations of the Lorentz group, DAN **176**, 502–505 (with V. A. Ponomarev). MR **36** # 2739.
= SMD **8**, 1114–1117.

[156] Cohomology of Lie groups with real coefficients, DAN **176**, 24–27 (with D. B. Fuks). MR **37** # 2252a.
= SMD **8**, 1031–1034.

[157] Topology of non-compact Lie groups, FAP **1**:4, 33–45 (with D. B. Fuks). MR **37** # 2253.
= FAA **1**, 285–295.

[158] Representations of the quaternion group over a disconnected locally compact field, DAN **177**, 17–20 (with M. I. Graev). MR **36** # 2742.
= SMD **8**, 1346–1349.

[159] Functional organization of afferent connections of Purkinje cells of the paramedian lobe of the cerebellum, DAN **177**, 732–753 (with Yu. I. Arshavskii, M. B. Berkinblit and V. S. Yakobson).

[160] Topological invariants of non-compact Lie groups connected with infinite-dimensional representations, DAN **177**, 763–766 (with D. B. Fuks). MR **37** # 2252b.
= SMD **8**, 1483–1486.

[161] The theory of non-linear oscillation of electron plasma, Zh. Vychisl. Mat. i Mat. Fiz. 7, 322–347 (with N. M. Zueva, V. S. Imshennik, O. V. Lokytsievskii, V. S. Ryaben'kii and L. G. Khazin).

[162] Irreducible representations of the Lie algebras of the group $U(p, q)$, in the collection "*High energy physics and the theory of elementary particles*", Izdat. Naukova Dumka, Kiev, 216–226 (with M. I. Graev). MR **37** # 3814.

1968

[163] Complexes of k-dimensional planes and Plancherel's formula for $GL(n, C)$, DAN **179**, 522–525 (with M. I. Graev). MR **37** # 4764.

= SMD **9**, 394–398.
[164] Admissible line complexes in the space $\mathbf{C}^n$, FAP **2**:3, 39–52 (with M. I. Graev). MR **38** # 6522.
= FAA **2**, 219–229.
[165] Indecomposable representations of the Lorentz group, UMN **23**:2, 3–60 (with V. A. Ponomarev). MR **37** # 5325.
= RMS **23**:2, 1–58.
[166] Surface changes disturbing intracellular homeostasis as a factor inducing cell growth and division, Currents in modern biology, no. 2, 43–55 (in English) (with Yu. M. Vasil'ev).
[167] The structure of the quotient skew field of the enveloping algebra of a semisimple Lie algebra, DAN **180**, 775–777 (with A. A. Kirillov). MR **37** # 5260.
= SMD **9**, 669–671.
[168] Classifying spaces for principal bundles with Hausdorff bases, DAN **181**, 515–518 (with D. B. Fuks). MR **38** # 716.
= SMD **9**, 851–854.
[169] Cohomology of Lie algebras of vector fields on a circle, FAP **2**:4, 92–93 (with D. B. Fuks). MR **39** # 6348a.
= FAA **2**, 342–343.
[170] Change of cellular surface – the basis of biological singularities of a tumour cell, Vestnik Akad. Med. Nauk SSSR, no. 3, 45–49 (with Yu. M. Vasil'ev).
[171] Representations of quaternion groups over locally compact and functional fields, FAP **2**:1 20–35 (with M. I. Graev). MR **38** # 4611.
= FAA **2**, 19–33.
[172] Wound healing in cell cultures, Experimental Cell Research **54**, 83–93 (in English) (with Yu. M. Vasil'ev, L. V. Domnin and R. I. Rapoport).
[173] The control of certain types of movement, in the collection: *Material of the international symposium* IFAK *on technical and biological problems of control* (with V. S. Gurfinkel, M. L. Shik, G. N. Orlovskii, F. V. Severin and A. G. Fel'dman).
[174] Factors controlling the proliferation of normal and tumour cells, in the book *Connective tissue in normal and pathological conditions*, Izdat. Nauka, Novosibirsk, 212–215 (with Yu. M. Vasil'ev, L. V. Erofeeva, O. Yu. Ivanova, I. L. Slavnaya and A. A. Yaskovets).

1969

[175] The structure of a Lie skew field connected with a semisimple decomposable Lie algebra, FAP **3**:1, 7–26 (with A. A. Kirillov). MR **39** # 2827.
= FAA **3**, 6–21.
[176] Differential forms and integral geometry, FAP **3**:2, 24–40 (with M. I. Graev and Z. Ya. Shapiro). MR **39** # 6232.
= FAA **3**, 101–114.
[177] Cohomology of the Lie algebra of vector fields on a manifold, FAP **3**:2, 87 (with D. B. Fuks). MR **39** # 6348b.
= FAA **3**, 155.
[178] Cohomology of the Lie algebra of tangent vector fields of a smooth manifold, FAP **3**:3, 32–52 (with D. B. Fuks). MR **41** # 1067.
= FAA **3**, 194–210.

[179] Quadratic forms over commutative group rings and *K*-theory, FAP **3**:4, 28–33 (with A. S. Mishchenko). MR **41** #9243.
= FAA **3**, 277–281.

[180] Remarks on the classification of a pair of commutating linear transformations in a finite-dimensional space, FAP **3**:4, 81–82 (with V. A. Ponomarev). MR **40** # 7279.
= FAA **3**, 325–329.

[181] Two types of granular cell in the cortex of the cerebellum, Neirofiziologiya **1**, 167–176 (with Yu. I. Arshavskii, M. B. Berkinblit and V. S. Yakobson).

[182] Stimulation of synthesis of DNA in mouse embryo fibroblastlike cells in vitro by factors of different character, DAN **187**, 913–915 (with Yu. M. Vasil'ev, V. I. Gel'shtein and E. K. Fetisova).

[183] Interaction in biological systems, Priroda, no. 6, 13–21; no. 7, 24–33 (with Yu. M. Vasil'ev, Sh. A. Guberman and M. L. Shik).

1970

[184] Integral geometry in projective space, FAP **4**:1, 14–32 (with M. I. Graev and Z. Ya. Shapiro). MR **43** #6856.
= FAA **4**, 12–28.

[185] Problems of integral geometry connected with a pair of Grassman manifolds, DAN **193**, 259–262 (with M. I. Graev and Z. Ya. Shapiro). MR **42** #3728.
= SMD **11**, 892–896.

[186] Classification of linear representations of the group SL(2, C), DAN **194**, 1002–1005 (with M. I. Graev and V. A. Ponomarev). MR **43** #2162.
= SMD **11**, 1319–1323.

[187] Cohomology of the Lie algebra of smooth vector fields, DAN **190**, 1267–1270 (with D. B. Fuks). MR **44** #2247.
= SMD **11**, 268–271.

[188] Cohomology of the Lie algebra of formal vector fields, Izv. Akad. Nauk Ser. Mat. **34**, 322–337 (with D. B. Fuks). MR **42** #1103.

[189] Cohomology of the Lie algebra of tangent vector fields of a smooth manifold, II, FAP **4**:4, 23–31 (with D. B. Fuks). MR **44** #2248.
= FAA **4**, 110–116.

[190] Cohomology of the Lie algebra of vector fields with non-trivial coefficients, FAP **4**:3, 10–25 (with D. B. Fuks). MR **45** #7752.
= FAA **4**, 181–192.

[191] The cohomology of infinite-dimensional Lie algebras, some questions of integral geometry, Internat. Congress Mathematicians, Nice 1970, **1**, 95–111 (in English).

[192] Differential operators on the fundamental affine space, DAN **195**, 1255–1258 (with I. N. Bernshtein and S. I. Gel'fand). MR **43** #3402.
= SMD **11**, 1646–1649.

[193] Stimulation of DNA synthesis in cultures of mouse embyro fibroblastlike cells, J.Cell Physiol. **75**, 305–313 (in English) (with Yu. M. Vasil'ev, V. I. Gel'shtein and E. K. Fetisova).

[194] Organization of afferent connections of intercalary neurons in the paramedian lobe of the cerebellum of a cat, DAN **193**, 250–253 (with Yu. I. Arshavskii, B. M. Berkinblit and V. S. Yakobson).

[195] Cycles representing cohomology classes of the Lie algebra of formal vector fields,

UMN **25**:5, 239–240 (with D. B. Fuks). MR **45** # 2737.
[196] Upper bounds for the cohomology of infinite-dimensional Lie algebras, FAP **4**:4, 70–71 (with D. B. Fuks). MR **44** # 4792.
= FAA **4**, 323–324.
[197] Mathematical prognosis of the outcome of haemorrhages with the aim of determining evidence for surgical treatment, Nevropatologii i psikhiatrii **2**, 177–181 (with M. L. Izvekova, Sh. A. Guberman, E. I. Kandel', N. M. Chebotareva, D. K. Lunev, I. F. Nikolaeva and N. V. Lebedeva).
[198] Features of the influence of the lateral reticular nucleus of medulla oblongata on the cortex of the cerebellum, Neirofiziologiya **2**, 581–586 (with Yu. I. Arshavskii, M. B. Berkenblit, O. I. Fukson and V. S. Yakobson).
[199] Effect of colcemid on the locomotory behaviour of fibroblasts, J. Embryology and Experimental Morphology **24**, 625–640 (in English) (with Yu. M. Vasil'ev, L. V. Domnin, O. Yu. Ivanova, S. G. Kommom and L. V. Ol'shevskaya).
[200] Problems of linear algebra and classification of quadruples of subspaces in a finite-dimensional vector space, Colloq. Math. soc. Janos Bolyai 5. Hilbert space operators, Tihany, Hungary (in English) (with V. A. Ponomarev).
[201] Crystal structure of paraoxyacetophenon, DAN **195**, 341–344 (with G. M. Lobanova, S. L. Ginzburg, M. G. Neigauz, L. A. Novakovskaya and G. V. Gurskaya).
= Soviet Physics Dokl. **15**, 999–1002.

1971

[202] Quadruples of subspaces of a finite-dimensional vector space, DAN **197**, 762–765 (with V. A. Ponomarev). MR **44** # 2762.
= SMD **12**, 535–539.
[203] The structure of representations generated by vectors of dominant weight, FAP **5**:1, 1–9 (with I. N. Bernshtein and S. I. Gel'fand). MR **45** # 298.
= FAA **5**, 1–8.
[204] Initiation of DNA synthesis in cultures of mouse fibroblastlike cells under the action of substances that disturb the formation of microtubes, DAN **197**, 1425–1428 (with Yu. M. Vasil'ev and V. I. Gel'shtein).
[205] Organization of projections of somatic nerves in different regions of the cortex of the cerebellum of a cat, Neirofiziologiya 3:2 (with Yu. I. Arshavskii, M. B. Berkinblit and O. I. Fukson).
[206] Background activity of Purkinje cells in intact and deafferentized frontal lobes of the cortex of the cerebellum of a cat, Biofizika **16**, 684–691 (with Yu. I. Arshavskii, M. B. Berkinblit, I. A. Keder-Stepanova, E. M. Smelyanskaya and V. S. Yakobson).
[207] Afferent connections and interaction of neurons of the cortex of the cerebellum, in the collection "*Structural and functional organization of the cerebellum*". Izdat. Nauka, Moscow, 40–47 (with Yu. I. Arshavskii, M. B. Berkinblit, O. I. Fuks and V. S. Yakobson).
[208] The reticular afferent system of the cerebellum and its functional significance, Izv. Akad. Nauk. SSSR Ser. Biol., no. 3, 375–383 (with Yu. I. Arshavskii, M. B. Berkinblit, O. I. Fukson and V. S. Yakobson).
[209] A quantitative estimate of the form and orientation of cell nuclei in a culture, Ontogenez **2**, 138–144 (with Yu. M. Vasil'ev, L. B. Margolis and V. I. Samoilov).
[210] An estimate of the pressure in the pulmonary artery from electro- and phonocardio-

graphical data under a defect of the intraventricular partition, Kardiologiya, no. 5, 84–87 (with G. G. Gel'shtein, Sh. A. Guberman, I. M. Rotvain and T. A. Fain).

[211] Functional role of the reticular afferent system of the cerebellum, Preprint of the IPM Akad. Nauk SSSR (with Yu. I. Arshavskii, M. B. Berkinblit, O. I. Fukson and V. S. Yakobson).

[212] Initiation of DNA synthesis in cell cultures by colcemid, Proc. Nat. Acad. Sci. USA **68**, 977–979 (in English) (with Yu. M. Vasil'ev and V. I. Gel'shtein).

[213] The kinetics of proliferation in cultures of mouse embryo fibroblastlike cells, Tsitologiya **13**, 1362–1377 (with V. Ya. Brodskii, Yu. M. Vasil'ev, V. I. Gel'shtein, L. V. Domnin, L. B. Klempner and T. L. Marshak).

[214] Representations of the group $GL(n, K)$, where K is a local field, Preprint 71, IPM Akad. Nauk SSSR (with D. A. Kazhdan).

[215] Some questions of differential geometry and the computation of the cohomology of Lie algebras of vector fields, DAN **200**, 269–272 (with D. A. Kazhdan). MR **44** # 4770.
= SMD **12**, 1367–1370.

[216] Recognition of the degree of pulmonary hypertonia under a defect of the intraventricular partition with the aid of the EVM, Krovoobrashchenie, no. 6 (with G. G. Gel'shtein, Sh. A. Guberman, I. M. Rotvain, V. A. Silin, V. K. Sukhov and T. L. Fain).

1972

[217] The action of metaphase inhibitors on the form and movement of interphase fibroblasts in a culture, Tsitologiya **14**, 80–88 (with Yu. M. Vasil'ev, L. V. Domnin and O. Yu. Ivanova).

[218] Suppression of reactions of Purkinje cells under preceding activation of the reticulo cerebellar path, Fiziol. Zh. SSSR **58**, 208–214 (with Yu. I. Arshavskii, M. B. Berkinblat, O. I. Fukson and V. S. Yakobson).

[219] Differential operators on a cubic cone, UMN **27**:1, 185–190 (with I. N. Bernshtein and S. I. Gel'fand).
= RMS **27**:1, 169–174.

[220] Actions of infinite-dimensional Lie algebras, FAP **6**:1, 10–15 (with D. A. Kazhdan and D. B. Fuks). MR **46** #922.
= FAA **6**, 9–13.

[221] Differential operators on the fundamental affine space and investigation of 𝔊-modules, Preprint no. 77, IPM Akad. Nauk SSSR (with I. N. Bernshtein and S. I. Gel'fand).

[222] Defective formation of the lamellar cytoplasmy neoplastic fibroblasts, Proc. Nat Acad. Sci. USA **69**, 248–252 (in English) (with L. V. Dominin, O. Yu. Ivanova, L. B. Margolis, Yu. M. Vasil'ev, L. V. Ol'shevskaya and Yu. A. Rovenskii).

[223] Criteria of high seismicity, determined by pattern reçognition, Proc. final Symp. Upper mantle project 13 (in English) (with Sh. A. Guberman, V. I. Keilis-Borok, M. L. Izvekova and E. Ya. Rantsman).

[224] Activity of the neurons of the dorsal spino-cerebellar tract under locomotion, Biofizika **17**, 487–494 (with Yu. I. Arshavskii, M. B. Berkinblit, G. N, Orlovskii and O. I. Fukson).

[225] Activity of the neurons of the ventral spino-cerebellar tract under locomotion, Biofizika **17**, 883–896 (with Yu. A. Arshavskii, M. B. Bertinblit, G. N. Orlovskii and O. I. Fukson).

[226] Activity of the neurons of the ventral spino-cerebellar tract under locomotion of cats with deafferentized hind legs, Biofizika **17**, 1113–1119 (with Yu. I. Arshavskii, M. B. Berkinblit, G. N. Orlovskii and O. I. Fukson).

[227] Criteria of high seismicity, DAN **202**, 1317–1320 (with Sh. A. Guberman, V. I. Keilis-Borok, M. L. Izvekova and E. Ya. Rantsman).

[228] Recordings of neurons of the dorsal spinocerebellar tract during evoked locomotion, Brain Research **43**, 272–275 (in English) (with Yu. I. Arshavskii, M. B. Berkinblit, G. N. Orlovskii and O. I. Fukson).

[229] Origin of modulation in neurons of the ventral spinocerebellar tract during locomotion, Brain Research **43**, 276–279 (in English) (with Yu. I. Arshavskii, M. B. Berkinblit, G. N. Orlovskii and O. I. Fukson).

[230] Cohomology of the Lie algebra of Hamiltonian formal vector fields, FAP **6**:3, 25–29 (with D. I. Kalinin and D. B. Fuks). MR **47** #1088.
= FAA **6**, 193–196.

[231] Representations of the group $GL(n, K)$, where K is a local field, FAP **6**:4, 73–74 (with D. A. Kazhdan).
= FAA **6**, 315–317.

[232] An attempt to carry over criteria of high seismicity from Central Asia to Anatolia and adjoining regions, DAN **210**, 327–330 (with Sh. A. Guberman, M. P. Zhidkov, M. S. Kaletskaya, V. I. Keilis-Borok and E. Ya. Rantsman).

[233] Regulation of the behaviour of connective tissue cells in multicell systems, in the book "histophysiology of connective tissue, Novosibirsk, **1**, 31–36 (with Yu. M. Vasil'ev, V. I. Gel'shtein and L. V. Domnin).

[234] The action of metaphase inhibitors on the form and movement of fibroblasts in a culture, Tsitologiya **14**, 80–88 (with Yu. M. Vasil'ev, L. V. Domnin, O. Yu. Ivanova, S. G. Kommom and L. V. Ol'shevskaya).

1973

[235] Recognition of places of possible origin of powerful earthquakes (in Eastern Central Asia), in the collection "*Computational Seismology*", no. 6 (with Sh. A. Guberman, M. L. Izvekova, V. I. Keilis-Borok and E. Ya. Rantsman).

[236] Activity of neurons of the cuneo-cerebellar tract under locomotion, Biofizika **18**, 126–131 (with Yu. I. Arshevskii, M. V. Berkinblit, G. N. Orlovskii and O. I. Fukson).

[237] Interactions of normal and neoplastic fibroblasts with the substratum, Ciba Foundation Symposium on Cell Locomotion, 312–331 (in English) (with Yu. M. Vasil'ev).

[238] Coxeter functors and Gabriel's theorem, UMN **28**:2, 19–33 (with V. A. Ponomarev, and I. N. Bernshtein).
= RMS **28**:2, 17–32.

[239] Schubert cells and cohomology of flag spaces, FAP **7**:1, 64–65 (with I. N. Bernshtein and S. I. Gel'fand). MR **47** # 6713.
= FAA 7, 56–57.

[240] Intracellular interaction in cultures of transformed fibroblasts of strain L and normal mouse fibroblasts, Tsitologiya **15**, 1024–1028 (with Yu. M. Vasil'ev, L. V. Domnin, O. Yu. Ivanova and L. B. Margolis).

[241] Schubert cells and cohomology of spaces G/P, UMN **28**:3, 3–26 (with I. N. Bernshtein and S. I. Gel'fand).
= RMS **28**:3, 1–26.

[242] Representations of the group $SL(2, \mathbf{R})$, where $\mathbf{R}$ is a ring of functions, UMN **28**:5, 83–128 (with M. I. Graev and A. M. Vershik).

= RMS **28**:5, 87–132.

[243] Disturbance of morphogenetic reactions of cells under tumourous transformation, Vestnik Akad. Med. Nauk SSSR, no. 4, 61–69 (with Yu. M. Vasil'ev).

1961[1)]

[244] The principle of non-local search in automatic optimization systems, DAN **137**, 295–298 (with M. L. Tzetlin). MR 23# B2155.
= Soviet Physics Dokl. **6**, 192–194.

1969[2)]

[245] Prognostic matematic alevelutiei ictusorilor hemorogice in scepul precizavii indicatiiler tratementului chirurgical, "K.Accidouteler vasculare cerebrale", Bucuresti, 44–45 (with Sh. A. Guberman, M. L. Izvekova, E. I. Kandel', T. V. Lebedeva, D.K. Lunev, I. F. Nikolaeva and N. M. Chebotareva).

[246] Computer prognosis of spontaneous intracerebral haemorrhage for the purpose of its surgical treatment, IV. Internat. Congress Neurosurgery, Excerpta Med., no. 32 (with Sh. A. Guberman, M. L. Izvekova, E. I. Kandel, T. V. Lebedeva, D. K. Lunev, I. F. Nikolaeva and N. M. Chebotareva).

Translated by
E. J. R. Primrose

1) An article omitted from the list of Gel'fand's published works in UMN **19**:3 (1964) = RMS **19**:3 (1964).
2) Added in proof.

Izrail Moiseevich Gelfand

(On his seventieth birthday)

Usp. Mat. Nauk SSSR **38** (6) (1983) 137–152
[Russ. Math. Surv. **38** (6) (1983) 145–163]

On 2 September 1983 Izrail' Moiseevich Gel'fand, Corresponding Member of the Academy of Sciences of the USSR, celebrated his seventieth birthday.

I.M. Gel'fand was born on 2 September 1913 in a small town Krasnye Okny in the Odessa district. He has devoted more than fifty years of his life to mathematics. His formal record of service to mathematics began in 1932, when he became a research student, having already acquired some experience of teaching mathematics, although he had had no secondary school education.

He himself said that the men who had the greatest influence on his scientific interests were Lavrent'ev, Lyusternik, Plesner, Shnirel'man, and especially his supervisor Kolmogorov.

For the major part of his life Gel'fand has worked in the field of functional analysis, although the breadth of his scientific interests still causes astonishment today, and it is easy to cite examples of his "sorties" into very distant fields of mathematics. His total allegiance to the methods of infinite-dimensional analysis became apparent when he turned to topology, and when he constructed representations of finite groups. His research began within the framework of functional analysis, which had been traditional for 30 years—the theory of Banach spaces, and for over 50 years Gel'fand actively influenced and to a considerable degree determined the shape of the contemporary problems of functional analysis. Important and characteristic in this connection are Gel'fand's papers and books, his survey articles and lectures, containing propositions and discussions of contemporary mathematical problems (in particular, his lectures at the International Congresses of Mathematicians in Amsterdam and Nice and the published material of a lecture at Edinburgh). The problems he raised repeatedly served as the starting point for important mathematical research (for example, the index problem).

This year marks the 40th anniversary of Gel'fand's world-famous seminar on functional analysis, whose topics reflect the leader's broad interpretation of this subject. Since 1967 he has been editor of the journal "*Funktsional'nyi analiz i evo prilozheniya*" (Functional analysis and its applications).

Gel'fand's studies on normed rings, infinite-dimensional representations of semisimple Lie groups, the inverse Sturm-Liouville problem, and generalized functions have already become classic. His papers on integral geometry, the cohomology of infinite-dimensional Lie algebras, linear algebra and mathematical physics enjoy equally wide recognition.

Gel'fand has constantly thought about problems of computational and applied mathematics. For almost a quarter of a century he has been concerned with the physiology of motion and the biology of cells, and indeed with biology, but not with the applications of mathematics to biology.

Teaching and work in the social sciences play an important part in Gel'fand's life. He has many direct and indirect pupils, for many years he has been a Professor at the Moscow State University and a member of the editorial boards of a number of Soviet and foreign scientific journals. Twenty years ago, on Gel'fand's initiative an All-Union Correspondence School of Mathematics was founded at the Moscow State University, and thanks to it young scholars in distant cities and villages in the country are given the chance to study mathematics in depth. From 1968–1970 he was President of the Moscow Mathematical Society, and was elected an Honorary Member of the Society.

Gel'fand's services to science have been recognized internationally by numerous awards. He is a Foreign Member of the National Academy of Science of the USA, of the American Academy of Arts and Science, of the Swedish Royal Academy of Sciences, the Royal Irish Academy, the London Mathematical Society, the Royal Society (London), the Institut de France, Académie des Sciences, and has received Honorary Doctorates from the Universities of Oxford, Harvard, Paris, and Uppsala; he was awarded the International Wolf Prize and the Wigner Gold Medal.

For his services to the advancement of science and the training of students Gel'fand was awarded three Orders of Lenin, two Orders of Labour of the Red Banner, two Orders of Friendship of Nations, and the "Badge of Honour". He was awarded a Lenin Prize and two State Prizes.

Gel'fand reaches his seventieth birthday full of original ideas and plans. We wish him many years of fruitful creative work.

We do not give here a survey of his work of many years[(1)], but we restrict ourselves to a discussion of some of his results of the last ten years.

Representation Theory.

Gel'fand's work on representation theory and its applications began more than forty years ago and had a decisive influence on the development of this whole branch of functional analysis. In the last ten years Gel'fand has continued to work on representation theory and has found new and striking results.

First of all, we mention a series of papers on models of the representations of finite and compact groups. The representation theory of these groups has

[(1)]See Uspekhi Mat. Nauk **19**:3 (1964), 187–205; **29**:1 (1974), 193–246. = Russian Math. Surveys **19**:3 (1964), 163–180; **29**:1 (1974), 1–61.

already become a classic area, but there are still many interesting unsolved problems. One of them is the construction for a given group G of an explicit model, that is, of a G-module in which every (or almost every in a certain sense) representation occurs precisely once. He succeeded in constructing such a model for the semisimple algebraic groups over a finite field [258] and the simply-connected compact Lie groups [274], [275].

He also continued his work on the representation of infinite-dimensional Lie groups. In [288] he constructed and studied a class of irreducible unitary representations of the groups of diffeomorphisms of a smooth manifold X. These representations are constructed in the space of functions on the set B_X of finite and the set Γ_X of infinite configurations in X (the latter endowed with a system of Poisson measures). He succeeded in computing completely the structure of the ring of representations of the group Diff X that are realized in this way.

In [326], [327], and [377] he studies the group G^X of smooth maps of a manifold X on a compact semisimple group G. This infinite-dimensional group arises naturally in the theory of gauge fields, and a knowledge of its unitary representations is important in mathematical physics. In the papers referred to he constructs an embedding of G^X in the group of motions of a Hilbert space H, by means of the so-called Maurer-Cartan cocycle. This makes it possible to construct a series of representations of G^X in the Fok space Exp X. It turned out that the representations thus constructed are irreducible only for dim $X \geqslant 3$. In the case dim $X = 1$ the representations are reducible, and in the case dim $X = 2$ the answer is not known.

Gel'fand also continued his work on purely algebraic questions of representation theory. A series of joint papers with Ponomareev [296], [353], [375] continues the study begun in [265] of linear representations of free modular structures. Of special interest is the study of perfect elements in modular structures. (An element is called perfect if under any indecomposable linear representation of the structure, it goes over into 0 or 1.) It turns out that in a free modular structure with r generators one can construct a countable collection of substructures $B^{\pm}(l)$, ($l = 1, 2, \ldots$), isomorphic to the structure of vertices of an r-dimensional cube.

Finally, in [292], [334], [335] he continued his study of the category O and the category of Harish-Chandra modules, which has been extensively developed in recent years. The idea behind these papers consists in translating the problem of classifying the representation with some properties or others into the language of linear algebra, that is, in stating it as a problem of the classification of collections of linear spaces and maps between them, satisfying certain commutation relations. In [335] he considers representations of specific Lie algebras; in [334] he indicates a general method for studying Harish-Chandra modules over any semisimple Lie algebra. In [292] analogous questions are studied for the category O.

In [340] he considered the reduction of the problem on the classification of algebraic vector bundles on projective spaces to a problem of linear algebra. It turned out that vector bundles on $\mathbf{P}^n$ correspond to representations of the exterior (Grassman) algebra of $n+1$ variables, that is, in essence the problem of the classification of algebraic vector bundles on a projective space is also a problem of linear algebra.

In [325] holomorphic discrete series of unitary representations of a real semisimple Lie group G are connected with the boundary values on G of holomorphic vector-valued functions on certain semigroups in the complex group G_C, which are Stein manifolds (G is the skeleton for them).

Integral Geometry.
Integral geometry as Gel'fand understands it originated in the interpretation of certain facts from the representation theory of semisimple Lie groups. It has been emphasized repeatedly that one of the basic problems in integral geometry is going beyond the limits of the geometry of homogeneous manifolds, so that the corresponding results on representations are a special case of essentially more general propositions. In recent years some results in this direction have been obtained.

Great difficulties in integral geometry stem from real problems. In complex integral geometry the discussion is confined so far to local problems, that is, to problems in which the inversion formulae contain only differential operators. Undoubtedly, in real integral geometry we cannot confine ourselves to such problems. The simplest example: the inversion formula for the Radon transformation in an even-dimensional space contains a non-local operator. Other examples are real semisimple Lie groups.

In [323] some problems connected with integration over planes in $\mathbf{R}^n$ for which the inversion formulae are non-local are studied. Here objects are introduced that in a certain sense are non-local analogues of differential forms. [382] contains a projective version of these formulae.

In [352] Gel'fand studies general problems of integral geometry connected with integration along curves and having local inversion formulae. Families of curves for which this condition is satisfied are called admissible. It is shown that in the inversion formulae for admissible families of curves there always appears an integration of a certain universal differential 1-form on the manifold of curves. It had been shown earlier that the inversion formulae for problems connected with integration over planes in $\mathbf{C}^n$ have a similar structure.

Admissible 2-parameter families of plane curves are described: they are always families of solutions of the equation $y'' = F(x, y, y')$, where F is a polynomial of degree 3 in y'. In [362] this description is generalized to n-parametric families in $\mathbf{R}^n$. Connections are found between admissible families of curves in the sense of integral geometry and Penrose's twistor approach to the construction of autodual solutions of the Einstein equation.

Admissible families of submanifolds of dimension greater than 1 are studied in [398]. The crucial step is this separation of a special infinitesimal structure on such families.

We draw attention to a comprehensive survey of integral geometry [364] in which, in particular, there is a detailed discussion of the problem of the integration of differential k-forms over planes of dimension p, $p < k$. The image is described by systems of differential equations generalizing classical equations of mathematical physics with zero mass.

Formal geometry and the study of combinatorial characteristic classes. As in earlier years, Gel'fand repeatedly turned to geometry in the last decade. All the diverse themes of his papers on geometry are linked by a common aim, which can be summed up as the "localization of topological invariants". Under the name of "formal geometry" this programme is outlined in Gel'fand's lecture at the International Congress of Mathematicians in 1970.

As is known, manifolds are homogeneous: the neighbourhood of every point of a manifold is constructed exactly as that of any other point. Let us put on a manifold any structure that destroys this homogeneity, say, a triangulation or a Riemannian metric. How can we now determine what contribution each piece of the manifold makes to one topological invariant or another? One of the most important problems of this kind is the search for "combinatorial formulae" for the characteristic classes of the manifold. Such formulae are well known for the Euler and Stiefel-Whitney classes, but for 40 years this list remained static. In [270], [276], [277], and [290] a combinatorial formula was found for the first rational Pontryagin class. This work gave the impetus for numerous studies on various areas of mathematics. We mention the exciting study of the geometry of Grassmanian manifolds, written jointly by Gel'fand and the American mathematician Macpherson [395]; and this study in turn served as the starting point for work by Atiyah, Bott and others, devoted to the "moment map" in symplectic geometry. It is worth noting that the combinatorial formula for the Pontryagin class contains deep mysteries, the solution of which may possibly carry combinatorial geometry to a new stage. One of the intriguing peculiarities of the formula is the presence of a dilogarithm—a classical non-elementary function which has the strange habit of turning up in key points of the most diverse areas of mathematics (such as the distribution of primes in number theory, the theory of tissues in algebraic geometry, the formula for the volume of a tetrahedron in Lobachevskii geometry, the construction of a regulating map for K_2 in algebraic K-theory, etc.).

We mention also other papers on the localization of topological invariants [253], [264], [287], [293] and closely related work on the cohomology of infinite-dimensional Lie algebras (the principal papers of this series belong to the end of the 60's and the beginning of the 70's, but some important results were obtained in the last decade—see [257], [341]).

The Hamiltonian formalism of completely integrable systems.
In the 60's there emerged the so-called "method of the inverse scattering problem" for finding solitary wave packets—solitons, that is, solutions of the famous Korteweg-de Vries equation in the class of sufficiently rapidly decreasing functions. From the very first this method made intensive use of notable results of Gel'fand, Levitan, Marchenko, and others, dating back to the early 50's, on the inverse scattering problem for the Schrödinger operator. During the first half of the 70's the method was extended to a number of other physically important evolution systems, a connection with the theory of classical completely integrable Hamiltonian systems was discovered, and a similar method was developed for periodic boundary conditions.

In 1975 Gel'fand began to work actively in this field and brought to it a number of important new ideas. The main trend of his research related to working out various aspects of the Hamiltonian formalism of systems of Lax type. We mention, in particular, the important class of Poisson brackets he discovered in [291], which made it possible to prove that systems of Lax type $\dot{L} = [A, L]$, where L is an operator of arbitrary order, are Hamiltonian (this was subsequently given a deep algebraic interpretation); it also led to a study of systems with a bundle of Poisson brackets, a link with infinite-dimensional Lie algebras, a number of important aspects of the formal calculus of variations, and to important applications of the technique to operators of fractional degree [278], [294], [295], [322], [339], [351], [376], and [387].

Medical diagnostic.
As regards applied problems, in the last decade Gel'fand paid much attention to the problems of medical diagnostic. A group led by him did original work by means of programs of image recognition. But gradually the limitations of this method became clear, above all in conjunction with the fact that in medicine one has to deal with small choices. The question arose of the optimal use of the experience of a highly-qualified clinical doctor, of the formalization of the diagnostic attitude of the doctor. For this purpose a special method was developed, which was called "diagnostic games" and was successfully applied to a number of problems in cardiology, gastroenterology, neurology, and nephrology ([300], [305], [306], [309]–[311], [342], [368], [372], [383], [385], [392], etc.).

The nerve mechanisms controlling movement.
Under Gel'fand's leadership the department of mathematical methods in biology of the Belozerskii Interfaculty Laboratory for the scientific study of problems in molecular biology and bioorganic chemistry at the Moscow State University developed a method of recording the activity of single nerve cells in various motor centres during rhythmic movements. With the help of this method they collected an enormous amount of experimental data on

the working of neural mechanisms of the spinal cord, of the brain stem and cerebellum, that control the movements of the extremities. It was shown that in response to a call from the locomotory or itching reflex in the spinal cord the central mechanisms for generating rhythmic activity underlying these movements, "close up". When the spinal mechanisms are closed and rhythmic movements are performed, signals are received via the spinal cord pathways in the cerebellum about the functioning of the spinal blood in the control of movements. It was observed that two types of signals arrive at the cerebellum: signals about the working of the operative motor apparatus and signals about that of the spinal generator of rhythmic activity. The signals arriving via the spinal cord pathways provoke rhythmic modulation of the activity of the neurons of the cerebellum, which in turn modulate the activity of the neurons of the descending paths originating in the brain stem. The signals travelling along the descending paths to the spinal cord regulate the level of muscular activity during movement. The decisive role in the formation of signals emerging from the cerebellum and going to the brain stem is played by the incoming signals concerning the working of the spinal generator of rhythmical activity. By provoking rhythmical modulations of the activity of neurons on the descending main pathways, the cerebellum regulates the transmission of signals from different parts of the cerebrum and from different receptors to the spinal cord corresponding to the phase of movement of the extremities.

The experimental results obtained made it possible to formulate some general ideas on the role of the cerebellum in controlling movement. According to these ideas, the motor behaviour of living creatures is made up of an organic collection of motor synergies (programs). The coordination of the work of the different motor synergies and the adaptation of the work of synergies to the changing conditions of the external surroundings are carried out by the cerebellum, which receives detailed information on the working of the various synergies and on the external surroundings, selects from the information received the essential information about the work of the synergies and the external surroundings, and influences effectively the transmission of signals from some sections of the cerebrum to others.

Biology of the cell.

Over the last ten years Gel'fand has continued his work on the biology of the normal and of the tumourous cell, which he had begun in the 60's. He carried out this work with a team of colleagues at the Moscow State University and the All-Union Oncological Research Centre AMN USSR (see the monographs [371] and [373]). The main trend of this work was, as before, the study of the reactions of cells to the surrounding micro-environment, in particular, to the morphogenetic processes that control the shape of cells and the creation from separate cells of complex multicellular systems—tissues and organs. The subjects of this research were normal and tumorous cells (fibroblasts and epithelia) grown in culture; in addition to methods already traditional for this team (scanning and transmitting electron microscope, microcinematography), in recent years immunological methods have been more widely used. In these studies the team succeeded in singling out common cellular reactions that underlie a number of outwardly diverse processes: the reaction of attachment of cells to the surface on which they grow, their extension, the growth of long cellular appendages (for example, on nerve cells), the movement of cells and the formation of contacts between cells, and phagocytes. They were called fundamental morphogenetic reactions. Two groups of such reactions were distinguished: pseudopodial reactions of attachment and processes of stabilization. By extending and attaching pseudopodia, the cell constantly "feels" the surrounding environment, while with the help of stabilization processes the cell sums up and remembers results of preceding pseudopodial reactions and decides the place of future such reactions.

An experimental analysis of pseudopodial reactions showed that the surface of extended pseudopodia has special properties: only this part of the surface is adhesive and capable of transferring groups of membranous albuminous receptors directly to the centre of the cell. Apparently, the molecular basis of these peculiarities of pseudopodia is the ability of a given part of the cell to "anchor" receptors of the membrane: to combine groups of receptors from within with actinic microfilaments. The selection and study of the fundamental morphogenetic reactions made it possible to look in a new way at many physiological and pathological processes. One of these is the formation of blood-clots in blood vessels. At the basis of blood-clot formation lie the processes of attachment of small non-nucleated cells circulating in the blood, thrombocytes, to the damaged vascular wall. Until then it was not clear why the thrombocytes do not attach to the undamaged vascular wall. Thrombocytes adhere to different surfaces by means of pseudopodial reactions. The surface of the platelets of the endothelium covering the inner vascular wall is not capable of forming pseudopodia. According to a hupothesis put forward by Vasil'ev and Gel'fand, this absence of pseudopodia is the reason why blood-clots do not form in the undamaged vessel.

Another very important problem connected with the study of morphogenetic reactions is the investigation of the mechanisms of the behaviour of tumorous cells. The inhibition of the normal morphogenetic capacity is characteristic of tumorous cells in an organism and in a culture. An experimental analysis of the inhibition of morphogenetic reactions in tumorous cells in culture, made by Gel'fand and his collaborators, showed that in tumorous transformations pseudopodial reactions of attachment are first shifted, while the normal capacity for processes of stabilization is usually preserved.

A number of data obtained by Gel'fand's group makes it plausible that the inhibition of the "anchoring" of groups of receptors of a membrane with actinic microfilaments is the molecular base of the changes in pseudopodial attachment.

Recently Gel'fand and Vasil'ev put forward a hypothesis according to which the inhibitions in the pseudopodial reactions of attachment connected with changes in "anchoring" are the cause of the irregular reproduction of tumorous cells. Thus, possible common mechanisms are outlined which determine the two main features of a tumorous cell, that are so dangerous for the organism: irregular reproduction and impaired capacity for morphogenesis.

N.N. Bogolyubov, S.G. Gindikin, A.A. Kirillov, A.N. Kolmogorov, S.P. Novikov, L.D. Faddeev

List of I.M. Gel'fand's published works[(1)]

[247] Contact inhibition of movement in the cultures of transformed cells, Proc. Nat. Acad. Sci. USA, **70** (1973), 2011–2014 (with Yu.M. Vasil'ev, V.I. Gel'stein, O.Yu. Ivanova, and L.B. Margolis) (in English).

[248] The structure of the lamellar cytoplasm of normal and tumorous fibroblasts, Papers from a Soviet-French symposium, in: *Ul'trastruktura rakovykh kletok* (Ultrastructure of cancerous cells), Nauka, Moscow 1973, 49–71 (with Yu.M. Vasil'ev, L.V. Domnina, E.E. Krivitska, L.V. Ol'shevskaya, Yu.A. Rovenskii, and A.P. Chern).

[249] Comparative study of density dependent inhibition of growth in the cultures of normal and neoplastic fibroblast-like cells. Abstracts 6th meeting of the European study group for cell proliferation, Nauka, Moscow 1973, 15 (with Yu.M. Vasil'ev, L.V. Domnina, O.Yu. Pletyushkina, and E.K. Fetisova) (in English).

[250] Factors inducing DNA synthesis and mitosis in normal and neoplastic cell culture. Abstracts 6th meeting of the European study group for cell proliferation, Nauka, Moscow 1973, 61 (with Yu.M. Vasil'ev (in English).

[251] Determination of criteria of high seismism by means of recognition algorithms, Vestnik Moscow Gos. Univ. **1973**, no. 5, 78–83 (with Sh.A. Guberman, M.I. Zhidkov, V.I. Keilis-Borok, E.Ya. Rantsman, and I.M. Rotvain).

[252] Interactions of normal and neoplastic cells with various surfaces. Neoplasma, **20** (1973), 583–585 (with A.D. Bershadskii, V.I. Gel'stein, L.V. Domnina, O.Yu. Ivanova, S.G. Komm, L.B. Margolis, and Yu.M. Vasil'ev) (in English).

[253] *PL*-foliations, Funktsional. Analiz i Prilozhen. 7:4 (1973), 29–37 (with D.B. Fuks). MR **49** # 3958, Zbl. **294** # 57016.
= Functional Anal. Appl. **7** (1973), 278–284.

[254] Some peculiarities of the organization of afferent links of the cerebellum, in: *IV Mezhdunar. biofiz. kongress Dokl. simp. Pushchino.* (4th International Biophysical congress, Pushchino symp.) 1973, vol. 3, 327–346 (with Yu.I. Arshavskii, M.B. Berkinblit, G.N. Orlovskii, O.I. Fukson, and B.S. Yakobson).

[(1)]The beginning of this list is published in Uspekhi Mat. Nauk, **19**:3 (1964), 199–205 (= Russian Math. Surveys **19**:3 (1964), 163–180) and **29**:1 (1974), 193–246 (= Russian Math. Surveys **29**:1 (1974), 1–61).

[255] Activity of the neurons of the cuneo-cerebellar tract for locomotion, Biofizika, **18** (1973), 126-131 (with Yu.I. Arshavskii, O.I. Fukson, and G.N. Orlovskii).
[256] Irreducible representations of the group G^X and cohomology, Funktsional. Analiz, i Prilozhen 8:2 (1974), 67-69 (with A.M. Vershik and M.I. Graev). MR **50** # 530. Zbl. **299** # 22004.
= Functional Anal. Appl. **8** (1974), 151-153.
[257] Cohomology of the Lie algebra of formal vector fields with coefficients in its dual space and variations of characteristic classes of foliations, Funktsional. Analiz i Prilozhen. 8:2 (1974), 13-29 (with B.L. Feigin and D.B. Fuks). MR **50** # 8553. Zbl. **298** # 57011.
= Functional Anal. Appl. **8** (1974), 99-112.
[258] A new model of representations of finite semisimple algebraic groups, Uspekhi Mat. Nauk **29**:3 (1974), 185-186 (with I.N. Bernstein and S.I. Gel'fand). MR **53** # 5760.
[259] Fonov activity of Pourkin cells of the paramedian part of the cortex of the cerebellum of a cat, Biofizika **19** (1974), 903-907 (with Yu.N. Arshavskii, M.B. Berkinblit, E.M. Smelyanskii, and V.S. Yakobson).
[260] Differences in the working of spino-cerebral tracts in artificial irritation and locomotion, in: *Mekhanizmy ob"edineniya neironov v nervnom tsentre* (Mechanisms of the union of neurons in the nerve centre) Nauka, Leningrad 1974, 99-105 (with Yu.I. Arshavskii, M.B. Berkinblit, G.N. Orlovskii, and O.I. Fukson).
[261] Recognition of places where strong earthquakes are possible. II, Four regions of Asia Minor and South-East Europe, *Vychislitel'naya seismologiya* (Computational seismology) Nauka, Moscow **1974**, no. 7, 3-39 (with Sh.A. Guberman, M.P. Zhidkov, V.I. Keilis-Borok, M.S. Kaletska, E.Ya. Rantsman, and I.M. Rotvain).
[262] Recognition of places where strong earthquakes are possible. III, The case when the boundaries of disjunctive nodes are not known, *Vychislitel'naya seismologiya* (Computational seismology) Nauka, Moscow **1974**, no. 7, 41-62 (with Sh.A. Guberman, M.P. Zhidkov, V.I. Keilis-Borok, E.Ya. Rantsman, and I.M. Rotvain).
[263] The results of intercellular impacts in cultures of normal and transformed fibroblasts, *Tsitologiya* (Cytology), **16** (1974), 752-756 (with Yu.M. Vasil'ev, B.I. Gel'stein, L.B. Margolis, O.Yu. Ivanova, and S.G. Komm).
[264] *PL*-foliations. II, Funktsional. Anal. i Prilozhen **8**:3 (1974), 7-11 (with D.B. Fuks). MR **54** # 6159.
= Functional Anal. Appl. 8 (1974), 197-200.
[265] Free modular lattices and their representations, Uspekhi Mat. Nauk **29**:6 (1974), 3-58 (with V.A. Ponomarev). MR **53** # 5393.
= Russian Math. Surveys **29**:6 (1974), 1-56.
[266] Orientation of mitosis of fibroblasts is determined in the interphase, Proc. Nat. Acad. Sci. USA **71** (1974), 2032 (with O.Yu. Ivanova, L.B. Margolis, and Yu.M. Vasil'ev) (in English).
[267] Peculiarities of information entering the cortex of the cerebellum via different afferent paths, structural and functional organization of the cerebellum, Naukova dumka, Kiev 1974, 34-41 (with Yu.I. Arshavskii, M.B. Berkinblit, G.N. Orlovskii, and O.I. Fukson).
[268] Contact inhibition of phagocytosis in epithelial sheets: alterations of cell surface properties induced by cell-cell contacts, Proc. Nat. Acad. Sci. USA **72** (1975), 719-722 (with Yu.M. Vasil'ev, L.V. Domnina, O.S. Zakharova, and A.V. Lyubimov) (in English).

[269] Spreading of normal and transformed fibroblasts in dense cultures, Exper. Cell Research **90** (1975), 317-327 (with A.P. Chern and Yu.M. Vasil'ev) (in English).
[270] The combinatorial computation of characteristic classes, Funkst. analiz. **9**:1 (1975), 54-55 (with A.M. Gabrielov and M.V. Losik). MR **51** # 1839.
= Functional Anal. Appl. **9** (1975), 48-49.
[271] Quantitative evaluation of cell orientation in culture, J. Cell Sci. **17** (1975), 1-10 (in English).
[272] Representation of the group of diffeomorphisms connected with infinite configurations, Preprint Inst. Appl. Math. **1975**, no. 46, 1-62 (with A.M. Vershik and M.I. Graev).
[273] The square root of a quasiregular representation of the group $SL(2, k)$, Funktsional. Anal. i Prilozhen. **9**:2 (1975), 64-66 (with A.M. Vershik and M.I. Graev). MR **51** # 8338.
= Functional Anal. Appl. **9** (1975), 146-148.
[274] Models of representations of compact Lie groups, Funktsional. Anal. i Prilozhen. **9**:4 (1975), 61-62 (with I.N. Bernstein and S.I. Gel'fand). MR **54** # 2884.
= Functional Anal. Appl. **9** (1975), 322-324.
[275] Models of representations of Lie groups, Proc. Petrovskii Sem. **2** (1975), 3-21 (with I.N. Bernstein and S.I. Gel'fand).
[276] Combinatorial computation of characteristic classes, Funkstional. Anal. i Prilozhen. **9**:2 (1975), 12-28 (with A.M. Gabrielov and M.V. Losik). MR **53** # 14504a.
= Functional Anal. Appl. **9** (1975), 103-115.
[277] Combinatorial computation of characteristic classes, Funkstional. Anal. i Prilozhen. **9**:3 (1975), 5-26 (with A.M. Gabrielov and M.V. Losik). MR **53** # 14504a.
= Functional Anal. Appl. **9** (1975), 186-202.
[278] The asymptotic behaviour of the resolvent of the Sturm-Liouville equations and the algebra of the Korteweg-de Vries equations, Uspekhi Mat. Nauk **30**:5 (1975), 67-100 (with L.A. Dikii). MR **58** # 22746. Zbl. **461** # 35072.
= Russian Math. Surveys **30**:5 (1975), 77-113.
[279] Processes determining the changes of shape of a cell after its separation from the epigastrium, *Tsitologiya* (Cytology) **1975**, no. 5, 633-638 (with Yu.M. Vasil'ev and I.S. Tint).
[280] Neoplastic fibroblasts sensitive to growth inhibition by parent normal cells, British J. Cancer **31** (1975), 535-543 (with Yu.M. Vasil'ev and O.Yu. Pletyushkina) (in English).
[281] Methods of measuring the orientation of cells, Ontogenez **6**:1 (1975), 105-110 (with Yu.M. Vasil'ev, L.B. Margolis, and V.I. Samoidov).
[282] Insensibility of dense cultures of transformed mice fibroblasts to the action of agents, stimulating the synthesis of DNA in cultures of normal cells, *Tsitologiya* (Cytology) **17** (1975), 442-446 (with Yu.M. Vasil'ev, O.Yu. Pletyushkin, and E.K. Fetisoba).
[283] The activity of neurons of the ventral spino-cerebral tract in "fictitious scratching", Biofizika **20** (1975), 748-749 (with Yu.I. Arshavskii, G.N. Orlovskii, and G.A. Pavlova).
[284] Origin of the modulation of the activity of vestibular-spinal neurons in scratching, Biofizika **20** (1975), 946-947 (with Yu.I. Arshavskii, G.N. Orlovskii, and G.A. Pavlova).
[285] On the role of the cerebellum in regulating some rhythmic movements (locomotion, scratching), Summaries 12th meeting of the All-Union Physiological Society, Tbilisi 1975, 15-16 (with Yu.I. Arshavskii, M.B. Berkinblit, T.G. Delyagina, G.N. Orlovskii, G.A. Pavlova, A.G. Fel'dman, and O.I. Fukson).

[286] On the role of the central program and afferent inflow in generation of scratching movements in the cat, Brain Research **100** (1975), 297-313 (with T.G. Delyagina, G.N. Orlovskii, and A.G. Fel'dman) (in English).

[287] The Gauss-Bonnet theorem and the Atiyah-Patodi-Singer functionals for the characteristic classes of foliations, Topology **15** (1975), 165-188 (with A.M. Gabrielov and D.B. Fuks) (in English). MR **55** # 4201.

[288] Representation of the group of diffeomorphisms, Uspekhi Mat. Nauk **30**:6 (1975), 3-50 (with A.M. Vershik and M.I. Graev). MR **53** # 3188.
= Russian Math. Surveys **30**:6 (1975), 1-50.

[289] Prognosis of a place where strong earthquakes occur, as a problem of recognition, in: *Modelirovanie obycheniya i povedeniya* (Modelling of training and behaviour), Nauka, Moscow 1975, 18-25 (with Sh.A. Guberman, L.P. Zhidkov, M.S. Kaletska, V.I. Keilis-Borok, I.M. Rotvain, and E.Ya. Rantsman).

[290] A local combinatorial formula for the first Pontryagin class, Funktsional. Anal. i Prilozhen. **10**:1 (1976), 14-17 (with A.M. Gabrielov and M.V. Losik). MR **53** # 14504b.
= Functional Anal. Appl. **10** (1976), 12-15.

[291] A Lie algebra structure in the formal calculus of variations, Funktsional. Anal. i Prilozhen. **10**:1 (1976), 1-8 (with L.A. Dikii). MR **57** # 7670.
= Functional Anal. Appl. **10** (1976), 16-22.

[292] A certain category of *G*-modules, Funktsional. Anal. i Prilozhen. **10**:2 (1976), 1-8 (with I.N. Bernstein and S.I. Gel'fand). MR **53** # 10880.
= Functional Anal. Appl. **10** (1976), 87-92.

[293] Atiyah-Patodi-Singer functionals for the characteristic classes of a tangent bundle, Funktsional. Anal. i Prilozhen. **10**:2 (1976), 13-28 (with A.M. Gabrielov and M.V. Losik). MR **54** # 1245.
= Functional Anal. Appl. **10** (1976), 95-107.

[294] Fractional powers of operators and Hamiltonian systems, Funktsional. Anal. i Prilozhen. **10**:4 (1976), 13-29 (with L.A. Dikii). MR **55** # 6484.
= Functional Anal. Appl. **10** (1976), 259-273.

[295] Poisson brackets and the kernel of the variational derivative in the formal calculus of variations, Funktsional. Anal. i Prilozhen. **10**:4 (1976), 30-34 (with Yu.I. Manin and M.A. Shubin). MR **55** # 13486.
= Functional Anal. Appl. **10** (1976), 274-278.

[296] Lattices, representations, and the algebras connected with them. I, Uspekhi Mat. Nauk **31**:5 (1976), 71-88 (with V.A. Ponomarev). MR **58** # 16779a.
= Russian Math. Surveys **31**:5 (1976), 67-85.

[297] Calculation of characteristic classes of combinatorial vector bundles, Preprint Inst. Appl. Mat. **1976**, no. 99 (with M.V. Losik).

[298] Conditions for the occurence of strong earthquakes (California and some other areas), in: *Vychislitel'naya seismologiya* (Computational seismology), Nauka, Moscow **1976**, no. 9, 3-92 (with Sh.A. Guberman, V.I. Keilis-Borok, F. Press, L. Knopov, I.M. Rotvain, A.M. Sadovskii, and E.Ya. Rantsman).

[299] Pattern recognition applied to earthquake epicenters in California, Phys. Earth and Planetary Interiors **1976**, no. 11, 277-283 (with Sh.A. Guberman, V.I. Keilis-Borok, L. Knopov, F.Press, E.Ya. Rantsman, I.M. Rotvain, and A.M. Sadovskii) (in English).

[300] First results of the prognostication of the effect of transmural (large focal) myocardial infarcts, in: *Aktyal'nye voprosy kardiologii. Otdalennye rezul'taty lecheniya elokachestvennykh opukholei* (Current problems in cardiology. Remote results of the treatment of malignant tumours), Nauka, Moscow 1976, 19-24 (with M.A. Alekseevskaya, I.V. Martynov, and V.M. Sablin).

[301] Formation of bundles of microfilaments during spreading of fibroblasts on the substratum, Exper. Cell Research **97** (1976), 241-248 (with E.E. Bragina and Yu.M. Vasil'ev) (in English).

[302] Serum dependence of expression of the transformed phenotype experiments with subline of mouse *L* fibroblasts adapted to growth in serum-free medium, Internat. J. Cancer **1976**, 83-92 (with A.D. Bershadskii, V.I. Gel'fand, V.I. Gel'stein, and Yu.M. Vasil'ev) (in English).

[303] Effect of colcemid on the spreading of fibroblasts in culture, Exper. Cell Research **101** (1976), 207-219 (with O.Yu. Ibanova, L.B. Margolis, and Yu.M. Vasil'ev) (in English).

[304] Active cell edge and movements of concanavalin *A* receptors on the surface of epithelial and fibroblastic cells, Proc. Nat. Acad. Sci. USA **73** (1976), 4085-4089 (with Yu.M. Vasil'ev, L.V. Domnina, N.A. Dorfman, and O.Yu. Pletyushkina) (in English).

[305] A computer study of prognosis of cerebral haemorrhage for choosing optimal treatment, European Congr. Neurosurgery, Edinburgh 1976, 71-72 (with Sh.A. Guberman, M.L. Izvekova, E.I. Kandel, D.K. Lunev, N.V. Lebedeva, I.F. Nikolaeva, N.M. Chebotareva, and E.V. Shmidt) (in English).

[306] Prognostication of the results of surgical treatment of haemorrhaging lesions by means of a computer, *Voprosy neirokhirurgii* (Problems of neurosurgery), **1976**, no. 3, 20-23 (with Sh.A. Guberman, M.L. Izvekova, E.I. Kandel, N.M. Chebotareva, and E.V. Shmidt).

[307] Effect of colcemid on the spreading of fibroblast in culture, Exper. Cell Research **101** (1976), 207-219 (with O.Yu. Ivanova, Yu.M. Vasil'ev, and L.B. Margolis) (in English).

[308] Effect of colcemid on morphogenetic processes and locomotion of fibroblasts, Cell mobility **3** (1976), 279-304 (with Yu.M. Vasil'ev) (in English).

[309] Prognostication of the result of a large focal myocardial infarct by means of a learning program, *Kardiologiya* (Cardiology) **17** (1977), 26-31 (with V.M.Sablin, M.A.Alekseevskaya, Sh.A.Guberman, I.V.Martynov, and I.M.Rotvain).

[310] Prognostication of the result of a myocardial infarct by means of the program "Cortex-3", *Kardiologiya* (Cardiology) **17**:6 (1977), 13-23 (with M.A. Alekseevskaya, L.D. Goloviya, Sh.A. Guberman, M.L. Izvekova, and A.L. Syrkin).

[311] Prognostication of the healing of duodenal ulcers, *Aktualnye voprosy gastroenterologii* (Current problems of gastroenterology) **1977**, no. 10, 42-51 (with M.Yu. Melikova, S.G. Gindikin, and M.L. Izvekova).

[312] A general guide-line or a general method for creating one (On ways of applying mathematical methods in medicine), Summaries of lectures at the All-Union Conf. on the theory and practice of automatic electrocardiological and clinical investigations, Kaunas 1977, 3-5 (with M.A. Alekseevskaya, M.L. Izvekova, L.D. Golovnya, I.V. Martynov, V.M. Sablin, and A.L. Syrkin).

[313] On the methodology of creating a formalized description of the patient (using the example of prognostication of remote results of electro-impulsive treatment of a constant form of flickering arrhythmy), Summaries of lectures at the All-Union Conf. on the theory and practice of automatic electrocardiological and clinical investigations, Kaunas 1977, 5-8 (with M.A. Alekseevskaya, E.S. Klyushin, A.V. Nedostup, and A.L. Syrkin).

[314] Effects of antitubilins on redistribution of cross-linked receptors on the surface of fibroblasts and epithelial cells, Proc. Nat. Acad. Sci. USA **74** (1977), 2865–2868 (with L.V. Domnina, O.Yu. Pletyushkin, and Yu.M. Vasil'ev) (in English).

[315] The influence of colcemid on the polarization of cells on narrow strips of the adhesive substratum, *Tsitologiya* (Cytology) **19** (1977), 357–360 (with O.Yu. Ivanova, S.G. Komm, L.B. Margolis, and Yu.M. Vasil'ev).

[316] The influence of serum on the development of cell transformation, Vestnik Akad. Mech. Nauk **3** (1977), 55–59 (with A.D. Bershadskii, V.I. Gel'fand, and Yu.M. Vasil'ev).

[317] Contact interaction of cell surfaces. Lectures at the Soviet-Italian symposium "Tissue proteinases in normal and pathological state" Moscow 22–27 September 1977 (with A.D. Bershadskii, Yu.M. Vasil'ev, A.D. Lyubimov, and Yu.A. Rovenskii).

[318] Mechanisms of morphogenesis in cell cultures, Internat. Rev. Cytology **50** (1977), 159–274 (with Yu.M. Vasil'ev) (in English).

[319] A formula for the analysis of histograms of intercellular intervals of Pourkine cells, Biofizika **22** (1977), 1072–1080 (with Yu.I. Arshavskii, M.B. Berkinblit, and V.S. Yakobson).

[320] Morphological division of mountainous countries by formalized criteria, *Vychislitel'naya seismologiya* (Computational seismology), **1977**, no. 10, 33–79 (with M.A. Alekseevskaya, A.M. Gabrielov, A.D. Gvishiani, and E.Ya. Rantsman).

[321] Formal morphostructural zoning at mountain territories, J. Geophysics **43** (1977), 227–235 (with M.A. Alekseevskaya, A.M. Gabrielov, A.D. Gvishiani, and E.Ya. Rantsman) (in English).

[322] The resolvent and Hamiltonian systems, Funktsional. Anal. i Prilozhen. **11**:2 (1977), 11–27 (with L.A. Dikii). MR **56** # 1359. Zbl. **412** # 58018. = Functional Anal. Appl. **11** (1977), 93–105.

[323] Non-local inversion formulae in real integral geometry, Funktsional. Anal. i Prilozhen. **11**:3 (1977), 12–19 (with S.G. Gindikin). MR **56** # 16265. = Functional Anal. Appl. **11** (1977).

[324] Representation lattices and the algebras connected with them, Uspekhi Mat. Nauk **32**:1 (1977), 55–107 (with V.A. Ponomarev). = Russian Math. Surveys **32**:1 (1977), 91–114.

[325] Complex manifolds whose skeletons are semisimple real Lie groups, and analytic discrete series of representations, Funktsional. Anal. i Prilozhen. **11**:4 (1977), 20–28 (with S.G. Gindikin). MR **58** # 11230. Zbl. **444** # 22006. = Functional Anal. Appl. **11** (1978), 258–265.

[326] Representations of the groups of smooth maps from a manifold X into a compact Lie group, Dokl. Akad. Nauk **323** (1977), 745–748 (with A.M. Vershik and M.I. Graev). MR **55** # 10602. = Soviet Math. Dokl. **18** (1977), 118–121.

[327] Representations of the group of smooth maps of a manifold X into a compact Lie group, Compositio Mat. **35** (1977), 299–334 (with M.I. Graev and A.M. Vershik). MR **58** # 28257 (in English).

[328] Activity of neurons of the lateral reticular nucleus in scratching, Biofizika **22**:1 (1977) (with Yu.I. Arshavskii, G.N. Orlovskii, and G.A. Pavlova).

[329] Influence of agents, destroying microtublets, on the distribution of receptors of the surface of cultured cells, *Tsitologiya* (Cytology) **20** (1978), 796–801 (with L.V. Domnina, O.Yu. Pletyushkina, and Yu.M. Vasil'ev).

[330] Messages conveyed by spino-cerebellar pathways during scratching in the cat. I, Activity of neurons of lateral reticular nucleus, Brain Research **151** (1978), 479–491 (with Yu.I. Arshavskii, G.N. Orlovskii, and G.A. Pavlova) (in English).

[331] Messages conveyed by spino-cerebellar pathways during scratching in the cat. 2, Activity of neurons of the ventral spino-cerebellar tract, Brain Research **151** (1978), 493-506 (with Yu.I. Arshavskii, G.N. Orlovskii, and G.A. Pavlova) (in English).

[332] Generation of scratching. 1, Activity of spinal interneurons during scratching, J. Neurophysiology **41** (1978), 1040-1057 (with M.B. Berkinblit, T.G. Delyagina, A.G. Fel'dman, and G.N. Orlovskii) (in English).

[333] Generation of scratching. 2, Non-regular regimes of generation, J. Neurophysiology **41** (1978), 1058-1069 (with M.B. Berkinblit, T.G. Delyagina, A.G. Fel'dman, and G.N. Orlovskii) (in English).

[334] Structure locale de la catégorie des modules de Harish-Chandra. I, C.R. Acad. Sci. Paris **A286** (1978), 435-437 (with I.N. Bernshtein and S.I. Gel'fand). Zbl. **416** # 22018; **431** # 22012. MR **58** # 16966.

[335] Structure locale de la catégorie des modules de Harish-Chandra. II, C.R. Acad. Sci. Paris **A286** (1978), 495-497 (with I.N. Bernshtein and S.I. Gel'fand). Zbl. **431** # 22013. MR **81e**:22026.

[336] Mechanisms of non-adhesiveness of endothelial and epithelial surfaces, Nature **275** (1978), 710-711 (with Yu.M. Vasil'ev) (in English).

[337] Microtubular system in cultured mouse epithelial cells, Cell Biol. Internat. Rep. **2** (1978), 345-351 (with Yu.M. Vasil'ev, A.D. Bershadskii, V.A. Rozenblat, and I.S. Tint) (in English).

[338] Calculus of jets and non-linear Hamiltonian systems, Funktsional. Anal. i Prilozhen. **12**:2 (1978), 8-23 (with L.A. Dikii). MR **58** # 18561. Zbl. **405** # 58027. = Functional Anal. Appl. **12** (1978), 81-94.

[339] The family of Hamiltonian structures connected with integrable non-linear differential equations (on the second Hamiltonian structure), Preprint Inst. Appl. Mat. **1978**, no. 136 (with L.A. Dikii). MR **81**:58027.

[340] Algebraic vector bundles on $\mathbf{P}^n$ and problems of linear algebra, Funktsional. Anal. i Prilozhen. **12**:3 (1978), 66-68 (with I.N. Bernshtein and S.I. Gel'fand). MR **80c**:14010a. Zbl. **402** # 14005. = Functional Anal. Appl. **12** (1979), 212-214.

[341] Cohomology of infinite-dimensional Lie algebras and Laplace operators, Funktsional. Anal. i Prilozhen. **12**:4 (1978), 1-5 (with B.L. Feigin and D.B. Fuks). MR **80i**:58050. Zbl. **404** # 17008. = Functional Anal. Appl. **12** (1978), 243-247.

[342] On one approach to formalization of the diagnostic attitude of a doctor (using the prognosis of the healing of duodenal ulcers), Summaries of lectures at the All-Union Conf. biological and medical cybernetics, vol. 2 (1978), 27-31 (with S.G. Gindikin, M.L. Izvekova, and M.Yu. Melikova).

[343] On some questions of mathematical diagnostics: examples of problems from gastroenterology, Summaries of lectures at the Second All-Union Congr. on Gastroenterology, vol. 2, Nauka, Moscow-Leningrad 1978, 57-58 (with S.G. Gindikin, M.L. Izvekova, and M.Yu. Melikova).

[344] Dualità, Enciclopedia Einaudi, Einaudi, Torino, vol. 5 (1978), 126-178 (with Yu.U. Manin).

[345] Messages conveyed by descending tract during scratching in the cat. I, Activity of vestibulospinal neurons, Brain Research **159** (1978), 88-110 (with Yu.I. Arshavskii, G.N. Orlovskii, and G.A. Pavlova) (in English).

[346] A new approach to the problem of the choice of information and formalization of the description of the patient for the solution of medical problems on a computer, Preprint Inst. Appl. Math. **1978**, no. 144 (with M.A. Alekseevskaya, E.S. Klyushin, A.V. Nedostup, and A.L. Syrkin).

[347] The gathering of medical information for processing on a computer (manual), Inst. Appl. Math. 1979 (with M.A. Alekseevskaya, E.S. Klyushin, and A.V. Nedostup).
[348] One method of formalizing the diagnostic attitude of a doctor (examples of prognosis of the healing of a duodenal ulcer), Preprint Acad. Sci. USSR, Sci. Committee on the complex problem "Cybernetics", Moscow 1979 (with S.G. Gindikin, M.L. Izvekovaya, and M.Yu. Melikova).
[349] Upper surfaces of epithelial sheets and of fluid lipid films are non-adhesive for platelets, Proc. Nat. Acad. Sci. USA **76** (1979), 2303-2305 (with L.B. Margolis, E.Yu. Vasil'eva, and Yu.M. Vasil'ev) (in English).
[350] Morphology of microtubular systems in epithelial cells in the kidney of a mouse, Ontogenez **10** (1979), 231-235 (with A.D. Bershadskii, I.S. Tint, V.I. Gel'fand, V.A. Rozenblat, and Yu.M. Vasil'ev).
[351] Integrable non-linear equations and Liouville's theorem, Funktsional. Anal. i Prilozhen. **13**:1 (1979), 8-20 (with L.A. Dikii). MR **80i**:58027. Zbl. **423** # 34003.
= Functional Anal. Appl. **13** (1979), 6-15. Zbl. **436** # 34002.
[352] A local problem of integral geometry in a space of curves, Funktsional. Anal. i Prilozhen. **13**:2 (1979), 11-31 (with S.G. Gindikin and Z.Ya. Shapiro). MR **80k**:53100. Zbl. **415** # 53046.
= Functional Anal. Appl. **13** (1980), 248-262. Zbl. **437** # 53057.
[353] Model algebras and representations of graphs, Funktsional. Anal. i Prilozhen. **13**:3 (1979), 1-12 (with V.A. Ponomarev).
= Functional Anal. Appl. **13** (1980), 157-166. Zbl. **437** # 16020.
[354] Hamiltonian operators and algebraic structures associated with them, Funktsional. Anal. i Prilozhen. **13**:4 (1979), 13-31 (with I.Ya. Dorfman). MR **81c**:58035.
= Functional Anal. Appl. **13** (1980), 248-262. Zbl. **437** # 58009.
[355] Problems of integral geometry in $\mathbf{RP}^n$ connected with the integration of differential forms, Funktsional. Anal. i Prilozhen. **13**:4 (1979), 64-66 (with S.G. Gindikin and M.I. Graev). MR **83a**:43006.
= Functional Anal. Appl. **13** (1980), 288-290. Zbl. **441** # 58002.
[356] The role of the brain stem and cerebellum in the regulation of rhythmic movements, Proc. 13 Congr. Pavlov Physiological Soc., vol. 1, Nauka, Leningrad 1979, 474-475 (with Yu.I. Arshavskii).
[357] The significance of signals passing along the various spino-cerebral pathways for the work of locomotive centres of the brain stem in scratching, in: *Neironnye mekhanizmy integrativnoi deyatel'nosti mozzhechka* (Neural mechanisms of integrated activity of the brain stem), Erevan 1979, 88-91 (with M.B. Berkinblit, G.N. Orlovskii, and Yu.I. Arshavskii).
[358] Signalling mechanisms of the scratching reflex and their interaction with the cerebellum, in: *Neironnye mekhanizmy integrativnoi deyatel'nosti mozzhechka* (Neural mechanisms of integrated activity of the brain stem), Erevan 1979, 92-96 (with M.B. Berkinblit, G.N. Orlovskii, and Yu.I. Arshavskii).
[359] The problem of classifying glomerule kidneys, Preprint Acad. Sci. USSR, Sci. Committee on the complex problem "Cybernetics", Moscow 1980 (with M.Ya. Ratner, B.I. Rozenfel'd, and V.V. Serov).
[360] Representation of graphs, Perfect sub-representations, Funktsional. Anal. i Prilozhen. **14**:3 (1980), 14-31 (with V.A. Ponomarev). MR **83c**:05113. Zbl. **453** # 05027.
= Functional Anal. Appl. **14** (1980), 177-190. Zbl. **464** # 05033.

[361] The Schouten bracket and Hamiltonian operators, Funktsional. Anal. i Prilozhen. **14**:3 (1980), 71-74 (with I.Ya. Dorfman). MR **82e**:58039. Zbl. **444** # 58010. = Functional Anal. Appl. **14** (1980), 223-226.

[362] Admissible n-dimensional complexes of curves in $\mathbf{R}^n$, Funktsional. Anal. i Prilozhen. **14**:4 (1980), 36-44 (with M.I. Graev). MR **82e**:53013. Zbl. **454** # 53042.
= Functional Anal. Appl. **14** (1980), 274-281. Zbl. **458** # 53041.

[363] Integral geometry for one-dimensional fibrations of general form over $\mathbf{RP}^n$, Preprint Inst. Appl. Math. **1980**, no. 60, 1-24 (with S.G. Gindikin and M.I. Graev). MR **82g**:53081.

[364] Integral geometry in affine and projective spaces, "*Sovremennye problemy matematiki*" (Current problems in mathematics), VINITI Moscow **16** (1980), 53-226 (with S.G. Gindikin and M.I. Graev). MR **82m**:43017. Zbl. **474** # 52005. = J. Soviet Math. **18** (1982), 39-167. Zbl. **474** # 52010.

[365] A formalized differentiated description of the motor of the stomach and the duodenal intestine, Preprint Acad. Sci. USSR, Sci. Committee on the complex problem "Cybernetics" 1980, 1-34 (with S.A. Chernyakevich and I.V. Cherednik).

[366] Study of the correlation between electrocardiograph and coronary data, Preprint Acad. Sci. USSR, Sci. Committee on the complex problem "Cybernetics" 1980, 1-28 (with G.G. Gel'shtein, I.P. Lukashevich, and M.A. Shifrin).

[367] Spreading of fibroblasts in a medium containing cytochalasin B: Formation of lamellar cytoplesma as a combination of several functionally different processes, Proc. Nat. Acad. Sci. USA **77** (1980), 5919-5922 (with Zh.L. Bliokh, L.V. Domnina, O.Yu. Ivanova, O.Yu. Pletyushkina, T.S. Svitkina, V.V. Smolyaninov, and Yu.M. Vasil'ev) (in English).

[368] Prognostication of complications and classification of patients with severe myocardial infarction, Summaries of lectures at the second All-Union Conf. "Theory and practice of automation of electrocardiological and clinical studies", Kaunas 1981, 274-276 (with A.L. Syrkin, V.L. Vakhlyaev, E.V. Pomerantsev, V.M. Sablin, M.N. Starkova, and V.A. Sulimova).

[369] Expressibility of electrocardiograph changes in severe disease of the coronary artery in patients with chronic ischemic heart disease, Summaries of lectures at the second All-Union Conf. "Theory and practice of automation of electrocardiological and clinical studies", Kaunas 1981, 304-307 (with G.G. Gel'shtein, L.S. Zingerman, I.P. Lukashevich, and M.A. Shifrin).

[370] Choice of information for the classification of patients with myocardial infarction and choice of medical tactics, Summaries of lectures at the second All-Union Conf. "Theory and practice of automation of electrocardiological and clinical studies", Kaunas 1981, 276-278 (with V.L. Vakhlyaev, E.V. Pomerantsev, B.I. Rozenfel'd, V.A. Sulimov, A.L. Syrkin, and S.M. Khoroshkin).

[371] Neoplastic and normal cells in culture, Cambridge University Press, London-Sydney 1981 (with Yu.M. Vasil'ev) (in English).

[372] The immediate prognosis for healing of duodenal ulcers (control of classification), Trans. Second Moscow Med. Inst. Ser. "Surgery" **1981**, no. 32, 73-80 (with M.Yu. Melikova, S.G. Gindikin, and M.L. Izvekova).

[373] *Vzaimodeistvie normal'nykh i opukholevykh kletok so sredoi* (Interaction of normal and tumorous cells with the medium), Nauka, Moscow 1981 (with Yu.M. Vasil'ev).

[374] Multi-purpose chart of a patient with myocardial infarction (for setting up a data bank in a computer), Preprint Acad. Sci. USSR Sci. Committee on the complex problem "Cybernetics" 1981 (with V.M. Alekseev, M.A. Alekseeskaya, E.E. Gogin, L.D. Goloviya, R.M. Zaslavska, M.L. Izvekova, E.S. Klyushin, I.V. Martynov, I.V. Sablin, A.L. Syrkin, and V.A. Ponomarev).

[375] Gabriel's theorem is also true for representations of graphs endowed with relations, Funktsional. Anal. i Prilozhen. **15**:2 (1981), 71-72 (with V.A. Ponomarev).
= Functional Anal. Appl. **15** (1981), 132-133. Zbl. **479** # 18003.

[376] Hamiltonian operators and infinite-dimensional Lie algebras, Funktsional. Anal. i Prilozhen. **15**:3 (1981), 23-40 (with I.Ya. Dorfman). MR **82j**:58045.
Zbl. **478** # 58013.
= Functional Anal. Appl. **15** (1982), 173-187. Zbl. **487** # 58008.

[377] Representations of the group of functions taking values in a compact Lie group, Compositio Mat. **42** (1981), 217-243 (with M.I. Graev and A.M. Vershik).
MR **83g**:22002. Zbl. **449** # 22019 (in English).

[378] Simmetria, Enciclopedia Einaudi, Einaudi, Torino 1981, vol. 12, 916-943 (with Yu.I. Manin).

[379] Mechanisms of morphological reactions determining the shape and movement of normal and transformed cells in culture, in: *Nemyshechnie sistemy* (Non-muscular systems), Nauka, Moscow 1981, 65-75 (with A.D. Bershadskii, Yu.M. Vasil'ev, Zh.L. Bliokh, L.V. Domnina, O.Yu. Ivanova, T.M. Svitkina, I.S. Tint, and V.V. Smolyaninov).

[380] Stabilization independent of micropipelets of the cell surface of normal and transformed connective tissue cells, *Tsitologiya* (Cytology), **23** (1981), 62-65 (with O.Yu. Ivanova, S.G. Komm, and Yu.M. Vasil'ev).

[381] Integral transformations connected with two remarkable complexes in projective space, Preprint Inst. Appl. Math. **1982**, no. 93 (with M.I. Graev).

[382] Non-local inversion formulae in a problem of integral geometry connected with p-dimensional planes in real projective space, Funktsional. Anal. i Prilozhen. **16**:3 (1982), 49-51 (with M.I. Graev and R. Roshy).
= Functional Anal. Appl. **16** (1982), 196-198.

[383] Classification of patients and prognosis of healing in myocardial infarction, Preprint Acad. Sci. USSR Sci. Committee on the complex problem "Cybernetics" 1982 (with M.N. Starkova and A.L. Syrkin).

[384] The methodology of comparing material from two hospitals and the construction of a single guide-line for the prognosis of the effect of a strong focal myocardial infarction, Preprint Acad. Sci. USSR Sci. Committee on the complex problem "Cybernetics" 1982 (with M.L. Izvekova, M.N. Starkova, and A.L. Syrkin).

[385] Determination of a morphological picture of glomerule kidney from clinical-functional data (by means of a formal scheme modelling the diagnostis of kidney consultants), Preprint Acad. Sci. USSR Sci. Committee on the complex problem "Cybernetics" **1982** (with M.A. Brodskii, V.A. Varshavskii, M.Ya. Ratner, B.I. Rozenfel'd, V.V. Serov, and I.I. Stenina).

[386] Selection of information for the classification of patients with myocardial infarction and choice of medical tactics, Preprint Acad. Sci. USSR Sci. Committee on the complex problem "Cybernetics" 1982 (with V.D. Vakhlyaev, B.I. Rozenfel'd, V.A. Sulimov, A.L. Syrkin, and S.M. Khorshkin).

[387] Hamiltonian operators and the classical Yang-Baxter equation, Funktsional. Anal. i Prilozhen. **16**:4 (1982), 1-9 (with I.Ya. Dorfman).
= Functional Anal. Appl. **16** (1982), 241-248.

[388] The role of the cerebellum in guiding rhythmic movement, Lecture at the First All-Union Biophysical Congress, Moscow 1982 (with Yu.I. Arshavskii and G.N. Orlovskii).

[389] Neural mechanisms in the generation of nutritional rhythmics in molluscs, Lecture at the First All-Union Biophysical Congress, Moscow 1982 (with Yu.I. Arshavskii, I.N. Beloozerova, Yu.V. Panchin, and G.N. Orlovskii).

[390] Possible common mechanism of morphological and growth-related alterations accompanying neoplastic transformation, Proc. Nat. Acad. Sci. USA **79** (1982), 2594–2597 (with Yu.M. Vasil'ev) (in English).

[391] A commutative model of the basic representation of the group $SL(2, \mathbf{R})^X$ connected with a unipotent subgroup, Preprint Inst. Appl. Math. **1982**, no. 169 (with A.M. Vershik and M.I. Graev).

[392] Structural organisation of data in problems of medical diagnosis and prognosis, Preprint Acad. Sci. USSR Sci. Committee on the complex problem "Cybernetics" 1982 (with B.I. Rozenfel'd and M.A. Shifrin).

[393] The Plancherel formula for the integral transformation connected with a complex of lines intersecting an algebraic straight line in $\mathbf{C}^3$ and $\mathbf{CP}^3$, Dokl. Akad. Nauk Arm. SSR **75**:1 (1982), 9–15 (with M.I. Graev, R.G. Airapetyan, and G.R. Oganesyan). Zbl. **504** # 43009.

[394] Effects of small doses of cytochalasins on fibroblasts: preferential changes of active edges and focal contacts, Proc. Nat. Acad. Sci. USA **79** (1982), 7754–7757 (with L.V. Domnina, V.I. Gel'fand, O.Yu. Ivanova, O.Yu. Pletyushkina, and Yu.M. Vasil'ev) (in English).

[395] Geometry in Grassmanians and a generalization of the dilogarithm, Advances in Maths. **44** (1982), 279–312 (with R. Macpherson) (in English). Zbl. **504** # 57021.

[396] An abstract Hamiltonian formalism for the classical Yang-Baxter bundles, Uspekhi Mat. Nauk **38**:3 (1983), 3–21 (with I.V. Cherednik).
= Russian Math. Surveys **38**:3 (1983), 1–22.

[397] Plancherel's theorem for the integral transformation connected with a complex of p-dimensional planes in $\mathbf{CP}^n$, Dokl. Akad. Nauk SSSR **268** (1983), 265–268 (with R.G. Airapetyan, M.I. Graev, and G.R. Oganesyan).
= Soviet Math. Dokl. **27** (1983), 47–50.

[398] Geometrical structures of double fibrations and their connection with some problems of integral geometry, Funktsional. Anal. i Prilozhen. **17**:2 (1983), 7–22 (with G.S. Shmelev).
= Functional. Anal. Appl. **17** (1983), 84–96.

[399] A commutative model of representations of the group of flows $SL(2, \mathbf{R})^X$ connected with a unipotent subgroup, Funktsional. Anal. i Prilozhen. **17**:2 (1983), 70–72 (with A.M. Vershik and M.I. Graev).
= Functional. Anal. Appl. **17** (1983), 137–139.

[400] Retrospective estimate of non-stable cardiac angina in various forms of myocardial infarction, *Klinicheskaya meditsina* (Clinical medicine), **1983**, no. 3, 28–31 (with A.Yu. Lyuiko, M.N. Starkova, and A.L. Syrkin).

[401] Prognosis of recidive haemorrhaging in patients with ulcerous disease of the stomach and duodenal intestine, *Vestnik khirurgii* (Bulletin of surgery), Grekov **130**:4 (1983), 21–24 (with A.A. Grinberg, M.L. Izvekova, and V.P. Lakhtina).

Translated by A. Lofthouse

Some remarks on I. M. Gelfand's works

In these comments, we will attempt to provide a brief overview of the material contained in the first of these series of articles. We should like to mention at the outset that our comments is not intended as a substitute for the two excellent survey articles on the work of Gelfand, published in Uspekhi on the occasion of his 50th and 60th birthdays, and republished in this volume. We recommend these articles to those of our readers who would like to have a more detailed introduction to Gelfand's work than we are able to supply in these few pages.

The papers of Gelfand written prior to 1960 have by now become so assimilated into the mathematical mainstream that it is difficult for analysts and geometers of our generation (i.e., of the 60's and 70's) to assess the impact they had when they first appeared. Some attempt will be made to do so, however, in the first couple of pages of this review in order to put the rest of the remarks we will make into perspective. The situation is quite different with the more recent papers. The Gelfand-Fuks papers triggered an enormous outburst of mathematical activity in the West in which we ourselves and our colleagues at Harvard and M.I.T. played a modest part; and the same can be said of the developments in integral geometry inspired by the seminal Gelfand-Graev paper [Vol. II, part 2, item 14], and the developments in ordinary differential equations inspired by the Gelfand-Dikij papers. Our problem with assessing these papers will be that we have become *too* involved with the developments related to them to put them into a historical perspective. Nevertheless we will try. When we discuss the Gelfand-Dikij papers, by the way, we will have to probe the pre-1960's literature a little more deeply since they address issues raised by the earlier papers of Gelfand-Levitan. For the most part, however, we will focus on the post-1960's literature.

1. The papers of the 1940's and 1950's

The theory of commutative normed rings, created by Gelfand in the late 1930's, has become today one of the most active areas of functional analysis. The key idea in Gelfand's theory – that maximal ideals are the underlying "points" of a commutative normed ring – not only revolutionized harmonic analysis but had an enormous impact in algebraic geometry. (One need only look at the development of the concept of the spectrum of a commutative ring and the concept of scheme in the algebraic geometry of the 1960's and

1970's to see how far beyond the borders of functional analysis Gelfand's ideas penetrated.)

After dealing brilliantly with the theory of *commutative* normed rings, Gelfand in collaboration with Najmark proved that *every* normed ring with an involution can be realized as a ring of linear operators in a Hilbert space with its natural involution. This represented a crucial step in the "coming of age" of the theory of C^* algebras: By 1950 C^* algebras had become an essential tool in the arsenal of functional analysis and, in theoretical physics, were to dominate for the next twenty years the approach to quantum field theory from the axiomatic point of view. In Gelfand's hands this result also served as a link between the theory of normed rings and the theory of representations of non-compact groups which was central to Gelfand's research interests in the late 1940's and early 1950's. It might be an interesting object of study for the historian of mathematics of the twenty-first century to compare the approach to normed rings by Gelfand and his collaborators with the approach of his contemporaries, von Neumann et al., in the United States. In the hands of Gelfand the theory was a prelude to the luxuriant development of representations of semi-simple Lie groups which is today a vibrant area of interaction between geometry and analysis. An analogous case-study might be made, by the way, of the history of the theory of generalized functions: With roots in the work of Hadamard, Riesz and Sobolev, this theory was developed more or less simultaneously by Schwartz in France and Gelfand and his collaborators in Russia. In Gelfand's hands this theory became a rich collection of geometrical examples whose properties are being investigated to this day, the theory of holonomic systems, homogeneous distributions and integral operators being some of its biproducts. The five volumes of Gelfand on the theory of generalized functions became instantaneous classics when they came out in the fifties and are still essential texts in the education of every analyst.

Concerning his other papers of the 50's his paper with Fomin on geodesic flows on surfaces of negative curvature brought this area back into the mathematical mainstream and paved the way for the deep investigations of Anosov and others on hyperbolicity in dynamical systems. About his work with Levitan on inverse spectral theory we will have more to say in § 4. As for his papers on generalized random processes and their relation with quantum field theory, we will confine our remarks to the observation that Gelfand's approach contained many ingredients of what is today the reigning viewpoint in this subject.

Let us now turn to a summary of the papers included here which reflect Gelfand's work after 1960.

2. *Index theory*

It is safe to say that Gelfand's short paper on the index theorem had the greatest repercussions on the mathematics of the 60's and 70's of all his major papers. In this paper he makes the simple (and profound) observation that the index of an elliptic operator is a homotopy invariant of the leading symbol, an observation which inspired Seeley, Atiyah-Singer, Fedosov and others to find a concrete expression for the index, and initiated an interest in index problems which has not abated to this day (vide the application of index theory to the problem of anomalies in gauge-field theory.)

3. Gelfand-Fuks cohomology

Another paper of Gelfand whose repercussions have been out of proportion to its modest length is [Vol. III, part 2, item 3]. Let $\mathscr{A}(M)$ be the Lie algebra of all smooth vector fields on a manifold M. It is a natural question to ask what the cohomology is of this Lie algebra and whether it describes any interesting topological features of M itself. It is difficult to understand, in retrospect, why this question was never raised before Gelfand raised it in the article cited above. Perhaps one expected that the answer would be either trivial or far too complicated to be interesting. However, in this article, Gelfand and Fuks computed the answer for S^1 and showed that *neither* is the case. Shortly afterwards, in [loc. cit, item 5], they worked out the cohomology of the algebra, W_n, of formal power series vector fields in n variables. In their result the topology of a certain CW complex puts in a surprising appearance. This led Bott to his well-known conjecture on the final answer for $H^*(\mathscr{A}(M), \mathbb{R})$ for arbitrary compact n-dimensional M: namely it is essentially the cohomology of the space of all sections of a certain fiber bundle over M with the CW complex mentioned above as typical fiber. Thus one consequence of this article is a completely unforseen tie-in between Gelfand-Fuks cohomology and the topology of function spaces. Another great surprise was a tie-in between Gelfand-Fuks cohomology and topological invariants of foliations: In 1971 Godbillon and Vey discovered a "non-classical" characteristic class in $H^3(M, \mathbb{R})$ associated to a codimension one foliation of M. Soon after a number of generalized Godbillon Vey classes were discovered and linked by Bott, Haefliger and others to Gelfand-Fuks cohomology by means of a universal map $H^*(W_n, \mathbb{R}) \to$ Godbillon-Vey classes of M_n.

4. Ordinary differential equations

The Gelfand-Levitan theory concerns itself with recapturing the potential term, q, in the ODE

$$\left(\frac{d^2}{dx^2} + q(x)\right) f = \lambda f$$

from the spectral data associated with this equation (such as eigenfunctions and norming constants or transmission and reflection coefficients.) The possibility of doing this had been observed prior to their work by Borg and Levinson (with Bargmann calling attention to the importance of norming coefficients as inverse spectral data.) However, what the Gelfand-Levitan theory provides is a *concrete* reconstruction scheme: To summarize the main ideas of the theory in a few lines, they noticed that the generalized eigenfunctions of the equation above can be obtained by applying a kind of Gram-Schmidt process to functions of the form, $A(\lambda)\, e^{i\lambda x}$, $-\infty < \lambda < \infty$. This leads to a Tricomi equation relating the spectral data to q in a very concrete way and enables one to view the transform

(*) 2nd order ODE's → Spectral data

as a kind of non-linear Fourier transform. (In the 70's one was to see that the analogy between (*) and the usual Fourier transform was not too far-fetched. Just as the usual Fourier transform reduces the standard diffusion equation, $\frac{du}{dt}=\frac{d^2u}{dx^2}$, to diagonal form, the transform (*) reduces its non-linear counterpart, the Korteweg-De Vries equation, to diagonal form.)

Gelfand became involved again in inverse spectral problems in the mid 70's in his joint work with Dikij. We would like to call attention to one result mentioned in these papers since it was later to play a very important role in the work of Adler and others: it was Gelfand's description of the Lax operator generating the various isospectral flows associated with the Hill's equation: Namely the k-th Lax operator is just

$$(D^2+q)^s_+, \quad s=\frac{2k+1}{2},$$

D^2+q being the Hill's equation, $(D^2+q)^s$ its s-th complex power (in the sense of Seeley) and $(D^2+q)^s_+$ the positive part of its pseudodifferential expansion in D.

More generally, in these papers, Gelfand and Dikij developed a theory of formal variational calculus which shed immense light on the Hamiltonian character of the N soliton equations for many non-linear partial differential equations in addition to providing formal means for algebraically computing their integrals.

5. *Integral geometry and intertwining operators*

Thanks to the work of Gelfand, Najmark, Harish-Chandra and Langlands one is very close to understanding what *all* the irreducible unitary representations are for the classical Lie groups. What is still mysterious, however, is the amazing number of redundant ways in which these representations occur in nature. For this reason we feel that the imminent classification of all irreducible unitary representations of the classical groups does not, like the classification of the finite simple groups a couple years ago, signal the end of an era. An optimist would say that, the preliminary work having been accomplished, one can now tackle the interesting problem: explaining the redundancies. This, of course, comes down to inventing procedures for "intertwining" any two geometrically dissimilar descriptions of a given irreducible on the Langlands list. It is not clear, a priori, that there are any general procedures. (Maybe the best one can hope for is a large arsenal of "ad hoc" procedures.) However, Gelfand and his collaborators make a strong case in Volume 5 of "Generalized Functions" for the contention that there are, at least, general *principles* for constructing intertwining operators, involving techniques from integral geometry. We will illustrate what we mean by an example: Let P^n be complex projective space and G_k the Grassmannian of k-dimensional projective subspaces of P^n. If we impose a Hermitian structure on $\mathbb{C}^{n+1}$ then each k-dimensional subspace of P^n acquires a Fubini-Study measure and so we can define an integral transform

$$(**)\quad f\in C^\infty(P^n)\to\hat{f}\in C^\infty(G_k)$$

by defining $\hat{f}(p)$ to be the integral of f over the k-plane represented by P. Special cases of this transform are well-known to classical geometers, and it is easy to verify its basic properties, e.g. that it is injective, that its range is characterized by a system of *P.D.E.*'s, etc.

It also has a relatively uninteresting intertwining property: it intertwines the representations of $SU(n+1)$ on the L^2 spaces of P^n and G_k. In [Vol III, part I, item 6] Gelfand and Graev observe that, in fact, it has a much more interesting intertwining property: with a little care one can define (**) without the choice of a Hermitian form on $\mathbb{C}^{n+1}$. It then becomes a (manifestly) $GL(n+1,\mathbb{C})$-invariant operator. What is more surprising is that one can construct a manifestly $GL(n+1,\mathbb{C})$-invariant inverse operator: Given a function on P^n one is able to recover its value at a point, x, by integrating over a appropriate algebraic cycle a $2k$-form, ω^x, on G_{k+1}, ω^x being constructed in a $GL(n+1,\mathbb{C})$-invariant fashion from the integral data, $\hat{f}$. For details of this beautiful inversion theorem, see the article cited above.

6. *Integral geometry and admissibility*

The integral transform described above is overdetermined for $1\leqq k<n-1$: one can construct f from its integrals over k planes, but, as is easy to see by examples, not all integrals over k planes are needed to reconstruct. In Volume 5 of "Generalized Functions" Gelfand raised for the first time the question: Can one characterize those n-dimensional subvarieties, W, of G_k with the property that the integral of f over k-planes corresponding to points of W is sufficient to reconstruct f, he called such W's *admissible* and he and Graev obtained, for $k=1$ and $n=3$ a complete classification of admissible W's. Perhaps the most surprising feature of their classification is that admissibility is *not* a generic property. In [loc. cit. item 4] Gelfand and M. Graev settled the admissibility question for $k=1$ and n arbitrary, and recently Gelfand-Gindinkin [loc. cit. item 8] and Gelfand-Graev-Rocu [loc. cit. item 13] have succeeded in shedding some light on the much harder admissibility question for real Grassmannians.

7. *Concluding remarks*

Unfortunately this survey is woefully inadequate as a comprehensive overview of Gelfand's work over five decades. What it has omitted, in addition to a report on papers which we felt unqualified to discuss (e.g. those on Verma modules and the so-called BGG papers) is a discussion of many ideas which haven't found their way into his publications at all: By this we mean remarks which have circulated widely as marginal comments on his work and that of others. We will mention two examples:

1) In his address to the International Congress in Stockholm in 1962, he called attention to the remarkable similarities between the zeta function of a homogeneous space and the Heisenberg S-matrix. The importance of this observation is now clear in light of the work of Fadeev-Pavlov and Lax-Phillips on the theory of scattering for Fuchsian groups.

2) Our second example has yet, we feel, to be properly appreciated. In his Harvard lectures in June of 1976 he pointed out that there are at present two kinds of invariants associated with differentiable structures on compact manifolds: one kind, like the index and heat invariants, are expressible as integrals of *local* data connected with the structure. The second kind, like the eta invariant, Ray-Singer torsion and the Chern-Simons invariants, don't have this property but their first variations do. Might this be the tip of the iceberg? Perhaps the next generation of global invariants will be non-local, have non-local first variation and have *local* second variation.

In fact, in the nebulous area between the world of smooth and of PL phenomena there may exist a whole hierarchy of such invariants, each class of invariants in the hierarchy being one order less "local" than the preceding!

V.W. Guillemin and S. Sternberg

Tentative table of contents

Volume II

Part I. General problems of representation theory

1. Irreducible unitary representations of locally bicompact groups (with D.A. Rajkov)
2. Unitary representations of the group of linear transformations of the straight line (with M.A. Najmark)
3. The centre of the infinitesimal group ring
4. Spherical functions on symmetric Riemannian spaces

Part II. Infinite-dimensional representations of semisimple Lie groups

1. Unitary representations of the Lorentz group (with M.A. Najmark)
2. Unitary representations of the Lorentz group (with M.A. Najmark)
3. On unitary representations of the complex unimodular group (with M.A. Najmark)
4. The fundamental series of irreducible representations of the complex unimodular group (with M.A. Najmark)
5. Complementary and degenerate series of representations of the complex unimodular group (with M.A. Najmark)
6. The trace in principal and complementary series representations of the complex unimodular group (with M.A. Najmark)
7. On the connection between representations of complex semisimple Lie groups and its maximal compact subgroup (with M.A. Najmark)
8. An analogue of the Plancherel formula for the complex unimodular group (with M.A. Najmark)
9. General relativistically invariant equations and infinite-dimensional representations of the Lorentz group (with A.M. Yaglom)
10. On the structure of the ring of rapidly decreasing functions on a Lie group
11. Unitary representations of classical groups. Introduction. 9. Spherical functions. 18. Transitivity classes for the set of pairs. Another way of describing representations of the complementary series (with M.A. Najmark)
12. Unitary representations of the real simple Lie groups (with M.I. Graev)

13. Unitary representations of real unimodular groups (Principal non-degenerate series) (with M.I. Graev)
14. On a general method for decomposition of the regular representation of a Lie group into irreducible representations (with M.I. Graev)

Part III. Geometry of homogeneous spaces; spherical functions; automorphic functions

1. Some remarks on the theory of spherical functions on symmetric Riemannian manifolds (with F.A. Berezin)
2. Geodesic flows on manifolds of constant negative curvature (with S.V. Fomin)
3. Geometry of homogeneous spaces, representations of groups in homogeneous spaces and related questions of integral geometry (with M.I. Graev)
4. Unitary representations in homogeneous spaces with discrete stationary groups (with I.I. Piatetski-Shapiro)
5. Unitary representations in a space G/Γ, where G is a group of n-by-n matrices and Γ is a subgroup of integer matrices (with I.I. Piatetski-Shapiro)

Part IV. Models of representations; representations of groups over various fields

1. Categories of group representations and the problem of classifying irreducible representations (with M.I. Graev)
2. Construction of irreducible representations of simple algebraic groups over a finite field (with M.I. Graev)
3. Representations of quaternion groups over locally compact and functional fields (with M.I. Graev)
4. A new model for representations of finite semisimple algebraic groups (with I.N. Bernshtejn and S.I. Gelfand)
5. Irreducible unitary representations of the group of unimodular second-order matrices with elements from a locally compact field (with M.I. Graev)
6. Plancherel's formula for the group of the unimodular second order matrices with elements in a locally compact field (with M.I. Graev)
7. On the representation of the group $GL(n,k)$, where K is a local field (with D.A. Kazhdan)
8 Models of representations of Lie groups (with I.N. Bernshtejn and S.I. Gelfand)
9. Complex manifolds whose skeletons are semisimple real Lie groups, and analytic discrete series of representations (with S.G. Gindikin)
10. Models of representations of classical groups and their hidden symmetries (with A.V. Zelevinskij)

Part V. Verma modules; resolutions of finite-dimensional representations

1. Differential operators on the principal affine space and investigation of G-modules (with I.N. Bernshtejn and S.I. Gelfand)

2. Structure of representations generated by vectors of heighest weight (with I.N. Bernshtejn and S.I. Gelfand)
3. Differential operators on a cubic cone (with I.N. Bernshtejn and S.I. Gelfand)
4. Schubert cells and cohomology of the spaces G/P (with I.N. Bernshtejn and S.I. Gelfand)
5. Category of G-modules (with I.N. Bernshtejn and S.I. Gelfand)
6. Structure locale de la catégorie des modules de Harish-Chandra I (avec I.N. Bernshtejn and S.I. Gelfand)
7. Structure locale de la catégorie des modules de Harish-Chandra II (avec I.N. Bernshtejn and S.I. Gelfand)
8. Algebraic bundles over P^n and problems of linear algebra (with I.N. Bernshtejn and S.I. Gelfand)

Part VI. Enveloping algebras and their quotient skew-fields

1. Fields associated with enveloping algebras of Lie algebras (with A.A. Kirillov)
2. Sur les corps liés aux algèbres enveloppantes des algèbres de Lie (avec A.A. Kirillov)
3. On the structure of the field of quotients of the enveloping algebra of a semisimple Lie algebra (with A.A. Kirillov)
4. The structure of the Lie field connected with a split semisimple Lie algebra (with A.A. Kirillov)

Part VII. Finite-dimensional representations

1. Finite-dimensional representations of the group of unimodular matrices (with M.L. Tsetlin)
2. Finite-dimensional representations of the group of orthogonal matrices (with M.L. Tsetlin)
3. Finite-dimensional irreducible representations of the unitary and the full linear groups and related special functions (with M.I. Graev)

Part VIII. Indecomposable representations of semisimple Lie groups and of finite-dimensional algebras; problems of linear algebra

1. Indecomposable representations of the Lorentz group (with V.A. Ponomarev)
2. Remarks on the classification of a pair of commuting linear transformations in a finite-dimensional space (with V.A. Ponomarev)
3. The classification of the linear representations of the group $SL(2, C)$ (with M.I. Graev and V.A. Ponomarev)
4. Quadruples of subspaces of a finite-dimensional vector space (with V.A. Ponomarev)

5. Coxeter functors and Gabriel's theorem (with I.N. Bernshtejn and S.I. Gelfand)
6. Model algebras and representations of graphs (with V.A. Ponomarev)

Part IX. Representations of infinite-dimensional groups

1. Representations of the group $SL(2, R)$, where R is a ring of functions (with M.I. Graev and A.M. Vershik)
2. Representations of the group of smooth mappings of a manifold X into a compact Lie group (with M.I. Graev and A.M. Vershik)
3. Irreducible representations of the group G^x and cohomologies (with M.I. Graev and A.M. Vershik)
4. Representations of the group of diffeomorphisms (with M.I. Graev and A.M. Vershik)
5. Representations of the group of functions taking values in a compact Lie group (with M.I. Graev and A.M. Vershik)
6. A commutative model of representation of the group of flows $SL(2, R)^x$ that is connected with a unipotent subgroup (with M.I. Graev and A.M. Vershik)

Volume III

Part I. Integral geometry

1. Integral transformations connected with straight line complexes in a complex affine space (with M.I. Graev)
2. Integral geometry on k-dimensional planes (with M.I. Graev and Z.Ya. Shapiro)
3. Complexes of k-dimensional planes in the space $\mathbf{C}^n$ and Plancherel's formula for the group $GL(n, C)$ (with M.I. Graev)
4. Complexes of straight lines in the space $\mathbf{C}^n$ (with M.I. Graev)
5. Differential forms and integral geometry (with M.I. Graev and Z.Ya. Shapiro)
6. Integral geometry in projective space (with M.I. Graev and Z.Ya. Shapiro)
7. A problem of integral geometry connected with a pair of Grassmann manifolds (with M.I. Graev and Z.Ya. Shapiro)
8. Nonlocal inversion formulas in real integral geometry (with S.G. Gindikin)
9. A local problem of integral geometry in a space of curves (with S.G. Gindikin and Z.Ya. Shapiro)
10. A problem of integral geometry in RP^n connected with the integration of differential forms (with S.G. Gindikin and M.I. Graev)
11. Integral geometry in affine and projective spaces (with S.G. Gindikin and M.I. Graev)

12. Geometric structures of double bundles and their relation to certain problems in integral geometry (with G.S. Shmelev)
13. The problem of integral geometry and intertwining operators for a pair of real Grassmannian manifolds (with M.I. Graev and R. Rosu)

Part II. Cohomology and characteristic classes

1. On classifying spaces for principal fiberings with Hausdorff bases (with D.B. Fuks)
2. Topology of noncompact Lie groups (with D.B. Fuks)
3. The cohomologies of the Lie algebra of the vector fields in a circle (with D.B. Fuks)
4. Cohomologies of the Lie algebra of tangential vector fields of a smooth manifold (with D.B. Fuks)
5. Cohomology of the Lie algebra of formal vector fields (with D.B. Fuks)
6. Cohomology of Lie algebra of tangential vector fields II (with D.B. Fuks)
7. Cohomology of Lie algebra of vector fields with nontrivial coefficients (with D.B. Fuks)
8. Quadratic forms over commutative group rings and the *K*-theory (with A.S. Mishchenko)
9. Upper bounds for cohomology of infinite-dimensional Lie algebras (with D.B. Fuks)
10. The action of infinite-dimensional Lie algebras (with D.B. Fuks and D.A. Kazhdan)
11. Cohomology of the Lie algebra of Hamiltonian formal vector fields (with D.B. Fuks and D.I. Kalinin)
12. *PL*-foliations (with D.B. Fuks)
13. Cohomology of the Lie algebra of formal vector fields with coefficients in its adjoint space and variations of characteristic classes of foliations (with B.L. Feigin and D.B. Fuks)
14. The Gauss-Bonnet theorem and the Atiyah-Patodi-Singer functionals for the characteristic classes of foliations (with D.B. Fuks and A.M. Gabrielov)
15. *PL*-foliations II (with D.B. Fuks)
16. Combinatorial calculus of characteristic classes (with A.M. Gabrielov and M.V. Losik)
17. Combinatorial calculus of characteristic classes (with A.M. Gabrielov and M.V. Losik)
18. A local combinatorial formula for the first class of Pontryagin (with A.M. Gabrielov and M.V. Losik)
19. Atiyah-Patodi-Singer functionals for characteristic functionals for tangent bundles (with A.M. Gabrielov and M.V. Losik)
20. Calculation of characteristic classes of combinatorial vector bundles (with M.V. Losik)
21. Cohomology of infinite-dimensional Lie algebras and Laplace operators (with B.L. Feigin and D.B. Fuks)
22. Geometry in Grassmannians and a generalization of the dilogarithm (with R. MacPherson)

Part III. Functional integration; probability; information theory

1. Generalized random processes
2. Integration in functional spaces and its applications to quantum physics (with A.M. Yaglom)
3. On the general definition of the amount of information (with A.N. Kolmogorov and A.M. Yaglom)
4. On the computation of the amount of information about a random function contained in another such function (with A.M. Yaglom)

Part IV. Mathematics of computation; cybernetics; biology

1. On the application of random tests (the Monte Carlo method) for the solution of the kinetic equation (with N.N. Chentsov, S.M. Fejnberg, and A.S. Frolov)
2. On difference-schemes for the solution of the heat conduction equation. The 'double sweep' method for the solution of difference equations (with O.V. Lokytsievskij)
3. Some methods of control for complex systems (with M.L. Tsetlin)
4. Determination of crystal structure by the method of non-local search (with Yu.G. Fedorov and I.I. Piatetski-Shapiro)
5. The determination of crystal structures by the method of *R*-factor minimization (with Yu.G. Fedorov, R.A. Kayushina, and B.K. Vainshtejn)
6. On certain classes of games and automata games (with I.I. Piatetski-Shapiro and M.L. Tsetlin)
7. Some questions in the investigation of movement (with V.S. Gurfinkel and M.L. Tsetlin)
8. Recording of neurons of the dorsal spinocerebellar tract during evoked locomotion (with Yu.I. Arshavskij, M.B. Berkenblit, O.I. Fukson, and G.N. Orlovskij)
9. Origin of modulation in neurons of the vertral spinocerebellar tract during locomotion (with Yu.I. Arshavskij, M.B. Berkenblit, O.I. Fukson, and G.N. Orlovskij)
10. Generation of scratching 1. Activity of spinal interneurons during scratching (with M.B. Berkinblit, T.G. Delyagina, A.G. Fe'ldman, and G.N. Orlovskij)
11. The cerebellum and control of rhythmical movements (with Yu.I. Arshavskij and G.N. Orlovskij)
12. Interrelationships of contacting cells in the cell complexes of mouse ascites hepatoma (with V.I. Guelshtejn, A.G. Malenkov, and Yu.M. Vasil'ev)
13. Initiation of DNA synthesis in cell cultures by colcemid (with V.I. Guelshtejn and Yu.M. Vasil'ev)
14. Mechanisms of morphogenesis in cell cultures (with Yu.M. Vasil'ev)
15. Possible common mechanism of morphological and growth-related alterations accompanying neoplastic transformation (with Yu.M. Vasil'ev)

Bibliography

The bibliography in these *Collected Papers* is a revised and updated version of the "List of I.M. Gelfand's Publications" published in the birthday addresses which can be found at the end of this volume. Hence the numbering of this Bibliography does not match that of these original 'Lists'.

The articles and monographs are listed in chronological order. The numbers in the right-hand column indicate where an article can be found in these Collected Papers, for example I.II.1 means volume I, part II, article 1.

From 1933–1947 the *Doklady* were published both in Russian and in a foreign language edition entitled *Comptes Rendus (Doklady) de l'Académie de l'URSS.*

Articles marked with an asterisk (*) were originally published in Russian and translated especially for this publication.

References to the reviews published in Mathematical Reviews (MR) and Zentralblatt für Mathematik (Zbl.) have been given as far as could be ascertained.

1936

1. Sur un lemme de la théorie des espaces linéaires. Izv. Nauchno-Issled. Inst. Mat. Khar'kov Univ., Ser. 4, **13** (1936) 35–40. Zbl. **14**:162 I.II.1

1937

2. Zur Theorie abstrakter Funktionen. Dokl. Akad. Nauk SSSR **17** (1937) 243–245. Zbl. **18**:71
3. Operatoren und abstrakte Funktionen. Dokl. Akad. Nauk SSSR **17** (1937) 245–248. Zbl. **18**:72

1938

4. Abstrakte Funktionen und lineare Operatoren. Mat. Sb., Nov. Ser. **4** (46) (1938) 235–284. Zbl. **20**:367 I.II.2

1939

5. (with A.N. Kolmogorov) On rings of continuous functions. Dokl. Akad. Nauk SSSR **22** (1939) 11–15. Zbl. **21**:411

6. On normed rings. Dokl. Akad. Nauk SSSR **23** (1939) 430–432. Zbl. **21**:294 I.II.4

7. To the theory of normed rings. II. On absolutely convergent trigonometrical series and integrals. Dokl. Akad. Nauk SSSR **25** (1939) 570–572. Zbl. **22**:357 I.II.5

8. To the theory of normed rings. III. On the ring of almost periodic functions. Dokl. Akad. Nauk SSSR **25** (1939) 573–574. Zbl. **22**:357 I.II.6

9. On one-parametrical groups of operators in a normed space. Dokl. Akad. Nauk SSSR **25** (1939) 713–718. Zbl. **22**:358 I.II.3

1940

10. (with D.A. Rajkov) On the theory of characters of commutative topological groups. Dokl. Akad. Nauk SSSR **28** (1940) 195–198. Zbl. **24**:120 I.II.7

11. Normierte Ringe. Mat. Sb., Nov. Ser. **9** (51) (1941) 3–23. Zbl. **24**:320 I.II.8

12. (with G.E. Shilov) Über verschiedene Methoden der Einführung der Topologie in die Menge der maximalen Ideale eines normierten Ringes. Mat. Sb., Nov. Ser. **9** (51) (1941) 25–38. Zbl. **24**:321 I.II.9

13. Ideale und primäre Ideale in normierten Ringen. Mat. Sb., Nov. Ser. **9** (51) (1941) 41–48. Zbl. **24**:322 I.II.10

14. Zur Theorie der Charaktere der Abelschen topologischen Gruppen. Mat. Sb., Nov. Ser. **9** (51) (1941) 49–50. Zbl. **24**:323 I.II.11

15. Über absolut konvergente trigonometrische Reihen und Integrale. Mat. Sb., Nov. Ser. **9** (51) (1941) 51–66. Zbl. **24**:323 I.II.12

1942

16. (with M.A. Najmark) On the embedding of normed rings into the ring of operators in Hilbert space. Mat. Sb., Nov. Ser. **12** (54) (1942) 197–213. Zbl. **60**:270 I.II.13

17. (with D.A. Rajkov) Irreducible unitary representations of locally bicompact groups. Mat. Sb. 13 (55), 301–316 (1942) [Transl., II. Ser., Am. Math. Soc. **36** (1964) 1–15]. Zbl. **166**:401 II.I.1

1943

18. (with D.A. Rajkov) Irreducible unitary representations of locally compact groups. Dokl. Akad. Nauk SSSR **42** (1943) 199–201. Zbl. **61**:253

1946

19. (with M.A. Najmark) Unitary representations of the Lorentz group. J. Phys., Acad. Sci. USSR **10** (1946) 93–94. Zbl. **61**:253 II.II.1

20. (with M.A. Najmark) On unitary representations of the complex unimodular group. Dokl. Akad. Nauk SSSR **54** (1946) 195–198. Zbl. **29**:5 II.II.3

21. (with D.A. Rajkov and G.E. Shilov) Commutative normed rings. Usp. Mat. Nauk **1**, 2 (1946) 48–146 [Transl., II. Ser., Am. Math. Soc. 5 (1957) 115–220]. Zbl. **201**:457 I.II.14
22. (with M.A. Najmark) Unitary representations of semisimple Lie groups I. Mat. Sb., Nov. Ser. **21** (63) (1946) 405–434 [Transl., II. Ser., Am. Math. Soc. (reprinted) **9** (1962) 1–14]. Zbl. **38**:17
*23. (with M.A. Najmark) Complementary and degenerate series of representations of the complex unimodular group. Dokl. Akad. Nauk SSSR **58** (1946) 1577–1580. Zbl. **37**:304 II.II.5

1947

24. (with M.A. Najmark) Unitary representations of the group of linear transformations of the straight line. Dokl. Akad. Nauk SSSR **55** (1947) 567–570. Zbl. **29**:5 II.I.2
25. (with M.A. Najmark) The principal series of irreducible representations of the complex unimodular group. Dokl. Akad. Nauk SSSR **56** (1947) 3–4. Zbl. **29**:5 II.II.4
*26. (with M.A. Najmark) Unitary representations of the Lorentz group. Izv. Akad. Nauk SSSR, Ser. Mat. **11** (1947) 411–504. Zbl. **37**:153 II.II.2

1948

27. Lectures on linear algebra. Moscow-Leningrad: Gostekhizdat 1948 (in Russian). Zbl. **38**:156
28. (with M.A. Najmark) Normed rings with an involution and their representations. Izv. Akad. Nauk SSSR, Ser. Mat. **12**, 445–480 (1948) (English translation in: Commutative normed rings, I.M. Gelfand, D.A. Rajkov and G.E. Shilov, pp. 240–274. Chelsea 1964). Zbl. **31**:34 I.II.15
29. (with A.M. Yaglom) General relativistically invariant equations and infinite-dimensional representations of the Lorentz group. Dokl. Akad. Nauk SSSR **59** (1948) 655–659 (in Russian). Zbl. **37**:127
30. Integral equations. Article in the Great Soviet Encyclopedia (1948) (in Russian)
*31. (with A.M. Yaglom) General relativistically invariant equations and infinite-dimensional representations of the Lorentz group. Zh. Ehksp. Teor. Fiz. **18** (1948) 703–733 II.II.9
*32. (with M.A. Najmark) The trace in principal and complementary series representations of the complex unimodular group. Dokl. Akad. Nauk SSSR **61** (1948) 9–11. Zbl. **35**:299 II.II.6
33. (with A.M. Yaglom) Relativistically invariant equations corresponding to a definite charge and a definite energy. Dokl. Akad. Nauk SSSR **63** (1948) 371–374 (in Russian). Zbl. **31**:95

34. (with A.M. Yaglom) Pauli's theorem for general relativistically invariant equations. Zh. Ehksp. Teor. Fiz. **18** (1948) 1096–1104 (in Russian)

35. (with A.M. Yaglom) Charge conjugacy for general relativistically invariant equations. Zh. Ehksp. Teor. Fiz. **18** (1948) 1105–1111 (in Russian)

*36. (with M.A. Najmark) On the connection between representations of complex semisimple Lie groups and its maximal compact subgroup. Dokl. Akad. Nauk SSSR **63** (1948) 225–228. Zbl. **35**:15 II.II.7

*37. (with M.A. Najmark) An analogue of the Plancherel formula for the complex unimodular group. Dokl. Akad. Nauk SSSR **63** (1948) 609–612. Zbl. **38**:18 II.II.8

1950

*38. The centre of the infinitesimal group ring. Mat. Sb., Nov. Ser. **26** (68) (1950) 103–112. Zbl. **35**:300 II.I.3

39. (with M.A. Najmark) The connexion between the unitary representations of the complex unimodular group and those of its unitary subgroup. Izv. Akad. Nauk SSSR, Ser. Mat. **14** (1950) 239–260 (in Russian). Zbl. **37**:15

40. Spherical functions on symmetric Riemannian spaces. Dokl. Akad. Nauk SSSR 70, (1950) 5–8 [Transl., II. Ser., Am. Math. Soc. **37** (1964) 39–43]. Zbl. **38**:274 II.I.4

*41. (with M.L. Tsetlin) Finite-dimensional representations of the group of unimodular matrices. Dokl. Akad. Nauk SSSR **71** (1950) 825–828. Zbl. **37**:153 II.VII.1

*42. (with M.L. Tsetlin) Finite-dimensional representations of the group of orthogonal matrices. Dokl. Akad. Nauk SSSR **71** (1950) 1017–1020. Zbl. **37**:153 II.VII.2

*43. Eigenfunction expansions for equations with periodic coefficients. Dokl. Akad. Nauk SSSR **73** (1950) 1117–1120. Zbl. **37**:345 I.III.1

44. (with M.A. Najmark) Unitary representations of classical groups. Tr. Mat. Inst. Steklova **36** (1950) 1–288 (in Russian). Zbl. **41**:362

(English transl. of the Introduction, Chap. 9 'Spherical functions' and Chap 18 'Transitivity classes for the set of pairs. Another way of describing representations of the complementary series' can be found in volume II) II.II.11

1951

45. (with S.V. Fomin) Unitary representations of Lie groups and geodesic flows on surfaces of constant negative curvature. Dokl. Akad. Nauk SSSR **76** (1951) 771–774 (in Russian). Zbl. **45**:388

46. Lectures on linear algebra, 2nd ed. Moscow-Leningrad: Gostekhizdat 1951 (English transl.: New York: Interscience 1961). Zbl. **98**:11

47. (with B.M. Levitan) On the determination of a differential equation from its spectral function. Izv. Akad. Nauk SSSR, Ser. Mat. **15** (1951) 309–361 [Transl., II. Ser., Am. Math. Soc. **1** (1955) 253–304]. Zbl. **44**:93 I.III.2

48. Remark on N.K. Bari's paper 'Biorthogonal system and bases in a Hilbert space'. Uch. Zap. Mosk. Gos. Univ., Ser. Mat. **140** (4) (1951) 224–225 (in Russian)

1952

49. (with Z. Ya. Shapiro) Representations of the group of rotations of 3-dimensional space and their applications. Usp. Mat. Nauk **7** (1) (1952) 3–117 (in Russian). Zbl. **49**:157

50. (with S.V. Fomin) Geodesic flows on manifolds of constant negative curvature. Usp. Mat. Nauk **7** (1) (1952) 118–137 [Transl., II. Ser., Am. Math. Soc. **1** (1955) 49–65]. Zbl. **66**:361 II.III.2

51. (with M.A. Najmark) Unitary representations of the unimodular group that contain an identity representation of the unitary subgroup. Tr. Mosk. Mat. O.-va **1** (1952) 423–473 (in Russian). Zbl. **49**:358

*52. (with M.I. Graev) Unitary representations of the real simple Lie groups. Dokl. Akad. Nauk SSSR **86** (1952) 461–463. Zbl. **49**:358 II.II.12

53. On the spectrum of non-selfadjoint differential operators. Usp. Mat. Nauk **7** (6) (1952) 183–184 (in Russian). Zbl. **48**:96

1953

*54. (with B.M. Levitan) On a simple identity for eigenvalues of second order differential operator. Dokl. Akad. Nauk SSSR **88** (1953) 593–596. Zbl. **53**:60 I.III.3

*55. (with M.I. Graev) Unitary representations of real unimodular groups (Principal non-degenerate series). Izv. Akad. Nauk SSSR, Ser. Mat. **17** (1952) 189–249. Zbl. **52**:341 II.II.13

*56. (with M.I. Graev) On a general method for decomposition of the regular representation of a Lie group into irreducible representations. Dokl. Akad. Nauk SSSR **92** (1952) 221–224. Zbl. **53**:15 II.II.14

57. (with M.I. Graev) The analogue of Plancherel's theorem for real unimodular groups. Dokl. Akad. Nauk SSSR **92** (1952) 461–464 (in Russian). Zbl. **53**:15

58. (with G.E. Shilov) Fourier transforms of rapidly increasing functions and questions of the uniqueness of the solution of Cauchy's problem. Usp. Mat. Nauk **8** (6) (1952) 3–54 [Transl., II. Ser., Am. Math. Soc. **5** (1957) 221–274]. Zbl. **52**:116

1954

*59. (with R.A. Minlos) Solution of quantum field equations. Dokl. Akad. Nauk SSSR **97** (1954) 209–212. Zbl. **58**:232 I.III.4

60. (with V.B. Lidskij) On the structure of the regions of stability of linear canonical systems of differential equations with periodic coefficients. Usp. Mat. Nauk **10** (1) (1955) 3–40 [Transl., II. Ser., Am. Math. Soc. **8** (1958) 143–181]. Zbl. **64**:89 I.III.5

*61. Generalized random processes. Dokl. Akad. Nauk SSSR **100** (1955) 853–856. Zbl. **68**:112 III.III.1

62. (with M.I. Graev) The traces of unitary representations of the real unimodular group. Dokl. Akad. Nauk SSSR **100** (1955) 1037–1040 (in Russian). Zbl. **64**:111

63. (with M.I. Graev) The analogue of Plancherel's formula for the classical groups. Tr. Mosk. Mat. O.-va **4** (1955) 375–404 [Transl., II. Ser., Am. Math. Soc. **9** (1958) 123–154. Zbl. **66**:20

64. (with Z.Ya. Shapiro) Homogeneous functions and their applications. Usp. Mat. Nauk **10** (3) (1955) 3–70 [Transl., II. Ser., Am. Math. Soc. **8** (1958) 21–85]. Zbl. **65**:101

65. (with G.E. Shilov) On a new method in theorems concerning the uniqueness of the solution of Cauchy's problem for systems of linear partial differential equations. Dokl. Akad. Nauk SSSR **102** (1955) 1065–1068 (in Russian). Zbl. **67**:72

1956

*66. (with A.G. Kostyuchenko) Eigenfunction expansions for differential and other operators. Dokl. Akad. Nauk SSSR **103** (1955) 349–352. Zbl. **65**:104 I.III.6

67. (with A.M. Yaglom) Integration in functional spaces and its applications to quantum physics. Usp. Mat. Nauk **11** (1) (1956) 77–114 [J. Math. Phys. **1** (1960) 48–69]. Zbl. **92**:451 III.III.2

*68. On identities for eigenvalues of a second order differential operator. Usp. Mat. Nauk **11** (1) (1956) 191–198. Zbl. **70**:83 I.III.7

69. (with F.A. Berezin) Some remarks on the theory of spherical functions on symmetric Riemannian manifolds. Tr. Mosk. Mat. O.-va **5** (1956) 311–351 [Transl., II. Ser., Am. Math. Soc. **21** (1962) 193–238]. Zbl. **72**:17 II.III.1

70. (with G.E. Shilov) Quelques applications de la théorie des fonctions généralisées. J. Math. Pures et Appl., IX. Ser. **35** (1956) 383–413. Zbl. **75**:285

71. (with A.N. Kolmogorov and A.M. Yaglom) On the general definition of the amount of information. Dokl. Akad. Nauk SSSR **111** (1956) 745–748 (German transl. in: Arbeiten zur Informationstheorie II, Mathematische Forschungsberichte. Berlin: VEB Deutscher Verlag der Wissenschaften 1958). Zbl. **71**:345 III.III.3

72. On some problems of functional analysis. Usp. Mat. Nauk **11** (6) (1956) 3–12 [Transl., II. Ser., Am. Math. Soc. **16** (1960) 315–324]. Zbl. **100**: 321 I.I.2

73. (with F.A. Berezin, M.I. Graev, and M.A. Najmark) Group representations. Usp. Mat. Nauk **11** (6) (1956) 13–40 [Transl., II. Ser., Am.Math. Soc. **16** (1960) 325–353]. Zbl. **74**: 103

74. (with N.N. Chentsov) On the numerical evaluation of continuous integrals. Zh. Ehksp. Teor. Fiz. **31** (1956) 1106–1107 (in Russian)

75. (with M.L. Tsetlin) On quantities with anomalous parity. Zh. Ehksp. Teor. Fiz. **31** (1956) 1107–1109 (in Russian)

1957

76. (with A.M. Yaglom) On the computation of the amount of information about a random function contained in another such function. Usp. Mat. Nauk **12** (1) (1957) 3–52 (German transl. in: Arbeiten zur Informationstheorie II, Mathematische Forschungsberichte. Berlin: VEB Deutscher Verlag der Wissenschaften 1958). Zbl. **78**: 322 III.III.4

77. On the subrings of the ring of continuous functions. Usp. Mat. Nauk **12** (1) (1957) 247–251 [Transl., II. Ser., Am. Math. Soc. **16** (1957) 477–479]. Zbl. **100**: 322

78. Some aspects of functional analysis and algebra. Proc. Int. Congr. Math. 1954, Amsterdam **1** (1957) 253–276. Zbl. **79**: 326 I.I.1

1958

79. (with G.E. Shilov) Generalized functions 1. Properties and operations. Moscow: Fizmatgiz 1958 (English transl.: New York: Academic Press 1964; German transl.: Berlin: VEB Deutscher Verlag der Wissenschaften 1960; French transl.: Paris: Dunod 1962). Zbl. **91**: 111

80. (with G.E. Shilov) Generalized functions 2. Spaces of fundamental and generalized functions. Moscow: Fizmatgiz 1958 (English transl.: New York: Academic Press 1968; German transl.: Berlin: VEB Deutscher Verlag der Wissenschaften 1960; French transl.: Paris: Dunod 1964). Zbl. **91**: 111

81. (with G.E. Shilov) Generalized functions 3. Theory of differential equations. Moscow: Fizmatgiz 1958 (English transl.: New York: Academic Press 1967; German transl.: Berlin: VEB Deutscher Verlag der Wissenschaften 1964; French transl.: Paris: Dunod 1964). Zbl. **91**: 111

82. (with K.I. Babenko) Some observations on hyperbolic systems. Nauchn. Dokl. Vyssh. Shk. **1** (1958) 12–18 (in Russian). Zbl. **144**: 139

83. (with R.A. Minlos and Z.Ya. Shapiro) Representations of the rotation and Lorentz groups. Moscow: Fizmatgiz 1958 (English transl.: London: Pergamon Press 1963). Zbl. **108**:220
84. (with R.A. Minlos and A.M. Yaglom) Path integrals. Proceedings of the Third All-Union Mathematics Congress **3** (1958) 521–531 (in Russian)
85. (with F.A. Berezin, M.I. Graev and M.A. Najmark) Representations of Lie groups. Proceedings of the Third All-Union Mathematics Congress **3** (1958) 246–254 (in Russian). Zbl. **97**:109
86. (with A.N. Kolmogorov and A.M. Yaglom) The amount of information and entropy for continuous distributions. Proceedings of the Third All-Union Mathematics Congress **3** (1958) 300–320 (in Russian). Zbl. **92**:340
87. (with I.G. Petrovskij and G.E. Shilov) The theory of systems of partial differential equations. Proceedings of the Third All-Union Mathematics Congress **3** (1958) 65–72 (in Russian). Zbl. **107**:74
88. (with S.I. Braginskij and R.P. Fedorenko) The theory of the compression and pulsation of a plasma column under a powerful pulse discharge. Fiz. Plazmy Probl. Upr. Termoyad. Reakts. **4** (1958) 201–222 (in Russian)
89. (with N.N. Chentsov and A.S. Frolov) The computation of continuous integrals by the Monte Carlo method. Izv. Vyssh. Uchebn. Zaved., Mat. **5** (6) (1958) 32–45 (in Russian). Zbl. **139**:323

1959

90. Some problems in the theory of quasilinear equations. Usp. Mat. Nauk **14** (2) (1959) 87–158 [Transl., II. Ser., Am. Math. Soc. **29** (1963) 295–381]. Zbl. **96**:66 I.III.8
91. (with I.I. Piatetski-Shapiro) Theory of representations and theory of automorphic functions. Usp. Mat. Nauk **14** (2) (1959) 171–194 [Transl., II. Ser., Am. Math. Soc. **26** (1963) 173–200]. Zbl. **121**:306
92. Some questions of analysis and differential equations. Usp. Mat. Nauk **14** (3) (1959) 3–19 [Transl., II. Ser., Am. Math. Soc. **26** (1963) 201–219]. Zbl. **91**:88 I.I.3
93. (with M.I. Graev) Geometry of homogeneous spaces, representations of groups in homogeneous spaces and related questions of integral geometry. Tr. Mosk. Mat. O.-va **88** (1959) 321–390 [Transl., II. Ser., Am. Math. Soc. **37** (1964) 351–429]. Zbl. **136**:434 II.III.3
94. (with M.I. Graev) The decomposition into irreducible components of representations of the Lorentz group in the spaces

of functions defined on symmetric spaces. Dokl. Akad. Nauk SSSR **127** (1959) 250–253 (in Russian). Zbl. **99**:321

*95. (with I.I. Piatetski-Shapiro) On a theorem of Poincaré. Dokl. Akad. Nauk SSSR **127** (3) (1959) 490–493. Zbl. **107**:171 I.III.9

*96. On the structure of the ring of rapidly decreasing functions on a Lie group. Dokl. Akad. Nauk SSSR **124** (1959) 19–21. Zbl. **103**:336 II.II.10

1960

97. (with N.N. Chentsov, S.M. Fejnberg, and A.S. Frolov) On the application of random tests (the Monte Carlo method) for the solution of the kinetic equation. Proceedings of the Second International Conference on the Peaceful Use of Atomic Energy. (Geneva, 1958), **2** (1960) 628–683 III.IV.1

98. (with Sya Do-Shin) On positive definite distributions. Usp. Mat. Nauk **15** (1) (1960) 185–190 (in Russian). Zbl. **97**:314

99. Integral geometry and its relation to the theory of group representations. Usp. Mat. Nauk **15** (2) (1960) 155–164 [Russ. Math. Surv. **15** (2) (1960) 143–151]. Zbl. **119**:177 I.I.4

100. (with M.I. Graev) Fourier transforms of rapidly decreasing functions on complex semisimple groups. Dokl. Akad. Nauk SSSR **131** (1960) 496–499 (in Russian). Zbl. **103**:337

101. (with M.L. Tsetlin) On continuous models of control systems. Dokl. Akad. Nauk SSSR **131** (1960) 1242–1245 (in Russian)

102. On elliptic equations. Usp. Mat. Nauk **15** (3) (1960) 121–132 [Russ. Math. Surv. **15** (1960) 113–123]. Zbl. **95**:78 I.I.5

103. On a paper by K. Hoffmann and I.M. Singer. Usp. Mat. Nauk **15** (3) (1960) 239–240 (in Russian). Zbl. **154**:386

104. (with D.A. Rajkov and G.E. Shilov) Commutative normed rings. Moscow: Fizmatgiz 1960 (English transl.: New York: Chelsea 1964; German transl.: Berlin: VEB Deutscher Verlag der Wissenschaften 1964; French transl.: Paris: Gauthier-Villars 1964). Zbl. **134**:321

105. (with M.I. Graev) Integrals over hyperplanes of fundamental and generalized functions. Dokl. Akad. Nauk SSSR **135** (1960) 1307–1310 [Sov. Math., Dokl. **1** (1960) 1369–1372]. Zbl. **108**:296

1961

106. (with N.Ya. Vilenkin) Generalized functions 4. Applications of harmonic analysis. Moscow: Fizmatgiz 1961 (English transl.: New York: Academic Press 1964; German transl.: Berlin: VEB Deutscher Verlag der Wissenschaften 1960; French transl.: Paris: Dunod 1967). Zbl. **136**:112

107. (with S.V. Fomin) Calculus of Variations. Moscow: Fizmatgiz (1961) (English transl.: Englewood Cliffs, N.J.: Prentice Hall 1963). Zbl. **127**:54

108. (with V.A. Borovikov, A.F. Grashin, and I.Ya. Pomeranchuk) Phase-shift analysis of *pp*-scattering at 95 *Mev*. Zh. Ehksp. Teor. Fiz. **40** (1961) 1106–111 [Soviet Physics **13** (4) (1961) 780–784]

109. (with A.F. Grashin and L.N. Ivanova) Phase-shift analysis of *pp*-scattering at an energy of 150 Mev. Zh. Ehksp. Teor. Fiz. **40** (5) (1961) 1338–1342 (in Russian)

110. (with V.S. Gurfinkel and M.L. Tsetlin) Some considerations on the tactics of making movements. Dokl. Akad. Nauk SSSR **139** (1961) 1250–1253 (in Russian)

111. (with M.I. Graev et al.) Magnetic surfaces of the 3-path helical magnetic field excited by a crimped field. Zh. Tekh. Fiz. **31** (1961) 1164–1169 (in Russian)

112. (with M.I. Graev) Integral transformations connected with straight line complexes in a complex affine space. Dokl. Akad. Nauk SSSR **138** (1961) 1266–1269 [Sov. Math., Dokl. **2** (1961) 809–812]. Zbl. **109**:151 III.I.1

113. (with M.L. Tsetlin) The principle of non-local search in automatic optimization systems. Dokl. Akad. Nauk SSSR **137** (1961) 295–298 [Sov. Phys., Dokl. **6** (1961) 192–194]

1962

114. (with M.L. Tsetlin) Some methods of control for complex systems. Usp. Mat. Nauk **17** (1) (1962) 3–25 [Russ. Math. Surv. **17** (1) (1962) 95–117]. Zbl. **107**:299 III.IV.3

115. (with A.I. Morozov, N.M. Zueva et al.) An example of the theoretical determination of a magnetic field that does not have magnetic surfaces. Dokl. Akad. Nauk SSSR **143** (1962) 81–83 (in Russian)

116. (with M.I. Graev) An application of the horysphere method to the spectral analysis of functions in real and imaginary Lobachevskii space. Tr. Mosk. Mat. O.-va **11** (1962) 243–308 (in Russian). Zbl. **176**:443

117. (with M.I. Graev and N.Ya. Vilenkin) Generalized functions 5. Integral geometry and representation theory. Moscow: Fizmatgiz 1962 (English transl.: New York: Academic Press 1966; French transl.: Paris: Dunod 1970). Zbl. **115**:167

118. (with M.I. Graev) Categories of group representations and the problem of classifying irreducible representations. Dokl. Akad. Nauk SSSR **146** (1962) 757–760 [Sov. Math., Dokl. **3** (1962) 1378–1381] II.IV.1

119. (with I.I. Piatetski-Shapiro) Unitary representations in homogeneous spaces with discrete stationary groups. Dokl. Akad. Nauk SSSR **147** (1962) 17–20 [Sov. Math., Dokl. **3** (1962) 1528–1531]. Zbl. **119**:270 II.III.4

120. (with I.I. Patetski-Shapiro) Unitary representations in a space G/Γ, where G is a group of n-by-n matrices and Γ is a subgroup of integer matrices. Dokl. Akad. Nauk SSSR **147** (1962) 275–278 [Sov. Math., Dokl. **3** (1962) 1574–1577]. Zbl. **119**:271 II.III.5

121. (with M.I. Graev) Construction of irreducible representations of simple algebraic groups over a finite field. Dokl. Akad. Nauk SSSR **147** (1962) 529–532 [Sov. Math., Dokl. **3** (1962) 1646–1649]. Zbl. **119**:269 II.IV.2

122. (with V.S. Gurfinkel and M.L. Tsetlin) On the techniques of control of complex systems and their relation to physiology. Symposium 'Biological aspects of cybernetics', pp. 66–73. Moscow: Publ. Akad. Nauk SSSR (1962) [Transl., II. Ser., Am. Math. Soc. **111** (1978) 213–219]

*123. (with O.V. Lokytsievskij) On difference-schemes for the solution of the heat conduction equation. The 'double sweep' method for the solution of difference equations. Supplements I and II to 'Introduction to the theory of difference-schemes' by S.K. Godunov and V.S. Ryaben'kij. Moscow: Fizmatgiz 1962. Zbl. **106**:319 III.IV.2

1963

124. (with L.M. Chailakhyan and S.A. Kovalev) Intracellular irritation of the various compartments of a frog's heart. Dokl. Akad. Nauk SSSR **148** (1963) 973–976 (in Russian)

125. (with M.I. Graev, A.I. Morozov et al.) On the structure of a toroidal magnetic field that does not have magnetic surfaces. Dokl. Akad. Nauk SSSR **148** (1963) 1286–1289 (in Russian)

126. (with M.I. Graev) Irreducible unitary representations of the group of unimodular second-order matrices with elements from a locally compact field. Dokl. Akad. Nauk SSSR **149** (1963) 499–502 [Sov. Math., Dokl. **4** (1963) 397–400]. Zbl. **119**:270 II.IV.5

127. (with I.I. Piatetski-Shapiro) Automorphic functions and the theory of representations. Tr. Mosk. Mat. O.-va **12** (1963) 389–412 [Trans. Mosc. Math. Soc. **12** (1965)]. Zbl. **136**:73

128. Automorphic functions and the theory of representations. Proc. Int. Congr. Math. Stockholm (1962) 74–85. Zbl. **138**:71 I.I.6

129. (with M.I. Graev) Plancherel's formula for the groups of the unimodular second order matrices with elements in a locally compact field. Dokl. Akad. Nauk SSSR **151** (1963) 262–264 [Sov. Math., Dokl. **4** (1963) 397–400]. Zbl. **204**:141 II.IV.6

130. (with M.I. Graev) Representations of a group of second order matrices with elements from a locally compact field and special functions on locally compact fields. Usp. Mat. Nauk **18** (4) (1963) 29–99 [Russ. Math. Surv. **18** (4) (1963) 29–100]. Zbl. **166**:402

131. (with A.Ya. Fridenshtejn) On the possible mechanism of change of immunological tolerance. Usp. Sov. Biol. **55** (1963) 428–429 (in Russian)

132. (with V.S. Gurfinkel, Ya.M. Kots, M.L. Shik, and M.L. Tsetlin) On the synchronization of motor units and some related ideas. Biofizika **8** (1963) 475–488 (in Russian)

133. (with Yu.G. Fedorov and I.I. Piatetski-Shapiro) Determination of crystal structure by the method of non-local search. Dokl. Akad. Nauk SSSR **152** (1963) 1045–1048 [Sov. Math. Dokl. **4** (1963) 1487–1490] III.IV.4

*134. (with Yu.G. Fedorov, R.A. Kayushina, and B.K. Vainshtejn) The determination of crystal structures by the method of *R*-factor minimization. Dokl. Akad. Nauk SSSR **153** (1963) 93–96 III.IV.5

135. (with I.I. Piatetski-Shapiro and M.L. Tsetlin) On certain classes of games and automata games. Dokl. Akad. Nauk SSSR **152** (1963) 845–848. [Transl., II. Ser., Am. Math. Soc. **87** (1970) 275–280]. MR **28**:1068. Zbl. **137**:143 III.IV.6

136. (with M.I. Graev) The structure of the ring of finite functions on the group of second-order unimodular matrices with elements from a disconnected locally compact field. Dokl. Akad. Nauk SSSR **153** (1963) 512–515 [Sov. Math. Dokl. **4** (1963) 1679–1700]. MR **33**:4183. Zbl. **199**:200

137. (with G.E. Shilov) Categories of finite-dimensional spaces. Vestn. Mosk. Univ., Ser. I. **4** (1963) 27–48 (in Russian). MR **28**:1223. Zbl. **161**:27

1964

138. (with V.I. Guelshtejn, A.G. Malenkov, and Yu.M. Vasil'ev) Characteristics of cell complexes of ascitic mouse hepatoma 22. Dokl. Akad. Nauk SSSR **156** (1964) 168–170 (in Russian)

139. (with M.I. Graev and I.I. Piatetski-Shapiro) Representations of adèle groups. Dokl. Akad. Nauk SSSR **156** (1964) 487–490. [Sov. Math., Dokl. (1964) 657–661]. MR **29**:2237. Zbl. **133**:294

140. (with V.S. Gurfinkel) Investigation of recognition activity. Biofizika **9** (1964) 710–717 (in Russian)

141. (with V.I. Bryzgalov, V.S. Gurfinkel, and M.L. Tsetlin) Homogeneous automata games and their simulation on digital

computers. Avtom. Telemekh. **25** (1964) 1572–1580 (in Russian). MR **30**:1897. Zbl. **141**:339

142. (with E.G. Glagoleva and A.A. Kirillov) The coordinate method. Moscow: Nauka 1964 (English transl.: New York: Gordon & Breach 1967; German transl.: Leipzig: Teubner 1968; Czech. Transl.: Bratislava: ALFA 1976)

1965

143. (with M.I. Graev) Finite-dimensional irreducible representations of the unitary and the full linear groups and related special functions. Izv. Akad. Nauk SSSR, Ser. Mat. **29** (1965) 1329–1356 [Transl., II. Ser., Am. Math. Soc. **64** (1965) 116–146]. MR **34**:1450. Zbl. **139**:307 II.VII.3
144. (with V.I. Guelshtejn, A.G. Malenkov, and Yu.M. Vasil'ev) Cell complexes in ascitic hepatomata of mice and rats, in the collection "Cell differentiation and induction mechanisms". Moscow: Nauka (1965) 220–232 (in Russian)
145. (with E.G. Glagoleva and E.E. Schnol) Functions and their graphs. Moscow: Nauka 1965 (English transl.: New York: Gordon & Breach 1967; German transl.: Leipzig: Teubner 1971). Zbl. **129**:267

1966

146. (with M.I. Graev and I.I. Piatetski-Shapiro) Theory of representations and automorphic functions. Moskau: Nauka 1966. (English transl.: Philadelphia London Toronto: Saunders 1969). MR **36**:3725. Zbl. **138**:72
147. (with A.A. Kirillov) Fields associated with enveloping algebras of Lie algebras. Dokl. Akad. Nauk SSSR **167** (1966) 503–505 [Sov. Math., Dokl. **7** (1966) 407–409]. Zbl. **149**:29 II.VI.1
148. (avec A.A. Kirillov) Sur les corps liés aux algèbres enveloppantes des algèbres de Lie., Publ. Math., Inst. Hautes Etud. Sci. **31** (1966) 509–523. MR **33**:7731 II.VI.2
149. (with V.I. Guelshtejn, A.G. Malenkov, and Yu.M. Vasil'ev) Local interactions of cells in cell complexes of ascitic hepatoma 22. Dokl. Akad. Nauk SSSR **167** (1966) 437–439 (in Russian)
150. (with V.I. Guelshtejn, A.G. Malenkov, and Yu.M. Vasil'ev) Interrelationships of contacting cells in the cell complexes of mouse ascites hepatoma. Int. J. Cancer **1** (1966) 451–462 III.IV.12
151. (with M.I. Graev and E.Ya. Shapiro) Integral geometry on a manifold of *k*-dimensional planes. Dokl. Akad. Nauk SSSR **168** (1966) 1236–1238 [Sov. Math., Dokl. **7** (1966) 801–804]. Zbl. **168**:201
152. (with L.V. Erofeeva, and Yu.M. Vasil'ev) The behaviour of fibroblasts of a cell culture on removal of part of the mono-

layer. Dokl. Akad. Nauk SSSR **171** (1966) 721–724 (in Russian)

153. (with M.L. Tsetlin) Mathematical simulation of mechanisms of the central nervous system. In the collection: Models of structure-functional organisation of certain biological systems, pp. 9–26. Moscow: Nauka 1966 (in Russian)

*154. (with V.S. Gurfinkel and M.L. Tsetlin) Some questions in the investigation of movement. In the collection: Models of structure-functional organisation of certain biological systems, pp. 264–276, Moscow: Nauka 1966 III.IV.7

155. (with Yu.G. Fedorov, S.L. Ginzburg, and E.B. Vul) The ravine method in problems of X-ray structure analysis, pp. 1–77, Moscow: Nauka 1966 (in Russian)

156. Lectures on linear algebra, pp. 1–280. Moscow: Nauka 1966 (in Russian). MR **34**:4274. Zbl. **158**:297

1967

157. (with M.I. Graev and Z.Ya. Shapiro) Integral geometry on *k*-dimensional planes. Funkts. Anal. Prilozh. **1** (1) (1967) 15–31 [Funct. Anal. Appl. **1** (1967) 14–27]. MR **35**:3620. Zbl. **164**:231 III.I.2

158. (with V.A. Ponomarev) Categories of Harish-Chandra models over the Lie algebra of the Lorentz group. Dokl. Akad. Nauk SSSR **176** (1967) 243–246 [Sov. Math., Dokl. **8** (1967) 1065–1068]. MR **36**:6552. Zbl. **241**:22025

159. (with V.A. Ponomarev) Classification of indecomposable infinitesimal representations of the Lorentz group. Dokl. Akad. Nauk SSSR **176** (1967) 502–505 [Sov. Math., Dokl. **8** (1967) 114–1117]. MR **36**:2739. Zbl. **246**:22013

160. (with D.B. Fuks) Cohomology of Lie groups with real coefficients. Dokl. Akad. Nauk SSSR **176** (1967) 24–27 [Sov. Math., Dokl. **8** (1967) 1031–1034]. MR **37**:2252a. Zbl. **169**:547

161. (with D.B. Fuks) Topology of noncompact Lie groups. Funkts. Anal. Prilozh. **1** (4) (1967) 33–45 [Funct. Anal. Appl. **1** (1967) 285–295]. MR **37**:2253. Zbl. **169**:547 III.II.2

162. (with M.I. Graev) Representations of the quaternion group over a disconnected locally compact continuous field. Dokl. Akad. Nauk SSSR **177** (1967) 17–20 [Sov. Math., Dokl. **8** (1967) 1346–1349]. MR **36**:2742. Zbl. **225**:22009

163. (with Yu.I. Arshavskij, M.B. Berkinblit, and V.S. Yakobson) Functional organization of afferent connections of Purkinje cells of the paramedian lobe of the cerebellum. Dokl. Akad. Nauk SSSR **177** (1967) 732–753 (in Russian)

164. (with D.B. Fuks) Topological invariants of non-compact Lie groups connected with infinite-dimensional representations.

Dokl. Akad. Nauk SSSR **177** (1967) 763–766 [Sov. Math., Dokl. **8** (1967) 1483–1486]. MR **37**:2252b. Zbl. **169**:548

165. (with V.S. Imshennik, L.G. Khazin, O.V. Lokytsievskij, V.S. Ryaben'kij, and N.M. Zueva) The theory of non-linear oscillation of electron plasma. Zh. Vychisl. Mat. Mat. Fiz. **7** (1967) 322–347 (in Russian). Zbl. **181**:575

166. (with M.I. Graev) Irreducible representations of the Lie algebras of the group $U(p, q)$. In the collection: High energy physics and the theory of elementary particles, pp. 216–226. Kiev: Naukova Dumka 1967 (in Russian). MR **37**:3814

1968

167. (with M.I. Graev) Complexes of k-dimensional planes in the space $\mathbf{C}^n$ and Plancherel's formula for the group $GL(n, \mathbf{C})$. Dokl. Akad. Nauk SSSR **179** (1968) 522–525 [Sov. Math., Dokl. **9** (1968) 394–398]. MR **37**:4764. Zbl. **198**:271 III.I.3

168. (with M.I. Graev) Complexes of straight lines in the space $\mathbf{C}^n$. Funkts. Anal. Prilozh. **2**(3) (1968) 39–52 [Funct. Anal. Appl. **2** (1968) 219–229]. MR **38**:6522. Zbl. **179**:509 III.I.4

169. (with V.A. Ponomarev) Indecomposable representations of the Lorentz group. Usp. Mat. Nauk **23** (2) 3–60 (1968) [Russ. Math. Surv. **23** (2) (1968) 1–58]. MR **38**:5325. Zbl. **236**:22012 II.VIII.1

170. (with Yu.M. Vasil'ev) Surface changes disturbing intracellular homeostasis as a factor inducing cell growth and division. Curr. Mod. Biol. **2** (1968) 43–55

171. (with A.A. Kirillov) On the structure of the field of quotients of the enveloping algebra of a semisimple Lie algebra. Dokl. Akad. Nauk SSSR **180** (1968) 775–777 [Sov. Math., Dokl. **9** (1968) 669–671]. MR **37**:5260. Zbl. **244**:17006 II.VI.3

172. (with D.B. Fuks) On classifying spaces for principal fiberings with Hausdorff bases. Dokl. Akad. Nauk SSSR **181** (1968) 515–518 [Sov. Math., Dokl. **9** (1968) 851–854]. MR **38**:716. Zbl. **181**:266 III.II.1

173. (with D.B. Fuks) The cohomologies of the Lie algebra of the vector fields in a circle. Funkts. Anal. Prilozh. **2** (4) (1968) 92–93 [Funct. Anal. Appl. **2** (1968) 342–343]. MR **39**:6348a. Zbl. **176**:115 III.II.3

174. (with Yu.M. Vasil'ev) Change of cellular surface – the basis of biological singularities of a tumor cell. Vestn. Akad. Med. Nauk SSSR **3** (1968) 45–49 (in Russian)

175. (with M.I. Graev) Representations of quaternion groups over locally compact and functional fields. Funkts. Anal. Prilozh. **2** (1) (1968) 20–35 [Funct. Anal. Appl. **2** (1968) 19–33]. MR **38**:4611. Zbl. **233**:20016 II.IV.3

176. (with L.V. Domnina, R.I. Rapoport, and Yu.E.M. Vasil'ev) Wound healing in cell cultures. Exp. Cell Res. **54** (1968) 83–93

177. (with A.B. Fel'dman, V.S. Gurfinkel, G.N. Orlovskij, F.V. Severin, and M.L. Shik) The control of certain types of movement. In the collection: Material of the international symposium IFAK on technical and biological problems of control (1968) (in Russian)

178. (with L.V. Erofeeva, O.Yu. Ivanova, I.L. Slavnaya, Yu.M. Vasil'ev, and A.A. Yaskovets) Factors controlling the proliferation of normal and tumour cells. In: Connective tissue in normal and pathological conditions, pp. 212–215. Novosibirsk: Nauka 1969 (in Russian)

1969

179. (with A.A. Kirillov) The structure of the Lie field connected with a split semisimple Lie algebra. Funkts. Anal. Prilozh. **3** (1) (1969) 7–26 [Funct. Anal. Appl. **3** (1969) 6–21]. MR **39**:2827. Zbl. **244**:17007 II.VI.4

180. (with M.I. Graev and Z.Ya. Shapiro) Differential forms and integral geometry. Funkts. Anal. Prilozh. **3** (2) (1969) 24–40 [Funct. Anal. Appl. **3** (1969) 101–114]. MR **39**:6232. Zbl. **191**:528 III.I.5

181. (with D.B. Fuks) Cohomology of the Lie algebra of vector fields on a manifold. Funkts. Anal. Prilozh. **3** (2) (1969) 87 [Funct. Anal. Appl. **3** (1969) 155]. MR **39**:6348 b. Zbl. **194**:246

182. (with D.B. Fuks) Cohomologies of the Lie algebra of tangential vector fields of a smooth manifold. Funkts. Anal. Prilozh. **3** (3) (1969) 32–52 [Funct. Anal. Appl. **3** (1969) 194–210]. MR **41**:1067. Zbl. **216**:203 III.II.4

183. (with A.S. Mishchenko) Quadratic forms over commutative group rings and the *K*-theory. Funkts. Anal. Prilozh. **3** (4) (1969) 28–33 [Funct. Anal. Appl. **3** (1969) 277–281]. MR **41**:9243. Zbl. **239**:55004 III.II.8

184. (with V.A. Ponomarev) Remarks on the classification of a pair of commuting linear transformations in a finite-dimensional space. Funkts. Anal. Prilozh. **3** (4) (1969) 81–82 [Funct. Anal. Appl. **3** (1969) 325–329]. MR **40**:7279 II.VIII.2

185. (with Yu.I. Arshavskij, M.B. Berkinblit, and V.S. Yakobson) Two types of granular cell in the cortex of the cerebellum. Nejrofiziologiya **1** (1969) 167–176 (in Russian)

186. (with E.K. Fetisova, V.I. Guelshtejn, and Yu.M. Vasil'ev) Stimulation of synthesis of DNA in mouse embryo fibroblast-like cells in vitro by factors of different character. Dokl. Akad. Nauk SSSR **187** (1969) 913–915 (in Russian)

187. (with Sh.A. Guberman, M.L. Shik, and Yu.M. Vasil'ev) Interaction in biological systems. Priroda **6**, 13–21; **7**, 24–33 (1969) (in Russian)

188. (with N.M. Chebotareva, Sh.A. Guberman, M.L. Izvekova, E.I. Kandel, T.V. Lebedeva, D.K. Luhev, and I.F. Nikolaeva) Prognostic matematic alevelutiei ictusorilor hemorogice in scepul preciazavii indicatiiler tratementului chirurgical. K. Accidouteler vasculare cerebrale, pp. 44–45, Bucuresti (1969)

189. (with N.M. Chebotareva, Sh.A. Guberman, M.L. Izvekova, E.I. Kandel, T.V. Lebedeva, D.K. Lunev, and I.F. Nikolaeva) Computer prognosis of spontaneous intracerebral haemorrhage for the purpose of its surgical treatment. IV. Int. Congr. Neurosurg., Excerpta Med. **32** (1969)

1970

190. (with M.I. Graev and Z.Ya. Shapiro) Integral geometry in projective space. Funkts. Anal. Prilozh. **4** (1) (1970) 14–32 [Funct. Anal. Appl. **4** (1970) 12–28]. MR **43**:6856 III.I.6

191. (with M.I. Graev and Z.Ya. Shapiro) A problem of integral geometry connected with a pair of Grassmann manifolds. Dokl. Akad. Nauk SSSR **193** (1970) 259–262 [Sov. Math., Dokl. **11** (1970) 892–896]. MR **42**:3728 III.I.7

192. (with M.I. Graev and V.A. Ponomarev) The classification of the linear representations of the group SL (2, *C*). Dokl. Akad. Nauk SSSR **194** (1970) 1002–1005 [Sov. Math., Dokl. **11** (1970) 1319–1323]. MR **43**:2162. Zbl. **229**:22024 IV.VIII.3

193. (with D.B. Fuks) Cohomology of the Lie algebra of formal vector fields. Dokl. Akad. Nauk SSSR **190** (1970) 1267–1270 [Sov. Math., Dokl. **11** (1970) 268–271]. MR **44**:2247. Zbl. **264**:17005

194. (with D.B. Fuks) Cohomology of the Lie algebra of formal vector fields. Izv. Akad. Nauk SSSR, Ser. Mat. **34** (1970) 322–337 [Math. USSR, Izv. **34** (1970) 327–342]. MR **44**:1103. Zbl. **216**:203 III.II.5

195. (with D.B. Fuks) Cohomology of Lie algebra of tangential vector fields II. Funkts. Anal. Prilozh. **4** (4) (1970) 23–31 [Funct. Anal. Appl. **4** (1970) 110–116]. MR **44**:2248. Zbl. **208**:514 III.II.6

196. (with D.B. Fuks) Cohomology of Lie algebra of vector fields with nontrivial coefficients. Funkts. Anal. Prilozh. **4** (3) (1970) 10–25 [Funct. Anal. Appl. **4** (1970) 181–192]. MR **44**:7752. Zbl. **222**:58001 III.II.7

197. The cohomology of infinite dimensional Lie algebras; some questions of integral geometry. Int. Congr. Math., Nice 1970, **1** (1970) 95–111. Zbl. **239**:58004 I.I.7

198. (with I.N. Bernshtejn and S.I. Gelfand) Differential operators on the base affine space. Dokl. Akad. Nauk SSSR **195** (1970) 1255–1258 [Sov. Math., Dokl. **11** (1970) 1646–1649]. MR **43**:3402. Zbl. **217**:369
199. (with E.K. Fetisova, V.I. Guelshtejn, and J.M. Vasil'ev) Stimulation of DNA synthesis in cultures of mouse embryo fibroblastlike cells. J. Cell Physiol. **75** (1970) 305–313
200. (with Yu.I. Arshavskij, B.M. Berkinblit, and V.S. Yakobson) Organization of afferent connections of intercalary neurons in the paramedian lobe of the cerebellum of a cat. Dokl. Akad. Nauk SSSR **193** (1970) 250–253 (in Russian)
201. (with D.B. Fuks) Cycles representing cohomology classes of the Lie algebra of formal vector fields. Usp. Mat. Nauk **25** (5) (1970) 239–240. MR **45**:2737 (in Russian). Zbl. **216**:204
202. (with D.B. Fuks) Upper bounds for cohomology of infinite-dimensional Lie algebras. Funkts. Anal. Prilozh. **4** (4) (1970) 70–71 [Funct. Anal. Appl. **4** (1970) 323–324]. MR **44**:4792. Zbl. **224**:18013 III.II.9
203. (with N.M. Chebotareva, Sh.A. Guberman, M.L. Izvekova, E.I. Kandel', N.V. Lebedeva, D.K. Lunev, and I.F. Nikolaeva) Mathematical prognosis of the outcome of haemorrhages with the aim of determining evidence for surgical treatment. Zh. Nevropatol. Psikhiatr. **2** (1970) 177–181 (in Russian)
204. (with Yu.I. Arshavskij, M.B. Berkenblit, O.I. Fukson, and V.S. Yakobson) Features of the influence of the lateral reticular nucleus of medulla oblongata on the cortex of the cerebellum. Nejrofiziologiya **2** (1970) 581–586 (in Russian)
205. (with L.V. Domnina, O.Yu. Ivanova, S.G. Komm, L.V. Ol'shevskaya, and Yu.M. Vasil'ev) Effect of colcemid on the locomotory behaviour of fibroblasts. J. Embryol. Exp. Morphol. **24** (1970) 625–640
206. (with V.A. Ponomarev) Problems of linear algebra and classification of quadruples of subspaces in a finite-dimensional vector space. Colloq. Math. Soc. Janos Bolyai **5**. Hilbert space operators. Tihany, Hungary 1970 (1972). Zbl. **294**:15002
207. (with S.L.Ginzburg, G.V. Gurskaya, G.M. Lobanova, M.G. Nejgauz, and L.A. Novakovskaya) Crystal structure of paroxyacetophenon. Dokl. Akad. Nauk SSSR **195** (1970) 341–344 [Sov. Phys., Dokl. **15** (1970) 999–1002]

1971

208. (with V.A. Ponomarev) Quadruples of subspaces of a finite-dimensional vector space. Dokl. Akad. Nauk SSSR **197** (1971)

762–765 [Sov. Math., Dokl. **12** (1971) 535–539]. MR **44**:2762. Zbl. **294**:15001 II.VIII.4

209. (with I.N. Bernshtejn and S.I. Gelfand) Structure of representations generated by vectors of heighest weight. Funkts. Anal. Prilozh. **5** (1) (1971) 1–9 [Funct. Anal. Appl. **5** (1971) 1–8]. MR **45**:298. Zbl. **246**:17008 II.V.2

210. (with V.I. Guelshtejn and Yu.M. Vasil'ev) Initiation of DNA synthesis in cultures of mouse fibroblastlike cells under the action of substances that disturb the formation of microtubes. Dokl. Akad. Nauk SSSR **197** (1971) 1425–1428 (in Russian)

211. (with Yu.I. Arshavskij, M.B. Berkinblit, and O.I. Fukson) Organization of projections of somatic nerves in different regions of the cortex of the cerebellum of a cat. Nejrofiziologiya **3** (2) (1971) (in Russian)

212. (with Yu.I. Arshavskij, M.B. Berkinblit, I.A. Keder-Stepanova, E.M. Smelyanskaya, and V.S. Yakobson) Background activity of Purkinje cells in intact and deafferentized frontal lobes of the cortex of the cerebellum of a cat. Biofizika **16** (1971) 684–691 (in Russian)

213. (with Yu.I. Arshavskij, M.B. Berkinblit, O.I. Fukson, and V.S. Yakobson) Afferent connections and interaction of neurons of the cortex of the cerebellum. In the collection: Structural and functional organization of the cerebellum, pp. 40–47. Moscow: Nauka (1971) (in Russian)

214. (with Yu.I. Arshavskij, M.B. Berkinblit, O.I. Fukson, and V.S. Yakobson) The reticular afferent system of the cerebellum and its functional significance. Izv. Akad. Nauk SSSR, Ser. Biol. **3** (1971) 375–383 (in Russian)

215. (with L.B. Margolis, V.I. Samojlov, and Yu.M. Vasil'ev) A quantitative estimate of the form and orientation of cell nuclei in a culture. Ontogenez **2** (1971) 138–144 (in Russian)

216. (with T.A. Fajn, Sh.A. Guberman, G.G. Guelshtejn, and I.M. Rotvajn) An estimate of the pressure in the pulmonary artery from electro- and phonocardiographical data under a defect of the intraventricular partition. Kardiologiya **5** (1971) 84–87 (in Russian)

217. (with Yu.I. Arshavskij, M.B. Berkinblit, O.I. Fukson, and V.S. Yakobson) Functional role of the reticular afferent system of the cerebellum. Prepr. IPM Akad. Nauk SSSR (1971) (in Russian)

218. (with V.I. Guelshtejn and Yu.M. Vasil'ev) Initiation of DNA synthesis in cell cultures by colcemid. Proc. Natl. Acad. Sci. USA **68** (1971) 977–979 III.IV.13

219. (with V.Ya. Brodskij, L.V. Domnina, V.I. Guelshtejn, L.B. Klempner, T.L. Marshak, and Yu.M. Vasil'ev) The kinetics

of proliferation in cultures of mouse embryo fibroblastlike cells. Tsitologiya **13** (1971) 1362–1377 (in Russian)

220. (with D.A. Kazhdan) Representations of the group $GL(n, K)$, where K is a local field. Prepr. **71**, IPM Akad. Nauk SSSR (1971) (English transl. in: Lie groups and their representations. Proc. Summer School in Group Representations. Bolyai Janos Math. Soc., Budapest 1971, pp. 95–118. New York: Halsted 1975). Zbl. **348**:22011
221. (with D.A. Kazhdan) Some questions of differential geometry and the computation of the cohomology of Lie algebras of vector fields. Dokl. Akad. Nauk SSSR **200** (1971) 269–272 [Sov. Math., Dokl. **12** (1971) 1367–1370]. MR **44**:4770. Zbl. **238**:58001
222. (with T.L. Fajn, Sh.A. Guberman, G.G. Guelshtejn, I.M. Rotvajn, V.A. Silin, and V.K. Sukhov) Recognition of the degree of pulmonary hypertonia under a defect of the intraventricular partition with the aid of the EVM. Krovoobrashchenie **6** (1971) (in Russian)

1972

223. (with L.V. Domnina, O.Yu. Ivanova, and Yu.M. Vasil'ev) The action of metaphase inhibitors on the form and movement of interphase fibroblasts in a culture. Tsitologiya **14** (1972) 80–88 (in Russian)
224. (with Yu.I. Arshavskij, M.B. Berkinblit, O.I. Fukson, and V.S. Yakobson) Suppression of reactions of Purkinje cells under preceding activation of the reticulo-cerebellar path. Fiziol. Zh. SSSR Im. I.M. Sechenova **58** (1972) 208–214 (in Russian)
225. (with I.N. Bernshtejn and S.I. Gel'fand) Differential operators on a cubic cone. Usp. Mat. Nauk **27** (1) (1972) 185–190 [Russ. Math. Surv. **27** (1) (1972) 185–190. Zbl. **257**:58010 II.V.3
226. (with D.B. Fuks and D.A. Kazhdan) The action of infinite-dimensional Lie algebras. Funkts. Anal. Prilozh. **6** (1) (1972) 10–15 [Funct. Anal. Appl. **6** (1972) 9–13]. MR **46**:922. Zbl. **267**:18023 III.II.10
227. (with I.N. Bernshtejn and S.I. Gel'fand) Differential operators on the principal affine space and investigation of G-modules. Prepr. **77**, IPM Akad. Nauk SSSR (1972) (English transl. in: Lie groups and their representations. Proc. Summer School in Group Representations. Bolyai Janos Math. Soc., Budapest 1971, pp. 21–64. New York: Halsted 1975). Zbl. **338**:58019 II.V.1
228. (with L.V. Domnina, O.Yu. Ivanova, L.B. Margolis, L.V. Ol'shevskaya, Yu.A. Rovenskij, and Yu.M. Vasil'ev) Defective formation of the lamellar cytoplasm in neoplastic fibroblasts. Proc. Natl. Acad. Sci. USA **69** (1972) 248–252

229. (with Sh.A. Guberman, M.L. Izvekova, V.J. Kejlis-Borok, and E.Ya. Rantsman) Criteria of high seismicity, determined by pattern recognition. Proc. Final Symp. Upper Mantle Project **13** (1972)

230. (with Yu.I. Arshavskij, M.B. Berkinblit, O.I. Fukson, and G.N. Orlovskij) Activity of the neurons of the dorsal spino-cerebellar tract under locomotion. Biofizika **17** (1972) 487–494 (in Russian)

231. (with Yu.A. Arshavskij, M.B. Berkinblit, O.I. Fukson, and G.N. Orlovskij) Activity of the neurons of the ventral spino-cerebellar tract under locomotion. Biofizika **17** (1972) 883–896 (in Russian)

232. (with Yu.I. Arshavskij, M.B. Berkenblit, O.I. Fukson, and G.N. Orlovskij) Activity of the neurons of the ventral spino-cerebellar tract under locomotion of cats with deafferentized hind legs. Biofizika **17** (1972) 1113–1119 (in Russian)

233. (with Sh.A. Guberman, M.L. Izvekova, V.I. Kejlis-Borok, E.Ya. Rantsman) Criteria of high seismicity. Dokl. Akad. Nauk SSSR **202** (1972) 1317–1320 (in Russian)

234. (with Yu.I. Arshavskij, M.B. Berkenblit, O.I. Fukson, and G.N. Orlovskij) Recordings of neurons of the dorsal spinocer-ebellar tract during evoked locomotion. Brain Res. **43** (1972) 272–275 III.IV.8

235. (with Yu.I. Arshavskij, M.B. Berkenblit, O.I. Fukson, and G.N. Orlovskij) Origin of modulation in neurons of the ventral spinocerebellar tract during locomotion. Brain Res. **43** (1972) 276–279 III.IV.9

236. (with D.B. Fuks and D.I. Kalinin) Cohomology of the Lie algebra of Hamiltonian formal vector fields. Funkts. Anal. Prilozh. **6** (63) (1972) 25–29 [Funct. Anal. Appl. **6** (1972) 193–196]. MR **47**:1088. Zbl. **259**:57023 III.II.11

237. (with D.A. Kazhdan) On the representation of the group $GL(n, k)$, where K is a local field. Funkts. Anal. Prilozh. **6** (4) (1972) 73–74 [Funct. Anal. Appl. **6** (1972) 315–317]. Zbl. **288**:22024 II.IV.7

238. (with Sh.A. Guberman, M.S. Kaletskaya, V.I. Kejlis-Borok, E.Ya. Rantsman, and M.P. Zhidkov) An attempt to carry over criteria of heigh seismicity from Central asia to Anatolia and adjoining regions. Dokl. Akad. Nauk SSSR **210** (1972) 327–330 (in Russian)

239. (with L.V. Domnina, V.I. Guelshtejn, and Yu.M. Vasil'ev) Regulation of the behaviour of connective tissue cells in multi-cell systems. In: Histophysiology of connective tissue, vol. **1**, pp. 31–36. Novosibirsk 1972

240. (with L.V. Domnina, O.Yu. Ivanova, S.G. Komm, L.V. Ol'shevskaya, and Yu.M. Vasil'ev) The action of metaphase inhibitors on the form and movement of fibroblasts in a culture. Tsitologiya **14** (1972) 80–88 (in Russian)

1973

241. (with Sh.A. Guberman, M.L. Izvekova, V.I. Kejlis-Borok, and E.Ya. Rantsman) Recognition of places of possible origin of powerful earthquakes (in Eastern Central Asia). Vychisl. Seismol. **6** (1973) (in Russian)
242. (with Yu.I. Arshavskij, M.V. Berkenblit, O.I. Fukson, and G.N. Orlovskij) Activity of neurons of the cuneo-cerebellar tract under locomotion. Biofizika **18** (1973) 126–131 (in Russian)
243. (with Yu.M. Vasil'ev) Interactions of normal and neoplastic fibroblasts with the substratum. Ciba Foundation Symposium on Cell Locomotion (1973) 312–331
244. (with I.N. Bernshtejn and V.A. Ponomarev) Coxeter functors and Gabriel's theorem. Usp. Mat. Nauk **28** (2) (1973) 19–33 [Russ. Math. Surv. **28** (2) (1973) 17–32]. Zbl. **269**:08001 II.VIII.5
245. (with I.N. Bernshtejn and S.I. Gel'fand) Schubert cells and cohomology of flag spaces. Funkts. Anal. Prilozh. **7** (1) (1973) 64–65 [Funct. Anal. Appl. **7** (1973) 53–55]. MR **47**:6713. Zbl. **282**:20035
246. (with L.V. Domnina, O.Yu. Ivanova, L.B. Margolis, and Yu.M. Vasil'ev) Intracellular interaction in cultures of transformed fibroblasts of strain L and normal mouse fibroblasts.Tsitologiya **15** (1973) 1024–1028 (in Russian)
247. (with I.N. Bernshtejn and S.I. Gelfand) Schubert cells and cohomology of the spaces G/P. Usp. Mat. Nauk **28** (3) (1973) 3–26 [Russ. Math. Surv. **28** (3) (1973) 1–26]. Zbl. **289**:57024 II.V.4
248. (with M.I. Graev and A.M. Vershik) Representations of the group $SL(2, R)$, where R is a ring of functions. Usp. Mat. Nauk. **28** (5) (1973) 82–128 [Russ. Math. Surv. **28** (5) (1973) 87–132]. Zbl. **297**:22003 II.IX.1
249. (with Yu.M. Vasil'ev) Disturbance of morphogenetic reactions of cells under tumorous transformation. Vestn. Akad. Med. Nauk SSSR **4** (1973) 61–69 (in Russian)
250. (with V.I. Guelshtejn, O.Yu. Ivanova, L.B. Margolis, and Yu.M. Vasil'ev) Contact inhibition of movement in the cultures of transformed cells. Proc. Natl. Acad. Sci. USA **70** (1973) 2011–2014
251. (with L.V. Domnina, E.E. Krivitska, L.V. Ol'shevskaya, Yu.A. Rovenskij, and Yu.M. Vasil'ev) The structure of the lamellar cytoplasm of normal and tumorous fibroblasts.

Papers from a Soviet-French symposium. In: Ultrastructure of cancerous cells, pp. 49–71. Moscow: Nauka 1973 (in Russian)

252. (with L.V. Domnina, E.K. Fetisova, O.Yu. Pletyushkina, and Yu.M. Vasil'ev) Comparative study of density dependent inhibition of growth in the cultures of normal and neoplastic fibroblast-like cells. Abstracts 6th meeting of the European study group for cell proliferation, p. 15. Moscow: Nauka 1973

253. (with Yu.M. Vasil'ev) Factors inducing DNA synthesis and mitosis in normal and neoplastic cell culture. Abstracts 6th meeting of the European study group for cell proliferation, p. 61. Moscow: Nauka 1973

254. (with Sh.A. Guberman, V.I. Kejlis-Borok, E.Ya. Rantsman, I.M. Rotvajn, and M.I. Zhidkov) Determination of criteria of high seismism by means of recognition algorithms. Vestn. Mosk. Gos. Univ. **5** (1973) 78–83 (in Russian)

255. (with A.D. Bershadskii, L.V. Domnina, V.I. Guelshtejn, O.Yu. Ivanova, S.G. Komm, L.B. Margolis, and Yu.M. Vasil'ev) Interactions of normal and neoplastic cells with various surfaces. Neoplasma **20** (1973) 583–585

256. (with D.B. Fuks) *PL*-foliations. Funkts. Anal. Prilozh. **7** (4) (1973) 29–37 [Funct. Anal. Appl. **7** (1973) 278–284]. MR **49**:3958. Zbl. **294**:57016. III.II.12

257. (with Yu.I. Arshavskij, M.B. Berkinblit, O.I. Fukson, G.N. Orlovskij, and B.S. Yakobson) Some peculiarities of the organization of afferent links of the cerebellum. In: 4th International Biophysical Congress, Pushchino Symp. **3** (1973) 327–346 (in Russian)

258. (with Yu.I. Arshavskij, O.I. Fukson, and G.N. Orlovskij) Activity of the neurons of the cuneo-cerebellar tract for locomotion. Biofizika **18** (1973) 126–131 (in Russian)

1974

259. (with M.I. Graev and A.M. Vershik) Irreducible representations of the group G^X and cohomologies. Funkts. Anal. Prilozh. **8** (2) (1974) 67–69 [Funct. Anal. Appl. **8** (1974) 151–153]. MR **50**:530. Zbl. **299**:22004. II.IX.3

260. (with B.L. Feigin and D.B. Fuks) Cohomology of the Lie algebra of formal vector fields with coefficients in its adjoint space and variations of characteristic classes of foliations. Funkts. Anal. Prilozh. **8** (2) (1974) 13–29 [Funct. Anal. Appl. **8** (1974) 99–112]. MR **50**:8553. Zbl. **298**:57011. III.II.13

*261. (with I.N. Bernshtejn and S.I. Gelfand) A new model for representations of finite semisimple algebraic groups. Usp. Mat. Nauk **29** (3) (1974) 185–186. MR 53:5760. Zbl. **354**:20031 II.IV.4

262. (with Yu.N. Arshavskij, M.B. Berkinblit, A.M. Smelyanskij, and V.S. Yakobson) Background activity of Pourkynje cells of the paramedian part of the cortex of the cerebellum of a cat. Biofizika **19** (1974) 903–907 (in Russian)

263. (with Yu.I. Arshavskij, M.B. Berkinblit, O.I. Fukson, and G.N. Orlovskij) Differences in the working of spino-cerebral tracts in artificial irritation and locomotion. In: Mechanisms of the union of neurons in the nerve centre, pp. 99–105. Leningrad: Nauka 1974 (in Russian)

264. (with Sh.A. Guberman, M.S. Kaletska, V.I. Kejlis-Borok, E.Ya. Rantsman, I.M. Rotvajn, and M.P. Zhidkov) Recognition of places where strong earthquakes are possible. II. Four regions of Asia Minor and South-East Europe. Vychisl. Seismol. **7** (1974) 3–39 (in Russian)

265. (with Sh.A. Guberman, V.I. Kejlis-Borok, E.Ya. Rantsman, I.M. Rotvajn, and M.P. Zhidkov) Recognition of places where strong earthquakes are possible. III. The case when the boundaries of disjunctive nodes are not known. Vychisl. Seismol. **7** (1974) 41–62 (in Russian)

266. (with V.I. Guelshtejn, O.Yu. Ivanova, S.G. Komm, and L.B. Margolis, and Yu.M. Vasil'ev) The results of intercellular impacts in cultures of normal and transformed fibroblasts. Tsitologiya **16** (1974) 752–756 (in Russian)

267. (with D.B. Fuks) *PL*-foliations. II. Funkts. Anal. Prilozh. **8** (3) (1974) 7–11 [Funct. Anal. Appl. **8** (1974) 197–200]. MR **54**:6159. Zbl. **316**:57010 III.II.15

268. (with V.A. Ponomarev) Free modular lattices and their representations. Usp. Mat. Nauk **29** (6) (1974) 3–58 [Russ. Math. Surv. **29** (6) (1974) 1–56]. MR **53**:5393. Zbl. **314**:15003

269. (with O.Yu. Ivanova, L.B. Margolis, and Yu.M. Vasil'ev) Orientation of mitosis of fibroblasts is determined in the interphase. Proc. Natl. Acad. Sci. USA **71** (1974) 2032

270. (with Yu.I. Arshavskij, M.B. Berkinblit, O.I. Fukson, and G.N. Orlovskij) Peculiarities of information entering the cortex of the cerebellum via different afferent paths, structural and functional organization of the cerebellum, pp. 34–41. Kiev: Naukova Dumka 1974 (in Russian)

1975

271. (with L.V. Domnina, A.V. Lyubimov, Yu.M. Vasil'ev, and O.S. Zakharova) Contact inhibition of phagocytosis in epithelial sheets: alterations of cell surface properties induced by cell-cell contacts. Proc. Natl. Acad. Sci. USA **72** (1975) 719–722

272. (with A.P. Chern and Yu.M. Vasil'ev) Spreading of normal and transformed fibroblasts in dense cultures. Exp. Cell Res. **90** (1975) 317–327

273. (with A.M. Gabrielov and M.V. Losik) The combinatorial computation of characteristic classes. Funkts. Anal. Prilozh. **9** (1975) 54–55 [Funct. Anal. Appl. **9** (1975) 48–49]. MR **51**:1839. Zbl. **312**:57016

274. Quantitative evaluation of cell orientation in culture. J. Cell. Sci. **17** (1975) 1–10

275. (with M.I. Graev and A.M. Vershik) Representations of the group of diffeomorphisms connected with infinite configurations. Prepr. Inst. Appl. Math. **46** (1975) 1–62 (in Russian)

276. (with M.I. Graev and A.M. Vershik) The square roots of quasiregular representations of the group $SL(2, k)$. Funkts. Anal. Prilozh. **9** (2) (1975) 64–66 [Funct. Anal. Appl. **9** (1975) 146–148]. MR **51**:8338. Zbl. **398**:22010

277. (with I.N. Bernshtejn and S.I. Gel'fand) Models of representations of compact Lie groups. Funkts. Anal. Prilozh. **9** (4) (1975) 61–62 [Funct. Anal. Appl. **9** (1975) 322–324]. MR **54**:2884. Zbl. **339**:22009

278. (with I.N. Bernshtejn and S.I. Gel'fand) Models of representations of Lie groups. Proc. Petrovskij Semin. **2** (1976) 3–21. [Sel. Math. Sov. **1** (2) (1981) 121–142] Zbl. **499**:22004 II.IV.8

279. (with A.M. Gabrielov and M.V. Losik) Combinatorial calculus of characteristic classes. Funkts. Anal. Prilozh. **9** (2) (1975) 12–28 [Funct. Anal. Appl. **9** (1975) 103–115]. MR **53**:14504a. Zbl. **312**:57016 III.II.16

280. (with A.M. Gabrielov and M.V. Losik) Combinatorial calculus of characteristic classes. Funkts. Anal. Prilozh. **9** (3) (1975) 5–26 [Funct. Anal. Appl. **9** (1975) 186–202]. MR **53**:14504a. Zbl. **341**:57017 III.II.17

281. (with L.A. Dikij) Asymptotic behaviour of the resolvent of Sturm-Liouville equations and the algebra of the Korteweg-de Vries equations. Usp. Mat. Nauk **30** (5) (1975) 67–100 [Russ. Math. Surv. **30** (5) (1975) 77–113]. MR **58**:22746. Zbl. **461**:35072. I.III.12

282. (with I.S. Tint and Yu.M. Vasil'ev) Processes determining the changes of shape of a cell after its separation from the epigastrium. Tsitologiya **5** (1975) 633–638 (in Russian)

283. (with O.Yu. Pletyushkina and Yu.M. Vasil'ev) Neoplastic fibroblasts sensitive to growth inhibition by parent normal cells. Br. J. Cancer **31** (1975) 535–543

284. (with L.B. Margolis, V.I. Samojlov, and Yu.M. Vasil'ev) Methods of measuring the orientation of cells. Ontogenez **6** (1) (1975) 105–110 (in Russian)

285. (with E.K. Fetisoba, O.Yu. Pletyushkina, and Yu.M. Vasil'ev) Insensibility of dense cultures of transformed mice fibroblasts to the action of agents, stimulating the synthesis of DNA in cultures of normal cells. Tsitologiya **17** (1975) 442–446 (in Russian)

286. (with Yu.I. Arshavskij, G.N. Orlovskij, and G.A. Pavlova) The activity of neurons of the ventral spino-cerebral tract in "fictitious scratching". Biofizika **20** (1975) 748–749 (in Russian)

287. (with Yu.I. Arshavskij, G.N. Orlovskij, and G.A. Pavlova) Origin of the modulation of the activity of vestibular-spinal neurons in scratching. Biofizika **20** (1975) 946–947 (in Russian)

288. (with Yu.I. Arshavskij, M.B. Berkinblit, T.G. Delyagina, A.G. Fel'dman, O.I. Fukson, G.N. Orlovskij, and G.A. Pavlova) On the role of the cerebellum in regulating some rhythmic movements (locomotion, scratching). Summaries 12th meeting of the All-Union Physiological Society, pp. 15–16. Tbilisi 1975 (in Russian)

289. (with T.G. Delyagina, A.G. Fel'dman, and G.N. Orlovskij) On the role of the central program and afferent inflow in generation of scratching movements in the cat. Brain Res. **100** (1975) 297–313

290. (with M.I. Graev and A.M. Vershik) Representations of the group of diffeomorphisms. Usp. Mat. Nauk **30** (6) (1975) 3–50 [Russ. Math. Surv. **30** (6) (1975) 1–50]. MR **53**:3188. Zbl. **317**:58009 II.IX.4

291. (with Sh.A. Guberman, M.S. Kaletska, V.I. Kejlis-Borok, E.Ya. Rantsman, I.M. Rotvajn, and L.P. Zhidkov) Prognosis of a place where strong earthquakes occur, as a problem of recognition. In: Modelling of training and behaviour, pp. 18–25. Moscow: Nauka 1975 (in Russian)

1976

292. (with D.B. Fuks and A.M. Gabrielov) The Gauss-Bonnet theorem and the Atiyah-Patodi-Singer functionals for the characteristic classes of foliations. Topology **15** (1976) 165–188. MR **55**:4201. Zbl. **347**:57009 III.II.14

293. (with A.M. Gabrielov and M.V. Losik) A local combinatorial formula for the first class of Pontrayagin. Funkts. Anal. Prilozh. **10** (1) (1976) 14–17 [Funct. Anal. Appl. **10** (1976) 12–15]. MR **53**:14504b. Zbl. **328**:57006 III.II.18

294. (with L.A. Dikij) A Lie algebra structure in a formal variational calculation. Funkts. Anal. Prilozh. **10** (1) (1976) 1–8

[Funct. Anal. Appl. **10** (1976) 16–22]. MR **57**:7670. Zbl. **347**:49023

295. (with I.N. Bernshtejn and S.I. Gel'fand) Category of *G*-modules. Funkts. Anal. Prilozh. **10** (2) (1976) 1–8 [Funct. Anal. Appl. **10** (1976) 87–92]. MR **53**:10880. Zbl. **353**:18013 II.V.5

296. (with A.M. Gabrielov and M.V. Losik) Atiyah-Patodi-Singer functionals for characteristic functionals for tangent bundles. Funkts. Anal. Prilozh. **10** (2) (1976) 13–28 [Funct. Anal. Appl. **10** (1976) 95–107]. MR **54**:1245. Zbl. **344**:57008 III.II.19

297. (with L.A. Dikij) Fractional powers of operators and Hamiltonian systems. Funkts. Anal. Prilozh. **10** (4) (1976) 13–29 [Funct. Anal. Appl. **10** (1976) 259–273]. MR **55**:6484. Zbl. **346**:35085 I.III.10

298. (with Yu.I. Manin and M.A. Shubin) Poisson brackets and the kernel of the variational derivative in the formal calculus of variations. Funkts. Anal. Prilozh. **10** (4) (1976) 30–34 [Funct. Anal. Appl. **10** (1976) 274–278]. MR **55**:13486. Zbl. **395**:58005

299. (with V.A. Ponomarev) Lattices, representations, and algebras connected with them. I. Usp. Mat. Nauk **31** (5) (1976) 71–88 [Russ. Math. Surv. **31** (5) (1976) 67–85]. MR **58**:16779a. Zbl. **358**:06020

*300. (with M.V. Losik) Calculation of characteristic classes of combinatorial vector bundles. Prepr. Inst. Appl. Mat. **99** (1976) III.II.20

301. (with Sh.A. Guberman, V.I. Kejlis-Borok, L. Knopov, E. Press, E.Ya. Rantsman, I.M. Rotvajn, and A.M. Sadovskij) Conditions for the occurence of strong earthquakes (California and some other areas). Vychisl. Seismol. **9** (1976) 3–92 (in Russian)

302. (with Sh.A. Guberman, V.I. Kejlis-Borok, L. Knopov, F. Press, E.Ya. Rantsman, I.M. Rotvajn, and A.M. Sadovskij) Pattern recognition applied to earthquake epicenters in California. Phys. Earth Planet. Inter. **11** (1976) 277–283

303. (with M.A. Alekseevskaya, I.V. Martynov, and V.M. Sablin) First results of the prognostication of the effect of transmural (large focal) myocardial infarcts. Aktual'nye voprosy kardiologii, Otdelennye rezul'taty lecheniya elokachestvennykh opukholej, 19–24. Moscow: Nauka 1976 (in Russian)

304. (with E.E. Bragina and Yu.M. Vasil'ev) Formation of bundles of microfilaments during spreading of fibroblasts on the substratum. Exp. Cell Res. **97** (1976) 241–248

305. (with A.D. Bershadskij, V.I. Gel'fand, V.I. Guelshtejn, and Yu.M. Vasil'ev) Serum dependence of expression of the transformed phenotype experiments with subline of mouse *L* fibro-

blasts adapted to growth in serum-free medium. Int. J. Cancer (1976) 84–92

306. (with O.Yu. Ivanova, L.B. Margolis, and Yu.M. Vasil'ev) Effect of colcemid on the spreading of fibroblasts in culture. Exp. Cell Res. **101** (1976) 207–219
307. (with L.V. Domnina, N.A. Dorfman, O.Yu. Pletyushkina, and Yu.M. Vasil'ev) Active cell edge and movements of concanavalin *A* receptors on the surface of epithelial and fibroblastic cells. Proc. Natl. Acad. Sci. USA **73** (1976) 4085–4089
308. (with N.M. Chebotareva, Sh.A. Guberman, M.L. Izvekova, E.I. Kandel, N.V. Lebedeva, D.K. Lunev, I.F. Nikolaeva, and E.V. Shmidt) A computer study of prognosis of cerebral haemorrhage for choosing optimal treatment. Eur. Congr. Neurosurg., pp. 71–72. Edinburgh (1976)
309. (with N.M. Chebotareva, Sh.A. Guberman, M.L. Izvekova, E.I. Kandel, and E.V. Shmidt) Prognostication of the results of surgical treatment of haemorrhaging lesions by means of a computer. Vopr. Nejrokhir. **3** (1976) 20–23 (in Russian)
310. (with O.Yu. Ivanova, L.B. Margolis, and Yu.M. Vasil'ev) Effect of colcemid on spreading of fibroblast in culture. Exp. Cell Res. **101** (1976) 207–219
311. (with Yu.M. Vasil'ev) Effect of colcemid on morphogenetic processes and locomotion of fibroblasts. Cell Motility **3** (1976) 279–304

1977

312. (with M.A. Alekseevskaya, Sh.A. Guberman, I.V. Martynov, I.M. Rotvajn, and V.M. Sablin) Prognostication of the result of a large focal myocardial infarct by means of learning program. Kardiologiya **17** (1977) 26–31 (in Russian)
313. (with M.A. Alekseevskaya, L.D. Golovnya, Sh.A. Gubermann, M.L. Izvekova, and A.L. Syrkin) Prognostication of the result of a myocardial infarct by means of the program "Cortex-3". Kardiologiya **17** (6) (1977) 13–23 (in Russian)
314. (with M.Yu. Melikova, S.G. Gindikin, and M.L. Izvekova) Prognostication of the healing of duodenal ulcers. Aktual. Vopr. Gastroenterol. **10** (1977) 42–51 (in Russian)
315. (with M.A. Alekseevskaya, L.D. Golovnya, M.L. Izvekova, I.V. Martynov, V.M. Sablin, and A.L. Syrkin) A general guide-line or a general method for creating one (On ways of applying mathematical methods in medicine). Summaries of lectures at the All-Union Conf. on the theory and practice of automatic electrocardiological and clinical investigations, pp. 3–5. Kaunas (1977) (in Russian)

316. (with M.A. Alekseevskaya, E.S. Klyushin, A.V. Nedostup and A.L. Syrkin) On the methodology of creating a formalized description of the patient (using the example of prognostication of remote results of electro-impulsive treatment of a constant form of flickering arrhythmy). Summaries of lectures at the All-Union Conf. on the theory and practice of automatic electrocardiological and clinical investigations, pp. 5–8. Kaunas (1977) (in Russian)

317. (with L.V. Domnina, O.Yu. Pletyushkina, and Yu.M. Vasil'ev) Effects of antitubilins on redistribution of cross-linked receptors on the surface of fibroblasts and epithelial cells. Proc. Natl. Acad. Sci. USA **74** (1977) 2865–2868

318. (with O.Yu. Ivanova, S.G. Komm, L.B. Margolis, and Yu.M. Vasil'ev) The influence of colcemid on the polarization of cells on narrow strips of the adhesive substratum. Tsitologiya **19** (1977) 357–360 (in Russian)

319. (with A.D. Bershadskij, V.I. Gel'fand, and Yu.M. Vasil'ev) The influence of serum on the development of cell transformation. Vestn. Akad. Mech. Nauk **3** (1977) 55–59 (in Russian)

320. (with A.D. Bershadskij, A.D. Lyubimov, Yu.A. Rovenskij, and Yu.M. Vasil'ev) Contact interaction of cell surfaces. Lectures at the Soviet-Italian symposium "Tissue proteinases in normal and pathological state", Moscow 22–27 September 1977 (in Russian)

321. (with Yu.M. Vasil'ev) Mechanisms of morphogenesis in cell cultures. Int. Rev. Cytol. **50** (1977) 159–274 III.IV.14

322. (with Yu.I. Arshavskij, M.B. Berkinblit, and V.S. Yakobson) A formula for the analysis of histograms of intercellular intervals of Pourkine cells. Biofizika **22** (1977) (in Russian)

323. (with M.A. Alekseevskaya, A.M.Gabrielov, A.D. Gvishiani, and E.Ya. Rantsman) Morphological division of mountainous countries by formalized criteria. Vychisl. Seismol. **10** (1977) 33–79 (in Russian)

324. (with M.A. Alekseevskaya, A.M. Gabrielov, A.D. Gvishiani, and E.Ya. Rantsman) Formal morphostructural zoning at mountain territories. J. Geophys. **43** (1977) 227–235

325. (with L.A. Dikij) The Resolvent and Hamiltonian systems. Funkts. Anal. Prilozh. **11** (2) (1977) 11–27 [Funct. Anal. Appl. **11** (1977) 93–105]. MR **56**:1359. Zbl. **357**:58005. I.III.13

326. (with S.G. Gindikin) Nonlocal inversion formulas in real integral geometry. Funkts. Anal. Prilozh. **11** (3) (1977) 12–19 [Funct. Anal. Appl. **11** (1977) 173–179]. MR **56**:16265. Zbl. **385**:53056 III.I.8

327. (with V.A. Ponomarev) Representation lattices and the algebras connected with them. Usp. Mat. Nauk **32** (1) (1977)

85–107 [Russ. Math. Surv. **32** (1) (1977) 91–114]. Zbl. **358**:06021

328. (with S.G. Gindikin) Complex manifolds whose skeletons are semisimple real Lie groups, and analytic discrete series of representations. Funkts. Anal. Prilozh. **11** (4) (1977) 20–28 [Funct. Anal. Appl. **11** (1977) 258–265]. MR **58**:11230. Zbl. **444**:22006. II.IV.9
329. (with M.I. Graev and A.M. Vershik) Representations of the group of smooth mappings from a manifold *X* into a compact Lie group. Dokl. Akad. Nauk SSSR **323** (1977) 745–748 [Sov. Math., Dokl. **18** (1977) 118–121]. MR **55**:10602. Zbl. **393**:22012
330. (with M.I. Graev and A.M. Vershik) Representations of the group of smooth mappings of a manifold *X* into a compact Lie group. Compos. Math. **35** (1977) 299–334. MR **58**:28257. Zbl. **368**:53034 II.IX.2
331. (with Yu.I. Arshavskij, G.N. Orlovskij, and G.A. Pavlova) Activity of neurons of the lateral reticular nucleus in scratching. Biofizika **22** (1) (1977) (in Russian)

1978

332. (with L.V. Domnina, O.Yu. Pletyushkina, and Yu.M. Vasil'ev) Influence of agents, destroying microtubules, on the distribution of receptors of the surface of cultured cells. Tsitologiya **20** (1978) 796–801 (in Russian)
333. (with Yu.I. Arshavskij, G.N. Orlovskij, and G.A. Pavlova) Messages conveyed by spino-cerebellar pathways during scratching in the cat. 1. Activity of neurons of lateral reticular nucleus. Brain Res. **151** (1978) 479–491
334. (with Yu.I. Arshavskij, G.N. Orlovskij, and G.A. Pavlova) Messages conveyed by spino-cerebellar pathways during scratching in the cat. 2. Activity of neurons of the ventral spino-cerebellar tract. Brain Res. **151** (1978) 493–506
335. (with M.B. Berkinblit, T.G. Delyagina, A.G. Fel'dman, and G.N. Orlovskij) Generation of scratching. 1. Activity of spinal interneurons during scratching. J. Neurophysiol. **41** (1978) 1040–1057 III.IV.10
336. (with M.B. Berkinblit, T.G. Delyagina, A.G. Fel'dman, and G.N. Orlovskij) Generation of scratching. 2. Non-regular regimes of generation. J. Neurophysiol. **41** (1978) 1058–1069
337. (with I.N. Bernshtejn and S.I. Gel'fand) Structure locale de la catégorie des modules de Harish-Chandra I. C.R. Acad. Sci., Paris, Ser. A **286** (1978) 435–437. MR **58**:16966, Zbl. **416**:22018 II.V.6

338. (with I.N. Bernshtejn and S.I. Gel'fand) Structure locale de la catégorie des modules de Harish-Chandra II. C.R. Acad. Sci., Paris, Ser. A **286** (1978) 495–497. MR **81e**:22026. Zbl. **431**:22013 II.V.7

339. (with Yu.M. Vasil'ev) Mechanisms of non-adhesiveness of endothelial and epithelial surfaces. Nature **275** (1978) 710–711

340. (with Yu.M. Vasil'ev, A.D. Bershadskij, V.A. Rozenblat, and I.S. Tint) Microtubular system in cultured mouse epithelial cells. Cell Biol. Int. Rep. **2** (1978) 345–351

341. (with L.A. Dikij) Variational calculus and the Korteweg-de Vries equations. Partial differential equations. Proc. All-Union Conf., Moscow 1976, dedic. I.G. Petrovskij pp. 81–83 (1978) (in Russian). Zbl. **498**:35074

342. (with L.A. Dikij) Calculus of jets and non-linear Hamiltonian systems. Funkts. Anal. Prilozh. **12** (2) (1978) 8–23 [Funct. Anal. Appl. **12** (1978) 81–94]. MR **58**:18561, Zbl. **388**:58009.

*343. (with L.A. Dikij) A family of Hamiltonian structures related to nonlinear integrable differential equations. Prepr. Inst. Appl. Mat. **136** (1978). MR **81**:58027 I.III.11

344. (with I.N. Bernshtejn and S.I. Gel'fand) Algebraic bundles over P^n and problems of linear algebra. Funkts. Anal. Prilozh. **12** (3) (1978) 66–68 [Funct. Anal. Appl. **12** (1978) 212–214]. MR **80c**:14010a, Zbl. **402**:14005. II.V.8

345. (with B.L. Feigin and D.B. Fuks) Cohomology of infinite-dimensional Lie algebras and Laplace operators. Funkts. Anal. Prilozh. **12** (4) (1978) 1–5 [Funct. Anal. Appl. **12** (1978) 243–247]. MR **80i**:58050, Zbl. **396**:17008. III.II.21

346. (with S.G. Gindikin, M.L. Izvekova, and M.Yu. Melikova) On one approach to formalization of the diagnostic attitude of a doctor (using the prognosis of the healing of duodenal ulcers). Summaries of lectures at the All-Union Conf. biological and medical cybernetics, vol. 2, pp. 27–31 (1978) (in Russian)

347. (with S.G. Gindikin, M.L. Izvekova, and M.Yu. Melikova) On some questions of mathematical diagnostics: examples of problems from gastroenterology. Summaries of lectures at the Second All-Union Cong. on Gastroenterology, vol. 2, pp. 57–58, Moscow-Leningrad: Nauka (1978) (in Russian)

348. (with Yu.I. Manin) Dualità, Enciclopedia Einaudi, Vol. 5, pp. 126–178. Einaudi, Torino (1978)

349. (with Yu.I. Arshavskij, G.N. Orlovskij, and G.A. Pavlova) Messages conveyed by descending tract during scratching in the cat. I. Activity of vestibulospinal neurons. Brain Res. **159** (1978) 88–110

350. (with M.A. Alekseevskaya, E.S. Klyushin, A.V. Nedostup, and A.L. Syrkin) A new approach to the problem of the choice of information and formalization of the description of the patient for the solution of medical problems on a computer. Prepr. Inst. Appl. Math. **144** (1978) (in Russian)

1979

351. (with M.A. Alekseevskaya, E.S. Klyushin, and A.V. Nedostup) The gathering of medical information for processing on a computer (manual). Inst. Appl. Math. (1979) (in Russian)
352. (with S.G. Gindikin, M.L. Izvekova, and M.Yu. Melikova) One method of formalizing the diagnostic attitude of a doctor (examples of prognosis of the healing of a duodenal ulcer). Prepr. Acad. Sci. USSR Sci. Committee on the complex problem "Cybernetics" (1979) (in Russian)
353. (with L.B. Margolis, E.J. Vasil'eva, and Yu.M. Vasil'ev) Upper surfaces of epithelial sheets and of fluid lipid films are non-adhesive for platelets. Proc. Natl. Acad. Sci. USA **76** (1979) 2303–2305
354. (with A.D. Bershadskij, V.I. Gel'fand, V.A. Rozenblat, I.S. Tint, and Yu.M. Vasil'ev) Morphology of microtubular systems in epithelial cells in the kidney of a mouse. Ontogenez **10** (1979) 231–235 (in Russian)
355. (with L.A. Dikij) Integrable nonlinear equations and the Liouville theorem. Funkts. Anal. Prilozh. **13** (1) (1979) 8–20 [Funct. Anal. Appl. **13** (1979) 6–15]. MR **80i**:58027. Zbl. **423**:34003. I.III.14
356. (with S.G. Gindikin and Z.Ya. Shapiro) A local problem of integral geometry in a space of curves. Funkts. Anal. Prilozh. **13** (2) (1979) 11–31 [Funct. Anal. Appl. **13** (1980) 87–102]. MR **80k**:53100. Zbl. **415**:53046. III.I.9
357. (with V.A. Ponomarev) Model algebras and representations of graphs. Funkts. Anal. Prilozh. **13** (3) (1979) 1–12. [Funct. Anal. Appl. **13** (1980) 157–166]. Zbl. **437**:16020 II.VIII.6
358. (with I.Ya. Dorfman) Hamiltonian operators and algebraic structures related to them. Funkts. Anal. Prilozh. **13** (4) (1979) 13–31 [Funct. Anal. Appl. **13** (1980) 248–262]. MR **81c**:58035. Zbl. **428**:58009 I.III.15
359. (with S.G. Gindikin and M.I. Graev) A problem of integral geometry in RP^n connected with the integration of differential forms. Funkts. Anal. Prilozh. **13** (4) (1979) 64–67 [Funct. Anal. Appl. **13** (1980) 288–290]. MR **83a**:43006. Zbl. **423**:58001 III.I.10
360. (with Yu.I. Arshavskij) The role of the brain stem and cerebellum in the regulation of rhythmic movements. Proc. 13 Congr. Pavlov Physiol. Soc., vol. 1, pp. 474–475. Leningrad: Nauka 1979 (in Russian)

361. (with Yu.I. Arshavskij, M.B. Berkinblit, and G.N. Orlovskij) The significance of signals passing along the various spino-cerebral pathways for the work of locomotive centres of the brain stem in scratching. In: Nejronnye mekhanizmy integrativnoj deyatel'nosti mozzhechka, pp. 88–91. Erevan 1979 (in Russian)

362. (with Yu.I. Arshavskij, M.B. Berkinblit, G.N. Orlovskij) Signalling mechanisms of the scratching reflex and their interaction with the cerebellum. In: Nejronnye mekhanizmy integrativnoj deyatel'nosti mozzhechka, pp. 92–96. Erevan 1979 (in Russian)

1980

363. (with M.Ya. Ratner, B.I. Rozenfel'd, and V.V. Serov) The problem of classifying glomerule kidneys. Prepr. Acad. Sci. USSR Sci. Committee on the complex problem "Cybernetics" (1980) (in Russian)

364. (with V.A. Ponomarev) Representations of graphs. Perfect sub-representations. Funkts. Anal. Prilozh. **14** (3) (1980) 14–31 [Funct. Anal. Appl. **14** (1980) 177–190]. MR **83c**:05113. Zbl. **453**:05027

365. (with I.Ya. Dorfman) The Schouten bracket and Hamiltonian operators. Funkts. Anal. Prilozh. **14** (3) (1980) 71–74 [Funct. Anal. Appl. **14** (1980) 223–226]. MR **82e**:58039. Zbl. **444**:58010. I.III.16

366. (with M.I. Graev) Admissible n-dimensional complexes of curves in $\mathbf{R}^n$. Funkts. Anal. Prilozh. **14** (4) (1980) 36–44 [Funct. Anal. Appl. **14** (1980) 274–281]. MR **82**:53013. Zbl. **454**:53042.

367. (with S.G. Gindikin and M.I. Graev) Integral geometry for one-dimensional fibrations of general form over $\mathbf{RP}^n$. Prepr. Inst. Appl. Math. **60** (1980) 1–24 (in Russian). MR **82g**:53081

368. (with S.G. Gindikin and M.I. Graev) Integral geometry in affine and projective spaces. Itogi Nauki Tekh., Ser. Sovrem. Probl. Mat. 16, 53–226, Moscow: VINITI (1980) [J. Sov. Math. **18** (1980) 39–167]. MR **82m**:43017. Zbl. **465**:52005. III.I.11

369. (with I.V. Cherednik and S.A. Chernyakevich) A formalized differentiated description of the motor of the stomach and the duodenal intestine. Prepr. Acad. Sci. USSR Sci. Committee on the complex problem "Cybernetics", pp. 1–34 (1980) (in Russian)

370. (with G.G. Guelshtejn, I.P. Lukashevich, and M.A. Shifrin) Study of the correlation between electrocardiograph and coronary data. Prepr. Acad. Sci. USSR Sci. Committee on the complex problem "Cybernetics", pp. 1–28 (1980) (in Russian)

371. (with Zh.L. Bliokh, L.V. Domnina, O.Yu. Ivanova, O.Yu. Pletyushkina, T.S. Svitkina, V.V. Smolyaninov, and Yu.M. Vasil'ev) Spreading of fibroblasts in a medium containing cytochalasin *B*: Formation of lamellar cytoplasm as a combination of several functionally different processes. Proc. Natl. Acad. Sci. USA **77** (1980) 5919–5922

1981

372. (with E.V. Pomerantsev, V.M. Sablin, M.N.Starkova, V.A. Sulimova, A.L. Syrkin, and V.L. Vakhlyaev) Prognostication of complications and classification of patients with severe myocardial infarcation. Summaries of lectures at the second All-Union Conf. "Theory and practice of automation of electrocardiological and clinical studies", pp. 274–276. Kaunas 1981 (in Russian)
373. (with G.G. Guelshtejn, I.P. Lukashhevich, M.A. Shifrin, and L.S. Zingerman) Expressibility of electrocardiograph changes in severe disease of the coronary artery in patients with chronic ischemic heart disease. Summaries of lectures at the second All-Union Conf. "Theory and practice of automation of electron-cardiological and clinical studies", pp. 304–307. Kaunas 1981 (in Russian)
374. (with S.M. Khoroshkin, E.V. Pomerantsev, B.I. Rozenfel'd, V.A. Sulimov, A.L. Syrkin, and V.L. Vaklyaev) Choice of information for the classification of patients with myocardial infarcation and choice of medical tactics. Summaries of lectures at the second All-Union Conf. "Theory and practice of automation of electrocardiological and clinical studies", pp. 267–278. Kaunas 1981 (in Russian)
375. (with Yu.M. Vasil'ev) Neoplastic and normal cells in culture. London-Sydney: Cambridge University Press 1981
376. (with S.G. Gindikin, M.L. Izvekova, and M.Yu. Melikova) The immediate prognosis for healing of duodenal ulcers (control of classification). Trans. Second Moscow Med. Inst. Ser. "Surgery" **32** (1981) 73–80 (in Russian)
377. (with Yu.M. Vasil'ev) Interaction of normal and tumorous cells with the medium. Moscow: Nauka 1981 (in Russian)
378. (with V.M. Alekseev, M.A. Alekseevskaya, E.E. Gogin, L.D. Golovnya, M.L. Izvekova, E.S. Klyushin, I.V. Martynov, V.A. Ponomarev, I.V. Sablin, A.L. Syrkin, and R.M. Zaslavska) Multi-purpose chart of a patient with myocardial infarction (for setting up a data bank in a computer). Prepr. Acad. Sci. USSR Sci. Committee on the complex problem "Cybernetics" (1981) (in Russian)
379. (with V.A. Ponomarev) Gabriel's theorem is also true for representations of graphs endowed with relations. Funkts. Anal.

Prilozh. **15** (2) (1981) 71–22 [Funct. Anal. Appl. **15** (1981) 132–133]. Zbl. **479**:18003

380. (with I.Ya. Dorfman) Hamiltonian operators and infinite-dimensional Lie algebras. Funkts. Anal. Prilozh. **15** (3) (1981) 23–40 [Funct. Anal. Appl. **15** (1982) 173–187]. MR **82j**:58045. Zbl. **478**:58013
381. (with Yu.I. Manin) Simmetria, Enciclopedia Einaudi, Vol. 12, pp. 916–943. Einaudi, Torino 1981
382. (with A.D. Bershadskij, Zh.L. Bliokh, L.V. Domnina, O.Yu. Ivanova, V.V. Smolyahinov, T.M. Svitkina, I.S. Tint, and Yu.M. Vasil'ev) Mechanisms of morphological reactions determining the shape and movement of normal and transformed cells in culture. In: Nemyshechnie sistemy, pp. 65–75. Moscow: Nauka 1981 (in Russian)
383. (with O.Yu. Ivanova, S.G., Komm, and Yu.M. Vasil'ev) Stabilization independent of micropipelets of the cell surface of normal and transformed connective tissue cells. Tsitologiya **23** (1981) 62–65 (in Russian)
384. (with M.I. Graev and A.M. Vershik) Representations of the group of functions taking values in a compact Lie group. Compos. Math. **42**, 217–243 (1981). MR **83g**:22002. Zbl. **449**:22019. II.IX.5

1982

385. (with M.I. Graev) Integral transformations connected with two remarkable complexes in projective space. Prepr. Inst. Appl. Math. **93** (1982) (in Russian)
386. (with M.I. Graev and R. Rosu) Non-local inversion formulae in a problem of integral geometry connected with *p*-dimensional planes in real projective space. Funkts. Anal. Prilozh. **16** (3) (1982) 49–51 [Funct. Anal. Appl. **16** (1982) 196–198]. Zbl. **511**:53072
387. (with M.N. Starkova and A.L. Syrkin) Classification of patients and prognosis of healing in myocardial infarction. Prepr. Acad. Sci. USSR Sci. Committee on the complex problem "Cybernetics" (1982) (in Russian)
388. (with M.L. Izvekova, M.N. Starkova, and A.L. Syrkin) The methodology of comparing material from two hospitals and the construction of a single guide-line for the prognosis of the effect of a strong focal myocardial infarction. Prepr. Acad. Sci. USSR Sci. Committee on the complex problem "Cybernetics" (1982) (in Russian)
389. (with M.A. Brodskij, M.Ya. Ratner, B.I. Rozenfel'd, V.V. Serov, I.I. Stenina, and V.A. Varshavskij) Determination of a morphological picture of glomerule kidney from clinical-

functional data (by means of a formal scheme modelling the diagnosis of kidney consultants). Prepr. Acad. Sci. USSR Sci. Committee on the complex problem "Cybernatics" (1982) (in Russian)

390. (with S.M. Khorshkin, B.I. Rozenfel'd, V.A. Sulimov, A.L. Syrkin, and V.D. Vakhlyaev) Selection of information for the classification of patients with myocardial infarcation and choice of medical tactics. Prepr. Acad. Sci. USSR Sci. Committee on the complex problem "Cybernetics" (1982) (in Russian)

391. (with I.Ya. Dorfman) Hamiltonian operators and the classical Yang-Baxter equation. Funkts. Anal. Prilozh. **16**(4) (1982) 1–9 [Funct. Anal. Appl. **16** (1982) 241–248]. Zbl. **527**:58018 I.III.17

392. (with Yu.I. Arshavskij and G.N. Orlovskij) The cerebellum and control of rhythmical movements. Trends Neurosci. **6** (10) (1983) 417–422 III.IV.11

393. (with Yu.I. Arshavskij, I.N. Beloozerova, G.N. Orlovskij, and Yu.V. Panchin) Neural mechanisms in the generation of nutritional rhythmics in molluscs. Lecture at the First All-Union Biophysical Congress. Moscow 1982 (in Russian)

394. (with Yu.M. Vasil'ev) Possible common mechanism of morphological and growth-related alterations accompanying neoplastic transformation. Proc. Natl. Acad. Sci. USA **79** (1982) 2594–2597 III.IV.15

395. (with M.I. Graev and A.M. Vershik) A commutative model of the basic representation of the group $SL(2, R)X$ connected with a unipotent subgroup. Prepr. Inst. Appl. Math. **169** (1982) (in Russian)

396. (with B.I. Rozenfel'd and M.A. Shifrin) Structural organisation of data in problems of medical diagnosis and prognosis. Prepr. Acad. Sci. USSR Sci. Committee on the complex problem "Cybernetics" (1982) (in Russian)

397. (with R.G. Ajrapetyan, M.I. Graev, and G.R. Oganesyan) The Plancherel formula for the integral transformation connected with a complex of lines intersecting an algebraic straight line in $\mathbf{C}^3$ and $\mathbf{CF}^3$. Dokl., Akad. Nauk Arm. SSR **75** (1) (1982) 9–15 (in Russian). Zbl. **504**:43009

398. (with L.V. Domnina, V.I. Gel'fand, O.Yu. Ivanova, O.Yu. Pletyushkina, and Yu.M. Vasil'ev) Effects of small doses of cytochalasins on fibroblasts: preferential changes of active edges and focal contacts. Proc. Natl. Acad. Sci. USA **79** (1982) 7754–7757

399. (with R. MacPherson) Geometry in Grassmannians and a generalization of the dilogarithm. Adv. Math. **44** (1982) 279–312. Zbl. **504**:57021 III.II.22

1983

400. (with I.V. Cherednik) An abstract Hamiltonian formalism for the classical Yang-Baxter bundles. Usp. Mat. Nauk **38** (3) (1983) 3–21 [Russ. Math. Surv. **38** (3) (1983) 1–22]. Zbl. **536**:58006
401. (with R.G. Ajrapetyan, M.I. Graev, and G.R. Oganesyan) Plancherel theorem for the integral transformation connected with a complex of p-dimensional planes in $\mathbf{CF}^n$. Dokl. Akad. Nauk SSSR **268**, 265–268 (1983) [Sov. Math., Dokl. **27** (1983) 47–50]. Zbl. **527**:53045
402. (with G.S. Shmelev) Geometric structures of double bundles and their relation to certain problems in integral geometry. Funkts. Anal. Prilozh. **17** (2) (1983) 7–22 [Funct. Anal. Appl. **17** (1983) 84–96]. Zbl. **519**:53058 III.I.12
403. (with M.I. Graev and A.M. Vershik) A commutative model of representation of the group of flows $SL(2, R)^X$ that is connected with a unipotent subgroup. Funkts. Anal. Prilozh. **17** (2) (1983) 70–72 [Funct. Anal. Appl. **17** (1983) 137–139]. Zbl. **536**:22008 II.IX.6
404. (with A.Yu. Lyuiko, M.N. Starkova, and A.L. Syrkin) Retrospective estimate of non-stable cardiac angina in various forms of myocardial infarction. Klin. Med. **3** (1983) 28–31 (in Russian)
405. (with A.A. Grinberg, M.L. Izvekova, and V.P. Lakhtina) Prognosis of recidive haemorrhaging in patients with ulcerous disease of the stomach and duodenal intestine. Vestn. Khir., Grekov **130** (4) (1983) 21–24 (in Russian)

1984

406. (with A.V. Zelevinskij) Models of representations of classical groups and their hidden symmetries. Funkts. Anal. Prilozh. **18** (3) (1984) 14–31 [Funct. Anal. Appl. **18** (1984) 183–198]. Zbl. **556**:22003 II.IV.10
407. (Yu.L. Daletskij) Some formal differential structures related to Lie superalgebras. Prepr. Inst. Math. **85** (1984) (in Russian)
408. (with R.G. Ajrapetyan, M.I. Graev, and G.R. Oganesyan) The Plancherel theorem for the integral transformation connected to a pair of Grassmannians. Izv. Akad. Nauk Arm. SSR, Mat. **19** (6) (1984) 467–483 [Sov. J. Contemp. Math. Anal., Arm. Acad. Sci. **18** (4) (1983) 21–32]. MR **86c**:53046. Zbl. **577**:44002
409. (with M.I. Graev and R. Rosu). The problem of integral geometry and intertwining operators for a pair of real Grassmannian manifolds. J. Oper. Theory **12** (2) (1984) 359–383. MR **86c**:22016. Zbl. **551**:53034 III.I.13

410. (with M.A. Brodskij, M.Ya. Ratner, B.I. Rozenfel'd, I.I. Stenina, and V.A. Varshavskij) Morphological-clinical variants of chronic glomerulonephritis and their role in evaluation of serenity of disease. Arkh. Patol. **11** (1984) 46–52 (in Russian)

411. Functions of the cerebellum for the control of rhythmic movements. Today's views on the function of the cerebellum, pp. 181–188. Erevan 1984 (in Russian)

412. (with Yu.I. Arshavskij and G.N. Orlovskij) The cerebellum and the control of rhythmic movements. Moscow: Nauka 1984 (in Russian)

413. (with Yu.I. Arshavskij, G.N. Orlovskij, G.A. Pavlova, and L.B. Popova) Origin of signals convate by the ventral spinocerebellar tract and spino-reticulo-cerebellar pathway. Exp. Brain Res. **54** (3) (1984) 426–431

414. (with L.V. Domnina, O.Yu. Ivanova, O.Yu. Pletyushkina, T.M. Svitkina, and Yu.M. Vasil'ev) Formation of processes in the spreading of fibroblasts in a medium with Cytochalasin B in vitro. Ontogenez **15** (3) (1984) 275–282 (in Russian)

415. (with Yu.M. Vasil'ev) Membrane-cytoskeleton interactions during cell spreading on non-cellular surfaces. 16th meeting of the Federation of European Biochemical Societies, Abstracts, p. 34 (1984) (in Russian)

416. (with A.D. Bershadskij, V.I. Gelfand, L.A. Lyass, A.S. Serpinskaya, and Yu.M. Vasil'ev) Multinucleation induced improvements of the spreading of the transformed cells on the substratum. Proc. Natl. Acad. Sci. USA **81** (1984) 3098–3102

1985

417. (with M.I. Graev) On some families of irreducible unitary representations of the group $U(\infty)$. Prepr. Inst. Appl. Math. **51** (1985) (in Russian)

418. (with A.V. Zelevinskij) Polyhedra in the scheme space and the canonical basis for irreducible representations of gl_3. Funkts. Anal. Prilozh. **19**(2) (1985) 72–75 [Funct. Anal. Appl. **19** (1985) 141–144]. Zbl. **606**:17006

419. (with A.V. Zelevinskij) The canonical basis in irreducible representations of gl_3 and applications. In: Proc. III. Int. Semin. on Group-Theoretical Methods in Physics, Yurmala, 1985 (in Russian)

420. (with A.V. Zelevinskij) Multiplicities and good bases for gl_n. In: Proc. III. Int. Semin. on Group-Theoretical Methods in Physics, Yurmala, 1985 (in Russian)

421. (with B.I. Rozenfel'd and M.A. Shifrin) Structural organization of data in medical diagnostics and prognosis. In: Problems of medical diagnostics and prognosis from the point of

view of a mathematician, I.M. Gelfand (ed.). Vopr. Kibern., Mosk. 112 (1985) 5–64 (in Russian)

422. (with N.M. Chebotareva, S.G. Gindikin, Sh.A. Guberman, M.L. Izvekova, E.I. Kandel, M.Yu. Melikova, B.I. Rozenfel'd, M.N. Starkova, and A.L. Syrkin) Some problems of classification and prognosis from various area of the medicin. In: Problems of medical diagnostics and prognosis from the point of view of a mathematician, I.M. Gelfand (ed.). Vopr. Kibern., Mosk. **112** (1985) 65–127 (in Russian)
423. (with G.I. Dzuba, Sh.A. Guberman, and L.V. Kuznetsov) Applications of the global approach to the discrimination of objects in the automatized analysis of chest x-rays. In: Problems of medical diagnostic and prognosis from the point of view of mathematician. I.M. Gelfand (ed.). Vopr. Kibern., Mosk. **112** (1985) 148–171 (in Russian)
424. (with B.I. Rozenfel'd, and M.A. Shifrin) "Diagnostic games" in medical diagnostics and prognosis. Psikhol. Zh. **5** (1985) (in Russian)
425. (with Yu.I. Arshavskij, G.N. Orlovskij, Yu.V. Panchin, G.A. Pavlova, and L.B. Popova) Regeneration of neurons in pedal ganglia of pteropodial mollusc Clione limacina. Neurophysiologiya **17** (4) (1985) 449–455 (in Russian)

1986

426. (with A.V. Zelevinskij) Representation models for classical groups and their higher symmetries. In: Elie Cartan et les mathématiques d'aujourdhui. The mathematical heritage of Elie Cartan, Sémin. Lyon 1984, Astérisque, No. Hors Sér., 117–128 (1985). Zbl. **594**:22007
427. (with G.N. Orlovskij and M.L. Shik) Locomotion and stratching in tetrapods. In: Neural control of rhythmic movements, J. Wiley, N.Y., 1986
428. General theory of hypergeometric functions. Dokl. Akad. Nauk SSSR **288** (1) (1986) 14–18 (in Russian)
429. (with S.I. Gelfand) Generalized hypergeometric equations. Dokl. Akad. Nauk SSSR **288** (2) (1986) 279–283 (in Russian)
430. (with M.I. Graev) Duality theorem for general hypergeometric functions. Dokl. Akad. Nauk SSSR **289** (1) (1986) 19–23 (in Russian)
431. (with A.B. Goncharov) A characterization of the Grassmannians. Dokl. Akad. Nauk SSSR **289** (5) (1986) 1047–1052 (in Russian)
432. (with A.V. Zelevinskij) Algebraic and combinatorial aspects of the general theory of hypergeometric functions. Funkts. Anal. Prilozh. **20** (3) (1986) 17–34 (in Russian)

1987

433. (with M. Goresky, R. MacPherson, V.V. Serganova) Combinatorial geometries, convex polyhedra, and Schubert cells. Adv. Math. **63** (3) (1987) 301–316
434. (with V.V. Serganova) On the notion of a general (W.P.)-matroid. Dokl. Akad. Nauk SSSR **292** (1) (1987) 15–20 (in Russian)
435. (with V.V. Serganova) Strata of maximal torus in a compact homogeneous space. Dokl. Akad. Nauk SSSR **292** (3) (1987) 524–528 (in Russian)
436. (with V.V. Serganova) Combinatorial geometries and strata of the torus on homogeneous compact manifolds. Usp. Mat. Nauk **42** (2) (1987) 107–133 (in Russian)
437. (with V.A. Vasil'ev, A.V. Zelevinskij) General hypergeometric functions on complex Grassmannians. Funkts. Anal. Prilozh. **21** (1) (1987) 23–38 (in Russian)
438. (with M.I. Graev) General hypergeometric functions on the Grassmannian $G_{3,6}$. Preprint IFM **123** (1987) 27pp. (in Russian)
439. (with M.I. Graev) Strata in $G_{3,6}$ and corresponding hypergeometric functions. Preprint IFM **127** (1987) 25pp. (in Russian)
440. (with M.I. Graev, A.V.Zelevinskij) Holonomic systems and series of hypergeometric type. Dokl. Akad. Nauk SSSR **295** (1) (1987) 14–19
441. (with A.N. Varchenko) On the Heaviside functions of the hyperplane's configuration. Funkts. Anal. Prilozh. **21** (4) (1987) 43pp.
442. (with T.V. Alekseevskaya, A.V. Zelevinskij) Arrangement of real hyperplanes and related, partition function. Dokl. Akad. Nauk SSSR (1987) (in Russian)
443. (with V.S. Retakh, V.V. Serganova) Generalized Airy functions, Schubert cells and Jordan groups. Dokl. Akad. Nauk SSSR (1987) (in Russian)
444. (with V.B. Dugina, T.M. Svitkina, J.M. Vasil'ev) New type of morphological reorganization induced by phorbol ester: reversible partition of cell into motile and stable domains. Proc. Natl. Acad. Sci. USA **84** (1987) 4122–4125

Acknowledgements

We would like to thank the original publishers of I.M. Gelfand's papers for granting permission to reprint them here.

The numbers following each source correspond to the numbering of the article in this volume.

Reprinted from *Actes du Congr. Intern. des Mathématiciens.* Nice 1970, I. © Gauthier-Villars: I.7

Reprinted from *Commutative Normed Rings,* Chelsea: New York, 1964. © Chelsea Publishing Company: II.15

Reprinted from *Dokl. Adad. Nauk. SSSR*: II.3, II.4, II.5, II.6, II.7

Reprinted from *Funct. Anal. Appl.* © Consultants Bureau: III.10, III.13, III.14, III.15, III.16, III.17

Reprinted from *Izv. of the Math. Research Inst.,* Kharkov Univ. Ser. 4 13;1: II.1

Reprinted from *Mat. Sb.*: II.2, II.8, II.9, II.10, II.11, II.12, II.13

Reprinted from *Proc. Intern. Cong. Math. 1954, Amsterdam I.* © North Holland Publishing Co.: I.1

Reprinted from *Proc. Intern. Cong. Math. 1962, Stockholm.* © Institut Mittag-Leffler: I.6

Reprinted from *Russ. Math. Surv.* © British Library: I.4, I.5, III.12, Essays in honour of Izrail M. Gelfand (pp. 735–830 in this volume)

Reprinted from *Transl., II. Ser., Am. Math. Soc.* © American Mathematical Society: I.2, I.3, II.14, III.2, III.5, III.8

Printed in the United States
By Bookmasters